Lecture Notes in Mechanical Engineering

Lecture Notes in Mechanical Engineering (LNME) publishes the latest developments in Mechanical Engineering—quickly, informally and with high quality. Original research reported in proceedings and post-proceedings represents the core of LNME. Volumes published in LNME embrace all aspects, subfields and new challenges of mechanical engineering. Topics in the series include:

- Engineering Design
- Machinery and Machine Elements
- Mechanical Structures and Stress Analysis
- Automotive Engineering
- Engine Technology
- Aerospace Technology and Astronautics
- Nanotechnology and Microengineering
- Control, Robotics, Mechatronics
- MEMS
- Theoretical and Applied Mechanics
- Dynamical Systems, Control
- Fluid Mechanics
- Engineering Thermodynamics, Heat and Mass Transfer
- Manufacturing
- Precision Engineering, Instrumentation, Measurement
- Materials Engineering
- Tribology and Surface Technology

To submit a proposal or request further information, please contact the Springer Editor of your location:

China: Dr. Mengchu Huang at mengchu.huang@springer.com
India: Priya Vyas at priya.vyas@springer.com
Rest of Asia, Australia, New Zealand: Swati Meherishi at swati.meherishi@springer.com
All other countries: Dr. Leontina Di Cecco at Leontina.dicecco@springer.com

To submit a proposal for a monograph, please check our Springer Tracts in Mechanical Engineering at http://www.springer.com/series/11693 or contact Leontina.dicecco@springer.com

Indexed by SCOPUS. The books of the series are submitted for indexing to Web of Science.

More information about this series at http://www.springer.com/series/11236

A. Arockiarajan · M. Duraiselvam ·
Ramesh Raju
Editors

Advances in Industrial Automation and Smart Manufacturing

Select Proceedings of ICAIASM 2019

Springer

Editors
A. Arockiarajan
Indian Institute of Technology Madras
Chennai, India

M. Duraiselvam
National Institute of Technology
Tiruchirappalli
Tiruchirappalli, India

Ramesh Raju
Santhiram Engineering College
Nandyal, India

ISSN 2195-4356 ISSN 2195-4364 (electronic)
Lecture Notes in Mechanical Engineering
ISBN 978-981-15-4738-6 ISBN 978-981-15-4739-3 (eBook)
https://doi.org/10.1007/978-981-15-4739-3

This Springer imprint is published by the registered company Springer Nature Singapore Pte Ltd.
The registered company address is: 152 Beach Road, #21-01/04 Gateway East, Singapore 189721, Singapore

Organizing Committee

Chief Patron

Dr. M. Santhiramudu, Chairman, Santhiram Engineering College, Nandyal

Patron

Shri. M. Sivaram, Managing Director, Santhiram Engineering College, Nandyal

Convener

Dr. M. V. Subramanyam, Principal, Santhiram Engineering College, Nandyal

Co-Convener and Organizing Secretary

Dr. Ramesh Raju, Professor, Department of Mechanical Engineering, SREC, Nandyal

International Advisory Committee

Dr. Suwas Krishna Nikumb, NRC Automotive & Surface Transportation, Canada
Dr. Nachiappan Subramanian, University of Susse, United Kingdom (UK)
Dr. Che-Hua Yang, Institute of Manufacturing Engineering, National Taipei Univ. of Tech., Taiwan
Dr. A. Senthil Kumar, National University of Singapore, Singapore
Dr. Manoj Gupta, National University of Singapore, Singapore
Dr. Tan Koon Tatt, MAHSA University, Malaysia
Dr. Rika Ampuh Hadiguna, University of Andalas, Indonesia
Dr. Baskar K, Shinas College of Technology, Sultanate of Oman

National Advisory Committee

Chairman

Shri. Jegadeesh T.J - ADS PRO-Shield- Chennai – Managing Director

Advisory

Dr. Soundarapandian, IIT Madras
Dr. Aravindan, IIT Delhi
Dr. Sahana Agarwal, NIT Raipur
Dr. K. Partrhiban, NIT Tiruchirappalli
Dr. T. Ramesh, NIT Tiruchirappalli
Dr. Ardhendu Saha, NIT Agartala
Dr. Ram Naresh Rai, NIT Agartala
Dr. Mainak Mallik, NIT Arunachala Pradesh
Dr. P. V. VARDE Senior Professor, Bhabha Atomic Research Centre, Mumbai
Dr. S. S. Panwar, Scientist, DRDL, Hyderabad

List of Contributors and Reviewers

Dr. I. A. Palani, IIT Indore
Dr. Shiva. S, IIT Jammu
Dr. Om Prakash, NIT Patna
Dr. S. Vinodh, NIT Trichy
Dr. N. M. Sivaram, NIT Puducherry
Dr. Sateeshkumar. V, NIT Trichy
Dr. Karpagaraj, NIT Patna
Dr. Vimal KEK, NIT Patna
Dr. Aravind Gali, NIT Jamshedpur
Dr. Sujith Kumar C. S, NIT Calicut
Dr. Kuntal Maji, NIT Patna
Dr. V. Saraswathy, CSIR - CERI
Dr. V. Dillibabu, GTRE, DRDO Bangalore
Dr. Ashwin C. Gowda, VTU, Belagavi
Dr. N. Manikandan, SVEC, Tirupati
Dr. D. Palanisamy, Adhi College, Chennai
Dr. P. S. Sivakumar, Mahsa University/Malaysia
Dr. N. Jeyaprakash, NTUT, Taiwan
Dr. Vinothkumar. S, Shandong University/China

Foreword

The present issue of lecture notes in Mechanical Engineering from Springer is the outcome of the International Conference on Advances in Industrial Automation and Smart Manufacturing (ICAIASM 2019) being held at Santhiram Engineering College, Nandyal 518501, Kurnool District, Andhra Pradesh, India, on July 26 and 27, 2019, as a parallel event with ICAEMS 2019, an International Conference on Advances in Engineering, Management and Sciences.

This edition of the ICAIASM Conference served as a platform to deliver a current and future to be products, technologies, goods and information. It gave overview of the standards and technologies used to implement Industry 4.0 and Industrial IoT-based environment for smart manufacturing. This book covers diversified topics of smart manufacturing technology.

I am also grateful to Chairman Dr. M. Santhiramudu; Managing Director Shri. M. Sivaram and Principal Dr. M. V. Subramanyam, Santhiram Engineering College, Nandyal; Members of the ICAIASM 2019 Committee; and the researchers who have actively participated in the elaboration of this book. A special thanks to Dr. A. Arockiarajan, Professor, IIT Madras, and Dr.-Ing. M. Duraiselvam, Professor, National Institute of Technology, Tiruchirappalli, who took the responsibility of being the editors of this book.

Nandyal, India

Dr. Ramesh Raju
(Organizing Secretary – ICAIASM 2019)

Preface

Santhiram Engineering College (SREC), Nandyal organized International Conference on Advances in Industrial Automation and Smart Manufacturing (ICAIASM 2019) on 26th and 27th July 2019. ICAIASM 2019 held for the second successive time in SREC, Nandyal. The aim of ICAIASM 2019 conference was to bridge the gap between academics, research centers and industry in the field of Smart Manufacturing Technology. ICAIASM 2019 included oral, poster and tutorial sessions given by experts in state-of-the-art topics.

The ICAIASM 2019 Conference helped to support and develop technologies for right understanding and implementation of concepts of smart manufacturing. The conference discussed about many industrial concepts like industrial IoT and cyber physical systems, advanced simulation and digital twin, wireless instrumentation, rapid prototyping and tooling, augmented reality, analytics and manufacturing operations management and so on.

Sessions empowered and enabled the participants to understand the changes in global manufacturing scenario. It also projected the global competitiveness of the Indian goods through smart manufacturing. This Conference ICAIASM is planned to be held once in two years, and will reflect the changing scenario in manufacturing and will empower the future growth in a sustainable manner.

Nandyal, India

Dr. A. Arockiarajan
Dr.-Ing. M. Duraiselvam
Dr. Ramesh Raju

Contents

About the Editors

Dr. A. Arockiarajan is currently working as a professor at Solid Mechanics Division, Department of Applied Mechanics, Indian Institute of Technology Madras, Chennai. He obtained his B.E. from M. S. University, Tirunelveli, M.E. from Bharathiar University, Coimbatore, India and Ph.D. from the University of Kaiserslautern, Germany. His areas of research include smart materials and structures, composites and bio-materials. He has published nearly 100 research articles in reputed international and national journals.

Dr. M. Duraiselvam is currently working as a professor at the Department of Production Engineering, National Institute of Technology Trichy. He graduated from Coimbatore Institute of Technology with a Bachelor's degree in Mechanical Engineering in 1996. He completed his Master's degree in Manufacturing Technology from National Institute of Technology, Tiruchirappalli in 1998. He also has an MBA from Madurai Kamaraj University specialized in operations management in 2003. He completed his Ph.D. in the area of laser cladding of turbine blades from Technical University of Clausthal, Germany in 2006 under DAAD Fellowship. Dr. Durai was also awarded Young Scientist by Department of Science and Technology (DST), Government of India for a fast track project. He has published nearly 80 research articles in reputed international and national journals.

Dr. Ramesh Raju is currently working as a professor in the Department of Mechanical Engineering, Santhiram Engineering College, Nandyal, Andhra Pradesh. He graduated from Shanmugha College of Engineering, Thanjavur, Bharathidasan University and obtained his post graduate degree from Thiagarajar College of Engineering Madurai, Anna University. He obtained his Ph.D. from National Institute of Technology Trichy. He has 5 years of industrial experience, 9 years of teaching experience and 7 years of research experience. His areas of interest are laser material processing, surface engineering and refractory materials.

Dr. Raju is a fellow of Indian Institute of Production Engineers (India) and a member of the Indian Society for Technical Education of India. He has published more than 40 research articles in respected international and national publications.

An Overview of Industry 4.0 Technologies and Benefits and Challenges That Incurred While Adopting It

V. M. Gobinath

Abstract The so-called Industry 4.0 is expected to be the Fourth Industrial Revolution and is to have a huge impact on the manufacturing industry. The ultimate aim of Industry 4.0 is to improve the manufacturing process by combining the virtual and physical world which can help us achieve fully automated manufacturing with adaptability while improving the product quality. The thesis also talks about the establishment of connectivity between various devices in an environment (factory) and also between humans and machines which helps us in collecting and analysing the data to come up with a latent solution for a complicated problem. By this way, customized products can be manufactured easily within the short time. However, challenges such as difficulty in adaptability of the current industries to this revolution are also mentioned in the following paper.

Keywords Internet of things (IOT) · Internet of services (IOS) · Cyber-physical system · Cloud manufacturing

1 Introduction

The First Industrial Revolution which included the transition of muscle power to the establishment of machines that was sourced by steam power was a big leap for the man kind. Followed by the Industry 1.0, the second revolution occurred after the invention of combustion engines and electricity. Third Industrial Revolution occurred after the invention of computers and its ability to minimize the time for an operation and also other advancement such as social media and artificial intelligence. Finally, we are in the verge for the initiation of Industry 4.0. The focus of Industry 4.0 depends on two technologies, which includes human–machine interface (HMI) and connectivity. HMI is a link between humans and machines which helps in guiding the operator towards the solution of the problem [1]. Connectivity includes the communication between the machines and tools or between a machine and another. The ultimate aim

---

V. M. Gobinath (✉)
Department of Mechanical, Industrial and Aerospace Engineering, Concordia University, 1455 De Maisonneuve Blvd. W, Montreal, QC, Canada

A. Arockiarajan et al. (eds.), *Advances in Industrial Automation and Smart Manufacturing*, Lecture Notes in Mechanical Engineering,
https://doi.org/10.1007/978-981-15-4739-3_1

of Industry 4.0 is to improve quality, cost efficiency, delivery time, flexibility and ergonomics. The fourth revolution is mainly based on the benefits of effective data acquisition. The Fourth Industrial Revolution is expected to bring more sophisticated process into manufacturing by the concept of digitalization.

2 Cyber-Physical System (Cps)

CPS is the base for the evolution of Industry 4.0. Cyber-physical system is a concept of creating a digital replica of the physical resources and then linking them together for augmentation of the efficiency of the output of the physical resources [2, 3]. In other words, CPS can be defined as the technology that helps the physical resources via digitalization. CPS can also be implemented in supply chains management to optimize the processes based on the real-time situations. CPS also enables for the self-con'uration and self-optimization of the machines based on the changing environmental conditions (Fig. 1).

The cyber-physical system consists of five levels:

- The first level is smart connection which includes the collection of data from the machines and devices through controllers, sensors and other secondary components.
- The second level involves data conversion where the raw data are converted into meaningful information.
- The third level is the cyber level where all the acquired information is taken to a common platform where it can be compared with other acquired information to identify any deviations and then to establish a countermeasure to correct the deviation.
- The fourth level is the cognition level in which interaction between the machines and humans takes place. In this level, various suggestions are brought up by the

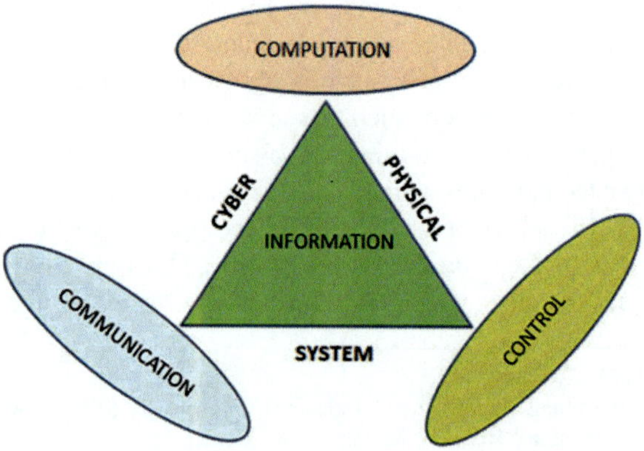

Fig. 1 Structure of cyber-physical system

machine for the humans to perform accordingly so as to increase the efficiency of the manufacturing process and also to improve the quality of the product.

- The fifth level of CPS includes self-configuration of the system based on the environment in such a way it remains as optimized as possible.

3 Vital Concepts and Technologies in Industry 4.0

As discussed earlier, the two main concepts that Industry 4.0 focusses on HMI and connectivity. These two bundles of technologies do not work alone but it works as a combination of technologies.

3.1 Connectivity

This includes linkage of machine with another machine/s for efficient collection and monitoring of the data. Efficient communication of information is the foundation for Industry 4.0. Connectivity plays a dominant role in the industries where the quality and efficiency of the product are vital in the production logic. The degree of connectivity can be of three types, namely one-to-one, one-to-many and many-to-many.

ONE-TO-ONE: Individual usage of machine data to monitor, control and optimize individual devices

ONE-TO-MANY: Networked machine parts allow collective usage of machine data to monitor, control and optimize all devices

MANY-TO-MANY: Integrated usage of machine data connecting machine data to processes.

The primary job of the connectivity concept is to communicate the data and information (both transmit and receive). Communication of information can be done using wired and wireless modes. Implementation of wired communication system is more powerful and sturdy but it is not flexible if the factory has to undergo any chance in infrastructure. However, wireless communication is easy to install but it lacks stability when it comes to large data transfer. The main theme of the connectivity between the machines is cyber-physical systems (CPS) [2]. Internet of things (IoT) is a network of devices that is linked with the CPS. IoT helps the machines and units to communicate with each other and its supply chain and thereby providing clear, accurate and timely data transfer to achieve co-ordination in supply chain. The data and information generated by the machines are collected and processed in the cloud where the performance and risk factors are calculated in real time. As a result of this, cloud computing will become an emerging concept in Internet.

Fig. 2 Structure of Internet of Things

3.1.1 Internet of Things (Iot) and Internet of Services

One of the recent developments of IT is IoT [4]. This technology has completely changed the face of supply chain management by improving the connectivity between the humans and machines [4, 5]. To summarize, IoT refers to a network of digitally connected devices that has the capability to interact within the factory and also between the factory and the supply chain. IoS helps in the enhancement of the product based on the market demand. This technology helps the manufacturing firms to communicate closely with the customers and understand their emotional needs through the products. Hence, a lot of time and resource are reduced that is spent on aggressive market research before the product reaches the market (Fig. 2).

3.1.2 Machine to Machine Interface (M2m)

As the name suggests, we can understand that this technology involves the creation of link between the machines so that data can be communicated among them [6]. One of the examples of M2M interface is the creation of digital link between the machines involved in product development and the machines involved in manufacturing process. Any changes in the product are captured in the database, and it is then transferred to the manufacturing strategy. This provides more flexibility in the creation of mass customized production. In future, robots are expected to be the complete serving entity in the factory system.

3.1.3 Big Data Analytics (BD)

Big data analytics is the glue that holds the concepts of connectivity and HMI. Big data refers to the collection, storage and analysis of a very large amount of data that is acquired from various devices [7]. Since this revolution is mainly driven by automation, a large amount of data is transferred, stored and processed. In order to do this, big data technology is very helpful. The main characteristics of big data include 4Vs: volume, variety, velocity, veracity. Volume refers to the amount of data generated, stored and processed. This characteristic determines whether the data generated can be considered as big data or not. Variety represents the type of data generated. It can be a text, drawing, audio or a visual representation. Data velocity denotes the speed at which the data are generated and analysed. Veracity represents how accurate is the data generated. This characteristic is directly linked to the quality of the generated data.

3.1.4 Cloud Computing

A cloud represents the common space in the Internet where the information can be stored and can be accessed anytime we want. Therefore, any data in the cloud can be accessed by any devices connected to the Internet [8]. The cloud data can be private or public. A public cloud involves storage and access of data and applications via Internet by the public (anyone can have access). Gmail is one of the best examples of public cloud computing where the applications and services launched by Google can be accessed by anyone through any device via Internet. In case of private cloud, data and services within the cloud are secured by a firewall. Private clouds are usually used by corporate and ITs to store, transfer and access data without any security breach. Microsoft Exchange is one of the examples of private cloud as it can be accessed only via VPNs (Fig. 3).

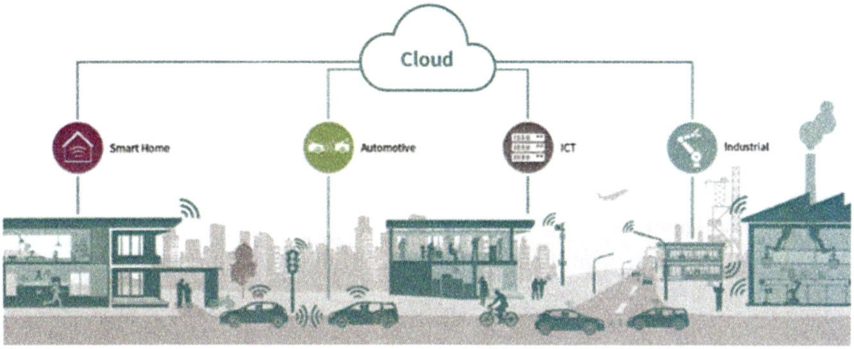

Fig. 3 Cloud storage

3.1.5 Cloud Manufacturing

Cloud manufacturing is a technology that developed after the advancement of cloud computing, IoT and several service-based technologies [9]. The aim of cloud manufacturing is to share the manufacturing resources and capabilities conveniently in order to provide suitable services for the required consumers. Here, all the resources and tools required for manufacturing are stored in the cloud. These resources can be accessed anytime from the cloud. Cloud manufacturing is broadly classified into two categories. The first category involves the storage of manufacturing software into the cloud. The second category involves the integration of real-time devices such as computers, CNC and other machines to the cloud. When all the physical resources in a factory are connected to the cloud, the unused or rarely used machines in the factory can be used for the benefits of other consumers as they will be aware about the availability of the machines they require via the data in cloud space. By this way, the unused machines can be utilized more effectively, and the initial cost of machines can be reduced.

3.2 Human–Machine Interface

As the name indicates, this technological bundle involves the interaction between the humans (Operators) and machines [1]. Though the whole concept of Industry 4.0 is to completely automate the manufacturing process via self- optimization, there are few challenges that are to be met. To overcome these challenges, human interaction with the machines is required now and then. HMI technology is usually preferred in the firms where flexibility is a key characteristic. Here, the operator is the flexible factor in the factory set-up who interacts with the machine manually. As they will be the primary problem-solvers and decision-makers, they must be provided with suffice data and information through devices such as mobiles and context sensitive devices. Through Industry 4.0, it is believed that the concept of automation brings us to a point called "mixed automation" where both humans and machine (robots) work together in the factory environment without any partition between them [1]. Various robot manufacturing firms have pointed out that collaborative robots have the capacity to interact and guide the operators rather than just to stop when the operators get close to them.

3.2.1 Augmented Reality (AR)

Augmented reality is a technique used to combine the actual scene viewed by the user in real time with the virtual scene generated by the computer [10]. This technique is very useful in easy interpretation of data from the machines, especially during human–machine interface (Fig. 4).

Fig. 4 Augmented reality via tablet

3.2.2 Collaborative Robots

Collaborative robots or cobots are the future serving entities in the factories in Industry 4.0 [11]. The key functions of these robots include autonomous operation in a collaborative workspace with humans and establishment of safety design features to protect humans from injury in factory environment. Collaborative robots are also capable of the concept of hand guiding where humans can guide the robots with hand-operated devices. These robots are provided with highly sensitive sensors which help them to stop moving if they find any operator close to the safety level [11, 12]. However, these robots are not very powerful nor very quick when compared to traditional factory robots and they need not be as their primary task is to work along with the humans. Therefore, they work at the speed and magnitude of an average human being (Fig. 5).

3.2.3 Exoskeletons

Exoskeletons are the mechanical gadgets that are worn by the operator which can act as a limb or muscle and can work in co-ordination with the movements given by the operator [13]. They are commonly used as an amplifying and assisting device. The concept of exoskeleton initially originated in defence department, however it is recognized that it might add significant value in manufacturing industries (Fig. 6).

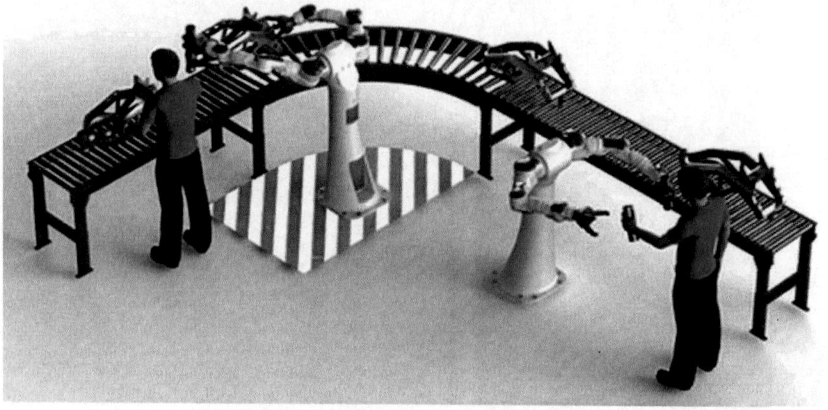

Fig. 5 Collaborative robots in action with humans. *Source* Yaskawa.eu.com

4 Smart Factory and Smart Production

Smart factory and smart production are the two vital areas in Industry 4.0. Smart factory represents the factory that is accompanied with technologies such as CPS, IoT, big data analytics, augmented reality and artificial intelligence. It mainly comprises the physical resources integrated with the local industrial network which in turn is connected to the cloud space. This cloud space is provided with terminals to access, control and supervise the manufacturing system. The physical resources include the real-time appliances and machines. These machines are actuated or operated from the cloud via the industrial networks to which they are connected. The cloud is the store house of all the data that is acquired from the physical resources. These data and information in the cloud could be accessed by the people via the terminals to which the cloud data is connected.

5 Key Features of Industry 4.0

The key feature for the successful establishment of Industry 4.0 is to integrate various domains in the supply chains. This integration of domains can be horizontal or vertical or end-to-end integration [14]. Horizontal integration involves the integration of two firms that are in the same stage and produces same or similar product/output in the supply chain. Through this integration communication between same stages of the supply chains within the firm or with other firms is possible. Vertical integration involves the integration of two firms that are at the different stages in the supply chain. This integration helps us to have better communication through all the domains in all the hierarchical flow. Vertical integration can either be forward or backward integration. When a company in the initial stage of supply chain controls or owns a company in the farther stages along the supply chain, then it is called forward integration. In contrast, when a company in the farther stage of the supply chain owns or controls a company in the beginning stage of the supply chain, then it is called as backward integration. The combination of the horizontal and vertical integration results in the formation of end-to-end integration that has the benefits of both types of integration.

6 Benefits of Industry 4.0

The benefits and outcomes of Industry 4.0 can be classified into three main categories

- New business models
- Efficiency
- Customization.

6.1 New Business Models

A business model is a template of all the processes happening within a firm. The new business model involves more interaction with the customers when compared to the current models. The technical advancement in Industry 4.0 enables increase in efficiency in the way by which data can be created and captured.

6.2 Efficiency of Operation

New technologies such as big data analytics and CPS help us to improve the operational capabilities through predictive and preventive maintenance, self-optimization of the machines, improved visualization of supply chains and quick decision-making. Due to this, the downtime of the machine can be cut or eliminated; hence, the operational efficiency of the firm will be improved.

6.3 Customization

The third benefit of Industry 4.0 is improved customer satisfaction. New technologies of Industry 4.0 help to learn more about their customers and receive real-time feedback. Through the feedback from the customers, mass personalized production is possible. Direct input and feedback from the customers will help the companies to design and produce more customized products for which shorter cycle time and low cost are sufficient when compared to standardized mass production. By this, both the producer and the customer will share the new value that is created.

7 Challenges in Adopting Industry 4.0

7.1 Cyber-Physical System

The lack of complete framework and physical resources is one of the big challenges in the implementation of CPS in current factories. Other external factors such as noise and vibrations also influence in the measurement of inputs from the devices which interrupts the entire cycle of data transfer. CPS may also face issues related to scalability, robustness and its complexity due to its need to fetch real-time data. Still studies have been carried out on how to error free interface between physical and digital world.

7.2 Internet of Things and Internet of Services

This involves the integration of various technologies from different domains which is the primary challenge. The main challenges in implementation of IoT and IoS are the lack of scalability, robustness, security, privacy breach and quality [5]. The existence of various working platforms for various devices makes it difficult for the researchers to integrate all the devices into one. For example, both Android and iPhone have different working platform. Under this case, the combination of these two devices to access a common data from the cloud is difficult as each device platform has its own privacy. Scalability is another challenge during the implantation of Industry 4.0 via IoT. As the number of device integrated to each other increases, large amount of resources for data storage and management is required. Security breach is one of the main challenges in the IoT as the potential attackers of the secured data increase due to the availability of global access to Internet; therefore, anyone can have access to devices connected to IoT.

7.3 Big Data Analytics

Since the whole concept of Industry 4.0 is digitalization and automation, a large amount of data reception and storage and processing is required. In order to do so, large amount of resources are required to maintain and process the data. Since these data are stored in a common space called as cloud storage, the question of how secured the data will be maintained will be raised. This storage of large amount of data with high confidentiality is again a challenge for adoption of Industry 4.0.

7.4 Cloud Manufacturing

Cloud manufacturing is a big challenge for the small and medium-scale enterprises (SME). The primary challenge that SME faces is the development of complex design via cloud manufacturing as it involves integration of highly advanced software, application and machines. The second challenge is the lack of follow-up actions as most of the company do not have the ability to offer that. The third challenge is the cost of the technology. Since cloud manufacturing involves advanced techniques and machines, the cost of the overall technology is high which is difficult for the small and medium-scale firms to adopt [9].

7.5 Augmented Reality

Augmented reality faces four main challenges. The first challenge is the exact overlay of the real-time information with the virtual environment. Very high accuracy in positioning and co-ordination of the reality and virtual environment are necessary for the efficient design and manufacturing process. The second challenge faced by AR system is registration. There are two possible errors in this concept, namely static and dynamic error [3]. The static inaccuracy is caused due to the errors present in the sensing devices, algorithms and other secondary devices. Dynamic errors are caused due to delays in synchronization and computing process. Dynamic errors are mainly due to latency problems.

7.6 Collaborative Robots

The primary challenge of installation of cobots is that they require separate working area. Though they are programmed to work alongside humans, the concept of human safety hinders the complete elimination of boundary between the robots and operators [12]. As these robots are collaborative, they always require human's assistance to work.

8 Conclusion

The primary aim of this paper is to bring awareness about the technology that will be used in Industry4.0. Through this paper, it is understood that Industry 4.0 will have a huge impact on the manufacturing industry that will drive us to the most advanced future where digitalized world will work in parallel with the physical world. Though this adaptation might be a challenge, we will be forced to adapt it due to the emerging technology and increased demands from the consumers.

References

1. Ansari F, Erol S, Sihn W (2018) Rethinking human-machine learning in industry 4.0: how does the paradigm shift treat the role of human learning? Procedia Manuf 23:122–177
2. Lee J, Bagheri B, Kao HA (2015) A cyber-physical system architecture for industry 4.0 based manufacturing sytems. Manuf lett 3:18–23
3. László L (2014) Cyber-physical production systems: roots, expectations and R&D challenges. Procedia CIRP 17:9–13
4. Gubbi J, Buyya R, Marusic S, Palaniswami M (2013) Internet of Things (IoT): a vision, architectural elements and future directions. Future Gener Comput Syst 29:1645–1660
5. Gigli M, Koo S (2011) Internet of Things: services and applications categorization. Adv Internet Things 1:27–31
6. Balamuralidhar P, Misra P, Pal A (2013) Software platforms for Internet of Things and M2M. J Indian Inst Sci 93:487–498
7. Elgendy N, Elragal A (2014) Big data analytics: a literature review paper. advances in data mining. Appl Theor Aspects 8557:214–227
8. Buyya R, Yeo CS, Venugopal S, Broberg J, Brandic I (2008) Cloud computing and emerging IT platforms: vision, hype and reality for delivering computing as the 5th utility. Future Gener Comput Syst 25:599–616
9. Ren L, Zhang L, Tao F, Zhao C, Chai X, Zhao X (2013) Cloud manfacturing: from concept to practise. Enterprise Inf Syst 9:186–209
10. Ling H (2017) Augmented reality in reality. IEEE Multimedia 24:10–15
11. Ranky PG (2003) Collaborative, synchronous robots serving machines and cells. Indus Robot Int J 30:213–217
12. Peshkin M, Colgate JE (1999) Cobots. Indus Robot Int J 26:335–341
13. Bogue R (2009) Exoskeletons and robotic prosthetics: a review of recent developments. Indus Robot Int J 36:421–427
14. Garvey GT (1995) Why reputation favors joint ventures over vertical and horizontal integration a simple model. J Econ Behav organ 28:387–397

Design, Fabrication and Testing of Magnetorheological Damper System for Machine Tool Application

Shreedhar Kolekar, V. Dhinakaran, T. Jagadeesha, and Choi Seung Bok

Abstract Magnetorheological (MR) damper makes use of magnetorheological fluids (MRFs) as a working fluid. MRFs are smart fluids which can change their properties under the influence of external stimulus. They can change behaviour from a liquid state to that of semi-solid with yield stress when exposed to external magnetic fields and back to liquid in absence of magnetic field. This change occurs in fractions of seconds. MRFs are suspensions of micron-sized magnetizable particles disperse in a nonmagnetic carrier fluid. These particles align themselves in the direction of the applied magnetic field forming a chain like structure, thereby resisting the fluid flow. In this project, we are optimizing the composition for a silicon oil-based MR fluid in the laboratory and then test it using a magnetic rheometer to measure different rheological properties such as viscosity, shear stress and yield stress, and then design and fabricate a damper-based on the same MR fluid. The damper is designed based on a mathematical model and further optimized using MATLAB. The damping coefficient varies in accordance with the intensity of the magnetic field, thus providing the required damping force. A test rig was designed for testing out the performance of the designed damper.

Keywords Magnetorheological fluid · Yield strength · Off-state fluid viscosity

---

S. Kolekar
Satara College of Engineering and Management, Satara 415015, India

V. Dhinakaran
Chennai Institute of Technology, Kundrathur, Chennai 600069, India

T. Jagadeesha (✉)
National Institute of Technology Calicut, Calicut 673601, India
e-mail: jagdishsg@nitc.ac.in

C. S. Bok
Inha University, Incheon 22212, South Korea

A. Arockiarajan et al. (eds.), *Advances in Industrial Automation and Smart Manufacturing*, Lecture Notes in Mechanical Engineering,
https://doi.org/10.1007/978-981-15-4739-3_2

1 Introduction

Magnetorheological (MR) fluids are fluids that respond to an applied magnetic field with a dramatic change in rheological behaviour. MR fluid dampers are new-type vibration control elements, having the advantage of rapid damping and stiffness changing in the presence of an applied magnetic field. Typical MR fluid has the advantages of higher yield stress (up to 50–100 kPa) insensitive to general contaminants, using only 12–24 V low voltage, with a relatively broad working temperature range (40–150 °C), so MR fluid damper has the characteristics of large damping force, low-power consumption, easy to control, etc [1–3].

Since MR fluid has high yield stress and may solidify at strong magnetic field, the MR dampers in a twin tube damper system may become rigid and may cause the system instability. The stability of MR damper is important for the effective and safe use of the damper, and almost no literatures till now focus on the stability analysis on twin-tube damper system. A twin-tube MR damper has similar construction to that of a common twin-tube damper; only difference is that it has MR fluid acts as working fluid and has magnetic coils to control or regulate the flow of fluid. The magnetic field changes the viscosity of the MR fluid in the orifices, thus regulating the flow across the chambers of the damper. When the flow varies, the damping effect of the damper changes accordingly [4–7].

Selection criteria of MR fluid components play very critical role as the change in one or more components or in their properties influences the MR effect. The commonly used carrier fluids are silicone oil, mineral oil and synthetic oil. Silicone oil is mostly preferred because of good heat-transfer characteristics, good high flash points, and good temperature stability, and there is little change in physical properties over a wide temperature range with a relative flat viscosity temperature and serviceability ranging from −40 to 204 °C. Magnetic particles used are of micrometer size, the particle size distribution ranging from 1 to 10 μm. The stability of the fluid increases with decreased size of magnetic particles. Also, the stabilizers are added to the fluid to ensure particle suspension in the carrier fluid [8, 9].

The main objective of this paper is to synthesize a silicone oil-based stable MR fluid, design and fabricate of twin-tube magnetorheological damper test rig using linear motor as actuator and study behaviour of the MR damper with different current supply to the magnetic coils.

2 Mathematical Modelling

In this section, mathematical modelling of twin-tube damper is discussed. In this model, it is assumed that the height of the gap is much smaller than its length and the width; therefore, the flow is considered as a flow between parallel plates, and thus, the valve mode simplification is reasonable. Figure 1 shows the schematic diagram of double-tube damper.

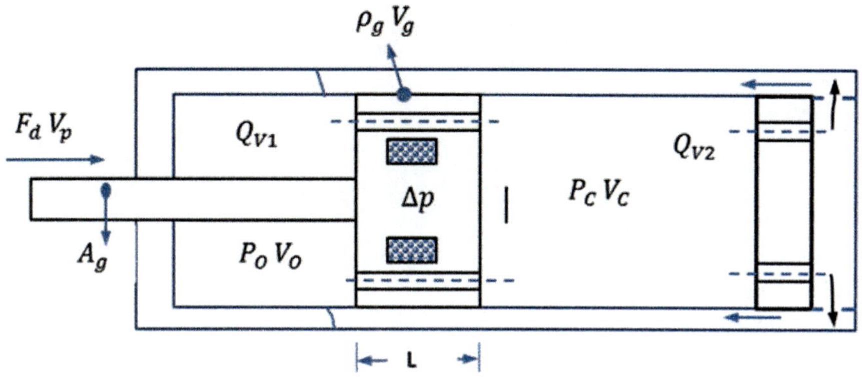

Fig. 1 Relation between rheology and particle polarization

Let Q_{v1} be the flow through annular gap in piston. The annular path has a length L and cross-sectional area A_g of and Q_{v2} be the net flow through foot valve and pressure drop across piston.

The net flow through foot valve and pressure drop across piston is given by

$$\Delta p = P_{\mathrm{r}} - P_{\mathrm{c}} \tag{1}$$

where P_{r} rebound chamber is pressure and P_{c} is compressor chamber pressure.

The bulk modules of fluid are given by

$$\beta_f = -V\frac{\mathrm{d}p}{\mathrm{d}V} \tag{2}$$

Isothermal compressibility can be written as

$$c_f = -\frac{1}{V}\frac{\mathrm{d}V}{\mathrm{d}p}$$

Rearranging we get

$$-\frac{\mathrm{d}V}{\mathrm{d}t} = c_f V\frac{\mathrm{d}P}{\mathrm{d}t}$$

Let volume of rebound chamber and compression chamber at any instant be V_{r} and V_{c}, respectively. Now change in volume cab written as

$$\frac{\mathrm{d}V}{\mathrm{d}t} = \sum Q_{\mathrm{in}} - Q_{\mathrm{out}}$$

Substituting value of $\frac{\mathrm{d}V}{\mathrm{d}t}$, we get

$$c_f V \frac{\mathrm{d}V}{\mathrm{d}t} = \sum Q_{\mathrm{in}} - Q_{\mathrm{out}}$$

Hence,

$$\frac{\mathrm{d}P}{\mathrm{d}t} = \frac{\beta_f}{V}\left(\sum Q_{\mathrm{in}} - Q_{\mathrm{out}}\right)$$

Here, β_{f} is influenced by stiffness of container β_{c}, i.e. expansion cylinder walls with pressure.

Combining fluid bulk modulus with volumetric effect, following relation was assembled.

$$\frac{1}{\beta} = \frac{1}{\beta_{\mathrm{f}}} + \frac{1}{\beta_{\mathrm{c}}}$$

Considering compliance of cylinder material gives

$$\frac{1}{\beta_{\mathrm{c}}} = \frac{12}{E_s}\left(v - \frac{D_0^2 + D_{\mathrm{p}}^2}{D_{\mathrm{i}}^2 - D_{\mathrm{p}}^2}\right) \tag{5}$$

where

E_{s} Young's modulus of elasticity for cylinder material
v Poisson's ratio
D_{p} piston outside diameter
D_{i} cylinder inside diameter
D_0 orifice diameter.

Also, compressibility of the gas is given by

$$\beta_f(P) = \beta_o\left(\frac{1 + k\left(\frac{P_{\mathrm{a}}}{P_{\mathrm{a}}+P}\right)^{\frac{1}{n}}}{1 + k\frac{P_{\mathrm{a}}^{\frac{1}{n}}}{n(P_{\mathrm{a}}+P)^{\frac{1}{n}}}}\right) \tag{6}$$

where

β_o bulk modulus of pure fluid
k relative gas coefficient
P_{a} atmospheric pressure
n adiabatic gas constant.

Let volume of rebound chamber and compression chamber at any instant be V_{r} and V_{c}, respectively; then,

$$\begin{aligned} V_{\mathrm{r}} &= V_{\mathrm{r0}} + \left(A_{\mathrm{p}} - A_{\mathrm{r}}\right)x_{\mathrm{p}} \\ V_{\mathrm{c}} &= V_{\mathrm{c0}} - A_{\mathrm{p}}x_{\mathrm{p}} \end{aligned} \tag{7}$$

where

V_{r0}, V_{c0} initial volume of respective chamber
A_p area of piston
A_r area of rod
x_p piston displacement.

From Eq. (3), pressure changes in rebound chamber can be written as

$$\dot{P}_r = \frac{\beta}{V_r}\left(A_p - A_r\right)V_p - Q_{v1} \tag{8}$$

And, pressure changes in compression chamber is given by

$$\dot{P}_c = \frac{\beta}{V_c}\left(-A_p V_p + Q_{v1} - Q_{v2}\right) \tag{9}$$

where

V_p piston velocity.

Cavitation effect can be accounted by imposing constraints on $\dot{P}_r$ and $\dot{P}_c$, i.e. $\dot{P}_r \geq P_V$ and $\dot{P}_c \geq P_V$, P_r is vapour pressure of fluid.

Now, considering inertia of mass of fluid inside annular gap inside damper is in motion

$$\Delta p \times A = \rho L \dot{Q}$$

So, flow through valve holes is given by

$$\dot{Q}_{V1} = \frac{A_g}{\rho L}(P_r - P_c - \Delta P_a) \tag{10}$$

where

A_g size of annular gap in piston
ρ density of fluid
ΔP_a pressure loss across annular path
L length of fluid.

Now assuming an adiabatic process pressure due to gas

$$P_g = P_{g0}\left(\frac{V_{g0}}{V_{g0} - \int Q_{v1}\,dt}\right)^n \tag{11}$$

where

P_{g0} initial gas pressure
V_{g0} initial gas volume

n adiabatic gas constant.

Discharge through foot valve is given by

$$Q = C_{\mathrm{d}} A \sqrt{\frac{2\Delta \mathrm{P}}{\rho\left(1 - \frac{D_{\mathrm{g}}}{D_{\mathrm{p}}}\right)}}$$

where

$$\Delta \mathrm{P} = P_{\mathrm{r}} - P_{\mathrm{c}} \text{ and } \frac{D_{\mathrm{g}}}{D_{\mathrm{p}}} \cong 0$$

So,

$$Q_{v2} = C_{\mathrm{d}} A \sqrt{\frac{2|P_{\mathrm{r}} - P_{\mathrm{c}}|}{\rho}} \quad (12)$$

and the damping force is given by

$$F_{\mathrm{d}} = \left(A_{\mathrm{p}} - A_{\mathrm{r}}\right) P_{\mathrm{r}} - A_{\mathrm{P}} P_{\mathrm{c}} + F_{\mathrm{f}} \quad (13)$$

where

F_{d} force due to friction at various interfaces.

Further, for mathematical modelling of shear stress equation, Bingham model is considered (Fig. 2).

As the MR fluid satisfies Bingham model, the shear stress is given by

$$\tau = \tau_0 + \mu \frac{\mathrm{d}u}{\mathrm{d}y} \quad (14)$$

where

τ shear stress

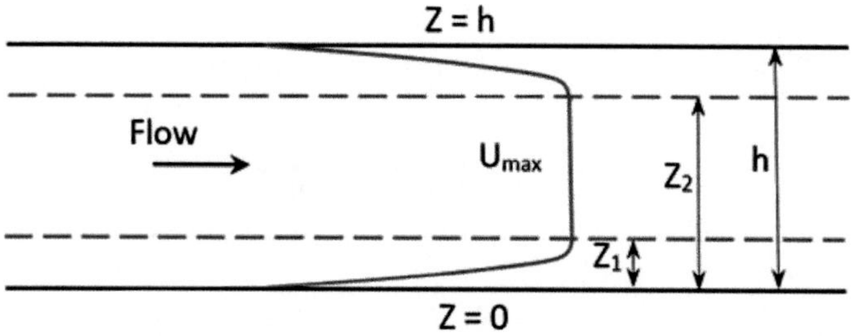

Fig. 2 Flow between two parallel plates

τ_0 shear stress due to magnetic field
μ viscosity of fluid
$\frac{du}{dy}$ velocity gradient.

Since, size of annular flow path is very small compared to size of piston area. So, the flow inside annulus can be approximated to flow between two parallel plates.

Now, from Navier Stokes' equation of momentum, we have

$$\rho\frac{\partial u}{\partial t}+\rho\frac{\partial(uv)}{\partial x}=-\frac{\partial p}{\partial x}+\frac{\partial\tau_{xy}}{\partial y} \tag{15}$$

where

τ is given by Eq. (14)
ρ density of fluid
u, v velocity component along x and y
p pressure.

Now, for a fully developed, steady, one-dimensional flow, Eq. (15) is simplified to give

$$\frac{\partial\tau}{\partial y}=\frac{\partial p}{\partial x}$$

On integrating both sides of above equation, we get

$$\tau = P_x Y + C$$

At boundary condition, i.e. at mid-section, the flow shear stress is zero, i.e.

$$z=\frac{h}{2}\text{ and }\tau=0$$

Now, applying above boundary condition to Eq. (14), we get

$$C=\frac{-P_x h}{2},$$

Hence, $\tau = P_x\left(Y-\frac{h}{2}\right)$, $\tau = 0$ for, Z1 < Z < Z2

Optimization of MR Valve

'Valve ratio' and 'controllable force' are two important parameters for evaluating the performance of an MR damper. The force exerted by a damper can be divided into uncontrolled force (F_τ) due to governable yield stress and controlled force (F_η). Uncontrolled force is due to the viscous force in the absence of applied magnetic field or off-state viscosity (Fig. 3).

Now, pressure difference across valve holes due to controlled force is given by

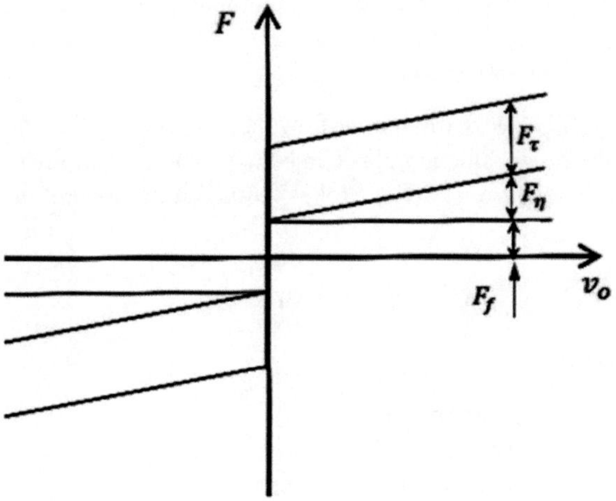

Fig. 3 Decomposition of force in MR damper

$$\Delta P_{A,\tau} = 2c\frac{t_f \tau_y}{t_g} \tag{16}$$

And pressure difference across valves holes due to uncontrolled force is given by

$$\Delta P_{A,\eta} = 6\frac{\eta L}{\pi t_g^3 R_d} \tag{17}$$

where

$$R_d = R - t_h - 0.5t_g$$

$c \approx 2.07 + 1/(1 + 0.4\Gamma)$ restricted in between the value of 2.07–3.07 (Fig. 4).

Valve ratio (λ) is the ratio of pressure difference across valve holes due to uncontrolled force to pressure difference across valve holes due to controlled force.

$$\text{So, } \lambda = \frac{\Delta P_{A,\eta}}{\Delta P_{A,\tau}} = \frac{3\eta L Q}{\pi c t_g^2 R_d t_f \tau_y} \tag{18}$$

It is required that valve ratio takes a small value, less than one. The force is inversely proportional to valve hole size; the controlled force must as large as possible to optimize the effectiveness of MR damper. Hence, valve holes of small sizes are required.

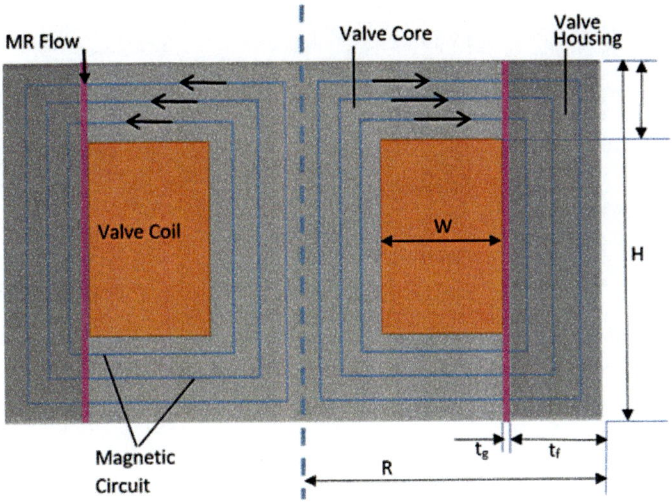

Fig. 4 Piston structure with magnetic field lines

3 Synthesis of MR Fluid

The major constituents of MR fluid are carrier fluid, magnetizable particles and additives. Carrier fluid is the major constituent of MR fluid by 50–80% by volume fraction. The commonly used carrier liquids are mineral oil, synthetic oil and silicone oil. Silicone oil is used as carrier fluid in our experiments.

The size of magnetic particles is of the order of 1–10 μm. The sedimentation increases with an increase in the size of magnetic particles. Carbonyl iron powders of sizes 2–10 micron is used in our experiments. The stabilizers serve to keep the particles suspended in the fluid. Vacuum grease and lithium grease are used as additives in our experiments; the MR fluid has been prepared using mechanical stirrer.

Initially, the mixture of carrier fluid and additive is prepared and stirred for 3 h. The amount of carbonyl iron particles as mentioned in table is then added into the prepared mixture and stirred for another 4 h. The magnetorheological fluid, developed in house, is used in the prototype damper. The fluid is a suspension of a 5- to 40-micron diameter sized magnetically susceptible particles mixed in silicone oil carrier fluid (Table 1).

Table 1 Properties of synthesized MR fluid

Constituents	Sample
Appearance	Dark grey liquid
Carried fluid	Silicone oil
Suspended particles (weight)	30%
Stabilizer (weight)	Lithium white grease, 12%
Surfactant (weight)	Triton X-100, 3%
Operating temperature	−30 to 150 °C
Off-state viscosity	422.39 mPa s at 40 °C 165.76 mPa s at 100 °C

4 Design of Damper System

4.1 Design of Mechanical Parts

Considering a cylinder of the radius $R = 35$ mm and valve hole length $L = 50$ mm. The post-yield viscosity of the MRF is assumed to be constant, $\eta = 0.\ 092$ Pa s and the flow rate of the MR valves are $Q = 3 \times 10^{-4}\ \mathrm{m}^3/\mathrm{s}$. The following parameters were varied for optimality of MR damper:

Number of holes (n)
Diameter of pitch circle for valve holes (t_h)
Flange thickness (t_f)
Diameter of the valve holes (t_g).

Here, t_h is variable accounting for pitch circle; it is the distance of pitch circles from periphery of piston.

(i) Number of holes: Number of holes is determined by taking, $t_h = 0.0025$ m, $t_f = 0.005$ m and $t_g = 0.003$ m.
From Fig. 5, valve ratio decreases with increase in number of valve holes; this is because with increase in number of holes, more of MR fluid comes under the influence of magnetic field in the valve holes, also because of decrease in the pressure difference across the chambers. But the number of holes cannot be increased beyond certain limit due design considerations.

(ii) Diameter of pitch circle: Diameter of pitch circle is determined by taking $n = 0.003$ m, $t_f = 0.005$ m and $t_g = 0.003$ m. From Fig. 6, minor changes in the diameter of pitch circle do not have significant effect on the valve ratio. So, it decided taking strength of design and wear into account.

(iii) Flange thickness: Flange thickness is calculated by taking $n = 0.003$ m, $t_h = 0.0025$ m and $t_g = 0.003$ m.
The plot in Fig. 7 shows that valve ratio decreases with the increase in flange thickness, because with increase in flange thickness area of MR fluid, flow length under magnetic influence increases.

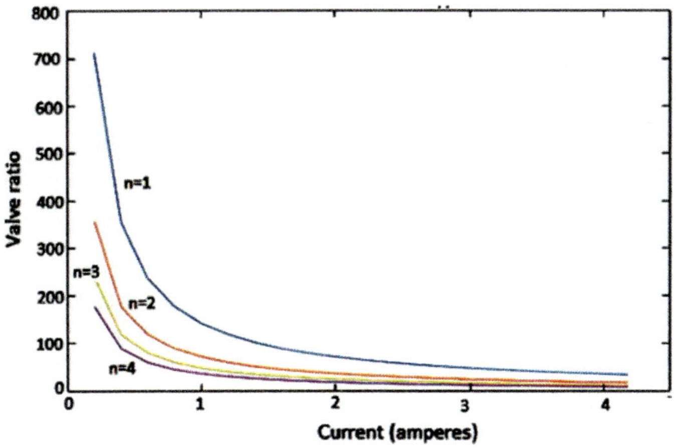

Fig. 5 Variation of valve ratio and applied current with different number of holes

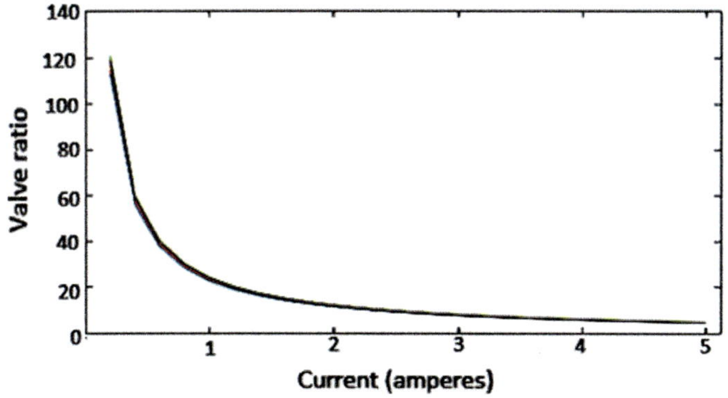

Fig. 6 Variation of valve ratio and applied current with different pitch circle diameter

(iv) Diameter of valve holes is calculated by taking $n = 0.003$ m, $t_h = 0.0025$ m and $t_f = 0.005$ m.

It is seen that with the increase in diameter of holes, the valve ratio decreases, because the pressure difference across the valve holes due to uncontrolled force decreases, thus decreasing the valve ratio. Hence, the final dimensions are fixed as given in Table 2.

4.2 Calculations for Armature Winding

The magnetic flux density is given by

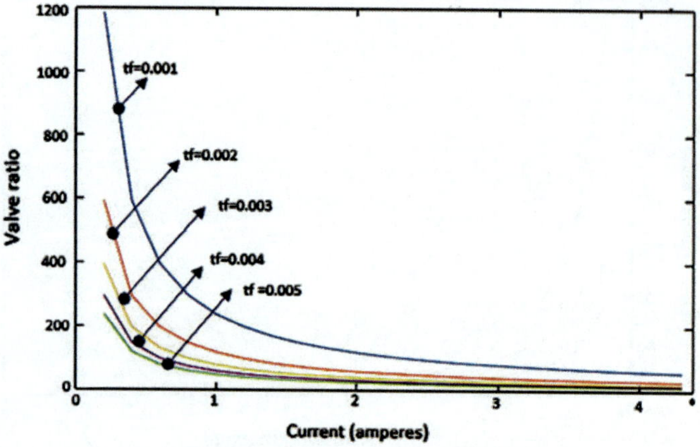

Fig. 7 Variation of valve ratio and applied current with different flange thickness

Table 2 Final dimensions

S. No.	Part	Dimension
1	Number of holes	3
2	Pitch circle distance	3.5 mm
3	Flange thickness	5 mm
4	Diameter of valve holes	3 mm

$$B = \mu_0 N I \tag{19}$$

$$B = \mu_0 \frac{N}{L} I \tag{20}$$

where

μ_0 permeability of air
N number of turns
L length of winding
I current supplied.

So, the number of turns is given by

$$N = \frac{BL}{\mu_0 I}$$

Now, assuming $B = 1$ tesla, $L = 5 \times 10^{-2}$ m and $I = 5$ A and taking 60% tangential efficiency; the number of turns required would be

$$N = \frac{(1\,\mathrm{T})\left(5 \times 10^{-2}\,\mathrm{m}\right)}{\left(4\pi \times 10^{-7}\,\mathrm{H}\right)(5\,\mathrm{A})}(1.6) = 12732 \text{ turns} \tag{21}$$

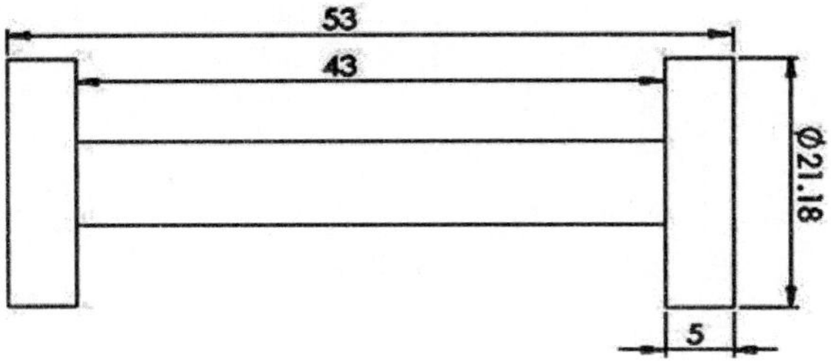

Fig. 8 Magnetic field intensity distribution for 70 mm core

Considering, a TWG rated wire of specifications diameter = 0.25 mm and current rating = 5 A maximum (Fig. 8).

Now, according to the geometry of designed armature, core number of turns in one plane or layer of winding can found by dividing the length of armature core by thickness of winding wire.

$$\text{No. of turns in each layer} = \frac{43\,\text{mm}}{0.25\,\text{mm}} = 172\ \text{turns}$$

And, number of layers of winding can be found by dividing the available radius for winding by thickness of winding wire.

$$\text{No. of layers of winding} = \frac{10\,\text{mm}}{0.25\,\text{mm}} = 40\ \text{layers}$$

Thus, possible number of turns on the armature core = 172 × 40 = 6880 turns, which is approximately 54% of required number of windings to obtain 1 T at 5 A; so, the designed armature core would be able to produce approx. 0.54 T at 5 A.

4.3 *Magnetic Analysis of Designed Armature*

The following sections discuss a 2D axis-symmetric analysis of MR damper. Using certain value of current, we calculate the magnetic flux density at the armature core, piston and MR fluid. All dimensions are in SI unit. The number of coil windings is 6000 and is responsible for magnetic flux field that energies the MR fluid. Since the functioning of MR dampers is based on magnetic flux density, current through the coils is changed to get variations in flux density (Fig. 9).

The considered model is assumed to have negligible flux leakage such that there is no saturation of material. The above assumptions simplify the model analysis.

For a static (DC) current, ANSYS requires the current to be input in the form of current density (current over the area of the coil). This boundary condition is used for models in which the flux is contained in an iron circuit. In post-processing, the forces are summarized for the engine, MR fluid and the engine housing, using a Maxwell stress tensor and a virtual work calculation. Flux density also is displayed.

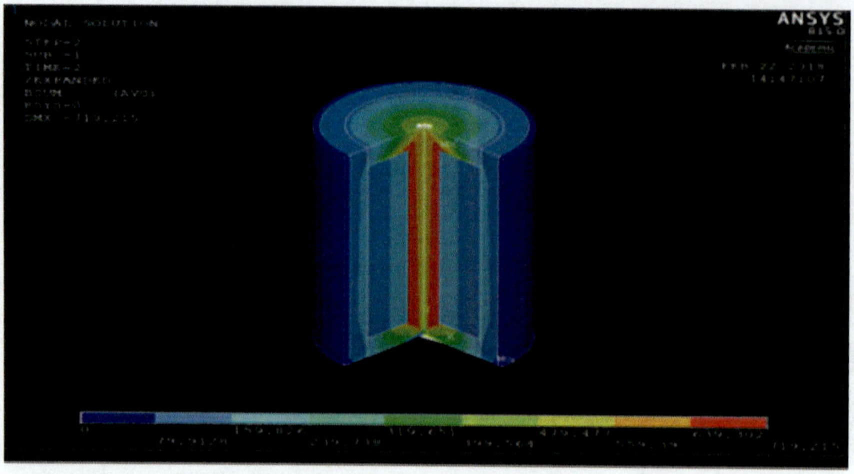

Fig. 9 Magnetic field intensity distribution for 70 mm core

The final post-processing operation computes the terminal parameters including coil inductance. Figure 10 shows the direction of the magnetic field lines across the piston. The space between the lines is directly proportional to the magnetic field intensity.

The closely spaced lines represent greater intensity. The flow of the fluid through the piston is normal to the direction of the field lines.

4.4 Fabrication of MR Damper

The MR damper consists of various parts such as outer shock tube, inner shock tube, piston cup and cap, piston rod, armature core, foot valve, inner cylinder cap, and upper and lower end caps which are assembled and filled with synthesized silicon-based MR fluid. Work regime of the MR fluid inside the damper is called the valve mode. In this model, the flow through valve holes is treated as a flow between two parallel still plates. The resistance to flow of a liquid through the valve holes is controlled by the change in the magnetic field, whose vector is normal to the path of the flow. The adjustment of the magnetic field is performed by the change of the current in the coil winding mounted on the piston. Various components of MR damper fabricated and the assembled MR damper are shown in Fig. 11.

5 Test Rig for Damper Testing

A novel test rig is designed and fabricated. The reciprocating motion for damper is provided by the tubular linear actuator by attaching the top part of tubular linear

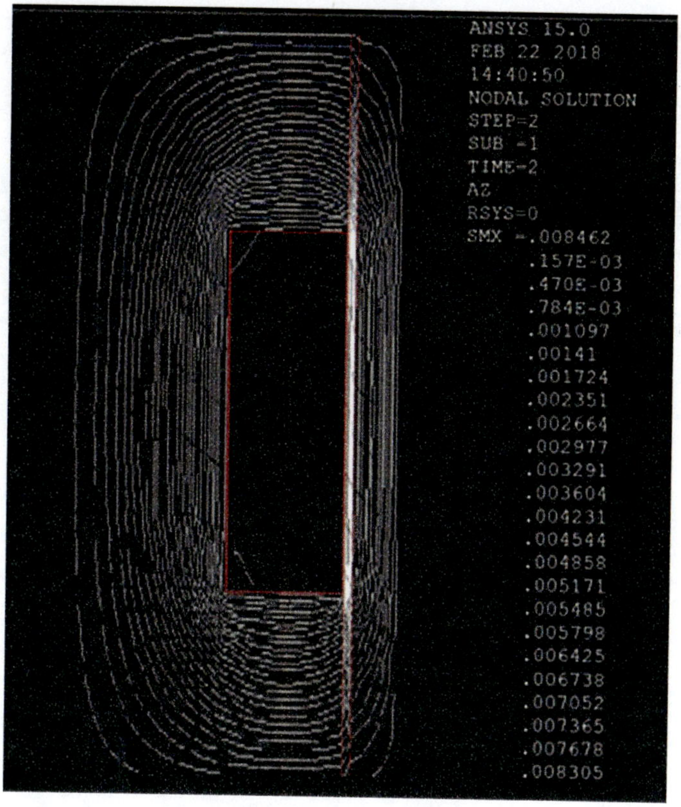

Fig. 10 Axisymmetric of MR damper in ANSYS

Fig. 11 Complete assembly of damper

actuator to the bottom part of the cylinder, and the test rig is designed in such a way that the total length remains constant. Two IR sensors are placed at a fixed distance on the GI pipes, and an IR emitter is placed at the middle of the cylinder. When the emitter crosses the first sensor and emits radiation, the time $t1$ for the respective fixed position of sensor 1 is noted. Similarly, when it moves upwards and crosses the second sensor, time $t2$ for the fixed position of sensor 2 is noted down. In this way, we get the change in position with respect to time and velocity is calculated. The purpose of the load cell is to get load time history. Firstly, the values of the load time graphs are obtained in LABVIEW, and later, they are exported to MATLAB. The final test rig assembly is shown in Fig. 12, and the electrical connections are shown in Fig. 13.

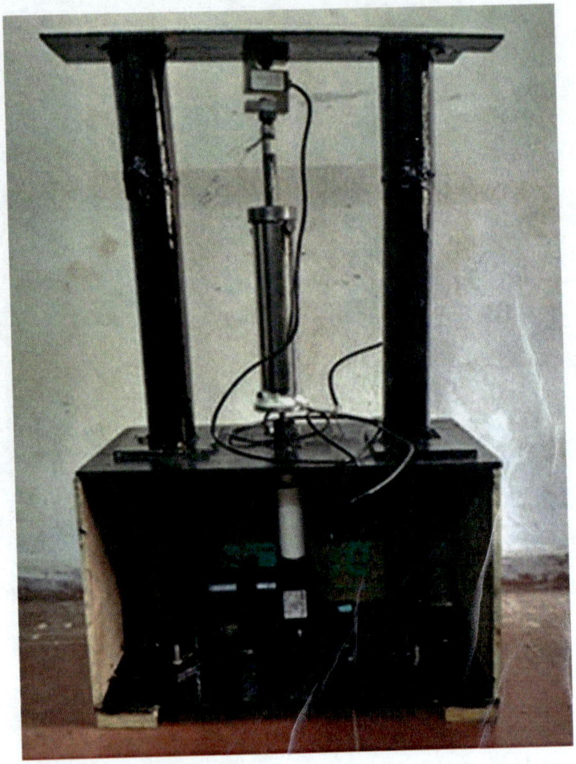

Fig. 12 Test rig set-up for MR damper

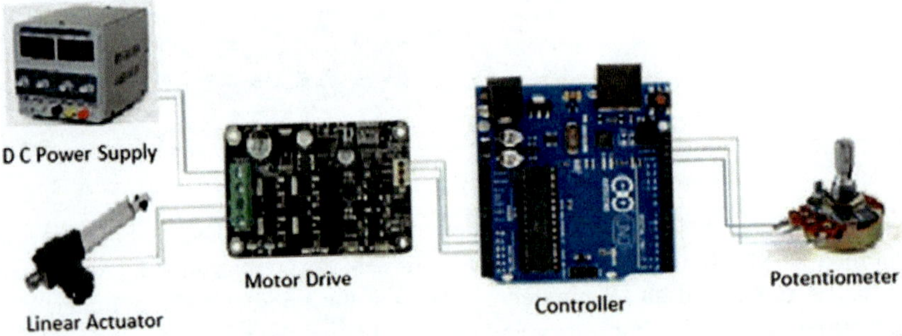

Fig. 13 Electrical circuit for controlling linear actuator

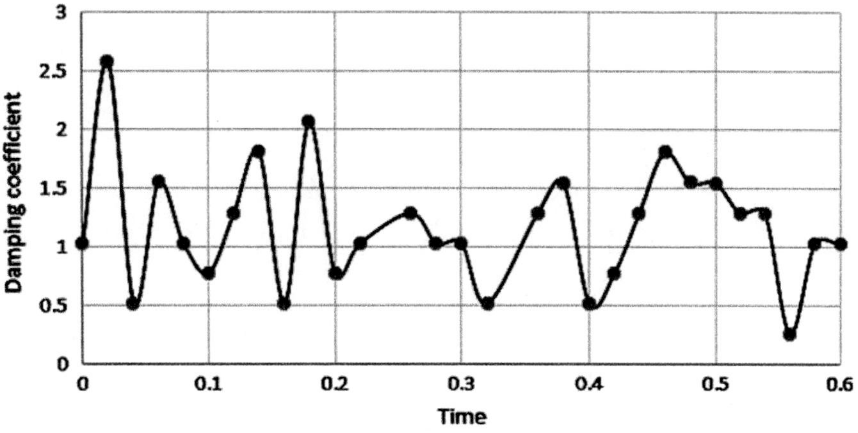

Fig. 14 Variation of damping coefficient with time at 1 A current

6 Results and Analysis

After fabrication of the components, they are assembled and electrical connections are made. When the current is increased, the MR fluids change its state from fluid to semi-solid. The amount of damping is function of state of the fluid velocity is obtained by knowing the distance travelled and the time taken to travel the distance. Force data are read from the DAQ output, and damping coefficient is calculated as function of the time.

In MR dampers, the damping coefficient will not be constant but the function of the field current. Figures 14, 15 and 16 show the variation of damping coefficient as a function of time.

7 Conclusions

The synthesized silicon oil-based MR fluid was tested in fabricated MR damper with test rig. The damping coefficient varies in accordance with the intensity of the magnetic field, thus providing the required damping force. For a MR fluid damper, vibration or damping of the system can be controlled just by controlling the magnetic field strength. As magnetic field strength value completely depends on the current supplied, keeping all the properties constant flexibility in vibration or damping of the system can be obtained just by controlling the current.

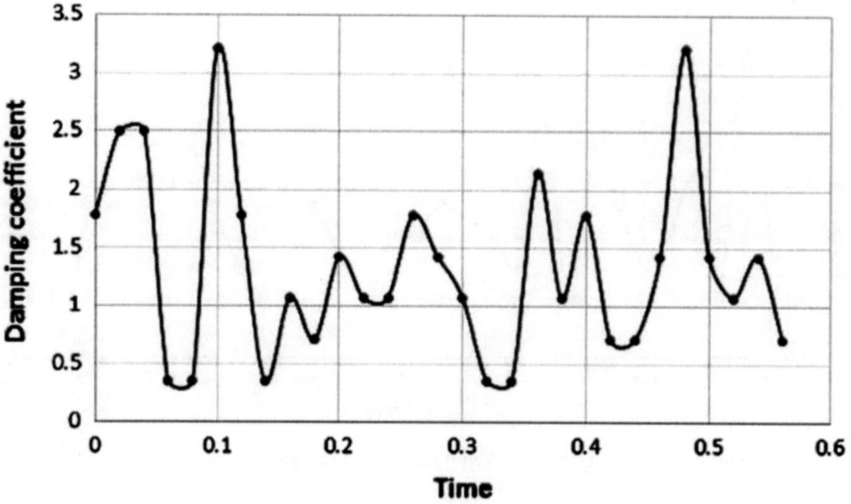

Fig. 15 Variation of damping coefficient with time at 1.2 A current

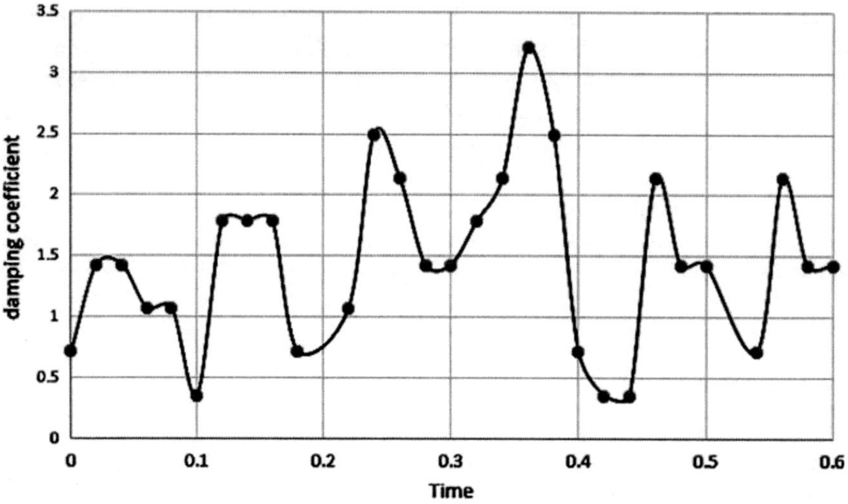

Fig. 16 Variation of damping coefficient with time at 1.4 A current

References

1. Lijesh KP, Kumar D, Hirani H (2017) Effect of disc hardness on MR brake performance. Eng Fail Anal 74:228–238
2. Jolly MR, Carlson JD, Munoz BC (1996) A model of the behaviour of magnetorheological materials. Smart Mater Struct 5(5):607
3. Poznić A, Zelić A, Szabó L (2012) Magnetorheological fluid brake–basic performances testing with magnetic field efficiency improvement proposal. Hung J Ind Chem 40(2):107–111

4. Khan SA, Suresh A, Seetharamaiah N (2014) Principles, characteristics and applications of magneto rheological fluid damper in flow and shear mode. Procedia Mater Sci 6:1547–1556
5. Kim MS, Liu YD, Park BJ, You CY, Choi HJ (2012) Carbonyl iron particles dispersed in a polymer solution and their rheological characteristics under applied magnetic field. J Ind Eng Chem 18(2):664–667
6. Wang D, Zi B, Zeng Y, Hou Y, Meng Q (2014) Temperature-dependent material properties of the components of magnetorheological fluids. J Mater Sci 49(24):8459–8470
7. Kumbhar BK, Patil SR (2014) A study on properties and selection criteria for magneto-rheological (MR) fluid components. Int J ChemTech Res 6:3303–3306
8. Bier R, Huey H, Associates V, Alto P (1984) Design considerations for magnetorheological brakes, Computer (Long Beach Calif) 19(5):834–837
9. Sarkar C, Hirani H (2013) Theoretical and experimental studies on a magnetorheological brake operating under compression plus shear mode. Smart Mater Struct 22(11):115032

Design and Fabrication of a Test Rig for Performance Analysis of a Pneumatic Muscle Actuator

M. Yashas, Antonio Dylan Do Rosario Carvalho, P. Navin Karanth, and Vijay Desai

Abstract Artificial pneumatic muscle actuator (PMA) is used to convert pneumatic power to mechanical force. These actuators are like biological muscles, hence its name. PMAs have a wide range of applications ranging from robotics, industrial to the medical applications, as a result of its high power-to-weight ratio, nonhazardous, and compliant nature. The PMA, like any other actuator, has to be tested and validated before using it in any real-world application. An experimental procedure is set up to allow the designer to predict, analyze, and optimize PMA performance before its use in sensitive applications. To achieve this, a rigid experimental test rig needs to be designed to test properties of PMA such as displacement, force, and load capacity. Result of the experiment conducted on the standard PMA by using developed test rig has 5% variation with the values of the manufacturer's datasheet.

Keywords Pneumatic artificial muscle · Actuator · Test rig · Festo fluidic muscle

1 Introduction

PMAs differ from traditional pneumatic actuators and thus can be used in a wide range of applications such as mobile, anthropomorphic, bionic and humanoid robots; rehabilitation and physiotherapeutic applications; as well as for automation of manufacturing processes. The principle of operation makes PMA as a single-acting pneumatic actuator. Other than PMAs, it has other names which are fluidic muscle [1], pneumatic artificial muscle [2], air muscle [3], axially contractible actuator [4], tension actuator [5, 6], fluid actuator [7].

The PMA has a long history with few accomplishments in different applications. McKibben was the first to develop rubber-tube-type PMA for an artificial limb. McKibben muscle made from an elastic rubber bladder as the inner material and the outer material is a nonextendable double-helix-woven sheath or called braided mesh

M. Yashas (✉) · A. D. Do Rosario Carvalho · P. Navin Karanth · V. Desai
Mechanical Engineering Department, National Institute of Technology Karnataka, Surathkal, India
e-mail: yashas.173mc004@nitk.edu.in

A. Arockiarajan et al. (eds.), *Advances in Industrial Automation and Smart Manufacturing*, Lecture Notes in Mechanical Engineering,
https://doi.org/10.1007/978-981-15-4739-3_3

shell. The PMA was redesigned in the 1980s and ended up being even more suitable with improvised capability. Bridgestone Pvt. Ltd. made use of PMA for painting applications as a part of the industry and a couple of other forms as helping disabled individuals [12].

Pneumatic muscles are preferred compared to pneumatic cylinders due to their high power-to-weight ratio, elastic structure, variable ability to install, least mechanical wear flexible dimensions, low cost, and high reliability of humans. PMA has its application in various fields: lifting load assists, tab punching, robotic prosthesis, wearable exoskeleton robot, or jumping robot (See Fig. 1).

The PMAs made of a thin rubber tube (bladder) are wrapped around with a braided mesh, and both the ends are completely sealed, one end is for inlet connection, and another one is for load attachment. When pressurized air is applied to the inlet of the PMA, the internal bladder will make a way to increase its volume against the inextensible natured braided mesh shell causing the actuator to shorten in its length and producing a pulling action when a load is attached. A new design of PMAs has been introduced to the market by Festo which is wrapped with an airtight, adjustable

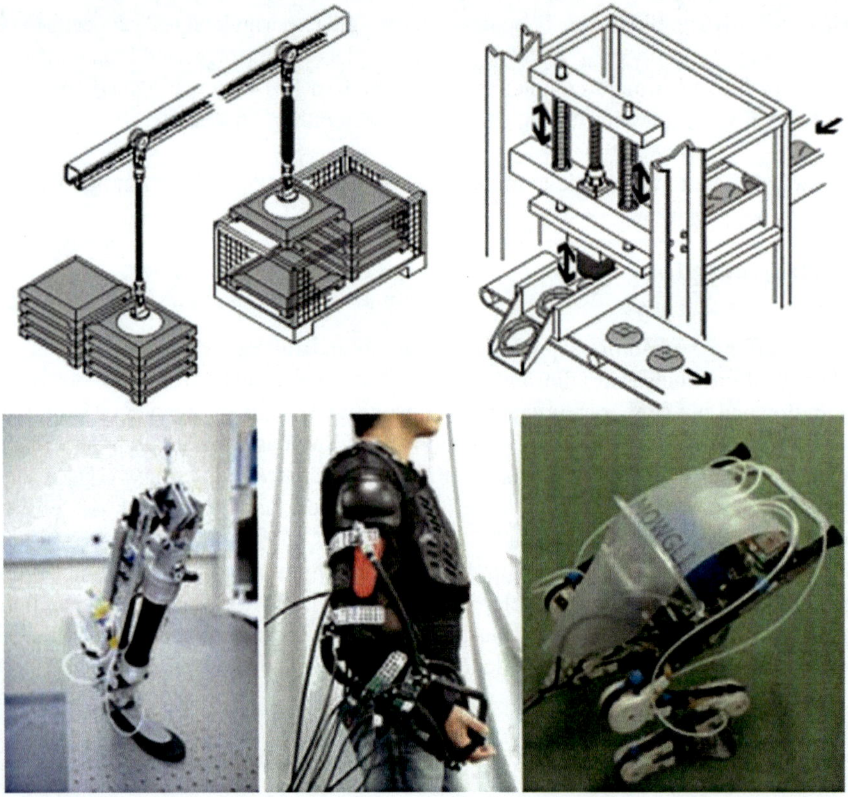

Fig. 1 Various applications of PMAs [25]

hose with a rhomboidally arranged nonelastic fibers. The three-dimensional patterned result and the grid pattern are deformed when the pressurized air sent inside the PMA. A pulling action takes place in the axial direction; this results in the shortening of the PMA as the pressure is increased in the inlet supply [15].

A pneumatic muscle is fabricated, and the static characteristics of it were tested and verified with the standard parameters. They have used a very compact control mechanism for the actuator, which does not even contain external sensors, pneumatic regulator, or direction control valves [8]. PMAs are generally used and mainly considered as active actuators, but their use in the passive application is very less. The improvement and validation of the stiffness model were done to help in applying PMA as a passive component on assisting devices without taking internal muscle pressure and muscle geometric parameters into consideration [9]. A novel smart PMA with a pressure sensing capability was proposed and developed. The fiber sleeve used in the actuator with conductive properties has been an essential element of this work. Therefore, the fiber sleeve works as both a sensor for testing displacement and also as an actuator element [10]. The considering requirements of a flexible actuation system and applied to the untethered robots. The braided pneumatic actuator is wholly studied, and experimental testing was performed to show the similar traditional benefits matching the pneumatic cylinder [11].

McKibben artificial PMA is the first developed, tested, and modeled PMA. It is made up of an elastic bladder wrapped around a braided cord and with its end fitting for the air inlet. The results were compared with human muscle properties to evaluate the suitability of McKibben actuators for human muscle emulation in biologically based robot arms [12]. The design of a traditional McKibben muscle was modified to include an inductive coil surrounding the muscle fibers. These new design measures coil inductance change to determine the contraction force and displacement, the application of this process to the variety of existing McKibben actuator designs [13]. A comparative study on performance is particularly from a dynamic and energetic point of view when using water and air as actuation means for the control of the fluidic muscle. Theoretical and experimental analyses on a dedicated test rig were carried out to assess the assumptions [14]. A designed test stand was used to know the static—isobaric, isotonic, and isometric characteristics of two pneumatic muscles: a Festo Fluidic Muscle MAS and a Shadow Air Muscle by using different methods and finally experimental properties compared with the simulated ones [15]. PMA was fabricated and tested for its mesh suitability, the effect of force on actuator contraction. Initial experimental results are used to verify the contraction percentage for various loads [16].

A flexible deflection measurement sensor utilizes a conductive fiber that is installed on the surface of the artificial muscle [17]. The potentiometer is directly inserted into the rubber tube for the purpose of measuring the deflection [18]. For measuring the expansion in the axial and radial direction of the muscles, McKibben artificial muscle is incorporated with a photo displacement sensor [19]. Elastomer strain sensor is used for detecting artificial muscle expansion quantity [20]. On an artificial muscle, a strain gauge is mounted, which can be used for detecting the muscle deformation [21]. To measure the axial deformation of the artificial muscle, a flexible electro-conductive rubber sensor was developed [22]. A rubber sensor with a conductive film in it is mounted on the McKibben artificial muscle, by the measurement of electrical resistance which gives the expansion quantity of the muscle [23]. Dielectric elastomer sensor with a contact with the inner surface is incorporated in the development of self-sensing artificial muscle for the deflection measurement [24]. This paper consists of four sections. After the introduction, in Sect. 2 methodology, the experimental test rig and measurements will be discussed. Section 3 presents the experimental outcomes and comparisons of fluidic muscles. Finally, Sect. 4 gives you the conclusion of the experimental work.

2 Methodology

All the investigation and experimental measurements are performed on the PMAs for analyzing the static characteristics of it like force and displacement during the contraction of the muscles. The test rig is designed mainly for checking the force and displacement characteristics of the muscle. The PMA that we are going to investigate is by Festo (DMSP-10-100 N–RM–RM) where the first number defines the inside diameter in mm, the second number defines the nominal length in mm, and RM means radial pneumatic connection at both ends.

2.1 Experimental Test Rig Setup for Displacement Measurement

For the displacement measurement, Fig. 2 shows that fluidic muscle is mounted vertically in the test rig during the experiment. One end of the pneumatic muscle is fixed while the other end is allowed to move freely with the linear bearing and the shaft rod.

Experimental test rig setup for displacement measurement consists of the following components and devices:

- 24 V DC regulated power supply.
- FLR unit.
- Air compressor–with 8 bar pressure supply.
- Laser sensor by KEYENCE IL series—IL200.

The test rig visually looks like a cuboid in shape with a length of 450 mm, the width of 190 mm, and height of 110 mm. It is made up of aluminum sections. Keyence IL200 series laser sensor has 280 mm sensing range at a distance of 300 mm from the target with a 160 mm dead zone range [26]. The top portion of the test rig is where the laser sensor is mounted. This laser gives the accurate measurement of displacement of the contracted PMA.

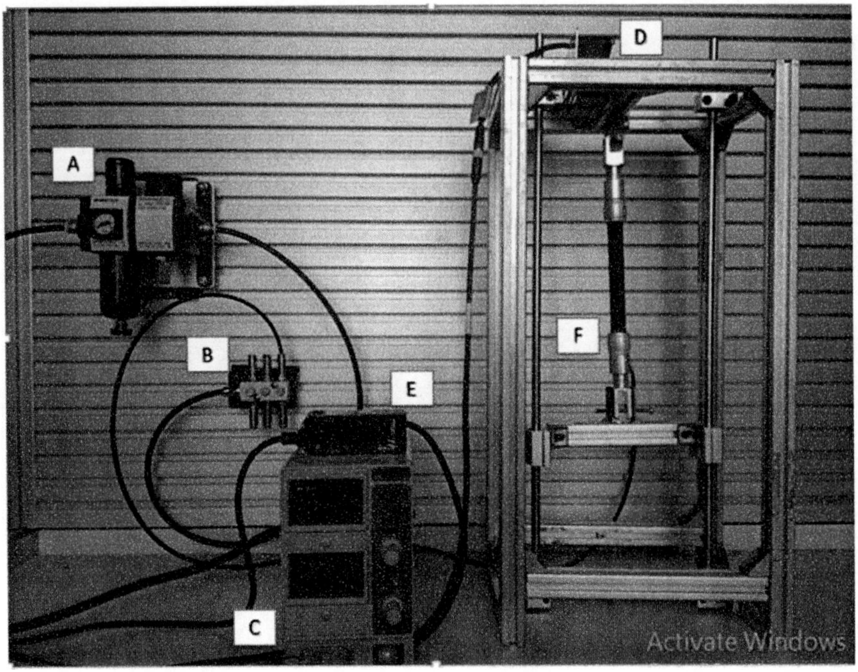

Fig. 2 Experimental setup for force measurement (A: FLR unit, B: distributer unit, C: 24 V DC supply, D: laser displacement sensor, E: IL 1000 display unit, F: pneumatic muscle by Festo)

Experimental procedure for the Displacement measurement. Primarily, we need to make sure that sufficient pressure is there in the compressor by verifying with the help of pressure gauge, completely release the pressure from the muscle, and record the distance indicated by the laser sensor. If it is not zero initially or you can shift it to zero, then start. Gradual increment of pressure in steps of 0.5 bar and distance are recorded till 7.5 bar and then decreased in steps of 0.5 bar till zero. The displacement measurement involved five experimental trials performed two commercially available muscles from Festo, both are of the same specification as mentioned above. Another five sets of displacement measurement were recorded by hanging a mass of 1027 gms. Results are then plotted and compared with the values of the manufacturer's datasheet, and a comparative analysis was conducted to verify whether both muscles are behaving the same. Contraction percentage can be achieved by using formula as shown in Eq. 1.

$$\text{Contraction }\% = \frac{\text{Displacement}}{\text{Original length}} \times 100\% \quad (1)$$

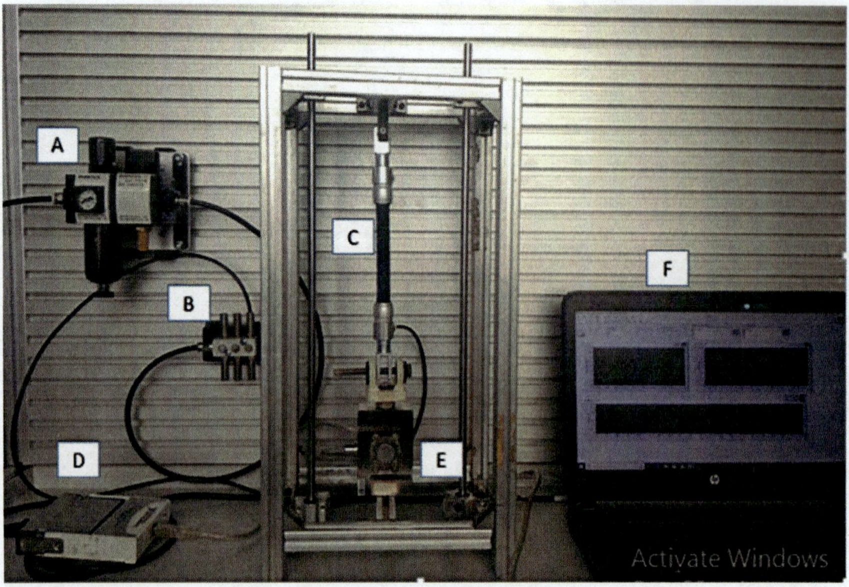

Fig. 3 Experimental setup for force measurement (A: FLR unit, B: distributer unit, C: pneumatic muscle by Festo, D: NI 9237 DAC, E: HBM S9M force transducer, F: PC with LabView)

2.2 *Experimental Test Rig Setup for Force Measurement*

For the force measurement, Fig. 3 shows that fluidic muscle is mounted in a similar vertical way as mounted for displacement measurement. One end of the muscle is made fixed, but another end is not allowed to slide anymore; it is connected to the force transducer which has knuckle joint kind of arrangement at the top of it.

Experimental test rig setup for displacement measurement consists of the following components and devices:

- FLR unit.
- Air compressor—with 8 bar pressure supply.
- Load cell by HBM (S9M/2KN).
- NI Hi-speed USB carrier.
- NI 9237 DAC.

For the force measurement of the PMAs, same test rig has been used with minor changes in the design. We have used HBM S9M series force transducer with 2KN capacity with a sensitivity of 2 mV/V. The mounting of the force transducer is between the base of the rig and the free end of the PMA. Data from the force transducer will be shown in the front panel of the LabView program through NI9237AC. Figure 4 shows the front panel of the LabView program. When neglecting the membrane's material deformation and the low inertial muscle properties, the generated force is expressed as (Eq. 2).

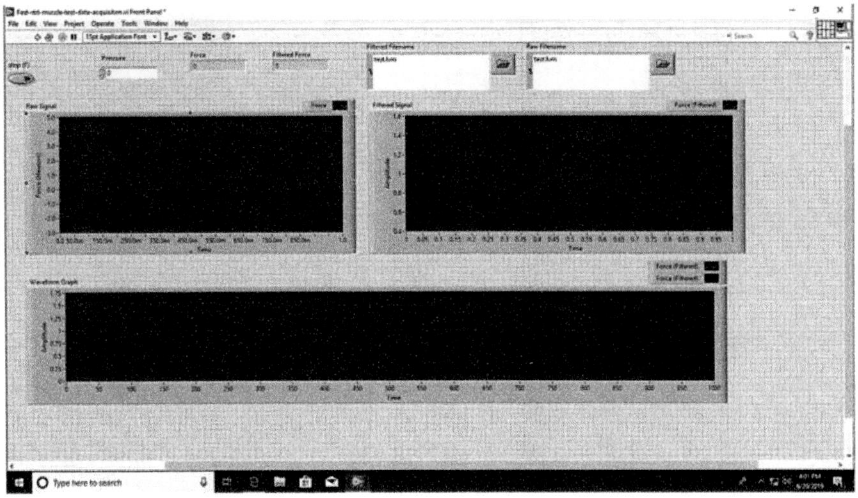

Fig. 4 Front panel of the LabView program

$$F = -P\frac{\mathrm{d}V}{\mathrm{d}l} \tag{2}$$

with 'p' the gauge pressure inside the muscle, dV the enclosed muscle volume changes, and dl the actuator length changes.

Experimental procedure for the Force Measurement. The following of same initial steps for the displacement measurement is done here, gradual increase in the pressure in steps of 0.5 bar and force is noted from the front panel of the LabView Program till 7.5 bar and then decreased in steps of 0.5 bar till zero. Five sets of force measurement were recorded for the two standard muscles from Festo. Results are then plotted and compared with the values of the manufacturer's datasheet, and comparison among the two muscles is to verify whether both muscles are behaving the same.

3 Experimental Results and Discussion

To study the characteristics such as force and displacement (contraction), two standard pneumatic muscles from Festo (DMSP-10-100 N-RM-RM) of 10 mm diameter and 100 mm length were used. As the pneumatic muscle actuator is nonconventional, this study aims at comparing two standard muscles of similar properties (such as length and diameter) and studying the error percentage for various inputs. As per the discussion in the previous section, the force experiment involved five trials. In each trial, the pressure is gradually increased in steps of 0.5 bar from 0 to 7.5 bar and gradually decreased from 7.5 to 0 bar.

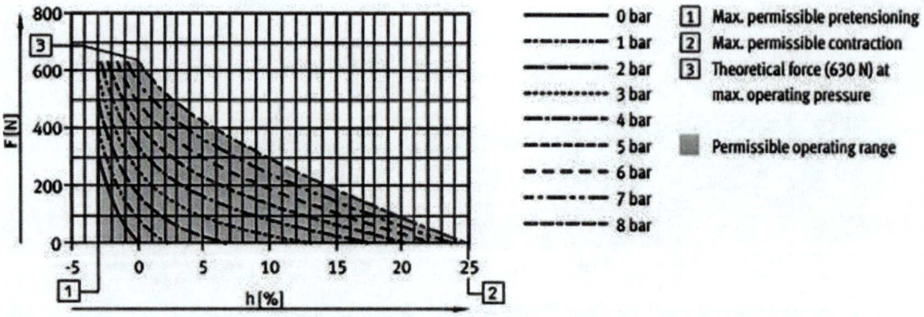

Fig. 5 Force–contraction function of DMSP-10-100 N [27]

The displacement experiment also involved five trials but was conducted for load and no load conditions (1027 gms), with each trial involving a gradual increment in pressure. As Fig. 5 shows standard data collected from the Festo catalog, it has both force and contraction percentage (h %) plotted for 0 to 8 bar.

Figure 6a, b shows average values of the displacement of the PMA plotted against pressure. From the plots, we notice the difference in the displacement for a specific value of pressure during contraction and retraction. It is evident that for a particular pressure, there is a small deviation in the displacement values during contraction and retraction cycles. In muscle 1, maximum deviation of displacement occurred at 3 bar with a 3.7 mm error, and minimum deviation of displacement occurred at 7.5 bar with a 0.3 mm error; and in muscle 2, maximum deviations of displacement occurred at 3 bar with a 4.46 mm error, and minimum deviation of displacement occurred at 7.5 bar with a 0.58 mm error.

Figure 7a, b shows average repeatability error percentage plotted against the pressure values. From the plots, we notice that the error value in both the muscles spikes suddenly at 1 bar pressure and then starts decreasing as the pressure increases. This error is a result of large deviations at lower pressures. In muscle 1, the maximum

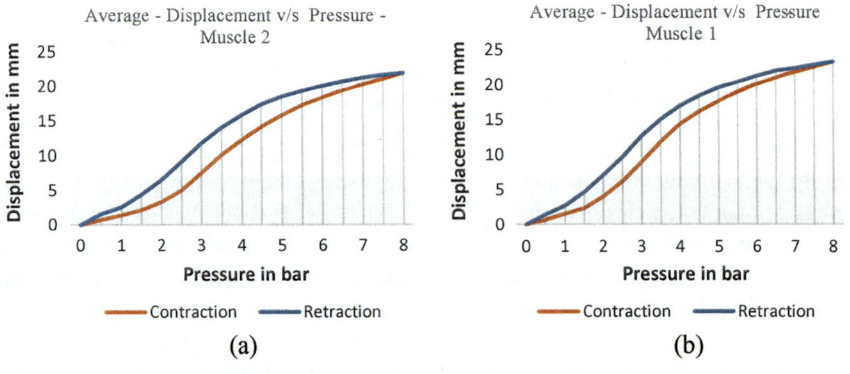

Fig. 6 Average displacement versus pressure plot (without weight) **a** muscle 2 **b** muscle 1

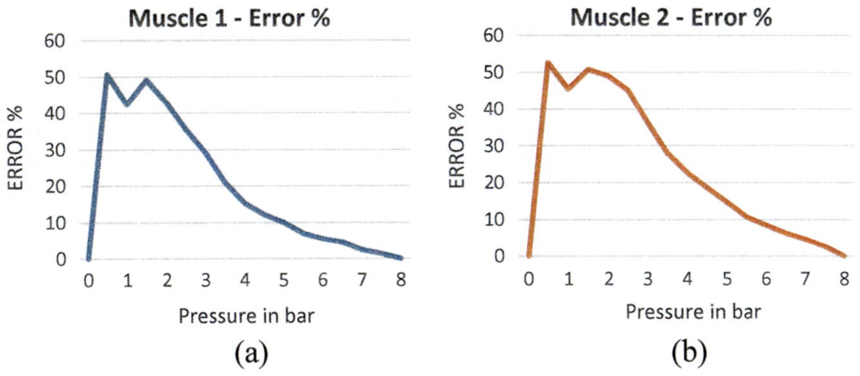

Fig. 7 Average repeatability error % versus pressure plot (without weight) (a) muscle 1 (b) muscle 2

percentage of error is at 0.5 bar (50.6%), and minimum percentage of an error is at 7.5 bar (1.3%). In muscle 2, maximum percentage of an error occurs at 0.5 bar (52.5%), and minimum percentage of an error occurs at 7.5 bar (2.6%). From the figures, we can analyze that slope of error drop in muscle 1 is steeper when compared to muscle 2, but initial neck portion of the graphs looks similar to each other.

Figure 8 shows all the displacement values during contraction and retraction of both the muscles with a weight of 1027gms. The plot depicts that both the muscles behave in the same way. Repeatability errors can be a result of the difference in the displacement value for the same pressure while contracting and retracting the actuators. Muscle 1 shows you the maximum 3.54 mm difference in displacement at 3 bar pressure and a minimum 0.26 mm difference in displacement at 7.5 bar pressure, while muscle 2 shows a maximum difference of 3.86 mm displacement at

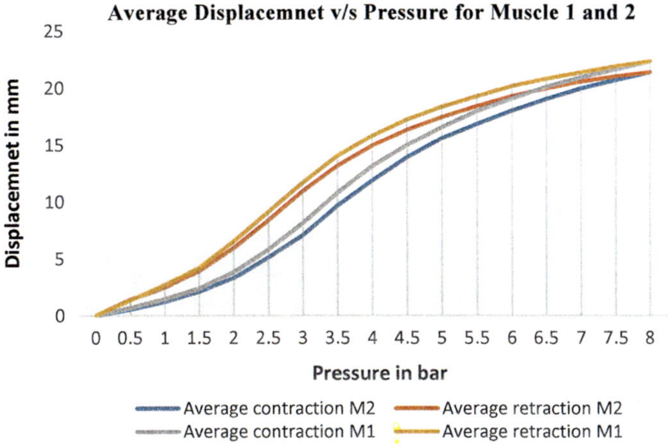

Fig. 8 Average displacement versus pressure plot (with weight)

3 bar pressure and a minimum of 0.26 mm difference in displacement at 0.32 bar pressure.

Figure 9 shows the repeatability error percentage in both the muscles. When we compare the behavior of the muscles, the maximum error percentage is seen in muscle 2 throughout the pressure range (0–8 bar). Muscle 2 has a maximum error of 60.8% at 0.5 bars and a minimum error of 1.5% at 7.5 bars, while muscle 1 has a maximum error of 48.4% at 0.5 bars and a minimum error of 1.1% at 7.5 bars. The error percentage plot of the muscle without load indicates that muscle 1 is showing good results with an average error reduction of 4.2%, while muscle 2 is showing an average 2.8% increase in error.

Figure 10a, b depicts the average force generated for different pressure conditions. Five sets of trails were conducted on both the muscles in steps of 0.5 bar from 0 to 7.5 bars. Muscle 1 contraction curve is slightly smoother than the retraction curve, but in the muscle 2, contraction curve is initially smoother than the retraction curve.

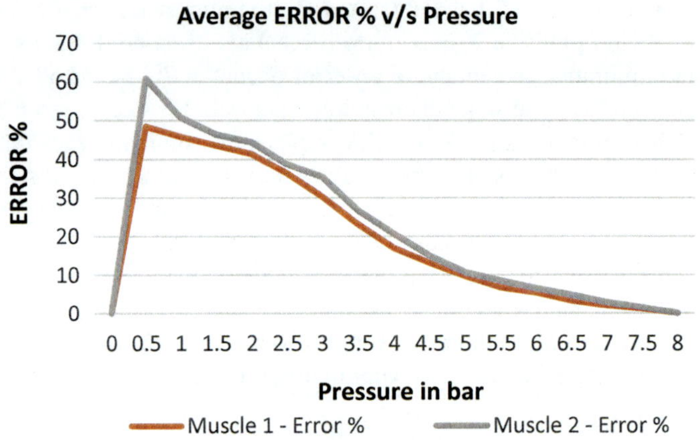

Fig. 9 Average repeatability error % versus pressure plot (with weight)

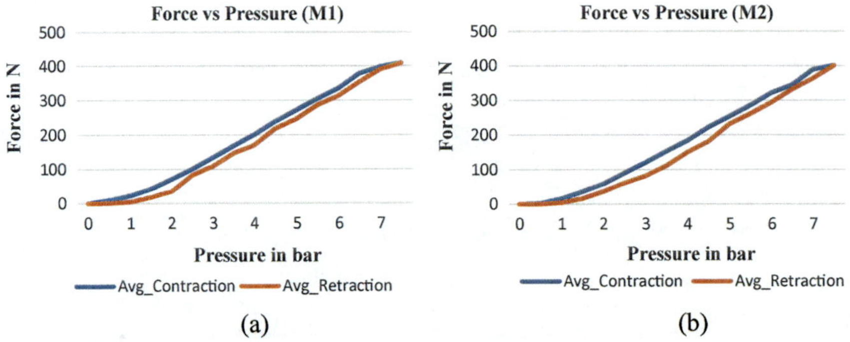

Fig. 10 Average force versus pressure plot **a** muscle 1 **b** muscle 2

As the pressure increases, the smoothness of the curve degrades (at 6 bar). There is a lack of repeatability of force value for a specific pressure in both muscles during contraction and retraction cycles. A deviation in force is seen throughout. Muscle 1 shows a maximum force deviation of 30.6 N at 4 bar, and muscle 2 shows a maximum force deviation of 41.7 N at 3.5 bar. Muscle 1 shows an average force deviation of 20.2 N, and muscle 2 shows an average force deviation of 23.5 N.

From Fig. 11a, we notice a sudden spike of force error reading at the initial stage of pressure rise and gradually decrease as the pressure increases further. The plot shows 75.8% (maximum) force error at 0.5 bar pressure and 2.17% (minimum) force error at 7.5 bar pressure. In Fig. 11b, the same kind of error spikes can be seen initially during retraction of muscles, but sudden fall of error percentage to zero can be depicted here in this plot. The plot shows 68.46% (maximum) force error at 0.5 bar pressure and 2.17% (minimum) force error at 7.5 bar pressure. Figure 11c shows the percentage of error in muscle 1 during its contraction and retraction, the same kind of behavior is seen as mentioned above about Fig. 11a, the slope of this is steeper, and then, plot is shown in Fig. 11d. A maximum force error of 89.8% at 0.5 bar

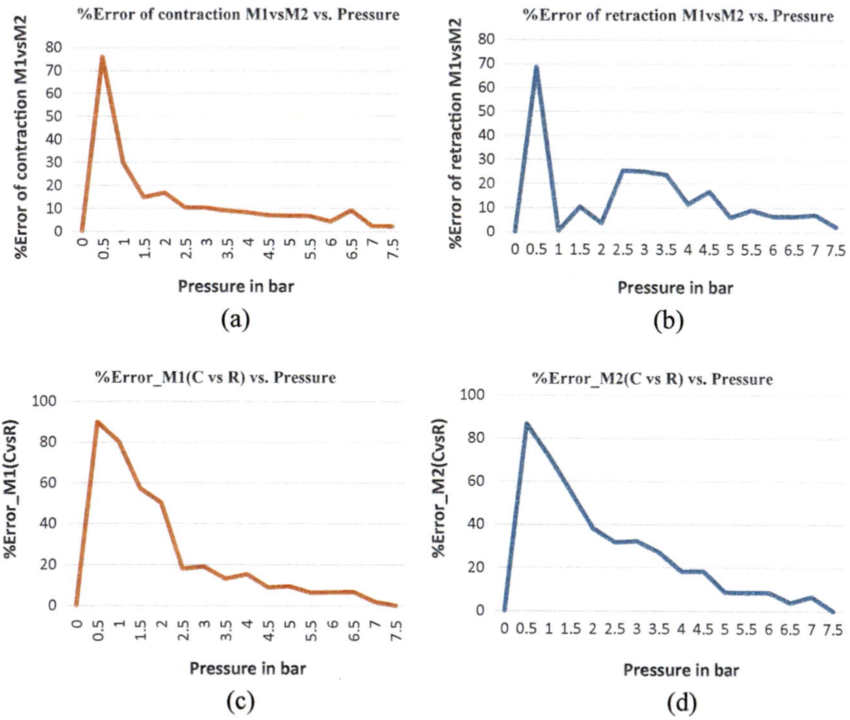

Fig. 11 Average force error percentage vs pressure **a** comparison between muscles for force error percentage generated during contraction, **b** comparison between muscles for force error percentage generated during contraction, **c** comparison of force error percentage generated during contraction and retraction in muscle M1, **d** comparison of force error percentage generated during contraction and retraction in muscle M2

pressure and 1.83% (minimum) force error is seen at 7.5 bar pressure (See Fig. 11c). The plot in Fig. 11d shows 86.72% (maximum) force error at 0.5 bar pressure and 6.63%(minimum) force error at 7.5 bar pressure.

4 Conclusion

The significance of PMA in certain situations is greater than standard pneumatic cylinders. The applications of PMA are widely spread from a biomedical to an industrial automation sector. Testing and evaluation of PMA need(s) to be performed before we use it in any practical applications.

This paper focuses on the fabricated test rig used for testing static characteristics of two commercially available muscles of the same specifications. The main reason for choosing two muscles of the same model was to study its repeatability characteristics.

- From the results of the experiments conducted, we can infer that the values of force and displacement at various pressures have a slight deviation during the contraction and retraction cycles, thus indicating an error of repeatability in PMA.
- The value of force and deflection at a particular pressure for muscle M1 and M2 also differs, thus indicating a repeatability error between two standard muscles of the same specification. From the study, it is also evident that the error percentage is high at a lower value of pressure and decreases gradually with increasing pressure.

Lastly, we can conclude that the test rig designed was very flexible to conduct a number of successful experiments on PMA for their static properties.

References

1. Andrikopoulos G, Nikolakopoulos G, Manesis S (2011) A survey on applications of pneumatic artificial muscles. In: 2011 19th mediterranean conference on control & automation (MED), IEEE, pp 1439–46, Jun 2011
2. Caldwell DG, Razak A, Goodwin M (1993) Braided pneumatic muscle actuators. IFAC Proc Vol 26(1):522–527
3. Chou CP, Hannaford B (1996) Measurement and modeling of McKibben pneumatic artificial muscles. IEEE Trans Robot Autom 12(1):90–102
4. Daerden F, Lefeber D (2002) Pneumatic artificial muscles: actuators for robotics and automation. European J Mech Env Eng 47(1):11–21
5. Daerden F, Lefeber D, Verrelst B, Van Ham R (2001) Pleated pneumatic artificial muscles: compliant robotic actuators. In: Proceedings on IEEE/RSJ International Conference on Intelligent Robots and Systems, IEEE, vol 4, pp 1958–1963
6. Doumit M, Leclair J (2017) Development and testing of stiffness model for pneumatic artificial muscle. Int J Mech Sci 120:30–41
7. Erin O, Pol N, Valle L, Park YL (2016) Design of a bio-inspired pneumatic artificial muscle with self-contained sensing. In: 2016 38th Annual International Conference of the IEEE Engineering in Medicine and Biology Society (EMBC), IEEE, pp 2115–2119

8. Focchi M, Guglielmino E, Semini C, Parmiggiani A, Tsagarakis N, Vanderborght B, Caldwell DG (2010) Water/air performance analysis of a fluidic muscle. In: 2010 IEEE/RSJ International Conference on Intelligent Robots and Systems, IEEE, pp 2194–2199
9. FestoHomepage https://www.festo.com/cat/enus_us/data/doc_enus/PDF/US/DMSPMAS_ENUS.PDFFluidicMuscleDMSP/MAS.Festoproductcatalog, p 36, Last accessed 20 Jun 2019
10. Gonzalez C (2015) What's the difference between pneumatic, hydraulic, and electrical actuators? Machine design
11. Hamamoto I, Akagi T, Dohta S (2006) Development of a flexible displacement sensor using nylon string coated with carbon and its application for McKibben actuator. In: Proceedings on SICE-ICASE international joint conference, pp 1943–1946
12. Jobbágy B, Šimšík D, Karch J, Onofrejová D (2014) Robotic arm with artificial muscles in rehabilitation. Procedia Eng 96:195–202
13. KeyenceHomepage https://www.controldesign.com/assets/10WPpdf/100208_Keyence_LaserSensor. Last accessed 20 Jun 2019
14. Kang BS, Kothera CS, Woods BKS, Wereley NM (2009) Dynamic modeling of Mckibben pneumatic artificial muscles for antagonistic actuation. In: Proceedings on IEEE international conference on robotics and automation Kobe Japan, pp 182–87
15. Oki K, Saga N, Nagase J (2010) Smart actuator using a pneumatic artificial muscle and a strain gage type bend sensor. In: Proceedings on International Conference Applied Bionics Biomechanics, pp 1–5
16. Laksanacharoen S (2004) Artificial muscle construction using natural rubber latex in Thailand. In: The 3rd Thailand and material science and technology conference, pp 10–11
17. Coulbourne NC, Son S, Fox JW (2007) Self-sensing McKibben actuators using dielectric elastomer sensors. In: Proceedings on Electroactive Polymer Actuators Development
18. Najmuddin WSWA, Mustaffa MT (2017) A study on contraction of pneumatic artificial muscle (PAM) for load-lifting. J Phys Conf Series, IOP Publishing 908(1):012036
19. Olympus Corporation, Japan Patent Kokai, 2010-155283, 2010-07-15 (in Japanese)
20. Kuriyama S, Ding M, Kurita Y, Ogasawara T, Ueda J (2009) Flexible sensor for McKibben pneumatic actuator. In: Proceedings on IEEE Sensors, pp 520–525
21. Wakimoto S, Suzumori K, Kanda T (2005) Development of intelligent McKibben actuator. Proc Intell Robot Syst 487–492
22. Sárosi J (2012) New approximation algorithm for the force of fluidic muscles. In: 2012 7th IEEE international symposium on applied computational intelligence and informatics (SACI), IEEE, pp 229–233
23. Takosoglu JE, Laski PA, Blasiak S, Bracha G, Pietrala D (2016) Determining the static characteristics of pneumatic muscles. Meas Control 49(2):62–71
24. Akagi T, Dohta S, Kenmotsu Y, Zhao F, Yoneda M (2012) Development of smart inner diameter sensor for position control of McKibben artificial muscle. Proc Int Symp Robot Intell Sens 10:105–112
25. Wang H, Totaro M, Beccai L (2018) Toward perceptive soft robots: progress and challenges. Adv Sci 5(9):1800541
26. Wakimoto S, Misumi J, Suzumori K (2016) New concept and fundamental experiments of a smart pneumatic artificial muscle with a conductive fiber. Sens Actuators A 250:48–54
27. Park YL, Chen B, Majidi C, Wood RJ, Nagpal R, Goldfield E (2012) Active modular elastomer sleeve for soft wearable assistance robots. In: Proceedings on International Conference on Intelligent Robotic Systems, pp 1595–1602

Fracture Analysis of C-Stringer and Hat Stringer on the Load Carrying Vehicle

B. Stalin, V. Dhinakaran, M. Ravichandran, K. Sathiya Moorthi, and J. Vairamuthu

Abstract In this paper, chassis frame stringer's fracture analysis in TATA 2516 TC truck is sited. To save the shock, twist and other loads, the chassis must be a high strengthened one. After that load, the chassis should hold out against fracture behaviour. The work performed on chassis stringer's fracture analysis with the properties of stiffness, strength by use of Ansys workbench software. Reports are showing that hat stringer has more strength than C-shaped stringer during fracture conditions.

Keywords Mechanical properties · FEA · Stringer · Fracture

1 Introduction

The chassis is a major part of the vehicle to compensate the entire load of the vehicle and goods to be supported. The chassis is supported by cross bars which is one type of stringer component. The transient dynamic analysis of the frame is also analyzed to bring on the frequency extraction of the frame. The mechanical properties variability of the aluminium alloys is studied under various parameters like the content of alloying element, the type of treatment, welding and test parameters. A more economical material partial factor value and a uniform reliability factor were

B. Stalin (✉) · K. Sathiya Moorthi
Department of Mechanical Engineering, Anna University, Regional Campus Madurai, Madurai, Tamil Nadu 625 019, India
e-mail: stalin1312@gmail.com

V. Dhinakaran
Department of Mechanical Engineering, Chennai Institute of Technology, Kundrathur, Chennai, Tamil Nadu, India

M. Ravichandran
Department of Mechanical Engineering, K. Ramakrishnan College of Engineering, Samayapuram, Trichy, Tamil Nadu 621 112, India

J. Vairamuthu
Department of Mechanical Engineering, Sethu Institute of Technology, Kariapatti, Virudhunagar Dist., Tamil Nadu, India

A. Arockiarajan et al. (eds.), *Advances in Industrial Automation and Smart Manufacturing*, Lecture Notes in Mechanical Engineering,
https://doi.org/10.1007/978-981-15-4739-3_4

provided a new value of partial values [1]. By use of FEA software, the analysis is done on stayed columns 3D critical buckling loads. The sliding stays modelling need shell elements and friction introduction at the saddle stay interface [2]. Liu et al. [3] have done an experiment that the new type high strength magnesium alloy composition Mg–Zn–Ce–Y–Zr by extrusion and direct ageing. The high strength value in yielding behaviour is 407 Mpa and in tensile is 421 Mpa, respectively. Puttaratorn Ekapun et al. [4] have designed and performed the analysis of an electromagnetic tricycle on the airport. The material chosen for the analysis is aluminium 5182 series (AL51), aluminium 6061 series (AL60) and nickel aluminium bronze 6300 series (C63). The force–displacement graph is developed for the static and dynamic analysis. The stress analysis is carried out on the Pro-Mechanica software and calculated the max stress generated on the chassis. The displacement analysis is also done and the calculated deformation value is in safe condition. The maximum strain value is also found in the analysis. Finally, the shear stress developed on the TC 2516 chassis is also retrieved from the analysis [5]. Ingole et al. [6] have estimated the tractor trailer stress analysis on the exiting trailer with the new modelled trailer configurations with respect to cross-section area. Then, the stress analysis was carried out to the chassis of a tractor trailer. Von-mises stress analysis is done to the chassis with three boundary conditions on the front axle and rear axle. The location of the maximum stress distribution is also found out and maximum deformation is also located [7]. Mordike et al. [8] have studied the properties and the application of the magnesium alloy materials. The specific strength of the magnesium alloy is improved by the Mg–Al alloy fabrication by the die casting process. From that die casting process, the high productivity, high precision and high quality surface can be attained.

2 Methodology

A stringer represents the load withstanding component on the frame. Here, the chassis frame cross member is analyzed in TATA 2516 TC truck against fracture properties. The component taken from the chassis is a C-stringer. We substitute for C-shaped cross member by hat-shaped cross members. The hat stringer under 9.53 mm fillet, 20 mm fillet, 30 mm fillet, 40 mm fillet is drawn by the use of CAD package and analyzed by using Ansys Workbench.

The C-stringer cross section is mentioned in form of 262 mm × 65 mm. The length of the stringer is 2440 mm. The fillet between the edges is 9.53 mm from Fig. 1. The material of the component is 9345 standard S37 steels. The properties of the S37 steel material are attained from the literature Patel et al. [5]. The hat stringer model developed by the cross-section dimensions is height 262 mm, top phase 132 mm and inside edge of 65 mm. The hat stringer length is 2440 mm. The fillet on the edges is 9.53, 20, 30 and 40 mm. The truck capacity is 25 tons and model number is TATA 2516TC. The weight to be applied on the chassis is 245250 N. The capacity factor is 1.25 and then the weight will be 306562 N. These load characteristics are obtained from the literature paper done by Patel et al. [5].

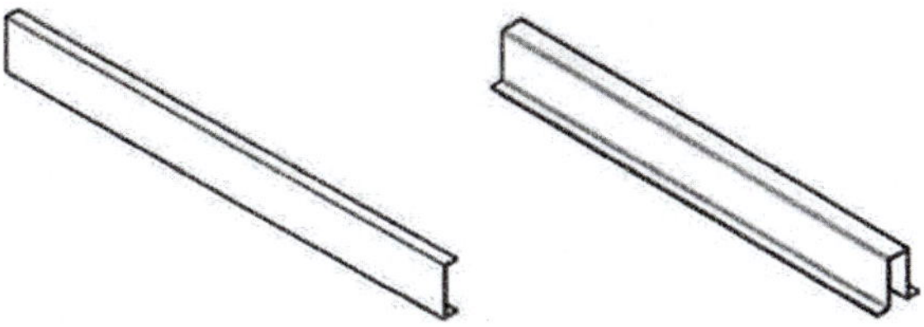

Fig. 1 C-stringer and hat-shaped stringer

3 Results and Discussion

The obtained result parameters are stress intensity factor at three modes and J-integral. The stress concentration factors are stress state near the tip of the crack in opening mode (K_1), sliding mode (K_2), tearing mode (K_3) and the J-integral means that a way to measure the strain energy rate or work per unit fracture area in a material.

Stress intensity factor in opening mode (K_1). The stress intensity value of opening mode for the C-stringer is −180 N/mm in minimum and −1419 N/mm at maximum. These values are obtained in the contour 6 results Fig. 2a. From Fig. 2b, we have observed the stress intensity factor of opening mode for the hat stringer with 9.53 mm fillet as 542.26 N/mm in maximum and 19.9 N/mm at minimum. Figure 2c is representing the stress intensity factor of hat stringer with 20 mm fillet on opening mode as 496.3 N/mm in high and 56.203 N/mm in low. 30 mm fillet hat stringer is getting the stress intensity factor value of the opening made as 394.7 N/mm in peak and 64.34 N/mm in the bottom, which is referred from Fig. 2d. Figure 2e is the solution image of the fracture analysis. From the figure, we have shown that the value of stress intensity factor in opening mode is 284.66 N/mm in large and 55.9 N/mm in small. From the solution images Fig. 2 and the graphically represented in Fig. 3, we

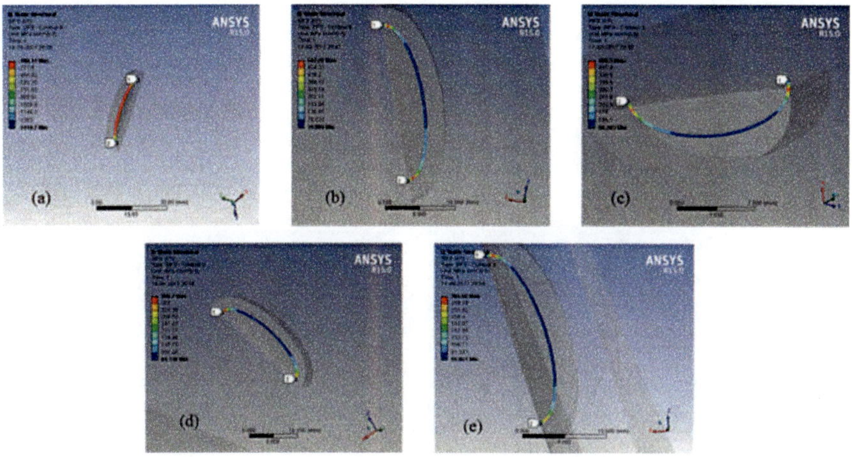

Fig. 2 K_I values for five structures: **a** C-stringer **b** hat stringer with 9.53 mm fillet **c** hat stringer with 20 mm fillet **d** hat stringer with 30 mm Fillet, and **e** hat stringer with 40 mm fillet

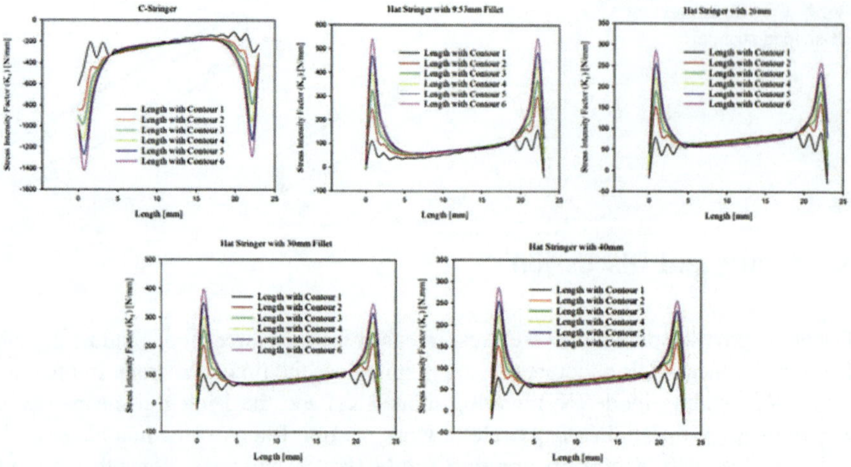

Fig. 3 Stress intensity factor (K_1) for six contours in every model

conclude that the hat stringer with 40 mm fillet is having the lower stress intensity factor values and in close range of zenith and nadir values for the opening mode fracture of analysis.

Stress intensity factor in sliding mode (K_2). The stress intensity value of sliding mode for the C-stringer is 22.85 N/mm in maximum and −340 N/mm at minimum. These values are obtained from the contour 6 results of Fig. 4a. From Fig. 4b, we have observed the stress intensity factor of sliding mode for the hat stringer with 9.53 mm fillet as 6.62 N/mm in maximum and −119.9 N/mm at minimum. Figure 4c

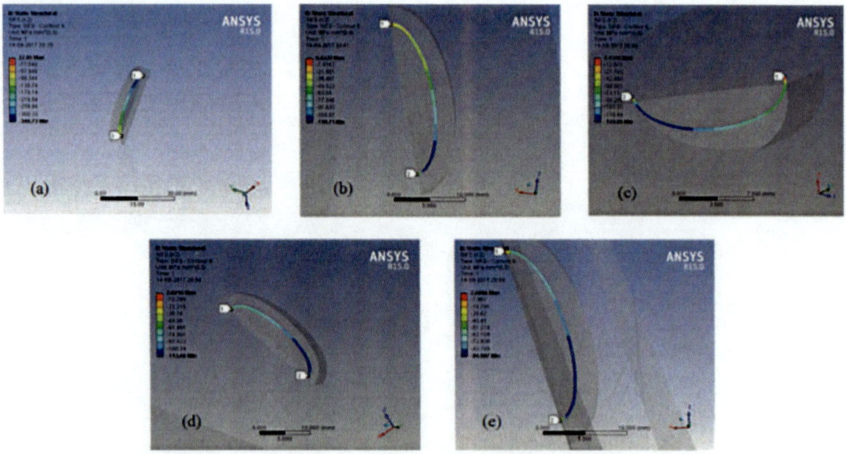

Fig. 4 K_2 values for five structures: **a** C-stringer, **b** hat stringer with 9.53 mm fillet, **c** hat stringer with 20 mm fillet, **d** hat stringer with 30 mm fillet, **e** hat stringer with 40 mm fillet

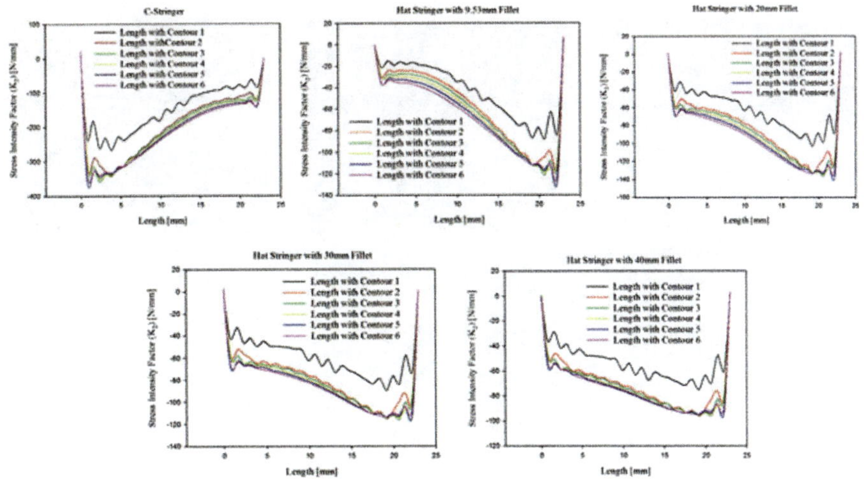

Fig. 5 Stress intensity factor (K_2) for six contours in every model

is representing the stress intensity factor of hat stringer with 20 mm fillet on sliding mode as 2.43 N/mm in high and −133.65 N/mm in low.

Hat stringer with 30 mm fillet is getting the stress intensity factor value of sliding mode as 2.62 N/mm in peak and −113.66 N/mm in the bottom which is referred from Fig. 4d. Figure 4e is the solution image of the fracture analysis. From the figure, we have shown that the value of stress intensity factor in sliding mode is 2.86 N/mm in large and −94.6 N/mm in small. From the solution results Fig. 4 and the graphically represented in Fig. 5, we conclude that the hat stringer with 40 mm fillet is having the lower stress intensity factor values and in close range of zenith and nadir values of contour 6 for the sliding mode of fracture analysis.

Stress intensity factor in tearing mode (K_3). The stress intensity value of tearing mode for the C-stringer is 658.63 N/mm in maximum and −640.15 N/mm at minimum. These values are obtained from the contour 6 results of Fig. 6a. From Fig. 6b, we have observed the stress intensity factor of tearing mode for the hat stringer with 9.53 mm fillet as 263.93 N/mm in maximum and −242.38 N/mm at minimum. Figure 6c is representing the stress intensity factor of hat stringer with 20 mm fillet on tearing mode as 235.76 N/mm in high and −198.16 N/mm in low. Hat stringer with 30 mm fillet is getting the stress intensity factor value of tearing mode as 184.15 N/mm in peak and −145.1 N/mm in the bottom which is referred from Fig. 6d. Figure 6e is the solution image of the fracture analysis. From the figure, we have shown that the value of stress intensity factor in tearing mode is 129.21 N/mm in large and −104.87 N/mm in small. From the solution images Fig. 6 and the graphically represented in Fig. 7, we conclude that the hat stringer with 40 mm fillet is having the lower stress intensity factor values and in close range of zenith and nadir values of contour 6 for the tearing mode of fracture analysis.

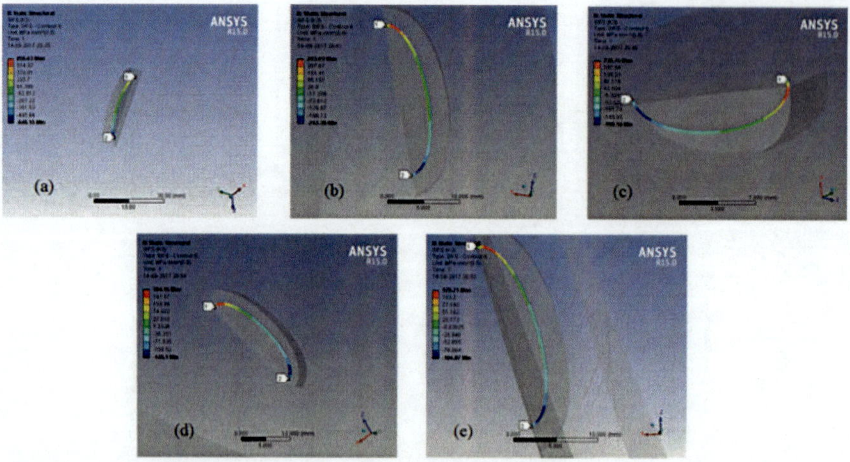

Fig. 6 K_3 Values for five stringer: **a** C-stringer **b** hat stringer with 9.53 mm fillet **c** hat stringer with 20 mm fillet **d** hat stringer with 30 mm fillet, and **e** hat stringer with 40 mm fillet

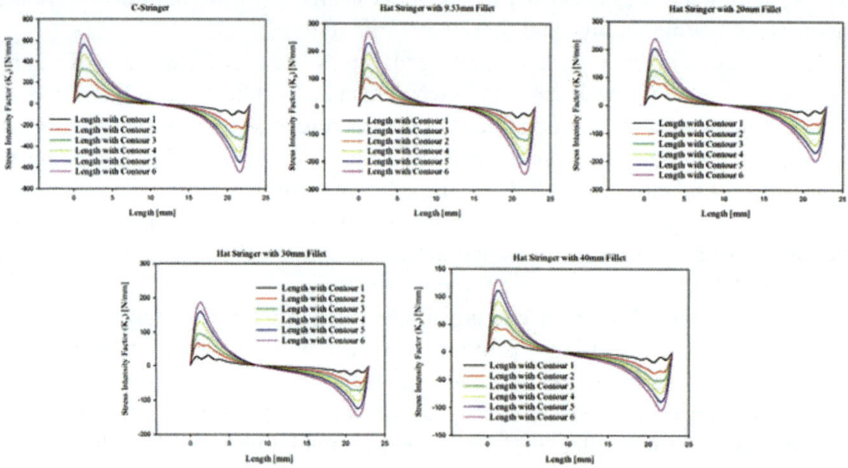

Fig. 7 Stress intensity factor (K_3) for six contours in every model

J-integral values. The J-integral on the crack for the C-stringer is 3.49 N/mm in maximum and −0.09 N/mm at minimum. These values are obtained from Fig. 8a. From Fig. 8b, we have observed the J-integral on the crack for the hat stringer with 9.53 mm fillet as −0.016 N/mm in maximum and −0.45 N/mm at minimum. Figure 8c is representing the J-integral of hat stringer with 20 mm fillet as 0.019 N/mm in high and −0.5 N/mm in low. Hat stringer with 30 mm fillet is getting the J-integral on the crack as 0.02 N/mm in peak and −0.5 N/mm at least which is referred from Fig. 8d. From Fig. 8e, we have been shown that the value of the J-integral is

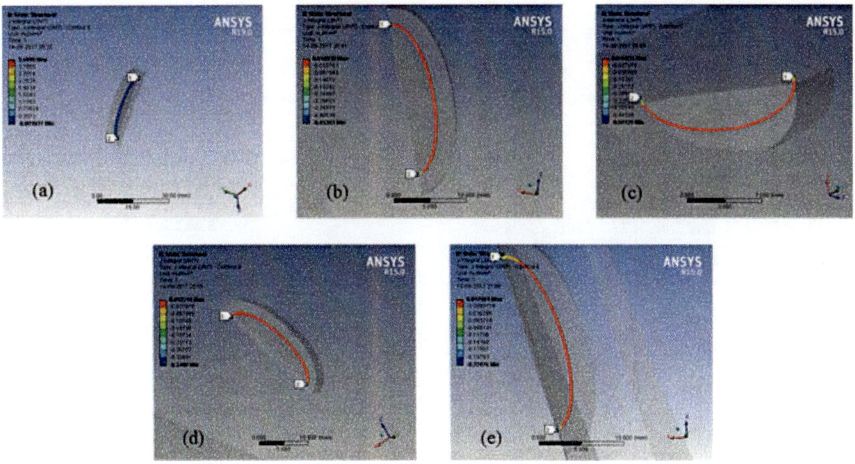

Fig. 8 J-Integral values for five structures: **a** C-stringer **b** hat stringer with 9.53 mm fillet **c** hat stringer with 20 mm fillet **d** hat stringer with 30 mm fillet, and **e** hat stringer with 40 mm fillet

0.012 N/mm in large and -0.35 N/mm in small. From the solution Fig. 8 and the graphically represented in Fig. 9, we conclude that the hat stringer with 40 mm fillet is having the lower J-integral values and in close range of zenith and nadir values of contour 6 in fracture analysis. The comparison to the fracture parameters is shown in Fig. 10.

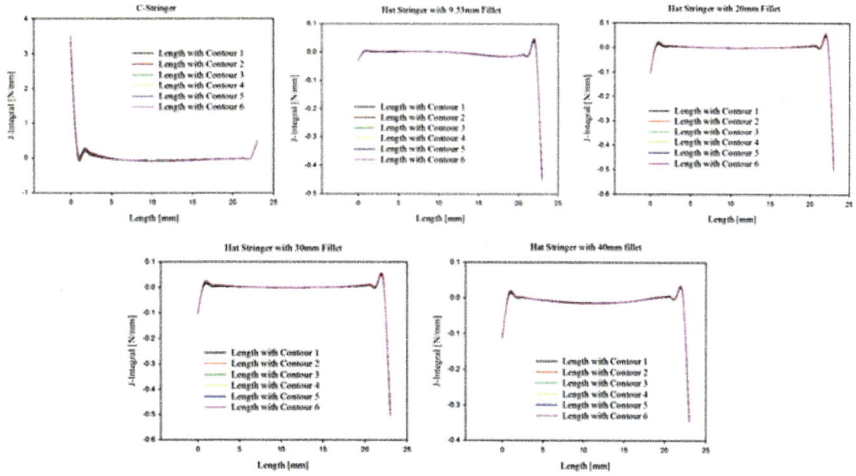

Fig. 9 J-integral values for six contours in every model

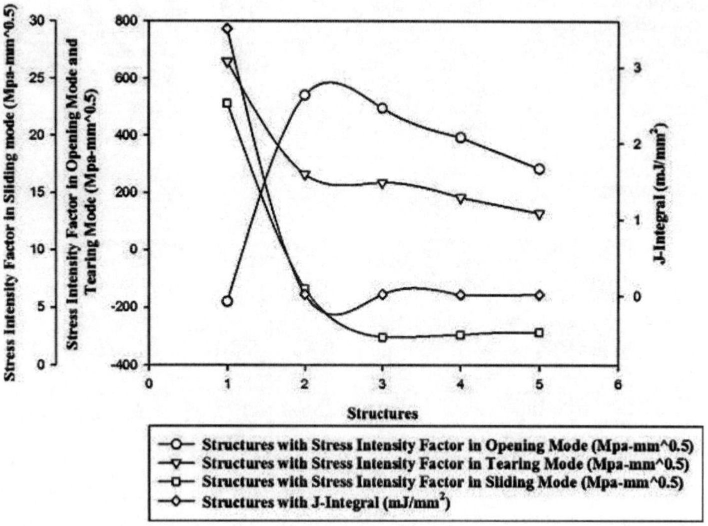

Fig. 10 Comparison of fracture parameters

4 Conclusion

This paper concludes that stringers fracture analysis on chassis frame of TATA 2516 TC truck under fracture mode behaviour. In fracture, stress concentration factors and J-integral values are good for the 40 mm fillet hat stringer. Hence, the best model for the replacement of the chassis C-stringer on TATA-2516 TC truck is a 40 mm fillet hat stringer.

References

1. Doksanovic T, Dzeba I, Markulak D (2017) Variability of structural aluminium alloys mechanical properties. Struct Saf 67:11–26
2. Pichal R, Machacek J (2017) Buckling and post-buckling of prestressed stainless steel stayed columns. Eng Struct Technol 9(2):63–69
3. Liu L, Chen X, Pan F, Gao S, Zhao C (2016) A new high-strength Mg–Zn–Ce–Y–Zr magnesium alloy. J Alloy Compd 688:537–541
4. Ekapun P, Pang TY (2015) Design and performance analysis of an electromagnetic tricycle operated in an airport. Procedia Eng 99:1330–1338
5. Patel RL, Gawande KR, Morabiya DB (2014) Design and analysis of chassis frame of TATA 2516TC. Int J Res Appl Sci Eng Technol 2(III)
6. Ingole NK, Bhope DV (2011) Stress analysis of tractor trailer chassis for self-weight reduction. Int J Eng Sci Technol 3(9):137–144

7. Kurdi O, Abdul Rahman R, Tamin MN (2008) Stress analysis of heavy duty truck chassis using finite element method. In: 2nd regional conference on vehicle engineering & technology on proceedings, Kuala Lumpur, vol 23(6), pp 292–297
8. Mordike BL, Ebert T (2001) Magnesium: properties, applications potential. Mater Sci Eng 302(1):37–45

Prioritization of Factors Influencing Sustainable Product Design in the Context of Green Consumer Behavior Using Hybrid AHP–ELECTRE II: A Case Study

Jeevan Kishore Reddy, Kandasamy Jayakrishna, and Sakthivel Aravind Raj

Abstract The purpose of this paper is to deal with the tradeoff that exists between the green consumer behavior and sustainable product features. The relation between them is complex and less explored. The mutual relationship between them involves various concepts, so this study makes use of the existing tool called hybrid multi-criterion decision-making (MCDM) method to prioritize the factors influencing sustainable product design in the context of green consumer behavior. The hybrid MCDM method AHP–ELECTRE II technique was used in this study to investigate the relationship between the green consumer behavior and sustainable product features. The practical applicability of this method was also proven in this paper using a case study done in an automotive manufacturing organization. Further the study reveals the most important determinants that a green consumer would look for while purchasing any product or service.

Keywords Sustainable product design · Green consumer behavior · AHP–ELECTRE II

1 Introduction

Consumers and organizations are now showing inclination toward purchasing of the sustainable products and services, and this change has demanded a thorough enquiry in the sustainable products features and their relationship with the consumer decision making. The modern organizations are well aware of their responsibility in persevering the society and environment directly or indirectly, helping them to

J. K. Reddy · K. Jayakrishna (✉) · S. Aravind Raj
School of Mechanical Engineering, Vellore Institute of Technology, Vellore 632014, India
e-mail: mail2jaikrish@gmail.com

J. K. Reddy
e-mail: jeevankishore98k@gmail.com

S. Aravind Raj
e-mail: aravindsakthivel@hotmail.com

A. Arockiarajan et al. (eds.), *Advances in Industrial Automation and Smart Manufacturing*, Lecture Notes in Mechanical Engineering,
https://doi.org/10.1007/978-981-15-4739-3_5

maintain their status or brand tag for their products in the global market. The tradeoff that exists between the green consumer behavior and sustainable product features is complex, remains less explored and marks the uniqueness of this study. The reason and justification for using the hybrid MCDM techniques like AHP and ELECTRE II are also explained in the later parts of this paper. The literature reviews on key topics like sustainable product design, green consumer behavior, AHP and ELECTRE II are done along with their applications.

2 Literature Review

The literature review has been done with the objective of bi-dimensional approach of product design and green consumer behavior. The various aspects of the literature deal with the multifaceted product designs and complex behavior of consumer decision making.

2.1 Review on Sustainable Product Design

Kasarda et al. [1] explained the new methodology design for adaptation (DFAD) for achieving sustainable design based on the hypothesis that a product is unable to adapt to change. In DFAD, products are conceptualized and modeled as dynamic systems based on classical control theory along with feedback control strategies to respond or adapt. The authors concluded that changing performance requirements are based on physical, cultural, environmental and/or economic considerations. Kaebernick et al. [2] developed the concept of an approach to product development, based on sustainable manufacturing. The author discussed the four examples of methodologies and decision tools which were the most important sources of environmental impacts of a product. The authors revealed that the importance of this study lies on the integrating concept rather than on the details of the methodologies. Howarth and Hadfield [3] examined the risks and benefits of the sustainable development using a concept model and a Bournemouth University practical model site in a structured manner. The authors evaluated the major social, economic and environmental risks and benefits in order to modify the design and to improve the sustainable development aspects. The methodology can be illustrated using the motor car as the product. Gehin et al. [4] combined the end-of-life (EoL) strategies with concurrent engineering principles to develop design aids which permits designers to compare their products to 'remanufacturable product profiles.' The authors aim was to why and how to integrate EoL strategies in the early design phases. Byggeth et al. [5] developed a method for sustainable product development (MSPD) by integrating social and ecological aspects of sustainability with a strategic business perspective in product development. The authors concluded that the MSPD promotes a 'bird's eye' perspective which aids the society in transformation toward sustainability by implementing in Swedish companies.

2.2 Review on Green Consumer Behavior

Gallastegui [6] studied the relevant literature and presented a review, summary of the most relevant on the eco-label issues. He divided the study into three areas such as the study of demand, the study of supply and the market and trade impacts of labeling programs. The author revealed the lack of proper and conclusive research to date and the complexity of the topic of research. He concluded that the ISO legislation and the need to regulate the use of words such as 'green' and 'bio' induce the consumer to error. Zhu et al. [7] evaluated and described the GSCM drivers, practices and performance among various Chinese manufacturing organizations using factor analysis. Using literature and industry expert input, the survey questionnaire was designed. The authors found that due to regulatory, competitive and marketing pressures and drivers, Chinese enterprises have increased their environmental awareness. For further development, efforts have been made by Chinese enterprises together with the Chinese government. Laroche et al. [8] investigated the demographic, psychological and behavioral profiles of consumers, who are willing to pay more for environmentally friendly products using various statistical analyses. The authors reported that today's ecological problems are severe that corporations do not act conscientiously toward the environment. They cited high importance on security and warm relationships with others, and they often consider ecological issues when making a purchase. Alba and Hutchinson [9] provided a conceptual organization for this diverse literature using two fundamental distinctions. First, consumer expertise is distinguished from product-related experience. Second, five distinct aspects, or dimensions, of expertise are identified: cognitive effort, cognitive structure, analysis, elaboration, and memory. Awan and Raza [10] examined about the important of environmental issues in consumer decision-making process and the important factors that affecting the consumers while taking decision toward electricity providing company. Self-administrated questionnaires and quantitative research methods are used to collect data. The author proved that when consumer makes decision, he does not consider only the product but he also keeps other factors in mind.

2.3 Review on Applications of AHP

Reza et al. [11] evaluated the sustainability analysis best flooring systems using MCDM technique, AHP. This study uses the triple bottom line criteria to bring out the sustainable flooring system among concrete, clay and expanded polystyrene (EPS). The three criteria taken into consideration are economic concerns, environmental concerns and socio-political concerns, and these are divided into 13 sub-criteria. Using life cycle analysis, it was found that EPS is the best sustainable flooring system. Hambali et al. [12] explained the application of AHP in the conceptual design stage of an automotive composite bumper beam. Using analytical hierarchical process, they selected a concept selection model, namely concurrent design concept selection and materials selection (CDCSMS), as the most appropriate selection model. Appropriate design concept was determined with the application of sensitivity analysis among eight design concepts. Badri [13] conducted the study on previous results of service quality attributes and identified 'sets of quality measure'. Priority weights for quality attributes are calculated using AHP and are then incorporated in a goal-programming model to select the best a set of quality control instruments. Yavuz et al. [14] selected the optimum support design for the main transport road, which was planned for deep coal seam panels of Western Lignite Corporation (WLC) Tuncbilek in Turkiye using AHP. The authors considered eight main objectives, namely four different displacement values for determined history locations, factor of safety (FOS), cost, labor and applicability factor for the selection of support design. After carrying out several numerical models for different support design, AHP method was incorporated to evaluate these support designs according to the pre-determined criteria. The result showed that such AHP application helped the engineers to evaluate the system alternatives for an underground mine. Yupu et al. [15] evaluated the sustainable development of coal mining cities in Heilongjiang province, based on the AHP method along with operational and scientific principles. A comprehensive evaluation is made on the indices of the four major coal mining cities of Heilongjiang using fuzzy comprehensive evaluation method. The result showed that the economic development and environment quality are the most important indices of the targeted layer influencing the sustainable development of coal mining cities of Heilongjiang.

2.4 Review on Applications of ELECTRE Method

Shanian and Savadogo [16] applied ELimination and Choice Expressing the REality (ELECTRE) and method to find out the most appropriate material for a thermally loaded conductor and formed the logical ranking for selecting the best material using material selection decision matrix and criteria sensitivity analysis. The authors also developed a computer program, Mathematica to facilitate the application of the method to other types of material selection problems. Kaya and Kahraman [17] proposed an environmental impact assessment methodology based on an integrated fuzzy AHP–ELECTRE approach for urban industrial planning. Using fuzzy AHP, the criteria weights are generated and fuzzy ELECTRE was used to assess the environmental impact generated. A fuzzy dominance relation (FDR) methodology was used to rank the alternatives from the most risky to the least along with the sensitivity analysis. Condor et al. [18] proposed ELECTRE TRI model for assessing synergies between the Rio Conventions at the project level in the forest sector to found a compromise solution to a problem with multidimensional and conflicting criteria including social, economic and environmental features. The authors showed that the procedures of the ELECTRE TRI model are an advantage of dealing with incomparability between the profiles and alternatives. Salimen et al. [19] have compared three multi-criterion decision-making analysis such as Preference Ranking Organisation Method for Enrichment Evaluations (PROMETHEE), ELECTRE and Simple Multi-Attribute Rating Technique (SMART) in the environmental context. They concluded that the results obtained by these methods will greatly vary and they recommended ELECTRE method to obtain the preferred ranking.

2.5 *Research Gap*

Despite the fact that many researchers have focused upon sustainable product design and green consumer behavior with application of MCDM techniques, only fewer efforts have been made to integrate the design requirements of a sustainable product and consumer behavior. In this context, this study explores the application of hybrid AHP-ELECTRE II methods for the prioritization of factors affecting consumer behavior in sustainable design context with respect to the case product.

3 Case Study

The case study has been developed in an automotive manufacturing organization situated in Bangalore, India (hereafter designated as XYZ). XYZ is the manufacturer of automotive seating. A cross-functional team comprising experts from consumer, stakeholder, design, manufacturing and quality was formulated. The cross-functional team framed and weighted the factors influencing the sustainable product design and the decision matrix.

4 Identification of Factors Influencing Sustainable Product Design and Green Consumer Behavior Determinants

Based upon the exhaustive literature review and the cross-functional team advice, the factors influencing sustainable product design identified include high-performance rate (C_1), aesthetics (C_2), ergonomically fit (C_3), safe to use (C_4), good EoL disposal (C_5), ease of disassembly (C_6), low cost (C_7) and use of renewable resources (C_8) and the green consumer behavior determinants include demographic (GCB_1), socio-cultural (GCB_2), cost (GCB_3), eco-literacy (GCB_4), perception of values (GCB_5) and availability (GCB_6). The factors influencing sustainable product design and green consumer determinants are presented in Table 1.

Table 1 Factors influencing sustainable product design and green consumer determinants

Sl. No.	Factors influencing sustainable product design	Green consumer behavior determinants
1	High-performance rate (C_1)	Demographic (GCB_1)
2	Aesthetics (C_2)	Socio-cultural (GCB_2)
3	Ergonomically fit (C_3)	Cost (GCB_3)
4	Safe to use (C_4)	Eco-literacy (GCB_4)
5	Good EoL disposal (C_5)	Perception of values (GCB_5)
6	Ease of disassembly (C_6)	Availability (GCB_6)
7	Low cost (C_7)	
8	Use of renewable resources (C_8)	

5 Computation of Weights Using AHP

Weighting the factors influencing sustainable product design, table values are obtained by using AHP measurement Linkert's scale (one to nine). Comparing high-performance rate and aesthetics is '3' and the reverse comparison is reciprocal of '0.333', this 0.333. The ΠCi is the multiplication of scales along each row, and the normalized weights are obtained by dividing each row value of (ΠCi)1/4 by its summation of column. Table 2 shows the computation of weights using AHP.

6 Steps in ELECTRE II

Step I. Formation of Decision and Normalization matrix
The decision matrix values portray the relational importance between each factor influencing sustainable product design and the green consumer behavior determinants. The decision value of every green consumer behavior determinant is named as GCB_1 to GCB_6. The decision matrix formulated by cross-functional team is shown in Table 3. The normalized decision matrix is framed by dividing each decision value by the summation of all the decision value along each row. The normalized decision value of every green consumer behavior determinant is named as gcb1 to gcb6 (Table 4).

Step II. Computation of concordance matrix and discordance matrix
The concordance index and discordance index in ELECTRE II include two tremendous conflicting affiliations, strong and weak, whereby strong-ranking and weak-ranking are deduced to obtain the final ranking [20]. The concordance index quantifies the degree of ascendancy of one green consumer behavior determinant over the other [21], and it is summed up using the formula

Table 2 Computation of weights using AHP

Criteria	C_1	C_2	C_3	C_4	C_5	C_6	C_7	C_8	ΠC_i	$(\Pi C_i)^{1/4}$	$[(\Pi C_i)^{1/4}]/\Sigma[(\Pi C_i)^{1/4}]$
C_1	1.000	5.000	3.000	3.000	3.000	5.000	5.000	5.000	16875.000	11.398	0.458
C_2	0.200	1.000	3.000	7.000	7.000	7.000	3.000	5.000	3087.000	7.454	0.299
C_3	0.333	0.333	1.000	3.000	7.000	3.000	3.000	3.000	63.000	2.817	0.113
C_4	0.333	0.143	0.333	1.000	5.000	5.000	7.000	7.000	19.444	2.100	0.084
C_5	0.333	0.143	0.143	0.200	1.000	7.000	5.000	3.000	0.143	0.615	0.025
C_6	0.200	0.143	0.333	0.200	0.143	1.000	7.000	3.000	0.006	0.275	0.011
C_7	0.200	0.333	0.333	0.143	0.200	0.143	1.000	7.000	0.001	0.159	0.006
C_8	0.200	0.200	0.333	0.143	0.333	0.333	0.143	1.000	0.000	0.074	0.003
Total										24.891	1.000

Table 3 Decision matrix

	Decision matrix					
	GCB_1	GCB_2	GCB_3	GCB_4	GCB_5	GCB_6
C_1	8	5	4	6	8	9
C_2	45	34	62	26	73	11
C_3	8	5	6	7	3	6
C_4	65	87	45	35	63	90
C_5	7	9	5	8	6	7
C_6	5	6	3	8	9	6
C_7	53	67	43	36	57	82
C_8	424	325	230	525	440	240

Table 4 Normalization matrix

	Normalized values					
	gcb_1	gcb_2	gcb_3	gcb_4	gcb_5	gcb_6
C_1	0.20	0.13	0.10	0.15	0.20	0.23
C_2	0.18	0.14	0.25	0.10	0.29	0.04
C_3	0.23	0.14	0.17	0.20	0.09	0.17
C_4	0.17	0.23	0.12	0.09	0.16	0.23
C_5	0.17	0.21	0.12	0.19	0.14	0.17
C_6	0.14	0.16	0.08	0.22	0.24	0.16
C_7	0.16	0.20	0.13	0.11	0.17	0.24
C_8	0.19	0.15	0.11	0.24	0.20	0.11

$$C(\mathrm{gcb}_i, \mathrm{gcb}_j) = \frac{W^+ + 0.5W^=}{W^+ + W^= + W^-}$$

where W^+ is the sum of the weight of factors influencing the sustainable product design in which one green consumer behavior determinant gcb_i is preferred to gcb_j, W^- is the sum of the weight of factors influencing the sustainable product design in which one green consumer behavior determinant gcb_j is preferred to gcb_i, $W^=$ is the sum of the weight of factors influencing the sustainable product design where both gcb_i and gcb_j are indifferent. The concordance index computed is shown in Table 5.

The discordance index quantifies the degree of limitation of one green consumer behavior determinant over the other [22], and it is summed up using the formula

$$d(\mathrm{gcb}_i, \mathrm{gcb}_j) = \frac{\max\limits_{\mathrm{gcb}_i > \mathrm{gcb}_j} [\mathrm{gcb}_i - \mathrm{gcb}_j]}{\max(|\mathrm{gcb}_i - \mathrm{gcb}_j|)}$$

The discordance index computed is shown in Table 6.

Table 5 Concordance index

CON	gcb_1	gcb_2	gcb_3	gcb_4	gcb_5	gcb_6
gcb_1		0.874	0.701	0.961	0.451	0.428
gcb_2	0.126		0.587	0.415	0.229	0.333
gcb_3	0.299	0.413		0.390	0.113	0.356
gcb_4	0.039	0.585	0.610		0.141	0.451
gcb_5	0.549	0.771	0.887	0.859		0.313
gcb_6	0.572	0.667	0.644	0.549	0.687	

Table 6 Discordance index

DISCON	gcb_1	gcb_2	gcb_3	gcb_4	gcb_5	gcb_6
gcb_1		1.000	1.283	0.961	1.000	1.000
gcb_2	0.667		0.978	1.000	0.460	0.916
gcb_3	0.677	1.000		1.000	0.529	1.000
gcb_4	1.000	0.678	0.942		0.610	0.913
gcb_5	0.781	1.000	1.000	1.000		1.000
gcb_6	0.633	1.000	0.615	1.000	0.347	

Step III. Computation of pure concordance and pure discordance matrix
To find out the full ranking, pure concordance and pure discordance indices were computed using the following formulas

$$\text{Pure concordance index } (C_i) = \sum_{j=1}^{n} c\left(\text{gcb}_i, \text{gcb}_j\right) - \sum_{i=1}^{n} c\left(\text{gcb}_j, \text{gcb}_i\right) \ \&$$

$$\text{Pure discordance index } (D_i) = \sum_{j=1}^{n} D\left(\text{gcb}_i, \text{gcb}_j\right) - \sum_{i=1}^{n} D\left(\text{gcb}_j, \text{gcb}_i\right)$$

respectively.

Where c (gcb_i, gcb_j) and d (gcb_i, gcb_j) are the concordance and discordance indices between each green consumer behavior determinants. The pure concordance and pure discordance indices calculated are listed in Tables 7 and 8, respectively.

The pure concordance index is ranked in the descending order of values, and the pure discordance index is ranked in the ascending order of values. The averages of both these ranks are calculated at the end to determine the final ranking of the green consumer behavior determinants.

Table 7 Pure concordance index

Determinants	c_i		Rank
gcb_1	c_1	1.829	6
gcb_2	c_2	−1.620	2
gcb_3	c_3	−1.857	1
gcb_4	c_4	−1.348	3
gcb_5	c_5	1.759	5
gcb_6	c_6	1.237	4

Table 8 Pure discordance index

Determinants	d_i		Rank
gcb_1	d_1	1.486	5
gcb_2	d_2	−0.657	3
gcb_3	d_3	−0.613	2
gcb_4	d_4	−0.817	4
gcb_5	d_5	1.835	6
gcb_6	d_6	−1.234	1

7 Results and Conclusion

The final ranking computed was in the order of gcb3 < gcb2 < gcb6 < gcb4 < gcb5 < gcb1. The results of this study reveal that the cultural (GCB3), socio-cultural (GCB2) and availability (GCB4) as the most important determinants of green consumer behavior. The prioritization of the green consumer behavior determinants helped the organization to identify the consumer's thoughts of buying the sustainable products and thereby improving their profit [23]. Understanding the customers forms the first and foremost task of any organization to withstand in the global market [24]. The methodology adopted in this study proves its practical applicability in capturing the consumer persuasion about sustainable product design and its need. Though the case study produced potential outcomes following the research, there are some limitations that could be improved in future research. In terms of factors influencing sustainable product design, this case study only took cross-functional team advice and is limited to papers reviewed. This could be broadened to wider range of factors and determinants from different organizations.

References

1. Mary E, Kasarda JP, Terpennya DI, Precoda KR, Jelesko J, Sahin A, Parkf J (2007) Design for adaptability (DFAD)—a new concept for achieving sustainable design. Robot Comput-Integr Manuf 23:727–734

2. Kaebernick H, Kara S, Sun M (2003) Sustainable product development and manufacturing by considering environmental requirements. Robot Comput Integr Manuf 19:461–468
3. Howarth G, Hadfield M (2006) A sustainable product design model. Mater Des 27:1128–1133
4. Gehin A, Zwolinski P, Brissaud D (2008) A tool to implement sustainable end-of-life strategies in the product development phase. J Clean Prod 16:566–576
5. Byggeth S, Broman G, Robert K-H (2007) A method for sustainable product development based on a modular system of guiding questions. J Clean Prod 15:1–11
6. Galarraga Gallastegui I (2002) The use of eco-labels: a review of the literature. Eur Env 12:316–331
7. Zhu Q, Sarkis J, Geng Y (2004) Green supply chain management in China: pressures, practices and performance. Int J Oper Prod Manage 25(5):449–468
8. Laroche M, Bergeron J, Barbaro-Forleo G (2001) Targeting consumers who are willing to pay more for environmentally friendly products. J Consum Mark 18(6):503–552
9. Alba JW, Hutchinson JW (1987) Dimensions of consumer expertise. J Consum Res 13(4):411–454
10. Awan YSAMA, Raza MA (2012) Green consumer behavior: empirical study of swedish consumer behavior. Rec Res Econ., pp 89–104. ISBN: 978-1-61804-061-9
11. Reza B, Sadiq R, Hewage K (2010) Sustainability assessment of flooring systems in the city of Tehran: An AHP-based life cycle analysis. Constr Build Mater 25:2053–2066
12. Hambali A, Sapuan SM, Ismail N, Nukman Y (2009) Application of analytical hierarchy process in the design concept selection of automotive composite bumper beam during the conceptual design stage. Sci Res Essay 4(4):198–211
13. Badri MA (2001) A combined AHP and GP model for quality control systems. Int J Prod Econ 72:27–40
14. Yavuz M, Iphar M, Once G (2008) The optimum support design selection by using AHP method for the main haulage road in WLC Tuncbilek colliery. Tunn Undergr Space Technol 23:111–119
15. Zhang Y, Sun Y, Qin J (2012) Sustainable development of coal cities in Heilongjiang province based on AHP method. Int J Mining Sci Technol 22:133–137
16. Shanian A, Savadogo O (2006) A material selection model based on the concept of multiple attribute decision making. Mater Des 27:329–337
17. Kaya T, Kahraman C (2011) An integrated fuzzy AHP–ELECTRE methodology for environmental impact assessment. Expert Syst Appl 38:8553–8562
18. RD Condor, Scarelli A, Valentini R (2011) Multicriteria decision aid to support multilateral environmental agreements in assessing international forestry projects. Int Environ Agreements 11:117–137
19. Salminen P, Hokkanen J, Lahdelma R (1998) Comparing multi-criteria methods in thecontext of environmental problems. Eur J Oper Res 104:485–496
20. Martin R, Michael B (1998) A new system for weighting environmental criteria for use within ELECTRE III. Eur J Oper Res 107:552–563
21. Roy B (1991) The outranking approach and the foundations of ELECTRE methods. Theor Dec 31(1):49–73
22. Roy B, Bertier P (1973) Lamethode ELECTREII—Une application au media—planning. In: Ross M (ed) OR'72. Amsterdam, North Holland
23. Soonthonsmai V (2007) Environmental or green marketing as global competitive edge: concept, synthesis, and implication. In: EABR (Business) and ETLC (Teaching) Conference Proceeding, Venice, Italy
24. Wasik JF (1996) Green marketing and management: a global perspective. Blackwell Publishers Inc., Cambridge, Mass

Evaluation of Material Properties and Abrasive Resistance of Tantalum Carbide-Based Hardox Steel for Construction Purpose

P. S. Senthil Kumar, S. Marichamy, C. Sivakandhan, B. Stalin, V. Dhinakaran, and I. Satyanarayana

Abstract In recent days, the demand of high strength and wear resistant steel has been increased. The material properties such as hardness and toughness are plays an important role of wear resistant. In this research, the tantalum carbide-based hardox steel is prepared by stir casting process. To obtain high hardness and toughness, 2, 4, and 6 weighted percentage of tantalum carbide particles are added to the hardox steel. The material properties such as impact strength, tensile strength, yield strength, hardness, percentage of elongation, and density are found on the stir casted steel specimens. The dry sand rubber abrasion apparatus is used to measure the abrasive resistance. In this abrasion test, the various input parameters such as disc speed, types of abrasive, and sliding distance are considered. This modified tantalum carbide-based hardox steel has recommended for civil engineering construction and other applications. The abrasive resistance rate is found by varying the input parameters. The most influential parameter for the abrasive resistance rate is found out through analysis of variance (ANOVA).

P. S. Senthil Kumar
Department of Civil Engineering, Sri Indu College of Engineering and Technology, Hyderabad, Telangana, India

S. Marichamy
Department of Mechanical Engineering, Sri Indu College of Engineering and Technology, Hyderabad, Telangana, India

C. Sivakandhan · I. Satyanarayana
Department of Mechanical Engineering, Sri Indu Institute of Engineering and Technology, Hyderabad, Telangana 501510, India

B. Stalin (✉)
Department of Mechanical Engineering, Anna University, Regional Campus Madurai, Madurai, Tamil Nadu 625019, India
e-mail: stalin1312@gmail.com

V. Dhinakaran
Department of Mechanical Engineering,
Chennai Institute of Technology, Kundrathur, Chennai, Tamil Nadu 600069, India

A. Arockiarajan et al. (eds.), *Advances in Industrial Automation and Smart Manufacturing*, Lecture Notes in Mechanical Engineering,
https://doi.org/10.1007/978-981-15-4739-3_6

Keywords Hardox steel · Tantalum carbide particles · Dry sand rubber abrasion · Analysis of variance · Stir casting process

1 Introduction

Normally, hardox steel has high strength and abrasion resistance. It has high toughness and impact strength. These material properties are further increased by the addition of tantalum carbide particles. The tantalum carbide (TaC_X)-based hardox steel is used in production of truck bodies, material handling system, crushers, and civil engineering applications. In any industrial applications, the impact and abrasive action can be considered [1]. The various material structures and abrasive resistance have been studied in the high strength low alloy steel [2]. An examination of mechanical property and three-body impact abrasive wear behavior has been studied in 0.27% carbon dual phase steel [3]. The abrasion resistance has been increased when the grain size was decreased [4]. The mechanical properties were mainly depending on the microstructure and grain size of the materials [5–7]. The mechanical properties and microstructures of Ti and Mo micro-alloyed medium Mn steel have been studied [8]. Wear resistant and martensite structure have been analyzed in hardox steel [9, 10].

The present work deals with fabrication and evaluation of material properties of tantalum carbide-based hardox steel. The abrasive resistance test was conducted and most influential parameter was found out through analysis of variance. The high amount of load withstanding capacity was confirmed through the deflection test.

2 Materials and Methodology

2.1 Stir Casting Process

The tantalum carbide (TaC_X)-based hardox steel was fabricated through stir casting method. The raw material of hardox steel was placed in a crucible furnace. The furnace was operating at 1230 °C temperature. The stirrer was applied at 600 rpm. After achieving a liquid state, the furnace was stopped. The preheated tantalum carbide particles were slowly added to hardox steel. Again the furnace temperature was raised and uniform stirrer was applied for 45 min. After that, furnace was stopped and immediately molten metal was transferred to the die unit. The same procedure was repeated for each specimen. The preheated tantalum carbide particles were added at the rate of weight percentage basis like as 2%, 4%, and 6%.

Table 1 Material properties

S. No.	Properties	2% of TaC_X	4% of TaC_X	6% of TaC_X
1	Hardness (BHW)	370	420	450
2	Tensile strength (N/mm^2)	920	1050	1300
3	Yield strength (N/mm^2)	800	920	1000
4	Impact strength (J)	25	35	50
5	% of elongation	10	12	15
6	Density (g/cc)	7.43	7.83	8.23

2.2 Properties of Hardox Steel

The material properties, namely hardness, tensile strength, impact strength, yield strength, percentage of elongation, and density, were evaluated and shown in Table 1 for three specimens with different composition of reinforcement (2%, 4%, and 6%) of TaC_X.

Brinell hardness tester with tungsten indenter, universal testing machine, and charpy testing machine was used to measure the mechanical properties. Archimedes principle was used to measure the material density. From Table 1, it can be observed that all values were gradually increased after the addition of TaC_X particles. Alloying elements such as nickel, silicon, aluminum, niobium, and chromium and percentage of composition of reinforcement materials play an important role in material properties.

3 Results and Discussion

3.1 Abrasive Resistance Test

The schematic diagram of dry sand rubber abrasion apparatus was shown in Fig. 1. The unit consists of rotating disc, abrasive feed system, specimen holder, and counter weight. The rotating wheel was made of rubber which is connected through an electric motor. The specimen to be tested was placed between the disc and the work holder. The specimen was rigidly fixed in the work holder. The abrasive particles like as foundry sand, silicon carbide, and tungsten carbide particles flow between the work piece and disc.

The weight of the work piece has been measured before and after the experiment. The abrasive resistance rate (ARR) was determined by the ratio between the weight differences to the initial weight of the specimen. The various input parameters such as disc speed (200, 400, and 600 rpm), types of abrasive such as dolomite powder, silicon carbide, and titanium carbide particles, and sliding distance (400, 600, and

Fig. 1 Schematic diagram of dry sand rubber abrasion apparatus

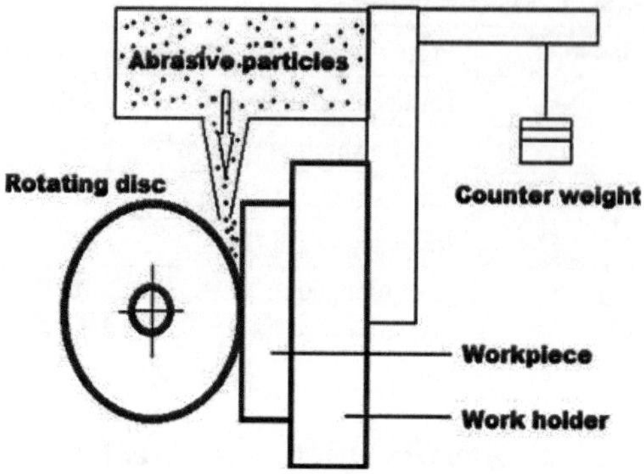

700 m) are considered. The experimental results for abrasion test were shown in Table 2.

Analysis of variance for abrasion test was shown in Table 3. The types of abrasive were main contribution parameter (70.37%) which affects abrasive resistance. The least contribution (2.6%) parameter was sliding distance which affects abrasive resistance.

Based on the parameters, a regression model was generated which was used to forecast the response. Here, *A*, *B*, and *C* are disc speed, types of abrasive, and sliding distance, respectively.

Table 2 Experimental results for abrasion test

Exp. No.	A: Disc speed (rpm)	B: Types of abrasive	C: Sliding distance (m)	Abrasive resistance rate (10^{-6})
1	200	Dolomite powder	400	0.114
2	200	Silicon carbide	600	0.125
3	200	Titanium carbide	700	0.167
4	400	Dolomite powder	600	0.121
5	400	Silicon carbide	700	0.143
6	400	Titanium carbide	400	0.176
7	600	Dolomite powder	700	0.134
8	600	Silicon carbide	400	0.154
9	600	Titanium carbide	600	0.234

Table 3 Analysis of variance for abrasion test

Source	DF	Adj. SS	Adj. MS	*F*-value	*P*-value	Percentage (%)
Disc speed (rpm)	2	0.002371	0.001185	3.82	0.207	21.42
Types of abrasive	2	0.007789	0.003894	12.55	0.074	70.37
Sliding distance (m)	2	0.000288	0.000144	0.46	0.683	02.60
Error	2	0.000621	0.000310	–	–	05.61
Total	8	0.011068	–	–	–	100
Model summary	$S = 0.0176163$		R-sq = 94.39%		R-sq(adj) = 77.57%	

$$\begin{aligned}\text{ARR} = {} & 0.152 - 0.016A - 200 - 0.005A - 400 \\ & + 0.022A - 600 - 0.029B - 0.011B \\ & + 0.040B - 0.004C - 400 + 0.008C \\ & - 600 - 0.004C - 700 \end{aligned} \tag{1}$$

4 Surface Plot Analysis

Figure 2 shows the surface plot analysis for disc speed, abrasive resistance rate, and types of abrasives. The abrasive rate was maximum when the discs speed is between 400 and 600 rpm. In this range, silicon carbide particle plays an vital role.

The surface plot analysis for disc speed, abrasive resistance rate, and sliding distance was shown in Fig. 3. The abrasive resistance (0.10–0.20) gradually increased

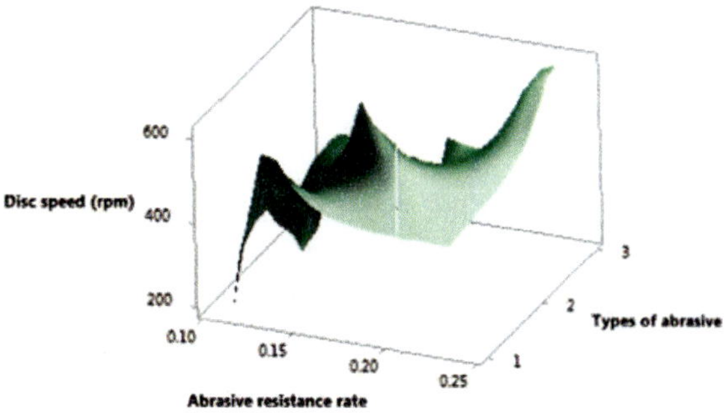

Fig. 2 Surface plot analysis for disc speed, abrasive resistance rate and types of abrasives

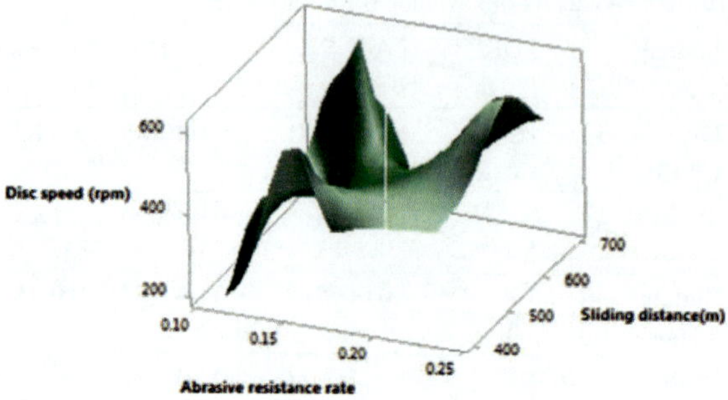

Fig. 3 Surface plot analysis for disc speed, abrasive resistance rate, and sliding distance

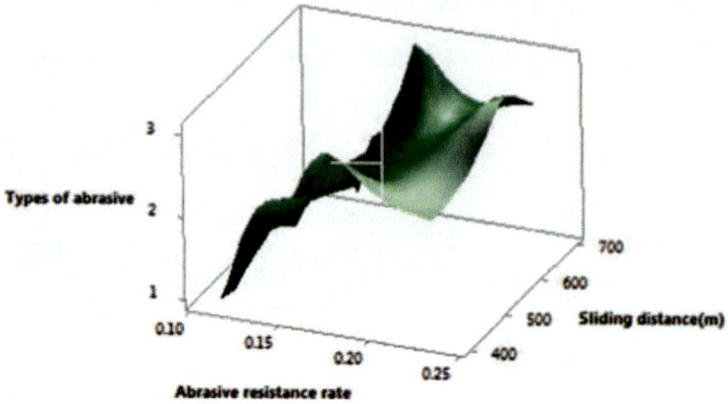

Fig. 4 Surface plot analysis for types of abrasive, abrasive resistance rate, and sliding distance

for the initial stage, i.e., disc speed is 200–400 rpm and sliding distance is 400–500 m. After that, it was decreased and then increased due to higher values of input parameters.

The surface plot analysis for disc speed, abrasive resistance rate, and sliding distance was shown in Fig. 4. The abrasive resistance rate was gradually increased from initial stage to final stage. The maximum abrasive resistance (0.25) rate was obtained at a sliding distance of 700 m. During this stage, titanium carbide plays an important role.

5 Deflection Test

The simply supported and the cantilever beam deflection test was conducted on tantalum carbide (TaC_X)-based hardox steel. The deflection was recorded through dial gauge. The specimen size of 1000 × 25 × 6 mm is allowed to test the deflection. The experimental results for deflection test under simply supported and cantilever beam were shown in Table 4.

From Table 4, it can be concluded that zero deflection was observed for any load (0–20 kg). Due to excellent properties especially high strength and hardness, it can withstand high amount of load without any deflection. Hence, it can be recommended for construction purpose.

Table 4 Experimental results for deflection test

S. No.	Load (kg)	Distance (mm)	Deflection
1	0	0	0
2	1	300	0
3	3	300	0
4	5	400	0
5	8	400	0
7	12	500	0
8	14	500	0
9	17	600	0
10	20	600	0

6 Conclusions

- The tantalum carbide (TaC_X)-based hardox steel was successfully prepared by stir casting process.
- The material properties, namely hardness, impact strength, tensile strength, yield strength, percentage of elongation, and density, were evaluated.
- It can be observed that all experimental values of material properties were gradually increased after the addition of TaC_X particles.
- The abrasive resistance test was conducted on tantalum carbide-based hardox steel and it can be concluded that from analysis of variance, the types of abrasive were main contribution parameter (70.37%) which affect abrasive resistance.
- From the deflection test, it can withstand high amounts of load without any deflection. Due to its excellent material properties and withstand high amounts of load, it can be recommended for civil engineering applications especially in construction purpose.

References

1. Sundstrom A, Rendnón J, Olsson M (2001) Wear behavior of some low alloyed steels under combined impact/abrasion contact conditions. Wear 250:744–754
2. Jha AK, Prasad BK, Modi OP, Das S, Yegneswaran AH (2003) Correlating microstructural features and mechanical properties with abrasion resistance of a high strength low alloy steel. Wear 254:120–128
3. Deng X, Wang Z, Tian Y, Fu T, Wang G (2013) An investigation of mechanical property and three-body impact abrasive wear behavior of a 0.27% C dual phase steel. Mater Des 49:220–225
4. Xu X, Xu W, Ederveen FH, Zwaag S (2013) design of low hardness abrasion resistant steels. Wear 301:89–93
5. Grajcar A, Kilarski A, Kozlowska A (2018) Microstructure-properties relationships in thermo mechanically processed medium-Mn steels with high Al content. Metals 8:929

6. Kim SJ, Lee CG, Choi I, Lee SE (2001) Effects of heat treatment and alloying elements on the microstructures and mechanical properties of 0.15% C transformation induced plasticity-aided cold-rolled steel sheets. Metall Mater Trans 32:505–514
7. Grajcar A, Kwasny W, Zalecki W (2015) Microstructure-property relationships in TRIP aided medium-C steel with lamellar retained austenite. Mater Sci Technol 31:781–794
8. Jeyaprakash N, Duraiselvam M, Raju R (2018) Modelling of Cr 3 C 2–25% NiCr laser alloyed cast iron in high temperature sliding wear condition using response surface methodology. Arch Metall Mater 63(3):1303–1315
9. Lee D, Kim JK, Lee S, Lee K, De Cooman BC (2017) Microstructures and mechanical properties of Ti and Mo micro-alloyed medium Mn steel. Mater Sci Eng A 706:1–14
10. Konovalov SV (2016) Formation wear resistant coatings on martensite steel hardox 450 by welding methods. Mater Sci Eng 142:2–5

Development of Portable Tabletop Equipments for Micromanufacturing System

T. T. M. Kannan and R. Elangovan

Abstract In the manufacturing industry, smart and beneficial evaluations are expected for the new century. Microfactory is a small dimensioned factory for recommendation of small production system. It is a future manufacturing concept which was initiated by Japanese research and development. Smaller-scale manufacture advancements have been relentlessly progressing as of late. Microfactory concepts are adopted in watch mechanism, fuel injector, nozzle and microelectronics. Tabletop microprocessing plant is the name predominantly smaller than usual assembling framework which creates and amasses of little segments without compromising machining tolerances. In this work, development of miniaturized machinery of portable lathe, milling, drilling and surface grinding machine has been fabricated as per requirements, and corresponding position errors are calculated, and forecast exactness was sufficient for utilization in the reasonable structure stages. Microfactory is small dimensioned production system which is used for manufacturing of miniature of components requiring high accuracy, high throughput and low creation cost. Scaling down of machine apparatuses leads to huge sparing in vitality, space, time and all resources.

Keywords Micromachines · Microfactory · TPI · Error analysis · Micromanufacturing

T. T. M. Kannan (✉)
Department of Mechanical Engineering, PRIST Deemed to be University, Thanjavur 613403, India
e-mail: ttmkresaerch@gmail.com

R. Elangovan
Department of Mechanical Engineering, Mookambigai College of Engineering, Pudukkottai 622502, India

A. Arockiarajan et al. (eds.), *Advances in Industrial Automation and Smart Manufacturing*, Lecture Notes in Mechanical Engineering,
https://doi.org/10.1007/978-981-15-4739-3_7

1 Introduction

Small-scale industrial facility is a name of an ultra smaller than expected assembling framework that comprises little of machine devices and controllers. These ultra-precision machines commonly require costly and specific structure which highlights to accomplish the ideal dimension of precision. Numerous wellsprings of blunder present in machine device scale good with scaling down permitting of configuration to meet the precision necessities in a more affordable machine instruments. Reasoning of improvement of tabletop types of gear for small-scale industrial facility to downscaling the creation framework and procedures closer the size of the item merchandize. Littler machines for littler parts' creation framework are the fundamental idea for manufacture of miniaturized scale machines. Miniaturized scale industrial facility alludes to little dimensioned manufacturing plant ready to deliver little dimensioned items. An expected advantage of the microfactory would reduce environmental impact and cost of miniature mechanical fabrication especially for "diverse types and small quantity production system. Tabletop small machines enable flexible layout changes; it can control the increase of cost when the product design has been modified. The expression "smaller-scale machining" is commonly used to characterize the act of material evacuation for creation of parts having measurement which lies somewhere in the range of 1 and 999 miniaturized scale meters. It was the first big success; the further steps into concentration on process physics of micro-machining include materials and microstructural efforts, machine tools, workpiece, tooling, design issues, software and simulation tools. Small of segment is required for a wide scope of uses from aviation to biomedical industries. The late quantities of examinations have been accounted for on the machining of miniaturized scale highlights and segments. The creators portrayed about estimation of situating mistakes between the item and cutting apparatus of each small-scale machines. The examination of estimated and determined blunders demonstrates that the situating mistakes of machine apparatuses are generally unsurprising by the proposed plan instrument. There was a proposition of a list to assess all-out execution of items by considering item esteems, costs and natural effects through item life cycle. Downsizing of manufacturing system can lead to smart solutions improving space utilization factor, reducing time, price and energy including environmental condition such as humidity, temperature and cleanliness. Desktop factories are small-size production system suitable for fabricating and assembling small parts and products without compromising machine tolerances. The authors introduced the structure and assessment of smaller than usual of machine instruments that are fit for accomplishing fundamentally higher cutting rates, delivering three-dimensional highlights in materials. The micro- and tabletop factories can bring enormous benefits against the regular factories in terms of their sustainability perspective such as ecological, economic and social which are not highlighted in the previous work.

2 Literature Survey

Furuta [1] have described the concept of microfactory system for saving of energy. This work conducted at mechanical engineering laboratory Japan demonstrated the significance and importance of microfactory for sustainable manufacturing system for environmental, social and economic impact. Michel et al. [2] have stated the building of minifactory from a technology construction kit for microcomponents in micromanufacturing system. They proposed the theory of small machines for small components. It consumes less power and time for the applications of biomedical and aeronautical industries. They developed the concept of a set of cooperating machines in tabletop containment over larger traditional machines such as reduced cost and space. Mekid [3] has discussed the development of meso-scale machine tool for micromachining. Small-scaled features are widely required in many industries including biomedical, consumer electronics, aerospace and defense. They also proposed a new microfactory architecture development that enables flexibility, modularity, automation and portability. Werkmeister and Slocum [4] have constructed a meso-mill for fabrication of miniature of components. They proved that microscale cutting mechanics phenomenon known as minimum chip thickness effect and micro-metal chips during machining of aluminum alloy using five-axis micro-milling machine. In addition, the meso-scale milling machine includes workpiece measurement and chip removal process using measurement censor. Tanaka [5] have defined the design evaluation of miniature of machine tool using microfactory concept. By using the form shaping theory, it is possible to evaluate the design parameter of micromachines. The design tool stimulates the performance of miniature of machine tools using five-axis multipurpose machines. Yuichi [6] have proposed a methodology of addressing product robustness of micromachine tool. They mentioned that prediction of accuracy, metal removal rate, cutting force and shear angle of micromachine tool is suitable for small production system. They also represented the measurement of positioning error using analytical results and experimental procedures which is dependent on some assumptions. Park and Malekian [7] have stated the mechanics of machining at microscale in microfactory concept. They described the machine tool structure loop stiffness, vibrations, precision axis of motion and high damping forces during machining of microcomponents. Orthogonal and oblique cutting of nonferrous material produced high wear rate and shear formation. Srinivasa and Shunmugam [8] have proposed the concept of factories of twenty-twenty in 2020. They described that small-size machine units and tabletop micromachines have been developed for different purposes like micro-drilling, micro-slat, micro-turning and micro-forming of aluminum alloy seeds. Micro- and desktop factories can bring multiple benefits of microfactory system against the conventional factories. Most of researcher constructs a microfactory in a common table, and all the micromachines are erected in a fixed layout of desktop, which is difficult to operate. The microfactory is evaluated only by performance index and efficiency of plant. Only a little work has been carried out for development of desktop factory and determination to analyze of micromachines. After 2013, there is no development of desktop factory due to implementation of MEMS devices for miniature of machines [9–15].

3 Objectives

- To develop a portable microfactory for miniature of components.
- To run the portable tabletop microfactory for small production system.
- To find efficiency of portable tabletop microfactory
- To reduce labor cost through proficient generation of tweaked green items on the spot.
- To find error developed in portable tabletop micromachinery during machining process
- To find total performance index (TPI) of portable tabletop microfactory.
- To reduce heat, emission, vibration, noise, humidity and temperature, and it should be eco-friendly.
- To enormous saving of space, energy, time, material and other resources.

4 Portable Tabletop MicroFactory

The developed microfactory enables considerable saving in terms of resources, time, energy, space and material. The main benefits of microfactories are smaller investment and less operating cost. Micromachining is a particular system connected to small-scale parts. Micromachining is a specific technique applied to microscale parts. Micromachining refers to amount of material removal from 1 micron to 999 microns. They can be referred as microstructures, microsystems, mechatronics and microstructure innovation. Micromanufacturing is the manufacturing of products in small quantities using small manufacturing facilities. Micromanufacturing in general and mass content micromanufacturing processes (microforming, microcasting and microwelding) in particular have extensive applications in coming future due to inherent advantages of miniaturization. Miniaturized scale manufacture is the way toward creating little structures of micrometer scales and littler. Miniaturized scale manufacture is the way toward creating little structures of micrometer scales and littler. Miniaturized machine tools such as portable lathe, milling, drilling and surface grinding machine have been fabricated as per requirements, and corresponding position errors are calculated, and accuracy of each minimachine tool was predicted. The portable microfactory was designed to be packaged in big-size suitcase to demonstrate its portability. Figure 1 shows the portable microfactory which has external dimensions of 900 × 600 mm and weight about 15.5 kg. It can be placed in the truck of an automobile and transported with mounted casters. It functions entirely

Fig. 1 Portable tabletop microfactory

by single-phase connection with 230 v AC power source, and its power consumption is 1300 watts during plant operation. A normal bit of the smaller-scale industrial facility was that the miniaturized scale plant would lessen ecological effect and expenses of little mechanical manufacture, particularly for "differing types-and-little amount creation." Since the littleness of the machines empowers adaptable format transforms, it can control the expansion of the costs when the item structures have been altered. The optimum plant layout may increase the productivity of micro-components. Consequently, smaller machines have reduced vibrations and inertia effects and elastic forces when compared to the conventional machineries. Further, the thermal deformations were found to be decreased linearly in small dimensioned machines. The miniaturization of machine tool is also found to be increased in degree of freedom of product design, thereby modifying the system design. Microfactory elements are portable, flexible in plant layout, less electrical power consumption, need less skilled operator, minimum space and time which are the superior benefits of the system.

5 Total Performance Index

TPI is an index that can be used to theoretically evaluate the efficiency of manufacturing system.

$$\mathrm{TPI} = \mathrm{UV}/\sqrt{\mathrm{LCC}}\sqrt{}\mathrm{LCE}$$

TPI = Total performance index, UV = Utility value of product, LCC = Life cycle cost of product

LCF = Lifecycle environmental impact of product.

Figure 2 shows that the number of hands is plotted on X-axis and efficiency of system is plotted on Y-axis. It represents the effectiveness of framework expanded by number of hands expanded. The straightforward file to assess the framework proficiency of a small-scale plant was proposed by thinking about throughput, machine cost, work cost and outflow created by machine devices, material and electrical machines. The aftereffects of analysis of the test creation procedure of small-scale industrial facility had the option to demonstrate that the framework had some appropriate setups. It will be a decent answer for "different types-and-little amount creation.

1. **Advantages of Microfactory**

- Eco-friendly and user friendly
- Portable, compact and tabletop
- Improving space utilization factor
- Low investment cost
- Improved production rate
- Lightweight
- Facility investment
- Increase flexibility of plant layout
- Increased position accuracy
- Sustainability such as social, environmental and economics
- Greater saving of energy, power space, material and time
- Reduced emission of heat, vibration and noise.

2. **Disadvantages of Microfactories**

- Problem associated during product handling
- Vibration developed during machining process
- Suitable for small production system
- Need smaller size jig, fixtures and other clamping devices
- Need magnetic table for erecting all machineries.

3. **Main Features of Micromachines**

- High exactness pivot of movement and control with level of accuracy
- Small space, assets and vitality effective
- Machine apparatus structure with high circle firmness and great damping capacity
- High thermal solidness and mechanical stability with low vibrations
- Variable shaft speed and feed can be accomplished
- Portable and low power utilization

6 Sustainability of MicroFactory

The cutting-edge generation framework is relied upon to limit the ecological burdens the framework causes during its lifetime. This sets necessities particularly for the vitality and asset utilization, outflow and wastage just as reusability and transfer of creation framework and its segments. The smaller-scale manufacturing plant stages bargain of little estimated creation gadgets when contrasted with conventional bigger industrial facilities. They require less production line floor space, devour less vitality and crude materials and make less waste and emanations. The miniaturized scale processing plant cannot improve the assembling workplace yet additionally give better support of end client. The little size of smaller-scale production line arrangements enables them to be carried nearer to the end client. Indeed, even to the point of scale, it guarantees the quicker and move tweaked administration and fulfilled client. The little size smaller-scale processing plants can be adaptability situated to the most advantageous area and enable simple adoptability to various requests. This supportive "on-the-spot" assembling and certainty that miniaturized scale industrial facilities and greater condition amicable contrasted with enormous processing plants. From the social perspective, it is imperative to limit dangerous workplaces, improve ergonomics of the workplace and to demonstrate the effectiveness innovativeness and strength of laborers. All tabletop types of gear are non-programmed and reasonable for small-scale machining. Spindle speed is directed by voltage controller, feed and profundity of cut which is balanced by smaller-scale screws. Less introductory cost, less talented administrator, compact and works by less electrical power are the primary centrality of tabletop small-scale machines. All tabletop supplies are non-programmed and reasonable for small-scale machining. Spindle speed is controlled by voltage controller, feed and profundity of cut which is balanced by miniaturized scale screws. Less introductory cost, less gifted administrator, versatile and works by less electrical power are the fundamental essentialness of tabletop smaller-scale machines. The supportability of scaled-down creation framework was talked about from three points of view—environmental, economic and social.

1. **Environmental perspective**

A. Greater saving of energy and resources for machining, building, stores and operation such as power sources, lighting and other systems.
B. Reduce heat, emission, vibration, noise, humidity and temperature from inside and outside of factories. It is really eco-efficient.

2. **Economic perspective**

A. Improved equipment portability for transport, installation, storage, replacement dismantle, reuse, recycling and scalability for changes in production volume.
B. The natural attention to buyers is always expanding, and environmental impression of items begins to move and increasingly noteworthy factor directing the buy choice.

3. Social

In social point of view, "microfactories" are minimized hazardous work environments. It improves ergonomics of work environment to pursue the efficiency.

7 Conclusion

The developed microfactory enables considerable saving in terms of resources, time, energy, space and material. The main benefits of microfactories are smaller investment and less operating cost. Miniaturized machinery of portable lathe, milling, drilling and surface grinding machine has been fabricated as per requirements, and corresponding position errors are calculated, and expectation exactness was sufficient for use in the theoretical structure stages. Littler machines have diminished vibrations and inactivity impacts contrast with the customary hardware. The scaling down of machine apparatus is likewise observed to be increment in level of opportunity of item plan, thereby adjusting the framework structure. The little size smaller-scale factories can be adaptably situated to the most advantageous areas and great answer for "assorted sorts and-little amount generation framework." Tabletop microfactory is the name superior miniature manufacturing system which fabricates and assembles the small components without compromising accuracy. From the social perspective, it is essential to limit perilous workplace and pursue effectiveness. The maintainability of scaled-down generation framework was talked about from three points of view—economic, ecological and social.

References

1. Furuta K (2000) Microfactory system in Japanese national R&D project. Singapore–Japan Forum on MEMS, Singapore. pp 14–19, Nov 2000
2. Michel P, Vogler MP, Liu X, Kapoor SG, DeVor RE, Ehmann KF (2002) Development of meso-scale machine tool (mMT) systems. Technical Paper—Society of Manufacturing Engineers. MS, (MS02-181), pp 1–9, Jan 2002
3. Mekid S (2004) High speed desktop ultra Precision CNC micro and meso machine. Adv Mater Res 739:640–646
4. Werkmeister J, Slocum A (2003) Design and fabrication of the meso-mill: a five-axis milling machine for meso-scaled parts. In: Winter topical meeting, machines and processes for micro scale and meso scale fabrication, metrology and assembly, pp 79–82, Apr 2003
5. Tanaka M (2001) Development of desktop machining micro factories Machine tools for the Micro factory. In: Proceedings of international workshop on micro mechanical fabrication techniques, pp 137–140, Apr 2001
6. Yuichi O (2004) Microfactory: concept, history, and developments. J Manuf Sci Eng 126:837–844
7. Park SS, Malekian M (2009) Mechanistic modeling and accurate measurement of micro end milling forces 58:49–52

8. Srinivasa YV, Shunmugam MS (2013) Mechanistic model for prediction of cutting forces in micro end-milling and experimental comparison. Int J Mach Tools Manuf 67:18–27
9. Codourey A, Perroud S, Mussard Y (2006) Miniature of reconfigurable assembly line for small products In: Proceedings of third International seminar on precision assembly, pp 19–21, Feb 2006
10. Mekid S, Ryu HS (2007) Rapid vision-based dimensional precision inspection of mesoscale artefacts. Proc Inst Mech Eng Part B J Eng Manuf 4:659–672
11. Keong Ng C, Melkote SN, Rahman M, Senthil Kumar A (2006) Experimental study of micro- and nano-scale cutting of aluminum 7075-T6. Int J Mach Tools Manuf 46:929–936
12. Brecher C, Utsch P, Klar R, Wenzel C (2010) Compact design for high precision machine tools. Int J Mach Tools Manuf 50:328–334
13. Ogedengbe T (2010) Contribution to the design and operation of a micro milling machine. PhD thesis, The University of Manchester
14. Razali AR, Qin Y (2012) A review on micro manufacturing, micro forming and their key issues. In: Malaysian technical university conference on engineering and technology
15. Nee AYC, Ong SK (2013) Virtual and augmented reality applications in manufacturing. In: 7 th international conference on manufacturing, modeling, management and control, pp 15–26

Analyzing the Manufacturing Operations and Identifying the Bottlenecks in Food Processing Industry

Chetan Prakash Chhalani, Piyushh Bhutoria, Yash Agarwal, and S. Aravind Raj

Abstract The prime focus of any industry is to consume the energy efficiently and identify the bottlenecks in the system that will enhance the efficiency of process and the system. The companies should not omit the problems those are considerably influencing the potential of the production processes. System and people accountable for those right chains of manufacturing process try to find ways to remove the bottlenecks and minimize the waiting time per process at the mechanical system. The speed of the production process is confined by the process which has the least efficiency. The inferred results and outputs of the experimental analysis display the chances of the bottlenecks in the system. The theory of constraints is used to identify the bottlenecks by taking the input data over a period of the time and analyzing it. Its data includes the average complete production time required for each individual product to be manufactured. The bottlenecks of the machine or system are found which further helps in increasing the efficiency of the system with efficient use of resources opted. The foremost aim of this paper is to spot the probability of implementing a preferred simulation tool and analyzing production process bottlenecks. The elemental approach of analysis utilized for the study was literature studies with simulation.

Keywords Bottleneck · Productivity · Production process · Food industry · Production line

1 Introduction

As per the theory of constraints (TOC), the main task of efficient production process management is to seek out and reduce the overall impact of the bottleneck for a particular firm by restructuring the rest of the resources around it. A bottleneck is a process with lowest output in a system that can cause parts to build up between processes

C. P. Chhalani · P. Bhutoria · Y. Agarwal · S. Aravind Raj (✉)
Department of Manufacturing Engineering, School of Mechanical Engineering, Vellore Institute of Technology, Vellore, Tamil Nadu 632014, India
e-mail: aravindsakthivel@hotmail.com

A. Arockiarajan et al. (eds.), *Advances in Industrial Automation and Smart Manufacturing*, Lecture Notes in Mechanical Engineering,
https://doi.org/10.1007/978-981-15-4739-3_8

and can also cause other operations to wait and be unproductive. It highlights the capability of production system as a whole.

The capability of production system may be a critical part of producing firms' competitive advantage, its associates in nursing influence on ecological problems and company's prices [1].

Production management may be divided into the following taking the prevalence of bottlenecks into consideration [2]:

- In a very state of affairs with zero bottlenecks within the production process method.
- In a very state of affairs with only one bottleneck within the production process method.
- In a very state of affairs with several bottlenecks within the production process method.

For the production process in case of zero bottlenecks, productivity is sufficient to fulfill the forecasting sales, whereas in case of one bottleneck it should be determined if it is the only process available or any alternative process is available. In this case, the contribution margin is a useful tool for analyzing each combination of processes and products. In the case of more than one bottlenecks parallel in the production process, decision becomes more complex. In any such case, the solutions to the problem are ought to be found using linear programming methods.

2 Literature Review

In the literature, the bottleneck is defined in several ways but many of them cannot be practically useful in general. In line with one among the definitions touching on production process, a bottleneck is associated key element of a production method, wherever each particular resource that has got to be accustomed maximizes production, it is used in 100%. A 100% use of the assembly capability for a given particular machine brings forth a considerable danger in the functioning of the production process used. The machines' bottlenecks are characterized majorly based on mishandling of the system, but which further tend toward failure of the system. Most of the time, completion date with respect to the production system incurs the additional issue.

The production potency and capability of the complete method are outlined as a limiting factor of the bottleneck. The smallest co-contributing parameters of a production system among the set of the contributing parameters are at the lowest level of any production cell or at the company's workstation level. This may cause a problem as such where machine before reaching the state of the bottlenecks finishes the processing of the system, so it cannot send material further, as a machine will

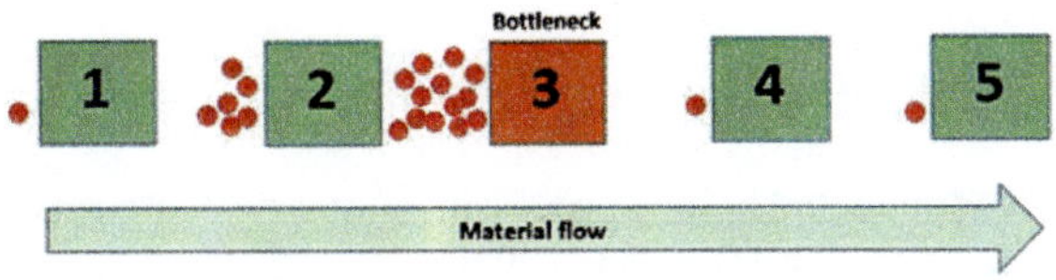

Fig. 1 General idea of bottleneck. *Source* Elaborated by the author based on [3, 4]

come after that, being the following method's bottlenecks, remains pinned in process previous orders.

Bottlenecks may be also increasing the time of the standoff within the production processes resulting at a sequent stage, and waiting time of the further future orders is increased drastically. Bottlenecks put forward the levels in the whole production methods. The existing definitions square measures steady in particularly one way—impacts on the potency of the manufacturing systems are gained majorly by bottlenecks, and the material flow in the production system is moreover as even loading in the machines. Figure presents the thought of the bottlenecks (Fig. 1).

Improving the working of the machines that is late, the assembly could be concerned problem. However, it ought to be kept in mind that before pursuing any act targeted toward developing the system further to the machines outlined for the bottlenecks, the particular locations are to be spotted.

In the theory of constraints, the first step of managing the level of the constraints in production line is to find the bottleneck. It associates limiting the process limitations. There are more furthermore levels of managing the constraints which even can be used for the simulation of the computer-aided simulation tools and models:

- Subordinating all alternatives operations to a selection made into stage two;
- Removing the process's bottlenecks;
- Coming back to stage one and reducing the limit output of inertia;
- Making a call on the way of employing a bottleneck;

As each process or systems have to have constraints, the finding and removing of a number of them lead to the circumstance of the new developed limitations that will change the previous already there. Five-stage process management procedure is the consecutive process.

The existing literature finds several strategies in sleuthing production bottlenecks. However, quiet few units have no swollen embellishments, and further case studies imply that the digital simulation models can be used in the respect. Therefore, the put forward issue needs more studies applying various packages.

The developing simulation tools with increasing the capabilities area unit associate degree in progress analysis challenge.

3 Methodology

The case study was conducted in food processing industry located at Guwahati, India. The data is collected for period from May 2018 to June 2018 (2 months). Discussions were done with internal experts in quality and production department. The sequence of work is done in following flowchart.

The TOC was formulated by Goldratt (1970s). To enhance the functioning of business companies, he applied the methods of exact sciences. In this study, the theory of constraints is used to focus only on production process of the company, but it can also be used for the intention of managing whole supply chain or whole company [5, 6].

Cyplik [7] implemented TOC in production area for pharmaceutical industry. TOC lays down a set of tools used to manage and control bottlenecks and, increase profit defined by TOC, consequently. The five focusing steps method is one of the basic tools in TOC. Continuous perfection of the system is its essential property (Fig. 2).

Step one identifies the system constraint. The process with least output efficiency in the production line is usually the constraint to the manufacturing process. Next, we decide how to exploit the constraint. In this step, the decision is made to streamline the use of critical resource and reducing unwanted activities. Next, we subordinate everything else. In this step, we subject every decisions made to work and the constraint's productivity. Next, we elevate the constraint. This step focuses on improving the overall productivity of the constraint, and finally, going back to step one. If the constraint is eliminated, we go back to the top, i.e., the first step.

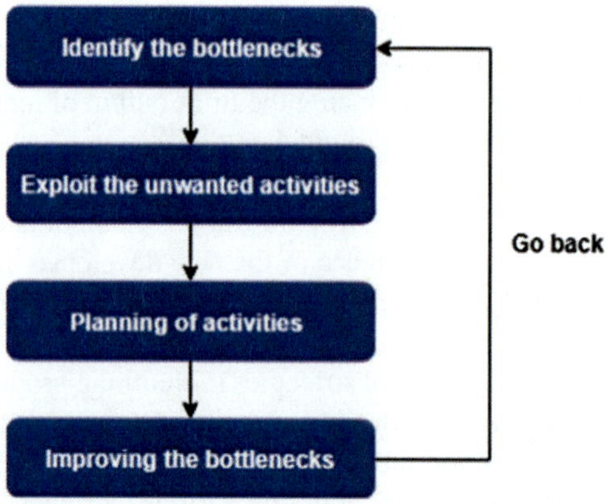

Fig. 2 Basic TOC tool

4 Experimentation/Modeling

The perused and examined production line system consists of two production lines each with eight machines in order to reach the production quantity. In production line 1, all the machines are fully automated and are directed to give out product S1, and each machine is allotted with an employee who is solely accountable for the end product and sending it for packaging. Process layout is the layout followed by that particular industry. The perused and examined production company data is stated as below:

- T_{av}—Machine availability time (480 min.),
- T_{ct}—Machining time,
- T_{st}—Machine preparation time,
- T_{u}—Periodic intervals or breaks.

The perused and examined line 1 gives out product S1 in shift one out of all the shifts performed in that particular day, usually there are three shifts in that particular company. Table 1 states the company's operational data in regards with the individual machinery.

All the machineries of the system in production line 1 possess different machining time, facilitating with the lease efficient machinery providing overall productivity of the system machinery.

The experimental modeling starts with analyzing and perusing the systems bottlenecks in the production line 1, to know the discrete productivity of the machinery. The mathematical expression of the productivity used of machinery is given below:

$$P_{E0} = \frac{(t_{av} - t_{st} - t_u)}{t_{ct}}$$

The productivity of operation on M1 is therefore

Table 1 Individual machine data line 1

Machine	T_{ct} (in sec)	T_{st} (in sec)	T_u (in sec)
M1	0.869	900	600
M2	0.895	900	480
M3	0.9375	900	1020
M4	1.395	900	1200
M5	1.7142	900	1800
M6	1.666	900	1260
M7	0.8	900	420
M8	1.395	900	1500

$$P_{\mathrm{Eo}} = \frac{(480 * 60) - 900 - 600}{0.869 * 60} = 524$$

The remaining machinery data is obtained similarly. Table 2 holds the data of the productivity of systems discrete operation in line 1.

Examining the discrete productivity of each machinery and pre-holding the mathematical tool with its assumptions used for the analyzing, i.e., TOC which states that the productivity of the weakest link in the line is adequate to the systems whole productivity, it is needed to be expressed in the form of productivity for line 1, i.e., 254 pieces from every shift are obtained. So, it is required for all the systems machinery operations to tool with respect to the discrete bottlenecks' productivity. Therefore, the potency of each discrete machinery is not obtained.

The degree of machine usability is known from the ratio of quotient of productivity of the entire manufacturing line (254 pieces) to the productivity of each machine. For M1, usage is

$$P_{\mathrm{M1}} = \frac{254}{524} = 48.5\%$$

Table 2 Line 1 discrete machine productivity

Machine	Productivity (pcs/shift)
M1	524
M2	511
M3	478
M4	319
M5	254
M6	267
M7	573
M8	315

Table 3 Line 1 degree of using each machine

Machine	Productivity	Degree of using individual stations (%)
M1	524	48.5
M2	511	49.7
M3	478	53.1
M4	319	79.6
M5	**254**	**100**
M6	267	95.1
M7	573	44.3
M8	315	80.6

Note The above obtained/formulated data has shown that machinery M5 had reached a one hundred percent of its usage which gives no revelation as their still exist bottlenecks in line 1 which restrict us to achieve the production capability fully

For the remaining machines, the usage is calculated similarly. Table 3 contains the degree of using individual machines in production line 1.

The above obtained/formulated data has shown that machinery M5 had reached a one hundred percent of its usage which gives no revelation as their still exist bottlenecks in line 1 which restrict us to achieve the production capability fully. As achieved discrete operations (i.e., M1) amount usage has given us an unsatisfactory result, which holds that increasing in the amount of the usage will cause the pilling of the stock which will further lead in increasing of the production value of the product. The weakest line (i.e., M7) has the least amount of production with the least usage of the machinery in the operations cause's bottlenecks.

Line 1 production line is not discrete but has additionally various machinery with the specific options as shown and presented in Table 1. Systems production process is classically formed as to allotment of one worker with each discrete machinery due to variability in machining time. Line 2 produces similarly as line 1 but different product (i.e., S2). The customer order is facilitated on regular basics which keeps on varying in entire year or in wide spectrum. Information and data of the amount of periodic orders are known, and machining time is given in Table 4.

The initial step of experimenting the bottlenecks for this kind of a line consists of verifying that if the systems production line is known to be capable of achieving its all of the orders. Machining time for each discrete machinery is therefore multiplied by the volume of the orders obtained for particular products. Knowing that system has three-shift production ($8\text{ h} \times 3 \times 60 \times 60 = 86400\text{ s}$), it is checked and known if all of the discrete machinery is potent to deliver production with the compliance of the time available for the production.

Daily M1 usage time is

$$M1 = 395 \times 0.83 + 223 \times 0.55 = 450.5\,\text{min}$$

Table 4 Line 2 machine operational data (in sec)

Product	S1	S2
Forecast number of orders	395	223
Margin	600	350
M1	0.83	0.55
M2	0.83	0.45
M3	0.83	0.33
M4	0.97	0.45
M5	0.76	0.33
M6	0.86	0.41
M7	0.68	0.33
M8	0.48	0.35

Table 5 Bottlenecks recognition

	Product S1	Product S2	Total
Forecast no. of orders	395	223	618
Margin	600	350	–
M1	327.85	122.65	450.5
M2	327.85	100.35	428.2
M3	327.85	73.59	401.4
M4	383.15	100.35	483.5
M5	300.2	73.59	373.79
M6	339.7	91.43	431.13
M7	268.6	73.59	342.19
M8	189.6	78.05	267.65

For other machineries, the indicator is known in the similar steps as in the above. Table 5 shows the recognition of the bottlenecks in line 2.

Table 5 brings us to conclude that M4 goes beyond the time available (480 min) on each shift. It means that it is ineffective of producing product to meet the daily demand of individual product. Therefore, consecutive steps of analyzing bottlenecks will consist of reducing the quantity of products produced so as to make the most use of the weakest link in the whole process. To this end, RCM should be determined as follows:

$$\text{Relative contribution margin (RCM)} = \frac{\text{contribution margin per piece}}{\text{time of bottleneck usage}}$$

The profitableness of the systems given goods produced is shown by the value of RCM. According to the factor, as increase in the worth of the systems indicator increases the production of the discrete good to achieve more cost effectiveness for that particular systems product. Value of RCM is obtained as follows:

$$\text{RCM}_{\text{UC}} = \frac{600}{0.97} = 618.55$$

For other discrete products, the RCM value is obtained in the same as shown above. Table 6 has the RCM value for every discrete product in the system.

From Table 6, we understood that product S2 is the most cost efficient and therefore arrive at the conclusion, whereas the product S1 is least economic. Thus, if it is compulsory to remove any part of production process, the part with least cost efficient needs to be given up. Consistent with the TOC, the solely optimized is known with the outlined bottlenecks.

So, the time consumed by M4 needs to be minimized, and time of producing the given systems product (i.e., S1) needs to be reduced with 3.5 min. (483.5−480 = 3.5). Working time of the machinery M4 should also be reduced by 3.5 min. As

Table 6 Each product's relative contribution margin

Product	S1	S2
Forecast No. of Orders	395	223
Margin	600	350
M4 (Sec)	0.97	0.45
RCM	618.55	777.77

there is decrease in working time, machinery M4 requires 0.97 min to produce each product and is shown in Table 4.

The best obtained solution in the present circumstances is

- Maximum volume of order obtained for producing product S1 is 223 pcs.
- Maximum volume of order obtained for producing product S2 is 395 pcs.

It is possible to determine the revenue by knowing the volume of orders for every product:

$$P = 395 \times 600 + 223 \times 350 = 315050$$

5 Results and Discussions

The scrutiny of the given data says that the bottleneck in the said example is the machine M7, with 97.6% productivity [8], whereas the leftover time period is given to stops in the production process that are prepared in advance. It is seen that the machine M7 impacts appreciable standoff at the machine M6, touching 57.6% of the simulation time. The found bottleneck impact increased in between operation stocks with further blocking of the machine M6. It also imposes the period of waiting time for the semifinished products at the machine M8 (Table 7).

6 Conclusion

The reason for many unfavorable circumstances at a company, resulting in significant monetary losses, is a bottleneck. Therefore, the whole production process needs to be strictly monitored. It permits early recognition of a bottleneck, hence limiting its effects to the lowest level.

The theory of constraints targets on enlarged output of constraint which, consequently, permits the rise within the output of the whole manufacturing unit. Thus, the foremost necessary component of theory of constraint is the speed at which

Table 7 Detailed time utilization of machines

Machine	Working time (in %)	Waiting time (in %)	Blocked time (in %)	Break time (in %)
M1	82.4	–	16.3	1.3
M2	86	5.7	7	1.3
M3	73.1	3.8	21.8	1.3
M4	54.4	44.3	–	1.3
M5	36.3	9.7	52.7	1.3
M6	41.1	–	57.6	1.3
M7	97.6	1.1	–	1.3
M8	47.5	42.53	8.67	1.3

a company produces and sells its products, receiving money in return. The exercise of applying the theory of constraints implies where one's attention needs to be centered, on what needs to be changed and not on what needs to be restricted. Therefore, corporates need to fully focus on the activities which increase the output.

The recommended solutions are associated with organizational behavior, as per the criterion of selection. The future analysis is directed toward developing additional elaborated alternative of determination of corresponding contribution margin within the manufacturing process.

The issue of the bottleneck is key core problem encountered by production companies while improving or formulating their manufacturing process. During this case study, key issue was to figure out a manufacturing process blockage with the assistance of a simulation model.

References

1. Kolinski A (2017) The impact of eco-efficiency in production on availability of machines and equipment. In: Golinska-Dawson P, Kolinski A (eds) Efficiency in sustainable supply chain. Springer International Publishing, Berlin
2. Kolinski A, Tomkowiak A (2010) Wykorzystanie koncepji analizy waskich gardel w zarzadzaniu produkcja, Gospodarka Materialowa i Logistyka, No 9.
3. Betterton CE, Silver SJ (2012) Detecting bottlenecks in serial production lines—A focus on interdeparture time variance. International Journal of Production Research 50(15): 4158–4174
4. Hsiao Y-C, Lin Y, Huang Y-K (2010) Optimal multi-stage logistic and inventory policies with production bottleneck in a serial supply chain. International Journal of Production Economics 124(2): 408–413
5. Cyplik P, Hadas L, Adamczak M, Domanski R, Kupczyk M, Pruska Z (2014) Measuring the level of integration in a sustainable supply chain. IFAC Proc Vol 47(3):4465–4470
6. Hadas L, Cyplik P, Adamczak M (2015) Dimensions for developing supply chain integration scenarios, business logistics in modern management
7. Cyplik P, Hadas L, Domanski R (2009) Implementation of the theory of constraints in the area of stock management within the supply chain-a case study. LogForum 5(3):1–12

8. Kolińska J, Domański R (2017) The analysis of production lines bottlenecks identification and ways of management. In: 17th international scientific conference business logistics in modern management, Osijek, Croatia, 12–13 Oct

Design and Optimization of Constrained Damping Layer Thickness of Aluminium Plate Structure at Various Wave Modes of Vibration

Rama Rao Thatigiri and Meera Saheb Koppanati

Abstract Different types of dynamic loads will originate from the external sources like aerodynamic forces in case of airborne applications and road unevenness in case of automotive applications. Aircraft structures, aerospace vehicle structures, automotive structures, etc., are manufactured by maximum utilization of plate-like structures only. Suppression of effect of dynamic loads on plate structures is most important at the stage of design of any product. Application of constrained damping layer in plate structure without any changes in structural dimensions particularly thickness of plate is the best method to control the dynamic load distribution in the plate structures. The constrained damping layer thickness varies quantitatively for different modes of vibration of plate. The determination of exact or optimized damping layer thickness requirement for the aluminium sandwiched plate structure is considered as the key factor to overcome the maximum effect of dynamic loads in plate-like structures. As part of this work, natural frequency of the plate along with loss factor, and damping layer thickness are computed on iteration methodology, and at convergence condition these values are taken as optimum for a particular mode of vibration. The proposed analytical method will be converted in the form of a general-purpose computer program in MATLAB. This code will be applied to any dimension of the plate at multiple vibration mode patterns. The outcome of this work would be an analytical method that can be used by any designer to find the optimum damping layer thickness of a plate-like structure in order to reduce the vibration response.

Keywords Structures · Aluminium plate · Damping layer thickness · Design and optimization · Constrained · MATLAB · And natural frequency

R. R. Thatigiri (✉) · M. S. Koppanati
Jawaharlal Nehru Technological University, Kakinada 533003, Andhra Pradesh, India
e-mail: ramaraohal@gmail.com

M. S. Koppanati
e-mail: meera.aec@gmail.com

A. Arockiarajan et al. (eds.), *Advances in Industrial Automation and Smart Manufacturing*, Lecture Notes in Mechanical Engineering,
https://doi.org/10.1007/978-981-15-4739-3_9

1 Introduction

Plate structures are often used in many applications like automotive, aerospace, aircraft, etc., as load-bearing platform like brackets. Dynamic loads like vibration originated from the external sources like road roughness in case of automobiles and aerodynamic pressure in case of aerospace vehicles. This will transmit to the subsystem which is mounted on plate through plate itself. Subsystems are supposed to deliver the intended functions safely without getting influenced by vibration. If the influence of vibration on subsystems is high, there are three ways to reduce the effect of vibration. First one is reducing the vibration at the source which is practically not possible. Second one is attempting on the subsystem which is also not possible as the design of the subsystem will be frozen. The only possible way is to modify the path, i.e. interface supporting structure (plate). The distinguishing features of a polymer, the damping layer, which make it useful for controlling the vibration, are its resilience and its high-energy dissipation capacity. In the part of the interface supporting structure, the vibration response reduced by sandwiching the constrained damping layer (viscoelastic material) between the two different metallic plates; here, the usage/requirement of damping layer is not defined [1, 2]. The structural design of the supporting structure is to be fine-tuned in such a way that the amplification in vibration between the supporting structure and the subsystem is to be minimized. Damping is a measure of rate at which vibration energy dissipates in any structure. In order to protect any structure from vibration, damping needs to be enhanced. Vibration characteristics, the resonance frequency, the loss factor, and the dynamic Young's modulus of structural materials, polymers, and composites vary markedly with temperature and loading rate [3]. Damping exists in many forms like material damping, dry friction damping, etc. Damping is even influenced by addition of damping layer to the plate in sandwiched configuration. During a flexural vibration, the damping layer is subjected to a large shear deformation and the shear damping is likely to exceed the extensional damping. The base layer and the constraining layer are assumed to be non-dissipative, and most of the energy is dissipated due to the shear deformation of the viscoelastic layer. However, for a given plate on which vibration response is to be minimized, damping layer is to be designed and optimum thickness of damping layer maximizes the damping factor. Design and optimization of damping layer for enhancing the damping of an aluminium plate is and thereby reducing vibration response is taken up in this work. To begin with theoretical formulation will be established for design and optimization of damping layer using which optimum damping layer thickness will be arrived at. Formulation will be extended to predict the vibration response of the plate without and with damping layer. This project aims at establishing theoretical formulae for design and optimization of damping layer for plate structure at different vibration modes of random vibration in MATLAB.

2 Design and Optimization of Constrained Damping Layer Thickness

Design input modal parameters of 1.5-mm thickness sandwiched three-layered rectangular four-side supported plate simply supported aluminium plate for the study:

1. Length of plate, $a = 1000$ mm
2. Width of plate, $b = 500$ mm
3. Thickness of plate, $h_1 = 1.5$ mm
4. Wave number in x-direction, m
5. Wave number in y-direction, n
6. Wave speed, C_L (aluminium) = 5150
7. Young's modulus, $e^3 = 7 \times 10^{10}$ Pa
8. Poisons ratio, $v = 0.3$
9. Damping factor for aluminium = 0.005 (0.5%).

The configuration of plate structure and dimensional notations of sandwiched three-layered rectangular four-side supported plate is shown in Fig. 1.

h_1 = Thickness of layer 1 (parent material)

h_2 = Thickness of damping layer (core material thickness)

h_3 = Thickness of layer 3 (parent material).

Damping layer, a viscoelastic material used as core material, the effect of high-damping material on the damping of the whole structure is influenced by the material stiffness as well as by its damping. These two properties are conveniently quantified by the complex Young's modulus $E^{\mathrm{I}} + iE^{\mathrm{II}}$ or the complex shear modulus $G^{\mathrm{I}} + iG^{\mathrm{II}}$. Both of these will be usually presented in the complex forms which embody the loss factors η_E and η_G, i.e. $E^* = E^{\mathrm{I}}(1 + i\,\eta_E)$ and $G^* = G^{\mathrm{I}}(1 + i\,\eta_G)$. η_E and η_G are usually assumed to be equal for a given material. When the material is subjected to

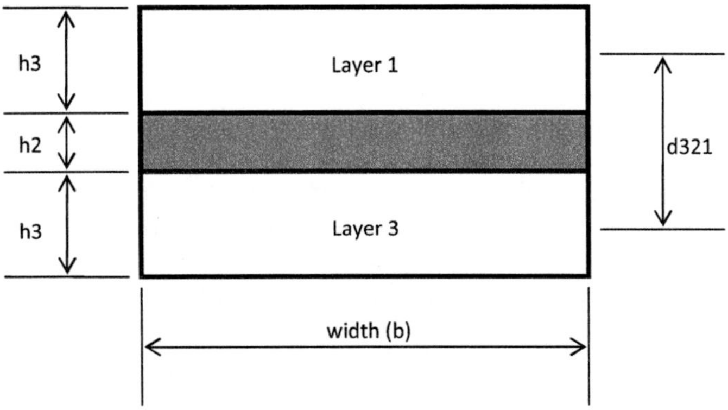

Fig. 1 Three-layered rectangular four-side supported plate

cyclic stress and strain with amplitudes σ_0 and ε_0, the maximum energy stored and energy dissipated per cycle in a unit volume are as follows:

$$\text{Maximum energy stored} = \delta U = \sigma_0^2/2E^{\mathrm{I}} = E^{\mathrm{I}}\varepsilon_0^2/2 \tag{1}$$

$$\text{Energy dissipated per cycle} = \delta W = 2\pi\,\eta_E\sigma_0^2/2E^{\mathrm{I}} = \pi\,E^{\mathrm{II}}\varepsilon_0^2 \tag{2}$$

For a given strain amplitude, the material which dissipates most energy is therefore the one with the highest loss modulus, E^{II}.

3 Formulae

Optimum damping layer thickness, loss factor, and natural frequency have been calculated for rectangular four-side supported plate structure. The governed general bending deflection equation of plate structure is

$$EI_t\frac{\partial^4 W}{\partial X^4} + m_t\frac{\partial^2 W}{\partial t^2} = p(x,t) \tag{3}$$

Layer 2 is now the damping layer and will be referred to as the core. It is assumed that damping layer will have negligible direct stiffness in the x-y plane, but to undergo no direct strain in the z-direction. When the whole plate bends, it therefore constrains the top and bottom plates to deflect in the z-direction by equal amounts $w(x, y, t)$. Its complex shear modulus is $G_2{}^* = G_2^{\mathrm{I}}(1 + i\,\eta_s)$, η_s being the shear loss factor of the material.

When the finite shear stiffness of the core is properly taken into account, the homogeneous differential equation with $\omega^2(1 + i\eta_{\mathrm{DC}})$ as a complex eigenvalue for $w(x, t)$ is of sixth order and can be shown to be

$$\frac{\mathrm{d}^6 w}{\mathrm{d}x^6} - g^*(1+Y)\frac{\mathrm{d}^4 w}{\mathrm{d}x} - \frac{\omega^2(1+i\eta_{\mathrm{DC}})m_t}{EI_t}\left(\frac{\mathrm{d}^2 w}{\mathrm{d}x^2} - g^* w\right) = 0 \tag{4}$$

The normalized imaginary part, η_{DC}, is not only the constant of proportionality between $p(x, t)$ and the local inertia loading. It is also the configuration loss factor of the modes $w(x)$ which satisfy the equation and has a different value for each mode. These modes are damped forced normal modes and can be used in a computational modal analysis of the response of the plate. The real part of the complex eigenvalue corresponds to the frequency at which the mode resonates when the above form of external harmonic loading is applied. In practice, it can be regarded as one of the resonance frequencies of the plate under any reasonable form of harmonic excitation.

$1 + Y$ is a non-dimensional measure of the flexural stiffness of the whole layered plate. From the earliest studies of constrained layers, Y has been called as the geometric parameter, although it is actually a function of both the geometry and the elastic moduli of the layers. The terminology will nevertheless be retained. In terms of the EAs and EIs of the two plate layers, Y is given by

$$d_{321} = \frac{(h_1 + h_3 + 2h_2)}{2} \tag{5}$$

$$Y = \frac{d_{321}^2}{EI_t}\left[\frac{E_1A_1E_3A_3}{E_1A_1 + E_3A_3}\right] \tag{6}$$

The non-dimensional stiffness and damping of the three-layered plate are expressed in as stiffness ratio due to added damping configuration, $K_{\mathrm{DC}},$ loss factor due to added damping configuration, $\eta_{\mathrm{DC}},$ and their product, the resonant displacement index, $K\eta_{\mathrm{DC}}$.

$$K_{\mathrm{DC}} = 1 + Y\left(\frac{g\left[1 + g\left(1 + \eta_{\mathrm{s}}^2\right)\right]}{1 + 2g + g^2(1 + \eta_{\mathrm{s}}^2)}\right) \tag{7}$$

$$\eta_{\mathrm{DC}} = \frac{g\eta_s Y}{1 + g(2 + Y) + g^2(1 + \eta_s^2)[1 + Y]} \tag{8}$$

$$K\eta_{\mathrm{DC}} = \frac{g\eta_s Y}{1 + 2g + g^2(1 + \eta_s^2)} \tag{9}$$

The above equations can be seen to depend on just three non-dimensional parameters: Y, shear parameter of plate structure (g), and the loss factor (η_s) of the damping layer. The shear modulus (G_2^1) of the core and the wave numbers (or mode numbers m and n) are combined together within g. Clearly, the stiffness ratio and modal loss factor both depend on gm and hence on m and so are both mode-dependent. They also vary from mode to mode, because different modes have different resonance frequencies, so G_s^{I} and η_s for the frequency-dependent core material have different values for different modes.

4 Procedure for Design and Optimization of Constrained Damping Layer Thickness

Necessary equations for design and optimization of constrained damping layer thickness for damping enhancement of plate are arranged in the form of sequence. The optimized natural frequency of a particular vibration mode of plate, maximum damping layer thickness, and loss factor can be obtained after repeating sequences of steps several times by maximizing the loss factor (2 times of the damping factor) of the plate. This procedure can be applied for any dimension of the plate to find the modal frequencies and damping factors of constrained damping layer plates for N number of vibration modes by considering some geometric parameters.

However in order to execute the above-said procedure to arrive at optimized damping layer thickness which maximizes the loss factor, MATLAB code is written as part of this work.

4.1 Development of Program in MATLAB

This work aims at design and optimization of constrained damping layer for enhancement of damping in an aluminium plate.

MATLAB code is given below.

```
% Beginning of code
clear
clf
m = 3; % mode number along length
n = 3; % mode number along width
a = 1; % Length of plate
b = 0.5; % Width of plate
h1 = 1.5/2000; % Thickness of plate
h3 = h1;
c1 = 5150; % Wave speed
v = 0.3; % Poisson's ratio
e3 = 7e10; % Young's modulus
y = 3; % Geometric parameter (Starting value)
j = 20; % Number of iterations
f = zeros(j,1); % Frequency
ndcmax = zeros(j,1); % Maximum loss factor (2 × damping factor)
h21opt = zeros(j,1); % Optimum thickness ratio of damping layer
dampthick = zeros(j,1); % Optimum thickness of damping layer
for i = 1:j;
kf2 = (pi^2*((m/a)^2 + (n/b)^2)); % Wave number
.......
.......
.......
results = [f h21opt dampthick ndcmax]
close; %closes all figures opened by default
%code for ploting
% plot(kdctemp,results(:,1),'-*')
% grid on
% xlabel('Stiffness (kdc)')
% ylabel('Natural frequency, Hz (f)')
% title('Stiffness of plate vs. Natural frequency')
```

```
% plot(kdctemp,results(:,3),'-*')
% grid on
% xlabel('Stiffness (kdc)')
% ylabel('damping Layer thickness, mm')
% title('Stiffness of plate vs. Damping layer thickness')
plot(kdctemp,results(:,4),'-*')
grid on
xlabel('Stiffness (kdc)')
ylabel('Maximum loss factor(ndcmax)')
title('Stiffness of plate vs. Maximum loss factor')
```

5 Results and Graphs

The results pertaining to design and optimization of constrained damping layer thickness which are obtained using formulae implemented in MATLAB are given below:

5.1 CASE 1: Vibration Mode Pattern of Plate (M = *3,* N = *3)*

1. Optimized natural frequency = 161.09 Hz
2. Optimized damping layer thickness = 0.8384 mm
3. Maximum loss factor = 0.4937.

Observations from Figs. 2, 3, and 4: The results of fundamental natural frequency, damping layer thickness, and maximum loss factor are increased directly proportional to the flexural stiffness of the plate structure and are optimized at the value of stiffness approximately 3.8.

CASE 2: Vibration Pattern of Plate ($m = 4, n = 3$)

1. Optimized natural frequency = 165.134 Hz
2. Optimized damping layer thickness = 0.4711
3. Maximum loss factor = 0.4165.

Observations from Figs. 5, 6 and 7: The results of fundamental natural frequency, damping layer thickness, and maximum loss factor are increased directly proportional to the flexural stiffness of the plate structure and are optimized at the value of stiffness approximately 3.0.

CASE 3: Vibration Pattern of plate ($m = 5, n = 4$)

1. Optimized fundamental natural frequency = 244.702 Hz.
2. Optimized damping layer thickness = 0.1193

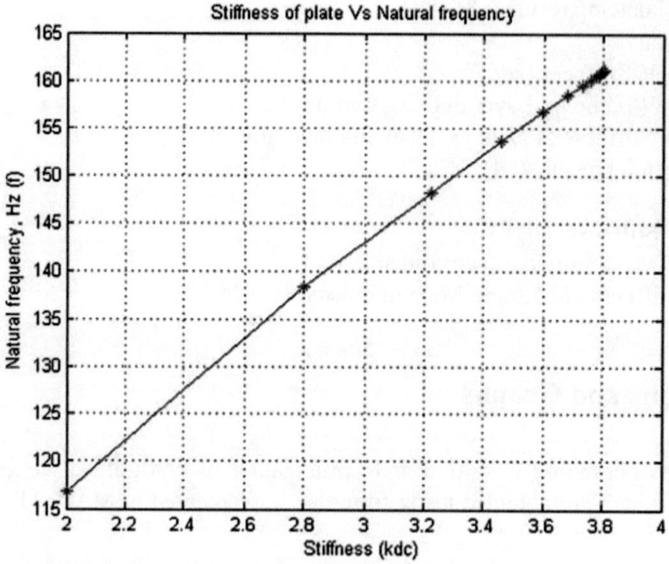

Fig. 2 Modal stiffness versus modal natural frequency of plate structure

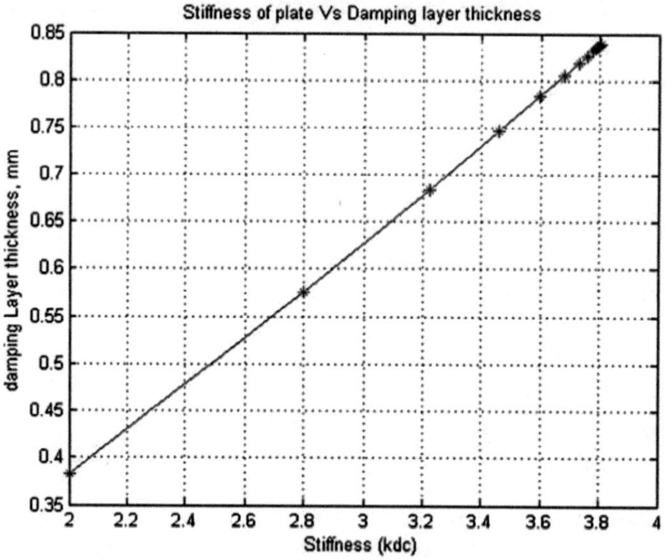

Fig. 3 Modal stiffness versus modal damping layer thickness of plate structure

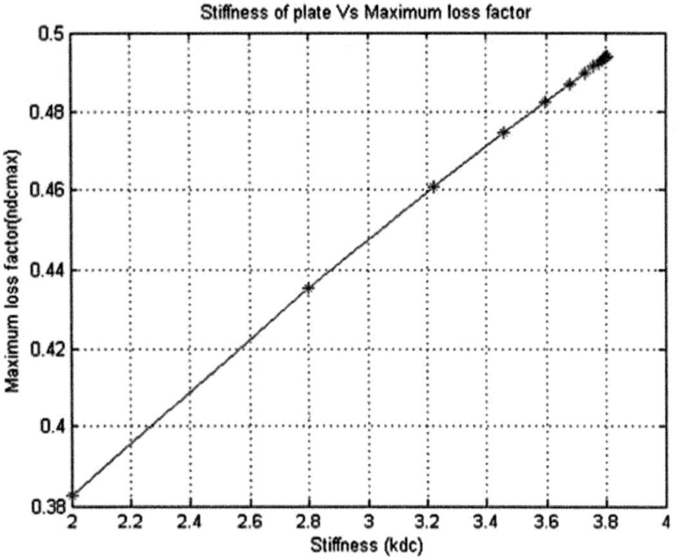

Fig. 4 Modal stiffness versus modal maximum loss factor

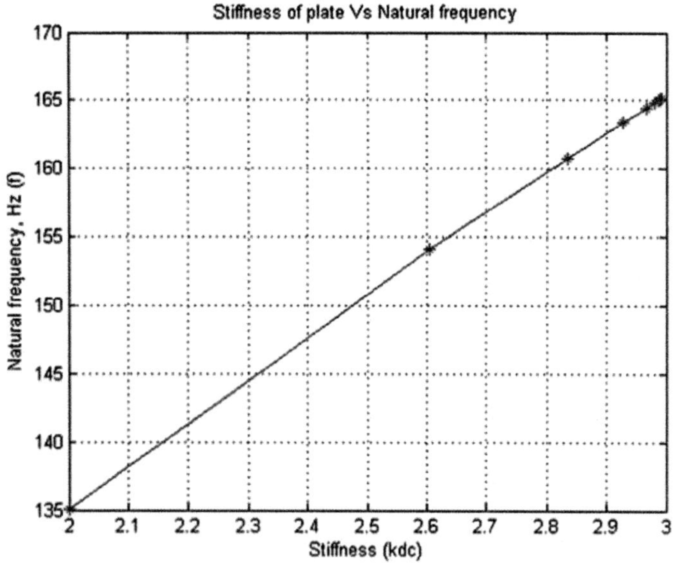

Fig. 5 Modal stiffness versus modal natural frequency of plate structure

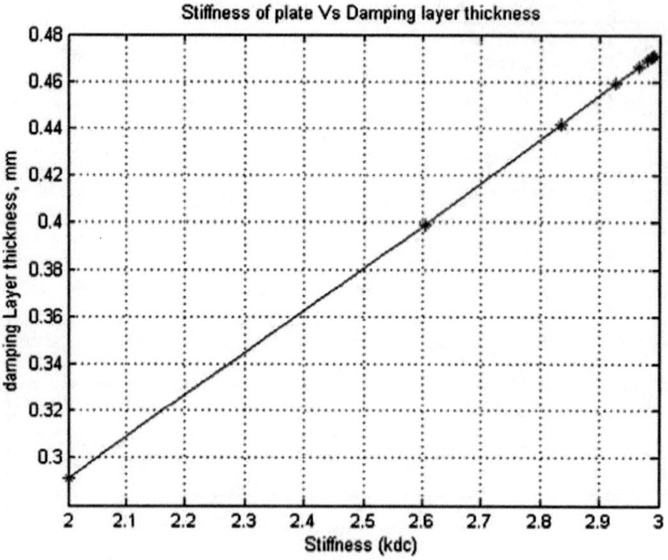

Fig. 6 Modal stiffness versus modal damping layer thickness of plate structure

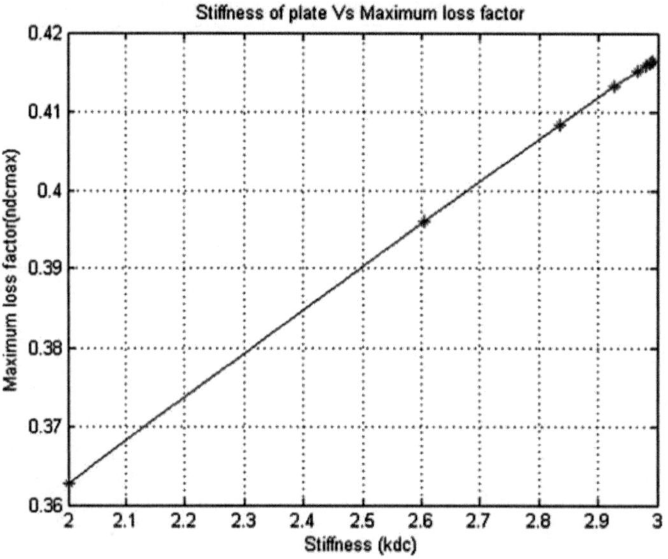

Fig. 7 Modal stiffness versus modal maximum loss factor

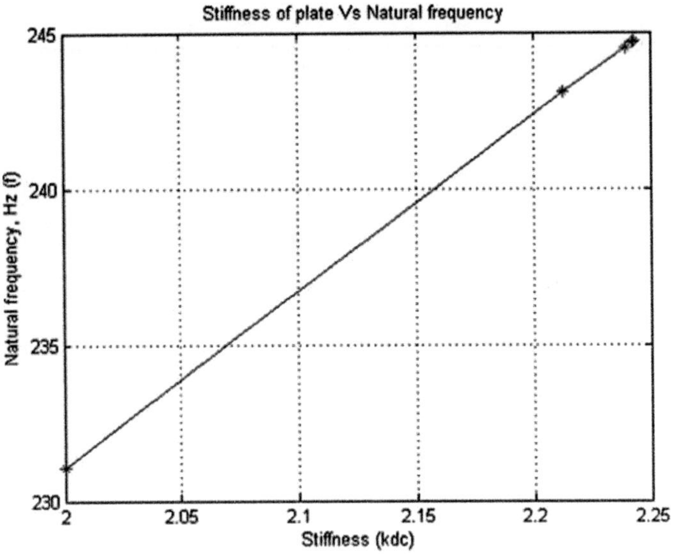

Fig. 8 Modal stiffness versus modal natural frequency of plate structure

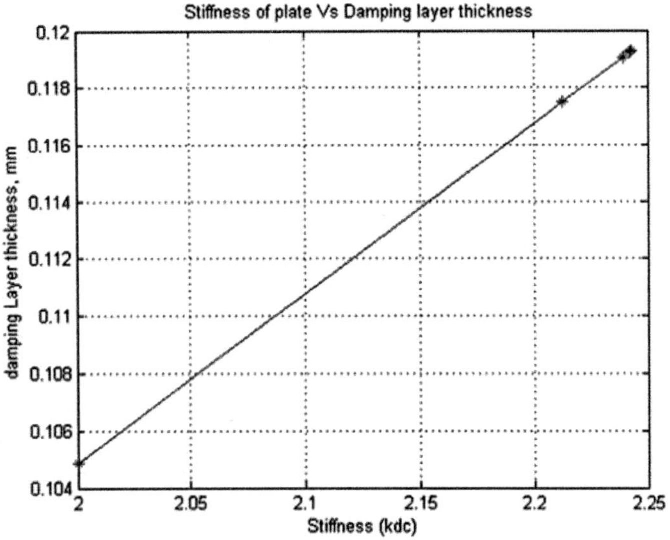

Fig. 9 Modal stiffness versus modal damping layer thickness of plate structure

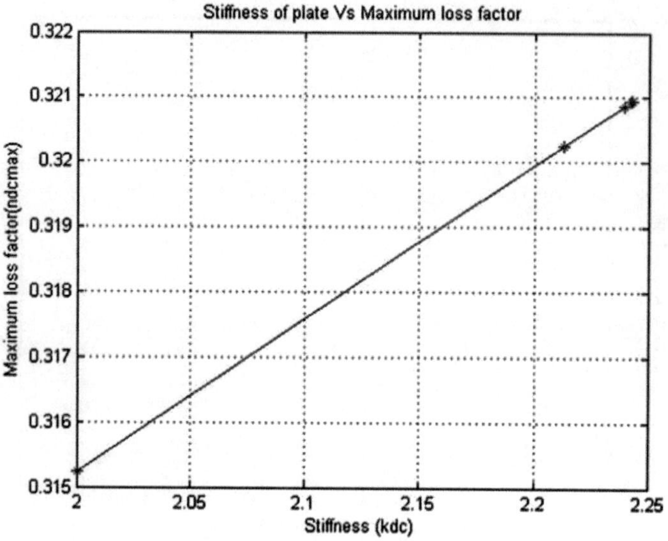

Fig. 10 Modal stiffness versus modal maximum loss factor

3. Maximum loss factor = 0.321.

Observations from Figs. 8, 9 and 10: The results of fundamental natural frequency, damping layer thickness, and maximum loss factor are increased directly proportional to the flexural stiffness of the plate structure and are optimized at the value of stiffness approximately 2.25.

CASE 4: Vibration pattern of plate ($m = 5, n = 5$)

1. Optimized fundamental natural frequency = 333.92 Hz
2. Optimized damping layer thickness = 0.0581 mm
3. Maximum loss factor = 0.3015.

Observations from Figs. 11, 12, and 13: The results of fundamental natural frequency, damping layer thickness, and maximum loss factor are increased directly proportional to the flexural stiffness of the plate structure and are optimized at the value of stiffness approximately 2.12.

6 Conclusion

The different iterations are performed on the optimized design of three-layered aluminium plate structure by varying the constrained damping layer thickness and flexural stiffness of the whole plate structure at various wave modes of vibration. Thus, we have obtained the optimized value of damping layer thickness at different

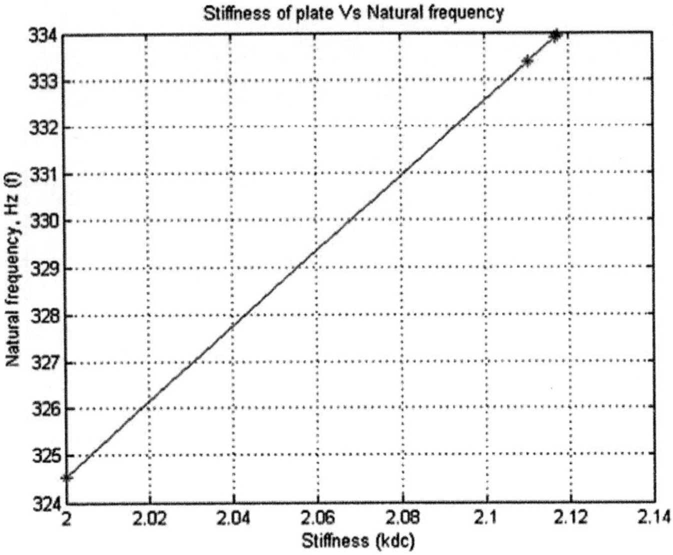

Fig. 11 Modal stiffness versus modal natural frequency of plate structure

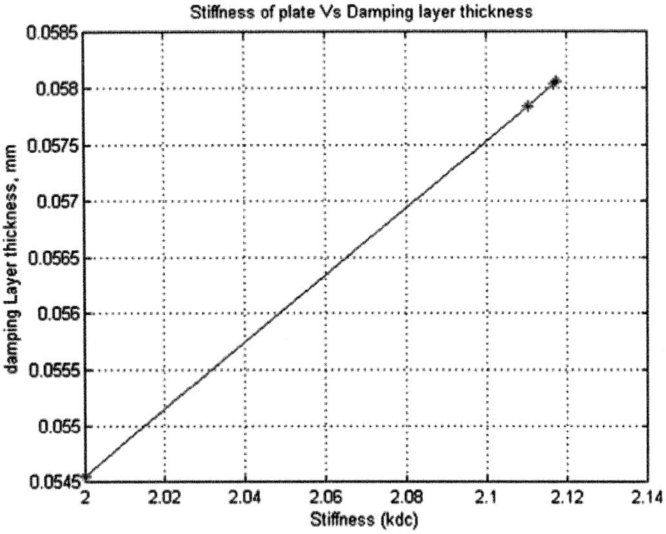

Fig. 12 Modal stiffness versus modal damping layer thickness of plate structure

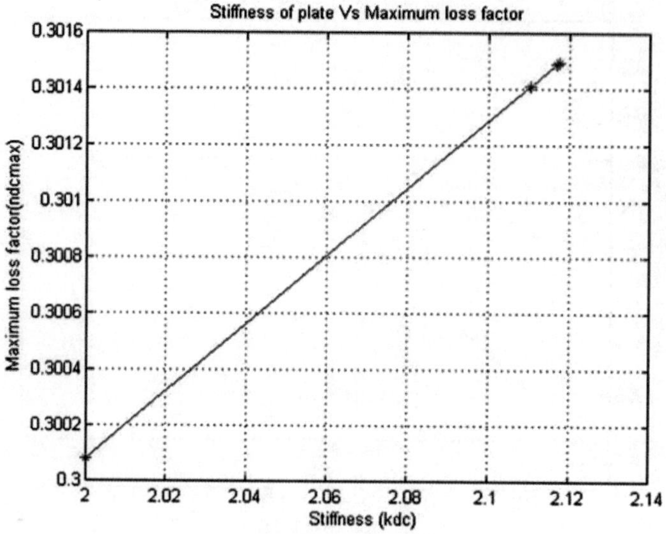

Fig. 13 Modal stiffness versus modal maximum loss factor

wave modes of vibration. The optimized damping layer thickness is more at lower modes of vibration, and thickness becomes less at higher modes of vibration for given plate structure. Hence, the determination of optimized damping layer thickness at various levels of vibration becomes useful for designing any product.

References

1. Patil V, Talole M, Kharche PP, Mahesh MS (2016) Analysis of constrained layer treatment for damping in skin panels of aircraft. Int J Trend Res Dev 3(3). ISSN 2394–9333
2. Pandit A, Oza N, Nimbhorkar S (2016) Investigation of partial constrained layer damping treatment effect on vibration analysis. Int J Sci Eng Res 7(10)
3. Gallimore C, Kochersberger K, DeVita R. Virginia Tech. Unmanned Systems Laboratory, Blacksburg, VA 24061

Design and Investigation of a Go e-Kart Frame by Utilizing CAD Tools

Syed Kwaja Moinuddin, K. Nagaraju, Turpunati Shahinsha, and K. L. Srinivasulu

Abstract This report says about demonstrating, playing out the static examining of a go e-kart body comprising of round pillars. Displaying is performed exploitation demonstrating programming framework, for example, NX computer-aided design and investigation on ANSYS [1]. The most redirection is set by performing basic investigation. The skeleton is planned such that it needs less material and, in addition, it is ground-breaking enough to oppose the various effects consequently [2]. Quality and lightweight were our essential idea all through the arranging of the suspension of the kart. Subsequently, PVC was assigned as partner in nursing adequate material for style with properties such as lightweight and high strength. Every effects and stresses pondered the extreme working conditions; then, the arranging was dissected inside the examination programming framework. Well-ordered adjustments in style were made as discovered fundamental and as investigated on the product framework [3]. When the entire examination and endorsement of style is investigated everything by arranging to create the criteria of stresses which are burden incited.

Keywords Chassis · NX CAD · PVC · ANSYS

S. K. Moinuddin (✉) · K. Nagaraju · K. L. Srinivasulu
Department of Mechanical Engineering, Santhiram Engineering College, Nandyal, Andhra Pradesh, India
e-mail: khnmoin5@gmail.com

K. Nagaraju
e-mail: rajumgm321@gmail.com

K. L. Srinivasulu
e-mail: k.l.srinu.123@gmail.com

T. Shahinsha
Department of Mechanical Engineering, Mahaveer Institute of Science and Technology, Hyderabad, Telangana, India
e-mail: shahinsha.cool@gmail.com

A. Arockiarajan et al. (eds.), *Advances in Industrial Automation and Smart Manufacturing*, Lecture Notes in Mechanical Engineering,
https://doi.org/10.1007/978-981-15-4739-3_10

1 Introduction

The task of the vehicle frame is to hold all the parts while riding and then to transmit perpendicular and directional abundance via the tires on the frame induced by accelerations. Many learners in engineering have an accompanying knowledge of stresses, and torsional stiffness lengthy till they navigate it. Many individuals pressure complete of fabric selection, but it is the gateway to a nice storage frame once you are acquainted with it. Since this will improve the performance, more particular concepts can still benefit. A design chapter can talk a lot about these things. On the 3D design tool, we built a frame design software pattern. To use this mode of machine code, its team has been able to evaluate the appearance in the 3D region and reduce manufacturing mistakes. Its primary objective throughout the structure of the frame was to obtain perfect blend between a wide and durable rider region with simple leakage and evacuation and flexible sizes to accomplish the necessary weight and rigidity requirements. Regarding the above in the event of a rollover situation, the sizes needed for the desired inspections were approximately set using a simulated structure.

The final design was chosen after a series of style modifications and subsequent estimations. The vehicle's layout process is continuous and is based on countless design activities.

The layout mechanism therefore concentrates on the following goals:

- Safety
- Serviceability
- Strength
- Ruggedness
- Standardization
- Cost
- Durability
- Lightweight
- High performance.

1.1 Literature Survey

A frame consists of an internal frame promoting the construction and use of a manmade product [4]. It is similar to the skeleton of an animal. The under part of an automobile is an example of a chassis, composed of the frame (on which the body is installed). When the operating equipment such as wheels and transmission is included, and often the rider's position, the setup will be defined as a floating structure. Its truss requires operator, engine, brake system, carburetor, and driving system load, but truss must have sufficient power to safeguard the operator in case of an effect. The rider's compartment must be able to withstand most of the stresses that are placed on it. It could be accomplished either by utilizing substance of elevated intensity or by better protecting components against the transmitted load. Frame must be made of ASME (AMERICAN SOCIETY OF MECHANICAL ENGINEERS) metal ducting of lowest possible demands for dimensions or even stability.

To achieve desire design, reading 3D vector graphics like firm works is crucial. Model assessment defines its constraints in the structure that performs a significant part in the system. The assessment will be assumed either structure is secure or not, it is also possible to see the curvature and force changing in the kart. Gautama Yadav et al. investigated that perhaps the X model they are using is many flexible and has a safety factor of 2.5 in particular. Our first concern was passenger safety as well as elevated model stability that enables the racing car throughout rotating. Thus, the strong structure or the powerful framework of a roll cage maximizes the job required and gives the required outcome. The structure weight is also lower. The length of the path would be the metric from the middle of the tire to the middle of the tire.

Measurement is done from the middle with twin tires. It will have a major impact on a go kart's acceleration and braking conduct.

2 Design of Chassis

These are a major component of a go e-kart. Either an empty channel or a supertubular region would be preferable if the vehicles were to become durable and stunned. Skeleton configuration ought to be to such an extent that it ought not be exposed to wind during sharp turns; in this way, it must have adequate pliable and flexible enough to oppose impacts of divergent powers.

2.1 Chassis Modification

The frame has been changed according to the structure imperatives set forth in the prior area. The frame produced alongside the motor's darting intent and distinct frill needed a strong foundation in which to mount its drive train frame. The generated case was then installed on to the foundation wall next to the tightening point and then solidly welded to it.

2.2 Design Consideration

The product has made it possible for our team to imagine a 3D plan and reduce factory errors. The key basis of the skeleton setup was perfect harmony between a large and ergonomic driver area with easy entry and departure and conservative measurement in order to meet the necessary weight and torsional inflexibility requirements. The last undercarriage setup was resolved after advancement of the configuration modifications, with the resulting restricted fundamental research using ANSYS-16 programming.

2.3 Design Destinations

1. Include complete rider certainty by obtaining necessary quality as well as torque unbending nature although structural rigidity through constant determination of pvc.

- Architecture of construction, and decrease of expenses in order to guarantee aggressive fabric costs and other go e-karts installation expenses. Strengthen user convenience by increasing horizontal room in the passenger area.
- Maximize simplicity in workability by making sure that individuals do not interfere with separate subsystems.

2. Decide on cost performance as far as large-scale manufacturing is concerned. Calculation of the car's Pressures under various stacking conditions. For instance, price, drivability, support, easy use, and safety, the product can prove to be skilled in every perspective.

2.4 3-D Modeling

2.4.1 Importance of Modeling

In the context of increasing worldwide market rivalry, a paradigm of offering goods with increasing diversity, reduced batches and higher quality can be described in the modern production setting. Unless better quality, reduced costs, and shorter fresh products are introduced, the industries cannot resist global competition. There is intense international rivalry, and the supply of skilled workers is falling. Technicians are now using computer-aided design (CAD), computer-aided manufacture (CAM), and computer-aided engineering (CAE) systems to automate their procedures by significantly improving computing energy and increasing the accessible design and manufacturing of software instruments. The PC framework is the use of NX DESING PLM programming in order to help create, adjust, examine, or enhance a scheme. NX DESIGN programming is used to increase the originator's profitability, enhance the configuration nature, enhance correspondence through paperwork, and create a database for assembly. NX DESIGN is an important contemporary manufacturing technology widely used in countless applications, including automobiles and aircraft. For the 3-D suspension program, our team used a 3-D modeling program named NX PLM.

3-D demonstration was finished by utilizing NX-CAD programming as appeared in Fig. 1 (Fig. 2).

2.4.2 Introduction to NX Unigraphics

NX is the sophisticated and closely integrated product development solution in the World. If offers enormous value to companies of all sizes, spanning the entire variety of product growth. It became complicated product designs and thus speeds up the process of marketing products. Sophisticated analysis, simulated graphic and engineering concurrent design, industry design, geometric designs, advanced graphics simulations, industrial designs, advanced analysis, and simultaneous engineering. Concepts in a knowledge-based way. The software has powerful hybrid modeling

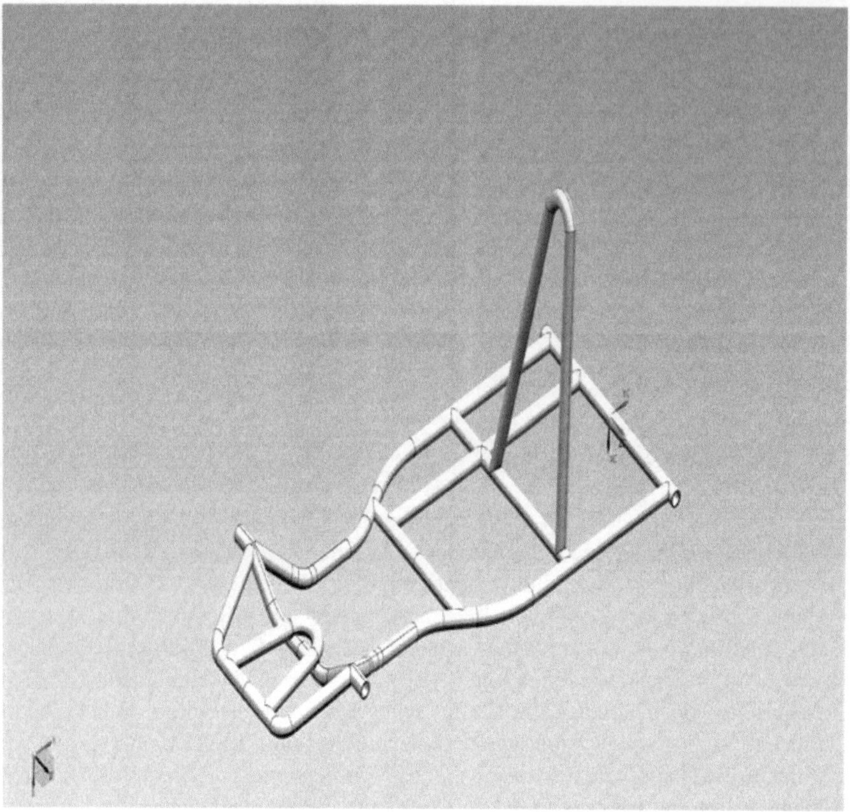

Fig. 1 3D CAD model of go e-kart hollow chassis

capabilities by integrating constraint based function modeling and explicit geometric modeling. It allows the user to design complex compounds in relation to modeling ordinary geometry parts.

3 Investigation of Hollow Frame in Static Stacking

The static burden style of undercarriage includes style of car once it is very still [5].

Static stacking on body:

- Driver alongside seat and frill.
- Roll confine.
- Engine.
- Steering framework.

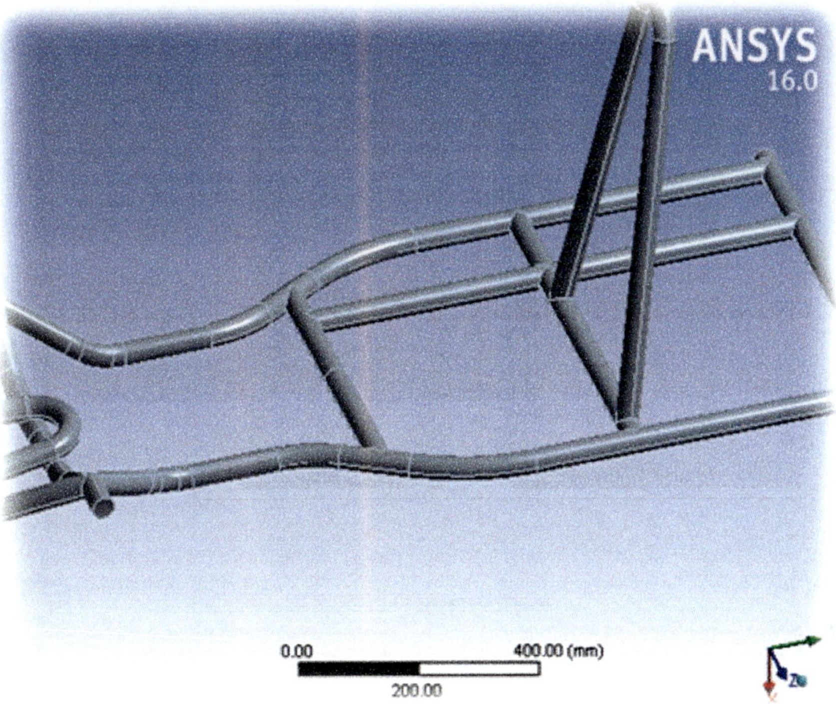

Fig. 2 3D CAD model of go e-kart solid chassis

Heap of rider and rider seat were considered while heap of guiding structure and so on are small when contrasted with above sections it tends to be ignored henceforth.

Static Loads on Hallow Chassis Total Deformation with Static Loads

See Figs. 3, 4, 5, 6, 7, 8, 9 and 10.

4 Analysis of Solid Chassis in Static Loading

See Figs. 11, 12, 13, 14, 15, 16, 17 and 18.

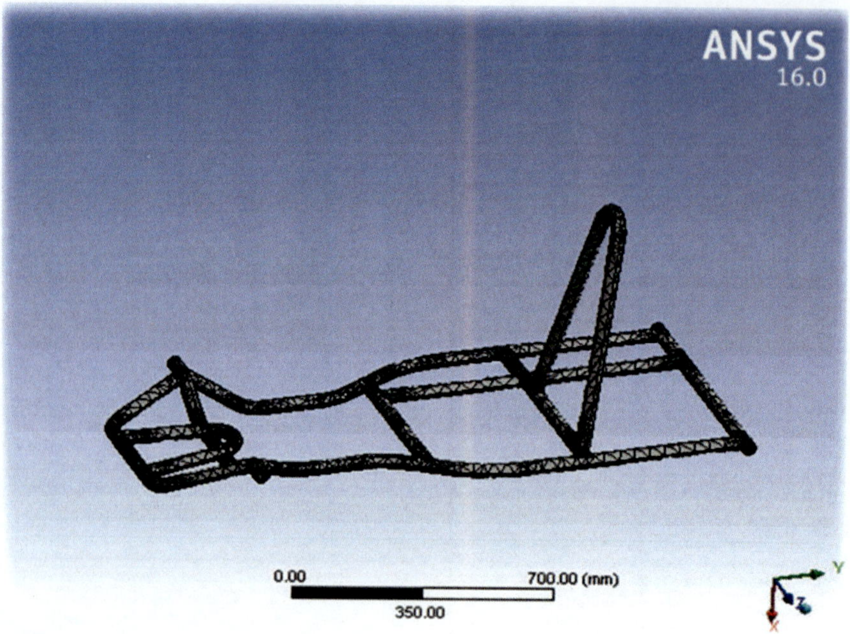

Fig. 3 Meshing of frame hollow chassis)

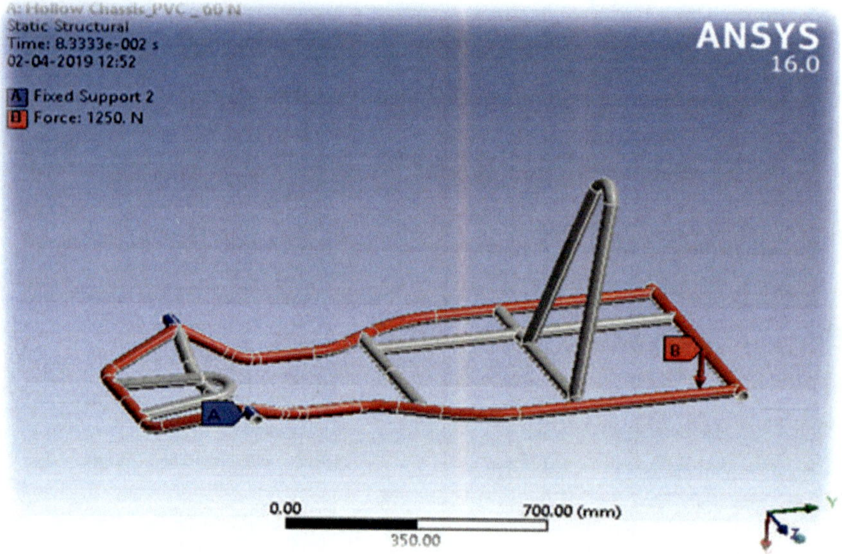

Fig. 4 Load under 2000 N of hollow chassis

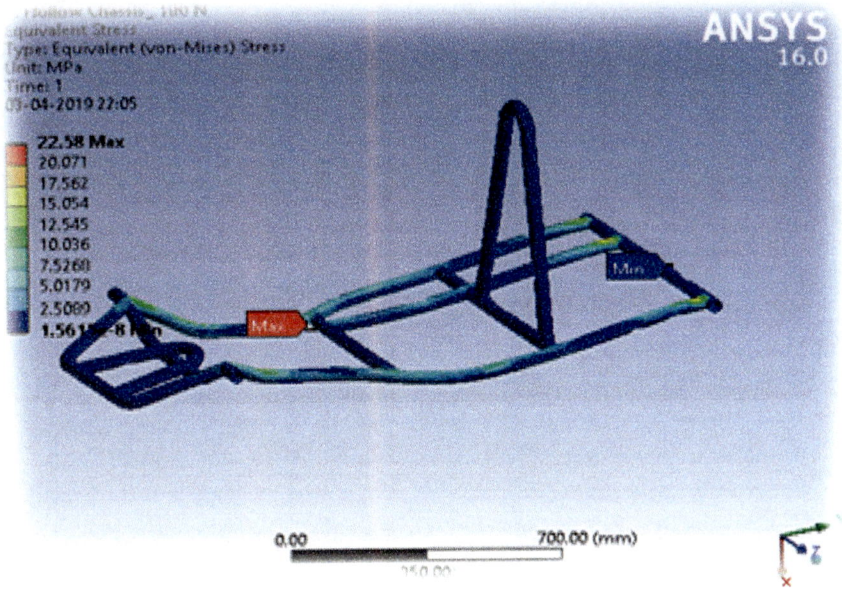

Fig. 5 Equivalent stress of hallow chassis under load condition of 1500 N

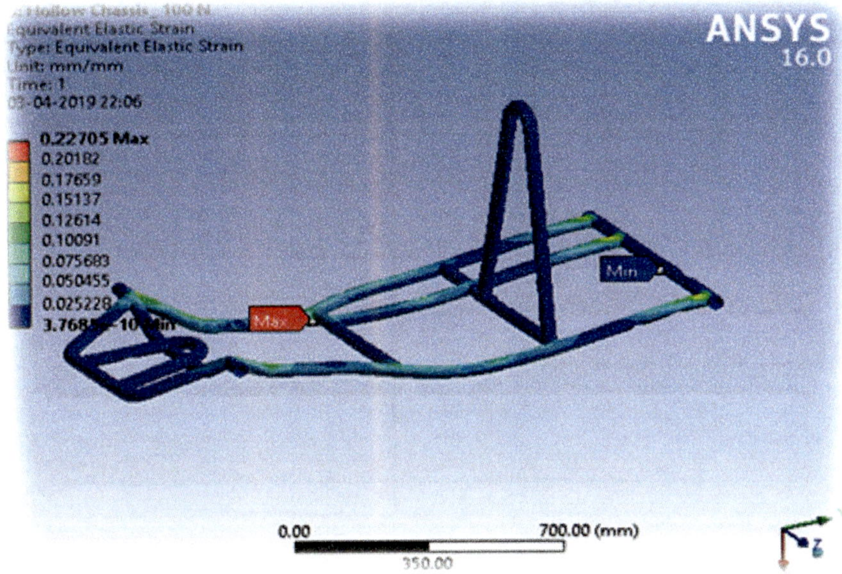

Fig. 6 Equivalent strain of hallow chassis under load condition of 1500 N

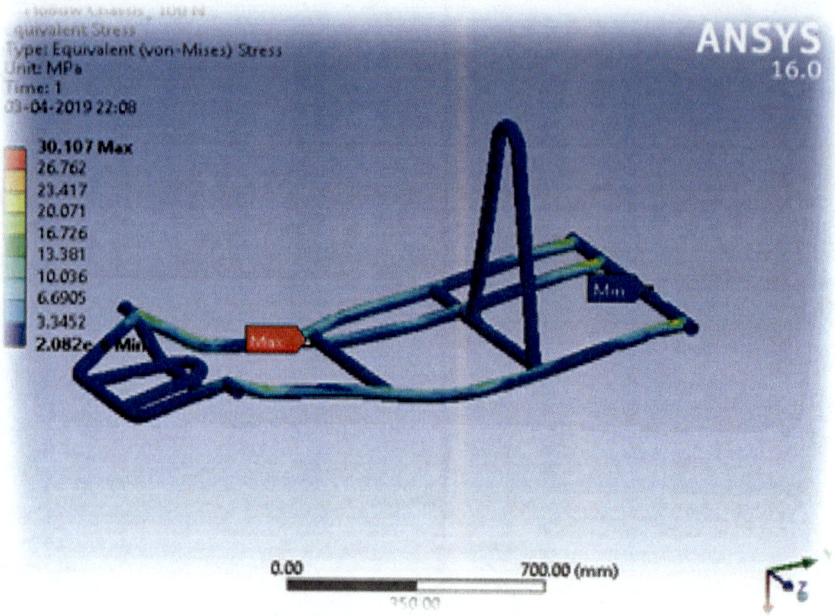

Fig. 7 Equivalent stresses of hollow chassis (PVC) under load condition of 2000 N

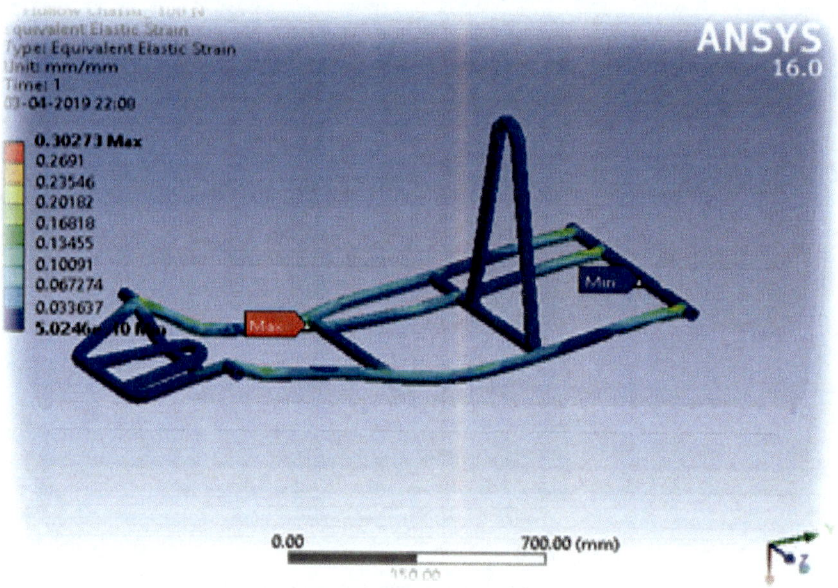

Fig. 8 Equivalent strain of hollow chassis (PVC) under load condition of 2000 N

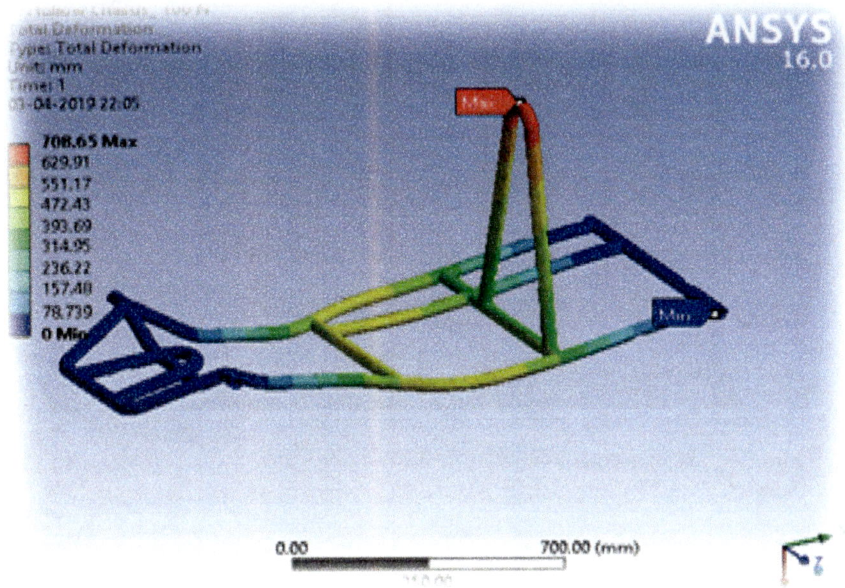

Fig. 9 Total deformation of hollow chassis (PVC) under load condition of 1500 N

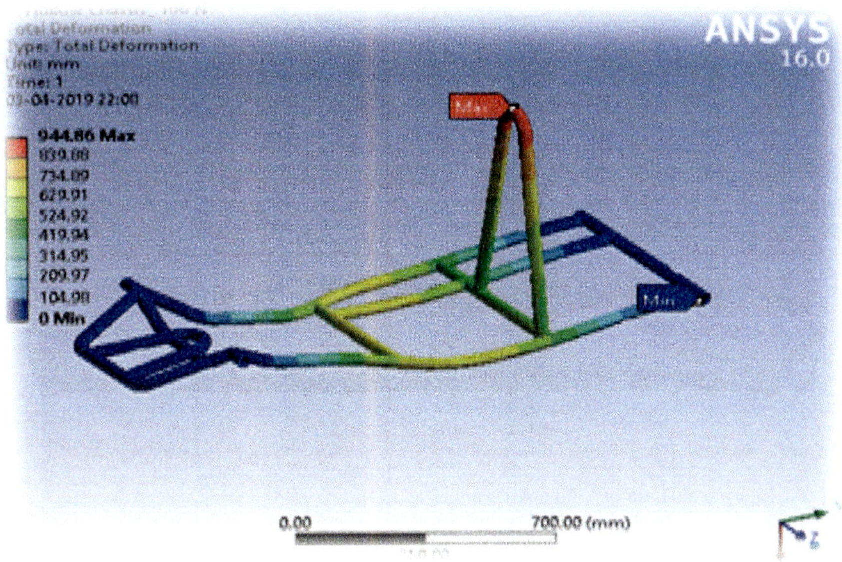

Fig. 10 Total deformation of hollow chassis (PVC) under load condition of 2000 N

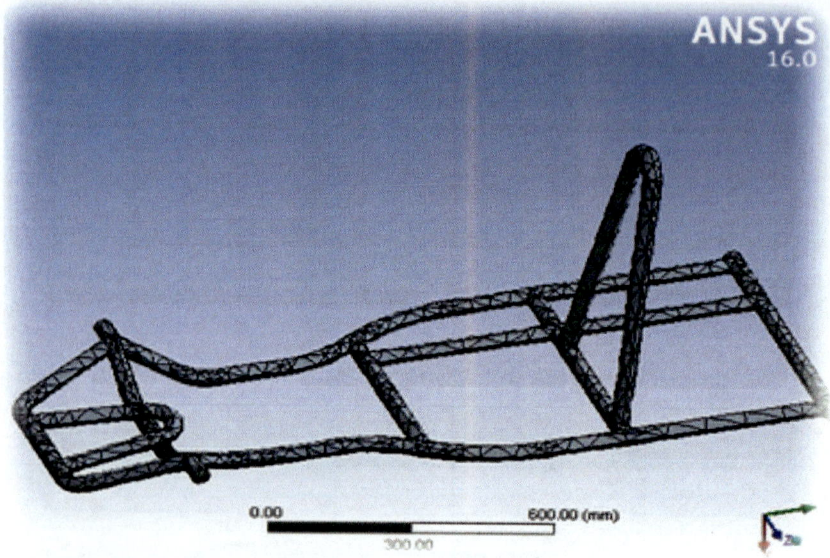

Fig. 11 Meshing of solid chassis under 2000 N

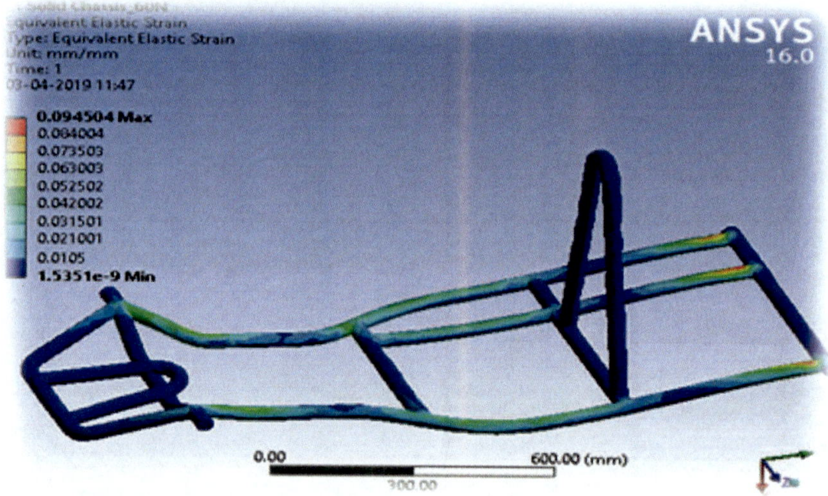

Fig. 12 Equivalent strain of solid chassis under 1500 N

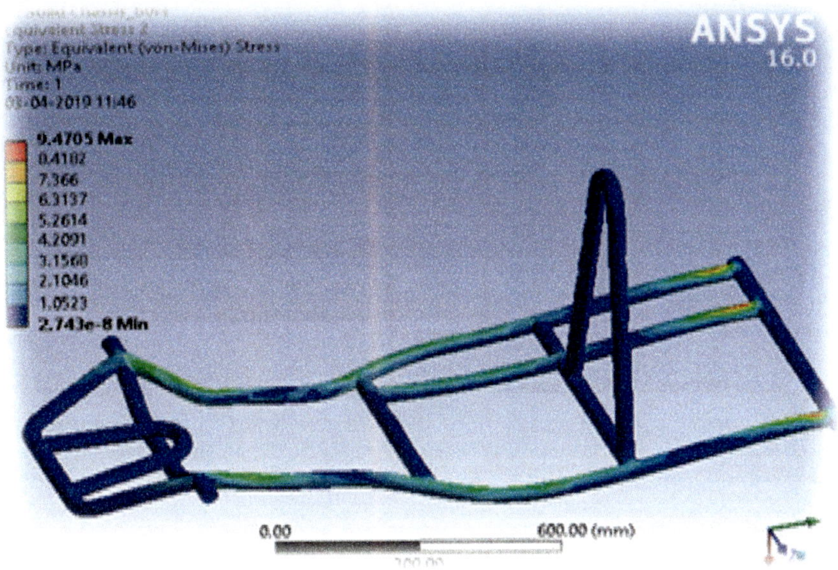

Fig. 13 Equivalent stress of solid chassis under 1500 N load

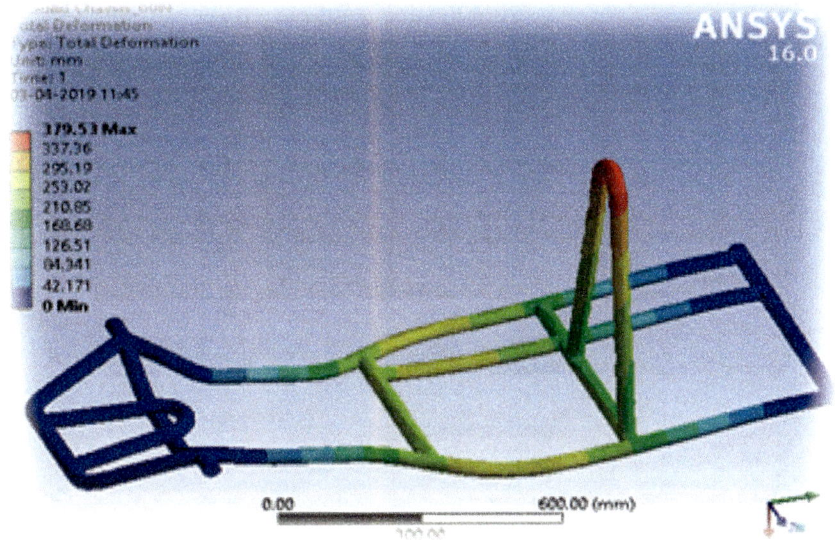

Fig. 14 Total deformation of solid chassis under 1500 N load

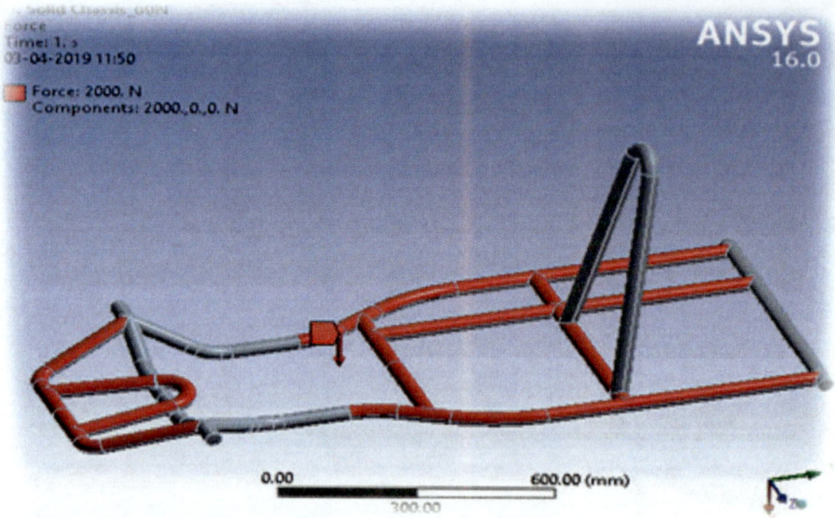

Fig. 15 Load under 2000 N of solid chassis under 2000 N load

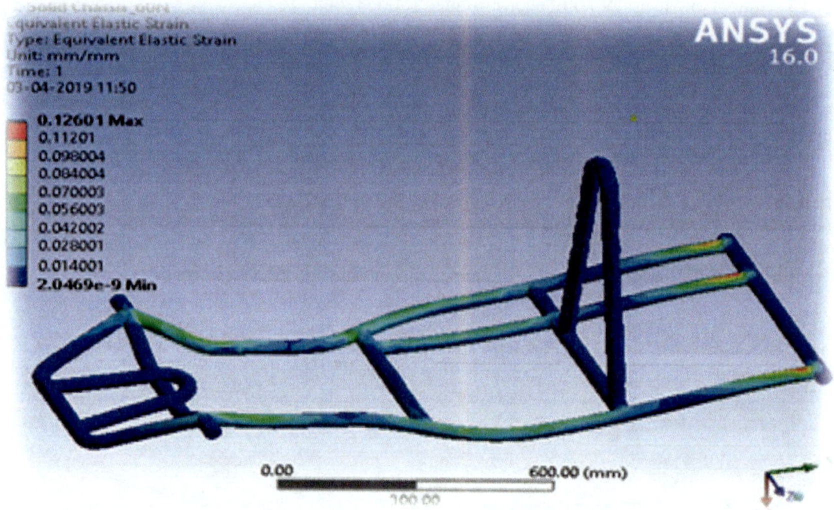

Fig. 16 Equivalent strain of solid chassis

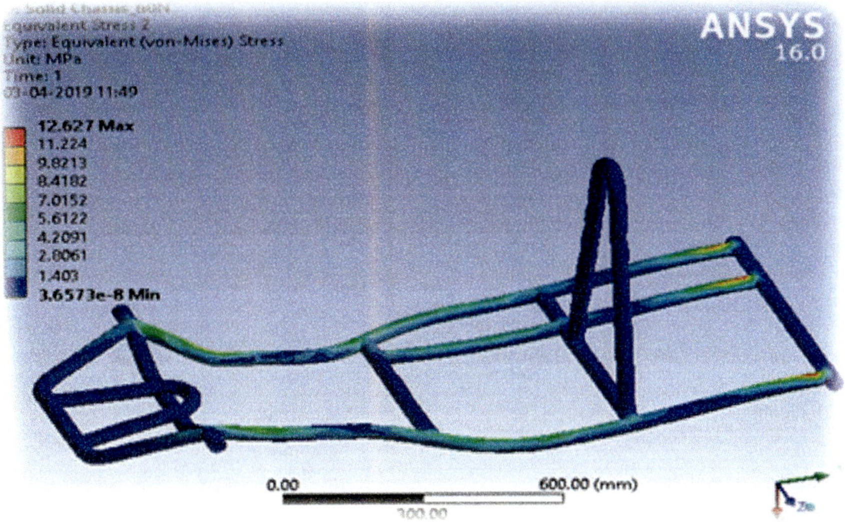

Fig. 17 Equivalent stress of solid chassis under 2000 N load

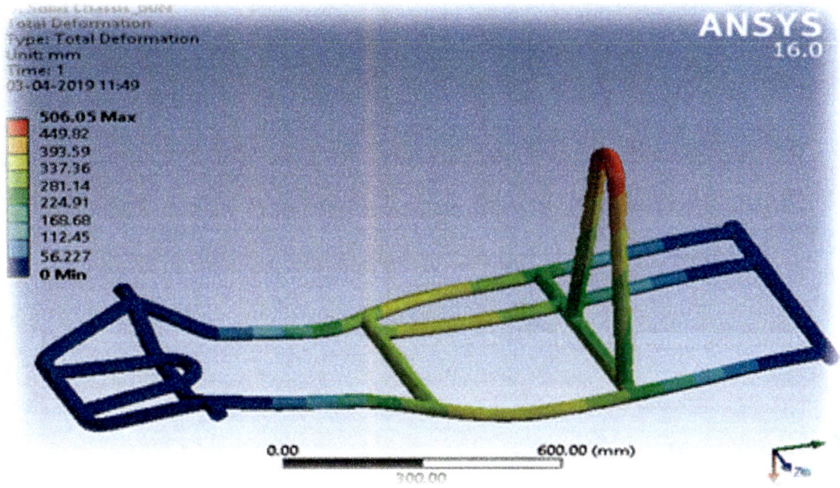

Fig. 18 Total deformation of solid chassis under 2000 N load

Table 1 Material properties

Mechanical properties of PVC	Values
Tensile strength	7300 Mpa
Tensile modulus of elasticity	410,000 Gpa
Flexural strength	15,000 N/P
Flexural modulus	420,000
Izod impact	0.9

5 Vehicle Specifications

Vehicle model	Make value
1.1 inch pipe	33.40 mm
Watts	250 W
Total length	1090 mm
Total width	520 mm
Most extreme speed	300 RPM
Generally speaking weight	Up to 200 kg
Material	PVC

6 Material Selection

The frame is comprised of PVC. This material was chosen because of its great mix of the majority of the run of the mill qualities of steel—quality, flexibility, and similar simplicity of machining (Table 1).

7 Result and Conclusion

The way to reasonable case style is that the any mass is reserved from the nonpartisan pivot the ton of furrowed it will be [6]. This one sentence is that the premise of car case style. This investigation endeavored to break down weight on the undercarriage style misuse limited segment examination. This is significant on the grounds that the incitement information is valuable for further plan improvement and thusly prompts cost viability. The final responds underneath shows the consequences done during investigation applied effectively to the situation of go e-kart (Tables 2 and 3).

Table 2 Hollow chassis of final result

Load condition	Equivalent stress (Mpa)	Equivalent strain	Total deformation
150 kg Min value	1.5615e−8	3.7685e−10	0
150 kg Max value	22.58	0.22705	708.5
200 kg Min value	2.082e−8	5.0246e−10	0
200 kg Max value	30.107	0.30273	944.06

Table 3 Solid chassis final result

Load condition	Equivalent stress (Mpa)	Equivalent strain	Total deformation
150 kg Min value	9.4705e−8	1.5351e−9	0
150 kg Max value	2.7433	0.09450	379.53
200 kg Min value	3.6573e-5	2.0469e-9	0
200 kg Max value	12.627	0012601	506.05

8 Conclusion

Static investigation utilizing limited component technique was effectively completed to decide most extreme redirection and its area on body structure. The consequences of examination are uncovered that the area of most extreme avoidance concurs well with hypothetical greatest area of basic bar. The investigation discovers the contrast among empty and strong case (PVC pipe) results. So, solid PVC with HDP is preferable in chassis development.

9 Future Scope

Therefore, in the future, go karts can be used as a mover for individuals, which has become comfortable and providing high convenience. For example, KTM ariel and KTM X-bow have generated ariel particle.

Therefore, go e-karts could be used as a mover of people in the future that is safer and more convenient.

References

Book

1. Adams H. Chassis engineering
2. Rajput RK, Strength of materials
3. Bhandari VB, Design of machine elements

Journal Paper

4. Sharma PK, Parekh NJ, Naik D (2014) Optimization and stress analysis in chassis in TATA turbo truck SE1613. IJEAT, no. 181–187
5. Kumar V, Patel V, Patel RI (2012) Structural analysis OF ladder chassis frame. WJST
6. Sane SS, Jadhav G, Anandraj H (1955) Stress analysis of light commercial vehicle chassis by FEM. Piaggio Vehicle Pte. Ltd, Pune

Design and Fabrication of an Automated Laptop Stand

Mona Sahu, Kondru Gnana Sundari, and Emerald Ninolin Stephen

Abstract In the recent past, laptops are becoming smaller, lighter, cheaper, and more powerful. Laptop usage is very common among IT professionals, academicians, students, and at home. Due to their portability feature, it takes time to position the laptop correctly to maintain correct posture to avoid pain during usage. Ideally, laptop users should use a laptop stand to avoid pain in neck. This study aimed to design and fabricate an automated laptop stand to improve the neck posture of a person. This laptop stand consists of a scissor mechanism, a linear actuator to automate the vertical linear motion and a Raspberry Pi Camera as the feedback system. The camera detects the eye coordinates of the user and sends the signal to the microcontroller to stop the actuator movement once the eye reaches the center of the frame. The fabrication of the laptop stand was successfully completed. The results showed that top line of the screen was adjusted close to the user's eye level.

Keywords Laptops · Posture · Ergonomics

1 Introduction

In this day and age, it is very common to own a laptop, starting from a school student to an IT employee [1]. The increased use of laptops has also increased the concern to analyze the musculoskeletal disorders (MSDs) associated with their usage. According to a study in 2008, almost 82% of the undergraduates were reported to own only a laptop instead of a desktop [2].

The prolonged usage of laptops without proper ergonomic concerns will lead to musculoskeletal disorders, especially in the neck region [3]. An average IT employee may spend 6-8 h of the workday on the computer [4]. With the rise of electronic commerce, people are likely to spend 2–4 additional hours per day on their personal computer. As more number of hours are spent working on a computer each day (both

M. Sahu (✉) · K. Gnana Sundari · E. N. Stephen
Karunya Institute of Technology and Sciences, Coimbatore 641114, India
e-mail: monasahu@karunya.edu

A. Arockiarajan et al. (eds.), *Advances in Industrial Automation and Smart Manufacturing*, Lecture Notes in Mechanical Engineering,
https://doi.org/10.1007/978-981-15-4739-3_11

at work and at home), it is important to take necessary steps to prevent the risk of MSDs caused.

According to ergonomic principles, the top of the laptop screen should be at the eye level and one arm distance from the user [5]. Due to the different anthropometric dimensions of each person, it is difficult to adjust the position of the laptop screen on a standard table. Hence, the user tends to flex his/her neck to view the laptop screen. Working in an unhealthy posture for a long time can lead to neck related MSDs [6].

The existing laptop stand available in the market can be manually adjusted. Most of the users are either ignorant or lazy to adjust the position of the laptop screen as per the ergonomic principles [7]. The height adjustment of the laptop screen in incorrect postures increases the risk for muscular skeletal disorder. Hence, the aim of this study was to design and fabricate an automated laptop stand to improve the neck posture of a person while using a laptop.

2 Selection of the Range of Movement of the Laptop

The range of the vertical movement of the stand is selected based on the Indian anthropometric data given by Chakrabarti [8]. The seated eye level from the floor of the 25th female percentile value was taken as lowest position of the laptop stand. The seated eye level from the floor of the 95th male percentile value was taken as highest position of the laptop stand.

Total height of a Person sitting on a chair = Popliteal + Height from seat pan to eye

Total height of a female (25th percentile) seated on a chair = 386 + 628 = 1014 mm

Total height of a male (95th percentile) seated on a chair = 471 + 805 = 1276 mm.

Hence, with respect to the Indian anthropometric details, the height range of the laptop stand was taken as 262 mm.

3 Components Used

3.1 Frame of the Laptop Stand

A scissor mechanism is used to provide the upward motion of the laptop stand. This mechanism consists of a four bar and slider crank mechanism. The links form a parallel kinematic mechanism as shown in Fig. 1 and permit a vertical movement of load. It contains 2 links crisscrossed with each other known as 'scissor arm'. The upward motion is produced by the application of force by an actuator, to the supports

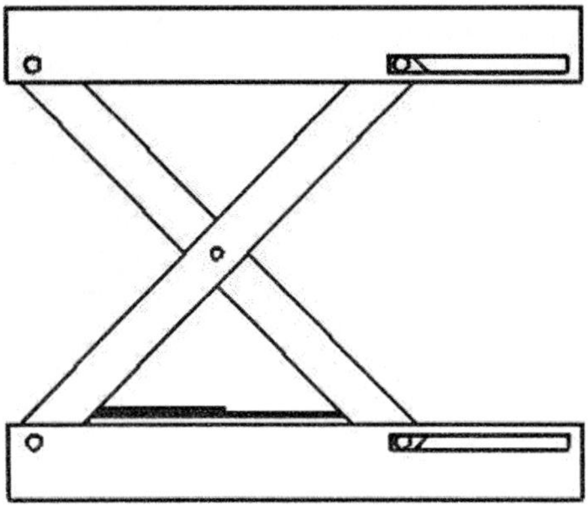

Fig. 1 The schematic diagram of the scissor mechanism

of the crisscross (X) pattern, which elongates the crossing pattern, pushing the load vertically upward. The degree of freedom for this mechanism is 1 as shown below:

Number of Links = $n_L = 4$
Number of Joints = $n_J = 5$
Degree of Freedom, $F = 3(n_L - 1) - 3n_J$

$$\text{Degree of Freedom, } \begin{aligned} F &= 3(n_L - 1) - 3n_J + \sum_{i=1}^{n_J} f_i \\ F &= 3(4-1) - 3(5) + 7 = 9 - 15 + 7 = 1 \end{aligned}.$$

3.2 *Linear Actuator*

The linear actuator is used to provide the force to lift the stand (Fig. 2). The actuator is connected via stainless steel rods, and the horizontal displacement of the actuator is converted by the scissor mechanism into vertical displacement. The actuator is powered by a 12 V DC battery. It is connected to the Raspberry Pi through a motor driver (L298) circuit. The specifications of the linear actuator are given below:

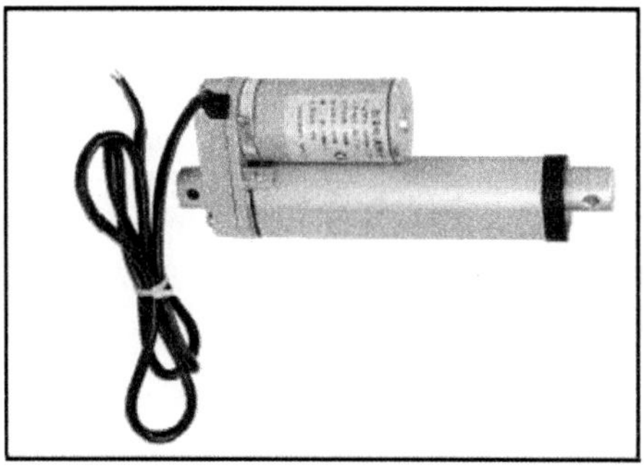

Fig. 2 Linear actuator

Fig. 3 Raspberry Pi camera

Material: aluminum alloy
Input voltage: 12 V DC
Speed: 3–40 mm/s
Load capacity: 100 N
Stroke length: 100 mm
Retracted length: 200 mm.

3.3 Feedback System

A Raspberry Pi Camera v2, a high-quality 5 megapixel Sony IMX219 image sensor is used as the feedback system. It is a custom-designed add-on board for Raspberry Pi, featuring a fixed focus lens. The Raspberry Camera is used to detect the eye and is programmed to send signal when the eye is positioned in the center of the camera's focus. The Raspberry Pi Camera is shown in Fig. 3.

4 Working Principle

The linear actuator, Raspberry Pi, and Raspberry Pi Camera v2 are connected with the motor driver L298N. The algorithm of the program for the automation of the laptop stand is shown in Fig. 4. The actuator and the camera are first switched on. The Raspberry Camera detects the eye of the user seated in front of the laptop screen and records its coordinates. The eye position in the camera frame moves up or down as the actuator lifts the laptop stand. The camera is programmed to send a signal to the controller to stop once the eye comes in the middle of the camera's focus. The controller provides the signal to the actuator to stop once the Raspberry Camera reaches the eye level.

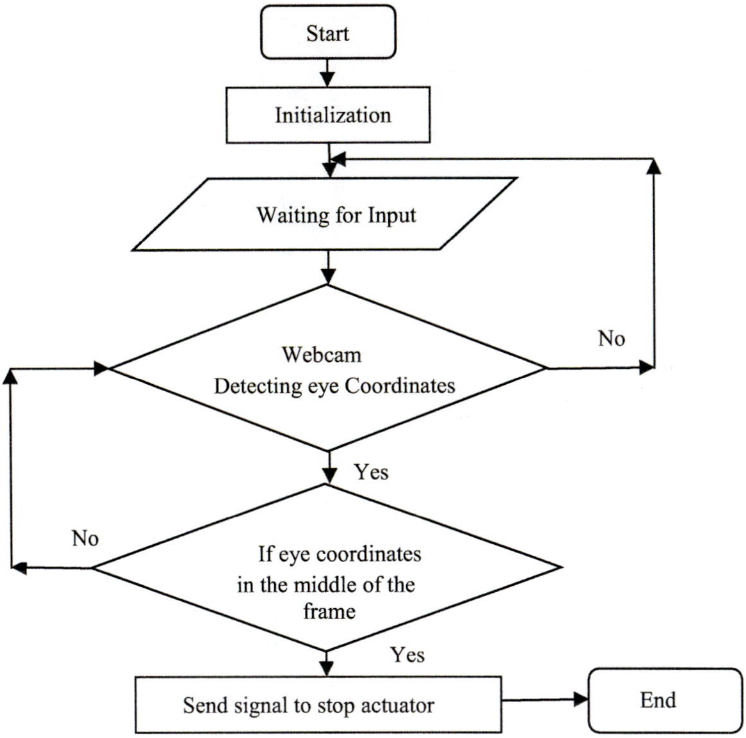

Fig. 4 Algorithm

Fig. 5 CAD model of the automated laptop stand

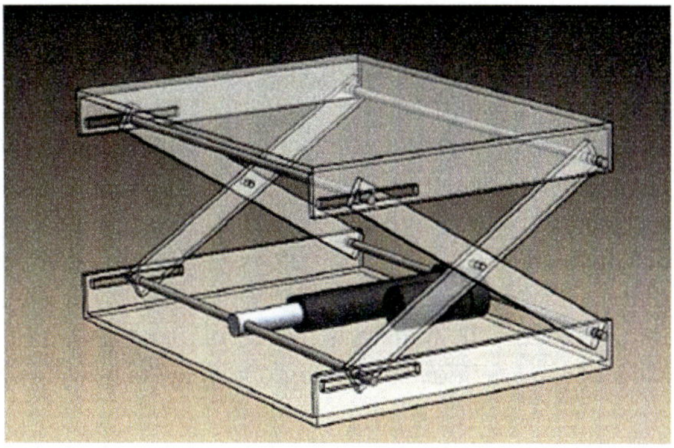

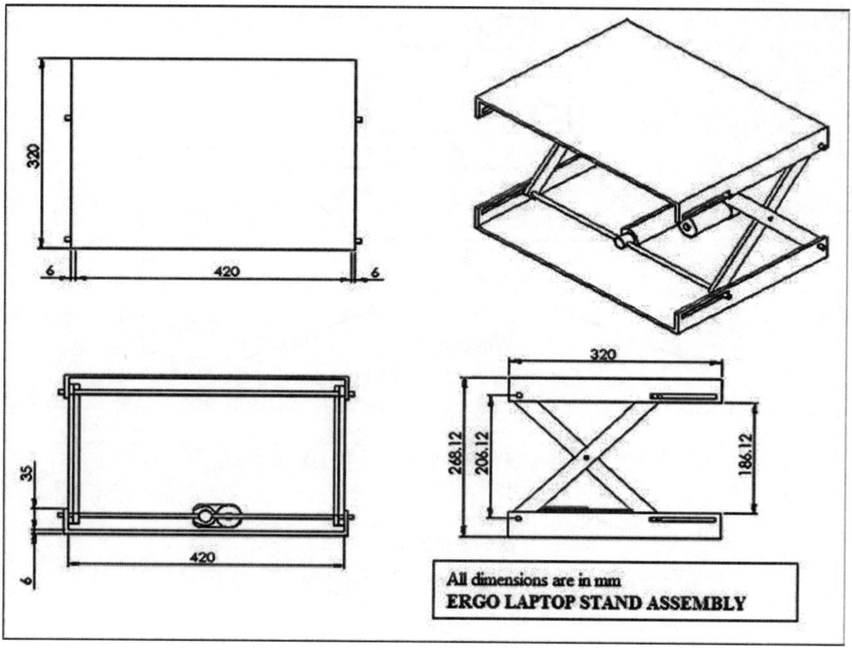

Fig. 6 Schematic diagram of the laptop stand

5 Fabrication

The computer-aided design (CAD) model of the laptop stand was designed in SolidWorks 2016. The isometric view of the model is shown in Fig. 5. The schematic diagram of the part and the assembly are shown in Fig. 6.

The design of the laptop stand is portable and easy to use. The frame is made of acrylic plastic which is cheap and light in weight. The sheet of acrylic plastic is cut to the required shape and size by the laser cutting process. The linkage is also made of acrylic material. A linear actuator is assembled using stainless steel rods on the base plate of the frame. The Raspberry Pi is placed at the back of the laptop screen, and the Raspberry Pi Camera is placed on the top line of the screen. The photographic view of the assembly of the laptop stand is shown in Fig. 7.

6 Testing of the Laptop Stand

Ten subjects of height varying from 25th percentile to 95th percentile male were selected for testing the functionality of the automated laptop stand. A no-objection statement was obtained from all the subjects for publication in revealing the human identity. Prior permissions were obtained from the Institutional Human Ethical

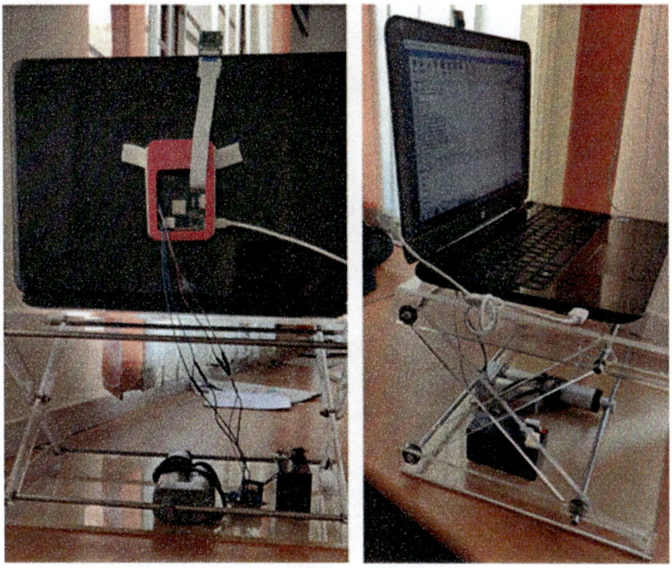

Fig. 7 The photographic view of the assembly of the laptop stand

Committee, for conducting laboratory experiments on human beings. The subjects were asked to sit on the computer chair and place their laptop on the automated laptop stand. The laptop stand was placed directly in front of the user such that the laptop screen was one arm distance from the subject. They were also asked to sit with their trunk straight and neck straight. The Raspberry Pi was placed at the back of the laptop screen, and the Raspberry Pi Camera was placed on the top line of the screen. The camera and the electronic circuit were switched on to detect the eye coordinates of the subject as shown in Fig. 8. As soon as the eye coordinates reach the middle of the frame, the actuator stops. After the automated adjustment of the laptop, the neck posture of the subjects was measured using a manual goniometer and the reading was recorded. Three trials were taken for each subject.

7 Results and Discussion

Figure 9 shows the neck angle recorded for the 10 subjects while viewing the top visible line of the laptop screen. An average of 20.9° neck extension was observed for the subjects. The testing of the automated laptop stand showed that there was a slight delay in receiving the signal to stop the actuator. It was also noticed that the position of the camera was one of the factors for the laptop stand to stop above the required height; i.e., the height of that the top visible line of the laptop screen should be 15° below the eye level.

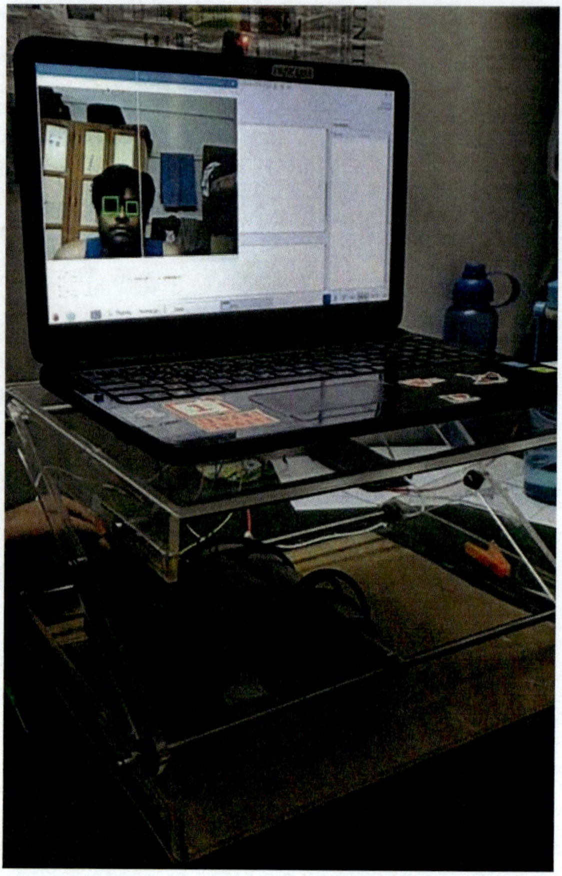

Fig. 8 Detection of eye coordinates

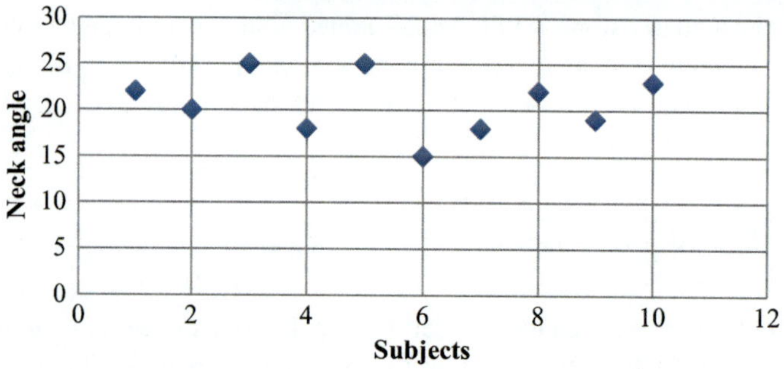

Fig. 9 Measured neck angle of subjects while looking at the top visible line of the laptop screen

The present research work leaves a wide scope for improvization and further investigations to explore design changes for the laptop stand to incorporate the desired ergonomic neck posture. Some recommendations for further improvizations include the following: to use a stepper motor instead of a linear actuator which is light in weight, to accurately automate the height adjustment as per ergonomic guidelines, and to automate the position of the laptop to maintain the required distance from the user.

8 Conclusion

This research was aimed to improve the neck posture while using laptops. The design of the laptop stand is attractive and esthetic. The design and fabrication were successfully completed. As per the results, the mean neck angle for the 10 subjects was 20°. The results suggested that with some further improvements in the design, the automated laptop stand can improve and maintain the height of the top visible line of the laptop screen correctly at 15° below eye level. It can be used by people who prefer working for long hours on laptops.

References

1. Smith S, Salaway G, Borreson Caruso J (2009) The ECAR study of undergraduate students and information technology. In: Research study, vol. 6. EDUCAUSE Center for Applied Research, Boulder, CO
2. Chang C, Amick B, Menedez CC, Robertson M, del Pino RJ, Dinnerlein JT (2008) Where and how college students use their laptop computers. In: Proceedings of the Human Factors and Ergonomics Society 52nd annual meeting, pp 629–633
3. Andersen JH, Fallentin N, Thomsen JF, Mikkelsen S (2011) Risk factors for neck and upper extremity disorders among computers users and the effect of interventions: an overview of systematic reviews. PLoS One 6(5)
4. de Looze MP, Kuijt-Evers LFM, van Dieën JH, Sitting comfort and discomfort and the relationships with objective measures. Ergonomics 46:985–998
5. ANSI/HFES (2007) Human factors engineering of computer workstations (ANSI/HFES 100–2007). Human Factors and Ergonomics Society, Santa Monica, CA
6. Price J, Dowell W (1998) Laptop configurations in offices: effects on posture and discomfort. In: Proceedings of the Human Factors and Ergonomics Society 42nd annual meeting, pp 629–633
7. Vink P, Porcar-Seder R, de Pozo AP, Krause (2015) Office chairs are often not adjusted by end-users. In: Proceedings of the Human Factors and Ergonomics Society 51st annual meeting, pp 1015–1019
8. Chakrabarti D (1997) Indian anthropometric dimensions for ergonomic design practice. National Institute of Design, Ahmedabad. Retrieved from http://books.google.com/books?id=koyAAAAAMAAJ

Design and Analysis of Progressive Die for Manufacturing the Gasket Part

Saurabh Priyadarshi, Premanand S. Chauhan, and Rajesh Pratap Singh

Abstract A progressive die is a set of tools performing series or several sheet-metal operations with a single tool. It includes two or more workstations for the production of products. In every pressing stroke, one operation is performed, and strip stock moves forward through the die strip. The main functional advantage of progressive die for the manufacturing industry is that it takes less time and gives a high volume of production. It can build Precision tools with less time-consumption. From the designing point of view, the CAD/CAM technique can be used to develop the part. This simulation application gives the designer, to perform or test and freedom to take risk-free decisions. In this project experiment, the main work is to design the gasket parts, design progressive die, and analyze the progressive die or gasket part whether there is any defect in production process analyses. Required solid modeling is done on SOLIDWORKS.

Keywords Progressive die design · Design analysis · Materials · Effect on production using simple die

1 Introduction

The rapidly growing automobile industry uses a wide range of sheet-metal gasket products, and the automobile industry manufactures a wide variety of gasket products at a very high volume. Nowadays, these industries are leading in producing new technologies. Different types of sheet-metal tools are used in the automobile industry to manufacture the gasket product at low cost with mass production of the product. Due to the very large requirement of the product, in minimum time, we can use a progressive die to reduce manufacturing time and cost. The gasket manufacturing industry makes very complicated geometry, and the stamping process executes several operations until the final product is formed. This process sequence for gasket development

S. Priyadarshi (✉) · P. S. Chauhan · R. P. Singh
Department of Mechanical Engineering, I.P.S. College of Technology & Management, Gwalior, Madhya Pradesh 474001, India
e-mail: 543saurabh@gmail.com

A. Arockiarajan et al. (eds.), *Advances in Industrial Automation and Smart Manufacturing*, Lecture Notes in Mechanical Engineering,
https://doi.org/10.1007/978-981-15-4739-3_12

is the best on-die geometry, the blank shape of the product, and process parameters. Each and every work station or stage is designed based on past experiences, Trials and Errors.

The development process of Die may be time-consuming. Therefore, we are using computer-aided design and simulation for planning and study the design process sequence of the progressive die. The progressive tooling for the stamping process is defined as a metal-forming process. It is widely used for manufacturing metal forming parts for automotive industries and electronics industries. The progressive die for stamping process consists of the various independent workstation. Each and every workstation performs one or two different operations. The product is transferred from one workstation to another workstation by the strip-stock or punching stock and the strip cut out in the final workstation. The decision to design or manufacturing the progressive tool is dependent on the complexity of the product and volume of the product and production cost or production time of the product. The use of progressive stamping dies to produce different types of parts reduces the product-related cost as possible. The main important or demanding benefit of the progressive die is to met precise work and durability.As the manufacturing of progressive dies is difficult, but some important points are considered to reduce the manufacturing time and cost, always consider the blank position, blank-boundary, part-edge-deformation, pilot, holes.

The benefits of using a progressive stamping die are to increase product productivity and reduce product cost with a high volume of manufacturing.

2 Existing Design for the Manufacture of Gasket Part Using Four Different Types of the Die, See Fig. 1

In this four-type of die, we manufacture the product, but the rejection ratio is too high hence to eliminate the rejection, we develop progressives die see Fig. 2.

2.1 Specification of Component for Manufacturing the Gasket Part

Material: Stainless steel 304

Overall OD: 60 mm

Overall ID: 48.8 mm

Overall length: 50.6 mm

Thickness of strip: 0.8 mm (Fig. 3).

Fig. 1 Simple die. **a** Blanking die, **b** deep drawing die, **c** post-forming die, **d** edge trimming die

Fig. 2 Rejection part

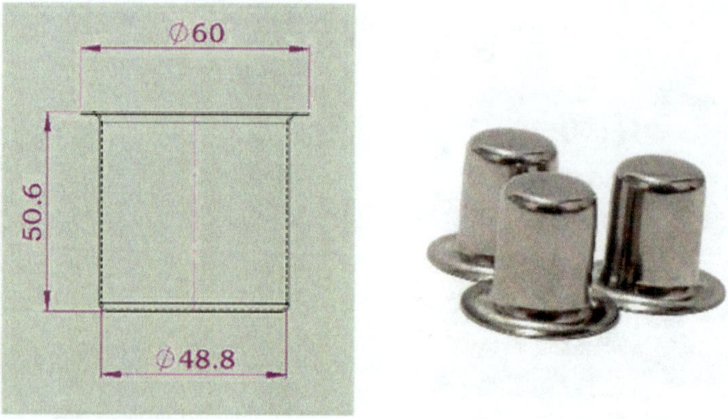

Fig. 3 Final product

2.2 *Design of Progressive Die*

Manufacturing Process of Progressive Die design (Fig. 4)

3 The Main Components of the Progressive Die

See Table 1.

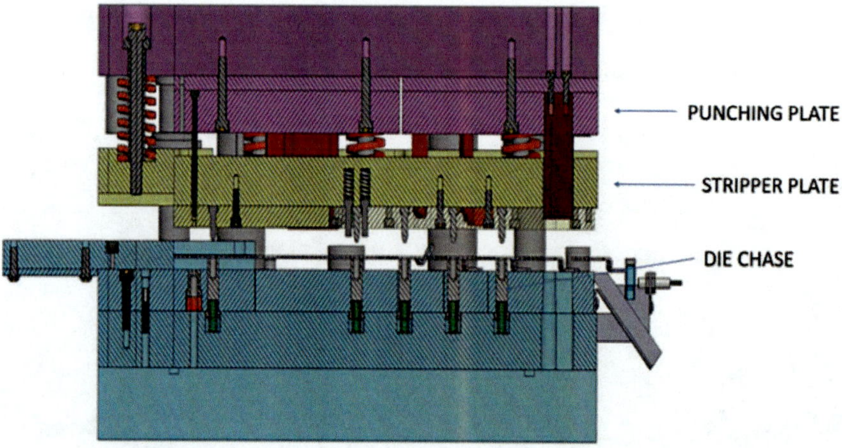

Fig. 4 Progressive die

Table 1 Main part of the progressive die

S. No.	Part name	Material
1	Punching plate	D2
2	Stripper plate	45C8
3	Die chase	FGM

3.1 Punching Plate

The punching plate may be a part of the progressive die to get rid of the finite volume of the sheet. Punch plate helps to hold and support the piercing operation, notching operation, and cut-off punches operation. However, the product of stainless steel is left softly and works as a prime grade punch. The material utilized in producing is AISI-D2. The length of the punching block for the punch plate is given, when the block is shanked or equal in diameter to the ram hole or slot. The shank is fastened in position with a screw. The half is the dowel pin along the side of the highest plate, retains or holds the punches. The center position or distance area that is unit picked up or transferred from the hardened die block to eliminate the likelihood of placement of punches and die openings, to dimensional changes through the heat treatment. All holes receive the body of punch in the unit area given H7 suitable vacant and lightweight press-fitting. The material for the punching plate is AISI D2 as a result of D2 steel and work as harder material.

The AISI D2 material properties are high-carbon steel, and the hardness is 55-62 HRC (Table 2).

3.1.1 The Design and Analysis of the Punching Plate on SOLIDWORKS Are as Follows

- **3D drawing and 2D drawing** (Fig. 5).
- **Load analysis on punching plate by SOLIDWORKS simulation application** (Fig. 6).

3.1.2 The Manufacturing Process of Punching Plate is as Follows

- **EDM (wire cut)** (Fig. 7)
- **Machining (VMC)** (Fig. 8)

Table 2 The AISI D2 steel composition

AISI D2 material composition percentage (%)				
C	SI	Cr	Mo	V
01.50	00.30	12.00	00.80	00.90

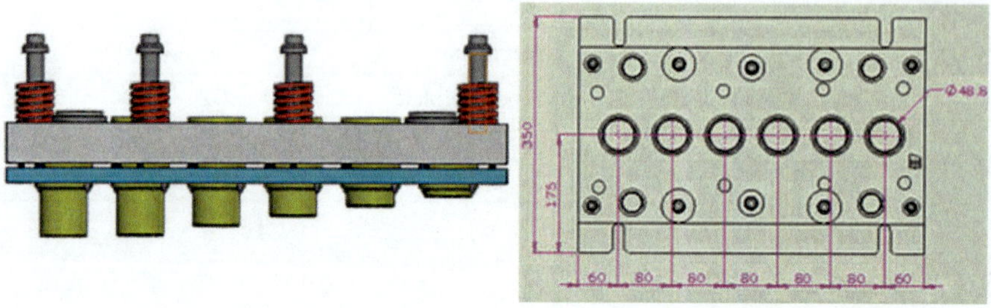

Fig. 5 Punching plate 3D and 2D drawing

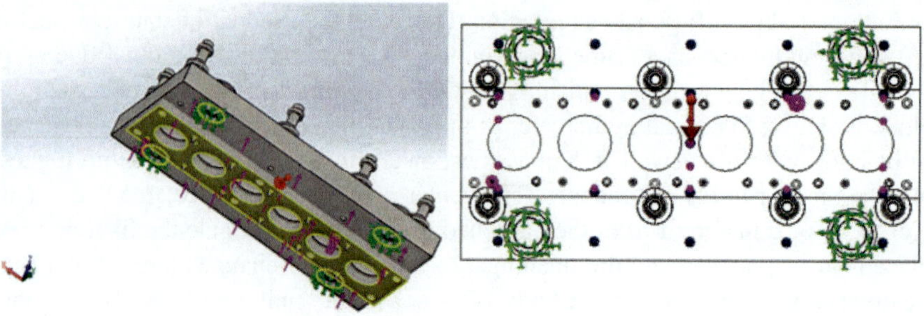

Fig. 6 Punching plate load analysis

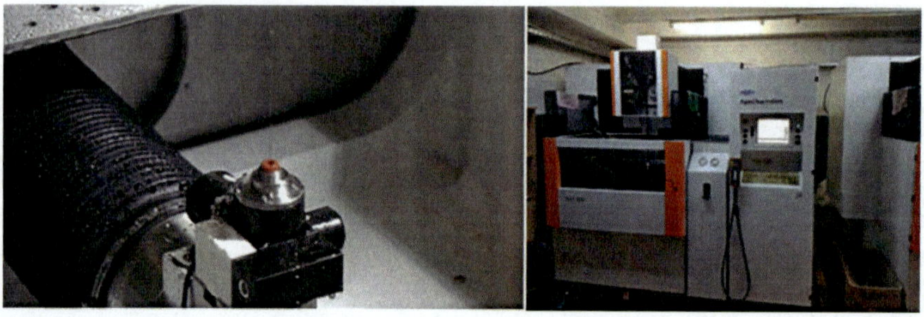

Fig. 7 EDM (wire-cut) machine

- **Hardening and tempering**
- **Surface grinding** (Table 3).

3.2 *Stripper Plate*

A stripper plate holds the metal tightly with the die throughout cutting or piercing, for preventing the metal from slipping or moving on the die. It conjointly helps forestall,

Fig. 8 Machining (VMC machine)

Table 3 AISI 45C8 chemical composition

45C8 steel material composition percentage (%)				
C	Mn	P	S	Si
0.36–0.44	0.60–1.00	5.00	0.50	0.10–0.40

the half from distorting, and permits the cutting punches to withdraw or "strip" from the metal coming by press stroke. Material for the stripper plate. The material for the stripper plate is AISI 45C8. As a result, to select 45C8 steel is typically equipped untreated, however, it is ready to be equipped in order within the normalized or finally heats treated. It is adequate for a large variety of applications.

3.2.1 The Design and Analysis of Stripper Plate on Solidworks Are as Follows

- 2D drawing.

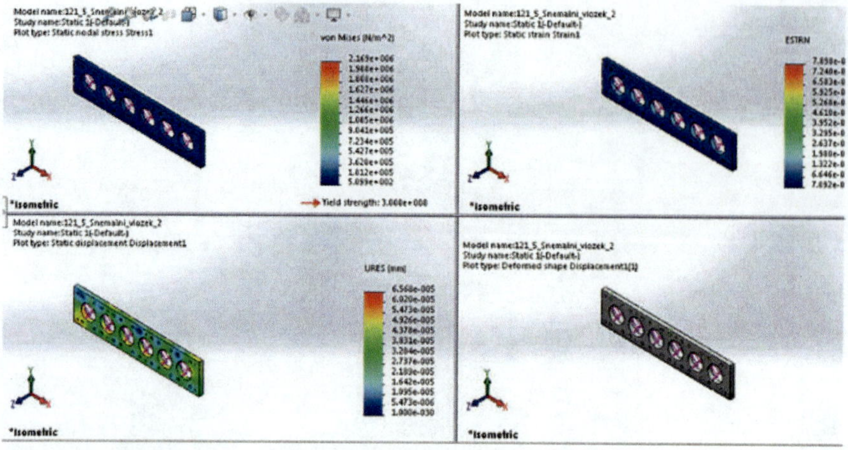

Fig. 9 Analysis report

- Analysis report image for crystal structure
- Analysis report image for static and dynamic loading strength (Fig. 9).

3.2.2 The Manufacturing Process of Punching Plate Is as Follows

- **EDM (wire cut)**
- **Machining (VMC)**
- **Hardening and tempering**

  Tempering (heat of between 550 and 660 °C then cool in oil or water).
  Normalizing (830–860 °C then it's cooled in air).

- **Surface grinding**

3.3 Die Chase Plate

The die chase plate is in the form of a plate and its work is as a gap filler, and it gives angular clearance to allow the escape of excellent blank profile. The waste skelton of the stock strip, from that blanks, is cut and recovered the salvaged material.

The material for the die chase plate is FGM material as a result functionally hierarchal materials could also be characterized by the variation in composition and structure bit by bit over volume, leading to corresponding changes within the properties of the fabric. The materials are often designed for specific performance and applications.

3.3.1 The Manufacturing Process of Die Chase Plate is as Follows

- **VMC machining process**
- **Surface grinding process**

The mechanical properties of cylindrical structures material manufactured from functionally hierarchic material (FGM) unendingly in microscopical sense from one surface to the opposite. The main important advantage is that it is well-designed structure and stand up to very high-temperature gradient environments, whereas maintaining their structural integrity. All material properties expressed as a '*P*'.

$$P = \{P0(P - 1(T - 1) + 1 + (PT) + (P2T2) + (P3T3)\}$$

where $P0, (P - 1), P, P2$, and $P3$ are constants within the cuboid or the fabric property. With the on top of the equation, the higher-order result of the temperature on the fabric is captured. The constant values for a few materials, namely semiconducting material chemical compound silicon nitride (Si3N4), unsullied or stainless steel (SUS304), zirconium dioxide (ZrO2), and Ti–6Al–4V.

3.4 The Progressive Die Design on SOLIDWORKS

SOLIDWORKS is a solid or surface modeling technique based on CAD "computer-aided design" and computer-aided engineering, the PC application that runs on a Windows system and develops the planning as per demand. The Dassault system develops or produces printed SOLIDWORKS software package. SOLIDWORKS application is solid modeling software package; it permits you to style the half and assembly the half within the three-dimensional house. The technique is typically operating for style complete product coming up with, for planning the half. Begin for sketching second profiles then use extruding, the extrude-cut and lofting command for solid form, when completed the half style, assemble the half and being the simulation method as per simulation, resulted half is changed. This method, once more and once more, simplifies and build a final product.

4 Result

The experimental production results are evaluated by the help of the production table or on the press machine experimentally, and the calculation work is done on SolidWorks software (Fig. 10).

Finite element analysis (FEA) is done for the punching plate, stripper plate, or die chase from this it is seen that the results are in an acceptable range. The analytical and finite element analysis (FEA) results are nearly equal and both are in the acceptable

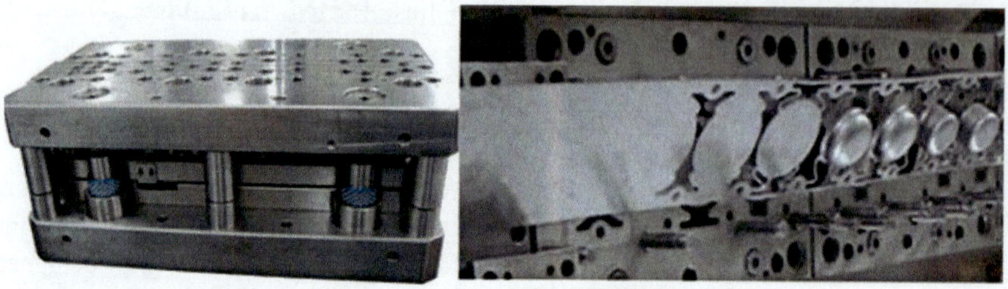

Fig. 10 Progressive die

Fig. 11 Final sheet-metal gasket parts

range. The total punching force required for punching plate or manufacturing the gasket part is 112 tones including 25 percent of the factor of safety. The calculation is done on SOLIDWORKS. So our designed safe is under given load conditions (Fig. 11).

It is concluded that all parameters for the manufacturing and implementation of progressive die for gasket manufacturing by the help of computer design process using SOLIDWORKS, the manufacturing cost and time are also reduced-drastically than the traditional/simple tooling method. Any sheet-metal manufacturing industry uses 100, 150, 200 tones press machine are easily available for manufacturing and testing purposes. For practical work in the press machine available in Fleet gasket India Pvt. Ltd. Sheet metal gasket part under one-month observation or analysis testing is done on the blanking plate, punching plate and bending punch to determine the strength of the progressive die. It shows the results, and the gasket part manufacturing rate is too high than the respective previous production rate (Figs. 12,13,14,15 and 16).

After the camper the chat-list for simple die or progressive die. The progressive die production rate is very high at very less part reject.

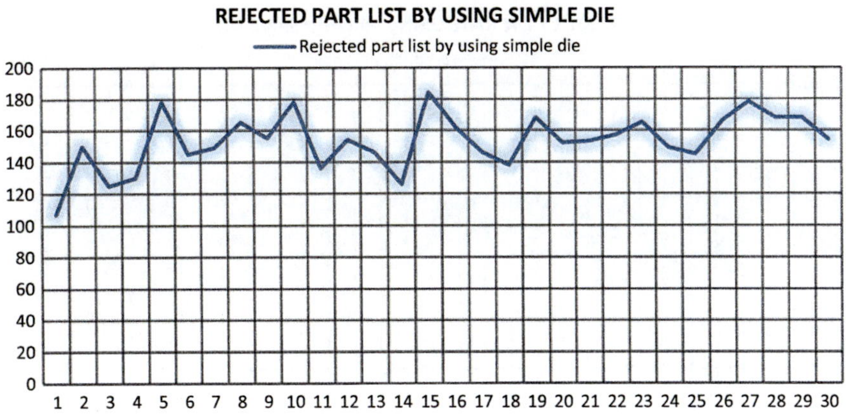

Fig. 12 One-month production report using simple die

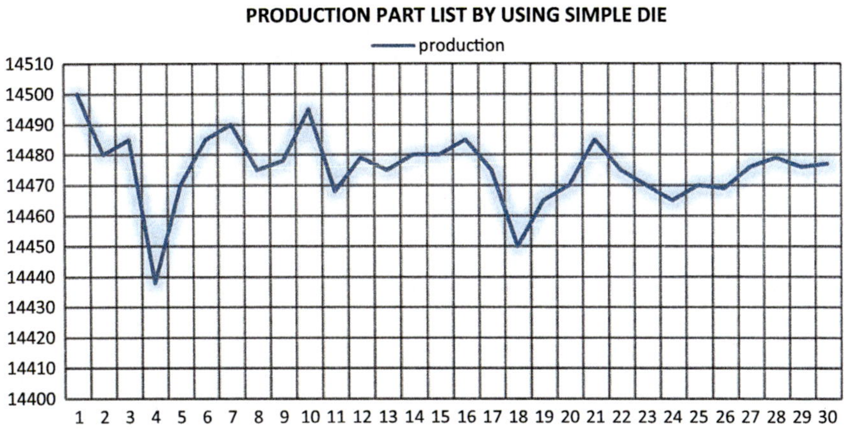

Fig. 13 One-month production report using simple die

5 Conclusion

The main focus of this paper was on the optimization of productivity of gasket parts manufacturing unit. This work is beneficial for the research and development departments of the automobile gasket manufacturing unit. The progressive die is very economical with suitable characteristics including strength and wears resistance. This analysis work may be beneficial for the sheet-metal work research industry, automobile industry, electrical industry, aerospace industry. The progressive die is very economical and produces sheet-metal gasket part from the metal sheet with suitable characteristics including strength and wear resistance. The implementation of progressive die for manufacturing process replaces the traditional method (simple

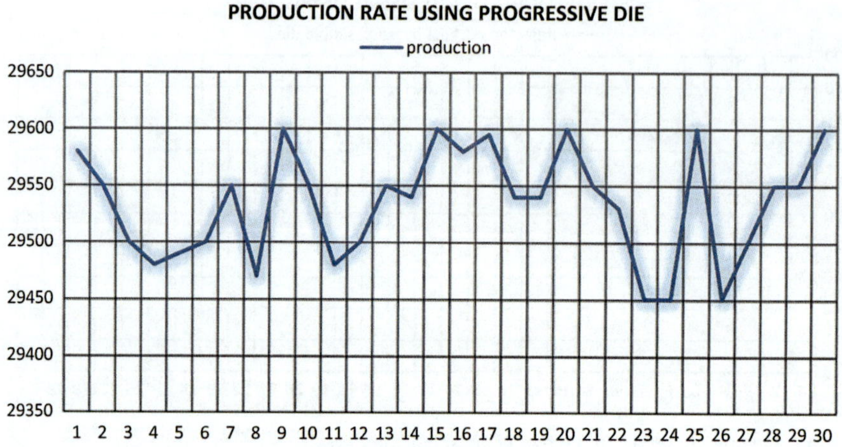

Fig. 14 One-month production report using progressive die

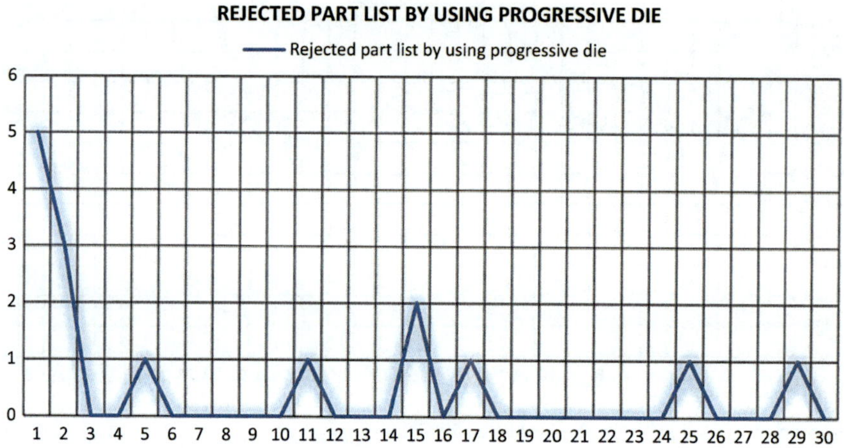

Fig. 15 One-month production report using progressive die

die) because the traditional method is a time-consuming process. This research deals with the production report for manufacturing gasket parts using a progressive die or simple die manufacturing. The simple die is also used for manufacturing, but it includes two or more die used for one-single product manufacturing (blanking die, deep drawing die, post-forming die, edge trimming die); therefore, the manufacturing cost of blanking dies, deep drawing dies, post-forming die, edge trimming die is highly expensive and the mass production rate is low compared to progressive die. In the manufacturing industry, 200 tones press machine is easily available, so the press machine is used for manufacturing gasket part. The progressive die is generally used in industries for developing sheet metal parts for manufacturing the high

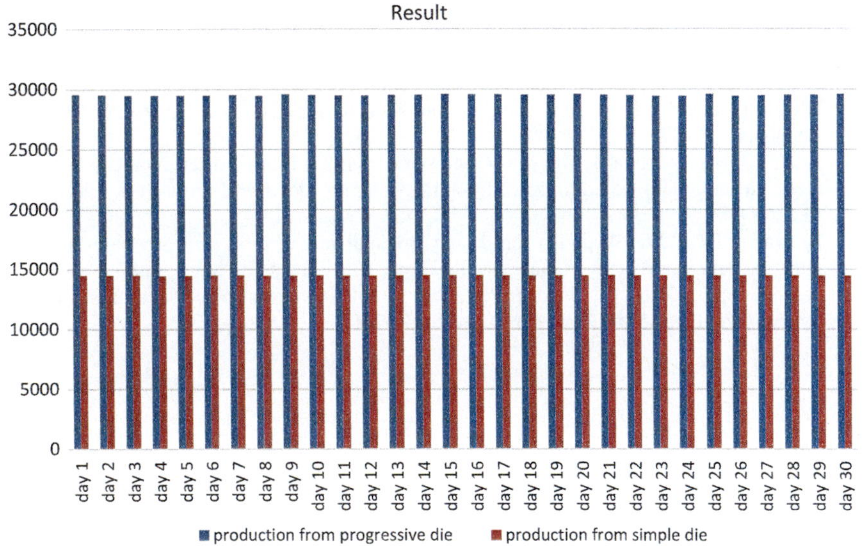

Fig. 16 One-month production report using progressive die verses simple die

volume of products. Hence, the designing of progressive die works perfectly and it includes 25 percent of the safety factor that improves its life.

Further work can be done on other segments of automobile components manufacturing industries as well as the industries where the simple die is used to manufacture components. The work can also be done to control progressive die with Artificial Intelligence (IoT).

References

1. Vinay Kumar AV, Ramegowda D, Design of progressive press tool for an alpha meter component. Int J Res Eng Technol. eISSN: 2319-1163, pISSN: 2321-7308
2. Ameresh H, Hari Shankar P (2013) Progressive tool design and analysis for 49, lever 5 stage tools. Int J Comput Trends Technol (IJCTT) 4:197. ISSN: 2231-2803 http://www.ijcttjournal.org. Graff KF (2015) Micromanufacturing engineering and technology, 2nd edn.
3. Planning of Progressive manufacturing Dies, https://www.autoform.com/en/newsevents/news/autoform-plus-r5-streamlining-sheet-metal-forming/
4. Young (1964) Synthetic structure of industrial plastics (Book style with paper title and editor). In: Peters J (ed) Plastics, 2nd edn. vol 3. McGraw-Hill, New York, pp 15–64
5. (2007) Int J Mach Tools Manuf 47:1088
6. Vogler MP, DeVor RE, Kapoor SG (2008) J Manuf Sci Eng 126:685. (2004) J Mechatron Manuf Syst 1:23
7. Aramcharoen A, Mativenga PT, Yang S, Cooke KE, Teer DG (2008) Int J Mach Tools Manuf 48:1578
8. Sinan F, Luke X, Weiss Lee E, Ozdoganlar OB (2008) Int J Mach Tools Manuf 48:459. J Mater Process Technol 190

9. Howell LJ (2000) Innovation in the automobile industry: a new era. 30(16):21
10. P. Cardoso and J.P. Davim // Materials and Manufacturing Processes 25 (2010) 1125
11. O'Donovan B, Eckert C, Clarkson J, Browning TR (2005) Design planning and modelling. In: Clarkson J, Eckert C (eds) Design process improvement: a review of current practice. Springer, London, pp 60–87
12. Adler PS, Mandelbaum A, Nguyen V, Schwerer E (1995) From project to process management: an empirically-based framework for analyzing product development time. Manage Sci 413:458–484
13. Browning TR (2001) Applying the design structure matrix to system decomposition and integration problems: a review and new directions. IEEE Trans Eng Manage 483:292–306
14. O'Donovan B, Eckert C, Clarkson PJ (2004) Simulating design processes to assist design process planning, ASME Paper No. DETC2004-57612

Design of Industrial Safety Helmets with Improved Stiffness Through Finite Element Analysis

R. Surendran, A. Rahul Kumar, A. Suresh, K. P. Manoj Kumar, V. Dhinakaran, and K. Karthikeyan

Abstract An industrial safety helmet is used to protect our head from the objects that are falling on the helmet by absorbing mechanical energy. The number of workers getting head injuries in the construction and industries are increasing every year. The risk of injuries can be reduced significantly by wearing an appropriate safety helmet. At present, the helmet used in the industry has low stiffness because of improper filling of material and uneven pressure distribution. In this paper, an effort has been taken to improve the stiffness of the industrial safety helmet by modifying the existing design. The stiffness of the industrial safety helmet is determined by doing the finite element analysis in ABAQUS. Acrylonitrile butadiene styrene (ABS) and natural fiber are preferred as the materials for the helmet. CATIA V5 is used for modeling the helmet, and the analysis is carried out using ABAQUS software. The standard design and modified design are analyzed, and the results are noted and also the suggestions are given for changing the design to improve the stiffness. The standard designed helmet is modified, and the analysis is done for the modified designed helmet. At last, the results of both the helmet are compared, and it is found that the newly designed helmet has better stiffness than the standard designed helmet.

Keywords Industrial safety helmet · Stiffness · Analysis · Abaqus

1 Introduction

The safety of workers has always been a very important issue in all industrial activities. The accidents mainly involve the head injury due to falling objects. The head is totally encased in bone and requires protection because the head injury can even lead to death. According to an international report, around 48,000 labors have died

R. Surendran (✉) · A. Rahul Kumar · A. Suresh · K. P. Manoj Kumar · V. Dhinakaran · K. Karthikeyan
Centre for Applied Research, Chennai Institute of Technology, Kundrathur, Chennai 600069, India
e-mail: surendranravimathi@gmail.com

A. Arockiarajan et al. (eds.), *Advances in Industrial Automation and Smart Manufacturing*, Lecture Notes in Mechanical Engineering,
https://doi.org/10.1007/978-981-15-4739-3_13

in the country because of accidents that are caused due to improper usage of the helmet. As per the review of the International Labor Organization and the British Safety Council, it is stated that nearly 48,000 people in the country die per year due to work-related hazards. The deaths in India are 20 times higher than in Britain that is related to work accidents. The Indian population of 1.35 billion, nearly 465 million, are serving in the industry, but barely 20% concerning that are embraced under the present health and safety legal structure. Hence, the hard hats or safety helmets are used to guard the workers against these injuries and serve as the primary source of defense against brain injury. Thus, the risk of brain injury is reduced by safety helmets which save lives. High-performing industrial safety helmets must be provided to the workers with good shock absorption properties and high stiffness to provide safety for workers. They should absorb the maximum energy from the impact and therefore minimize the forces transmitted onto the user's head. If the helmet is properly selected, most head injuries can be avoided. A lot of research has been carried out in the field for improving the quality of the industrial safety helmet to improve the shock-absorbing capacity. Rajasekar et al. [1] had proposed a finite element analysis of the helmet. The materials selected for the helmet are natural fiber and polypropylene. They have designed it by using Creo software. A comparison study is made by taking into account the deflection and stress.

Kostopoulos at el. [2] had inscribed in his work the dynamic response of the helmet against impact load. The helmet materials chosen are carbon, glass, and Kevlar fiber. The study is done by LS-DYNA 3D. In his work, the helmet consists of two-layered shells, and Kevlar fiber is used to make the outer layer, while glass fiber is chosen for the inner layer. He inferred that the Kevlar fiber exhibits a more suited response than other materials. Arunprasath et al. [3] had done his work in the field of the safety helmet. They had fabricated the outer shell of the helmet by using a hand layup technique. The preferred materials for the manufacture of the outer shell are coconut shell powder, coconut leaf midrib, and glass fiber-reinforced composite. LS Dyna software is used for the analysis, and the results are compared. The authors suggested that these composites can be used as an alternative material for manufacturing the helmet due to their excellent properties. Ariff had fabricated the helmet by coconut fiber. The main focus of their work involved the change in the inner shell of the helmet. The modeling is done in CATIA, and the analysis is performed using ANSYS to compare the mechanical properties. They had also discussed the cost analysis to lessen the price of the safety helmet. They had concluded that these composites exhibit excellent shock-absorbing ability. Hence, the composites showed the equivalent properties as exhibited by the conventional material used for the helmet and weight are similar. Thus, these composites can be used as an alternative material for the manufacturing of the helmet [4].

Aly et al. [5] had examined the composite case material for the helmet. The trial of the helmet is done by the drop test. The composites materials are correlated with acrylonitrile butadiene styrene (ABS). The drop test is done to determine its impact strength. In their work, they had arrived at the conclusion that the composites such as glass fiber woven fabric furnish more reliable properties than ABS. Thus, these composite materials are an alternative material that can be used instead of ABS plastic for the production of the helmet. The work is done by Johnson et al. [6] for the development of the hybrid bio-composites. Jute bio-fibers blend with glass fibers and in the form of pultruded sheets to prepare the bio-composite. They had fabricated the helmet by using these bio-composites. They concluded that these bio-composites can be developed as the material for the making of the helmet because these are economically advantageous.

Murali et al. [7] had performed his work in the field of hybrid composites for the helmet. They have fabricated the helmet by using the hybrid composites. The hybrid composites selected are banana, sisal, and jute. These hybrid composites not only result in the mass reduction of the helmet almost 50% but also display outstanding durability. Thus, these hybrid composites can be used as a replacement for the plastics used for the production of the helmet. Rodríguez-Millána et al. [8] had made a study on the effect of critical loading on the helmet. A serious problem in industrial situations is the injury of the head. They had concentrated on the design adjustment of the helmet to protect the head against blast fulminations. A computational study is done by them to verify the response against the load and concluded with the best design in response to blast loading.

2 Material Used

Based on the literature survey, we have selected the materials for the analysis of the safety helmet which are

- **ABS** -Acrylonitrile Butadiene Styrene
- **Natural Fiber**
  - → Banana Fiber
  - → Coconut Fiber

The material properties of the above-mentioned materials are listed in Table 1. These materials are widely used for the manufacture of the safety helmets. We have attempted to improve the stiffness of the safety helmet by modifying the design that can be easily manufactured by any of these materials.

Table 1 Material properties

Name of the material		Density (kg/m^3)	Young's modulus (GPa)	Poisson's ratio
Acrylonitrile butadiene styrene		1020	2	0.4
Natural fiber	Banana fiber	1101	7.7–20	0.3
	Coconut fiber	1288	4–6	0.3

3 Design Specification

The modeling of the helmet is done in CATIA with the help of standard dimensions. As CATIA is one of the excellent software for designing, it is recommended by various reputed organizations. The main design specifications of the industrial safety helmets are:

- The material chosen for the manufacture of the helmet should be sweat-resistant, non-irritant, and a high shock-absorbing capacity.
- The headband of the helmet should not be less than 30 mm in width and must be tractable.
- The headband must be fitted with anti-concussion strips whose width should not be less than 19 mm, and it must have at least four anchoring points, building the cradle.
- The straps must ensure clearance of at least 30 mm between the top of the wearers head and inside top of the helmet crown at the smallest size adjustment of the headband.
- The straps must have a maximum wearing height of not less than 80 mm.
- The helmet shall be provided either with chinstrap or nape strap.
- The chinstrap should be of at least 19 mm width with an adjustment device to maintain the tension.
- The nape strap shall be either integral or as an attachment to the headband and adjustable.
- The mass of the complete helmet should not exceed 400 g (Figs. 1 and 2).

Fig. 1 Standard helmet

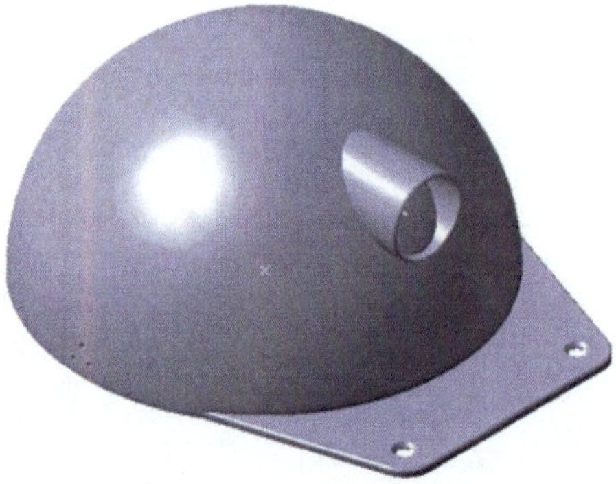

Fig. 2 Modified helmet

Table 2 Mesh details

Part name	Element type	Element shape	Geometric order	No. of elements	No. of nodes
Standard helmet	C3D10	Tetrahedral	Quadratic	93,583	165,397
Modified helmet	C3D10	Tetrahedral	Quadratic	132,473	220,628

The results noted for both the helmet and stiffness are calculated

4 Finite Element Analysis

The finite element analysis is performed by ABAQUS software. The part model is imported into the ABAQUS, and the mentioned properties are applied for the safety helmet. The helmet is analyzed for all the three materials, and the results are compared for the standard used helmet and the modified helmet. The static general analysis is done to determine the stiffness. The bottom case of the helmet is fixed. The load is given on the top of the helmet. The loads applied on the helmet are 100, 150, and 200 N. The elements used to mesh the part is tetrahedral. The feature of the mesh is furnished in Table 2.

5 Results and Discussions

The investigation of the helmet is performed in ABAQUS software. The literature study has been useful for selecting the materials for the helmet. The materials selected for the analysis are ABS and natural fiber such as banana and coconut fibers. These materials show excellent properties and are lightweight when compared with the conventionally used materials. The usage of the natural fiber-reinforced polymer composites not only improves the standard of the helmet but also reduces the burden on the environment due to their biodegradable behavior. The analysis is done to determine the stiffness of the helmet for all the three materials. The word 'stiffness' can be coined as the ability of the material to counter deformation in response to an applied load. The stiffness is obtained by referring the value of deflection in the ABAQUS for the applied load. The loads to be applied to the top of the helmet are 100, 150, and 200 N. Both the helmets are subjected to all the three load cases, and their response is recorded. The maximum deflection obtained from the analysis is tabulated, and stiffness is calculated by using the formula given below

$$\text{Stiffness} = \text{Load}/\text{Deflection}$$
$$K = F/\delta(\text{N}/\text{mm})$$

First, the material selected for analysis is acrylonitrile butadiene styrene (ABS). The deflection caused by the applied load noted for both the helmet and stiffness is calculated. The results are compared by plotting a graph between the applied load (*N*) on *Y*-axis and the deflection (mm) on *X*-axis.

Load case: 1 100 N load is applied at the top of the safety helmet (Fig. 3).

Load case: 2 150 N load is applied at the top of the safety helmet (Fig. 4).

Load case: 3 200 N load is applied at the top of the safety helmet (Fig. 5).

From Fig. 6, it is inferred that the resistivity to deformation for the applied load of the modified helmet is more than the standard helmet. Hence, the stiffness of the modified helmet has been improved due to change in design. After completing the

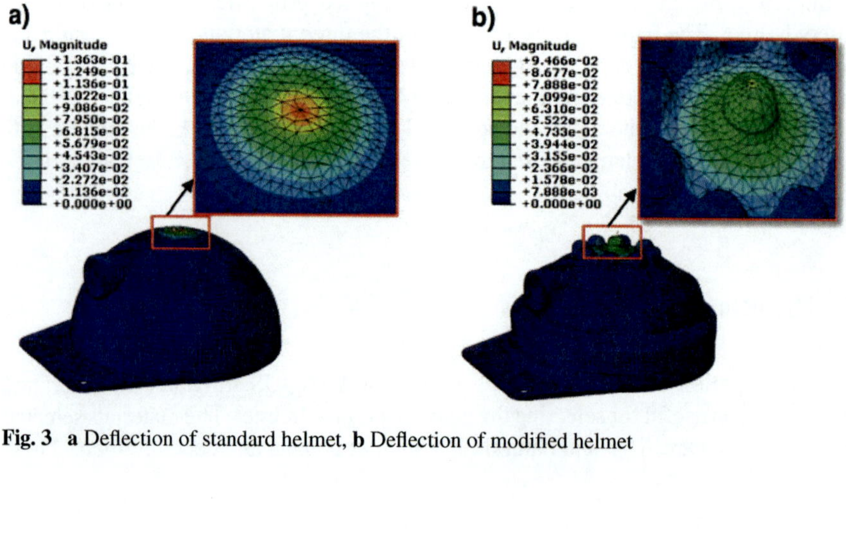

Fig. 3 **a** Deflection of standard helmet, **b** Deflection of modified helmet

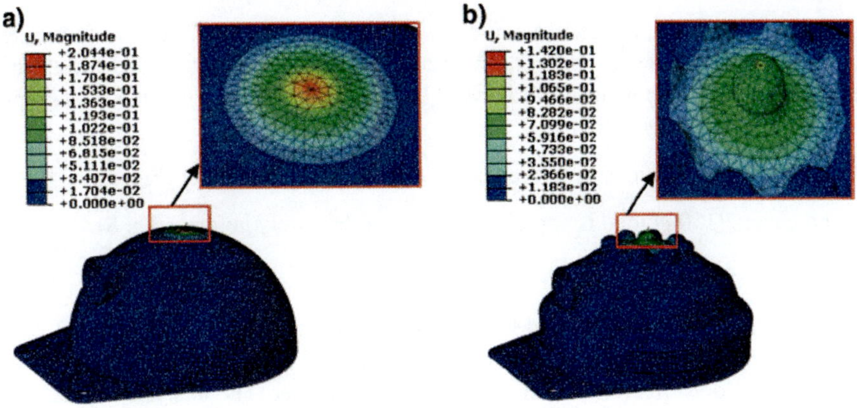

Fig. 4 **a** Deflection of standard helmet, **b** Deflection of modified helmet

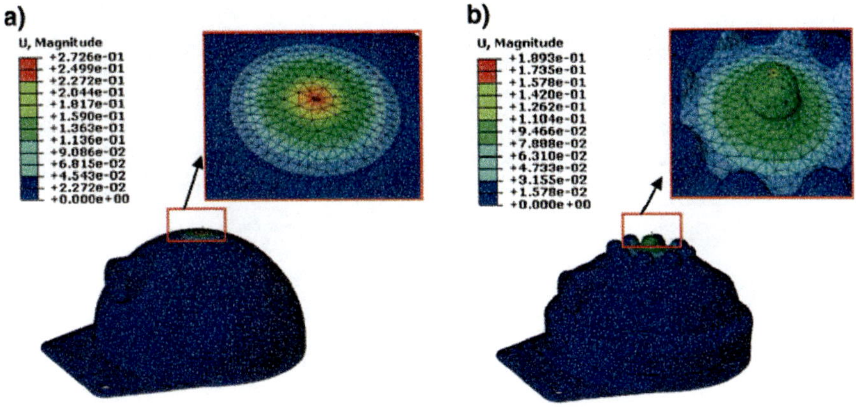

Fig. 5 **a** Deflection of standard helmet, **b** Deflection of modified helmet

Fig. 6 Load versus deflection for ABS

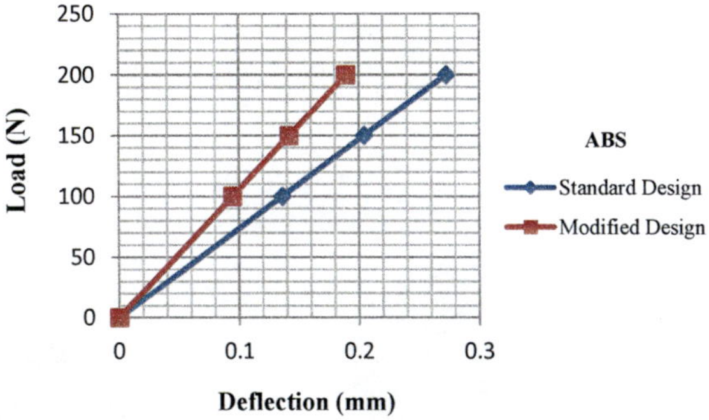

analysis using ABS, the analysis is done by using natural fiber, i.e., coconut fiber and followed by banana fiber. The deflection of the helmet is recorded for all the load cases, and the conclusion is derived by plotting a stiffness graph, i.e., a graph between load and deflection.

Load case: 1 100 N load is applied at the top of the safety helmet (Fig. 7).

Load case: 2 150 N load is applied at the top of the safety helmet (Fig. 8).

Load case: 3 200 N load is applied at the top of the safety helmet (Fig. 9).

The stiffness of the modified helmets has been increased when compared with the standard design. From Fig. 10, it is inferred that the modified helmet has higher resistivity to deformation than the standard used helmet. The analysis is followed by using natural fiber (banana fiber) and after the completion of coconut fiber. The

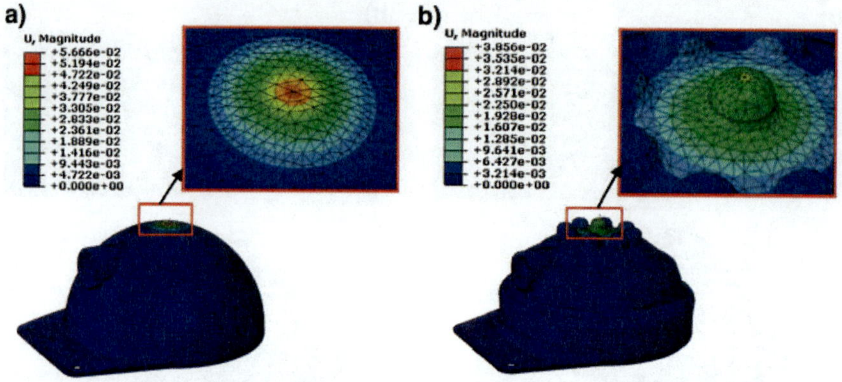

Fig. 7 **a** Deflection of standard helmet, **b** Deflection of modified helmet

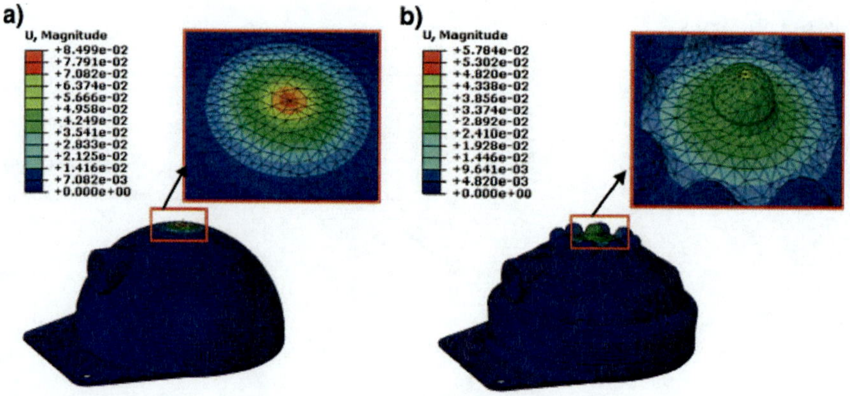

Fig. 8 **a** Deflection of standard helmet, **b** Deflection of modified helmet

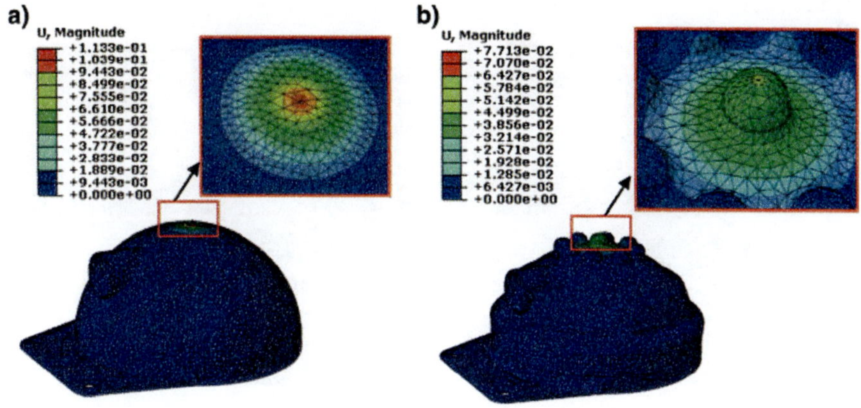

Fig. 9 **a** Deflection of standard helmet, **b** Deflection of modified helmet

Fig. 10 Load versus deflection for coconut fiber

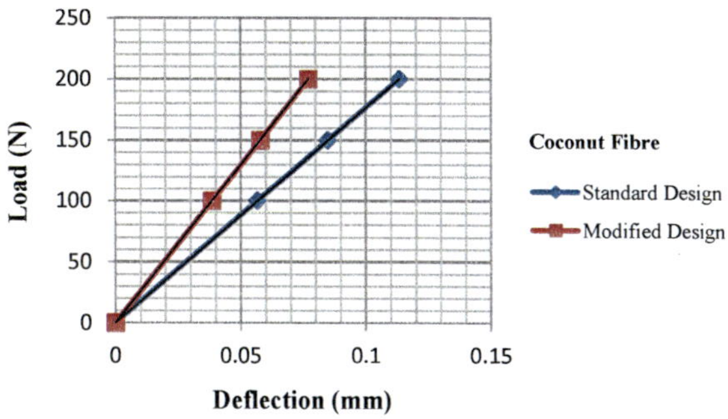

similar procedure is repeated for doing the analysis, and the stiffness is compared by taking deflection into account for all the load cases.

Load case: 1 100 N load is applied at the top of the safety helmet (Fig. 11).

Load case: 2 150 N load is applied at the top of the safety helmet (Fig. 12).

Load case: 3 200 N load is applied at the top of the safety helmet (Fig. 13).

Figure 14 shows that the modified helmet stiffness has been improved by modifying the design. After the completion of all the analysis, the deflection of both the modified helmet and the standard helmet in response to all the load cases are tabulated. From Fig. 15, the stiffness for modified is higher than the standard used helmet irrespective of the material used. The modified designed helmet can be manufactured by any of the material (Table 3).

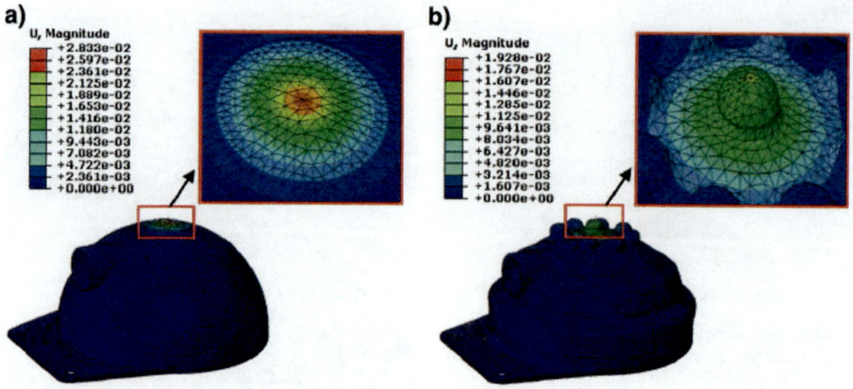

Fig. 11 **a** Deflection of standard helmet, **b** Deflection of modified helmet

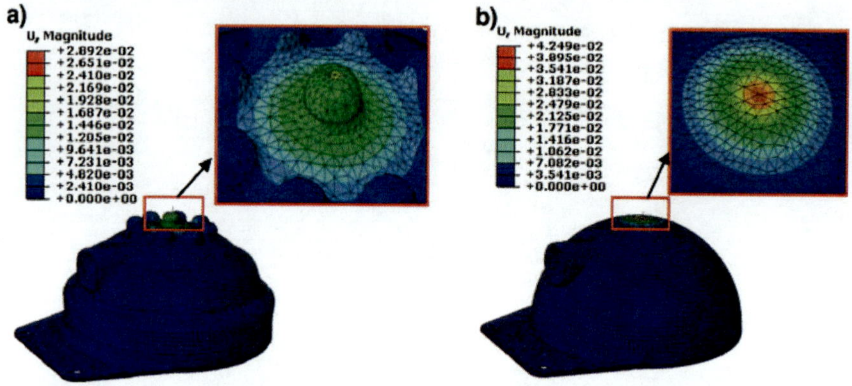

Fig. 12 **a** Deflection of standard helmet, **b** Deflection of modified helmet

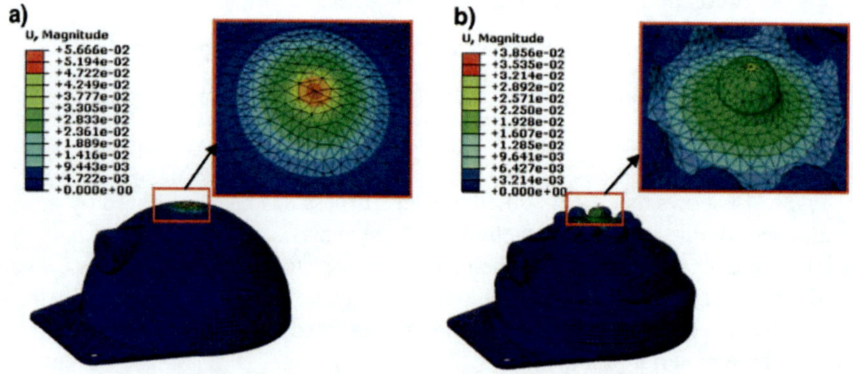

Fig. 13 **a** Deflection of standard helmet, **b** Deflection of modified helmet

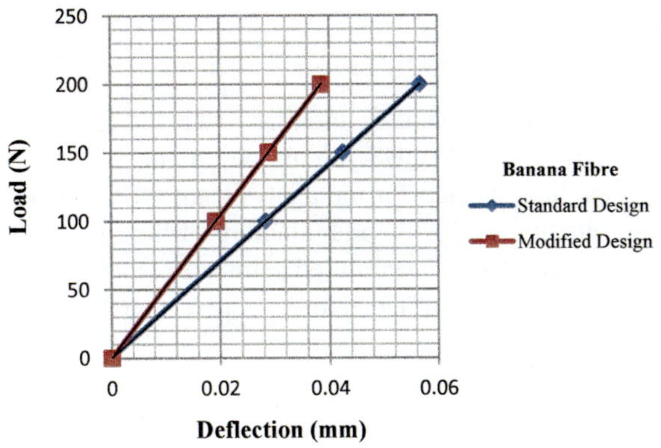

Fig. 14 Load versus deflection for banana fiber

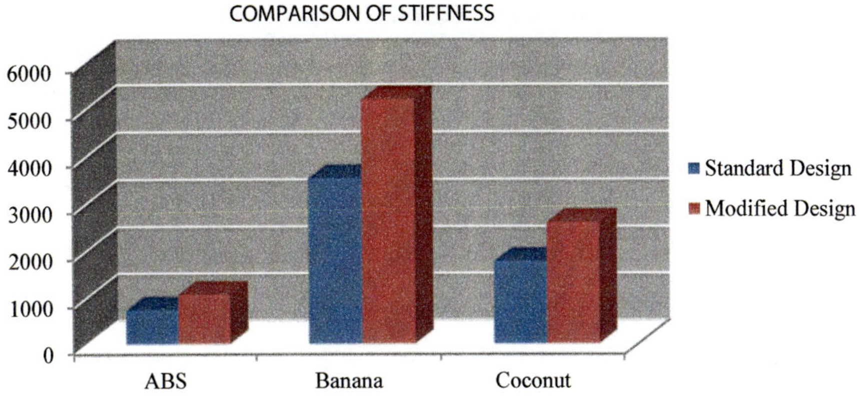

Fig. 15 Load versus deflection for ABS

Table 3 Stiffness of the helmets

Material	Load (N)	Deflection (mm)		Stiffness (N/mm)	
		Standard design	Modified design	Standard design	Modified design
ABS	100	0.1363	0.09466	733.675715	1056.41242
	150	0.2044	0.1420	733.855186	1056.33803
	200	0.2726	0.1893	733.675715	1056.52404
Banana fiber	100	0.02833	0.01928	3529.82704	5186.72199
	150	0.04249	0.02892	3530.24241	5186.72199
	200	0.05666	0.03856	3529.82704	5186.72199
Coconut fiber	100	0.05666	0.03856	1764.91352	2593.3610
	150	0.08499	0.05784	1764.91352	2593.3610
	200	0.1133	0.07713	1765.22507	2593.02476

6 Conclusion

This study indicates the validation of the modified designed helmet with the standard helmet by doing analysis in ABAQUS, when both the helmets are subjected to the same load. The materials selected for the analysis are ABS, natural fiber (banana, coconut) that are widely used for the manufacture of the industrial safety helmet. The analysis results showed that the modified helmet provides more resistant to load. Hence, it is concluded that the modified helmet is stiffer than the standard used helmet irrespective of the material used.

References

1. Rajasekar K, Ashokkumar K, Narayanan L (2015) Design and analysis of industrial safety helmet using natural fibers. Int J Innov Eng Technol (IJIET) 5(3)
2. Kostopoulos V, Markopoulos YP, Giannopoulos G, Vlachos DE (2002) Finite element analysis of impact damage response of composite motorcycle safety helmets. Compos B Eng 33(2):99–107
3. Arunprasath B, Narasiman E, Sivakumar BG (2015) Fabrication and analysis of flax preperg natural fiber reinforced epoxy composite material for helmet outer shell. Int J Innov Res Sci Eng Technol 4(9):8500–8506
4. Ariff TF, Jalil ME, Bahar R (2014) Design improvements in the inner shell of a motorbike helmet using coconut fiber composite. Abstr Emerg Trends Scie Res 2
5. Aly NM, Ali Marwa A et al, Evaluation of the performance of new laminated composite shells for motorcycle helmets. Int J Eng Technol (IJET)
6. Johnson S, Kang L, Akil HM (2016) Mechanical behavior of jute hybrid bio-composites. Compos Part B: Eng 91:83–93
7. Murali, B., D. Chandramohan, S. K. Nagoor Vali, and B. Mohan. "Fabrication of industrial safety helmet by using hybrid composite materials." Journal of Middle East Applied Science and Technology (JMEAST) 15 (2014): 584–587
8. Rodríguez-Millán M, Tan LB, Tse KM, Lee HP, Miguelez MH (2017) Effect of full helmet systems on human head responses under blast loading. Mater Des 117

Development of Injection Moulding and Its Statically Method Experimental Study During the Manufacturing Process

M. Usha Rani, Y. Ramamohan Reddy, V. Viswanatha Chari, and G. Srinivas Kumar

Abstract Infusion shaping is an assembling procedure for creating part for both thermoplastic and thermosetting plastic material. Goal of this undertaking is to decide the parameter impact to the war page and shrinkage of the hand telephone packaging and to decide the enhancement parameter for lessen the war page and shrinkage the most critical issues underway of plastic parts utilizing infusion forming. In this examination, impact of infusion shaping parameters on the shrinkage in polypropylene (PP) and polystyrene (PS) is explored. The connection among information and yield of the procedure is examined utilizing relapse strategy and analysis of variance method. The chosen input parameters are dissolving temperature, infusion weight, pressing weight, and pressing time. Impact of these parameters on the shrinkage of previously mentioned materials is considered utilizing scientific displaying. Thus, efficiency will be diminished while the endeavors are channelized to upgrade quality. To guarantee high caliber and efficiency, it is important to enhance machining parameters. Different reactions of nature of infusion forming process have been examined based on execution parameters and strategies. This paper means to show plastic infusion shaping procedure conditions. The handling conditions fulfilled quality-based item fabricating.

Keywords Molding · Tensile test

1 Introduction

These days, focused market expects makers to deliver fantastic parts, with bring down cost at all conceivable time. Infusion shaping is known as a compelling procedure for large-scale manufacturing of plastic parts with convoluted structures and high-dimensional accuracy. In this technique, high weight liquid polymer is infused to the hole with wanted shape. Next, under high weight, liquid sets. Amid the procedure, plastic materials are under high weight and temperature. Materials are cooled to get

M. Usha Rani (✉) · Y. Ramamohan Reddy · V. Viswanatha Chari · G. Srinivas Kumar
Srinivasa Ramanujan Institute of Technology, Anantapur, India
e-mail: usharani.mec@srit.ac.in

A. Arockiarajan et al. (eds.), *Advances in Industrial Automation and Smart Manufacturing*, Lecture Notes in Mechanical Engineering,
https://doi.org/10.1007/978-981-15-4739-3_14

wanted shape. Infusion shaping procedure can be isolated into four phases: plasticization, infusion, pressing, and cooling the plan of plastic segments three kinds of fashioners ordinarily interface in the advancement of item Typically connect in the improvement of the item, Industrial architects and specialists in ergonomics and feel, create quality that specifically collaborate with the client and give generally speaking structure. Mechanical specialists build up the segments that make up the item. These parts satisfy the capacities that the clients indicate. Generation design includes and alters highlights essential for part produce. The outline of items expects specialists to be educated about essential quality. In infusion forming, exceptionally confused parts can be produced and sizes may extend from little to expansive. Infusion Molding is a cyclic make up the item. These parts satisfy the capacities that the clients indicate. Generation design includes and alters highlights essential for part produce. The outline of items expects specialists to be educated about essential quality. In infusion forming, exceptionally confused parts can be produced, and sizes may extend from little to expansive. Infusion molding is a cyclic procedure for delivering copy articles from a form and is the most broadly utilized for polymer preparing. The primary favorable position of this procedure is the limit of monotonously manufacturing parts having complex shapes and geometries at high generation rates with least cost. Most extreme polymers might be infusion shaped, as thermoplastics, fiber-strengthened thermo plastics, thermosetting plastics, and elastomers (Fig. 1).

The infusion forming process has a progression of tasks that are consecutively done that prompt the change of plastic pellets into a shaped part. Cooling makes the plastic to set and turn out to be dimensionally steady before expulsion. Warmth that has been exchanged to the shape by the liquid plastic is diverted by a coolant that courses through cored entries in the form. Coolant temperature and stream rate decide the effectiveness of warmth expulsion. In any case, factors like thickness of the divider between the shape hole and coolant chamber and the material of the form will be researched. Cooling the shaped parts consistently may mean either, cooling the

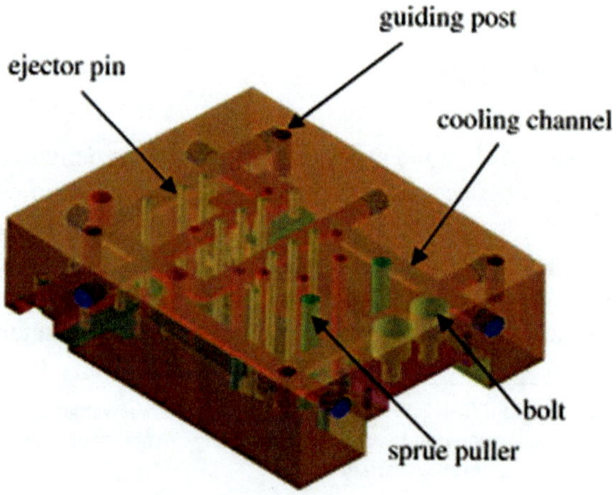

Fig. 1 An injection mold cooling system

shape with various stream rates of cooling medium in various territories or, utilizing a similar stream rate all through the shape however with various temperatures of cooling medium.

Injection Molding

All infusion machines have a type of security interlock framework that counteracts access to the molds amid the clasping and infusion stages when the machine is working semi-consequently Injection shaping machine while the particular destinations of the exploration work are to outline and develop a little infusion forming machine, and testing. The extent of the work is to outline and develop a financially savvy and ecologically cordial little infusion forming machine for the creation of little plastic articles. The exploration work will include outline idea, tasks, plan examination that will involve outline of infusion plunger, engine determination, outline of the handle, and the use on the handle of the machine. Development of little infusion shaping machine for framing little plastic articles in little scale businesses was a result of the way that most infusion forming machines were of enormous size and most little scale enterprises in creating nations could not abstain from getting them because of their expenses. In taking care of this issue, there is a need to outline little infusion shaping machine that is avoidable by little scale enterprises for generation of little plastic articles; this is the method of reasoning behind this work.

Problem Statement

Form filling chiefly relies upon the temperature upkeep stream inside the shape depressions. The imperfections principally happen with the stream rate thought and the infusion weights. Symmetrical dissemination of temperature at the corner areas of pit can be seen in single hole molds, while the multi-pit shape with complicity turns into a major assignment for the ventures. By centering the relative deficiencies happen in the form planning, a nearby advancement with stream examination is required question look into By considering the above elements considered the temperature territories must be reproduced alongside the thickness of parts to keep up consistent levels of temperature stream of any protest with variable thickness considered as an issue for investigate.

Objectives

- To plan single-frame base for multipurpose use with the distinction in insert.
- To dismember the imbuement stream examination to be viewed and entertainments will make with different temperatures and weights.
- To measure the complexities between hunk inserts and hot sprinters.
- To separate the stream examination for multi gap shape by changing sprinter and entryway diagrams.

Scope of Work

The temperature field was two-dimensional—one orchestrate in the stream course and the other in the thickness heading, inciting the so-called 1½D approach. The imbued polymer was believed to be a Newtonian fluid, and restricted difference frameworks were used to numerically light up the course of action of modify conditions. Performed one-dimensional stream examination was combined with a sparkle change condition for a rectangular opening. Recalling the genuine target to build up the past ways to deal with oversees more sensible geometries, conformal mapping or disintegrating of complex shape distresses in various fundamental parts which were utilized to stretch out the 1½D way to deal with deal with all the all the more confounding stream conditions. Regardless, these techniques require satisfactory accord to be worthy, and the course of action precision solidly depends upon how the geometry is distributed, sharp judgment from the customer.

2 Literature Review

Saturday and Emmanuel [1] discovered plastic items valuable to empower capacity, bundling, transportation and offers of their items. Numerous works have been done in plastic creation on virgin materials, reused materials and also fiber-strengthened plastics yet should has not been done on the embellishment framework to break down the impact of an infusion shaping framework as it influence item qualities through satisfactory material determination, clasping power, infusion or pouring temperature, water cooling framework and temperature, thickness of materials in the cooling chamber as they influence the execution of the infusion form.

Peponi et al. [2] the warming framework as it influences the measure of warmth vitality required to liquefy virgin plastic materials or reused material is resolved on the limit of the trim machine as a unit. Be that as it may, warming component size and consistent high voltage supply could improve the warmth dissemination rate in the groups. The infusion framework with an unpredictable pitched screw transport can be improved by controlling the speed of its prime mover; in any case, encounter has demonstrated that the speed is coordinated with the limit of the warming framework groups divided doing the sink lodging the machining infusion framework.

Sahputra [3] displayed an improved wax model of gas turbine cutting edge Their investigation on the form filling condition in the venture throwing process, they clarify that significant strides in speculation throwing forms are infusion embellishment of Wax design, fired covering, expelling wax, drying and material packaging. In the shape producing they think about infusion temperature and holding time as preparing factors They observed that holding time to be more predominant than that of infusion temperature.

The examination uncovers to us that the physical assorted among laminar and violent stream is that laminar stream of a fluid when each particle of the fluid takes after a smooth way and never intrude with each other in trim. One delayed consequence of laminar stream is that the speed of the fluid is predictable whenever in the fluid. Meanwhile in tempestuous stream, a sporadic stream that is portrayed by little whirlpool areas. The speed of this fluid is undeniably not steady at each point [4–12].

3 Methodology

The examination and build up of an estimation demonstrate which encourages a specialist to enhance the estimation of assembling hours in the shape fabricating. That unsupported master estimation speaks to an extremely wide arrangement space. Infusion forming machine offers numerous preferences to choices fabricating strategies, including insignificant misfortunes from scrap (since scrap pieces can be liquefied and reused), and negligible completing necessities. Infusion shaping machine contrasts from metal kick the bucket throwing, in that liquid metals can basically be poured and plastic gums must be infused with compel. The procedure includes presenting crude materials in type of granules into one end of a warmed barrel, warming the materials in the warming chamber, and compelling the liquid metal into a shut form, where the last cementing of the liquid metal in type of the setup of the shape hole takes. The expecting infusion machine will be produced using mellow steel and medium carbon steel. It must be utilized for the generation of little parts, for example, key holder, bottle top, count, ruler, and garments peg. The mellow steel is utilized for the development of supporting plates, container, centralized server, form, and platens, handle, and tie bars. This is on account that they are not subjected to steady warmth. It is effortlessly weldable and has great functionality and however indicates poor reaction to warm treatment.

Developments for Injection Molding Machines

Improvement of little infusion shaping machine for framing little plastic articles in little scale enterprises was contemplated. This work which involved outline, development, and test little infusion shaping machine was equipped for framing little plastic articles by infusing liquid gums into a shut, cooled form, where it sets to give the coveted items produced. The machine was outlined and built to function as a model for creating little plastic parts. Outline idea, task, and get-together of segments parts were made. Additionally, working illustrations and materials determination were made in view of computations of the width of infusion plunger, number of teeth required for the plunger rack and goad adapt, the rakish speed, number of upheaval, torque and power acquired from the electric engine chose and the use on the handle of the machine.

Injection Molding Machines Operations

Infusion-forming machine molds can be affixed in either an even or vertical position. The larger part of machines are on a level plane arranged, yet vertical machines are utilized in some specialty applications, for example, embed forming, enabling the machine to exploit gravity. Some vertical machines additionally do not require the shape to be attached. There are numerous approaches to affix the instruments to the platens, the most well-known being manual clips (the two parts are dashed to the platens); be that as it may, water-driven braces (chocks are utilized to hold the device setup) and attractive cinches are additionally utilized. The attractive and water-driven clips are utilized where quick instrument changes are required (Fig. 2).

Fig. 2 Injection molding machine

Manufacturing Techniques
Infusion forming is utilized more widely in the assembling field to make different items for meeting necessities. With a specific end goal to deliver amazing infusion items monetarily, we should consider the issue about infusion shape exhaustively. The nature of infusion form straightforwardly influences the effectiveness, quality, and cost of shaped items. It assumes a vital part in the framing of infusion items. Not just the surface quality and accuracy of plastic items are totally controlled by the shape, yet in addition the inside quality and productivity of framing items are likewise influenced by the form. A component-based 3D design has been set up for the extraction of focus, and pit with single gloom shape makes easy to plot. Parts with unclear layout of the case which is having items the sprinter arrangement will change. In the present work, the model has been created in solid works, and shape base arrangement was set up in the same sensitive item.

Phases of Injection Molding
Stream development is stressed over the lead of plastics in the midst of the shape filling process. A plastic part's properties depend upon how the part is shaped. Two areas having vague estimations and delivered utilizing a comparable material yet formed under different conditions will have unmistakable uneasiness and shrinkage levels and will bear on differently in the field, inferring that they are eventually two one of a kind parts. The manner in which the plastic streams into the frame is of focal

criticalness in choosing the idea of the part. The route toward filling the shape can be especially destitute down with the ability to predict weight, temperature, and stress shape is an adjusted and machined steel plate having gaps which are made of plastic leave mixed for game plan of a segment. It contains two sections which are a bit of formed cut. The mix of surfaces of molded fragments is machined definitely so there can be no spillage of plastic spilt line, and if spillage happens, it will be exorbitant to remove.

Injection Temperature Monitoring
For it tends to be contrasted and plastic infusion process, temperature checking strategy is broadly utilized in equipment as the principle screen of conveyed observing, or depend on the administrator's experience judgment, temperature estimation accuracy is low, the perception isn't helpful, auspiciousness, frequently prompts temperature alteration in a working cycle, diminish the infusion the achievement rate of cost increments, and as a result of the test question, the temperature parameters in the numerical interim unique, regularly caused by the instrument checking framework is made out of unit there are a considerable measure of burden to the ongoing testing and examination. What's more, elastic infusion process, continuous and exactness of temperature checking is all the more requesting, subsequently can't screen the conventional technique to address the issues.

4 Results

Infusion forming is maybe the most widely recognized and vital of all plastic preparing forms. The procedure is to a great degree flexible and can deliver extremely complex formed parts, with the utilization of multi-sided molds. Indeed, even parts with metal supplements can be created. Injection forming is an imperative assembling process for the mass delivering of plastic items in complex shapes and sizes with high accuracy. Lately, the creation of infusion-shaped plastic items has expanded quickly on the grounds that plastic items are light in weight, low in cost, and quick for shape framing, in this way making them more prevalent than the metallic ones. Because of developing plastics applications, expanding client request and fast development of the worldwide commercial center, the quality prerequisites of infusion shaped items have turned out to be more stringent. Since, the item quality prerequisites turn out to be more stringent in the plastic business, the assurance of the ideal infusion forming process parameters for advancement of new items and the change of existing item quality has turned into a functioning examination region. The setting of process parameters and their enhancement are perceived as essential ways to deal with enhance the nature of the formed items at no extra cost for shape repairs. These embellishment parameters may influence the nature of the formed items. Little changes of trim parameters may give a critical effect to the plastic material's attributes. Numerous trial works were completed to explore the impact

Fig. 3 Injection molding manufacturing machine

of the infusion shaping parameters on the nature of shaped items and the event of embellishment surrenders (Fig. 3).

The execution of the shape cost depends upon the glow transferability of fluid metal through frame discouragement, and it will streamline the thing cost system of warmth trade when fluid metal experiencing the pit with consistent temperature and to keep up until the point that the moment that hole filling will give the suitability of significant worth shape. For achieving the quality shape, change is required for multi-pit molds plan. The cooling time must be enhanced before embarking to the shape preparation to avoid war page and under fill. Frame-impacting materials must be moved up to its mechanical and warm properties to fulfill the glow trade essentials for predictable stream of temperature in the instrument geometry.

Table 1 elucidates the quality extending and flexibility level of the shape in the midst of it subjected to malleable, flexural, weight, and Izod influence quality and assorted regards.

Table 1 Strength elongation elasticity percentages

Recycled number		0	1	2	3	5
Tensile	Strength	100	98	97	97	96
	Elongation	100	97	98	95	95
	Modulus	100	106	102	103	104
Flexural	Strength	100	99	98	96	97
	Modulus	100	99	98	96	97
Compression	Strength	100	97	10	96	98
	Modulus	100	98	101	96	98
Izod impact strength		100	83	93	92	91

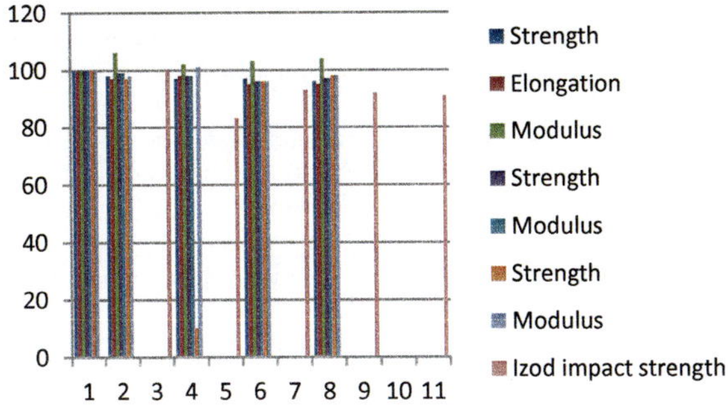

The above chart elucidates the quality expansion and flexibility level of the frame in the midst of it is subjected to tractable, flexural, weight, and Izod influence quality and different regards.

5 Conclusions

The innovation, then again is the primary procedure parameters during the time spent infusion, study, temperature and weight consistency, stream infusion, impact of hole size and entryway measure and different factors on the diverse level of infusion shaping the procedure of both. In the outline and utilization of elastic infusion machine, ought to completely think about the effect of different variables. The most widely recognized imperfections in infusion shaping incorporate rankles, consume marks, streak, sink marks, short shot, weld line and distorting. Infusion forming is one of the normal procedures associated with plastic industry. In any case, infusion forming is a mind boggling process because of numerous changes required, for example, the part configuration, shape configuration, machine execution, and process parameter setting. These changes are important to create great quality plastic part. It is notable that procedure parameter setting is the primary intuitive remedial activities that ought to be performed to achieve quality prerequisites. Inability to set the fitting parameters will result in an expanded cost and decreased quality and profitability of infusion forming items. Deciding the procedure parameter settings for plastic infusion forming significantly influences the nature of the plastic infusion-shaped item.

References

1. Saturday A, Emmanuel A (2016) Parametric analysis of an injection molding system performance for PET products production. Scholars J Eng Technol (SJET) 4(9):441–458. ISSN: 2321-435X
2. Peponi L, Puglia D, Torre L et al (2014) Processing of nanostructured polymers and advanced polymeric based Nano composites. Mater Sci Eng: R: Rep 85:1–46
3. Sahputra IH (2007) Comparison of two flow analysis software for injection moulding tool design. In: IEEE international conference on industrial engineering and engineering management, pp 607– 611
4. Sadeghi BHM (2010) A BP-neural network predictor model for plastic injection
5. Dominick V et al (2003) Replication characterization of micro ribs fabricated by combining ultraprecision machining and microinjection molding. Polym Eng Sci 156(45):257. ISSN: 2347-6710
6. Sazaki et al (2000) Modeling and simulation for micro injection molding process. Comput Fluid Dynam Technol Appl 42(10):410–413
7. Yeh, Wu JS (2015) Injection molding of polymeric LIGA HARMs. Microsyst Technol 76(14):1. ISSN 1466-8033. Molding process. J Mater Process Technol 103(3):411–416
8. Ringertz (2005) An investigation into hesitation effects in oscillating flows. Proc Soc Plast Eng Ann Tech Conf 98(12):36–42
9. Chen W et al (2004) Injection embellishment of polymeric LIGA HARMs. Micro Syst Technol 10(14):241-251. ISSN: 1432-881. Minh PS et al (2012) Injection molding of polymer micro- and submicron structures with high aspect ratios 155(14):1–14. ISSN: 2249-7846
10. Pye R (2009) Replication characterization of microribs fabricated by combining ultraprecision machining and microinjection molding. Polym Eng Sci 98(8):36–42. ISSN: 015533-1658
11. Reifschneider (2001) The micro injection molding process for polymeric components manufacturing. 8(283):1. ISSN: 0097–6326
12. Rawin et al (1997) Injection molding of polymer micro- and submicron structures with high aspect ratios. 26(278P:1. ISSN: 0036-8075

Buckling Analysis of C-Stringer and Hat Stringer on the Load Carrying Vehicle

B. Stalin, V. Dhinakaran, M. Ravichandran, K. Sathiya Moorthi, and J. Vairamuthu

Abstract In this paper, the stringer's buckling analysis is analyzed and reported. To block the shock, twist, and other stresses, the chassis should be in very high strength. More than that of strength, the chassis should tolerate the buckling condition of the frame. Here, the work is done toward stringer analysis in chassis with the properties of buckling behavior, strength by use of FEA software (Ansys Workbench). The result shows that hat stringer is stronger than C-stringer during the stringer on buckling conditions.

Keywords Strength · FEA · Stringer bar · Buckling

1 Introduction

The chassis is a major component of the load carrying vehicles to withstand the entire load of the vehicle and goods to be travelled. The chassis is supported by cross bars which is one type of stringer component. The stringer types are in shape of C-stringer, I-stringer, hat stringer, L-stringer, and so on. Ren et al. [1] has resolved the finite element analysis on the frame of SX360dump trucks to analyze the vibration mode and optimization of the frame in safe condition. Lanzutti et al. [2] has analyzed

B. Stalin (✉) · K. Sathiya Moorthi
Department of Mechanical Engineering, Anna University, Regional Campus Madurai, Madurai 625 019, Tamil Nadu, India
e-mail: stalin1312@gmail.com

V. Dhinakaran
Department of Mechanical Engineering, Chennai Institute of Technology, Kundrathur, Chennai, Tamil Nadu, India

M. Ravichandran
Department of Mechanical Engineering, K. Ramakrishnan College of Engineering, Samayapuram, Trichy, Tamil Nadu 621 112, India

J. Vairamuthu
Department of Mechanical Engineering, Sethu Institute of Technology, Kariapatti, Virudhunagar, Tamil Nadu, India

A. Arockiarajan et al. (eds.), *Advances in Industrial Automation and Smart Manufacturing*, Lecture Notes in Mechanical Engineering,
https://doi.org/10.1007/978-981-15-4739-3_15

the failure of the press that works continuously on the process plant. Mo et al. [3] has investigated hat stringer buckling and postbuckling response subjected to axial compression. The SEM images and XRD patterns are examined. Raghuvanshi et al. [4] has studied the design and development of the kart chassis. The finite element analysis of the automotive coach and the experimental analysis of the coach is carried out, and the results are compared. The finite element model of the chassis is developed and analyzed in six orders on FEA software (Han et al. [5]). The high-strength magnesium alloy was fabricated by an ingot metallurgical process with hot extrusion followed by aging. The composition of the materials in an alloy were Mg, 1.8% of Gd, 1.8% of Y, 0.7% of Zn, and 0.2% of Zr. Then, the electron back scatter diffraction (EBSD) analysis and transmission electron microscopy (TEM) analysis are carried out to the sample of the manufactured alloy (Homma et al. [6]). Karaoglu et al. [7] has resolved the stress analysis of truck chassis with riveted joints which is performed in Finite element analysis software. First of all, the wheel forces are calculated, and then, wheel moments are found (Conle et al. [8]).

2 Methodology

A stringer is a longitudinal structural part of the chassis frame. Here, the chassis frame cross bar is analyzed in the vehicle TATA 2516 TC truck. The current part which is presented in the chassis is C-stringer. Now, we have changed the C-stringer to hat stringer. The hat stringer with 9.53 mm fillet, 20 mm fillet, 30 mm fillet, and 40 mm fillet are made by using a CAD package (AutoCAD 3D) and analyzed by using FEA software (Ansys Workbench).

The C-stringer cross section is 262 mm × 65 mm. The length of stringer bar is 2440 mm. The fillet between the edges is 9.53 mm from Fig. 1. The material of the component is 9345 standard S37 steels. The properties of the S37 steel material are attained from the literature Raghuvanshi et al. [4]. The hat stringer bar cross section dimensions are 262 mm height, 132 mm top phase, and inside edge of 65 mm. The hat stringer length is 2440 mm. The edge fillets are 9.53 mm, 20 mm, 30 mm, and 40 mm.

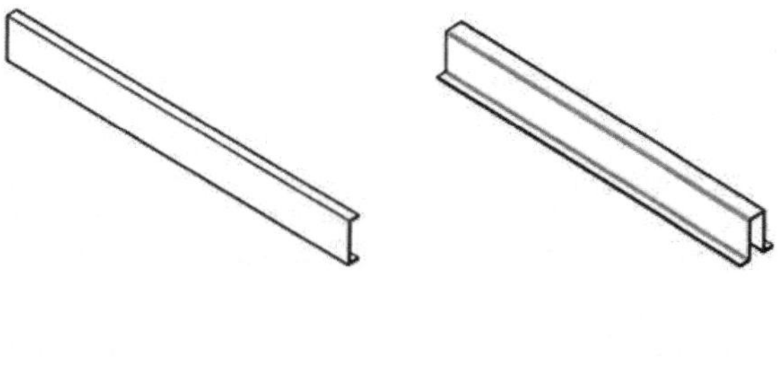

Fig. 1 C-shaped stringer and hat-shaped stringer

3 Results and Discussion

In buckling analysis, two types of buckling action to be carried out. The first criteria is loading on the axis of the stringer, and the another criteria is loading on the center of the stringer top region. The load applied on three axis condition is 1 MN.

The stringer component should resist the maximum amount of buckling load in the chassis because it should carry all parts weight without bending from its position and shape. Here, the buckling analysis is carried out in two types of loading. They are centrally loaded analysis of stringer and axially loaded analysis of stringer (*X*-axis, *Y*-axis, and *Z*-axis).

Load on *X*-axis. From Fig. 2a, the load multiplier of the C-stringer on mode 1 is 0.095 of applied load for maximum deformation of 1 mm which is gathered. The load multiplier value is 0.83 of applied load which is obtained from mode 2. The load multiplier value is 1.03 of applied load to the deformation of 1.2 mm on mode 3. The load multiplier values for the three modes of hat stringer with 9.53 mm fillet are 0.477, 2.51, and 3.08 of applied load which are obtained from Fig. 2b. These values are obtained for the deformation of 1 mm in each mode. From Fig. 2c, the load multiplier values of the three modes are 0.48, 2.5, and 3.24 of applied loads. These values are retrieved from the deformation of 1 mm in each mode of the stringer. The buckling load multiplier value for hat stringer with fillet 20 mm is 3.24 of applied load. The load multiplier values for hat stringer with 30 mm fillet are 0.49, 2.49, and 3.34 of applied load which are achieved from Fig. 2d. Load multiplier value 3.34 of applied load while buckling is the high value of the three modes. The buckling load

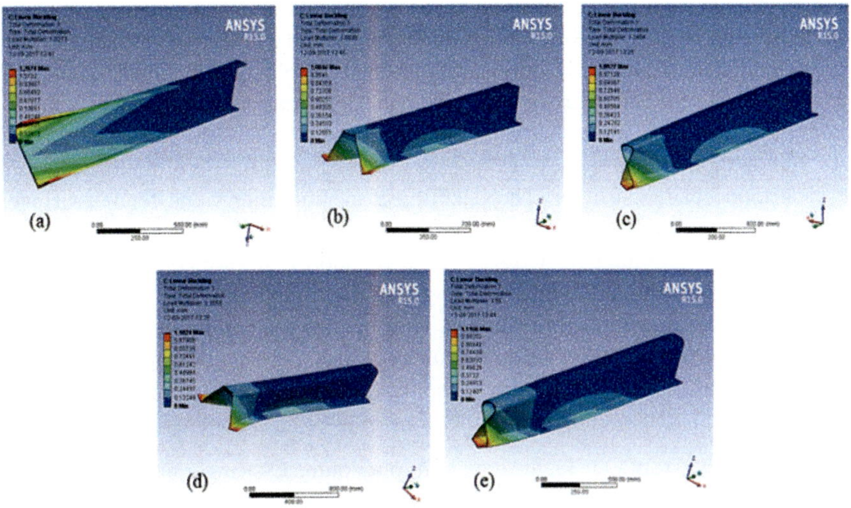

Fig. 2 Load multiplier values of five structures loaded in *X*-axis: **a** C-stringer, **b** hat stringer with 9.53 mm fillet, **c** hat stringer with 20 mm fillet, **d** hat stringer with 30 mm fillet, and **e** hat stringer with 40 mm fillet

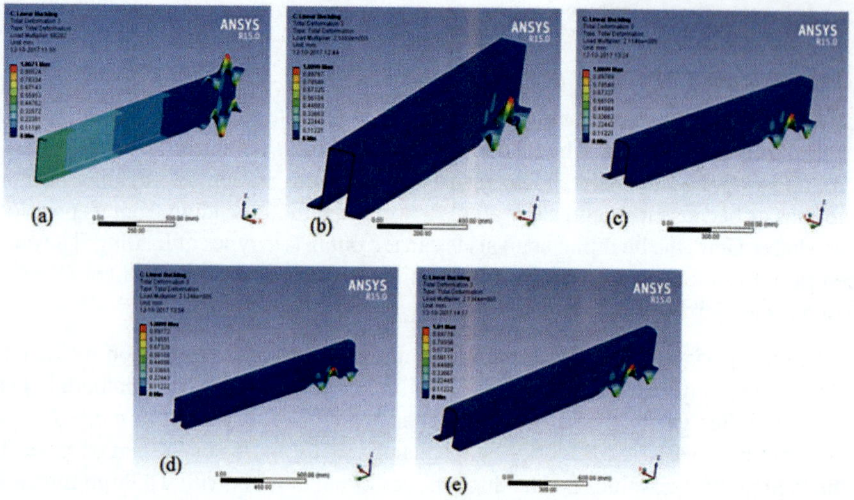

Fig. 3 Load multiplier values of five structures loaded in *Y*-axis: **a** C-stringer, **b** hat stringer with 9.53 mm fillet, **c** hat stringer with 20 mm fillet, **d** hat stringer with 30 mm fillet, and **e** hat stringer with 40 mm fillet

multiplier values for the hat stringer with 40 mm fillet are 0.499, 2.48, and 3.56 of applied loads on the three modes of total deformation. The maximum value of the load multiplier is 3.56 which is retrieved from Fig. 2e.

Load on *Y*-axis. From Fig. 3a, the load multiplier of the C-stringer on mode 1 is 37,608, 44,523, and 68,282 of applied loads for maximum deformation of 1 mm which is gathered. The load multiplier values for the three modes of hat stringer with 9.53 mm fillet are 124,070, 167,500, and 210,080 of applied load which are obtained from Fig. 3b. These values are obtained from the deformation of 1 mm in each mode. From Fig. 3c, the load multiplier values of the three modes are 140,930, 183,050, and 211,460 of applied loads. These values are retrieved from the deformation of 1 mm in each mode of the stringer. The buckling load multiplier value for hat stringer with fillet 20 mm is 211,460 of applied loads. The load multiplier values for hat stringer with 30 mm fillet are 167,420, 196,390, and 212,460 of applied load which are achieved from Fig. 3d, while buckling load multiplier value 212,460 of applied load is the high value of the three modes. The buckling load multiplier values for the hat stringer with 40 mm fillet are 172,380, 198,450, and 213,440 of applied loads on the three modes of total deformation. The maximum value of the load multiplier is 213,440 which is retrieved from Fig. 3e.

Load on *Z*-axis. From Fig. 4a, the load multiplier of the C-stringer on mode 1, 2, and 3 are 193,720, 257,830, and 304,930 of applied loads for maximum deformation of 1 mm which is gathered. The load multiplier values for the three modes of hat stringer with 9.53 mm, 20 mm, 30 mm, and 40 mm fillet are 477,560, 548,050, and 622,490 of applied loads; 487,900, 556,740, and 642,360 of applied loads; 509,780, 569,320,

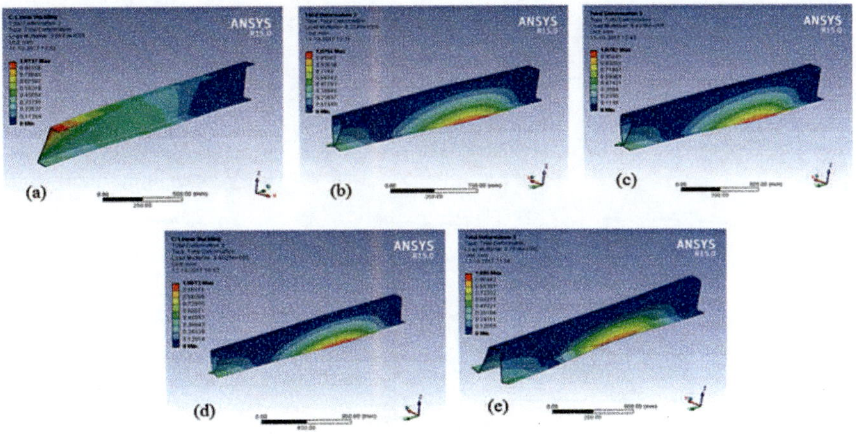

Fig. 4 Load multiplier values of five structures loaded in *Z*-axis: **a** C-stringer, **b** hat stringer with 9.53 mm fillet, **c** hat stringer with 20 mm fillet, **d** hat stringer with 30 mm fillet, **e** hat stringer with 40 mm fillet

and 660,250 of applied loads; and 514,980, 580,560, and 679,360 of applied loads which are obtained from Fig. 4b–e. These values are obtained from the deformation of 1 mm in each mode.

Centrally loaded analysis. The buckling analysis of the stringer by applying the center load on the stringer top region. The load conditions are point load of 4050 N and pressure load of 17.69 Mpa. This load condition is applied to all stringers, and the deformation and the load multiplier are calculated. The buckling load multiplier values for the C-stringer on three modes are 0.76, 1.21, and 1.24 of applied loads. The high value of load multiplier is 3.24 of applied load which can be gathered from Fig. 5a. From Fig. 5b–e, the buckling load multiplier for the various modes of the hat stringer with 9.53, 20, 30, and 40 mm fillet is 2.03, 2.71, and 2.72 of applied loads; 2.5, 3.51, and 3.53 of applied loads; 3.67, 4.68, and 4.74 of applied loads; and 4.64, 6.85, and 7.12 of applied loads.

In buckling, the load multiplier values in *X*-axis, *Y*-axis, and *Z*-axis are more in hat stringer with 40 mm fillet while comparing to other stringer. The load multiplier value of center loading is very high as compared to the other structures in Fig. 6.

4 Conclusion

This paper concludes that buckling analysis of the chassis frame stringer on TATA 2516 TC truck. In buckling, the load rating values of all axis and center loading are very high on the 40 mm fillet hat stringer as compared to remaining structures. Hence, the suitable model for alternative of the chassis C-stringer on TATA-2516 TC truck is a hat stringer 40 mm fillet for buckling behavior.

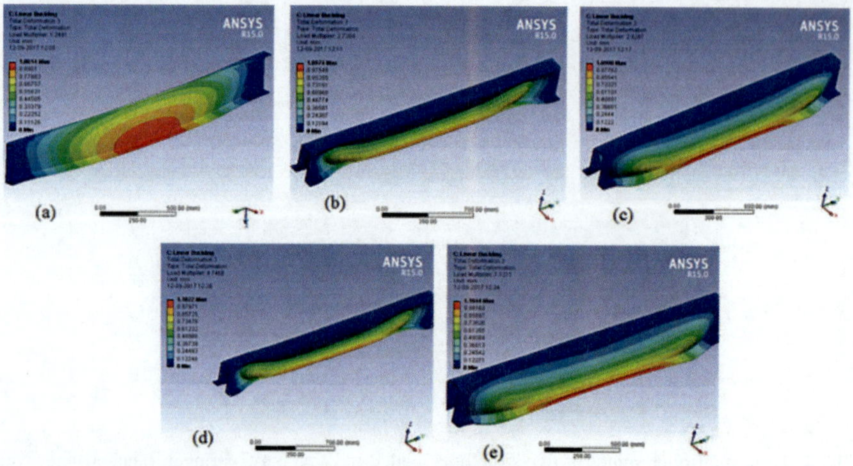

Fig. 5 Load multiplier values of five structures loaded in the center: **a** C-stringer, **b** hat stringer with 9.53 mm fillet, **c** hat stringer with 20 mm fillet, **d** hat stringer with 30 mm fillet, and **e** hat stringer with 40 mm fillet

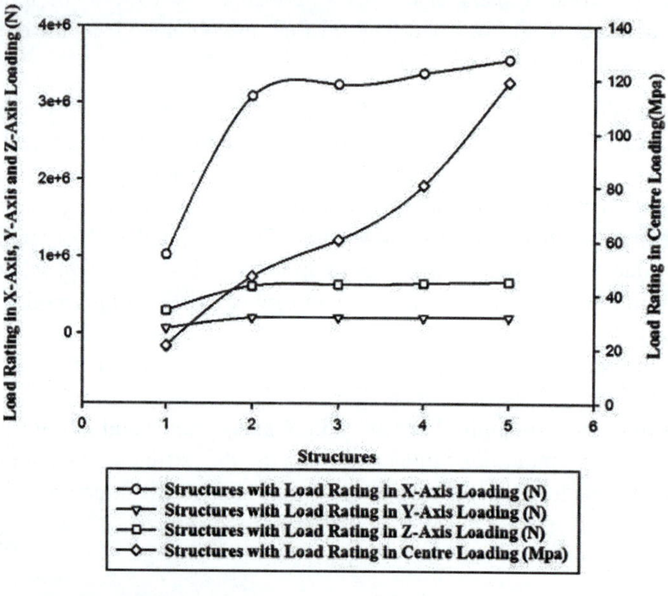

Fig. 6 Comparison of buckling parameters

References

1. Ren Y, Yongchang Y, Zhao B, Fan C, Li H (2017) Finite element analysis and optimal design for the frame of SX360 dump trucks. Procedia Eng 174:638–647
2. Lanzutti A, Andreatta F, Raffaelli A, Magnana M, Zuliani L, Fantoni M, Fedrizzi L (2017) Failure analysis of a continuous press component in MDF production plant. Eng Fail Anal 82:493–500
3. Mo Y, Ge D, He B (2016) Experiment and optimization of the hat-stringer-stiffened composite panels under axial compression. Compos: Eng Part. B 84:285–293
4. Raghuvanshi AC, Srivastav T, Mishra RK (2015) Design and development of foldable kart chassis. Mater Today 2(4–5):1707–1713
5. Han T, Huang C, Tan ACC (2014) Experimental and finite element analysis to identify the source of vibration of a coach. Eng Fail Anal 44:100–109
6. Homma T, Kunito N, Kamado S (2009) Fabrication of extraordinary high-strength magnesium alloy by hot extrusion. Scripta Mater 61(6):644–647
7. Karaoglu C, Kuralay NS (2002) Stress analysis of a truck chassis with riveted joints. Finite Elem Anal Des 38(12):1115–1130
8. Conle FA, Mousseau CW (1991) Using vehicle dynamics simulations and finite-element results to generate fatigue life contours for chassis components. Int J Fatigue 13(3):195–205

Design and Development of IC Engine Fuel Injector Nozzle Holder Body Using Hydraulic Fixture

M. Purusothaman, K. Prudhvi, T. N. Valarmathi, J. Hemanadh, and S. Ganesan

Abstract In automotive industries, cost reduction is the major challenge for engineer. Every year they need some percentage reduction in manufacturing cost at the same time they need quality improvement. The automotive industries are facing tough environment to reduce the total manufacturing cost, reduce the waste and finally improve quality. Generally, various methods are followed to overcome these. The main cause for waste is the dimension criticality in the production to satisfy the customer and to meet the competitive world. Engine is the heart of any automobile vehicle, and the performance of the engine depends on the nozzle delivery pressure. Dimension of the nozzle part determines the performance and efficiency. This work minimizes the cost and effort involved in the production of nozzle holder body by eliminating the rejection, rework, loading/unloading/surface damages and operator fatigue. The production rate is increasing with improved quality using an effective and automated hydraulic fixture system.

Keywords IC engine · Fuel injector nozzle · Hydraulic fixture · Performance improvement

1 Introduction

Nozzle holder body is a one of the major parts in injector assembly. Variation in the dimension NHB leads to rework or rejection in the production of nozzle holder body (NHB). The main objective or aim of the project is to eliminate the rejection and rework in the production of nozzle holder body using hydraulic fixture. The purpose of this project is to minimize the cost and effort involved by eliminating the rejection, rework, loading damages, operator fatigue and also by increasing the production rate using an effective and automated hydraulic fixture system. The suitable modifications are to be incorporated in the automobile radiator, and CFD analysis is being carried

M. Purusothaman (✉) · K. Prudhvi · T. N. Valarmathi · J. Hemanadh · S. Ganesan
School of Mechanical Engineering, Sathyabama Institute of Science and Technology, Chennai 600119, India
e-mail: purusothmani@gmail.com

A. Arockiarajan et al. (eds.), *Advances in Industrial Automation and Smart Manufacturing*, Lecture Notes in Mechanical Engineering,
https://doi.org/10.1007/978-981-15-4739-3_16

out to fulfill the requirements to improve the performance of IC engine [1]. When a vehicle is parked in open sky leads to sudden rise of inside cabin temperature, it may damage the property as well as harm to the children or pets when left in the vehicle itself. Analysis is being carried out to reduce the cabin temperature by introducing PCM setup on the roof of the vehicle [2]. In this study, the small modification is to be carried out in the exhaust manifold to improve the overall engine performance [3]. The complication on this challenging boundary problem is discussed, and the integrative nature of the solution ideology is emphasized [4]. This paper reveals the design of jigs and fixtures for hydraulic press machine in the manufacturing sectors. The present-day problem in industry is facing the usage of hydraulic press machine when the demand has increased which occurs on the gripping or holding the workpiece safely [5]. This work projects the issues of construction, functions and special characteristics of the proposed computer-aided design system. Sample fixture design drawings for boring fixtures of bracket parts were provided as example [6]. This paper expresses that general and comprehensive automatic fixture design with thumb rule approach developed the total system [7]. This paper executes the reduction in fixtures operation time, increases productivity and high quality of operation is possible [8].

2 Materials and Methods

Based on cycle time and productivity hours, the number of component targeted per day is 197 nos. Out of 197, an average of 25–28 components gets rejected. The main cause for the rejection is due to the depth variation during solenoid hole drilling operation as shown in Fig. 1. The depth of the solenoid hole is 29.4 mm with the tolerance of 40 microns. Since the tolerance limit is more critical, a small variation in holding the NHB while machining causes depth variation. Achieving the required dimension with the existing equipment and process becomes undesirable as given in Table 1.

2.1 *Power Pack Assembly*

The pump, motor, pressure regulator, strainer, pressure relief valve, pressure gauge, etc., are mounted on the skeleton of the power pack. The assembled power pack is shown in Fig. 2. The motor is driven by the electrical energy. The motor drives the pump through coupling and creates a pressure difference in the system. Due to the pressure difference, the pump delivers the oil from tank to pressure control valve at high pressure. The pressure regulator is set to the required delivery pressure by adjusting the regulator screw. The excess pressure developed in the power pack is returned to the tank through a return circuit. The exact required pressure is sent to the four-way direction control valve.

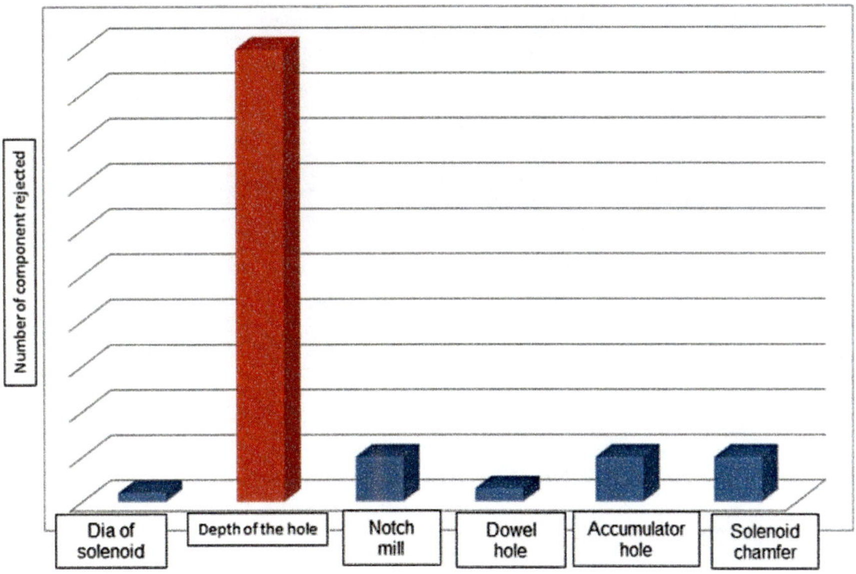

Fig. 1 Rejection causes

Table 1 Drilling operation dimensions

S. No.	Operations	Dia. in mm	Depth in mm
1	Solenoid hole enlarging	10.5 ± 0.05	**29.4 ± 0.04**
2	Dowel hole (2 no's)	1.75/1.92	
3	Accumulator hole	2	12 ± 0.01
4	Notch mill	1.3	
5	Solenoid hole chamfering	10.5	

2.2 *Sequence of Operation*

When cradle at 0° operator loads the workpiece in the collet through guide pin. When the operation starts the cycle, the cradle is tilted to 180°. Then the component de-clamps for 2 s. The hydraulic cylinder is actuated by power pack, and the piston moves up (extends). With an 35 bar pressure, the workpiece is rigidly clamped between the stopper and the hydraulic cylinder. After that the collet clamps the workpiece again. To ensure that there is no movement on workpiece while machining. The drilling cycle stars, and after completion of drilling holes the cycle ends. After the cycle ends, the pressure applied by the cylinder is removed with the assistance of direction control valve and the cradle rotates back reaches 0°. Then the collet releases the pressure and allows the workpiece to unload. And through the direction control valve,

Fig. 2 Power pack assembly

the hydraulic actuates the self-ejection system. The operator unloads the machined workpiece from the machine.

2.3 *Fixture with Workpiece*

Figure 3 shows the fixture at 0° with workpiece clamped on it. Figure 4 shows the top view of fixture after 180° of rotation. The component is rigidly clamped in between the hydraulic cylinder and stopper. After the cycle ends, the pressure applied by the cylinder is removed with the assistance of direction control valve. Now the cradle rotates back to 0°. Then the collet releases the pressure and allows the workpiece to unload.

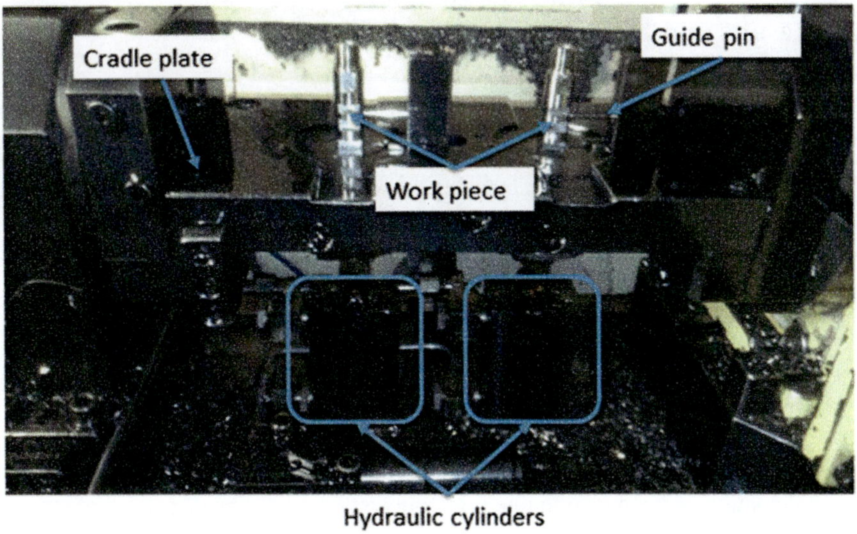

Fig. 3 Fixture with workpiece tilted to 0°

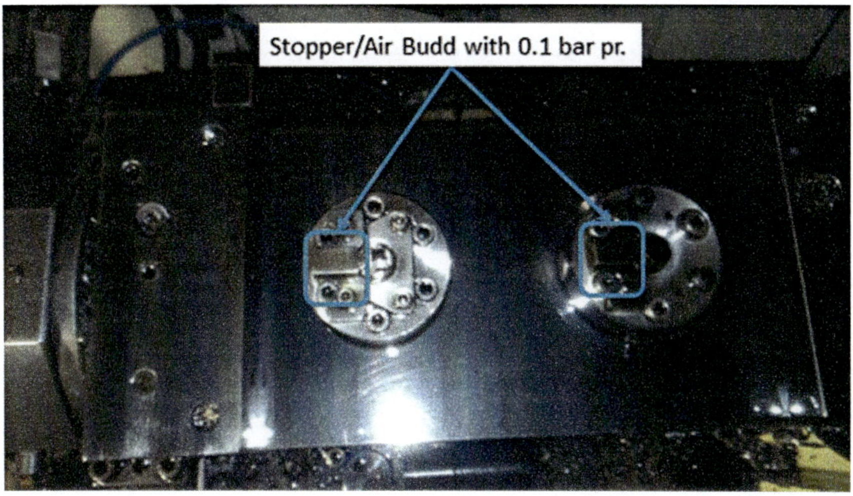

Fig. 4 Fixture with workpiece tilted to 180°

3 Results and Discussion

3.1 Measurement of the Job Movement

The below measurement is taken by using the work offset option in VMC before and after machining. Initially after clamping the workpiece, the work offset will be set to zero. But after the machining while checking the work offset, we found that the component moves about axially during machining process.

3.2 Calculation of Rejection Rate

The productivity hours and non-productivity hours in a shift are shown in Table 2 and calculation of the rejection rate is as follows.

$$\begin{aligned}\text{Total productivity time per shift} &= \text{Total shift time} \\ &\quad - \text{Non - productivity time} \\ &= 510 - 75 \\ &= 435\text{min}\end{aligned}$$

$$\begin{aligned}\text{No of component targeted per shift} &= \text{Total productivity} \\ &\quad \text{time per shift/cycle time} \\ &= 435/2.2(\text{min}) \\ &= 197\ \text{Nos.}(\text{approx} = 200)\end{aligned}$$

Actual no of component produced = 190 (±3) No's.

No. of rejections per shift = 25–28 No's

Total rejection rate = 28/190 = 14.7%.

Table 2 Productivity hours

Shift timing		Non-productivity hours	
Starting time	7:45	Tea break	15 min
Ending time	4:15	Lunch break	45 min
Total hours	8:30	Tea break	15 min
Total shift time	510 min	Non-productive time	75 min

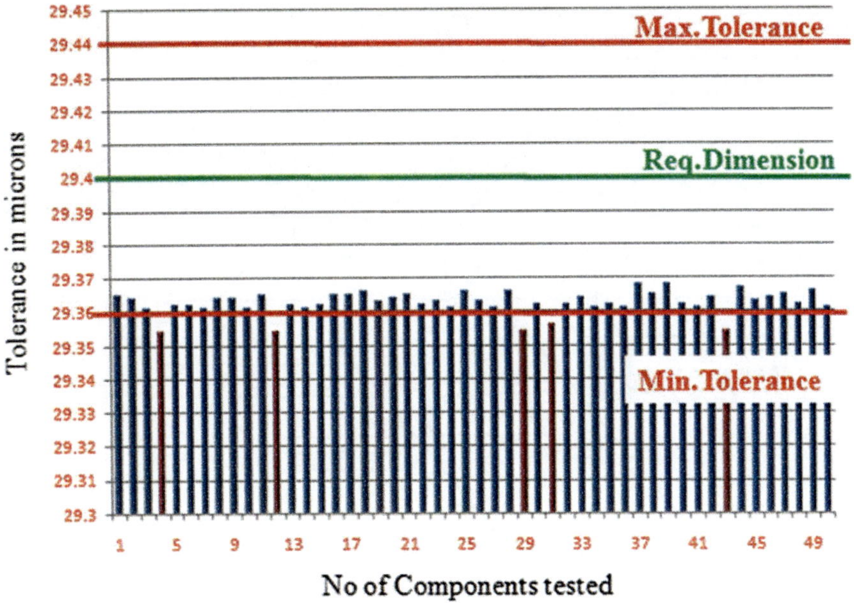

Fig. 5 Solenoid hole depth variation

3.3 Solenoid Hole Depth Variation Before Implementation

In a random selection, fifty components were picked for the depth measurement. Figure 5 shows the depth variation for the selected fifty parts. Out of fifty components, five components were not within the tolerance limit dimension.

No. of component tested = 50 No's.

No. of rejections = 5 No's.

Total rejection rate = 5/50 = 10%.

Therefore, the rejection rate lies in between 10 and 15% because of depth variation.

3.4 Root Cause Analysis

There are plenty of causes which can result into dimension variation. The below fishbone in Fig. 6 will give a brief idea about possible root causes for the depth variation. Based on the root cause analysis (fishbone diagram) as shown in Fig. 6, the following check has been made as shown in Table 3. The issue found is the job moment along axially while machining. The basic approach when a job moves down while machining is to provide an extra support in the opposite direction of its movement. But since the cradle plate rotates 180°, we cannot implement a fixed

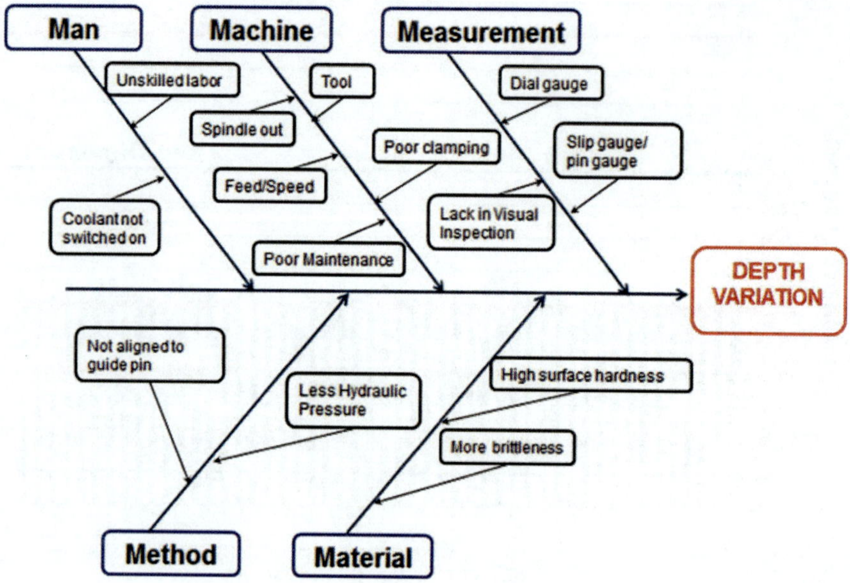

Fig. 6 Root cause analysis

Table 3 Possible causes

Category	Possible causes	Validation	Result
Man	Unskilled operator	Checked trained operator	Insignificant
	Wrong loading	Checked loading	Insignificant
Machine	Play in chuck	Checked OK	Insignificant
	Improper maintenance	Checked OK	Insignificant
Measurement	Lack in calibration	Calibration report verified	Insignificant
	Improper gauge	Gauge verified OK	Insignificant
Clamping	Hydraulic pressure	No variation	Insignificant
	Component sliding	Additional support needed	Significant
	Metal chips in holding area	Checked OK	Insignificant
Material	Defective input RM	Report verified OK	Insignificant
	RM dimensional variation	Checked OK	Insignificant
Coolant	Coolant on/off	Checked	Insignificant

block or a fixed support. We need to implement a fixture that should be fixed out of the radius of cradle plate. And also it should have some kind of extension so that it supports the job from the bottom after the cradle plate turns about 180°.

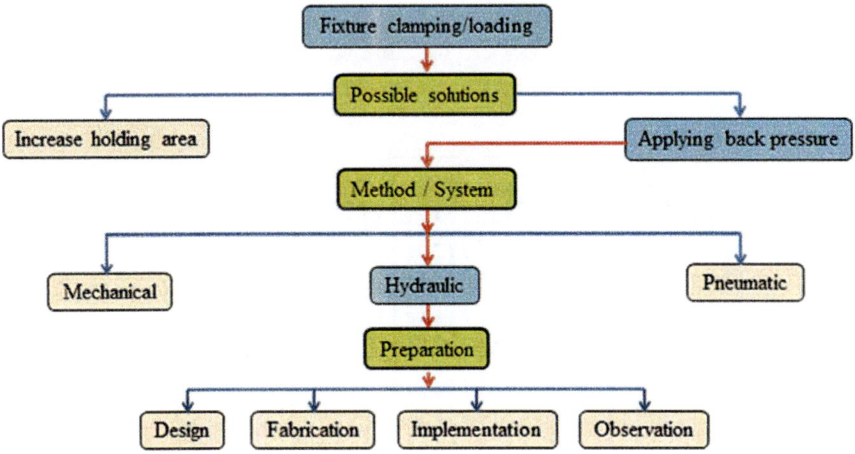

Fig. 7 Possible solution

3.5 *Prime Cause Validation*

The possible cause for the depth variation and checks has been tabulated. The prime cause or the significant cause for the depth variation is identified in Table 3.

As stated in the above prime cause validation, we found that the issue is with the clamping which leads to movement of NHB during drilling operation. The solution is shown Fig. 7. Increasing clamping area is not possible since the job is pre-machined so the total clamping area is limited to 35 mm. So applying back pressure is the only way to fix this issue. Applying back pressure can be done by several methods. It can be mechanical or hydraulic or pneumatic system.

Since movement of mechanical parts and pneumatic system are noisy, heavy construction, automation complexities, poor repeatability, speed, not reliable, difficulty in control and space constraints are in the mechanical system and pneumatic system is not selected. In the hydraulic system, pressure can be varied by adjusting the pressure regulator, less noisy, wide range of speed, load controls and deliver high pressure with more precise movement the hydraulic system is selected in this project to solve this issue.

3.6 *Depth Variation After Implementation of New Fixture*

As the same way in a randam selection, fifty components were picked for the depth measurement. Figure 8 shows the depth variation for the selected fifty parts. Out of fifty components, all the fifty components are within the tolerance limit. Though after implementing the fixture, there are some deviations in the dimension because of influence by other parameters, and they are within the tolerance limit. Hence, the

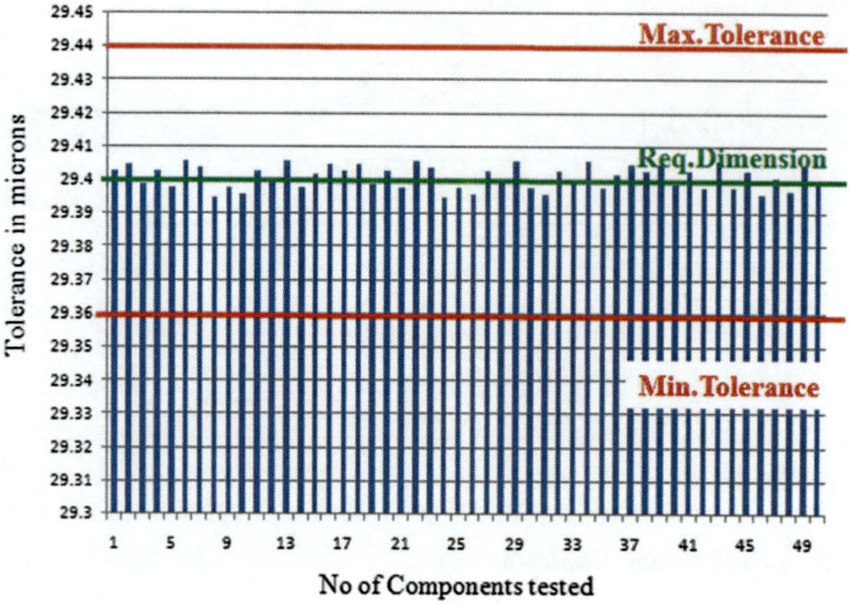

Fig. 8 Solenoid hole depth variation

observation data clearly shows the successful implementation of the fixture which eliminates the rejection and rework. From the depth variation its conclude that the rejection rate becomes zero percentage after implementation of the new fixture.

4 Conclusion

The purpose of this project is to reduce the rejection and rework in the production of nozzle holder body. According to the trial experiments and data observed, the deviation in the dimension of the depth is reduced. The rework and rejection rate in the production of the nozzle holder body are reduced. The hydraulic fixture implementation helped to achieved the required dimension within the tolerance. The rejection rate is eliminated from 15 to 0%. It also avoids rework time and wastage of pre-machining cost. This fixture enhances the production rate, reduces the operator fatigue and a high degree of dimensional accuracy is achieved with high quality and customer satisfaction. Concluded that the elimination of rejection in the production of nozzle holder body using the hydraulic fixture is successfully implemented.

References

1. Senthilkumar G, Ramachandran S, Purusothaman M (2010) Indigenous development of automobile radiator using CFD. IEEE Front Autom Mec Eng (FAME) 373–376
2. Purusothaman M, Sam Cornilius C, Siva R (2017) Experimental investigation of thermal performance in a vehicle cabin test setup with pcm in the roof IOP Conf Ser: Mater Sci Eng 197(1). https://doi.org/10.1088/1757-899x/197/1/012073
3. Valarmathi TN, Sekar S, Purusothaman M, Saravanan J, Balan KN, Sekar SD, Mothilal T (2018) Design and thermal analysis of coated and uncoated exhaust manifold. Int J Ambient Energy
4. Gandhi MV, Thompson BS (1986) Automated design of modular fixtures for flexible manufacturing systems. J Manuf Syst 5(4):243–252
5. Hirmanto RS, Taufik S, Sivarao Hambali A, Tajul AA (2012) Design of jigs and fixtures for hydraulic press machine. Ind Eng J 1:19–24
6. Jiang W, Wang Z, Cai Y, Wang KK (1988) Computer-aided group fixture design. CIRP Ann-Manuf Technol 37(1):145–148
7. Trappey JC, Liu CR (1990) A literature survey of fixture design automation. Int J Adv Manuf Technol 5:240–255
8. Peshatwar SV, Raut LP (2013) Design and development of fixture for eccentric shaft—a review. Int J Eng Res Appl 3(1):1591–96

Topology Optimization of Steering Knuckle

V. Dhinakaran, A. Rahul Kumar, Rishiekesh Ramgopal, Surendar Kannan, B. Stalin, and T. Jagadeesha

Abstract One of the essential parts of an automobile is the steering knuckle. The suspension system, wheel hub, steering system, and braking system are associated with the guidance of the steering knuckle. As it is a critical component in the automobile, it experiences the diverse loads constrained to various conditions. With the advancement of technology, works are going to produce great comfort which involves astonishing safety characteristics. As a consequence of the progression of the vehicle in the field of comfort, the weight of the vehicle gets increased. This addition in weight not only issues in declining fuel efficiency but also degrades the overall performance of the vehicle. Thus, the demand for today's automotive manufacturers is the weight reduction of the vehicle to enhance its performance. In this project work, an attempt has been taken for the topology optimization of the steering knuckle. The foremost intention of our task is to decrease weight while maintaining stiffness. The modeling of the knuckle is completed using Catia software as per the design restraints, and the stiffness analysis is done in OptiStruct (Hyper Works). After the completion of the modeling, the stiffness analysis is carried out, design modifications are suggested, and topology optimization is performed on the steering knuckle for lowering of the weight.

Keywords Steering knuckle · Optimization · Stiffness

V. Dhinakaran (✉) · A. R. Kumar · S. Kannan
Centre for Applied Research, Chennai Institute of Technology, Chennai 600069, India
e-mail: dhinakaranv@citchennai.net

R. Ramgopal
Department of Mechanical Engineering, National Institute of Technology, Tiruchirappalli, Tamil Nadu, India

B. Stalin
Department of Mechanical Engineering, Anna University, Regional Campus Madurai, Madurai 625 019, Tamil Nadu, India

T. Jagadeesha
Department of Mechanical Engineering, National Institute of Technology, Calicut, Kerala, India

A. Arockiarajan et al. (eds.), *Advances in Industrial Automation and Smart Manufacturing*, Lecture Notes in Mechanical Engineering,
https://doi.org/10.1007/978-981-15-4739-3_17

1 Introduction

Optimization is enhancing as one of the developing technologies in the automotive industry. The reduction of weight in the autofield is to promote vehicle performance. The process of reducing weight has continued by advancements in materials, revised design, fabrication means, and optimization methods. But of all the process, optimization is found to be superior and is implanted more to reach the merest weight by highest or reasonable performance. The link that supports the steering unit to turn the front wheels is the steering knuckle. The stub axle, tie rod, and axle housing are connected with the knuckle. One terminal of the knuckle is united to the axle casing by kingpin, while another terminal is attached to the tie rod. A bearing is utilized in settling the wheel hub with the knuckle. The tie rod is employed in producing the linear motion, which is converted into the angular motion with the help of the steering knuckle. The major working systems of the vehicle such as the suspension system, steering system, and the braking system are inter-related with the knuckle. The main breakdown of the steering knuckle is a fatigued failure, which results from time-varying loads during its operating conditions. The major part that accounts for weight reduction in the automotive field is the steering knuckle. During the topology optimization of the knuckle, the weight of the part should be reduced while constraining the various features such as the strength and stiffness. The optimization should not result in the reduction of either stiffness or the strength of the components.

The design is said to be an ideal design if it delivers the desired tasks and is completely trustworthy under the utmost performing circumstances. The ideal design should be easy for the manufacturer and the material preferred must be cost-effective. The easy way to predict the characteristics of any component is by doing a finite element analysis. This helps in observing the actual operation of the components under different circumstances and added provides information about the failure. Thus, the analysis can be used as a tool for the optimization process. The weight reduction of any parts of the automobile will result in decreasing the overall weight of the vehicle. The mass reduction results in the enhancement of the efficiency of the auto. The material used for the production of the vehicle will be saved if the weight is minimized. The weight of the vehicle is directly influenced by the consumption of fuel and emissions that are coming out as exhaust. Thus, automobile enterprises are interested in working in different ways to improve the standard of the vehicle (Fig. 1).

The innovation process in the field of producing light alloys for the vehicle is to reduce the weight and also focuses on the design outlines to find newly designed parts or modify the existing parts which result in increasing the efficiency. In the year 2001, Chang et al. [1] had presented their work in the field of shape optimization of the structural components. The main propaganda of the works is the process of manufacturing of the vehicle components by advance technology with the cost being a constraint. The author inferred that the parts of the automobile can be produced by advance machining processes such as solid freeform fabrication, virtual manufacturing, and computer numerical control. Roy et al. [2] had concentrated in their work

Fig. 1 Wheel assembly's split view

to develop a new approach for manual optimization. The author had also discussed the difficulties encountered by the engineering fellows. The author had suggested two ways for the problem of the optimization. One is the evaluation of the design, while the other is degrees of freedom. They concluded with a summary of various methods used in design optimization. They have identified the problem of scalability as a major issue in the field of design optimization. Jhala (2009) has worked on the comparison of the fatigue characteristics of the steering knuckle. They have selected three different materials for the manufacturing of the steering knuckle. The materials chosen are cast iron, forged steel, and cast aluminum. Further, the work is carried out by the Finite element analysis. The FEA results in the predictions of the stress patterns in each part. They had also experimented with all the components. The experimental result is correlated with the FEA result, and the fatigue responses of all the materials are compared. They arrive at the conclusion that the knuckle made of forged steel exhibits better fatigue performance than the remaining two materials [3].

The work presented by Vijayarangan et al. [4] had published his work in the field of optimization. They had considered the advanced materials such as metal matrix composites (MMCs) for the manufacture of the knuckle. The ANSYS software is used for the structural investigation of the steering knuckle. The FEA results show that the strut region deflects more and stress is highest at this location. They had also conducted the experiment to note the actual behavior of the knuckle. At last, both the results are compared. They concluded that the MMCs result in a 55% less weight than the conventionally used materials for the production of the steering knuckle. In the year 2013, as per the report published by the United States, government revealed that they are working on the improvement of the vehicle. They have set an objective to decrease the weight of the chassis and suspension system by one-fourth at the end of 2020 [5].

In 2013, the process of designing a knuckle by the usage of the scratch is proposed by Rajendran et al. [6]. They have used Opti Struct for the optimization of the knuckle. They had makes the utilization of the hyper works. Nirala et al. [7] had performed topology optimization in his work. The T clamp cylinder is optimized to decrease the mass by using computational analysis. After the optimization of the cylinder, the analysis is carried out on the cylinder to check whether it is acceptable or not. They concluded that the optimized part exhibits a similar strength as the existing model.

In this paper, an attempt has been done to minimize the weight of the steering knuckle. The weight reduction is done by topology optimization. From the literature survey, we concluded that the weight reduction of the parts of the vehicle results in improving fuel effectiveness and thus enhancing the overall performance. The optimization of any components involves two different processes. One includes the alternative material instead of the existing material, while the other way is improving the design standards with the same material. We have done the topology optimization of the steering knuckle by taking the second case.

2 Methodology

The application of the design is to withstand the static as well as dynamic loads or forces occurring in the product during its operation. The following assumptions and data are considered while doing analysis

- The model is imported and the material property is assigned.
- The element size and properties are defined.
- The load conditions are evaluated with the optimized factor of safety.
- Boundary conditions and constraints are applied to the part.
- Stress and deformation values are obtained analytically.
- The results are compared with the permissible limit.
- Optimization is done to design.
- Then, the design is evaluated again.
- The result is compared and the design is finalized.

3 Designing a Cad Model

The part of the knuckle is designed in CATIA software. During the designing, the standard specification of the steering knuckle is maintained (Fig. 2).

4 Finite Element Analysis

Computational mechanics has advanced immensely over the last few decades, and today, it is the need of the hour for researchers. Structural modeling and analysis are playing a vital role in determining the deflection, induced stress, in response to the applied load and identifying the regions where failures are most likely to occur in the model. The knowledge of the behavior of a structural analysis assists in improving

Fig. 2 Knuckle joint

the performance and increasing the efficiency of the design. The material preferred for the knuckle is Aluminum 2011 T3 Alloy (Table 1).

The CAD design of the steering knuckle is imported into the Analysis software. The part is examined for the geometry error because the geometry error results in the mesh complication. The meshing is done by using hyper mesh. The mesh details are provided in Table 2.

Loading and Boundary Conditions Boundary conditions are structural constraints applied to the object on evaluation of design. For testing the design, the design application parameters such as the application and action of forces, hinge and pivot points, and action of forces are evaluated for the study of stress and other linear parameters. The knuckle is constrained at wheel center for all the degree of freedom. The different loading condition is described below:

1. At the struct location, 1 N load is applied about TX, TY, and TZ.

Table 1 Material properties

Density (Kg/m^3)	Young's modulus (MPa)	Poisson's ratio	Yield strength (MPa)	Ultimate tensile strength (MPa)
2770	69,000	0.33	280	310

Table 2 Mesh details

Element	No. of elements	No. of nodes
TRIA	167,259	261,566

Fig. 3 Boundary conditions

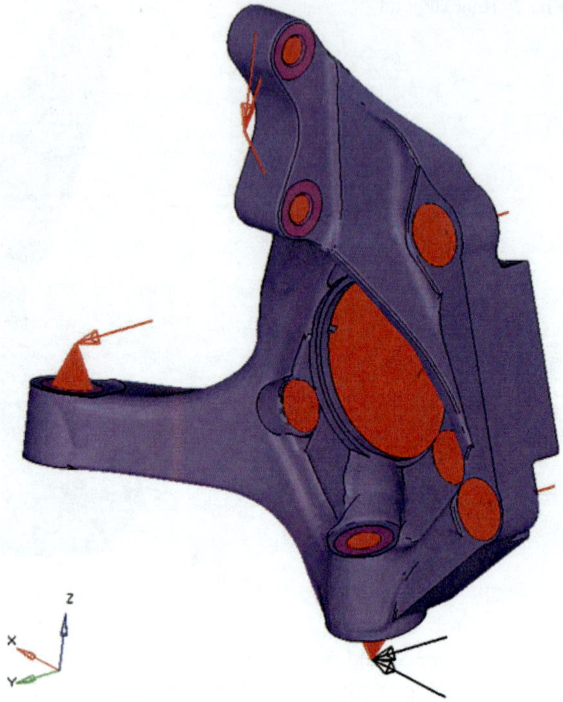

2. At the tie rod location, 1 N load is applied about TY.
3. At the lower ball joint location, 1 N load is applied about TX, TY.
4. At the upper and lower caliper ear, 1 N load is applied about TY (Fig. 3).

5 Observation

The FEA results in the prediction of the deflection and the stress-induced in the components for different loading conditions. The stress pattern helps in determining the region to eliminate the material from the region where there are low-stress concentrations. The topology optimization is done on the steering knuckle. After the completion of the optimization, the optimized part is analyzed with the same procedure being followed before. The contour plot of displacement for different loading is shown below.

Load Case 1
At the lower ball joint location, 1 N load is applied about TX, TY (Fig. 4).

Load Case 2 At the tie rod location, 1 N load is applied about TY (Fig. 5).

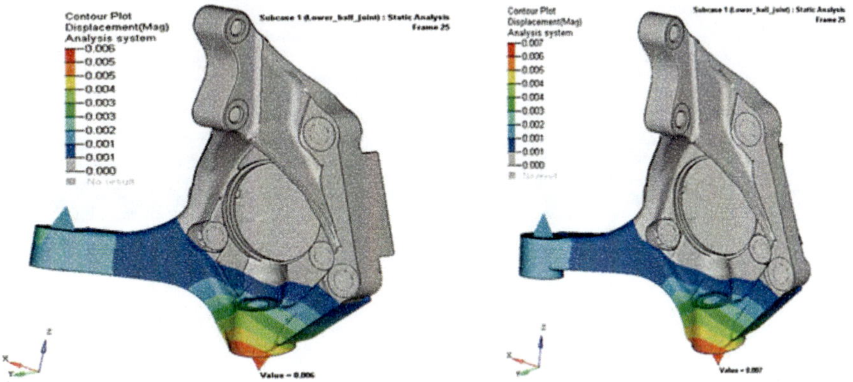

Fig. 4 Displacement for load case 1

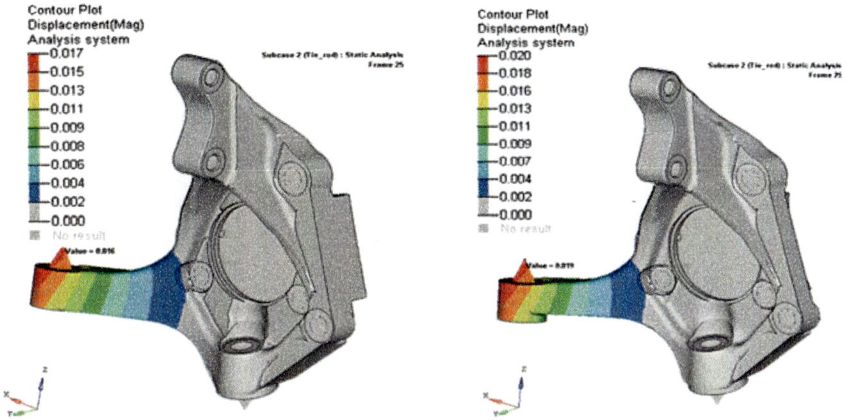

Fig. 5 Displacement for load case 2

Load Case 3
At the struct location, 1 N load is applied about TX, TY, and TZ (Fig. 6).

Load Case 4
At the upper caliper ear, 1 N load is applied about TY (Fig. 7).

Load Case 5
At the lower caliper ear, 1 N load is applied about TY (Fig. 8).

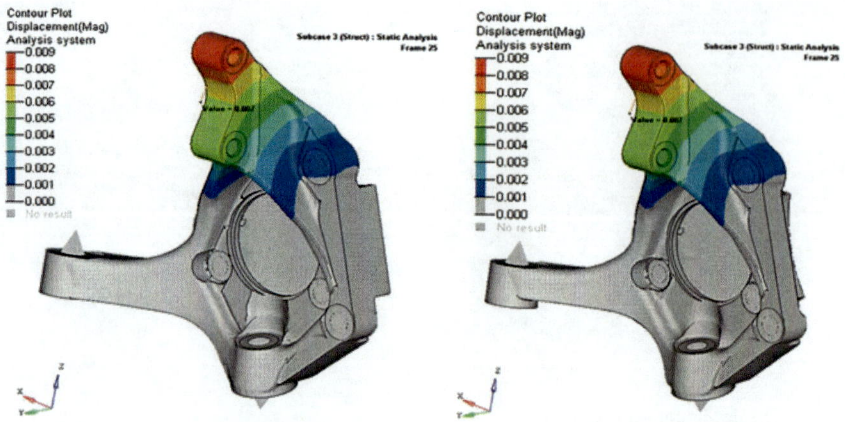

Fig. 6 Displacement for load case 3

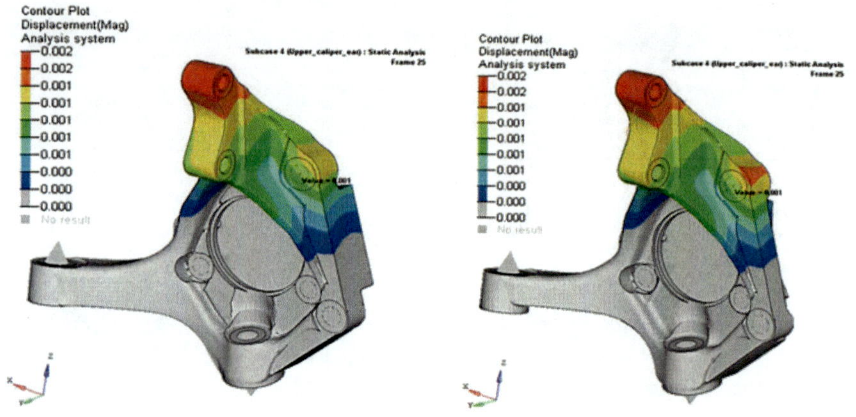

Fig. 7 Displacement for load case 4

6 Result

The main purpose of the topology optimization is to decrease the weight of the parts, and also, stiffness values have been maintained within the range. We have taken care of all the manufacturing feasibility and constrain, while design modification is done on the components (Table 3).

After performing the optimization, the results are recorded. It is understood that the optimized part consists of an 8.35% mass reduction than the initial part (Table 4).

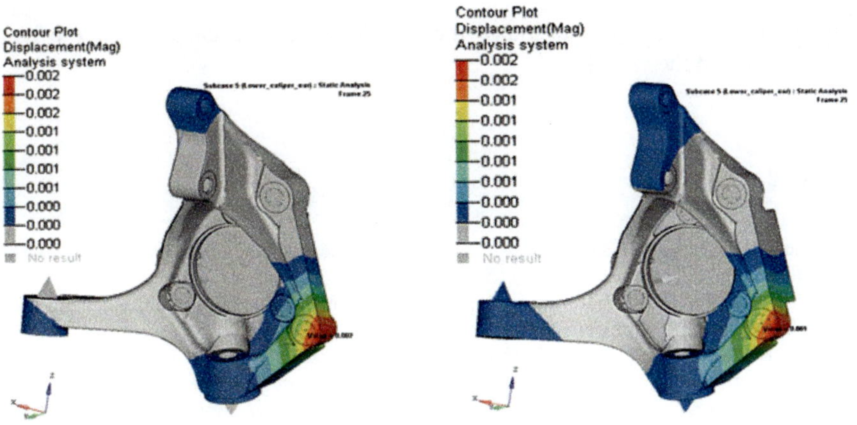

Fig. 8 Displacement for load case 5

Table 3 Stiffness of the component

Load case	Load (N)	Displacement (mm)		Stiffness (KN/mm)	
		Initial design	Optimized design	Initial design	Optimized design
Lower ball joint	1000	0.006	0.007	167	143
Tie rod	1000	0.016	0.019	63	53
Struct	1000	0.007	0.007	143	143
Upper caliper ear	1000	0.001	0.001	1000	1000
Lower caliper ear	1000	0.001	0.002	1000	500

Table 4 Mass comparison

	Initial design	Optimized design	Reduction %
Mass	9.394 kg	8.670 kg	8.35

7 Conclusion

The motto of our work is to minimize the weight of the steering knuckle by performing topology optimization. The optimization is done by maintaining the stiffness of the knuckle as a constraint. The FEA is done on the initial design, and the region which shows less stress concentration is focused to remove the material. The optimization is done, and the optimized part is again analyzed with the same load and boundary condition. The maximum mass that can be reduced from the components by keeping the stiffness as a constraint is nearly 8%. Hence, we concluded that the weight reduction of the component will result in material saving and cost reduction. Thus, the optimization will help in improving fuel efficiency and enhancing the overall performance of the vehicle.

References

1. Chang KH, Tang PS (2001) Integration of design and manufacturing for structural shape optimization. Adv Eng Softw 32(7):555–567
2. Roy R, H S, Teti R (2008) Recent advances in engineering design optimisation: Challenges and future trends. CIRP Ann 57(2):697–715
3. Jhala RL, Kothari KD (2009) Component fatigue behaviors and life predictions of a steering knuckle using finite element analysis
4. Vijayarangan S, Rajamanickam N, Sivananth V (2013) Evaluation of metal matrix composite to replace spheroidal graphite iron for a critical component, steering knuckle. Mater Des 43:532–541
5. US department of energy Vehicle Technology Office "WORKSHOP REPORT: Light-Duty Vehicles Technical Requirements and Gaps for Lightweight and Propulsion Materials" Feb 2013
6. Rajendran RS, Sudalaimuthu S, Sixth M (2013) Knuckle development process with the help of optimization techniques. Altair Technol Conf India
7. Niral P, Chauhan M (2013) FEA and topology optimization of 1000T clamp cylinder for injection molding machine. Procedia Eng 51:617–623

Recent Ergonomic Interventions and Evaluations on Laptop, Smartphones and Desktop Computer Users

Mona Sahu, Kondru Gnana Sundari, and Abhishek David

Abstract The use of computers has been rapidly rising over the recent decade. In the recent past along with the usage of desktops, laptops and smartphones are widely used in most of the offices. With the advancement in technology in smartphones, people spend a lot of time using a smartphone for the basic need of networking. It was noticed that laptops and smartphones users tend to attain incorrect postures when compared with desktop users. It is the prime concern for the organizations to improve their work environment in order to optimize the health, safety, comfort and effectiveness of their employees. This paper is aimed to study the ergonomic interventions carried out to reduce pain and to improve the posture while working on desktop, laptops and smartphones. The ergonomic evaluation methods to assess the computer workstations are also laid out. This review will be helpful for the users to adjust their workstation correctly and designers to design the workstation to include better usability feature as per ergonomic guidelines during laptop, smartphone and desktop usage.

Keywords Computer workstation · Posture · Ergonomics

1 Introduction

Laptops are widely used over desktops in most of the companies, and almost all the students in schools and colleges use their own personal laptops. The portability and ease of use of laptops made them more preferable over desktops. In 2006, around 71% of the UG students used a desktop. In the year 2009, it was found that the usage of desktops declined drastically to 44% [1]. According to a study in 2008, almost 82% of the undergraduates were reported to own only a laptop instead of a desktop [2]. Furthermore, they spent an average of 21.3 h a week on their computers [1]. The increase in the use of laptops has increased the need for analysing the musculoskeletal disorders associated with them. The increase of physical discomfort on the upper

M. Sahu (✉) · K. Gnana Sundari · A. David
Karunya Institute of Technology and Sciences, Coimbatore 641114, India
e-mail: monasahu@karunya.edu

A. Arockiarajan et al. (eds.), *Advances in Industrial Automation and Smart Manufacturing*, Lecture Notes in Mechanical Engineering,
https://doi.org/10.1007/978-981-15-4739-3_18

extremity of the body is mainly related to the prolonged use of laptops in non-neutral postures [3, 4]. It is vital to use arm support to reduce the muscle activities of the shoulder in order to prevent discomfort among laptop users. Inbuilt keyboard and mouse pad of the laptop tend to cause discomfort in the wrist and elbow during long working hours, in such cases the user is recommended to position the data entry devices like keyboard in line with the elbow height. The position of the computer's display is also a factor which influences strain in the user's eye and leads pain in the body. Hence, the top of display can be placed at or near the eye level, and the viewing distance can be maintained at around 0.5 m [5].

Smartphone has played an irreplaceable role in human life. The number of smartphone users has increased a lot over the years. The number of smartphone users has increased from about 10° of the world's population in 2011 to 36° in 2018 [6]. Smartphone has become multifunctional over the years, and people prefer smartphone for many of their daily tasks. Due to the low cost and convenience of the mobile handled devices, almost everybody these days from child to adult of every age owns at least one type of mobile device. In the past years after the first release of smartphone, the touch screen has dominated the markets and the daily habits of usage of such devices have been considerably affected. A survey in Canada among staff and students shows that the 4.65 h are spent on the smartphone a day [7]. The keypad interface of a smartphone is convenient and user-friendly, but the posture adapted while typing is non-neutral and leads to pain. Therefore, there is a need to improve the performance of the keyboard typing. While using a smartphone, the thumb is mainly used for every operating task. This considerable use of mobile phone has raised concern about musculoskeletal problems.

In various places such as industries, offices, schools and colleges, people are often required to work for long hours on computers. This requires them to remain in certain positions like sitting continuously and to do monotonous work. Long hours of computer usage in incorrect posture, results in physical discomfort, pain in different body regions, poor posture and also reduces the efficiency. In order to avoid this, ergonomically designed computer workstations are essential to reduce pain and improve productivity.

Based on various studies, it can be inferred that one of mostly reported musculoskeletal occupational injuries is carpal tunnel syndrome (CTS). Computer usage is a low load repetitive task. Prolonged use of computers is a major risk factor for the development of CTS. While working on the keyboard and mouse, non-neutral forearm and wrist postures lead to sustained loading on the muscles of the forearm and wrist. This is another risk factor of CTS. The study done by Doohee You et al. concluded that the CTS affected approximately 5 million workers in the USA, and the cost of medical care had been expected to be over $2 billion annually [8, 9]. Several medical and personal factors can increase the risk of CTS. Studies have shown the relation between occupational factors and CTS. These factors include the mechanical work done at the workplace. It has been concluded in recent reviews that there is a relationship between CTS and repetitive movements, bending\twisting of wrist and forceful manual exertion in the workplace [10].

This review is intended to examine the existing risk factors for pain during desktop computer, laptop and smartphone usage, ergonomic evaluation techniques used and ergonomic interventions carried out to reduce pain in office employees.

2 Methods

All the relevant articles published between 2010 and 2018 were searched by using various electronic databases such as Science Direct, Scopus and Google Scholar. The searches were made by using the following keywords: computer workstation, keyboard, laptop, smartphones, ergonomics, carpal tunnel syndrome and wrist posture.

3 Laptops Workstations

3.1 Challenges Faced While Using a Laptop

The portability and size of laptops make it more convenient and easy to use. Due to the portability feature of a laptop, the users normally tend to sit or stand or lie down in inconvenient postures. The prolonged usage in such unhealthy postures may lead to discomfort and hence injury [11]. It was also found by Rafael et al. that many users are not aware of the ergonomic practices that are to be adopted while working on a laptop. Hence, to reduce the health issues for laptop users, the need arises to provide adequate ergonomic training and information on the ideal practices to be followed. It is also essential to find the incorrect postures adapted by laptop users so that the musculoskeletal disorders can be reduced. Therefore, this overview is an attempt to present the incorrect postures adapted while using a laptop and the occurrence of pain in different body regions which leads to musculoskeletal disorders (MSDs).

3.2 Risk Factors for Pain

Pain in the neck region. According to the research done in various parts of the world, the usage of laptop computers causes larger neck flexion angles [12–15]. Larger neck flexion angles are mainly due to the reduced range of movement which could lead to strain in the neck [16]. The inbuilt keyboard and mouse in laptops cannot be adjusted as per the user's convenience. The non-adjustment feature of the placement of the input devices is the major cause for the adapted incorrect working postures and further causing pain. According to International Association for the Study of Pain, it is observed that neck pain affects around 30%–50% of the general population annually, among which 15% will experience long-lasting neck pain (over 3 months) [16]. Korhonen et al. showed that 34.4% of the employees had neck pain while working with laptops, and the risk of neck pain is due to the poor placement of keyboards [17]. Cagnie et al. studied a 12-month prevalence of neck pain among office workers and found that 45.5% of the employees had neck pain while working with laptops [18]. The findings by Cagnie et al. are also consistent with Korhonen et al. that is there is a high correlation between neck pain and posture of the neck bent forward for a long period of time.

Pain in the eyes. The prolonged usage of laptops and other VDUs lead to computer vision syndrome (CVS) also referred to as digital eye strain. With an increase in the duration of digital screen usage, the discomfort level appeared to increase. Most of the usual symptoms related with CVS include eyestrain, dry eyes, red eyes, blurred vision, headaches and neck and shoulder pain. The main causes for these symptoms are glare on a digital screen, inappropriate illumination, incorrect viewing distances, non-neutral seating posture and a combination of these risk factors.

3.3 Ergonomic Interventions to Reduce Pain

Ergonomic interventions to reduce pain while using laptops include workstation adjustments as well as ergonomic training to the users. A study found that laptop users can improve head and neck postures with increasing angle of the laptop support surface. However, wrist extension increases with increased angle while using internal input devices [19]. Supporting the forearm on the work surface may increase comfort and decrease muscular load on the neck and shoulders [20]. According to a study, a common ergonomic problem among students was not using tables and chairs while using laptops. Using tables and chairs while working with laptops, using external keyboards to reduce stress on the shoulders, applying the right ergonomic methods and doing short exercises were recommended to prevent musculoskeletal disorders [21].

Lizette Mowatt found that if the device is placed below eye level, it would reduce the neck pain and visual symptoms [22]. Among college going students, there exists a high prevalence of symptoms of CVS, in particular eye strain, burning and neck pain. This can be improved through improved ergonomic practices. Figure 1 shows the standard ergonomic practice to be followed while using a laptop.

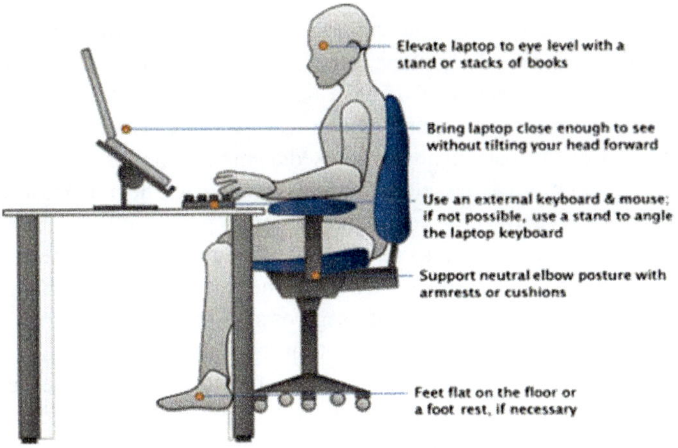

Fig. 1 Standard ergonomic practice during laptop usage [50]

4 Literature Review on Smartphone Usage

4.1 Challenges Faced While Using a Smartphone

The prolonged usage of the smartphone is directly related to the reported pain symptom in the neck, shoulder and thumb of a person. The research done by Gustafsson et al. clearly indicates that people mostly use one thumb rather than two hands while texting on a smartphone and moreover the greater muscle activities detected in the muscle abductor pollicis longus (APL) of the thumb while keying at a quicker speed confirmed the same [23]. Boucaut 2018 et al. used the Rapid Upper Limb Assessment (RULA) method to find the risk factors related to pain while using a smartphone [24]. It was found that muscle use and posture are the two main risk factors for pain and additional research needs to be done to solve the problems related to smartphone users [24]. In Korea, 55.8% of smartphone users experienced musculoskeletal symptoms in at least one body part, especially in the neck [25]. Neck pain is considered to be the most usual musculoskeletal disorder in smartphone users. Similarly, in Thailand, 26.3–60% of the mobile users suffered with shoulder and neck symptoms [26]. Awkward postures like neck flexion and forward bending of the trunk can lead to strain in the muscles, tendon and ligaments and can also compress the nerves leading to numbness and irritation [27].

4.2 Risk Factors for Pain

Pain in neck region. Most of the smartphone tasks like texting, emailing, browsing and gaming involve the user holding the device with one or two hands with the eyes looking down, neck flexed and using the thumb to control the device. An acute pain arises in the neck and shoulder muscles in such postures for a long period of time. The daily time spent in using a hand-held mobile device was found to be directly correlated with reporting recent neck pain [7]. The neck flexion angles were measured from 18 participants who used a smartphone for common tasks like mailing, texting, web browsing, etc., and the average neck flexion angle was found to be 33°–45° [28]. There are several studies stating prolonged neck flexion as the major risk factor behind neck pain [29–31]. Neck pain may originate from any of the pain-sensitive structures in the neck such as the vertebral bones, ligaments, nerve roots and particular muscles [32]. The usage of smartphone also causes headaches and numbness or tingling into the upper extremity of the body.

Hansraj (2014) created cervical spine model with realistic values in Cosmos works for smartphone users, and he calculated the forces on the cervical spine with increasing degrees of neck flexion [33]. The results showed a rise in force on the cervical spine from neutral to 60° of flexion. Figure 2 shows the rise in force on the cervical spine and the angle of flexion of the neck. From his findings, it is inferred that a normal smartphone user may experience up to a fourfold increase in load on the cervical spine compared to the neutral posture. The average neck flexion angle is found to be 33°–45° [28].

Pain in thumb region. The tapping of the thumb is mainly done while texting on a smartphone. While texting, the thumb needs to be flexed or extended, adducted or abducted [34]. Hence, the thumb movements are complex involving the muscles of the hand and forearm. Two simple designs of smartphones are available in the market. One type of phones has an input physical keyboard in the bottom half of the phone and the screen on the top half. Another type of phones with virtual touch screen keyboard available as desired. In the present day, phones with touch screen are more convenient and hence are mostly and widely used by the users. Kietrys et al. compared the muscle activity while texting on a keypad phone and a touch

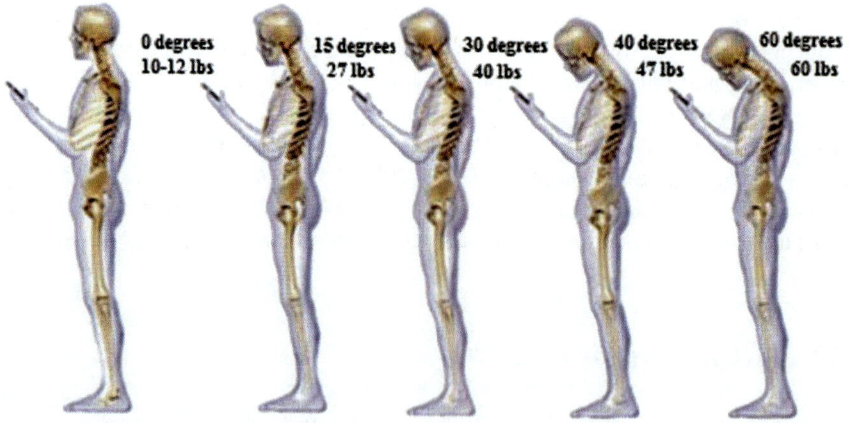

Fig. 2 Increase in force on the cervical spine and the angle of flexion [33]

screen phone. It was found that the usage of keypad phone leads to greater muscle activity in thumb and forearm muscles when compared with touch screen phones [35]. Although a larger display is suitable for reading a range of contents, from an ergonomic perspective a medium-sized display is preferred as it would provide better hand grip, considering most of the smartphones are used with one hand grip. Increasing the display size would reduce the grip comfort. A study on the effects of screen size on muscle activities found that there was an increase in muscle activities according to increasing screen size of the smartphone [35]. Furthermore, an increase in the screen size demands more extension in the thumb muscles causing fatigue and pain in the thumb.

4.3 Ergonomic Interventions to Reduce Pain

The ergonomic interventions to prevent or reduce neck-related MSDs in video display terminal (VDT) users include rest breaks [36], training on ergonomics [37], using arm support [38] and pointing devices [38, 39]. However, there are also studies that showed that rest breaks [40], training on ergonomics [41] and using arm supports [42] found no effect on neck-related MSDs. Workstation adjustments were also found to have no effect on neck MSDs [43, 44].

Two systematic reviews further examined these interventions and could not find any strong evidence that any of these specific ergonomic interventions had positive effects on MSDs [45, 46]. The findings showed that the effect of the interventions on neck-related MSDs among VDU users is equivocal. Further investigation of these interventions should be needed. The ergonomic interventions among smartphone usage are very few in number.

An awareness system was developed that would give signal displayed on smartphone to alert users for any bad posture in the middle of a every 5 min task [47]. The results of the study showed that poor postures of the neck and upper extremity improved right after the signal. However, the poor postures continued as time elapsed. The authors suggested that a much more effective warning system should be developed [47].

Study was conducted among smartphone users with MSDs in neck and/or upper extremities with those without symptoms and found that a greater number of the latter group used back and forearm supports which gave a more neutral head position as compared to the former group [48]. Based on the results, back and forearm supports were recommended to be used for better comfort and avoid sitting with neck flexion to prevent MSDs during smartphone use.

Tang came up with a significant intervention of using prism glass, which would turn the vision to a downward 90° angle [49]. It seemed to offer an effective way to eliminate neck flexion among smartphone users. Their study on 14 males and 11 females found that the prism view generated much lower neck muscle activity, more neutral postures and less neck discomfort, compared to the direct view. Though the typing speed in prism view was significantly lower compared to the direct view,

Fig. 3 Postures with and without prism glasses [49]

the accuracy was the same. However, this intervention needs further development to address major concerns like glasses being too heavy and viewing area being too small. Figure 3 shows the postures with and without prism glasses.

5 Desktop Workstations

5.1 *Ergonomic Evaluation*

Lynn C. Onyebeke et al. aimed to examine the effect of the six arm support conditions on the posture and muscle activity of the wrist, forearm, upper arm and shoulder. The applied forces on the supports were also measured. It was found that the shoulder muscle activity and torque were considerably reduced while working on the forearm which was consistent with the research done by Nag et al. [51]. Among the six types of supports, raised palm reduced extension of the wrist and also the applied forces to the mouse pad. Forearm support reduced shoulder flexion torque. The subjects experienced and reported less discomfort while suing a support [52]. Figure 4 shows the six types of supports.

Various factors leading work-related musculoskeletal disorders in office were investigated by M. Matos et al. Rapid Assessment Office Strain (ROSA) method was used for the evaluation. The sections of the computer workstation, i.e. chair, monitor, mouse, keyboard and telephone were evaluated. The results indicated that the ROSA score does not only indicate the availability of poor workstation but also that the users are unaware of how to use the workstation available correctly. It was also noticed that the main affected area of the upper extremity was the shoulder and the cervical segments. This was mainly due to the incorrect sitting posture all through the day giving rise to increase in muscular activity in the shoulder and cervical area [53]. J. Liebregts et al. have successfully established the reliability of the ROSA method in conducting assessments [54].

A majority of the call centre workers according to Worawan Poochadaa were more prone to higher risk of pain, and it was suggested that proper ergonomic training should be given to office workers to prevent MSDs [55]. Cross-sectional studies done

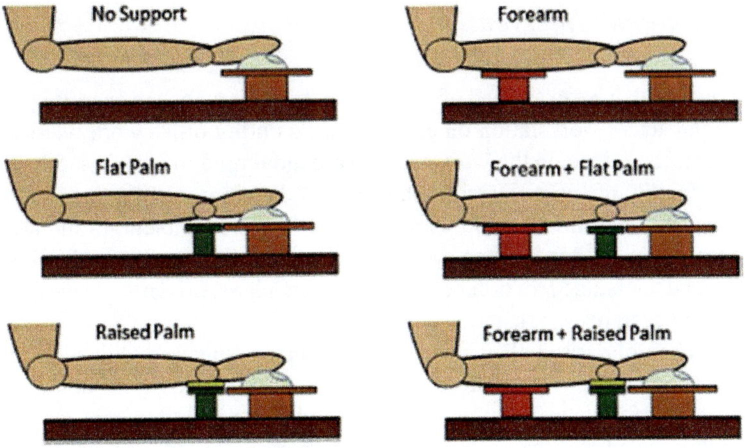

Fig. 4 Six types of supports while using mouse [52]

by Carole James et al. among computer users showed that the neck, lower back and shoulder were the most affected parts of the body [56].

Davis K. G. et al. evaluated various postures experienced when using a keyboard and a touch screen while in seated or in standing position [57]. Using an angled touch screen input device when seated at the lower work surface and standing at the high work surface was considered to be the most suitable postural condition. More than using a keyboard, touch screens input device reduced the discomfort. It is advised that call centre workers whose work demands them to sit and stand for different operations are benefitted with sit–stand tables. The change in upper extremity posture and the muscle activity of the user when using a sit–stand workstation were researched by Michael Y. Lin [58]. The differences between working on a standing and sitting workstations were studied. It was found that both the workstations induced different wrist, shoulder postures and muscle effort. Thomas Karakolis et al. have also tested the effectiveness of the sit–stand workstations. It was concluded that while using sit–stand workstations, discomfort was reduced and the productivity of the user remained unchanged [59].

It was found by David Ben Kumah that computer user's lack of proper information and ergonomic training led to discomfort [60]. Therefore, proper training and engineering interventions should be incorporated to reduce sustained non-neutral postures. The research done by Sevim Celik concluded that the working environment must be arranged ergonomically, and proper measures should be taken so that office workers do not suffer from musculoskeletal disorders [61].

Luís Lavina et al. have analysed the interaction between human and the computer, and they have summarized information and data in order to develop new solutions for the computer users [62]. Leena Korpinen et al. have found that people who use computers and mobile phones suffer from a high risk of numbness and pain in their lower back and hip [63].

5.2 Ergonomic Intervention

Katja Koren et al. have examined office workers for their work and exercise [64]. The effects of an active workstation on error and time during office work were studied. Cycling exercises done on the workstation were monitored. It was found that typing time significantly increased. Results showed that cognitive performance and typing error remained the same. It was concluded that the task completion time increased when exercises were done on the cycling workstation during office tasks. The error rate remained unchanged. In order to achieve the standards for daily physical activity, such suitable designed workstation is required.

Dianne Commissaris et al. evaluated performance of the user while working with dynamic workstations over a short period of time [65]. Tests were done to check whether the performance parameters differed while using a dynamic workstation. The results showed that the performance parameters were affected while using the dynamic workstations in comparison with the sit condition. It was found that while working on the dynamic workstation, performance parameters, such as reading, were not affected. It was concluded that health effects were caused due to inadequate physical activity. The leading cause of health effects was found to be sedentary work with little physical exercise. Performances of common office works, such as cognitive tasks and reading, were not much affected when using a dynamic workstation. Research done by Justine Leavy et al. suggested that adjustable workstations can be used to reduce extended sitting time [66].

Sisay A. Workineh et al. developed an ergonomic computer workstation to evaluate multiple working positions of the user [67]. The four different types of the positions were lean forward, lean back, upright and zero-gravity position. The comfort of the user in the four different positions was compared. It was found that the zero-gravity and lean back position gave more comfort in the lower back as compared to the other positions. The upright position was not preferred for most of the tasks.

Abigail Werth et al. have tested posture differences, muscle activity and performance between three portable computing devices: slate computer, netbook and laptop and desk and computer during typing tasks [68]. The forearm and neck postural analysis was done. The muscle activity of neck, wrist and forearm was captured. Out of the four workstations evaluated, typing on slate computer leads to more extended neck, wrist and elbow postures. Moreover, the performance also was affected. Hence, using compact computers causes the user to adapt more incorrect postures which in turn increases the injury of the neck and upper extremity. The limitations of the study were that the readings were captured in a short period of 30 min, because typing works are generally done for longer periods of time. Other limitation was that not all muscles were selected for this study; only few commonly assessed muscles were selected for the study.

Ardalan Shariat et al. have done research to check the effectiveness of the combination of ergonomic modifications and exercise and also individual components on the cause of pain in office workers [69]. The results were assessed by a questionnaire. Cornell Musculoskeletal Disorders Questionnaire was used to analyse the results between the three treatment groups. The results were analysed after 2 months, 4 months and 6 months' intervention. The results indicated that after the 4 months' intervention, there was no difference between the treatment groups. It can be concluded that there was improvement in all the three groups. But, after 6 month period intervention, there was significant improvement in exercise group. In conclusion, it was suggested that for long-term treatment, exercise training was more beneficial compared to just modifying the ergonomic conditions.

Sonya Goostrey et al. have examined the impact on the muscle activity and movement of the neck while placing documents in three positions which are a vertical laterally placed document holder level with the computer screen, flat on the table to the left of the keyboard and in line above the keyboard [70]. Figure 5 shows the positions of placing document holder. It was noted that when documents were placed in level to the computer screen, the neck flexion and muscle activity were the less than the rest two positions. The greatest neck movement was produced by the desktop position. Further studies and testing of other holders have to be done to get the best use of this commonly used office equipment.

Annina B. Schmid et al. researched the correlation between ergonomic devices and reducing carpel tunnel pressure in users affected with CTS [71]. It is also noted that the usage of ergonomically designed devices did not reduce carpal tunnel pressure. The study concluded that it was not recommended for or against any specific ergonomically designed device and that the selection of these devices depends mainly on the preference of the user.

Arvin Fathel et al. had designed software for indicating the user to maintain the corrected and normal body posture. Diagnostic images analyser captured the posture and the position of the user, sitting on the computer chair [72]. The software identified the posture of the user. If the posture was found to be wrong, the software warns

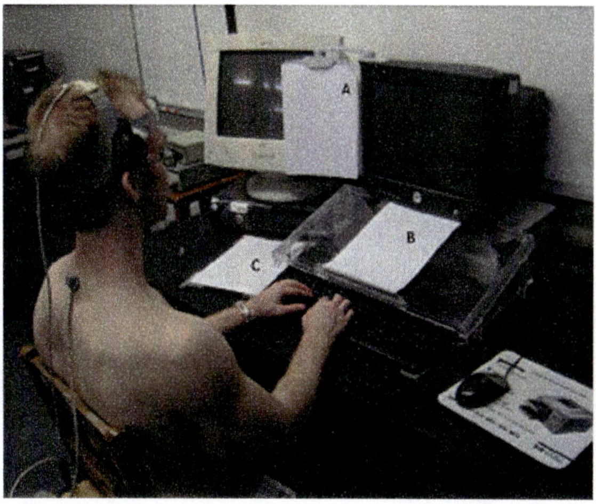

Fig. 5 Three positions of placing document holder: A. vertical laterally placed B. in line C. flat on the table to the left of the keyboard [70]

Table 1 Ergonomic interventions done to reduce pain among computer users from year 2013–2018

S. No.	Year	Authors	Ergonomic Interventions	Results
1	2018	Wu et al.	ActiveErgo automatically adjusts the monitor height to the seated height of the user [73]	Significantly improves the posture compared to a manual configuration in both speed and accuracy and helps significantly more users to fully meet ergonomics guidelines
2	2017	Ardalan Shariat et al.	Modifying existing workstation, exercise and ergonomic training	Long-term treatment, exercise training is more beneficial compared to just modifying the ergonomic conditions
3	2016	Katja Koren et al.	Cycling on the computer workstation	The typing time is significantly increased. The task completion time increased when exercises were done on the cycling workstation during office tasks The error rate remained unchanged
4	2015	Dianne A. C. M Commissaris et al.	Performance of the user while working with dynamic workstations over a short period of time	The health effects were caused due to inadequate physical activity The leading cause of health effects was found to be sedentary work with little physical exercise Performances of common office works, such as cognitive tasks and reading, were not much affected when using a dynamic workstation
5	2015	Sisay A. Workineh et al.	Evaluated multiple working positions of the user	The zero-gravity and lean back position gave more comfort in the lower back as compared to the other positions The upright position was not preferred for most of the tasks
6	2015	Arvin Fathel et al.	Software for indicating the user to maintain the correct and normal body posture	The use of this software constantly modifies the user's position and prevention of wrong posture

(continued)

Table 1 (continued)

S. No.	Year	Authors	Ergonomic Interventions	Results
7	2014	Abigail Werth et al.	The posture differences, muscle activity and performance between three portable computing devices: slate computer, netbook and laptop and desk and computer during typing tasks were compared	The slate computer was found to result in the lowest muscle activation
8	2014	Annina B. Schmid et al.	Correlation between ergonomic devices and reducing carpel tunnel pressure in users affected with CTS	The usage of ergonomically designed devices did not reduce carpal tunnel pressure. The selection of these devices depends mainly on the preference of the user
9	2013	Sonya Goostrey et al.	The impact on the muscle activity and movement of the neck while placing documents in three positions	The greatest neck movement was produced by the desktop position. Documents placed in level to the computer screen led to less neck flexion and muscle activity than the rest two positions

the user. This study showed that the use of this software constantly modifies the user's position and prevention of wrong posture. Table 1 summarises the ergo.nomic interventions done to reduce pain among computer users from year 2013–2018.

6 Conclusion

Laptops and smartphones have become irreplaceable in our day-to-day life. The various risk factors associated with the wrong postures adapted while using laptops and smartphones are studied. Also the ergonomic interventions to avoid/prevent musculoskeletal disorders due to back/awkward postures were discussed. It is observed that the ergonomic intervention to enhance good posture during smartphone usage was very limited. Hence, further research and engineering practices need to be carried out to address the problem.

Ergonomic design for computer workstations has caused considerable concern to maintain good health in employees. It is understood from the literature review that, in the recent past, most of the companies have adjustable workstations. It was also found that not adequate ergonomic training and knowledge are provided in most of the offices. The employees who have received ergonomic training have reported that they are unable to adjust the workstation as per ergonomic guidelines due to factors such as work pressure and time pressure. Hence, though ergonomic standards and guidelines are presented, it is vital to make these guidelines.

With the increase in technology and usage of computers in every field, the main ergonomic intervention made thus far was software designed to show the right posture on the user's computer screen in every one hour. This, to an extent, helped the user to adjust their posture correctly. Another successful intervention was to have automated adjustable workstations.

References

1. Smith S, Salaway G, Borreson Caruso J (2009) The ECAR study of undergraduate students and information technology. In: Research study, vol 6. EDUCAUSE Center for Applied Research, Boulder, CO
2. Chang C, Amick B, Menedez CC, Robertson M, del Pino RJ, Dinnerlein JT (2008) Where and how college students use their laptop computers. In: Proceedings of the human factors and ergonomics society 52nd annual meeting, pp 629–633
3. Burgess-Limerick R, Plooy A, Ankrum DR (1998) The effect of imposed and self-selected computer monitor height on posture and gaze angle. Clin Biomech 13(8):584–592
4. Gerr F, Marcus M, Monteilh C (2004) Epidemiology of musculoskeletal disorders among computer users: lesson learned from the role of posture and keyboard use. J Electromyogr Kinesiol 14(1):25–31
5. ANSI/HFES (2007) Human factors engineering of computer workstations (ANSI/HFES 100-2007). Human Factors and Ergonomics Society, Santa Monica, CA

6. STATISTA 2018. Number of smartphone users worldwide from 2014 to 2020 (in billions). Homepage: https://www.statista.com/statistics/330695/number-of-smartphone-users-worldwide
7. Berolo S, Wells RP, Amick BC 3rd (2011) Musculoskeletal symptoms among mobile hand-held device users and their relationship to device use: a preliminary study in a Canadian university population. Appl Ergon 42(2):371–378
8. You D, Smith AH, Rempel D (2014) Meta-Analysis: Association between Wrist Posture and Carpal Tunnel syndrome among workers. Safety and Health at work 5, 27–31
9. Whoo HC, White P, Lai C (2017) Ergonomics standards and guidelines for computer workstation design and the impact on user's health—a review. Ergonomics 59(3):464–475
10. Agarwal S, Rajajeyakumar (2018) Early diagnosis and prevention of repetitive strain injury induced carpal tunnel syndrome among computer users. Clin Exp Psychol 4(1):188
11. Rafael M, Tatiana G (2007) A survey of ergonomic issues associated with a university laptop program. J Educ Human Dev 1(2):1–15
12. Price J, Dowell W (1998) Laptop configurations in offices: effects on posture and discomfort. In: Proceedings of the human factors and ergonomics society 42nd annual meeting, pp 629–633
13. Straker L, Jones KJ, Miller J (1997) A comparison of the postures assumed when using laptop computers and desktop computers. Appl Ergon 28:263–268
14. Sommerich CM, Starr H, Smith CA, Shivers C (2002) Effects of notebook computer configuration and task on user biomechanics, productivity, and comfort. Int J Ind Ergon 30:7–31
15. Seghers J, Jochem A, Spaepen A (2003) Posture, muscle activity and muscle fatigue in prolonged VDT work at different screen height settings. Ergonomics 46:714–730
16. IASP (2009) International association for the study of pain
17. Korhonen T, Ketola R, Toivonen R, Luukkonen R, Hakkanen M, Viikari-Juntura E (2003) Work related and individual predictors for incident neck pain among office employees working with video display units. Occup Environ Med 60(7):475–482
18. Cagnie B, Danneels L, Van Tiggelen D, De Loose V, Cambier D (2007) Individual and work related risk factors for neck pain among office workers: a cross sectional study. Eur Spine J 16(5):679–686
19. Asundi K, Odell D, Luce A, Dennerlein JT (2012) Changes in posture through the use of simple inclines with notebook computers placed on a standard desk. Appl Ergon 400–407
20. Cook C, Burgess-Limerick R, Papalia S (2014) The effect of upper extremity support on upper extremity posture and muscle activity during keyboard use. Appl Ergon 35(3):285–292
21. Shokri S, Ghalenoei M, Taban E, Ahmadi O, Kouhnavard B (2015) Evaluation of prevalence of musculoskeletal disorders among students using portable computer in faculty of health. Qazvin University of Medical Sciences
22. Mowatt L, Gordon C, Babu A, Santosh R, Jones T (2017) To determine the prevalence of computer vision syndrome (CVS) and ergonomic practices among students in the Faculty of Medical Sciences at The University of the West Indies (UWI), Jamaica
23. Gustafsson E, Johnson PW, Hagberg M (2010) Thumb postures & physical loads during mobile phone use—a comparison of young adults with and without musculoskeletal symptoms. J Electromyogr Kinesiol 20(1):127–135
24. Namwongsa S, Puntumetakul R, Neubert MS, Chaiklieng S, Boucaut R, Ergonomic risk assessment of smartphone users using the Rapid Upper Limb Assessment (RULA) tool. PLoS ONE 13(8):e0203394. https://doi.org/10.1371/journal.pone.0203394
25. Kim HJ, Kim JS (2015) The relationship between smartphone use and subjective musculoskeletal symptoms and university students. Journal of Physical Therapy Science 27(3):575–579
26. Namwongsa S, Puntumetakul R, Swangnetr M (2017) Prevalence of musculoskeletal disorders of smartphone users in Khon Kaen University students, Thailand. In: Dare to desire: the 2nd national ergonomics conference
27. Chaiklieng S, Pannak A, Duangprom N (2016) The assessment of ergonomics risk on upper limb disorder among electronic workers. Srinagarind Med J 31(2):202–209

28. Lee S, Kang H, Shin G (2015) Head flexion angle while using a smartphone. Ergonomics 58(2):220–226
29. Jonsson B (1988) The static load component in muscle work. Eur J Appl Physiol Occup Physiol 57(3):305–310
30. Ohlsson K, Attewell RG, Palsson B, Karlsson B, Balogh I, Johnsson B, Ahlm A, Skerfving S (1995) Repetitive industrial work and neck and upper limb disorders in females. Am J Ind Med 27(5):731–747
31. Andersen JH, Fallentin N, Thomsen JF, Mikkelsen S (2011) Risk factors for neck and upper extremity disorders among computers users and the effect of interventions: an overview of systematic reviews. PLoS One 6(5)
32. Delisa J, Garie D, Gans B, Gatens P, Leonard J (1998) Rehabilitation medicine principle and practice. Lippincott Company, Philadelphia
33. Hansraj KK (2014) Assessment of stresses in the cervical spine caused by posture and position of the head. Surg Technol Int 25:277–279
34. Greene WB, Heckman JD (1994) The clinical measurement of joint motion. American Academy of Orthopaedic Surgeons, Rosemont
35. Kietrys DM, Gerg MJ, Dropkin J, Gold JE (2015) Mobile input device type, texting style and screen size influence upper extremity and trapezius muscle activity, and cervical posture while texting. Appl Ergon 50:98–104
36. Galinsky TL, Swanson NG, Sauter SL, Hurrell JJ, Schleifer LM (2000) A field study of supplementary rest breaks for data-entry operators. Ergonomics 43(5):622–638
37. Peper E, Gibney KH, Wilson VE (2004) Group training with healthy computing practices to prevent repetitive strain injury (rsi): a preliminary study. Appl Psychophysiol Biofeedback 29(4):279–287
38. Rempel DM, Krause N, Goldberg R, Benner D, Hudes M (2006) A randomised controlled trial evaluating the effects of two workstation interventions on upper body pain and incident musculoskeletal disorders among computer operators. Occup Environ Med (London, England) 63(5):300–306
39. Aaras A, Ro O, Thoresen M (1999) Can a more neutral position of the forearm when operating a computer mouse reduce the pain level for visual display unit operators? A prospective epidemiological intervention study. Int J Human-Comput Interact 11(2):79–94
40. Mclean L, Tingley M, Scott RN, Rickards J (2001) Computer terminal work and the benefit of microbreaks. Appl Ergon 32(3):225–237
41. Greene BL, Dejoy DM, Olejnik S (2005) Effects of an active ergonomics training program on risk exposure, worker beliefs, and symptoms in computer users. Work 24(1):41–52
42. Lintula M, Nevala-Puranen N, Louhevaara V (2001) Effects of ergorest arm supports on muscle strain and wrist positions during the use of the mouse and keyboard in work with visual display units: a work site intervention. Int J Occup Saf Ergon 7(1):103–116
43. Ketola R, Toivonen R, Hakkanen M, Luukkonen R, Takala EP, Viikari-Juntura E, Expert Group In E (2002) Effects of ergonomic intervention in work with video display units. Scand J Work Environ Health 28(1):18–24
44. Cook C, Burgess-Limerick R (2004) The effect of forearm support on musculoskeletal discomfort during call centre work. Appl Ergon 35(4):337–342
45. Brewer S, Van Eerd D, Amick BC 3rd, Irvin E, Daum KM, Gerr F, Moore JS, Cullen K, Rempel D (2006) Workplace interventions to prevent musculoskeletal and visual symptoms and disorders among computer users: a systematic review. J Occup Rehabil 16(3):325–358
46. Andersen JH, Fallentin N, Thomsen JF, Mikkelsen S (2011) Risk factors for neck and upper extremity disorders among computers users and the effect of interventions: an overview of systematic reviews. PLoS ONE 6(5):e19691
47. Hashimoto D, Kotani K, Suzuki S, Asao T (2015) Influence of posture awareness signals to the upper extremity posture and muscular loads during mobile device use. In: Proceedings 19th triennial congress of the IEA, Melbourne 9–14 Aug 2015
48. Gustafsson E, Johnson PW, Lindegard A, Hagberg M (2011) Technique, muscle activity and kinematic differences in young adults texting on mobile phones. Ergonomics 54(5):77–87

49. Tang M (2016) An investigation of an ergonomic intervention on neck biomechanics and pain due to smartphone use
50. UCDAVIS Occupational Health services. Homepage: https://safetyservices.ucdavis.edu/article/laptop-ergonomics
51. Nag PK, Pal S, Nag A, Vyas H (2009) Influence of arm and wrist support on forearm and back muscle activity in computer keyboard operation. Appl Ergon 40:286–291
52. Onyebeke LC, Young JG, Trudeau MB, Dennerlein JT (2013) Effects of forearm and palm supports on the upper extremity during computer mouse use. Appl Ergon 45:564–570
53. Matos M, Arezesb PM (2015) Ergonomic evaluation of office workplaces with rapid office strain assessment (ROSA). Procedia Manuf 3:4689–4694
54. Liebregts J, Sonne M, Potvin JR (2015) Photograph-based ergonomic evaluations using the rapid office strain assessment (ROSA). Appl Ergon 52:317–324
55. Poochadaa W, Chaiklieng S (2015) Ergonomic risk assessment among call center workers. Procedia Manuf 3:4613–4620
56. James C, James D, Nie V, Schumacher T, Guest M, Tessier J, Marley J, Bohatko-Naismith J, Snodgrass S (2018) Musculoskeletal discomfort and use of computers in the university environment. Appl Ergon 69:128–135
57. Davis KG, Hammer MJ, Kotowski SE, Bhattacharya A (2014) An ergonomic comparison of data entry work using a keyboard vs. touch screen input device while standing and sitting. J Ergon S4(1)
58. Lin MY, Barbir A, Dennerlein JT (2017) Evaluating biomechanics of use-selected sitting and standing computer workstation. Appl Ergon 65:382–388
59. Karakolis T, Callaghan JP (2013) The impact of sit-stand office workstations on worker discomfort and productivity: a review. Appl Ergon 45(3):799–806
60. Kumah DB (2016) Ergonomic challenges of employees using computers at work in a tertiary institution in Ghana. Optometry: Open Access 1(2)
61. Celik S, Celik K, Dirmese E, Tasdemir N, Arik T, Buyukkara I (2016) Determination of pain in musculoskeletal system reported by office workers and the pain risk factors. Int J Occup Med Environ Health 31(1):91–111
62. Lavina L, Filgueirasa E, Figueiredo C, Desouzart G (2015) Understanding the interaction of design students with portable computers in the classroom. Procedia Manuf 3:6274–6281
63. Korpinen L, Paakkonen R, Gobba F (2015) Self-reported ache, pain, or numbness in hip and lower back and use of computers and cell phones amongst Finns aged 18–65. Int J Ind Ergon 48:70–76
64. Koren K, Pisot R, Simunic B (2015) Active workstation allows office workers to work efficiently while exercising moderately. Appl Ergon 54:83–89
65. Commissaris DACM, K€onemann R, van Mastrigt SH, Burford E-M, Botter J, Douwes M, Ellegast RP (2014) Effects of a standing and three dynamic workstations on computer task performance and cognitive function tests. Appl Ergon 45:1570–1578
66. Leavy J, Jancey J (2016) Stand by me: qualitative insights into the ease of use of adjustable workstations. AIMS Public Health 3(3):633–662
67. Workineh SA, Yamaura H (2015) Multi-position ergonomic computer workstation design to increase comfort of computer work. Int J Ind Ergon 53:1–9
68. Werth A, Babski-Reeves K (2014) Effects of portable computing devices on posture, muscle activation levels and efficiency. Appl Ergon 45:1603–1609
69. Shariat A, Cleland JA, Danaeed M, Kargarfardb M, Sangelaji B, Tamrin SBM (2017) Effects of stretching exercise training and ergonomic modifications on musculoskeletal discomforts of office workers: a randomized controlled trial. Braz J Phys Ther 22(2):144–153
70. Goostrey S, Treleaven J, Johnston V (2013) Evaluation of document location during computer use in terms of neck muscle activity and neck movement. Appl Ergon 45:767–772
71. Schmid AB, Kubler PA, Johnston V, Coppieters MW (2014) A vertical mouse and ergonomic mouse pads alter wrist position but do not reduce carpal tunnel pressure in patients with carpal tunnel syndrome. Appl Ergon 47:151–156

72. Fathel A (2015) Control software designation of postural and ergonomic variables during working with computer users. J Inf Technol Softw Eng 5(3)
73. Wu Y, Wu T, Taele P, Wang B, Liu J, Ku P, Chen MY (2018) ActiveErgo : automatic and personalized ergonomics using self-actuating furniture. In: Proceedings of the SIGCHI conference on human factors in computing systems—CHI'18. https://doi.org/10.1145/3173574.3174132

A Review on Pre-installing Investigations of Earth Air Tube Heat Exchanger (EATHE)

Saif Nawaz Ahmad and Om Prakash

Abstract Use of conventional energy for heating/cooling is increasing day by day, since this conventional energy has produced by fossil fuels which are limited source. So this increasing demand of energy may be reduced using renewable sources. There are so many passive technologies available for heating/cooling space that utilizing renewable energy. Earth air tube heat exchanger (EATHE) is one of the viable alternatives for heating/cooling applications that utilizing geothermal energy (a type of renewable energy). EATHE extracts earth's stored energy (heat) for heating and releases heat to earth for cooling, thereby makes earth as source and sinks, respectively. The fundamental goal of this paper is to review the previous work concerning pre-installation investigations of EATHE system. Hence, in this article, a literature research has been performed regarding various parameters which affect the design and experimental results of EATHE system that must be taken into account before installing the system. Geological survey, climatic conditions, thermo-physical properties of soil and pipes, heat estimation are first and the most important parameters that must be considered before designing and installing the EATHE system. So this work provides useful information to researchers who willing to research in this field.

Keywords Earth air tube heat exchanger (EATHE) · Source · Sink · Geothermal energy

S. N. Ahmad (✉) · O. Prakash
National Institute of Technology, Patna 800005, India
e-mail: saif.nawaz.ahmad@gmail.com

O. Prakash
e-mail: om.prakash@nitp.ac.in

A. Arockiarajan et al. (eds.), *Advances in Industrial Automation and Smart Manufacturing*, Lecture Notes in Mechanical Engineering,
https://doi.org/10.1007/978-981-15-4739-3_19

1 Introduction

Energy contributes one of the essential ingredients for the existence of human activities and it is the foundation stone for financial growth of any nation. In the recent decades, energy consumption reached high in magnitudes so there is a need to implementing energy conservative and energy management techniques to meet the energy requirement. For space cooling and heating, the conventional systems are used that are top drivers of global electricity demand and is expected to be triple by 2050 [1]. The present consumption of energy for heating and cooling of space for building sector is approximately 32–33% [2]. Many passive systems are being invented and used nowadays to meet the cooling and heating requirements as well as reducing the dependency on primary energy consumption. The most popular passive system among them is EATHE system that works on renewable technique and has found applications in residential and commercial buildings, greenhouses, etc. The EATHE system utilizing the soil that situated beneath the surface of ground as heat source or sink and air that flow inside the tube as heat-carrying medium. It has been often shown that the temperature of soil (underground) being constant at depth approximately 1.5–2 m around the year for a particular locality [3]. Earth air tube heat exchanger performance depends on dispersal of temperature around soil, moisture and seasonal varying ambient temperature [4]. The categorization of geological properties of particular location is also considered regarding installation of EATHE system. The designer must have information regarding thermal and physical properties of soil before designing the EATHE system [5]. The impact of optimum length and individual earth's surface treatments on thermal performance of pipes that are used for heating/cooling of buildings has been conducted for hot-dry, composite and cold climates stands for Jodhpur, Delhi and Leh, respectively [6].

1.1 Working Principle of EATHE System

In an EATHE system, the pipes of designed dimension and materials are covered in the earth at given fixed depth. The soil release the thermal energy to air that flowing inside pipe in winter season for space heating when the outdoor temperature is lesser than underground temperature, and in this way earth work as source, while in summer season for space cooling the soil absorbs thermal energy from air through pipe whenever outdoor ambient temperature is greater than the underground temperature and hence earth works as sink. Thus, heat is moved to or from the bounding soil by pipe and flowing air through conduction and convection, respectively [7]. Fans, blower or any passive system are attached with EATHE for creating pressure difference by which air to be flow continuously within the system. Many researchers have studied and described the earth air tube heat exchanger that integrated into building and greenhouses are effective technique for air conditioning system.

1.2 *Classification of EATHE System*

There are so many types of this system available around the globe that are categorized on different basis. Among them two major types are categorized named open loop ground heat exchangers as shown in Fig. 1a and closed loop ground heat exchanger as shown in Fig. 1b. In open loop system category, the ground may be works directly for heating/cooling a medium while in closed loop ground may be used indirectly for conditioning purposes. There is advantage of closed loop on open loop that the closed loop system is more conducive than open loop system and also in this system the problem of humidity reduced drastically.

2 Literature Review

Before installation of EATHE, the very important parameter which is to be kept in mind is the climatic and geographical condition of particular locations. There is some technical feasibility which is to be identified by observing the weather data such as minimum and maximum ambient air temperature, humidity ratio, relative humidity as well the underground temperature at certain depth. A theoretical model of an earth air tube heat exchanger (EATHE) has been developed by Al-ajmi et al. [8] for hot, arid climate, and thereby predicted outlet temperature and cooling load potential. Their results reveal that the energy demand for cooling of a typical house is reduced by 30% in a peak summer season and temperature reduction of 2.81 °C also occurs. Chel and Tiwari [9] develop a thermal model of EATHE for anticipating its performance and also the life cycle cost analysis under composite climate of New Delhi in India. Bansal et al. [10] carried out year round hourly analysis of EATHE for hot and dry climatic conditions integrated with evaporative cooler. According to

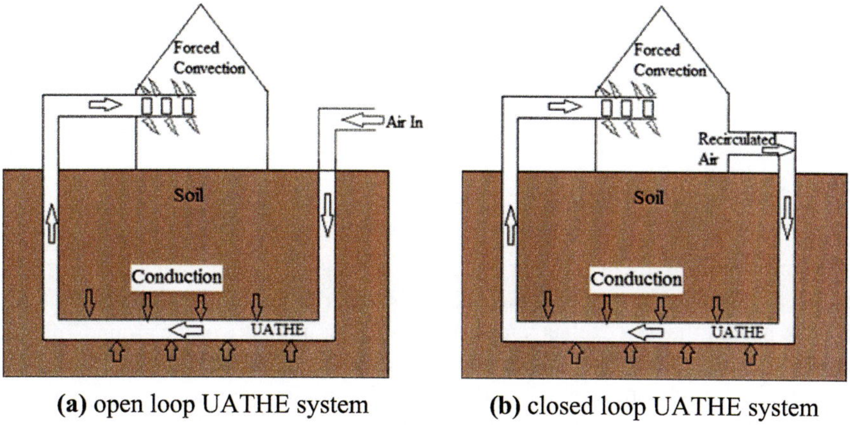

Fig. 1 **a** Open loop UATHE system. **b** Closed loop UATHE system

their results, it was observed that for hot and dry climate the EATHE system would be more useful when it is coupled to evaporative cooler. Sanusi et al. [11] predicting the performance of EATHE by performing an experimental investigation under hot and humid climate of university campus of Malaysia. Their experimental result shows that for humid climate the temperature was reduced by of 6.4 °C while for hot season it was significantly reduced by 6.9 °C. Gan [12] studied the dynamic interaction between soil and atmosphere for EATHE by developing a three-dimensional numerical model. Benhammou et al. [13] presented numerical method based on transient study for passive mode of cooling for buildings under hot and arid climates. They also examine the different designed parameters which affects the performance of EATHE that integrated with wind tower.

Khabbaz et al. [14] concluded that EATHE system is more effective for refreshing the air under semi-arid climate like in the region of Marrekeh, after analyzing the results of both the study, such as numerical as well as experimental. Belatrache [15] perform the numerical study of EATHE system at different operating parameters under arid climate of south Algeria. Fazlikhani et al. [16] developed the steady-state model of EATHE to examine and compare the performance and energy-saving potential for two climates of Iran such as cold and hot arid. They concluded that EATHE is more conducive in hot arid climate than cold climate. Li et al. [17] perform feasibility study of an EATHE for critically cold regions in order to preheating the fresh air. After performing their experimental investigation, they conclude that the fresh air get heated sufficiently for given cold regions by aforesaid system. Rodrigues et al. [18] presented numerical model for investigating its performance under different types of soils for coastal area, in the southern Brazilian city of Rio Grande. They obtained the geotechnical profiles of the soil through standard penetration tests (SPT) for thermal performance evaluations.

Alrobaian [19] performs experimental investigation of EATHE integrated with PV panel under hot climate of Egypt for cooling system. It was concluded after analyzing his study that this model was able to reduce temperature at given rate of air flow as compared to other reference model.

3 Pre-installing Factors Affecting the EATHE System

3.1 Soil Temperature

The temperature gradient between heat exchanger and ground is a major factor of heat transfer between them. Hence, this should be determined by the soil temperatures that are nearby the heat exchanger. Therefore, the temperature of the ground directly affects the efficiency of heating or cooling as well as the effectiveness of the given system. The temperature of soil varies according to both day to day and seasonal cycles. The day to day cycles vanishes within every 10 of cm and the seasonal cycles vanishes at larger depths, but the temperature remains approximately constant and

Table 1 Thermo-physical properties of different soils [24]

Types of soil	Density (kg/m^3)	Thermal conductivity (W/mK)	Thermal diffusivity (m^2/s)	Specific heat (J/kg K)
Wet clay	1800	1.49	6.18×10^{-7}	1340
Dry clay	1650	2.3	4.89×10^{-7}	2850
Limestone	1670	0.71	1.91×10^{-7}	2230
Sand	1520	1.24	6.94×10^{-7}	1650

equals to the average temperature of air at depths more than 15 m Rybach [20]. Due to thermal diffusivity of soil the function of air temperature taken annually is being transmitted down Keery et al. [21]. For increasing the temperature of subsurface, the surface of earth should be glazed or blackened while the earth's subsurface would be painted white or can be shaded or wetted by spraying the water for decreasing the temperature of subsurface Peretti et al. [22].

3.2 Thermal Properties of Soil

The heat transfer rate from the soil, or to the soil, is mainly calculated by some of its thermal properties such as thermal conductivity and thermal diffusivity. As we know that the thermal conductivity of any material is the capacity to conduct or transfer heat through it, while thermal diffusivity shows the rate of transfer of heat through any medium or materials. Mathur et al. [23] considered soil of three distinct thermal diffusivities such as 1.37×10^{-7} m^2/s, 4.37×10^{-7} m^2/s and 9.69×10^{-7} m^2/s and investigating the thermal performance of EATHE. Their study reveals that the higher thermal diffusivity contained soil transfer heat at faster rate and more heat content. The various ground thermo-physical properties are compiled in Table 1.

3.3 Ground and Geological Conditions

Whenever this system is being installed there are so many primary ground engineering exposure would need to be kept in mind, such as thickness and the nature of deposits on ground, geological weathered bedrock depth and strength as well as any hazardous ground conditions. The geological and climatic conditions of the site would be identified and predicted before installing the EATHE system. The ground condition may also directly influenced by bedrock strength, thickness of deposits and depth of weathered bedrock. The method that used for trenching and/or drilling and their associated cost also effect on installation of EATHE system. For best result of heat exchange, the soil must be packed tightly surrounding the pipe, therefore, a clay with compacted its particle or sand is approved by Ascione et al. [25]. Li et al. [26] found that the performance of this system is significantly influenced by the climate and geographical conditions of any particular locations and region. Chiesa [24] explained the climatic potential of EATHE system for cooling as well as heating mode of applications. They explained this potential using two indexes named cooling degree days (CDD) and heating degree days (HDD). The inlet air temperature also affects the performance of this system.

3.4 Groundwater

The conditions of groundwater may also affect the efficiency and effectiveness of EATHE system. As we know that the saturation levels of sediments and rocks affect its thermal properties. The mudstones that are low permeable rocks get saturated after going few meters inside the surface of ground irrespective of the elevations where it situated. The thickness of zone of instauration may be large generally at high elevations for more permeable limestones and sandstones and it may vary seasonally. The seasonal water level fluctuations are also affected the system as well.

4 Conclusions

The performance of this system depends on environmental and geological conditions as well as local site conditions. EATHE system may be established in many types of climates such as humid subtropical, oceanic, hot desert, hot arid, Mediterranean and composite climates. It is found from various contemplation that for wet season the drop of temperature was 6.4 °C while for hot-dry season the significantly drop in temperature was 6.9 °C there is about 30% reduction in cooling energy demand for typical house in a peak summer season. It has been also concluded that the locations with high variable diurnal as well as seasonal surrounding air temperature were more desirable for the EATHE system because it was found that if there is small variation in ambient air temperature, then system has low performance. This system can be designed and developed both for cool climates and warm climates. Hence, the familiarity about temperature of grounds and the geological conditions may lead to the performance while designing the system and it may also help in planning and costing of installation of the EATHE system.

References

1. International Energy Agency. http://www.iea.org
2. Nejat P, Jomehzadeh F, Mahdi M, Gohari M (2015) A global review of energy consumption, CO_2 emissions and policy in the residential sector (with an overview of the top ten CO_2 emitting countries). Renew Sustain Energy Rev 43:843–862
3. Bisoniya TS, Kumar A (2014) Heating potential evaluation of earth—air heat exchanger system for winter season. J Building Phy 1–19
4. Kumar R, Kaushik SC, Garg SN (2006) Heating and cooling potential of an earth-to-air heat exchanger using artificial neural network. Renew Energy 31(8):1139–1155
5. Bisoniya TS, Kumar A, Baredar P (2014) Study on calculation models of earth-air heat exchanger systems. J Energy 2014:1–15. https://doi.org/10.1155/2014/859286
6. Sodha MS, Buddhi D, Sawhney RL (1991) Thermal performance of underground air pipe: different earth surface treatments. Energy Convers Manage 31(1):95–104
7. Sehli A, Hasni A, Tamali M (2012) The potential of earth-air heat exchangers for low energy cooling of buildings in South Algeria. Energy Procedia 18:496–506
8. Al-ajmi F, Loveday DL, Hanby VI (2006) The cooling potential of earth – air heat exchangers for domestic buildings in a desert climate. Build Environ 41:235–244
9. Chel A, Tiwari GN (2009) Performance evaluation and life cycle cost analysis of earth to air heat exchanger integrated with adobe building for New Delhi composite climate. Energy Build 41(1):56–66
10. Bansal V, Mishra R, Agarwal GD, Mathur J (2012) Performance analysis of integrated earth air-tunnel-evaporative cooling system in hot and dry climate. Energy and Build 47:525–532
11. Sanusi ANZ, Shao L, Ibrahim N (2013) Passive ground cooling system for low energy buildings in Malaysia (hot and humid climates). Renew Energy 49:193–196
12. Gan G (2015) Simulation of dynamic interactions of the earth—air heat exchanger with soil and atmosphere for preheating of ventilation air. Appl Energy 158:118–132
13. Benhammou M, Draoui B, Zerrouki M, Marif Y (2015) Performance analysis of an earth-to-air heat exchanger assisted by a wind tower for passive cooling of buildings in arid and hot climate. Energy Convers Manag 91:1–11
14. Khabbaz M, Benhamou B, Limam K, Hollmuller P, Hamdi H, Bennouna A (2016) Experimental and numerical study of an earth-to-air heat exchanger for air cooling in a residential building in hot semi-arid climate. Energy Build 125:109–121
15. Belatrache D (2017) Numerical analysis of earth air heat exchangers at operating conditions in arid climates. Int J Hydrogen Energy 42:8898–8904
16. Fazlikhani F, Goudarzi H, Solgi E (2017) Numerical analysis of the efficiency of earth to air heat exchange systems in cold and hot-arid climates. Energy Convers Manag 148:78–89
17. Li H, Ni L, Liu G, Zhao Z, Yao Y (2018) Feasibility study on applications of an Earth-air Heat Exchanger (EATHE) for preheating fresh air in severe cold regions. Renew Energy 133:1268–1284
18. Rodrigues MK, Coswig FS, Camargo KR, Isoldi LA, Brum RS, Ramalho JVA et al (2018) Thermal performance simulations of Earth-Air Heat Exchangers for different soils of a coastal city using in-situ data. Sustain Energy Technol Assessments 30:224–229
19. Alrobaian AA (2019) Performance of PV panel coupled with geothermal air cooling system subjected to hot climatic. Appl Therm Eng 148:1–9
20. Rybach L (2010) Ground-source heat pump systems. GHC Bulletin 16–26
21. Keery J, Binley A, Crook N, Smith JWN (2007) Temporal and spatial variability of groundwater-surface water fluxes: development and application of an analytical method using temperature time series. J Hydrol 336:1–16
22. Peretti C, Zarrella A, Carli MD, Zecchin R (2013) The design and environmental evaluation of earth-to-air heat exchangers (EATHE). A literature review. Renew Sustain Energy Rev 28:107–116

23. Mathur A, Mathur S, Agrawal GD, Mathur J (2017) Comparative study of straight and spiral earth air tunnel heat exchanger system operated in cooling and heating modes. Renew Energy 108:474–487
24. Chiesa G (2017) Integration of renewable energy in the built environment (Electricity, Heating and Coling). Energy Procedia 122:517–522
25. Ascione F, Bellia L, Minichiello F (2011) Earth-to-air heat exchangers for Italian climates. Renew Energy 36:2177–2188
26. Li Z, Zhu W, Bai T, Zheng M (2009) Experimental study of a ground sink direct cooling system in cold areas. Energy Build 41:1233–1237

Assessment of Green Sustainability Index of a Machining Process Using Multigrade Fuzzy Approach

S. Dhanalakshmi and T. Rameshbabu

Abstract This paper focuses on assessing the environmental sustainability of a machining process. Environmental sustainability involves production of products by eco-friendly methods assuring a green environment. A sustainability assessment model is presented considering the green characteristics of a machining operation in three orientations: productivity, environmental performance, and economic performance using multigrade fuzzy approach. The area which needs improvement in the sustainability level of a process is identified, and proposals are given which assists the industrialists to remain competitive ensuring a green environment. The usage of this model will accelerate the green practices in machining process.

Keywords Sustainability assessment · Multigrade fuzzy approach · Green sustainability index

1 Introduction

With the growing interest of the end users for greener products and production technologies, issues related to environmental considerations, sustainable manufacturing is becoming a source for the manufacturers to remain competitive. It insists the industrialists to focus on assessing sustainability level of machining processes. Sustainability assessment is a methodology whose prime objective is to give a clear picture about the interactions of a manufacturing process with the environment ensuring safety to employees and environment and sounds economical. Sustainability assessment is perceived as a vital tool to support to move toward sustainability. Sustainable production is an approach to enhance the performance of a manufacturing process. Hence, it becomes essential for the manufacturing firms to implement sustainability to

S. Dhanalakshmi (✉)
Department of Mechanical Engineering, T.P.E.V.R. Government Polytechnic College, Vellore 632002, India
e-mail: sdltpevr@gmail.com

T. Rameshbabu
Department of Industrial Engineering, Anna University, Chennai 600025, India

A. Arockiarajan et al. (eds.), *Advances in Industrial Automation and Smart Manufacturing*, Lecture Notes in Mechanical Engineering,
https://doi.org/10.1007/978-981-15-4739-3_20

Table 1 Thesaurus of study

Key area field	**Manufacturing paradigm** Sustainable manufacturing Green manufacturing	**Framework** Equity Environment Economy	**Machining process** Turning, milling, facing, drilling, shaping, electrical discharge machining
Stage and development	**Activity** Assessment	**Design** Methodology, framework,	**Assessment** Multigrade fuzzy approach
Productivity	Surface roughness, dimensional accuracy, vibration, tool wear, chip mass, machining time, material removal rate, equipment maintenance, operator skill		
Environment	Water consumed, cutting fluid consumed, coolant loss, material consumption, labor engaged, toxicity, disposal, airborne emissions, randicity, radiation, acidification		
Economy	Recycling, cooling lubricant processing, energy consumption, frequency of tool change, startup time, idle time, setup cost, energy cost, processing cost, machine cost, tool cost		
Results	**Sustainability level** Green sustainability index		**Improvement proposals**

survive in the competitive world. There arises a need to evaluate the level of sustainability emphasizing greener system. Taking this into consideration, an attempt is made to assess the sustainability by considering the following objectives.

- Development of a conceptual framework which shows causes of the agencies influencing the machining system in productivity, environment, and economic perspectives.
- Applying multigrade fuzzy methodology for sustainability assessment.
- Finding the drag factors and suggestions for areas to enhance sustainable production. The details are given in Table 1.

2 Methodology

Figure 1 shows the methodology adopted, and it describes the way of attaining the proposed models.

3 Literature Review

This literature review is presented by taking into account sustainability assessment in machining process and criteria considered in the study as perspectives. Munda [1] developed a multicriterion network to assess sustainability. Narita et al. [2] and

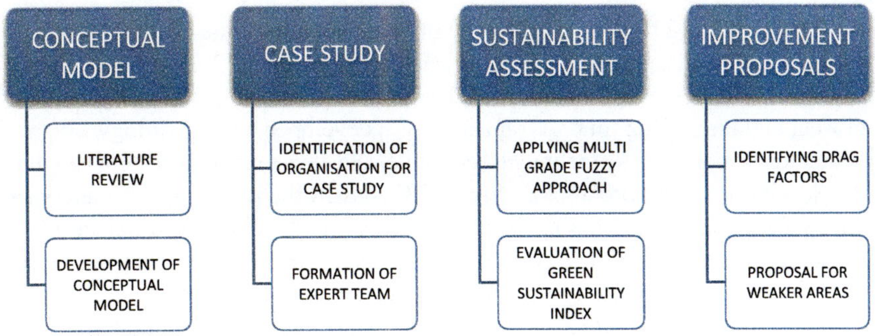

Fig. 1 Methodology

Narita and Fujimoto [3] verified environmental burden by calculating the impact in dry, wet, and MQL machining operations. The main pillars to achieve sustainability in machining are identified by Kopac and Pusavec [4] by comparing and evaluating machining systems at different cutting environments. Reduction of energy consumed improves the sustainability of the machining process proposed by Rajemi [5]. Bhanot et al. [6] analyzed the most sustainable parameters and indicators in three dimensions for milling and turning processes. Li et al. [7] proposed eco-efficiency model to evaluate energy and resource utilization for a grinding process. Shao et al. [8] generated an application neural model for representing manufacturing processes to optimize sustainability performance using sustainable process analytics formalism.

Yan et al. [9] evaluated sustainability performance of face milling operation. Kadam [10] proved in turning of Inconel 718 that medium MRR using water vapor as coolant seems to be sustainable cleaner machining technique. Singh and Madan [11] assessed the most sustainable turning process from three alternate machining technologies from the overall sustainability score. Vinodh [12] performed sustainability assessment on process perspective considering 12 criteria and 37 attributes. A framework presented to determine the sustainability of process orientation and identified the obstacles to enable sustainability improvement [13] (Feng et al. [14]; Ghandehariun et al. [15]; Hegab et al. [16]; Gocksen bas et al. [17]). Katna et al. [18] developed a non-hazardous biodegradable cutting fluid which is safe to environment and humans, saving disposal cost. Wang and Hu [19] suggested that reduced vibration during machining can optimize the machining process. Gunay et al. [20] evaluated the necessity of sustainability by investigating optimized production efficiency, minimized environmental impact, and maintaining social equity. Rotella et al. [21] addressed the impacts of sustainability elements such as cost, energy intake, waste reduction, and environment for a crankshaft oil-hole drilling process. The environmental impact of a turning process in terms of machine tools was presented by Campatelli and Scippa [22]. Shokrani et al. [23] reviewed different materials, cutting fluids, and their drawbacks considering health and environment concerns associated with machining. The environmental impact of a machining process depends on the material, and design of the product is demonstrated with the production of a simple jig by Harun et al. [24]. With a prototype of NC machining time estimation system, machining time and sequences are optimized by Borker et al. [25]. Kubler et al. [26] analyzed the resource optimization of a multipass turning operation by reducing tool wear and machining time. Avram et al. [27] developed a methodology for green manufacturing considering technical, economic, and environmental criteria and illustrated it with a milling process. Lu et al. [28] studied the planning parameters and their contribution toward the environmental, cost, and social performance of the machining process. Schultheiss et al. [29] evaluated the tool's life cycle and effective usage of resources, energy to improve quality and to minimize the cost, and environmental impact. Table 2 shows the sustainable enable for productivity.

A novel approach to attain environmental sustainability in manufacturing was proposed by Adler et al. [31]. Sokovic and Mijanovic [32] modeled the environmental effect of cutting fluid by quantifying the rate of cutting fluid leaving the machining area. Winter et al. [33] developed an empirical model to determine the eco-efficiency of grinding process. Chuang and Yang [34] evaluated the performance of a green manufacturing system with ten strategic subjects and 74 assessment factors. Kafara et al. [35] showed that additive manufacturing has lower environmental impact than the conventional machining process. Table 3 shows the sustainable enable for environment.

Even though many researches on sustainability assessment are performed on machining process in the past, only a few systematic approaches were used. The selection of criteria and their relative importance to rank the sustainability level

Table 2 Sustainable Enabler: Productivity

Sustainable criterion	References
Job fidelity $\mathbf{SC_{11}}$	Bas et al. [17], Bhanot et al. [6] katna et al. [18], Rogov et al. [30]
Tool prominence $\mathbf{SC_{12}}$	Gunay et al. [20], Lu and Rotella [21], Bas et al. [17], Campatelli and Scippa[22], Bhanot et al. [6], Shokrani et al. [23]
Machining impact $\mathbf{SC_{13}}$	Lu and Rotella [21], Harun [24], Borkar et al. [25] Kübler et al. [26]
Quality conformance $\mathbf{SC_{14}}$	Lu and Rotella [21], Pusavac et al. [4]

Table 3 Sustainable Enabler: Environment

Sustainable criterion	References
Utility factors $\mathbf{SC_{21}}$	Avram et al. [27], Lu et al. [28], Shokrani et al. [23], Pandian et al. [36], Kerbrat et al. [37], Filleti et al. [38], Yan et al. [7], Lee et al. [39], Shin et al. [40]
Source components $\mathbf{SC_{22}}$	Schlosser [41], Ingarao et al. [42], Rao [43]
Hazard stimulators $\mathbf{SC_{23}}$	Bell et al. [44], Adler et al. [31], Yucel and Gunay [20], Hermann et al. [45], Pineda-Henson et al. [46],
Global warming indicators $\mathbf{SC_{24}}$	Hermann et al. [45], Kafara et al. [35], Oliveira [47], Salonitis et al. [48]

Table 4 Sustainable enabler: economy

Sustainable criterion	References
Renewable elements $\mathbf{SC_{31}}$	Chuang and Yang [34], Pineda-Henson et al. [46], Lee et al. [39], Campatelli and Scippa [22], Hermann et al. [45], Li et al. [7]
Time constraints $\mathbf{SC_{32}}$	Rogov et al. [30], Zhu et al. [49], Li et al. [7]
Stake contributors $\mathbf{SC_{33}}$	Campatelli and Scippa [22], Rao [43], Musca et al. [50]
Overheads $\mathbf{SC_{34}}$	Pusavec and Kopac [4], Filleti et al. [38], Lu and Rotella [21], Kübler et al. [26], Singh and Madan [11]

provide an accurate evaluation of the machining process. The uniqueness of the work is that the path leading to sustainability is traced with some new attributes which are not yet discussed in the context of machining process. Table 4 shows the sustainable enable for economy.

4 Conceptual Model

The conceptual framework for sustainability assessment is presented in Table 5. The proposed framework for sustainability assessment has been described from three orientations namely productivity, environmental performance, and economic performance. This framework is a three-tier model. First tier indicates sustainability enablers, second tier shows sustainability 12 criteria, and third tier has 31 sustainability attributes. The productivity enabler has criteria namely job fidelity, tool prominence, machining impact, and quality conformance. The environmental enabler consists of criteria such as utility factors, source components, hazard stimulators, and global warming indicators. The criteria pertaining to economic enabler include criteria like renewable elements, time constraints, stake contributors, and overheads. Criteria with their aspects are exhibited in Table 6.

5 Profile of Case Study Organization

The case study is conducted in an automobile parts manufacturing industry located at Chennai, Tamil Nadu, India. The organization is a leading manufacturer of automobile parts. It has 250 employees working in the organization. It supplies the parts to automobile manufacturers. It has implemented quality management system and environmental management system. The products are produced by environmentally friendly process, and it is ensured at each and every stages of production. The employees are motivated with rewards and incentives with their performance. It has a well-established management information system to store data related to product and process. Effective training and awareness programs are being conducted

Table 5 Conceptual framework for sustainability assessment

Sustainability enabler	Sustainability criteria	Sustainability attributes
Productivity $\mathbf{SE_1}$	Job fidelity $\mathbf{SC_{11}}$	• Surface roughness of the part $\mathbf{SA_{111}}$ • Dimensional accuracy of the part $\mathbf{SA_{112}}$
	Tool prominence $\mathbf{SC_{12}}$	• Tool wear SA_{121} • Chip mass $\mathbf{SA_{122}}$
	Machining impact $\mathbf{SC_{13}}$	• Machining time of the part $\mathbf{SA_{131}}$ • Material removal rate $\mathbf{SA_{132}}$
	Quality Conformance $\mathbf{SC_{14}}$	• Equipment maintenance $\mathbf{SA_{141}}$ • Operator skill $\mathbf{SA_{142}}$
Environmental performance $\mathbf{SE_2}$	Utility factors $\mathbf{SC_{21}}$	• Water consumed during machining $\mathbf{SA_{211}}$ • Cutting fluid used for cutting $\mathbf{SA_{212}}$ • Coolant used while machining $\mathbf{SA_{213}}$
	Source components $\mathbf{SC_{22}}$	• Material consumed for part production $\mathbf{SA_{221}}$ • Number of labors engaged $\mathbf{SA_{222}}$
	Hazard stimulators $\mathbf{SC_{23}}$	• Human toxicity $\mathbf{SA_{231}}$ • Disposal of material $\mathbf{SA_{232}}$ • Airborne emissions $\mathbf{SA_{234}}$
	Global warming indicators $\mathbf{SC_{24}}$	• Radiation $\mathbf{SA}$ $_{\mathbf{242}}$ • Acidification $\mathbf{SA_{241}}$
Economic performance $\mathbf{SE_3}$	Renewable elements $\mathbf{SC_{31}}$	• Recycling of materials $\mathbf{SA_{311}}$ • Cooling lubricant processing $\mathbf{SA_{312}}$
	Time constraints $\mathbf{SC_{32}}$	• Frequency of tool change $\mathbf{SA_{321}}$ • Startup time $\mathbf{SA_{322}}$
	Stake contributors $\mathbf{SC_{41}}$	• Machine cost $\mathbf{SA_{411}}$ • Tool cost $\mathbf{SA_{412}}$
	Overheads $\mathbf{SC_{42}}$	• Setup cost $\mathbf{SA_{421}}$ • Processing cost $\mathbf{SA_{422}}$

to employees to realize their responsibilities. Safe working environment is ensured, and the workers are insisted to follow the safety practices in all operations.

Table 6 Criteria with their aspects

Sustainability criteria	Meaning
Job fidelity	The precision of the job in terms of surface texture
Tool prominence	The major role of tool in machining
Machining impact	The rate of output of machining
Quality conformance	The main contributor for the quality of the product
Source components Global warming indicators Renewable elements	The rate of exploitation of resources Root cause for global warming from machining The reusability, recyclability of auxiliary substances
Time constraints	Time lapsed other than machining
Overheads	The indirect expense spent for a process

5.1 Need of This Study

All organization even though they are good in market need to retain their position in the competitive world. The only way to identify their position in the market is assessing their sustainability level. An approach is essential to calculate the GS level and to determine the drag factors which are hindrance to sustainability. There arises a need to develop a conceptual framework with process perspective.

5.2 Expert Team

Six experts involved in discussion for the assessment purpose. The experts are heads of design, production, quality, research and development, and marketing departments who possess rich knowledge and experience on the machining process in the organization. By interacting with the experts, they are made familiar about the assessment work and are asked to give weights for performance rating.

6 Assessment of Green Sustainability Index by Multigrade Fuzzy Approach

The green sustainability index of an industry is calculated by GSI. It is the product of overall assessment factor R and overall weight W. At the initial level, single factor vector and weights from expert team were used. Sum of the weights awarded by the experts should satisfy Eq. (1). Using Eq. (2), single factor vector for the enabler was obtained.

$$\sum_{i=1}^{I} w_i = 1, \tag{1}$$

$$I_{ij} = \begin{bmatrix} R_{11} & \dots & R_{1n} \\ \vdots & \ddots & \vdots \\ R_{k1} & \dots & R_{kn} \end{bmatrix} \times \begin{bmatrix} W_1 \\ \vdots \\ W_k \end{bmatrix} = \left[\left(\sum_{k=1}^{k} R_{k1} \times W_k \right) \dots \left(\sum_{k=1}^{k} R_{kn} \times W_k \right) \right] \tag{2}$$

where w_i indicates weights awarded by the expert team. R_{kn} performance rating for kth attribute obtained from nth expert, $[R_{k1} \dots R_{kn}]$, single factor vector obtained from expert term. In the next level, with the single factor vector, green sustainability index of the organization was obtained. It is the product of overall single factor assessment vector and the overall weight of the expert team. The equation for calculating overall sustainability index is given by

$$\text{GSI} = R_i \times W_i \tag{3}$$

The expert team is given a scale of five grades to assess the sustainability level for the criteria and attributes identified. GSI = {10,8,6,4,2}. Here, 8–10 represents 'greatly sustainable', 6–8 represents sustainable, 4–6 represents generally sustainable, 2–4 represents not sustainable, and less than 2 represents greatly unsustainable. Table 7 shows the assessment table with weights given by experts.

Weights related to job fidelity criterion $W_{11} = (0.5, 0.3, 0.2)$.

Assessment vector of job fidelity criterion is given by

$$R_{11} = \begin{bmatrix} 8 & 9 & 8 & 9 & 10 & 8 \\ 6 & 7 & 7 & 6 & 8 & 7 \\ 5 & 4 & 4 & 5 & 7 & 6 \end{bmatrix}$$

Index of job fidelity criterion is given by

$$I_{11} = W_{11} \times R_{11} : I_{11} = (6.8,\ 7.4,\ 6.9,\ 7.3,\ 8.8,\ 7.3)$$

Table 7 Weights of experts for criterion assessment calculation

I_i	I_{ij}	I_{ijk}	E_1	E_2	E_3	E_4	E_5	E_6	W_{ij}	W_i	W
I_1	I_{11}	I_{111}	8	9	8	9	10	8	0.5	0.3	0.35
		I_{112}	6	7	7	6	8	7	0.3		
		I_{113}	5	4	4	5	7	6	0.2		
	I_{12}	I_{121}	7	6	7	8	7	8	0.5	0.2	
		I_{122}	5	6	4	5	4	4	0.5		
	I_{13}	I_{131}	5	4	5	7	6	6	0.5	0.2	
		I_{132}	8	7	6	7	8	6	0.5		
	I_{14}	I_{141}	5	7	8	5	7	6	0.5	0.3	
		I_{142}	7	8	10	8	7	8	0.5		
I_2	I_{21}	I_{211}	7	8	9	8	8	9	0.4	0.2	0.4
		I_{212}	5	5	6	5	7	5	0.3		
		I_{213}	4	5	4	5	5	4	0.3		
	I_{22}	I_{221}	5	4	6	5	4	5	0.5	0.1	
		I_{222}	8	7	7	9	8	6	0.5		
	I_{23}	I_{231}	10	9	9	8	8	8	0.5	0.55	
		I_{232}	7	8	7	8	7	7	0.3		
		I_{233}	6	5	5	7	6	7	0.1		
		I_{234}	5	4	4	5	5	6	0.1		
	I_{24}	I_{241}	6	4	6	7	5	6	0.5	0.15	
		I_{242}	5	4	6	6	5	5	0.5		
I_3	I_{31}	I_{311}	8	9	9	8	10	8	0.35	0.3	0.25
		I_{312}	5	4	5	4	5	6	0.3		
		I_{313}	4	6	6	6	5	6	0.35		
	I_{32}	I_{321}	5	4	5	4	4	5	0.3		
		I_{322}	6	7	6	6	5	6	0.3	0.2	
		I_{323}	9	10	8	8	9	8	0.4		
	I_{33}	I_{331}	4	5	4	5	4	6	0.5	0.25	
		I_{332}	7	8	9	7	7	8	0.5		
	I_{34}	I_{341}	4	5	4	4	5	4	0.3	0.25	
		I_{342}	5	5	6	6	7	5	0.35		
		I_{343}	8	9	7	8	8	6	0.35		

Using the same principle, the index related to other defined criteria has been arrived.

6.1 Enabler Assessment Calculation

Weights obtained for productivity enabler are $W_1 = (0.3, 0.2, 0.2, 0.3)$.

Assessment vector of productivity enabler is given by

$$R_1 = \begin{bmatrix} 6.8\ 7.4\ 6.9\ 7.3\ 8.8\ 7.3 \\ 6.0\ 6.0\ 5.5\ 6.5\ 5.5\ 6.0 \\ 6.5\ 5.5\ 5.5\ 7.0\ 7.0\ 6.0 \\ 6.0\ 7.5\ 9.0\ 6.5\ 7.0\ 7.0 \end{bmatrix}$$

Index related to productivity enabler $I_1 = W_1 \times R_1 : I_1 = (6.3,6.8,7.0,6.8,7.2,6.7)$.

In the same way, the indices are obtained for the environmental and economic sustainability enablers.

$I_2 = (7.1,6.7,7.1,7.1,6.8,6.7)$: $I_3 = (5.9,6.6,6.4,6.1,6.4,6.8)$.

6.2 Assessment of Green Sustainability Index

The green sustainability index is calculated and given below:

$$\text{Overall weight } W = (0.35,\ 0.4,\ 0.25)$$

$$\text{Overall assessment vector } R = \begin{bmatrix} 6.3\ 6.8\ 7.0\ 6.8\ 7.2\ 6.7 \\ 7.1\ 6.7\ 7.1\ 7.1\ 6.8\ 6.7 \\ 5.9\ 6.6\ 6.4\ 6.1\ 6.4\ 6.8 \end{bmatrix}$$

Green sustainability index $\mathrm{GSI} = W_i \times R_i$

$\mathrm{GSI} = (6.8,6.7,6.9,6.8,6.8,6.7)$; $\mathrm{GSI} = \frac{1}{S}\,(6.8,6.7,6.9,6.8,6.8,6.7) = 6.7 \in (6,8)$.

7 Results and Discussion

The green sustainability index calculated using multigrade fuzzy method is 6.7, and it is within the range (6–8). Hence, it is obvious that the organization is sustainable. In order to trace the obstacles to enhance the sustainability level, a gap analysis was done. Using the equation given below, green sustainability importance index (GSII_{ij}) value for each criterion was computed. A green sustainability, which combines the performance rating and weight for each criterion, was used to find the drag factors. Then, using equation GSI, the calculated GSII_{ij} is converted into a number. When the value of GSII is low for a criterion, the degree of contribution for this factor is low. A value less than 6 was set as threshold by the management to identify the drag criteria. It was found that three criteria possess a score less than the threshold value.

$$\text{GSII}_{ij} = \begin{bmatrix} R_{11} & \dots & R_{1n} \\ \vdots & \ddots & \vdots \\ R_{k1} & \dots & R_{kn} \end{bmatrix} \times \begin{bmatrix} W_1 \\ \vdots \\ W_k \end{bmatrix}$$
$$= \left[\left(\sum_{k=1}^{k} R_{k1} \times W_k \right) \dots \left(\sum_{k=1}^{k} R_{kn} \times W_k \right) \right] \quad (4)$$

Practical Implications

The identified drag factors are tool prominence, hazard stimulators, and stake contributors. A team was formed to give suggestions to improve the weaker areas. An interaction session was arranged to make the team members aware of sustainable manufacturing. The proposals for improvement were coined in consultation with the expert team.

Managerial Implications

This approach helps the managers not only to assess the sustainability level of their organization but also to periodically trace the drag factors which are the hindrance to the improvement. By adopting this methodology, the managers can assure an environmental friendly manufacturing system prevailing in their organization (Fig. 2 and Table 8).

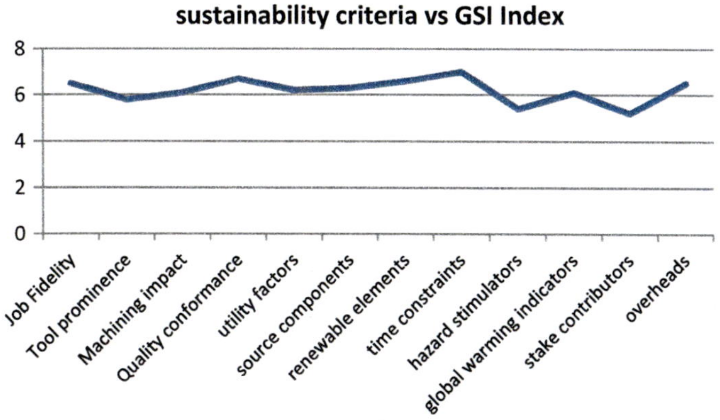

Fig. 2 Drag Factors Identification

Table 8 Improvement proposals

S. No.	Weaker area	Improvement proposals
1	Tool prominence	Training can be conducted for the workers emphasizing the efficient use of tools for machining
2	Hazard stimulators	Awareness program may be conducted among employees to use safety and health prevention devices
3	Stake contributors	By maintaining a proper inventory control system, the effective utilization of capital can be assured

8 Conclusion

Sustainability in manufacturing process is gaining more importance as it directly affects the overall performance of the organization. To stay competitive and to survive, it becomes essential for every organization to evaluate its sustainability level. The approach presented in the study will guide the managers to assess the GSI of their organization. Sustainable manufacturing is implementing environmental friendly machining practices ensuring safety and reducing adverse effects of machining. By using the MGF approach, it is easier to assess the GSI level. By computing the GSII, the drag factors can be identified and by implementing the improvement proposals the GSI level can be improved significantly.

References

1. Munda G (2005) "Measuring sustainability": a multi-criterion framework. Environ Dev Sustain 7(1):117–134
2. Narita H, Desmira N, Fujimoto H (2008) Environmental burden analysis for machining operation using LCA method. In: Manufacturing systems and technologies for the new frontier. Springer, London, pp 65–68
3. Narita H, Fujimoto H (2009) Analysis of environmental impact due to machine tool operation. Int J Autom Technol 3(1):49–55
4. Pusavec F, Kopac J (2009) Achieving and implementation of sustainability principles in machining processes. J Adv Prod Eng Manag 3:58–69
5. Rajemi MF, Mativenga PT, Aramcharoen A (2010) Sustainable machining: selection of optimum turning conditions based on minimum energy considerations. J Clean Prod 18(10–11):1059–1065
6. Bhanot N, Rao PV, Deshmukh SG (2015) Sustainable manufacturing: an interaction analysis for machining parameters using graph theory. Procedia-Soc Behav Sci 189:57–63
7. Yan J, Feng C, Li L (2014) Sustainability assessment of machining process based on extension theory and entropy weight approach. Int J Adv Manuf Technol 71(5–8):1419–1431
8. Shao H, Wang HL, Zhao XM (2004) A cutting power model for tool wear monitoring in milling. Int J Mach Tools Manuf 44(14):1503–1509
9. Yan J, Li L (2013) Multi-objective optimization of milling parameters–the trade-offs between energy, production rate and cutting quality. J Clean Prod 52:462–471
10. Kadam GS, Pawade RS (2017) Surface integrity and sustainability assessment in high-speed machining of Inconel 718—an eco-friendly green approach. J Clean Prod 147:273–283

11. Singh K, Madan J (2015) Sustainability Performance Assessment of Alternate Machining Technologies. In: ASME 2015 international manufacturing science and engineering conference. American Society of Mechanical Engineers, pp V002T05A006–V002T05A006
12. Vinodh S (2010) Improvement of agility and sustainability: a case study in an Indian rotary switches manufacturing organisation. J Clean Prod 18(10–11):1015–1020
13. Vimal KEK, Vinodh S, Muralidharan R (2015) An approach for evaluation of process sustainability using multi-grade fuzzy method. Int J Sustain Eng 8(1):40–54
14. Feng SC, Joung CB, Li G (2010, May) Development overview of sustainable manufacturing metrics. In Proceedings of the 17th CIRP international conference on life cycle engineering (Vol. 6). Hefei, PRC
15. Ghandehariun A, Nazzal Y, Kishawy H, Al-Arifi NS (2015) Investigation of sustainability in machining processes: exergy analysis of turning operations. Int J Exergy 17(1):1–16
16. Hegab HA, Darras B, Kishawy HA (2018) Towards sustainability assessment of machining processes. J Clean Prod 170:694–703
17. Bas G, Stoev L, Durakbasa NM (2015) Assessment of the production quality in machining by integrating a system of high precision measurement. Procedia Eng 100:1616–1624
18. Katna R, Singh K, Agrawal N, Jain S (2017) Green manufacturing—performance of a biodegradable cutting fluid. Mater Manuf Processes 32(13):1522–1527
19. Wang J, Hu H (2006) Vibration-based fault diagnosis of pump using fuzzy technique. Measurement 39(2):176–185
20. Yucel E, Gunay M (2013) An evaluation on machining processes for sustainable manufacturing. Gazi Univ J Sci 26(2):241–252
21. Lu T, Rotella G, Feng SC, Badurdeen F, Dillon OW, Rouch K, Jawahir IS (2012) Metrics-based sustainability assessment of a drilling process. In: Sustainable manufacturing. Springer, Berlin, Heidelberg, pp 59–64
22. Campatelli G, Scippa A (2016) Environmental impact reduction for a turning process: comparative analysis of lubrication and cutting inserts substitution strategies. Procedia CIRP 55:200–205
23. Shokrani A, Dhokia V, Newman ST (2012) Environmentally conscious machining of difficult-to-machine materials with regard to cutting fluids. Int J Mach Tools Manuf 57:83–101
24. Harun MHS, Taha Z, Salaam HA (2013) Sustainable manufacturing: effect of material selection and design on the environmental impact in the manufacturing process. IOP Conf Ser Mater Sci Eng 50(1):012060
25. Borkar BR, Puri YM, Kuthe AM, Deshpande PS (2014) Optimization of machining time using feature based process planning. In: 5th international & 26th all India manufacturing technology, design and research conference (AIMTDR 2014)
26. Kübler F, Böhner J, Steinhilper R (2015) Resource efficiency optimization of manufacturing processes using evolutionary computation: a turning case. Procedia CIRP 29:822–827
27. Avram O, Stroud I, Xirouchakis P (2011) A multi-criteria decision method for sustainability assessment of the use phase of machine tool systems. Int J Adv Manuf Technol 53(5–8):811–828
28. Lu T, Gupta A, Jayal AD, Badurdeen F, Feng SC, Dillon Jr OW, Jawahir IS (2011) A framework of product and process metrics for sustainable manufacturing. In: Advances in Sustainable Manufacturing. Springer, Berlin, Heidelberg, pp 333–338
29. Schultheiss F, Zhou J, Gröntoft E, Ståhl JE (2013) Sustainable machining through increasing the cutting tool utilization. J Clean Prod 59:298–307
30. Rogov VA, Ghorbani S, Popikov AN, Polushin NI (2017) Improvement of cutting tool performance during machining process by using different shim. Arch Civ Mech Eng 17(3):694–710
31. Adler DP, Hii WS, Michalek DJ, Sutherland JW (2006) Examining the role of cutting fluids in machining and efforts to address associated environmental/health concerns. Mach Sci Technol 10(1):23–58
32. Soković M, Mijanović K (2001) Ecological aspects of the cutting fluids and its influence on quantifiable parameters of the cutting processes. J Mat Process Technol 109(1–2):181–189

33. Li W, Winter M, Kara S, Herrmann C (2012) Eco-efficiency of manufacturing processes: a grinding case. CIRP Ann Manuf Technol 61(1):59–62
34. Chuang SP, Yang CL (2014) Key success factors when implementing a green-manufacturing system. Prod Plann Control 25(11):923–937
35. Kafara M, Süchting M, Kemnitzer J, Westermann HH, Steinhilper R (2017) Comparative life cycle assessment of conventional and additive manufacturing in mold core making for CFRP production. Procedia Manuf 8:2
36. Pandian GS, Jawahar N, Nachiappan SP (2013) Questionnaire for the development of environmental regulation management system for Indian manufacturing industries. Manag Arts 56:13330–13343
37. Kerbrat O, Le Bourhis F, Mognol P, Hascoët JY (2016) Environmental impact assessment studies in additive manufacturing. In: Handbook of sustainability in additive manufacturing. Springer, Singapore, pp 31–63
38. Filleti RA, Silva DA, Silva EJ, Ometto AR (2014) Dynamic system for life cycle inventory and impact assessment of manufacturing processes. Procedia CIRP 15:531–536
39. Lee CM, Choi YH, Ha JH, Woo WS (2017) Eco-friendly technology for recycling of cutting fluids and metal chips: a review. Int J Precis Eng Manuf-Green Technol 4(4):457–468
40. Shin SJ, Suh SH, Stroud I (2015) A green productivity based process planning system for a machining process. Int J Prod Res 53(17):5085–5105
41. Schlosser R (2013) Sustainability of energy and material consumption within manufacturing processes. https://doi.org/10.14279/depositonce-3753
42. Ingarao G, Di Lorenzo R, Micari F (2011) Sustainability issues in sheet metal forming processes: an overview. J Clean Prod 19(4):337–347
43. Rao RV (2008) Environmental impact assessment of manufacturing processes using a combinatorial mathematics based decision making method. Int J Sustain Eng 1(1):42–50
44. Bell DD, Chou J, Nowag L, Liang SY (1999) Modeling of the environmental effect of cutting fluid©. Tribol Trans 42(1):168–173
45. Hermann BG, Kroeze C, Jawjit W (2007) Assessing environmental performance by combining life cycle assessment, multi-criteria analysis and environmental performance indicators. J Clean Prod 15(18):1787–1796
46. Pineda-Henson R, Culaba AB, Mendoza GA (2002) Evaluating environmental performance of pulp and paper manufacturing using the analytic hierarchy process and life-cycle assessment. J Ind Ecol 6(1):15–28
47. Oliveira M (2012) Towards Benign metal-forming: the assessment of the environmental performance of metal-sheet forming processes. In: Metal forming-process, tools, design. InTech
48. Salonitis K, Chondros T, Chryssolouris G (2008) Grinding wheel effect in the grind-hardening process. Int J Adv Manuf Technol 38(1–2):48–58
49. Zhu XR, Wang GP (2014) The analysis and improvement measures on green impact of CNC machine tool spindle system. In Applied Mechanics and Materials. Trans Tech Publications, (Vol. 529, pp. 207–212)
50. Musca G, Mihalache A, Tabacaru L (2016). Increase productivity and cost optimization in CNC manufacturing. IOP Conf. Ser. Mater. Sci. Eng. 161(1):012019

Multiobjective Optimization of End Milling Process on Monel Using Grey Relational Analysis

Muhammed Shihan, J. Chandradass, and T. T. M. Kannan

Abstract The present examination manages the machining of Monel K-400 to enhance the processing procedure parameters to limit the surface roughness and machining time and to amplify the material removal rate. L_{16} symmetrical exhibit utilizing Taguchi's procedure with 4 levels and 3 factors are considered in end milling process. In this work, spindle speed, feed and depth of cut are taken as information process parameters of end milling process. Surface roughness, machining time and material removal rate are taken as response parameters. CBN-coated end milling cutter is utilized in this work. Grey relational analysis is used to solve the end milling process with multiple performance characteristics. Analysis of variance was additionally applied to identify the most significant factor of machining process on Monel K-400. Experimental results shown that machining performance in the milling can be improved effectively through this Taguchi-based grey relational analysis approach.

Keywords End milling · Monel K-400 · DOE · Taguchi · Grey relational analysis

1 Introduction

End milling is one of most popular machining processes to produce slots, keyways and geometric surfaces on the parts. Surface roughness greatly influences the performance of mechanical parts. It has a great impact on mechanical behaviour and corrosion resistance. Since only proper selection process parameters are required to obtain surface roughness, the effect of cutting parameters is reflected on surface roughness, surface texture and dimensional deviation of product. The three factors are very important in any milling process namely spindle speed, feed and depth of cut. Monel K-400 is a popular grade of super alloy material due to their excellent mechanical

M. Shihan (✉) · T. T. M. Kannan
Department of Mechanical Engineering, Prist University, Thanjavur 613403, Tamilnadu, India
e-mail: mhdshihan@gmail.com

J. Chandradass
Centre for Automotive Materials, Department of Automobile Engineering, SRM University, Kattankulathur, Chennai 603203, Tamilnadu, India

A. Arockiarajan et al. (eds.), *Advances in Industrial Automation and Smart Manufacturing*, Lecture Notes in Mechanical Engineering,
https://doi.org/10.1007/978-981-15-4739-3_21

and chemical properties. Monel is a nickel–copper alloy that is resistant to corrosion in many environments. It consists of two crystalline solids that form a single new solid. Monel 400 contains at least 63% nickel, between 29 and 34% copper, between 2 and 2.5% iron and between 1.5 and 2% manganese. Monel 405 adds no more than 0.5% silicon, and Monel K-500 adds between 2.3 and 3.15% aluminium and between 0.35 and 0.85% titanium. These and other variations all are valued for their resistance to attack by acids and alkalis, as well as for their high mechanical strength and good ductility. Recent design cooling jackets are solid copper with Monel 500 tube embedded inside. The tube is bent into many turns to maximize heat transfer from the solid copper to water flowing in the Monel tube. Other applications sometimes include chemical plants, including environments using sulphuric acid and hydrofluoric acid. There are numerous coating techniques available in the market. The industry segment is constantly in search and development of new coatings and enhanced properties. CBN-coated tools play vital role in machining hard and very hard material. It removes the material with high accuracy and surface finish. Chahal et al. [1] studied the effect of input parameters in end milling process parameters such as spindle speed, feed and depth of cut on H-13 die steel material using L_9 orthogonal array used as design of experiment and concluded that high spindle speed is a dominating parameter for achieving lower surface roughness of hard steel machining process. Akhilsoman et al. [2] conducted experiments with Monel K-400 in face milling process. They found that lower surface roughness is obtained in high spindle speed and higher material removal rate is obtained in high value of depth of cut. They also reported that coated milling cutter can be used in machining of Monel K-400 material. Lu et al. [3] found larger material removal rate in face milling operation. They conducted experiments based on L_{18} orthogonal array of design of experiment concept. Feed rate is the most dominating process parameter in face milling operation and produces good surface quality and more material removal rate. Lohithaksha et al. [4] announced procedure parameter advancement of end-processing task on Inconel 718 super combination with multi reaction criteria dependent on Taguchi's symmetrical cluster with dark social examination. Aggarwal et al. [5] led probe end-processing procedure of AISI H11 steel dependent on Taguchi's dim social investigation. They finished up axle speed, which is an overwhelming parameter in end-processing procedure and gave better surface harshness great surface integrity. Kuran et al. [6] talked about a use of Taguchi's trial strategy for incorporating end-processing process parameters of AISI 304 hardened steel. They spoke to that unite is an affecting parameter for delivering surface nature of processed surfaces. Upinderkumar et al. [7] examined ideal cutting parameters in fast machining of H 13 dry condition. Taguchi's L9 symmetrical array and broke down by ANOVA. Liaoa et al. [8] examined conduct of end-processing procedure of Inconel 718 super compound utilizing solidified carbide instrument. They anticipate high axle speed and produce great surface trustworthiness and keep up lower surface unpleasantness and furthermore notice solidified carbide apparatus is appropriate for super compound material. Chang and Lu [9] examined on expectation of surface harshness in side processing activities utilizing the diverse polynomial systems. They directed arrangement of investigations to think about the impacts of the different cutting parameters on surface unpleasantness. Alam et al. [10] created statistical expectation model of normal surface unpleasantness utilizing three-level full factorial structure of investigations on a vertical machining centre (VMC) with a rapid processing connection (HES 510), utilizing cutting velocity, profundity of cut and feed as machining factors. Cakir et al. [11] analysed the impacts of cutting velocity, feed rate and profundity of cut onto the surface harshness utilizing the scientific model and performed arrangement of going tests to assess the impact of two understood covering layers onto the surface unpleasantness. Yang et al. [12] connected the structures of tests (DOE) to enhance parameters of a CNC in end-processing for high-virtue graphite under dry machining and concentrated the dimensional exactness of furrow width and the plane of end processing. Arranging trial depended on a Taguchi symmetrical exhibit table, and ANOVA was adjusted to break down the most powerful factors on the CNC end-processing process. The end-processing procedure is one of the most broadly utilized material expulsion forms among different machining forms in industry. At last plants, the cutting tasks are basic as on account of face processing performed on the highest point of a level surface with an inflexible shaper, a processing of exceptionally complex parts. Kareem et al. [13] advanced processing parameters utilizing Taguchi strategy for limiting surface unpleasantness and directed analyses in a CNC processing machine with cutting profundity, cutting pace and feed rate. Zhang et al. [14] considered the Taguchi structure application to improve surface quality in a CNC face processing task with feed rate, axle speed and profundity of cut as control factors and the working chamber temperature and the use of various device embeds as the commotion factors. ANOVA is connected to dissect the huge elements influencing surface unpleasantness. El-Sonbaty et al. [15] created artificial neural systems (ANNs) models for the expectation of the connection between the cutting conditions and the comparing fractal parameters of machined surfaces in face-processing activity with rotational speed, feed, profundity of cut, pre-instrument flank wear and vibration level as the information parameters and the yield are the fractal parameters determined.

A few research works have been done utilizing nickel-based super composites to contemplate the impact of cutting conditions in machining [16–21]. The determination of machining parameters assumes an essential job and poor choice prompts wearing and breaking of cutting apparatuses and may bring about efficient misfortunes, for example harmed workpiece and poor surface quality [22–25].

After the basic audit of above machining issues, the fundamental point of the present work is to examine the effect of various machining parameters, for example shaft speed, feed and depth of cut on end processing of Monel K-400 compound. Taguchi structure system is used for exploratory arranging during end processing of Monel K-400 composite. The outcomes are investigated to accomplish ideal surface unpleasantness and material evacuation rate. Dark social investigation is executed to join the different reactions into single numerical score; these scores are positioned to decide the ideal machine parameter settings. Affirmation tests were performed by utilizing tests. ANOVA is connected to explore the high effect parameters on the numerous performance attributes.

2 Experimental Results

A. Materials and processes

The test study was completed in dry conditions on a CNC vertical processing machine furnished with a greatest shaft speed of 7000 rpm, feed rate of 10 m/min and a 3.8-kW drive engine. CNC part programs for instrument ways were made. The workpiece material utilized was Monel K-400 as a 75 mm (length) 50 mm (width) 12 mm (height) square having hardness of 45 HRc. The workpiece material is mounted onto the machine table to give most extreme unbending nature. The exploratory arrangement of the work piece for end plant is appeared in Fig. 1. The point-by-point data on compound structure and mechanical properties of this Monel K-400 amalgam is given in Tables 1 and 2 individually. The apparatus utilized for performing end processing activity is CBN covered carbide (10 mm width, 16—woodwinds). The machined surface was estimated at three distinct positions utilizing a surf test (Make—Mahr

Fig. 1 Experimental set-up for end milling operation

Table 1 Chemical composition of Monel K-400

Ni	Cu	Al	Ti	C	Mn	Fe	Si
70	24	2.0	0.8	0.25	1.04	2.0	0.50

Table 2 Mechanical properties of Monel K-400

Tensile strength	Elongation	Hardness
(Ksi)	(%)	(HRc)
180	30	35alS

Table 3 Process parameters and their levels

Parameter	Unit	Level 1	Level 2	Level 3	Level 4
spindle speed (s)	Rpm	1400	2100	2800	3500
Feed (f)	Mm/tooth	0.03	0.06	0.09	0.12
Depth of cut (d)	mm	0.75	1.00	1.25	1.50

surf test) estimating instrument with the cut-off length 2.5 mm, and the normal surface unpleasantness (Ra) esteem is recorded in microns. Material removal rate (MRR) is utilized as another exhibition measure to assess a machining execution. Material removal rate is communicated as the measure of material expelled under a time of machining time and is communicated in mm^3/sec. cBN (cubic boron nitride) sintered composite is second hardest material by precious stone and multiple times harder than tungsten carbide, additionally having highlights of solid warmth obstruction and high warm conductivity. Notwithstanding, less extreme normal for cBN frequently causes chipping of hardware edge effectively; cBN is recommended for finishing of hard materials with less cutting load on the tool edge, which guarantees extra-long tool life.

3 Plan of Experiments

The system of Taguchi for three components at four levels is utilized for the implementation of the arrangement of analyses. Taguchi's L_{16} symmetrical cluster is utilized to characterize the 16 preliminary conditions. Just the fundamental impacts are of intrigue, and factor communications are not contemplated. The procedure parameters and levels are recorded in Table 3. Everyone of the 16 preliminaries and the normal reaction esteems are utilized for the investigation. Table 4 demonstrates the trial design and comparing normal test outcomes. The info parameters of end milling are spindle speed, feed and depth of cut, and responses are material removal rate, machining time and surface roughness.

4 Result and Discussion

4.1 Machining Time

Prediction of the machining time of cutting tool during machining process is very important in order to obtain high precision parts. Machining time is closely related with production and productivity. The machining time is measured by digital stop watch and tabulated as given below (Table 5).

Table 4 Experimental layout using an L_{16} orthogonal array

Expt. No.	Process parameters		
	Spindle speed	Feed	Depth of cut
1	1	1	1
2	1	2	2
3	1	3	3
4	1	4	4
5	2	1	2
6	2	2	1
7	2	3	4
8	2	4	3
9	3	1	3
10	3	2	4
11	3	3	1
12	3	4	2
13	4	1	4
14	4	2	3
15	4	3	2
16	4	4	1

Table 5 Milling parameters of machining time

Spindle speed	Feed	DOC	Machining time	S-N ratio
1	1	1	32	−30.37
1	2	2	27	−28.94
1	3	3	23	−27.60
1	4	4	21	−26.62
2	1	2	22	−25.57
2	2	1	19	−22.92
2	3	4	18	−26.44
2	4	3	13	−25.10
3	1	3	20	−21.58
3	2	4	17	−24.08
3	3	1	11	−22.27
3	4	2	15	−21.58
4	1	4	12	−24.08
4	2	3	11	−22.27
4	3	2	14	−28.52
4	4	1	10	−20.00

Figure 2 shows that main effect plot for machining time of end milling process on Monel k-400. The higher value of signal-to-noise ratio indicates the favourable parameter of machining time and 1200 rpm of spindle speed and 0.06 mm/rev of feed rate and 0.6 mm of depth of cut. It also indicates the lower machining time is achieved in higher spindle speed and feed.

Table 6 shows that response table for adjusted signal-to-noise ratio of machining time of end milling process on Monel k-400. It represent spindle speed is an influencing parameter of end milling process. The lower machining time is achieved by high spindle speed of machining process based on signal-to-noise ratio.

Table 7 shows that ANOVA value of machining time of end milling process of Monel k-400. It represents higher values of *F* indicate the dominating parameter of

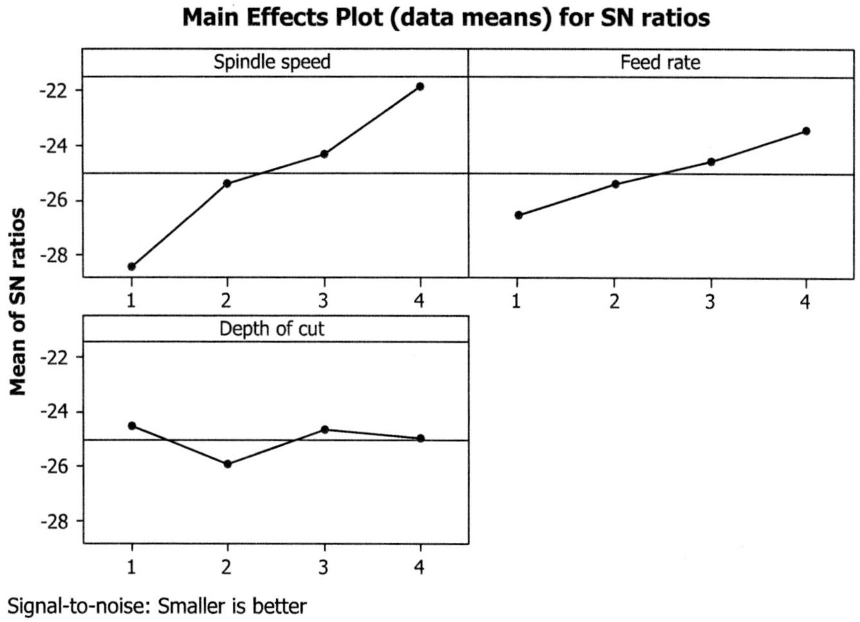

Fig. 2 Main effect plot for machining time

Table 6 Response table for machining time

Level	Spindle speed	Feed rate	Depth of cut
1	−28.44	−26.58	−24.49
2	−25.44	−25.41	−25.95
3	−24.30	−24.57	−24.64
4	−21.85	−23.46	−24.95
Delta	6.60	3.12	1.45
Rank	1	2	3

Table 7 Analysis of variance for machining time

Source	DF	Seq SS	Adj SS	Adj SS	F	P
Spindle speed	3	428.	428	142	19.9	0.00
Feed rate	3	107	107	35.6	4.98	0.04
Depth of cut	3	18	18.5	6.1	0.86	0.51
Error	6	43	43.0	7.1	–	–
Total	15	597	–	–	–	–

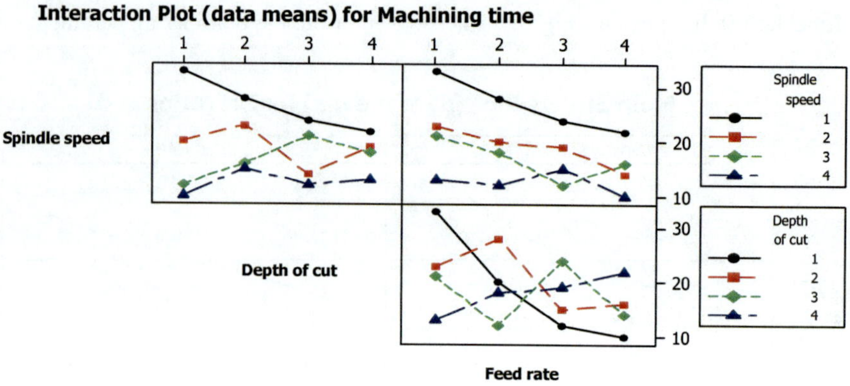

Fig. 3 Interaction plot for machining time

end milling process. The machining time influence the machining process for better productivity.

Figure 3 shows that interaction plot for machining time of end milling process on Monel k-400. It indicates the spindle speed, feed rate and depth of cut are the dependant parameters of end milling process for achieving lower machining time. The dependant parameters of end milling process produce good surface finish and productivity.

4.2 *Material Removal Rate*

Material removal rate is defined as volume of material removed divided by machining time. Material removal rate is the rate of which the cross section area of material being removed moves through the work piece. Material removal rate is calculated from the difference in volume of workpiece before and after each experiment and weight of workpiece is measured by digital scale. The material removal rate of each experiment is tabulated in Table 8.

Figure 4 shows that main effect plot for material removal rate of end milling process on Monel k-400. The higher value of signal-to-noise ratio indicates the

Table 8 Milling parameters of MRR

Spindle speed	Feed	DOC	MRR	S-N ratio
1	1	1	33	−30.37
1	2	2	28	−28.94
1	3	3	24	−27.60
1	4	4	22	−26.62
2	1	2	23	−25.57
2	2	1	20	−22.92
2	3	4	19	−26.44
2	4	3	14	−25.10
3	1	3	21	−21.58
3	2	4	18	−24.08
3	3	1	12	−22.27
3	4	2	16	−21.58
4	1	4	13	−24.08
4	2	3	12	−22.27
4	3	2	15	−28.52
4	4	1	10	−20.00

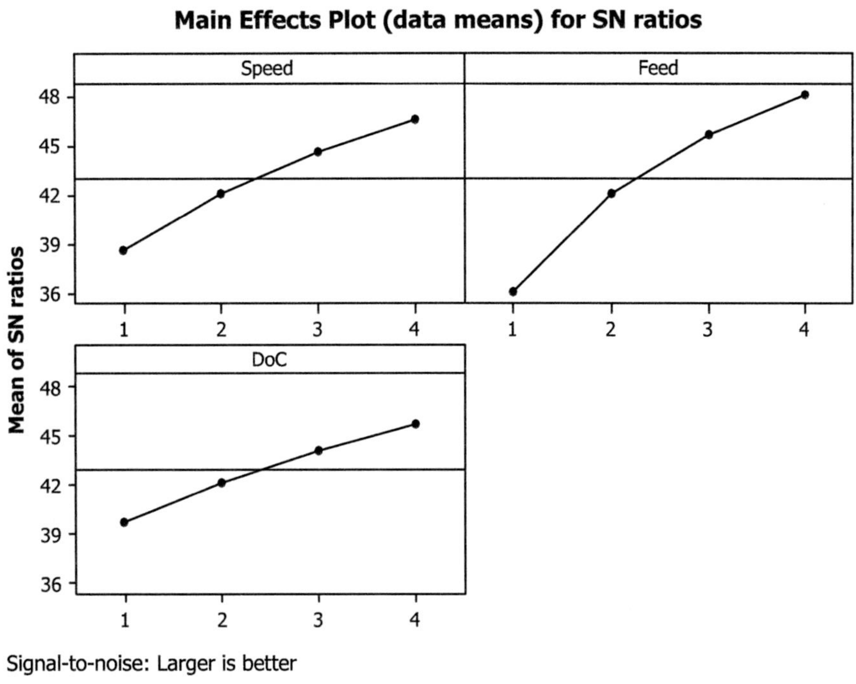

Fig. 4 Main effect plot for MRR

Table 9 Response table for MRR

Level	Spindle speed	Feed rate	Depth of cut
1	38.57	36.04	39.70
2	42.09	42.06	42.20
3	44.59	45.59	44.14
4	46.53	48.08	45.73
Delta	46.53	12.04	6.03
Rank	2	1	3

Table 10 ANOVA for MRR

Source	DF	Seq SS	Adj SS	Adj SS	F	P
Spindle speed	3	25748	25748	858	8.5	0.014
Feed rate	3	70781	70781	23594	23.6	0.001
Depth of cut	3	9034	9034	301	3.01	0.116
Error	6	5996	5996	999	–	–
Total	15	–	–	–	–	–

favourable parameter of material removal rate and 600 rpm of spindle speed and 0.02 mm/rev of feed rate and 0.2 mm of depth of cut. It also indicates the higher rate of material removal rate achieved in higher feed rate.

Table 9 shows the response table for adjusted signal-to-noise ratio of material removal rate of end milling process on Monel k-500. It represents feed rate is an influencing parameter of end milling process. The higher material removal rate is achieved by medium of feed rate of machining process based on signal-to-noise ratio.

Table 10 shows the ANOVA value of material removal rate of end milling process of Monel k-400. It represents higher values of F indicate the dominating parameter of end milling process. The material removal rate influences the machining process for better productivity.

Figure 5 demonstrates that interaction plot for material evacuation rate of end milling process on Monel k-400. It shows the spindle speed, feed rate and depth of cut are the dependant parameters of end milling process for accomplishing lower machining time. The dependant parameters of end milling process produce good surface finish and productivity.

4.3 *Surface Roughness*

Surface roughness is one of the most critical prerequisites in machining process. A sensibly decent surface completion is wanted to improve the mechanical properties, for example fatigue, sway, consumption opposition and aesthetical intrigue of the

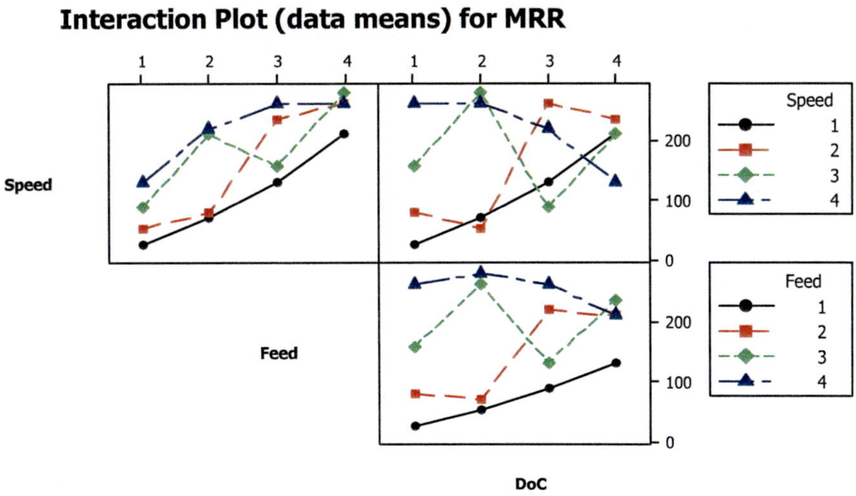

Fig. 5 Interaction plot for MRR

item. Get better surface completion of processed items; the ideal machining parameters and instrument geometry are to be chosen. Surface harshness has gotten serious consideration for a long time, and it is a key procedure to asses nature of a particular product. The surface unpleasantness affects mechaincal properties, and it likewise influences other practical characteristics of parts like grating, wear, light reflection, heat transmission, grease and electrical conductivity. Surface harshness was measure utilizing unpleasantness metre with a precision of 0.01 Âμm after the end processing procedure of work piece. Stylus profilers are contact profilers in that they physically contact the surface being estimated, ordinarily with a precious stone tip on a gently adjusted arm. The stylus is hauled in contact with the surface, and as the precious stone tip experiences crests and valleys, the tip is raised and brought down from a generally straight line. The sum to which the stylus is raised or brought down at some random point is recorded. The result from a stylus measurement is regularly communicated as a solitary parameter (Ra); however, this number is calculated from a direct informational collection, which is typically diagrammed as a line in a profile plot. Surface roughness fundamentally relies on feed rate, spindle speed and depth of cut is appeared in Table 12. The most good estimation of end milling process on the grounds that surface roughness has direct connection with quality S/N proportion was assessed dependent on the following condition.

Table 11 shows that signal-to-noise ratio data and control factors. It indicates the degree of predictable performance of a product and process in the presence of noise factor. Consequently, the level that has higher value determines the optimal level of each factor 1200 rpm of spindle speed, 0.06 mm/rev of feed rate and 0.4 mm of depth of cut are the optimal value for achieving lower surface roughness of end milling process on Monel k-400 using coated carbide end milling cutter.

Table 11 Milling parameters of surface roughness

Spindle speed	Feed	DOC	Surface roughness	S-N ratio
1	1	1	0.559	5.0518
1	2	2	0.391	8.1565
1	3	3	0.384	8.3134
1	4	4	0.407	7.8081
2	1	2	0.454	6.8589
3	4	2	0.470	6.5580
4	1	4	0.365	8.7541
4	2	3	0.384	8.3134
4	3	2	0.247	12.1461
4	4	1	0.372	8.5891

Figure 6 shows the main effect plot for surface roughness of end milling process on Monel k-400. The higher value of signal-to-noise ratio indicates the favourable parameter of machining time and 1200 rpm of spindle speed and 0.06 mm/rev of feed

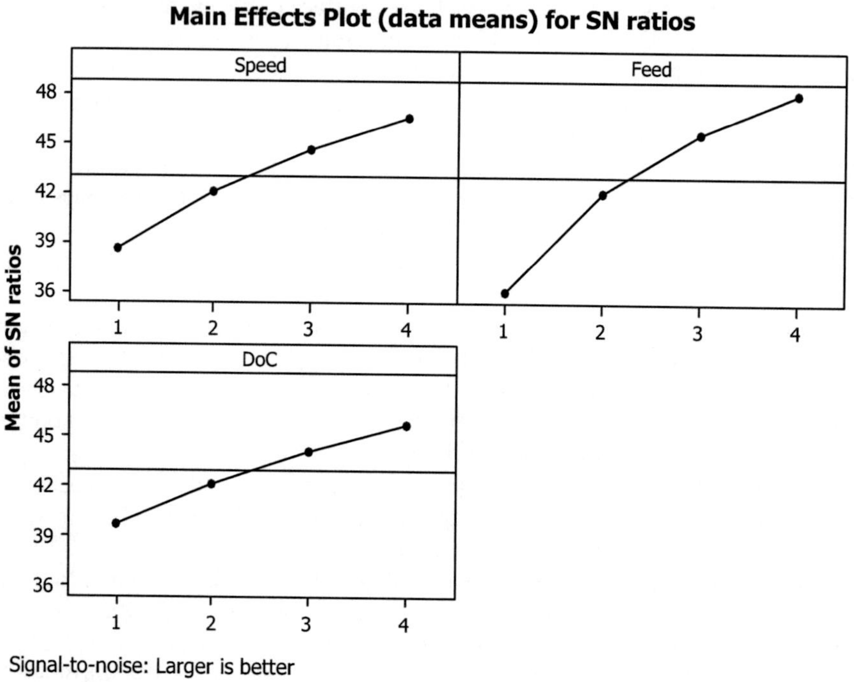

Fig. 6 Main effect plot for SR

Table 12 Response table for surface roughness

Level	Spindle speed	Feed rate	Depth of cut
1	7.332	10.019	10.046
2	9.083	10.184	8.430
3	10.778	8.983	10.729
4	9.451	7.458	9.440
Delta	3.445	2.724	3.290
Rank	1	3	2

Table 13 ANOVA for MRR

Source	DF	Seq SS	Adj SS	Adj SS	F	P
Spindle speed	3	0.0107	0.0107	0.0035	2.07	0.25
Feed rate	3	0.00536	0.00536	0.001788	1.04	0.44
Depth of cut	3	0.0154	0.0154	0.00516	3.00	0.11
Error	6	0.0103	0.0103	0.010329	–	–
Total	15	0.0418	–	–	–	–

rate and 0.4 mm of depth of cut. It also indicates that the lower surface roughness is achieved in higher spindle speed and feed.

Table 12 shows the response table for adjusted signal-to-noise ratio of surface roughness of end milling process on Monel k-400. It represents spindle speed is an influencing parameter of end milling process. The lower surface roughness is achieved by high spindle speed of surface roughness based on signal-to-noise ratio.

Table 13 shows the ANOVA value of surface roughness of end milling process of Monel k-400. It represents higher values of F indicate the dominating parameter of end milling process. The surface roughness influences the machining process for better productivity.

4.4 Grey Relational Analysis

In grey relational analysis, the capacity of components is ignored in circumstances where the scope of the grouping is bigger and the standard worth is huge. Dark social generation is an information procedure which moving the first succession to a comparable grouping; for this reason, the trial result is standardized in the range between zero and one; the standardization should be possible from three distinctive approaches.

(1) **The higher is better**

$$x_i(k) = \frac{x_i(k) - \min x_i(k)}{\max x_i(k) - \min x_i(k)} \quad (1)$$

(2) **The smaller is better**

$$x_i(k) = \frac{\max x_i(k) - x_i(k)}{\max x_i(k) - \min x_i(k)} \quad (2)$$

4.5 *Grey Relational Analysis of Process Parameter*

The grey grade of machining time (Table 14), surface roughness (Table 15), and material removal rate (Table 15) are mentioned below. The grey relational analysis of end milling process is also mentioned in Tables 16 and 17.

Overall grey relational grade is found by averaging the dim social coefficient corresponding to choose reactions. This system changes over a various reaction.

Methodology progression process advancement issue into a solitary response streamlining circumstance with the target capacity is in general dark social grade. The ideal parametric mix is then assessed by amplifying the overall dark social evaluation.

The direct effects of machining parameters on surface roughness are analysed and discussed in the graphical form. This helps to determine which parameter

Table 14 Grey grade of machining time

S. No	Machining time	SNR (machining time)	NSR (machining time)
1	33	−30.3730	1
2	28	−28.9432	0.8623
3	24	−27.6042	0.7332
4	22	−26.8485	0.6603
5	23	−27.2346	0.6976
6	20	−26.0206	0.5805
7	19	−25.5751	0.5376
8	14	−22.9226	0.2818
9	21	−26.4444	0.6214
10	18	−25.1055	0.4923
11	12	−21.5836	0.1527
12	16	−24.0824	0.3936
13	13	−22.2789	0.2197
14	12	−21.5836	0.1527
15	15	−23.5218	0.3396
16	10	−20.0000	0

Table 15 Grey grade of surface roughness

S. No	Surface roughness	SNR (surface roughness)	NSR (surface roughness)
1	0.559	5.0518	1
2	0.391	8.1565	0.7838
3	0.384	8.3134	0.7728
4	0.407	7.8081	0.8080
5	0.454	6.8589	0.8741
6	0.144	16.8328	0.1796
7	0.515	5.7639	0.9504
8	0.453	6.8780	0.8728
9	0.107	19.4123	0
10	0.425	7.4322	0.8342
11	0.327	9.7090	0.6756
12	0.470	6.5580	0.8951
13	0.365	8.7541	0.7421
14	0.380	8.3134	0.7728
15	0.247	12.1461	0.5059
16	0.372	8.5891	0.7536

Table 16 Grey grade for material removal rate

S. No	Material removal rate	SNR (material removal rate)	NSR (material removal rate)
1	26.38	28.4255	0
2	70.37	36.9478	0.4144
3	131.94	42.4075	0.6799
4	211.11	46.4902	0.8784
5	52.78	34.4494	0.2929
6	79.17	37.9712	0.4642
7	237.50	47.5133	0.9282
8	263.89	48.4285	0.9727
9	87.96	38.8857	0.5086
10	211.11	46.4902	0.8784
11	158.33	43.9913	0.7569
12	281.48	48.9890	1
13	131.94	42.4075	0.6799
14	219.91	46.8449	0.8957
15	263.89	48.4285	0.9727
16	263.89	48.4285	0.9727

Table 17 Grey relational analysis for end milling process

Input parameters			Output parameters			Taguchi outputs			Gray realational outputs			Results	
Speed	Feed	DoC	MT	MRR	SR	SNRA (MT)	SNRA (MRR)	SNRA (SR)	NRS (MT)	NRS (MRR)	NRS(SR)	AVG	RANK
1400	0.03	0.75	33	26.38	0.55	−30.3703	28.4	5.05	1	0	1	0.66	8
1400	0.06	1.00	28	70.37	0.39	−28.9432	36.9	8.15	0.86	0.4144	0.7838	0.686	7
1400	0.09	1.25	24	131.9	0.38	−27.6042	42.4	8.31	0.73	0.6799	0.7728	0.72	5
1400	0.12	1.50	22	211.1	0.40	−26.8485	46.4	7.80	0.66	0.8784	0.8080	0.78	2
2100	0.03	1.00	23	52.78	0.45	−27.2346	34.4	6.85	0.69	0.2929	0.8741	0.62	9
2100	0.06	0.75	20	79.17	0.14	−26.0206	37.9	16.8	0.58	0.4642	0.1796	0.40	15
2100	0.09	1.50	16	237.5	0.51	−25.5751	47.5	5.76	0.53	0.9282	0.9504	0.80	1
2100	0.12	1.25	14	263.8	0.45	−22.9226	48.4	6.87	0.28	0.9727	0.8728	0.70	6
2800	0.03	1.25	21	87.96	0.10	−26.4444	38.8	19.4	0.62	0.5086	0	0.37	16
2800	0.06	1.50	18	211.1	0.42	−25.1055	46.4	7.43	0.49	0.8784	0.8342	0.73	4
2800	0.09	0.75	12	158.3	0.32	−21.5836	43.9	9.70	0.15	0.7569	0.6756	0.52	14
2800	0.12	1.00	16	281.4	0.47	−24.0824	48.9	6.55	0.39	1	0.8951	0.76	3
3500	0.03	1.50	13	131.9	0.36	−22.2789	42.4	8.75	0.21	0.6799	0.7421	0.54	13
3500	0.06	1.25	12	219.9	0.38	−21.5836	46.8	8.31	0.15	0.8957	0.7728	0.60	10
3500	0.09	1.00	15	263.8	0.24	−23.5218	48.4	12.1	0.33	0.9727	0.5059	0.60	11
3500	0.12	0.75	10	263.8	0.37	−20.0000	48.4	8.58	0	0.9727	0.7536	0.57	12

are statistically significant in improving the machining performance in end-milling process.

5 Confirmation Experiment

After selecting the optimal level of machining parameters, the next step is the prediction and verification of the improvement in performance characteristics with the help of the selected optimal level of the machining parameters. The grey relational grade is estimated as

$$\hat{\varphi} = \varphi_m + \sum_{i=1}^{n} \bar{\varphi}_j - \varphi_m \quad (3)$$

where *m* is the total mean of the grey relational grade, *j* is the mean of the grey relational grade at the optimum level, and *n* is the number of machining parameters that influences the multiple performance characteristics. Based on Eq (3), the grey relational grade is estimated using the optimal machining parameters.

Table 18 shows the results of the confirmation experiment utilizing the ideal machining parameters. The surface harshness (SR) is improved from 0.21 to 0.19 μm, and the material expulsion rate (MRR) is significantly expanded from 4.308 to 7.100 mm^3/sec. It is unmistakably demonstrated that various presentation qualities at last milling of Monel k-400 are incredibly improved through this investigation.

Table 18 Results of machining performance using initial and optimal machining parameters

	Initial machining parameters	Optimal machining parameters	
		Prediction	Experiment
Setting level	A1B1C1	(A1B4C1)	A1B4C1
Machining time (MT)	28.33	26.42	26.75
Surface roughness (SR)	0.444	0.45825	0.43525
Material removal rate (MRR)	76.23	219.8521	219.9075
Grey relational grade	0.3647	0.8522	
Improvement in grey relational grade = 0.4575			

6 Conclusion

The present work has successfully displayed the use of Taguchi based grey relational for multireaction enhancement of procedure parameters of end milling of Monel k-400. The significant ends are given underneath

- The ideal cutting parameter for machining procedure of Monel k-400 is 3500 rpm of axle speed, 0.12 mm/rev of feed rate and 1.5 mm of depth of cut for accomplishing lower surface harshness.
- The ideal cutting parameters for machining procedure of Monel k-400 3500 rpm of axle speed, 0.12 mm/rev of feed rate and 0.75 mm of depth of cut for accomplishing the most minimal machining time.
- The ideal cutting parameters for machining procedure of Monel k-400 1400 rpm of axle speed, 0.03 mm/rev of feed rate and 0.75 mm of depth of cut for accomplishing higher material expulsion rate.
- Multiresponse optimization represents 2100 rpm of shaft speed, 0.09 mm/rev of feed rate and 1.5 mm of depth of cut is the ideal parameters for fulfilling bring down the best and higher the best ideas of end milling process of Monel k-400.
- Taguchi's grey relational analysis does not include any entangled scientifically and, in this way, can be utilized by specialists without a strong statistical background.

References

1. Chahal, M., (2013) Investigations of machining parameters on surface roughness in CNC milling using Taguchi technique. Innovative Syst Des Eng 4(7):5–10
2. Akhilsoman, Vignesh S, Shanbhag, Sakthivel M (2015) Influence of MQL application on Ra, cutting force and tool wear for milling of Monel k-400. Int J Mech Prod Eng 3(5):118–123
3. Lu HS, Chang CK, Hwang NC, Chung CT (2009) Greyrelationalanalysiscoupledwithprincipal-componentanalysisforoptimizationdesignofthe cutting parameters in high-speed end milling. J Mater Process Technol 209:3808–3817
4. Maiyara LM, Ramanujamb R, Venkatesanc K, Jeraldd J (2013) Optimization of mackhining parameters for end milling of inconel 718 super alloy using taguchi based grey relational analysis. Procedia Eng 64:1276–1282
5. Aggarwal N, Sharma SK (2014) Optimization of machining parameters in end milling of AISI H11 steel alloy by Taguchi based grey relational analysis. Int J Current Eng Technol 4(4):2797–2803
6. Kuram E, Simsek BT, Ozcelik B, Demirbas E, Askin S (2010) Optimization of the cutting fluids and parameters using Taguchi and ANOVA in milling. In: Proceedings of the world congress on engineering 2010 Vol II WCE 2010, June 30 - July 2, 2010, London, UK
7. Upinder K, Deepak N (2013) Optimization of cutting parameters in high speed turning by grey relational analysis. Int J Eng Res Appl (IJERA) 3(1):832–839
8. Liaoa YS, Linb HM, Wang JH (2008) BehaviorsofendmillingInconel718 super alloy by cemented carbide tools. J Mater Proc Technol 201(1–3):460–465
9. Chang C-K, Lu HS (2006) Study on the prediction model of surface roughness for side milling operations. Int J Adv Manuf Technol 29:867–878

10. Alam S, Nurul Amin AKM, Patwari AU (2008) Surface roughness prediction model in high speed end-milling of Ti-6Al-4V. Competitive Manuf Proc 2nd Intl & 23rd AIMTDR conf
11. Cakir MC, Ensarioglu C, Demirayak, I (2009) Mathematical modeling of surface roughness for evaluating the effects of cutting parameters and coating material. J Mater Process Technol 209:102–109
12. Yang YK, Chuang MT, Lin SS (2009) Optimization of dry machining parameters for high-purity graphite in end milling process via design of experiments methods. J Mater Process Technol 209:4395–4400
13. Fadheel KI, Tariq M (2014) Optimization of end milling parameters of AISI 1055 by Taguchi method. Int J Adv Res Eng Technol (IJARET) 5(3):09–20
14. Zhang JZ, Chen JC, Kirby ED (2007) Surface roughness optimization in an end-milling operation using the Taguchi design method. J Mat Processing Technol 187(4):233–239
15. El-Sonbaty IA, Khashaba UA, Selmy AL, Ali IA (2008) Prediction of surface roughness profiles for milled surfaces using an artificial neural network and fractal geometry approach. J Mater Process Technol 200:271–278
16. Dudzinski D, Devillez A, Moufki A, Larrouque're D, Zerrouki V, Vigneau J (2004) A review of developments towards dry and high speed machining of Inconel 718 alloy. Int J Mach Tools Manuf 44:439–456
17. Raju R, Manikandan N, Palanisamy D, Arulkirubakaran D, Sambathkumar S, Bhanu Prakash P (2018) Optimization of process parameters in electrical discharge machining of haste alloy C276 using Taguchi's method. Mater Today: Proc 5(6):14432–14439
18. Kose E, Kurt A, Seker U (2008) The effects of the feed rate on the cutting tool stresses in machining of Inconel 718. J Mater Process Technol 196(1–3) 165–173
19. Ezugwu EO (2005) Key improvements in the machining of difficult-to-cut aerospace superalloys. Int J Mach Tools Manuf 45(12–13):1353–1367
20. Thirugnanasambantham KG, Raju R, Sankaramoorthy T, Velmurugan P, Kannagi A, Reddy MCK, Chandra VR (2018) Degradation mechanism for high-temperature sliding wear in surface-modified In718 superalloy. Cogent Eng 5(1):1–11
21. Olovsjö S, Hammersberg P, Ståhl JE, Avdovic P, Nyborg L (2012) Methodology for evaluating effects of material characteristics on machinability—theory and statistics-based modeling applied on Alloy 718. Int J Adv Manuf Technol 59:55–66
22. Zhang JZ, Chen JC, Kirby ED (2007) Surface roughness optimization in an end-milling operation using the Taguchi design method. J Mater Process Technol 184:233–239
23. Nalbant M, Altin A, Gokkaya H (2007) The effect of coating material and geometry of cutting tool and cutting speed on machinability properties of Inconel 718 super alloys. Mater Des 28:1719–1724
24. Sivalingam V, Sun J, Yang B, Liu K, Raju R (2018) Machining performance and tool wear analysis on cryogenic treated insert during end milling of Ti-6Al-4V alloy. J Manuf Process 36:188–196
25. Darwish SM (2018) The impact of tool material and cutting parameters on surface roughness of supermet 718 nickel super alloy. J Mater Process Technol 97:10–18

Investigations on Wire Electrical Discharge Machining of Nickel-Based Superalloy Using Taguchi's Approach

N. Manikandan, J. S. Binoj, K. C. Varaprasad, P. Thejasree, and Ramesh Raju

Abstract Incoloy 800HT is one of the hard-to-machine materials and extensively employed in high-temperature applications. It possess better strength and lower thermal diffusion that results in poor machinability by conventional machining methods. To overcome such kind of demerits, unconventional methods of material removal process have been evolved and assumed to be an appropriate alternative method for conventional methods of machining. Wire electrical discharge machining (WEDM) is one amongst the variant evolved from the perception of electrical discharge machining that is predominantly adopted for machining of hard materials and also for making complicated shaped components. In this exploration, an experimental study has been done on WEDM of nickel-based superalloy and mainly focusing on optimizing the process variables for machining of Incoloy 800HT by using Taguchi's response analysis. The experimental runs are planned by Taguchi's method. The experimentation has been done by considering the independent process variables such as pulse on time, pulse off time and peak current at three different levels. Material removal rate (MRR) and surface roughness (SR) are the preferred as performance characteristics in this analysis. From this experimental analysis, the impact of each individual process parameters on the desired response parameters has been studied.

Keywords WEDM · Superalloy · Incoloy 800HT · Taguchi's method · S/N ratio analysis · Single objective optimization

N. Manikandan (✉) · J. S. Binoj · K. C. Varaprasad · P. Thejasree
Micro Machining Research Centre, Sree Vidyanikethan Engineering College (Autonomous), Tirupati 517102, Andhra Pradesh, India
e-mail: thuraiyurmani@gmail.com

R. Raju
Department of Mechanical Engineering, Santhiram Engineering College, Nandyal 518501, Andhra Pradesh, India

A. Arockiarajan et al. (eds.), *Advances in Industrial Automation and Smart Manufacturing*, Lecture Notes in Mechanical Engineering,
https://doi.org/10.1007/978-981-15-4739-3_22

1 Introduction

Wire spark-erosion machining or Wire Electrical Discharge Machining (WEDM) is one amongst the contemporary method of material removal and most familiar method used in various manufacturing industries for making the components such as turbine blades of aircraft engines, nozzles for fuel injectors and dies. WEDM has been employed for machining harder, corrosive and wear-resistant electrically conductive materials and also for hard-to-machine work materials. The concept of WEDM has been developed from the electrical discharge machining (EDM) and mainly adopted for making complicated shapes on the work materials which is not possible by the traditional method of material removal process [1–5]. In WEDM process, the removal of material takes place based on the electrodischarge erosion effect of electrical sparks occurred in amongst the electrode wire and the work material that are separated with the help of a dielectric fluid as shown in Fig. 1.

The voltage has been supplied in amid the electrode wire and selected work material with the existence of dielectric fluid for melting the work material surface by the discharge of sparks on the surface. Because of the running of electrode wire from one end to another end, continuously the material removal takes place [6–8].

Nickel-based superalloys are considered as predominant work materials in the manufacturing industries because of its inherent properties and advantages than titanium-based alloys. Heat-resistant and retaining the desired properties at high temperatures, high resistance to corrosion and high melting temperature are some imperative advantages of nickel-based superalloys [9–11]. Incoloy 800HT is comprehensively adopted in heat treating components such as baskets, fixtures and trays because of its better corrosion resistance properties. Furthermore, this material has the applications in petroleum processing industries as the material for heat exchanger [12]. Nickel-based superalloys execute a highly important role in aerospace applications, engines for rockets, nuclear reactors, thermal power plants, submarines and high temperature applications. Since these alloys are difficult to machine with

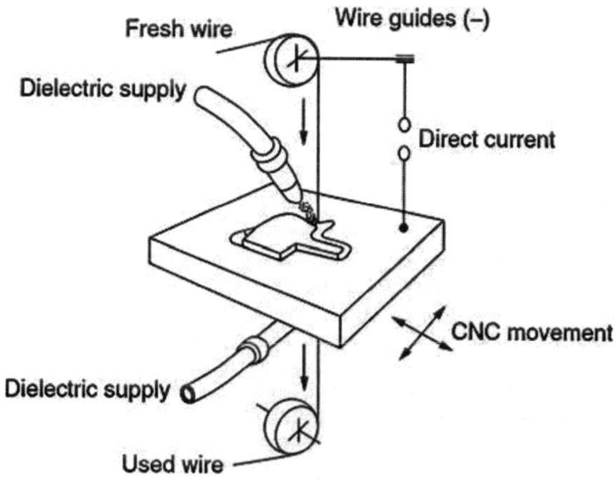

Fig. 1 Schematic of wire EDM (Ref: El-Hofy [7])

the help of traditional methods, the nontraditional methods have been proposed to machine these alloys. The complication of machining of these kinds of hard-to-machine materials by conventional process leads to the development of advanced machining process such as WEDM and also helps to make the materials that are not possible by conventional methods of material removal [13, 14].

An experimental investigation has been performed on WEDM process for analysing the machining performance using Inconel as work material. Taguchi's approach has been adopted for designing the experiments and also for single objective optimization. A multiperformance index has been derived by using appropriate method of multiple objective optimization [15]. Taguchi method has been employed for attaining better surface finish in WEDM of tool steel, and it is revealed from the investigation that the adopted method provides better improvement in the desired objective and the WEDM process has been modified for obtaining better machinability during machining H.S.S, titanium Nimonic and aluminium alloys as work material [16–19].

It is surmised from the available literature, the importance and machining of Incoloy 800HT by WEDM process desires attention. In this current exploration, an exertion was taken for establishing the prominence of process parameter for the preferred performance attributes named as material removal rate and surface roughness of the machined surface. Taguchi's approach has been engaged for designing the experiment trials and also for single objective optimization of desired performance measures.

2 Materials and Methods

Incoloy 800HT is a nickel-based superalloy that possesses exceptional temperature and corrosion resistance properties. Because of its outstanding properties, the work material has broader applications especially in heat treatment equipments, heat exchangers and super heater tubes in boilers. Incoloy 800HT has been opted as work material (size of 25 mm diameter and 150 mm length) in this proposed exploration, and it is clamped inner side of the machining chamber as illustrated in Fig. 2. WEDM machine (Concord Make—Model DK—7732) has been employed for experimentation purpose. Reusable molybdenum wire has been used as tool electrode, and de-ionized water is used as dielectric fluid in this present investigation.

In conventional approach of designing the experiments, there is a need of conducting more number of experimental trials with opted process parameters and levels. Such kind of issues are resolved by adopting the Taguchi's experimental design approach. An unique design of layout has been proposed by Taguchi for conducting the experiments that is called as orthogonal array (OA) and also for inspecting the independent process variables with the help of minimum set of experimental trials. Pulse on time (μs), Pulse off time (μs) and peak current (A) are chosen as independent variables based upon the available literature. The material removal rate (MRR) and surface roughness (SR) are opted as performance characteristics in

Fig. 2 Experimental set-up

Table 1 Machining process parameters and levels

Symbols	Variables	Levels		
		1	2	3
A	Pulse on time (μs)	10	20	30
B	Pulse off time (μs)	5	10	15
C	Peak current (A)	1	2	3

this present experimentation. The nominated independent process parameters, levels and the ranges are depicted in Table 1, and as per these values, an L_{27} OA has been chosen for wire EDM of Incoloy 800HT.

3 Results and Discussion

The experimental runs have been performed as per L_{27} OA for exploring the prominence of independent parameters on desired performance characteristics in wire EDM of Incoloy 800HT. An exertion was made to establish the best possible process variables for accomplishing the effective and capable wire EDM process. In the wire EDM, greater removal rate of material and lesser values of surface roughness are the indicator of superior performance measures. So MRR is considered as larger the better criterion and surface roughness is considered as smaller the better criterion.

3.1 Influence of Process Parameters on Material Removal Rate

Figure 3 illustrates the response graph for rate of material removal during machining of Incoloy 800HT. From the response graph, it is perceived that the rate of removal of material is getting increased with an improvement of values in pulse-on time, and also it is getting reduced with an increment in pulse-off-time. Also, it is witnessed that the pulse-on-time is utmost predominant process variable for the rate of material removal rate. With increasing values in pulse-on-time, the discharge energy given to the machining chamber will be more that consequences in an influential explosion that has the possibility of leading to an increment in the rate of material removal. Increasing in the values of pulse-on-time has the capability of getting increase in the amount of electrons that impacting on the work material, therefore it starts eroding more quantity of material from the surface of the work piece per spark discharge.

Taguchi's analysis for MRR has been accomplished, and the outcomes are presented in Table 2. The best set of process variables for attaining an improved material removal rate is $A3B1C3$. It means that the optimal process parameters combination for improved performance is 'P_{on}' – 30 μs; 'P_{off}' – 5 μs; peak current is 3 A. Also, it is perceived from the exploration that the 'P_{on}' is the momentous process variable, and then, it is trailed by 'P_{off}' and peak current.

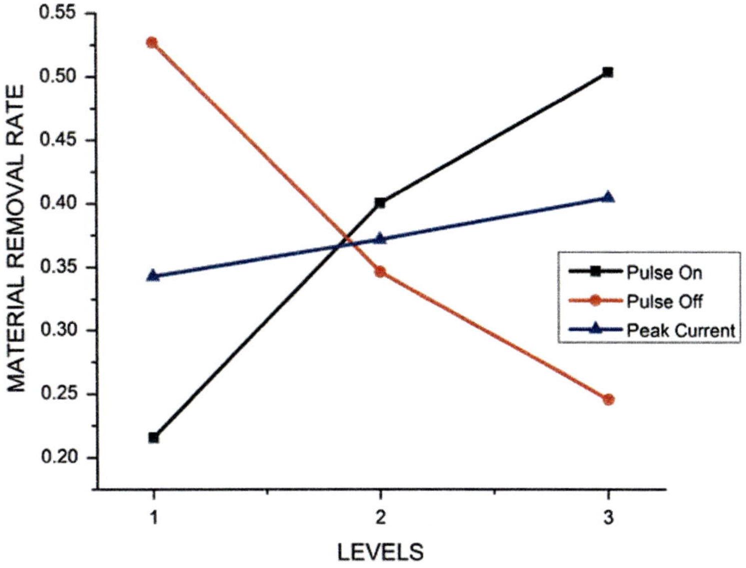

Fig. 3 Response analysis for material removal rate

Table 2 Response table for MRR—WEDM of Incoloy 800HT

Levels	Means			S/N Ratio		
	A	B	C	A	B	C
1	0.2157	0.5270	0.3428	−14.018	−6.025	−10.398
2	0.3999	0.3460	0.3714	−8.458	−9.834	−9.491
3	0.5028	0.2454	0.4042	−6.332	−12.948	−8.920
Delta	0.2871	0.2816	0.0614	7.687	6.923	1.478
Rank	1	2	3	1	2	3

3.2 *Influence of Process Parameters on Surface Roughness*

The graphical illustration of response graph for surface roughness during wire EDM of Incoloy 800HT is depicted in Fig. 4. It is perceived from the graph that the values of roughness of the electrically discharge machined surface getting increased with an increase of 'P_{on}' and the same has been getting diminished with an increment of 'P_{off}'. The superior energy of discharge provided crater in a greater manner that causes more roughness at the machined surface. The pulse on time is deemed as the supreme persuasive process variable for roughness of the machined surface.

Taguchi analysis for surface roughness has been executed, and the results are presented in Table 3. The best set of process variables for attaining better surface roughness is *A1B3C3*. It means that the optimal process variable combination for improved performance is 'P_{on}' − 10 μs; 'P_{off}' − 15 μs; peak current is 3 A. Also,

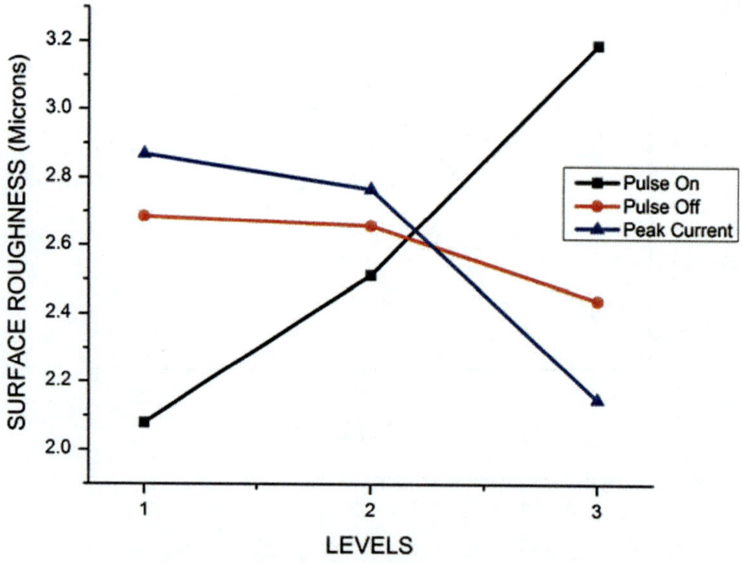

Fig. 4 Response graph for surface roughness

Table 3 Response table for surface roughness for Incoloy 800HT

Levels	MEANS			S/N RATIO		
	A	B	C	A	B	C
1	2.080	2.686	2.869	−6.123	−8.417	−8.908
2	2.513	2.658	2.766	−7.861	−8.275	−8.632
3	3.190	2.440	2.149	−9.958	−7.342	−6.493
Delta	1.11	0.246	0.72	3.745	1.075	2.415
Rank	1	3	2	1	3	2

it is witnessed from the analysis that the 'P_{on}' is the major variable, and then, it is trailed by peak current and pulse off time.

4 Conclusions

This current experimentation details a single aspect optimization problem of Wire-EDM of Incoloy 800HT with the help of Taguchi analysis. In this investigation, the rate of material removal (MRR) and roughness of machined surface (SR) were deemed as the performance characteristics and the conclusion attained are discussed as below:

- Taguchi's approach has been adopted for designing the experimental runs, and also, it has been adopted for optimizing the individual variables for accomplishing better performance in machining.
- The prevalence of deemed independent parameters on the preferred performance features was made known by Taguchi's analysis. Pulse on time is the major process variable for the better and improved performance characteristics.
- The detailed Taguchi's approach in this present exploration is greatly appropriate for establishing the best set of independent process parameters for attaining the improved performance in any advanced machining process.
- The outcome attained from this exploration will be a wide-ranging support to the manufacturers for enhancing the rate of production and quality of products made with an assistance of WEDM process.

References

1. Kanlayasiri K, Boonmung S (2007) Effects of wire-EDM machining variables on surface roughness of newly developed DC 53 die steel: design of experiments and regression model. J Mater Process Technol 192–193:459–464

2. Davim JP (2008) Machining fundamentals and recent advances. Springer-Verlag London Limited, British Library Cataloguing in Publication Data. https://doi.org/10.1007/978-1-84800-213-5
3. Patil N, Brahmankar PK (2010) Determination of material removal rate in wire electro-discharge machining of metal matrix composites using dimensional analysis. Int J Adv Manuf Technol 51(5–8):599–610. https://doi.org/10.1007/s00170-010-2633-3
4. Huang Y, Ming W, Guo J, Zhang Z, Liu G, Li M, Zhang G (2013) Optimization of cutting conditions of YG15 on rough and finish cutting in WEDM based on statistical analyses. Int J Adv Manuf Technol 69(5–8):993–1008. https://doi.org/10.1007/s00170-013-5037-3
5. Saha P, Tarafdar D, Pal S, Saha P, Srivastava A, Das K (2009) Modeling of wire electro-discharge machining of TiC/Fe in situ metal matrix composite using normalized RBFN with enhanced k means clustering technique. Int J Adv Manuf Technol 43(1–2):107–116. https://doi.org/10.1007/s00170-008-1679-y
6. Moulton DB (1999) Wire EDM the fundamentals. Sugar Grove, IL: EDM network (www.notebookmanuals.bestmanualguide.com)
7. El-Hofy H (2005) Advanced machining processes. McGraw-Hill. https://doi.org/10.1036/0071466940
8. Sommer C, Sommer S (2005) Complete EDM handbook. Advance Pub
9. Wu Q (2007) Serrated chip formation and tool-edge wear in high-speed machining of advanced aerospace materials. Utah State University, Logan, Utah
10. Saoubi RM, Outeiro JC, Chandrasekaran H, Dillon OW Jr, Jawahir IS (2008) A review of surface integrity in machining and its impact on functional performance and life of machined products. Int J Sustain Manuf 1:203–236
11. Guo YB, Li W, Jawahir IS (2009) Surface integrity characterization and prediction in machining of hardened and difficult-to-machine alloys; a state-of-the-art research review and analysis. Mach Sci Technol 13:437–470
12. Yilbas BS, Khaled M, Gondal MA, Ourfelli M, Khan Z (1999) Nano-second pulse laser treatment of Incoloy 800 HT alloy-corrosion properties. Opt Lasers Eng 32:157–172
13. Takayama Y, Makino Y, Niu Y, Uchida H (2016) The latest technology of Wire-cut EDM. Procedia CIRP 42:623–626
14. Manikandan N, Arulkirubakaran D, Palanisamy D, Raju R (2019) Influence of wire-EDM textured conventional tungsten carbide inserts in machining of aerospace materials (Ti–6Al–4 V alloy). Mater Manuf Processes 34(1):103–111
15. Yang CB, Lin CG, Chiang HL, Chen CC (2017) Single and multiobjective optimization of Inconel 718 nickel-based superalloy in the wire electrical discharge machining. Int J Adv Manuf Technol 93(9–12):3075–3084
16. Sudhakara D, Prasanthi G (2014) Application of Taguchi method for determining optimum surface roughness in wire electric discharge machining of P/M cold worked tool steel (Vanadis-4E). Procedia Eng 97:1565–1576
17. Ramesh NN, Harinarayana K, Naik BB (2014) Machining characteristics of HSS & Titanium using electro discharge sawing and wire-electrodischarge machining. Procedia Mater Sci 6:1253–1259
18. Goswami A, Kumar J (2014) Optimization in wire-cut EDM of Nimonic-80A using Taguchi's approach and utility concept. Eng Sci Technol Int J 17(4):236–246
19. Manikandan N, Binoj JS, Varaprasad KC, Sabari SS, Raju R (2019) Investigations on wire spark erosion machining of aluminum-based metal matrix composites. In: Advances in manufacturing technology. Springer, Singapore, pp 361–369

Comparison of Duplex Stainless Steel (2205) Spur Gears Cut by Wire Electrodischarge Machining (WEDM) and Hobbing Under Dry Condition

Mukesh Kumar Choudhary, V. Dhinakaran, and T. Jagadeesha

Abstract The present work deals with the influence of machining process on surface roughness and tool wear performances of spur gears under dry conditions, that is without any lubrication. Spur gears profiles are mathematically modelled and generated in CAD software. Using super-duplex stainless steel, spur gears has fabricated using hobbing and wire cut EDM processes. An experimental rig has developed to test the gear under dry operation conditions and variation of teeth wear and roughness has measured under dry conditions (without lubricants). Teeth of wire cut EDM gears are stronger compared to hobbed gear for same dimensions. Teeth surface roughness of the WEDM gears is much higher compared to hobbed gears. Initially (at lower RPM), wear rate of the WEDM gears teeth is little higher compared to hobbed gears because of higher surface roughness of WEDM gears.

Keywords Spur gears · WEDM · Hobbing · Dry conditions · Wear life

1 Introduction

In search of more corrosion resistant and mechanically superior materials for highly aggressive environments such as hot chlorinated seawater and highly acidic, chloride containing media, duplex stainless steels (DSSs) have emerged as the best suited materials. DSS 2205 (ASME Standard A240—UNS S32205) is a highly alloyed super-duplex stainless steel having high tensile strength, high fatigue strength, high toughness, high critical pitting temperature and corrosion resistance which sets it apart from other generations of DSSs. Hence, it is used in high end applications such as desalination equipment, chemical industries, pressure vessels, piping and

M. K. Choudhary · T. Jagadeesha (✉)
National Institute of Technology, Calicut, Kerala, India
e-mail: jagdishsg@nitc.ac.in

V. Dhinakaran
Department of Mechanical Engineering,
Chennai Institute of Technology, Kundrathur, Chennai 600069, India

A. Arockiarajan et al. (eds.), *Advances in Industrial Automation and Smart Manufacturing*, Lecture Notes in Mechanical Engineering,
https://doi.org/10.1007/978-981-15-4739-3_23

Table 1 Mechanical properties of duplex stainless steel 2205

Tensile strength (MPa)	Yield strength (MPa)	Hardness (HRC)
620	450	31

heat exchangers, tropical marine usages, offshore oil and gas exploration and petrochemical processing. Wire electrodischarge machining (WEDM) is a manufacturing system for high precision cutting of complex shapes from hard metals or those which are difficult to cut using conventional machining methods. Hobbing is one of the common methods employed in industries to manufacture gears. Electrical current and pulse on time are two process parameters in WEDM. Several researchers suggested manufacturing of spur gears using WEDM processes. Jose Luis et al. manufactured spur tooth gear made of Ti6Al4V alloy using WEDM processes [1]. A geometrical model of the gear was developed using MATLAB code by them. Equation of geometrical conversion of involute curve for straight line segments demonstrated by Lee and Lee [2]. Yanzhongwang [3] optimized the spray parameters of oil jet lubricated spur gears. When the gear rotating with high speed, lubricating oil cannot reach to meshing zone effectively [4–6]. Mohapatra et al. carried out modelling and analysis of WEDM cutting processes. [7]. Thermophysical model for WEDM was developed by Shahane and Pande [8]. Kahraman [9] developed time-dependent mixed elastohydrodynamic lubrication system for spur gears. Very hard and difficult to cut spur gears can be manufactured using WEDM processes [10–17]. From the literature survey, it is observed that there exists a research opportunities in modelling, manufacturing and performance analysis of duplex stainless steel gears.

2 Experimentation

2.1 Material Selection

Super duplex stainless steel (SDSS) G2205 is used as the workpiece material for the gear manufacture. Properties of SDSS are given in Table 1.

2.2 Wire EDM Machine Selection and Process Details

DK7735 CNC wire cut EDM machine is used to manufacture spur gear which is shown in Fig. 1a. Gears are produced by controlled erosion of electrically conductive materials such as duplex stainless steel by the initiation of rapid and repetitive spark discharge between the electrode tool and the workpiece separated by a small gap (0.02 mm). A voltage (80 V) is applied between the tool and the workpiece which are separated by dielectric fluid. Electrons from tool collide with the dielectric molecules

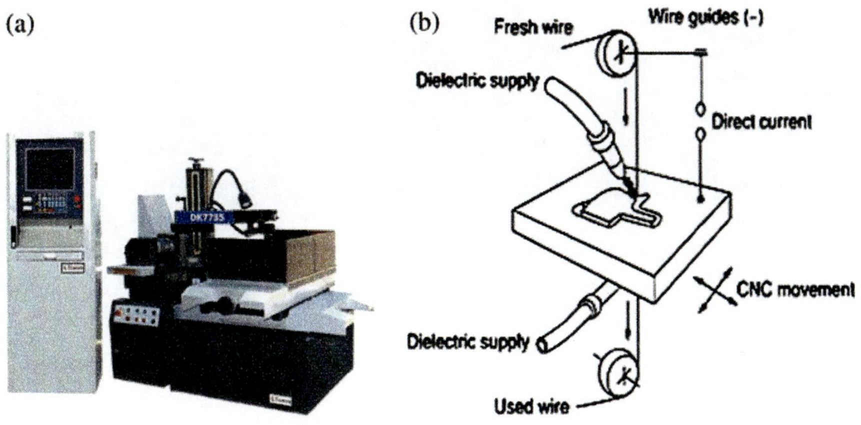

Fig. 1 **a** CNC wire cut EDM machine. **b** WEDM process

which results in formation of positive ions and electrons. These electron flows create compression shock on work surface which results in the rise in temperature up to 10000 °C. Heat generated melts and vaporizes the material. Low viscosity deionized water is used as dielectric. Machine consists of 4 axes movement. Workpiece can be moved along x and y direction. The wire guide can be moved u and v direction. The desired taper angle is achieved by simultaneous control of movement of XY table and UV table with the help of controller as shown in Fig. 1b. Adaptive control system is used for maintaining constant gap between wire and work.

2.3 *Specification of Gears*

For fabrication of WEDM gears, molybdenum wire of diameter 0.18 mm has used with the spark gap of 0.02 mm. Pulse on and off time is 20 and 7 microseconds, respectively. Voltage and current are 80 V and 2.7 amperes, respectively. Wire tension is 900 g, and wire moving speed is 11.5 m/s. Hob cutter of high carbon steel has used for fabrication of hobbed spur gear. Spur gears are manufactured in hobbing and CNC wire cut EDM machine. Typical gears are shown in Fig. 2a, b, respectively. Addendum and dedendum diameters are 64 mm and 56.38 mm, respectively. Module of the gear is 2 mm; pressure angle is 20 deg; and number of teeth is 30. Tooth thickness for hobbed gear is 2.94 mm, and that of the WEDM is 3.14 mm. Two pairs of spur gear have fabricated. One pair of WEDM gears is marked as 1(driving) and 2(driven), and other pair hobbed gears has also marked as 1 (driving) and 2 (driven).

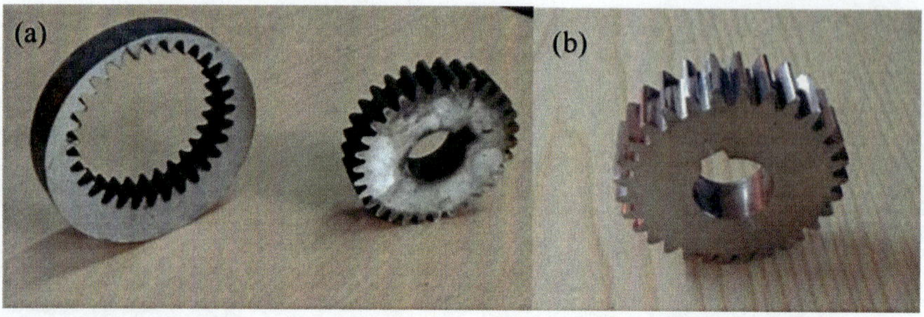

Fig. 2 **a** Fabricated WEDM spur gear **b** fabricated hobbed spur gear

2.4 Experimental Set-Up

Experimental set-up is developed and is shown in Fig. 3. 0.5 HP three-phase AC motor has used with variable frequency controller for controlling the speed of motor. Two parallel shafts are supported on bearing pedestal via love-jaw coupling. Spur gears are mounted on two solid shafts. The speed of the motor can be controlled by using variable frequency controller. Speeds used in these experiments are 100, 200, 300, 400, 500, 600, 700 and 800 RPMs.

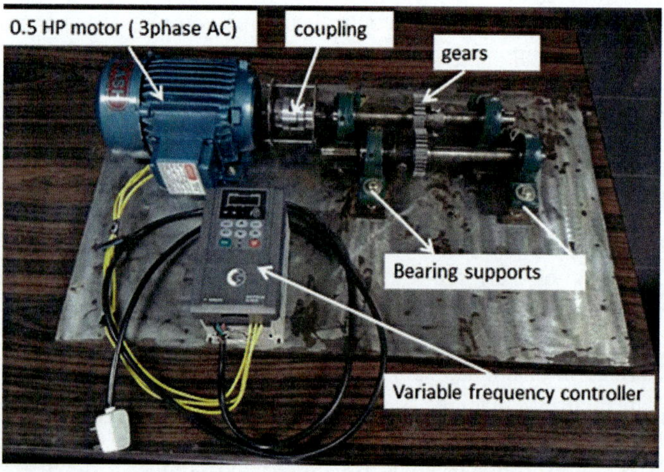

Fig. 3 Schematic representation of experimental set-up

3 Results and Discussion

3.1 Roughness and Tooth Thickness of WEDM Gears Under Dry Condition

Two WEDM gears were run for six hours at 100, 200, 300, 400, 500, 600, 700 and 800 RPMs. At the end of each six hours, roughness and tooth thickness are measured. Table 2 gives the roughness and tooth thickness. Averages of 2 readings at two spots are taken in each iteration. Ra (1) and Ra (2) indicate the average roughness values for gear 1 and 2, respectively. T (1) and T (2) are the tooth thickness for gear 1 (driving) and 2 (driven), respectively. It is found that as the RPM increases, the average roughness decreases in dry condition. The teeth thickness decreases as the RPM increases.

3.2 Roughness and Tooth Thickness of Hobbed Gears Under Dry Condition

Two hobbed gears are run for six hours at 100, 200, 300, 400, 500, 600, 700 and 800 RPMs. At the end of each six hours, roughness and tooth thickness are measured. Table 3 gives the roughness and tooth thickness. Averages of 2 readings at two spots have taken in each iteration. Ra (1) and Ra (2) indicate the average roughness values for gear 1 and 2, respectively. *T* (1) and *T* (2) are the tooth thickness for gear 1(driving) and 2 (driven), respectively. It is found that as the RPM increases, the average roughness increases in dry condition. The tooth thickness decreases as the RPM increases.

3.3 Tooth Profiles at the Beginning and End of the Test in Dry Condition

Tooth profiles at the end of each RPM test are taken using ALICONA. Figure 4 shows the tooth profile after 100 RPM test and after 800 RPM test. Figure 4a shows the tooth profile of WEDM driving gear after 100 RPM test (after 6 h), Fig. 4b shows the tooth profile of WEDM driving gear after 800 RPM test (after 48 h), Fig. 4c shows the tooth profile of hobbed gears after 100 RPM test (after 6 h), and Fig. 4d shows the tooth profile of hobbed gears after 800 RPM test (after 48 h). Some pit marks are observed (at face of gears) after 800 RPM test (i.e. after 48 h of running). In both WEDM gears and hobbed gears, pit marks are observed to be at the same spot.

Table 2 Roughness and teeth thickness of WEDM gears at different RPMs

No. of iteration	RPM	Ra (1)	Avg. Ra (1)	T (1)	T_a (1)	Ra (2)	Avg. Ra (2)	T (2)	T_a (2)
0	000	6.977, 6.897	6.937	3.14, 3.14	3.14	6.988, 6.869	6.942	3.14, 3.14	3.14
1	100	5.716, 5.819	5.768	3.14, 3.14	3.14	6.100, 5.810	5.955	3.14, 3.12	3.13
2	200	5.416, 5.819	5.618	3.12, 3.14	3.13	5.400, 5.510	5.455	3.14, 3.12	3.13
3	300	5.210, 5.110	5.160	3.12, 3.14	3.13	5.110, 5.170	5.140	3.12, 3.12	3.12
4	400	4.516, 4.550	4.990	3.12, 3.12	3.12	4.880, 4.918	4.899	3.10, 3.12	3.11
5	500	4.770, 4.520	4.533	3.10, 3.12	3.11	4.660, 4.513	4.587	3.10, 3.10	3.10
6	600	4.770, 4.520	4.645	3.10, 3.08	3.09	4.670, 4.440	4.555	3.08, 3.08	3.08
7	700	3.990, 3.976	3.983	3.08, 3.08	3.08	3.918, 3.880	3.899	3.06, 3.08	3.07
8	800	3.440, 3.142	3.291	3.04, 3.04	3.04	3.480, 4.000	3.480	3.04, 3.04	3.04

Table 3 Roughness and teeth thickness of hobbed gears at different RPMs

No. of iteration	RPM	Ra (1)	Avg. Ra (1)	*T* (1)	T_a (1)	Ra (2)	Avg. Ra(2)	*T* (2)	T_a (2)
0	000	0.225, 0.234	0.230	2.94, 2.94	2.94	0.230, 0.228	0.229	2.94, 2.94	2.94
1	100	0.240, 0.244	0.242	2.94, 2.94	2.94	0.242, 0.240	0.241	2.94, 2.94	2.94
2	200	0.266, 0.260	0.263	2.94, 2.92	2.93	0.260, 0.262	0.261	2.94, 2.94	2.94
3	300	0.298, 0.300	0.299	2.92, 2.94	2.93	0.302, 0.306	0.304	2.94, 2.92	2.93
4	400	0.398, 0.390	0.349	2.90, 2.90	2.90	0.399, 0.370	0.385	2.92, 2.94	2.93
5	500	0.467, 0.468	0.468	2.90, 2.90	2.90	0.498, 0.460	0.479	2.90, 2.88	2.89
6	600	0.499, 0.560	0.530	2.88, 2.90	2.89	0.542, 0.560	0.551	2.86, 2.86	2.86
7	700	0.706, 0.799	0.754	2.88, 2.86	2.87	0.709, 0.810	0.760	2.84, 2.86	2.85
8	800	0.860, 0.880	0.870	2.86, 2.84	2.85	0.854, 0.752	0.803	2.84, 2.84	2.84

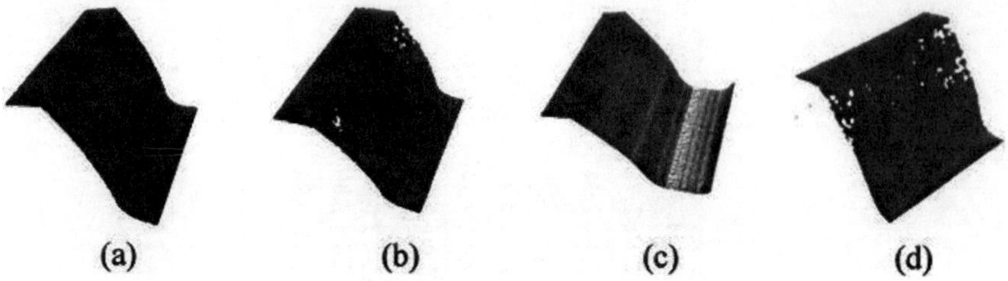

Fig. 4 Teeth profile representation after 100 and 800 RPMs in dry condition

Table 4 Wear rates of WEDM and hobbed (machined) gears under dry condition

		Wear rate of gear teeth in mm			
No. of iteration	Gear speed in (RPM)	WEDM (1)	Machined (1)	WEDM (2)	Machined (2)
0	000	0	0	0	0
1	100	0	0	0.01	0
2	200	0.01	0.01	0.01	0
3	300	0.01	0.01	0.02	0.01
4	400	0.02	0.04	0.03	0.01
5	500	0.03	0.04	0.04	0.05
6	600	0.05	0.05	0.06	0.08
7	700	0.06	0.07	0.07	0.09
8	800	0.1	0.09	0.1	0.1

3.4 Wear Test Under Dry Condition

At the end of the each run, gear is dismantled and tooth thickness is measured. Wear is calculated based on the difference in tooth thickness. Table 4 shows the wear rate of machined and WEDM gears. (1) indicates driver gear 1, and (2) indicates driven gear 2. It can be observed that after running at 100 RPMs, loss of the tooth thickness for both gears is negligible. Loss of teeth thickness of the machined gear is higher at 700 RPMs, and behaviour of wear for both gears is non-linear at various RPM values.

4 Conclusions

The influence of two machining processed gears on roughness and wear performances is investigated in this work. Three-dimensional spur gears are modelled in solid works and converted into 2D CAD model to give an input for CNC wire cut EDM machine. Using Super-duplex stainless steel spur gears are cut using hobbing and wire cut

EDM processes. An experimental set-up has developed to test the gear under various operation conditions, and variation of wear rate and roughness has measured under dry conditions. Teeth of wire cut EDM gears are stronger compared to hobbed gears for same dimensions. Teeth surface roughness of the WEDM gears is much higher than hobbed gears. Wear of the WEDM gears is lower compared to hobbed gears at higher RPM because of more accurate involute profile. For same module and pitch circle radius, WEDM gears have higher teeth thickness. Therefore, these gears are stronger and longer life compared to hobbed gears.

References

1. Talón JL, Ortega JC, Gómez CL, Sancho ER, Olmos EF (2010) Manufacture of a spur tooth gear in Ti–6Al–4V alloy by electrical discharge. Comput Aided Des 42(3):221–230
2. Lee RS, Lee JN (2001) A new tool-path generation method using a cylindrical end mill for 5-axis machining of a spatial cam with a conical meshing element. Int J Adv Manuf Technol 18(9):615–623
3. Lugt PM, Morales-Espejel GE (2011) A review of elasto-hydrodynamic lubrication theory. Tribol Trans 54(3):470–496
4. Biboulet N, Colin F, Lubrecht AA (2013) Friction in starved hydrodynamically lubricated line contacts. Tribol Int 58:1–6
5. Mishra SP, Polycarpou AA (2011) Tribological studies of unpolished laser surface textures under starved lubrication conditions for use in air-conditioning and refrigeration compressors. Tribol Int 44(12):1890–1901
6. Mohapatra KD, Shaibu VB, Sahoo SK (2018) Modeling and analysis of Wire EDM in a Gear Cutting process for a 2D Model. Mater Today: Proceedings 5(2):4793–4802
7. Shahane S, Pande SS (2016) Development of a thermo-physical model for multi-spark wire EDM process. Procedia Manuf 5:205–219
8. Li S, Kahraman A (2010) A transient mixed elastohydrodynamic lubrication model for spur gear pairs. J Tribol 132(1):011501
9. Gamage JR, DeSilva AK (2016) Effect of wire breakage on the process energy utilisation of EDM. Procedia Cirp 42:586–590
10. Joshi SN, Pande SS (2009) Development of an intelligent process model for EDM. Int J Adv Manuf Technol 45(3–4):300
11. Li L, Li ZY, Wei XT, Cheng X (2015) Machining characteristics of Inconel 718 by sinking-EDM and wire-EDM. Mater Manuf Process 30(8):968–973
12. Mandal A, Dixit AR, Das AK, Mandal N (2016) Modeling and optimization of machining nimonic C-263 superalloy using multicut strategy in WEDM. Mater Manuf Process 31(7):860–868
13. Manikandan N, Arulkirubakaran D, Palanisamy D, Raju R (2019) Influence of wire-EDM textured conventional tungsten carbide inserts in machining of aerospace materials (Ti–6Al–4V alloy). Mater Manuf Process 34(1):103–111
14. Manikandan N, Binoj JS, Varaprasad KC, Sabari SS, Raju R (2019) Investigations on wire spark erosion machining of aluminum-based metal matrix composites. Lecture notes in mechanical engineering. In: Advances in Manufacturing Technology, Springer, Singapore, pp 361–369
15. Doğan O, Karpat F (2019) Crack detection for spur gears with asymmetric teeth based on the dynamic transmission error. Mech Mach Theory 133:417–431
16. Hammami M, Fernandes CM, Martins R, Abbes MS, Haddar M, Seabra J (2019) Torque loss in FZG-A10 gears lubricated with axle oils. Tribol Int 131:112–127
17. Wen Q, Du Q, Zhai X (2018) A new analytical model to calculate the maximum tooth root stress and critical section location of spur gear. Mech Mach Theory 128:275–286

Modeling and Optimization of Machining Parameters for Turning of Mild Steel Using Single-Point Cutting Tool Made of P20 Tool Steel

S. Dinesh, V. Vijayan, A. Parthiban, C. Saravanan, and B. Suresh Kumar

Abstract The work deals with optimization of the feasible parameters and identification of the most feasible parameter for turning operation using the response surface methodology, Grey–Taguchi method, and ANOVA. Mild steel was turned using a single-point cutting tool made of P20 tool steel by varying the parameters such as depth of cut, feed rate, and rake angle. L27 array was used to conduct the experiments. The responses such as material removal rate, surface roughness, and cutting force were measured. The optimized parameter is identified using Grey–Taguchi method. RSM is used to carry interaction study, and the most critical parameter is identified for ANOVA.

Keywords P20 steel · Response surface methodology · Grey–Taguchi method · ANOVA

1 Introduction

A secondary manufacturing process that is used to achieve the required shape through material removal from the workpiece diametrically is called turning. Single-point cutting tools made of HSS and inserts of various materials have been used to carry out turning operation. In certain conditions, the profile of the tools has been changed to improve the process. This change in tool profiles has improved the performance of the process by reducing the stress concentration at the cutting edges and reduced the

---

S. Dinesh (✉) · V. Vijayan · B. Suresh Kumar
Department of Mechanical Engineering, K. Ramakrishnan College of Technology, Kariyamanikam Road, Samayapuram, Tiruchirappalli, Tamil Nadu 621112, India
e-mail: dineshmechian@gmail.com

A. Parthiban
Department of Mechanical Engineering, Vels Institute of Technology & Advanced Studies, Chennai, Tamil Nadu 600117, India

C. Saravanan
Department of Mechanical Engineering, University College of Engineering (BIT Campus), Tiruchirappalli 620024, India

A. Arockiarajan et al. (eds.), *Advances in Industrial Automation and Smart Manufacturing*, Lecture Notes in Mechanical Engineering,
https://doi.org/10.1007/978-981-15-4739-3_24

Table 1 Different levels and parameters

Level/parameters	Level 1	Level 2	Level 3
Feed (rev/min)	0.01	0.015	0.02
Rake angle (°)	7	8	9
Depth of cut (mm)	0.5	1	1.5

tool wear up to an extend [1]. Materials such as HSS, PCBN, cemented carbide, and ceramics have been used for tool materials [2]. The pattern of the microtexture and the optimal tool material selection reduced tool wear, cutting force, and surface finish [3]. Surface quality improvement and cutting force reduction can be accomplished by proper selection of controlling parameters. The parameters were optimized for MQL to reduce the cutting force and improve surface quality [4]. The cutting force and cutting temperature are reduced up to 56.1 and 27.1% by cutting microgrooves of higher thickness in the tool [5]. Toll wear was reduced with increased hardness at flank on Haynes 282 superalloy [6]. The effect of input parameters over output can be determined using techniques such as Arbitrary Lagrangian–Eulerian (ALE) formulation approach, ANOVA, etc. [7–9].

2 Materials, Methods, and Experimentation

The tool material selected for this turning process is steel P20 material. The steel P20 material is a high tensile material with high chromium content. The dimension of the material is 60 mm in length and 24 mm in diameter. Center lathe was used for machining the workpiece. The tool material was quenched and ground using tool grinder. The input parameters are tabulated in Table 1. The responses such as material removal rate, surface roughness, and cutting force were measured. Cutting force was measured using lathe tool dynamometer and SJ210 surface roughness tester was used for measuring surface roughness. The experimental results are tabulated in Table 2. Figure 1 shows the hardened tool made of P20 tool steel.

3 Interpretation of Results

Response surface methodology is used for developing the prediction model and interaction study of the critical parameters that govern the responses [10–17]. Regarding SR, an increase in SR is observed with the increase in depth of cut with the least SR being achieved at minimum feed rate. The increase in SR is linear in nature. A considerable decrease in SR has been encountered with increase in rake angle. The minimum SR is achieved at minimum depth of cut, higher rake angle, and minimal feed rate as shown in Fig. 2a–c.

Table 2 Experimental results

Input control parameters				Output responses		
Trial No.	Feed rate (mm/rev)	Rake angle (°)	Depth of cut (mm)	Surface roughness (microns)	Cutting force (kgf)	MRR (gm/min)
1	0.01	9	0.5	1.57	11.6	0.003
2	0.01	9	1	1.77	18.5	0.004
3	0.01	9	1.5	2.38	35.7	0.005
4	0.01	8	0.5	1.54	13.9	0.006
5	0.01	8	1	1.83	18.4	0.008
6	0.01	8	1.5	2.18	19.0	0.017
7	0.01	7	0.5	1.59	10.4	0.013
8	0.01	7	1	1.69	16.4	0.012
9	0.01	7	1.5	2.44	20.0	0.019
10	0.015	7	0.5	3.74	11.6	0.012
11	0.015	7	1	3.56	18.5	0.018
12	0.015	7	1.5	2.15	36.0	0.020
13	0.015	8	0.5	3.68	16.3	0.004
14	0.015	8	1	4	30.5	0.019
15	0.015	8	1.5	2.99	26.8	0.019
16	0.015	9	0.5	3.1	14.2	0.008
17	0.015	9	1	6.26	15.0	0.011
18	0.015	9	1.5	2.87	25.3	0.008
19	0.02	9	0.5	2.08	18.4	0.005
20	0.02	9	1	1.96	24.6	0.014
21	0.02	9	1.5	2.6	28.4	0.018
22	0.02	8	0.5	2.14	11.7	0.008
23	0.02	8	1	1.9	19.8	0.007
24	0.02	8	1.5	2.91	28.0	0.017
25	0.02	7	0.5	1.69	13.2	0.003
26	0.02	7	1	1.85	20.5	0.007
27	0.02	7	1.5	1.74	24.9	0.018

The cutting force has varied in hyperbolic fashion which shows that the cutting force is minimal at the mid-span of the rake angle, feed rate, and depth of cut. A small exponential increase in cutting force is observed until the mid-span of depth of cut and cutting force increases at faster rate exponentially thereafter as shown in Fig. 3a–c. MRR is maximum at minimal feed rate, maximum depth of cut, and rake angle. An exponential increase in MRR is achieved with increasing feed rate, and MRR has decreased with increasing rake angle as shown in Fig. 4a–c.

Fig. 1 Hardened P20 toll

ANOVA is used to pick the most contributing parameters for MRR, SR, and cutting forces [17]. Depth of cut is identified as the most controllable parameter MRR and cutting force. The models developed for SR and cutting force had a p value of <0.0001, and for model developed for MRR, it was found to be 0.0002. The models developed were significant with a 5.7, 12 and 5.34 for MRR, SR, and cutting force. The secondary control parameter which had significant effect over MRR was rake angle and feed rate for SR and cutting force. The R-squared value for MRR and cutting force was 0.77 and 0.81 which is lesser than 0.91 for SR. This proves the developed model shall be used for predicting the desired responses with the degree of accuracy being higher for SR and lowest for MRR. The adjusted R squared value also follows the same trend as R-squared values for all the responses. The ANOVA table for MRR, SR and Cutting force are shown in Tables 3, 4 and 5.

The feasible combination of parameters was identified using Grey–Taguchi optimization technique. The Grey relational grade was calculated based on the S/N ratio obtained for the responses. Plot between GRG and the levels of various parameters is shown in Fig. 5. This infers that maximum productivity with minimal SR and cutting force can be achieved by posting a feed rate of 0.015 mm/rev, rake angle of 7°, and 1.5 mm of depth of cut. Table 6 infers the percentage of contribution of the input parameters in enhancement of the responses. The major contributor has been identified as the depth of cut followed by feed rate and the least contributor being rake angle.

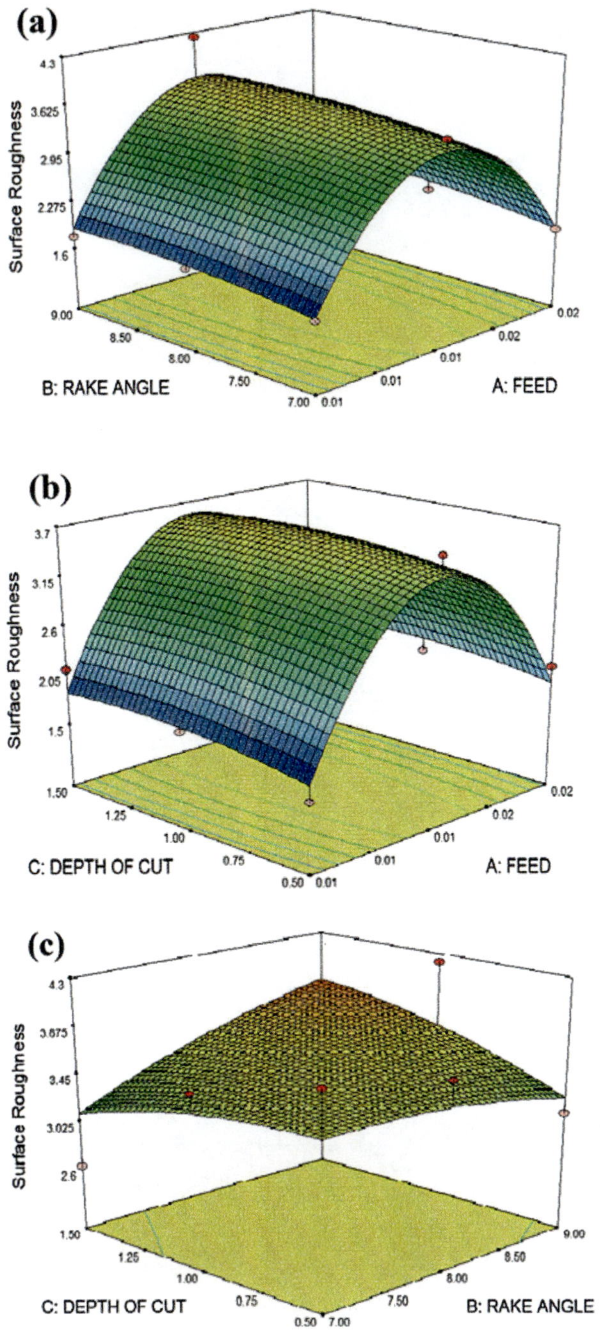

Fig. 2 Effect of controlling parameters over surface roughness

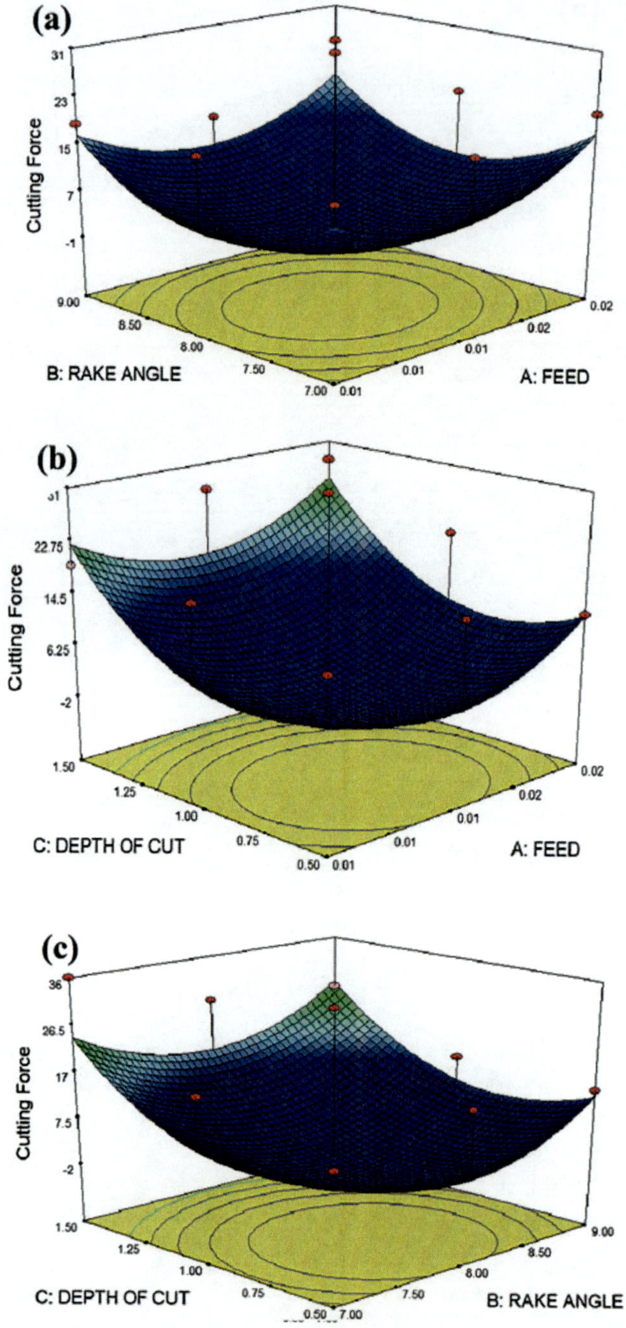

Fig. 3 Effect of controlling parameters over cutting force

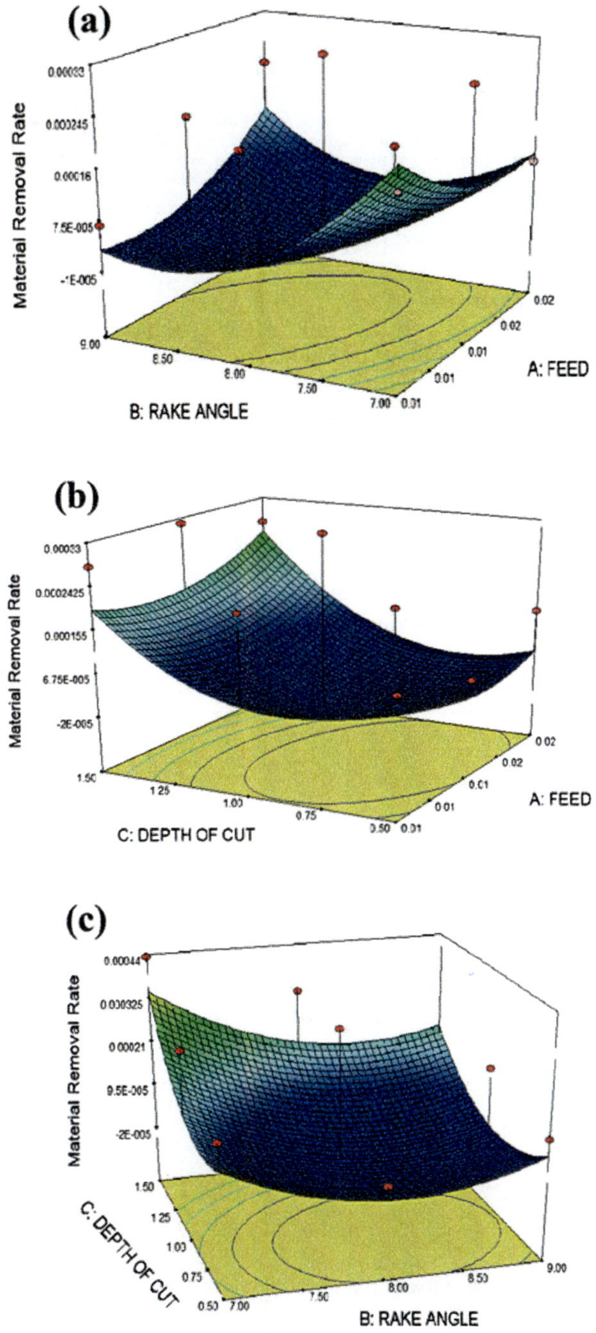

Fig. 4 Effect of controlling parameters over MRR

Table 3 ANOVA for material removal rate

Source	Sum of squares	df	Mean square	*F* Value	*p*-value Prob > *F*	Source
Model	1E−06	9	1.14E−07	7.05	0.0002	Significant
A-Feed rate	3E−10	1	2.73E−10	0.02	0.90	
B-Rake angle	4E−08	1	4.13E−08	2.55	0.13	
C-Depth of cut	1E−07	1	1.06E−07	6.55	0.02	
AB	4E−08	1	4.06E−08	2.51	0.13	
AC	7E−09	1	7.27E−09	0.45	0.51	
BC	9E−09	1	8.63E−09	0.53	0.47	
A^2	2E−08	1	1.82E−08	1.12	0.30	
B^2	7E−08	1	6.93E−08	4.28	0.05	
C^2	9E−08	1	8.99E−08	5.55	0.03	
Residual	3E−07	18	1.62E−08			
Total	1E−06	27				

Table 4 ANOVA for surface roughness

Source	Sum of squares	df	Mean square	*F* value	*p*-value Prob > *F*	Source
Model	16.850	9	1.87	20.1	<0.0001	Significant
A-Feed rate	0.213	1	0.21	2.3	0.1487	
B-Rake angle	0.289	1	0.29	3.1	0.0964	
C-Depth of cut	0.061	1	0.06	0.7	0.4289	
AB	0.018	1	0.02	0.2	0.6691	
AC	0.003	1	0.00	0.0	0.8523	
BC	0.745	1	0.75	8.0	0.0116	
A^2	15.424	1	15.42	165.4	<0.0001	
B^2	0.038	1	0.04	0.4	0.5296	
C^2	0.058	1	0.06	0.6	0.4411	
Residual	1.585	17	0.09			
Total	18.435	26				

4 Conclusion

L 27 full factorial array was used to conduct experimentation on mild steel rod with newly developed single point cutting tool made if P20 tool steel with varying rake angles. The results were studied with response surface methodology, ANOVA, and Grey–Taguchi's method using which the following findings were recorded. Depth of cut was identified as the most contributing parameter for MRR and cutting force. Feed

Table 5 ANOVA for cutting force

Source	Sum of squares	df	Mean square	*F* value	*p*-value Prob > *F*	Source
Model	10,200.9	9	1133.4	9.09	<0.0001	Significant
A-Feed rate	36.9	1	36.9	0.30	0.593	
B-Rake angle	22.7	1	22.7	0.18	0.6745	
C-Depth of cut	838.5	1	838.5	6.73	0.0184	
AB	3.3	1	3.3	0.03	0.8724	
AC	0.1	1	0.1	0.00	0.9798	
BC	0.0	1	0.0	0.00	0.9879	
A^2	458.8	1	458.8	3.68	0.0711	
B^2	680.6	1	680.6	5.46	0.0312	
C^2	736.4	1	736.4	5.91	0.0258	
Residual	2244.0	18	124.7			
Total	12,444.9	27				

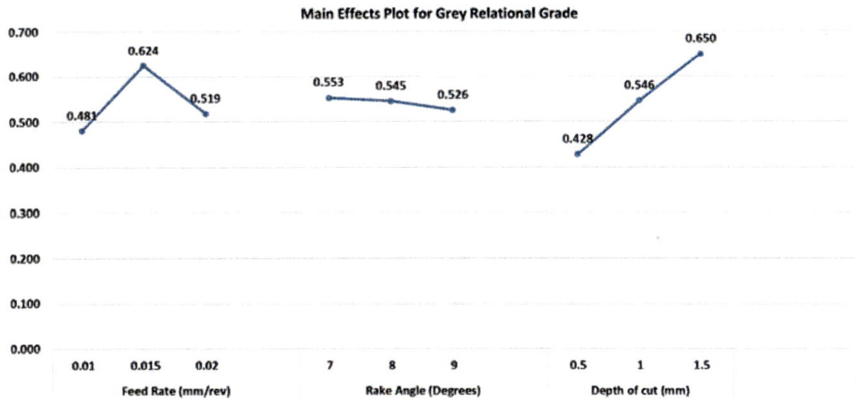

Fig. 5 Main effect plot for Grey relational grade

Table 6 Contribution of controlling parameters

Factors	DoF	SS	MS	% Contribution
Feed rate	2	0.0994	0.0497	22.53
Rake angle	2	0.0035	0.0018	0.80
Depth of cut	2	0.2227	0.1114	50.51
Error	20	0.1153	0.0058	26.16
Total	26	0.4410	0.0170	100.00

rate was identified to have effective control over surface roughness. The secondary control parameter which had significant effect over MRR was rake angle and feed rate for SR and cutting force. The developed models which were significant can be used for predicting the responses, but the model being more effective of predicting SR followed by cutting force and MRR. The optimal conditions for enhancement of productivity and quality were found to be the lower rake angle of 7 degree, higher depth of cut of 1.5 mm, and 0.015 mm of feed. Upon analysis of the results based on S/N ratio, major contributor has been identified as the depth of cut followed by feed rate and the least contributor being rake angle.

References

1. Liu G, Huang C, Rui S, Ozel T, Liu Y, Xu L (2019) 3D FEM simulation of the turning process of stainless steel 17-4PH with differently texturized cutting tools. Int J Mech Sci. https://doi.org/10.1016/j.ijmecsci.2019.03.016
2. Sirtuli LJ, Boing D, Schroeter RB (2019) Evaluation of layer adhered on PCBN tools during turning of AISI D2 steel. Int J Refract Met Hard Mater. https://doi.org/10.1016/j.ijrmhm.2019.104977
3. Cui X, Jiao F, Ming P, Guo J (2017) Reliability analysis of ceramic cutting tools in continuous and interrupted hard turning. Ceram Int 43(13):10109–10122. https://doi.org/10.1016/j.ceramint.2017.05.032
4. Patole PB, Kulkarni VV (2015) Optimization of process parameters based on surface roughness and cutting force in MQL turning of AISI 4340 using nano fluid. Procedia Manuf 1:454–465
5. Suárez A, Veiga F, de Lacalle LL, Polvorosa R, Wretland A (2019) An investigation of cutting forces and tool wear in turning of Haynes 282. J Manuf Processes 37:529–540
6. Liang X, Liu Z, Yao G, Wang B, Ren X (2019) Investigation of surface topography and its deterioration resulting from tool wear evolution when dry turning of titanium alloy Ti-6Al-4V. Tribol Int. https://doi.org/10.1016/j.triboint.2019.02.049
7. Sadílek M, Dubský J, Sadílková Z, Poruba Z () Cutting forces during turning with variable depth of cut, Perspect Sci 7:357—363 (2016)
8. Zhang G, Guo C (2016) Modeling of cutting force distribution on tool edge in turning process. Procedia Manuf 1:454–465
9. Kuntoğlu M, Sağlam H (2019) Investigation of progressive tool wear for determining of optimized machining parameters in turning. Measurement. https://doi.org/10.1016/j.measurement.2019.04.022
10. Yokeswaran R, Karuppusamy S, Arul S (2018) Experimental investigation and optimization of welding parameters of metal inert gas welding process in joining dissimilar metals. Int J Mech Prod Eng Res Dev 8(1):415–422
11. Dillibabu V, Duraiselvam M, Chandrasekhar U, Raju R (2017) Microstructural studies on laser dissimilar welded Ni and steel alloys for aeronautical turbine applications. Lasers Eng 37(4–6):247–260
12. Senthilkumar N, Ganapathy T, Tamizharasan T (2014) Optimisation of machining and geometrical parameters in turning process using Taguchi method. Aust J Mech Eng 12(2):233–246
13. Raju R, Sivalingam V, Sun J, Natarajan M, Zhao Y (2019) Experimental and Taguchi-based grey approach of laser metal deposition technique on nickel-based superalloy. Trans Indian Inst Met 72(1):205–214
14. Tamizharasan T, Senthil Kumar S, Selvkumar V, Dinesh S (2019) Taguchi's methodology of optimizing turning parameters over chip thickness ratio in machining PM AMMC. SN Appl Sci 1:160

15. Manikandan N, Raju R, Palanisamy D, Arulkirubakaran D, Kumar S (2018) Investigation on Ti6Al4V laser metal deposition using Taguchi based grey approach. Mater Today Proc 5(6):14375–14383
16. Dinesh S, Godwin Antony A, Rajaguru K, Vijayan V (2016) Experimental investigation and optimization of material removal rate and surface roughness in CNC turning of EN24 alloy steel. Asian J Res Soc Sci Humanit 6(8):849–863
17. Saravanan S, Ravichandran M, Stalin B, Saravanavel S, Sukumar S (2019) Optimization of process parameters of electrochemical machining of TiC-Reinforced AA6063 composites. Lecture Notes in Mechanical engineering. https://doi.org/10.1007/978-981-13-6374-0_33

Application of Entropy—Deng's Similarity Approach for Optimization of Single-Point Incremental Forming Process Parameters of Titanium Grade 2 Sheets

G. Yoganjaneyulu and C. Sathiya Narayanan

Abstract Incrementally forming sheet metals using a single ball-ended tool is one of steadfast areas in which industry is capable of focusing its attention in the future prospect appropriate to huge opportunities existing in automation. Because of low pace of production and action of spring back that effect the dimension, this process has not been adopted in industries. This technique can get better production rate by increasing the rotating spindle speed significantly. This present paper mainly reveals a multiple response optimization technique on process parameters of SPIF by using combination of entropy weight measurement method and similarity-based approach developed by Deng to find out the optimal parameters. Experiments for this study were designed using an orthogonal array, L27, according to Taguchi's method taking input factors namely the type of oil, feed rate, step depth, and tool diameter and spindle speed to obtain suitable responses. Optimum conditions were decided based on output response values and an experimental validation conducted using the above decided parameters. After validation, it has been found that there was an improvement of 9.80%.

Keywords Entropy weight measurement method · Deng's similarity-based approach · SEM

1 Introduction

The sheet metal which is majorly used in aerospace parts, marine components, orthopedic implants, and dental parts is titanium sheet metal. Since ratio between strength and weight is high for titanium and also it corrodes less than aluminum and steel, it

G. Yoganjaneyulu (✉)
Department of Mechanical Engineering, Vignan's Institute of Information Technology, Visakhapatnam, Andhra Pradesh, India
e-mail: yogi.08390@gmail.com

C. Sathiya Narayanan
Department of Production Engineering, National Institute of Technology, Tiruchirapalli, Tamil Nadu, India

A. Arockiarajan et al. (eds.), *Advances in Industrial Automation and Smart Manufacturing*, Lecture Notes in Mechanical Engineering,
https://doi.org/10.1007/978-981-15-4739-3_25

is being used majorly. Artificial ankle and other parts used in medical applications are customized and manufactured by incremental sheet metal forming. The same process helps in sheet exalting and forming shapes which was complicated [1, 2]. Narayanasamy et al. [3–6] revealed the forming, fracture, and corrosion behavior, and extend of forming, fracture, and also strain limit at wrinkling using limit diagrams for different thickness steel and titanium sheet materials. Fiorotto et al. [7] studied the thickness of strains experimentally through individual materials and conducted the SPIF process of sandwich panels. Fan et al. [8] reported that the Ti–6Al–4V alloy is well formed by electric hot incremental forming. Suitable lubricant, range of temperature, and way of lubrication are required for efficient process of forming.

Gupta et al. [9] used Taguchi technique with fuzzy logic for multiresponse optimization of high-speed computer numerical controlled machine turning process. The author also analyzed signal-to-noise ratio (S/N ratio), ANOVA, and suitable cutting parameters in high-speed turning parameters in cryogenic atmosphere. Echrif and Hrairi [10] studied surface roughness variation in a negative incremental sheet forming process. To overcome the effect of spindle speed, forming tool diameter, feed rate, and vertical step size were studied using Taguchi's method, DoE, and ANOVA. Raju and Narayanan [11] analyzed the combination of both Taguchi's grey relational analysis (TGRA) approach and response surface methodology (RSM) which are used to find the input variables that are optimal, namely diameter of tool, feed rate, spindle rotational speed, sheet numbers, and vertical step down was essential to attain suitable responses. Chinnaiyan and Jeevanantham [12] carried out the formation of SPIF on AA5052 alloy. In order to investigate the optimal process, parameters such as end type of tool, thickness of the sheet, feed rate, diameter of tool, vertical step-depth, spindle rotational speed, and type of lubricant were analyzed by using multi-objective responses namely grey relational analysis (GRA) and ANOVA. Taguchi method-based GRA combined with principle component analysis (PCA) was used to find the optimal forming factors. Rajesh et al. [13] introduced the multi-performance responses namely surface roughness and formability of the sheet were investigated with respect to the rolling direction. TGRA combined with PCA had been applied to achieve the optimal objectives. Deng [14] developed a novel approach to identify the best optimal alternative among the available design combination. A vast amount of earlier works were happened with Topsis method which has certain limitations that can be overcome by applying Deng's similarity method.

In this current attempt, experiments were conducted to investigate ascertain effects, such as time of forming, limiting wall angle, limiting strain, surface roughness, and spring back. Entropy weight measurement method with similarity-based approach proposed by Deng has been used in accomplishing this objective. The optimization of the forming parameters on the Taguchi-based entropy weight measurement method with coupled the above-mentioned Deng's approach is being the focus of this research work.

2 Materials and Methods

2.1 Materials Selection

The grade 2 titanium sheet with thickness of 1 mm was used in SPIF. The experimental setup, tool, and fixture used in SPIF as shown in Fig. 1, and deformed components are represented in Fig. 2. In this present study, the tool were made up of F6 tool steel and the various diameters of balls such as 8.5, 10.5, and 12.5 mm diameters were positioned below the shank. Lubricant which was used to achieve better surface finish on the deformed sheets was coconut oil. Different forming parameters namely type of lubricant, tool diameter, rate of feed, speed rate, and also step downs were selected which gives five factors, and three levels as represented in Tables 1 and 2.

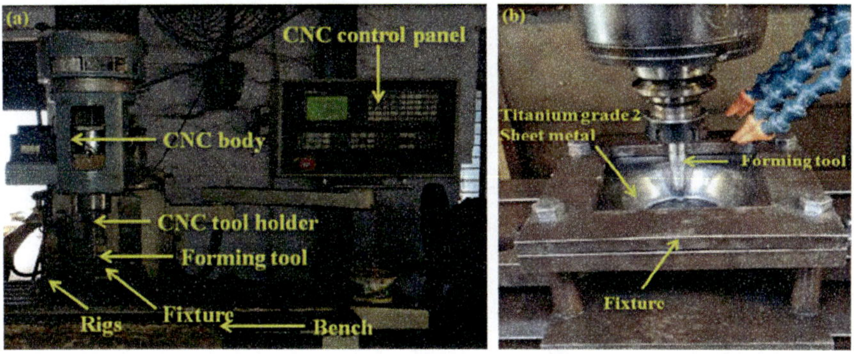

Fig. 1 **a** Experimental setup of CNC milling machine, **b** fixture and tool

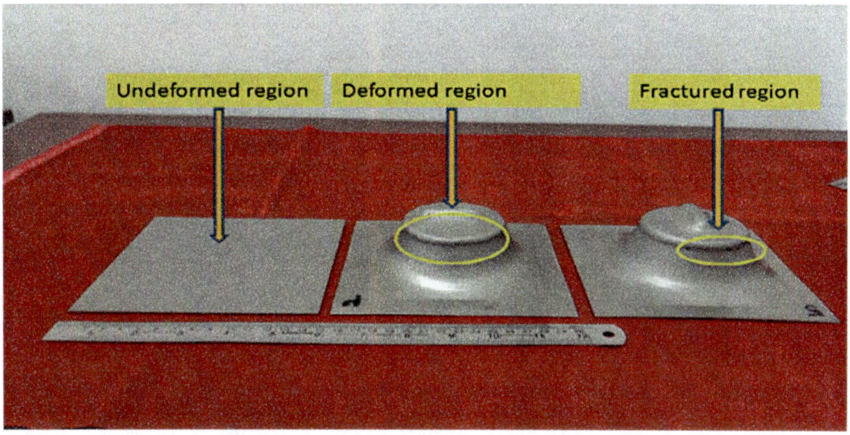

Fig. 2 Titanium grade 2 sheet components formed by SPIF process

Table 1 Factors and levels of the SPIF experiment

S. No.	Factors	Factors notation	Levels			Units
			1	2	3	
1	Type of lubricant	A	Oil	Grease	Dry	
2	Tool diameter	B	8.5	10.5	12.5	mm
3	Feed rate	C	300	600	900	mm/min
4	Speed	D	300	450	550	RPM
5	Vertical step depth	E	0.2	0.4	0.6	mm

Table 2 Experimental layout using an L27 orthogonal array

S. No.	Type of lubricant	Tool diameter (mm)	Feed (mm/min)	Speed (rpm)	Vertical step depth (mm)
1	Oil	8.5	300	350	0.2
2	Oil	8.5	300	350	0.4
3	Oil	8.5	300	350	0.6
4	Oil	10.5	600	450	0.2
5	Oil	10.5	600	450	0.4
6	Oil	10.5	600	450	0.6
7	Oil	12.5	900	550	0.2
8	Oil	12.5	900	550	0.4
9	Oil	12.5	900	550	0.6
10	Grease	8.5	600	550	0.2
11	Grease	8.5	600	550	0.4
12	Grease	8.5	600	550	0.6
13	Grease	10.5	900	350	0.2
14	Grease	10.5	900	350	0.4
15	Grease	10.5	900	350	0.6
16	Grease	12.5	300	450	0.2
17	Grease	12.5	300	450	0.4
18	Grease	12.5	300	450	0.6
19	Dry	8.5	900	450	0.2
20	Dry	8.5	900	450	0.4
21	Dry	8.5	900	450	0.6
22	Dry	10.5	300	550	0.2
23	Dry	10.5	300	550	0.4
24	Dry	10.5	300	550	0.6
25	Dry	12.5	600	350	0.2
26	Dry	12.5	600	350	0.4
27	Dry	12.5	600	350	0.6

The forming time is calculated from the vertical machining center itself by taking into the variation of starting time from end time of forming process into consideration. The deformed components of maximum limiting wall angle and maximum depth of deformed component were also measured with coordinate measuring machine (CMM) having an accuracy of 4 μm; for readings, the average values were determined. At fracture, the major and minor diameters were noted with the help of video measuring machine (VMM) with accuracy of 5 μm. The limiting major true strain and limiting minor true strain values can be determined as described by Raju et al. [11]. The surface roughness was noted using surface roughness tester with an accuracy of 2 μm, and the similar observations were made by [11]. The average formability, spring back, and the surface roughness were measured and calculated by Raju et al. [11]. The mechanism of fracture behavior for titanium grade 2 sheets and sheets was examined with scanning electron microscope (SEM) (model: JEOL–JSM 7500 F), which are in good agreements by Yoganjaneyulu et al. [5].

The experimental results of five output responses namely forming time, formability, wall angle, surface roughness, and spring back were measured and represented in Table 3.

3 Results and Discussions

3.1 Entropy Weight Method

The entropy weight method and Deng's similarity approach which is multiresponse decision creation by assimilating the entropy weight measurement method by using the Deng's method. In this Deng's similarity approach, best optimized parameters have been determined [15]. In this method of entropy weight measurement process variables, weights are determined with no consideration of decision. Basic concept of this method uses the concept that the more useful is the superior weight index value. This is in good agreement with each response [16].

Displayed equations are centered and set on a separate line.

$$W_i = \frac{1 - ej}{n - \sum_{j=1}^{n} eij} \tag{1}$$

3.2 Similarity-Based Approach Proposed by Deng's

After attaining the weightages of every response, Deng's technique is explained in the subsequent ladder, which is in good agreements by [14, 16].

Table 3 Experimental results for L27 orthogonal array layout

Exp. No.	Forming time (min)	Wall angle (°)	Formability	Surface roughness R_a (μm)	Spring back (mm)
1	60.03	56.525	0.439	1.99	4.098
2	45.07	64.049	0.805	2.07	3.033
3	19.51	60.447	0.930	2.14	2.756
4	37.14	60.474	0.691	1.3	2.451
5	20.27	63.730	0.62	2.09	4.145
6	7.36	55.702	0.976	2.58	3.735
7	23.13	59.823	0.795	1.04	3.26
8	15.55	57.660	0.894	2.09	3.564
9	5.30	63.193	1.040	2.82	2.956
10	36.03	58.498	0.802	2.53	3.712
11	23.31	58.378	0.760	1.79	4.598
12	10.22	60.765	0.978	2.33	2.17
13	24.57	62.256	1.088	2.33	2.133
14	12.27	61.810	1.038	2.15	2.859
15	6.21	63.741	0.628	2.32	2.187
16	60.31	54.273	1.714	2.2	2.385
17	34.33	57.98	1.02	1.85	2.42
18	16.51	66.158	0.612	1.38	3.243
19	22.22	58.650	0.615	2.33	4.463
20	18.55	64.958	0.990	1.99	4.258
21	6.11	64.451	0.589	2.08	2.035
22	60.7	55.968	0.637	2.4	5.396
23	35.24	60.861	1.229	2.48	2.534
24	20.40	62.991	0.808	2.33	3.505
25	31.24	57.729	1.002	2.6	3.759
26	17.26	61.878	0.603	1.71	2.956
27	8.56	61.811	0.761	1.83	2.432

- Determine decision matrix.
- Assign the weightages for SPIF outputs.
- Normalization of decision matrix.
- Find the performance matrix.
- Determine the negative ideal solution (NIS) and positive ideal solution (PIS).
- Determine the degree of conflict between every alternative Ai, PIS, and NIS.
- Obtain the similarity and conflict of between NIS and PIS solutions. The $|Ci|$ and S_i^{+-} denotes similarity in between the alternatives; *Ai* and *Aj* are the superior; and the same is represented in Table 6.

- Calculate the overall performance index, and similar observations were reported by elsewhere [15, 16].

$$P_i = \frac{S_i^+}{S_i^+ + S_i^-}, \quad i = 1, 2, \ldots, n \tag{2}$$

- Ranking as directed by Deng's method of similarity. Once after analyzing the SPIF output responses, the 9th experimental run reveals in ranking and best optimistic parameters of SPIF which is the bigger performance index value (Tables 4, 5 and 6).

It is observed from Table 7 that 9th experiment gives the maximum performance index (Pi). So among the 27 experiments conducted with input factors, the corresponding 9th experiment will give the best optimal solution. The analysis of the responses is conducted using Minitab17, and the optimal factors are obtained as tool diameter of 12.5 mm, speed of spindle rotational 450 rpm, rate of feed 900 mm/min, and vertical step down 0.6 mm.

The preferred optimum input-level parameters not only prognoses but similarly validates the confirmation test, which uses the ideal level of design parameters. Then the predictable S/N ratio, by using the following Eq. (3):

$$\widehat{\eta} = \eta_m + \sum_{i=1}^{q} \bar{\eta} - \eta_m \tag{3}$$

where η_m stands for average S/N ratio, $\bar{\eta}$ stands for means S/N ratio relating to the ith parameter on the jth level, and q stands for number of important parameters. Deng's similarity base approach rank is predicted using the Eq. (3) and best optimum combination parameters in SPIF related to different response characteristics revealed. The confirmation test is executed using SPIF, which verifies and improves the performance characteristic of titanium grade 2 sheets. It is observed from the main effect and S/N ratio graph highest values of Deng's can be achieved and represented in Table 8. The main effects plot, S/N ratio, and interaction plot of Deng's similarity-based Taguchi approach values are shown in Figs. 3, 4 and 5. It is observed from the interaction graph that there is a minimum interaction between diameter of tool and rate of feed. A maximum interaction is observed between the type of lubricant and vertical step down. Therefore, Deng's similarity based approach rank for multiperformance optimizations in SPIF process is found to be improving between 0.5917 and 0.6560 as results which are shown in Table 8.

3.3 Scanning Electron Microscope (SEM) Studies

Figure 6 presents the fractography in which failure infers the sheets of titanium grade 2 sheets are well seen under shear mode of fracture, the sheets that were deformed

Table 4 Normalize and decision matrix with 'n' alternatives and 'm' attributes

	'm' Attributes					
	SPIF Responses					
	Exp. No.	Forming time (min)	Wall angle (°)	Formability	Surface roughness R_a (μm)	Spring back (mm)
n' Alternatives	1	60.03	56.525	0.439	1.99	4.098
	2	45.07	64.049	0.805	2.07	3.033
	3	19.51	60.447	0.930	2.14	2.756
	4	37.14	60.474	0.691	1.3	2.451
	5	20.27	63.730	0.62	2.09	4.145
	6	7.36	55.702	0.976	2.58	3.735
	7	23.13	59.823	0.795	1.04	3.26
	8	15.55	57.660	0.894	2.09	3.564
	9	5.30	63.193	1.040	2.82	2.956
	10	36.03	58.498	0.802	2.53	3.712
	11	23.31	58.378	0.760	1.79	4.598
	12	10.22	60.765	0.978	2.33	2.17
	13	24.57	62.256	1.088	2.33	2.133
	14	12.27	61.810	1.038	2.15	2.859
	15	6.21	63.741	0.628	2.32	2.187
	16	60.31	54.273	1.714	2.2	2.385
	17	34.33	57.98	1.02	1.85	2.42
	18	16.51	66.158	0.612	1.38	3.243
	19	22.22	58.650	0.615	2.33	4.463
	20	18.55	64.958	0.990	1.99	4.258
	21	6.11	64.451	0.589	2.08	2.035
	22	60.7	55.968	0.637	2.4	5.396
	23	35.24	60.861	1.229	2.48	2.534
	24	20.40	62.991	0.808	2.33	3.505
	25	31.24	57.729	1.002	2.6	3.759
	26	17.26	61.878	0.603	1.71	2.956
	27	8.56	61.811	0.761	1.83	2.432
	'm' Attributes					
	Normalized responses					
	Exp. No.	Forming time (min)	Wall angle (°)	Formability	Surface roughness R_a (μm)	Spring back (mm)
'n' Alternatives	1	0.387034	0.179436	0.09481	0.17895	0.236288

(continued)

Table 4 (continued)

'm' Attributes					
Normalized responses					
Exp. No.	Forming time (min)	Wall angle (°)	Formability	Surface roughness R_a (μm)	Spring back (mm)
2	0.290582	0.20332	0.17385	0.18614	0.174881
3	0.125788	0.191886	0.20085	0.19243	0.158909
4	0.239454	0.191971	0.14923	0.11690	0.141323
5	0.130688	0.202307	0.13390	0.18794	0.238998
6	0.047452	0.176823	0.21078	0.23200	0.215358
7	0.149127	0.189905	0.17169	0.09352	0.187969
8	0.100256	0.183039	0.19307	0.18794	0.205498
9	0.034171	0.200603	0.22460	0.25358	0.170441
10	0.232298	0.185699	0.17320	0.22751	0.214032
11	0.150288	0.185318	0.16413	0.16096	0.265118
12	0.065892	0.192895	0.21121	0.20952	0.125121
13	0.158411	0.197628	0.23497	0.20952	0.122987
14	0.079109	0.196213	0.22417	0.19333	0.164848
15	0.040038	0.202342	0.13562	0.20862	0.126101
16	0.388839	0.172287	0.37017	0.19783	0.137518
17	0.221337	0.184054	0.22028	0.16636	0.139536
18	0.106446	0.210015	0.13217	0.12409	0.186989
19	0.14326	0.186181	0.132821	0.209525	0.257334
20	0.119598	0.206206	0.21380	0.17895	0.245514
21	0.039393	0.204596	0.12720	0.18704	0.117337
22	0.391354	0.177667	0.13757	0.21581	0.31113
23	0.227204	0.1932	0.26542	0.22301	0.146109
24	0.131526	0.199962	0.17450	0.20952	0.202096
25	0.201415	0.183258	0.21640	0.23380	0.216741
26	0.111281	0.196428	0.13022	0.15377	0.170441
27	0.055189	0.196216	0.16435	0.16456	0.140228

under a mode of transition from shear mode to inter crystalline separation; further, some of sheets fracture were faintly developed due to the inter crystalline fracture even later a significant extent of plastic deformation.

Table 5 Weighted performance matrix

Exp. No.	Forming time (min)	Wall angle (°)	Formability	Surface roughness R_a (μm)	Spring back (mm)
1	0.08286	0.03642	0.024472	0.032921	0.033286
2	0.062211	0.041268	0.044874	0.034245	0.024635
3	0.02693	0.038947	0.051842	0.035403	0.022386
4	0.051265	0.038964	0.038519	0.021506	0.019908
5	0.027979	0.041062	0.034561	0.034576	0.033668
6	0.010159	0.03589	0.054406	0.042682	0.030337
7	0.031927	0.038545	0.044316	0.017205	0.026479
8	0.021464	0.037151	0.049835	0.034576	0.028948
9	0.007316	0.040716	0.057974	0.046653	0.02401
10	0.049733	0.037691	0.044706	0.041855	0.030151
11	0.032175	0.037614	0.042365	0.029613	0.037347
12	0.014107	0.039152	0.054517	0.038546	0.017626
13	0.033914	0.040113	0.060649	0.038546	0.017325
14	0.016936	0.039825	0.057862	0.035568	0.023222
15	0.008572	0.041069	0.035007	0.038381	0.017764
16	0.083247	0.034969	0.095545	0.036396	0.019372
17	0.047386	0.037358	0.056859	0.030605	0.019656
18	0.022789	0.042627	0.034115	0.02283	0.026341
19	0.030671	0.037789	0.034282	0.038546	0.036251
20	0.025605	0.041854	0.055186	0.032921	0.034585
21	0.008434	0.041527	0.032833	0.03441	0.016529
22	0.083785	0.036061	0.035509	0.039704	0.043829
23	0.048642	0.039214	0.068509	0.041028	0.020582
24	0.028158	0.040586	0.045041	0.038546	0.028469
25	0.043121	0.037196	0.055855	0.043013	0.030532
26	0.023824	0.039869	0.033613	0.028289	0.02401
27	0.011815	0.039826	0.042421	0.030275	0.019754

4 Conclusions

This paper focused on a multiresponse optimization of entropy weight measurement method and Deng's similarity approach for optimizing the SPIF process parameters. The output response parameters to obtain realistic or superlative responses on titanium grade 2 sheets can be concluded as following:

Table 6 Degree of conflicts between alternatives

Exp. No.	Cos θ_i^+	Cos θ_i^-	C^+	C^-
1	0.497998	0.990666	0.05219	0.103821
2	0.720314	0.947508	0.069712	0.0917
3	0.885826	0.808587	0.072422	0.066107
4	0.740449	0.925058	0.059646	0.074517
5	0.769696	0.876112	0.059583	0.06782
6	0.88734	0.705461	0.074734	0.059416
7	0.86072	0.840743	0.063636	0.062159
8	0.88331	0.793699	0.070432	0.063287
9	0.896281	0.664286	0.079285	0.058763
10	0.749984	0.936583	0.069362	0.08662
11	0.807745	0.878438	0.065205	0.070912
12	0.915084	0.703944	0.073779	0.056756
13	0.892202	0.80328	0.080908	0.072844
14	0.925747	0.724218	0.077593	0.060702
15	0.822419	0.709972	0.056826	0.049056
16	0.821594	0.834803	0.113194	0.115014
17	0.84444	0.869765	0.076453	0.078746
18	0.822053	0.826328	0.056449	0.056742
19	0.74393	0.89594	0.059253	0.07136
20	0.886239	0.800888	0.077933	0.070427
21	0.824226	0.706942	0.054242	0.046524
22	0.565051	0.993431	0.06462	0.11361
23	0.864414	0.844567	0.089433	0.08738
24	0.833452	0.847245	0.068552	0.069686
25	0.824677	0.884852	0.078864	0.084618
26	0.812625	0.847685	0.055496	0.05789
27	0.894067	0.720739	0.062149	0.050101

- Entropy weight measurement method and similarity-based approach by Deng are habitually preferable method for multicriterion verdict assembly method in SPIF of titanium grade 2.
- To attain superlative SPIF responses, the optimal input process parameters are followed and obtained from Entropy weight measurement method and similarity-based approach by Deng: diameter of tool 12.5 mm, spindle rotational speed 450 rpm, rate of feed is 900 mm/min, and VSD is 0.6 mm; these experimental values show attainment of best optimal parameters.

Table 7 Similarity and overall performance index values of all alternatives

Exp. No.	S^+	S^-	P_i	Ranking
1	0.005613	0.01181	0.322183	27
2	0.007498	0.010431	0.41821	25
3	0.00779	0.00752	0.508813	11
4	0.006415	0.008476	0.430802	24
5	0.006409	0.007715	0.453763	21
6	0.008038	0.006759	0.543241	4
7	0.006845	0.007071	0.491877	12
8	0.007576	0.007199	0.512746	8
9	0.008528	0.006684	0.560591	1
10	0.00746	0.009853	0.430904	23
11	0.007013	0.008066	0.465088	20
12	0.007936	0.006456	0.551402	2
13	0.008702	0.008286	0.51225	9
14	0.008346	0.006905	0.547241	3
15	0.006112	0.00558	0.522746	7
16	0.012175	0.013083	0.482026	15
17	0.008223	0.008957	0.478631	17
18	0.006071	0.006454	0.484714	14
19	0.006373	0.008117	0.439817	22
20	0.008382	0.008011	0.51132	10
21	0.005834	0.005292	0.52436	6
22	0.00695	0.012923	0.34973	26
23	0.009619	0.009939	0.491813	13
24	0.007373	0.007927	0.48191	16
25	0.008482	0.009625	0.46844	19
26	0.005969	0.006585	0.475465	18
27	0.006685	0.005699	0.539797	5

- In the interaction graph, minimum interaction is observed between diameter of tool and rate of feed but a maximum interaction is observed between type of lubricant and vertical step down.
- The scanning electron microscope (SEM) analysis results show the fractured samples of titanium grade 2 sheet materials with moderate plastic deformation.
- The confirmation test is conducted with the best optimal parameter conditions, which shows an improvement of 9.80% in the Deng's similarity approach.
- Entropy weight measurement method and similarity-based approach by Deng are best superior method for multicriteria decision making method in SPIF of titanium grade 2.

Table 8 Confirmation test

Trial No.	Input parameters	Values best input parameters	Estimated Deng's similarity approach	Obtained values Deng's similarity	Improvement Deng's similarity approach
1	Type of lubricant	Grease	0.5917	0.6560	0.0643
	Tool diameter	12.5 mm			
	Feed	900 mm/min			
	Speed	450 rpm			
	VSD	0.6 mm			

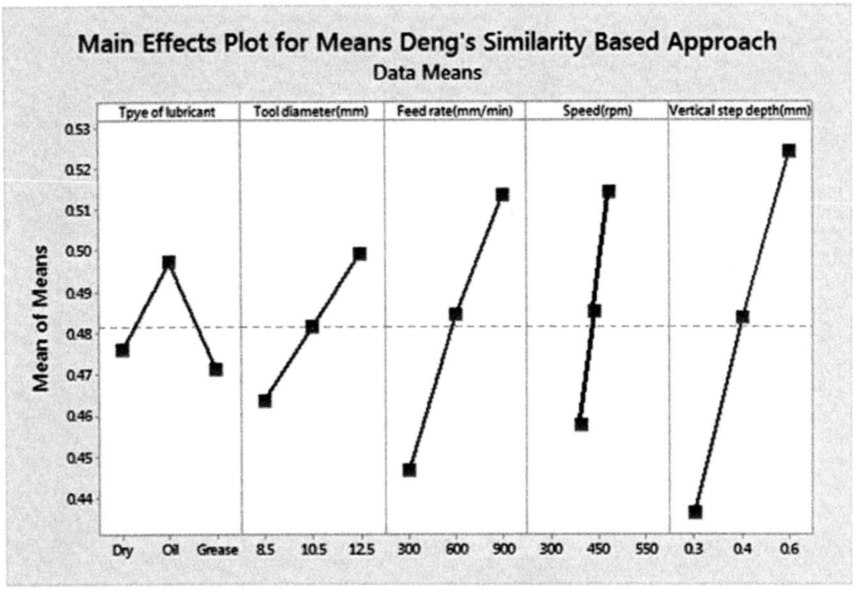

Fig. 3 The main effect plot for Deng's method showing optimal levels

- Hence, this study is convenient and successfully optimized with regard to the multiresponse forming problem in cooperation with researchers and industries to solve the vigorous problem. This newly optimization method has been effectively applied for further sheet metal forming industries.

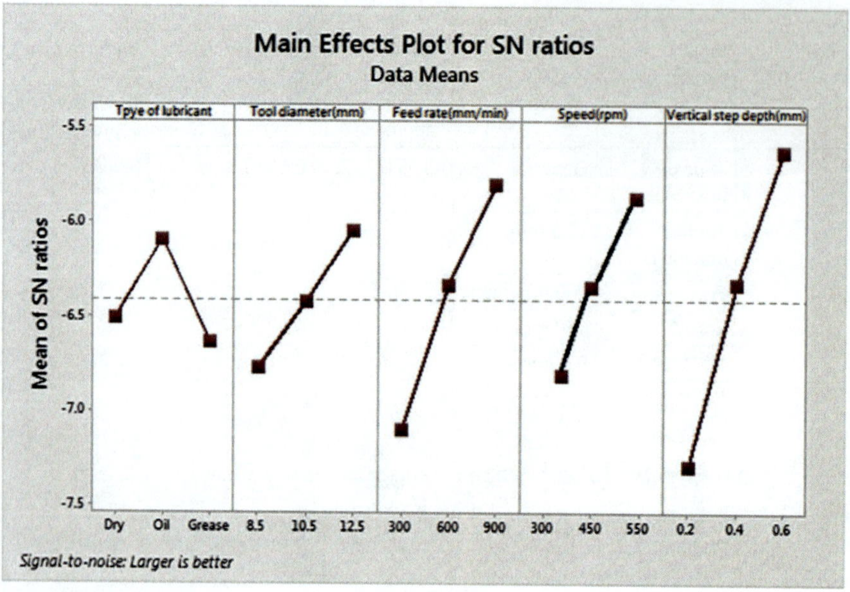

Fig. 4 The S/N ratios plot of best optimal parametric combination of S/N ratio using Deng's Taguchi approach

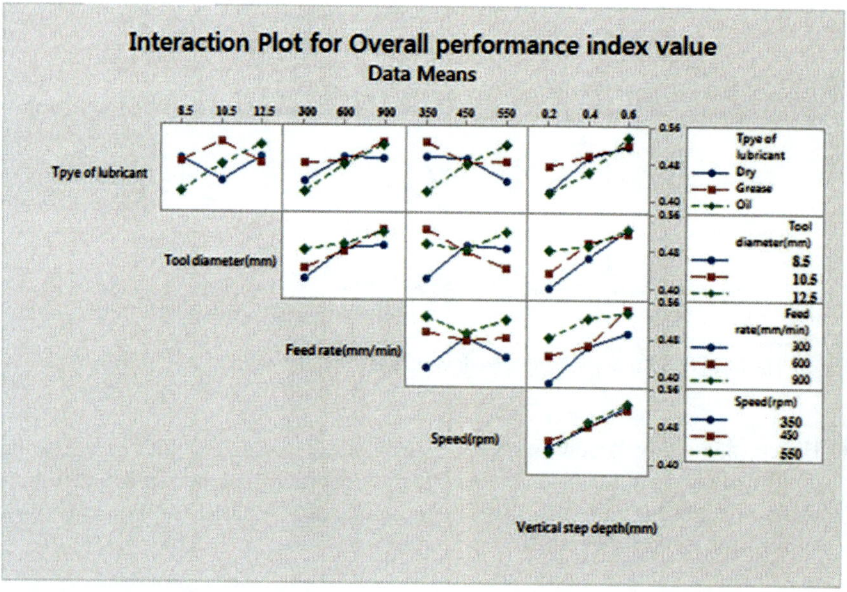

Fig. 5 The interaction plot of optimal parametric combination using Deng's approach

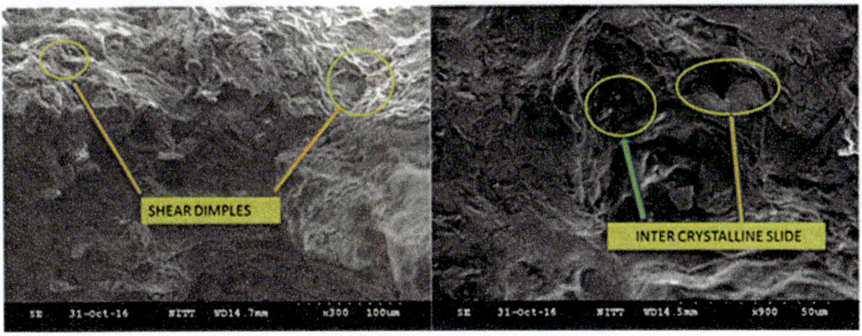

Fig. 6 SEM fractographs of deformed surface

References

1. Ambrogio G, Filice L, Gagliardi F (2012) Formability of lightweight alloys by hot incremental sheet forming. Mater Des 34:501–508
2. Ambrogio G, De Napoli L, Filice L, Gagliardi F, Muzzupappa M (2005) Application of incremental forming process for high customised medical product manufacturing. J Mater Process Technol 162:156–162
3. Narayanasamy R, Narayanan CS (2008) Forming, fracture and wrinkling limit diagram for if steel sheets of different thickness. Mater Des 29(7):1467–1475
4. Narayanasamy R, Parthasarathi NL, Narayanan CS, Venugopal T, Pradhan HT (2008) A study on fracture behaviour of three different high strength low alloy steel sheets during formation with different strain ratios. Mater Des 29(9):1868–1885
5. Yoganjaneyulu G, Narayanan CS, Narayanasamy R (2018) Investigation on the fracture behavior of titanium grade 2 sheets by using the single point incremental forming process. J Manuf Processes 35:197–204
6. Yoganjaneyulu G, Phaneendra Y, Ravikumar VV, Narayanan CS (2019) Investigations on the void coalescence and corrosion behaviour of titanium grade 4 sheets during single point incremental forming process. Department of Production Engineering, NIT Tiruchirappalli, India. Department of Aeronautical Engineering, Andhra University, Vishakhapatnam, India. Department of Metallurgical and Materials Engineering, NIT Tiruchirappalli, India
7. Fiorotto M, Sorgente M, Lucchetta G (2010) Preliminary studies on single point incremental forming for composite materials. IntJ Mater Form 3(1):951–954
8. Fan G, Sun F, Meng X, Gao L, Tong G (2010) Electric hot incremental forming of Ti-6Al-4V titanium sheet. Int J Adv Manuf Technol 49(9–12):941–947
9. Gupta A, Singh H, Aggarwal A (2011) Taguchi-fuzzy multi output optimization (MOO) in high speed CNC turning of AISI P-20 tool steel. Expert Syst Appl 38(6):6822–6828
10. Echrif SB, Hrairi M (2014) Significant parameters for the surface roughness in incremental forming process. Mater Manuf Processes 29(6):697–703
11. Raju C, Narayanan CS (2016) Application of a hybrid optimization technique in a multiple sheet single point incremental forming process. Measurement 78:296–308
12. Chinnaiyan P, Jeevanantham AK (2014) Multi-objective optimization of single point incremental sheet forming of AA5052 using Taguchi based gray relational analysis coupled with principal component analysis. Int J Precis Eng Manuf 15(11):2309–2316
13. Rajesh S, Devaraj D, Pandian RS, Rajakarunakaran S (2013) Multi-response optimization of machining parameters on red mud-based aluminum metal matrix composites in turning process. Int J Adv Manuf Technol 67(1–4):811–821
14. Deng H (2007) A similarity-based approach to ranking multicriteria alternatives. In: International conference on intelligent computing. Springer, Berlin, Heidelberg, pp 253–262

15. Palanisamy A, Selvaraj T (2018) Optimisation of turning parameters on heat treated INCOLOY 800H using cryogenically treated CVD tool with grey-based entropy method. Int J Mach Mach Mater 20(5):401–424
16. Sivaprasad PV, Haq NA (2019) An Entropy-Deng's similarity-based technique for modeling and optimization of process variables for laser micro drilling of alloy-X

Experimental Studies on Deep Cryo Treated Plus Tempered Tungsten Carbide Inserts in Turning Operation

K. Arunkarthikeyan and K. Balamurugan

Abstract Improving the tool life will increase the production by reducing the tool change time and tool cost. Various surface modification techniques are available such as commercial coatings method, cryogenic treatment, nitriding and cyaniding. The cryogenic treatment of the cutting tools enhances the micro-structural properties of the tool and used in increasing the tool life. In the present work, cryogenic treated tungsten carbide tool is used to examine the machinability characteristics on low-carbon AISI 1020 steel for turning operation. The experimental study on machining operations is conducted by Taguchi's L9 orthogonal array for three levels with three factors. The governing parameters of cutting velocity, feed rate and depth of cut on tool wear rate and surface finish on the machined surface. The effect of each parameter on output response is verified through ANOVA. In that, the depth of cut is the most significant factor compared to other factors. The optimum machining condition is predicted through grey relational analysis.

Keywords Cryogenic treatment · Tool wear rate · Grey relational analysis · AISI 1020 steel

1 Introduction

The sub-zero heat treatment or cryogenic treatment is being for a wide range of applications ferrous, nonferrous alloys, ceramics, carbides, plastics as well as the process and the coolant for high-temperature machine cutting applications [1]. Cryogenic treatment is a process, the material is subjected to below zero degree Celsius, in terms of Kelvin scale and the consequence of cryogenic heat treatment on cutting tools, enhances the mechanical properties like abrasion and wear resistance, toughness, hardness and electrical conductivity [2]. Based on applications and properties of tool material, cryogenic treatment is applied at various temperature ranges in between –

K. Arunkarthikeyan · K. Balamurugan (✉)
Department of Mechanical Engineering, VFSTR (Deemed to be University), Guntur, Andhra Pradesh 522213, India
e-mail: kbalan2000@gmail.com

A. Arockiarajan et al. (eds.), *Advances in Industrial Automation and Smart Manufacturing*, Lecture Notes in Mechanical Engineering,
https://doi.org/10.1007/978-981-15-4739-3_26

80 and −196 °C. It is also classified into shallow treated and deeply treated based on the cryogenic soaking duration of tool material [3]. Cryogenically treatment enables the mechanical properties of tool material with finely distributed carbides, augments better surface properties [4]. This type of nonconventional treatment is appreciable for different types of cutting tools, tool steels, alloy steels, plastics and ceramics [5]. The sub-zero cryo treatment has many advantages, the result of the treatment was martensite formation replaced by retained austenite. It is the reason behind that for cryogenically treated materials enhances the high wear resistance along with good surface finish for various machining conditions [6]. The researcher inspected and reported the performance of cryogenic heat treated tools for various machining operation. The cryo treatment increases the tool life and greatly influences on cutting force. Holding temperature (during cryo treatment and post-tempering treatment) and tempering time (cryogenic treatment time and post heat treatment time) with determine the tool stability as well as wear resistance rate of the tool material [7]. Investigators make the comparative performance investigation on cryogenic treated high speed steel (HSS) for various machining conditions. Their study reveals reasons for the tool life and stability of cryo treated tool inserts. From the earlier study confirms that, incomplete transformation of austenite to martensite structure after the cryogenic treatment with change in tempering time [8].

This work deals with the cryogenically soaked plus tempered (CSPT) tungsten carbide inserts subjected to turning operation by NAGMATI 175 semiautomatic lathe at different operating conditions. CSPT tungsten carbide inserts will have excessive amount of carbides formation on the surface, which will provide an excellent wear resistant with appreciable level of average surface roughness (Ra). The optimum machining condition is predicted through grey relational analysis (GRA). The cutting tool performance characteristics of CSPT tungsten carbide insert at different machining conditions for turning operation on low-carbon AISI 1020 steel is studied and reported.

2 Materials and Methods

2.1 Preparation of a Cryogenically Treated Tool

Cryogenic treated tungsten carbide insert has better machining performance compared to the untreated ones [9]. Cryogenic soaking temperature and time, post-tempering cycle and holding time are the variance is whether increase or decreases the performances of the tool with the effect of rearrangement of the grains in its surface [10]. Commercially available tungsten carbide insert is soaked in cryogenic environment for 72 h and post-tempered in induction furnace for a period of 2 h in 200 °C is shown in Fig. 1.

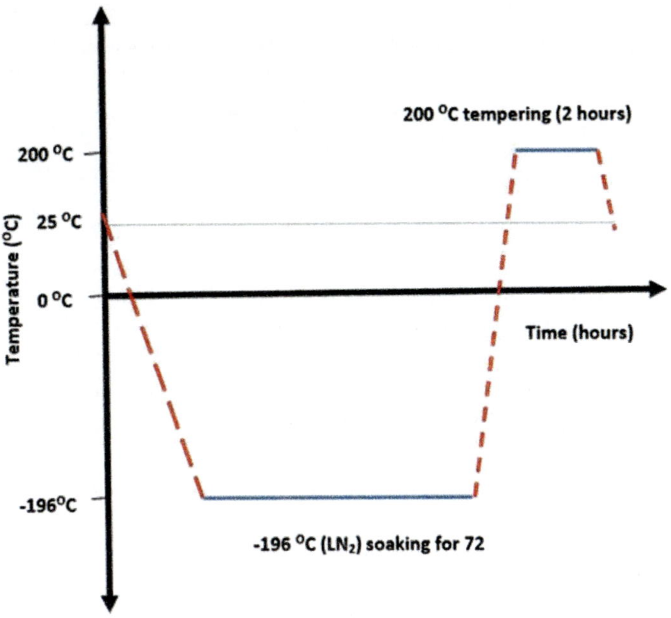

Fig. 1 Deep cryogenic treatment cycle procedure

2.2 *Workpiece Material*

A low-carbon AISI 1020 steel is taken as test material for analysing the tool performance of the CSPT tungsten carbide inserts. The workpiece material (low-carbon AISI 1020 steel) is used for various applications like gudgeon pin, machinery parts, spindle gear, worm gear, shafts, light duty gears, axles and ratchets. The work material hardness is 65 HRB in B scale. The workpiece material (AISI 1020 low-carbon steel) Ø50 × 100 mm dimensions and it kept constant throughout the turning operations. The effect of cryo teated insert on hardness was shown in Table 1.

The CSPT tungsten carbide cutting insert performance and untreated tungsten carbide cutting insert performance are observed by various trails of dry turning operation conducted for three factors with three levels (Table 2). The dimensions of the workpiece kept constant for all machining condition. New tool tip is used to take the experimental observations.

Table 1 Hardness of cryo treated and untreated cutting inserts

S. No	Treatment	Vicker's hardness (kgf/mm^2)
1	Cryogenically treated tungsten carbide inserts	1646.6
2	Untreated tungsten carbide inserts	1592.4

Table 2 Machine setting for turning operation

Parameters	Cutting velocity (m/min)	Feed rate (mm/rev)	Depth of cut (mm)
Levels	79	0.102	0.4
	121	0.143	0.6
	188	0.191	0.8

Subsequently, the completion of every dry turning operation on low-carbon AISI 1020 steel material, the surface roughness value of the turned surface is observed by handy type surface roughness tester, Japan made Mitutoyo SJ-210 of 2 μm stylus probe radius with linear interpolation for 1 mm turned surface. In that, the average surface roughness (Ra) values were determined from three dissimilar sample surfaces.

The tool wear rate is measured according to their weight loss after every cycle of turning operation. The turning operation is conducted for 2 min and its kept constant for all experiments. The electronic weight balance device is used to measure weight loss. Equation 1 is used to calculate the tool wear rate,

$$\text{TWR} = \frac{W_\text{b} - W_\text{a}}{M_\text{t}} \tag{1}$$

W_b Weight of the tool before machining in mg
W_a Weight of the tool after machining in mg
M_t Machining time in minutes.

3 Results and Discussion

The hardness for the CSPT and untreated tungsten carbide inserts was measured in Vickers's microhardness tester. The hardness is measured by applying a load 100 kg for 10 s. The hardness values are measured at various places of the inserts and the better value is tabulated in Table 1. The turning operation is conducted and their corresponding output responses (TWR and Ra) are tabulated in Table 3. The prediction of the CSPT tungsten carbide tool performance done by multi-response GRA for observed values. ANOVA is developed for determining the effects of individual input parameters to the output characteristics. The CSPT tungsten carbide insert and the untreated tungsten carbide insert display the poor 'Ra' value appearances at higher feed rate, high depth of cut and lower cutting speeds. The peak value of Ra obtained by the untreated tungsten carbide insert under dry turning operation is 4.203 μm at lowest speed (79 m/min), peak value of depth of cut (0.8 mm) and peak value of feed rate (0.191 mm/rev), and for cryogenically soaked plus tempered tungsten carbide Ra value is 3.106 μm. The lowest value of Ra detected is at moderate feed rate (0.143 mm/rev), the minimal value of depth of cut (0.4 mm) and higher cutting

Table 3 Ra and TWR observation for untreated and cryo treated tungsten carbide

Trail. No	Speed (rpm)	Cutting velocity (m/min)	Feed rate (mm/rev)	Depth of cut (mm)	Ra		TWR	
					Cryo treated inserts (μm)	Untreated insert (μm)	Cryo treated inserts (mg/min)	Untreated insert (mg/min)
1	502	79	0.102	0.4	2.506	3.574	6	13
2	502	79	0.143	0.6	3.050	3.884	16	24
3	502	79	0.191	0.8	3.502	4.203	29	35
4	770	121	0.102	0.6	2.507	3.104	15	19
5	770	121	0.143	0.8	3.025	3.963	28	35
6	770	121	0.191	0.4	3.064	4.106	10	14
7	1196	188	0.102	0.8	2.124	3.901	21	31
8	1196	188	0.143	0.4	2.021	2.982	13	14
9	1196	188	0.191	0.6	3.062	4.053	22	30

velocity (181 m/min). The results conclude the value of Ra obtained by CSPT tungsten carbide insert has superior surface behaviours than untreated tungsten carbide insert at greater speed and lesser feed rates.

The maximum wear rate is registered for experiment trail number ‘3’ for both (CSPT and untreated tungsten carbide) inserts and their values are 29 mg/min and 35 mg/min, respectively. The reason for the maximum TWR is peak value of depth of cut and cutting velocity. The maximum depth of cut effect to create more contact of tool on workpiece and its subjected to high cutting velocity, more amount of heat is generated on the tool-workpiece contact zone, which leads to creep wear. Meanwhile, low TWR is observed for CSPT and untreated inserts in experiment trail 1 and their values are 6 mg/min and 13 mg/min, respectively. It reveals that low level of cutting parameters (cutting velocity = 79 m/min, feed rate = 0.102 mm/rev and depth of cut = 0.4 mm) enhance the life of the tool.

Tungsten carbide inserts with the treatment of cryogenic soaked plus tempered inserts as a tool material preferred for machining of low-carbon steels than untreated tungsten carbide inserts.

3.1 Grey Relational Analysis

In this work, the output responses (Ra and TWR) were optimized through the Taguchi's grey relational analysis multi-response approach. By using GRA approach to quantify the results according to the scale limits of 0 to 1. Both responses were taken as a minimum the better condition and computed through the grey relational algorithm. The step by step procedure [11] for GRA algorithm applied for each observation and is tabulated in Table 4. From that, the third experimental trail has

Table 4 Taguchi's grey relational analysis for cryogenically treated tool

S. No	S/N ratio		Normalized S/N		Grey coefficient		Grey grade	Rank
	TWR	Ra	TWR	Ra	TWR	Ra		
1	−15.5630	-7.979	0	0.3913	0.5000	0.6216	0.5608	9
2	−24.0824	-9.686	0.6225	0.7486	0.7260	0.7991	0.7625	4
3	−29.2480	−10.886	1.0000	1.0000	1	1	1	1
4	−23.5218	−7.983	0.5815	0.3920	0.7050	0.6218	0.6634	7
5	−28.9432	−9.614	0.9777	0.7337	0.9782	0.7896	0.8839	2
6	−20.0000	−9.725	0.3242	0.7570	0.5967	0.8044	0.7006	5
7	−26.4444	−6.543	0.7951	0.0904	0.8300	0.5236	0.6768	6
8	−22.2789	−6.111	0.4907	0.0000	0.6626	0.5000	0.5813	8
9	−26.8485	−9.720	0.8246	0.7558	0.8508	0.8037	0.8273	3

Table 5 Response table for grey relational grade

Factors	Level-1	Level-2	Level-3
Cutting velocity	0.7744	0.7493	0.6951
Feed rate	0.6337	0.7425	0.8426
Depth of cut	0.6142	0.7511	0.8536

reached the maximum value. The first experimental trial has less significant trail compared to other trails. An ideal machine setting arrangement of cutting conditions for achieving the lowest Ra value and lowest TWR is determined from the maximum grey relational grade values are shown in Table 5, and maximum values are in each level is found to be the feed rate, 0.191 mm/rev, depth of cut, 0.8 mm, and cutting speed, 79 m/min, from the confirmed results. Figure 2 clearly shows the main effect plot for cutting input parameters according to the value of the grey relational grade. At the machining condition, first level of cutting velocity, third level of feed rate and depth of cut will yield the optimum yield condition.

3.2 *ANOVA for Grey Relational Grade*

ANOVA observation is shown in Table 6. The influence each cutting parameter affected the output response variables (maximum mean value in each case of inputs). Grey relational grade (GRG) is calculated and analysed to understand the effect of influencing parameters, i.e., cutting velocity, feed rate and depth of cut on the output responses. The depth of cut is the most influencing factor among all and its contribution on the output response is 52.04%. The feed rate and the cutting velocity are the factors influenced less, compared to the depth of cut and their contributions are 39.48% and 5.94%, respectively.

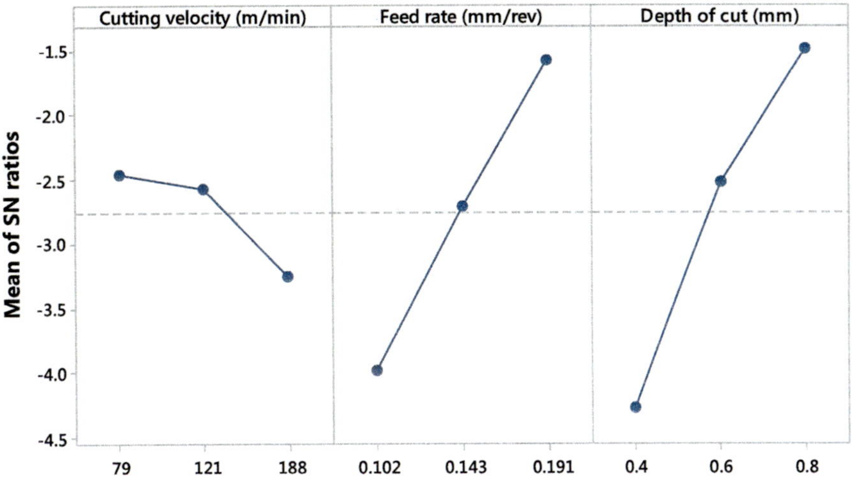

Fig. 2 Main effects plot for S/N (grey relational grade)

Table 6 ANOVA table for grey relational grade

Source	DOF	AdjSS	AdjMS	*F*-value	Contribution (%)
Cutting velocity	1	0.1609	0.0536	11.65	5.94
Feed rate	1	0.0094	0.0098	77.49	39.48
Depth of cut	1	0.0652	0.0652	102.14	52.04
Error	5	0.0859	0.0859		2.55
Total	8	0.0042			

3.3 Surface Plot Analysis for Ra and TWR

The experimental work is conducted and the output responses are observed. The 3D surface plot drawn for observed values of Ra and TWR is shown in Figs. 3 and 4. Figure 3b and c is the evident that, feed rate is significant factor in Ra increment during machining, followed by depth of cut. The depth of cut increment increases the Ra value is shown in Fig. 3a, c. While considering TWR increment, the depth of cut is the major contributor is shown in Fig. 4a, c. The feed rate is another significant factor for increasing the TWR after the depth of cut and it is confirmed from Fig. 4b, c.

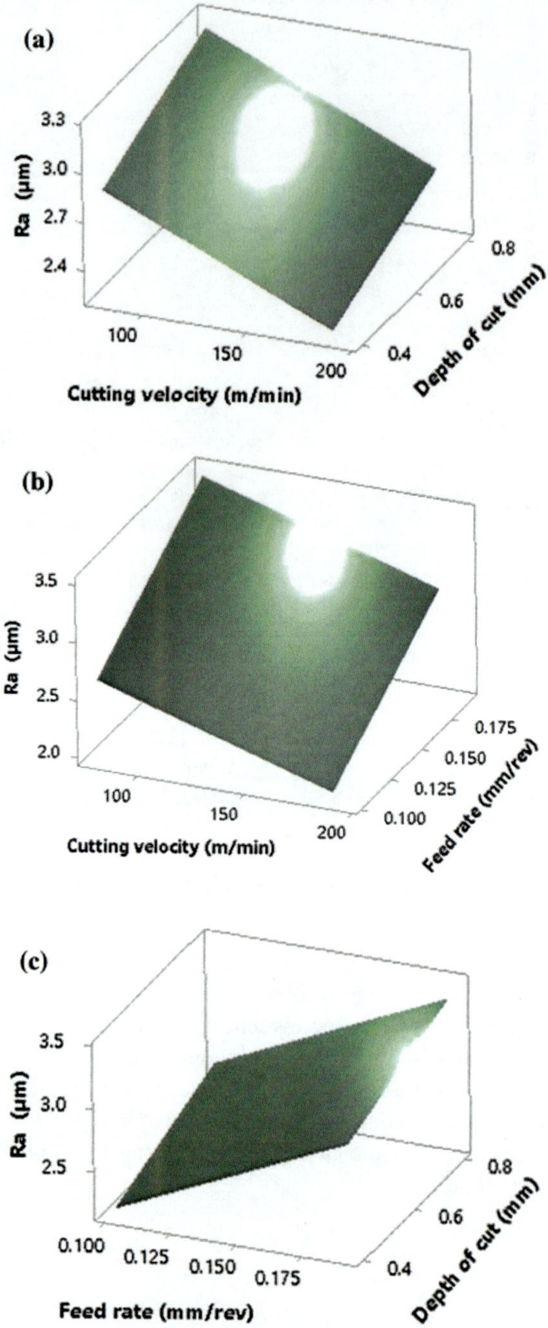

Fig. 3 Surface roughness characteristics of cryo treated tungsten carbide insert at various cutting parameters **a** cutting velocity and depth of cut, **b** cutting velocity and feed rate, **c** feed rate and depth of cut

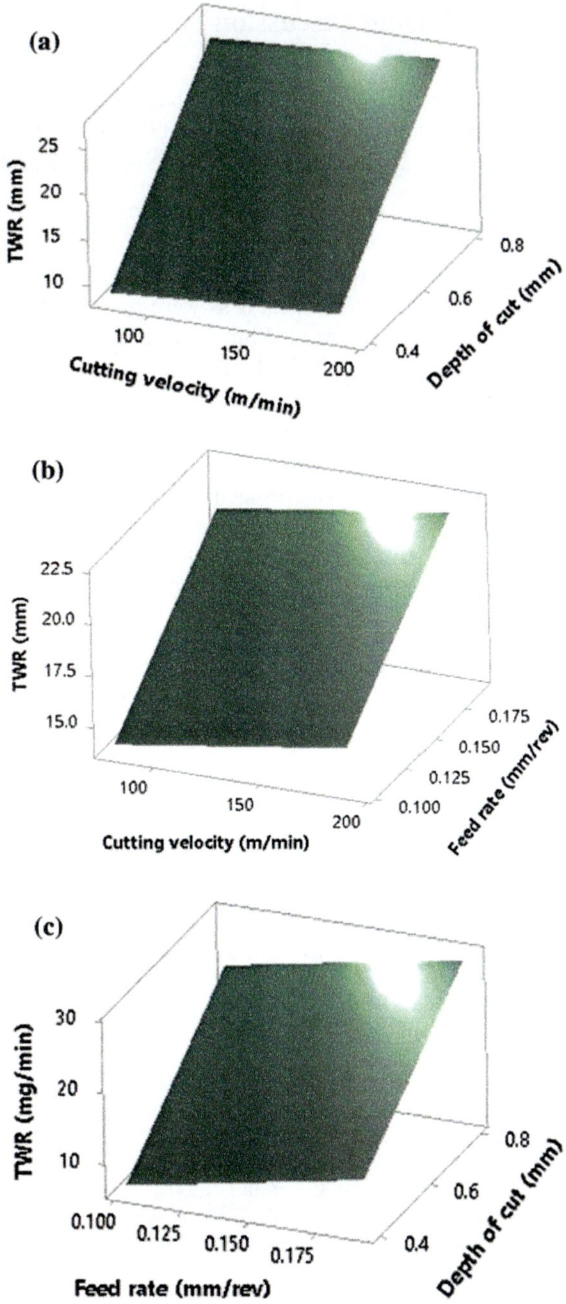

Fig. 4 Tool wear rate characteristics of cryo treated tungsten carbide insert at various cutting parameters **a** cutting velocity and depth of cut, **b** cutting velocity and feed rate, **c** feed rate and depth of cut

4 Results and Discussion

The experimental studies carried on the performance of tungsten carbide inserts with deep cryogenically soaked plus tempered tool inserts compared with the performance of conventionally used tungsten carbide inserts have the following conclusions:

- The conventionally used tungsten carbide insert is soaked at cryogenic environment for 72 h and tempered for 2 h in 200 °C environment.
- The turning operation is carried out according to Taguchi's L9 OA. Tool wear rate and average surface roughness were optimized through grey relation approach. From that, the optimized input machining condition is identified to be cutting velocity = 79 m/min, feed rate = 0.191 mm/rev and depth of cut = 0.8 mm.
- ANOVA study exhibits the input significant factors contribution over the output responses. From ANOVA, the depth of cut is found to be a dominating factor affect the output responses and its contribution is 52.04% hence on contribution of 39.48% and 5.94%, respectively.
- Deep cryogenic treated tungsten carbide inserts can be a suitable insert than conventionally available tungsten carbide inserts to get better machining characteristics for turning of AISI 1020 steel.

Acknowledgements The authors wish to acknowledge the support of the Mechanical Engineering department, VFSTR (Deemed to be University) to accomplish the work.

References

1. Gu K, Wang J, Zhou Y (2019) Effect of cryogenic treatment on wear resistance of Ti-6Al-4 V alloy for biomedical applications. J MechBehav Biomed 30:131–139
2. Reddy TVS, Sornakumar T, Reddy MV, Venkataram R, Senthilkumar A (2009) A turning studies of deep cryogenic treated p-40 tungsten carbide cutting tool inserts. Tech Commun sci-tech, 13:269–281
3. Reitz W, Pendray J (2001) Cryoprocessing of materials: a review of current status. Mater Manuf Process 16:829–840
4. Seah KHW, Rahman M, Yong KH (2003) Performance evaluation of cryogenically treated tungsten carbide cutting tool inserts. J Eng Manuf 217:29–43
5. Podgornik B, Leskovšek V, Vižintin J (2009) Influence of deep- cryogenic treatment on tribological properties of P/M high-speed steel. Mater Manuf Process 24:734–738
6. Gallagher AH, Agosti CD, Roth JT (2005) Effect of cryogenic treatments on tungsten carbide tool life: microstructural analysis. Trans North Am Manuf Res Inst SME 33:153–160
7. Gopal Krishna PV, Kishore K, Ramadevudu G, Ali S (2012) Performance evaluation of cryogenic treated tools in turning. Adv Product Eng Manage 3:187–194
8. Candane D, Alagumurthi N, Palaniraja K (2013) Effect of cryogenic treatment on microstructure and wear characteristics of AISI M35 HSS. Int J Mater Sci Appl 2:56–65
9. Raj Mohan BVS, Suresh RK (2015) Multi objective optimization of cutting parameters during turning of En31 alloy steel using ant colony optimization. Int J Mech Eng Technol 6:31–45

10. Yıldız Y, Nalbant M (2008) A review of cryogenic cooling in machining processes. Int J Mach Tools Manuf 48:947–964
11. Balamurugan K, Uthayakumar M, Sankar S, Hareesh US, Warrier KGK (2017) Mathematical modelling on multiple variables in machining LaPO4/Y2O3 composite by abrasive waterjet. Int J Mach Mach Mater 19:426–439

Dual Response Surface and Desirability Approach for Modeling and Optimization of Abrasive Water Jet Machining Process

G. Rajyalakshmi, K. Jayakrishna, and S. Aravind Raj

Abstract Earlier decades have observed an expeditious increase in the evolution of harder and complex to cut materials. Abrasive water jet machining is one of the novel nontraditional cutting processes that found diverse applications in machining different kinds of difficult-to-machine materials. It is a commercial process for heat sensible materials, which are difficult to machine by heat generated machining techniques. Process parameters play an important role in finding the economics of machining process at good quality. This research focused on the comparative prediction of responses during AWJ machining of granite using response surface methodology and desirability approach. MRR and surface roughness are considered as responses which will decide the quality of machining process. Experiments were carried out according to RSM design. RSM is adopted for prediction of responses in order to compare experimental values. Finally, optimizations have been done for determining the best set of machining parameters.

Keywords AWJM · Response surface methodology · Desirability analysis · Modeling · Optimization

1 Introduction

AWJ process is an advanced machining process for difficult to cut materials. Collision of abrasive material with water is used to erode the work piece material [1]. High pressure water mixes with abrasives in mixing chamber and this abrasive water jet coming through the nozzle cuts the work piece material. This process is a well-known advanced technique for cutting difficult-to-machine materials. It is a technique, which machines without heat generation and the machined surface is generated without any HAZ or residual stress virtually [2]. This process is inert toward properties of material with high versatile and flexible in machining. The main disadvantage of this technique is its vociferous and disordered working environment [3].

G. Rajyalakshmi (✉) · K. Jayakrishna · S. Aravind Raj
School of Mechanical Engineering, Vellore Institute of Technology, Vellore, Tamil Nadu, India
e-mail: rajyalakshmimed@gmail.com

A. Arockiarajan et al. (eds.), *Advances in Industrial Automation and Smart Manufacturing*, Lecture Notes in Mechanical Engineering,
https://doi.org/10.1007/978-981-15-4739-3_27

AWJ machining process is nonconformist yet totally a flexible process. Fret issue of high speed water jet combined with grating is used for metal removal. The intention of cutting by water jet is approved from universe which has been exhibiting the incident of gnawing of the material by a flow of water. For the improvement in AWJ technology, when the fluid flow is combined with sand particles, the pace of erosion is viewed to be quicker. High speed water jet is fabricated by crossing high tensioned water over a very tiny orifice during that time the abrasive is combined with high tensioned water in a blending cabin to produce abrasive jet. These complete cutting ends accommodated by water particles enlarge the percentage of gnawing, developing the machining proficiency of AWJ is very complex compared to pure WJ. The mechanism of material removal contains the following steps: micro-cutting, micro-fracture and plowing deformation. An imperious mechanism of metal cutting is detected by the features of influencing AW grains, viz. physical dimensions of the work piece, stiffness, the jet crash angle, feed rate. Micro-cutting and pushing deformation are allied with plastic materials and shoal crash angles; where micro-machining arise for spiny particles during which plowing deformation is allied with round-shaped particles. Micro-cracking observes when the grains crash about right to the metal piece. Micro-cutting is trusted to be the imperious mechanism for MRR in AWJ cutting of fiber-reinforced plastics [4, 5]. Figure 1 shows a simple schematic diagram of abrasive jet machining (AJM). Pressurized jet may comprehend of an intensifier, prime mover, controller, and an accumulator. Pressurized water at 200–400 Mpa is fed through high pressure jet tube to the module called cutting head. As described in Fig. 1, the flow of water then carries along with a tiny hole (of 0.2–0.3 mm diameter) at high pressure, to frame a velocity (200–300 m/s) WJ at very high rates. Water jet is then entered inside the stirring cabin to blend by abrasives equipped by abrasive furnishing system forming it which exit from vent nozzle of comparably huge diameter with 0.7–0.8 mm. CNC system is used for controlling the

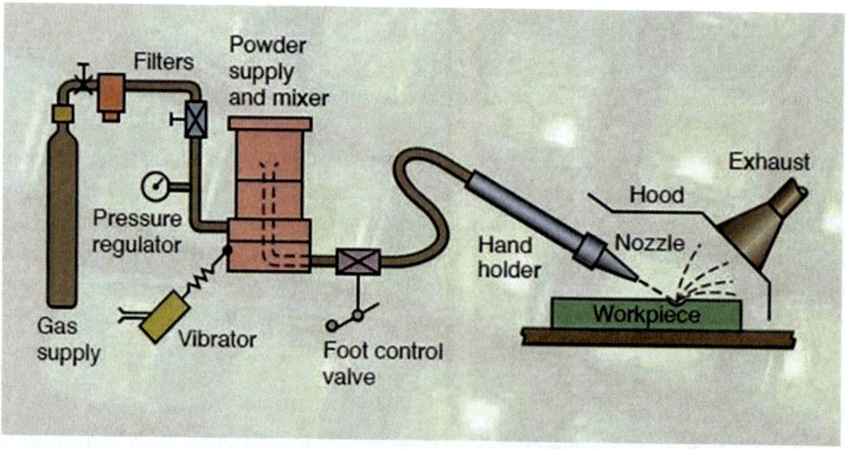

Fig. 1 A simple schematic diagram of Abrasive Jet Machining (AJM)

location and motion of the cutting head [6]. AWJ machining includes utterly a small number of influential parameters that can show the impact on performance of AWJ cutting. Figure 1 shows their general relationship. Moreover, main and easy-to-adopt parameters are considered approximately for obtaining the superior quality of the cut.

Due to its peculiar parameters and amusing features, such as invulnerability and encumbering to scratches, cracks, stains, spills, heat, cold, and moisture, granite has been extensively adopted as bounded stone in societal and economic applications in regular life [7]. AWJ is an advanced tool for penetrating rocks and rocklike metals. It is best suited for cutting, pre-weakening and drilling of tough rocks [8]. For production industries and also for other civil and mining engineering fields, this process is an assured tool due to its exceptional properties of inflexible cutting, and acceptable surface roughness, minor kerf widths, expanded tool life, trouble free-form cutting, flexible process, dust free, ergonomic working conditions, and environment. These highlights make the process an eco-friendly process compared to other conventional machining processes like sawing (circular) in raw stone cutting and figuring areas [9].

There are abundant parameters associated with this process. The connected parameters of AWH machining process are figured out in Fig. 2 [10, 11]. Many of the researchers have focused on understanding the effect of the most influential parameters on responses like material removal rate, surface roughness, kerf and microstructural properties. Very few researchers have been suggested techniques for enhancing

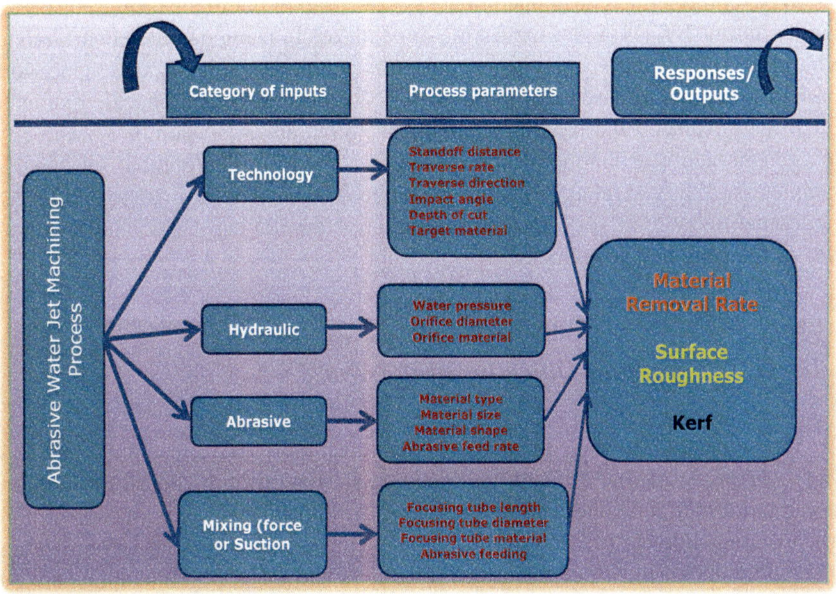

Fig. 2 Process parameters and responses in AWJ process

the surface finish and kerf quality [12, 13]. Based on a thorough knowledge of the AWJC metal removal mechanism, various methods such as polishing [14], turning [15], drilling [16], milling [17] and surface finishing [18] have been progressed to design and machine economically.

Earlier commented, AWJ machining process has about no boundaries. This process is highly preferable for 2D and 3D machining. There is no variation between cutting of cheap construction steel or stainless steel materials, both are equally machined well. If quality is highly demanded during machining process, the process parameters (Fig. 2) must be properly selected and the machining process needs to be ended before occurrence of the abrasion deformation section.

Parameter optimization for higher material removal and recommended surface roughness was done out by Nagdeve et al. [19]. Taguchi's integrated with ANOVA were adopted for optimization. The analysis exhibits that the standoff distance affects the MRR, whereas surface roughness was highly influenced by abrasive flow rate. Ramprasad et al. [20] investigated that the water pressure was the most influential factor for stainless steel 403 work material followed by standoff distance and abrasive flow rate.

Liu et al. [21] used response surface methodology (RSM) with Box–Behnken design (BBD). The outcomes concluded that the transverse speed is an influential factor along with water pressure, abrasive flow rate and tilt angle. AWJ process was optimized using DOE with statistical approach [22]. The results showed that the suitable set of process parameters was responsible for the enhanced quality of finishing, dimension accuracy and for higher productivity.

An ANN model was developed to guess the speed of cut to the promised surface quality during AWJM [23]. Forecasting of depth of cut using neural network model is build [24] published a paper on genetic neuro technique stating that neural network model to predict cut depth is introduced by considering focusing nozzle diameter, rate of jet traverse, water pressure and abrasive flow rate. ANN combined with GA, i.e., genetic neuro technique, is showcased for suggesting the optimal set of input parameters [25]. A start has been done by Ramesh Babu and Srinivasu to build ANN model and fuzzy logic for different figuring are a employing AWJC along with an idea of focusing nozzle diameter [26].

1.1 Motivation and Objective of the Work

The researchers have, even, donated toward exposure of ABWJ machining process. Moreover investigation is being continued still. This proposes the need of future research. Further and continued research may gush through several aspects of more crucial, predictive control and performance of the grinding process. Optimization is certainly one of the aspects. Surface roughness and material removal rate are the most important responses while considering AWJ machining process.

In addition, process variables are anticipated to influence surface roughness, as well as material removal rate during AWJ machining process. Taking into account the aspects, the current work has been done to analyze how the assorted input variables impact on surface integrity and MRR in AWJ machining on granite material. Multi-objective optimization has been attempted by considering simultaneous minimization of surface roughness and maximization of material removal rate. The observation has been done through experiments; data examined using RSM and desirability approach. Mathematical modeling has also been done, to determine the relation between the responses and process parameters. It is predicted that this work and future research in this area will finally direct to build a vigorous knowledge and database, this may facilitate proper selection of input variables more reliably and predictively for people in industry to achieve the desired performance in abrasive water jet machining of granite.

This work is targeted on predicting responses of AWJ process parameters for machining a granite material applying RSM and desirability and also to optimize the process for qualitative productivity.

2 Experimentation

Abrasive water jet machining has been carried out at various levels of process variables as per Box–Behnken design matrix, which is deliberated subsequently. An abrasive jet begin out the like a pure water jet. As the lean flow of water leaves the nozzle, abrasive is joined to the stream and blended. The beam of water precipitates abrasive particles to speeds higher in order to cut much harder materials. The blending of abrasive particles in water jet is in such a way that water jets momentum is shifted to the abrasives. The coherent, abrasive water jet that exists the AWJM nozzle can cut various materials, such as metals, glass, ceramics and composites. The work material (Fig. 3) used in this experiment is granite machined into 27 samples of 2 cm × 2 cm

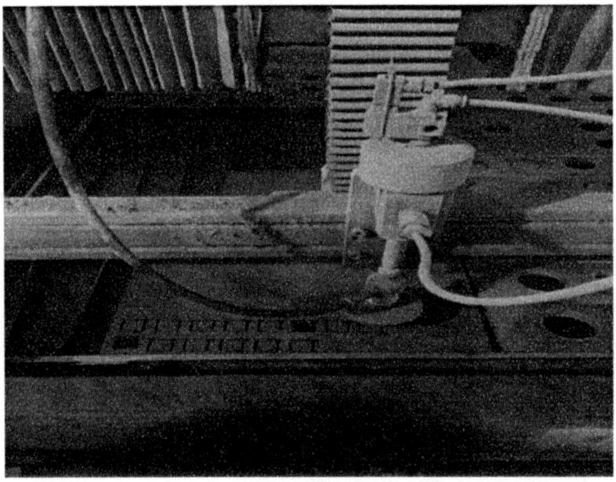

Fig. 3 Granite cutting using AWJM

dimension. Each experiment is repeated twice and average of response values are considered for analysis.

2.1 Experimental Design

2.1.1 Response Surface Methodology

A response surface design is a set of futuristic design of experiments (DOE) techniques that help you comprehend and optimize your response in a better manner. Response surface methodology is frequently applied to cultivate models after most influential factors have been determined using factorial designs.

If a response of interest is influenced by multiple variables RSM is one of the best techniques for modeling and analysis because it is a combination of both mathematical and statistical technique, and then the target is the optimization of output by simultaneous variation of input factors. RSM is one of the operative techniques for enhancing, improving and optimizing the machining process by joining various process parameters and to decide the impact of critical interactions on the achievement of the output variables. For adopting RSM, choice of range of process parameters and effective design of experiment is in need. In present work, model is built by adopting design of experiment with Box–Behnken design and regression analysis in addition with ANOVA for judging about the performance of the model. The responses were modeled to equip quadratic pattern depicted by the given equation.

$$\begin{aligned} Y = &\beta_0 + \beta_1 x_1 + \beta_2 x_2 + \beta_3 x_3 + \beta_{12} x_1 x_2 + \beta_{13} x_1 x_3 \\ &+ \beta_{23} x_2 x_3 + \beta_{11} x_1^2 + \beta_{22} x_2^2 + \beta_{33} x_3^2 \end{aligned}$$

Here, the notations with β are called approximating function parameters.Y is an output parameter and x is process parameter. In this work, all 27 experiments (Table 2) were planned based on Box–Behnken method. Four process parameters (Table 1) were considered for evaluating the performance of machining process. The responses considered for evaluation are material removal rate (MRR) and surface roughness (R_a). Experiments were conducted according to the order suggested by RSM methodology and the results were calculated as per the below formula and recorded in Table 3.

Table 1 Process parameters and their levels

Level	Pressure (bar)	Abrasive (gm/min)	Stand-off (mm)	Feed rate (mm/min)
-1	3000	500	2	350
0	3200	600	2.5	400
1	3600	800	3	495

Table 2 Box Behnken Experimental design

S.NO	PRESSURE (bar)	ABRASIVE (gm/min)	STAND-OFF (mm)	FEED RATE (mm/min)
1	0	−1	0	1
2	0	0	0	0
3	0	1	0	-1
4	−1	1	0	0
5	0	0	−1	−1
6	1	0	0	−1
7	1	0	0	1
8	1	0	−1	0
9	0	1	0	1
10	−1	0	0	−1
11	0	0	−1	1
12	0	1	1	0
13	0	−1	1	0
14	0	0	0	0
15	0	−1	0	−1
16	−1	−1	0	0
17	0	0	1	1
18	−1	0	−1	0
19	0	1	−1	0
20	1	1	0	0
21	0	−1	−1	0
22	0	0	1	−1
23	1	−1	0	0
24	1	0	1	0
25	−1	0	1	0
26	−1	0	0	1
27	0	0	0	0

$\text{MRR} = H_t \, \text{WD}_i$
H_t = depth of penetration
V_f = traverse speed of the abrasive water jet
W = width of the kerf
H_t = 16.5 mm
V_f = varies at three different levels
W = varies depending on the kerf value.

Kerf width is calculated using Vernier caliper. Surface roughness of the samples is found using roughness testing machine.

Table 3 Experimental Results

S.NO	PRESSURE (bar)	ABRASIVE (gm/min)	STAND-OFF DISTANCE (mm)	Traverse speed (mm/min)	KERF (mm)	SR (μm)	MRR (mm^3/min)
1	3200	500	2.5	495	0.17	6.3496	528
2	3200	600	2.5	400	0.43	5.2516	150.15
3	3200	800	2.5	350	0.1	6.0605	528
4	3000	800	2.5	400	0.3	7.9998	2376
5	3200	600	2	350	0.15	6.2907	3185.325
6	3600	600	2.5	350	0.026	4.9293	4290
7	3600	600	2.5	495	0.39	7.3537	1388.475
8	3600	600	2	400	0.08	6.9946	2838
9	3200	800	2.5	495	0.08	6.0177	1122
10	3000	600	2.5	350	0.01	6.5055	303.6
11	3200	600	2	495	0.09	5.6061	735.075
12	3200	800	3	400	0.1	6.0196	792
13	3200	500	3	400	0.09	5.8027	594
14	3200	600	2.5	400	0.046	7.0307	130.68
15	3200	500	2.5	350	0.27	6.1613	1559.25
16	3000	500	2.5	400	0.32	7.2511	577.5
17	3200	600	3	495	0.016	5.5828	653.4
18	3000	600	2	400	0.13	6.7500	660
19	3200	800	2	400	0.17	5.9860	866.25
20	3600	800	2.5	400	0.08	6.4728	693
21	3200	500	2	400	0.12	6.2216	924
22	3200	600	3	350	0.12	5.3928	198
23	3600	500	2.5	400	0.36	6.1543	81.675
24	3600	600	3	400	0.65	6.1904	57.75
25	3000	600	3	400	0.03	6.0323	1980
26	3000	600	2.5	495	0.01	6.4509	858
27	3200	600	2.5	400	0.14	6.4300	2112

2.2 *Empirical Model Formulation*

RSM has been implemented to calculate MRR and surface roughness of granite material during AWJ machining process. Box–Behnken design of 27 experiments is considered establishing the model for formulation of relationship between responses and process parameters. Quadratic model for surface roughness and material removal rate is framed based on Eqs. 1 and 2.

Following equations show the response surface equation for surface roughness and material removal rates in terms of process variables.

$$\begin{aligned} \text{MRR} = {} & 1355 + 481A - 275B + 139C + 189D + 277A*A + 27B*B \\ & - 332C*C - 506D*D - 429A*B + 1106A*C + 753A*D \\ & - 66B*C + 62B*D - 108C*D \end{aligned} \quad (1)$$

Regression equation of surface roughness

$$\begin{aligned} \text{SR} = {} & 62374 - 2412A + 513B - 2357C + 1684D + 4929A*A + 1694B*B \\ & - 2840C*C - 3051D*D - 1076A*B - 216A*C + 6198A*D \\ & + 1131B*C - 578B*D + 2186C*D \end{aligned} \quad (2)$$

3 Results and Discussion

Based on response surface analysis, the impact of process variables on responses is discussed in this section.

3.1 Effect on Process Parameters on Surface Roughness

The goodness-of-fit in regression and ANOVA is examined by residual plots. Studying residual plots assist to find whether the ordinary least squares assumptions are being rescued. If these assumptions are fulfilled, then ordinary least squares regression will generate unbiased coefficient estimates with the least possible deviation. The normal plot of residuals is used to check the assumption that the residuals are normally distributed. The plot of residuals versus fits is used to check the assumption that the residuals have a fixed variance.

It is clear that surface roughness (Fig. 4) is mainly influenced by pressure then by SOD and feed rate in the sequence. Surface roughness decreases with growth in standoff distance. High surface finish is obtained at low feed rate and high standoff distance. An optimal value of surface roughness of 5–6 μm is obtained at 3 mm standoff distance and at a feed rate of 350 mm/min. At low pressure and low abrasive flow rate, surface roughness is expected at very high rate.

Non-influential interaction terms have been eliminated. The accuracy of model in terms of the three tests, and analysis for lack of fit is verified by ANOVA. The standard of degree of fit is represented by coefficient of determination (*R*-sqr). The statistics of *R*-squared values clearly exhibits that, the pattern elucidates 92.2% of the total deviation. The obtained R^2 value after calibrated for measure of the model is 85.39%, smaller than permissible variation betwixt *R*-sqr and *R*-adj sqr. When comparing

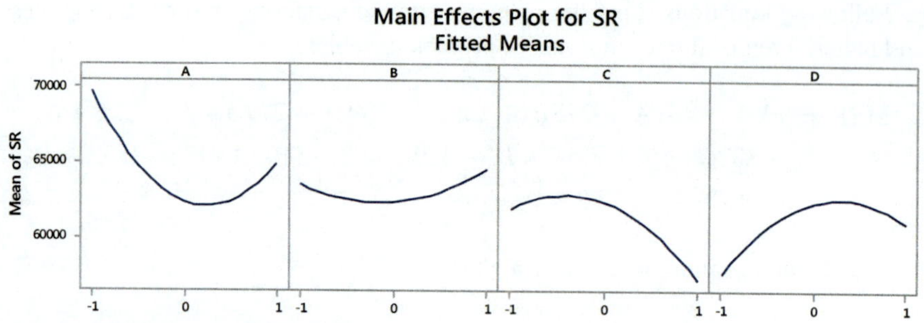

Fig. 4 Main effects plot for SR

$R^2_{\text{Adj}} = 0.8539$ with $R^2_{\text{pred}} = 0.4692$ explains both the terms are mutually covenant with one other and the pattern would be demanded to justify 46.93% variability in new data.

Fairness of the suggested model in RSM is to be inspected by residual reports. Plots of all residuals of surface roughness have represented by Fig. 5. It gives the information about errors and assumptions considered in this research. Closeness of points to the linear line indicates that expectations are not offended, because of normal and independent distribution of errors. Residual versus run number curve (Fig. 5) demonstrates that there is no intricate structure and pristine pattern, which discover that independent and fixed diverged assumptions are accepted and no relationship has been observed between residuals. Figure 5 represents the errors with normal

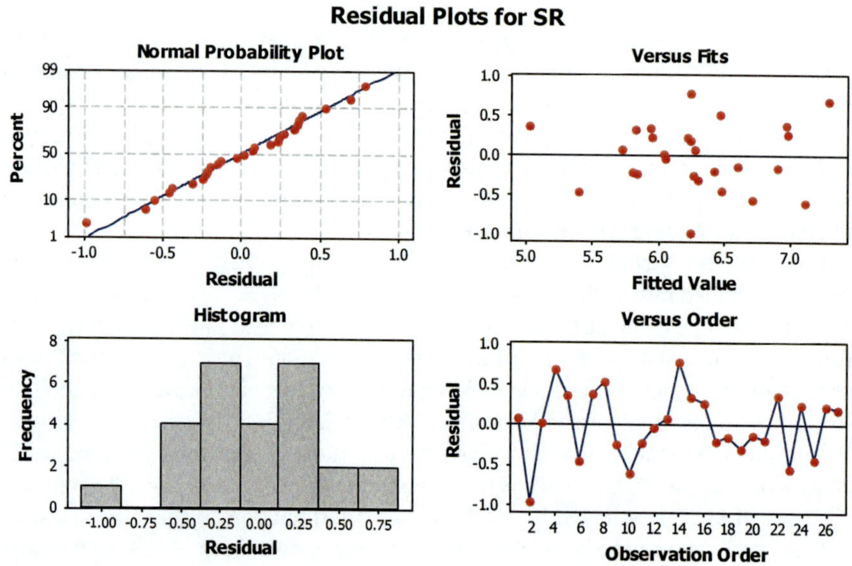

Fig. 5 Residual plots of SR

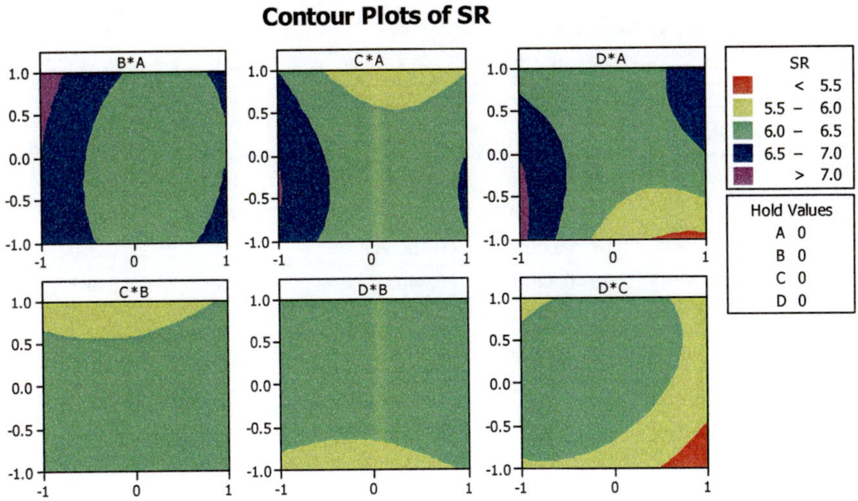

Fig. 6 Contour plots of SR

distribution, since actual and predicted values follow a straight line. It is concluded that the fairness of the suggested model based on above explanation. No cause was found to suspect any violator of independence or constant deviation assumptions. A contour plot provides (Fig. 6) a two-dimensional view of the surface where points that have the same response are connected to produce contour lines of constant responses. Contour plots are useful for establishing the response values and operating conditions (Table 4).

3.2 *Effect of Process Variables on MRR*

Figure 7 describes a linear increase of MRR as pressure increases. Abrasive flow rate does not have any significant impact on MRR. Both SOD and traverse speed has similar impact on MRR. Figure 7 shows that maximum material removal rate can be achieved at high pressure and low SOD, traverse speed.

Non-influential interaction terms have been eliminated. An accuracy of the model consists of the three tests, i.e., significance of regression model and significance of model coefficients, and tests for lack of fit is verified by ANOVA (Table 5). The standard of degree of fit is shown by coefficient of determination (*R*-squared). *R*-squared statistics clearly exhibit that the model explains 90.6% of the total deviation. The obtained R^2 value after calibrated for size (terms) of model is 88.23%, smaller than the permissible variation between *R*-squared and adjusted *R*-squared. Comparison of $R^2_{\text{Adj}} = 0.8826$ with $R^2_{\text{pred}} = 0.5273$ explains that both the terms are in mutual covenant with each other and the model would be demanded to justify 52.73% variability in new data.

Table 4 Analysis of Variance of SR

Source	DF	Adj SS	Adj MS	F-Value	P-Value
Regression	14	7.13283	0.50949	1.34	0.310
Linear	4	1.73670	0.43418	1.14	0.384
A	1	0.69818	0.69818	1.83	0.201
B	1	0.03160	0.03160	0.08	0.778
C	1	0.66665	0. 66665	1.75	0.210
D	1	0.34027	0.34027	0.89	0.363
Square	4	3.55586	0.88897	2.33	0.115
A*A	1	1.29556	1.29556	3.40	0.090
B*B	1	0.15305	1.5305478	0.40	0.538
C*C	1	0.43023	0.43023	1.13	0.309
D*D	1	0.49649	0.49649	1.30	0.276
Interaction		1.84027	0.30671	0.81	0.585
A*B	1	0.04627	0.04627	0.12	0.733
A*C	1	0.00187	0.00187	0.00	0.945
A*D	1	1.53636	1.53636	4.04	0.068
B*C	1	0.05119	0.05119	0.13	0.720
B*D	1	0.01335	0.01335	0.04	0.855
C*D	1	0.19123	0.19123	0.50	0.492
Residual Error	12	4.5688	0.38074		
Lack-of-Fit	10	2.93066	0.29307	0.38	0.891
Pure Error	2	16.3822	0.81911		
Total	26	11.7017			

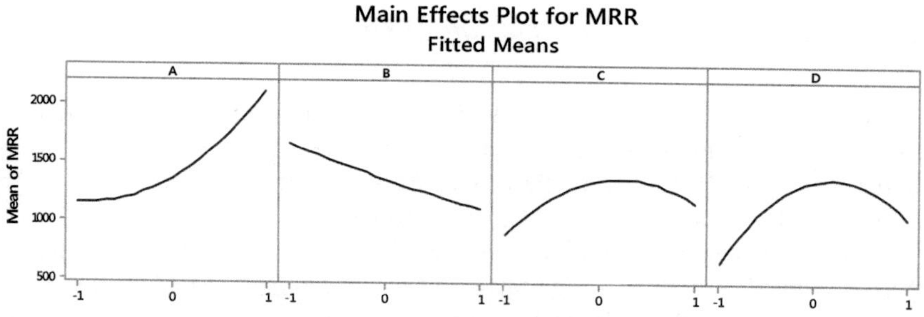

Fig. 7 Main effects plot for MRR

Table 5 Analysis of Variance of MRR

Source	DF	Adj SS	Adj MS	F-Value	P-Value
Regression	14	17757533	1268395	1.38	0.291
Linear	4	4864353	1268395	1.32	0.317
A	1	560650	560650	0.61	0.45
B	1	372002	372002	0.40	0.537
C	1	2028285	2028285	2.21	0.163
D	1	1903416	1903416	2.07	0.176
Square	4	2564554	641139	0.70	0.608
A*A	1	1110488	1110488	1.21	0.293
B*B	1	273571	273571	0.30	0.595
C*C	1	88453	88453	0.10	0.762
D*D	1	698701	1721437	0.76	0.400
Interaction	6	10328625	0.30671	1.87	0.167
A*B	1	352346	352346	0.38	0.547
A*C	1	4203013	4203013	4.57	0.054
A*D	1	2985854	2985854	3.25	0.097
B*C	1	16352	16352	0.02	0.896
B*D	1	660359	660359	0.72	0.413
C*D	1	2110700	2110700	2.30	0.156
Residual Error	12	11027763	918980		
Lack-of-Fit	10	8436142	8436142	0.65	0.738
Pure Error	2	2591621	1295811		
Total	26	28785296			

Fairness of the proposed model in RSM can be inspected by residual analysis. Plots of all residuals of surface roughness have represented by Fig. 8. It gives the information about errors and assumptions considered in this research. Closeness of points to the linear line indicates that expectations are not offended, because of normal and independent distribution of errors. Residual versus run number curve (Fig. 8) demonstrates that there is no intricate structure and pristine pattern, which discover that independent and fixed diverged assumptions are accepted and no relationship has been observed between residuals. Figure 5 represents the errors with normal distribution, since actual and predicted values follow a straight line. It is concluded that the fairness of the suggested model based on above explanation. No cause was found to suspect any violator of independence or constant deviation assumptions. It is resulted from counter plots (Fig. 9) that an optimum value of 2000–3000 mm^3/s of MRR can be achieved by setting pressure to 3600 bar, SOD 2 mm and traverse speed to 350 mm/min. MRR is observed to be minimum at low abrasive flow rate and high SOD, traverse speed.

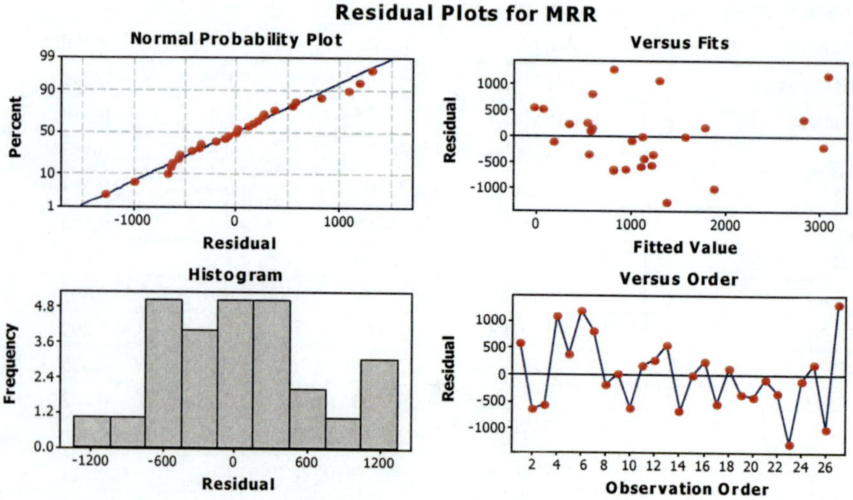

Fig. 8 Normal probability plots of MRR

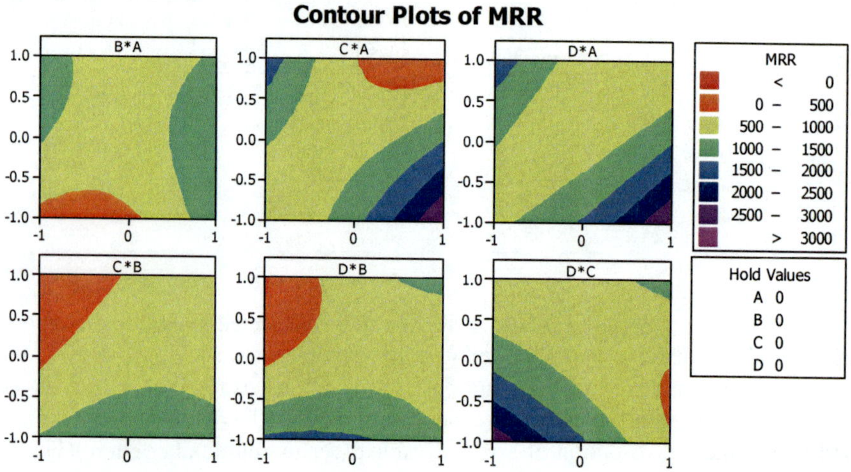

Fig. 9 Contour plots of MRR

4 Desirability Approach for Multi-objective Optimization

4.1 Optimization Formulation

The objective of this work is to obtain the maximum possible material removal rate (MRR) or productivity by maintaining the surface roughness at best favorable amount by optimizing the AWJ process. But, these two objects are in combat by nature, i.e., it is not possible to present single set of process parameter as the wholly optimal set of process parameters subjected to various experimental conditions. Otherwise, based on surface roughness, the current optimization problem should apart into various machining reign. An assumption is made that optimum process parameters which fulfill the constraints claimed by the experimental circumstances and met the requirements of the surface finish during the same time. Hence, the optimization of the AWJ machining process is formulated mathematically as follows:

$$\text{Max}: \; F(X) = \text{MRR}$$
$$\text{Subject to}: \; G(X) = R_a \leq R_{amax}$$

$$\begin{aligned}\text{Side constraints}: 3000 &\leq x_1 \leq 3600 \\ 500 &\leq x_2 \leq 800 \\ 2 &\leq x_3 \leq 3 \\ 350 &\leq x_4 \leq 495\end{aligned}$$

where x_1, x_2, x_3 and x_4 are process parameters (pressure, abrasive, standoff and feed rate, respectively). According to the importance of surface roughness value, the R_{amax} values are consigned and will be interpreted in detail during the optimization.

4.2 *Desirability Approach for Optimization*

Derringer and Suich [27] popularized this optimization technique. It is a search-based class of optimization making appliance of desirability functions for predicting optimal set of process parameters globally. As a whole, the method is primarily to transform each output y_i into a distinct desirability function d_i and that arrays over

$$0 \leq d_i \leq 1$$

where if the output y_i is at its target, then $d_i = 1$, and if the response exceeds an acceptable region, $d_i = 0$. For defining the shape of desirability function, weight factor (r), with a positive number, is associated with it.

Regarding the optimal or ambitious value, the sensitivity should be low for the desirability function, if the weight is chosen to be less than 1. Otherwise, the explored algorithm detects a point which is some way distant from the needed optimal value, and then the desirability function value will be falling to low value compared to its maximum amount. Considering a priority weightage with above one, has the contrary issue, and fixed it to one, assist an offset or mean sensitivity by the linear desirability function [28, 29]. The separate desirability functions are interpreted according to the target of optimization.

If the target or goal T_i for the output y_i is a larger value, then:

$$d_i = \left\{ \left(\frac{y - y_{\min}}{y_{\max} - y_{\min}} \right)^{\mathrm{wt}}, \left\{ \begin{array}{c} 0 \to y \le y_{\max} \\ y_{\min} \le y \le y_{\max} \\ 1 \to y \ge y_{\max} \end{array} \right\} \right\} \quad (3)$$

$$d_i = \left\{ \left(\frac{y - y_{\max}}{y_{\min} - y_{\max}} \right)^{\mathrm{wt}}, \left\{ \begin{array}{c} 1 \to y \le y_{\min} \\ y_{\min} \le y \le y_{\max} \\ 0 \to y \ge y_{\max} \end{array} \right\} \right\} \quad (4)$$

Here, d was a desirability function of y, $y_{\min}$ and $y_{\max}$ are boundaries (lower and upper) of response value of 'y', w_t was priority weightage lies in the range 0.1–10 to fit the outline of desirability function. An overall desirability function D(0 _ D _ 1) was interpreted as the geometric mean of separate desirability functions. The geometric mean of all responses of single objectives was given as multiple objective functions as shown in Eq. (5). The higher the D value leads for better desirability (Table 6) of the collective response levels.

$$D = (d_1 \times d_2 \times \cdots \times d_n)^{1/n} \quad (5)$$

Non-influential interaction terms have been eliminated. The model accuracy consists of the three tests (Fig. 10) i.e., significance of regression model, significance of model coefficients and tests for lack of fit is verified by ANOVA (Table 6). R-squared value represents the standard of degree of fit. R-squared figures clearly exhibits that the model shows 95.87% of the total deviation. The obtained R^2 value after calibrated for size (terms) of model is 90.64%, smaller than the permissible variation between R-sq and adjusted R-sq. Contrast of $R^2_{\mathrm{Adj}} = 0.9064$ with $R^2_{\mathrm{pred}} = 0.4082$ explains that both the terms are in mutual covenant with each other and the model would be demanded to justify 40.82% variability in new data. Based on contour plots (Fig. 11), a maximum value of desirability is achieved at lower conditions of SOD and feed rate. Pressure and abrasives are suggested to maintain at medium or at higher range for multi-objective optimization to achieve the set target.

From the generated quadratic mathematical responses, d_1 and d_2 are deliberated as the desirability functions for the MRR and Ra, respectively. Furthermore, the goals are set for maximization of MRR and minimization of R_a while being inferior to the accurate assigned value. 4 and 6 μm are two values, which have been selected for R_{amax}. The choice of these values is based on the span of surface roughness value attained during the experimentation (Table 3). At the end, unit weight factor ($r = 1$) was advised for each desirability function which indicates similar concern for both the responses during optimization.

Figure 12 represents the optimization results of AWJ process. In this figure, input variable is denoted by each column and responses are denoted by each row, i.e., output parameters. However, each cell indicates the variation in output with respect to the function of one of the input factors by maintaining the other variables constant. Current optimal parametric settings are represented by the cells inside the vertical lines, whereas present performance values are indicated by the dotted flat line. 1 and

Table 6 Desirability values of responses and Composite desirability

Ex.No	Desirability of SR	Desirability of MRR	Composite desirability	Rank
1.	0.732916	0.333333	0.494272	18
2.	0.946029	0.147758	0.373876	23
3.	0.794591	0.333333	0.514649	16
4.	0	0.740107	0	27
5.	0.745895	0.859643	0.800752	2
6.	1	1	1	1
7.	0.458212	0.560736	0.506889	17
8.	0.571821	0.810506	0.680782	5
9.	0.803319	0.50146	0.634691	7
10.	0.697397	0.241018	0.409982	21
11.	0.882865	0.400049	0.594297	9
12.	0.802934	0.41652	0.578307	11
13.	0.845804	0.355957	0.548699	14
14.	0.56144	0.131271	0.271478	24
15.	0.773645	0.595631	0.678828	6
16.	0.493349	0.350438	0.415798	20
17.	0.887155	0.375154	0.576905	12
18.	0.63772	0.377227	0.490474	19
19.	0.809724	0.437073	0.594902	8
20.	0.704995	0.387424	0.52262	15
21.	0.76084	0.452414	0.586698	10
22.	0.92139	0.18204	0.409548	22
23.	0.775118	0.075187	0.241409	25
24.	0.767492	0	0	26

-1 denote the high and small range of each process design variable, respectively. The critical part of this figure is the setting of optimal parameter necessitates gaining the determined objective standard, positioned at the center row at the middle of peak and deep row, represented in disciplined form. MRR at the maximized value is shown by the first column of each figure, accomplished along with the lowered values of Ra. The compound desirability (D) as well as specific desirability (d) is detected to be unity, which assures the survival of optimal points (global optimal) in each individual case [27, 29] (Table 7).

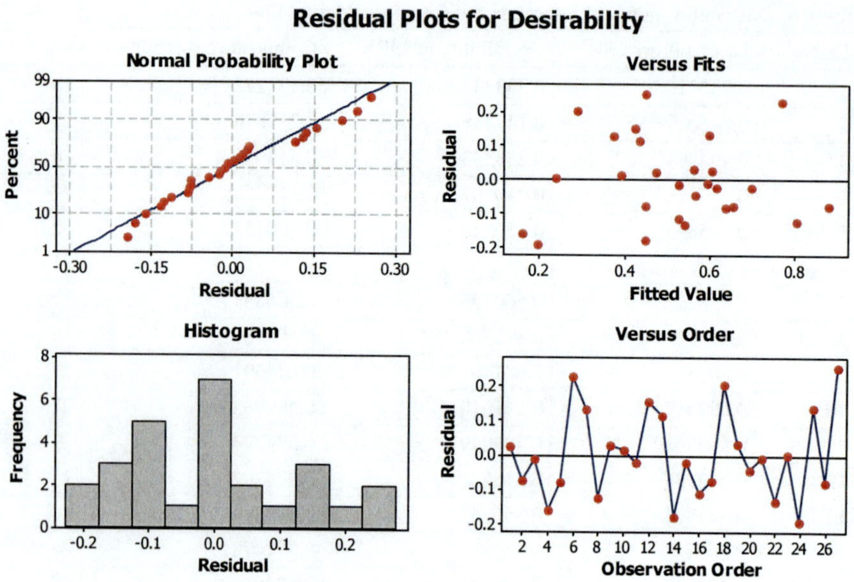

Fig. 10 Normal probability plot with MRR as response

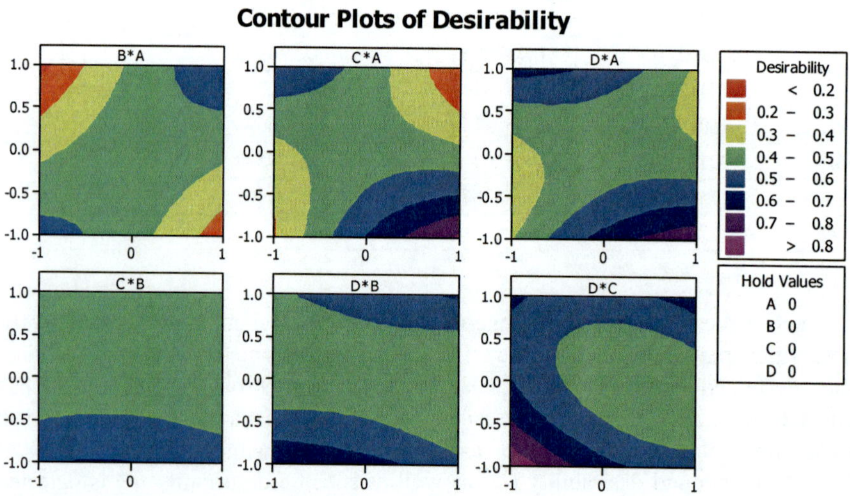

Fig. 11 Contour plots of desirability

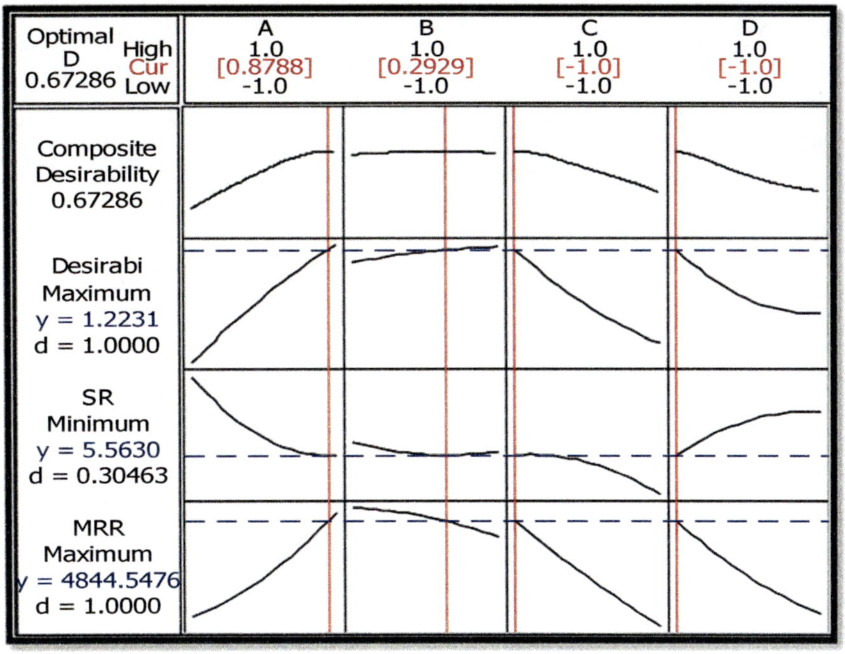

Fig. 12 Optimal results of AWJ process through desirability approach

5 Conclusions

The impact of AWJ machining variables on responses is reported in results and discussion section. From the analysis of experimental results, the below conclusions were listed.

- From results analysis, it was found that a linear increase of MRR as pressure increases. Rate of abrasive flow does not have any major impact on MRR. SOD and traverse speed have shown similar impact on MRR. Maximum material removal rate is achieved at peak pressure and small SOD, traverse speed.
- From contour plots, it is observed that an optimum value of 2000–3000 mm^3/s of MRR can be attained by setting pressure to 3600 bar, SOD 2 mm and traverse speed to 350 mm/min. MRR is observed to be minimum at low abrasive flow rate and high SOD, traverse speed.
- Pressure has shown a high influence on surface roughness followed by SOD and feed rate. Surface roughness decreases with rise in SOD.
- With minimum feed rate and maximum SOD high surface finish are achieved. An optimal value of surface roughness of 5–6 μm is obtained at 3 mm standoff distance and at a feed rate of 350 mm/min. At low pressure and low abrasive flow rate surface roughness expected at very high.

Table 7 Analysis of Variance for Desirability

Source	DF	Adj SS	Adj MS	F-Value	P-Value
Regression	14	0.81105	0.057932	1.72	0.177
Linear	4	0.095607	0.023902	0.71	0.602
A	1	0.009940	0.009940	0.29	0.597
B	1	0.001211	0.001211	0.04	0.853
C	1	0.067503	0.067503	2.00	0.183
D	1	0.016954	0.016954	0.50	0.492
Square	4	0.219948	0.054987	1.63	0.231
A*A	1	0.015664	0.015664	0.46	0.509
B*B	1	0.002722	0.002722	0.08	0.781
C*C	1	0.002722	0.002722	0.90	0.361
D*D	1	0.116993	0.116993	3.47	0.087
Interaction	6	0.495495	0.082582	2.45	0.088
A*B	1	0.121456	0.121456	3.60	0.082
A*C	1	0.213785	0.213785	6.33	0.027
A*D	1	0.102011	0.102011	3.02	0.108
B*C	1	0.000115	0.000115	0.00	0.955
B*D	1	0.023195	0.023195	0.69	0.423
C*D	1	0.034934	0.034934	1.03	0.329
Residual Error	12	0.405090	0.033758		
Lack-of-Fit	10	0.302045	0.030204	0.59	0.770
Pure Error	2	0.103046	0.051523		
Total	26	1.21614			

- Desirability approach is used for multi-objective optimization and it proved that, effective and versatile in finding optimal set of process parameters subjected to a defined level of surface roughness.
- Though AWJ process is distinct and complicated in nature, its performance can be investigated easily and deliberated through a unified approach by integrating desirability concept with RSM.
- After experimental verification, the attained optimal settings show maximum MRR as 4844.76 mm/m^3 and acceptable amount of surface roughness is 5.5630. The composite desirability is 0.67286, which are satisfactory from the point of view of engineering discipline.

References

1. Mishra PK (2005) Non-conventional machining. Narosa Publishing House, Third reprint
2. Khan AA, Hague MM (2007) Performance of different abrasive material during abrasive water jet machining of glass. J Mater Process Technol 191:404–407
3. Azmir MA, Ahsan AK (2009) A study of abrasive water jet machining process on glass/epoxy composite laminate. J Mater Process Technol 209:6168–6173
4. Folkes J (2009) Waterjet an innovative tool for manufacturing. J Mat Proc Tech 209:6181–6189
5. Patel SR (2010) Waterjet and abrasive waterjet Machining. A survey of capability a Machine in the Field. In: National Conference MEET'10, CD of Proceeding, Vardhman College of Engineering, Hyderabad, 4–5 June
6. Patel SR, Shaikh AA (2013) Control and measurement of abrasive flow rate in an abrasive waterjet machine. Int J Innovat Res Sci, Engg and Technol 2(11):7675–7679
7. Zhao C, Gong H, Feng FZ, Li ZJ (2013) Experimental study on the cutting force difference between rotary ultrasonic machining and conventional diamond grinding of K9 glass. Mach Sci Technol: An Internat J 17(1):129–144
8. Momber AW, Kovacevic R (1997) Test parameter analysis in abrasive water jet cutting of rocklike materials. Int J Rock Mech Min Sci 34:17–25
9. Vundavilli RP, Parappagoudar MB, Kodali SP, Benguluri S (2012) Fuzzy logic-based expert system for prediction of depth of cut in abrasive water jet machining process. Knowl Based Syst 27:456–464
10. John Rozario Jegaraj J, Babu NR (2007) A soft computing approach for controlling the quality of cut with abrasive waterjet cutting system experiencing orifice and focusing tube wear. J Mater Proc Technol 185(1–3):217–227
11. Shanmugam DK, Wang J, Liu H (2008) Minimization of kerf tapers in abrasive waterjet machining of alumina ceramics using a compensation technique. Int J Mach Tools Manuf 48:1527–1534
12. Shanmugam DK, Masood SH (2009) An investigation of kerf characteristics in abrasive waterjet cutting of layered composites. Int J Mater Proc Technol 209:3887–3893
13. Lemma E et al (2002) Optimising the AWJ cutting process of ductile materials using nozzle oscillation technique. Int J Mach Tools Manuf 42:781–789
14. Fenggang L, Geskin ES, Tismenetskiy L (1996) Feasibility study of abrasive waterjet polishing. In: 13th International Conference on Jetting Technology. Sardinia, pp 709–723
15. Hashish M (1987) Turning with abrasive-waterjets—a first investigation. Trans ASME J Eng Ind 109:281–290
16. Yong Z, Kovacevic R (1997) Modeling of jet flow drilling with consideration of the chaotic erosion histories of particles. Wear 209:284–291
17. Hashish M (1989) An investigation of milling with abrasive-waterjets. Trans ASME J Eng Ind 111(2):158–166
18. Borkowski P (2004) Theoretical and experimental basis of hydro-jet surface treatment. Koszalin
19. Nagdeve L, Chaturvedi V, Vimal J (2012) Implementation of Taguchi approach for optimization of abrasive water jet machining process parameters. Int J Instrum Control Autom (IJICA) 1(3–4):9–13
20. Ramprasad GU, Hassan K (2015) Optimization MRR of stainless steel 403 in abrasive water jet machining using anova and taguchi method. Int J Eng Res Appl 5(5):86–91
21. Liu D et al (2014) Modeling and optimization of operating parameters for abrasive waterjet turning alumina ceramics using response surface methodology combined with Box-Behnken design. Ceram Int 40(6):7899–7908
22. Ibraheem HMA, Iqbal A, Hashemipour M (2015) Numerical optimization of hole making in GFRP composite using abrasive water jet machining process. J Chin Inst Eng 38(1):66–76
23. Lu Y, Li X, Jiao B, Liao Y (2005) Application of artificial neural networks in abrasive water jet cutting process. In: International symposium on neural networks. Chongqing, China, pp 877–882

24. Srinivasu DS, Ramesh BN, Srinivasa YG, Louis H, Peter D, Versemann R (2005) Genetically evolved artificial neural networks built with sparse data for predicting depth of cut in abrasive water jet cutting. In: Proceedings American Water Jet Conference. Houston, Texas, pp 1–16
25. Srinivasu DS, Ramesh BNA (2008) Neuro-Genetic approach for selection of process parameters in abrasive water jet cutting considering variation in diameter of focusing nozzle. Int J Appl Soft Compt 8(1):809–819
26. Srinivasu DS, Ramesh BN, Srinivasa YG (2004) Neuro-Genetic approach for automatic selection of process parameters in abrasive water jet cutting considering the variation in bore diameter of focusing nozzle. In: Proceedings 21st all india manufacturing technology design and research conference, Vellore, India. December 2004, pp 887–893
27. Derringer G, Suich R (1980) Simultaneous optimization of several response variables. J Qual Technol 12(4):214–219
28. Myers RH, Montgomery DC (2002) Response surface methodology, 2nd edn. Wiley, New York. ISBN 0-471-41255-4
29. Castillo ED, Montgomery DC, Mc Carville DR (1996) Modified desirability functions for multiple response optimizations. J Qual Technol 28(3):337–345

Taguchi Optimization of AWJM Process Parameters on Aluminium Hybrid Composite

P. Ganesan, C. Sivakandhan, S. Marichamy, D. Madan, B. Stalin, and V. Dhinakaran

Abstract Metal matrix composites are used in the aircraft structural components and automobile sectors. Abrasive waterjet machining is most economic and effective non-traditional machining process for machine aluminium metal matrix composites. The fabrications have been carried out of AA2024-B_4C-TiC hybrid composites and the experiments conducted on a water jet machine. The effect of machining parameters for AWJM of hybrid AMMC using the Taguchi technique was carried out, and the optimal parameters for surface roughness, kerf angle and MRR were determined. The percentage of contribution for each input such as water jet pressure, traverse rate, and standoff has been found by ANOVA.

Keywords AA2024 · Hybrid composite · Taguchi technique · Analysis of variance · Abrasive water jet machining

P. Ganesan · S. Marichamy · D. Madan
Department of Mechanical Engineering, Sri Indu College of Engineering and Technology, Hyderabad, Telangana, India

C. Sivakandhan
Department of Mechanical Engineering, Sri Indu Institute of Engineering and Technology, Hyderabad, Telangana 501510, India

B. Stalin (✉)
Department of Mechanical Engineering, Anna University, Regional Campus Madurai, Madurai 625019, Tamil Nadu, India
e-mail: stalin1312@gmail.com

V. Dhinakaran
Department of Mechanical Engineering, Chennai Institute of Technology, Kundrathur, Tamil Nadu, India

A. Arockiarajan et al. (eds.), *Advances in Industrial Automation and Smart Manufacturing*, Lecture Notes in Mechanical Engineering,
https://doi.org/10.1007/978-981-15-4739-3_28

1 Introduction

The research and development work has been rapidly increasing in aluminium metal matrix composites, due to ease of preparation, low processing cost involved and adaptability to a wide range of properties, and cast aluminium alloy-based composites have been a subject of great interest to the researchers and technologists. Numbers of researchers have been conducted in the area of AMC, and some of them concentrated on the hybrid composites [1]. Sharma et al. [2] have studied that a wide variety of ceramic particulates such as SiC, Al_2O_3, B_4C, TiN, TiO_2 and TiC are used as reinforcement in aluminium metal matrix composites to get an improved mechanical and tribological properties. AA 2024 is used in valve fittings, hydraulic valve bodies, missile parts and munitions. Recent years, researchers concentrate on hybrid AMMCs having two or more reinforcement. Bhandare et al. [3] have fabricated MMC using stir casting processing. Stir casting is the most economical methods of processing of MMC. The properties of MMCs depend upon many processing parameters, and preparation of MMC with reinforcement of SiC, Al_2O_3, and graphite of varying proportion and wettability between all these materials are assured. Venkatachalam and Kumaravel [4] concluded that a numerous of fabrication methods are existing, and the stir casting process is a simplest fabrication route, low cost and most effective method. In case of mass production, stir casting is the best suitable method because of low cost and effectiveness and also greatest viable and prevalent method. It is considered to be a possible method for manufacturing application in terms of fabrication capacity and cost-effectiveness.

The machining of ceramic reinforced composites is difficult and non-viable using conventional machining because the addition of ceramic particles increases the hardness of composite and also the presence of ceramic particles reduces the tool life.

Kechagias et al. [5] discussed those operational parameters, abrasive parameters, hydraulic parameters and nozzle parameters. Patel and Tandon [6] have discussed the performance of response parameters such as the material removal rate (MRR), depth of cut, kerf width and surface roughness. Most of the previous research papers are based on abrasive waterjet machining to machine the different materials for different applications, but this study deals with abrasive waterjet machining for aluminium alloy-boron carbide-titanium hybrid composite [7].

Present work deals with fabrication of aluminium hybrid composite, abrasive waterjet machining process and Taguchi optimization.

2 Method of Fabrication

The material composition of AA2024 is given in Table 1. Boron carbide is one of the hardest and low-density materials, by-products in the production of metal borides and grey in colour, abundantly available and having hardness values of 2750 HV,

Table 1 Material composition of AA2024

Cu	Mg	Mn	Fe	Si	Zn	Cr	Ti	Al
4.85	1.31	0.667	0.254	0.11	0.079	0.033	0.008	Bal

Fig. 1 Stir casting set-up

which is greater than that of SiC. Titanium carbide is a tremendously hard ceramic material, and it has the appearance of black powder. It has a hardness value of 2470 HV, which is equal to that of SiC.

The stir casting set-up of composite preparation is shown in Fig. 1. The selected particles are B_4C (150 μm particle size and density of 2.5278 g/cm^3) and TiC (150 μm particle size and a density of 4.52 g/cm^3). The hybrid composites were produced through stir casting process by the two-step method. The first AA2024 alloy was melted in a resistance furnace to a temperature of 640 °C and then cooled to 490 °C. The alloy was in slurry state, and the slurry was stirred by a stirrer to create a vortex. While stirring the slurry, the two preheated ceramic particles were added in equal volume percentage to it. Now, the slurry with ceramic particles was again heated to 650 °C and stirring was carried out for 10 min with the average stirring rate of 300 rev/min. The molten mixture was poured into the dies and then cooled. The procedure was followed to prepare specimens of composites with particles of 5% (2.5% B_4C + 2.5% TiC), 10% (5% B_4C + 5% TiC), 15% (7.5% B_4C + 7.5% TiC) and 20% (10% B_4C + 10% TiC), to the size of 100 mm × 100 mm × 15 mm [8].

3 Machining

The machining was done on water jet cutter (Model DWJ1313-FB) with operating pressure of 300 MPa and traverse speed of 0–60 mm/min. The average 80 mesh size garnet was used as abrasive particles. The jet of water impinged vertically downward direction on the workpiece. The orifice assembly has a 0.7 mm diameter size carbide nozzle. For entire machining, mass flow rate of abrasive particles was maintained as 3.1 L/min. The selected parameters are water jet pressure, traverse speed and standoff distance selected to machine the hybrid composites. The average surface roughness was measured on the water jet machined surface using a surface tester (Mitutoyo Model SJ401). The kerf width at the top and bottom is measured by using an optical microscope, and the kerf angle is calculated using Eq. 1. The selected orthogonal array was L_{27} based on experiments conducted. The MRR is found using Eq. 2 [9].

$$\text{Kerf Angle} = \tan^{-1}(w_1 - w_2/2t) \tag{1}$$

$$\text{MRR} = (w_1 + w_2/2) \times \text{Thickness of work piece}(t) \times \text{Cutting velocity}(v) \tag{2}$$

4 Results and Discussion

Taguchi's optimization technique of selecting a limited number of experiments which creates the most information is identified as a partial fraction experiment [10]. To identify the optimum level of parameters, conducted confirmation experiments have been conducted and calculated the percentage improvements in the specified objectives for different applications, but this study deals with the optimization of machining parameters AA2024-B_4C-TiC hybrid composite. The experiments conducted as per the factors and their levels as given in Table 2.

Eighty-one experiments were conducted (27 experiments with 3 times) according to orthogonal array. Equations 3 and 4 show that the S/N ratio for smaller the better type and larger the better category [8].

$$\acute{\eta} = -10\log_{10}\left\{\frac{1}{n}\sum_{i=1}^{n} y_i^2\right\} \tag{3}$$

$$\acute{\eta} = -10\log_{10}\left(\frac{1}{n}\right)\sum_{i=1}^{n}\frac{1}{y_{ij}^2} \tag{4}$$

Table 2 Factors with their levels

	Level 1	Level 2	Level 3
Water jet pressure (JP, MPa)	220	240	260
Stand-off distance (SOD, mm)	1	2	3
Traverse speed (TS, mm/min)	20	30	40

Table 3 Minimization of SR for aluminium hybrid MMC

Level	A	B	C
1	−14.99	−14.63	−14.64
2	−14.78	−14.74	−14.75
3	−14.78	−14.94	−14.92
Delta	0.45	0.31	0.28
Rank	1	2	3

Table 4 Minimization of KA for aluminium hybrid MMC

Level	A	B	C
1	2.355	2.654	2.658
2	2.446	2.465	2.462
3	2.616	2.297	2.296
Delta	0.261	0.357	0.362
Rank	3	2	1

Table 5 Maximization of MRR for aluminium hybrid MMC

Level	A	B	C
1	41.73	41.77	41.98
2	42.08	42.08	42.08
3	42.40	42.37	42.16
Delta	0.67	0.60	0.19
Rank	1	2	3

Tables 3, 4 and 5 indicate the optimal setting levels for each factor resulting from S/N ratios. Figures 2, 3 and 4 show the main effect plot for S/N ratio of SR, KA and MRR. Table 3 indicates that the Taguchi analysis for SR delta value for water jet pressure is −14.99 and traverse speed is −14.94, and from the SR delta value, the water jet pressure is affected more on SR followed by traverse speed.

The optimum parameter for minimization of SR is water jet pressure as 220 MPa, traverse speed as 60 mm/min and standoff distance as 3 mm. Table 4 indicates that the Taguchi analysis of KA delta value for standoff distance is 2.296 and traverse speed is 2.297. It can be understood that the standoff distance has the strongest effect on KA followed by traverse speed. The optimum parameter for minimization of KA is water jet pressure as 220 MPa, traverse speed as 60 mm/min and standoff distance as 3 mm.

Table 5 indicates that the Taguchi analysis of MRR delta value for water jet pressure is 42.40 and traverse speed is 42.97, and from MRR delta value, the water jet pressure is affected more on MRR followed by traverse speed. The optimum

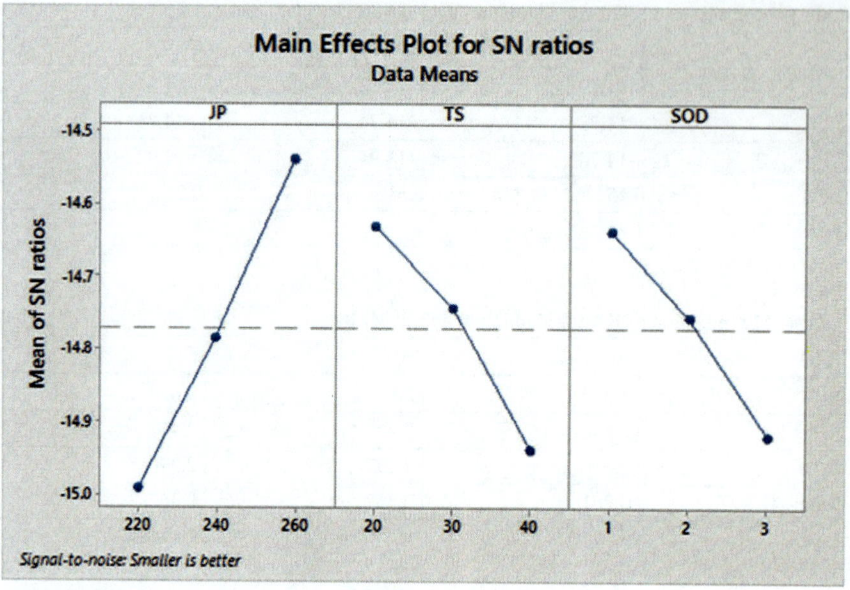

Fig. 2 Main effects plot of SR for aluminium hybrid MMC

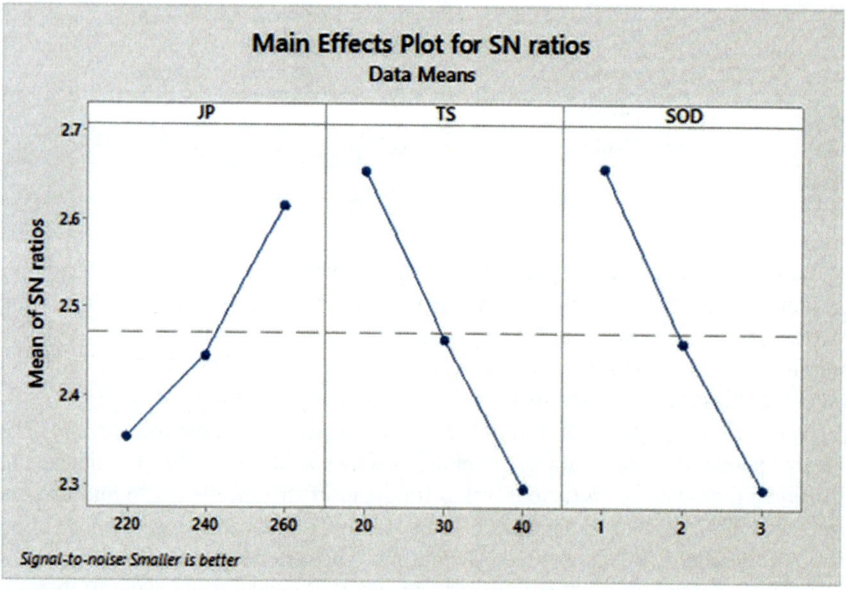

Fig. 3 Main effects plot of KA for aluminium hybrid MMC

parameter for minimization of MRR is water jet pressure as 260 MPa, traverse speed as 40 mm/min and standoff distance as 3 mm.

Most of the previous articles based on ANOVA, a great deal of variance, generally specify that there was a significant finding from the research. It is computationally well-designed and relatively robust against violations of its assumptions [11]. Table 6 indicates ANOVA for surface roughness and F-ratios. The water jet pressure influences more parameter on the surface roughness. The water jet pressure, jet traverse speed and standoff distance contributions in the surface roughness are 48.72%, 23.02% and 18.76%, respectively.

ANOVA general linear model of kerf angle is indicated in Table 7. The kerf angle is affected more by the standoff distance followed by traverse speed and water jet pressure. The water jet pressure, traverse speed and standoff distance percentage contributions are 20.79%, 37.81% and 38.91%, respectively.

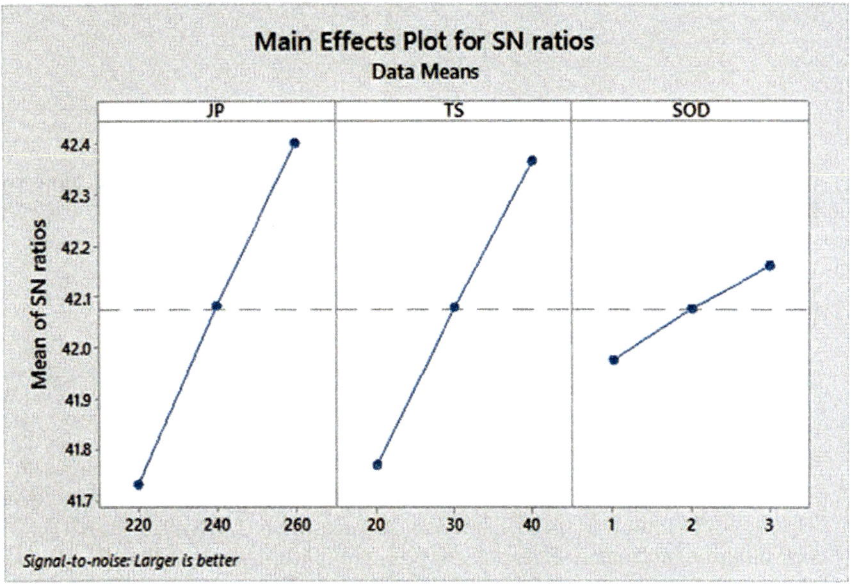

Fig. 4 Main effects plot of MRR for aluminium hybrid MMC

Table 6 Results of analysis of variance for surface roughness

Parameters	*df*	Adj SS	Adj MS	*F*-value	*P*-value	% Contribution
JP (Mpa)	2	0.36444	0.182219	51.00	0.0001	48.72
TS (mm/min)	2	0.17234	0.086172	24.12	0.0000	23.02
SOD (mm)	2	0.14048	0.070238	19.66	0.0007	18.76
Error	20	0.07146	0.003573			
Total	26	0.74871				

Table 7 Results of analysis of variance for kerf angle

Parameters	*df*	Adj SS	Adj MS	*F*-value	*P*-value	% Contribution
JP (Mpa)	2	0.002375	0.001188	83.78	0.0000	20.79
TS (mm/min)	2	0.004319	0.002159	152.34	0.0000	37.81
SOD (mm)	2	0.004445	0.002222	156.79	0.0005	38.91
Error	20	0.000284	0.000014			
Total	26	0.011423				

Table 8 Results of analysis of variance for MRR

Parameters	*df*	Adj SS	Adj MS	*F*-value	*P*-value	% Contribution
JP (Mpa)	2	434.36	217.178	364.03	0.0000	53.02
TS (mm/min)	2	338.93	169.467	284.06	0.0001	41.37
SOD (mm)	2	33.96	16.980	28.46	0.0003	4.15
Error	20	11.93	0.597			
Total	26	819.18				

Table 8 indicates ANOVA for material removal rate, and it is affected more by water jet pressure, traverse speed and minor effect of standoff distance. The water jet pressure, traverse speed and standoff distance percentage contributions are 53.02%, 41.37% and 4.15%, respectively.

5 Conclusions

- Taguchi analysis shows that the surface roughness delta value for water jet pressure is −14.99 and traverse speed is −14.94, and water jet pressure is affected more on SR followed by traverse speed. The water jet pressure is 220 MPa, traverse speed is 60 mm/min, and standoff distance is 3 mm to minimize the surface roughness.
- From the Taguchi analysis, the kerf angle delta value for standoff distance is 2.296 and traverse speed is 2.297, and standoff distance has the strongest effect on kerf angle followed by traverse speed. The water jet pressure is 220 MPa, traverse speed is 60 mm/min, and standoff distance is 3 mm to minimize the kerf angle.
- Taguchi analysis obviously shows the MRR delta value for water jet pressure is 42.40 and traverse speed is 42.97, and water jet pressure has the strongest effect on MRR followed by traverse speed. The water jet pressure is 260 MPa, traverse speed is 60 mm/min, and standoff distance is 3 mm to minimize the MRR.
- ANOVA of surface roughness, kerf angle and MRR has been determined, and water jet pressure is more influencing parameter on the surface roughness. The water jet pressure, jet traverse speed and standoff distance in the surface finish contributions are 48.72%, 23.02% and 18.76%, respectively.

- Kerf angle is affected more by the standoff distance followed by traverse speed and water jet pressure. The water jet pressure, traverse speed and standoff distance percentage contributions are 20.79%, 37.81% and 38.91%, respectively.
- MRR is affected more by water jet pressure and traverse speed, and there is a minor effect of standoff distance. The water jet pressure, traverse speed and standoff distance percentage contributions are 53.02%, 41.37% and 4.15%, respectively.

References

1. Ganesan P, Saravanan M (2012) Review on machining and tribological behaviors of aluminium hybrid composites. Procedia Eng 38:1399–1408
2. Sharma P, Khanduja D, Sharma S (2014) Tribological and mechanical behavior of particulate aluminum matrix composites'. J Reinf Plast Compos 33(23):2192–2202
3. Bhandare RG, Sonawane PM (2014) Preparation of aluminium matrix composite by using stir casting method and it's characterization. Int J Curr Eng Technol 3(3):2277–3106
4. Venkatachalam G, Kumaravel A (2017) Fabrication and characterization of A356-basalt ash-fly ash composites processed by stir casting method. Polym Polym Compos 25(3):209–213
5. Kechagias J, Petropoulos G, Vaxevanidis N (2012) Application of Taguchi design for quality characterization of abrasive waterjet machining of TRIP sheet steels. Int J Adv Manuf Technol 62(5):635–643
6. Patel D, Tandon P (2015) Experimental investigations of thermally enhanced abrasive waterjet machining of hard-to-machine metals. CIRP J Manuf Sci Technol 10(1):92–101
7. Manikandan N, Binoj JS, Varaprasad KC, Sabari SS, Raju R (2019) Investigations on wire spark erosion machining of aluminum-based metal matrix composites. Lecture Notes in Mechanical engineering, Advances in manufacturing technology. Springer, Singapore, pp 361–369
8. Saravanan M, Ganesan P (2017) Investigation of mechanical and corrosion properties of AA2024–B_4C–TiC hybrid metal matrix composites. Surf Rev Lett 25(8):1850109–1850117
9. Youssef HA, El-Hofy H (2008) Machining technology: machine tools and operations, 1st edn. CRC Press, Boca Raton
10. Pannerselvam R (2012) Design, and analysis of experiments. PHI Learning Pvt. Ltd., New Delhi
11. Sasikumar KS, Arulshri KP, Ponappa K, Uthayakumar M (2016) A study on kerf characteristics of hybrid aluminium 7075 metal matrix composites machined using abrasive waterjet machining technology. Proc Inst Mech Eng Part B: J Eng Manuf 16(1):1–15

ECM Machining and Its Process Optimization for AISI 304 Steel

D. Arulkirubakaran, Malkiya Ralsine Prince, D. Palanisamy, N. Manikandan, and R. Ramesh

Abstract In this present investigation, the experiments were conducted on AISI304 stainless steel by using ECM with hexagonal shaped copper electrode for determining desired performance measures. Mathematical/empirical relations are used for relating and analyzing the interaction of various output measures such as rate of material removed with respect to time, surface roughness, and overcut through the RSM approach based on the experimental data. Also, analysis of variance (ANOVA) has been adopted for determining the importance of the independent process variables on the desired output measures. From the analysis, it is observed that the concentration of electrolyte and voltage are the predominant machining variables for the material removal rate and surface roughness.

Keywords ECM · AISI 304 Steel · Electrolyte · MRR

1 Introduction

Electrochemical machining (ECM) process is an unconventional process works on the reversed electroplating method which is employed to cut different work materials such as titanium alloys, stainless steel, and super alloys. During the electroplating process, the high current is sent to the workpiece and the tool (cathode) through the

D. Arulkirubakaran (✉) · M. R. Prince
Department of Mechanical Engineering, Karunya Institute of Technology and Sciences, Coimbatore, India
e-mail: arulkirubakaran@karunya.edu

D. Palanisamy
Dr. APJ Abdul Kalam Research Center, Adhi College of Engineering and Technology, Kanchipuram, India

N. Manikandan
Micromachining Research Centre, Sree Vidyanikethan Engineering College, Tirupati, AP, India

R. Ramesh
Santhiram Engineering College, Kurnool, AP, India

A. Arockiarajan et al. (eds.), *Advances in Industrial Automation and Smart Manufacturing*, Lecture Notes in Mechanical Engineering,
https://doi.org/10.1007/978-981-15-4739-3_29

conductive medium (electrolyte). The required cavity or shape is achieved based on the replica of the tool geometry. Wei Han and Masanori Kunieda [1] developed the novel method to change the mode of machining from micro-ECM to micro-EDM using similar electrodes and same kind electrolyte. It was found that machining of tungsten material using both the techniques was more effective. Moreover, the through-holes were obtained in both the machining techniques. Also, the surface finish was found to be good in ECM machining than the micro-EDM. Wu et al. [2] attempted μECM using external vibration assisted method combined with polishing arrangement for the 3D component. Based on the experimental results, the vibration-assisted techniques gives better surface finish for curved and semi-curved surfaces.

Bhattacharyya et al. [3] investigated the micro-ECM machining characteristics. It was developed effectively to control MRR and accuracy of machining for meeting out the small-scale machining prerequisites. The authors found that machining trials were performed with the combination of various input process parameters (electrolyte concentration, amplitude, and micro-tool vibration). Neto et al. [4] demonstrated the ECM machining with respect to the process inputs such as voltage, rate of feed, electrolyte flow, and concentration of electrolyte. The various output measures identified are overcut (OC), material removal rate (MRR), and hardness. The input feed was the principal variable. It influences more with respect to the removal rate. Mukherjee et al. [5] investigated the role of NaCl solution in ECM machining on the iron workpiece. It was found that over-voltage and NaCl solution provide more penetration rate and the equilibrium gap during ECM machining process. Rajurkar et al. [6] investigated the merits of ECM process. The poor dimensional accuracy is highly influenced by higher MRR. Also, burr-free and smooth machined surface were noticed during the ECM machining. Moreover, it was noticed by the author that the simulation results good in correlation with experimental results.

Ahna et al. [7] investigated the rare use of ECM process in micro-manufacturing for the normal ground surface is not localized. For obtaining the localize zone, short pulses with tens nanosecond period are employed. The time duration of pulse, pulse frequency, and voltage on the localization distance have been evaluated measured using 8 μm diameter micro-hole, which was drilled on the 304 stainless steel foil [8, 9]. The objective is to optimize the surface roughness (Ra), overcut (OC), and MRR for the stainless steel (AISI304) with a Cu electrode. The experiments have been conducted using response surface methodology. The experiments are performed on the AISI 304 stainless steel workpiece material with the selected input process variables such as rate of feed, voltage, and electrolyte concentration with smooth work flow rate of electrolyte, electrolyte conductivity, and current across the work electrodes.

Table 1 Chemical composition of 304 stainless steel

Grade	Min	C	Mn	Si	P	S	Cr	Mo	Ni	N
		–	–	–	–	–	18.0	–	8.0	–
304	Max	0.08	2	0.75	0.045	0.030	20.0	–	10.5	0.10

2 Materials and Methods

The experimental runs are performed on ECM setup which contains of main subsystems such as machining chamber, panel for control, and circulation system for electrolyte. Tool feed is attached with DC servo motor; the voltage (V), current (C), time (T), and feed rate (F) are controlled by the control panel arrangement for various experimental runs. The electrolytic solution is pumped from the storage tank, and the passed electrolyte for the machining will be redirected to the storage tank by the circulating motor arrangement system. The sludge made after the circulation in the machining chamber will be dropped at the lower portion of the storage tank, and it can be cleaned in an easy manner with an aid of drain plug. The electrolyte flow control is controlled by flow control valve. In this present investigation, the copper electrode with hexagonal shape rod of 40 mm length and the dimensional gap between the electrode and work sample is 3 mm for passing the electrolyte to the machining area. The work sample material is used as AISI 304 stainless steel, and the chemical composition is shown in Table 1.

3 Results and Discussion

3.1 Effect on Material Removal Rate

The machinability of ECM process depends purely upon the concentration of electrolyte solution, feed rate of cathode, and supply voltage. The effect of different independent machining process variables on MRR (means) are demonstrated in Fig. 1. The material removal with respect time was gradually getting decreased with respect to the level of electrolyte concentration.

The material removal with respect to time is gradually getting increased with increasing in the voltage levels of 10–13.5; then, it diminishes with above 13.5 V. Moreover, MRR decreases on the feed rate with the range of 0.4–0.6, and then, it increases above the level of 0.6. The reason for the increment in MRR is due to lesser electrode gap that induces the current density in the gap with the constant anodic dissolution. For higher electrolytic concentration, better electrolyte conductivity between tool and work sample and higher ionic dissolution would be the responsible for the improvement in value of MRR. Table 2 shows the analysis of variance for the means of MRR. Table 3 shows the estimated regression coefficients for MRR.

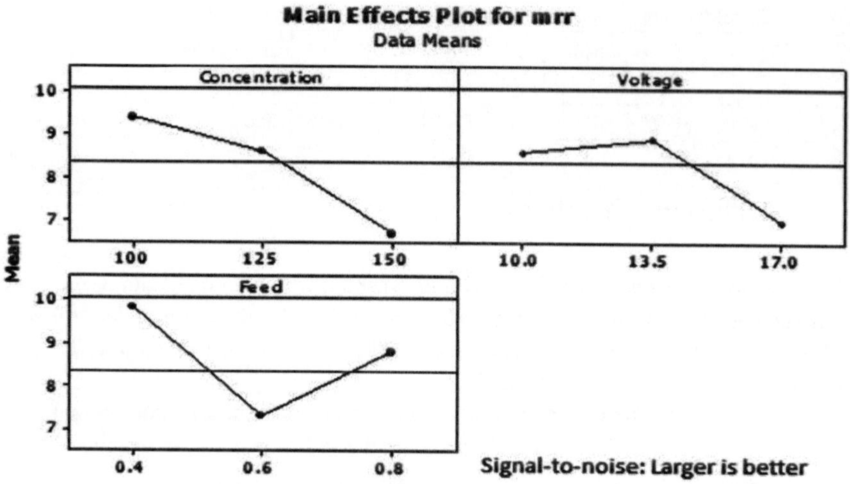

Fig. 1 Main effects of machining parameters on MRR

Table 2 Analysis of variance for the means of MRR

Source	DF	Seq SS	Adj SS	Adj MS	F	P
Regression	9	101.351	101.351	11.2612	6.50	0.004
Linear	3	28.039	28.039	9.3463	5.40	0.018
Square	3	69.574	69.574	23.1914	13.40	0.001
Interaction	3	3.738	3.738	1.2458	0.72	0.563
Lack-of-fit	5	13.027	13.027	2.6054	3.04	0.124
Pure error	5	4.286	4.286	0.8572		

Table 3 Estimated regression coefficients for MRR

Term	Coef	SE Coef	T	P	Remarks
Constant	8.230	0.4523	18.312	0.000	Significant
Concentration	−1.3652	0.4161	−3.281	0.008	Significant
Voltage	−0.8152	0.4161	−1.959	0.079	Non significant
Feed	−0.5247	0.4161	−1.261	0.236	Non significant
Concentration * concentration	−1.6630	0.7935	−2.096	0.063	Non significant
Voltage * voltage	−2.9880	0.7935	−3.766	0.004	Significant
Feed * feed	4.8020	0.7935	6.052	0.000	Significant
Concentration * voltage	0.2778	0.4652	0.597	0.564	Non significant
Concentration * feed	0.6122	0.4652	1.316	0.218	Non significant
Voltage * feed	0.1234	0.4652	0.265	0.796	Non significant

S = 1.31579; R-Sq = 85.41%; R-Sq(adj) = 72.28%

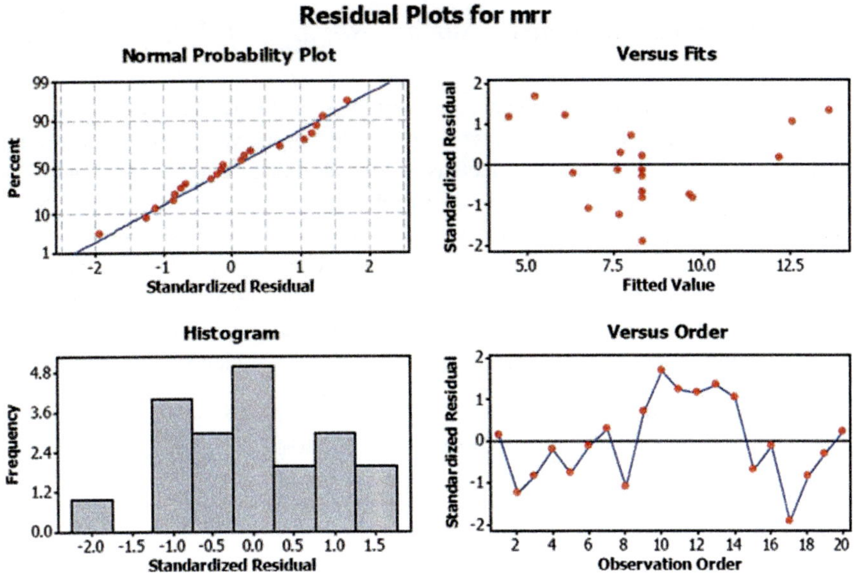

Fig. 2 Residual plots for MRR

$R^2 = 85.41\%$ specifies that the developed regression model is capable of envisaging the desired response with good accuracy.

The value of R^2 (adj) = 72.28%. The modeling error with respect standard deviation, S = 1.31579, electrolyte concentration (P = 0.008), is significant. Multiple voltage squares ($V * V$) and feed rate ($F * F$) are significant. The square multiples of electrolyte concentration ($C * C$) and interactions $C * V$ and $C * F$ are trivial solution. The MRR residual plot is illustrated in Fig. 2. The normal probability plot exhibits that the results attained from the experimentation are nearly normally disseminated, and the considered variables in this experimentation are significantly influencing the desired responses. The trained and tested residue ranges from −2 and 2 were used to fit values designate by the variance which is consistent, and a nonlinear connection is existing, and also no outliers are present in the data.

Graphical representation of histogram is the evidences that the data attained from the experimentation are very nearly in normal dispersed and this happens because the identified spots are found to very minimal. The residual graph shows the systematic influences in the experimental data.

From RSM, empirical connection among the desired responses and factors in coded forms are presented as $\text{MRR} = -7.01301 + 0.494256 \times \text{Concentration} + 0.243917 \times (\text{Voltage})^2 + 120.051 \times (\text{Feed})^2$.

3.2 Machining Outcome on Surface Roughness (SR)

The ascendency of different independent variables on SR (means) is exhibited in Fig. 3. The surface finish marginally is getting better for the variation concentrations with the range 100–125 and then decreases. Surface finish is found to be more for the larger voltage inputs. But surface finish diminishes for the variation in feed range from 0.4 to 0.6, and it get raise in the surface roughness values (Table 4).

R_a diminishes with the increment in the inter-electrode gap, and also, R_a enhances with the addition of electrolyte concentration, feed rate, and voltage. An increased electrolyte concentration might lead to the possibility of the passage of stray current to the machining periphery. The appraised regression coefficients values for SR are

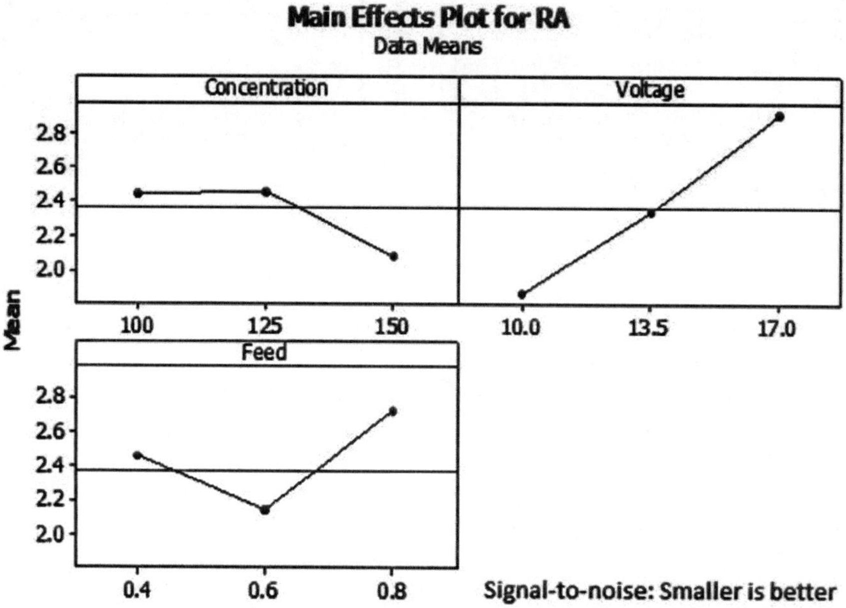

Fig. 3 Main effects of machining parameters on SR

Table 4 Analysis of variance for means of SR

Source	DF	Seq SS	Adj SS	Adj MS	F	P
Regression	9	7.06540	7.06540	0.78504	5.95	0.005
Linear	3	3.21680	3.21680	1.07227	8.13	0.005
Square	3	2.74440	2.74440	0.91480	6.94	0.008
Interaction	3	1.10420	1.10420	0.36807	2.79	0.095
Lack-of-fit	5	0.86988	0.86988	0.17398	1.94	0.243
Pure error	5	0.44880	0.44880	0.08976		

depicted in Table 5. R^2 = 84.27% shows that the model is capable of predicting the desired responses in an accurate manner. Adjusted R^2 is a modified R^2 which has been trialed of with present number of input variables in the machining process model and the R^2 (modified (adj)) = 70.12%.

The errors in terms of standard deviation in the modeling, S = 0.363136, parameter voltage (P = 0.001), are significant. The electrolyte concentration (P = 0.140) and feed (P = 0.277) are insignificant. The residual plot of SR is exhibited in Fig. 4.

Table 5 Estimated regression coefficients for SR

Term	Coeff	SE Coeff	T	P	Remarks
Constant	2.29600	0.1248	18.392	0.000	Significant
Concentration	−0.18400	0.1148	−1.602	0.140	Non significant
Voltage	0.52000	0.1148	4.528	0.001	Significant
Feed	0.13200	0.1148	1.149	0.277	Non significant
Concentration * concentration	−0.70000	0.2190	−3.197	0.010	Significant
Voltage * voltage	−0.08000	0.2190	−0.365	0.722	Non significant
Feed * feed	0.92000	0.2190	4.201	0.002	Significant
Concentration * voltage	0.08000	0.1284	0.623	0.547	Non significant
Concentration * feed	−0.18000	0.1284	−1.402	0.191	Non significant
Voltage * feed	0.31500	0.1284	2.453	0.034	Significant

S = 0.363136; R-Sq = 84.27%; R-Sq(adj) = 70.12%

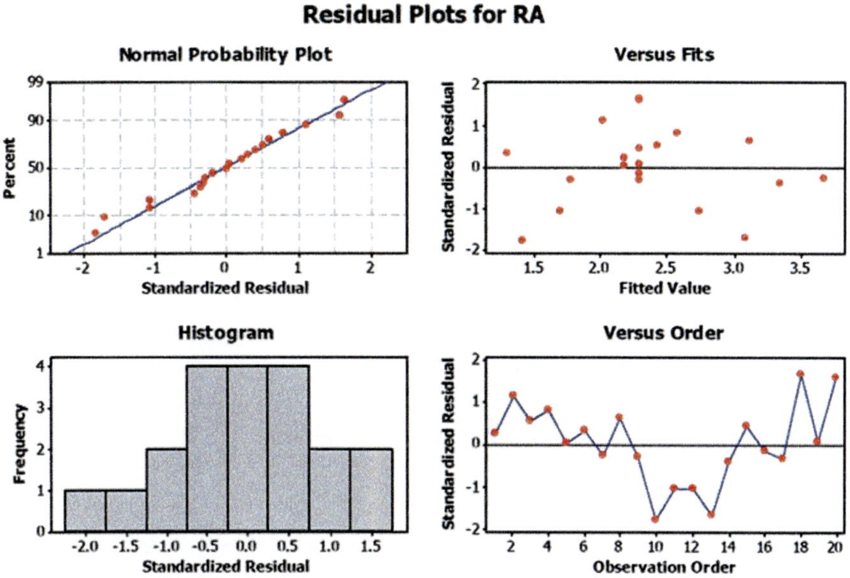

Fig. 4 Residual plots for SR

Standard occurrence (normal probability) plot illustrates that the data used for analysis are not normally dispersed and the independent process parameters considered in this experimentation are high impact on the desired responses.

The residue (Standardized) values vary from −2 and 2. It is found that the residual values show the consistent variation and a nonlinear relationship. Histogram evidences that the output data are closely in normal distribution with standard normal data, and it is due to the number of points presented and found very minimum. Residual plot shows that the nearly systematic effects were observed in the data with the standard data plot.

From RSM, empirical relationship between response and factors in coded forms is given as follows: $SR = -7.10806 - 0.0593878 \times \text{Voltage} - 0.00112000 \times (\text{Concentration})2 + 23.0000 \times (\text{Feed})2 + 0.450000 \times \text{Voltage} \times \text{Feed}$.

4 Conclusion

In this investigation, experiment on electrochemical machining evaluated the influence of machining variables on desired responses: overcut (OC), surface roughness (SR), and material removal rate (MRR) of the stainless steel AISI304 specimen using a copper electrode. The experimental runs were performed under various machining process parameter combination such as feed (F), voltage (V), and electrolyte concentration(C).

The conclusions drawn from the experimental investigations are as follows:

1. The removal of material is highly impacted by the interaction of rate of feed followed by voltage and then concentration. Also, MRR gradually getting decreased with an raise in the feed range (0.4–0.6) and electrolyte concentration. MRR getting increased with an augment in supply voltage up to the range of 10–13.5, and then, it diminishes.
2. The maximum MRR is achieved based on the optimum electrolyte concentration of 100 gm/L, voltage rate of 13.5 V, and feed rate of 0.6 mm/rate.
3. Parameters affecting surface finish are voltage and interaction of feed. Surface finish getting increased with increase in voltage.
4. The surface finish marginally getting increased with an amplification in concentration for the range up to 100–125 and it diminishes with argument in feed with the range of 0.4–0.6, and then, it diminishes.
5. The minimum surface finish is noticed with the concentration of electrolyte at 125 gm/L, voltage of 10 V and feed of 0.6 mm/min.

References

1. Han W, Kunieda M (2019) A novel method to switch machining mode between Micro-ECM and Micro-EDM using oxide film on surface of tungsten electrode. Precis Eng 56:455–465
2. Wu ZZ, Wu XY, Lei JG, Xu B, Jiang K, Zhong JM, Ruan SC (2018) Vibration-assisted micro-ECM combined with polishing to machine 3D microcavities by using an electrolyte with suspended B4C particles. J Mater Process Technol 255:275–284
3. Bhattacharyya B, Sorkhel SK (1999) Investigation for controlled electrochemical machining through response surface methodology-based approach. J Mater Process Technol 86:200–207
4. da Silva Neto JC, da Silva EM, da Silva MB (2006) Intervening variables in electrochemical machining. J Mater Process Technol 179:92–96
5. Mukherjee SK, Kumar S, Srivastava PK, Effect of electrolyte on the current- carrying process in electrochemical machining. In: Proceedings of I Mech E, vol 221 Part C: J Mech Eng Sci
6. Rajurkar KP, Wei B, Schnacker CL (1993) Monitoring and control of electrochemical machining (ECM). J Eng Ind 115/217
7. Ahn SH, Ryu SH, Choi DK, Chu CN (2004) Electro-chemical micro drilling using ultra short pulses. J Precis Eng 28:129–134
8. Raju R, Manikandan N, Palanisamy D, Arulkirubakaran D, Sambathkumar S, Bhanu Prakash P (2018) Optimization of process parameters in electrical discharge machining of haste alloy C276 using Taguchi's method. Mater Today: Proc 5(6):14432–14439
9. Manikandan N, Raju R, Palanisamy D, Arulkirubakaran D, Kumar S (2018) Investigation on Ti6Al4V laser metal deposition using Taguchi based grey approach. Mater Today: Proc 5(6):14375–14383

Investigation on Electrochemical Micromachining (EMM) of AA-MMC Using Acidified Sodium Nitrate Electrolyte

M. Soundarrajan, R. Thanigaivelan, and S. Maniraj

Abstract Machining of aluminum alloy (AA) -based composites is essential for the manufacturing sector due to its high strength, stability, and less weight. Accordingly, the outcome of metal matrix composites (MMC) needs to be understood for further processing in other application. Hence, in this paper an attempt made to study the process parameter of electrochemical micromachining (EMM) such as machining voltage (Mv), electrolyte concentration (Ec), duty cycle (Dc) on material removal rate (MRR), and overcut (OC) using acidified and non-acidified electrolyte. The AA-MMC (AA 6063, reinforced with 12% Sic and 5% Gr) is fabricated using stir casting method by weight fraction and considered as work material. Along with that the experiments are conducted in two electrolytes for enhancement of EMM performance, which are $NaNO_3$ (aqueous non-acidified sodium nitrate) and $NaNO_3$ + 10 ml of H_2SO_4 (aqueous acidified sodium nitrate). Therefore, 3.41 times better MRR is obtained in the aqueous acidified sodium nitrate electrolyte when compared to the aqueous non-acidified sodium nitrate electrolyte at the machining condition of 90% Dc, 30 g/l Ec and 15 V Mv. Additionally, scanning electron microscope (SEM) images are taken for the understanding of micro-hole surfaces and its intermolecular structure.

Keywords Electrochemical · Acidified · AA6063 · MMC · Electrolyte · H_2SO_4

M. Soundarrajan (✉) · R. Thanigaivelan
Department of Mechanical Engineering, Muthayammal Engineering College (Autonomous), Rasipuram, Namakkal 637408, India
e-mail: soundarrajan05@gmail.com

R. Thanigaivelan
e-mail: tvelan10@gmail.com

S. Maniraj
Department of Mechanical Engineering, Paavai Engineering College (Autonomous), Pachal, Namakkal 637018, India
e-mail: maniraj4016@gmail.com

A. Arockiarajan et al. (eds.), *Advances in Industrial Automation and Smart Manufacturing*, Lecture Notes in Mechanical Engineering,
https://doi.org/10.1007/978-981-15-4739-3_30

1 Introduction

Aluminum metal matrix (Al-MMC) composites are finding a variety of applications in aviation and automobile manufacturing sectors. A general reason behind, it has lightweight, high strength, high corrosion resistance, and high-temperature resistance, etc. Consequently, these properties are makes tough to machine in traditional method. Therefore, no-heat affected zone, non-contact machining surface such as EMM is suitable for machining the Al-MMC. To improve the EMM performance for precision manufacturing, various endeavors are carried out by the researchers in worldwide for the last decade on various alloy materials. In the way, Mingcheng et al. [1] have fabricated the square holes using EMM on SS 321 with 0.1 M of aqueous H_2SO_4 electrolyte. They have predicted the significant improvement in corner radius and machining rate with H_2SO_4 electrolyte. Geethapriyan et al. [2] have suspended the copper nanoparticles in the electrolyte to machining the SS 430 using EMM. They have noted significant change in machining performance using the nanoparticles suspended electrolyte. Also, they mentioned that the copper nanoparticles were increase the electrolyte conductivity and protect the stray current effect on machining zone. One of our previous research [3], we have studied the EMM on copper using ultraviolet heated electrolyte. We have noticed that the heated electrolyte produced the excellent material removal rate due to high conductivity of electrolyte. Yu el al. [4] have investigated the EMM on titanium using ethylene glycol-based NaCl electrolyte. They have found higher machining rate in EG-based NaCl electrolyte than the water-based NaCl electrolyte. Singh el al. [5] have tried the EMM on SS 304 and copper using different electrolytes such as $NaNO_3$, NaCl, and HCL. They have noted that among these three electrolytes, HCL produces the higher MRR. Thanigaivelan et al. [6] have used the infrared radiation to heating the electrolyte for EMM of copper material. They noticed the higher overcut due to heating of subsystem of EMM, whereas obtained 3 times better MRR. Wang et al. [7] have performed the micro-holes on laser treated inconel 718 alloy by EMM using annealed $NaNO_3$ electrolyte. They have noticed the smooth surface with no cracks formation on annealed electrolyte. Hackert et al. [8] have conducted the experiments on Al-Metal matrix composites using electrochemical machining with aqueous electrolytes such as $NaNO_3$ and NaCl. They have noted the higher passivation effects on $NaNO_3$ than the NaCl electrolyte. Singh et al. [9] have studied the EMM process parameters for inconel 825 alloy using NaCl electrolyte. They have found the notable increment in surface morphologies such as roughness, overcut, and material removal rate. From the above-literature survey, it is the evident that electrolyte plays a major role on EMM to deciding the machining performance. Generally, sulfuric acids are commonly used in the chemical industries to manufacture the other acids. Also, it is highly soluble acid in water. Hence, in this research the electrolyte (sodium nitrate) $NaNO_3$ is acidified with 10 ml of sulfuric acid (H_2SO_4) for machining high-strength AA-MMC (AA 6063, reinforced with 12% Sic and 5%Gr). The process parameters such as electrolyte concentration, machining voltage, and duty cycle are considered for evaluating response of such as material removal rate and overcut.

2 Experimental Method

In this research, an indigenously developed EMM setup has been used to fabricate the micro-hole which is shown in Fig. 1. AA-MMC 0.5-mm-thickness plate is fabricated by stir casting method using aluminum alloy (AA) 6063, 12% silicon (by weight fraction), and 5% graphite particles (by weight fraction) which are considered as work material [pradeep]. The experiments are conducted by the two types of electrolytes such as aqueous sodium nitrate ($NaNO_3$) and 10 ml of sulfuric acid (H_2SO_4) mixed on aqueous $NaNO_3$ for the different concentration levels. Generally, the H_2SO_4 is used in chemical industries to manufacturing other acids, surfactants, and metal oxides. Since it has high adhesion character with water, the diameter 460 μm stainless steel is considered as electrode and coated with epoxy resin to hinder the overcut.

The EMM unit consists of various components such as pulse rectifier, tool feeding system, and electrolyte filter. The completion of through micro-hole is identified by hydrogen bubble beneath the work material. The process parameters used to conduct the experiments are displayed in Table 1. The experiments are carried out through changing of one parameter at a time method which is displayed in Table 2. The MRR is calculated using weight differences between before and after machining. Overcut is calculated by the differences in diameter of hole and tool. The optical microscope has been used to verify the hole diameter.

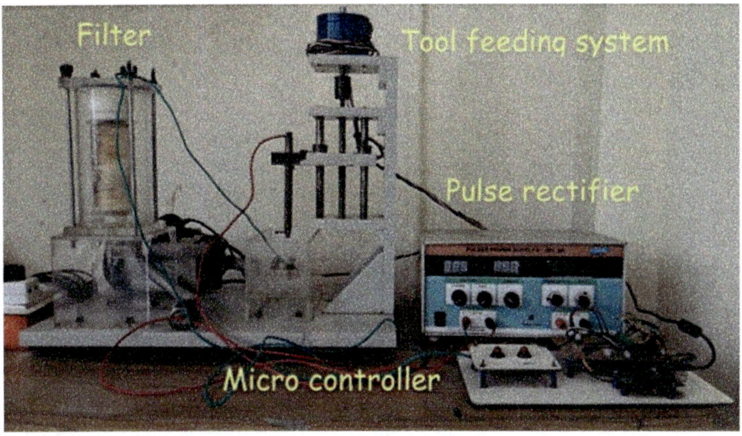

Fig. 1 EMM setup

Table 1 Levels of parameters

Symbol	Factors	Level 1
Ec	Electrolyte concentration in g/l	10–30
Mv	Machining voltage in V	7–15
Dc	Duty cycle in %	50–90
Sf	Sulfuric acid in ml	10–30

Table 2 Process planning for the experiments

Ex. No.	Ec in g/l	Mc in V	Dc in %
1	10	15	90
2	15	15	90
3	20	15	90
4	25	15	90
5	30	15	90
6	30	7	90
7	30	9	90
8	30	11	90
9	30	13	90
10	30	15	90
11	30	15	50
12	30	15	60
13	30	15	70
14	30	15	80
15	30	15	90

3 Result and Discussion

3.1 Results of Ec on MRR and OC

The results of Ec on MRR and OC are displayed as graph which is shown in Fig. 2. The figure is the evidences for acidified electrolyte produces higher material removal rate than non-acidified electrolyte. Since the ion dissociations are more in acidified electrolyte rather than the normal aqueous non-acidified electrolyte. The dissociation of ions is the most determining factor for the conductivity of electrolyte. Here the electrolyte sodium nitrate is still ionized by the 10 ml of sulfuric acid. Hence, the electrolyte increases the dissociation of ions highly in the acidified electrolyte. This high dissociation of ions increases the conductivity of electrolyte. The higher conductivity of electrolyte increases the localization effect on the machining zone [10]. Therefore, higher material removal rate is obtained in the acidified electrolyte than the

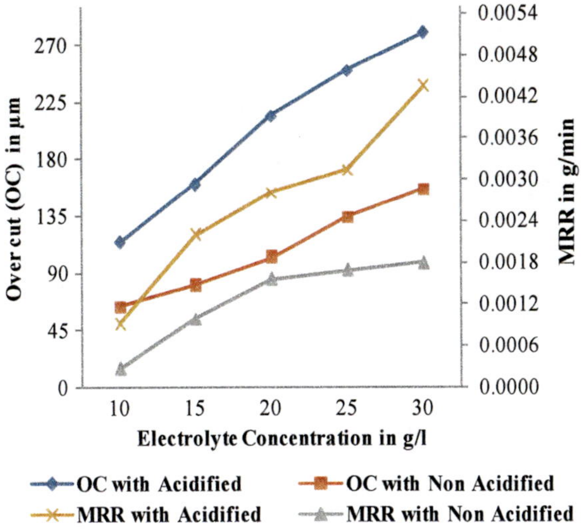

Fig. 2 Results of Ec on MRR on OC at 15 V Mv and 90% Dc for acidified electrolyte

non-acidified electrolyte. Also, it is obvious that higher parameter level will obtain the higher material removal rate and overcut. Therefore in acidified electrolyte, 1.80 times higher MRR are obtained in the machining condition that, 25 g/l electrolyte concentration, 15 v machining voltage and 90% duty cycle. And 2.52 times better MRR are obtained when compared with non-acidified electrolyte at the machining condition of 30 g/l electrolyte concentration, 15 v machining voltage, and 90% duty cycle. Generally, the sodium nitrate crates the higher sludge while mixing with the aqueous medium [11]. Furthermore, the chemical reaction of the aluminum and water is forming the slight passivation layer on metal surface which can trigger the sludge formation in aqueous sodium nitrate electrolyte [12]. Therefore, the sludge formation in non-acidified electrolyte creates the higher current instability on the machining zone, whereas the sludge formation in the acidified electrolyte is hindered significantly. Since the sodium nitrate is fully dissolved in sulfuric acid and forms the nitric acid and sodium bisulfate, therefore sludge formation has been hindered significantly in acidified electrolyte. Hence, the higher material removal rate has been attained in acidified electrolyte.

3.2 Results of Mv on MRR and OC

Figure 3 represents the results of machining voltage on MRR and OC. The increasing machining voltage in acidified electrolyte produces the 1.22 times higher MRR at the machining condition of 13 v machining voltage, 30 g/l electrolyte concentration, and 90% duty cycle than the non-acidified electrolyte. And 2.09 times better MRR are

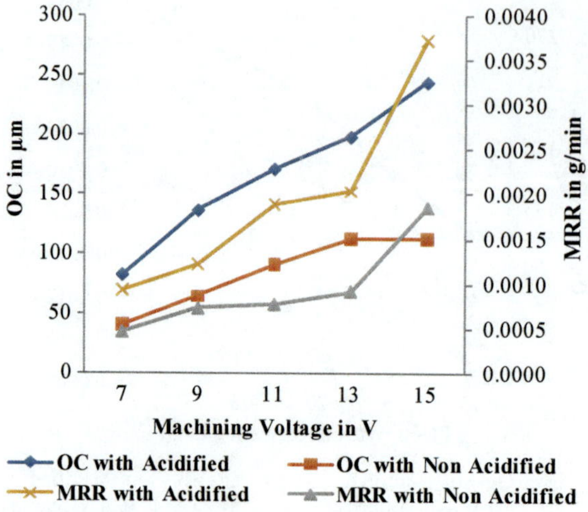

Fig. 3 Results of Mv on MRR on OC at 30 g/l Ec and 90% Dc for acidified electrolyte

obtained in 15 v machining voltage, 30 g/l electrolyte concentration, and 90% duty cycle of acidified electrolyte compared to the non-acidified electrolyte. Although Fig. 4a shows wobbled machining surfaces for the non-acidified electrolyte, Fig. 4b shows the over-etched surface for the acidified electrolyte. In that partial of micro-hole is found to be good and smooth profile in acidified electrolyte when compared to non-acidified electrolyte. Since it is obvious that AA-MMCs are found to be light and soft metal, therefore, the AA-MMC's can be etched easily when increasing voltage by the electrochemical reactions in acidified electrolyte.

3.3 *Results of Dc on MRR and OC*

In Fig. 5, the graph plotted for the results of duty cycle on MRR and OC. It shows that duty cycle plays the significant role on the machining performances. Since in acidified electrolyte, 3.12 times higher MRR are obtained than the non-acidified electrolyte. The parameter combination is used for the acidified electrolyte that is 80% duty cycle, 30 g/l electrolyte concentration, and 15 v machining voltage. And 3.41 times higher MRR are obtained at the machining combination of 90% duty cycle, 30 g/l electrolyte concentration, and 15 v machining voltage in acidified electrolyte. The duty cycle controls the current polarization effect on the machining zone. Due to this AA-MMC involved in overetching on its machining surface which can be identified using Fig. 6a, b for non-acidified and acidized electrolyte.

Figure 6b shows more intercrystalline etching on acidified than non-acidified electrolyte, which are found at the machining condition of 15 v machining voltage, 30 g/l electrolyte concentration, and 90% duty cycle. Also, Fig. 7a shows that, the

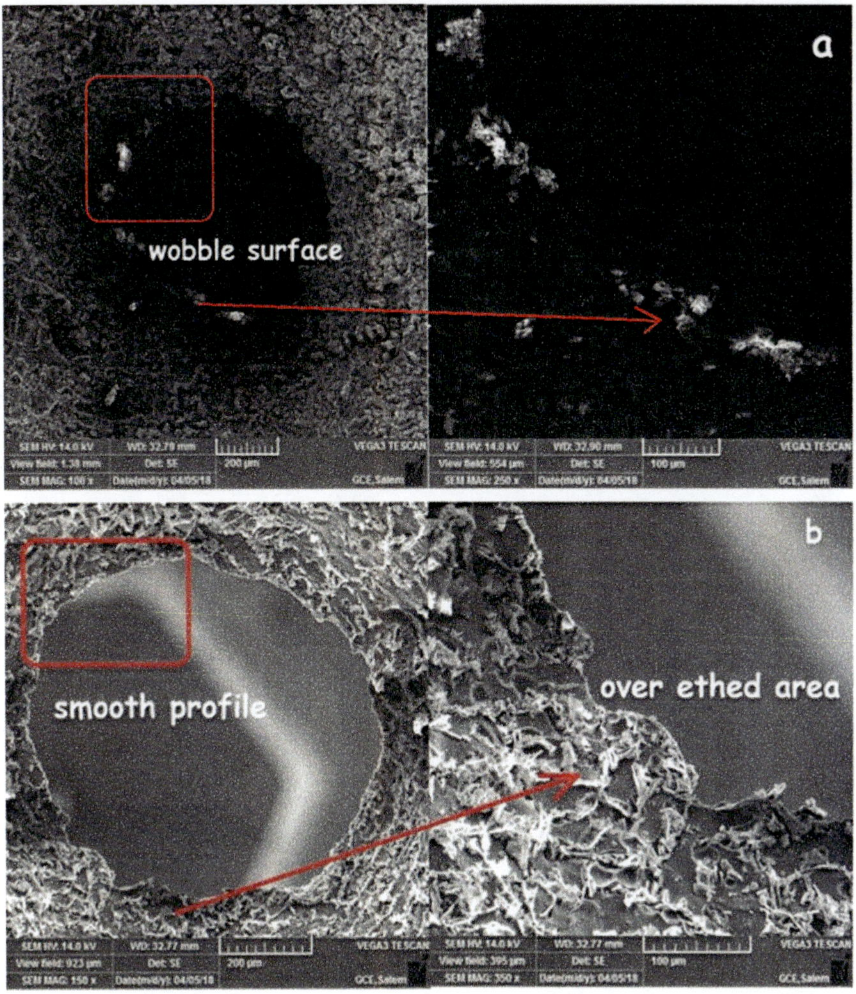

Fig. 4 SEM images of micro-hole machined at 7 V Mv, 30 g/l Ec and 90% Dc **a** non-acidified, **b** acidified electrolyte

non-acidified electrolyte affects the metal surface lighter, whereas Fig. 7b shows the high etched area on machining surface which can be identified by the presents of reinforced particles in metal surfaces. While compared to these two diagrams, the reinforced particles marks were highlighted high in acidified electrolyte than the non-acidified electrolyte. The overcut of micro-holes is found high in the acidified electrolyte, although the less amount of intercrystalline effect is found on machined surface in acidified electrolyte.

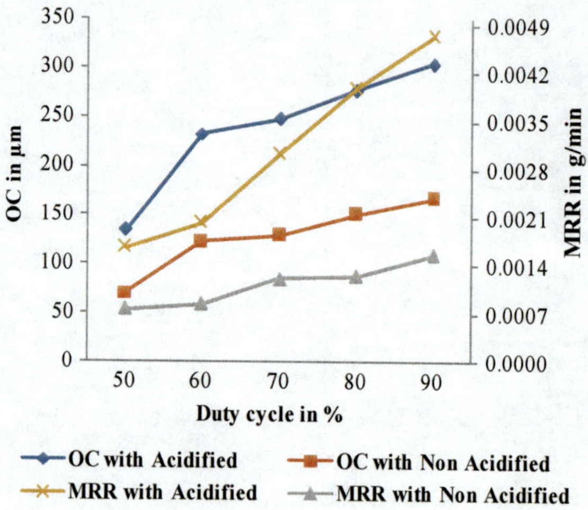

Fig. 5 Results of Dc on MRR on OC at 30 g/l Ec and 15 V Mv for acidified electrolyte

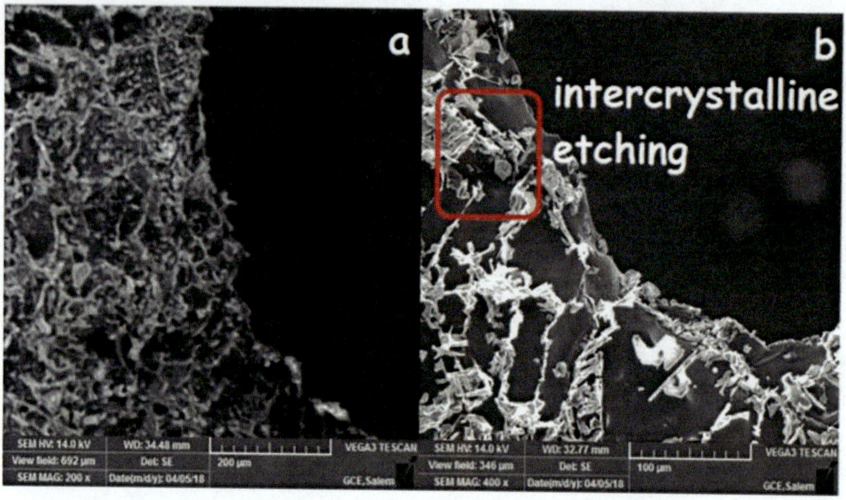

Fig. 6 SEM images of micro-hole circumference machined at 15 V Mv, 30 g/l Ec and 90% Dc **a** non-acidified, **b** acidified electrolyte

4 Conclusion

There are numerous applications available for AA-MMC's in manufacturing, automobile, aviation, and food industries. In this research made attempt on AA-MMC (AA 6063, reinforced with 12% Sic and 5% Gr) using acidified electrolyte to increase the EMM performance. The material removal rate is attained to be 3.12 times higher than the non-acidified electrolyte at the parameter combination of 80% duty cycle,

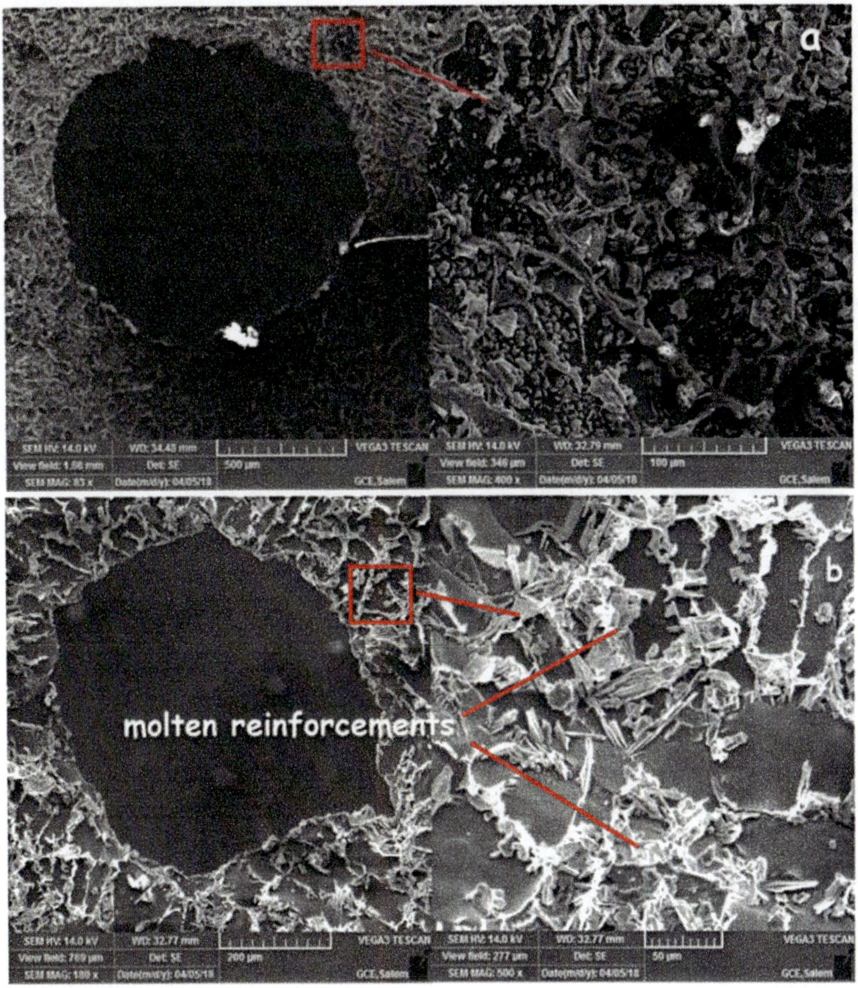

Fig. 7 SEM images of micro-hole surface machined at 15 V Mv, 30 g/l Ec and 90% Dc **a** non-acidified, **b** acidified electrolyte

30 g/l electrolyte concentration, and 15 v machining voltage. The next highest MRR obtained in acidified electrolyte that 3.41 times at the machining condition of 90% duty cycle, 30 g/l electrolyte concentration, and 15 v machining voltage. The acidified electrolyte shows a better MRR than the non-acidified normal electrolyte. Moreover, acidified electrolyte helps to hinder the sludge formation in the electrolyte. Due to the less sludge, the uniform micro-holes are obtained. Scanning electron microscope images are taken to understand the effect of acidified and non-acidified electrolyte on machining surface structure. Also, a uniform overcut is obtained in the acidified

electrolyte. Therefore, the acidified electrolyte would be suitable for machining AA-MMC's in EMM. Furthermore, experiments can conduct with EMM to hinder the intercrystalline attack on the machining surface.

Acknowledgements The authors thank the Government College of Engineering, Salem, for providing the SEM facilities. The authors thank the management of Muthayammal Engineering College, Rasipuram, Tamil Nadu, for the encouragement and support. The authors are grateful to the management of Sona College of Technology, Salem, for providing the optical microscope facilities to verify the overcut.

References

1. Mingcheng G, Yongbin Z, Lingchao M (2019) Electrochemical Micromachining of Square Holes in Stainless Steel in H2SO4. Int J Electrochem Sci 14:414–426
2. Geethapriyan T, Muthuramalingam T, Vasanth S, Thavamani J, Srinivasan VH (2019) Influence of nanoparticles-suspended electrolyte on machinability of stainless steel 430 using electrochemical micro-machining process. In: Advances in manufacturing processes. Springer, Singapore, pp 433–440
3. Soundarrajan M, Thanigaivelan R (2018) Investigation on electrochemical micromachining (ECMM) of copper inorganic material using UV heated electrolyte. Russ J Appl Chem 91(11):1805–1813
4. Yu Ning, Fang Xiaolong, Meng Lingchao, Zeng Yongbin, Zhu Di (2018) Electrochemical micromachining of titanium microstructures in an NaCl–ethylene glycol electrolyte. J Appl Electrochem 48(3):263–273
5. Singh Ramandeep (2018) To study the effect of different electrolytes and their concentrations on electrochemical micromachining. In: AIP conference proceedings, vol 1943, no 1. AIP Publishing, p 020046
6. Thanigaivelan R, Arunachalam RM, Kumar M, Dheeraj BP (2018) Performance of electrochemical micromachining of copper through infrared heated electrolyte. Mater Manuf Processes 33(4):383–389
7. Wang X, Ningsong Q, Guo P, Fang X, Lin X (2017) Electrochemical machining properties of the laser rapid formed Inconel 718 alloy in NaNO3 solution. J Electrochem Soc 164(14):E548–E559
8. Hackert-Oschätzchen M, Lehnert N, Martin A, Schubert A (2016) Jet electrochemical machining of particle reinforced aluminum matrix composites with different neutral electrolytes. In: IOP conference series: materials science and engineering, vol 118, no 1. IOP Publishing, p 012036
9. Singh A, Anandita S, Gangopadhyay S (2015) Microstructural analysis and multiresponse optimization during ECM of Inconel 825 using hybrid approach. Mater Manuf Processes 30(7):842–851
10. Thanigaivelan R, Arunachalam RM, Karthikeyan B, Loganathan P (2013) Electrochemical micromachining of stainless steel with acidified sodium nitrate electrolyte. Procedia CIRP 6:351–355
11. Soundarrajan M, Thanigaivelan R (2017) Intervening variables in electrochemical micro machining for copper. In: Proceedings of 10th international conference on precision, meso, micro and nano engineering (COPEN 10), Indian Institute of Technology Madras, Chennai-600036, India
12. Yoon S, Jang H-S, Kim S, Kim J, Cho KY (2017) Crater-like architectural aluminum current collectors with superior electrochemical performance for Li-ion batteries. J Electroanal Chem 797:37–41

Experimental Investigation of Nd:YAG Laser Welding of Inconel 625 Alloy Sheet

Sudhani Srikanth, A. Parthiban, V. Vijayan, S. Dinesh, and S. Sathish

Abstract The Nd:YAG laser welding is an advanced process for sheet metal welding. Inconel 625 alloy sheet is widely used in aircraft structure. The present paper is experimentally investigated about the Nd:YAG laser welding of Inconel 625 alloy sheet with improvement of geometrical accuracy and hardness of the welding area. The experiments carried out by using Box-Behnken design approach to predict the effect of Nd:YAG laser welding quality based on the welding parameters on laser power, welding speed, and frequency. In that, Box-Behnken design was used based on response surface methodology model, which is used on Design-Expert Software. The relationship between experimental value and RSM model is compared. Finally, based on the results, the proposed response surface models minimized welding bead width and hardness of the welding area for Nd:YAG laser welding of Inconel 625 alloy sheet.

Keywords Nd:YAG laser welding · Inconel 625 alloy · Box-Behnken design · Bead width · Hardness

S. Srikanth
Research Scholar, Department of Mechanical Engineering, Vels Institute of Science Technology & Advanced Studies, Chennai 600117, India

A. Parthiban (✉)
Associate Professor, Department of Mechanical Engineering, Vels Institute of Science Technology & Advanced Studies, Chennai 600117, India
e-mail: parthibana@yahoo.co.in

V. Vijayan
Professor, Department of Mechanical Engineering, K. Ramakrishnan College of Technology, Tiruchirappalli 621112, India

S. Dinesh
Assistant Professor, Department of Mechanical Engineering, K. Ramakrishnan College of Technology, Tiruchirappalli 621112, India

S. Sathish
Associate Professor, Department of Aeronautical Engineering, Hindustan Institute of Technology and Science, Chennai 603103, India

A. Arockiarajan et al. (eds.), *Advances in Industrial Automation and Smart Manufacturing*, Lecture Notes in Mechanical Engineering,
https://doi.org/10.1007/978-981-15-4739-3_31

1 Introduction

Laser welding is the advanced welding process used in sheet metal fabrication industries. Commonly, in industries, two types of laser welding are used: Nd:YAG and fiber, and broadly, Nd:YAG laser welding is used in most of the sheet metal industries. This work mainly focused on welding quality as well as hardness of Nd:YAG laser welding of Inconel 625 alloy sheet [1]. The Nd:YAG laser is a solid state laser; it is mainly working at a wavelength of 1.06 μm, and fiber laser is a gas-type laser; it has 10.6 μm of wavelength [2]. The Nd:YAG lasers have lower beam power and have worked continuously in pulsed mode; mainly welding the lower thickness of sheet metal at high accuracy [3]. Fiber laser welds all types of sheet metals at high speed because it delivers the maximum laser power and mass production [4]. At the same time, the maximum peak laser beam power at high welding speed of Nd:YAG laser welding to reflect a lesser degree by metal surfaces and this high absorptive of the Nd:YAG laser enable to process [5]. Nd:YAG laser welding of Inconel alloy material has the most important research for making of high accuracy of welding surfaces [6]. The quality of welding mainly depends on the input parameters such as laser beam power, welding speed, and frequency [7]. Most of the researchers concentrate on only geometrical analysis of welding area and also not consider about welding area hardness [8–11], so that the present work tries to concentrate on the response surface methodology-based Box-Behnken design approach. And the welding of Inconel 625 alloy sheet is used; the welding geometry and hardness of welding area are investigated [12–14].

2 Materials and Methods

In this work, Nd:YAG laser welding of Inconel 625 alloy sheet metal to cut curved profiles is considered.

2.1 *Experimental Procedure*

The experiments are carried out by ABB make Nd:YAG laser welding machine as shown in Fig. 1, and the specification of machine is shown in Table 1. The workpiece considered for this work is Inconel 625 alloys; 1.2-mm-thickness sheet is used as shown in Fig. 2. The input parameters considered are laser beam power, welding speed, and frequency. Three levels of each factor are considered; the ranges of input parameters are tabulated in Table 2. For conducting the experiments, the Box-Behnken design was used and 17 experimental runs were carried out as shown in Table 3 [12].

Fig. 1 Nd:YAG laser welding machine

Table 1 Specification of Nd:YAG laser welding

Model	ABB make
Laser type	Nd:YAG
Mode	Pulsed wave (PW)
Power	500 W
Focal length	127.5 mm
Focal position	−0.5
Optical lens	KCl lens
Gas nozzle	Cylindrical, flat topped, coaxial with a 2 mm orifice
Cutting position	Flat position

2.2 *ANOVA*

The analysis of variance the collection of mathematical data and to analysis the problems. Table 4 shows the result of bead width model value, and this model is most significant. Table 5 shows the result for hardness value; this model is most significant because the input parameters are most affected by bead width and hardness.

Fig. 2 Welding workpiece

Table 2 Ranges of welding parameters

Parameters	Units	Level1	Level2	Level3
Laser power	W	180	190	200
Welding speed	mm/min	10	15	20
Frequency	Hz	10	11	12

2.3 Response Surface Graph

Figure 3 response surface graph for bead width in this Fig. 3a the bead width is minimum at low beam power the beam power was increase bead width also increase, at the all level of welding speed bead width are constant. At the same time Fig. 3b at low frequency the bead width are increases at the same time minimum beam power bead width also decreases. Figure 3c the low frequency and low welding speed Hardness are minimum at the same time of interval high frequency low welding speed to achieve high hardness.

Figure 4 response surface graph for bead width in this Fig. 4a the Hardness is minimum at low beam power the beam power was increase bead width also increase, at the all level of welding speed Hardness are constant. At the same time Fig. 3b at low frequency the Hardness are increases at the same time minimum beam power Hardness also decreases. Figure 3c the low frequency and welding speed Hardness are low at the same time of interval high frequency low welding speed Hardness are maximum.

Table 6 shows comparison of experimental and RSM model values, and all level

Table 3 Experimental plan

Experimental number	Laser power (W)	Welding speed (mm/min)	Frequency (Hz)	Bead width (mm)	Hardness (HV)
1	190	20	10	1.449	228.4
2	190	15	11	1.416	227.5
3	200	15	12	1.535	239.8
4	190	15	11	1.416	228.9
5	190	15	11	1.433	226.3
6	190	15	11	1.436	225.9
7	190	10	12	1.465	225.1
8	190	15	11	1.413	229.1
9	180	15	12	1.321	213.2
10	180	10	11	1.331	216.5
11	190	20	12	1.456	224.1
12	180	20	11	1.342	217.9
13	200	15	10	1.51	236.9
14	190	10	10	1.419	226.9
15	200	20	11	1.491	238.9
16	180	15	10	1.355	216.5
17	200	10	11	1.581	234.4

Table 4 ANOVA table for bead width

Source	Sum of squares	DF	Mean square	F value	p-value
Model	0.0796	9	0.0088	26.8011	0.0001
A	0.0737	1	0.0737	223.4375	< 0.0001
B	0.0004	1	0.0004	1.2744	0.2961
C	0.0002	1	0.0002	0.7334	0.4201
AB	0.0026	1	0.0026	7.7287	0.0273
AC	0.0009	1	0.0009	2.6373	0.1484
BC	0.0004	1	0.0004	1.1524	0.3187
A	0.0000	1	0.0000	0.0402	0.8468
B	0.0010	1	0.0010	2.9578	0.1291
C	0.0004	1	0.0004	1.0859	0.3320
Residual	0.0023	7	0.0003		
Cor total	0.0819	16			

Table 5 ANOVA table for hardness

Source	Sum of squares	DF	Mean square	F value	p-value
Model	951.0817	9	105.6757	30.2926	<0.0001
A	922.3513	1	922.3513	264.3977	<0.0001
B	5.1200	1	5.1200	1.4677	0.2650
C	5.2813	1	5.2813	1.5139	0.2583
AB	2.4025	1	2.4025	0.6887	0.4340
AC	9.6100	1	9.6100	2.7548	0.1409
BC	1.5625	1	1.5625	0.4479	0.5248
A	0.0206	1	0.0206	0.0059	0.9409
B	1.2506	1	1.2506	0.3585	0.5682
C	3.1869	1	3.1869	0.9136	0.3710
Residual	24.4195	7	3.4885		
Cor Total	975.5012	16			

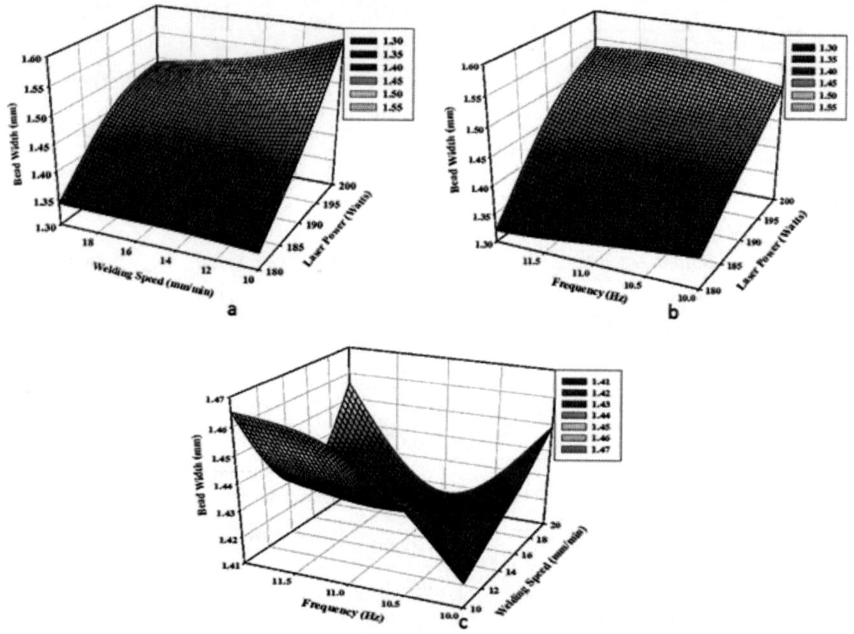

Fig. 3 Response surface graph for bead width

of conditions is experimental, and RSM models values are within the deviations [15].

Figure 5 shows the validation plots for bead width. That plots are plotted at all conditions of model values, so that from the plots, experimental values are nearer to the RSM model, satisfied at all level of conditions. Figure 6 also indicates the

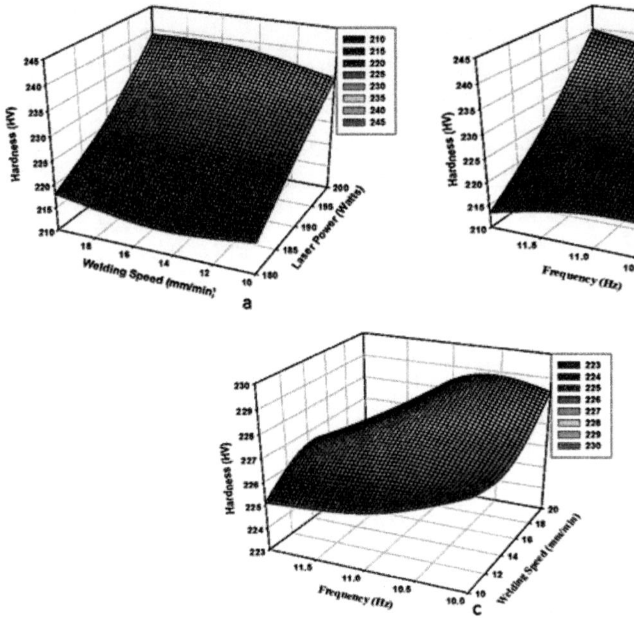

Fig. 4 Response surface graph for hardness

comparison and validation plots for hardness; it is satisfied at all experimental values of RSM models.

3 Conclusion

In this work, Box-Behnken design was used. The developed RSM model predicts the bead width and hardness to set the laser beam power, welding speed, and frequency of Nd:YAG laser welding of Inconel 625 alloy sheet. From this analysis, the following conclusions are made.

- The relationship between the bead width and hardness and the Nd:YAG laser welding parameters as well as interactions is considered.
- The welding speed is at lower level 10 mm/min, Laser beam power 190 watts and frequency 11 Hz is the bead width is minimum and hardness are high.
- The welding speed is at lower level 10 mm/min, Laser beam power 190 watts and frequency 11 Hz is lower value of bead width and hardness values.
- The welding speed is highest influence on the bead width and hardness followed by other parameters.

Table 6 Comparison of experimental and RSM model values

Experimental number	Laser power (W)	Welding speed (mm/min)	Frequency (Hz)	Bead width (mm)		Hardness (HV)	
				Experimental value	Model value	Experimental value	Model value
1	190	20	10	1.449	1.444	228.4	228.36
2	190	15	11	1.416	1.423	227.5	227.54
3	200	15	12	1.535	1.546	239.8	238.08
4	190	15	11	1.416	1.423	228.9	227.54
5	190	15	11	1.433	1.423	226.3	227.54
6	190	15	11	1.436	1.423	225.9	227.54
7	190	10	12	1.465	1.470	225.1	225.14
8	190	15	11	1.413	1.423	229.1	227.54
9	180	15	12	1.321	1.325	213.2	213.50
10	180	10	11	1.331	1.322	216.5	216.16
11	190	20	12	1.456	1.436	224.1	225.49
12	180	20	11	1.342	1.358	217.9	216.21
13	200	15	10	1.51	1.506	236.9	236.60
14	190	10	10	1.419	1.439	226.9	225.51
15	200	20	11	1.491	1.500	238.9	239.24
16	180	15	10	1.355	1.343	216.5	218.23
17	200	10	11	1.581	1.565	234.4	236.09

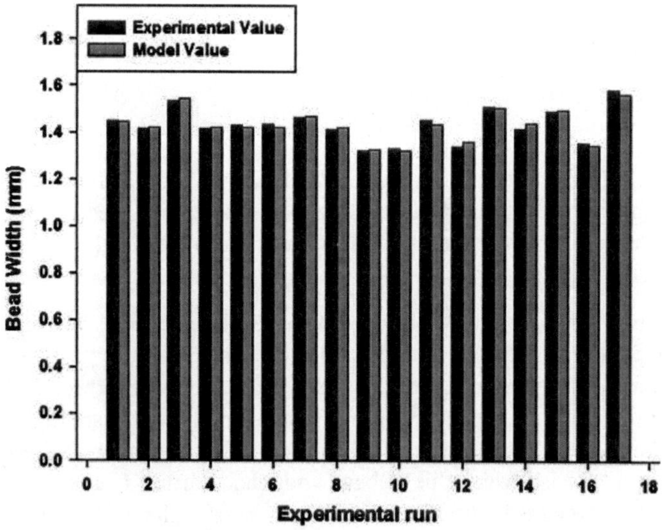

Fig. 5 Comparison and validation of result for bead width

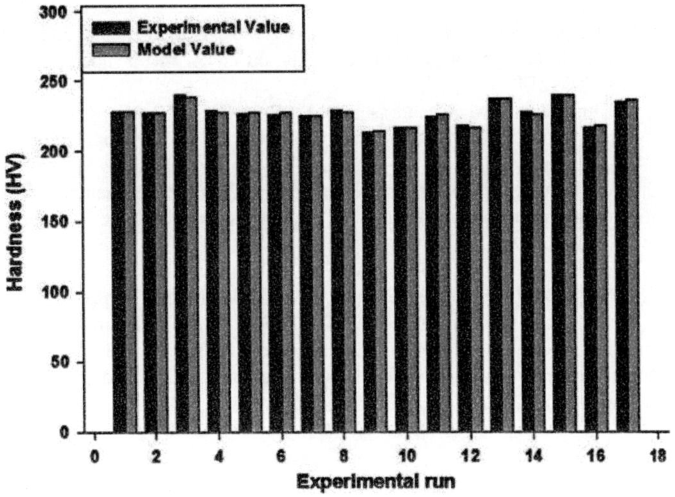

Fig. 6 Comparison and validation of result hardness

- The validation of result RSM model is satisfied at closer than experimental values. From the result the RSM model can be used successfully to predict the Hardness and Bead width in Nd:YAG laser welding of Inconel 625 alloy sheet.

References

1. Yadava V, Sharma A (2011) Optimization of kerf quality using robust design of experiments during Nd:YAG laser cutting of thin aluminum alloy sheet for straight profile. MIT Int J Mech Eng 1(1):1–8
2. Abdul Ghany K, Newishy M (2005) Cutting of 1.2 mm thick austenitic stainless steel sheet using pulsed and CW Nd: YAG Laser. J Mater Process Technol 168:438–447
3. Parthiban A, Ravikumar R, Suresh Kumar B, Baskar N (2015) Process performance with regards to surface roughness of the CO_2 laser cutting of AA6061–T6 aluminium alloy. Lasers Eng 32(3):327–341
4. Parthiban A, Ravikumar R, Abdul Zubar H, Duraiselvam M (2014) Experimental investigation of CO_2 laser cutting on AISI 316L sheet. J Sci Ind Res 73:387–393
5. Rajaram N, Sheikh-Ahamed J, Cheraghi SH (2003) CO_2 laser cut quality of 4130 steel. Int J Mach Tools Manuf 43:351–358
6. Srinivasan R, Vijayan V, Sridhar K (2017) Computational fluid dynamic analysis of missile with grid fins. J Appl Fluid Mech 10(Special Issue):33–39
7. Madic M, Radovanovic M, Nedic B (2012) Correlation between surface roughness characteristics in CO_2 laser cutting of mild steel. Tribol Ind 34(4):232–238
8. Manikandan N, Raju R, Palanisamy D, Arulkirubakaran D, Kumar S (2018) Investigation on Ti6Al4V laser metal deposition using Taguchi based grey approach. Mater Today: Proc 5(6):14375–14383
9. Parthiban A, Prasath JP, Vivek P, Pugazhenthi R (2018) Experimental investigation of plasma arc cutting for stainless steel sheet. Int J Mech Prod Eng Res Dev 8(1):907–914

10. Raju R, Sivalingam V, Sun J, Natarajan M, Zhao Y (2019) Experimental and Taguchi-based grey approach of laser metal deposition technique on nickel-based superalloy. Trans Indian Inst Met 72(1):205–214
11. Dillibabu V, Duraiselvam M, Chandrasekhar U, Raju R (2017) Microstructural studies on laser dissimilar welded Ni and steel alloys for aeronautical turbine applications. Lasers Eng 37(4-6):247–260
12. Dinesh S, Godwin Antony A, Rajaguru K, Vijayan V (2016) Experimental investigation and optimization of machining parameters in CNC turning operation of duplex stainless steel. Asian J Res Soc Sci Hum 6(10):179–195
13. Parthiban A, Sathish S, Chandrasekaran M (2017) Optimization of abrasive water jet cutting parameter for AISI 316L stainless steel sheet. J Appl Fluid Mech 10(Special Issue):15–22
14. Parthiban A, Sathish S, Chandrasekaran M (2017) Fuzzy logic based modeling of CO_2 laser cutting for stainless steel sheet. ARPN J Eng Appl Sci 12(6):1780–1784
15. Suresh Kumar B, Vijayan V, Baskar N (2016) Comparison of coated and uncoated carbide drill bits for drilling titanium grade 2 material. Mechanika 22(6):571–576

Prediction of Performance Measures in Wire Electrical Discharge Machining of Aluminum–Fly Ash Composites Using Regression Analysis

D. Palanisamy, N. Manikandan, Ramesh Raju, D. Arul Kirubakaran, and J. S. Binoj

Abstract Wire electrical discharge machining (WEDM) is one among the contemporary method of material removal process which is established from the perception of spark erosion machining. Non-traditional method of material removal processes has been evolved to overcome some difficulties in making of intricate shapes in any electrically conductive work material. Decision making is increasing fractious in the manufacturing domain because of its need for manufacturing the quality components with high precision and rapid changes in design. This present exploration deals with the evolvement of multiple regression equations for predicting the desired performance measures in wire EDM process. The experimental trials are devised with the help of Taguchi's experimental design approach, and the empirical equations are established using multiple regression models. Analysis of variance (ANOVA) is adopted for determining the importance of process variables which affects the desired performance measures. The evolved method for prediction of desired performance measures is flexible, competent, and accurate than existing models, and it scopes a better monitoring system. Finally, the evolved models are validated. The predicted results based on the developed models are in better covenant with the experimental outcomes.

Keywords WEDM · Aluminum alloy · Al 7075 · Aerospace materials · Taguchi's method · Single objective Optimization · Regression models

D. Palanisamy (✉)
Dr. APJ Abdul Kalam Research Centre, Adhi College of Engineering and Technology, Kancheepuram, India
e-mail: dpalanisamy.me@gmail.com

N. Manikandan · J. S. Binoj
Micro Machining Research Centre, Sree Vidyanikethan Engg College (Autonomous), Tirupati, Andhra Pradesh, India

R. Raju
Santhiram Engineering College, Nandyal, Andhra Pradesh, India

D. Arul Kirubakaran
School of Mechanical Engineering, Karunya University, Coimbatore, Tamil Nadu, India

A. Arockiarajan et al. (eds.), *Advances in Industrial Automation and Smart Manufacturing*, Lecture Notes in Mechanical Engineering,
https://doi.org/10.1007/978-981-15-4739-3_32

1 Introduction

The automobile and aerospace industries are looking for the development of lightweight materials for enhanced mechanical properties and strength to weight ratio. Metal matrix composites (MMCs) possesses the capability of spectacular thrifty in weight of the segments created associated great strength-to-weight proportion once contrasted with prevailing material and an outstanding choice for the metals used in grouping of segments adopted in vehicle businesses [1]. Because of the top quality and hardness of the reinforced materials gift in MMC's, they were exhausting for material removal process [2]. Hybrid MMC (HMMC) was factory-made utilizing quite one strengthening particulates, has the larger strength to stiffness proportions, and performs superior to compound and MMC even at a greater temperature environment. MMC is often factory-made by completely different methods, for instance, squeeze casting technique, stir casting technique, metallurgy, and fluid penetration. Among the different techniques indicated, blend throwing procedure is a straightforward and most prudent strategy for producing MMC [3–5]. Particulate strengthened aluminum framework composite is picking up significance in light of their ease with points of interest like isotropic properties and the likelihood of auxiliary handling encouraging manufacture of optional segments. Cast aluminum matrix molecule fortified composites that possess a greater specific strength, specific modulus, and great wear opposition when contrasted with unreinforced combinations and the machining of MMC is troublesome because of the exceedingly grating nature of clay fortifications when customary machining was utilized [6–9].

Wire EDM process is a method in which a thin metal wire associated with dielectric fluid permits the wire to move by the employment of heat from electrical sparks [10–12]. EDM includes the evacuation of metal by cathartic an electrical result a throbbing DC power provided between the electrode and the workpiece. It is most commonly utilized for the removal of material from hard materials [13]. Since MMC is typically innovative, the machinability analysis has not been targeted completely.

The impact of process parameters, for example, beat term, beat off-time, voltage, servo feed, servo voltage, dielectric flushing weight, wire feed, and wire strain on surface roughness of various hard materials was contemplated. It was discovered that the surface unpleasantness of hard materials was essentially influenced by servo voltage, wire feed, and wire tension [14–17]. Mahapatra and Patnaik [18] expressed that the impact of release flow, beat length, and dielectric stream rate was predominant when contrasted with that of heartbeat recurrence, wire feed, and wire pressure on roughness of machined D2 tool steel material. An investigation deliberated the impact of WEDM process variables on machining of WC-5.3% Co composites and expressed that a superior surface completion can be accomplished by setting the ideal wire counterbalance esteem. In this present examination, tests are led to discover the impact of WEDM variables on MRR and surface unpleasantness in aluminum–fly ash composites. The determination of this material was made considering its wide scope of uses in enterprises. The parameters are improved for better cutting conditions [19].

It is surmised from the available literature that there is an importance requires an investigational study on WEDM of aluminum–fly ash composites, also in development and prediction desired performance measures. In this present research article, an exertion has been considered for examining the influence of process variables on desired performance measure MRR and surface roughness by Taguchi's method and to develop the empirical relations for the desired performance measure with the help of regression methods.

2 Materials and Methods

Aluminum–fly ash composites were made by stir casting method. Economically accessible pure aluminum Al 7075 has been melted in an obstruction warmed mute heater and threw in a dirt graphite pot, and it has been degassed by progressively including hexachloroethane tablets. The fly ash remains particles that were preheated to expel the dampness. Commercially pure aluminum was liquefied by increasing its temperature, and it was degassed by cleansing hexachloroethane tablets. The stir casting setup with stirrer arrangement is shown in Fig. 1. The melt temperature was kept up at 953–993 K during the expansion of the particles. At that point, the melt was casted and the substance synthesis and smaller scale hardness investigation were accomplished for aluminum–fly ash composites. Figure 2 demonstrates the SEM micrograph of manufactured Aluminium Fly ash composite.

By combining all possible parameters, an orthogonal array L_{27} is opted for conducting the experimental runs by utilizing the full factorial experimental design.

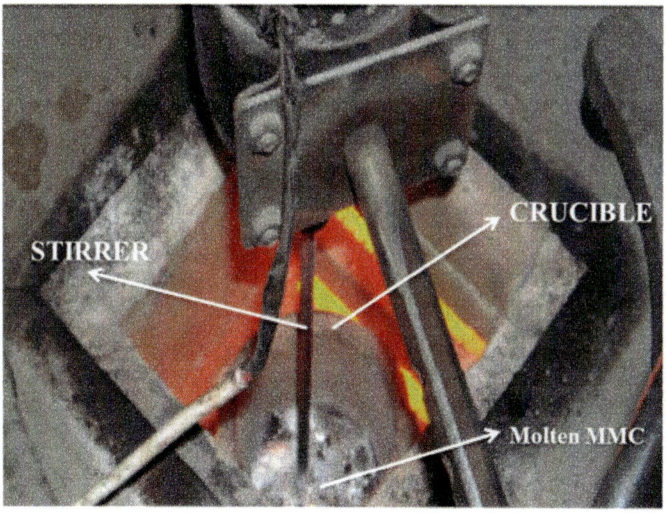

Fig. 1 Stir casting setup

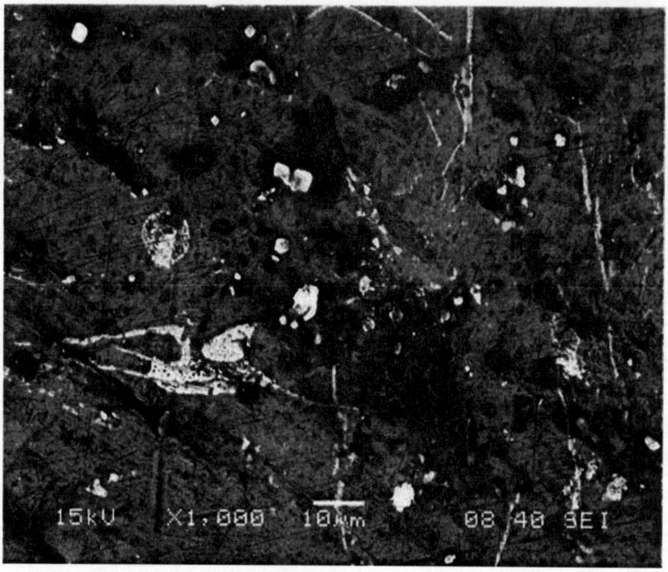

Fig. 2 SEM micrograph of Al-fly ash composite

In this investigation, Electronica Ultracut S1 wire EDM is employed to perform the experimental runs and the wire electrode of brass material with diameter of 0.25 mm has been employed for this investigation.

In this present research, the independent process variables are chosen as pulse on-time, pulse off-time, flushing pressure, feed of wire, and servo voltage and each at three levels have been considered on the desired-dependent measures such as MRR and surface roughness (Ra). The MRR was measured by a CONTECH precision weighing machine with 0.0001 g accuracy. The surface roughness of the surface was evaluated using a Mitutoyo SJ301 surface roughness tester. Each experimental run was performed twice, and the average reading is noted for evaluation. The machining process variables and the levels adopted for this experimentation are represented in Table 1.

Table 1 Process parameters and levels

Symbols	Parameters	Levels		
		1	2	3
A	Pulse on-time (μs)	105	108	111
B	Pulse off-time (μs)	54	57	60
C	Flushing pressure (kg/cm^2)	6	9	12
D	Wire feed (mm/min)	2	4	6
E	Servo voltage (V)	20	25	30

3 Results and Discussion

The experimental trials have been performed as per L_{27} OA for exploring the prominence of independent variables on desired performance characteristics in wire EDM of Al-Fly ash composite material. An exertion has been made to ascertain the best possible process variables for accomplishing the efficient wire EDM. In the wire EDM process, superior removal rate of material and lesser values of surface roughness are the indicator of superior performance measures. So MRR is considered as greater the better criterion, and surface roughness is considered as lesser the better criterion.

3.1 Influence of Process Variables on Material Removal Rate

The response graph for rate of material removal during machining of Al-fly ash composites is shown in Fig. 3. From the response graph, it is perceived that the rate of removal of material is getting increased with an increasing values of pulse on-time, flushing pressure, and wire feed. Also it is getting reduced with an increased values of pulse off-time and servo voltage. Also, it is witnessed that the pulse on-time is most significant variable for the MRR.

With increasing values in pulse-on-time, the discharge energy given to the machining chamber will be more that consequences in an influential explosion that has the possibility of leading to an increment in the rate of material removal.

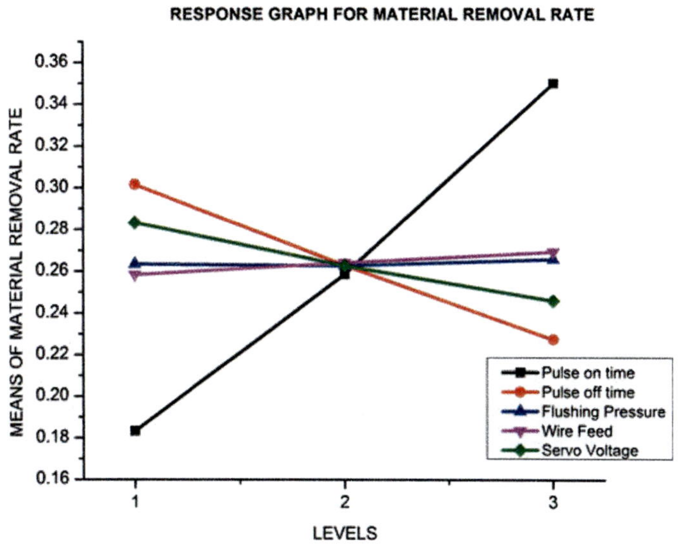

Fig. 3 Response analysis for material removal rate

Table 2 Taguchi's analysis for MRR—WEDM of Al-Fly ash composites

Levels	Means of material removal rate				
	A	B	C	D	E
1	0.1833	**0.3015**	0.2634	0.2584	**0.2832**
2	0.2585	0.2628	0.2626	0.2641	0.2626
3	**0.3499**	0.2275	**0.2658**	**0.2693**	0.2460
Delta	0.1665	0.0741	0.0032	0.0109	0.0371
Rank	1	2	5	4	3

Increasing the values of wire feed and flushing pressure makes possible the availability of favourable scenario for the easy removal of material from the work surface. Increasing in the values of pulse-on-time has the capability of getting increase in the amount of electrons that impacting on the work material, therefore it starts eroding more quantity of material from the surface of the work piece per spark discharge.

Taguchi's analysis for MRR has been accomplished, and the outcomes are presented in the response table as shown in Table 2. The best set of process parameters for attaining an improved material removal rate is $A_3B_1C_3D_3E_1$. It means that the optimum process variables combination for better performance is 'P_{on}'—111(μs); 'P_{off}'—54(μs); flushing pressure—12 kg/cm^2; wire feed—6 mm/min; servo voltage—20 V. Also, it is perceived from the analysis that the 'P_{on}' is the significant process variable, and then it is trailed by 'P_{off}', servo voltage, wire feed, and flushing pressure.

3.2 *Influence of Process Parameters on Surface Roughness*

The graphical illustration of response graph for roughness of machined surface roughness during Wire EDM of Al-Fly ash composites is depicted in Fig. 4.

It is perceived from the graph that the values of roughness of the electrically discharge machined surface getting increased with an increase of 'wire feed and flushing pressure,' and the same has been getting diminished with an increment of 'Pon', 'Poff', and applied voltage. The pulse on-time is deemed as the supreme persuasive process variable for roughness of the machined surface. Taguchi analysis for surface roughness has been executed, and the results are presented in the response table as exhibited in Table 3.

The best set of process parameters for attaining better surface roughness is $A_3B_3C_3D_1E_1$. It means that the optimum process variables combination for better performance is 'P_{on}'—111 μs; 'P_{off}'—60 μs; flushing pressure—12 kg/cm^2; wire feed—2 mm/min; servo voltage—20 V. Also, it is observed from the analysis that the 'P_{on}' is the major variable and then it is trailed by pulse off time, servo voltage, flushing pressure and wire feed.

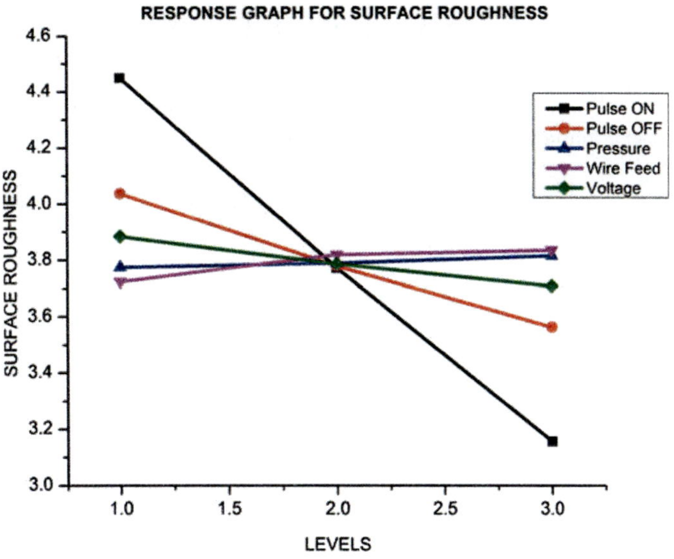

Fig. 4 Response graph for surface roughness

Table 3 Response table for Surface Roughness for WEDM of Al-Fly ash composites

Levels	Means of Surface Roughness				
	A	B	C	D	E
1	4.281	3.928	3.665	**3.718**	**3.668**
2	3.595	3.844	3.889	3.761	3.686
3	**3.412**	**3.516**	**3.734**	3.810	3.934
Delta	0.869	0.412	0.224	0.092	0.267
Rank	1	2	4	5	3

3.3 Development of Multiple Regression Models

Multiple regression method has been adopted to determine the correlation among the selected independent and dependent process variables for various advanced machining processes [20]. The regression models have different models such as linear, quadratic, interaction, and full model and have been exhibited in Eqs. (1–4) (for three input parameters):

Linear model:

$$y = \beta_0 + \beta_1 X_1 + \beta_2 X_2 + \beta_3 X_3 \tag{1}$$

Quadratic model:

$$y = \beta_0 + \beta_1 X_1 + \beta_2 X_2 + \beta_3 X_3 + \beta_4 (X_1)^2 + \beta_5 (X_2)^2 + \beta_6 (X_3)^2 \quad (2)$$

Interaction equation:

$$y = \beta_0 + \beta_1 X_1 + \beta_2 X_2 + \beta_3 X_3 + \beta_4 X_1 X_2 + \beta_5 X_1 X_3 + \beta_6 X_2 X_3 \quad (3)$$

Second-order model (Full model):

$$y = \beta_0 + \beta_1 X_1 + \beta_2 X_2 + \beta_3 X_3 + \beta_4 (X_1)^2 + \beta_5 (X_2)^2 + \beta_6 (X_3)^2 + \beta_7 X_1 X_2 + \beta_8 X_1 X_3 + \beta_9 X_2 X_3 \quad (4)$$

where 'y' is criterion variable and 'X_1', 'X_2', and 'X_3' are predictor variables. 'β_1', 'β_2', 'β_3', …, 'β_n' are regression coefficients.

3.3.1 Regression Model for MRR

The empirical equations for associating the relationship among the desired performance measure and the considered process variables has been obtained and the desired regression equation for material removal rate is exposed in Eq. (5).

$$\begin{aligned} \text{MRR} = {} & 6.16767 - 0.185654\,\text{Pulse ON} \\ & + 0.10883\,\text{Pulse OFF} - 0.0420963\,\text{Flushing Pressure} \\ & + 0.00140926\,\text{Pulse ON} * \text{Pulse ON} \\ & - 0.0016321\,\text{Pulse ON} * \text{Pulse OFF} \\ & + 0.000397531\,\text{Pulse ON} * \text{Flushing Pressure} \\ & + 0.000475309\,\text{Pulse OFF} * \text{Pulse OFF} \\ & - 7.90123\text{e} - 005\,\text{Flushing Pressure} * \text{Flushing Pressure}. \end{aligned} \quad (5)$$

3.3.2 Regression Model for Surface Roughness

The empirical equations for associating the relationship among the desired performance measure and the considered process variables has been obtained and the desired regression equation for surface roughness is exposed in Eq. (6).

$$\begin{aligned} \text{SR} = {} & 16.159 - 0.351593\,\text{Pulse ON} \\ & + 0.680685\,\text{Pulse OFF} + 0.269093\,\text{Flushing Pressure} \\ & + 0.00348148\,\text{Pulse ON} * \text{Pulse ON} \end{aligned}$$

$$
\begin{aligned}
&- 0.0105185 \text{ Pulse ON} * \text{Pulse OFF} \\
&- 0.00181481 \text{ Pulse ON} * \text{Flushing Pressure} \\
&+ 0.00327778 \text{ Pulse OFF} * \text{Pulse OFF} \\
&- 0.00457407 \text{ Flushing Pressure} * \text{Flushing Pressure}.
\end{aligned} \tag{6}
$$

4 Conclusions

This current experimentation details a single aspect optimization problem of wire EDM of Al-Fly ash composites with the help of Taguchi analysis. In this investigation, the rate of material removal rate (MRR) and surface roughness (SR) were deemed as the performance characteristics, and the conclusions attained are discussed as below:

- Taguchi's approach has been adopted for designing the experimental runs, and also, it has been adopted for optimizing the individual variables for accomplishing better performance in machining.
- The prevalence of deemed independent parameters on the preferred performance aspects was made known by Taguchi's analysis. Pulse on-time is the major process variable for a better and improved performance characteristics.
- The detailed Taguchi's approach in this present exploration is greatly appropriate for establishing the best set of process parameters for attaining the improved performance in any advanced machining process.
- The regression models were developed for desired performance measures, and from the regression coefficient values of developed empirical relations, the models were capable for predicting the desired performance measures.
- The outcome attained from this exploration will be a wide-ranging support to the manufacturers for enhancing the rate of production and quality of products made with an assistance of WEDM process.

References

1. Vijayarangan S, Rajamanickam N, Sivananth V (2013) Evaluation of metal matrix composite to replace spheroidal graphite iron for a critical component, steering knuckle. Mater Des 43:532–541
2. Chadwick G (1989) The machining of metal matrix composites with polycrystalline diamond tools. In: The materials revolution through the 90 s
3. Yigezu BS, Jha PK, Mahapatra MM (2013) The key attributes of synthesizing ceramic particulate reinforced Al-based matrix composites through stir casting process: a review. Mater Manuf Processes 28(9):969–979
4. Taha MA (2001) Practicalization of cast metal matrix composites (MMCCs). Mater Des 22(6):431–441

5. Hashim J, Looney L, Hashmi MSJ (1999) Metal matrix composites: production by the stir casting method. J Mater Process Technol 92:1–7
6. Yan BH, Wang CC (1993) Machinability of SiC particles reinforced Al alloy composite material. J Jpn Inst Light Met 43(4):187–192
7. Monaghan J, O'reilly P (1992) The drilling of an Al/SiC metal-matrix composite. J Mater Process Technol 33(4):469–480
8. Yue TM, Dai Y, Lau WS (1996) An examination of wire electrical discharge machining (WEDM) of Al2O3 particulate reinforced aluminum based composites. Mater Manuf Process 11(3):341–350
9. Ramulu M (1988) EDM sinker cutting of a ceramic particulate composite, SiC-TiB sub 2. Adv Ceram Mater (USA) 3(4)
10. Zhu Y, Liang T, Gu L, Zhao W (2017) Precision machining of high aspect-ratio rotational part with wire electro discharge machining. J Mech Sci Technol 31(3):1391–1399
11. Menzies I, Koshy P (2008) Assessment of abrasion-assisted material removal in wire EDM. CIRP Ann 57(1):195–198
12. Manikandan N, Arulkirubakaran D, Palanisamy D, Raju R (2019) Influence of wire-EDM textured conventional tungsten carbide inserts in machining of aerospace materials (Ti–6Al–4V alloy). Mater Manuf Processes 34(1):103–111
13. Uhlmann E, Piltz S, Oberschmidt D (2008) Machining of micro rotational parts by wire electrical discharge grinding. Prod Eng Res Devel 2(3):227–233
14. Alias A, Abdullah B, Abbas NM (2012) Influence of machine feed rate in WEDM of titanium Ti-6Al-4V with constant current (6A) using brass wire. Procedia Eng 41:1806–1811
15. Manikandan N, Binoj JS, Varaprasad KC, Sabari SS, Raju R (2019) Investigations on wire spark erosion machining of aluminum-based metal matrix composites. In: Advances in manufacturing technology. Springer, Singapore, pp 361–369
16. Kuriakose S, Mohan K, Shunmugam MS (2003) Data mining applied to wire-EDM process. J Mater Process Technol 142(1):182–189
17. Rahman MM, Khan M, Rahman A, Kadirgama K, Noor MM, Bakar RA (2011) Optimization of machining parameters on tool wear rate of Ti-6Al-4V through EDM using copper tungsten electrode: a statistical approach. In: Advanced materials research, vol 152. Trans Tech Publications, pp 1595–1602
18. Mahapatra SS, Patnaik A (2007) Optimization of wire electrical discharge machining (WEDM) process parameters using Taguchi method. Int J Adv Manuf Technol 34(9–10):911–925
19. Jangra KK (2015) An experimental study for multi-pass cutting operation in wire electrical discharge machining of WC-5.3% Co composite. Int J Adv Manuf Technol 76(5–8):971–982
20. Manikandan N, Kumanan S, Sathiyanarayanan C (2017) Optimisation of electrochemical drilling process using Taguchi method and regression analysis. Int J Mach Mach Mater 19(2):136–159

Wear Behaviour of Duplex Stainless Steel Spur Gear Produced by CNC Wire Electro-discharge Machining Under Wet Lubrication—An Experimental Approach

Mukesh Kumar Choudhary, V. Dhinakaran, and T. Jagadeesha

Abstract The present work deals with the influence of machining process on surface roughness and teeth wear performances of spur gears under wet condition (with lubrication). Spur gear profiles are mathematically modelled and generated in CAD software. Using super-duplex stainless steel, spur gears have fabricated using hobbing and wire-cut EDM processes. An experimental rig has developed to test the gear under various operating conditions, and variation of gear teeth wear and roughness has measured under wet lubrication conditions. Lubrication mechanism is modelled using ANSYS CFD, and volume fraction of oil availability during mating of gears teeth has evaluated under various speeds. Teeth of wire-cut EDM gears are stronger compared to hobbed gears for the same dimensions. Teeth surface roughness of the WEDM gears is much higher compared to hobbed gears. Wear of the WEDM gears is lower compared to hobbed gears.

Keywords Spur gears · WEDM · Hobbing · Gear lubrication · Wear rate

1 Introduction

In search of more corrosion resistant and mechanically superior materials for highly aggressive environments such as hot chlorinated seawater and highly acidic, chloride containing media, duplex stainless steels (DSSs) have emerged as the best-suited materials. DSS 2205 (ASME Standard A240—UNS S32205) is a highly alloyed super-duplex stainless steel having high tensile strength, high fatigue strength, high toughness, high critical pitting temperature, corrosion resistance which sets it apart from other generations of DSSs. Hence, it is used in high-end applications such as desalination equipment, chemical industries, pressure vessels, piping and heat

M. K. Choudhary · T. Jagadeesha (✉)
National Institute of Technology Calicut, Calicut, Kerala, India
e-mail: jagdishsg@nitc.ac.in

V. Dhinakaran
Department of Mechanical Engineering,
Chennai Institute of Technology, Kundrathur, Chennai 600069, India

A. Arockiarajan et al. (eds.), *Advances in Industrial Automation and Smart Manufacturing*, Lecture Notes in Mechanical Engineering,
https://doi.org/10.1007/978-981-15-4739-3_33

Table 1 Mechanical properties of duplex stainless steel 2205

Tensile strength (MPa)	Yield strength (MPa)	Hardness (HRC)
620	450	31

exchangers, tropical marine usages, offshore oil and gas exploration and petrochemical processing. Wire electro-discharge machining (WEDM) is a manufacturing system for high precision cutting of complex shapes from hard metals or those which are difficult to cut using conventional machining methods. Hobbing is one of the common methods employed in industries to manufacture gears. Electrical current and pulse on time are two process parameters in WEDM. Several researchers suggested manufacturing of spur gears using WEDM processes. Jose Luis et al. manufactured spur tooth gear made of Ti6Al4V alloy using WEDM processes [1]. A geometrical model of the gear was developed using MATLAB code by them. Equation of geometrical conversion of involute curve for straight-line segments demonstrated by Lee and Lee [2]. Yanzhongwang [3] optimized the spray parameters of oil jet lubricated spur gears. When the gear rotating with high speed, lubricating oil cannot reach to meshing zone effectively [4–8]. Mohapatra et al. carried out modelling and analysis of WEDM cutting processes. [9]. Thermophysical model for WEDM was developed by Shahane and Pande [10]. Kahraman [11] developed time-dependent mixed elastohydrodynamic lubrication system for spur gears. Very hard and difficult to cut spur gears can be manufactured using WEDM processes [12–21]. From the literature survey, it is observed that there exists research opportunities in modelling, manufacturing, and performance analysis of duplex stainless steel gears.

2 Experimentation

2.1 Material Selection

Super-duplex stainless steel (SDSS) G2205 is used as the workpiece material for the gear manufacture. Properties of SDSS are given in Table 1.

2.2 Machine Tool Selection

DK7735 CNC wire-cut machine (Fig. 1) and hobbing machine (Fig. 2) were used to fabrication of spur gear. For fabrication of WEDM gears, molybdenum wire of diameter 0.18 mm has been used with the spark gap of 0.02 mm. Pulse on and off time are 20 and 7 ms, respectively. Voltage and current are 80 V and 2.7 A, respectively. Wire tension is 900 g, and wire moving speed is 11.5 m/s. Hob cutter of high carbon steel has been used for the fabrication of hobbed spur gear.

Fig. 1 CNC wire-cut EDM machine

Fig. 2 Hobbing machine

2.3 Specification of Gears

Spur gears are manufactured in hobbing and CNC wire-cut EDM machine. Typical gear is shown in Figs. 3 and 4. Addendum diameter is 64 mm. Dedendum diameter is 56.38 mm. Module of the gear is 2 mm. Pressure angle is 20 deg. Number of teeth is 30. Tooth thickness for hobbed gear is 2.86 mm and that of the WEDM is 3.04 mm. Two pairs of spur gear have fabricated. One pair of WEDM gears which is marked as 1 (driving) and 2 (driven) and other pair hobbed gears which have also marked as 1 (driving) and 2 (driven).

Fig. 3 Fabricated WEDM spur gear

Fig. 4 Fabricated hobbed spur gear

2.4 Experimental Setup

Experimental setup is developed and is shown in Fig. 5. The three-phase 0.5 HP AC motor has used with variable frequency controller (VFC). Two parallel shafts are supported on bearing pedestal via love-jaw coupling. Spur gears are mounted on two

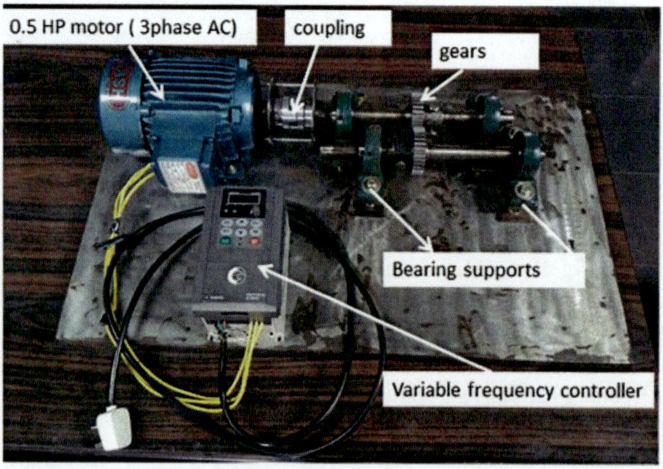

Fig. 5 Schematic representation of experimental setup

solid shafts. The speed of the motor can be controlled by using variable frequency controller. Speeds used in these experiments are 100, 200, 300, 400, 500, 600, 700 and 800 RPMs.

3 Results and Discussion

3.1 Roughness and Tooth Thickness of WEDM Gears Under Wet Condition

Two WEDM gears were run for six hours at 100, 200, 300, 400, 500, 600, 700 and 800 RPMs using lubricant Hydraulic Grade 46 is used as a lubricant in all the experiments. At the end of each six hours, roughness and tooth thickness are measured. Table 2 gives the roughness and tooth thickness. Averages of 2 readings at two spots are taken in each iteration. Ra (1) and Ra (2) indicate the average roughness values for gear 1 and 2, respectively. T (1) and T (2) are the tooth thickness for gear 1 (driving) and 2 (driven), respectively.

It is found that as the RPM increases the average roughness decreases in wet condition. The tooth thickness decreases as the RPM increase. Tooth thickness decreases drastically at higher RPM.

3.2 Roughness and Tooth Thickness of Hobbed Gears Under Wet Condition

Two hobbed gears were run for six hours at 100, 200, 300, 400, 500, 600, 700 and 800 RPMs using lubricant. At the end of each six hours, roughness and tooth thickness are measured. Table 3 gives the roughness and teeth thickness.

Averages of 2 readings at two spots have taken in each iteration. Ra(1) and Ra(2) indicate the average roughness values for gears 1 and 2, respectively. T(1) and T(2) are the tooth thickness for gear 1 (driving) and 2 (driven) gears, respectively.

3.3 Tooth Profiles at the Beginning and End of the Test in Wet Condition

Tooth profiles at the end of each RPM test have taken using ALICONA. Figure 6 shows the tooth profile after 100 RPM test and after 800 RPM test. Figure 6a shows the tooth profile of WEDM driving gear after 100 RPM test (after 6 h), and Fig. 6b shows the tooth profile of WEDM driving gear after 800 RPM test (after 48 h), and Fig. 6c shows the tooth profile of hobbed gears after 100 RPM test(after 6 h), and

Table 2 Roughness and teeth thickness of WEDM gears at different RPMs

No. of iteration	RPM	Ra(1)	Avg. Ra(1)	T (1)	$T_a(1)$	Ra(2)	Avg. Ra(2)	T(2)	$T_a(2)$
0	000	3.440, 3.142	3.291	3.04, 3.04	3.04	3.487, 4.000	3.480	3.04, 3.04	3.04
1	100	4.180, 3.400	3.790	3.04, 3.04	3.04	3.150, 3.490	3.320	3.04, 3.04	3.04
2	200	4.080, 3.990	3.990	3.04, 3.02	3.03	3.610, 3.570	3.590	3.04, 3.02	3.03
3	300	3.442, 3.450	3.440	3.04, 3.04	3.04	3.658, 3.449	3.554	3.04, 3.02	3.03
4	400	3.450, 3.140	3.295	3.02, 3.02	3.02	3.390, 3.987	3.689	3.02, 3.02	3.02
5	500	2.990, 2.820	2.905	3.02, 3.00	3.01	3.161, 3.982	3.572	3.02, 3.02	3.02
6	600	2.440, 2.928	2.684	3.00, 3.00	3.00	2.772, 2.670	2.721	3.00, 3.00	3.00
7	700	2.220, 1.982	2.101	3.00, 2.98	2.99	2.400, 2.418	2.409	3.00, 2.98	2.99
8	800	2.260, 2.450	2.355	2.98, 3.00	2.99	2.180, 2.880	2.530	2.98, 2.96	2.97

Table 3 Roughness and teeth thickness of hobbed gears at different RPMs

No. of iteration	RPM	Ra(1)	Avg. Ra(1)	T (1)	$T_a(1)$	Ra(2)	Avg. Ra(2)	T(2)	$T_a(2)$
0	000	0.860, 0.880	0.870	2.86, 2.86	2.86	0.854, 0.752	0.803	2.84, 2.86	2.85
1	100	0.842, 0.844	0.843	2.84, 2.86	2.85	0.822, 0.844	0.833	2.84, 2.84	2.84
2	200	0.800, 0.812	0.806	2.84, 2.84	2.84	0.810, 0.811	0.811	2.84, 2.84	2.84
3	300	0.815, 0.822	0.819	2.84, 2.84	2.84	0.820, 0.800	0.810	2.80, 2.84	2.82
4	400	0.858, 0.859	0.859	2.82, 2.84	2.83	0.810, 0.829	0.820	2.80, 2.82	2.81
5	500	0.881, 0.851	0.866	2.82, 2.82	2.82	0.823, 0.871	0.847	2.80, 2.80	2.80
6	600	0.884, 0.891	0.888	2.80, 2.80	2.80	0.845, 0.865	0.855	2.78, 2.80	2.79
7	700	0.872, 0.898	0.885	2.78, 2.80	2.79	0.849, 0.872	0.861	2.78, 2.78	2.78
8	800	0.898, 0.905	0.902	2.78, 2.78	2.78	0.896, 0.819	0.858	2.78, 2.76	2.77

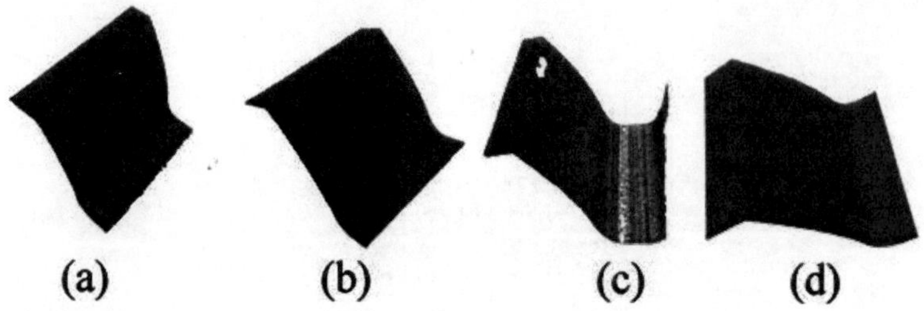

Fig. 6 Teeth profile representation after 100 and 800 RPMs in wet condition

Fig. 6d shows the tooth profile of hobbed gears after 800 RPM test(after 48 h). No pit marks are observed after 800 RPM test (i.e. after 48 h of running) as shown in wet conditions.

3.4 *Wear Test Under Wet Condition*

At the end of each run, gear is dismantled and tooth thickness is measured. Wear is calculated based on the difference in tooth thickness. Table 4 shows the wear rate of machined and WEDM gears. (1) indicates driver gear 1, and (2) indicates driven gear 2. It can be observed that after running at 100 RPMs, loss of the tooth thickness for both gears is negligible. Loss of the material at 200 and 800 RPMs is the same for both hobbed and WEDM gears. Behaviour of wear rate for both gears is nonlinear at various RPM values.

Table 4 Wear rates of WEDM and hobbed (machined) gears under wet condition

		Wear rate of gear teeth in mm			
No. of iteration	Gear speed in (RPM)	WEDM (1)	Machined (1)	WEDM (2)	Machined (2)
0	000	0	0	0	0
1	100	0	0	0	0
2	200	0.01	0.01	0.01	0
3	300	0.02	0.01	0.01	0.02
4	400	0.04	0.02	0.02	0.03
5	500	0.05	0.03	0.02	0.04
6	600	0.06	0.05	0.04	0.05
7	700	0.07	0.06	0.05	0.06
8	800	0.07	0.07	0.07	0.07

Table 5 Oil volume fraction at velocity of 1 m/s

No. of iteration	Speed of gear (RPM)	Oil volume fraction at V = 1 m/s
1	100	0.1867
2	200	0.1809
3	300	0.1712
4	400	0.1688
5	500	0.1612
6	600	0.1513
7	700	0.1490
8	800	0.1469

3.5 Lubrication Flow Analysis

There are two types of lubrication normally used in spur gears: into mesh and out of mesh lubrication method. Into mesh type of lubrication is generally used in gear lubrication. Three-dimensional CAD model of the spur gear pairs is used for CFD analysis. The density and viscosity of the lubricating oil are 871 kg/m^3 and 46.3 cSt, respectively.

Oil volume fraction at spray velocity of 1 m/s is given in Table 5. It can be observed that as the RPM increases, the volume of oil availability between the gear teeth meshing zone reduces.

4 Conclusions

Three-dimensional spur gears are modelled in SolidWorks and converted into 2D CAD model to give as an input for CNC wire-cut EDM machine. Using super-duplex stainless steel, spur gears are cut using hobbing and wire-cut EDM processes. An experimental setup has developed to test the gear under wet lubrication conditions and variation of gear teeth wear and roughness has measured under wet lubrication conditions. Lubrication mechanism is modelled using ANSYS CFD, and volume fraction of oil availability at gears meshing zone under various speeds has evaluated. As machined conditions, WEDM gears have higher surface roughness compared to hobbed gears. Due to the high initial surface roughness of WEDM gear, Ra value is always decreased in wet conditions. While for hobbed gears due to low initial surface roughness, the Ra value always increases in wet conditions at different RPM. Teeth thickness always decreases of both gear pairs in wet conditions. Wear rate of teeth thickness of both gear pairs in wet conditions is lower. As RPM increases, surface roughness value of machined gear pair is almost constant in wet conditions, while WEDM gear pair, Ra value slightly decreases. Life of WEDM gear is high irrespective of machined gear.

References

1. Talón JL, Ortega JC, Gómez CL, Sancho ER, Olmos EF (2010) Manufacture of a spur tooth gear in Ti–6Al–4V alloy by electrical discharge. Comput Aided Des 42(3):221–230
2. Lee RS, Lee JN (2001) A new tool-path generation method using a cylindrical end mill for 5-axis machining of a spatial cam with a conical meshing element. Int J Adv Manufacturing Technol 18(9):615–623
3. Lugt PM, Morales-Espejel GE (2011) A review of elasto-hydrodynamic lubrication theory. Tribol Trans 54(3):470–496
4. Biboulet N, Colin F, Lubrecht AA (2013) Friction in starved hydrodynamically lubricated line contacts. Tribol Int 58:1–6
5. Thirugnanasambantham KG, Raju R, Sankaramoorthy T, Velmurugan P, Kannagi A, Reddy MCK, Chandra VR (2018) Degradation mechanism for high-temperature sliding wear in surface-modified in 718 superalloy. Cogent Eng 5(1):1–11
6. Mishra SP, Polycarpou AA (2011) Tribological studies of unpolished laser surface textures under starved lubrication conditions for use in air-conditioning and refrigeration compressors. Tribol Int 44(12):1890–1901
7. Jeyaprakash N, Duraiselvam M, Raju R (2018) Modelling of Cr 3 C 2–25% NiCr laser alloyed cast iron in high temperature sliding wear condition using response surface methodology. Arch Metall Mater 63(3):1303–1315
8. Mohapatra KD, Shaibu VB, Sahoo SK (2018) Modeling and Analysis of Wire EDM in a Gear Cutting Process for a 2D Model. Mater Today: Proc 5(2):4793–4802
9. Shahane S, Pande SS (2016) Development of a thermo-physical model for multi-spark wire EDM process. Procedia Manuf 5:205–219
10. Li S, Kahraman A (2010) A transient mixed elastohydrodynamic lubrication model for spur gear pairs. J Tribol 132(1):011501
11. Gamage JR, DeSilva AK (2016) Effect of wire breakage on the process energy utilisation of EDM. Procedia Cirp 42:586–590
12. Joshi SN, Pande SS (2009) Development of an intelligent process model for EDM. Int J Adv Manuf Technol 45(3–4):300
13. Li L, Li ZY, Wei XT, Cheng X (2015) Machining characteristics of Inconel 718 by sinking-EDM and wire-EDM. Mater Manuf Process 30(8):968–973
14. Mandal A, Dixit AR, Das AK, Mandal N (2016) Modeling and optimization of machining nimonic C-263 superalloy using multicut strategy in WEDM. Mater Manuf Process 31(7):860–868
15. Doğan O, Karpat F (2019) Crack detection for spur gears with asymmetric teeth based on the dynamic transmission error. Mech Mach Theory 133:417–431
16. Manikandan N, Arulkirubakaran D, Palanisamy D, Raju R (2019) Influence of wire-EDM textured conventional tungsten carbide inserts in machining of aerospace materials (Ti–6Al–4V alloy). Mater Manuf Process 34(1):103–111
17. Hammami M, Fernandes CM, Martins R, Abbes MS, Haddar M, Seabra J (2019) Torque loss in FZG-A10 gears lubricated with axle oils. Tribol Int 131:112–127
18. Wen Q, Du Q, Zhai X (2018) A new analytical model to calculate the maximum tooth root stress and critical section location of spur gear. Mech Mach Theory 128:275–286
19. Wang Y, Song G, Niu W, Chen Y (2018) Influence of oil injection methods on the lubrication process of high speed spur gears. Tribol Int 121:180–189
20. Manikandan N, Binoj JS, Varaprasad KC, Sabari SS, Raju R (2019) Investigations on wire spark erosion machining of aluminum-based metal matrix composites. Lecture Notes in Mechanical engineering, Advances in Manufacturing technology. Springer, Singapore, pp 361–369
21. Wang Y, Song G, Niu W, Chen Y (2018) Optimized design of spray parameters of oil jet lubricated spur gears. Tribol Int 120:149–158

Application of Artificial Neural Network to Friction Stir Welding Process of AA7050 Aluminum Alloy

Aniket K. Dutt, K. Sindhuja, Siriyapureddy Vijaya Narasimha Reddy, and Pawan Kumar

Abstract In the present study, an artificial neural networks (ANN) model is developed for the analysis and correlation between friction stir welding (FSW) parameters, namely traverse speed and rotation rate with mechanical properties. The study focuses on FSW of precipitation strengthened AA7050 aluminum alloys. FSW generates enormous heat and strain, which modify the microstructure of AA7050 alloy. In AA7050 alloy, the precipitation of strengthened phase depends on peak temperature achieved during the FSW process and peak temperature depends on FSW parameters. The input for the ANN simulation is FSW parameters and output is the weld metal hardness and heat affected zone (HAZ) hardness, peak temperature of weld nugget, and peak temperature of HAZ. The simulated results showed agreement with the literature data.

Keywords FSW process · ANN · Aluminum alloy · Thermal cycle

1 Introduction

Friction stir welding (FSW) process is a solid-state welding process and used for various alloys including aluminum alloys, magnesium alloys, titanium alloys and steel [1]. Figure 1 represents the typical FSW process. Two materials are joined using the FSW process. Initially, materials are clamped properly on FSW platform followed by insertion of tool into the workpiece and traverse along the weld direction. The tool has two important features: pin and shoulder. The pin portion inserted completed into the workpiece and the shoulder area contacts to the workpiece surface. The heat is generated due to tool and material friction. The generated heat softens the material and tool rotation results in material flow. There are two main FSW process parameters, namely rotation rate and welding speed. The FSW process parameters modify the microstructure, which improves the mechanical properties. The invention

A. K. Dutt (✉) · K. Sindhuja · S. V. N. Reddy · P. Kumar
Department of Mechanical Engineering, Madanapalle Institute of Technology and Science, Madanapalle, Chittoor, Andhra Pradesh 517325, India
e-mail: aniketdutt@my.unt.edu

A. Arockiarajan et al. (eds.), *Advances in Industrial Automation and Smart Manufacturing*, Lecture Notes in Mechanical Engineering,
https://doi.org/10.1007/978-981-15-4739-3_34

Fig. 1 Typical FSW process

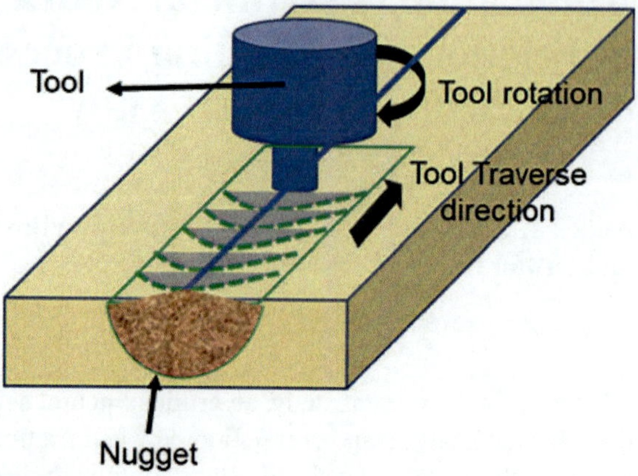

of the FSW process by TWI, UK, was done for solid-state joining [2]. The application of aluminum alloys is in the field of automobile, aircraft, and various other industries.

7XXX series are precipitation strengthened alloys and it is established that 7050-T7451 is strengthened due to the presence of coherent η' phase [3]. A general precipitation sequence in most of 7XXX series alloy is

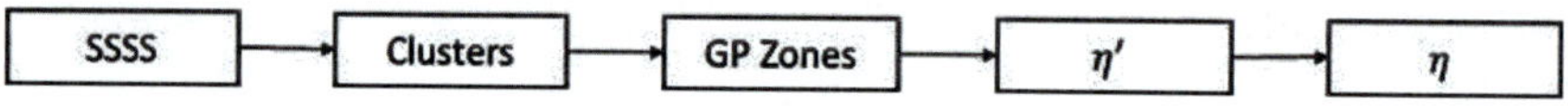

The precipitation of η' and η phases depends on the peak temperature attained during the FSW process [3]. Coherent η' phase strengthened the alloy more compared to incoherent η phase. The dissolution of strengthening phase starts at a temperature above 190 °C. The η phase precipitates between 215 and 250 °C. Above 250 °C, it starts coarsening and begin to dissolve at 320 °C. The hardness of heat affected zone of 7050 alloy is minimum as the temperature exist in the range of 300–350 °C. Hwang and Chou observed the minimum hardness of 7075 weld for thermal cycle, which peak temperature is 377 °C. This is attributed to dissolution of η phase. The strength or hardness of 7050 alloy is related to the size of phases and their fractions [4]. The η' phase possesses an HCP crystal structure. The size of η' phase is 3–4 nm in thickness and 5–10 nm in width, whereas η phase is incoherent and cigar-like structure.

The ANN has become very important research fields for engineering in last two decades. With the advancement of computer technology, it is becoming more popular [5, 6]. ANN possesses the ability to determine the complex relationships between inputs and outputs. In recent years, few authors performed ANN on friction stir welding on aluminum alloys [7–11]. Lakshminarayanan and Balasubramanian [10] performed ANN on AA7039 alloy to predict the tensile strength. Okuyucu et al. [7] performed ANN to FSW of Al plates to predict mechanical properties.

In the present study, application of ANN to the FSW of AA7050 alloys has been investigated to predict the hardness of heat affected zone (HAZ), the hardness of weld metal, the peak temperature of HAZ, and peak temperature of weld metal.

2 Materials and Method

The ANN is performed on the AA7050 alloy. The composition of AA7050 alloy is tabulated in Table 1.

The input data for ANN simulation has taken from the literature and shown in Figs. 2 and 3. The hardness of weld nugget and HAZ in terms of welding speed and rotating speed are as shown in Fig. 2. The peak temperature of the thermal cycle of weld nugget and HAZ in terms of welding speed and rotating speed is as shown in Fig. 3.

ANN solves complex problems using computational models. It has three layers, namely input, hidden, and outer layers. The input layers consist of input parameters.

Table 1 Weight percent of AA7050

Element	Zn	Mg	Cu	Zr	Al
wt%	6.2	2.35	2.3	0.1	Remaining

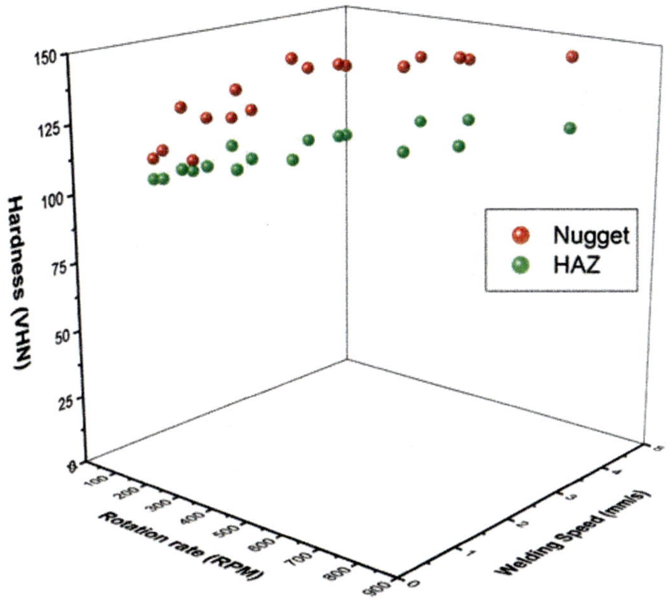

Fig. 2 Hardness of nugget and HAZ zone with respect to welding speed and rotation rate

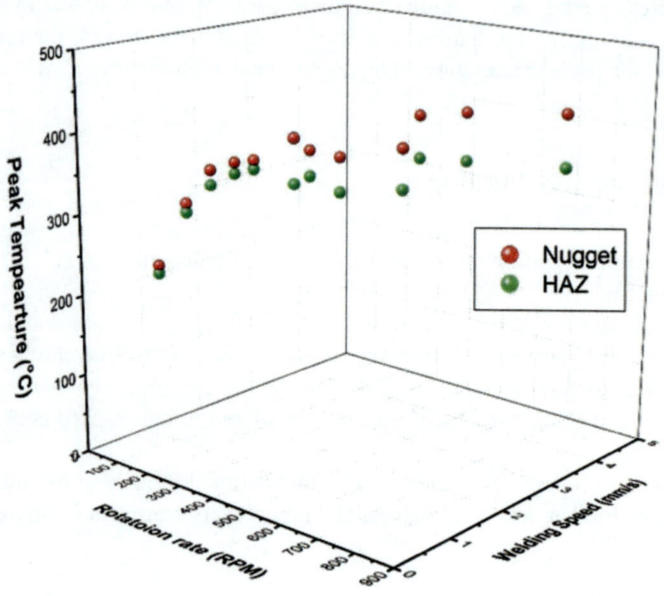

Fig. 3 Peak temperature of nugget and HAZ region with respect to welding speed and rotation rate

The FSW parameters are input parameters in the present study. The information of input layer is passed to the hidden layer following to output layers. A neuron is a basic unit of ANN and it relates to each other by synapses. Each synapse relates to weight. Different weights and bias are provided. A two-layer feed-forward network has applied with sigmoid hidden neurons and liner output neurons. The training, validation, and testing data are kept as 70%, 15%, and 15%, respectively. In the training step, data are provided to network for training and network is adjusted based on the error function. During the validation step, training is halted, and network generalization is measured. In the testing step, independent measure of network performance is performed during and after the training. Levenberg–Marquardt backpropagation algorithm is used to train network. The ANN is performed on three different numbers of hidden layers. These three neural networks are as follows: 2-1-1, 2-5-1, and 2-10-1. The neural network diagrams for these three networks are shown in Fig. 4.

3 Result and Discussions

The result of ANN simulation for the hardness of weld nugget with FSW parameters rotation rate and welding speed is shown in Fig. 5. Three networks 2-1-1, 2-5-1, and 2-10-1 are applied for simulation. The increase in hidden layer neurons improves the coefficient of correlation. The correlation coefficient of networks 2-1-1, 2-5-1, and 2-10-1 are 0.96,415, 0.96,198, and 0.97,831, respectively. The predicted hardness

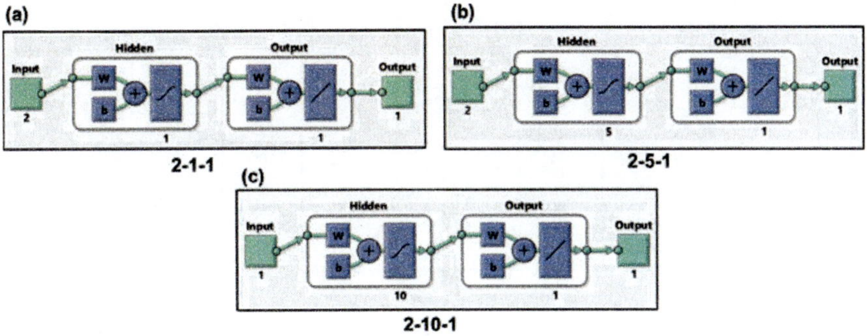

Fig. 4 Neural network diagrams for **a** 2-1-1 network, **b** 2-5-1 network, and **c** 2-10-1 network

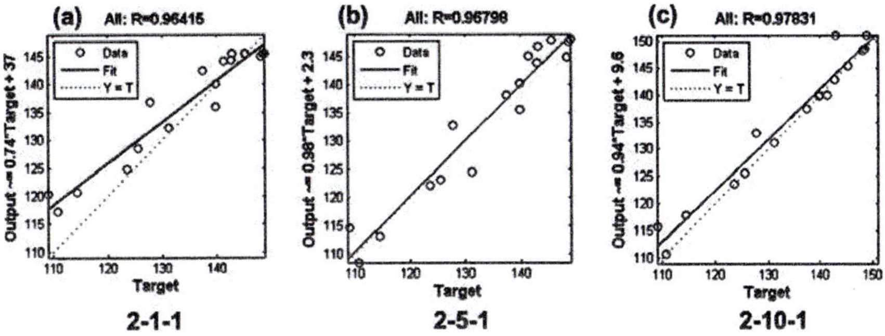

Fig. 5 The correlation between output and target value of hardness for weld nugget for networks: **a** 2-1-1, **b** 2-5-1, and **c** 2-10-1

value of weld nugget (output) showed a better correlation with the target values or literature values. The 2-10-1 network exhibited the best correlation.

The result of ANN simulation for the hardness of the HAZ with FSW parameters rotation rate and welding speed is shown in Fig. 6. The correlation coefficient of 2-1-1 and 2-5-1 networks are 0.95576 and 0.95361, respectively. There is no significant change in the correlation coefficient among 2-1-1 and 2-5-1 networks. Whereas a further increase in hidden layers neuron in 2-10-1 improves the coefficient of correlation to 0.96494. Thus, it is confirmed that the increase in hidden layer neurons improves the coefficient of correlation in general.

The result of ANN simulation for a peak temperature of weld nugget with FSW parameters rotation rate and welding speed is shown in Fig. 7. Here, three networks 2-1-1, 2-5-1, and 2-10-1 are applied for simulation, which showed the trend of improvement of the coefficient of correlation with the increase in hidden layer neurons. The correlation coefficient of networks 2-1-1, 2-5-1, and 2-10-1 are 0.8,531, 0.94,734, and 0.96,089, respectively. The predicted peak temperatures value of weld nugget (output) showed a better correlation with the target values. The 2-10-1 network exhibited the best correlation in comparison to 2-1-1 and 2-5-1 network.

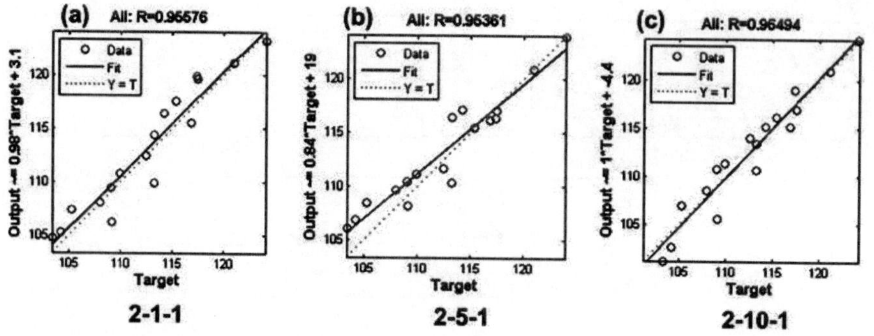

Fig. 6 Correlation between output and the target value of hardness for the HAZ for networks: **a** 2-1-1, **b** 2-5-1, and **c** 2-10-1

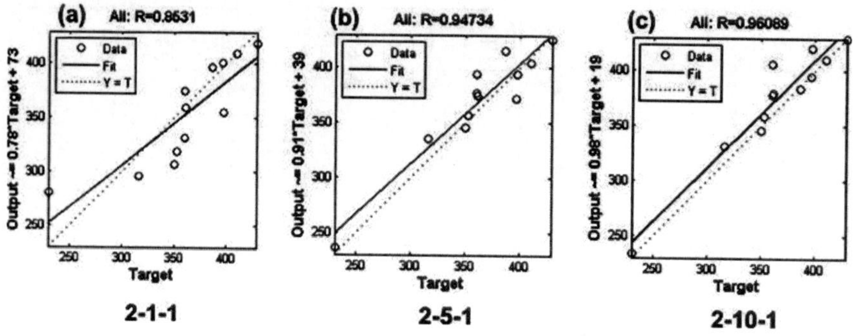

Fig. 7 Correlation between output and target value of peak temperature for weld nugget for networks: (a) 2-1-1, (b) 2-5-1, and (iii) 2-10-1

The result of ANN simulation for a peak temperature of HAZ with FSW parameters rotation rate and welding speed are shown in Fig. 8. The correlation coefficient of 2-1-1, 2-5-,1 and 2-10-1 networks are 0.8,531,0.94,734, and 0.96,089, respectively. The increase in hidden layer neurons improved the correlation coefficient significantly. The output values showed a better correlation with the target values and it is confirmed that the correlation improves with increase in hidden layer neurons.

The above ANNs results confirmed that an increase in hidden layer neurons improves the correlation coefficient for both hardness and peak temperature simulations. In the present study, ANN also performed for peak temperature apart from mechanical property hardness. 7050 alloy is precipitation strengthened alloy and strongly depend on the thermal cycle. The FSW process generates significant strain and enormous heat. The high strain produced during the FSW process reduces the grain size, which increases the strength. Although, the main strengthening mechanism in 7050 is precipitation strengthening. As discussed earlier, the η' phase is strengthening phase and the formation and dissolution depend on the thermal cycle, mainly, peak temperature. Another phase η which has less contribution to

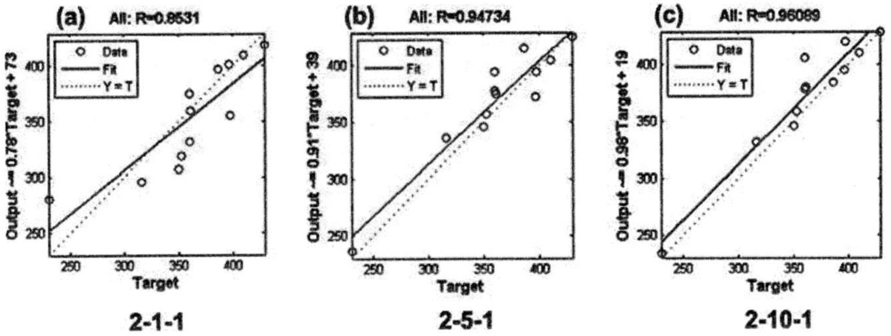

Fig. 8 Correlation between output and the target value of peak temperature for HAZ for networks: **a** 2-1-1, **b** 2-5-1, and **c** 2-10-1

the strengthening of alloy compared to η' phase starts coarsening above 250 °C and dissolve above 320 °C [4, 12–14]. The peak temperature observed during FSW process is significantly higher than 190 °C, dissolving temperature of strengthening η' phase [4, 12–14]. Thus, η' phase dissolved during the heating cycle. The FSW parameters which exhibited the highest peak temperature re-dissolved the η phase and subsequent cooling generates GP zones and thus has higher hardness than lower peak temperature. This confirms the higher hardness of alloy with increasing peak temperature.

4 Conclusions

The conclusions are:

1. ANNs are developed for the FSW process of AA7050 alloys.
2. The output parameters such as hardness of nugget of weld zone, the hardness of the HAZ, the peak temperature of nugget of weld zone, and HAZ showed a better correlation coefficient.
3. The correlation coefficient is improving with an increase in neurons in hidden layers. The neural network 2-10-1 showed a better correlation in comparison to 2-1-1 and 2-5-1.

References

1. Mishra RS, Ma Z (2005) Friction stir welding and processing. Mater Sci Eng: R: Rep 50:1–78
2. Thomas W, Nicholas E, Needham J, Church M, Templesmith P, Dawes C (1991) International patent PCT, GB92/02203
3. Dixit M, Mishra R, Sankaran K (2008) Structure–property correlations in Al 7050 and Al 7055 high-strength aluminum alloys. Mater Sci Eng, A 478:163–172

4. Reynolds AP, Tang W, Khandkar Z, Khan JA, Lindner K (2005) Relationships between weld parameters, hardness distribution and temperature history in alloy 7050 friction stir welds. Sci Technol Weld Join 10:190–199
5. Zeng Q, Zhang L, Xu Y, Cheng L, Yan X, Zu J, Dai G (2009) Designing expert system for in situ toughened Si3N4 based on adaptive neural fuzzy inference system and genetic algorithms. Mater Des 30:256–259
6. Larkiola J, Myllykoski P, Korhonen A, Cser L (1998) The role of neural networks in the optimisation of rolling processes. J Mater Process Technol 80:16–23
7. Okuyucu H, Kurt A, Arcaklioglu E (2007) Artificial neural network application to the friction stir welding of aluminum plates. Mater Des 28:78–84
8. Yousif Y, Daws K, Kazem B (2008) Prediction of friction stir welding characteristic using neural network. Jordan J Mech Ind Eng 2
9. Tansel IN, Demetgul M, Okuyucu H, Yapici A (2010) Optimizations of friction stir welding of aluminum alloy by using genetically optimized neural network. Int J Adv Manuf Technol 48:95–101
10. Lakshminarayanan A, Balasubramanian V (2009) Comparison of RSM with ANN in predicting tensile strength of friction stir welded AA7039 aluminium alloy joints. Trans Nonferr Met Soc China 19:9–18
11. Jiménez-Macías E, Sánchez-Roca A, Carvajal-Fals H, Blanco-Fernández J, Martínez-Cámara E (2014) Wavelets application in prediction of friction stir welding parameters of alloy joints from vibroacoustic ANN-based model
12. Jata K, Sankaran K, Ruschau J (2000) Friction-stir welding effects on microstructure and fatigue of aluminum alloy 7050-T7451. Metall Mater Trans A 31:2181–2192
13. Viana F, Pinto A, Santos H, Lopes A (1999) Retrogression and re-ageing of 7075 aluminium alloy: microstructural characterization. J Mater Process Technol 92:54–59
14. Archambault P, Godard D (2000) High temperature precipitation kinetics and TTT curve of a 7xxx alloy by in-situ electrical resistivity measurements and differential calorimetry [Time-Temperature-Transformation]. Scr Mater 42

Experimental Investigation on Velocity and Angle of Butt Weld Joint by Using TIG Welding

K. Nagaraju, Syed Kwaja Moinuddin, P. Moulali, and M. Ravichandra

Abstract Welding is the method of joining two metals and is neither same material nor different materials. The welding process is mostly used in manufacturing and thermal power plants, etc. In this paper, we find out tensile resistance of butt weld joint of v models with an angle of 30°, 40°, and 50° and these models made from mild steel. Nowadays, we are using different welding speeds, but when we take exact speed, then only to get the quality and strength of material, but here we used different types of welding speeds such has 0.5, 1, and 1.5 cm/s for v-groove angled models of 30°, 40°, and 50°. But actually v-groove angle uses up to 60°.

Keywords TIG welding machine mild steel · Taguchi method · ANNOVA

1 Introduction

1.1 TIG Welding

TIG welding stands for gas tungsten arc welding. The TIG welding consists of oxyacetylene cylinder, electric holder, tungsten electrode, welding machine, insulating and working piece. In the operation of welding machine one terminal of electrode is connected to gas cylinder and another terminal connected to welding machine. Due to this, heat is produced in electric holder that heat passes to filler material then finally with help of tungsten electrode molten metal passage at the joining of two metals. TIG welding process is different method when compared to other welding processes [1–3].

K. Nagaraju (✉) · S. K. Moinuddin · P. Moulali · M. Ravichandra
Department of Mechanical Engineering, Santhiram Engineering College, Nandyal, Kurnool, Andhra Pradesh 518501, India
e-mail: rajumgm321@gmail.com

S. K. Moinuddin
e-mail: khnmoin5@gmail.com

A. Arockiarajan et al. (eds.), *Advances in Industrial Automation and Smart Manufacturing*, Lecture Notes in Mechanical Engineering,
https://doi.org/10.1007/978-981-15-4739-3_35

1.2 About Welding Joints

Joining of more than two metal pieces is called as joint and these joints are prepared with exact geometry. From American society, we have five types of welding joints that are butt, corner, lap, edge, and T-joint. Butt weld joint of two pieces is to be joined in same plane [4, 5]. Mostly butt welding joints are used in thin metal sheets because we apply on one single pass of molten metal. These welded joints are prepared with skilled persons because of joints prepared in thin sheet metals [6].

1.3 The Taguchi Method-Robust Design

For development of new product, so many factors to be considered like input variables and design. Taguchi philosophy is the best method for making the optimum inputs to develop product by using design of experiments. This is a technique for finding precise concentrations of inputs and parameters to create a product of excellent quality. Also we will obtain a solid structure with this Taguchi technique.

2 Methodology

In this paper, we use mild steel material to prepare V-groove geometry models with various angles like 30°, 40°, 50°. A different welding speeds like 0.5, 1.0 and 1.5 cm/s use the TIG welding machine to prepare a welding assembly. Finally, to evaluate and analyzed the tensile strength of butt weld joint.

3 Experimental Procedure

3.1 Preparations of Weld Joints

In this experiment, we have used TIG welding machine to cut the steel material with angles of 30°, 40°, 50° and also used to join the materials with different speeds such as 0.5, 1, 1.5 cm/s, to check the ultimate tensile strength of prepared models (Figs. 1, 2 and 3) [7, 8].

Fig. 1 To prepare the models with different v-groove angle

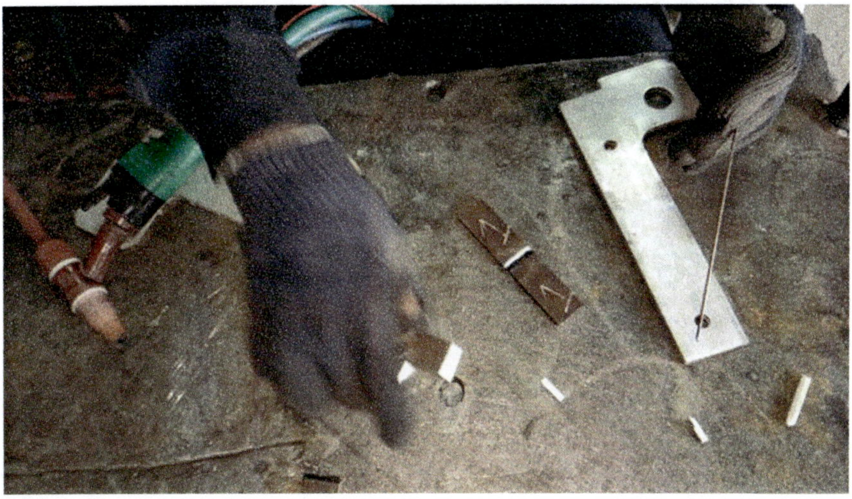

Fig. 2 To join the materials

Fig. 3 Welded V-groove models

3.2 Experimentation

Tensile test is done for different v-groove models and evaluated the ultimate tensile strength values. These values are optimized by using Minitab 18 software and which includes Taguchi orthogonal array and experiment parameters are selected from Table 1 and optimization array is shown in Table 2 (Table 3).

Table 1 Speed and angles of butt weld

Parameters	Level 1	Level 2	Level 3
Velocity (cm/s)	0.5	1.0	1.5
Angle (°)	30	40	50

Table 2 Optimization array

Groove angle (°)	Velocity (cm/s)
30	0.5
30	1.0
30	1.5
40	0.5
40	1.0
40	1.5
50	0.5
50	1.0
50	1.5

Table 3 Experimental values from different angles and velocity

Groove angle (°)	Welding velocity (cm/s)	Ultimate tensile strength (MPa)
30	0.5	385.2
30	1.0	425.6
30	1.5	460.197
40	0.5	410
40	1.0	462.581
40	1.5	370
50	0.5	380.287
50	1.0	365
50	1.5	370.25

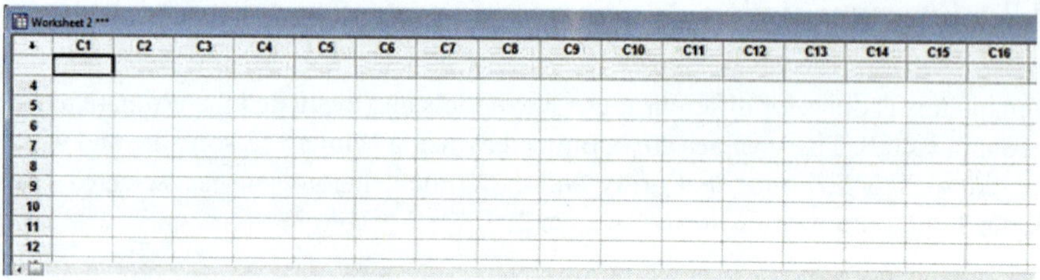

Fig. 4 Work sheet for design orthogonal array

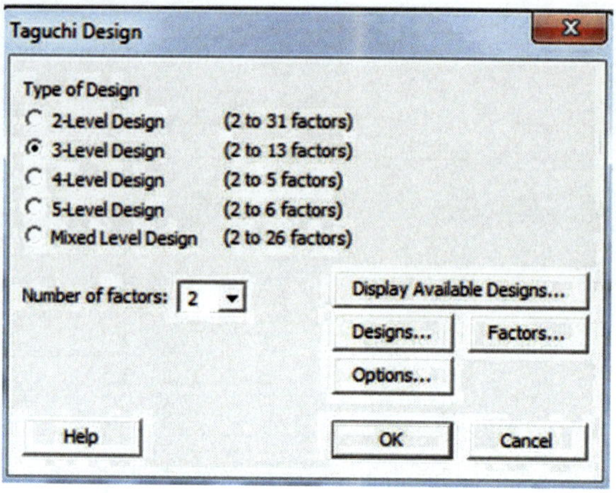

Fig. 5 Select factors for taguchi design

Ultimate tensile strength optimization with Minitab software18

Orthogonal array design and *S*/*N* ratio and mean values

Step-1 to design orthogonal array (Fig. 4)

Step-2 select factors (Figs. 5, 6, 7, 8, 9, 10, 11, 12, 13 and 14)

The aim of Taguchi method is find better response from the signal to noise (*S*/*N*). Because of which the distinctive differences of quality induced by uncontrollable parameters can be reduced [9]. The ultimate tensile strength is taken as a quality feature with the "bigger-the-better" idea. The larger-the-better for the *S*/*N* ratio is:

$$S/N = -10 * \log(\Sigma(1/n) * (1/Y2))$$

where

$n =$ number of test/row readings, here $n = 1$ and

$Y =$ Run/Row calculated value.

The *S*/*N* ratio values are calculated from equation and by using Minitab 18 software.

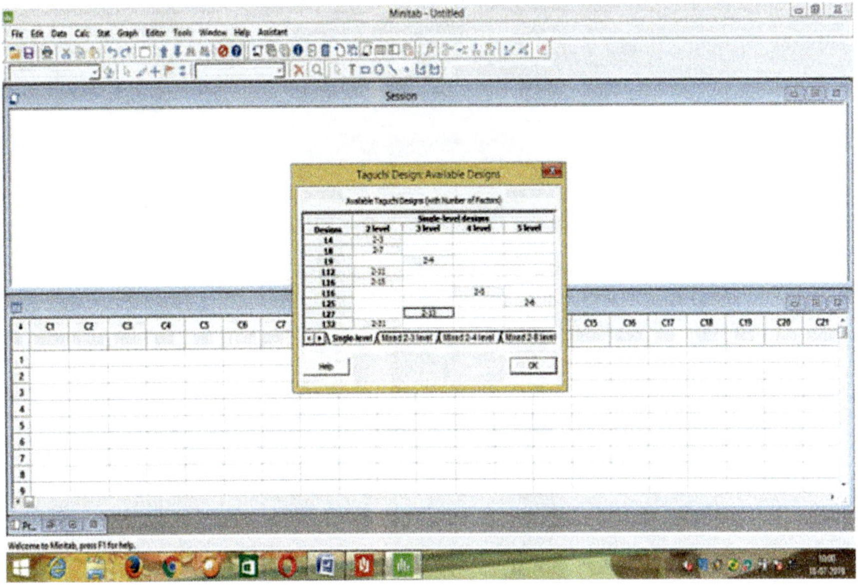

Fig. 6 Taguchi design available designs

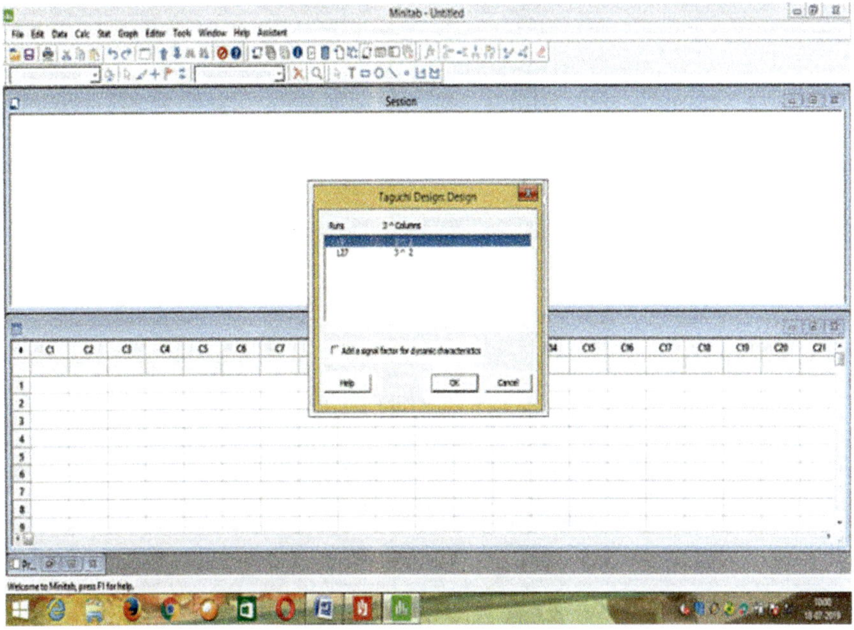

Fig. 7 Taguchi design: design

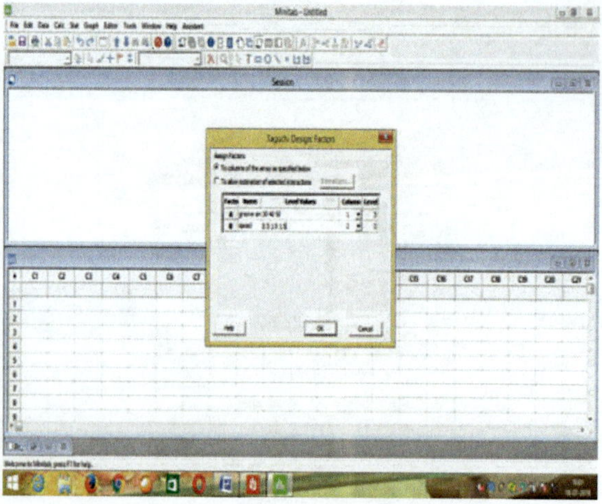

Fig. 8 Taguchi design factors

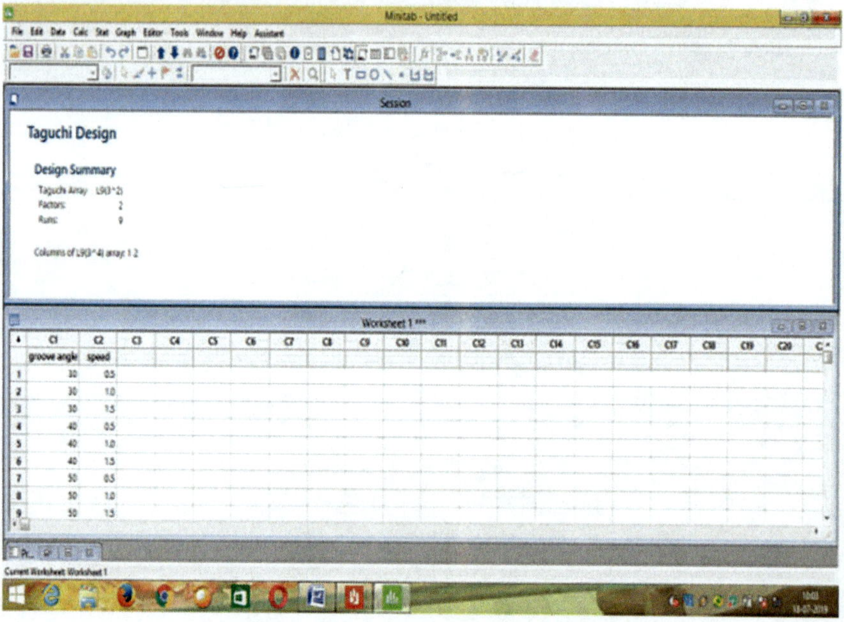

Fig. 9 Design summary

4 Result

The tensile strength values are evaluated from the experiment and from Taguchi software, values are shown in Table 4 (Figs. 15, 16 and 17).

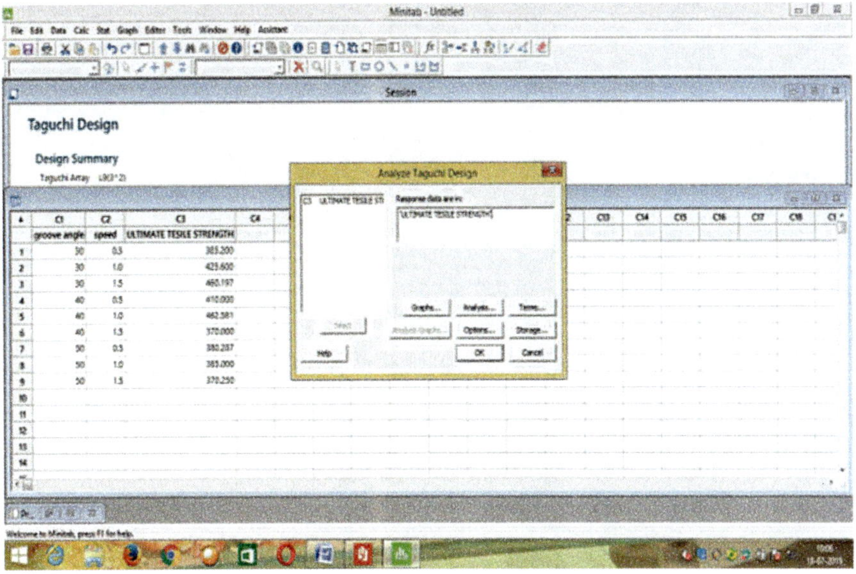

Fig. 10 Analyze taguchi design

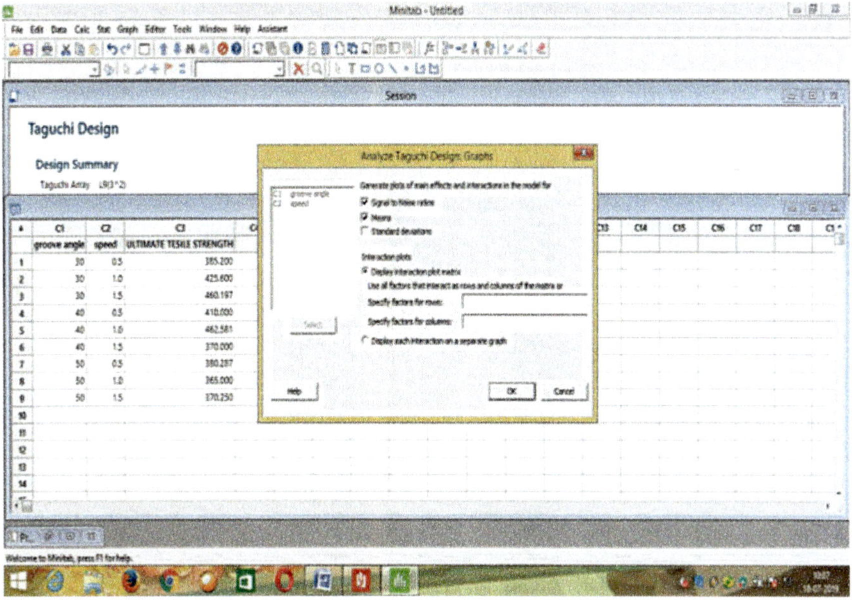

Fig. 11 Analyze taguchi design summary

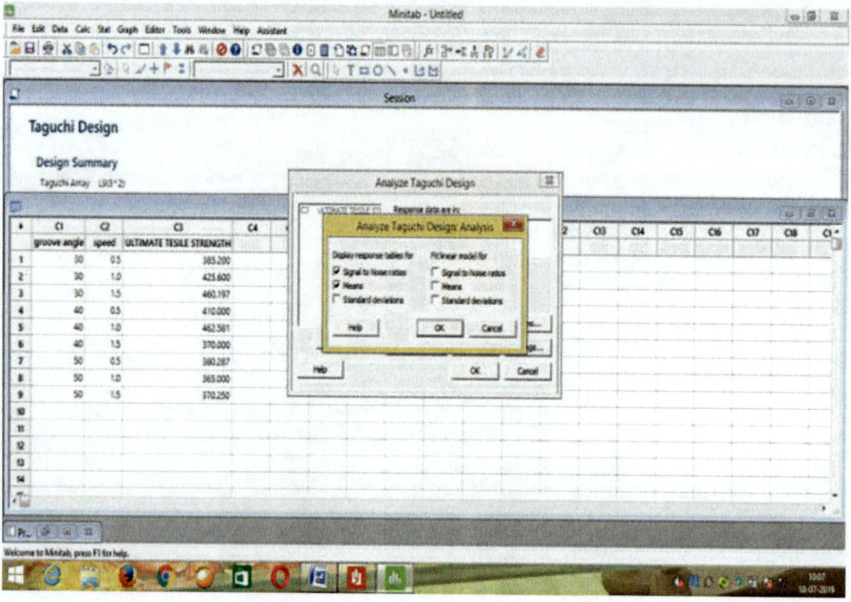

Fig. 12 Design summary for design Analysis

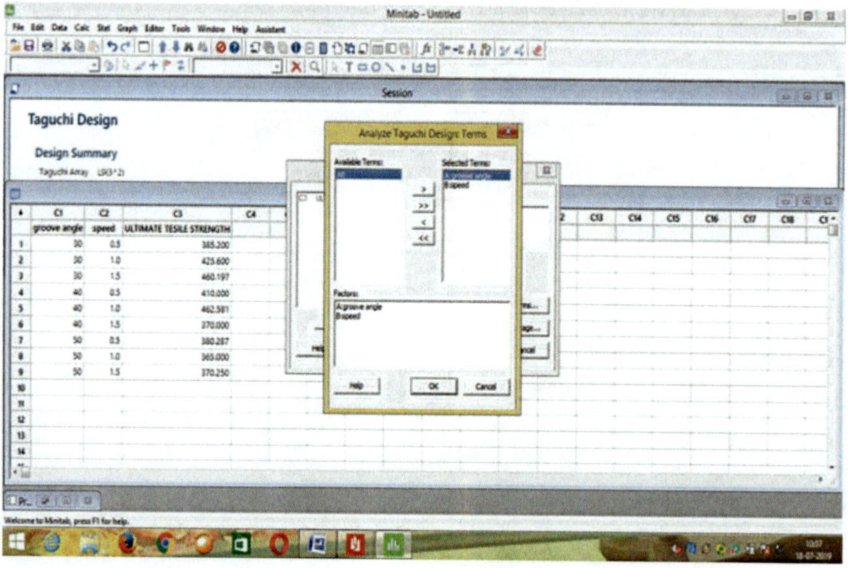

Fig. 13 Design terms

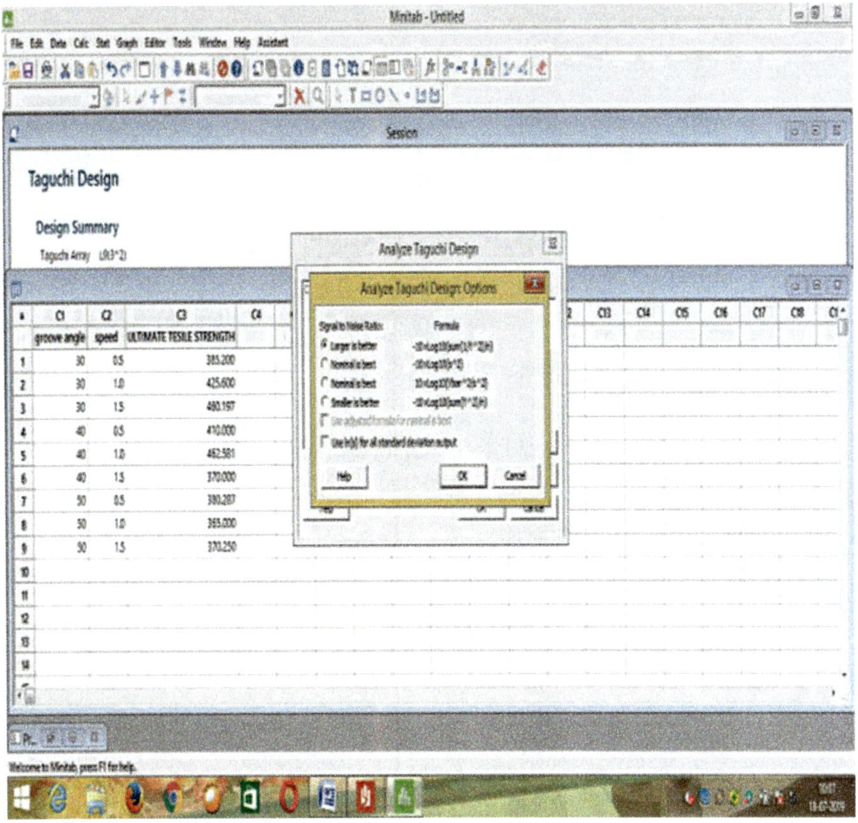

Fig. 14 Design options

Table 4 *S/N* ratio from higher the lower

Level number	Groove angle	Speed
Level 1	52.52	51.86
Level 2	52.31	52.0
Level 3	51.41	52.38

5 Conclusion

Taguchi method is fulfilling all experiment needs. Using Minitab 18 software, all feasible values are calculated. ANNOVA helps to determine the parameters significance. Lastly, the optimum value is to figure out using Minitab18 software and the welding factors are same as to groove angle and velocity for TIG work piece welding. In this paper, the optimal parameters of groove angles 30°, 40°, 50°, and speeds are 0.5, 1.0, and 1.5 cm/s. By consideration of above parameters, experimental work is

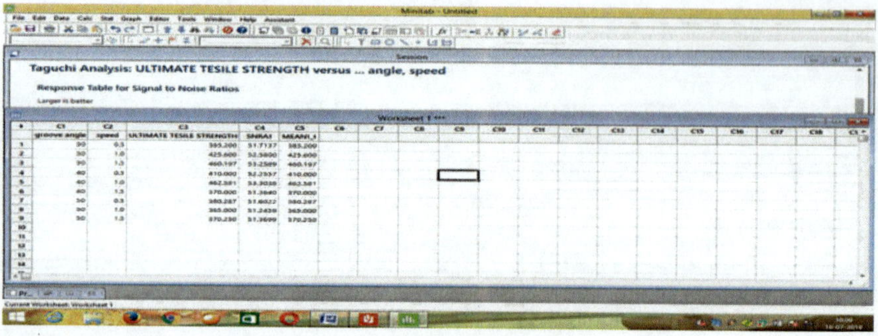

Fig. 15 Ultimate tensile strength Vs angle, speed

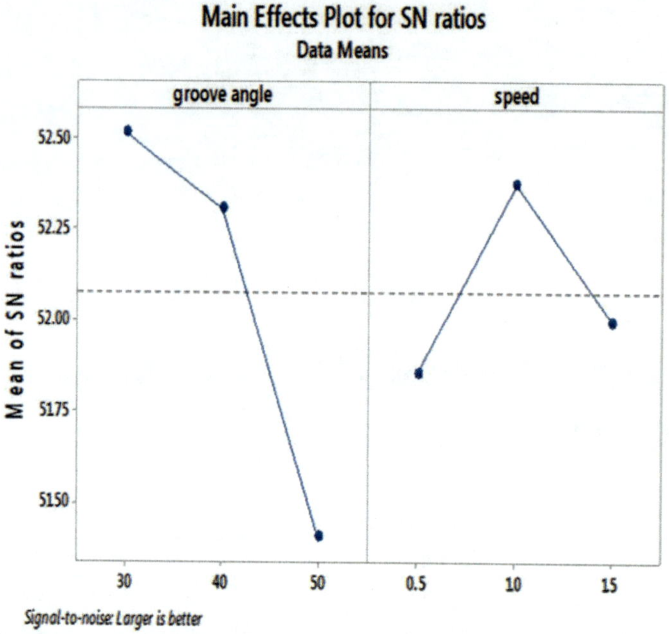

Fig. 16 Main effects plot for SN ratios

conducted and finally validated the ultimate tensile strength values. From the experimental results to optimize the result by using Taguchi method, and finally concluding that optimum tensile value at groove angle 30 and welding speed is 1.5 cm/s.

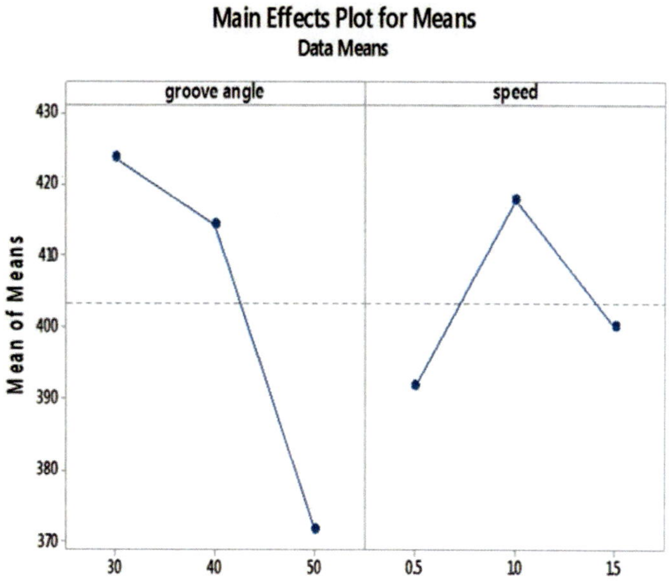

Fig. 17 Main effects plot for means

References

1. Rajkumar NS, Srihari (2001) A examine of impact of groove perspective on angular distortion & impact strength in butt weld. In: International convention on mechanical engineering
2. Ren N, Zan M (2012) Constructing effect of weld & warmness affected zone on deformation behaviour of welded tubes in numerical manipulate bending procedure. J Mater Process Era 43:259–269
3. Boniface RE (2012) Welding engineering. Mc Grow-Hill Book Company, New York
4. Sattari-Far I, Farahani MR (2009) Effect on weld groove form & bypass range on residual stresses 86:723–731
5. Tseng KH, Hsu CY (2011) Performance of activated TIG system in austenitic stainless steel welds. J Mater Process Technol 211(3):503–512
6. Akbari D, Sattari-Far I (2009) Effect of welding heat input on residual stresses in butt weld of assorted pipe joints. International magazine of stress vessel & piping 86:769–776
7. Hyde TH, Williams JA, Sun W (2002) Factors, defined from analysis, contributing to the creep performance of weld repairs. Creep performance of weld repairs, vol 1, Issue 3. OMMI
8. Hyde TH, Williams JA, Becker AA, Sun W (2003) A overview of the finite detail evaluation of repaired welds under creep conditions. Review of FE evaluation of repaired 79:654–663
9. Narang HK, Singh UP, Mahapatra MM, Jha PK (2011) Prediction of the weld pool welds, vol 2, Issue 2. OMMI

Predictive Models for Wire Spark Erosion Machining of AA 7075 Alloy Using Multiple Regression Analysis

N. Manikandan, J. S. Binoj, P. C. Krishnamachary, P. Thejasree, and D. Arul Kirubakaran

Abstract AA 7075 is an aluminium alloy that possesses exceptional mechanical characteristics such as better ductility and fatigue also. This material has extensively employed in making of aerospace components such as aircraft fittings, shafts, gears and also for manufacturing defence components. Apart from these, it has outstanding corrosion resistance property that makes this one as an important material for various engineering applications. Non-traditional method of material removal processes have been evolved to overcome some difficulties in making of intricate shapes. Wire Electrical Discharge Machining (WEDM) is one among the contemporary method of material removal and evolved from the concept of electrical discharge machining process that is predominantly adopted for making of intricate-shaped components. In this exploration, an experimental analysis has been done on WEDM of AA 7075 alloy using Taguchi's response analysis approach and mainly focusing on development of multiple regression models. The experimental runs are planned by Taguchi's experimental design technique. The experiments have been performed by taking into account, the process variables such as pulse on time, pulse off time and servo voltage at three different levels. Material removal rate and surface roughness are the preferred performance aspects deemed in this analysis. The multiple regression models have been evolved for predicting the desired performance measures.

Keywords WEDM · Aluminium alloy · Al 7075 · Aerospace materials · Taguchi's method · *S/N* ratio analysis · Optimization · Regression models

N. Manikandan (✉) · J. S. Binoj · P. C. Krishnamachary · P. Thejasree
Micro Machining Research Centre, Sree Vidyanikethan Engineering College (Autonomous), Tirupati, Andhra Pradesh 517102, India
e-mail: thuraiyurmani@gmail.com

D. Arul Kirubakaran
Department of Mechanical Engineering, Karunya Institute of Technology and Sciences, Coimbatore 641114, India

A. Arockiarajan et al. (eds.), *Advances in Industrial Automation and Smart Manufacturing*, Lecture Notes in Mechanical Engineering,
https://doi.org/10.1007/978-981-15-4739-3_36

1 Introduction

Wire spark erosion machining or Wire Electrical Discharge Machining (WEDM) is one amongst the contemporary method of material removal and most familiar method used in various manufacturing industries for making the components such as turbine blades of aircraft engines, nozzles for fuel injectors and dies. WEDM has been employed for machining harder, corrosive and wear resistant, electrically conductive material and also for hard to machine work materials. The concept of WEDM has been developed from the electrical discharge machining (EDM) and mainly adopted for making complicated shapes on the work materials which is not possible by the traditional method of material removal process [1–5]. In WEDM process, the removal of material takes place based on the electro-discharge erosion effect of electrical sparks occurred in amongst the electrode wire and the work material that are separated with the help of a dielectric fluid as shown in Fig. 1.

The voltage has been supplied in amid the electrode wire and selected work material with the existence of dielectric fluid for melting the work specimen surface by the discharge of sparks on the surface. Because of the running of electrode wire from one end to another end, continuously the material removal takes place [6–8]. Aluminium alloys are best suited for various engineering applications since it possess more specific strength, rigidity at elevated temperatures. However, the properties of material require to be improved while using for advanced structural applications. Also these alloys provide agreeable strength-to-weight ratio while comparing with other materials that are employed for various structural materials [9].

Aluminium alloys are considered as lightweight materials predominantly used work materials in the manufacturing industries because of its intrinsic properties such as better ductility and strength. The exceptional characteristics of these alloys makes this material as suitable for aircraft applications. Apart from these, AA 7075

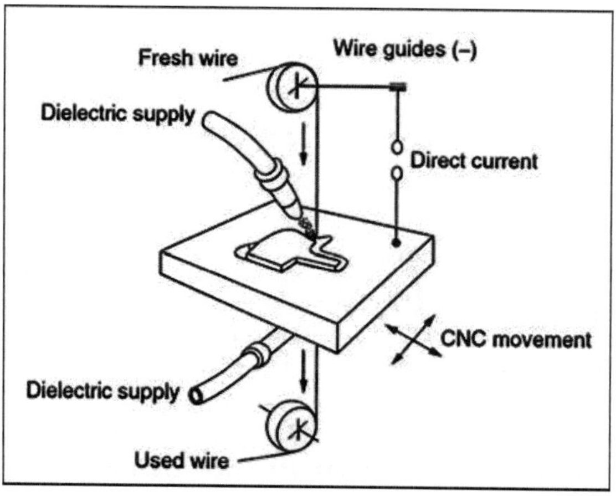

Fig. 1 Schematic of wire EDM [7]

alloy has better resistant to corrosion and more strength while comparing with remaining types of alloys. Various researchers revealed that AA 7075 alloys are mostly employed effectively for making the aircraft and their parts. An investigation has been performed on WEDM of AA 7075 by using RSM method. The significance of individual process parameters was ascertained with the help of aforementioned method [10–12]. The grey system theory has been adopted for optimizing the multi aspects in WEDM of SS 304 and from the investigation, it is perceived that the Taguchi's approach and ANOVA offer better performance for determining the significance of individual variables on the desired performance measures [13, 14]. The complication of machining of hard materials by conventional process leads to the development of advanced machining process such as WEDM and also helps to make the materials that are not possible by conventional methods of material removal [15]. An experimental investigation has been performed on WEDM process for analyzing the machining performance using Inconel as work material. Taguchi's approach has been adopted for designing the experiments and also for single-objective optimization. A multi-performance index has been derived by using appropriate method of multiple objective optimization [16]. Taguchi method has been employed for attaining better surface finish in WEDM of tool steel and it is revealed from the investigation that the adopted method provides better improvement in the desired objective and the WEDM process has been modified for obtaining better machinability during machining H.S.S, titanium nimonic and aluminium alloys as work material [17–20].

It is surmised from the available literature, the importance and machining of AA 7075 alloy by WEDM process needs attention. In this current exploration, an exertion has been taken for ascertaining the prominence of process variables for the preferred performance aspects named as material removal rate and surface roughness of the wire EDMed surface. Taguchi's approach has been engaged for designing the experiment trials and also for single-objective optimization of desired performance measures. Multiple regression analysis method has been involved for developing the predictive models for desired performance measures.

2 Materials and Methods

Aluminium 7075 alloy possesses exceptional strength and better resistance to corrosion. Due to its outstanding characteristics, the work material has broader applications especially in aircraft and other related applications. AA 7075 has been opted as work material (size of 25 mm diameter and 150 mm length) in this proposed exploration and it is clamped inner side of the machining chamber as illustrated in Fig. 2. WEDM machine (Concord Make-Model DK-7732) has been employed for experimentation purpose. Reusable molybdenum wire has been used as tool electrode and de-ionized water is used as dielectric fluid in this present investigation.

In conventional approach of designing the experiments, there is a need of conducting more number of experimental trials with opted process parameters and

Fig. 2 Experimental set-up

Table 1 Machining process parameters and levels

Symbols	Variables	Levels		
		1	2	3
A	Pulse on time (μs)	10	20	30
B	Pulse off time (μs)	5	10	15
C	Peak current (A)	1	2	3

levels. Such kind of issues are resolved by adopting the Taguchi's experimental design approach. An unique design of layout has been proposed by Taguchi for conducting the experiments, that is called as orthogonal array (OA) and also for inspecting the independent process variables with the help of minimum set of experimental trials. Pulse on time (μs), Pulse off time (μs) and peak current (A) are preferred as input variables based upon the available literature. The material removal rate (MRR) and surface roughness (SR) are opted as performance characteristics in this current experimentation. The nominated independent process variables, levels and the range of those values are depicted in Table 1 and as per these values an L_{27} OA has been opted for Wire EDM of AA 7075 alloy.

3 Results and Discussion

The experimental runs have been performed as per L_{27} OA for exploring the prominence of independent variables on desired performance characteristics in wire EDM of AA 7075 alloy. An exertion has been made to ascertain the best possible process variables for accomplishing the effectual and competent process of Wire EDM. In this process, superior removal rate of material and lesser values of surface roughness are the indicators of superior performance aspects. So MRR is considered as larger the better criterion and surface roughness is considered as smaller the better criterion.

3.1 Influence of Process Parameters on Material Removal Rate

The response graph for rate of removal of work surface material during machining of AA 7075 alloy is illustrated in Fig. 3. From the response graph, it is perceived that the rate of removal of material is getting increased with an improvement of values in pulse on duration, and furthermore it is getting diminished with an addition in pulse off time. Additionally, it is perceived that the pulse off time is most extreme process variable for the rate of material removal rate. With increasing values in pulse-on-time, the discharge energy given to the machining chamber will be more that consequences in an influential explosion that has the tendency of an increment in the rate of material removal. Increasing in the values of pulse-on-time has the capability of getting increase in the amount of electrons that impacting on the work material, therefore it starts eroding more quantity of material from the surface of the work piece per spark discharge.

Taguchi's analysis for MRR has been accomplished and the outcomes are presented in Table 2. The best set of process parameters for attaining an improved material removal rate is *A*3*B*1*C*3. It means that the optimal process variables combination for improved performance is 'P_{on}'—30 µs; 'P_{off}'—5 µs; and peak current is 3 A. Also it is witnessed from the investigation that the 'P_{off}' is the noteworthy process variable and then it is trailed by peak current and 'P_{on}'

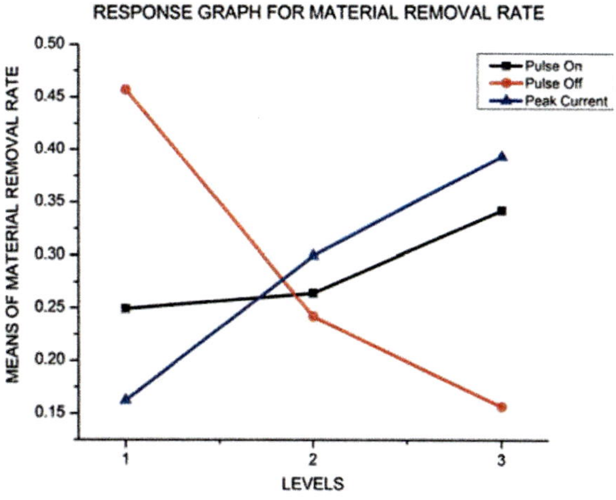

Fig. 3 Response graph for material removal rate

Table 2 Taguchi's analysis for MRR–WEDM of AA 7075 alloy

Levels	Means			*S/N* ratio		
	A	B	C	A	B	C
1	0.2492	0.4568	0.1623	−10.743	−7.47	−16.956
2	0.2641	0.2416	0.2994	−13.589	−13.053	−11.345
3	0.3417	0.1566	0.3933	−12.964	−16.773	−8.995
Delta	0.0925	0.3001	0.2309	2.846	9.303	7.961
Rank	3	1	2	3	1	2

3.2 *Influence of Process Parameters on Surface Roughness*

The graphical illustration of response graph for surface roughness during wire EDM of AA 7075 alloy is depicted in Fig. 4. It is perceived from the graph that the values of roughness of the electrically discharge machined surface getting increased with an increase of 'P_{on}' and the same has been getting diminished with an increment of peak current and 'P_{off}'. The superior energy of discharge provided crater in a greater manner that reasons more roughness at the machined surface. The 'P_{on}' time is deemed as the supreme persuasive process variable for roughness of the machined surface.

Taguchi analysis for surface roughness has been executed and the results are presented in in Table 3. The best combination of process variables for attaining better surface roughness is *A*1*B*3*C*3. It means that the optimal process parameter combination for improved performance is 'P_{on}'—10 μs; 'P_{off}'—15 μs; peak current is 3 A. Also it is witnessed from the analysis that the pulse on time is the major variable and then it is trailed by peak current and pulse off time.

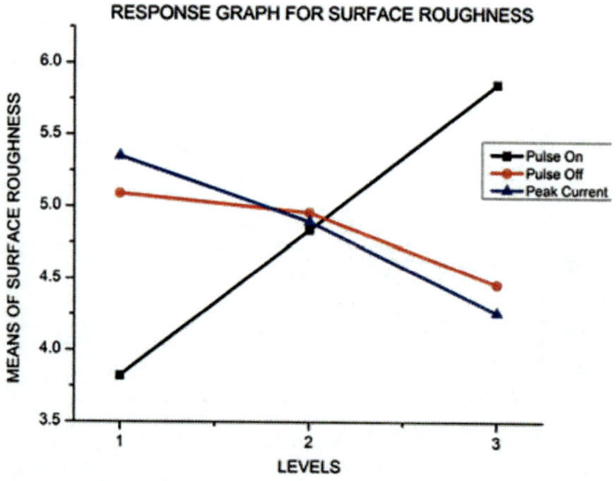

Fig. 4 Response graph for surface roughness

Table 3 Response table for surface roughness for AA 7075 alloy

Levels	Means			*S/N* ratio		
	A	B	C	A	B	C
1	3.822	5.090	5.352	−11.53	−13.94	−14.44
2	4.838	4.959	4.896	−13.63	−13.73	−13.66
3	5.853	4.464	4.266	−15.31	−12.8	−12.37
Delta	2.031	0.626	1.087	3.78	1.14	2.07
Rank	1	3	2	1	3	2

3.3 Evolution of Multiple Regression Models

Multiple regressions method has been adopted to determine the correlation amongst the selected independent and dependent process variables. The regression models have different models such as linear, quadratic, interaction and full model and have been exhibited in Eqs. (1–4) (for three input parameters):

Linear:

$$y = \beta_0 + \beta_1 X_1 + \beta_2 X_2 + \beta_3 X_3 \quad (1)$$

Quadratic equation:

$$y = \beta_0 + \beta_1 X_1 + \beta_2 X_2 + \beta_3 X_3 + \beta_4 (X_1)^2 + \beta_5 (X_2)^2 + \beta_6 (X_3)^2 \quad (2)$$

Interaction equation:

$$y = \beta_0 + \beta_1 X_1 + \beta_2 X_2 + \beta_3 X_3 + \beta_4 X_1 X_2 + \beta_5 X_1 X_3 + \beta_6 X_2 X_3 \quad (3)$$

Second order (Full model model):

$$\begin{aligned} y = \beta_0 + \beta_1 X_1 + \beta_2 X_2 + \beta_3 X_3 + \beta_4 (X_1)^2 + \beta_5 (X_2)^2 \\ + \beta_6 (X_3)^2 + \beta_7 X_1 X_2 + \beta_8 X_1 X_3 + \beta_9 X_2 X_3 \end{aligned} \quad (4)$$

Where 'y' is criterion variable and 'X_1', 'X_2' and 'X_3' are predictor variables. 'β_1', 'β_2', 'β_3', ….., 'β_n' are regression coefficients.

3.3.1 Regression Model for Material Removal Rate

The empirical equations for connecting the relationship among the desired performance measure and the considered process variables has been obtained and the desired regression equation for material removal rate is shown in Eq. (5)

$$\begin{aligned}\text{MRR} = {} & 0.644478 - 0.0313317 * (A) - 0.0715578 * (B) \\ & + 0.320122 * (C) + 0.0005375 * (A * A) \\ & + 0.000624833 * (A * B) + 0.002606 * (B * B) \\ & - 0.0115367 * (B * C) - 0.0001475 * (A * C) - 0.0215833(C * C)\end{aligned} \quad (5)$$

3.3.2 Regression Model for Surface Roughness

The empirical equations for connecting the relationship among the desired performance measure and the considered process variables has been obtained and the desired regression equation for surface roughness is shownin Eq. (6)

$$\begin{aligned}\text{SR} = {} & 0.496667 + 0.116556 * (A) + 0.147556 * (B) \\ & + 0.745 * (C) + 4.38847e\,018 * (A * A) + 3.33333e - 005(A * B) \\ & - 0.00726667(B * B) + 0.0298333 * (B * C) \\ & - 0.00766667 * (C * A) - 0.0866667 * (C * C)\end{aligned} \quad (6)$$

4 Conclusions

This current experimentation details a single aspect optimization problem of wire-EDM of AA 7075 alloy with the help of Taguchi analysis. In this investigation, the rate of removal material (MRR) and roughness of WEDMed surface (SR) were deemed as the performance characteristics and the conclusion attained is discussed as below:

- Taguchi's approach has been adopted for designing the experimental runs and also it has been adopted for optimizing the individual variables for accomplishing better performance in machining.
- The prevalence of deemed independent parameters on the preferred performance measures was revealed by Taguchi's analysis. The significant process parameter which helps to attain better and improved machining performance are determined by Taguchi's single response analysis.
- The detailed Taguchi's approach in this present exploration is greatly appropriate for concluding the best set of independent process parameters for attaining the improved performance in any advanced machining process.
- Mathematical models have been evolved to correlate the desired performance measure MRR and surface roughness with the various input machining variables. The correlation coefficient values of developed regression models shows that the models are fit for further effective and efficient prediction.

- The outcome attained from this exploration will be a wide-ranging support to the manufacturers for enhancing the rate of production and quality of products made with an assistance of WEDM process.

References

1. Kanlayasiri K, Boonmung S (2007) Effects of wire-EDM machining variables on surface roughness of newly developed DC 53 die steel: design of experiments and regression model. J Mater Process Technol 192–193:459–464
2. Davim JP (2008) Machining fundamentals and recent advances. Springer-Verlag London Limited, British Library Cataloguing in Publication Data. http://doi.org/10.1007/978-1-84800-213-5
3. Patil N, Brahmankar PK (2010) Determination of material removal rate in wire electro-discharge machining of metal matrix composites using dimensional analysis. Int J Adv Manuf Technol 51(5–8):599–610. https://doi.org/10.1007/s00170-010-2633-3
4. Huang Y, Ming W, Guo J, Zhang Z, Liu G, Li M, Zhang G (2013) Optimization of cutting conditions of YG15 on rough and finish cutting in WEDM based on statistical analyses. Int J Adv Manuf Technol 69(5–8):993–1008. https://doi.org/10.1007/s00170-013-5037-3
5. Saha P, Tarafdar D, Pal S, Saha P, Srivastava A, Das K (2009) Modeling of wire electro-discharge machining of TiC/Fe in situ metal matrix composite using normalized RBFN with enhanced k means clustering technique. Int J Adv Manuf Technol 43(1–2):107–116. https://doi.org/10.1007/s00170-008-1679-y
6. Moulton DB (1999) Wire EDM the fundamentals. EDM Network, Sugar Grove, IL. www.notebookmanuals.bestmanualguide.com
7. El-Hofy H (2005) Advanced machining processes. McGraw-Hill. https://doi.org/10.1036/0071466940
8. Sommer C, Sommer S (2005) Complete EDM handbook. Advance Publications, New York
9. Kumar SM, Pramod R, Kumar MS, Govindaraju HK (2014) Evaluation of fracture toughness and mechanical properties of aluminum alloy 7075, T6 with nickel coating. Procedia Eng 97:178–185
10. Dursun T, Soutis C (2014) Recent developments in advanced aircraft aluminium alloys. Mater Des 1980–2015(56):862–871
11. Rambabu PPNKV, Prasad NE, Kutumbarao VV, Wanhill RJH (2017) Aluminium alloys for aerospace applications. In: Aerospace materials and material technologies. Springer, Singapore, pp 29–52
12. Gopalakannan S, Senthilvelan T, Ranganathan S (2012) Modeling and optimization of EDM process parameters on machining of Al 7075-B4C MMC using RSM. Procedia Eng 38:685–690
13. Khan ZA, Siddiquee AN, Khan NZ, Khan U, Quadir GA (2014) Multi response optimization of wire electrical discharge machining process parameters using Taguchi based grey relational analysis. Procedia Mater Sci 6:1683–1695
14. Lajis MA, Radzi HCDM, Amin AKMN (2009) The implementation of Taguchi method on EDM process of tungsten carbide. Eur J Sci Res 26(4):609–617
15. Manikandan N, Arulkirubakaran D, Palanisamy D, Raju R (2019) Influence of wire-EDM textured conventional tungsten carbide inserts in machining of aerospace materials (Ti–6Al–4V alloy). Mater Manuf Process 34(1):103–111
16. Yang CB, Lin CG, Chiang HL, Chen CC (2017) Single and multi-objective optimization of Inconel 718 nickel-based superalloy in the wire electrical discharge machining. Int J Adv Manuf Technol 93(9–12):3075–3084
17. Sudhakara D, Prasanthi G (2014) Application of Taguchi method for determining optimum surface roughness in wire electric discharge machining of P/M cold worked tool steel (Vanadis-4E). Procedia Eng 97:1565–1576

18. Ramesh NN, Harinarayana K, Naik BB (2014) Machining characteristics of HSS & titanium using electro discharge sawing and wire-electro discharge machining. Procedia Mater Sci 6:1253–1259
19. Goswami A, Kumar J (2014) Optimization in wire-cut EDM of Nimonic-80A using Taguchi's approach and utility concept. Eng Sci Technol Int J 17(4):236–246
20. Manikandan N, Binoj JS, Varaprasad KC, Sabari SS, Raju R (2019) Investigations on wire spark erosion machining of aluminum-based metal matrix composites. In: Advances in manufacturing technology. Springer, Singapore, pp 361–369

Finite Element Modelling of Cutting Forces in Turning of Ti–6Al–4V Alloy

N. Subhash, V. Dhinakaran, and T. Jagadeesha

Abstract Titanium alloys are commonly used in aviation industries, and the reasons for using titanium alloys when compared to commonly available materials are high strength-to-weight ratio, mechanical properties and corrosion resistance. All these superior properties result in a major disadvantage, that is poor machinability. Excessive cutting forces and heat are generated during machining of titanium alloys so that induces poor tool life. Experimental studies of conventional machining on these materials are expensive and time-consuming. So, a dynamic 3D finite element model was developed in ABAQUS for predicting the cutting force in turning process considering the Lagrangian formulation for meshing and considering the tool as flexible. With the help of FE model, the process parameters such as cutting speed, feed and depth of cut can be optimized for cutting forces, which results in power savings as well as increased productivity in turning process of difficult to machine materials.

Keywords ABAQUS · Finite element model · Cutting forces · Ti6Al4V

1 Introduction

Ti–6Al–4V is an alloy which offers superior strength-to-weight ratio, excellent corrosion resistance and high toughness, and these titanium alloys are normally used in aerospace and biomedical applications for its unique properties. Ti–6Al–4V is considered to be a difficult to machine material due to its low thermal conductivity and high rigidity [1]. The cutting performance of the titanium alloy can be increased by improving the tool material and its coatings [2]. Due to the presence of coating, the nose radius will be affected, and it must be taken into account in the performance analysis of the tool. Mishra et al. [3] focused on the effect of textured tools on the

N. Subhash · T. Jagadeesha (✉)
National Institute of Technology Calicut, Calicut, Kerala 673601, India
e-mail: jagdishsg@nitc.ac.in

V. Dhinakaran
Department of Mechanical Engineering, Chennai Institute of Technology, Kundrathur, Chennai, Tamil Nadu 600069, India

A. Arockiarajan et al. (eds.), *Advances in Industrial Automation and Smart Manufacturing*, Lecture Notes in Mechanical Engineering,
https://doi.org/10.1007/978-981-15-4739-3_37

cutting forces through 2D finite element modelling of turning process. Simulation of turning with plain textured and TiAlN-coated textured tools were performed and compared it with plain and coated non-textured tools. Effectiveness of the textured tools was evaluated with respect to cutting force and average cutting coefficient. Ozel et al. [4] investigated the machining of Ti–6Al–4V alloy using cutting tools having multi-layer coating for cutting forces and tool wear. FE modelling was used to predict the force, tool wear, chip formation and temperature on these cutting inserts. It was observed that coated cutting insert designs showed some advantage on temperature distribution and tool wear contours [5, 6]. Mishra et al. [7] focused their study on residual stresses and temperature with plain textured and TiAlN-coated tools during turning of titanium alloy through 2D and 3D finite element modelling. The results of the simulation showed that lower temperature and residual stresses were produced during the turning process with textured TiAlN-coated tool. Ducobu et al. [8] suggested a 3D FE coupled Eulerian–Lagrangian (CEL) model, and it is compared with 2D orthogonal model. The simulation was focussed on the prediction of chip formation, cutting forces and lateral expansion of the chip. It was observed that the plain stain assumption can be closely approximated if the width is sufficiently large and cubic elements should be adopted for modelling the workpiece [2, 9, 10].

Most of the works were on the orthogonal simulation of the turning process. In the study, 3D FE simulation of turning of Ti–6Al–4V alloy was performed for predicting the cutting forces considering the effect of nose radius by incorporating the actual dimensions and angles of the cutting tools using ABAQUS/dynamic explicit. Both tool and the workpiece were considered to be flexible, and Lagrangian formulation was used for meshing generation. Johnson–Cook material model has been utilized for defining the flow stress. The simulation results were compared with the experimental values reported in the literature [3].

2 Finite Element Modelling

Finite element simulation for predicting cutting forces during turning of Ti–6Al–4V alloy using plane carbide insert was performed in ABAQUS/Explicit software package. Lagrangian formulation was used for the mesh generation of the workpiece and cutting insert. Four-noded hexahedral element was selected to model the workpiece and tetrahedral elements to model the cutting insert. The workpiece is partitioned based on the required so that mesh size can be made finer in the region where actual cutting takes place to increase the accuracy of the simulation results. The main reason for creating partition is to save the computational time. Mesh size varies from 40 to 1000 μm, resulting in 134,000 number of nodes in the workpiece as shown in Fig. 1b.

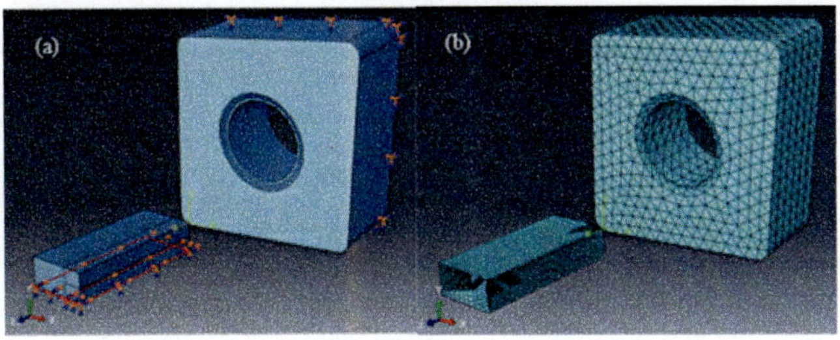

Fig. 1 **a** Boundary conditions and **b** meshing

2.1 Workpiece Material Model

Accuracy of the simulation results depends on the selection of the appropriate material model, material properties, interaction conditions and boundary conditions. In this study, Johnson–Cook material model was selected which is capable of incorporating the effect of various strain rates and temperature. Johnson–Cook (JC) equation which is relating the flow stress (σ_e) with plastic strain, strain rates and temperature is given in the equation.

$$\overline{\sigma}_e = \left[A + B\varepsilon^n\right]\left[1 + C\ln\left(\frac{\dot{\varepsilon}}{\dot{\varepsilon}_0}\right)\right]\left[1 - \left(\frac{T - T_0}{T_m - T_0}\right)\right] \tag{1}$$

where $\overline{\sigma}_e$ and $\dot{\varepsilon}_0$ are flow stress and reference plastic strain rate, respectively, and $1/s\varepsilon$ and $\dot{\varepsilon}$ are equivalent plastic strain and equivalent plastic strain rate, respectively. T, T_m and T_0 are the cutting, melting and ambient temperature, respectively. A, B, n, m and C are J-C constitutive material model constants. The mechanical properties and J-C material model constants are given in Tables 1 and 2, respectively.

Table 1 Mechanical and thermal properties of tool and workpiece [3]

Properties	Ti–6Al–4V	WC/C_0
Density (kg/m^3)	4436	14,900
Thermal expansion (°C^{-1})	9.1e6	4.8e6
Young's modulus (Pa)	1.13e11	6.45e11
Poisson's ratio	0.34	0.24
Thermal conductivity (Wm^{-1} C^{-1})	6.6	24
Specific heat (J kg^{-1} C^{-1})	670	206

Table 2 J-C material model constants for Ti–6Al–4V alloy [3]

A (MPa)	B (MPa)	n	m	C	T_{melt} (°C)	T_0 (°C)
782.7	498.4	0.28	0.8	0.028	1660	20

Table 3 J-C damage parameters for Ti–6Al–4V alloy [3]

D_1	D_2	D_3	D_4	D_5	Reference strain rate (s^{-1})
−0.09	0.25	−0.5	0.014	3.87	1

The contact was defined between the rake surface and nodes of the workpiece material. The interaction parameters such as coefficient of friction were taken as 0.6 in the present study to model the friction condition as the chip flows over the rake surface. Johnson–Cook fracture model was selected for defining the damage criteria of Ti–6Al–4V. Johnson–Cook (JC) equation which is relating the fracture strain with plastic strain, strain rates and temperature is given in the equation.

$$\varepsilon_f = \left[D_1 + D_2 \exp\left(D_3\sigma^*\right)\right]\left[1 + D_4 \ln\left(\frac{\dot{\varepsilon}}{\dot{\varepsilon}_p}\right)\right]\left[1 + D_5\left(\frac{T - T_0}{T_m - T_0}\right)\right] \quad (2)$$

where ε_f is fracture strain, $\sigma*$ is stress triaxiality and D_1, D_2, D_3, D_4 and D_5 are the J-C fracture model constants which are listed in Table 3.

In the 3D simulation model, the workpiece is modelled as a rectangular block with length 10 mm, width 3 mm and 2 mm thick, and simulation was performed for a length of 10 mm. A constant velocity is given to the workpiece which corresponds to the actual cutting speed, and the cutting insert is considered to be fixed as the boundary conditions as shown in Fig. 1a. Insert was modelled with geometry CNMA120408 in ABAQUS. Tool and the workpiece were assembled such that it results in the rake angle of −60, inclination angle of 00 and relief angle of 50. Chip formation and cutting force generated during machining is shown in Fig. 2.

3 Results and Discussion

The finite element simulation was developed for turning process of Ti–6Al–4V alloy using ABAQUS. Simulation was performed for various cutting speed and feeds and at a constant depth. Figure 3 shows the cutting forces generated during turning of Ti–6Al–4V from the 3D simulation developed at cutting speeds and feed of 100 m/min and 0.2 mm/rev, respectively, along the length of cut. It was observed that the cutting forces are increased considerably with feed and decreased with cutting speed. The simulation results are validated by the experimental results reported by Mishra et al. [3] where cutting experiments were performed using plane and coated tungsten carbide tool with geometry CNMA120408 which was PVD coated with TiAlN. Round bar was used with 50 mm diameter and 250 mm length for the experiment which was performed on CNC turning centre. The process parameter used for the experiment is given in Table 4. Forces were measured and recorded using the dynamometer (Kistler Make) with charge amplifier.

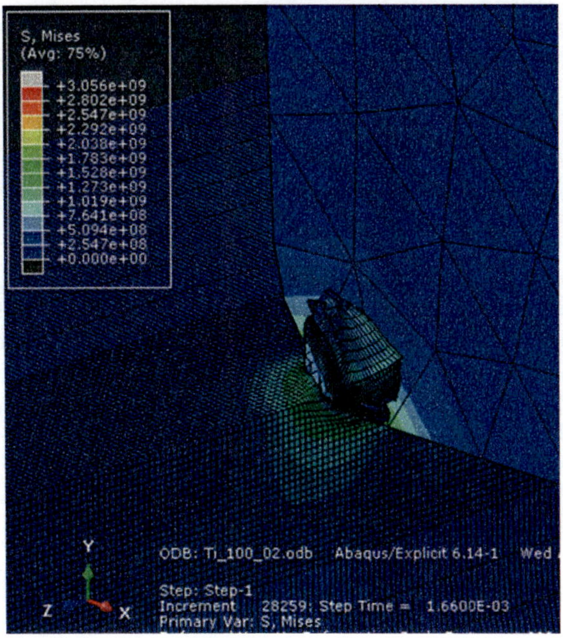

Fig. 2 Chip formation in the finite element model

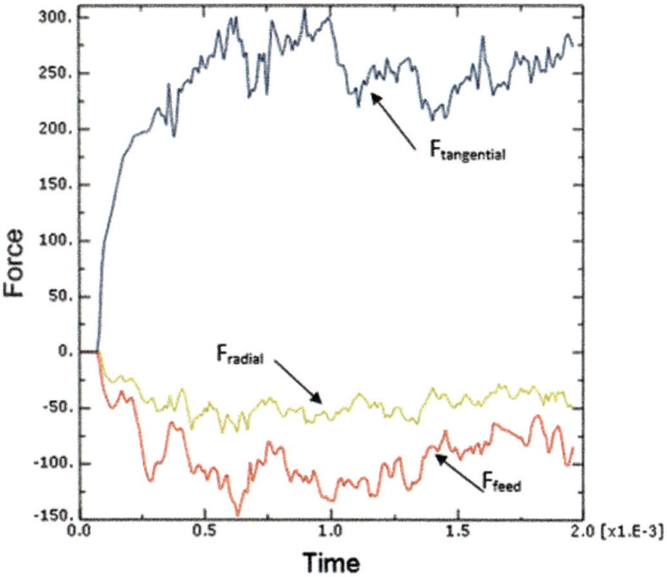

Fig. 3 Simulated cutting forces at $V = 100$ m/min, $f = 0.2$ mm/rev, $d = 0.5$ mm

Table 4 Process parameters used in the simulation

Process parameters	Value
Cutting speed (m/min)	100, 130, 160
Feed (mm/rev)	0.1, 0.2
Depth of cut (mm)	0.5
Machining condition	Dry machining

Figure 4 shows the experimental and simulation results for tangential cutting force at different process parameters. It is observed that as the feed increases the cutting force increases, this is due to the fact that as the feed increases, the area of the chip coming in contact with the cutting edge increases, and hence, cutting force increases significantly. But as the cutting speed is increased, the tendency of built-up edge (BUE) formation will reduce which results in reduced cutting forces in all three directions. The cutting forces were averaged over the cutting length in the simulation and compared with the experimental results. It was seen that the simulated results matches with the trend observed in the experiments with maximum error of 20%.

4 Conclusions

In the present work, FE model for oblique cutting was developed incorporating the flexibility of the tool for predicting the cutting forces during turning of Ti–6Al–4V alloy using ABAQUS/Explicit 6.18. This FE model was able to predict the variation of cutting forces generated during the machining operation for various cutting speeds and feed at a constant depth of cut. The simulated results were compared with the experimental values reported in the literature, and it was observed that the model was able to follow the pattern developed by the experimental results. The effect of coating was not incorporated in the model due to its complexity in modelling. But, the accuracy of the model can be increased by using the appropriate interaction parameters and heat partition between tool and workpiece material. Presence of coating on the insert results in variation of coefficient of friction between rake face and the chip formed during machining when compared to plain carbide tool.

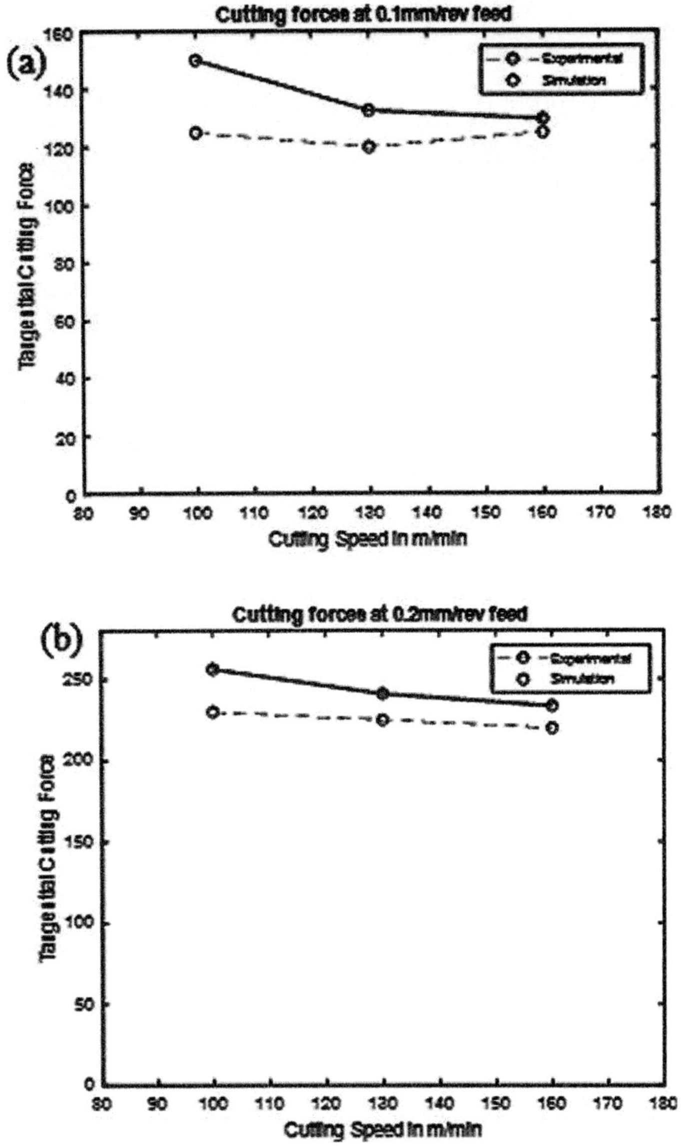

Fig. 4 Comparison between experimental and FE results at **a** feed = 0.1 mm/rev, **b** feed = 0.2 mm/rev

References

1. Byrne G, Dornfeld D, Denkena B (2003) Advancing cutting technology. CIRP Ann 52(2):483–507
2. Cotterell M, Byrne G (2008) Dynamics of chip formation during orthogonal cutting of titanium alloy Ti–6Al–4V. CIRP Ann 57(1):93–96
3. Mishra SK, Ghosh S, Aravindan S (2019) Performance of laser processed carbide tools for machining of Ti6Al4V alloys: a combined study on experimental and finite element analysis. Precis Eng 56:370–385
4. Özel T, Sima M, Srivastava AK, Kaftanoglu B (2010) Investigations on the effects of multi-layered coated inserts in machining Ti–6Al–4V alloy with experiments and finite element simulations. CIRP Ann 59(1):77–82
5. Sivalingam V, Sun J, Yang B, Liu K, Raju R (2018) Machining performance and tool wear analysis on cryogenic treated insert during end milling of Ti-6Al-4V alloy. J Manuf Process 36:188–196
6. Arulkirubakaran D, Senthilkumar V, Dinesh S, Velmurugan C, Manikandan N, Raju R (2018) Effect of textured tools on machining of Ti-6Al-4V alloy under lubricant condition. Mater Today Proc 5(6):14230–14236
7. Mishra SK, Ghosh S, Aravindan S (2017) Finite element investigations on temperature and residual stresses during machining Ti6Al4V alloy using TiAlN coated plain and textured tools. In: Proceedings of international conference on precision, meso, micro and nano engineering, COPEN, vol 10, pp 979–982
8. Ducobu F, Rivière-Lorphèvre E, Filippi E (2017) Finite element modelling of 3D orthogonal cutting experimental tests with the coupled Eulerian-Lagrangian (CEL) formulation. Finite Elem Anal Des 134:27–40
9. Komanduri R, Von Turkovich BF (1981) New observations on the mechanism of chip formation when machining titanium alloys. Wear 69(2):179–188
10. Ozel T, Sima M, Srivastava AK (2010) Finite element simulation of high speed machining Ti-6Ai-4V alloy using modified material models. In: 38th annual North American manufacturing research conference, NAMRC 38, pp 49–56

Analysis of Centerless Grinding Process Parameters in Machining SS 316 L Steel for Improved Productivity

P. Sakthivel, S. Dinesh, A. Godwin Antony, and V. Vijayan

Abstract Power losses in transmission affect the maximum speed of the vehicle and increase the fuel consumption. This is due to improper surface quality resulting from the intend to achieve maximum productivity. The work deals with analyzing the effect of each input parameter such as regulating wheel speed, regulating wheel angle and depth of cut on the response, and obtaining the most significant parameter that controls material removal rate by analysis of variance. The most feasible set of parameters are determined by response surface methodology.

Keywords Productivity · Material removal rate · Analysis by variance · Response surface methodology · Surface roughness (Ra)

1 Introduction

Centerless grinding process is used for grinding cylindrical parts such as shafts [1, 2]. The most commonly used type in industries is through-feed centerless grinding process. The grinding process is a very critical and tough process to handle as improper grinding will result in rejection of the completed job (or) might result is premature failure [3]. It is clear that the input parameters have a significant effect on the output parameters such as stack removal rate, surface roughness, and roundness of the workpiece. So, it is very important to optimize the input parameters so as to get improved surface finish and roundness at higher productivity rate [4]. Various researches have been conducted by changing the grinding wheel materials such as carbon nanotubes [5]. The surface roughness shall be reduced by electrolytic

P. Sakthivel (✉)
Department of Mechanical Engineering, Sri Krishna College of Technology, Coimbatore, India
e-mail: skt4design@gmail.com

S. Dinesh · A. Godwin Antony · V. Vijayan
Department of Mechanical Engineering, K. Ramakrishnan College of Technology, Trichy, Tamil Nadu, India

A. Arockiarajan et al. (eds.), *Advances in Industrial Automation and Smart Manufacturing*, Lecture Notes in Mechanical Engineering,
https://doi.org/10.1007/978-981-15-4739-3_38

Table 1 Process variables

Input parameters	Least point	Modest point	Utmost point
Regulating wheel speed (rpm)	12	25	46
Regulating wheel angle (°)	2	3	4
Depth of cut (mm)	0.1	0.16	0.2

in process dressing and electro-discharge turning process [6]. The decrease in waviness of the machined surface can be predicted with the frequency characteristics of grinding force [7]. The accuracy of the centerless grinding with maximum stock removal rate can be achieved at minimum feed rate in in-feed centerless grinding process [8]. The stability of the machine and chatter affects the performance of machine and will cause uneven distribution of grinding force resulting in uneven machined surface [9]. The roundness shall be improved by setting lower feed rate and higher speed without affecting the productivity [10].

The chatter stability shall be improved by through FE modeling of the process [11]. The experiments were decided based on design of experiments [12–14]. The most significant parameter that governs the response shall be identified by ANOVA analysis [15]. Response surface methodology can be used to study the interaction effect of input parameters and output responses [16–21]. Taguchi methodology has been a significant tool for predicting the responses based on the input parameters [16, 22–24]. Grey relational analysis is also used to optimize the centerless grinding process for improving productivity, surface finish, and roundness of machined workpiece [2].

2 Materials and Methods

Star make model centerless grinding machine was used for grinding the workpiece. MITUTOYO make tester was used to measure the surface roughness. The lateral and longitudinal dimensions of the workpiece were measured using vernier calliper ranging from 0 to 300 mm. Weight was measured using a weighing pan of capacity 400 gm with three decimal precision. Table 1 shows the experimental parameters considered for analysis, and Table 2 shows the experimental results.

3 Result Analysis

The impact of regulating wheel speed, depth of cut, and angle are shown in Figs. 1, 2 and 3. It is evident the MRR has increased in a hyperbolic fashion when there is increase in regulating wheel speed. Depending upon the increasing regulating wheel

Table 2 Experimental results

Experiment number	Regulating wheel speed (rpm)	Regulating wheel angle (°)	Depth of cut (mm)	Material removal rate (kg/s)
1	12	2	0.1	0.28
2	12	2	0.16	0.54
3	12	2	0.2	0.24
4	12	3	0.1	0.14
5	12	3	0.16	0.14
6	12	3	0.2	1.22
7	12	4	0.1	0.15
8	12	4	0.16	0.16
9	25	4	0.2	0.09
10	25	2	0.1	0.08
11	25	2	0.16	0.08
12	25	2	0.2	0.09
13	25	3	0.1	0.07
14	25	3	0.16	0.06
15	25	3	0.2	0.08
16	25	4	0.1	0.05
17	25	4	0.16	0.09
18	25	4	0.2	0.06
19	46	2	0.1	0.03
20	46	2	0.16	0.03
21	46	2	0.2	0.02
22	46	3	0.1	0.03
23	46	3	0.16	0.03
24	46	3	0.2	0.06
25	46	4	0.1	0.06
26	46	4	0.16	0.04
27	46	4	0.2	0.45

angle, there is considerable increase in MRR. The maximum MRR is obtained at minimum wheel speed. There is no increase (or) decrease in MRR with respect to depth of cut. So, it is evident that depth of cut is the least significant parameter.

Table 3 shows the analysis by variance and proves that the model is significant for predicting MRR. Regulating wheel speed is the most significant parameter for material removal rate followed by regulating wheel angle. Depth of cut has the least

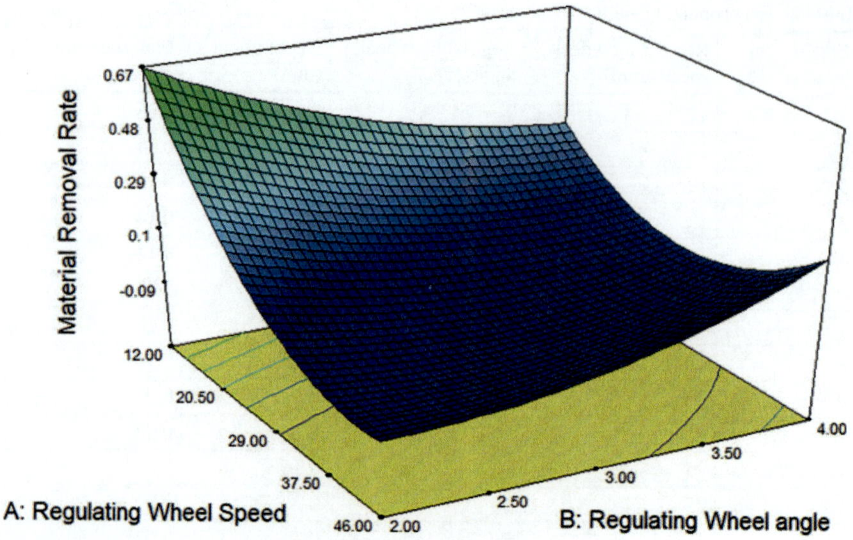

Fig. 1 Impact of regulating wheel speed and angle on MRR

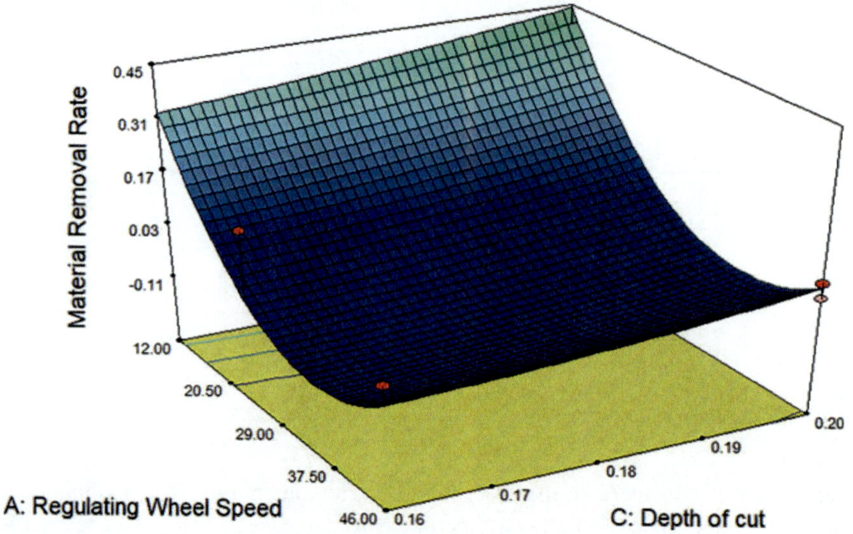

Fig. 2 Impact of regulating wheel speed and depth of cut on MRR

effect on MRR. Figure 4 shows the most feasible set of parameters. It is advisable to use minimum regulating wheel speed, regulating wheel angle, and maximum depth of cut. The desirabity value is 0.583.

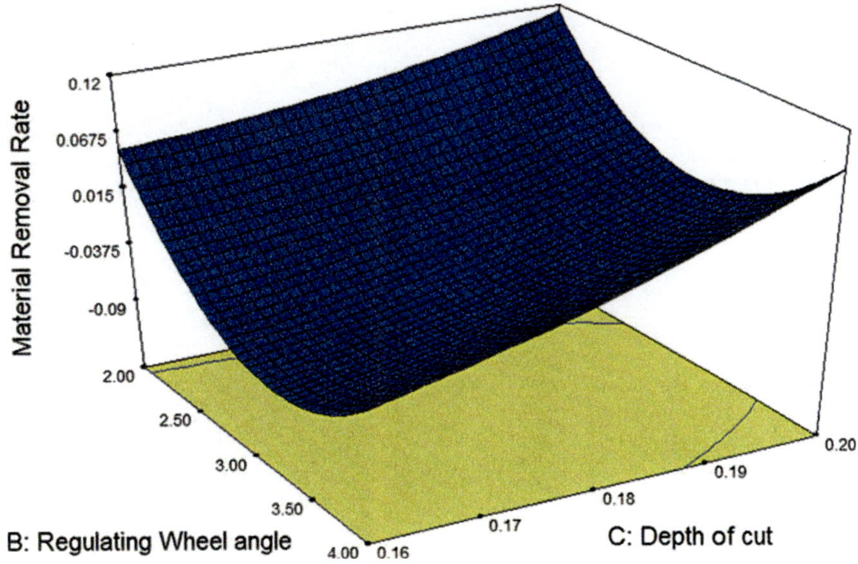

Fig. 3 Impact of regulating wheel angle and depth of cut on MRR

Table 3 ANOVA analysis for MRR

Source	Sum of squares	DOF	Mean square	*F*-value	*P*-value Prob > *F*
Model	1.029	9	0.114	3.599	0.011
A-Regulating wheel speed	0.316	1	0.316	9.931	0.006
B-Regulating wheel angle	0.009	1	0.009	0.293	0.595
C-Depth of cut	0.036	1	0.036	1.145	0.299
AB	0.267	1	0.267	8.409	0.010
AC	0.008	1	0.008	0.256	0.619
BC	0.011	1	0.011	0.36	0.556
A^2	0.268	1	0.268	8.439	0.010
B^2	0.033	1	0.033	1.024	0.326
C^2	0.021	1	0.021	0.648	0.432
Residual	0.540	17	0.032		
Lack of fit	0.531	11	0.048	30.73	0.000
Pure error	0.009	6	0.002		
Cor total	1.569	26			

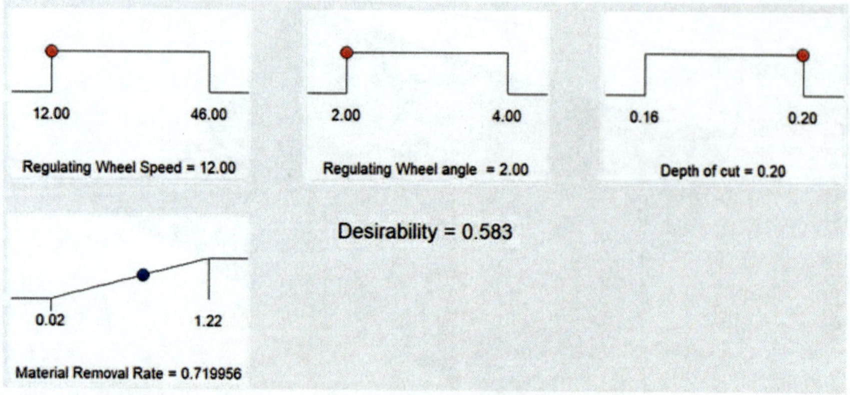

Fig. 4 Optimized parameters

4 Conclusion

Response surface methodology and ANOVA analysis was carried out for analysis of MRR in centerless grinding process. The results are as follows.

1. Regulating wheel speed is the most significant parameter for material removal rate followed by regulating wheel angle. MRR has increased in a hyperbolic fashion when there is increase in regulating wheel speed.
2. Increasing regulating wheel angle results in considerable increase in MRR. Depth of cut has the least effect on MRR.
3. The most feasible set of parameters are 12 rpm of regulating wheel speed, 2 degree of regulating wheel angle, and 0.2 mm of depth of cut.

References

1. Dinesh S, Rajkumar T, Muthukumarasamy S, Sathish Kumar G, Kajendirakumar V, Suresh Kumar B (2018) Analysis and optimization of machining parameters in through feed centerless grinding of high carbon steel. J Mech Eng Technol 9(13):431–44
2. Dinesh S, Godwin Antony A, Rajaguru K, Vijayan V (2017) Experimental investigation and optimization of material removal rate and surface roughness in centerless grinding of magnesium alloy using grey relational analysis. Mech Mech Eng 21(1):17–28
3. Krajnik P, Kopac J, Sluga A (2005) Design of grinding factors based on response surface methodology. J Mater Process Technol 629–636
4. Khoi PB, Trung DD, Cuong N, Man ND (2015) Research on optimization of plunge centerless grinding process using genetic algorithm and response surface method. Int J Scient Eng Technol 4(3):207–211
5. Sethuramalingam P, Vinayagam BK (2016) Multi objective optimization of multi wall carbon nanotube based nanogrinding wheel using grey relational and regression analysis. J Industr Eng, India Series C

6. Ohmori H, Li W, Makinouchi A, Bandyopadhyay BP (2001) Efficient and precision grinding of small hard and brittle cylindrical parts by the centerless grinding process combined with electro-discharge truing and electrolytic in-process dressing. J Mater Proc Technol 322–327
7. Siddiquee AN, Khan ZA, Mallick Z (2012) Grey relational analysis coupled with principal component analysis for optimisation design of the process parameters in in-feed centreless cylindrical grinding. Int J Adv Manufact Technol 983–992
8. Hashimoto F, Lahoti GD (2004) Optimization of set-up conditions for stability of the centerless grinding process. CIRP Ann Manufact Technol 53:271–274
9. Xu W, Wu Y, Sato T, Lin W (2010) Effects of process parameters on workpiece roundness in tangential-feed centerless grinding using a surface grinder. J Mater Proc Technol 210:759–766
10. Alvarez J, Barrenetxea D, Marquinez JI, Bediaga I, Gallego I (2014) Continuous variable feed rate: a novel method for improving infeed grinding processes, Int J Adv Manufact Technol 53–61
11. Rascalha A, Brandão LC, Filho SLMR (2013) Optimization of the dressing operation using load cells and the Taguchi method in the centerless grinding process. Int J Adv Manufact Technol 1103–1112
12. Gajalakshmi K, Senthilkumar N, Prabu B, Multi-response optimization of dry sliding wear parameters of AA6026 using hybrid grey relational analysis coupled with response surface method (GRSM). Measur Control. https://doi.org/10.1177/0020294019842603
13. Thirumalvalavan S, Senthilkumar N (2019) Evaluation of mechanical properties of aluminium alloy (LM25) reinforced with fused silica metal matrix composite. Ind J Eng Mater Sci 26(1):59–66
14. Tamizharasan T, Senthil Kumar N, Selvkumar V, Dinesh S (2019) Taguchis methodology of optimizing turning parameters over chip thickness ratio in machining PM AMMC, SN Appl Sci 1:160
15. Saravanan S, Ravichandran M, Stalin B, Saravanavel S, Sukumar S, Optimization of process parameters of electrochemical machining of TiC-reinforced AA6063 composites, Lecture Notes in Mechanical Engineering. https://doi.org/10.1007/978-981-13-6374-0_33
16. Dinesh S, Godwin Antony A, Rajaguru K, Parameshwaran P (2018) Comprehensive analysis of wire electric discharge machining process in machining high chromium high carbon steel. Int J Mechan Product Eng Res Dev 8(1):65–74
17. Raju R, Sivalingam V, Sun J, Natarajan M, Zhao Y (2019) Experimental and taguchi-based grey approach of laser metal deposition technique on nickel-based superalloy. Trans Indian Inst Met 72(1):205–214
18. Venkatesh R, Vijayan V (2016) Performance evaluation of multipurpose solar heating system. Mech Mech Eng 20(4):359–370
19. Raju R, Manikandan N, Palanisamy D, Arulkirubakaran D, Sambathkumar S, Bhanu Prakash P (2018) Optimization of process parameters in electrical discharge machining of haste alloy C276 using Taguchi's method. Mater Today: Proc 5(6):14432–14439
20. Dinesh S, Godwin Antony A, Rajaguru K, Vijayan V (2016) Experimental investigation and optimization of material removal rate and surface roughness in CNC turning of EN24 alloy steel. Mech Mech Eng 20(4):451–466
21. Manikandan N, Raju R, Palanisamy D, Arulkirubakaran D, Kumar S (2018) Investigation on Ti6Al4V laser metal deposition using Taguchi based grey approach. Mater Today: Proc 5(6):14375–14383
22. Parameswaran P, Godwin Antony A, Dinesh S, Radhakrishnan K (2018) Experimental study on mechanical and corrosion characteristics NAB alloy with the addition of chromium materials today proceedings 5:8089–8094
23. Xu W, Wu Y (2011) A new in-feed centerless grinding technique using a surface grinder. J Mater Process Technol 211:141–149
24. Yokeswaran R, Karuppusamy S, Arul S (2018) Experimental investigation and optimization of welding parameters of metal inert gas welding process in joining dissimilar metals. Int J Mech Product Eng Res Develop 8(1):415–422

Mechanical Properties and Microstructure of Nano-crystalline AlN Coating on Mild Steel

A. Rahul Kumar, P. Prince Packiaraj, A. Suresh, V. Dhinakaran, M. Vinayagamoorthy, and S. Sivaraj

Abstract Aluminum nitride (AlN) is a tetrahedral III-V compound which crystallizes under ambient conditions in the thermodynamically stable hexagonal quartzite structure. AlN thin films are of high importance because of their exceptional mechanical properties, thermal behavior, and optical characteristics. AlN thin films exhibit broad bandgap with a large refractive index and less absorption coefficient. They exhibit high hardness, high breakdown voltage, low dielectric constant, high electrical resistivity (~1013 Ω-cm), high piezoelectric coupling coefficient, thermal stability (2200 °C), and high thermal conductivity (~320 W/mK). It is the most used and essential material for the manufacture of piezoelectric sensors, actuators, and surface acoustic wave (SAW) devices. It also finds applications in optical devices, hard coatings, and protective coatings, etc. In this paper, an effort has been taken to determine the properties of AlN coated mild steel. The coating is done using DC magnetron sputtering. The properties such as mechanical, corrosion, and microstructure are investigated using X-ray diffraction (XRD), scanning electron microscopy (SEM), micro-indentation, and wear test, respectively

Keywords Microstructure · Coating · Mechanical

A. Rahul Kumar (✉) · V. Dhinakaran · M. Vinayagamoorthy · S. Sivaraj
Centre for Applied Research, Chennai Institute of Technology, Kundrathur, Chennai 600069, India
e-mail: arahulkumar14@gmail.com

P. Prince Packiaraj · A. Suresh
Department of Mechanical Engineering, Chennai Institute of Technology, Kundrathur, Chennai 600069, India

A. Arockiarajan et al. (eds.), *Advances in Industrial Automation and Smart Manufacturing*, Lecture Notes in Mechanical Engineering,
https://doi.org/10.1007/978-981-15-4739-3_39

1 Introduction

Aluminum nitride (AlN) is one of the widely used materials which belong to the III–V compound. AlN finds its application in various fields with the advancement of technology. This exhibits excellent mechanical and physical behavior. The versatility of AlN makes it an important material in the field of engineering, research, and material evolution. It is evolving as a promising future material for coating because of its great wear resistivity. This has excellent thermal conductivity and high hardness. The materials also resist corrosion which makes it more important for corrosion resisting purposes. The major use of the AlN coating involves heat tubs and as a thin film. Some of the important applications of AlN are:

- Heat sinks
- Thin-film resistors
- Tough coatings
- Optoelectronic application
- Transistors field effect
- Coating for dielectric
- In a device that needs a high-frequency surface acoustic wave
- Piezoelectric MEMS applications
- Resonators, microphones, and gyroscopes.

Since it finds its vast application in the engineering field as a coating material, we have selected AlN as a coating material for the investigations. The AlN coating can be done by various means such as chemical vapor deposition (CVD), pulsed laser ablation (PLA), and molecular beam epitaxy (MBE). At present, there have been very good competitions among the engineers, researchers, material scientists, and the industrialists to find the best and opted method for coating. The coating should adhere to the surface of the base metal. There is healthy research going into the filed for innovating new technologies for coating. Based on the study, we have opted for the DC magnetron sputtering method for coating. Aluminum nitride (AlN) has been used in many applications because it exhibits outstanding mechanical, chemical, and physical behavior. Buschow et al. [1] had discussed the major properties exhibited by AlN which includes high thermal conductivity and high hardness. The author had also suggested the corrosive resistivity of it. Strite S had performed a work in the field which determines the optical property of the AlN. The author had concluded that AlN exhibits excellent visual transparency. This can be noticed in visible as well as near-infrared ranges [2].

Wang et al. [3] had proposed his work in the field of coating of AlN by ion beam technology. The author had done a comparative study on the formation of multilayer AlN. They concluded that the AlN coating exhibits brilliant resistivity against corrosion and finds its main application in the field of heat sinks. Roman Y. G had done his work for aluminum nitride coatings which is produced by chemical vapor deposition. They had discussed all the properties of the coating done by CVD and laser, and concluded that these are used widely in the production of thin-film resistors

[4–9]. Chen et al. [10] had published his works on the preparation of aluminum nitride (AlN) coatings. The coating is done at room temperature by them around 800–1200 V. They had studied the inter-relationship between the argon ion beam voltages and the temperature of the coating at which it gets deposited. They had concluded with the microstructure of the AlN coating produced in the process. Feng et al. [11] worked in the field of determining the mechanical properties along with its microstructure change. They had worked on Inconel 718, a super alloy, by doing a laser weld. They have performed the testing of the plate to determine the mechanical properties of the super alloy.

Many techniques are available for thin-film deposition. Physical processes and chemical processes are two main techniques used to produce a thin film of metal oxides on suitable substrates. The physical processes are mainly based on evaporation and sputtering. First, the evaporation process involves the transformation of the material from a solid state or a liquid state to the vapor state by supplying thermal energy. The steam forms are consolidated form with the solid state on heating, and the unheated substrates undergo atomic level flaws. In evaporation, the difference in the vapor pressures poses the problem of stoichiometry. On the other hand, target poisoning and arcing are the main drawbacks of the reactive sputtering technique. The second methods, i.e., chemical methods, include sol-gel, plating, and chemical vapor deposition. The method of CVD means the coating deposition on the base metal. This method will result in the production of a condensed phase plus gaseous products. Temperatures that are high enough to activate the chemical reactions are needed. In chemical methods, the adhesion of the material with the substrate, porous nature, and impurity are the major drawbacks [12, 13].

An effort has been taken to study about the mechanical properties and microstructure of nano-crystalline AlN coating on mild steel. The coating method preferred is magnetron sputtering which is followed by different tests conducted on the specimen to determine its mechanical properties and microstructure.

2 Experimental Work

The base metal mild steel (MS) is made into the size of 25 × 25 mm. The surface of the mild steel part is ground with the help of SiC to eliminate the oxides and additional contaminants. The glossy substrates are degreased by using acetone. Then, the parts are electro-washed in alkali solution containing sodium hydroxide and sodium carbonate. The electro-cleaning is done for two minutes at 70 °C temperature. This is rinsed by soaking with triple distilled water (Table 1).

Table 1 Mild steel chemical composition

Elements	C	Si	Mn	Fe
Content (wt%)	0.37	0.28	0.66	98.69

The AlN coating is deposited on well-cleaned MS substrates. The AlN coating is done with the help of using a HIND HIVAC DC magnetron sputtering machine. Around 6–10 mbar base pressure is maintained, and the substrate temperature was varied. The remaining coating parameters are being held at optimum during the whole experiment. The substrates are etched for 15 min by maintaining a DC power of 50 W. Then, the high-purity argon is supplied inside the vacuum chamber for the plasma generation where the pressure is maintained constant at 3×10^{-3} mbar for all depositions. After the deposition of the AlN coating on mild steel, various tests are conducted on specimen to determine its mechanical properties and microstructure.

The unique X-ray diffraction pattern is generated by doing the XRD analysis that produces a unique "fingerprint" of the crystals (Fig. 1; Table 2).

$$\textbf{Braggslaw}: \mathbf{2d \sin\theta = n\lambda}$$

Fig. 1 DC magnetron sputtering unit

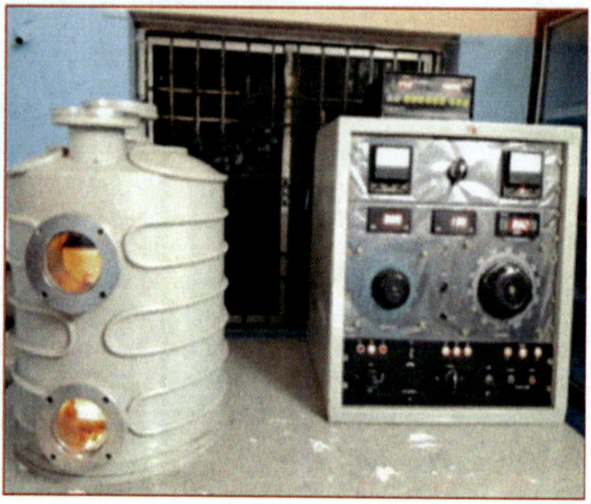

Table 2 Deposition conditions

RF power	100 W
Flow rate of argon	30 SCCM
Flow rate of nitrogen	15 SCCM
Substrate	MS
Substrate temperature	RT and 500 °C
Deposition time	1 h
Target	Aluminum
Base vacuum	3.0×10^{-6} mbar
Deposition pressure	3.0×10^{-2} mbar
Substrate to target distance	100 mm

where $\boldsymbol{n}$ is a positive integer and $\boldsymbol{\lambda}$ is the wavelength of incident wave.

The crystallite size can be determined from the broadening of corresponding X-ray peaks by Scherrer's equation

$$D = \frac{k\lambda}{\beta \cos \theta}$$

where

$\boldsymbol{D}$ is the crystallite size,
$\boldsymbol{\beta}$ $\sqrt{(B2 - b2)}$,
$\boldsymbol{\lambda}$ is the wavelength of X-ray radiation (CuKα1 = 1.5406 Å),
$\boldsymbol{\theta}$ is the diffraction angle, and
$\boldsymbol{k}$ is Scherrer's constant (assuming 0.9, since the particles are spherical)

Scanning electron microscope (SEM) is primarily used to study the surface morphology. A scanning electron microscope (SEM) is a type of electron microscope that produces images of a sample by scanning it with a focused beam of electrons. The electrons interact with atoms in the sample, producing various signals that contain information about the sample's surface topography and composition. Indentation techniques such as the Knoop and Vickers techniques are commonly used to measure micro-scale hardness of materials. The micro-hardness indentation measurement is a simple, reproducible, and cheap method, which is often used for the characterization of surface for mechanical applications. Pin-on-disk wear testing is a method of study the coefficient of friction, frictional force, and rate of wear between two materials. Dry sliding wear tests for different number of specimens were conducted by using a pin-on-disk machine. Wear is a process of removal of material from one or both of two solid surfaces in solid-state contact (Fig. 2).

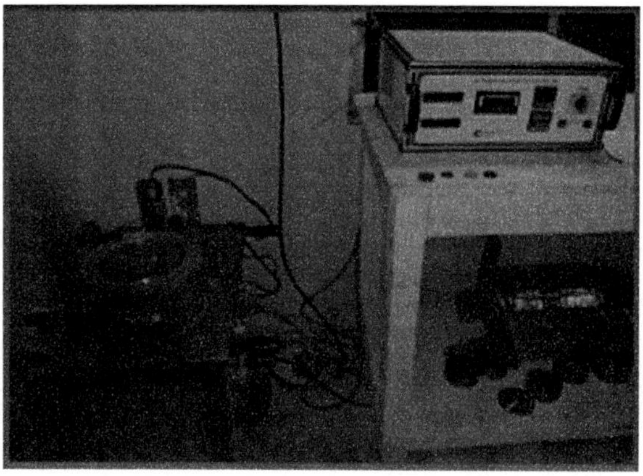

Fig. 2 Pin-on-disk wear test machine

Fig. 3 Before and after coating of AlN on mild steel

3 Results and Discussions

3.1 Thickness Measurement: Dektak Profilometer

The thickness of the films is measured by using Dektak profilometer (Dektak 6 M-Stylus profiler by Veeco, USA). A fine diamond tip consists of the stylus of the size of the radius of ~2 μm that is compressed on to the exterior by a pressure of 500 N/cm^2. The corresponding tip mass of only 0.1 g moved along the surface of the substrate. The vertical displacement of the tip is created by variations of the surface. These irregularities are transformed into an electrical signal. Then, this electric signal is amplified and recorded. The outcomes attained by the stylus are found to be in good agreement with other methods. It is capable of measuring step heights between 100 Å and ~50 μm with a resolution of 10 Å. The thickness of the AlN coatings deposited at different temperatures on silicon is measured by using the stylus profilometer and is found to be in the range (Fig. 3).

3.2 Structural Studies-XRD

The pattern of XRD of the AlN coated is shown in Fig. 4. The film-coated at room temperature has not shown any peaks. The film-coated at 500 °C had shown the peak at an angle of 70.2° corresponding to h-AlN (103) indicating the polycrystalline hexagonal structure. The crystalline particle size can be determined by using Scherrer's formula if the values of the incident X-ray wavelength (λ), full width at half maximum (FWHM) of the peaks (β), and diffraction angle (Θ) are provided. The crystallite size is found to be 12 nm.

$$D = \frac{k\lambda}{\beta \cos \theta}$$

Scherrer's constant $k = 0.9$,

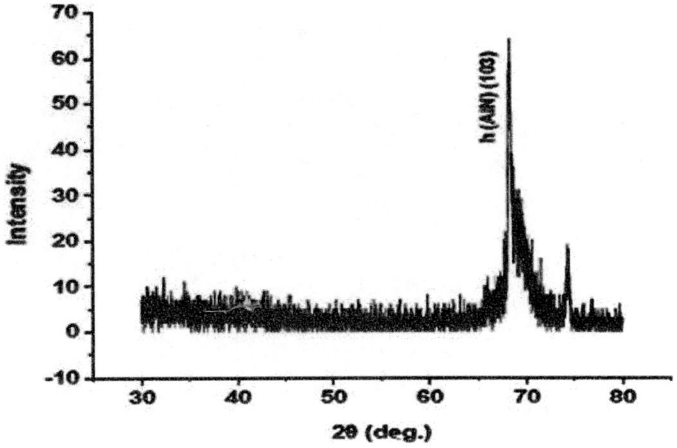

Fig. 4 XRD analysis of ALN coating

λ = 1.5418 × 10^{-10} m or 1.5418 Å
FWHM $\beta = 0.8820$
Angle of diffraction $\theta = 41.88$

$$D = \frac{0.9 \times 1.5418 \times 10^{-10}}{0.8820 \times \cos(41.88)}$$

$D = 121.08 \times 10^{-10}$
$D = 12$ nm.

3.3 Scanning Electron Microscopy

SEM image shows the smooth surface morphology with dense formation of grains. The grains are in uniform size and distributed homogeneously (Fig. 5).

4 Mechanical Properties

4.1 Pin on Disk

Figure 6 shows the coefficient of friction of the coated disk. It displays an accelerating drift with respect to time that lasts for 10 s (stage S1). The curve seized quite a consistent value (steady state, stage S2). The average of the coefficient of the friction for both the stage (S1 and S2) is 0.52. The initial value of the coefficient of the friction is determined by calculating the mean values for the first 10 s. The equivalent method

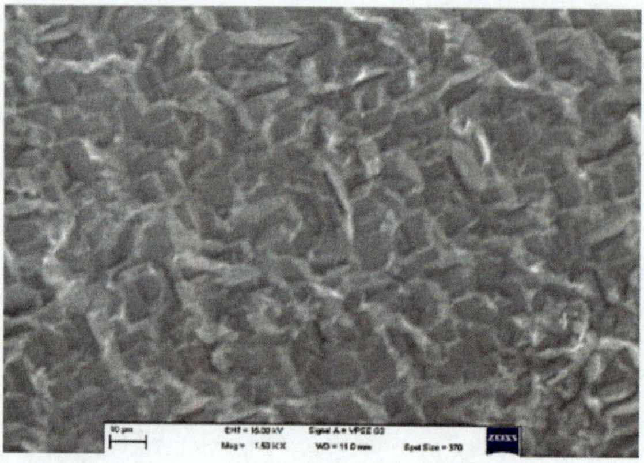

Fig. 5 SEM analysis of ALN coating

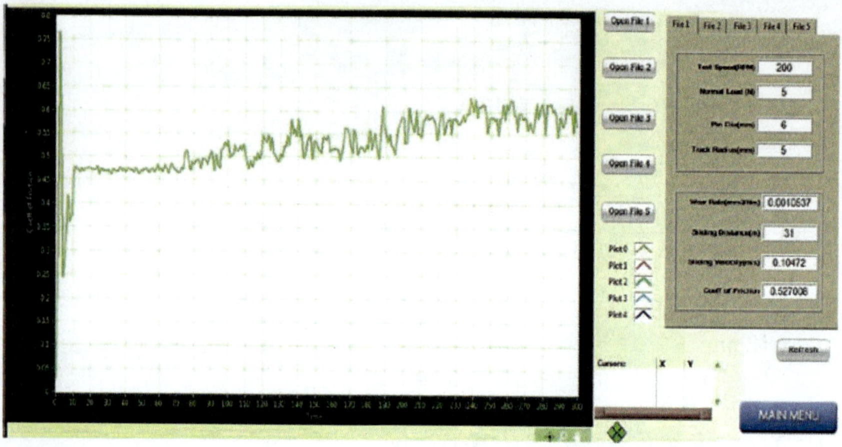

Fig. 6 Comparison of CoF versus time

is practiced to determine the steady-state coefficient of friction. The result exhibits the low coefficient of friction of the coated disk than the uncoated one.

Figure 7 showed the wear behavior of the mild steel sample. It showed that the wear is increased until 20 s. After that, the wear is constant with concerning time. This analysis indicated that the wear increased for a small interval of time, due to the particles, debris, or any contamination present on the disk or pin. After that, the wear in almost constant, and it showed the wear of 3 microns with time. This is due to the mild steel surface which is hardened due to the diffusion of boron atoms and carbon into the surface of the mild steel (Fig. 8).

The hardness value of **600.1 HV** is obtained for **0.100 g** load for the film deposited at RT.

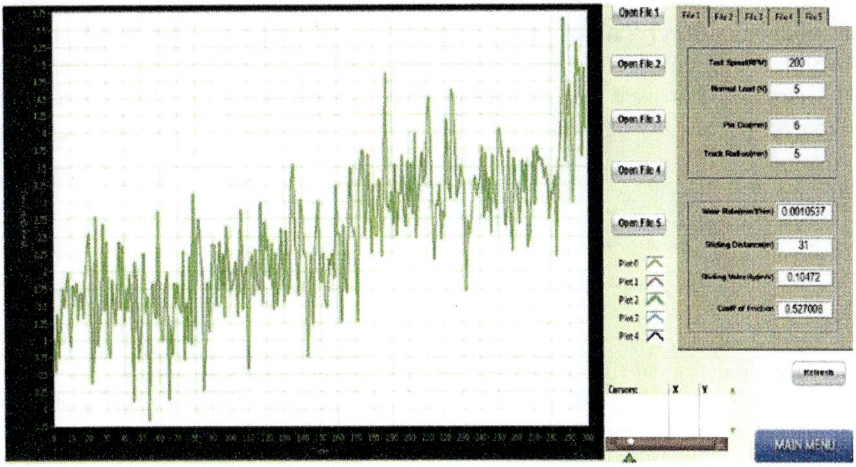

Fig. 7 Comparison of wear versus time

Fig. 8 Hardness measurement

5 Conclusion

AlN thin films are prepared successfully on low carbon steel and substrates of silicon at room temperature and 500 °C by DC magnetron sputtering method. The prepared films are characterized using XRD, SEM, micro-indentation, wear test to study the microstructure and mechanical properties of the films. The XRD analysis showed amorphous nature at room temperature, while the films deposited at 500 °C indicating the peaks corresponding to the hexagonal structure of the films with 12 nm being a crystallite size. SEM shows the smooth morphology of the films with uniform distribution of the crystallites. The hardness of the AlN film is found as 600 HV for the film deposited at RT, while the film at 500 °C shows the hardness of 355 HV. The hardness decreased with increasing temperature. Wear test on AlN thin film is prepared successfully on mild steel at RT, 500 °C, which indicates the peak value of 3 microns.

References

1. Buschow KHJ, Chan RW, Flemings MC, Ilschner B, Kramer EJ, Mahajan S (eds) (2001) Encyclopedia of Materials: Science and Technology, vol 1. Elsevier, Amsterdam
2. Strite S, Morkoç H (1992) GaN, AlN, and InN: a review. J Vac Sci Technol B: Microelectron Nanometer Struct Proc Meas Phenomena 10(4):1237–1266
3. Wang X, Kolitsch A, Prokert F, Möller W (1998) Ion beam assisted deposition of AlN monolithic films and Al/AlN multilayers: a comparative study. Surf Coat Technol 103:334–339
4. Roman YG, Adriaansen APM (1989) Aluminium nitride films made by low pressure chemical vapour deposition: Preparation and properties. Thin Solid Films 169(2):241–248
5. Jeyaprakash N, Duraiselvam M, Raju R (2018) Modelling of Cr 3 C 2–25% NiCr laser alloyed cast iron in high temperature sliding wear condition using response surface methodology. Arch Metall Mater 63(3):1303–1315
6. Reier T, Simson S, Schultze JW (1998) Corrosion and XPS investigations of nitrogen implanted aluminium. Electrochim Acta 43(1–2):149—158
7. Imamura, S, Fukui H, Shibata A, Omori N, Setoyama M (2007) Properties and cutting performance of AlTiCrN/TiSiCN bilayer coatings deposited by cathodic-arc ion plating. Surface and Coatings Technol 202(4–7):820–825
8. Subramanian B, Ananthakumar R, Jayachandran M (2010) Microstructural, mechanical and electrochemical corrosion properties of sputtered titanium–aluminum–nitride films for bio-implants. Vacuum 85(5):601–609
9. Engelmark F, Fucntes G, Katardjiev IV, Harsta A, Smith U, Berg S (2000) J Vac Sci Technol A, Vac Surf Films 18:1609
10. Chen, H-Y, Han S, Shih HC (2006) The characterization of aluminum nitride thin films prepared by dual ion beam sputtering. Surf Coat Technol 200(10):3326–3329
11. Qu, F, Liu X, Fei XING, Zhang K (2015) MATEC web of conferences, vol 21. EDP science, p 06005
12. Ahmadi A, Hadipour NL, Kamfiroozi M, Bagheri Z (2012) Theoretical study of aluminium nitride nanotubes for chemical sensing of formaldehyde. Sens Actuators B: Chem 161(1):1025–1029. https://doi.org/10.1016/j.snb.2011.12.001
13. Raju R, Sivalingam V, Sun J, Natarajan M, Zhao Y (2019) Experimental and taguchi-based grey approach of laser metal deposition technique on nickel-based superalloy. Trans Indian Inst Met 72(1):205–214

Material Characterization and Parametric Effect on Nickel-Coated Mild Steel Sheets by Electroplating Process

D. Pritima, P. Padmanabhan, S. Marichamy, C. Sivakandhan, B. Stalin, and V. Dhinakaran

Abstract In manufacturing industries, sheet metal plays an important role. To improve the properties, coating has been done on the sheet metal. In this work, nickel is coated on mild steel sheets by electroplating process. The hardness value is measured before and after the coating process. The morphology and material composition of the sheet metal were studied through a scanning electron microscope (SEM) and energy-dispersive analysis X-ray (EDAX). It is observed that nickel is uniformly dissolved on the mild steel surface. During the coating process, the input parameters such as magnitude of current, time and temperature are considered. Based on the input parameters, the coating thickness has been measured. The effect of parameter has been analyzed through analysis of variance.

Keywords Nickel-coated mild steel · Time · Temperature · Coating thickness · Analysis of variance

---

D. Pritima
Department of Mechanical Engineering, Vickram College of Engineering, Sivagangai, Tamil Nadu 630561, India

P. Padmanabhan
Department of Mechanical Engineering, V.V. College of Engineering, Tisaiyanvilai, Tamil Nadu 627657, India

S. Marichamy
Department of Mechanical Engineering, Sri Indu College of Engineering and Technology, Hyderabad, Telangana, India

C. Sivakandhan
Department of Mechanical Engineering, Sri Indu Institute of Engineering and Technology, Hyderabad, Telangana 501510, India

B. Stalin (✉)
Department of Mechanical Engineering, Anna University, Regional Campus Madurai, Madurai, Tamil Nadu 625019, India
e-mail: stalin1312@gmail.com

V. Dhinakaran
Department of Mechanical Engineering, Chennai Institute of Technology, Kundrathur, Chennai, Tamil Nadu 600069, India

A. Arockiarajan et al. (eds.), *Advances in Industrial Automation and Smart Manufacturing*, Lecture Notes in Mechanical Engineering,
https://doi.org/10.1007/978-981-15-4739-3_40

1 Introduction

To improve corrosion and wear resistance, nickel is coated on the mild steel sheet. The martensitic transformation behavior and applications were discussed in stainless steel sheet [1]. The favorable mechanical properties and surface texture were used to increase applications of stainless steel sheet [2, 3]. In the recent development of industries, the multi-functionality of clad sheet metals was introduced [4]. The ductility is one of the most important properties in any sheet metal. The stress and strain rates mainly depend on the ductility of the sheet metal [5]. The effect of electroplating parameters and microstructure has been analyzed in nanocrystalline-based nickel coating [6]. The operation parameters and its contribution were discussed in nickel plating process [7]. The quantitative analysis has been carried out on nickel-coated hard metal by electroplating process [8].

The present experimental investigation clearly describes about nickel-coated mild steel sheets by electroplating process. Through SEM and EDAX, sheet metal characterization was discussed. The sheet metal coating thickness and effect of parameter have been analyzed through analysis of variance.

2 Experimental Details

The mild steel sheet with 63 HRC was considered as the base metal. Before the electroplating process, the base metal sheet was cleaned. It was immersed in an electrolyte solution. The dissolved nickel ions were attracted by mild steel sheet which was connected to the cathode terminal. The input parameters such as magnitude of current, time and temperature are considered. According to the input parameters, the coating thickness has been found. The SEM-ZEISS machine was used to provide scanning electron microscopic analysis. The composition of the coated sheet was demonstrated by EDAX. After the coating process, sheet with 74 HRC has been achieved.

3 Results and Discussion

3.1 SEM and EDAX of Nickel-Coated Mild Steel Sheets

The morphology analysis was analyzed through SEM image. After electroplating process, the microstructure of the coated steel sheets was analyzed and it is shown in Fig. 1. The nickel was properly coated on the surface of the mild steel sheets, and nickel ions were dissolved in mild steel uniformly. In Fig. 2, EDAX shows the results of various weight percentages of elements after nickel coated of mild steel.

Fig. 1 SEM images of nickel-coated mild steel sheets

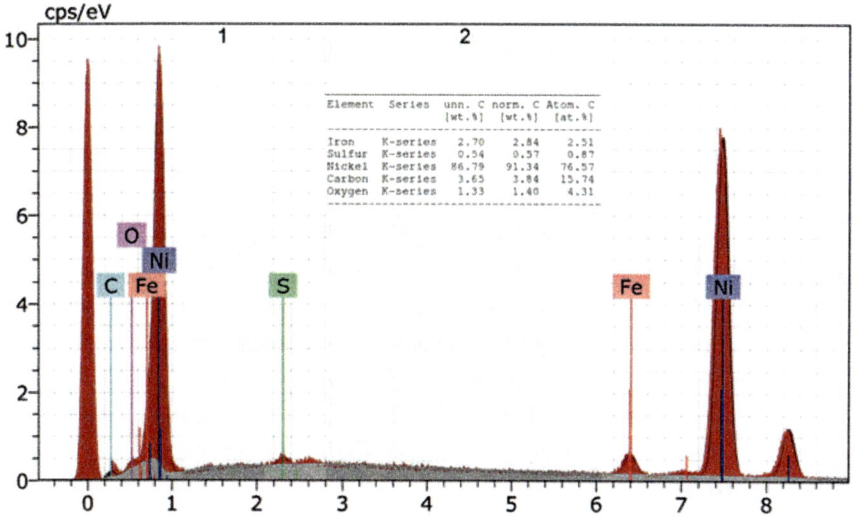

Fig. 2 EDAX of nickel-coated mild steel sheets

Table 1 Investigation results of electroplating process

S. No.	Magnitude of current (amps)	Temperature (°C)	Time (min)	Coating thickness (mm)
1	1	36	10	1.01
2	1	38	12	1.11
3	1	40	15	1.23
4	1.5	36	12	1.16
5	1.5	38	15	1.26
6	1.5	40	10	1.37
7	2	36	15	1.39
8	2	38	10	1.42
9	2	40	12	1.56

Table 2 Analysis of variance for electroplating process

Source	DF	Adj. SS	Adj. MS	*F*-value	*P*-value	Percentage (%)
Magnitude of current	2	0.174489	0.087244	280.43	0.004	73.54
Temperature	2	0.061089	0.030544	98.18	0.010	25.74
Time	2	0.001089	0.000544	1.75	0.364	00.46
Error	2	0.000622	0.000311	–	–	00.26
Total	8	0.237289	–	–	–	100

4 Machining

The aim of the electroplating process is to improve the properties of the sheet metal. The various input parameters such as magnitude of current, time and temperature are considered for this experimental work. The coating thickness was recorded when changing the input parameters. The investigation results of electroplating process are shown in Table 1.

Analysis of variance for electroplating process is shown in Table 2. The magnitude of current was the most influential parameter. Time was the least contribution parameter which affects the coating thickness.

5 Contour Plot Analysis

The contour plot of coating thickness against the magnitude of current and temperature is shown in Fig. 3. The high amount of coating thickness is obtained at current of 2 amps and temperature of 40 °C. During this period, more amounts of nickel ions are deposited on mild steel sheet.

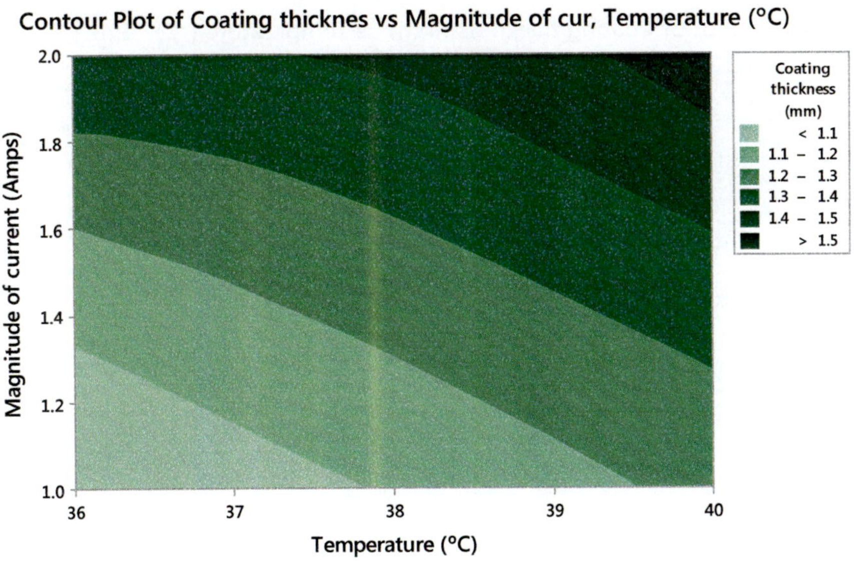

Fig. 3 Contour plot of coating thickness—magnitude of current and temperature

The coating thickness against the magnitude of current and time is shown in Fig. 4. The maximum of coating thickness has been achieved at current of 2 amps and time

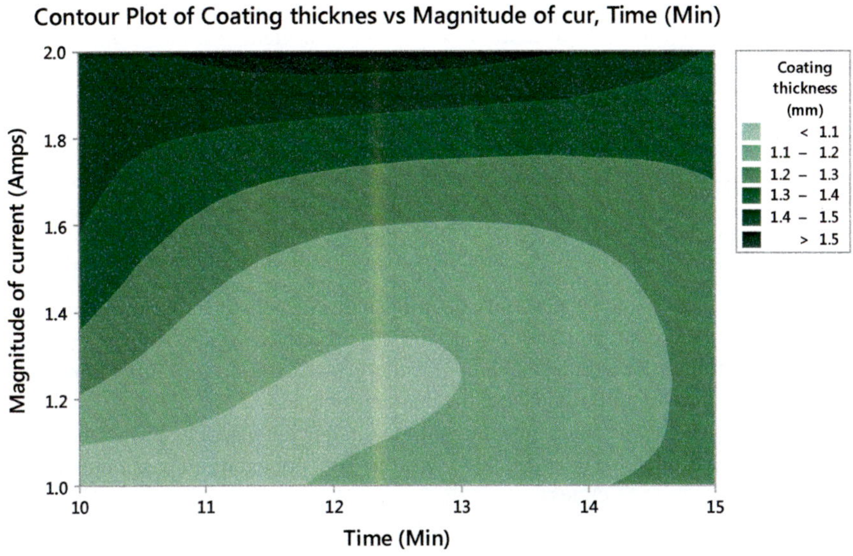

Fig. 4 Contour plot of coating thickness—magnitude of current and time

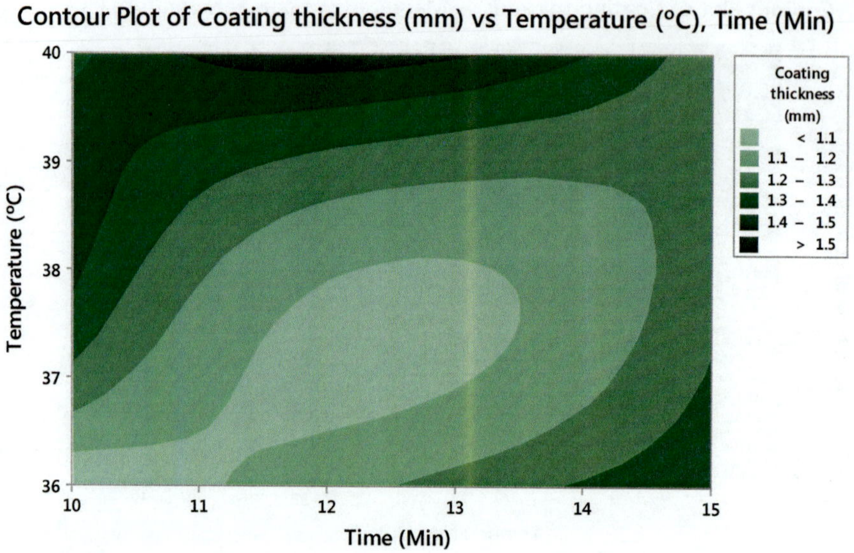

Fig. 5 Contour plot of coating thickness—temperature and time

of 11–14 min. The coating thickness mainly depends on the magnitude of current passed through electroplating process.

The coating thickness against the magnitude of temperature and time is shown in Fig. 5. The coating thickness was maximum at temperature of 40 °C and time of 11–13 min. During this temperature, more amounts of nickel ions are transferred to the mild steel sheet.

6 Conclusions

The following conclusions were made after the experimental investigation:

- The nickel was coated successfully on mild steel sheets by electroplating process.
- The sheet metal characterization was studied through SEM and EDAX.
- The magnitude of current (1–2 amps), time (10–15 min) and temperature (36–40 °C) was considered to measure the coated thickness.
- From analysis of variance, the magnitude of current (73.54%) was the most influential parameter. Time (73.54%) was the least contribution parameter which affects the coating thickness.
- The contour plot analysis provides interaction between coating thickness and variables.

References

1. Ishimaru E, Hamasaki H, Yoshida F (2015) Deformation-induced martensitic transformation behavior of type 304 stainless steel sheet in draw-bending process. J Mater Process Technol 223:34–38
2. Zhao O, Gardner L, Young B (2016) Structural performance of stainless steel circular hollow sections under combined axial load and bending—Part 1: Experiments and numerical modeling. Thin-Walled Struct 10:231–239
3. Mori K, Akita K, Abe Y (2007) Springback behaviour in bending of ultra-high-strength steel sheets using CNC servo press. Int J Mach Tools Manuf 47:321–325
4. Yilamu K, Hino R, Hamasaki H, Yoshida F (2010) Air bending and springback of stainless steel clad aluminum sheet. J Mater Process Technol 210:272–278
5. Oya T, Tiesler N, Kawanishi S, Yanagimoto J, Koseki T (2010) Experimental and numerical analysis of multilayered steel sheets upon bending. J Mater Process Technol 210:1926–1933
6. Rashidi AM, Amadeh A (2010) Effect of electroplating parameters on microstructure of nano crystalline nickel coatings. J Mater Sci Technol 26(1):82–86
7. Sadiku-Agboola O, Sadiku ER, Ojo OI, Akanji OL, Biotidara OF (2011) Influence of operation parameters on metal deposition in bright nickel-plating process. Portugaliae Electrochimica Acta 29(2):91–100
8. Wahab HA, Noordin MY, Izman S (2013) Denni Kurniawan: quantitative analysis of electroplated nickel coating on hard metal. Scient World J 631936:1–6

Evaluation of Mechanical Properties on Ni–Cr Alloy-Coated Marine Structures

A. Amala Mithin Minther Singh, P. Arul Franco, J. S. Binoj, and N. Manikandan

Abstract Surface coating has been used in abundance of engineering applications and components to protect and strengthen the surface of the material from corrosion which damages because of some atmospheric conditions and enhance their span of materials life. Corrosion or abrasion is caused naturally in which the materials destruct gradually using electrochemical or chemical reaction with their environment. This corrosion in metals occurs mainly due to air, water, and other environmental aspects. Steel structures have been used in many years of marine structural applications due to high fatigue strength, higher tensile strength, excellent dimensional stability, and thermal conductivity. The marine structure gets corroded due to the sea water and atmospheric conditions. The marine structures are made up of low carbon steel with various thickness from 6 to 20 mm, depends on the marine applications. The low carbon steel easily corrodes, so that the paint coating on the low carbon steel is in existence. This paint-coated surface directly reacts with marine atmosphere and forms a rust layer between 6 and 12 months. To prevent this rust layer formation, we need to bond coat with excellent corrosive resistive material under marine environment. These structures were partially immersed and floating in sea water. In fishing mechanized boat, the marine hull corrodes quickly due to marine atmosphere and heat lost from the engine room. This work focuses on protecting the marine structure subjected to corrosion and improving the mechanical strength and lifespan of marine structures with microparticle coating on the marine hull surface. The Ni–Cr powder coating was layered on the low carbon steel structure using plasma spray coating techniques with thickness of 100 and 150 μm. These prepared specimens was sized according to ASTM standards to perform the mechanical properties such as tensile, flexural and impact testing, and hardness tests like Rockwell, Brinell, Vickers, etc. The morphological structure also studied using SEM. The results show

A. Amala Mithin Minther Singh (✉) · P. Arul Franco
Department of Mechanical Engineering, University College of Engineering (Anna University Constituent College), Konam, Nagercoil, Tamil Nadu 629004, India
e-mail: mithinlites2011@gmail.com

J. S. Binoj · N. Manikandan
Micro Machining Research Centre, Department of Mechanical Engineering, Sree Vidyanikethan Engineering College (Autonomous), Tirupati, Andhra Pradesh 517102, India

A. Arockiarajan et al. (eds.), *Advances in Industrial Automation and Smart Manufacturing*, Lecture Notes in Mechanical Engineering,
https://doi.org/10.1007/978-981-15-4739-3_41

that the utilization of Ni–Cr micro-particle coating improves the mechanical strength of the mechanized boat hull from the paint coating in existence. The result reveals that coating improves the mechanical strength as well as its lifespan.

Keywords Low carbon steel · Ni–Cr · Plasma spray coating · Mechanical properties · Marine structure

1 Introduction

The work focuses on improving the corrosion resistive rate of mechanized boat hull which is made up of low carbon steel with Ni-based alloy coating by thermal spray coating techniques. In India, states like Tamilnadu and Kerala have long coastal area and occupation of many people is fishing. They are working with small crafts and mechanized boat. Mostly, they are using mechanized boat for fishing, because in small crafts, they cannot stay for few days in sea, in mechanized boat they can stay few days (i.e., 6–15 days and more). So they preferred mechanized boat, around more than 20,000 mechanized boats were used for fishing and 2 lakhs peoples are working in it. These mechanized boat hull structures are made up of mild steel structure due to its high strength with the thickness ranges from 6 to 20 mm, depends upon the usage in marine applications. It always floats in the marine water so the surface corrodes easily due to the sea water and atmospheric conditions.

Marine atmosphere has higher level of humidity and salinity and more corrosion, and these effects lead to failures of 30% on marine structure and other equipments. In worldwide, around \$50–80 billion has lost every year due to marine corrosion [1]. The low carbon steel easily corrodes, so that the paint coating on the low carbon steel is in existence. This paint-coated surface directly reacts with marine atmosphere and forms a rust layer between 6 and 12 months. To prevent this rust layer formation, we need to bond coat with excellent corrosive resistive material under marine environment. These structures were partially immersed and floating in sea water. In fishing mechanized boat, the marine hull corrodes quickly due to marine atmosphere and heat lost from the engine room (Figs. 1 and 2).

Fig. 1 Corroded hull structure

Fig. 2 Mechanized boat

Coating in the present scenario works well and has sufficient lifespan when they are applied as per the producer's specifications [2]. There are many forms of coatings which includes organic and metallic protective coatings to prevent and control corrosion [3].

This problem has challenged to the manufacturing industry for many years, but it needs to be proactive to prevent and control marine corrosion. The expertise shares their knowledge and research findings to develop modern engineering components to control the damaging effects of the marine atmosphere [4] (Figs. 3 and 4).

Active anticorrosive elements like metallic chromates passivate the surface by covering with an oxide film. Nickel-based alloys are usually employed as an anti-corrosive metal, preventing discharge of current from naval steel to electrolyte by electrochemical attachment with less passive anode, which covers sacrificial coating. [5].

Ni–Cr incorporated plasma spray coating with two different thicknesses 100 and 150 μm on the hull surface improves the strength of the hull structure, prevents corrosion, and enhances their mechanical properties and lifespan of the marine structures.

Fig. 3 Mechanized boat pro-E model

Fig. 4 Hull and deck house structure pro-E model

2 Materials and Methods

2.1 *Mild Steel*

Mild steel acts as a base metal for this work, and it is a form of low carbon steel and cheap in construction materials due to its low cost and high tensile strength. Alloy contains chromium, nickel, cobalt, titanium, tungsten, niobium, molybdenum, vanadium, or zirconium and adds any other element to obtain a desired alloying effect. This metal easily reacts with atmospheric oxygen and moisture leads to the degradation and rust formation of material properties, i.e., corrosion [6].

2.2 *Nichrome*

Nichrome is an alloy of 80% Ni and 20% Cr and widely used as corrosion resistance and electrical resistance applications. It is silvery-gray in color and resistance to oxidation (Table 1; Fig. 5).

Table 1 Nichrome property at STP

Material property	Values
Density	8400 kg/m^3
Modulus of elasticity	2.2×10^{11} N/m^2
Melting point	1400 °C
Specific heat	450 J/kgK
Thermal conductivity	11.3 W/mK
Thermal expansion	14×10^{-6} K^{-1}

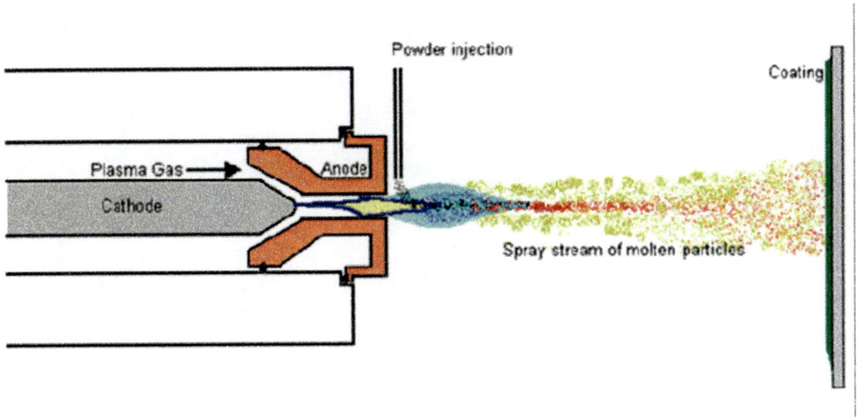

Fig. 5 Plasma spray process

2.3 Specimen Preparation

Base metal mild steel was cut into a size of 500 mm × 500 mm × 6 mm. It is washed with fresh water for three times and allowed it to dry. Then, it should be cleaned with acetone which has the ability to remove the presence of dust and impurities on the metal.

2.4 Powder Preparation

Nichrome metal is synthesized using heavy ball milling techniques, and 70 μm thickness nichrome powder was obtained (Fig. 6).

The nichrome powder was coated on the low carbon steel surface of 500 mm × 500 mm × 6 mm with two different thickness of 100 and 150 μm using the plasma spray method at 80–100 °C (Figs. 7 and 8; Table 2).

These samples were sized using the EDM CNC wire cut machine SODICK—AQ537L. The brass wire of diameter 0.25 mm was used to cut the samples, and mineral water was used to reject the heat from the metal and wire, which can be prevent from thermal effects. The wire feed is 470 mm/min and feed of current is 180–230 Amps.

Fig. 6 Nichrome powder 70 μm thickness

Fig. 7 100 μm coated plate

Fig. 8 150 μm coated plate

Table 2 Plasma spray coating parameter

Process parameters	Values/types
Gun	3 MB
Current	500Amps
Nozzle	GH
Powder feed	100–120 gm/min
Spray	4–6″
Gas	Hydrogen (pressure—50 PSI flow rate: 15–18 SCFH)
Voltage	65–75 Volts
Gas	Argon (pressure—100–120 PSI flow rate: 150–165 SCFH)

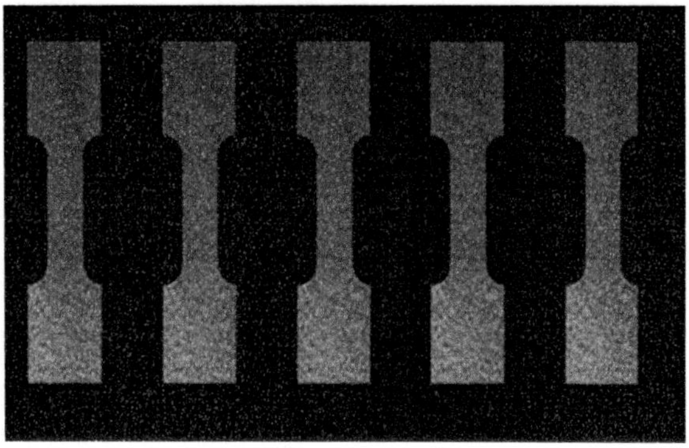

Fig. 9 Samples for tensile test

3 Experimental Tests

3.1 Mechanical Behavior

3.1.1 Tensile Test

This test was conducted using UTM according to ASTM D638. In this, 5 of each uncoated mild steel samples, 100 μm and 150 μm thickness coated samples totally 15 pieces of samples were tested (Fig. 9).

3.1.2 Flexural Test

This test was conducted using UTM according to ASTM D790. In this, 5 of each uncoated mild steel samples, 100 and 150 μm thickness coated samples; totally, 15 pieces of samples were tested (Fig. 10).

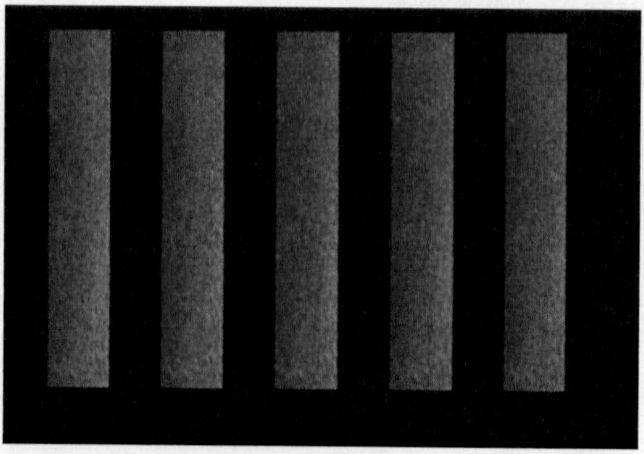

Fig. 10 Samples for flexural test

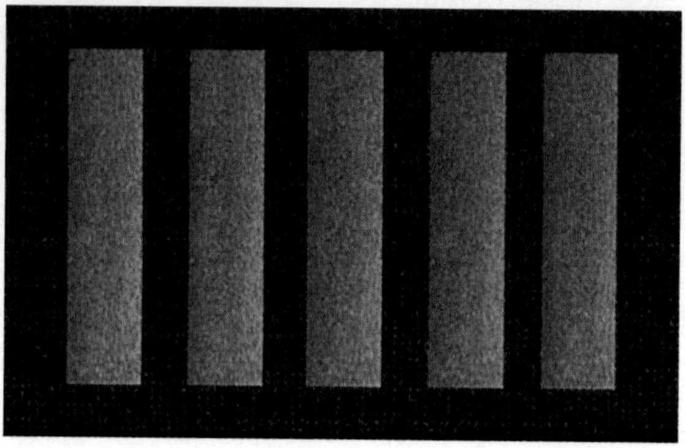

Fig. 11 Samples for impact test

3.1.3 Impact Test

This test was conducted using Charpy Impact Testing Machine according to ASTM D256. In this, 5 of each uncoated mild steel samples, 100 and 150 μm thickness coated samples; totally, 15 pieces of samples were tested (Fig. 11).

3.1.4 Hardness Test

This test was conducted using Rockwell Hardness TestMachine (Eq. 1), Brinell Harness Test Machine (Eq. 2), and Vickers Hardness Test Machine (Eqs. 3 and 4) according to ASTM E18, ASTM E10, and ASTM E384, respectively (Fig. 12).

$$\text{Rockwell Hardness Number (RHN)} = N - \frac{h}{s} \quad (1)$$

where N is number of specific Rockwell Hardness Scale (diamond indenter-100 units and steel ball indenter-130 units), h is permanent depth of indentation in mm under

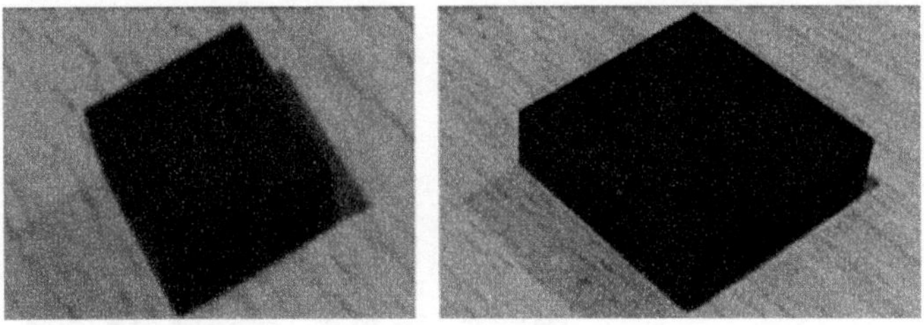

Fig. 12 Hardness test samples

removal of test force applied, and s is scale unit (0.002 mm for A, B, C, D, E, F, G, H, K and 0.001 mm for 15 N, 30 N, 45 N, 15T, 30T, 45T).

$$\text{Brinell Hardness Number (BHN)} = 0.102 \times \frac{2F}{\pi D(D - \sqrt{D^2 - d^2}} \tag{2}$$

where F is force applied in N, D is ball diameter in mm, and d is indentation mean diameter in mm.

$$\text{Vickers Hardness Number} = 0.0102 \times \frac{2F \sin 136/2}{d^2} \tag{3}$$

$$\text{Vickers Hardness Number} = 0.1891 \times F/d^2 \tag{4}$$

where F is force applied in N and d is indentation arithmetic mean length of two diagonals in mm.

3.2 Morphological Behavior

The surface morphology of the coated specimen was examined using SEM under various magnifications (500x & 200 μm) of COSLAB Model: SEG100. The specimens were mounted onto SEM holder using double-sided electrically conducting carbon adhesive tapes to prevent surface charge on the specimens when exposed to the electron beam. The prepared coated surfaces were sputter with gold prior as an overcoat to study their morphology which makes the conductive surface to avoid gathering of electron charge. SEM uses a focused electron probe to extract microstructural and chemical information presented point by point from a region of the sample. Due to higher spatial resolution, SEM is a powerful tool to characterize a wide range of specimen's morphology at nanometer to micrometer scale length.

For marine applications, a SEM is used to examine surface orientation, surface irregularities, or fracture areas of the specimens. A SEM is used to measure the depth of thin coatings. Initially, specimens have to be tested are sputtered and coated with gold and then placed in a vacuum chamber to view on the computer

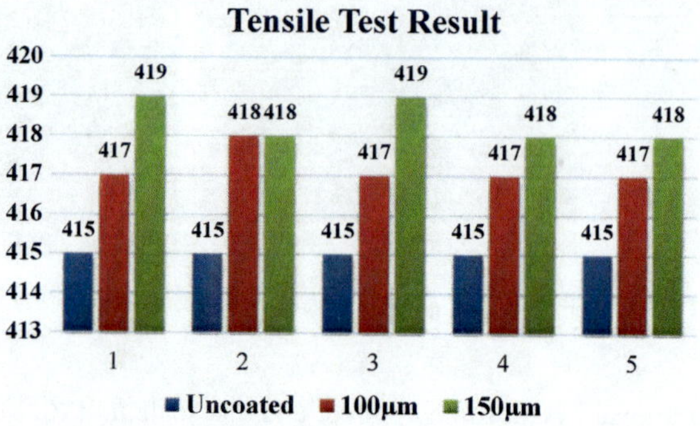

Fig. 13 Tensile properties of specimen

Table 3 Hardness values for coated specimens

Hardness	Uncoated	100 μm	150 μm
Rockwell (RHN)	71.5	71.7	72
Brinell (BHN)	127	127.4	127.9
Vickers (HV)	131	131.3	131.7

monitor/display screen at up to 10,000 × magnification. Polaroid photographs are taken for a permanent record of the specimen.

4 Results and Discussions

4.1 Mechanical Behavior

4.1.1 Tensile Test

From Fig. 13, the 150 μm thickness coated material has higher tensile strength than the 100 μm and uncoated specimen. Therefore, it clearly shows that coated nichrome with increase in thickness has more tensile strength and we can implement this for the building of marine hull structures (Table 3).

4.1.2 Flexural Test

From Fig. 14, the 150 μm thickness coated material has higher flexural strength than the 100 μm and uncoated specimen. Therefore, it clearly shows that coated nichrome with increase in thickness has more flexural strength and we can implement this for the building of marine hull structures.

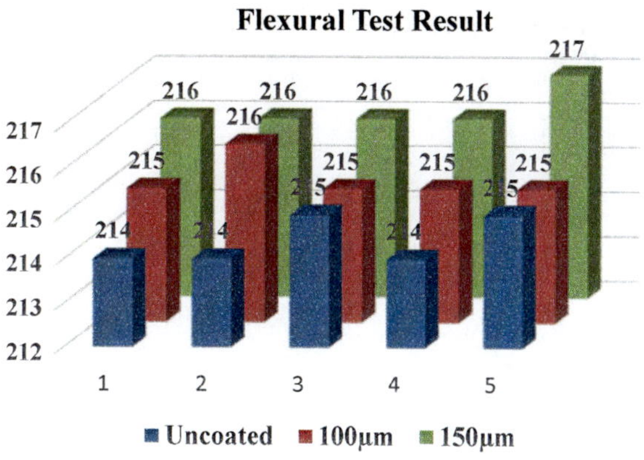

Fig. 14 Flexural properties of specimen

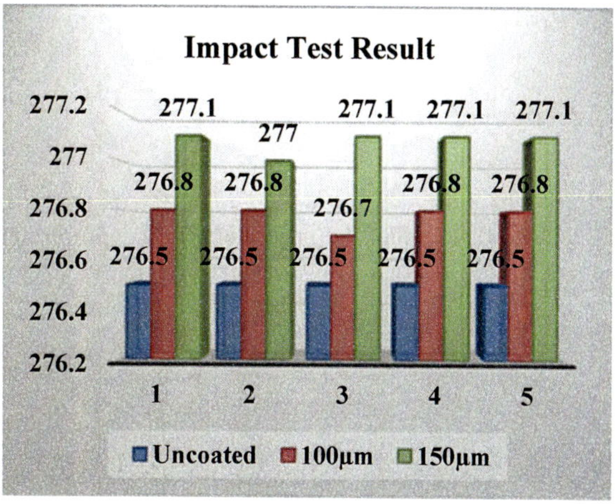

Fig. 15 Impact properties of specimen

4.1.3 Impact Test

From Fig. 15, the 150 µm thickness coated material has higher impact strength than the 100 µm and uncoated specimen. Therefore, it clearly shows that coated nichrome with increase in thickness has more impact strength and we can implement this for the building of marine hull structures.

4.1.4 Hardness Test

Table 4 shows that Rockwell Hardness, Brinell Hardness, and Vicker Hardness values are more in the 150 and 100 µm thickness than the uncoated values. It clearly shows that increase in thickness of coating gives higher hardness values, i.e., stronger

Table 4 Hardness test result comparison

	Parameters	Values	Hardness
Rockwell	N	130 (ball indenter)	72 RHN
	h	0.116 mm	
	S	0.002 mm	
Brinell	F	29,430 N	127.9 BHN
	D	10 mm (ball intender)	
	d	5.22 mm	
Vickers	F	981 N	131.7 HV
	D	1.18	

than the base specimen. As per the results obtained, we can implement this coated specimen for the marine hull applications.

4.2 *Morphological Structure*

Figures 16 and 17 show the scanning electron microscopic image of coated material. It clearly depicts that the base metal and coating metal have good bonding nature, and as in Fig. 17, it has porosity nature due to less coating thickness of the material.

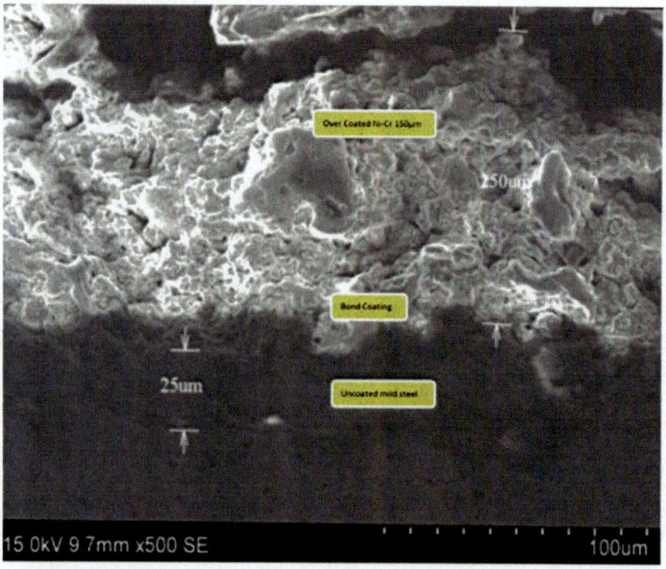

Fig. 16 SEM image of coated Ni–Cr 150 µm

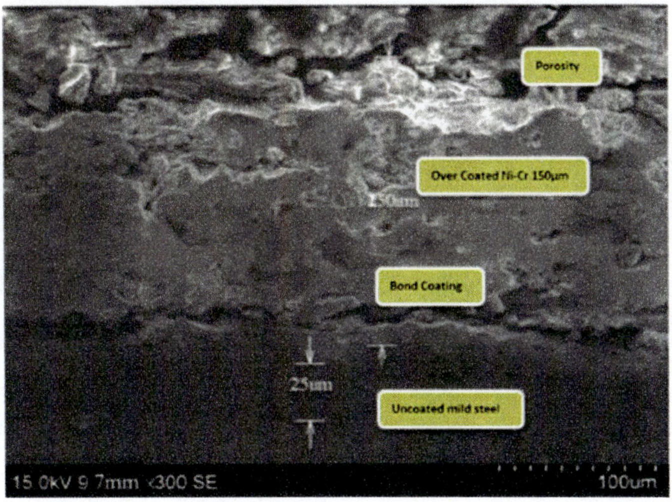

Fig. 17 SEM image of coated Ni–Cr 100 μm

Due to the bonding nature of the Ni–Cr and the substrate, coated specimens can be utilized to manufacture the mechanized boat hull.

5 Conclusions

The work focused on improving the mechanical strength of the coated material surface and the substrate. The above-mentioned experimental results show that tensile, flexural, and impact strength of the coated 150 μm thickness is higher than the uncoated mild steel. Rockwell, Brinell, and Vickers also give more hardness strength when compared with the uncoated low carbon steel. Mechanized boat hull coated with nichrome particle with increases in coating thickness gives excellent mechanical strength than the conventional uncoated low carbon steel structures. SEM suggests that the surface coated material has good bonding nature. Ni–Cr coating can be utilized on the low carbon steel mechanized boat hull structure, which reduces corrosion, increases mechanical strength, painting cost, fisherman wealth, and its lifespan. So we conclude that the Ni–Cr-coated specimen can be utilized to develop the mechanized boat hull.

References

1. NACE Int'l Co C Study (2001) Costs of corrosion control methods include services, R&D, education and training, implementation of corrosion prevention systems—A major expense for the owner/operator. National Oceanic and Atmospheric Administration (NOAA), Silver Spring
2. Oakley R, Qineti Q (2009) Corrosion testing in support of marine applications' technical presentations at the April 2009 Meeting, Marine Corrosion Forum (MCF). http://www.marinecorrosionforum.org/tpapr09.htm
3. GuedesSoares C, Garbatov Y, Zayed, A, Wang G (2009) Influence of environmental factors on corrosion of ship structures in marine atmosphere. Corros Scie 51:2014–2026. http://dx.doi.org/10.1016/j.corsci.2009.05.028
4. Al-Turaif H (2009) Surface coating properties of different shape and size pigment blends. Progress in OrganicCoatings 65:322–327
5. Deflorian F, Rossi S (2006) An EIS Study of ion diffusion through organic coatings. Electrochim Acta 51:1736–1744
6. Mohan S R, Priyadharshini S (2017) Investigation of corrosion behavior of mild steel using Al, Zn, Ni–Cr alloy coatings using plasma spray technique in IJLTEMAS, 6(4), April 2017 ISSN 2278245

Experimentation and Process Parametric Optimization of 3D Printing of ABS-Based Polymer Parts

Ramesh Raju, T. Arun Selvakumar, P. Mohammed Rizwan Ali, P. Satheesh Kumar, and D. Giridhar

Abstract The layer-by-layer addition of material technology is termed as additive or constructive manufacturing. It has gained a significant importance in latest era of manufacturing due to its simplicity. In this process, the procedure starts with generation of CAD model and ends with printing of CAD data in a 3D printer. As parts of complex and intricate geometries can be produced much easily when compared to traditional manufacturing practices, it has a wide variety of applications in the field of automotive, aerospace, consumer products and bio-medics. With an increased demand in various sectors, the consideration in terms of part quality and strength is also increasing day by day. The present work focuses on the single-objective optimization of the input parameter layer thickness, fill density and printing speed using Taguchi L9 optimization technique. The response factors are printing time, specimen thickness and surface roughness.

Keywords Fused deposition modeling · 3D printing · Polymer composite · Additive manufacturing · Taguchi's approach · Response analysis · Optimization

---

R. Raju (✉) · T. Arun Selvakumar
Department of Mechanical Engineering, Santhiram Engineering College, Nandyal, Andhra Pradesh 518501, India
e-mail: mrramesh2002@gmail.com

P. Mohammed Rizwan Ali
DesignTech Systems Ltd., Pune, India
e-mail: rizwanali.p@designtechsys.com

P. Satheesh Kumar
Defence Research and Development Organisation, Government of India, New Delhi, India

D. Giridhar
Erode Sengunthar Engineering College, Perundurai, Tamil Nadu, India
e-mail: giridhar@alumni.iim.ac.in

A. Arockiarajan et al. (eds.), *Advances in Industrial Automation and Smart Manufacturing*, Lecture Notes in Mechanical Engineering,
https://doi.org/10.1007/978-981-15-4739-3_42

1 Introduction

Additive manufacturing is the promising technique in recent times for manufacturing most complicated parts which are all used in different industrial sectors like aerospace, automobile, marine, dental and biomedical and many different sectors. The conventional techniques are not suitable to produce higher accuracy products as they always encountered with wear and tear and vibrations and so on [3, 13]. It further increases the complexities in machining and fabrication process [11]. Here in additive manufacturing methods, one should not bother about the coolants, cutting fluids, machinability and other metal working process.

The additive manufacturing can be performed using different techniques; 3D printing is one such method especially developed for printing 3D articles. This 3D printing-based additive manufacturing is performed in a progressive layer-by-layer approach, mainly through computerized controls. The geometry of the 3D article can be derived from any modeling tools and converted into electronic information source for further execution. All together, this 3D printer can be termed as modern industrial robot.

The 3D printing concepts were started evolving from 1970 onward; however, many of the options were limited and not matched by that time. Hence, the concept of rapid prototyping (RP) was utilized in the 3D printing. Based on the material type, the concept was named as stereolithography laser sintering and so on. The RP requires clear understanding on several constraints mainly the design and development aspect. Also, it requires additional information with respect to material data, experimental conditions and previous test analysis. The understanding of the process parameters is most important factor to achieve a best product: The layer thickness played such an imperative role in product quality during RP process [1].

Selective laser sintering (SLS) was successfully incorporated for printing of 3D articles using CO_2 laser [12]. The accuracy of the 3D-printed parts was selected as the major influencing factor: the shrinkage allowance factor considered one among the parameter in STL file and processed under SLS technology. The shrinkage factor had been investigated in response to process parameters and optimized for higher accuracy. The similar factor also influenced machining with fused deposition machining with FDM 1650 machining [2, 6, 10]. The Taguchi technique was made to do statistical analysis on resin-based polymerization techniques to attain higher accuracy [4].

The layer orientation played a vital role with respect to mechanical properties of the 3D-printed articles using RP techniques [5]. The article printed with zero degree layer orientation attained superior strength compared with other different layer orientations [14]. The ABS-based articles were printed using FDM-based 3D printers to achieve greater elastic property [7, 8]. The ANOVA-based optimization technique was implemented by the authors to perceive the optimal level for clear understanding. The mechanical properties of the 3D-printed nanocomposite material part were analyzed. The test details on 3D-printed articles revealed the existence of anisotropic compressive strength on most of the fabricated articles [9].

The previous study indicates that lot of work have been carried out toward optimization of 3D printer process parameter for ABS materials with respect to layer orientation and thickness. However, the work on fill density and printing speed was found very scarcely in present scenario. This paper addressed optimization of layer thickness, fill density and printing speed in a 3D printing of ABS material using Taguchi method. The experimented samples were analyzed with the response factors of printing time, specimen thickness and surface roughness.

2 Materials and Methods

Acrylonitrile butadiene styrene (ABS) is an opaque thermoplastic and amorphous polymer. "Thermoplastic" (as opposed to "thermoset") has to do with the way the material responds to heat. Thermoplastics become liquid (i.e., have a "glass transition") at a certain temperature (221 °F in the case of ABS plastic). They can be heated to their melting point, cooled and re-heated again without significant degradation. Instead of burning, thermoplastics like ABS liquefy which allows them to be easily injection molded and then subsequently recycled. By contrast, thermoset plastics can only be heated once (typically during the injection molding process). The first heating causes thermoset materials to set, resulting in a chemical change that cannot be reversed. If you tried to heat a thermoset plastic to a high temperature a second time, it would simply burn. This characteristic makes thermoset materials poor candidates for recycling. ABS is also an amorphous material meaning that it does not exhibit the ordered characteristics of crystalline solids.

2.1 Input Parameters

Varying input parameters are layer thickness, fill density and printing speed. These are varied in the Cura software. The values for the three input factors are given in Table 1. The minimum and maximum layer thickness is 0.06 and 0.4, respectively. The 3D printer we are using has 0.4-mm diameter nozzle. Fill density range is from 0 to 100%. And the range of printing speed is 20–150 mm/s. The standard L9 combinations for three input factors are given in Table 2.

Table 1 Input parameters

S. No.	Layer thickness (mm) (A)	Fill density (%) (B)	Printing speed (mm/s) (C)
1	0.1	35	30
2	0.15	50	50
3	0.2	65	70

Table 2 Standard L9 combinations for three input factors

Exp. No.	Combinations		
1	A1	B1	C1
2	A1	B2	C2
3	A1	B3	C3
4	A2	B1	C2
5	A2	B2	C3
6	A2	B3	C1
7	A3	B1	C3
8	A3	B2	C1
9	A3	B3	C2

For the above nine trails, nine ABS parts are printed, respectively, by varying the input parameters. Here, the extruder temperature is 240 C, bet temperature is 100 C and these values are constant for all the 9 experiments.

2.2 *Response Factors*

Printing Time: Time taken to print each component individually is called printing time, and it is calculated by stopwatch in seconds. Time starts when the nozzle begins from origin and ends when the part is 100% printed.

Prism Thickness: Printed part thickness is calculated by Vernier callipers of least count 0.1 mm, and the values are tabulated in a sequence manner.

Surface Roughness: It is defined as the deviations of peaks and troughs of varying heights, depths and spacing on the surfaces. If the deviations are large, the surface is rough; if they are small, the surface is smooth. Here, we have used a portable and digital roughness measuring instrument of area 0.8 mm * 5 mm.

3 Results and Discussion

3.1 *Response Factor Results*

The printed specimens and their response factors are indicated in Fig. 1 and Table 3, respectively.

Experiment 1: Printing time for this part is higher compared to all parts, because here we have given the least input values, i.e., layer thickness 0.1 mm, fill density 35%,

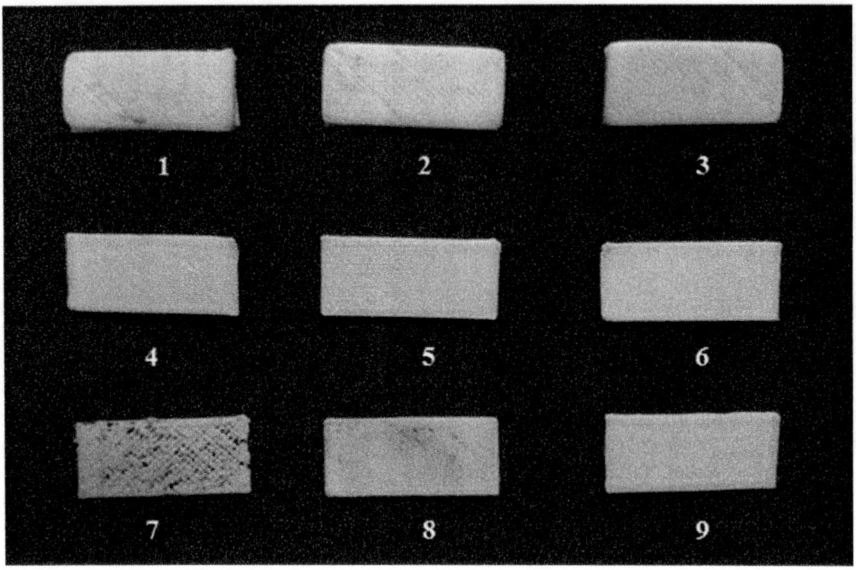

Fig. 1 ABS printed specimens

Table 3 Input and response factors

Exp. No.	Input			Response factors		
	Layer thickness (mm)	Fill density (%)	Printing speed (mm/s)	Printing time (s)	Part thickness (mm)	Surface roughness (μm)
1	0.1	35	30	513	1.9	0.004
2	0.1	50	50	399	1.8	0.003
3	0.1	65	70	341	1.7	0.004
4	0.15	35	50	302	1.9	0.005
5	0.15	50	70	268	2	0.005
6	0.15	65	30	408	1.9	0.005
7	0.2	35	70	202	1.2	0.006
8	0.2	50	50	310	2	0.005
9	0.2	65	30	257	2	0.005

printing speed 30 mm/s and the roughness 0.004. During printing, the part will get slightly peel from the hotbed of temperature 100 C.

Experiment 2: Here, again time and thickness decrease as speed increases. Again the part peels off from the bed.

Experiment 3: In this trail, thickness increases to 1.9 due to decrease in speed. Time taken to print is again decreased due to increase in layer thickness, and surface roughness increases to 0.005. No peeling in the part.

Experiment 4: Here, again time increases due to increase in speed, and part thickness increases to highest value. No peeling for this part also.

Experiment 5: In this experiment, printing time increases suddenly to 408 s due to decrease in printing speed. Part is peeling off moderately.

Experiment 6: In this experiment, printing time increases suddenly to 408 s due to decrease in printing speed. Part is peeling off moderately.

Experiment 7: This part is the thinnest of all the other due to increase in speed and decrease in fill density. Surface roughness is also high due to increase in layer thickness, i.e., 0.006. By observing this part, we can say that it is having lowest quality compared to other parts.

Experiment 8: Surface roughness decreases to 0.005, and printing time increases to 310 s due to decrease in printing speed.

Experiment 9: There is no change in roughness and thickness compared to previous part. Only the printing time decreases here. No peeling of the part from the bed.

3.2 *Taguchi Analysis: Printing Time*

Here, we analyze three input factors and one response factor at a time. In this, the response factor is printing time. It is calculated in seconds. The printing time always should be less. Hence, in options we select smaller is better. The model summary is listed in Table 4 and Fig. 2.

Regression Equation

$$\text{Printing Time (s)} = 756.4 - 1613 \text{ Layer Thickness (mm)} - 0.122 \text{ Fill Density (\%)} - 3.500 \text{ Printing Speed (mm/s)}$$

Table 4 Model summary for printing time

S	R-sq	R-sq (adj)	R-sq (pred)
22.9485	96.30%	94.07%	85.55%

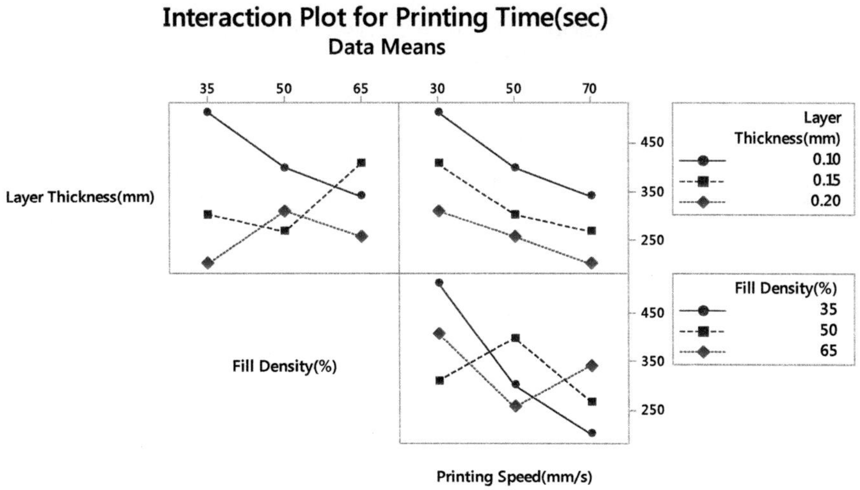

Fig. 2 Interaction plot for printing time

Table 5 Model summary for thickness

S	R-sq	R-sq (adj)	R-sq (pred)
0.250555	39.12%	2.59%	0.00%

3.3 Taguchi Analysis: Thickness

Response Table for Signal-to-Noise Ratios: For part thickness, printing speed has the highest values in S/N ratio and mean graphs having rank 1 followed by fill density and layer thickness. The optimal values are layer thickness 0.15 mm, fill density 50% and surface roughness 70 mm/s. The model summary is listed in Table 5 and Fig. 3.

Regression Equation:

$$\text{Thickness (mm)} = 1.964 - 0.67\ \text{Layer Thickness (mm)} + 0.00667\ \text{Fill Density (\%)} - 0.00750\ \text{Printing Speed (mm/s)}$$

3.4 Taguchi Analysis: Surface Roughness

Here, layer thickness has the highest values and the rank is 1 followed by fill density and surface roughness having the same rank 2.5. The optimal values are layer thickness 0.1 mm, fill density 50% and surface roughness 50 mm/s. The model summary is listed in Table 6 and Fig. 4.

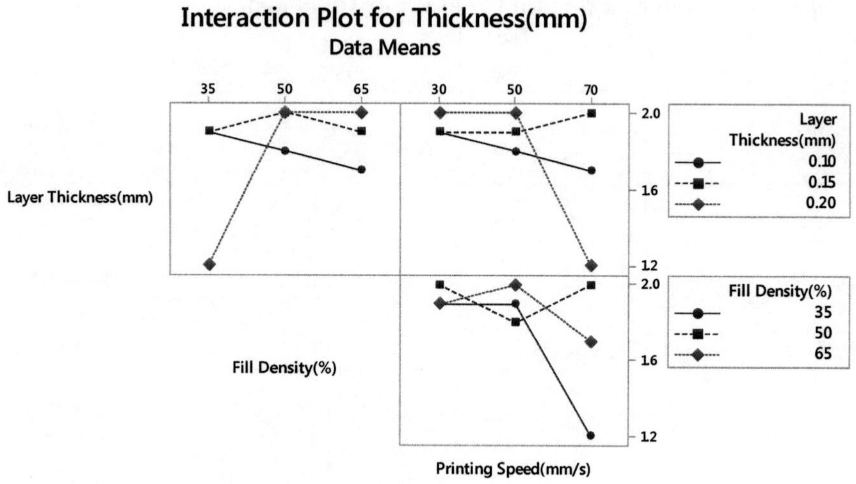

Fig. 3 Interaction plot for thickness

Table 6 Model summary for surface roughness

S	R-sq	R-sq (adj)	R-sq (pred)
0.0005477	75.00%	60.00%	30.85%

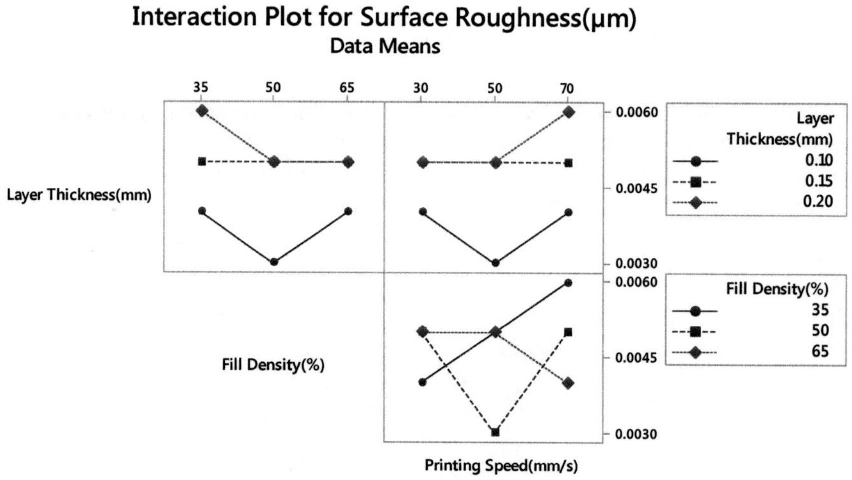

Fig. 4 Interaction plot for surface roughness

Regression Equation

$$\begin{aligned}\text{Surface Roughness } (\mu\text{m})\ \text{Surface Roughness } (\mu\text{m}) &= 0.00231\\ &+0.01667 \text{ Layer Thickness (mm)}\\ &- 0.000011 \text{ Fill Density } (\%)\\ &+ 0.000008 \text{PrintingSpeed(mm/s)}\end{aligned}$$

4 Conclusions

3D printing of ABS-based polymer material was obtained through fused deposition technique, and the process parameters were optimized using Taguchi-based technique. Based on the performance measures, it is concluded that

- Single-objective optimization for printing time: The most effecting input factor is layer thickness, and the optimal values are layer thickness 0.2 mm, fill density 35% and surface roughness 70 mm/s.
- Single-objective optimization for part thickness: The most effecting input factor is printing speed, and the optimal values are layer thickness 0.15 mm, fill density 50% and surface roughness 70 mm/s.
- Single-objective optimization for surface roughness: The most effecting input factor is layer thickness, and the optimal values are layer thickness 0.1 mm, fill density 50% and surface roughness 50 mm/s.
- The multi-objective optimization to predict the optimal process parameter for all three printing time, part thickness and surface roughness will be performed under the future scope of our project.

References

1. Anitha R, Arunachalam S, Radhakrishnan P (2001) Critical parameters influencing the quality of prototypes in fused deposition modelling. J Mater Process Technol 118(1–3):385–388. https://doi.org/10.1016/S0924-0136(01)00980-3
2. Bharath V, Dharma PN, Anshuman R, Henderson M (2000) Sensitivity of RP surface finish to process parameter variation in Solid free form fabrication proceedings. Solid free form fabrication proceedings. The University of Texas, Austin, pp 252–258
3. Bonifacio MER, Diniz AE (1994) Correlating tool wear, tool life, surface roughness and tool vibration in finish turning with coated carbide tools. Wear 173:137–144. https://doi.org/10.1016/0043-1648(94)90266-6
4. Campanelli SL, Cardano G, Giannoccaro R, Ludovic AD, Bohez ELJ (2007) Statistical analysis of stereolithographic process to improve the accuracy. Comput Aided Des 39(1):80–86. https://doi.org/10.1016/j.cad.2006.10.003

5. Es Said OS, Foyos J, Noorani R, Mandelson M, Marloth R, Pregger BA (2000) Effect of layer orientation on mechanical properties of rapid prototyped samples. Mater Manufact Process 15(1):107–122. https://doi.org/10.1080/10426910008912976
6. Górski F, Kuczko W, Wichniarek R (2013) Influence of process parameters on dimensional accuracy of parts manufactured using fused deposition modelling technology. Adv Sci Technol Res J 7(19):27–35
7. Galantucci LM, Lavecchia F, Percoco G (2008) Study of compression properties of topologically optimized FDM made structured parts. CIRP Ann 57(1):243–246. https://doi.org/10.1016/j.cirp.2008.03.009
8. Khan ZA, Lee BH, Abdullah J (2005) Optimization of rapid prototyping parameters for production of flexible ABS object. J Mater Process Technol 169(1):54–61. https://doi.org/10.1016/j.jmatprotec.2005.02.259
9. Lee CS, Kim SG, Kim HJ, Ahn SH (2007) Measurement of anisotropic compressive strength of rapid prototyping parts. J Mater Process Technol 187–188:630–637. https://doi.org/10.1016/j.jmatprotec.2006.11.095
10. Nancharaiah T, Ranga Raju D, Ramachandra Raju V (2010) An experimental investigation on surface quality and dimensional accuracy of FDM component. Int J Emerg Technol 1(2):106–111
11. Nouari M, List G, Girot F, Coupard D (2003) Experimental analysis and optimization of tool wear in dry machining of aluminium alloy. Wear 255(7–12):1359–1368. https://doi.org/10.1016/S0043-1648(03)00105-4
12. Pandey PM, Ragunath N (2007) Improving accuracy through shrinkage modelling by using Taguchi method in selective laser sintering. Int J Mach Tools Manuf 47(6):985–995. https://doi.org/10.1016/j.ijmachtools.2006.07.001
13. Rao SB (1986) Tool wear monitoring through the dynamics of stable turning. Journal of Engineering for Industry 108(3):184–190. https://doi.org/10.1115/1.3187062
14. Sood AK, Ohdar RK, Mahapatra SS (2009) Improving dimensional accuracy of processed part using grey taguchi method. J Mater Des 30(9):4243–4252

Experimental Analysis on Wire Electrical Discharge Machining of Inconel 718 Using Taguchi's Method

P. Thejasree, J. S. Binoj, P. C. Krishnamachary, N. Manikandan, and D. Palanisamy

Abstract Superalloy is a high-performance alloy exhibiting certain key features such as better mechanical and creep strength, thermal resistance, surface stability, oxidation and corrosion resistance. Inconel 718 is one among the superalloy possessing hard-to-machine material and mainly employed for greater temperature applications. However, in conventional machining methods owing to its peak strength and low thermal diffusivity superalloy results in poor machinability. Hence, to overcome the above deficiency, unconventional machining processes have been developed and presumed to be a suitable substitute technique for traditional machining methods. Among them, wire electrical discharge machining (WEDM) is a vibrant technique adopted from the principle of electrical discharge machining for machining harder-to-machine materials and also for creating complex-shaped components. The present work deals with the WEDM of nickel-based superalloy with main focus on optimizing the process parameters for machining of Inconel 718 by Taguchi's response analysis by considering independent process variables such as peak current, pulse on time and pulse off time at three different levels. Based on the results, it was noted that material removal rate (MRR) and kerf width (KW) are key performance indicators and through the experimental analysis influence on each individual process variables on the desired response parameters have been performed.

Keywords WEDM · Superalloy · Inconel 718 · Taguchi's method · Optimization

---

P. Thejasree · J. S. Binoj (✉) · P. C. Krishnamachary · N. Manikandan
Department of Mechanical Engineering, Micro Machining Research Centre,
Sree Vidyanikethan Engineering College (Autonomous), Tirupati 517102, Andhra Pradesh, India
e-mail: binojlaxman@gmail.com

D. Palanisamy
Department of Mechanical Engineering, Dr. APJ Abdul Kalam Research Centre,
Adhi College of Engineering and Technology, Kancheepuram 631605, Tamil Nadu, India

A. Arockiarajan et al. (eds.), *Advances in Industrial Automation and Smart Manufacturing*, Lecture Notes in Mechanical Engineering,
https://doi.org/10.1007/978-981-15-4739-3_43

1 Introduction

Wire electrical discharge machining (WEDM), also entitled wire spark erosion machining, is one of the supreme inventions influencing the tooling and machining industry. This is the most familiar method used to make extremely complicated parts such as turbine blades of aircraft engines, nozzles for fuel injectors and dies automatically and economically with appreciable accuracy, productivity and quality. WEDM has been employed for machining harder, corrosive and wear-resistant electrically conductive material and also for hard-to-machine work materials. The concept of WEDM is mainly adopted for making complicated shapes on the work materials which is not possible by the traditional method of material removal process and has been developed from the electrical discharge machining (EDM) [1–5]. As shown in Fig. 1, the material removal takes place in WEDM based on the electro-discharge destruction influence of electrical sparks amid electrode wire and work material that were separated with the help of a dielectric fluid.

The material removal takes place continuously because of the continuous running of electrode wire from one end to another end. The voltage has been supplied in the middle of the electrode wire and selected work material along with the existence of dielectric fluid for melting the work material surface by the discharge of sparks [6–8].

Nickel alloys are the extensively used materials than any other alloys for higher temperature claims owing to its intrinsic properties like high-strength, resisting ability to corrosion and oxidation, and their special magnetic and thermal expansion properties. Heat resistant and retaining the desired properties at high temperatures, high resistance to corrosion and high melting temperature are some imperative

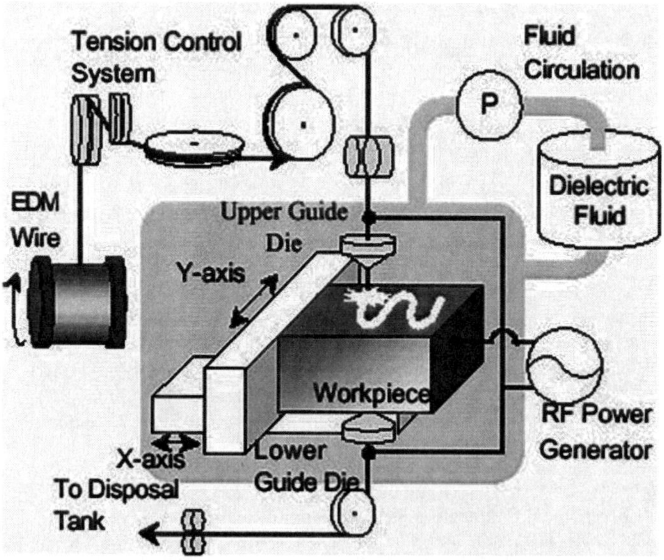

Fig. 1 Schematic of wire EDM

advantages of nickel-based superalloys [9–11]. Inconel 718 is comprehensively adopted in gas turbine blades and in chemical industries because of its better corrosion resistance properties. Furthermore, this material has the applications in petroleum processing industries as the material for heat exchanger [12]. Nickel-based superalloys accomplish a highly significant role in aerospace applications, engines for rockets, nuclear reactors, thermal power plants, submarines and high temperature applications. As these alloys are difficult to machine using the conventional methods, the nonconventional methods have been proposed to machine these alloys. The barrier of machining of these kinds of hard-to-machine materials by conventional process leads to the development of advanced machining process such as WEDM and also helps to make the materials that are not possible by conventional methods of material removal [13, 14].

Taguchi's approach has been implemented for designing the experiments and also for single-objective optimization. A multi-performance index has been derived by using appropriate method of multiple objective optimization [15]. Taguchi method has been employed for attaining better surface finish in WEDM of tool steel, it is revealed from the investigation that the adopted method provides better improvement in the desired objective, and the WEDM process has been modified for obtaining better machinability during machining H.S.S, titanium, Nimonic and aluminium alloys as work material [16–19]. In the current work, machinability studies were performed on WEDM process for evaluating the machining performance while machining Inconel 718. An effort was taken to set up the important process variables for attaining the performance individualities like material removal rate as well as kerf width of machined surfaces. Taguchi's approach has been engaged for designing the experiment trials and optimization of desired performance measures.

2 Materials and Methods

Inconel 718 alloy was notable for its specific properties like high strength and also known as corrosion-resistant nickel-based chromium material which may be employed for −423° to 1300 °F temperature applications. The age-hardenable alloy can be readily fabricated, even into complex parts that possess exceptional temperature and corrosion resistance properties. Because of its exceptional properties, the work material has broader applications in many areas. In the present study, Inconel 718 has been opted as work material (size of 25-mm diameter and 150-mm length) and it is fastened inner side of the machining chamber as illustrated in Fig. 2. Moreover, the chemical composition of Inconel 718 alloy is represented in Table 1 for highlighting the significance of the selected work material. WEDM machine (Concord Make—Model DK-7732) with reusable molybdenum wire is used in this present exploration.

By using an experimental design approach such as Taguchi, the issue of going for more number of experimental trials can be resolved. A unique design of layout has been proposed by Taguchi for conducting the experiments that is called as orthogonal array (OA) and also for examining independent process factors with the help

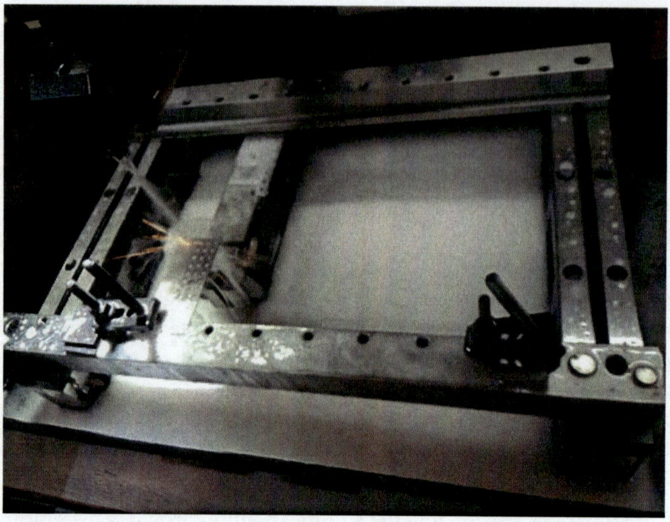

Fig. 2 Experimental set-up

Table 1 Chemical composition of Inconel 718 alloy

Element	C	Mn	Si	Ti	Al	Co	Mb	Cb	Fe	Cr	Ni
Composition (%)	0.08	0.35	0.35	0.6	0.8	1.0	3.0	5.0	17.0	19.0	52.82

Table 2 Process parameters for machining at different levels

Symbols	Variables	Levels		
		1	2	3
A	Pulse on time (μs)	10	20	30
B	Pulse off time (μs)	5	10	15
C	Peak current (A)	1	2	3

of minimum set of experimental trials. Pulse on time (μs), pulse off time (μs) and peak current are selected as input variables based on the available literature. Kerf width (KW) as well as material removal rate (MRR) are opted as performance measures in this present experimentation. The nominated independent process variables, levels and the range of those values are depicted in Table 2, and as per these values an L27 OA has been opted for wire EDM of Inconel 718.

3 Results and Discussion

For exploring the significance of input variables on desired performance characteristics of Inconel 718 in wire EDM, the experimental runs have been performed as per

L27 OA. An effort has been made to determine the best possible process variables for achieving the effective and proficient wire EDM process. In the wire EDM process, more removal rate of material and less values of kerf width are the indicators of greater performance measures. So, MRR can be treated as larger the better criterion and kerf width can be treated as smaller the better criterion.

3.1 Impact of Process Variables on Material Removal Rate

The response curve for removal of material during machining of Inconel 718 is shown in Fig. 3. From the graph, it is professed that the rate of removal of material is increasing with an enhancement of values in pulse on time and peak current as well as getting reduced with increase in pulse off time. Moreover, it is witnessed that the peak current was extremely leading process variable for removal of material. With increasing values in pulse on time and peak current, the discharge energy given to machining chamber was more which consequences in an influential bang that has the possibility of leading to increment in material removal rate. Increasing the values of pulse-on-time leads to increase for electrons directly affecting the material removal rate and thereby results in eroding more quantity of material from the surface of the sample per spark discharge.

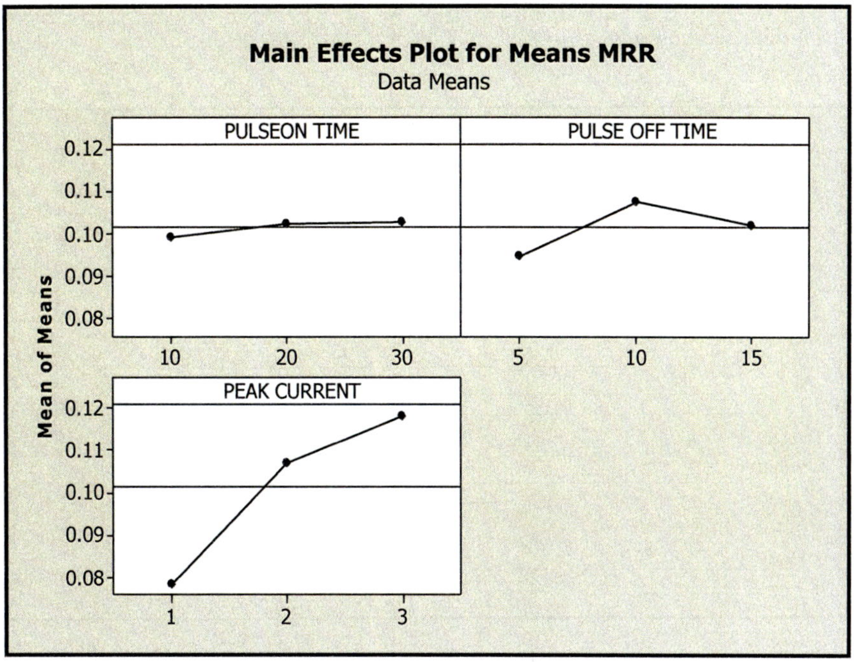

Fig. 3 Material removal rate (means) response curve

Table 3 Taguchi's response analysis for MRR—WEDM of Inconel 718

Levels	Means			S/N ratio		
	A	B	C	A	B	C
1	0.09898	0.09464	0.07835	−20.25	−20.69	−22.19
2	0.10241	0.10757	0.10733	−19.96	−19.49	−19.43
3	0.10285	0.10203	0.11856	−19.98	−19.98	−18.55
Delta	0.01293	0.01293	0.04021	0.30	1.19	3.64
Rank	3	2	1	3	2	1

Taguchi's response analysis for MRR has been accomplished, and results are presented in Table 3. The best process variables set for attaining improved material removal rate are *A*3*B*2*C*3. It means that the optimum process variable combination for better performance is 'P_{on}'—30 μs; 'P_{off}'— 10 μs; and peak current—3 A. Also, it is perceived from the investigation that the peak current is the significant process variable and then it is trailed by pulse off and pulse on time.

3.2 *Influence of Process Parameters on Kerf Width*

Graphical illustration of response graph for kerf width during wire EDM of Inconel 718 is depicted in Fig. 4.

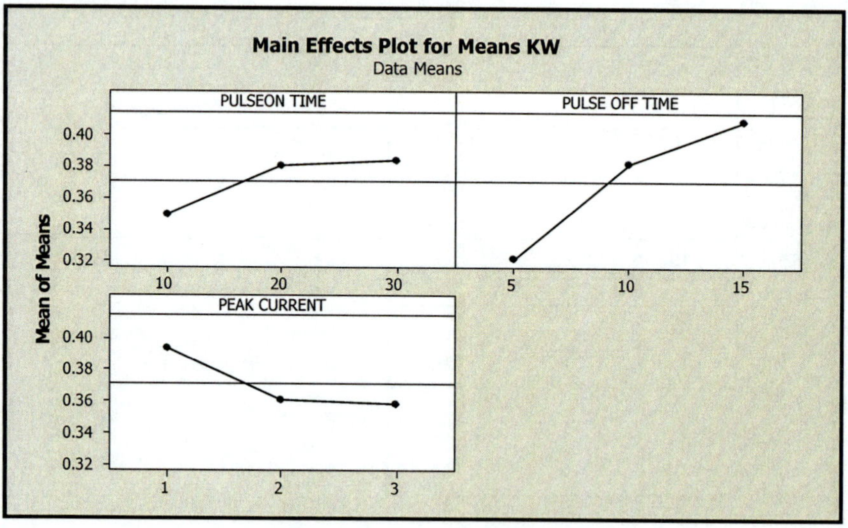

Fig. 4 Response graph for kerf width

Table 4 Response table for kerf width for Inconel 718

Levels	Means			S/N ratio		
	A	B	C	A	B	C
1	0.3489	0.3218	0.3940	9.347	10.015	8.163
2	0.3802	0.3822	0.3609	8.426	8.476	9.001
3	0.3840	0.4091	0.3582	8.523	7.804	9.132
Delta	0.0351	0.0873	0.0358	0.920	2.211	0.907
Rank	3	1	2	3	1	2

It is apparent from graph that the values of kerf width of machined surface were found increased with a rise of 'P_{on}' and 'P_{off}'. The superior energy of discharge provided crater in a greater manner that reasons for more kerf width at the machined surface. The pulse off time is deemed as the supreme persuasive process variable for kerf width.

Taguchi analysis for kerf width has been executed, and the results are presented in Table 4. The best combination of process parameters for attaining better kerf width is $A1B1C3$. It means that the optimal process factor combination for improved performance is 'Pon'—10 μs; 'Poff'—5 μs; and peak current—3 A. Also, it is witnessed from the analysis that the pulse off time is the major variable and then it is trailed by peak current and pulse on time.

4 Conclusions

Present investigation describes a single feature optimization problem of Inconel 718 while machining with wire EDM using Taguchi analysis. The kerf width (KW) and material removal rate (MRR) of machined surface were considered as the performance indicators, and the below conclusions were attained:

- The experimental runs were designed, and the individual variable's performance is optimized using Taguchi's approach.
- Effect of independent parameters on the desired performance indicators was made clear by Taguchi's analysis and Crucial indicator for attaining better performance characteristics has been determined for considered performance measures.
- The result achieved from this investigation will be used as a reference in the field of manufacturing of the products through WEDM to attain improved production rate and quality.
- The present study carried out by using Taguchi's approach is very much suitable to set up the best set of process variables to get enhanced performance in any relevant machining process.

References

1. Kanlayasiri K, Boonmung S (2007) Effects of wire-EDM machining variables on Kerf width of newly developed DC 53 die steel: design of experiments and regression model. J Mater Proc Technol 192–193:459–464
2. Davim JP (2008) Machining fundamentals and recent advances. Springer-Verlag London Limited, British Library Cataloguing in Publication Data. https://doi.org/10.1007/978-1-84800-213-5
3. Patil N, Brahmankar PK (2010) Determination of material removal rate in wire electro-discharge machining of metal matrix composites using dimensional analysis. Int J Adv Manufact Technol 51(5–8):599–610. https://doi.org/10.1007/s00170-010-2633-3
4. Huang Y, Ming W, Guo J, Zhang Z, Liu G, Li M, Zhang G (2013) Optimization of cutting conditions of YG15 on rough and finish cutting in WEDM based on statistical analyses. Int J Adv Manufact Technol 69(5–8):993–1008. https://doi.org/10.1007/s00170-013-5037-3
5. Saha P, Tarafdar D, Pal S, Saha P, Srivastava A, Das K (2009) Modeling of wire electro-discharge machining of TiC/Fe in situ metal matrix composite using normalized RBFN with enhanced k means clustering technique. Int J Adv Manufact Technol 43(1–2):107–116. https://doi.org/10.1007/s00170-008-1679-y
6. Moulton DB (1999) Wire EDM the fundamentals. EDM network, Sugar Grove, IL. www.notebookmanuals.bestmanualguide.com
7. El-Hofy H (2005) Advanced machining processes. McGraw-Hill. https://doi.org/10.1036/0071466940
8. Sommer C, Sommer S (2005) Complete EDM handbook. Advance Pub
9. Wu Q (2007) Serrated chip formation and tool-edge wear in high-speed machining of advanced aerospace materials. Utah State University, Logan, Utah
10. Saoubi RM, Outeiro JC, Chandrasekaran H, Dillon OW Jr, Jawahir IS (2008) A review of surface integrity in machining and its impact on functional performance and life of machined products. Int J Sustain Manufact 1:203–236
11. Guo YB, Li W, Jawahir IS (2009) Surface integrity characterization and prediction in machining of hardened and difficult-to-machine alloys; a state-of-the-art research review and analysis. Mach Sci Technology 13:437–470
12. Yilbas BS, Khaled M, Gondal MA, Ourfelli M, Khan Z (1999) Nano-second pulse laser treatment of Inconel 718 HT alloy-corrosion properties. Opt and Lasers Eng 32:157–172
13. Takayama Y, Makino Y, Niu Y, Uchida H (2016) The latest technology of Wire-cut EDM. Procedia CIRP 42:623–626
14. Manikandan N, Arulkirubakaran D, Palanisamy D, Raju R (2019) Influence of wire-EDM textured conventional tungsten carbide inserts in machining of aerospace materials (Ti–6Al–4 V alloy). Mater Manufact Proc 34(1):103–111
15. Yang CB, Lin CG, Chiang HL, Chen CC (2017) Single and multiobjective optimization of Inconel 718 nickel-based superalloy in the wire electrical discharge machining. Inter J Adv Manufact Technol 93(9–12):3075–3084
16. Sudhakara D, Prasanthi G (2014) Application of Taguchi method for determining optimum Kerf width in wire electric discharge machining of P/M cold worked tool steel (Vanadis-4E). Procedia Eng 97:1565–1576
17. Ramesh NN, Harinarayana K, Naik BB (2014) Machining characteristics of HSS & titanium using electro discharge sawing and wire-electrodischarge machining. Procedia Mater Sci 6:1253–1259
18. Goswami A, Kumar J (2014) Optimization in wire-cut EDM of Nimonic-80A using Taguchi's approach and utility concept. Eng Sci Technol An Int J 17(4):236–246
19. Manikandan N, Binoj JS, Varaprasad KC, Sabari SS, Raju R (2019) Investigations on wire spark erosion machining of aluminum-based metal matrix composites. In: Advances in manufacturing technology. Springer, Singapore, pp 361–369

Experimental Evaluation of Cutting Process Parameters in Cryogenic Machining of Duplex Stainless Steel

D. Narayanan, V. G. Salunkhe, V. Dhinakaran, and T. Jagadeesha

Abstract The present work deals with the influence of cryogenic coolants LN_2 delivered through holes made on flank surface and rake surface of tungsten carbide cutting tool inserts in turning of super duplex stainless steel (SDSS) using in-house developed cryogenic setup. Experiments were conducted with cryogenically treated tool, cryogenically treated tool with tempering, and cryogenic coolant directly passed through modified cutting tool insert. Results are compared with dry cutting conditions. The cutting conditions are low feed rate/high depth of cut, medium feed rate/medium depth of cut, and high feed rate/low depth of cut. The material removal rate and cutting speed are kept constant under all three cutting conditions. Microstructural study of the tool as received and cryogenically treated is examined using SEM. Population of harder tungsten carbide phase (gamma phase) is found to be more in cryogenically treated tool. Due to tempering, hardness of insert is improved by 8% which in turn increased tool life. Direct supply of LN_2 through modified cutting tool increased tool life by 23%, more than the cryogenically tempered tool. There are no appreciable changes in temperature of cutting tool under dry cutting and cryogenically treated inserts. However, there is a large difference observed in temperature of cutting tool when LN_2 is supplied through modified insert directly, which in turn yielded high tool life.

Keywords Cryogenically treated inserts · Cryogenic machining · Tool wear · Cutting temperature

D. Narayanan · T. Jagadeesha (✉)
Department of Mechanical Engineering, National Institute of Technology Calicut, Calicut, Kerala 673601, India
e-mail: jagdishsg@nitc.ac.in

V. G. Salunkhe
Department of Mechanical Engineering, ADCET, Ashta, Maharashtra 416301, India

V. Dhinakaran
Department of Mechanical Engineering,
Chennai Institute of Technology, Kundrathur, Chennai 600069, India

A. Arockiarajan et al. (eds.), *Advances in Industrial Automation and Smart Manufacturing*, Lecture Notes in Mechanical Engineering,
https://doi.org/10.1007/978-981-15-4739-3_44

Table 1 Mechanical properties of duplex stainless steel 2507

Tensile strength (MPa)	Yield strength (MPa)	Hardness (HRC)
795	550	32

1 Introduction

Machining operation for super-duplex stainless steels requires higher cutting forces and thus is the cause for tool wear at a rapid speed. Process capability of machining is one of the ways to prolong the cutting tool life and it can be achieved with better tool design and cryogenics [1–3]. Favorable chip formation is achieved with the help of cryogenic cooling. [4–7]. While comparing it was found that wet machining gave prior works that have reported that cryogenic CO_2 has substantially reduced the cutting temperature and cutting forces when compared to LN_2 coolant (cryogenic), wet cutting, and dry cutting, respectively [8–10]. This current study focuses on finding the most optimum parameter combination which gives higher tool life.

2 Experimentation

2.1 *Material Selection*

Super-duplex stainless steel (SDSS) 2507 is used as the workpiece material (Ø 36 × 300 mm) for the experimentation. Properties of SDSS are given in Table 1.

2.2 *Tool Selection and Modification*

PVD-coated tungsten carbide inserts (CNMG 120408MT12) are used for the experimentation. Tools are purchased from Taeugh Tech, Coimbatore. ISO DCLNL 2525 M12 tool holder is selected to hold tungsten carbide inserts. Holes for LN_2 delivery is created in CNMG 120408MT12 tool by using electric discharge machining (Electro cut, NIT Calicut). The rake surface of the tool was modified by making a hole having diameter of 2 mm along with depth of 2.4 mm. Modified cutting tool insert is shown in Fig. 2.

Table 2 Experimental conditions investigated

Cutting velocity (V_c) in m/min	Feed (S_o) in rev/min	Depth of cut in mm	Material removal rate in cm^3/min
113	0.35	1.2	47.460
113	0.26	1.6	47.008
113	0.21	2.0	47.460

2.3 Experimental Setup

Schematic setup of the experimental setup is shown in Fig. 3, and in-house developed system is shown in Fig. 4a. LN_2 coolant from Dewar is pushed into the in-house designed transfer device using compressed air. Details of the transfer device are shown in Fig. 4b.

The cutting temperature was measured by using K-type thermocouple, and it was recorded in PC by using data acquisition system Agilent 34972A. The tool wear was measured using 3D profilometer. Based on the initial trials and manufacturers recommendation, the cutting conditions are chosen. The details of the cutting parameters are given in Table 2. During machining, the modified cutting tool insert provided the coolant LN_2 from the rake and flank at a pressure of 2 bars in order to splatter at the tool–chip interface to achieve maximum cooling effect. The two side holes guide a small quantity of LN_2 which would reach the minor and the major significant cutting edges to provide active cooling at the interface.

2.4 Cryogenically Treated Tool Inserts

Cryogenic treatment was given to the carbide tools, by soaking at −180 °C for time duration of 24 h and then tempered in electric muffle furnace at a temperature of 245 °C for 180 min followed by furnace cooling. Hardness is measured using microhardness tester (VMT-X7) with diamond indentor. Diameter and minor load are 1 mm and 1 kg, respectively.

3 Results and Discussion

3.1 Cutting Force

Main cutting forces and feed forces were measured during the machining by using strain gauges. Figure 1 shows the strain gauges arrangement on tool holder. The

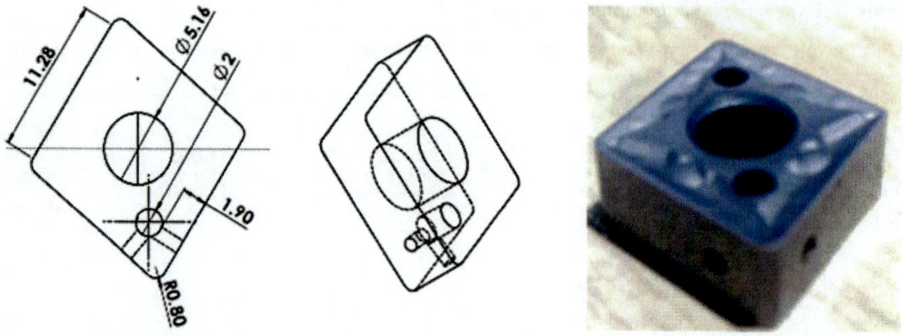

Fig. 1 Modified cutting tool insert

ascertained cutting forces depend on the shear strength of workpiece, chip shear angle, and friction between the tool and work material.

As can be seen in Table 1, various changes in cutting forces happened with a variation of depth of cut and feed rate during dry machining. As the depth of cut increased or feed rate decreased, the main cutting forces are decreased. Because the machining with high feed rate generates the high cutting zone temperature, the deduction in main cutting forces with increasing feed rate may be assigned the thermal softening. And also as decreases, the depth of the cut area of tool engagement on the primary flank is minimized. The chip–tool interface temperature for S_o = 0.35 mm/rev, t = 1.2 mm of dry machining and cryogenically treated inserts (without tempered, with tempered) was same but cutting forces are reduced in machining with cryogenically treated inserts because bond strength (sticking of particles one to another) between the particles is quite high so, high force is required to separate them their equilibrium position so, the sharpness of cutting tool still retain at high depth cut and low feed rate. So, the main cutting forces are reduced in machining with treated inserts and cryogenically treated with tempered shown effective reduction in main cutting forces compared to without tempered inserts because of tempering bond strength increased in with tempered inserts as shown in Fig. 2.

From Table 1, the S_o = 0.25 mm/rev, t = 1.6 mm and S_o = 0.25 mm/rev, t = 2 mm cutting condition of cryogenic machining generates the less cutting forces compared to dry, cryogenically treated inserts without tempered and with tempered inserts. Figures 3 and 4 show the influence of feed rate and depth of cut on main cutting forces and feed forces (Fig. 5; Table 3).

The basic aim of cryogenic (LN_2 as coolant) machining is not to cool the workpiece but to attempt to cool the chip–tool interface and flank face by delivering the cryogenic liquid nitrogen coolant through modified cutting tool inserts as shown in Fig. 1. The pressurized jet of liquid nitrogen increases the chip curl, and reducing the chip contact length yielding this reduces friction force. The reduced friction force does not hinder the flow of chip to that extent leading to less chip thickness and less chip reduction coefficient which leads to a reduction in all components of cutting forces at high depth of cut cutting condition. Further work is needed to clear up the effects of

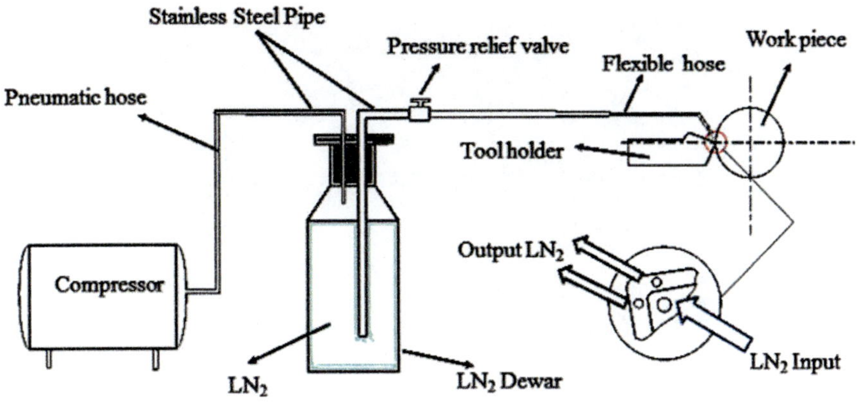

Fig. 2 Schematic representation of experimental setup

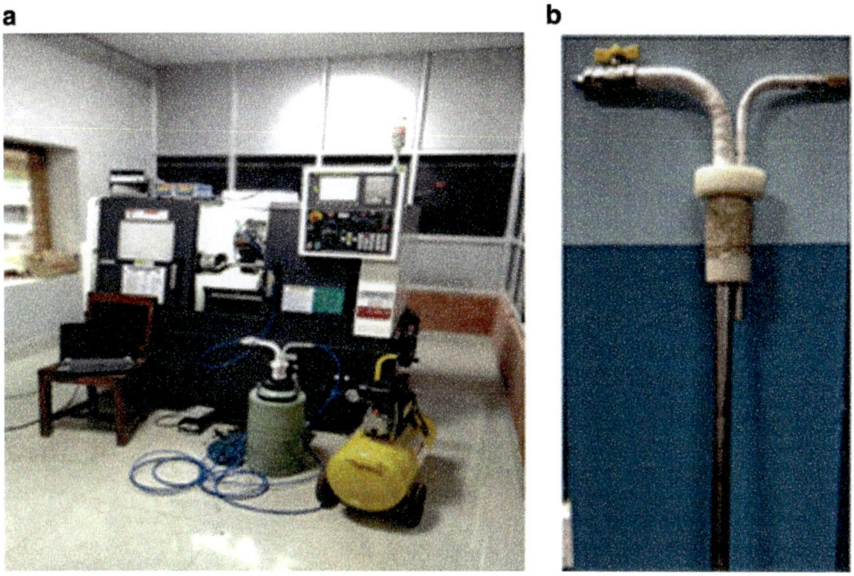

Fig. 3 **a** Experimental setup at NIT, Calicut and **b** transfer device

cryogenic LN_2 coolant on the microstructure and surface roughness of super duplex stainless steel 2507.

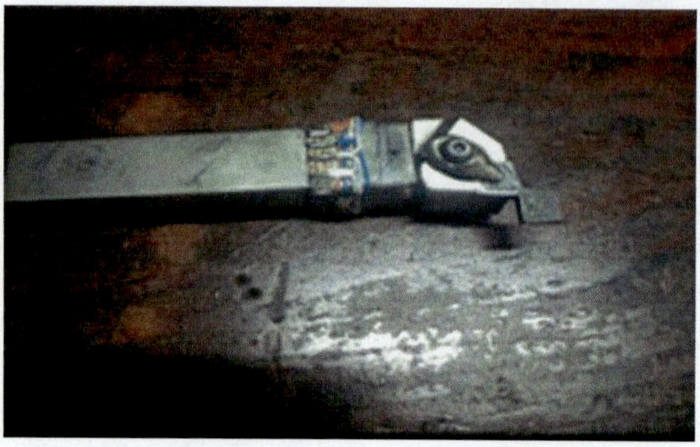

Fig. 4 Photograph showing the location of strain gauges arrangement

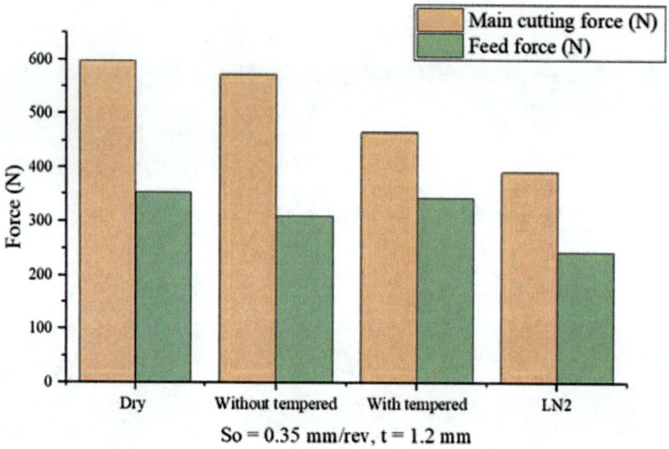

Fig. 5 Average cutting and feed forces for $N = 1000$ rpm, $S_o = 0.35$ mm/rev and $t = 1.2$ mm experimental condition

3.2 *Surface Roughness*

Any machined product's surface finishing significantly affects its quality and service life. For all machining conditions and parameters, the surface roughness (Ra) was measured for the machined surface. From the measurements, the median Ra was calculated. Figure 6 shows the impact of feed and depth of cut on roughness. The roughness of the SDSS improved dramatically by raising the feed rate. The roughness of the SDSS reduces as the depth of cut increases. When cryogenic LN_2 is applied, cutting forces are decreased more efficiently than dry cutting, cutting with cryogenically handled without tempered and tempered inserts (Fig. 7; Table 4).

The cryogenic coolant as LN_2 reaches the interface of the chip tool insert, efficiently decreasing the friction between the chip and the tool. The reduced cutting

Table 3 Main cutting force and feed force for each cutting and coolant condition

Environment conditions	Cutting speed = 1000 rpm, feed = 0.35 mm/rev, depth of cut = 1.2 mm		Cutting speed = 1000 rpm, feed = 0.25 mm/rev, depth of cut = 1.6 mm		Cutting speed = 1000 rpm, feed = 0.15 mm/rev, depth of cut = 2 mm	
	Main cutting force	Feed force	Main cutting force	Feed force	Main cutting force	Feed force
Dry machining	598.41	353.16	863.28	622.93	1152.67	961.38
Cryogenically treated without tempered	571.98	310.12	701.41	534.64	1064.38	873.09
Cryogenically treated with tempered	465.97	343.35	608.22	318.82	711.22	529.74
Cryogenic machining	392.40	245.25	480.69	284.49	544.45	259.96

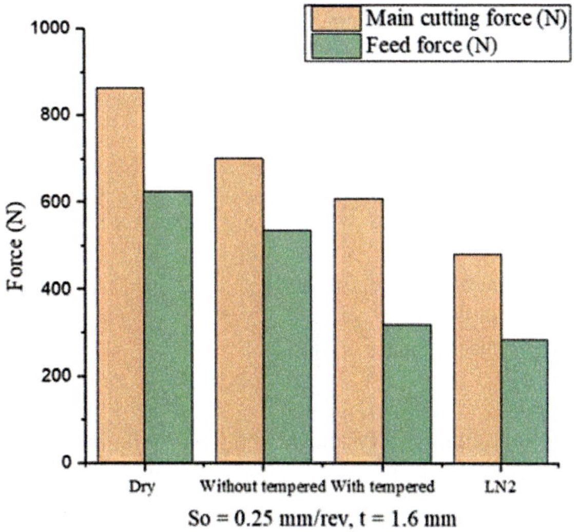

Fig. 6 Average cutting and feed forces for $N = 1000$ rpm, $S_o = 0.25$ mm/rev, and $t = 1.6$ mm experimental condition

force, therefore, results in minimal vibration resulting in less roughness values for the parts being machined. Figure 8 shows the surface roughness of super duplex stainless steel.

The surface finish is found to be better in cryogenic conditions, because the chip breakability is better during machining and less accumulation of chips near the cutting zone, and thereby frictional contact of the chips with the finished workpiece is avoided. The surface finish gets better in cryogenic LN_2 machining compared to dry

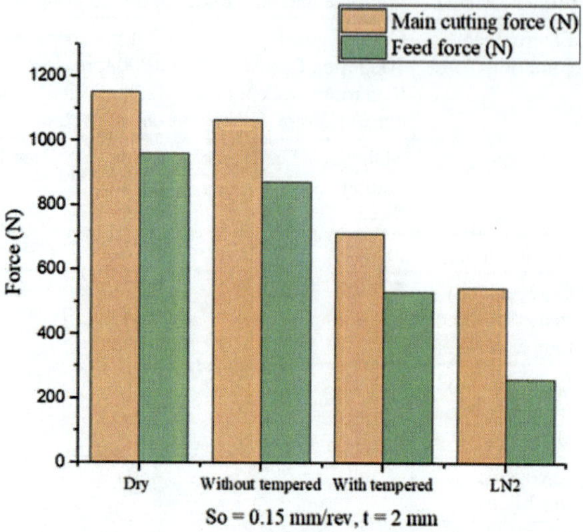

Fig. 7 Average cutting and feed forces for N = 1000 rpm, S_o = 0.15 mm/rev, and t = 2 mm experimental condition

Table 4 Surface roughness (Ra) of SDSS for each cutting and coolant condition

Surface roughness (Ra) in μm			
Environmental conditions	V_c = 114 m/min, S_o = 0.35 mm/rev, t = 1.2 mm	V_c = 114 m/min, S_o = 0.25 mm/rev, t = 1.6 mm	V_c = 114 m/min, S_o = 0.15 mm/rev, t = 2 mm
Dry machining	3.2504	2.6203	2.4560
Cryogenically treated without tempered	3.0520	2.4997	2.3645
Cryogenically treated with tempered	1.9950	1.4295	1.1336
Cryogenic machining	0.6231	0.3562	0.2939

cutting, cutting cryogenically treated without tempered and with tempered inserts around 81%, 75%, and 69%, respectively.

3.3 *Influence of Cryogenic Cooling on Chip Morphology*

One of the most important elements of machining is chip control. The chip shape generated significantly affects any machining industry's efficiency. The development of chips and their fracture in machining is very essential as it impacts the surface, the precision of the job piece, and the life of the tool. It is therefore vital to produce an appropriate type of chip with excellent chip control in machining.

In Fig. 10a, the fracture of chips is quite high because at S_o = 0.35 mm/rev and t = 1.2 mm, tool temperature is high due to high temperature and the friction between

Fig. 8 Surface roughness of SDSS at end of cutting at $N = 1000$ rpm, $S_o = 0.35$ mm/rev, and $t =$ 1.2 mm under **a** dry machining, **b** cryogenically treated without tempered, **c** with tempered, and **d** cryogenic machining

chip and tool interface is high; this results in high friction force and less in tool life. From Fig. 10b, segmented chips are formed during the cryogenic machining. The fracture of chips is less in cryogenic machining. In cryogenic machining, the LN_2 cryogenic is used as the coolant. It is more effective in removing the heat from cutting zone. This results less in cutting zone temperature and friction between chip and tool interface.

As feed rate decreases and depth of cut increases, the length of the chips increases, and from Figs. 8 and 9, continuous chips are formed during dry cutting. These continuous chips have entangled the finished part and spoiled the finished surface. In cryogenic machining, even at low feed rate discontinuous chips are formed because the pressurized liquid nitrogen is delivered through rake surface (modified tool insert) and then chips curl will increase thus causes discontinuous chips. And also, in cryogenic machining due to rapid cooling of chips, they become brittle. So, in cryogenic machining, breakability of chips is high and this causes less friction between chip–tool interface and good surface finish (Figs. 10, 11 and 12).

4 Conclusions

In-house experimental setup to machine super-duplex stainless steel is developed. Machining was done with cryogenically treated tool with/without tempering. Feed rate has more influence on cutting tool temperature compared to depth of cut under all cutting conditions. The pressurized jet of liquid nitrogen increases the chip curl, and reducing the chip contact length yielding reduces friction force. The reduced friction force does not hinder the flow of chip to that extent leading to less chip

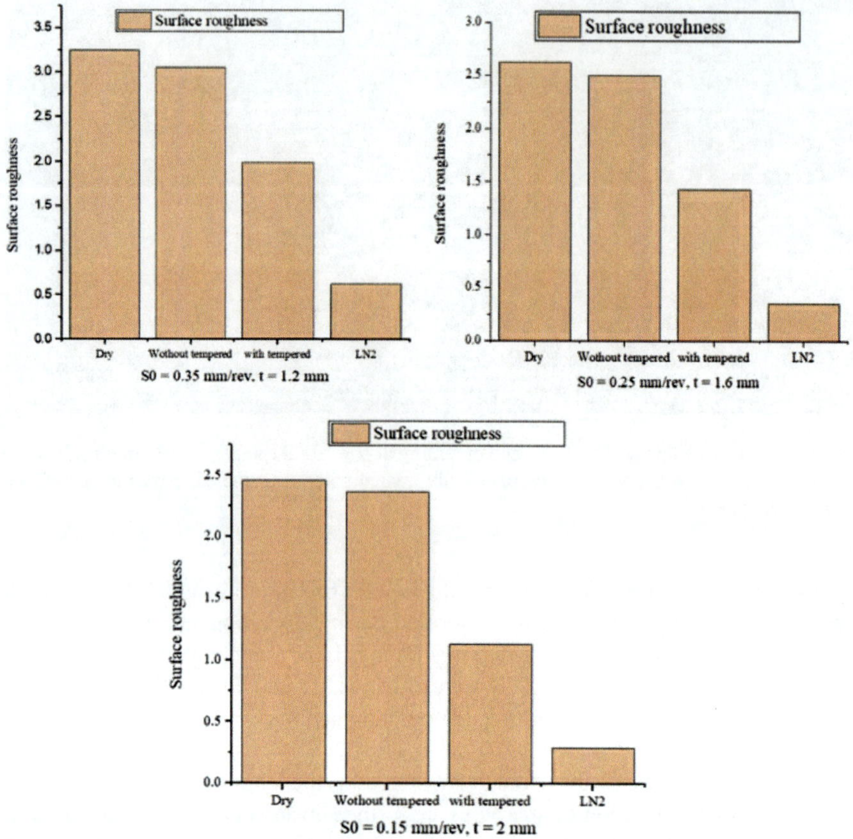

Fig. 9 Average surface roughness of SDSS for each experimental condition

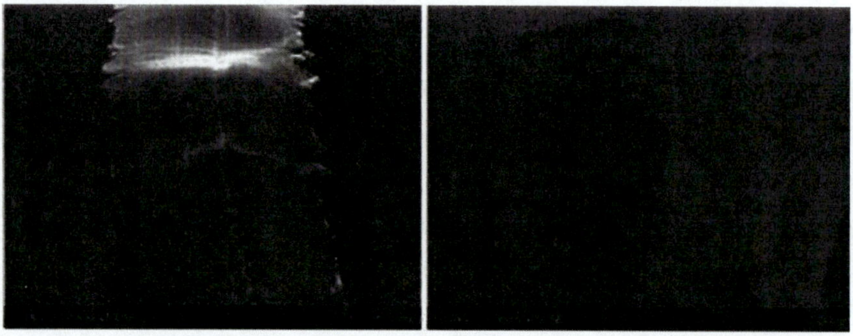

Fig. 10 Chip morphology of SDSS at end of cutting at $N = 1000$ rpm, $S_o = 0.35$ mm/rev, and $t =$ 1.2 mm under **a** dry cutting and **b** cryogenic machining

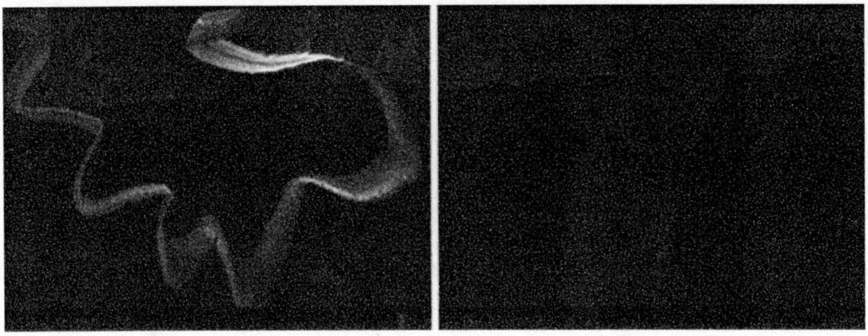

Fig. 11 Chip morphology of SDSS at end of cutting at $N = 1000$ rpm, $S_o = 0.25$ mm/rev, and $t =$ 1.6 mm under **a** dry cutting and **b** cryogenic machining

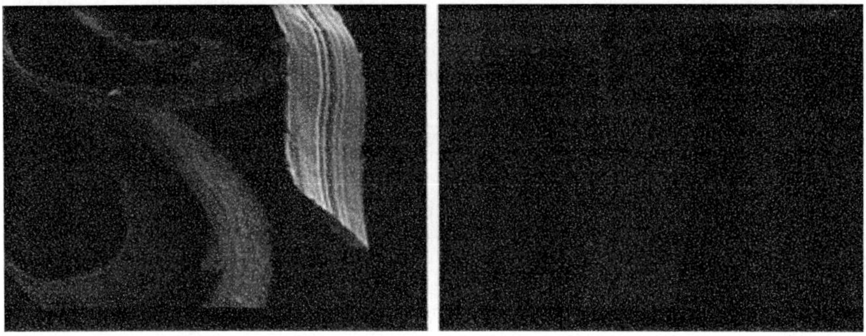

Fig. 12 Chip morphology of SDSS at end of cutting at $N = 1000$ rpm, $S_o = 0.15$ mm/rev, and $t =$ 2.0 mm under **a** dry cutting and **b** cryogenic machining

thickness and less chip reduction coefficient which leads to a reduction in all components of cutting forces at high depth of cut cutting condition. The surface finish gets better in cryogenic LN_2 machining compared to dry cutting, cutting cryogenically treated without tempered and with tempered inserts around 81%, 75%, and 69%, respectively. In cryogenic machining, breakability of chips is high and this causes less friction between chip–tool interface and good surface finish compared to dry cutting.

Acknowledgements This work has been funded by the Kerala State Council for Science, Technology, and Environment, Sanction No. KSCSTE/1452/2018-TDAP dated October 16, 2018.

References

1. Deshpande RG, Venugopal KA (2018) Machining with cryogenically treated carbide cutting tool inserts. Mater Today: Proc 5(1):1872–1878
2. Reddy TS, Ajaykumar BS, Reddy MV, Venkataram R (2007) Improvement of tool life of cryogenically treated P-30 tools. In: Proceedings of international conference on advanced materials and composites (ICAMC) at the national institute for interdisciplinary science and technology, CSIR, Trivandrum, pp 457–460
3. Yong AYL, Seah KHW, Rahman M (2007) Performance of cryogenically treated tungsten carbide tools in turning. Int J Adv Manuf Technol 46:2051–2056
4. Isik Y (2016) Using internally cooled cutting tools in the machining of difficult-to-cut materials based on Waspaloy. Adv Mech Eng 8(5):1687814016647888
5. Paul S, Dhar NR, Chattopadhyay AB (2001) Beneficial effects of cryogenic cooling over dry and wet machining on tool wear and surface finish in turning AISI 1060 steel. J Mater Process Technol 116(1):44–48
6. Dhar NR, Paul S, Chattopadhyay AB (2002) Role of cryogenic cooling on cutting temperature in turning steel. J Manuf Sci Eng 124(1):146–154
7. Bermingham MJ, Kirsch J, Sun S, Palanisamy S, Dargusch MS (2011) New observations on tool life, cutting forces and chip morphology in cryogenic machining Ti-6Al-4 V. Int J Mach Tools Manuf 51(6):500–511
8. Jerold BD, Kumar MP (2013) The influence of cryogenic coolants in machining of Ti–6Al–4 V. J Manuf Sci Eng 135(3):031005
9. Sivalingam V, Sun J, Yang B, Liu K, Raju R (2018) Machining performance and tool wear analysis on cryogenic treated insert during end milling of Ti-6Al-4 V alloy. J Manuf Proc 36:188–196
10. Arulkirubakaran D, Senthilkumar V, Dinesh S, Velmurugan C, Manikandan N, Raju R (2018) Effect of textured tools on machining of Ti-6Al-4V alloy under lubricant condition. Mater Today: Proc 5(6):14230–14236

Evaluation of Tensile Strength of Friction Welded Monel and ETP Copper Dissimilar Joints

S. Marimuthu, K. R. Balasubramanian, and T. T. M. Kannan

Abstract Friction welding is a solid-state joining process where heat is produced by scouring of two faying surfaces of materials. The dissimilar joints of Monel and ETP copper are used in boiler industry for the production of stop valves, pipe couplings, and heat exchangers. Monel is a nickel alloy having greater strength and hardness. Normally, it is difficult to weld by fusion methods due to its associated properties of heating under controlled. So that micro-microscopic particles Ni_3 (Al, Ti) are precipitated throughout matrix. Such problem can be alleviated by friction joining process. Friction welding process parameters play an important role in making quality weld joints. In this research work, friction welding of ETP copper and Monel is analyzed using design of experiments. The process parameters such as friction pressure, upset pressure, and friction time are designed by Taguchi's L_{16} orthogonal array. The significance of quality of welding process was found by analysis of variance (ANOVA). The empirical relationship is established to predict the higher ultimate tensile strength of welded joints of dissimilar materials.

Keywords Friction welding · Monel · Copper · Orthogonal array · Analysis of variance

1 Introduction

Friction welding is a strong state welding process which suggests that in this technique, warmth is not given from outside the system, and no fluid state of metal is accessible in this methodology. In this welding technique, warmth is created by precisely incited disintegration by sliding over the piece. After some time, it builds high temperature which is called plastic stage in which it is set up to be

S. Marimuthu (✉) · K. R. Balasubramanian
Department of Mechanical Engineering, National Institute of Technology, Trichy 620015, India
e-mail: smarimuthuu@yahoo.com

T. T. M. Kannan
Department of Mechanical Engineering, PRIST Deemed to Be University, Thanjavur, Tamil Nadu 613403, India

A. Arockiarajan et al. (eds.), *Advances in Industrial Automation and Smart Manufacturing*, Lecture Notes in Mechanical Engineering,
https://doi.org/10.1007/978-981-15-4739-3_45

join by applying satisfactory and basic weight on both bit of work piece. As no dissolving occurs in grating welding process, so it is excluded in combination welding process yet for the most part included in manufactured welding process. Monel is a nickel amalgams, essentially made out of nickel (up to 67%) and copper, with little extents of iron, manganese, carbon, and silicon, more grounded than encouraged nickel. Monel composites are impenetrable to utilization by various administrators, including rapidly spilling ocean water. The specific properties of Monel are high consumption protection from acids and soluble bases, high mechanical quality, great malleability, protection from salts, moderately ease, accessibility in various structures including hot and cold-moved sheets, plates, bars, and tubes. Electrolytic tough pitch copper (ETP) is an industrially unadulterated high conductivity evaluation of copper refined by electrolytic testimony which is then dissolved and oxidized to the "tough pitch" condition with controlled low oxygen content. This is the most generally utilised copper on account of its blend of electrical and heat conductivity, erosion prevention, usefulness and stylish appearance. Hot working temperature, hot formability, excellent cold working, great machinability and stress relieving. Tensile testing is a fundamental test in which a sample is subjected to a controlled tension until failure. Properties that are specifically estimated by means of a malleable test are extreme rigidity, breaking quality, most extreme prolongation and decrease in area. From these estimations, the accompanying properties can likewise be resolved.

2 Literature Survey

Mumin [1] investigated the friction welding parameters of dissimilar materials of SS and Cu. They predict the ultimate tensile strength using resource surface methodology and concluded that frictional press/frictional time have the greatest influence on the tensile strength of the joints.

Ambroziak [2] investigated the friction welded joints of niobium to steel/titanium to tungsten using interlayered behavior. They found that the use of interlayers ensures the absence of microcracking in the joint and further explained about steel + niobium and titanium + niobium using copper as interlayer and finally concluded that hard intermetallic layer is formed by optimum weld parameters.

Muralidharan et al. [3] evaluated microstructures and mechanical properties of aluminum + copper dissimilar materials by friction welding and found AA6082-T6 ALLOY and COPPER can be friction welded with optimum parameter such as forging pressure of 160Mpa and friction time of 4 s. The friction welded joints have achieved a higher tensile strength compared to the base metal.

Sathiya et al. [4] experimentally investigated friction welded joints of austenitic and ferritic stainless steels and found that the friction welded joints exhibited better properties than fusion processed joints and further found that the joint strength decreased with an increase in the friction time and hardness increases with increase in friction time.

Pavendhan et al. [5] optimized the rubbing welding parameters of disparate welding of carbon steel and tempered steel utilizing resource surface methodology (RSM); the elasticity of the joint relies upon parameters, for example, contact pressure, fashion weight, erosion time, and produce time and further discovered that the greatest rigidity of 543 MPa was gotten utilizing the ideal welding parameters, for example, grinding weight 90 MPa, upset weight 90 MPa, grating time 6 s, and manufacture time 6 s.

Sathiya et al. [6] advanced the contact welding parameters of AISI 3 0 4 by utilizing non-regular methods and artificial neural network (ANN). The relationship between information parameters, for example, warming weight, warming time, upset pressure, agitated time with the yield parameters such as elasticity and metal misfortune through ANN to advance the welding parameters GA, SA, and PSO systems were utilized in that the GA strategies beat well for the contact welding process.

Vishnu and Joseph [7] advanced the rubbing welding parameters of medium carbon steel utilizing GA for amplifying the rigidity and found that the procedure parameters significantly affect elasticity and turn speed was to have a more prominent surmising on elasticity of the joints pursued by bombshell weight and friction weight.

Sathiya et al. [8] optimized the erosion welding parameters utilizing SA to create the upgraded blaze width, streak stature, and glimmer thickness and furthermore discovered that the rate variety between the genuine and anticipated.

Anand et al. [9] anticipated machining parameters, for example, warming weight, heating time, upset pressure, and upset time, are improved with the thought of multi-reactions utilizing composite attractive quality worth, and the ideal degrees of parameters have been distinguished, and noteworthy, commitment of parameters has been controlled by ANOVA. They investigated that furious weight commitment is most extreme pursued by warming weight, warming time, and upset time.

Pandya et al. [10] examined the impact of parameters on the copy-off of erosion welded Inconel 718 and treated steel 304 utilizing Taguchi L25 exhibit method and found that rotational speed is the fundamental factor influencing the consume off length (BoL).

Anand et al. [9] predicted the erosion welding process parameters of Incoloy 800H utilizing artificial neural network (ANN) studies. ANN with GA may give great capacity to foresee the grinding welded process parameter. Distinctive ANNs, batch back propagation (BBP), quick propagation (QP), Levenburg–Marquardt back propagation (LM), and genetic algorithm (GANN) have been assessed regarding their consistency in both forward and turn around headings, and advancement of procedure parameters was completed effectively.

Shanjeevi et al. (2013) assessed the mechanical and metallurgical properties of dissimilar materials (austenitic SS + COPPER) by erosion welding through Taguchi strategy L27. They anticipated that the elasticity of the joints was fluctuated with various rotational rates 500, 1000 and 2000 RPM and further discovered that the utilization of higher grinding weight with low vexed weight expands the rigidity though with lower rubbing weight results decline in elasticity.

Mercan et al. [11] studied the impact of welding parameters on the weariness properties of AISI2205 – AISI1025 by rubbing welding and found that high RPM, low erosion weight, and low grinding occasions may expand the weakness quality.

Radoslow Winiczenko [12] presented the impact of grinding welding parameters on the elasticity and miniaturized scale auxiliary properties of unique AISI1020-ASTMA536 joints utilizing half and half RSM and hereditary calculation (GA)-based method. They found that contact time is a more grounded determinant in changing elasticity pursued by miracle power and rubbing power. Elasticity is found to diminish with an expansion in rubbing time, upset power as a negative impact on the rigidity. As agitated power builds, the rigidity diminishes.

Akbarimousavi and Goharikia [13] have explored on the mechanical properties and microstructure of divergent CP-titanium and AISI 316L austenitic treated steel constant rubbing welded joints. They found that the honesty of the weld was accomplished by small-scale hardness and tractable test and furthermore discovered that extent of rigidity of unique welded joints was beneath that of the titanium base material. This examination demonstrates that preheating of the AISI 316 stainless steel to about 700 °C is required to join the CP-titanium to AISI 316L stainless steel further demonstrates that preheating would create fragile oxide layers on the AISI 316 stainless steel. The weak layers avert the great joint interface among titanium and AISI 316LSS; anyway by smoothing the surfaces and applying the ideal produced press, the oxide layers can be expelled from the joint with blaze, and consequently, the higher quality joint can be gotten by preheating and applying satisfactory fashioned pressure.

Dey et al. [14] contemplated the joining of titanium to 304 L stainless steel by rubbing welding to create joints that are more grounded than the titanium base material as affirmed by ductile tests, and elastic disappointment happened in the titanium base material. They discovered miniaturized scale hardness profile and affirmed the event of strain solidifying of Ti close to the joint interface in the as welded condition and decrease in the impacts of strain solidifying by PWHT. They further found in the erosion test in bubbling nitric corrosive according to ASTM A-262 practice affirmed that the normal consumption rate of the joint happen.

Sahin et al. [15] dissected the rubbing welding process in connection to the welding of copper and steel bars. They led analyses under various welding conditions by methods for which metallurgical and microprobe examination of the weld cross-segments was completed. They found that the temperature variety at the interface in the spiral bearings assumes the key job on the dissemination procedure and the advancement of a heat-affected zone, which thusly influences the nature of the weld. The temperature attains its maximum value a long way from the inside yet not at the free surface, noticing that the warmth move co-effective decides the separation of the area of this max esteem from the surface. Then again, the width of the HAZ is influenced by the temperature. The HAZ is more extensive in the instance of a higher warm diffusivity area than that relating to a lower thermal diffusivity district. Further, they found that the dissemination happens because of nearby liquefying and physical blending and it is more prominent around the area where higher temps are gotten.

Ananthapadmanaban et al. [16] hypothetically and tentatively concentrated the mechanical properties of erosion welded MS to SS joints. They found the quality of the joints acquired was great and malleability was sensible. Substance structure of the steels utilized for welding assumes a significant job in choosing the properties of the weld. Likewise, discovered nearness of sulfur and phosphorous in the steel could have come about in low estimations of malleability in the weld.

Kimura et al. [17] explored crack of rubbing welded joint between unadulterated nickel and unadulterated aluminum with PWHT. Discovered joints autogenously fracture from the adjacent bit of moderate layer (interlayer) comprising of intermetallic compound (IMC) on the weld surface at high temperature, during the cooling procedure after PWHT and warmed at warming temperature for long warming time as IMC interlayer developed among Ni and Al base metals and found that the break of the joint happened at the weld interface between NiAl layer and Al base metal. They inferred that one of the fundamental purposes behind the crack joint could be as remarkable diminishing of the holding quality between NiAl interlayer and Al base metal, which was delivered with PWHT.

Sarsilmaz et al. [18] talked about the smaller scale structure and mechanical properties of Armor 500/AISI 2205 steel by grinding welding at various welding parameters, for example, grating weight and erosion times. Reinforcement and duplex SS were success completely joined by grating welding process. The utilization of higher grinding weight and friction time expands the elasticity of contact welded joints. The most astounding elasticity of grating welded couples of AISI 2205 and protection 500 was estimated at 1020 Mpa, and it was 116% of AISI 2205 base rigidity. The majority of the rubbing welded joints have higher elasticity than AISI 2205 base material. The break occurred outside transitional locale in AISI 2205 side and the cross-sectional constriction neck down to a point crack, demonstrating approximately 70% decrease in region. It is "further found that from EDS results, chromium and Fe have a high level of dissemination because of their high limit grinding co-efficient and low thermal conductivity compared to high thickness components".

All past research were performed rubbing welding process on various dissimilar material joints of copper, aluminum, steel, and composite material and furthermore led mechanical properties of welded joints [19]. But there is no previous research work on dissimilar material joints of super alloy family of Monel and copper with tensile strength analysis. This work reduces the production cycle time of weld process, cost economic, and martial loss during weld process.

3 Experimental Procedure

Unique materials of Monel K-500 (UNS NO5500) and ETP copper (ASTM C11000) with round and hollow bars of 16 mm breadth were acquired. The chemical structure of disparate materials is given in Tables 1 and 2. The rubbing welded tests are machined to a measurement of 15 mm and 100 mm length. The welded part is all around cleaned and cleaned by utilizing the media of CH3)2CO. Investigations are

Table 1 Chemical composition of Monel K-500

Element	Ni	C	Mn	Fe	S	Si	Cu	Al	Ti
%	62	0.25	1.5	2.0	0.01	0.5	29	3.0	0.85

Table 2 Chemical composition of ETP copper

Element	Cu	Bi	O	Pb
%	99.90	0.005	0.04	0.005

led, dependent on Taguchi's L 16 orthogonal array with 4 levels and 3 components, which are shown in Tables 3 and 4. Welding investigations were done on KUKA friction welding machine (Fig. 1) with consistent rotational speed of 1500 rpm given to one of the workpiece during warming stage. The disparate materials of Monel and copper-welded joint can be seen from Fig. 2 that joints were acquired effectively

Table 3 Friction welding parameters and their levels

Sr. No	Weld parameters	Level 1	Level 2	Level 3	Level 4
1	Friction pressure (bar)	20	25	30	35
2	Upset pressure (bar)	15	20	25	30
3	Friction time (s)	5	7	9	11

Table 4 L_{16} orthogonal array of friction welded parameters

Test No. friction pressure (bar)	Upset pressure (bar)	Friction time (s)
1 20	15	5
2 20	20	7
3 20	25	9
4 20	30	11
5 25	15	7
6 25	20	5
7 25	25	11
8 25	30	9
9 30	15	9
10 30	20	11
11 30	25	5
12 30	30	7
13 35	15	11
14 35	20	9
15 35	25	7
16 35	30	5

Fig. 1 KUKA friction welding machine (WRI-BHEL—Trichy)

Fig. 2 Friction welded joints of Monel K-500 and ETP copper

and streak development was seen in all cases. In this work, super-combination and copper material unique welded joints are made and pliable test is directed according to ASTM D638 standard strategy.

4 Result and Discussion

In friction welding process, heat is created between finishes of workpiece by direct change of mechanical energy into thermal energy. During relative movement of end surfaces, a lot of warmth is scattered and shapes a glimmer and length of the material gets abbreviated utilizing diverse task parameters. The tensile quality of the weld joints is recorded in Table 5 to accomplish higher quality. Grating time ought to be held as shorter as could be expected under the circumstances. Among all examples made by erosion welding, tests 13 and 16 got most minimal 154 Mpa and most noteworthy 231 MPa of elasticity esteems, respectively (Fig. 3). Low vexed weight results in lacking time for the material to warmth up, and bond quality is decreased. The bond quality is practically identical to that of parent material in Monel k-500.

Table 6 shows that friction pressure is a dominating parameter of friction welded joints of Monel and ETP copper. Main objective of higher tensile strength of dissimilar welded joints of Monel and ETP copper is achieved by minimizing the standard deviation and maximizing the S/N ratio. Results increase the friction time, increases

Table 5 Friction welding parameters of dissimilar materials

Friction pressure (bar)	Upset pressure (bar)	Friction time (bar)	Tensile strength (MPa)	S/N ratio
20	15	5	215	46.64
20	20	7	220	46.84
20	25	9	221	46.88
20	30	11	204	46.19
25	15	7	212	46.52
25	20	5	215	46.64
25	25	11	185	45.34
25	30	9	217	46.72
30	15	9	205	46.19
30	20	11	198	45.93
30	25	5	219	46.80
30	30	7	222	46.92
35	15	11	154	43.75
35	20	9	213	46.56
35	25	7	216	46.68
35	30	5	231	47.27

Fig. 3 Tensile test samples of dissimilar joint of Monel and copper

Table 6 Response table for signal-to-noise ratio (larger the best)

Level	Friction pressure	Upset pressure	Friction time
1	46.64	45.78	46.84
2	46.31	46.50	46.75
3	46.47	46.43	46.59
4	46.07	46.78	45.30
Delta	0.57	1.00	1.54
Rank	3	2	1

heat generation at the interface, and promotes the formation of brittle phase at the interface.

ANOVA technique was utilized to check sufficiency of the creating experimental relationship. In this examination, the ideal degree of certainty was taken to be 95%. This relationship is viewed as sufficient whenever determined *F* estimation of created relationship surpasses the standard organized *R* esteem for the ideal degree of certainty.

Table 7 shows the analysis of variance for tensile strength of dissimilar joints. It represents that maximum tensile strength of friction welded joints was attributed

Table 7 Analysis of variance for tensile strength, using adjusted SS for tests

Source	DF	Seq SS	Adj SS	Adj MS	*F*	*P*	% of Contribution
Friction pressure	3	289.25	289.25	96.42	1.05	0.435	5.785
Upset pressure	3	1042.2	1042.25	347.42	3.80	0.077	20.846
Friction time	3	3119.2	3119.25	1039.7	11.3	0.007	62.38
Error	6	549.00	549.00	9150	–	–	10.982
Total	15	4999.7	–	–	–	–	100

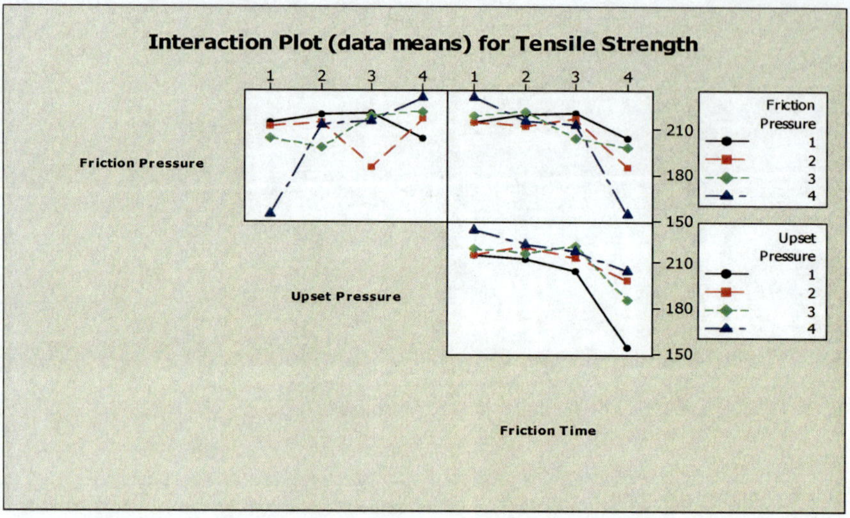

Fig. 4 Interaction plot of tensile strength of dissimilar welded joints

under the following welding conditions using F value in adjusted SS test, i.e., 35 bar of friction pressure, 30 bar of upset pressure, and 5 s of friction time for achieving higher tensile strength. During friction welding process, the strength of welds obtained with dissimilar material strongly depends upon the temperature attained by each substrate. ANOVA table represents friction time is the most dominant parameter based on percent of contribution to provide better tensile strength of dissimilar joints of Monel and ETP copper.

Figure 4 shows the interaction plot between friction welding process parameters of dissimilar materials of Monel and ETP copper. It represents that all friction welding process parameters are dependable and produce good weldability property. It also shows higher level of friction pressure, higher level of upset pressure, and low level of friction pressure.

5 Conclusion

The dissimilar materials combination of Monel K 500 and ETP copper joints are friction welded and tensile strength of weld joints are evaluated. Based on result analysis, the following conclusions are obtained.

- Friction welding has been successfully used to join Monel k-500 with ETP copper material. There is no melting, no solidification, and defects occurred. It is economically clean, no objectionable smoke, fumes, or gases generated that needed to be exhausted. Resulting joints are forged quality with 100% butt weld joints through the contact area as fast as other welding techniques.

- 35 bar of friction pressure, 30 bar of upset pressure, and 5 s of friction time are found to be the optimized parameter to obtain higher tensile strength. During friction welding process, the strength of welds obtained with dissimilar material strongly depends upon the friction time.
- All friction welding process parameters are dependable and produce good weldability property. It is obtained that higher level of friction pressure, higher level of upset pressure, and low level of friction pressure are the best option to achieve higher tensile strength between dissimilar welded joints.

References

1. Mumin S (2015) Optimizing the parameters for friction welding stainless steel to copper parts. Mater Technol 50:109–115
2. Ambroziak A (1996) Friction welding of dissimilar metal joints in niobium or titanium to steel or Tungsten. Weld Int 10(4):268–273
3. Muralidharan CH, Haribabu S, Hariprasadareddy Y, Muthupandi V, Siva Prasad K (2014) Evaluation of microstructures and Mechanical properties of dissimilar materials by friction welding. In: International conference on advances in manufacturing and materials engineering (AMME 2014), Procedia Mater Sci 5:1107–1113
4. Sathiya P, Aravindan S, Noorul Haq A (2008) Some experimental investigations on friction welded stainless steel joints. Mater Des 29:1099–1109
5. Pavendhan R, Lakshminarayanan PR, Balasubramanian V (2012) Optimization of friction weld ing process parameters for joining carbon steel and stainless steel. J Iron Steel Res, Int 29:1099–1109
6. Sathiya P, Aravindan S, Noorul Haq A, Panneerselvam K (2006) Optimization of friction welding parameters using simulated annealing. Indian J Eng Mater Sci 13:37–44
7. Vishnu PS, Joseph S (2014) Optimization of friction welding parameters for joining medium carbon steel using response surface methodology. Int J Eng Res Technol (IJERT) 3(10):338–345
8. Sathiya P, Aravindan S, Noorul Haq A, Panneerselvam K (2009) Optimaization of friction welding parameters using evolutionary computational techniques. J Mater Process Technol 209:2576–2584
9. Anand K, Barik BK, Tamil Mannan K, Sathiya P (2015) Artificial neural network modeling studies to predict the friction welding process parameters of Incoloy 800 H joints. Eng Sci Technol, Int J 1–14
10. Pandya PR, Bagesar A, Patel J (2014) Effect of welding parameters on burn-off length for friction welding of Inconel 718 and Stainless Steel 304 for production of bimetal poppet exhaust valve. Int J Sci Res Dev (IJSRD) 2(04):529–533
11. Mercan S, Aydin S, Ozdemir N (2015) Effect of welding parameters on the fatigue properties of dissimilar AISI 2205-AISI 1020 joined by friction welding. Int J Fatigue 81:78–90
12. Winiczenko R (2016) Effect of friction welding parameters on the tensile strength sand microstructural properties of dissimilar AISI 1020-ASTM A536 joints. Int J Adv Manuf Technol 84:941–955
13. Akbarimousavi SAA, Goharikia M (2011) Investigations on the mechanical properties and microstructure of dissimilar Cp-titanium and AISI 316L austenitic stainless steel continuous friction welds. Mater Des 32:3066–3075
14. Dey HC, Ashfaq M, Bhaduri AK, Prasad Rao K (2009) Joining of titanium to 304L stainless steel by friction welding. J Mater Process Technol 209:5862–5870
15. Sahin AZ, Yibas BS, Ahmed M, Nickel J (1998) Analysis of friction welding process in relation to the welding of copper and steel bars. J Mater Process Technol 82:127–136

16. Ananthapadmanaban D, Seshagiri Rao V, Abraham N, Prasad Rao K (2009) A study of mechanical properties of friction welded mild steel to stainless steel joints. Mater Des 30:2642–2646
17. Kimura M, Fuji A, Kono Y, Itoh S, Kim YC (2014) Investigations of facture for friction welded joint between pure nickel and pure aluminium with post-weld heat treatment. Mater Des 57:503–509
18. Sarsilmaz F, Kirik I, Bati S (2017) Microstructure and mechanical properties of armor 500/AISI 2205 steel joint by friction welding. J Manuf Process 28:131–136
19. Dillibabu V, Duraiselvam M, Chandrasekhar U, Raju R (2017) Microstructural studies on laser dissimilar welded Ni and steel alloys for aeronautical turbine applications. Lasers Eng 37(4–6):247–260

Micromechanical Modeling of Ferrite–Austenite Interphase of 23Cr-6Ni-3Mo Duplex Stainless Steel with an Initial Equiaxed Austenite Morphology Using Finite Element Methods

Pawan Kumar, Amit Roy Choudhury, Saroj Basantia, and Aniket Dutt

Abstract In the present work, two different regenerative volume elements (RVEs) were used for FEM analysis and modeling to study the effect of the austenite–ferrite interphase boundary on the strength of the 23Cr-6Ni-3Mo duplex stainless steel with an initial equiaxed austenite morphology. It is proposed that the method of assigning the "two noded beam elements" with suitable material properties along the ferrite–austenite interphase may provide an idea to simulate the mechanical response of duplex stainless steel within the elastic limit.

Keywords Micromechanical modeling · Duplex stainless steel · Stress distribution · Finite element method

1 Introduction

It is very difficult to perform the experiment to find out materials properties related to its microstructure. It is proposed that if it can be simulated through FEM then it will be easier to find out the mechanical response of a given microstructure. However, there is a huge challenge in simulation in case of duplex stainless steel having two distinct phases. In case of dual-phase steel (and stainless steel), the study of deformation behavior, stress and strain distributions, and failure initiation using finite element method based upon microstructure is reported by various researchers [1–4]. Some researchers also reported that duplex steel can be considered as a "composite material model" and the bulk mechanical behavior of such steel is a function of their individual phase properties like volume fractions, stiffness, and Poisson's ratio [5–8].

P. Kumar (✉) · A. Dutt
Madanapalle Institute of Technology & Science, Madanapalle, Andhra Pradesh, India
e-mail: pawanchauhan4444@gmail.com

A. R. Choudhury · S. Basantia
Indian Institute of Engineering Science and Technology, Shibpur, Shibpur, India

A. Arockiarajan et al. (eds.), *Advances in Industrial Automation and Smart Manufacturing*, Lecture Notes in Mechanical Engineering,
https://doi.org/10.1007/978-981-15-4739-3_46

However, it feels that a limited work has been done to address the deformation behavior of duplex stainless steel using finite element modeling. Schwarmet et al. have reported the deformation behavior and phase properties of austenite and ferrite phases of CF–3 and CF–8 duplex stainless steel using microstructure-based finite element models [9]. They have reported the greater mean elastic modulus of ferrite in comparison with austenite. They have also found that the higher von Mises stresses are located within ferrite, while higher von Mises strains are located in austenite (during deformation within elastic limit) [10]. In context to the above work, they have not reported the interphase properties of such steel during elastic deformation. In the present study, an initiative has been taken to find out the optimum mechanical properties along the interphase (ferrite–austenite) using FEM beam elements, which can more or less satisfy the effect of the interphase boundary on the strength of duplex stainless steel within elastic limit.

In the present work, two different regenerative volume elements (RVEs) have been used for FEM analysis and modeling to study the effect of the interphase boundary on the strength of the duplex stainless steel. However, it is suggested that a number of RVEs can be modeled and analyzed for better and more appropriate conclusion, and it will also impart the critical checking of its validation.

2 Tools and Techniques Used

2.1 *SimplewareScanIP*

It provides the processing of three-dimensional image data of magnetic resonance imaging, micro-CT, scanning electron microscopy image and optical microstructure, etc. It provides complete image visualization, its analysis and provides the quantification tools. It also has video recording features and option to export models/meshes from segmented data for CAD and 3D printing. In the present investigation, this software has been used to generate the regenerative volume elements (RVEs) for FEM analysis.

2.2 *Finite Element Method*

Finite element method of structural analysis was introduced in 1956, and over the years, it has established itself as a quite reliable method for investigating stress and strains generated within a system of bodies under various loading conditions including static and dynamic loads [9]. The analytical solutions to real-life problems, which are obtained by solving ordinary or partial differential equations, are often difficult to compute due to the complicated geometries, loadings, and material properties. Hence, numerical methods (FEA) are employed for acceptable approximate solutions. In FEA, instead of solving the problem for the entire body in one operation, the formulation of equations for each finite element is done and combines them to obtain the solution of the whole body. This method involves discretization of a continuum into a mesh of very small elements which are virtually connected at their nodes. Then, the overall stiffness matrix is computed for such a discretized body by a numerical technique. Finally, displacements at nodes are computed under the application of a load set. Stresses and strains are also computed from the displacement values by using the relationship of solid mechanics. There are certain steps in formulating a finite element analysis of a physical problem that is common to all such analyses. These steps (preprocessing, solution, and post-processing) are embodied in commercial finite element software packages and are implicitly incorporated in this work.

2.2.1 Concept of Node and Element

For analysis of a boundary value problem, a model is made graphically. Then, it is subdivided into multiple numbers of small geometric shapes. This procedure is called discretization, and the small geometric figures are called elements. An element is the basic building block of finite element analysis. There are several basic types of elements. The type of element for finite elements analysis that is used depends on the type of object that is to be modeled for finite element analysis and the type of analysis that is going to be performed. An element is a mathematical relation that defines how the degrees of freedom of a node related to the next. These elements can be lines (trusses or beams), areas (2-D or 3-D plates and membranes), or solids (bricks or tetrahedral). It also relates how the deflections create stresses. These elements are interconnected with each other at some points and some boundary and/or surfaces. These points are called nodes. Each element contains the material and geometrical properties. The material properties inside an element are assumed to be constant. The physical object can be modeled by choosing appropriate element such as frame element, plate element, shell element, and solid element.

2.3 ANSYS Software

The word ANSYS is originated from analysis system. It is an engineering simulation software. It is a comprehensive finite element (FE) analysis tool for structural analysis, including linear, nonlinear, and dynamic studies. The engineering simulation product provides a complete set of elements behavior, materials models, and equation solvers for a wide range of mechanical design problems. It also offers thermal and thermal-structural analysis. The analysis in ANSYS includes preprocessing, analysis, and post-processing of datas. In preprocessing of material using ANSYS, the nodes are selected and elements are defined. It is followed by putting the material properties and the adequate boundary conditions. The FEM analysis using ANSYS is nodal analysis, linear analysis, and deformation analysis, and the post-processing of data includes calculation of moments, forces, displacements, stresses at different contours, strain contours, and von Mises stress. In the present investigation, this software has been extensively used for computing the displacements, stresses, and strains.

3 Materials and Methods

The material under investigation was 23Cr-6Ni-3Mo duplex stainless steel with an initial equiaxed austenite morphology (Fig. 1). It was observed that the electron back-scattered diffraction (EBSD) microstructure was roughly 50% austenite and 50% ferrite. The initial microstructure was in EBSD format and it was imported to Simpleware software in order to convert it into RVEs. However, a required threshold value was applied to the phase map for capturing the individual phases (austenite and ferrite) present in the microstructure. The threshold value was applied in such a way that the phase fraction and microstructure remained unchanged. The phase fraction and morphology of individual phase remain unchanged by applying proper threshold value. The procedure was then followed by the re-dimensioning to the required dimension (6.7 mm of length and breadth) of the FE model. The size of the elements was controlled by using the software, which had helped in the conversion study of the model. This software was also used to allot the phases (as two different materials) so that at the time of FEM analysis, their respective material properties (for

Fig. 1 EBSD phase map of 23Cr-6Ni-3Mo duplex stainless steel with an initial equiaxed austenite morphology

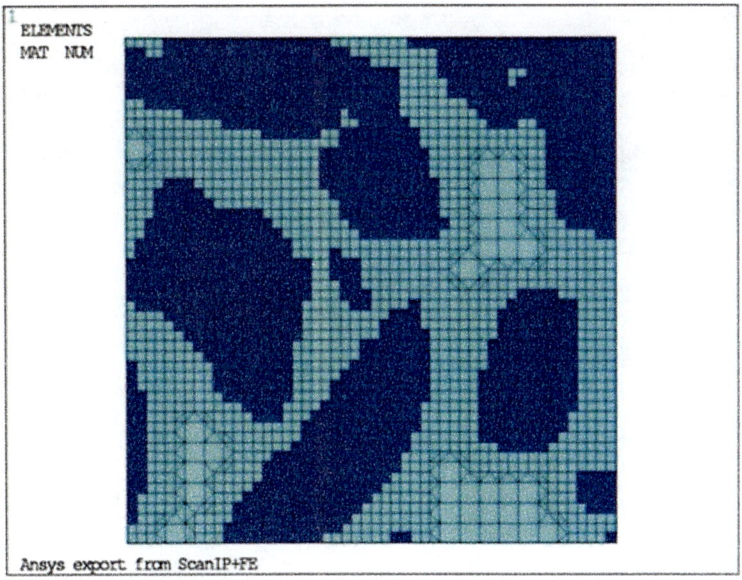

Fig. 2 Regenerative volume element-1 (RVE-1), created by using ScanIP 2016.09-SR1 image analysis software

two distinct phases) can be assigned. There were two regenerative volume elements (RVEs) in numbers, generated (with an austenite and ferrite volume % of 50 each) for present analysis as shown in Figs. 2 and 3.

The two RVEs (Figs. 2 and 3) were analyzed one by one. It is known that the Simpleware software is capable to separate out the two phases (austenite and ferrite) with a distinguished boundary and it has the capability to mesh it (to discretized RVEs into nodes and elements) for FEM analysis. Thereafter, these entire finite elements data (nodes and elements) was imported to FEM software package ANSYS. In ANSYS, static structural analysis type was chosen for analysis. When it was imported to ANSYS, we considered it as a thin slice, and the shell elements were used for deformation analysis. To model the interphase properties of duplex stainless steel, beam elements were assigned at common nodes along the interphase boundary. However, RVEs without beam have also been used for FEM analysis. When the RVEs were analyzed without the beam elements, it reciprocates the individual phase properties contrary to interphase properties. The shell elements were taken into consideration to avoid the possible mismatch of the degrees of freedom among the elements, shell, and beam (at the common nodes). A program in C language was written to find out the corresponding interphase nodes (of the two RVEs) and to generate the beam elements (having 2 nodes) along the interphase. The two noded beam elements with six degrees of freedom (three translational and three rotational) per nodes were used for stress analysis. The deformation analysis (within elastic limit) using ANSYS was performed for both the RVEs with the beam (along with the interphase) and without

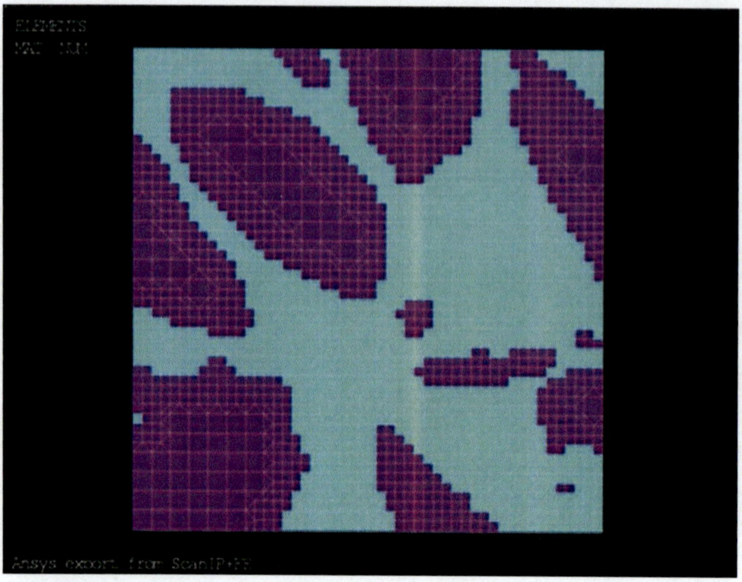

Fig. 3 Regenerative volume element-2 (RVE-2), created by using ScanIP 2016.09-SR1 image analysis software

beam. It is our proposed hypothesis that the increased yield strength for duplex steel (in comparison with single-phase austenitic or ferritic steel) can be modeled by using beam elements and assigning the definite material properties at those beam elements. For deformation analysis (within elastic limit) of the material, one end was fixed and forces (and strain) were applied at other ends. The trial and error approach has been used in the present deformation analysis. The magnitude of the force was chosen in such a way that it produced the strain of 0.2% (0.002) in RVEs (without beam). Thereafter, a greater force (of a magnitude of nearly two times the previous one) was applied to the same RVEs by assigning beam elements along the interphase. However, the mechanical properties were assigned to the beam elements in such a way that it still produced a strain of 0.2% (0.002) in the RVEs (with beam). The same procedure was repeated for strain analysis also. The FEM analysis was done and the von Mises stress and stress contours along ferrite–austenite interphase, grain boundary areas, and nodes of the geometry were observed.

4 Results and Discussion

In the present investigation, the magnitude of applied load was 500 N, and the strain applied was 0.2% (nearly 0.002) which was well within the elastic limit of the material under investigation. A stiffness of 2×10^{14} MPa was used for beam element and the thickness of the beam used was nearly 1/16th of the grain size. The stiffness of material without the beam was taken as 2×10^3 MPa. The SHELL 181 and BEAM 188 were used in present FEM modeling [11].

4.1 Von Mises Stress Distribution Using FEM in Load Control Mode

The von Mises elastic stress distribution contour plots using FEM analysis (with beam elements) showed that the higher stresses were observed close to irregular boundaries of the interphase; however, a lower magnitude of stress was associated within the interphase. As Young's modulus of beam material is considered much higher (E = 2E14 MPa) as well as due to irregularities of interphase line, stress concentration is taking place at sharp contours. On the other hand, inside the grain, Young's modulus is less than the interphase beam element. So, the lower stress was observed as shown in Figs. 4 and 5, respectively. However, the value of strain for RVE-1 and RVE-2 was 0.0019 and 0.0015, respectively (Figs. 4 and 5).

In context to the above analysis, when the same load and boundary conditions were applied on above RVEs (RVE-1 and RVE-2) without beam elements, the von Mises stress distribution was approximately uniform and the value of strain was 0.0033 for RVE-1 and 0.0028 for RVE-2 (Figs. 6 and 7, respectively). Hence, it can

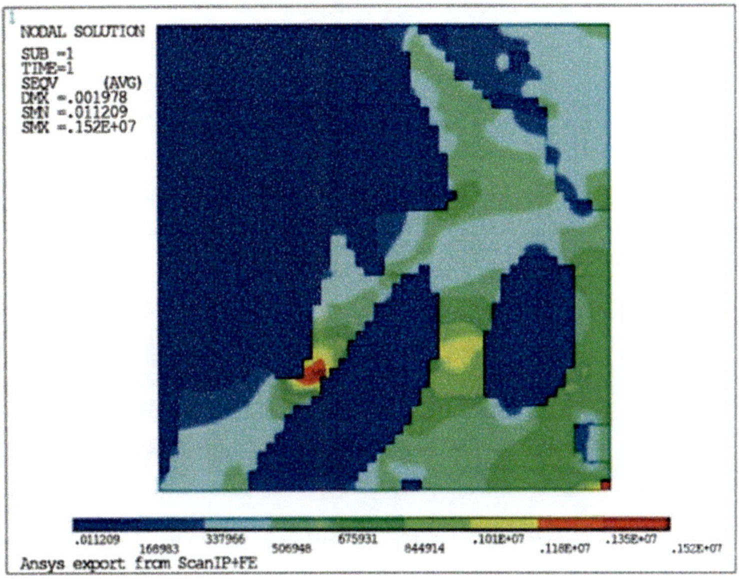

Fig. 4 Contour plots of FEM results of von Mises elastic stress (for RVE-1 with beam element) in load control mode

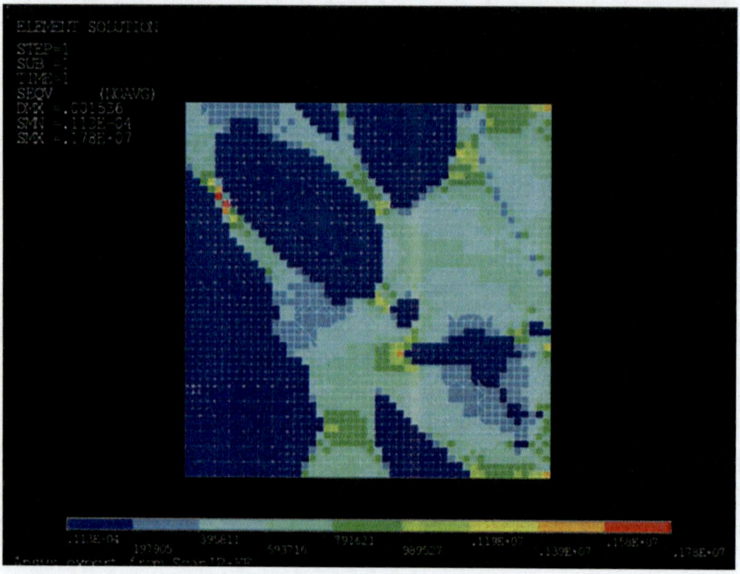

Fig. 5 Contour plots of FEM results of von Mises elastic stress (for RVE-2 with beam element) in load control mode

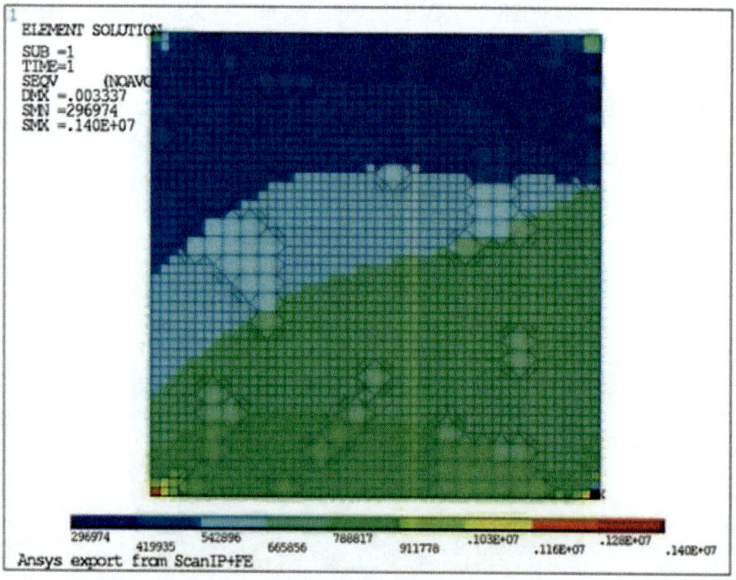

Fig. 6 Contour plots of FEM results of von Mises elastic stress (for RVE-1 without beam element) in load control mode

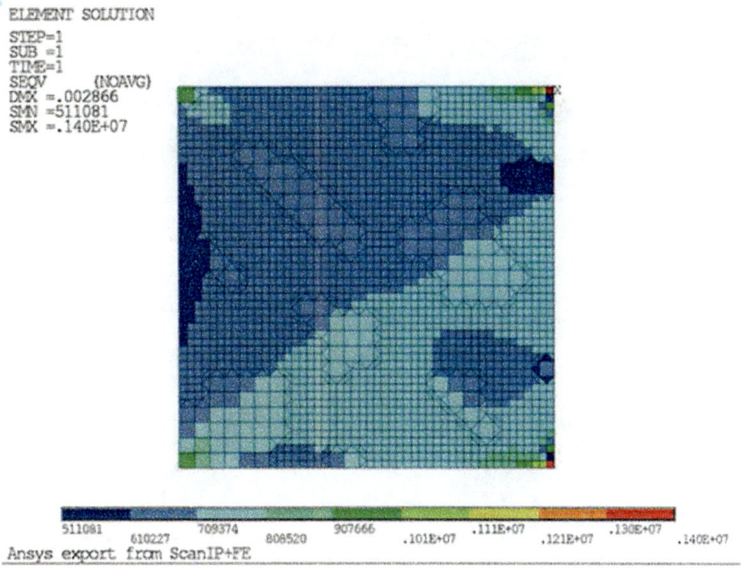

Fig. 7 Contour plots of FEM results of von Mises elastic stress (for RVE-2 without beam element) in load control mode

be said that the strain was reduced to approximately half in case of RVEs with beam element as compared to RVEs without beam element.

4.2 Von Mises Stress Distribution Using FEM in Displacement Control Mode

In case of von Mises elastic stress distribution contour plots using FEM analysis (with beam element) in displacement control, the figures depict that there is a huge stress variation location to location in the model even that varies from 0.005 to 2.22 MPa for RVE-1 (Fig. 8) and 0.01 MPa-4.5 MPa for RVE-2 (Fig. 9).

However, when the FEM analysis was done without the beam elements keeping the boundary as same as with beam, it was observed that the variation of von Mises stress distribution was minimum along most of the areas. The difference of stresses at the different location is much less and almost very close to average calculated stress value. (Stress = Force/Area). The stress variation observed from 0.6 MPa to 0.8 MPa for RVE-1 (Fig. 10) and 0.3 MPa to 1 MPa for RVE-2 (Fig. 11), only. These small variations occur due to unequal distribution of nodes, etc. But there is a considerable amount of difference in stress and strain that was observed at the locations where boundary conditions (Fixed) are put and forces were applied.

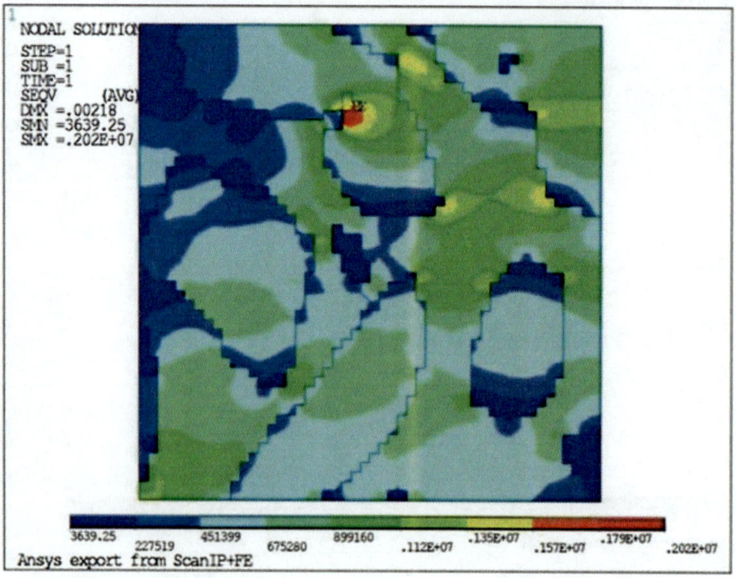

Fig. 8 Contour plots of FEM results of von Mises elastic stress (for RVE-1 with beam element) in displacement control mode

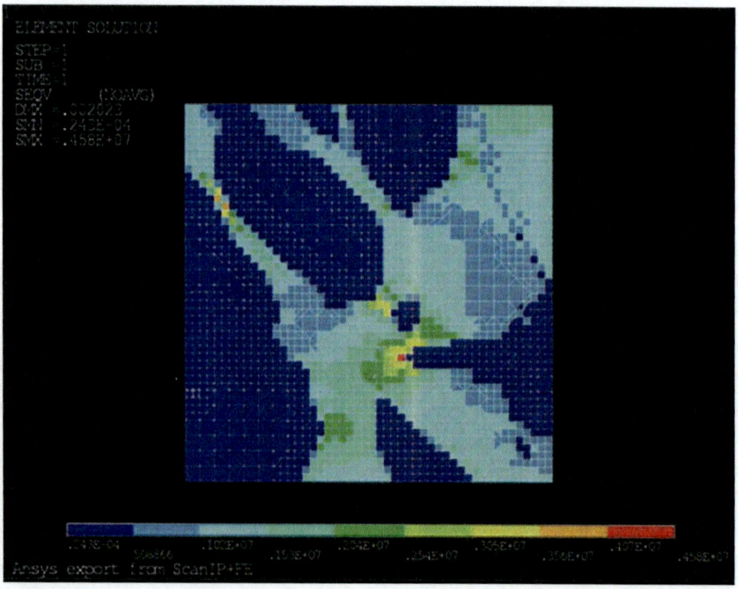

Fig. 9 Contour plots of FEM results of von Mises elastic stress (for RVE-2 with beam element) in displacement control mode

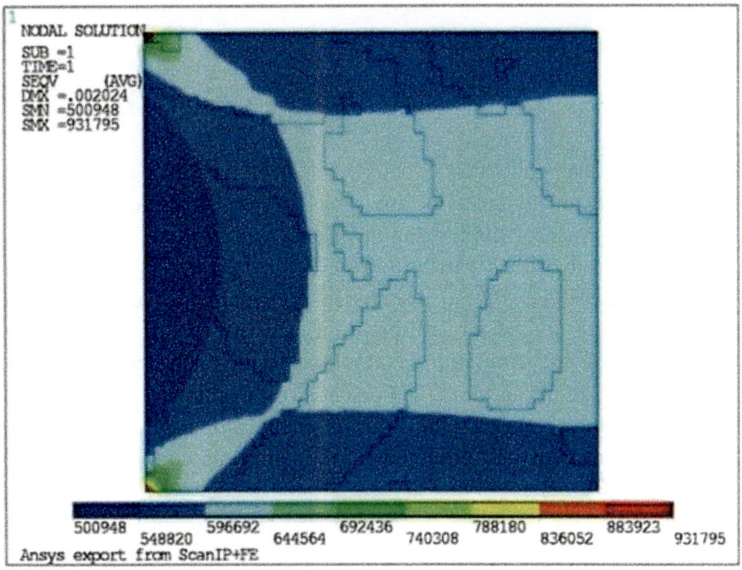

Fig. 10 Contour plots of FEM results of von Mises elastic stress (for RVE-1 without beam element) in displacement control mode

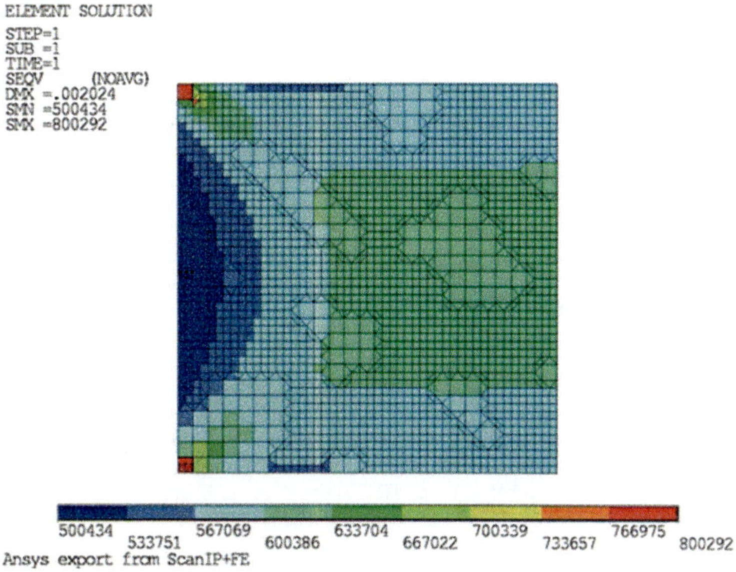

Fig. 11 Contour plots of FEM results of von Mises elastic stress (for RVE-2 without beam element) in displacement control mode

5 Conclusion

The principle premise of the present research yields specific results, which are concluded as:

The hypothesis of assigning the "two noded beam elements" with suitable material properties along the ferrite–austenite interphase may provide an idea to simulate the mechanical response of duplex stainless steel within the elastic limit. Couple of analyses, which have been carried out here, showed a very close outcome to experimental results. However, it seems that the more number of regenerative volume elements, considering numbers of different cases, can be performed to check the proper validation of FEM simulation methodology which is proposed in the present work. Therefore, it can be said that the present work could provide an idea in order to study the effect of interphase on the deformation behavior of duplex stainless steel within the elastic limits using finite element modeling methodology with modified boundary condition.

References

1. Sun X, Choi KS, Soulami A, Liu WN, Khaleel MA (2009) On key factors influencing ductile fractures of dual phase (DP) steels. Mater Sci Eng, A 526(1–2):140–149
2. Paul SK (2012) Micromechanics based modeling of dual phase steels: prediction of ductility and failure modes. Comput Mater Sci 1(56):34–42
3. Choi KS, Liu WN, Sun X, Khaleel MA (2009) Microstructure-based constitutive modeling of TRIP steel: Prediction of ductility and failure modes under different loading conditions. Acta Mater 57(8):2592–2604
4. Jeong CU, Heo YU, Choi JY, Woo W, Choi SH (2015) A study on the micromechanical behaviors of duplex stainless steel under uniaxial tension using ex-situ experimentation and the crystal plasticity finite element method. Int J Plast 1(75):22–38
5. Guo EY, Singh SS, Xie HX, Williams JJ, Jing T, Chawla N (2014) Microstructure-based modeling of deformation in steels based on constitutive relationships from micropillar compression. Steel Res Int 85(6):946–953
6. Ankem S, Margolin H, Greene CA, Neuberger BW, Oberson PG (2006) Mechanical properties of alloys consisting of two ductile phases. Prog Mater Sci 51(5):632–709
7. Jeyaprakash N, Duraiselvam M, Raju R (2018) Modelling of Cr 3 C 2–25% NiCr laser alloyed cast iron in high temperature sliding wear condition using response surface methodology. Arch Metall Mater 3(63):1303–1315
8. Choi KS, Soulami A, Liu WN, Sun X, Khaleel MA (2010) Influence of various material design parameters on deformation behaviors of TRIP steels. Comput Mater Sci 50(2):720–730
9. Schwarm SC, Kolli RP, Aydogan E, Mburu S, Ankem S (2017) Characterization of phase properties and deformation in ferritic-austenitic duplex stainless steels by nanoindentation and finite element method. Mater Sci Eng, A 5(680):359–367
10. Turner MJ (1956) Stiffness and deflection analysis of complex structures. J Aeronaut Sci 23(9):805–823
11. Documentation AN, Mechanical AP (2013) Element reference I. Element library solid, 186

Investigation on Slurry Pot Erosion Wear Behaviour of AA5083 Aluminium Alloy

K. Sasidhar Reddy, B. Sasi Prasad, E. Sai Kiran Gowd, A. Sekhar Babu, B. Santhosh Kumar, R. Manoj Kumar, and S. Baskaran

Abstract In this study, the erosion wear behaviour of AA5083 aluminium alloy is investigated by slurry pot erosion method. Various sizes of silica sand particles are used to evaluate the erosion wear behaviour of AA5083 alloy by measuring the mass loss. By employing Taguchi L9 orthogonal array, experiments were conducted and optimum process parameters are identified by analysing results of erosion test. Based on the ANOVA, the significant contribution of input parameters is identified. The dominant material removal mechanisms are observed through an optical microscope images, and reasons are discussed in detail.

Keywords Slurry erosion · AA5083 aluminium alloy · Taguchi analysis · ANOVA · Wear mechanisms

---

K. Sasidhar Reddy · B. Sasi Prasad · E. Sai Kiran Gowd · A. Sekhar Babu · B. Santhosh Kumar · R. Manoj Kumar · S. Baskaran (✉)
Department of Mechanical Engineering, Madanapalle Institute of Technology & Science, Madanapalle, AP, India
e-mail: baskaranslvm@gmail.com

K. Sasidhar Reddy
e-mail: sasi.konga@gmail.com

B. Sasi Prasad
e-mail: sasi210898@gmail.com

E. Sai Kiran Gowd
e-mail: saikiraneediga@gmail.com

A. Sekhar Babu
e-mail: babusekhar212@gmail.com

B. Santhosh Kumar
e-mail: santhoshkumarsant9@gmail.com

R. Manoj Kumar
e-mail: iitrmanoj@gmail.com

A. Arockiarajan et al. (eds.), *Advances in Industrial Automation and Smart Manufacturing*, Lecture Notes in Mechanical Engineering,
https://doi.org/10.1007/978-981-15-4739-3_47

1 Introduction

AA5083 aluminium alloy finds many applications in shipbuilding, rail cars, vehicle bodies, tip truck bodies, military vehicles, high performance vessels, mine skips and cages because of light weight, low density, good formability, better strength, high weldability and high corrosion resistance. Moreover, it is well known for cryogenic applications such as bridges, marine structure, chemical processing equipment, superconducting machinery, storage and transport tanks for cryogenic fluids, due to its exceptional performance in extreme environments [1, 2]. Even though it has good applications and better properties, AA5083 has very poor tribological behaviour which restricts the usage of the alloy in many areas. Therefore, enhancing the wear properties during sliding, reciprocating, rolling and eroding situations is very important to increase the service life of the components made by AA5083 alloy. In recent years, significant attention paid towards an improvement of erosion wear resistance due to its severity of the problems caused by this phenomenon. Kishor et al. [3] investigated the slurry erosion wear behaviour of 13Cr4Ni stainless steel after thermo-mechanical processing at 950 °C using a strain rate of 0.001 s^{-1}. They reported that thermo-mechanically processed stainless steel specimens exhibited higher slurry erosion wear resistance compare to as-received stainless steel due to the formation of refined austenite and martensite lath packets. The same authors [4] have studied the similar kind of investigation on 16Cr5Ni stainless steel using silica sand particles. The slurry erosion wear test is performed on thermo mechanically processed specimens for 24 h. They found that the erosion wear occurred by fatigue-based material removal mechanism.

In another study, Nair et al. [5] studied the slurry erosion wear behaviour of single phase $Al_{0.1}$CoCrFeNi high entropy alloy and compared the results with two structural materials such as mild steel and SS316L under similar test conditions. High entropy alloy showed that better erosion wear resistance compared to mild steel and dominant erosion wear mechanisms is found as ploughing, microcutting and severe plastic deformation for high entropy alloy, mild steel and SS316L stainless steel, respectively. Ojala et al. [6] studied the slurry erosion wear performance of wear resistant steels at 45° and 90° sample angles by employing granite and quartz as abrasives. They investigated the erosion wear performance with respect to hardness and microstructure of the steels. High-stress abrasive wear is observed as dominant mechanism when the size of the abrasive particles exceeds 1–2 mm. Lindgren and Perolainen [7] investigated the effect of erodent characteristics on titanium (ASTM Grade 2) by using different erodent such as quartz, ore, matte, concentrate and tailings through slurry pot erosion. The results are indicated that material loss by erosion occurred due to kinetic energy, size and shape of the erodents. In a similar way, the same authors [8] have investigated the erosion wear on austenitic and duplex stainless steel grades. Among the austenitic stainless steel grades tested, 316L has shown higher erosion resistance compared to 904L due to higher surface hardness which prevents shear localization. Naz et al. [9] studied surface topography of the mild steel after performing sand slurry test. The parameters such as sand size, dry sand velocity and solution stirring rate were significantly affect the surface of the mild steel. The large size sand particles damaged the surface higher compared to fine sand particles. Nguyen [10] performed the erosion experiments on stainless steel to study the effects of erodent size on erosion pattern, erosion rate, erosion mechanism and eroded profile. The results of both numerical simulation and experiments are correlated well and showed that "W" shape erodent pattern in smaller particle size and "U" shape pattern in large particle size. By considering all the recent studies, it is identified that very limited work has been published related to erosion wear behaviour non-ferrous alloys such as aluminium alloy. Therefore, the present study aimed to investigate the slurry erosion wear behaviour of AA5083 alloy using Taguchi orthogonal array method.

2 Experimental Work

AA5083 aluminium alloy is selected as a working material to investigate its erosion wear behaviour, and slurry pot erosion setup is used to carry out the experiments. Erodent particle size, rotational speed and time are selected as an important input process parameter. Based on three different values of each parameters (shown in Table 1), Taguchi L9 orthogonal array is selected to conduct the experiments. The erosion wear behaviour of AA5083 alloy is measured in terms of mass loss. Silica sand is used as an erodent with different particle sizes of 100–300 (μm), 300–600 (μm) and 600–1000 (μm). Table 1 depicts the experimental plan with results of slurry pot erosion under different experimental conditions. According to Taguchi analysis, the S/N ratio values are calculated based on smaller the better characteristics for the output response of mass loss. The minimum value of mass loss indicated that higher erosion wear resistance of the AA5083 alloy.

Figure 1 shows the complete experimental set-up of slurry pot erosion tester. Figure 1a shows the silica sand particles used in this study. The various sand particles

Table 1 Experimental plan with mass loss

E. No.	Particle size (A) (μm)	Speed (B) (rpm)	Time (C) (h)	Mass loss (g)	S/N ratio
1	100–300	2000	1	0.0157	36.0379
2	100–300	2400	2	0.0924	20.6866
3	100–300	2800	3	0.1039	19.6635
4	300–600	2000	2	0.1178	18.5756
5	300–600	2400	3	0.1498	16.4892
6	300–600	2800	1	0.0403	27.8939
7	600–1000	2000	3	0.0675	23.4139
8	600–1000	2400	1	0.0120	38.4164
9	600–1000	2800	2	0.0141	37.0156

Fig. 1 Experimental set-up **a** silica sand **b** slurry pot with baffle plates **c** stirrer with specimen holder **d** slurry pot erosion tester

size distribution is obtained by dry sieve analysis process. Figure 1b shows the slurry pot container with four vertical baffle plates which are equally spaced. The baffle plates are connected by welding and located inside of the pot container to restrict the rotational flow of slurry. The main function of the slurry pot container is to prepare the homogeneous mixing of sand and water for different particle sizes with same concentration. The slurry concentration kept constant for all the experiments are 1:3 (i.e. one part of silica sand and three parts of water). For each experimental run, fresh slurry is used to test the specimen. Figure 1c shows the stirrer with specimen holder. In order to ensure the homogeneous mixing of sand and water for different experimental combinations, the pot container has a stirrer which is rotated with the help of an electric motor. The rotational speed of a motor can be controlled by speed regulator. The specimen holder is equally spaced, and specimens are fitted at 90° sample angle to the holder by adhesive bonding. Figure 1d shows the slurry pot erosion tester. The specimens were weighed before and after for each experiment with an electronic weighing balance machine having an accuracy of 0.1 mg.

Finally, after performing the erosion test as per Table 1, the specimens are removed from the holder, cleaned in acetone and dried before weighing to calculate the mass

Table 2 Response table

Level	Particle size (μm)	Speed (rpm)	Time (h)
1	0.07071	0.06703	0.02269
2	0.10264	0.08474	0.07477
3	0.03120	0.05278	0.10709
Delta	0.07144	0.03195	0.08439
Rank	2	3	1

loss accurately. The eroded surfaces are analysed using optical microscope (OM) to identify the dominant erosion mechanism.

3 Results and Discussion

3.1 Response Table

Minitab 17 software is used to analyse the results of slurry erosion test. From the response table, the influential input parameters on output response can be identified easily by delta value and rank. Based on Table 2, it is identified that time be the most influencing parameter among three input parameters followed by particle size and rotational speed.

3.2 Main Effect Plot

In Taguchi analysis, based on selected quality characteristics such as larger the better, smaller the better and nominal the better, the optimum process parameters to achieve the better output response can be identified easily. According to erosion wear study, the mass loss should be minimum and smaller the better characteristics is chosen to calculate S/N ratio. The main effect plot for the input parameters is shown in Fig. 2. In order to obtain the minimum mass loss, the optimum experimental condition is identified by selecting the lowest level points in the main effect plot. The optimum parameter combination is noted as particle size should be high (600–1000 μm), rotational speed should be high (2800 rpm), and time should be low (1 h). From the plot, it is observed that particle size and speed follow the same pattern on mass loss (i.e. initially the values are increased with mass loss then decreased). The time has linear effect on the mass loss; i.e., irrespective of the particle size and speed, mass loss increased with increasing time [11].

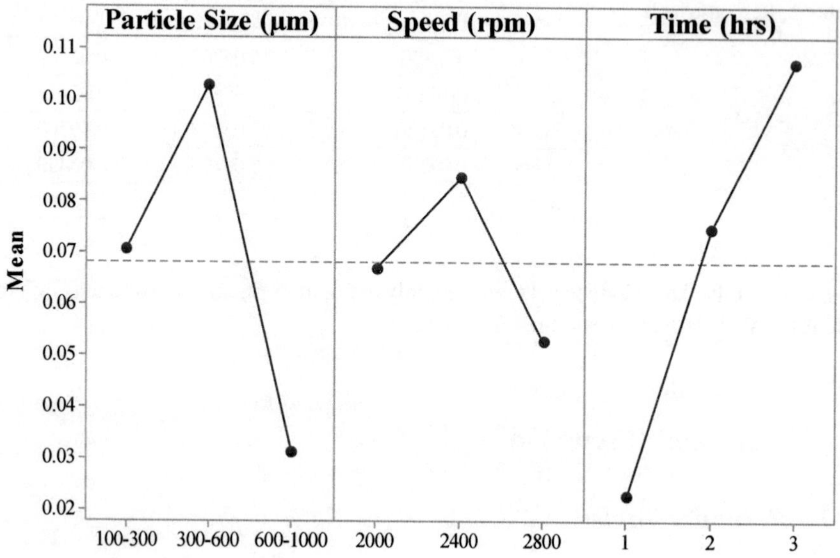

Fig. 2 Main effect plot

3.3 Interaction Plot

Figure 3 depicts the interaction effect of input parameters on the mass loss. The crossed lines are indicated that there is an interaction effect between the parameters. The interaction between particle size and speed showed that irrespective of different

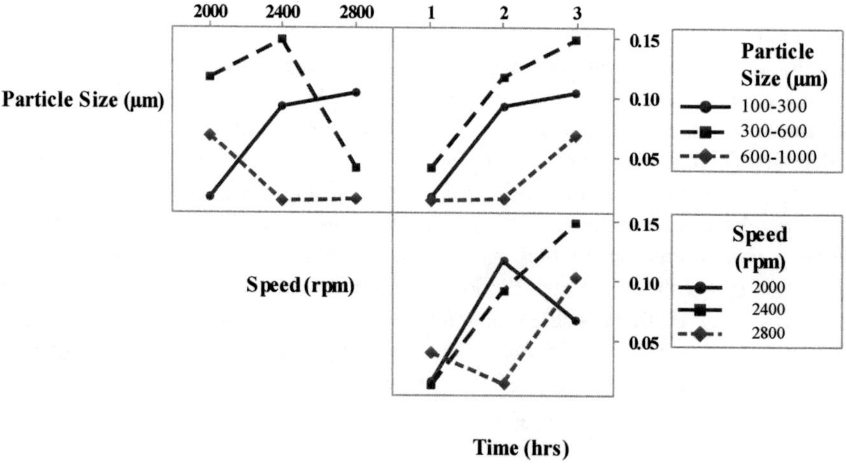

Fig. 3 Interaction plot

speeds, there is no interaction between the particle sizes of 300–600 (μm) and 600–1000 (μm). But, a significant interaction is observed between particle sizes of 100–300 (μm) and 600–1000 (μm) at lower speed. In a similar way, there is an interaction between the particle sizes of 100–300 (μm) and 300–600 (μm) at higher speed. No interaction is observed between particle size and time in all the three levels (i.e. no crossed line); both the parameters have the linear effect on the mass loss. Much significant interaction is observed between rotational speed and time at low and high levels. At low level of 1 h time, there is much interaction in all the speeds, and in the same way, significant interaction is observed at higher level of 2–3 h time.

3.4 Probability Plot

Figure 4 shows the probability plot of mass loss drawn at 95% confidence level. All the values of mass loss nearly fall on the 45° straight line which is indicated that values are normally distributed. In order to predict the mass loss (ML) within the specified level values, the regression equation is developed by using MINITAB 17 software. The plus and minus signs indicated in the equation showed the positive and negative effect on the mass loss, respectively. The multiple regression equation is as follows.

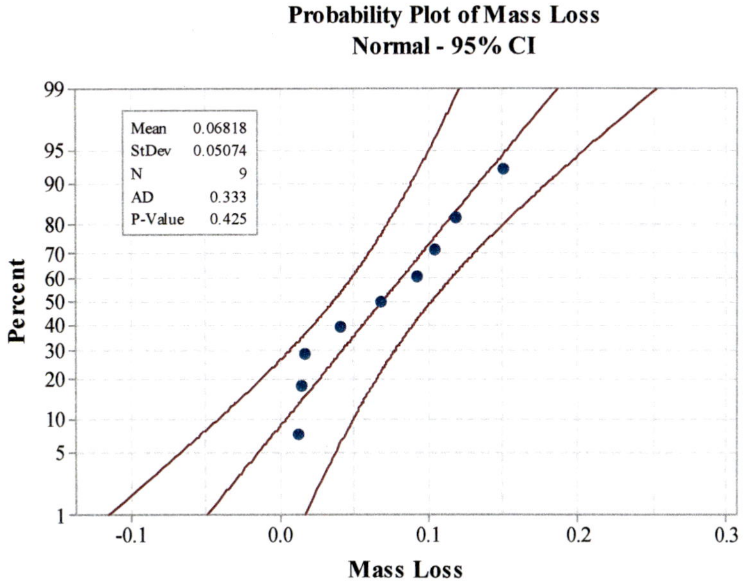

Fig. 4 Probability plot

Table 3 ANOVA

Source	Df[a]	AdjSS[b]	AdjMS[c]	F	P[d]	P (%)
A	2	0.007685	0.003842	15.47	0.061	37.31
B	2	0.001537	0.000769	3.09	0.244	7.46
C	2	0.010879	0.005439	21.89	0.044	52.82
Error	2	0.000497	0.000248			2.41
Total	8	0.020598				100

S = 0.0157626, R-sq = 97.59%, R-sq(adj) = 90.35%, where [a]degrees of freedom, [b]adjusted sums of squares, [c]adjusted mean squares and [d]probability

ML = 0.06818 + 0.00253 Particle Size (μm)_100–300 + 0.03446 Particle Size (μm)_300–600 − 0.03698 Particle Size (μm)_600–1000 − 0.00115 Speed (rpm)_2000 + 0.01655 Speed (rpm)_2400 − 0.01540 Speed (rpm)_2800 − 0.04549 Time (hrs)_1 + 0.00659 Time (hrs)_2 + 0.03890 Time (hrs)_3

3.5 ANOVA

Analysis of variance (ANOVA) results are shown in Table 3. The results are indicated that all the process parameters are significantly contributed on the mass loss. Among the three parameters, time has much significant contribution of 52.82% followed by particle size of 37.31% and rotational speed of 7.46% on the mass loss. R-sq value and error are obtained as 97.59% and 2.41%, respectively. The significant parameters are identified by observing the percentage of contribution which is more than error percentage. According to ANOVA, percentage of contribution for all the process parameters is identified as significant.

3.6 Eroded Surface Analysis

All the eroded surfaces were characterized by optical microscopy (OM) to identify the dominant material removal mechanism. Figure 5 shows the OM image of all the experimental conditions at 100× magnification. Figure 5a shows the image of eroded surface at 100–300 μm, 2000 rpm and 1 h experimental combination. The eroded surface showed very fine micro-indentations throughout surface. This is due to the silica particle size of 100–300 μm which indent surface. Few places show lips and fragments. Other than this, the eroded surface is observed smooth which yields the mass loss of 0.0157 g. Figure 5b shows eroded surface at 100–300 μm, 2400 rpm, and 2 h experimental combination. The same size of silica particles removes the material by ploughing and microcutting mechanism when increasing the speed and time to 2400 rpm and 2 h, respectively. This is attributed to the effect hardness of silica sand on AA5083 alloy. During continuous impact of the sand particles on the surface of AA5083, the sharp edges of the particles may penetrate on the specimen surface; due to this phenomenon, some places of eroded surfaces are observed by ploughing and microcutting mechanisms [3, 4, 10].

Figure 5c shows that eroded surface at 100–300 μm, 2800 rpm and 3 h. At higher rotational speed of 2800 rpm with more time (3 h), the eroded surface damaged little bit more amount of lips in the form of plastic deformation and microcutting mechanisms. In addition to that, few places shown that multiple cracks on the surface. Over the time, crack could propagate and join together and finally detach from the surface. Compared to the lower speed and lower time (i.e. 2000 rpm and 1 h), the silica sand particle size of 100–300 μm produced more loss of 0.1039 g at higher speed and more time. Figure 5d shows eroded surface at 300–600 μm, 2000 rpm and 2 h experimental combination. Deeper pit-like indentations along with micro-indentations are observed on the eroded surface of AA5083 alloy which indicates that penetration of hard silica particles mostly at normal to the surface. In few places, due to the heavy impact of the particles, deeper spherical craters also formed. Figure 5e shows that eroded surface at 300–600 μm, 2400 rpm and 3 h experimental combination. The observed material removal mechanisms are lips, fragments, craters and microcrack.

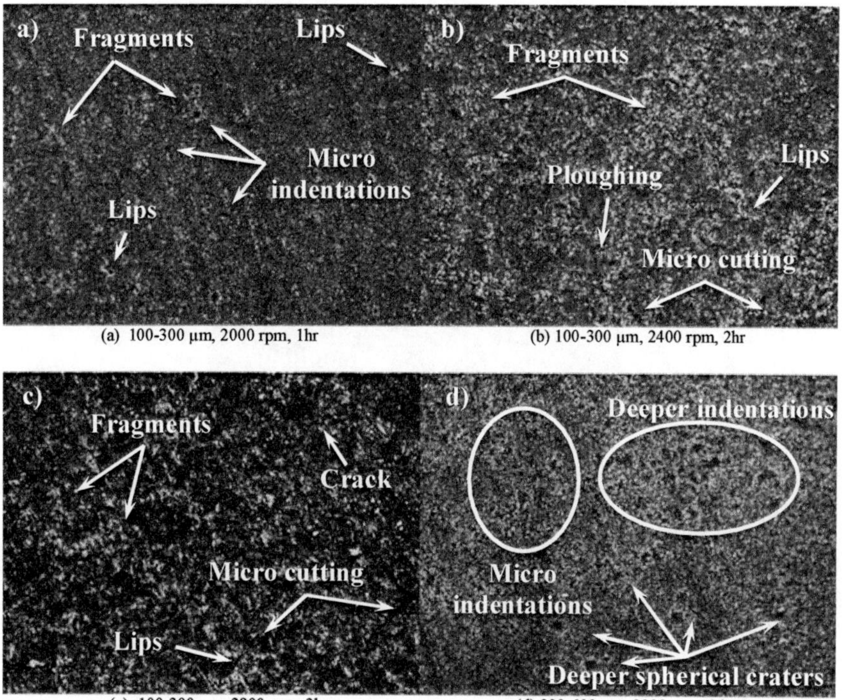

Fig. 5 OM image of eroded surfaces at different experimental conditions

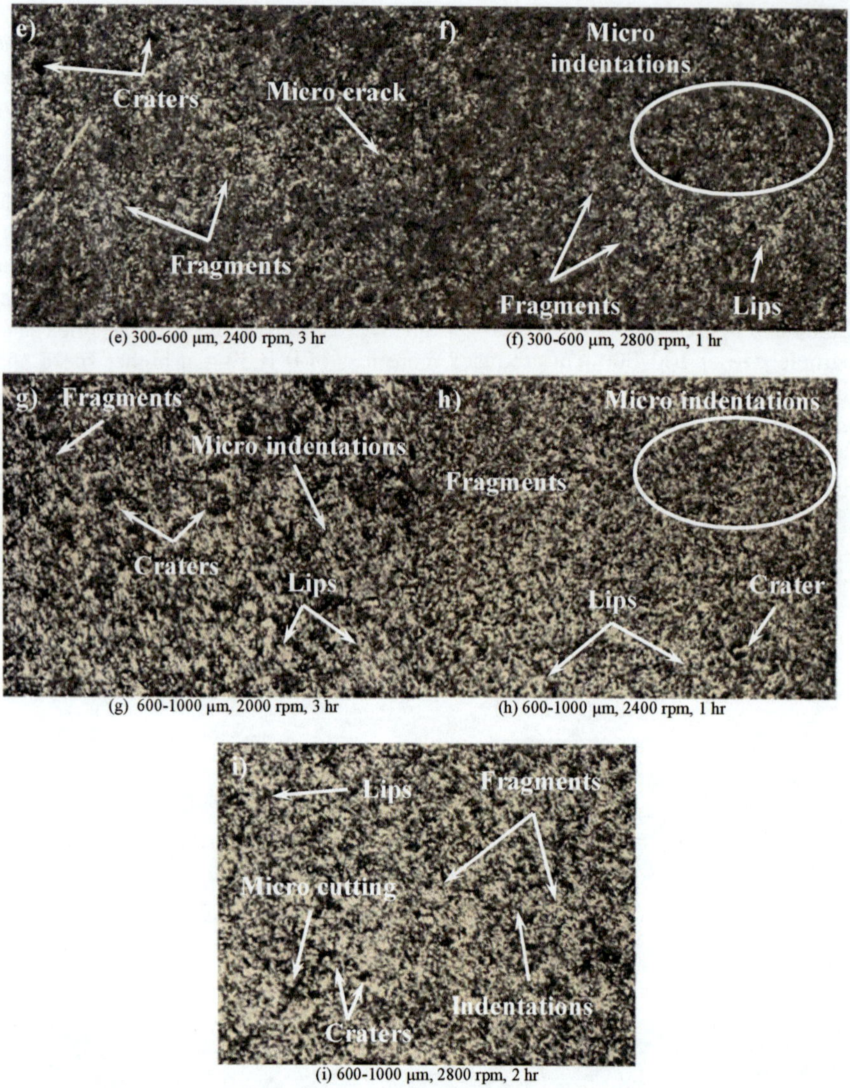

(e) 300-600 μm, 2400 rpm, 3 hr

(f) 300-600 μm, 2800 rpm, 1 hr

(g) 600-1000 μm, 2000 rpm, 3 hr

(h) 600-1000 μm, 2400 rpm, 1 hr

(i) 600-1000 μm, 2800 rpm, 2 hr

Fig. 5 (continued)

No much difference is observed on eroded surface (shown in Fig. 5f) at 300–600 μm, 2800 rpm and 1 h experimental combination compared to Fig. 5e.

Only very few differences are observed in the eroded surfaces of 600–1000 μm size silica particle which is shown in Fig. 5g–i. All the eroded surfaces are heavily affected by lips, indentations, fragments and craters. Continuous impact of the large size (600–1000 μm) sand particles resulted in surface protrusions (lips), which confirms the occurrence of plastic deformation. This phenomenon mainly depends on the AA5083 alloy properties such as hardness, strength and wear resistance [5, 6]. AA5083 has low hardness and moderate strength, but good formability compared to silica sand. Due to the properties of the alloy and the particle trajectory during the rotation, the surface is affected by the impact of the particles which causes the indentations and deeper craters on the surface [7, 8]. By observing all the eroded surfaces, the dominant erosion wear mechanisms are identified as plastic deformation, indentations, craters, ploughing, microcutting and microcracks.

4 Conclusion

The slurry erosion wear behaviour of AA5083 alloy was investigated by employing Taguchi method. From the main effect plot, the optimum process parameter was identified to obtain minimum mass loss as particle size (600–1000 μm), rotational speed (2800 rpm) and time (1 h). Among the three input parameters, time has the strongest influence on mass loss and the percentage of contribution was calculated as 52.82% by ANOVA. By observing all the eroded surfaces, the dominant erosion wear mechanisms were identified as plastic deformation, indentations, craters, ploughing, microcutting and microcracks.

References

1. Huang C, Wu Z, Huang R, Wang W, Li L (2017) Mechanical properties of AA5083 in different tempers at low temperatures. In: IOP conference series: materials science and engineering, vol 279, no 1, p 012002. IOP Publishing
2. Jun L, Tan M-J, Castagne S (2010) Formability in AA5083 and AA6061 alloys for light weight applications. Mater Des 31:S66–S70
3. Kishor Brij, Chaudhari GP, Nath SK (2016) Slurry erosion of thermo-mechanically processed 13Cr4Ni stainless steel. Tribol Int 93:50–57
4. Kishor Brij, Chaudhari GP, Nath SK (2018) Slurry erosion behaviour of thermomechanically treated 16Cr5Ni stainless steel. Tribol Int 119:411–418
5. Nair RB, Karthikeyan Selvam, Arora HS, Sundeep Mukherjee, Singh H, Grewal HS (2017) Slurry erosion behavior of high entropy alloys. Wear 386:230–238
6. Niko O, Valtonen K, Antikainen A, Kemppainen A, Minkkinen J, Oja O, Kuokkala V-T (2016) Wear performance of quenched wear resistant steels in abrasive slurry erosion. Wear 354:21–31
7. Lindgren M, Perolainen J (2014) Slurry pot investigation of the influence of erodant characteristics on the erosion resistance of titanium. Wear 321:64–69
8. Lindgren M, Perolainen J (2014) Slurry pot investigation of the influence of erodent characteristics on the erosion resistance of austenitic and duplex stainless steel grades. Wear 319:38–48
9. Naz MY, Sulaiman SA, Shukrullah S, Ghaffar A, Ibrahim KA, AbdEl-Salam NM (2017) Development of erosion-corrosion mechanisms for the study of steel surface behavior in a sand slurry. Measurement 106:203–210

10. Nguyen VB, Nguyen QB, Zhang YW, Lim CYH, Khoo BC (2016) Effect of particle size on erosion characteristics. Wear 348:126–137
11. Dube Narendra M, Anirudh Dube, Veeregowda DH, Iyer SB (2009) Experimental technique to analyse the slurry erosion wear due to turbulence. Wear 267:259–263

Tribological Behaviour and Electric Discharge Drilling of Duplex Silicon Metal Matrix

T. Vishnu Vardhan, S. Marichamy, B. Stalin, J. Vairamuthu, and V. Dhinakaran

Abstract Silicon metal matrix composites are extensively used in many industries such as automobile, aerospace, marine and mineral processing industries due to their enhanced wear properties. This experimental investigation deals with tribological behaviour and electric discharge drilling (EDD) of the silicon metal matrix. The duplex silicon metal matrix is synthesized by stir casting process and characterization was studied through a scanning electron microscope (SEM) and energy dispersive analysis of X-rays (EDAX), respectively. Mechanical properties such as hardness, tensile strength and impact strength were estimated after adding the weight percentage of titanium carbide. Drilling holes were formed on the surface of the metal matrix using electric discharge machining (EDM). Wear characteristics are carried out by pin on disc apparatus. Applied load, sliding velocity and weight percentage of reinforcement were considered as the input parameters. Wear rate is the response parameter for this experiment. After wear test, morphology of the worn out surfaces was analysed by SEM. It was clearly observed that mechanical and wear resistance were improved by increasing the weight percentage of reinforcement material. The

T. Vishnu Vardhan
Department of Mechanical Engineering, CMR Institute of Technology, Hyderabad, Telangana 501401, India

S. Marichamy (✉)
Department of Mechanical Engineering, Sri Indu College of Engineering and Technology, Hyderabad, Telangana 501510, India
e-mail: smari2k5@gmail.com

B. Stalin
Department of Mechanical Engineering, Anna University, Regional Campus Madurai, Madurai, Tamil Nadu 625019, India

J. Vairamuthu
Department of Mechanical Engineering, Sethu Institute of Technology, Pulloor, Kariapatti, Tamil Nadu 626115, India

V. Dhinakaran
Department of Mechanical Engineering, Chennai Institute of Technology, Kundrathur, Tamil Nadu 600069, India

A. Arockiarajan et al. (eds.), *Advances in Industrial Automation and Smart Manufacturing*, Lecture Notes in Mechanical Engineering,
https://doi.org/10.1007/978-981-15-4739-3_48

influential parameter and its effect have been found out by analysis of variance (ANOVA).

Keywords Duplex silicon metal matrix · Scanning electron microscope (SEM) · Analysis of variance (ANOVA) · Wear rate · Tribological behaviour · Electric discharge drilling (EDD)

1 Introduction

In modern days, demand of metal matrix has been increased in every field because of its excellent properties such as strength, hardness, toughness, ductility and malleability. The duplex silicon metal matrix is used in various applications such as automobiles, aerospace, healthcare products, computer and microelectronics industries. Wear plays an important role and mainly depends on the composition of the alloys. Thermal energy was used in EDM process. EDM is used for machining very hard materials, and it is used for machining of dies, cavities and precision parts. In this process, tool and workpiece are submerged into the dielectric fluid and spark was produced between the tool and workpiece [1]. The material properties such as hardness, toughness and thermal conductivity were increased due to the addition of titanium carbide to the alloys [2, 3]. The parameter interactions were studied on titanium carbide composite by EDM process [4]. Responses such as metal removal rate and the tool wear rate are found out during electrical discharge machining of Al–4Cu–6Si alloy reinforced with SiC composites [5]. Al-SiC composite was drilled by powder-mixed EDM, and the machining effect was compared with conventional EDM [6]. Material properties mainly depend on the nature of reinforcement added to the alloy [7]. Inconel metallurgical characteristics were studied during the electrical discharge machining process with copper electrode [8]. The material removal rate has been improved by using powder-mixed EDM [9]. The current was the most influential parameter while machining of aluminium boron carbide ($Al–B_4C$) composite [10]. Wear properties of the material mainly depend on the microstructural characteristics such as reinforcement particle size, processing method, volume fraction and distribution of the particles [11, 12]. Wear properties of copper-based aluminium composite have been studied under dry lubrication conditions [13]. Wear resistance, wear debris and particle agglomeration were studied on WC particle-reinforced copper matrix [14]. Reinforcement of metal matrix has better wear resistance than other materials [15]. Wear rate depends on the percentage volume of reinforcement addition to the alloys [16]. Wear behaviour of aluminium metal matrix composite was studied under different reinforcements [17]. The tool wear rate has been analysed in tungsten powder-mixed electric discharge machining of AA6061/10% SiC composite [18]. The most influential parameter was found by ANOVA during machining [19–26]. The predicted model and effect of parameters were analysed during electrical discharge machining of TiC mixed ceramic [27].

Present work deals with material synthesis, characterization, tribological behaviour and electric discharge drilling of the duplex silicon metal matrix.

2 Fabrication of Materials

The duplex silicon metal matrix was produced by stir casting technique. The billet of silicon, copper, was placed in a furnace crucible and heated at 1250 °C. The preheated aluminium, magnesium and titanium carbide were mixed in the crucible. The continuous stirrer was applied at 800 rpm. Initially, the temperature has been raised to the liquidus level and then allowed for solidification. The molten metal was poured into the mould after uniform mixture was obtained. The slab with shape and size 100 × 50 × 5 mm was produced. After fabrication of metal matrix, the various material properties were evaluated and shown in Table 1. The density of the duplex silicon metal matrix found to be 9.11 g/cm^3.

2.1 Material Characterization

The morphology of the material was studied through an SEM image shown in Fig. 1. From the image, the white region indicates the silicon particles and the black region

Table 1 Mechanical properties

Equipment used	Properties	Results
Brinell hardness tester	Hardness	450 BHN
Universal testing machine	Tensile strength	730 Mpa
Izod impact tester	Impact strength	16 J
Universal testing machine	Percentage of elongation	13%

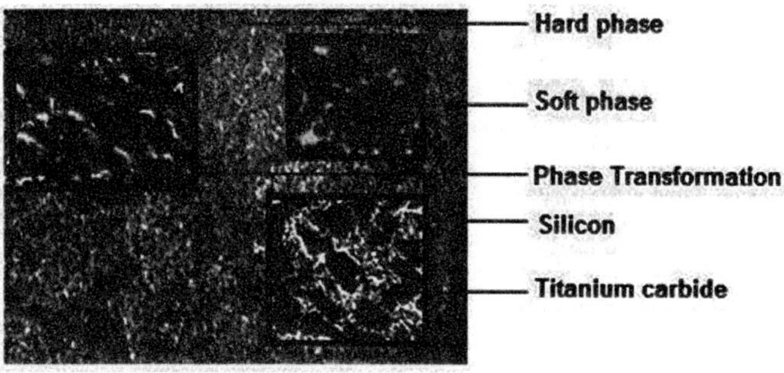

Fig. 1 SEM image of the duplex silicon metal matrix

shows the titanium carbide particles. Alloy elements are uniformly spread over the material structure. High melted particles and bonding strength are also shown in SEM. In the stir casting process, the temperature was raised to liquidus level and allowed for solidification. Due to transition of temperature and rate of cooling, it consists of soft and hard phase structure. Hence, this type of material structure was called as a duplex silicon metal matrix.

The chemical composition was confirmed through EDAX image as shown in Fig. 2. This MMC consists of silicon, copper, aluminium, magnesium and titanium carbide (Fig. 3).

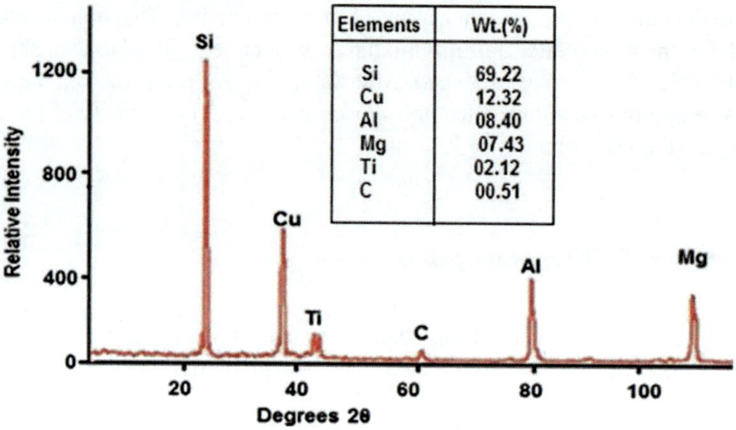

Fig. 2 EDAX image of the duplex silicon metal matrix

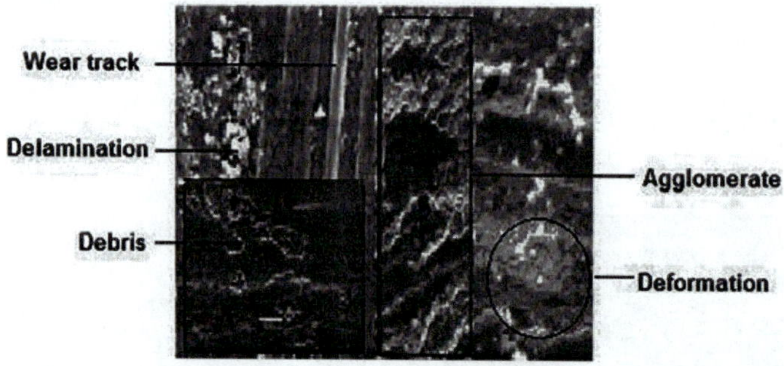

Fig. 3 Worn-out surface of the duplex silicon metal matrix

3 Electric Discharge Drilling

Due to inherent phase structure, silicon metal matrix was drilled by conventional process, which is very difficult. Hence, it was machined by electrical discharge machining (EDM) process. In this process, the material was removed by a powerful spark which is produced between the electrode and the workpiece. The more amount of spark was produced when increasing the current intensity. Hence, EDM process is also called as a spark erosion machining process. In this experimental investigation, silver tungsten is considered a hallow electrode material and kerosene is considered as the dielectric fluid. 6 mm-diameter holes were produced on the synthesized metal matrix. Current, voltage, pulse on time and flushing pressure are considered as the input parameters. Material removal rate and tool wear rate are considered as the output parameters. Initially, EDM experimental observations were recorded without using a SiC abrasive mixture in the flushing unit while machining of the metal matrix. After that, the SiC abrasive mixture is used in the flushing unit. The changes of the response were recorded with and without using SiC abrasive mixtures. The EDM experimental observations without SiC abrasive mixture are shown in Table 2.

By improving the material abrasion, silicon carbide abrasives are used to flow through the hollow electrode. The EDM experimental observations with SiC abrasive mixture are shown in Table 3.

By improving the material abrasion, silicon carbide abrasives are used to flow through the hollow electrode. The EDM experimental observations with SiC abrasive mixture are shown in Table 3.

The differences recorded with and without silicon carbide abrasives were used in EDM process and shown in Table 4.

From Table 4, high amount of material removal rate and minimum tool wear rate was obtained using with SiC abrasive mixture. The material removal rate was increased with increasing current, pulse on time and SiC abrasive mixture.

Table 2 The EDM experimental observations without SiC abrasive mixture

Current (A)	Pulse ON time (μs)	Voltage (V)	MRR (mm^3/min)	TWR (mm^3/min)
6	80	40	5.59	0.14
6	160	50	2.53	0.18
6	220	60	3.17	0.21
10	80	50	9.23	0.43
10	160	60	12.26	0.33
10	220	40	8.76	0.67
16	80	60	19.34	0.96
16	160	40	15.45	0.52
16	220	50	20.33	0.93

Table 3 The EDM experimental observations with SiC abrasive mixture

Current (A) A	Pulse ON time (μs) B	Voltage (V) C	MRR (mm^3/min)	TWR (mm^3/min)
6	80	40	7.29	0.15
6	160	50	3.12	0.11
6	220	60	4.53	0.19
10	80	50	10.13	0.22
10	160	60	14.24	0.17
10	220	40	10.66	0.43
16	80	60	21.13	0.76
16	160	40	19.11	0.34
16	220	50	22.14	0.63

Table 4 The EDM experimental observations without SiC abrasive mixture

Current (A) A	Pulse ON time (μs) B	Voltage (V) C	MRR (mm^3/min)	TWR (mm^3/min)
6	80	40	7.29	0.15
6	160	50	3.12	0.11
6	220	60	4.53	0.19
10	80	50	10.13	0.22
10	160	60	14.24	0.17
10	220	40	10.66	0.43
16	80	60	21.13	0.76
16	160	40	19.11	0.34
16	220	50	22.14	0.63

4 Evaluation of Frictional Properties

Wear is an important role to design any components. The excessive wear may lead to failure of components due to friction is produced between the rotating elements. Pin is made up of a duplex silicon metal matrix with cylindrical shape and size of 9 mm diameter and 30 mm height. The disc material is made up of EN-31. During wear test, the various control parameters such as applied load, sliding distance and disc speed were considered, whereas wear rate is the response parameter. The track diameter 100 mm is maintained for this experiment. The experimental results for frictional wear are shown in Table 5.

Table 5 Experimental results for frictional wear

Applied load (N) A	Sliding distance (m) B	Disc speed (rpm) C	Wear rate (mm^3/m)
10	500	400	0.0021
10	800	600	0.0031
10	1000	800	0.0011
20	500	600	0.0027
20	800	800	0.0069
20	1000	400	0.0002
30	500	800	0.0076
30	800	400	0.0097
30	1000	600	0.0002

4.1 Analysis of Wear Worn-Out Surface

The worn-out structure of the pin is analysed by scanning electron microscope (SEM). From this image, it clearly shows that plastic shearing occurred on the worn-out surface. The reinforcements perform as an abrasive between the pin and the disc during wear test. At higher applied load, grooves and wear debris were formed on the surface due to plastic deformation. The fracture was produced when the induced stress exceeds the fracture strength. The rate of wear debris grooves and delamination depends upon the fractured particles at higher applied load. Wear debris can be produced which depends on the adhesive and the plastic shearing of the material [28].

5 Analysis of Variance

The contribution parameter was found out by analysis of variance (ANOVA). It is an objective decision-based statistical tool to test the significance of each parameter and their interactions. Analysis of variance for MRR, TWR and wear rate are shown in Tables 6, 7 and 8. ANOVA is used to interpret the experimental results, and it is used to found the contribution levels [29].

From Table 6, current has provided 94.09% contribution effect on the material removal rate. It is clearly observed that the material removal rate increases with increase of current.

The mathematical models were used to predict the response, and they are shown in Eqs. (1)–(3), where A, B and C are the control parameters such as current, pulse on time and voltage.

$$\begin{aligned}\text{MRR}\left(\text{mm}^3/\text{min}\right) = & 12.48 - 7.50\,A - 6 - 0.81\,A - 10 \\ & + 8.31\,A - 16 + 0.37\,B - 80 - 0.33\,B - 160\end{aligned}$$

Table 6 Analysis of variance for MRR

Source	DF	Adj SS	Adj MS	F-Value	P-Value	Percentage %
Current	2	378.020	189.010	19.33	0.049	94.09
Pulse on time	2	0.728	0.364	0.04	0.964	0.19
Voltage	2	3.466	1.733	0.18	0.849	0.86
Error	2	19.561	9.780			4.86
Total	8	401.77				100

S = 3.12736; R-sq = 95.13%; R-sq (adj) = 86.53%

Table 7 Analysis of variance for SR

Source	DF	Adj SS	Adj MS	F-Value	P-Value	Percentage %
Current	2	0.289267	0.144633	5.60	0.152	68.40
Pulse on time	2	0.074600	0.037300	1.44	0.409	17.63
Voltage	2	0.007467	0.003733	0.14	0.874	1.76
Error	2	0.051667	0.025833			12.21
Total	8	0.423000				100

$S = 0.160728$; R-sq = 87.79%; R-sq (adj) = 88.33%

Table 8 Analysis of variance for wear rate

Source	DF	Adj SS	Adj MS	F-Value	P-Value	Percentage %
Applied load (N)	2	0.000022	0.000011	7.00	0.125	22.68
Sliding distance (m)	2	0.000056	0.000028	17.89	0.053	57.74
Disc speed (rpm)	2	0.000016	0.000008	5.01	0.166	16.49
Error	2	0.000003	0.000002			3.09
Total	8	0.000097				100

$S = 0.0012503$; R-sq = 96.76%; R-sq(adj) = 87.06%

$$\begin{aligned} &- 0.04\,B - 220 - 0.13\,C - 40 - 0.69\,C - 50 \\ &+ 0.82\,C - 60 \end{aligned} \tag{1}$$

From Table 7, current has provided 68.40% contribution effect on the tool wear rate. It is clearly observed that tool wear rate decreases with increase of current.

$$\begin{aligned} \text{TWR}\left(\text{mm}^3/\text{min}\right) = {} & 0.3333 - 0.1833A - 6 - 0.0600\,A \\ & - 10 + 0.2433A - 16 + 0.0433B - 80 - 0.1267\,B \\ & - 160 + 0.0833B - 220 - 0.0267C - 40 - 0.0133C \\ & - 50 + 0.0400C - 60 \end{aligned} \tag{2}$$

From Table 7, sliding distance has provided 57.74% contribution effect on wear rate. It is clearly observed that the wear rate mainly depends on sliding distance, where A, B and C are the input parameters of the wear test such as applied load, sliding distance and disc speed.

$$\begin{aligned}\text{Wear rate}\left(\text{mm}^3/\text{m}\right) = {} & 0.003733 - 0.001633\,A - 10 - 0.000467\,A - 20 \\ & + 0.002100\,A - 30 + 0.000400\,B - 500 + 0.002833\,B \\ & - 800 - 0.003233\,B - 1000 + 0.000267\,C - 400 \\ & - 0.001733\,C - 600 + 0.001467\,C - 800 \end{aligned} \tag{3}$$

6 Conclusion

The duplex silicon metal matrix was synthesized by stir casting route successfully. Material characterization was investigated through SEM and EDAX. Electrical discharge drilling experiments were conducted on the metal matrix. The tribological behaviour of the duplex silicon metal matrix was studied through pin on disc apparatus. From analysis of variance, current has provided 94.09% contribution effect on the material removal rate and it has provided 68.40% contribution effect on the tool wear rate. Sliding distance has provided 57.74% contribution effect on wear rate.

References

1. Jahan MP, Ragman M, Wong YS (2011) A review on the conventional and micro-electro discharge machining of Tungsten Carbide. Int J Mach Tools Manuf 51(12):837
2. Deng J (2001) Friction and wear behavior of Al_2O_3 ceramic composites at temperatures up to 800°C. Ceramics Int 27:135
3. Deng J, Tongkun C, Junlong S (2005) Microstructure and mechanical properties of hot-pressed Al_2O_3/TiC Ceramic composites with the additions of solid lubricants. Ceramics Int 31(1):249
4. Patel KM, Pandey PM, Rao PV (2010) Optimization of process parameters for multi-performance characteristics in EDM of Al_2O_3 Ceramic composite. Int J Adv Manuf Technol 47(9):1137
5. Dhar S, Purohit R, Saini N, Sharma A, Kumar H (2007) Mathematical modelling of electric discharge machining of cast Al–4Cu–6Si Alloy–10 wt. % SiC composites. J Mater Process Technol 194(1):24
6. Hu FQ, Cao FY, Song BY, Hou PJ, Zhang Y (2013) Surface properties of SiCp/Al composite by powder-mixed EDM. Proc CIRP 6:101
7. Yan BH, Wang C, Huang L (2000) Machining characteristics ofAl2O3/6061Al composite using rotary EDM with a disk like electrode. Int J Adv Manuf Technol 16:322
8. Rahul Datta S, Biswal BB, Mahapatra SS (2017) Electrical discharge machining of Inconel 825 using cryogenically treated copper electrode: emphasis on surface integrity and metallurgical characteristics. J Manuf Process 26:188
9. Singh B (2015) Influences of process parameters on MRR improvement in simple and powder-mixed EDM of AA6061/10%SiC composite. Mater Manuf Process 30(3):102

10. Kumar P (2016) Experimental Investigation and optimization of EDM process parameters for machining of Aluminum Boron Carbide composite. Mach Sci Technol 20:2
11. Kumar R, Mohanasundaram KM, Arumaikkannu S (2012) Analysis of parameters influencing wear and frictional behavior of Aluminum-fly ash composites. Tribol Trans 55(6):723
12. Gaitonde VN, Karnik Jayaprakash MS (2012) Some studies on wear and corrosion properties of Al5083/Al_2O_3/Graphite hybrid composites. J Minerals Mater Characterization Eng 11(7):695–703
13. Siddhalingeshwar I, Mitra DC (2011) Sliding wear behavior of in situ Al–4.5Cu–5TiB2 composite processed by single and multiple roll passes in mushy state. Wear 271(5–6):748–759
14. Deshpande L (2006) Wear resistance of WC particle reinforced copper matrix composites and the effect of porosity. Mater Sci Eng, A 418(1–2):137–145
15. Mallikarjuna S, Reddy H, Ahamadsab H (2017) Evaluation of impact properties of fly ash and S-glass reinforced Al-4046 hybrid metal matrix composites. Mater Today Proc 4:12285–12290
16. Reddy S, Kesavan VR (2016) Investigation of mechanical properties of Aluminium6061-Silicon Carbide Boron Carbide metal matrix composite. Silicon 1:1–8
17. Khan D (2017) Comparative study on erosive wear response of SiC reinforced and fly ash reinforced aluminium based metal matrix composite. Mater Today Proc 4:10093–10098
18. Singh K (2016) Investigation of the tool wear rate in Tungsten powder-mixed electric discharge machining of AA6061/10% SiC composite. Mater Manuf Process 31(4):456–466
19. Shandilya JPK, Jain, NK (2012) Parametric optimization during wire electrical discharge machining using response surface methodology. Proc Eng 38:2371–2377
20. Raju R, Sivalingam V, Sun J, Natarajan M, Zhao Y (2019) Experimental and Taguchi-based grey approach of laser metal deposition technique on Nickel-based superalloy. Trans Indian Inst Metals 72(1):205–214
21. Raju R, Manikandan N, Palanisamy D, Arulkirubakaran D, Sambathkumar S, Bhanu Prakash P (2018) Optimization of process parameters in electrical discharge machining of haste alloy C276 using Taguchi's method, Mater Today Proc 5(6):14432–14439
22. Manikandan N, Raju R, Palanisamy D, Arulkirubakaran D, Kumar S (2018) Investigation on Ti6Al4 V laser metal deposition using Taguchi based grey approach. Mater Today Proc 5(6):14375–14383
23. Thirugnanasambantham KG, Raju R, Sankaramoorthy T, Velmurugan P, Kannagi A, Reddy MCK, Chandra VR (2018) Degradation mechanism for high-temperature sliding wear in surface-modified In718 superalloy. Cogent Eng 5(1):1–11
24. Jeyaprakash N, Duraiselvam M, Raju R (2018) Modelling of Cr 3 C 2–25% NiCr laser alloyed cast iron in high temperature sliding wear condition using response surface methodology. Arch Metall Mater 63(3):1303–1315
25. Dillibabu V, Duraiselvam M, Chandrasekhar U, Raju R (2017) Microstructural studies on laser dissimilar welded Ni and steel alloys for aeronautical turbine applications. Lasers Eng 37(4–6):247–260
26. Sivalingam V, Sun J, Yang B, Liu K, Raju R (2018) Machining performance and tool wear analysis on cryogenic treated insert during end milling of Ti-6Al-4V alloy. J Manuf Process 36:188–196
27. Chiang KT (2008) Modeling and analysis of the effects of machining parameters on the performance characteristics in the EDM process of Al_2O_3 + TiC mixed ceramic. Int J Adv Manuf Technol 37:5233
28. Veeresh Kumar GB, Rao CSP, Selvaraj N (2012) Studies on mechanical and dry sliding wear of Al6061–SiC composites. Compos Part B Eng 43(3):1185–1191
29. Raghunath N, Pandey PM (2007) Improving accuracy through shrinkage modelling by using Taguchi method in selective sintering. Int J Mach Tools Manuf 47(1):985

Tensile and Hardness Behavior of RRA Treated Aluminum 7075 Alloy

K. Sunil Kumar and P. L. Srinivasamurthy

Abstract Hardness plays major role on metals to understand the behavior of the resistance to plastic deformation with respect to force. In this movement, I utilized all around naturally known as two-stage heat treatment called RRA is connected on AL 7075 composite to inspect the corrosion behavior.This procedure incorporates the different phases of warmth medicines incorporates, strengthening, high temperature preprecipitation, counterfeit maturing (T6),retrogression and re-maturing. The examples are at first aged to 473° and then extinguished in water medium.The examples are additionally aged to 18 h, 22 h, 26 h separately. After this examples are warmed 230°, 250°, 270°, 290° with various time interims 7 and 14 min individually to improve the quality of the material this procedure, we call it as retrogression and re-aging. After this directed test and came to realize that by utilizing RRA strategy, we can expand the quality of the material to most extreme degree. With this RRA methodology, we can increase the tensile strength and hardness of al alloy.

Keywords Tensile · Hardness · RRA treated · Aluminum 7075

1 Introduction

In materials, the aluminum is one the most important materials in resources. The main ore is bauxite is the source for aluminum. Its chemistry defines the material goods such as density electrical conductivity, corrosion resistance and strength [1–3].

The difficulties made on materials for enhanced overall performance are so countless and different that no one material can fulfill them. Aluminum alloy is the best for the certain requirements like thermal conductivities, high ductility, low cost, high strength, corrosion resistance, etc. [4–8].

K. Sunil Kumar (✉)
Department of Mechanical Engineering, RLJIT, Doddaballapur, Karnataka, India
e-mail: agnisai111@gmail.com

P. L. Srinivasamurthy
Department of Mechanical Engineering, MSRIT, Bangalore, Karnataka, India

A. Arockiarajan et al. (eds.), *Advances in Industrial Automation and Smart Manufacturing*, Lecture Notes in Mechanical Engineering,
https://doi.org/10.1007/978-981-15-4739-3_49

The capacity of any material is to resist from the impact it depends on the hardness as well as a ductility of materials. The different hardness may be obtained in metals by tempering and heat treatment processes [9].

The one more properties like tensile strength as we know that it is the capability of a material to resist a drawing force. Strength requires undergoing through the elastic to plastic bend. It is known to express as tensile force for the particular unit area of the material to divide [10–14].

Corrosion resistance and elastic modulus are the main parameters in engineering materials; these are mainly applicable in mechanical devices and structures. The productiveness of the hardness is scratch and rebound with the indentation with respect to indenter on the specimen [15–18].

2 Methodology

2.1 Material Selected and Its Composition-Aluminum 7075 Alloy

See Fig. 1.

Fig. 1 Elements composition

Element	Weight %
Zn	5.690
Mg	2.074
Cu	1.507
Mn	0.069
Cr	0.244
Ti	0.080
Fe	0.222
Si	0.036
Al	Balance

2.2 Methodology Route Map

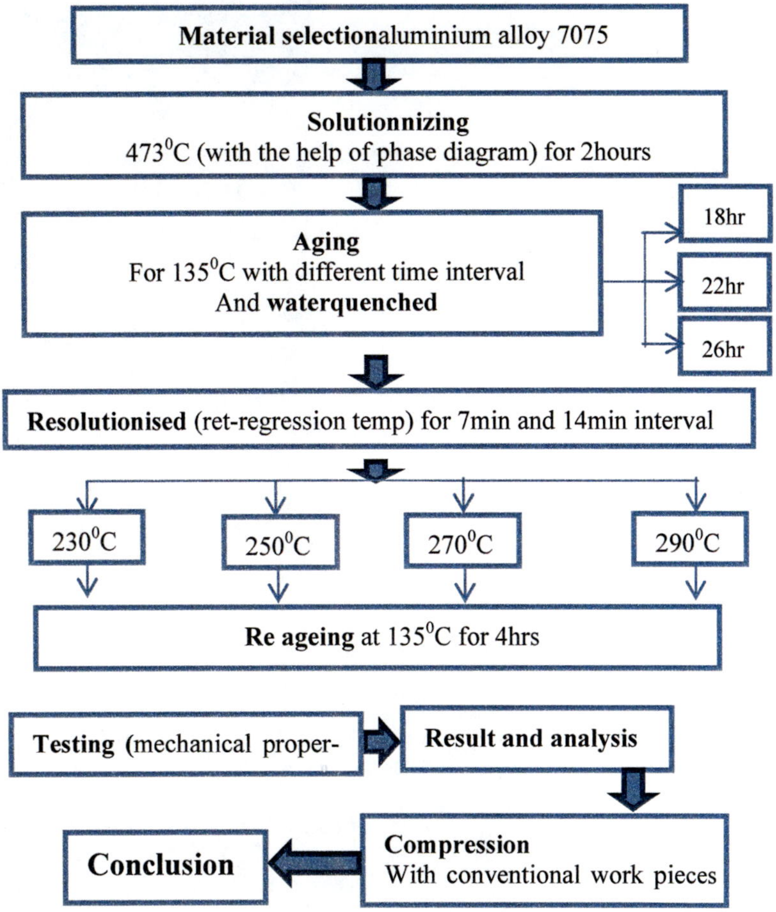

2.3 Furnace Setup and Specimens

See Figs. 2, 3, 4, 5 and 6.

Fig. 2 Hardness process setup

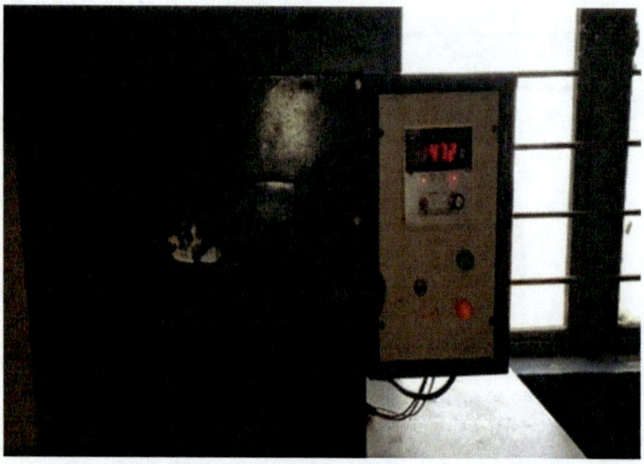

Fig. 3 Furnace

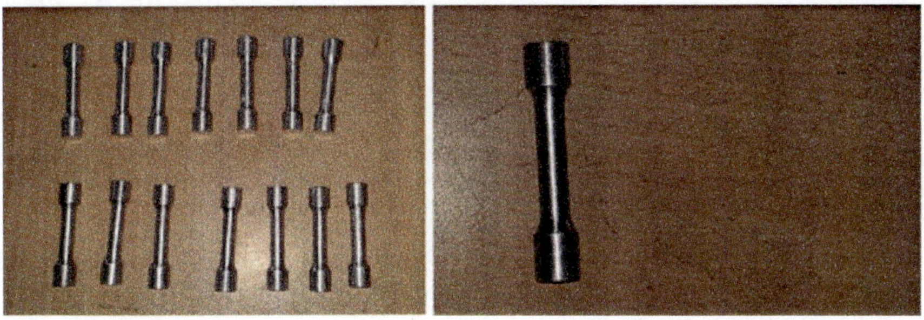

Fig. 4 Tensile Specimen

Fig. 5 RHN versus Age hardening time

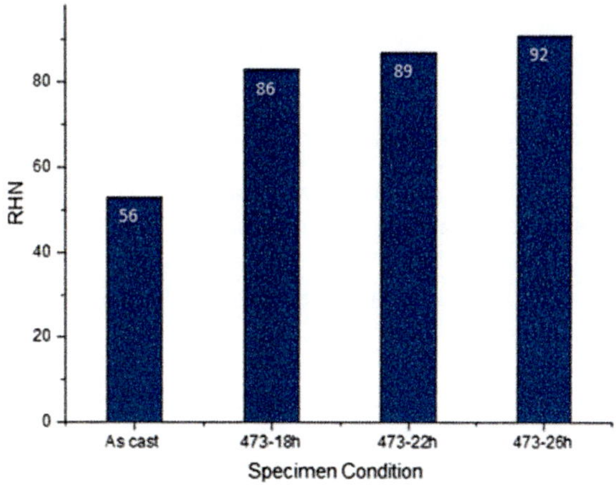

Fig. 6 Effect of hardness on RRA 473 °C-18 h specimens for the time 7 and 14 min duration

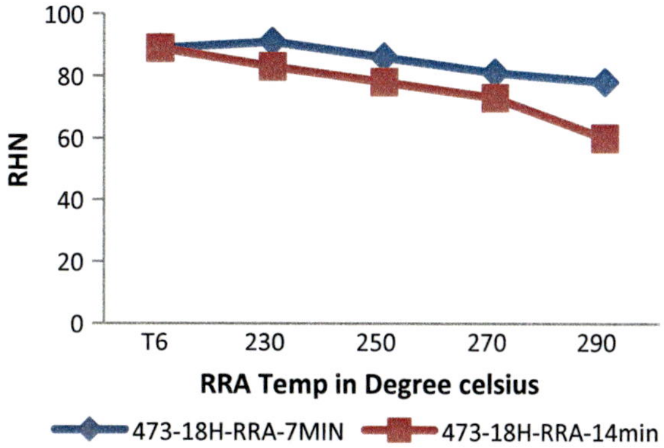

3 Experimental Details

3.1 Hardness Test

Effect of hardness with respect to age hardening of aluminum

In our experimentation, the test work pieces are solutionized to 473oC and aging for 18 h, 22 h and 26 h with T6 temper condition, respectively. To carry out hardness test, we have chosen rockwell hardness on scale (RB); it is performed on the surface of the test specimens. Proper covenant distance is taken for the between indentations. The four of average results are shown in Table 1.

Effect of time (aging) on hardness

See Fig. 7.

Discussion and Results

The hardness is comparatively increased with age hardening and solutionization, the atoms are solute atoms will be in the condition of superheated in the matrix and atoms take participation during quenching.

One of the reasons to increase the hardness is solute atoms are homogeneously precipitated in the grain boundaries.

Table 1 Hardness of as built, aged and solution treated

Alloy Name	Hardness(RHN)
As built(cast)	56
473 °C-18 h	86
473 °C-22 h	89
473 °C-26 h	92

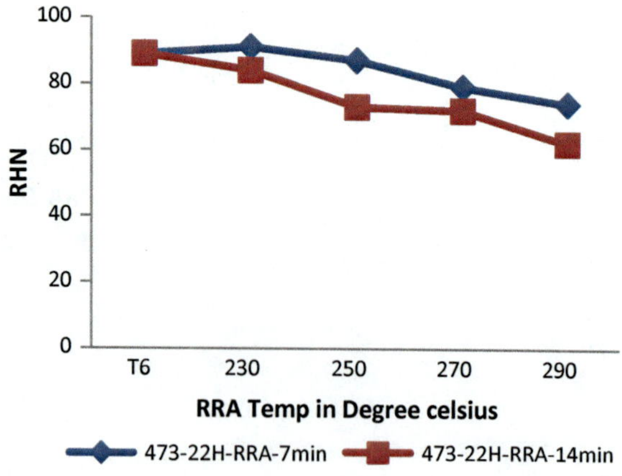

Fig. 7 Effect of hardness on RRA 473 °C-22 h specimens for the time 7 and 14 min duration

As age hardening increase, the hardness increased. The valves are 55.5%, 61.25% and 70.3%.

Effect of RRA treatment on hardness

Afterword the solutionization RRA treatment is carried out with 230, 250°C, 270 and 290 °C with a hold time of 7 and 14 min (Figs. 8, 9 and 10; Tables 2, 3 and 4).

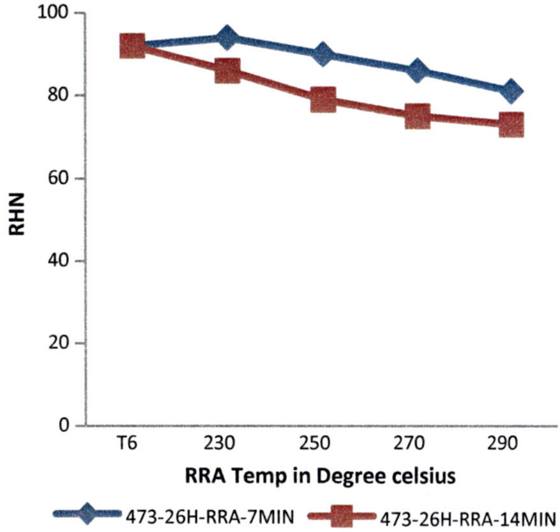

Fig. 8 Effect of hardness on RRA 473 °C-26 h specimens for the time 7 and 14 min duration

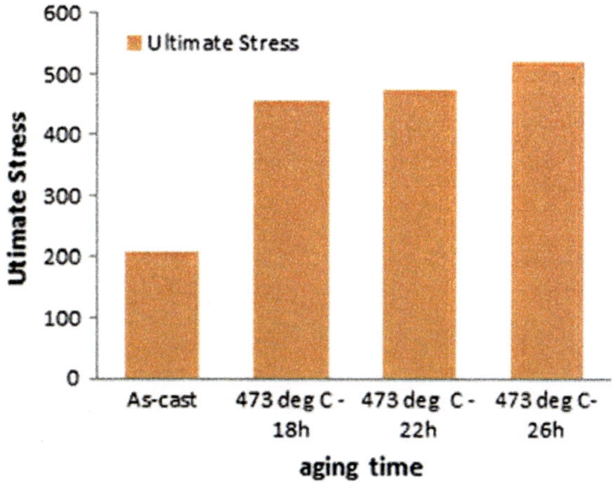

Fig. 9 Effect of aging versus UTS

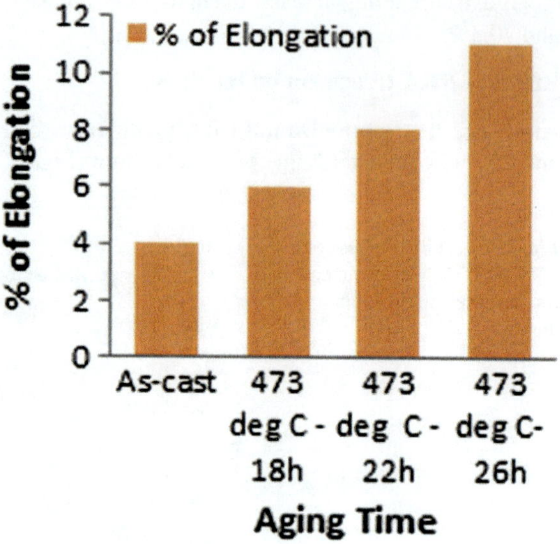

Fig. 10 Elongation versus Aging Time

Table 2 Hardness values of RRA work pieces (18 h)

S. No	Retrogression temperature in °C	473 °C-18 h-RRA-7 min	473 °C-18 h-RRA-14 min
1	T6	89	89
2	230	91	83
3	250	86	78
4	270	81	73
5	290	78	60

Table 3 Hardness values of RRA work pieces (22 h)

S. No	Retrogression temperature in °C	473 °C-221-RRA-7 min	473 °C-22 h-RRA-14 min
1	T6	89	89
2	230	91	84
3	250	87	73
4	270	79	72
5	290	74	62

Discussion and Results

The hardness of Al7075 alloy has noticeable increment in the beginning and with retrogation temperature at 230 °C for the hold of 7 min and then it decreased as a retrogation increased and it is increased to 14 min.

Table 4 Hardness values of RRA work pieces (26 h)

S. No	Retrogression temperature in °C	473 °C-26 L-RRA-7 min	473 °C-26 h-RRA-14 min
1	T6	92	92
2	230	94	86
3	250	90	79
4	270	86	75
5	290	81	73

It occurs stable phases by time of re-aging treatment. The parameters re-aging and retrogation lead to overaging and grain coarsening this leads to change in hardness is decreased.

By the graphs, we came to know that there is an increment of hardness 5% by RRA treatment followed by the temperature of 230 °C and time 7 min. And maximum decrease by 10% while the temperature is 473 °C-26 h and time is 14 min.

3.2 Tensile Test

The testing specimens are solutionized to 473 °C and aged for 18, 22 and 26 h, T6 temper condition, respectively. Tensile test is carried out and results are tabulated.

Effect of aging versus UTS

Discussion on Results

Graph 5 shows the increase in UTS with solutionization and the comparison with as cast 100 to 150% is increased (Fig. 11; Tables 5, 6 and 7).

Effect of aging time on % of Elongation

See Fig. 12.

3.3 Tensile Fracture Studies

Fracture studies on as cast, RRA treated and solution treated.

The above figure shows the SEM graphs after the tensile fracture. The surface fracture shows the non-uniform grains without treatment. It observed the elemental size is small in grains and evenly distributed 473 °C-22 h in solution treated. Further is decreased in size and distribution in RRA treated alloy (Figs. 13 and 14; Table 8).

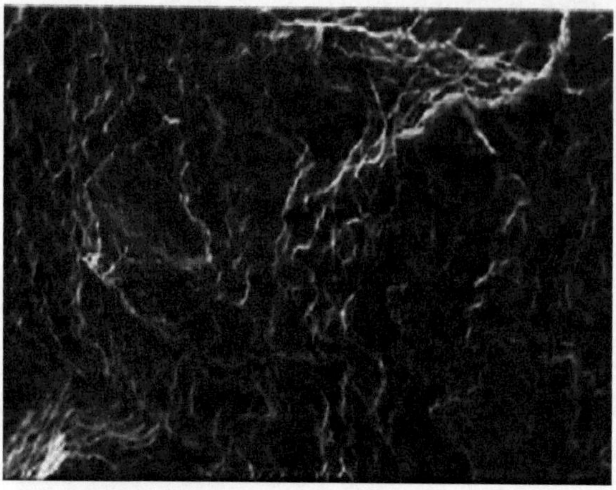

Fig. 11 Surface Fracture as cost specimen

Table 5 Tensile strength and elongation as cost and solutionzed specimen

S. No.	Specimen state	Ultimate stress MPa	% Elongation
1	As-cast	210	4
2	473 °C-18 h	454	6
3	473 °C-22 h	473	8
4	473 °C-26 h	520	11

Table 6 UTS and elongation of RRA treated-18 h

S.no	Specimen Condition	Ultimate stress in MPa	% of Elongation
1	473-18 h-230-7 min	475	7.9
2	473-18 h-230-14 min	428	8.8
3	473-18 h-250-7 min	444	9.0
4	473-18 h-250-14 min	359	11.8
5	473-18 h-270-7 min	401	10.6
6	473-18 h-270-14 min	320	10.4
7	473-18 h-290-7 min	361	11.3
8	473-18 h-290-14 min	206	12.1

Discussion on Results

Graph six appearances the effect of % elongation and time cast and solutionized alloy. It indicates the slight expansion in ductility with aging time. It starts from 4%–5% of 473-18 h alloy. Similarly, 7% for 473-22 h and 10% for 473-26 h as compared with cast one.

UTS and % of elongation are enhanced with aging treatment, as cost 4.01% elongation and UTS is 211 MPa. 473 °C-18 h results in 53.37% increase in UTS (treated), next one 473 °C-22 h-473 °C-26 h are 55.33% and 57.9% increased, respectively.

Table 7 UTS and elongation of RRA treated -22 h

S. No.	Specimen condition	Ultimate Stress in MPa	% of Elongation
1	473-22 h-230-7 min	508	9.2
2	473-22 h-230-14 min	429	10.4
3	473-22 h-250-7 min	480	10.9
4	473-22 h-250-14 min	396	11.8
5	473-22 h-270-7 min	408	10.6
6	473-22 h-270-14 min	376	12.2
7	473-22 h-290-7 min	390	12.9
8	473-22 h-290-14 min	348	13.3

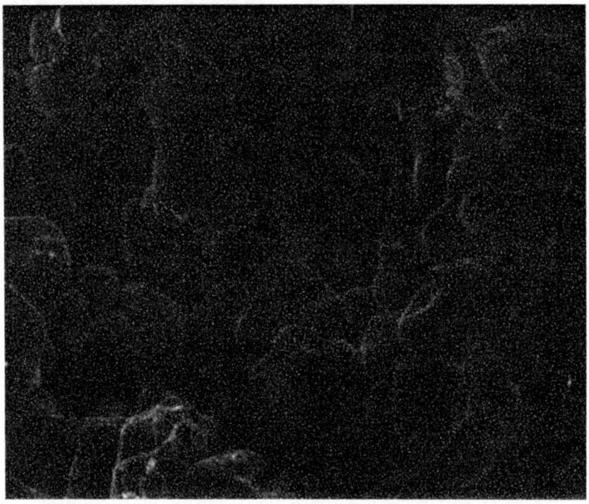

Fig. 12 Surface Fracture Treated 473 °C-18 h specimen

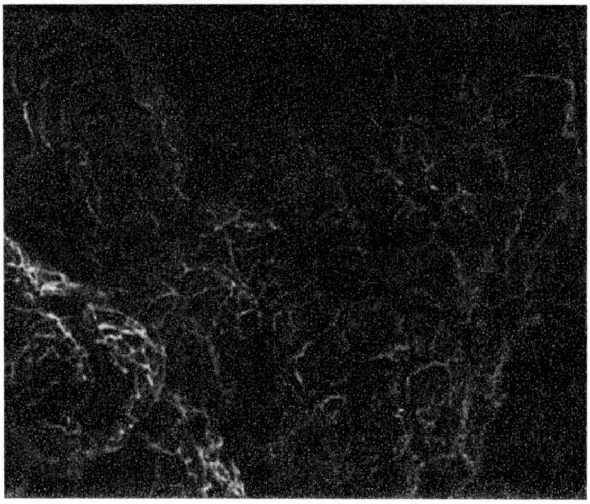

Fig. 13 Surface Fracture Treated 473 °C-22 h specimen

Fig. 14 Surface Fracture Treated 473 °C-26 h specimen

Table 8 UTS and elongation of RRA treated -26 h

.	Specimen condition	Ultimate Stress in MPa	% of elongation
1	473-22 h-230-7 min	518	11.4
2	473-22 h-230-14 min	481	11.9
3	473-22 h-250-7 min	488	11.8
4	473-22 h-250-14 min	409	13.4
5	473-22 h-270-7 min	419	12.6
6	473-22 h-270-14 min	390	13.8
7	473-22 h-290-7 min	419	12.9
8	473-22 h-290-14 min	363	14.2

In RRA treatment, we achieved maximum UTS of 506 MPa in 473 °C-26 h-230 °C-7 min specimens. It indicates increasing age time at lowest retrogression temperature at lowest retrogression time. The UTS has increased 70% by solutionized.

4 Conclusion

Al7075 specimens are prepared according to ASTM standards. The treatment process and process parameters having significant impact on the mechanical properties.

Aged specimens are improving its mechanical properties, and the tensile strength and resistance decrease with increasing aging temperature.

Tensile strength and hardness are improved after RRA heat treatment. In the results, we observed that the hardness and strength are increased with the increase in the aging (RRA) duration.

References

1. American society for metals—ASM Handbook, properties and selection, Nonferrous Alloys and Special Purpose Materials ASM International Handbook Committee (1990) vol 2, pp 137–138
2. Machler R, Uggowitzer PJ, Solenthaler C, Pedrazzoli RM, Spiedel MO (1991) Structure, mechanical properties, and stress corrosion behavior of high strength spray deposited 7000 series aluminum alloy. Mater Sci Technol 7:447–451
3. Chee FT, Mohamad RS (2009) Effect of hardness test on preceipitation hardening aluminum alloy 6061-T. Chiang Mai J Sci 36(3):276–286
4. Smallman RE (1985) Modern physical metallurgy, 4th edn. London Butterworths & Co
5. Lavernia EJ, Rai G, Grant NJ (1990) Rapid solidification processing of 7xxx aluminum alloys: a review. Mater Sci Eng 79:211–221
6. Sanctis MD (1991) Structure and properties of rapidly solidified ultrahigh strength Al-Zn-Mg-Cu alloys produced by spray deposition. Mater Sci Eng, A 141:103–121
7. Baradeswaran A, Elaya Perumal A (2014) Study on mechanical and wear properties of Al 7075/Al_2O_3/graphite hybrid composites. Compos Part B 56:464–471
8. Umanath K, Palanikumar K, Selvamani ST (2013) Analysis of dry sliding wear behavior of Al601/SiC/Al_2O_3 hybrid metal matrix composites. Compos Part B 53:159–168
9. Vijaya Ramnath B, Elanchezhian C, Jaivignesh M, Rajesh S, Parswajinan C, Ghias SA. Evaluation of mechanical properties of aluminum alloy-alumina–boron carbide metal matrix composites. Mater Des 58:332–338
10. Mishra SK, Biswas S, Satapathy A (2014) A study on processing, characterization and erosion wear behavior of silican carbide particles filled ZA-27 metal matrix composites. Mater Des 55:958–965
11. Huerta E, Corona JE, Olivia AI (2010) Univeral testing machine for mechanical properties of thin materials. Rev Mex Fis 56(4):317–322
12. Guerra M, Schmidt C, McClure JC, Murr LE, Nunes AC (2003) Flow patterns during friction stir welding. Mater Characterization 49:95–101
13. Li X, Zhang D, Qiu C, Wen Z (2012) Microstructure and mechanical properties of dissimilar pure copper/ 1350 aluminum alloy butt joints by friction stir welding. Trans Nonferrous Met Soc China 22:1298–1306
14. Shafiei-Zarghani A, Kashani-Bozorg SF, Zarei-Hanzaki A (2009) Microstructures and mechanical properties of Al/Al_2O_3 surface nano-composite layer produced by friction stir processing. Mater Sci Eng, A 500:84–91
15. Usman AM, Raji A, Waziri NH, Hassan MA (2014) Production and characterization of alluminum alloy—bagasse ash composites. IOSR J Mech Civil Eng 11(4):38–44
16. Baradeswaran A, Elaya Perumal A (2014) Wear and mechanical characterization of Al 7075/ graphite composites. Compos Part B Eng 56:72–76
17. Baradeswaran A, Elaya Perumal A (2014) Study on wear and mechanical properties of Al7075/ Al_2O_3/graphite hybrid composites. Compos Part B Eng 56:64–471
18. Ibrahim IA, Mohamed FA, Lavernia EJ (1991) Particulate reinforced metal matrix composites—a review. J Mater Sci 26(5):1137–1156

Investigations on Wire Spark Erosion Machining of AA 6061 Alloy Using Taguchi's Approach

J. S. Binoj, N. Manikandan, K. C. Varaprasad, P. Thejasree, and Ramesh Raju

Abstract AA 6061 is an aluminium alloy having good mechanical possessions such as high strength and ductility plus fatigue. Owing to its specific properties, AA 6061 is widely used in fabricating aerospace components such as aircraft fittings, shafts, gears and defence products. Moreover, its outstanding corrosion resistance property suits it as a significant material for many engineering applications. Non-traditional machining processes have been progressed to overwhelm some difficulties in making intricate shapes, and wire electrical discharge machining (WEDM) was one among that machining technique aimed at material removal established after the theory of electrical discharge machining process. In this present study, the experimental runs were planned by Taguchi's design approach and then the experimental analysis has been done on WEDM of AA 6061 alloy by Taguchi's analysis. The optimum process parameters have been determined for attaining improved machining performance with the help of Taguchi's response analysis.

Keywords WEDM · Aluminium alloy · AA 6061 · Taguchi's method · Single objective optimization

1 Introduction

One of the modern methods of material removal, commonly used in various manufacturing industries for making components like turbine blades of aircraft engines, nozzles for fuel injectors and dies is wire electrical discharge machining (WEDM).

J. S. Binoj (✉) · N. Manikandan · K. C. Varaprasad · P. Thejasree
Department of Mechanical Engineering, Micro Machining Research Centre,
Sree Vidyanikethan Engineering College (Autonomous),
Tirupati, Andhra Pradesh 517102, India
e-mail: binojlaxman@gmail.com

R. Raju
Department of Mechanical Engineering, Santhiram Engineering College, Nandyal,
Andhra Pradesh 518501, India

A. Arockiarajan et al. (eds.), *Advances in Industrial Automation and Smart Manufacturing*, Lecture Notes in Mechanical Engineering,
https://doi.org/10.1007/978-981-15-4739-3_50

WEDM has been employed for machining high hardness, corrosive and wear resistive electrically conductive materials. The principle of WEDM has been developed from the electrical discharge machining (EDM) and mainly implemented for making difficult shapes which is not possible by the conventional material removing process [1–5]. In WEDM process, the removal of material takes place based on the electro-discharge erosion effect of electrical sparks occurred in the electrode wire and work material separated with a dielectric fluid as shown in Fig. 1.

Supply of voltage in the middle of electrode and work material with existence of dielectric fluid for melting material surface is through spark erosion. Moreover, since the channel of electrode wire is from one side to other continuously, removal of material takes place [6–8]. Aluminium alloys are mostly suited for numerous engineering applications since it possesses more specific strength and rigidity at elevated temperatures. However, the properties of material are required to be enhanced while deploying it for advanced structural applications. Also, these alloys provide amenable strength to weight ratio while relating with other materials that are employed for various structural materials [9].

Aluminium alloys are considered predominantly as lightweight materials in manufacturing industries owing to its inherent properties like better ductility and strength. However, due to its exceptional characteristics, it is mainly used in aircraft structural applications. Apart from these, AA 6061 alloy has better corrosion resistance and strength compared to other alloys. Researchers had reported that aluminium alloys are mostly employed for effective making of aircraft and its parts. An investigation has been performed on WEDM of aluminium alloy by using RSM method and the significance of individual process parameters were ascertained [10–12]. The Grey relation analysis has been adopted in WEDM of SS 304 for optimizing the multi-aspects and perceived that Taguchi's approach and ANOVA promote better performance for

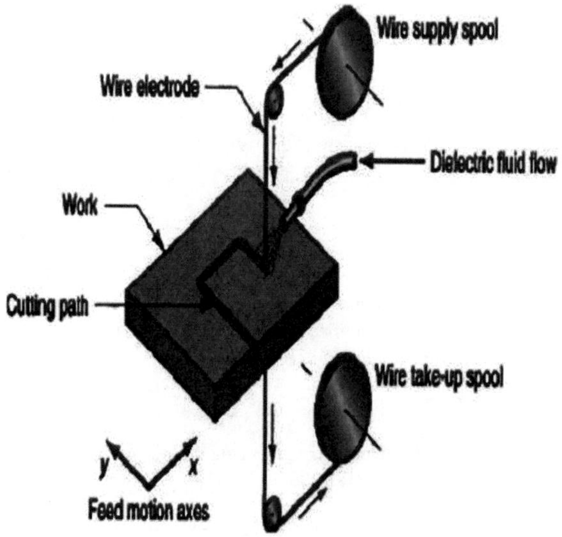

Fig. 1 Schematic of wire EDM

finding the importance of each variable over the desired performance assets [13, 14]. The problem of machining these kinds of hard materials by traditional process had lead to the development of advanced machining process like WEDM and altered the scenario by making the materials machined using the proposed methods [15].

An experimental investigation has been performed on WEDM process for analyzing the machining performance using Inconel as work material. Taguchi's approach has been implemented for designing the experiments and single-objective optimization. Also, multi-performance measures have been derived by a suitable method of multiple objective optimizations [16]. The Taguchi method has been employed for attaining better surface finish in WEDM of tool steel and is revealed from the examination that adopted technique provides a better improvement in the desired objective and WEDM process has been modified for obtaining better machinability during machining of H.S.S, titanium Nimonic and aluminium alloys [17–20]. From the available literature, it is perceived that the importance and machining of aluminium alloy by WEDM process needs further attention. Hence, in this present work, an exertion was taken for analyzing the importance of process variables to attain the desired performance features like Kerf width of the machined surface and material removal rate. Taguchi's approach has been engaged for designing the experiment trials and single-objective optimization of desired performance measures.

2 Materials and Methods

AA 6061 alloy possesses extraordinary strength and better resistance to corrosion, and because of its outstanding properties, it is widely used in aircraft and other applications. AA-6061 has been opted as work material (breadth—50 mm, thickness—6 mm and length—110 mm) in this work and is fixed in the inner side of machining chamber as shown in Fig. 2. WEDM machine (Concord Make-Model DK-7732) has been employed for experimentation purpose and reusable molybdenum wire is employed as tool electrode and the dielectric fluid used was de-ionized water.

Generally, in experimental design, conducting many experimental trials with opted process parameters and levels is needed. Such kind of issues are resolved by adopting Taguchi's experimental design approach, and a unique design layout has been proposed by Taguchi for conducting the experiments which is called as orthogonal array (OA) for examining independent process parameters using the help of minimum set of experimental trials. Pulse-on time (μs), pulse-off time (μs) and peak current (A) were designated as input parameter and Kerf width (KW), and material removal rate (MRR) is opted as output performance measures. The nominated independent process parameters, levels and the range of those values are depicted in Table 1, and as per these values, L27 OA has been opted for wire EDM of AA 6061 alloy.

Fig. 2 Experimental set-up

Table 1 Machinability process parameters and levels

Symbols	Variables	Levels		
		1	2	3
A	Pulse-on time (μs)	10	20	30
B	Pulse-off time (μs)	5	10	15
C	Peak current (A)	1	2	3

3 Results and Discussions

Experimental runs have been performed as per L27 OA for exploring the significance of input parameters on desired output characteristics in wire EDM of AA 6061 alloy. An exertion has been made to ascertain the best possible process parameters for accomplishing the operational and competent wire EDM process. In wire EDM process, greater material removal rate and lesser values of Kerf width are the indicator of superior performance measures. So, MRR is considered as larger the better criterion and Kerf width is considered as smaller the better criterion.

3.1 Effect of Material Removal Rate Over Process Variables

Response analysis curve for the removal of material during the machining of AA 6061 alloy is displayed in Fig. 3. It is observed from the response graph that the material removal rate was increased by way of enhancement of pulse-on time and peak current as well as getting decreased with increment in pulse-off time. By this, it was witnessed that the pulse-off time is the principal process variable for material removal rate. Moreover, increment of values in pulse-on-time discharged more energy to the machining chamber with significance influential explosion leading to increased values in the material removal rate. In addition, cumulative values of pulse-on time had raised capability of gaining amount of electrons impacting on workpiece, thereby it starts to wear away extra volume of material from the specimen surface per spark discharge.

Response analysis using Taguchi's approach for the MRR has been accomplished and the outcomes are shown in Table 2. The best set of process parameters for achieving a better material removal rate is *A*3*B*1*C*3 and the optimum process variables combination is 'P_{on}'—30 µs; 'P_{off}'—5 µs; Peak current is 3 A. Moreover, it is perceived from the examination that pulse-off time is the noteworthy process variable and then it is trailed by peak current and pulse-on time.

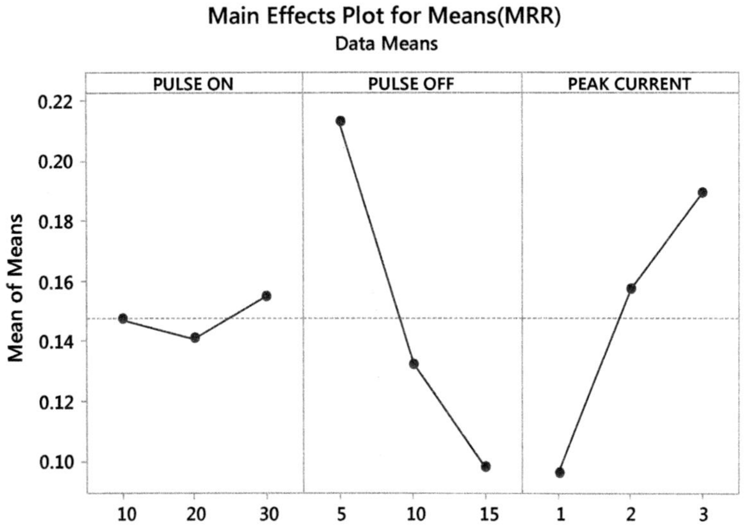

Fig. 3 Response analysis for material removal rate

Table 2 Taguchi's response analysis for MRR–WEDM of AA 6061 alloy

Level	Means			S/N ratio		
	A	B	C	A	B	C
1	0.14743	0.21302	0.09624	−17.35	−13.67	−21.27
2	0.14087	0.13220	0.15750	−18.15	−18.10	−16.54
3	0.15498	0.09806	0.18953	−17.08	−20.80	−14.77
Delta	0.01411	0.11497	0.09329	1.07	7.12	6.50
Rank	3	1	2	3	1	2

3.2 *Effect of Kerf Width Over Process Variables*

The graphical illustration of response graph for Kerf width during wire EDM of AA 6061 alloy is depicted in Fig. 4.

It is noted by the graph that value of Kerf width of electrically discharge machined surface is getting decreased with an increase of 'Pon' and similarly getting diminished with an increment of 'Poff'. The superior energy of discharge provided crater in a greater manner that roots additional Kerf width at the machined surface. The pulse-on time is believed as supreme persuasive process variable for Kerf width of the surface machined.

Taguchi analysis for Kerf width has been executed and the results are presented in Table 3. However, the best set of process parameters for attaining better Kerf width is *A3B3C3*. It means that the optimum process parameter combination for better

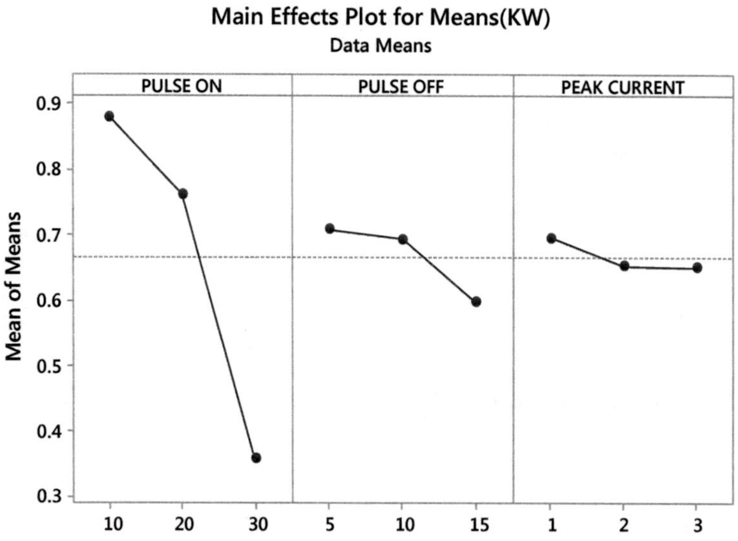

Fig. 4 Response graph for Kerf width

Table 3 Response table for Kerf width for AA 6061 alloy

Level	Means			S/N ratio		
	A	B	C	A	B	C
1	0.8800	0.7089	0.6956	1.116	3.601	3.791
2	0.7622	0.6933	0.6533	2.730	3.907	4.558
3	0.3578	0.5978	0.6511	8.969	5.306	4.466
Delta	0.5222	0.1111	0.0444	7.852	1.705	0.766
Rank	1	2	3	1	2	3

performance is 'Pon'—30 μs; 'Poff'—15 μs; and peak current is 3 A. Also, it is apparent from analysis that 'Pon' is major variable and then it is trailed by pulse-off time and peak current.

4 Conclusions

The current work explains a sole aspect optimization problem on wire EDM of AA 6061 alloy by the help of Taguchi's analysis. In this study, the variables like material removal rate (MRR) and Kerf width of machined surface (KW) were considered as the performance features and the following conclusions are drawn:

- Taguchi's approach has been adopted for designing the experimental runs and also it has been adopted for optimizing the individual variables for accomplishing better performance in machining.
- The prevalence of deemed independent parameters on the anticipated performance faces were made known by Taguchi's analysis. The best set of combinations for attaining better and improved machining performance were determined.
- The detailed Taguchi's approach in this present exploration is greatly appropriate to establish the best set of process variables for attaining the improved performance in any advanced machining process.
- The outcome attained from this exploration will be a wide-ranging support to the manufacturers for enhancing the rate of production and quality of products made with an assistance of WEDM process.

References

1. Kanlayasiri K, Boonmung S (2007) Effects of wire-EDM machining variables on Kerf width of newly developed DC 53 die steel: design of experiments and regression model. J Mater Process Technol 192–193:459–464
2. Davim JP (2008) Machining fundamentals and recent advances. Springer-Verlag London Limited, British Library Cataloguing in Publication Data. https://doi.org/10.1007/978-1-84800-213-5
3. Patil N, Brahmankar PK (2010) Determination of material removal rate in wire electro-discharge machining of metal matrix composites using dimensional analysis. Int J Adv Manuf Technol 51(5–8):599–610. https://doi.org/10.1007/s00170-010-2633-3
4. Huang Y, Ming W, Guo J, Zhang Z, Liu G, Li M, Zhang G (2013) Optimization of cutting conditions of YG15 on rough and finish cutting in WEDM based on statistical analyses. Int J Adv Manuf Technol 69(5–8):993–1008. https://doi.org/10.1007/s00170-013-5037-3
5. Saha P, Tarafdar D, Pal S, Saha P, Srivastava A, Das K (2009) Modeling of wire electro-discharge machining of TiC/Fe in situ metal matrix composite using normalized RBFN with enhanced k means clustering technique. Int J Adv Manuf Technol 43(1–2):107–116. https://doi.org/10.1007/s00170-008-1679-y
6. Moulton DB (1999) Wire EDM the fundamentals. EDM network, Sugar Grove, IL. www.notebookmanuals.bestmanualguide.com
7. El-Hofy H (2005) Advanced machining processes. McGraw-Hill. https://doi.org/10.1036/0071466940
8. Sommer C, Sommer S (2005) Complete EDM handbook. Advance Pub
9. Kumar SM, Pramod R, Kumar MS, Govindaraju HK (2014) Evaluation of fracture toughness and mechanical properties of aluminum alloy 7075, T6 with nickel coating. Proc Eng 97:178–185
10. Dursun T, Soutis C (2014) Recent developments in advanced aircraft aluminium alloys. Mater Des 1980–2015(56):862–871
11. Rambabu PPNKV, Prasad NE, Kutumbarao VV, Wanhill RJH (2017) Aluminium alloys for aerospace applications. In: Aerospace materials and material technologies, Springer, Singapore, pp 29–52
12. Gopalakannan S, Senthilvelan T, Ranganathan S (2012) Modeling and optimization of EDM process parameters on machining of Al 7075-B4C MMC using RSM. Procedia Eng 38:685–690
13. Khan ZA, Siddiquee AN, Khan NZ, Khan U, Quadir GA (2014) Multi response optimization of wire electrical discharge machining process parameters using Taguchi based grey relational analysis. Procedia Mater Sci 6:1683–1695
14. Lajis MA, Radzi HCDM, Amin AKMN (2009) The implementation of Taguchi method on EDM process of tungsten carbide. Eur J Sci Res 26(4):609–617
15. Manikandan N, Arulkirubakaran D, Palanisamy D, Raju R (2019) Influence of wire-EDM textured conventional tungsten carbide inserts in machining of aerospace materials (Ti–6Al–4 V alloy). Mater Manuf Proc 34(1):103–111
16. Yang CB, Lin CG, Chiang HL, Chen CC (2017) Single and multiobjective optimization of Inconel 718 nickel-based superalloy in the wire electrical discharge machining. Int J Adv Manufact Technol 93(9–12):3075–3084
17. Sudhakara D, Prasanthi G (2014) Application of Taguchi method for determining optimum Kerf width in wire electric discharge machining of P/M cold worked tool steel (Vanadis-4E). Proc Eng 97:1565–1576
18. Ramesh NN, Harinarayana K, Naik BB (2014) Machining characteristics of HSS & titanium using electro discharge sawing and wire-electrodischarge machining. Procedia Mater Sci 6:1253–1259

19. Goswami A, Kumar J (2014) Optimization in wire-cut EDM of Nimonic-80A using Taguchi's approach and utility concept. Eng Sci Technol Int J 17(4):236–246
20. Manikandan N, Binoj JS, Varaprasad KC, Sabari SS, Raju R (2019) Investigations on wire spark erosion machining of aluminum-based metal matrix composites. In: Advances in manufacturing technology. Springer, Singapore, pp 361–369

A FEA Model to Predict Mechanical Properties of Laminated Bamboo Composites

P. M. Bupathi Ram, V. Dhinakaran, K. P. Manoj Kumar, Surendar Kannan, and H. Mohit

Abstract Composites materials are widely used in many fields of applications for their unique properties. The increasing demand for composites materials is due to the advantages such as lightweight, high strength, high stiffness, superior fatigue life, tremendous corrosion resistance, and low-cost manufacturing. In this study, a finite element analysis (FEA) model was analyzed with the ABAQUS FEA software. The simulation was done to compare the simulation results with the experimental results of laminated bamboo composites in terms of tensile response. The maximum and minimum stress values for hypothetical layered laminated bamboo composites were obtained and were compared with the experimental results. The failure load, tensile strength, elastic behavior, and further properties of the laminated plates under axial load are determined from stress–strain curves. The simulation is performed with the material, having orientation angles of 0° for all plies. The simulation was successfully carried out and the results obtained from the simulation are compared with the results obtained from the experimental.

Keywords Composites · Tensile testing · Finite element analysis

1 Introduction

In today's modern technology, there is a huge need for materials with high mechanical performance especially in the field of aerospace application. Hence, in order to fulfill this requirement, a material known as composites has been found out. A composite material is generally made by integrating two or more materials usually reinforcement and a resin. The composite materials can combine the properties of two materials by which they are built which results in an increase in the mechanical properties

P. M. Bupathi Ram (✉) · V. Dhinakaran · K. P. Manoj Kumar · S. Kannan
Centre for Applied Research, Chennai Institute of Technology, Kundrathur, Chennai 600069, India
e-mail: bupathi1062001@gmail.com

H. Mohit
King Mongkut's University of Technology North Bangkok, Pune, Bang Sue, Bangkok 10800, Thailand

A. Arockiarajan et al. (eds.), *Advances in Industrial Automation and Smart Manufacturing*, Lecture Notes in Mechanical Engineering,
https://doi.org/10.1007/978-981-15-4739-3_51

of the composite material to meet the requirements for an application. The use of composites is increasing day by day in various field of applications, especially in aerospace applications such as in making frames, panels, interior components and in construction of buildings which requires properties such as low density, high strength, high stiffness, abrasion and impact resistant and also are not corroded easily [1–4].

Bamboo culms are a strong material having high mechanical properties such as high tensile strength which is good enough for structural purposes [5, 6]. In order to use them for structural purposes, they are converted into engineered bamboo timber which can be used as a substitute for wood. Bamboo timber is made by integrating two or more laminas with adhesives [7, 8].

While there are several publications for the experimental analysis of bamboo composites, only very few are there for finite element analysis (FEM) of the composites [9–16]. The tensile behavior of the composites is based on the lamina configuration. Each composite material is made of a particular lamina configuration [17]. The tensile test was done experimentally and it is compared with the simulation results. Moreover, this publication shows the comparison of experimental analysis and finite element analysis of the bamboo composites.

2 Simulation

In this analysis, the simulation was carried out in the framework SIMULIA ABAQUS FEA software. In ABAQUS, there are many ways for composite material modeling such as mixed modeling, microscopic modeling, macroscopic modeling, discrete reinforcement modeling, and sub-modeling. But mostly the composite materials are modeled as layered shells, layered-solids, stacked solid elements, and stacked or layered continuum shells. The aim of the analysis is to study the mechanical response of bamboo composites under axial loading and to compare the simulation results with the experimental result. The composite laminas are stacked up together with orientational angle 0° with lamina thickness of 1 mm [17]. The specimen is cut according to the ASTM D3039 with dimensions, 200 mm × 16 mm × 4 mm and a tab length of 50 mm. The anisotropic property is defined in ABAQUS using linear elastic property.

The values of E1, E2, ν12, G12, G13, and G23 are important for orthotropic materials, in which E1 is the longitudinal modulus, E2 is the transverse modulus, ν12 is the major Poisson's ratio and G12, G13, and G23 are the in-plane shear modulus. The elastic constants for the bamboo composite with linear elastic behavior are shown in Table 1. To carry out the simulation for isotropic lamina, the value of G23 is required, which is obtained by Eq. (1) [13].

$$\mathrm{G23} = \frac{\mathrm{E2}}{2}(1 + \nu 23) \tag{1}$$

Table 1 Elastic constants for laminate bamboo composites

Elastic modulus along fiber direction	E1 = 6.3 GPa
Elastic modulus across fiber direction	E2 = 2 GPa
Shear modulus	G12 = 0.672 GPa
Major Poisson's ratio	ν12 = 0.32

The step is created as static, general, and the boundary conditions (BC) are applied as symmetry/antisymmetry/encastre in the mechanical category and the ENCASTRE (U1 = U2 = U3 = UR1 = UR2 = UR3 = 0) option is selected which fixes the selected section firmly even when loads are acted upon it. The boundary conditions are applied as displacement/rotation during the simulation. Then, the values of displacement are changed one by one and the results are visualized in the visualization module. The results are generated from the simulation and are verified with the experimental results. The tensile properties of experimentally tested laminated bamboo composites (LLBCs) are shown in Table 2.

The simulation was carried out by either defining displacement or maximum load required for fracture. If displacement is defined, then the reaction force (RF) is noted and verified or else the reaction force (RF) is defined and the respective displacement is noted and verified. The results obtained from the simulation are noted in Table 3.

The tensile test is conducted with one of the ends of the specimen fixed and the other end is pulled apart so that the maximum load that the specimen can withstand can be determined. During the tensile test, the end grips are pulled apart from each other at a constant rate. The load on the specimen and the corresponding stress–strain

Table 2 Tensile test results of experimentally tested laminated bamboo composites (LLBCs)

Sample no.	Displacement on max. load (mm)	Max. load (kN)	Maximum stress (MPa)	Strain on max. load (mm/mm)
1	6.78	13.49	183.35	0.133
2	8.51	16.19	220.26	0.168
3	9.38	16.79	240.92	0.185
4	8.44	16.83	210.58	0.166
5	10.43	15	191.07	0.205
6	8.11	10.36	152.47	0.16
7	10.79	15.87	251.56	0.212
8	9.59	16.96	226.54	0.189
9	8.52	15.92	215.71	0.168
10	7.9	15.96	213.18	0.156
Average	8.85	15.34	210.56	0.1742
SD	1.21	2.03	28.83	0.0237
CV	13.7	13.26	13.69	13.648

Table 3 Tensile test results obtained from simulation

Sample no.	Displacement on max. load (mm)	Maximum stress (MPa)	Reaction force (kN)
1	6.78	196.7	8.74
2	8.51	246.9	10.94
3	9.38	272.1	12.1
4	8.44	244.8	10.89
5	10.43	302.6	13.46
6	8.11	235.3	10.47
7	10.79	313	13.92
8	9.59	278.2	12.38
9	8.52	247.2	10.99
10	7.9	229.2	10.19
Average	8.85	256.7	11.42
SD	1.21	35.1	1.56
CV	13.7	397.4	17.68

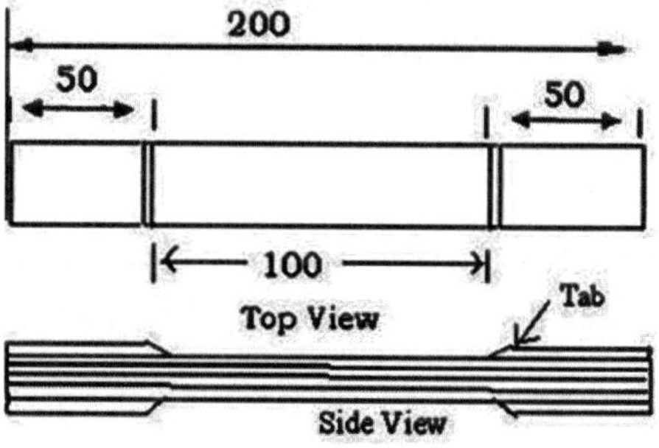

Fig. 1 Tensile test specimen [6]

behavior is continuously monitored until failure and plotted as a stress–strain curve as shown in Fig. 1.

3 Experimental Analysis

The experimental work carried out by Verma et al. [7, 8] has been used to avail the tensile results and data were used as the base for the simulation. The tensile testing is done by applying an axial load or tension to a test sample up to the point of fracture. The specimen is cut into dimensions of ASTM standards. The tensile testing shows the mechanical behavior of the material under the axial load. The tensile test generates a stress–strain data which can be used to determine the mechanical response. The data generated during the tensile test help in finding the mechanical properties

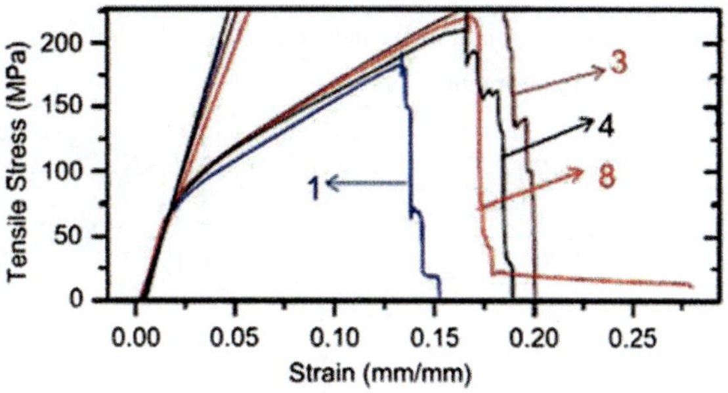

Fig. 2 Stress–strain curves for LLBCs [6]

of the material [7, 8]. From the stress–strain data, properties are ultimate tensile strength (UTS), yield strength, and measurements such as elongation, maximum strain, maximum stress, fracture point, and reduction in the area of the specimen.

The composite specimen is prepared through the fabrication process. The different layers of the composite materials with different orientational angles are prepared. Then, these layers are stacked together by using adhesives to combine them [18]. Then, a weight is kept over the fabricated composite for the better combining of the layers. If the specimen is well fabricated, accurate results can be obtained [19].

The specimen is fixed in the universal testing machine (UTM) with both end grips fixed in the machine. During the test, both the end grips are pulled apart at a constant rate by which axial tension acts on the specimen until a fracture occurs and changes in dimensions of the specimen occurs. The results are recorded and plotted as shown in Fig. 2. The strain gauge is also attached to the specimen with which strain data are obtained. The simulation results are matched with the simulation results.

The mechanical properties such as the measurements, ductility, and strength are calculated by the technician after the tensile test specimen has broken [20]. The broken test sample is again made to join together. The length and area of cross section after failure are measured and then compared with the original dimensions and the elongation of the broken test specimen is obtained.

4 Results and Discussion

The simulation was done on ten sample specimens of laminated bamboo composite material by varying the displacement as shown in Table 4. The magnitude of the displacement for each simulation is applied according to the data shown in Table 2. The corresponding results obtained are tabulated and is verified in Table 3. From the results, for a displacement of 6.78 mm, the maximum stress is 196.7 MPa and the reaction load is 8.74 kN as shown in Fig. 3a, b, whereas the experimental results are 183.25 MPa and 13.49 kN, respectively.

Then, the displacement applied is changed as 8.51 mm, the maximum stress is 246.9 MPa, and the reaction load is 10.94 kN, and the corresponding results are

Table 4 Comparison of experimental results and simulation results

Specimen	Experimental			Simulation		
	Displacement (mm)	Maximum stress (MPa)	Fracture load (kN)	Displacement (mm)	Maximum stress (MPa)	Fracture load (kN)
1	6.78	183.35	13.49	6.78	196.7	8.74
2	8.51	220.26	16.19	8.51	246.9	10.94
3	9.38	240.92	16.79	9.38	272.1	12.10
4	8.44	210.58	16.83	8.44	244.8	10.89
5	10.43	191.07	15.00	10.43	302.6	13.46
6	8.11	152.47	10.36	8.11	253.3	10.47
7	10.79	251.56	15.87	10.79	313.0	13.92
8	9.59	226.54	16.96	9.59	278.2	12.38
9	8.52	215.71	15.92	8.52	247.2	10.99
10	7.90	213.18	15.96	7.90	229.2	10.19
Average	8.85	210.56	15.34	8.85	256.7	11.42
CD	1.21	28.83	2.03	1.21	35.1	1.56
SD	13.70	13.69	13.26	13.70	397.4	17.68

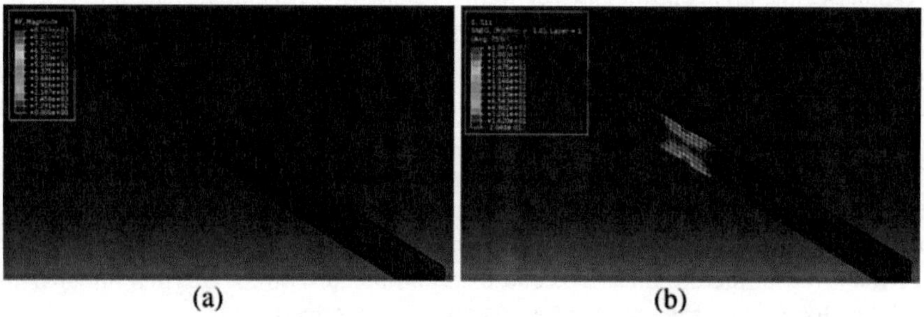

Fig. 3 Specimen no. 1 **a** contour plot of reaction force (RF) **b** contour plot of S11, stress

tabulated as shown in Fig. 4a, b, whereas the experimental results are 220.26 MPa and 16.19 kN, respectively.

Then, the displacement applied is changed as 9.38 mm, the maximum stress is 272.1 MPa, and the reaction load is 12.10 kN, and the corresponding results are tabulated as shown in Fig. 5a, b, whereas the experimental results are 240.92 MPa and 16.79 kN, respectively.

Then, the displacement applied is changed as 8.44 mm, the maximum stress is 244.8 MPa, and the reaction load is 10.89 KN, and the corresponding results are tabulated as shown in Fig. 6a, b, whereas the experimental results are 210.58 MPa and 16.83 kN, respectively.

Then, the displacement applied is changed as 10.43 mm, the maximum stress is 302.6 MPa, and the reaction load is 13.46 kN, and the corresponding results are

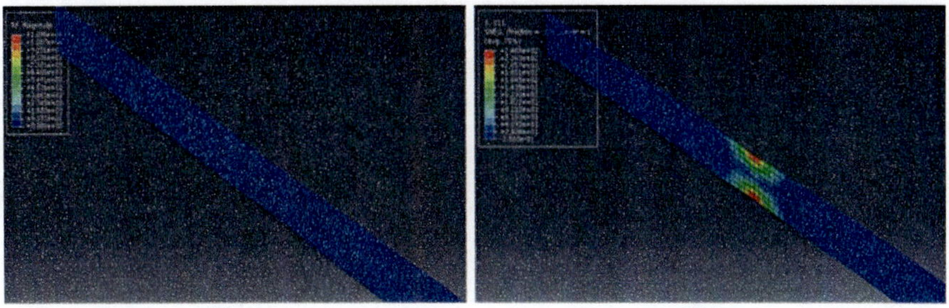

(a) (b)

Fig. 4 Specimen no. 2 **a** contour plot of reaction force (RF) **b** contour plot of S11, stress

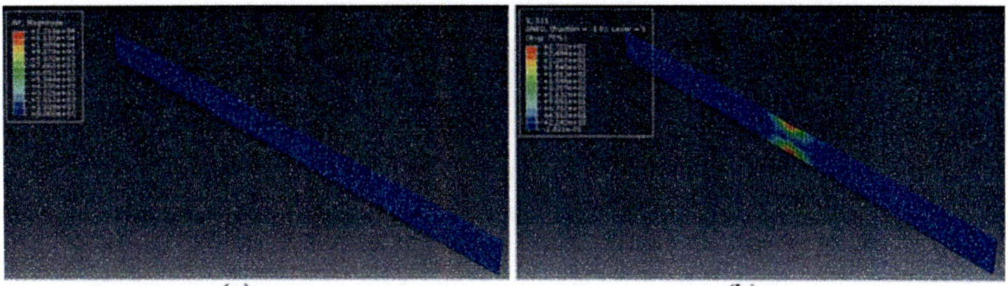

(a) (b)

Fig. 5 Specimen no. 3 **a** contour plot of reaction force (RF) **b** contour plot of S11, stress

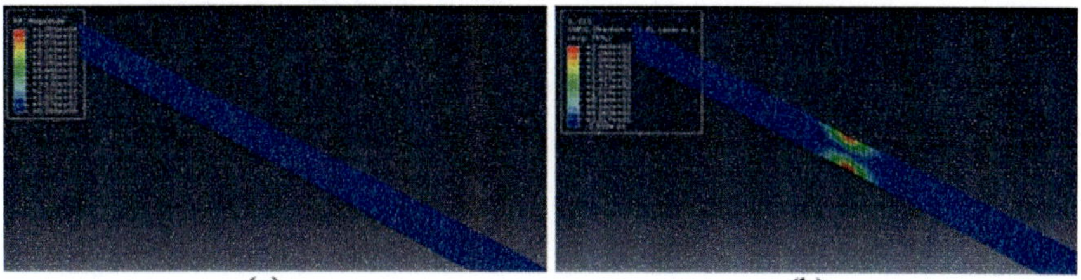

(a) (b)

Fig. 6 Specimen no. 4 **a** contour plot of reaction force (RF) **b** contour plot of S11, stress

tabulated as shown in Fig. 7a, b, whereas the experimental results are 191.07 MPa and 15.00 kN, respectively

Likewise, all the values of displacement obtained from the experiment are compared with the simulation results as described above.

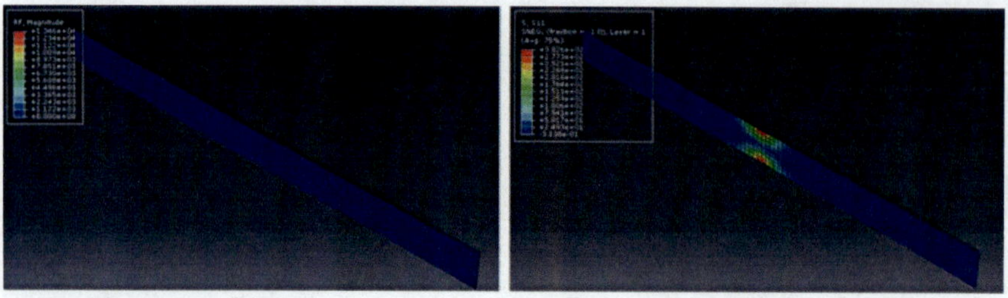

(a) (b)

Fig. 7 Specimen no.4 **a** contour plot of reaction force (RF) **b** contour plot of S11, stress

5 Conclusions

This paper compares the simulation results with the experimental results of layered laminated bamboo composites. A model for tensile testing is created and the simulation results were obtained and it is compared with the experimental results from the experiment. The material definitions are given according to the experimental analysis document.

The ABAQUS framework well suits to analyze the mechanical components, the part assemblies and to visualize the tensile behavior through the finite element analysis method. It well suits for the analysis of parameters such as stress, strain, deflection, displacement, and thermal response of materials. The analysis aims to study the mechanical response of laminated bamboo composites and to compare the simulation results with the experimental results.

To conclude that the analysis is correct, the comparison is made between the experimental and the simulation results. From the analysis, the simulation results almost matched the experimental results. The error percentage can be calculated as the ratio of the difference between the experimental results and the simulation results to the experimental results multiplied by 100. Equation (2) can be used to calculate the error percentage.

Therefore, it is concluded that the simulation results match the experimental results. Also, the simulation can be run for the experiments and the results can be validated. From the analysis, the error percentage between the experimental results and the simulation results is about 10% to 20%. However, for analysis, the user needs to know the theory of part designs, way of choosing the material properties, and the necessary boundary conditions (BC). If the user is well known about the material properties and their definitions, the model can provide accurate results with error percentage even less than 10%.

References

1. Nurhaniza M, Ariffin MKA, Aidy Ali, Mustapha F, Noraini AW (2010) Finite element analysis of composites materials for aerospace applications. In: IOP conference series: materials science and engineering, vol 11, no 1, p 012010. IOP Publishing
2. Mangalgiri PD (1999) Composite materials for aerospace applications. Bull Mater Sci 22(3):657–664
3. Manikandan N, Binoj JS, Varaprasad KC, Sabari SS, Raju R (2019) Investigations on wire spark erosion machining of aluminum-based metal matrix composites. Lecture Notes in Mechanical Engineering, In: Advances in manufacturing technology, pp 361–369
4. Ochoa OO, Reddy JN (1992) Finite element analysis of composite laminates. In: Finite element analysis of composite laminates. Springer, Dordrecht, pp 37–109
5. Chuma S, Hirohashi M, Ohgama T, Kasahara Y (1990) Composite structure and tensile properties of Mousou bamboo. Zairyou 39:847–851
6. Naik NK (2000) Report on mechanical and physic-chemical properties of bamboo. IIT Bombay
7. Verma CS, Chariar VM (2013) Stiffness and strength analysis of four layered laminate bamboo composite at macroscopic scale. Compos Part B Eng 45(1):369–376
8. Verma CS, Rajesh Purohit, Rana RS, Mohit H (2017) Mechanical properties of bamboo laminates with other composites. Mater Today Proc 4(2):3380–3386
9. Jain S, Rakesh Kumar, Jindal UC (1992) Mechanical behaviour of bamboo and bamboo composite. J Mater Sci 27(17):4598–4604
10. Okubo K, Fujii T, Yamamoto Y (2004) Development of bamboo-based polymer composites and their mechanical properties. Compos Part A Appl Sci Manuf 35(3):377–383
11. Thwe MM, Liao K (2003) Durability of bamboo-glass fiber reinforced polymer matrix hybrid composites. Compos Sci Technol 63(3–4):375–387
12. Kinoshita H, Kaizu K, Fukuda M, Tokunaga H, Koga K, Ikeda K (2009) Development of green composite consists of woodchips, bamboo fibers and biodegradable adhesive. Compos Part B Eng 40(7):607–612
13. Shibata S, Yongca Fukumoto I (2008) Flexural modulus of unidirectional and random composites made from biodegradable resin and bamboo and kanaffibers. Compos Part A, 9–15
14. Jindal UC (1986) Development and testing of bamboo-fibres reinforced plastic composites. J Compos Mater 20(1):19–29
15. Chen X, Guo Q, Mi Y (1998) Bamboo fiber-reinforced polypropylene composites: a study of the mechanical properties. J Appl Polym Sci 69(10):1891–1899
16. Rajulu A, Varada S, Allah Baksh G, Ramachandra Reddy, Narasimha Chary K (1998) Chemical resistance and tensile properties of short bamboo fiber reinforced epoxy composites. J Reinforced Plastics Compos 17(17):1507–1511
17. Soden PD, Hinton MJ, Kaddour AS (2004) Lamina properties, lay-up configurations and loading conditions for a range of fibre reinforced composite laminates. In: Failure criteria in fibre-reinforced-polymer composites, pp 30–51. Elsevier
18. Mili F, Necib B (2009) The effect of stacking sequence on the impact-induced damage in cross-ply E-glass/epoxy composite plates. Arch Appl Mech 79(11):1019–1031
19. Autio M, Parviainen H, Pramila A (1992) Accuracy of the finite element method in analyzing laminated plate and pipe structures. Mech Compos Mater 28(3):236–245
20. Oztan C, Karkkainen R, Fittipaldi M, Nygren G, Roberson L, Lane M, Celik E (2019) Microstructure and mechanical properties of three dimensional-printed continuous fiber composites. J Compos Mater 53(2):271–280

Contact Stress Evaluation of Micro-Grooving Process of Alumina Ceramic and Validation with Acoustic Emission Parameters

D. Giridhar and Ramesh Raju

Abstract Response of an engineering material to applied stress is normally evaluated by stress analysis to assess the zone of strength and weakness. Such an analysis will illustrate the possibility of crack-free machining. Grooving of ceramics with diamond indenter is associated with upsetting of the material drawn to the tool wedge and relaxation of the material trailing behind. The upsetting is normally associated with inducing of compressive stresses, while the tensile stress of trailing materials can lead to cracking. For a crack-free machining, effective control of grooving parameters is required. In the present study, the effect of grooving parameters, such as normal load, angle of inclination of the indenter, grooving speed and point radius of the indenter in micro-grooving process of alumina on the contact stress were taken into consideration for the study. It was observed that all the four parameters have significant effect on the contact stress along the width direction of the indenter, depth direction of the workpiece and shear direction of the indenter. The contact stresses at different directions were correlated with AE parameters such as RDC, number of events and RMS.

Keywords Contact stress · Micro-grooving process · Ceramic machining · Alumina · Acoustic emission

Nomenclature

B	Friction factor
$\underline{W}$	Distributed normal load per unit length
σ_x, σ_z	Stresses in x and z directions, N/m^2, respectively

D. Giridhar (✉)
Erode Sengunthar Engineering College, Perundurai, Tamilnadu, India
e-mail: giridhar@alumni.iim.ac.in

Ramesh Raju
Santhiram Engineering College, Nandyal, Kurnool, AP, India
e-mail: mrramesh2002@gmail.com

A. Arockiarajan et al. (eds.), *Advances in Industrial Automation and Smart Manufacturing*, Lecture Notes in Mechanical Engineering,
https://doi.org/10.1007/978-981-15-4739-3_52

σ_{zx}	Shear stress, N/m^2
τ_m	Maximum shear stress, N/m^2
σ_1, σ_2	Principal stress, N/m^2
α	Is the angle between the corresponding planes of the principal curvature,
υ_1 and υ_2	Are the Poisson's ratio of tool and workpiece
b	Is the half contact width, μm
E_1 and E_2	Are the modulus of elasticity of tool and workpiece, GPa
f	Distributed normal load per unit length, N/m
R	Radius of the indenter, μm
R_1, R'_1, R_2 and R'_2	Are the principal radii of curvature at the point of contact (Subscripts 1 and 2 refer to the two bodies making the contact)
W_y	Load at which yielding occurs, N
y	Yield strength of the material, N/m^2

1 Introduction

Ceramics are mostly machined by abrasive processing such as grinding to attain the desired shape tolerance and surface texture, while exposed to machining force and temperature depending on the machining conditions. The forces at the contact region of the abrasive grains and the machined surface give rise to severe plastic deformation. The occurrence of material removal mechanisms can be correlated with the fluctuation in the tangential force in scratch process [2]. Due to the plastic deformation, subsequent dislodgement of material removal occurs. The interaction of individual grains in the workpiece causes repeated compressive and tensile stressing of the workpiece as shown in Fig. 1. The tensile stresses can cause crack propagation in brittle crack-sensitive materials. This effect is most pronounced with large

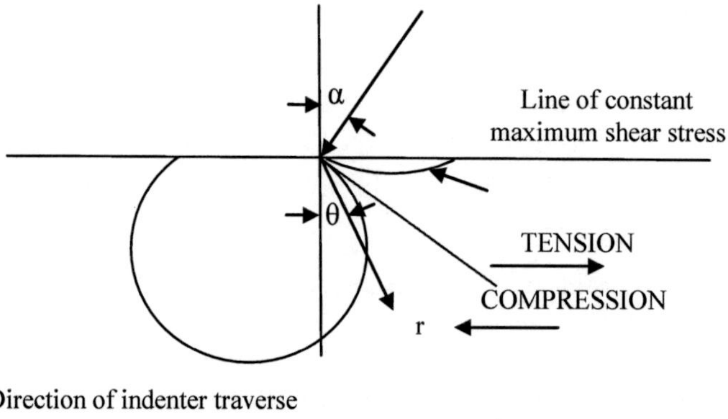

Fig. 1 Effect of point load (plain strain) inclined to a surface at an angle α [7]

ratio values of maximum uncut chip thickness and high values of force. In general compressive stress is infinite at the point of contact and reduces with increasing radius of contact. In practice, infinite stress is impossible since the material yields when contact gives rise to a finite contact region. The material becomes plastic whenever the stress reaches the yield point. The mode of material removal can be brittle or ductile depending on two criteria; resolved tensile stress on the cleavage plane and shear stress on the slip plane. Prominence of any one criterion depends on the size of the stress field caused by the particular machining process. The critical tensile stress, sensitive to pre-existing defects, showed a marked size effect, with the ductile mode being associated with small depths of cut [8]. There is no quantitative understanding of the way in which the sliding contact stress field is modified over stress-free surfaces. Until such information becomes available, it is impossible to quantify this effect. The above-response phenomenon has important implications on application of acoustic emission sensing in instrumented scratch tests [11]. Studies have been carried out on the scribing characteristics of brittle materials to identify the ductile to brittle region [12]. Micro-cracking and plastic deformation patterns were rationalized using elastic stress field by Boussinesq- and Cerruti-field solution on ZrB2–5 wt% SiC [4]. The study focuses on the effect of normal load, angle of inclination, grooving speed and point radius during the indentation process with respect to the stress field. It is observed in literature, AE has good correlation between strain and acoustic emission (AE) in a cyclic loading process [1]. AE has been used to characterize the material removal mechanisms in zirconia toughened alumina composite [5]. AE has been related to residual stresses [3] and frequency parameters have also been used in order to characterize the occurrence of microdamage in zirconia-based ceramics [9]. The study also aims to find the correlation between the AE parameters and contact stress values, in order to effectively characterize the process.

2 Experimental Procedure

The study on precision grooving of alumina ceramic (AD 998 grade) has been carried out to assess the possibility of ductile regime machining. The material removal mechanisms are studied by monitoring the machining dynamics.

As a part of the study, hinge-based scratch setup was designed and fabricated as shown in Fig. 2. In this study, hinged-based scratch setup used to simulate flexible abrasion process (i.e., resinoid bonded abrasion). The setup is mounted on aluminum base with anti-vibration mount with arrangement of two hinges cascaded in a horizontal frame. The set up is provided with a belt drive and rack and pinion arrangement for indenter traversing/grooving as shown in Fig. 2. The setup has provision for positioning of AE sensor below the workpiece, mounted on dynamometer for force monitoring; the signals were processed through ADLINK DAQ card (Model No.: 9812). Variation of the normal force and tangential force components were recorded. The experiments were conducted with levels as shown in Table 1. The stepper motor is controlled by speed and position using a controller interfaced with PC. The approach

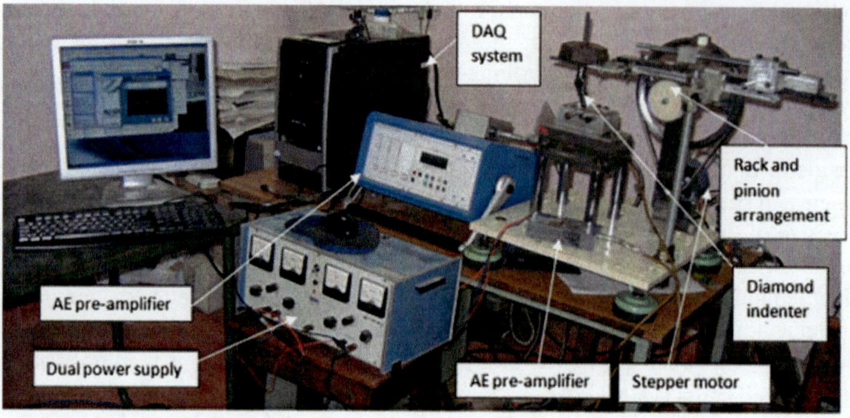

Fig. 2 Hinge-based scratch setup

Table 1 Levels of the experiments

Parameter	Levels
Normal load (N)	5, 10, 15
Angle of inclination (°) γ	−10, −5, 0, 5, 10
Grooving speed (mm/s)	16, 20
Point radius of the indenter (μm)	58, 45

angle is the angle between the axis of the indenter and the specimen normal as shown in Fig. 3. The direction of the tangential force component (F_y) is indicated in Fig. 3.

Fig. 3 Inclination of the indenter **a** positive, **b** negative

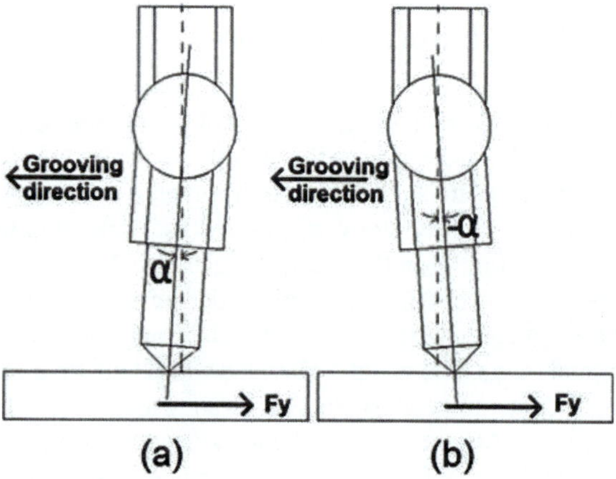

3 Contact Stress Evaluation

Stresses developed in the work-tool interface during the grooving process were evaluated for different grooving conditions (Table 1). Under high local stress, surface and sub-surface cracks can interact causing the detachment of small chips of material resulting in material dislodgement and surface roughness. Cracks nucleated close to the surface may also propagate further into the bulk, leading to macroscopic brittle failure [11].

Stress along different directions indicated in Fig. 4 of the indenter is given by the following equations [10]:

Stress along the width (groove) direction (σ_x):

$$\sigma_x = -\frac{b}{\pi\Delta}\left[\begin{array}{l} z\left(\frac{b^2+2z^2+2x^2}{b}\phi_1 - \frac{2\pi}{b} - 3x\phi_2\right) \\ +\beta\left\{\left(2x^2 - 2b^2 - 3z^2\right)\phi_2 + \frac{2\pi x}{b} + 2\left(b^2 - x^2 - z^2\right)\frac{x}{b}\phi_1\right\} \end{array}\right] \quad (1)$$

Stress along the depth direction (σ_z):

$$\sigma_z = -\frac{b}{\pi\Delta}\left\{z(b\phi_1 - x\phi_2) + \beta z^2\phi_2\right\} \quad (2)$$

Stress along the shear stress direction (σ_{xz}):

$$\sigma_{zx} = -\frac{b}{\pi\Delta}\left[z^2\phi_2 + \beta\left\{(b^2 + 2x^2 + 2z^2)\frac{z}{b}\phi_1 - 2\pi\frac{z}{b} - 3xz\phi_2\right\}\right] \quad (3)$$

where ϕ_1 and ϕ_2 are

$$\phi_1 = \frac{\pi(M+N)}{MN\sqrt{(2MN + 2x^2 + 2z^2 - 2b^2)}}$$

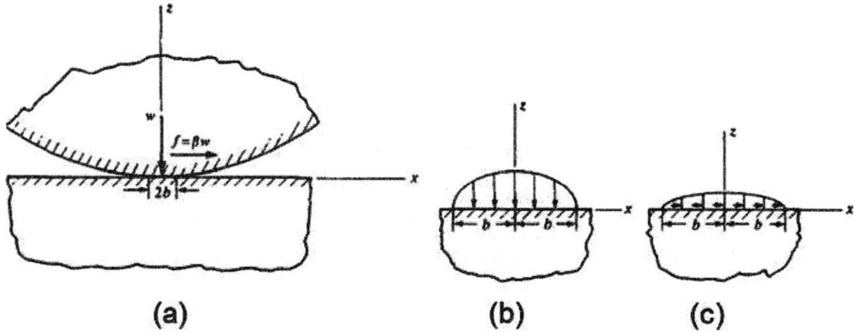

Fig. 4 **a** Sphere and flat surface contact simulating indenter point on flat surface, **b** stress acting along depth direction (*Z*), **c** stress acting along width direction (*X*)

$$\phi_2 = \frac{\pi(M-N)}{MN\sqrt{(2MN + 2x^2 + 2z^2 - 2b^2)}}$$

$$M = \sqrt{\{(b+x)^2 + z^2\}};$$

$$N = \sqrt{\{(b-x)^2 + z^2\}};$$

Δ---depends on the geometry of the bodies and material properties

$$\Delta = \frac{1}{A+B}\left(\frac{1-\upsilon_1^2}{E_1} + \frac{1-\upsilon_2^2}{E_2}\right)$$

$$A = \frac{1}{4}\left(\frac{1}{R_1} + \frac{1}{R_2} + \frac{1}{R_1'} + \frac{1}{R_2'}\right) + \frac{1}{4}\sqrt{\left[\left\{\left(\frac{1}{R_1} - \frac{1}{R_1'}\right) + \left(\frac{1}{R_2} - \frac{1}{R_2'}\right)\right\}^2 - 4\left(\frac{1}{R_1} - \frac{1}{R_1'}\right)\left(\frac{1}{R_2} - \frac{1}{R_2'}\right)\sin^2\alpha\right]}$$

$$B = \frac{1}{4}\left(\frac{1}{R_1} + \frac{1}{R_2} + \frac{1}{R_1'} + \frac{1}{R_2'}\right) - \frac{1}{4}\sqrt{\left[\left\{\left(\frac{1}{R_1} - \frac{1}{R_1'}\right) + \left(\frac{1}{R_2} - \frac{1}{R_2'}\right)\right\}^2 - 4\left(\frac{1}{R_1} - \frac{1}{R_1'}\right)\left(\frac{1}{R_2} - \frac{1}{R_2'}\right)\sin^2\alpha\right]}$$

$$b = \sqrt[3]{\frac{3W_yR}{4E_{sr}}}$$

$$W_y = 21.17(200e - 6)^2(y)\left(\frac{y}{E_{sr}}\right)^2$$

$$E_{sr} = \frac{1}{\left(\frac{1-\upsilon_1^2}{E_1} + \frac{1-\upsilon_2^2}{E_2}\right)}$$

3.1 Stress Along Width Direction of the Groove (σ_x)

Stress along the width direction was evaluated for different conditions of the grooving process. Typical profile of stress along the width of the groove is given in Fig. 5.

It is observed in Fig. 5, the stress distribution to be more compressive symmetric in nature around the center of the tip. As the depth increases, the stress changes from compressive mode to tensile mode. This may be due to side flow of material around the indenter.

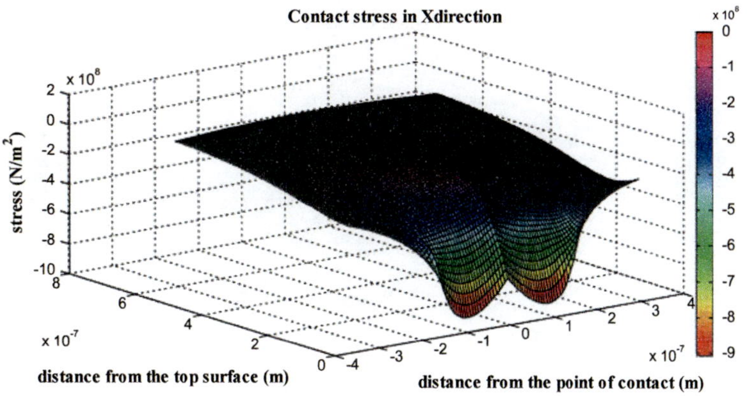

Fig. 5 Stress in width direction (*X*) of the groove

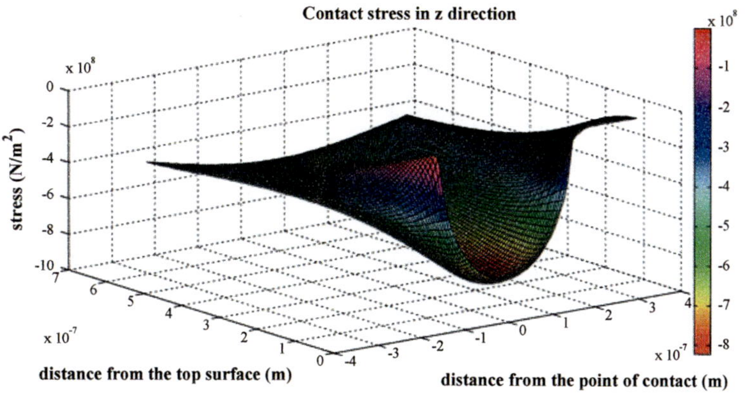

Fig. 6 Stress in depth direction (*Z*) of the groove

3.2 Stress Along Depth Direction of the Groove (σ_Z)

It is observed that the maximum stress occurs at the center of the tip and at very small distance from the top surface of the workpiece as shown in Fig. 6. The distribution is symmetric in nature. The stress is compressive in nature.

3.3 Stress Along the Shear Direction of the Groove (σ_{XZ})

It is observed that when coefficient of friction μ is less than 1, maximum tensile shear stress is to be maximum at small distance from the center of the contact point as shown in Fig. 7. It is also observed that at the center of the indenter, the stress

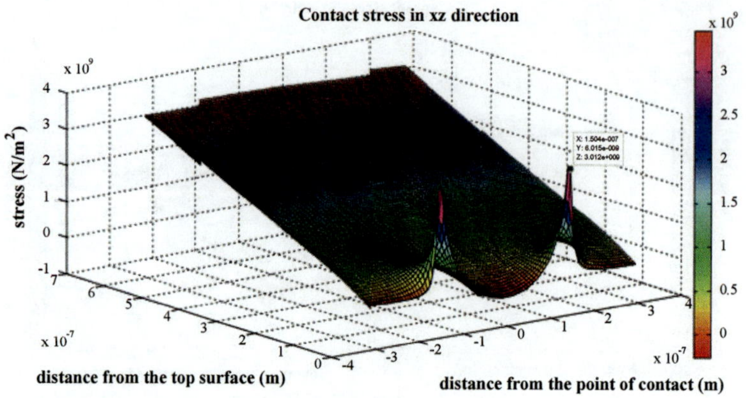

Fig. 7 Stress in shear direction (*XZ*) (coefficient of friction = 0.43)

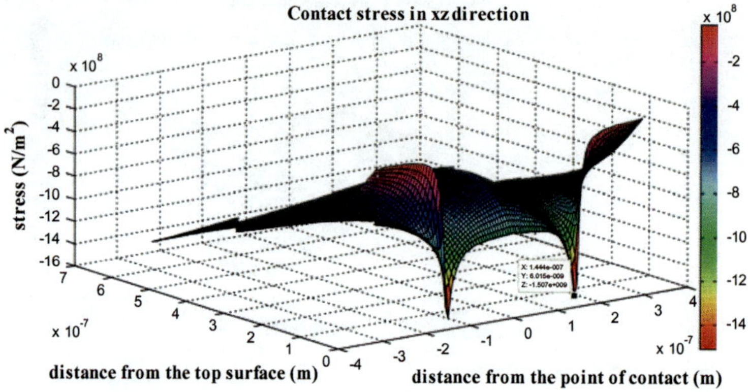

Fig. 8 Stress in shear direction (*XZ*) (coefficient of friction = 1.20)

tends to be compressive. Like other stresses the stress distribution is symmetric in nature. If $\mu < 1$ less resistance to traverse, i.e., less upsetting. If $\mu > 1$, more upsetting is observed.

If the coefficient of friction is greater than 1, maximum compressive shear stress is observed to occur at small distance from the center of the contact point as shown in Fig. 8. It is also observed that at the center of the indenter, the stress tends to be less compressive.

4 Parametric Effect on the Stress

Contact stress evaluated from the monitored grooving force for grooving speed of 20 mm/s is presented in the following illustrations. Peak stresses in the stress contour

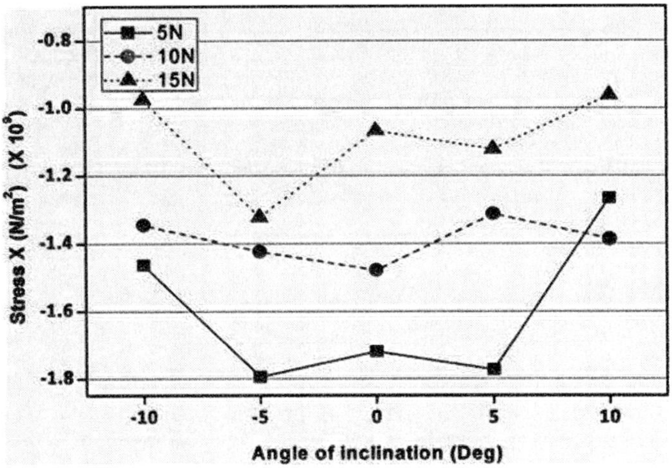

Fig. 9 Variation of stress along width (σ_x) with angle of inclination

are considered for evaluation. Stress along width direction is more compressive at 5 N loading condition, which may be due to increased side flow of the material during plowing dominant grooving as shown in Fig. 9. With 5 N loading condition, the stress along the width (X direction) decreases up to 0° inclination angle followed by an increase. With 10 N loading scheme, similar trend is observed with higher-order stress values more or less steady around 1.4 * 10^9 MPa. While with 15 N loading conditions, stress in *X* direction decreases up to −5° and increases with 0°. With positive inclination angles, contact stress increases which indicates mode change. Stress along the X direction increases at higher load. The reduced side flow with higher loads of indenter could have resulted in lower-order compressive stress (i.e., more tensile). It is also observed that for 5 and 10 N loading at higher inclination angle (negative inclination), stress along X direction tends to be more compressive, indicating ductile mode of grooving.

It is observed that stress along width direction (*X* direction), depth direction (*Z* direction) and the shear direction (*XZ* direction) increases at higher normal load as shown in Figs. 9, 10 and 11, respectively. It confirms that with higher normal load of indenter (up to 10 N), ductile mode of material removal occurs. For 5 N loading scheme, stress along *Z* direction tends to be more compressive with inclination angle up to 5°, above which the stress tends to be tensile. At −10°, indenter tends to bounce associated with unsteady grooving. With 10 N loading scheme, the response of the stress *Z* tends to be invariant except with higher inclination angle indicating more or less steady grooving. With 15 N loading condition, tends to be more compressive up to −5° above which the stress tends to be tensile, where mode change occurs. The stress tends to be less compressive (tensile) with increasing indentation load (normal load) resulting in less compressive stress, possibly due to occurrence of median cracking at higher loads.

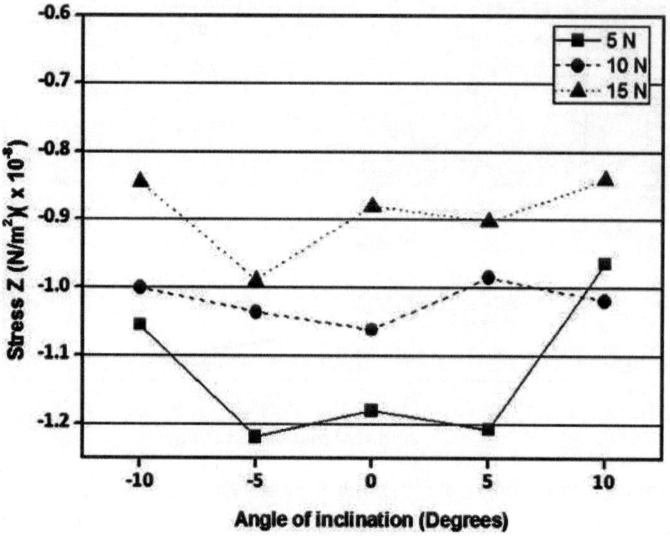

Fig. 10 Variation of stress along depth (σ_z) with angle of inclination

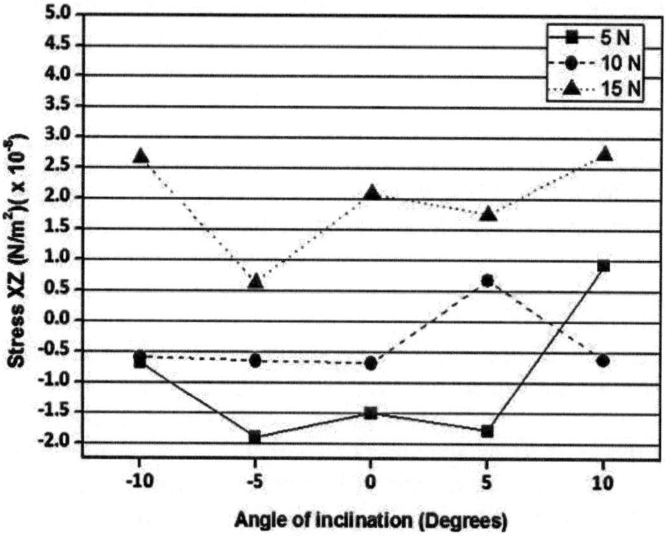

Fig. 11 Variation of stress along shear direction (σ_{xz}) with angle of inclination

Considering the stress along shear direction (σ_{xz}) at 5 N, (Fig. 11), it tends to be more compressive up to 5° angle of inclination and above which it changes to tensile indicating possible change in mode of grooving (partial ductile). With 10 N load mostly steady grooving is observed, whereas with 15 N loading condition, more tensile mode of grooving is observed, which leads to rupture of the material.

Observation of the stress along width, depth and shear directions indicate that stress along depth with 5 and 10 N loads tends to facilitate in ductile mode. It also shows that in 15 N loading conditions leads to brittle mode.

Referring to the illustration in Figs. 10, 11 and 12, it is seen that with smaller load of 5 N, the stresses along width, depth and shear direction tends to be more compressive, indicating dominant ductile (rupture free) dislodgement of material. With higher loads (15 N), dominant brittle mode can be inferred. Also, grooving in upsetting regime (−10°, −5° inclination angle) favor ductile condition. As a whole it can be summed up that with negative inclination angle (approach), 20 mm/s speed of grooving and 5–10 N loading will facilitate ductile mode of grooving.

Combined parametric influence on stress (σ_x, σ_z and σ_{xz}) is illustrated in Fig. 12. It can be inferred that mostly with 5 and 10 N loading dominant ductile mode occur, while with 15 N loading dominant brittle mode results.

Typical variation of the interface stress around width direction (σ_x) influenced by inclination angle and point radius of the indenter is shown in Fig. 13. It is seen that

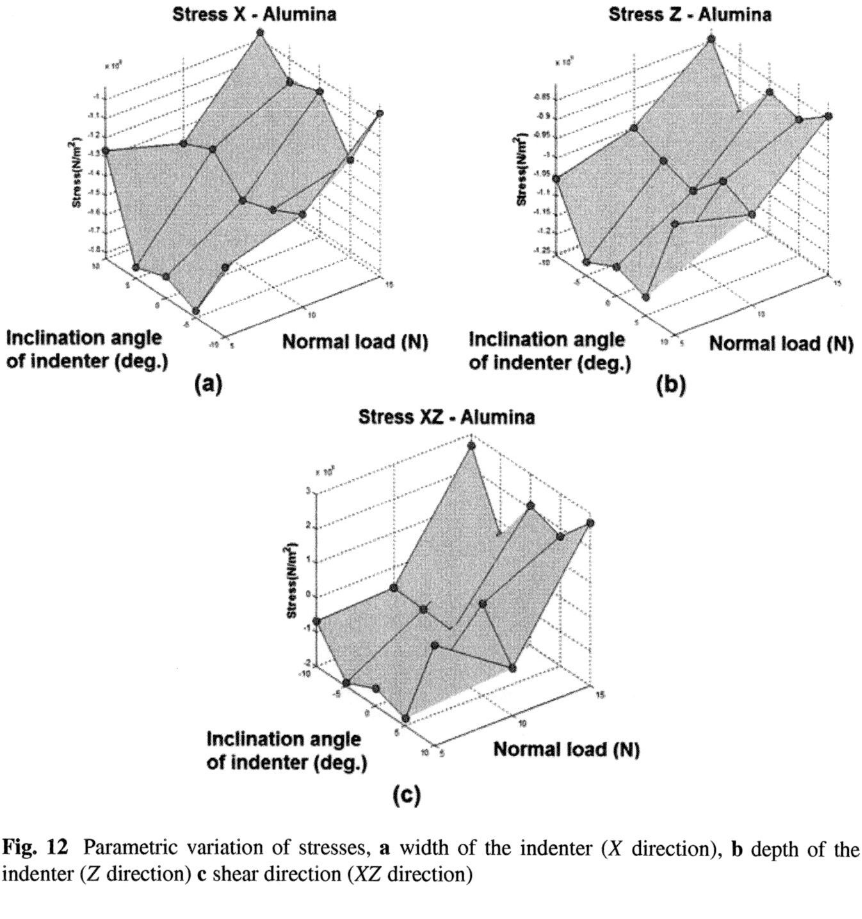

Fig. 12 Parametric variation of stresses, **a** width of the indenter (*X* direction), **b** depth of the indenter (*Z* direction) **c** shear direction (*XZ* direction)

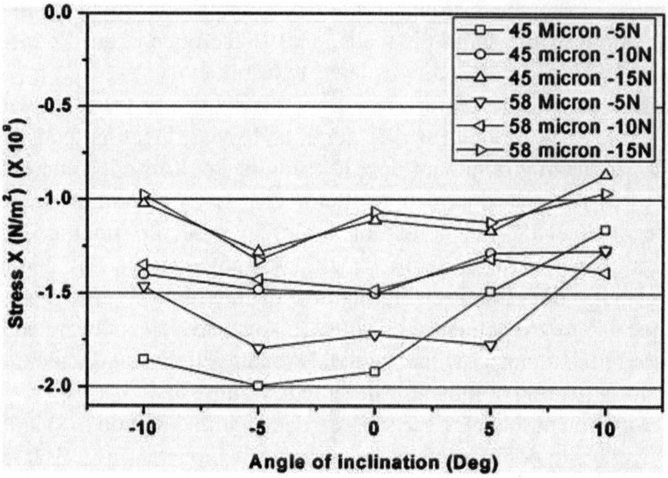

Fig. 13 Variation of stress along width at different indenter radii at 20 mm/s

barring 5 N loading, both the indenters exhibit only a marginal variation for 10 and 15 N loading. The stress along the width direction *X* tends to be less compressive with increasing (+ve) inclination angles indicating possible ductile brittle mode of grooving. Also grooving with indenter of point radius 45 μm exhibits more compressive stress. A similar trend can be seen with stress along *Z* direction as shown in Fig. 14. The observation on σ_{xz} (Fig. 15) indicates that mostly with 45 μm indenter tensile stress occurs. In 58 μm indenter, mostly compressive stress is observed at 5

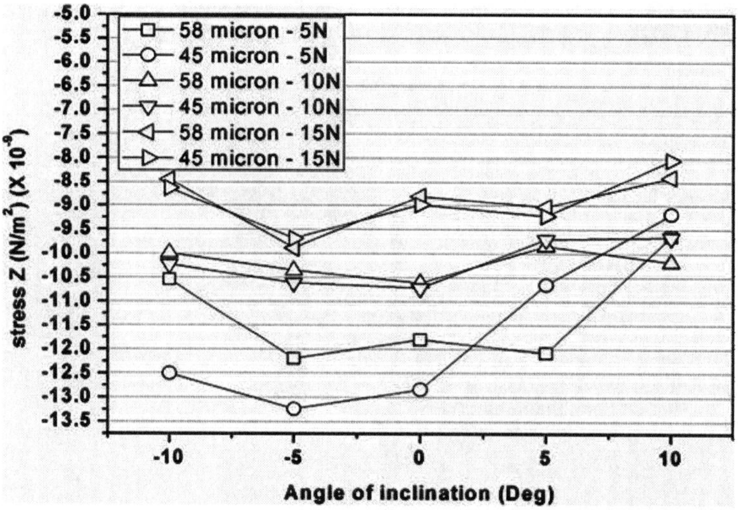

Fig. 14 Variation of stress along depth direction at different indenter radii at 20 mm/s

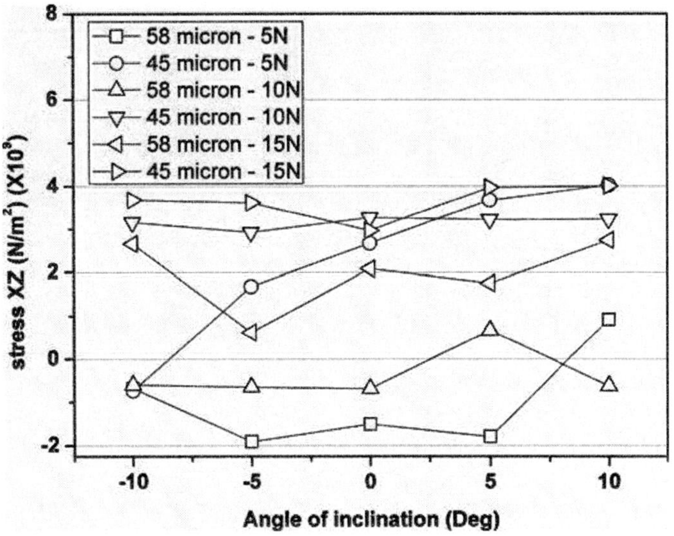

Fig. 15 Variation of stress along shear direction at different indenter radii at 20 mm/s

and 10 N loading conditions. It shows that 15 N loading exhibits dominant brittle mode of grooving.

In the upsetting regime (−10° to 0°), mostly higher-order component (σ_x) is seen with 45 μm indenter favoring ductile mode. In the deformation regime (0° to 10°), less compressive stress is observed indicating partial ductile to brittle mode of grooving. The relatively sharper indenter can facilitate less bouncing (steadier grooving) compared to 58 μm indenter. Figure 13 supplements the above observation. The use of a relatively sharper indenter with low to medium loads is advantageous for ductile mode of scratching [3].

Considering the variation of stress along width (X) at different speeds (Figs. 16, 17 and 18), it is observed that there is a reduction in stress with increase in speed of grooving. With increasing load, stress along width tends to be less compressive, whereas with increasing speed it tends to be more compressive as seen in Fig. 16. Mixed mode of grooving is observed for 16 mm/s. With inclination angles of −10° and −5° higher-order compressive stress occurs compared to 5° and 10° inclination angles.

Typical variation of σ_z with grooving parameter is shown in Fig. 17. For 58 μm point radius, it is seen that with increasing load of indenter less compressive (σ_z) stress occurs. Also with increasing speed of grooving the stresses along both x and z directions is relatively more compressive favoring ductile grooving. Also increasing load is associated with relatively less compressive stress. This implies that the mode of grooving with small to medium load and higher speed will be dominantly ductile, whereas it tends to be brittle at lower speed. It is seen that with 20 mm/s, mostly the observations on stress σ_x, σ_z and σ_{xz} indicate ductile mode of grooving, especially with −5°–0° inclination angle.

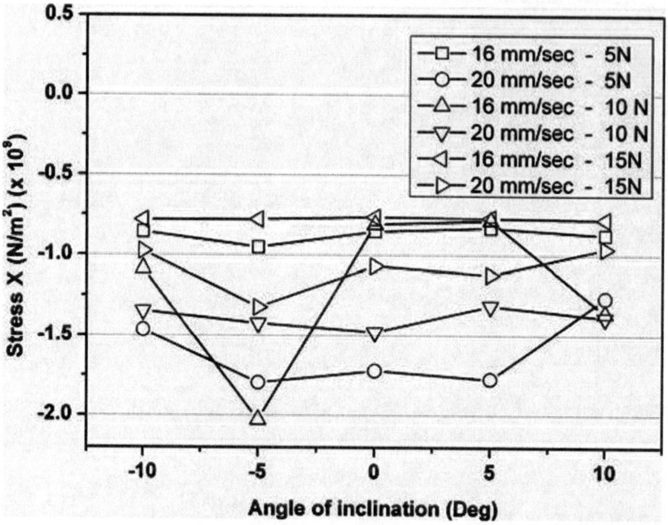

Fig. 16 Stress along width direction (X) at different grooving speed with 58 μm indenter

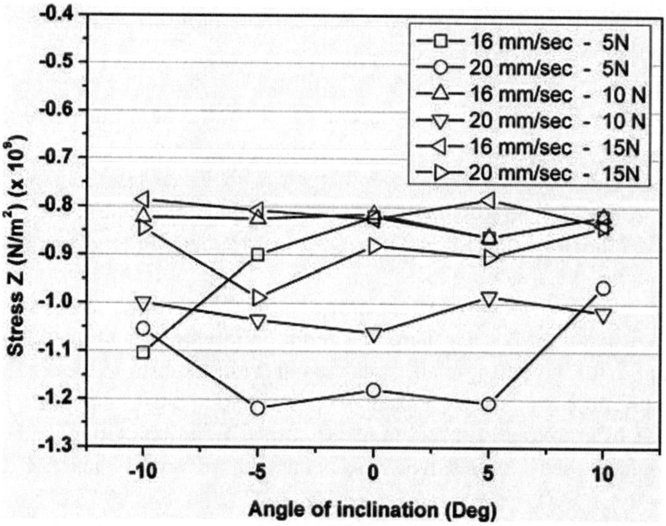

Fig. 17 Stress along depth direction (Z) at different grooving speed with 58 μm indenter

Typical observed variation of stress σ_{xz} influenced by grooving parameters is shown in Fig. 18. Stress along XZ direction tends to be relatively less tensile with increasing speed of grooving. Also with increasing inclination angle (more positive approach) the stress is more tensile. Relatively higher tensile stress can be seen with higher load (15 N). Referring to the illustrations on σ_x, σ_z and σ_{xz}, it can be inferred

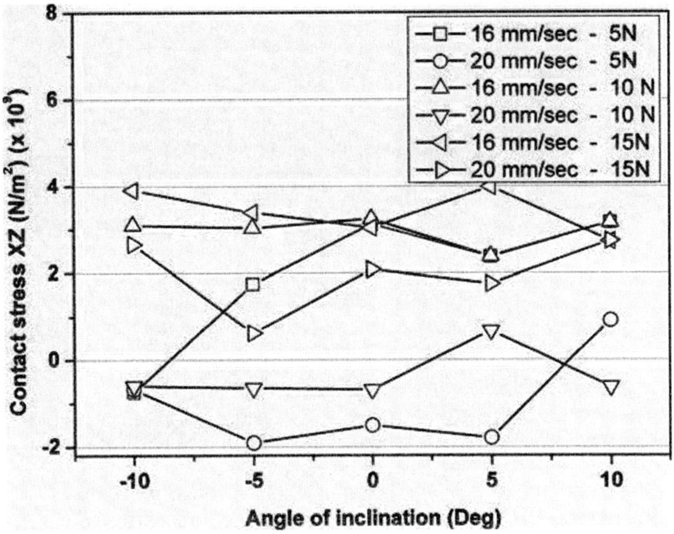

Fig. 18 Stress along shear direction (*XZ*) at different grooving speed at 58 μm indenter

that with 5 N loading dominant ductile mode is seen to occur up to 0° (upsetting regime) while with 10 N loading, mostly ductile mode is seen to occur, and 15 N loading facilitates dominant brittle mode of grooving.

5 Correlation with Stress Waves with AE Responses

For validating the contact stresses, acoustic emission (AE) parameters are taken into consideration. AE signals are basically elastic stress waves emanated from the material under stress during the grooving process. Hence, a correlation between the contact stress and consequent AE features can be used to advantage to monitor the acoustic emission as an indirect indicator for induced stresses during machining. The relaxation energy depends on the resilient energy or strain energy. In a cyclic load testing of barium osumilite ceramic matrix composite, the strain observed was found to have good correlation with AE hit (i.e., RDC) [7]. Some parameters of AE which are taken for the correlation are RMS, number of events and ring down counts (RDC) [6]. The features are illustrated in Fig. 19.

One of the important features of the monitored acoustic emission signal is its RMS value. AE RMS value indicates proportionally the energy release rate: hence, characterizing the monitored raw AE signal in terms of its RMS value can be a useful indicator. Accordingly the RMS of the monitored acoustic emission from the alumina during grooving has been correlated with evaluated contact stress components along the groove-width direction along the groove-depth direction and resultant shear direction. Grinding of ceramics is a tricky phenomenon, controlling the depth

Fig. 19 Parameters of AE

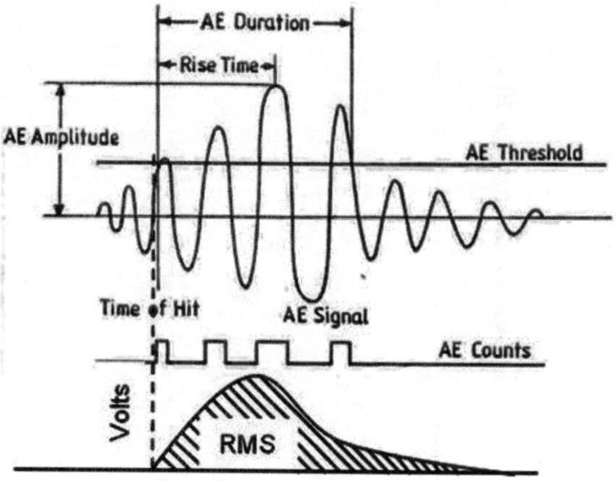

of cut, monitoring the AE parameters and correlating contact stress during machining process will help to identify the mode of grooving. Accordingly suitable decisions, such as change speed of grinding, depth of cut and orientation of indenter can be done.

Typical observed dependency of AE response on stress with RMS during grooving of alumina ceramic is presented in Fig. 20. From Fig. 20a, it is seen that for 5 N loading with increase in AE RMS up to certain contact stress value above which a reduction occurs (contact stress tending to be less compressive). Similarly, the AE RMS tends to drop down with increasing tensile stress along depth of the groove (σ_z) as shown in Fig. 20b. This is also reflected in Fig. 20c. It indicates ductile or partial mode of grooving. For 10 N loading condition, with increasing tensile load σ_x AE RMS tends to drop and rise while AE RMS increases with increasing stress along depth direction (σ_z). A proportional rise in AE RMS with increasing tensile (σ_{xz}) can be seen. This indicates ductile mode of grooving. For 15 N loading condition, a proportional rise in AE RMS with increasing stress along width of the groove (σ_x), stress along depth of the groove (σ_z) and stress along the shear direction (σ_{xz}). It indicates partial ductile or brittle mode of grooving at higher loading condition. Higher-order RMS is observed at 15 N loading condition indicating brittle rupture of the material.

Figure 21 shows the relationship between number of events and stress along width of the groove (σ_x), stress along depth of the groove (σ_z) and stress along the shear direction (σ_{xz}). An increase in the number of events occurs with increasing compressive stress along width direction (σ_x) and depth direction (σ_z). The decreasing trend for 5 N loading condition may be attributed to the plowing process. With 10 N loading, the number of events reduces with increasing compressive stress along the width of the groove (σ_x), depth of the groove (σ_z) and shear direction of the groove (σ_{xz}). In 15 N loading, the number of events tends to drop down and increases with stress along the width of the groove (σ_x), depth of the groove (σ_z) and shear direction of the groove (σ_{xz}). The decreasing trend in 10 and 15 N loading conditions may be attributed to ductile mode and micro-fracture mode of material, respectively.

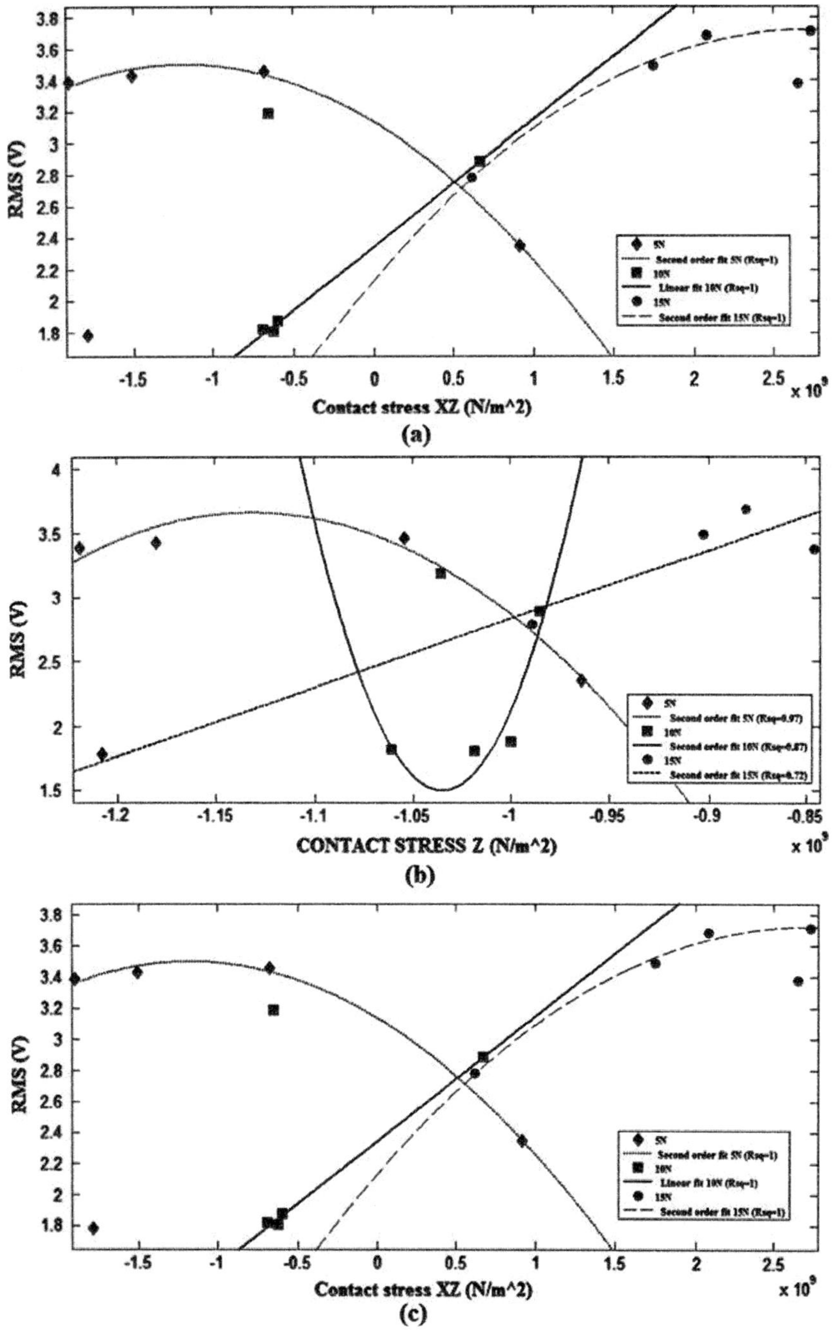

Fig. 20 Correlation of stress with cumulative RMS at different loading conditions **a** along width direction, **b** along depth direction, **c** along shear direction in 58 μm indenter at 20 mm/s

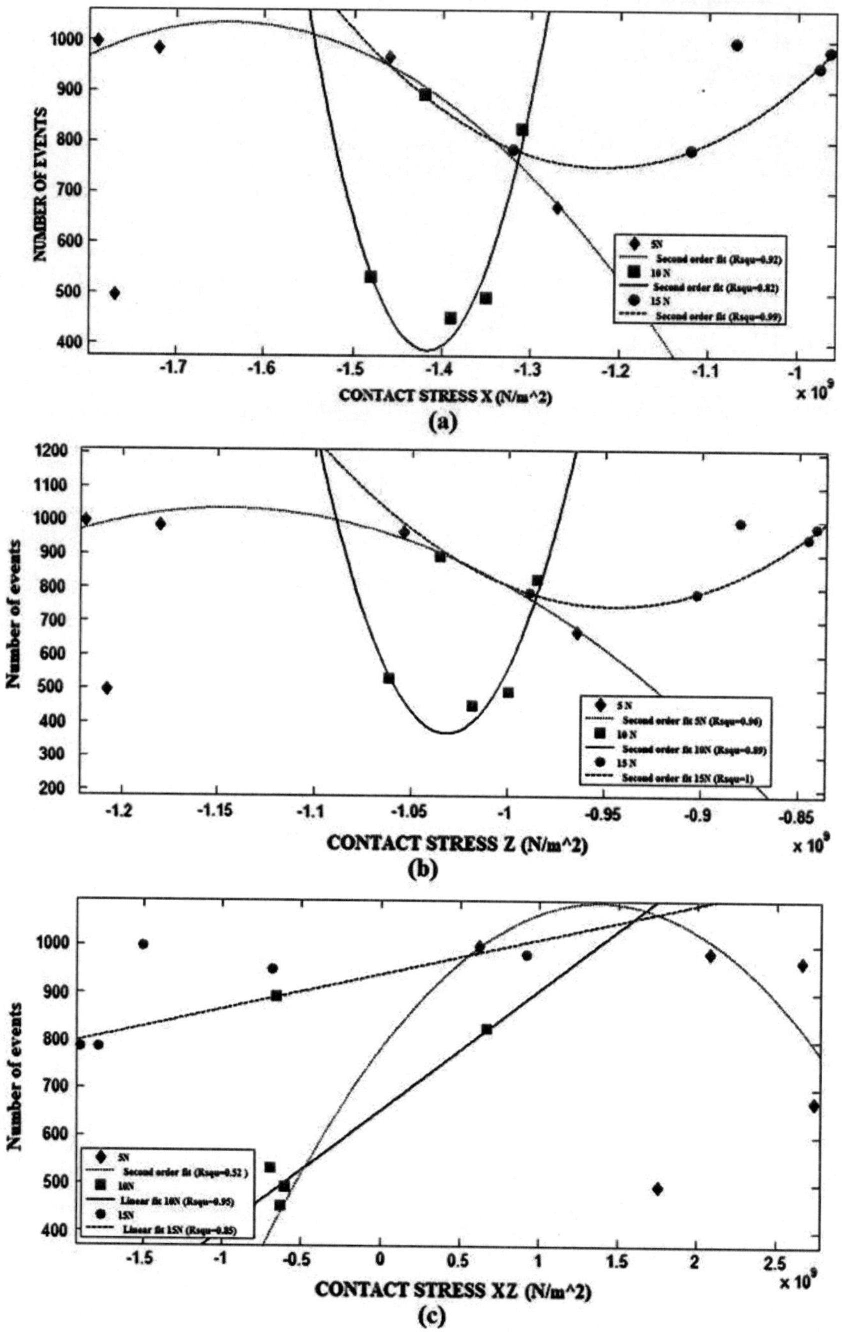

Fig. 21 Correlation of stress with number of events at different loading conditions **a** along width direction, **b** along depth direction, **c** along shear direction in 58 μm indenter at 20 mm/s

The observed relationships between RDC and stress components are presented in Fig. 22. Higher RDC are preferred for ductile mode of material removal which is clearly observed at compressive contact stresses. In all the loading conditions, RDC drops down with increase in tensile stress along the width of the groove (σ_x), depth of the groove (σ_z) and shear direction of the groove (σ_{xz}), which may be attributed to crack formation. More scatter is observed in 5 N loading condition compared to other loading conditions indicating plowing mode of material removal which is indicated in the correlation coefficient values. The illustrations in Fig. 22 clearly indicate correlation between the AE features and the stresses along depth, width and shear directions (i.e., R-square values for 5 N–0.68, 0.71, 0.6, for 10 N–0.80, 0.81, 0.95 and for 15 N–0.71, 0.92, 0.71).

6 Conclusions

The monitoring of grooving force and acoustic emission signal help to evaluate the contact stress and the corresponding material response in terms of acoustic emission, and thereby identify the mode of grooving.

(1) A significant variation in contact stress along the directions of groove width, depth and resultant shear direction from compressive to tensile mode with grooving conditions indicates change in mode of grooving.
(2) Stress along shear has maximum tensile value when coefficient of friction is less than 1, whereas it is compressive when coefficient of friction in more than 1.
(3) With variation of angle of inclination, stresses changes significantly. While stresses along width direction (σ_x), are of tensile nature at 0° inclination angle, stresses along the depth and shear directions are compressive; with extreme positive and negative inclination it is vice versa.
(4) It is also observed that at speed of 20 mm/s, the mode of material removal is ductile, whereas at speed of 16 mm/s, it is brittle.
(5) AE parameters like RMS, number of events and RDC of the signal has strong correlation with stress values.

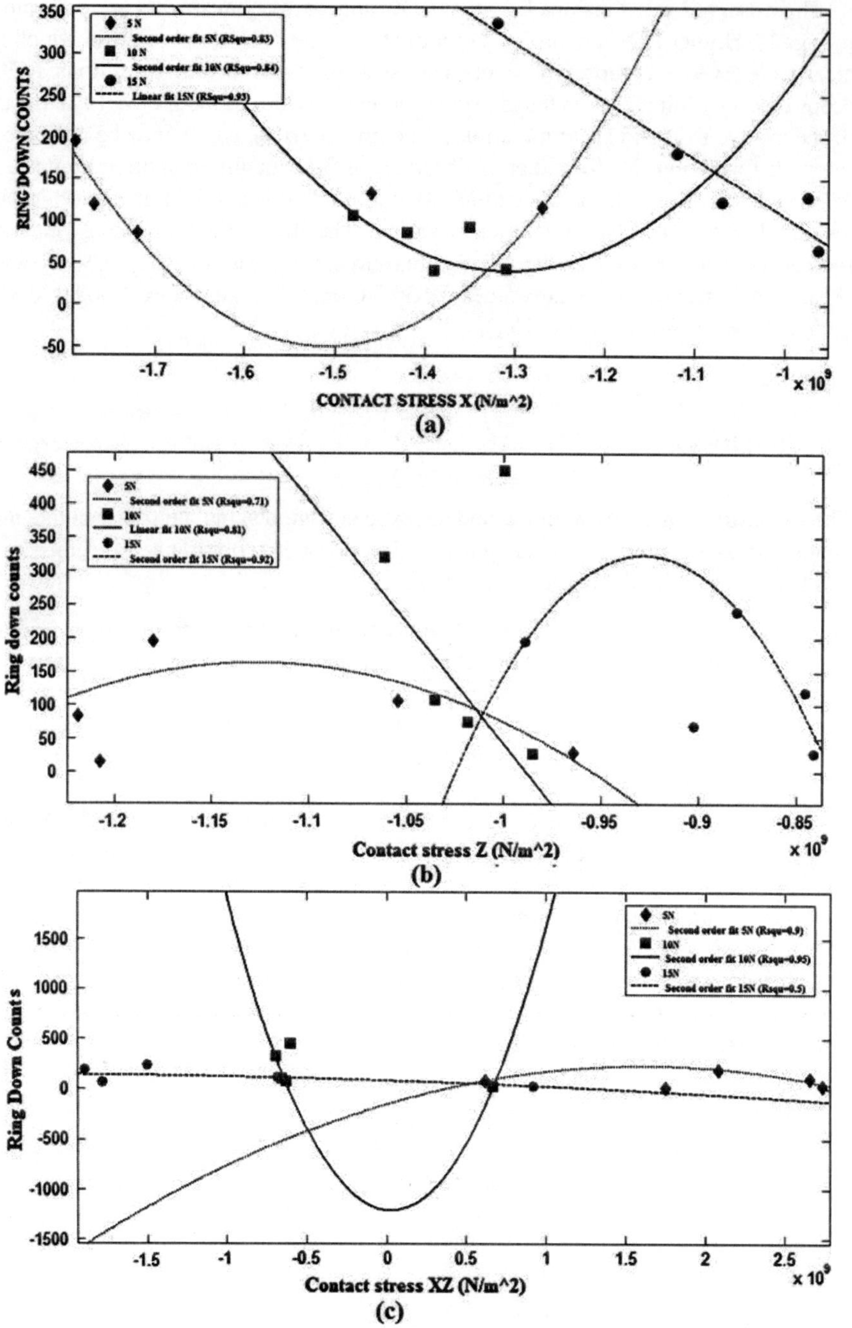

Fig. 22 Correlation of stress with ring down counts (RDC) at different loading conditions **a** along width direction, **b** along depth direction, **c** along shear direction in 58 μm indenter at 20 mm/s

References

1. Aggelis D, Dassios K, Kordatos E, Matikas T (2013) Damage accumulation in cyclically-loaded glass-ceramic matrix composites monitored by acoustic emission. Sci World J
2. Axen N, Kahlman L, Hutchings I (1997) Correlations between tangential force and damage mechanisms in the scratch testing of ceramics. Tribol Int 30(7):467–474
3. Evans A (1975) Residual stress measurement using acoustic emission. J Am Ceram Soc 58(5-6):239–243
4. Ghosh D, Subhash G, Radhakrishnan R, Sudarshan TS (2008) Scratch-induced microplasticity and microcracking in zirconium diboride-silicon carbide composite. Acta Mater 56(13):3011–3022
5. Giridhar D, Vijayaraghavan L, Krishnamurthy R (2012) Acoustic emission response of sintered alumina zirconia composite during grooving process. NDT E Int 46:55–62
6. Krishnamurthy R (1993) Acoustic emission for condition monitoring in manufacturing. In: Paper Presented at the Frontiers of Tribology and condition monitoring June 19, 1993. Chennai, India
7. Marinescu ID, Dimitrov B (2004) Tribology of abrasive machining processes. William Andrew Pub
8. Ngoi B, Sreejith P (2000) Ductile regime finish machining—a review. Int J Adv Manuf Technol 16(8):547–550
9. Patapy C, Proust A, Marlot D, Huger M, Chotard T (2010) Characterization by acoustic emission pattern recognition of microstructure evolution in a fused-cast refractory during high temperature cycling. J Eur Ceram Soc 30(15):3093–3101
10. Shukla A, Nigam H (1985) A numerical-experimental analysis of the contact stress problem. J Strain Anal Eng Des 20(4):241–245
11. Swindlehurst W, Wilshaw T (1976) Acoustic emisison from microcracks during sliding contact. J Mater Sci 11(6):1183–1186
12. Wang JJJ, Liao YY (2008) Critical depth of cut and specific cutting energy of a microscribing process for hard and brittle materials. J Eng Mater Technol 130(1):8–14

Numerical Analysis of Autofrettaged High-Pressure Aluminium Cylinder

Neelkant Patil, Shivarudraiah, and Kalmeshwar Ullegaddi

Abstract The main objective of this research article was to investigate the influence of autofrettaged cylindrical high-pressure vessels used for abrasive jet cutting machines and to Estimate the induced residual stresses during unloading conditions using finite element analysis. Autofrettaged cylinders were subjected to hydraulic pumps to study the bursting pressure along with different orientation (0°, 30°, 60°, 90°) of crack. In addition, the strength of the vessel was determined by using Brinell's hardness equipment along with the radial thickness of the cylinder. A numerical investigation on the autofrettaged cylinder was done by using Abaqus 6.13 with a 2D four-node bilinear solid plane element for nonlinear analysis.

Keywords Autofrettage · Mandrill · Aluminium cylinder · Residual stress · Brinell's hardness

1 Introduction

Autofrettage is the recently developed technology in which hydraulic pressure has been introducing at the one end of the cylinder specimen (another end is closed), due to which high pressure deformed the internal surface of the specimen beyond the elastic limit, as a result, residual stresses are formed due to a sudden release of working pressure. Residual stresses are at an inner portion of the cylindrical cross section, but the Bauschinger effect can decrease compressive residual hoop stresses near the bore of the cylinder. Therefore, components are widely used in nuclear power plants, chemical industries, aerospace, battle gun barrel and many other applications [1]. Many researchers investigated to understand the autofrettage concept and effect of autofrettage parameters on stress behaviour and pattern of distribution in pressure

N. Patil (✉) · Shivarudraiah
Department of Mechanical Engineering, UVCE, Bengaluru 560001, India
e-mail: patil.chambol@gmail.com

K. Ullegaddi
Department of Mechanical Engineering, JSS Academy of Technical Education, Bengaluru 560060, India

A. Arockiarajan et al. (eds.), *Advances in Industrial Automation and Smart Manufacturing*, Lecture Notes in Mechanical Engineering,
https://doi.org/10.1007/978-981-15-4739-3_53

cylinder/vessels. In this research work, the primary focus on the stress behaviour inside the wall of both autofrettaged cylinder and the non-autofrettaged cylinder was compared. Also, the relation between thick cylinder dimensional change and stress/strain distribution is formulated. It was shown that in the autofrettage process, no permanent deformation/ no damage is seen but stress distribution was investigated [2].

In the present investigation, the thick-walled cylinder is subjected to a uniform internal pressure of P_{i}. Outside pressure (atmospheric pressure) is negligible compared to internal pressure. The temperature distribution inside the wall is also assumed to be negligible. The complete geometrical test specimens were developing using cylindrical coordinates systems [3]. Wahi et al. [4] autofrettaged spherical pressure cylinders subject to distinctive autofrettage pressures assessed and the residual stress distribution were found. Results are collected by adding to an augmentation of variable material properties like Young's modulus and Poisson's ratio at different temperature conditions (VMP) methods and are confined with results from the finite element method. It is shown that utilizing the material model can cause a significant error in the assessment of residual hoop stress, particularly close to the inner surface of the cylinder. The optimum autofrettage pressure for making the desired residual stress state is presented and validated with experimental results. Farrahi et al. [5] studied on the residual stresses in an autofrettaged thick cylinder made of the functionally graded metal–ceramic composite are assessed by the variable material properties. They conclude that compressive residual stresses at the inner section to the outer radius have better fatigue life and load-bearing capacity of autofrettaged cylinder. Wire winding and autofrettage technique for the thick-walled cylinder were concentrated [6]. By investigating the impact of wire winding on an autofrettaged thick-walled cylinder utilizing variable material properties method, it was used to calculate the residual stress in the autofrettaged cylinder. The wire-wound autofrettaged vessel is validated by the finite element method, and it demonstrates that the residual hoop stress in a wire-wound autofrettaged cylinder has more stress distribution in the vessel. Autofrettage process is limited to metal materials and which is having high toughness. Autofrettage process is not possible to adopt in polymers and ceramic materials [7], metal plasticity behaviour several techniques address the limitation low-temperature autofrettage of the vessel shows higher compressive residual hoop stresses achieved by this technique [8]. The key reason is that stiffness and metal are higher at low temperature, and thermal strain accumulated during warm up to room temperature evaluates compressive stresses induced in the vessel components. Investigation of autofrettage process parameters on buckling behaviour of metallic cylinder is studied experimentally and numerically. Numerical analysis on autofrettaged cylinder is done in Abaqus 6.13, radial residual stresses and residual hoop stresses are correlated with the experimental results.

2 Numerical Studies

The 2D shell model was done in Abaqus 6.13, and simulation for the thick-walled cylinder was carried out on Abaqus CAE 6.13. The explicit analysis is used to study the behaviour of autofrettage on stress distribution among the geometrical model. A quarter of the geometry 2D model is considered in Fig. 1a, and in Fig. 1b the depth of the model in the z-direction is 1 mm. It assumed that this thickness is far enough from the two ends of the pipe; this dimension will not affect our studies.

In preprocessing, initially construct a 2D CAED model with Abaqus/CAE and follow steps to execute a given problem with input to deck preparation. The parameters need to be considered are the material properties, step time, meshing, loading, field output request and create the job in this manner to analyse the autofrettage process in the thick-walled cylinder.

The material properties obtained from testing, the engineering stress and engineering strain were plotted until the fracture of the specimen. From the engineering stress and engineering strain, the true stress and true strain are obtained from the given formulae as $\sigma_t = \sigma_e\ (1+\varepsilon_e)$ and $\varepsilon_t = \ln\ (1+\ \varepsilon_e)$, where σ_t is true stress, σ_e is engineering stress, ε_t is true strain, and ε_e is engineering strain. From the yield point of the material, the true stress and true strain were obtained. Figure 2a shows the elastic property (Young's modulus of 68,354.3 MPa) and Poisson's ratio of 0.33 input to the deck preparation. Figure 2b shows the plastic property of the material to the true stress versus true strain for the material Al 6351 T6. Figure 3a shows the global element size to discretize the model and curvature control to capture all areas of geometry. Figure 3b shows the meshing of the deformable body. Figure 4a, b shows the boundary condition of a body with *X*-symmetric and *Y*-symmetric, respectively.

Applying the internal pressure on a 2D planar model by restricted amplitude mode, the load applied in two steps is shown in Fig. 5a, b. In the initial stage, the

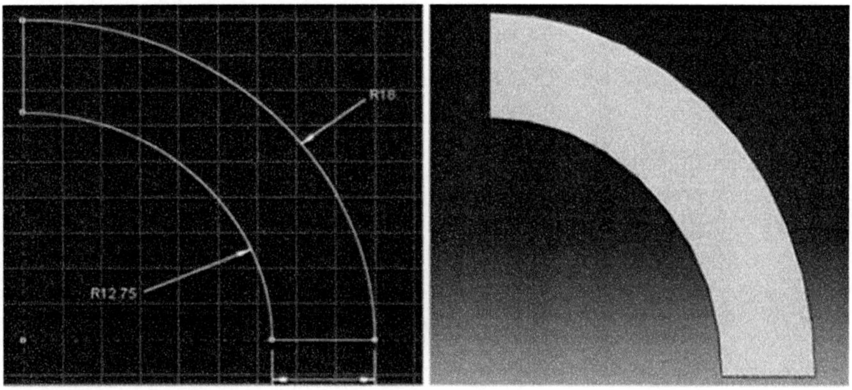

(a) (b)

Fig. 1 **a** Drawing thick-walled cylinder model in ABAQUS 6.13, **b** 2D planar cylindrical model in Abaqus 6.13

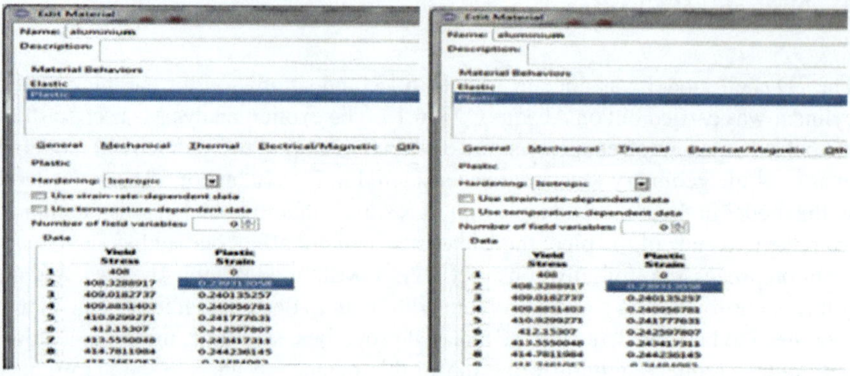

(a) (b)

Fig. 2 A Elastic property of aluminium 6351 T6

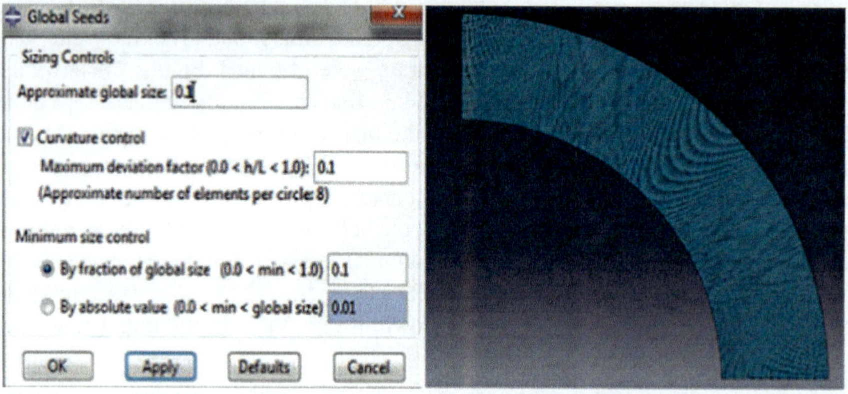

(a) (b)

Fig. 3 **a** Mesh size control **b** Meshing of 2D planar deformable body

thick cylinder subjected to internal pressure is then released. This process of loading and unloading pressure under the effect of internal pressure and application of step time, field output request and job creation of the deck model is shown in Fig. 6a, b.

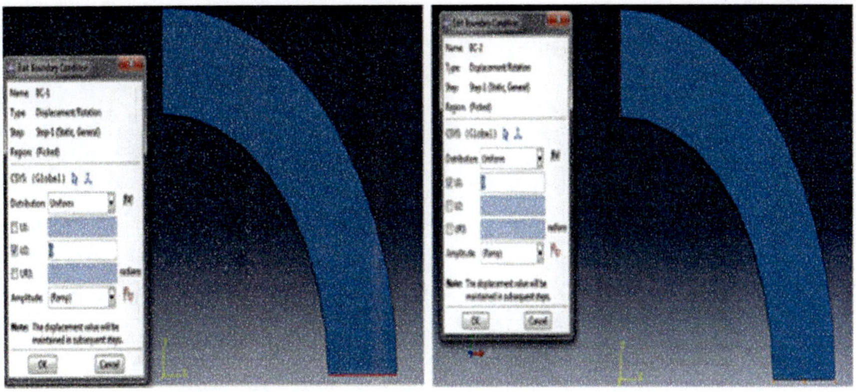

(a) (b)

Fig. 4 **a** Boundary condition *X*-symmetric **b** Boundary condition *Y*-symmetric

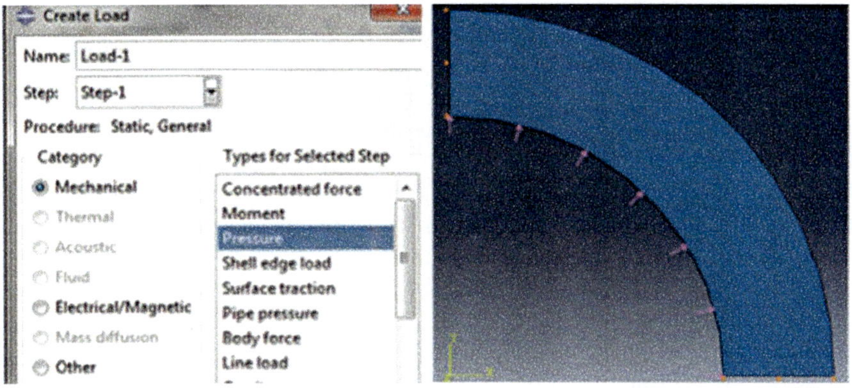

(a) (b)

Fig. 5 **a** Internal loading mechanical pressure **b** Internal loading on the body

3 Results and Discussion

3.1 Simulation in Abaqus

The residual stresses are obtained from finite element analysis by initially changing the coordinate system from the Cartesian coordinate (x, y and z) to the cylindrical coordinate system (r, θ and z). The variation of residual stress over the true length of the specimen at a thickness of cylindrical direction is shown in Fig. 8. The residual stress is compressive to the mid surface layer of the specimen and is tensile after that till a distance and converges to zero. Radial residual stresses are obtained due to the variation of the working pressure in such a way that plots the nature of the curve, as

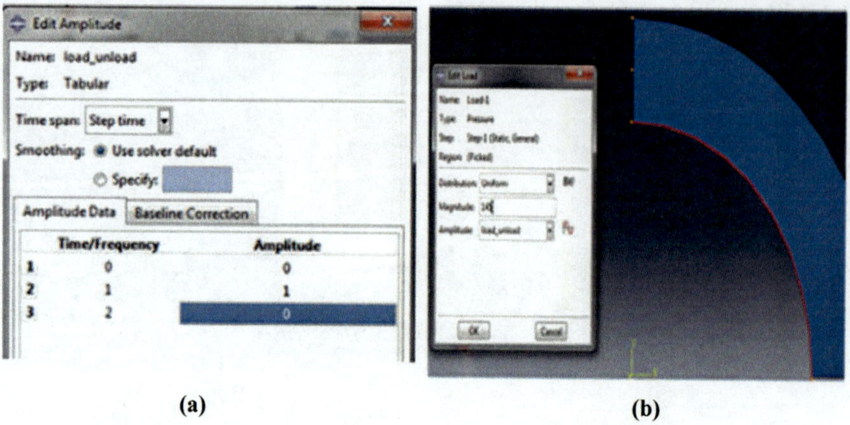

Fig. 6 **a** Internal amplitude on body **b** Internal pressure loading and unloading

shown in Fig. 7. Residual stresses are compressive and analysed at different pressure levels by studying the nature of the curve.

Figure 8 depicts the plot between radial residual stress and radial thickness for different autofrettage pressure. It can be easily inferred that at low autofrettage pressure, radial residual stresses are lesser than the high autofrettage pressure, hence high autofrettage pressure induces the higher residual stresses at the inner surface of the thick-walled cylinder, and it increases the strength and fatigue life of components. Thick cylinders were subjected to internal pressure results in the inner layer deformed beyond the elastic zone, and working pressure increases due to which radial residual stresses were increased, as shown in Figs. 8, 9 and 10. The following graph depicts autofrettage pressure increases from 120 MPa, 130 MPa, 135 MPa, 145 MPa, 150 MPa and 155 MPa, respectively, and the nature of the graph studied variation in radial residual stress.

Fig. 7 Radial residual stress distribution in the thick cylinder for various internal working pressures

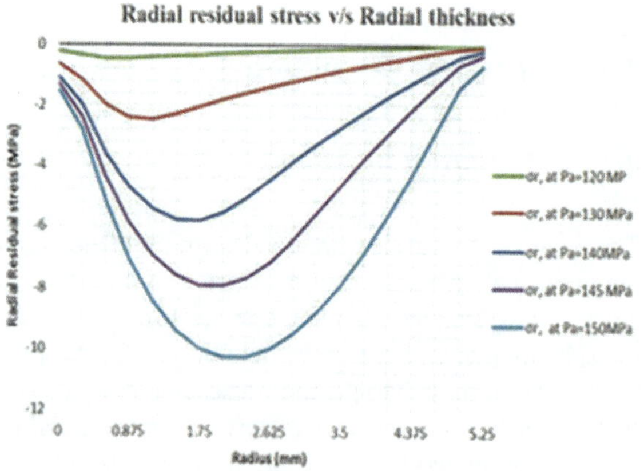

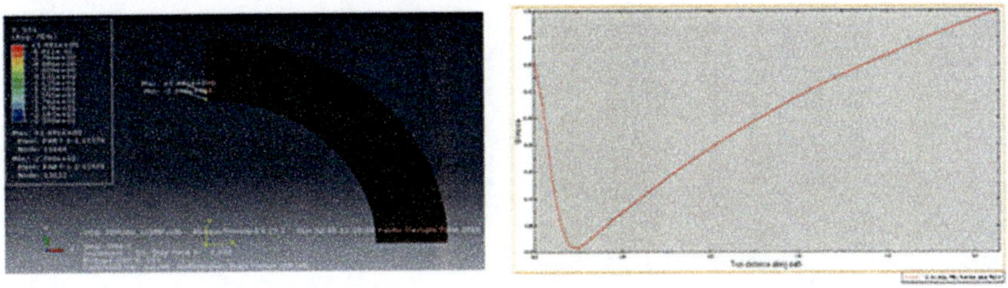

Fig. 8 Radial residual stress and nature of curve at PA = 120 MPa, mesh size 0.1 mm

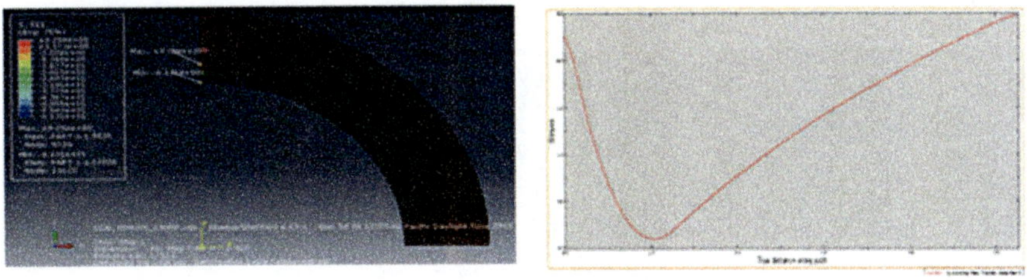

Fig. 9 Radial residual stress and nature of curve at PA = 130 MPa, mesh size 0.1 mm

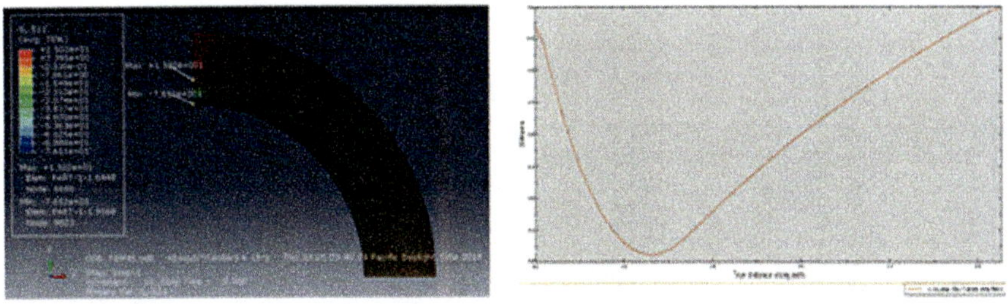

Fig. 10 Radial residual stress and nature of curve at PA = 135 MPa, mesh size 0.1 mm

In this study, the pressure applied on an aluminium cylinder with an increment of 1 MPa at each time step and it observed that increasing the pressure results in increasing residual stresses. Therefore, increasing residual stresses are useful for increasing the strengthening and fatigue life of cylindrical components. Radial residual stress distribution of thick cylinders is studied for different autofrettage pressure. Figure 8 shows radial residual stresses ranging from −23.99 MPa to 1.48 MPa, at P_i=120 MPa, Fig. 9 shows −61.31 MPa to 9.25 MPa, at P_i = 130 MPa, and Fig. 10 shows −71.51 MPa to 15.02 MPa, at P_i = 135 MPa. Figure 11 shows radial residual stresses from −117.2 MPa to 43.65 MPa, at internal pressure P_i = 145 MPa. Figure 12 shows radial residual stress from −136.6 MPa to 76.2 MPa, at P_i = 150 MPa, and Fig. 13 shows radial residual stress ranges from −90.74 MPa to 94.64 MPa, at P_i = 155 Mpa.

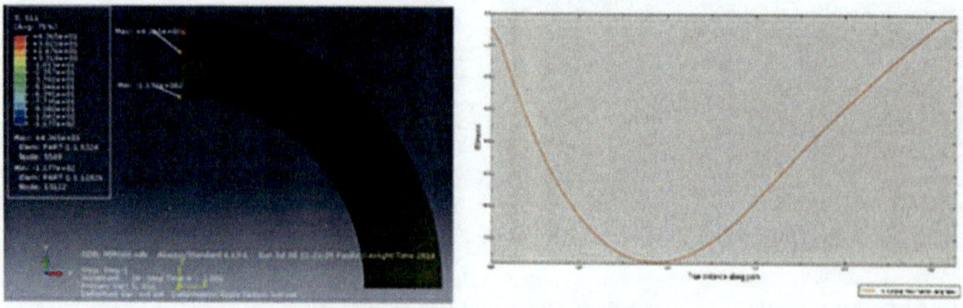

Fig. 11 Radial residual stress and nature of curve at PA = 145 MPa

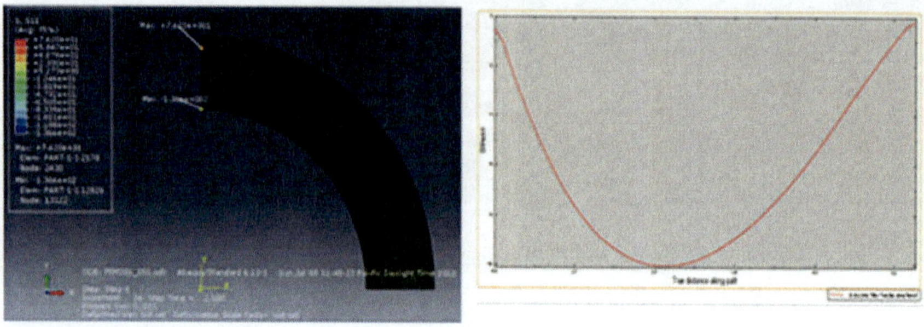

Fig. 12 Radial residual stress and nature of curve at PA = 150 MPa

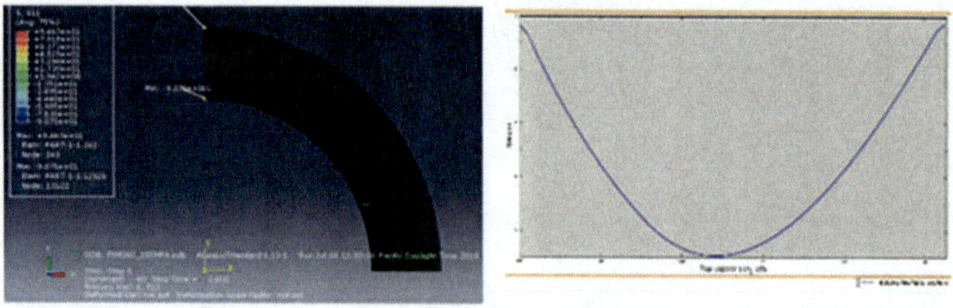

Fig. 13 Radial residual stress and nature of curve at PA = 155 MPa

3.2 *Hoop Residual Stress Distribution in a Thick Cylindrical Structure*

Figure 14 shows the plot between residual hoop stress and radial thickness for different autofrettage pressure. From the graph, it can be easily inferred that at low autofrettage pressure, residual hoop stresses are lesser than high autofrettage pressure. Therefore, high autofrettage pressure induces the higher residual hoop stresses in the thick-walled cylinder. Residual hoop stresses for different autofrettage pressure

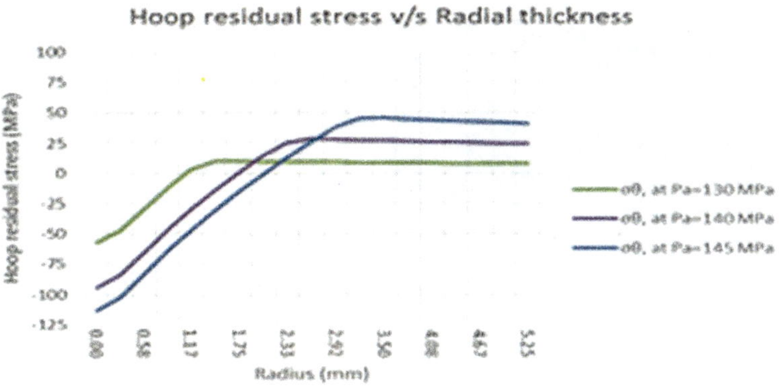

Fig. 14 Hoop residual stress distribution in a thick cylinder for various autofrettage pressures

are shown in the following figures. Figure 15 shows the residual hoop stress ranges vary from −57.01 MPa to 7.6 MPa, at PA = 130 MPa. Figure 16 shows hoop residual stresses vary from-94.16 MPa to 24.2 MPa, at PA = 140 MPa. Figure 17 shows hoop residual stresses vary from −112.85 MPa to 41.177 MPa, at PA = 145 MPa. Figure 18 shows hoop residual stresses vary from −87.15 MPa to 91.115 MPa, at PA = 155 MPa. It clearly depicts that as the autofrettage increases, residual hoop stress

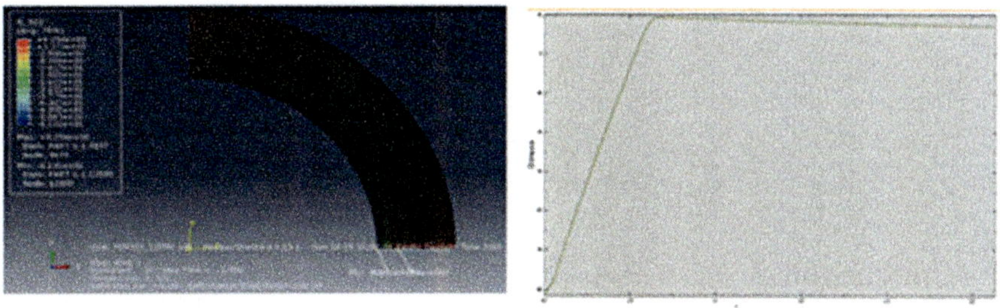

Fig. 15 Hoop residual stress distribution and nature of curve at PA =130 MPa

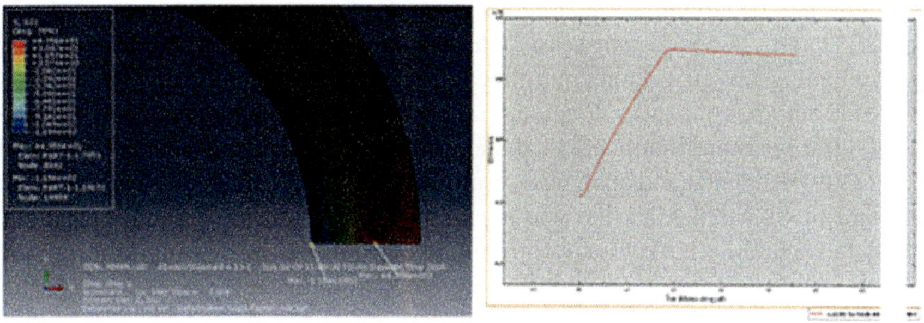

Fig. 16 Hoop residual stress distribution and nature of curve at PA =140 MPa

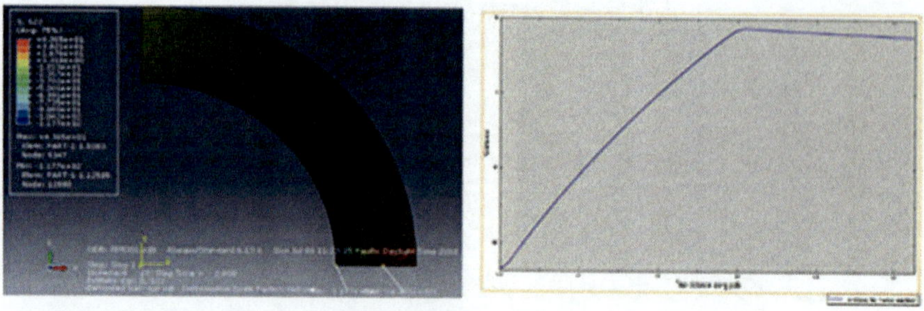

Fig. 17 Hoop residual stress distribution and nature of curve at PA = 145 MPa

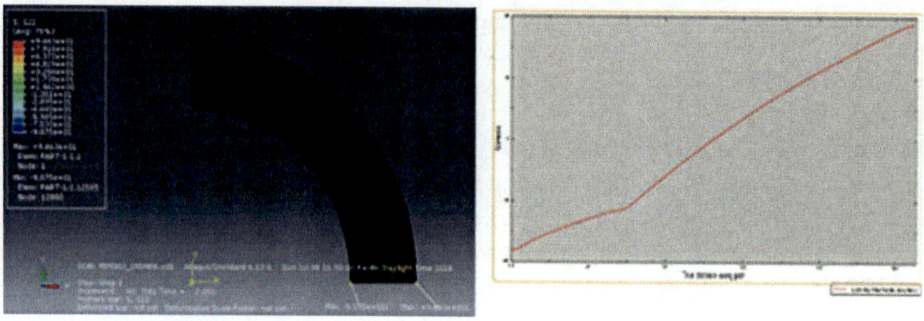

Fig. 18 Hoop residual stress distribution and nature of curve at PA = 155 MPa

also increases and results in increases in the bursting pressure of the thick cylindrical structure.

4 Conclusion

Autofrettage process increases the pressure bearing capacity as it decreases the hoop and radial stress at the inner surface of cylindrical structures. Autofrettage process on aluminium 6351 cylindrical structure changed the inner diameter from 25.5 mm to 27.84 mm by the percentage of 9.176. The thickness of the cylindrical structure reduced along the circumferential direction, and residual hoop stresses increased of the same radial thickness direction. Experimentally calculated hoop stress at the outer surface is more significant than hoop stress calculated from Abaqus 6.14, and it shows experimental and analytical results differed by 2% to 7%. The autofrettage pressure must be higher than the working pressure to increase the strength and fatigue life of the cylindrical structure. Controlling the residual stresses with respect to the pressure plays the leading role in the process of autofrettage.

References

1. Salzar RS (1999) Influence of autofrettage on metal matrix composite reinforced gun barrels. Composites: Part B 30:841–847
2. Haghpanah B, Ajdari A, Nayeb-Hashemi H, Vaziri A (2010) Autofrettage of layered and functionally graded metal-ceramic composite vessels. Compos Struct 92:1813–1822. https://doi.org/10.1016/j.compstruct.2010.01.019
3. Chen H-W, Sun H-K, Liu T-C (2009) Autofrettage analysis of a fibre-reinforced composite tube structure incorporated with a SMA. Compos Struct 89:497–508
4. Wahi N, Ayob A, Elbasheer MK (2011) Effect of Autofrettage on allowable pressure of thick-walled cylinders. IPCBEE 15:183–254
5. Farrahi G, Voyiadjis G, Hoseini S, Hosseinian E (2013) Residual stress analyses of re-autofrettaged thick-walled tubes. Int J Pressure Vessels Piping 98:57–64. https://doi.org/10.1016/j.ijpvp.2012.07.007
6. Zhong H, Puttaguntaa S (2009) Computer modeling of internal pressure autofrettage process of a thick-walled cylinder with the Bauschinger effect. Am Trans Eng Appl Sci 23:123–186
7. Jain A, Khanwelkar S, Saurav K, Landge A, Yadav U (2003) Design and performance of hydraulic autofrettage using universal testing machine. ISSN: vol-1, pp 2249-5762 (Online)
8. Teng H, Bate SK, Beardsmore DW (2009) Determination of residual stress profiles of pipe girth weld using a unified parametric function form.In: Proceedings of the ASME 2009 Pressure Vessels and Piping Conference. Volume 6: Materials and Fabrication, Parts A and B. Prague, Czech Republic. July 26–30, pp 381–388. ASME. https://doi.org/10.1115/PVP2009-77316

Investigation of Crack Detection Technique in a Rotating Shaft by Using Vibration Measurement

T. Jagadeesha, V. G. Salunkhe, R. G. Desavale, P. B. Patil, M. B. Kumbhar, and A. R. Koli

Abstract Cracks in the shafts are critical factor, as they are difficult to detect during operation. Operatives and maintenance personnel of critical plant machinery are particularly interested in early detection of symptoms that can lead to in-service failure of shafts. Many researchers have developed new diagnostic techniques, in order to detect cracks in shaft rotor system is, therefore, highly essential to avoid catastrophic failure. Vibration signals used as a diagnostic tool for smooth functioning and to avoid an undue stoppage. This paper incorporates experimental and numerical investigation of effect of crack in a shaft, on vibration amplitude levels, frequency spectrum and natural frequency. The experimental setup is developed to study the effect of crack on vibration amplitude. The vibration signals are measured with the help of state-of-the-art two-channel CEMB N500 FFT analyzer. ANSYS 14 is used to modal analysis for finding theoretical natural frequencies. Impact Hammer Testing is carried out with the help of DEWE-soft 43 platform along with data acquisition system and electronic instrumentation for finding experimental natural frequencies of a system. The results of analysis show that vibration analysis can be used as a tool to detect crack in shaft, before they cause any failure.

Keywords Vibration of shaft · Crack · Frequency spectrum · FFT analyzer · Impact hammer testing · DEWE-soft 43

T. Jagadeesha (✉)
Department of Mechanical Engineering, National Institute of Technology Calicut, Calicut, Kerala 673601, India
e-mail: jagdishsg@nitc.ac.in

V. G. Salunkhe · P. B. Patil
Department of Mechanical Engineering, ADCET, Ashta, Maharashtra 416301, India

R. G. Desavale
Department of Mechanical Engineering, RIT, Sakharale, Maharashtra 415414, India

M. B. Kumbhar
Faculty of Mechanical Engineering, STES, Lonavala, Maharashtra 410401, India

A. R. Koli
Faculty of Mechanical Engineering, TKIET, Warananagar, Maharashtra 416113, India

A. Arockiarajan et al. (eds.), *Advances in Industrial Automation and Smart Manufacturing*, Lecture Notes in Mechanical Engineering,
https://doi.org/10.1007/978-981-15-4739-3_54

1 Introduction

Rotating machineries are widely used in the modern industry. The application ranges from domestic appliances to power plants and aerospace equipment. In recent years, focus has been on safety of operation and increasing the service period beyond the original service life. The shafts in operation are sometimes susceptible to serious defects that develop without apparent warning. They are prime candidates for fatigue cracks because of the rapidly fluctuating nature of bending stresses, the presence of numerous stress raisers and possible design and manufacturing flaws. Cracks are defined as any unintentional discontinuities in the shaft material. Wide variations in temperature and environment during operation also contribute to conditions conducive to eventual fatigue failure. The total shaft failure can be catastrophic with enormous costs in down time, consequential damage to equipment and potential injury to personnel. Total failure occurs when the specimen has completely fractured into two or more parts. Safe and reliable operation of equipment relies on proactive maintenance aided by newly emerging diagnostic technologies. In ductile materials, such as low/medium-alloy steels used for turbo-machine shafts, cracks are initiated as tiny discontinuities that grow in size when the component is subjected to cyclic stresses. It is important to detect them before they reach the critical size and cause total failure of the shaft. Vibration signals can be used as a diagnostic tool for smooth functioning and to avoid an undue stoppage.

2 Literature Survey

Many investigators have studied and developed new methods for detection of crack in rotating shaft. Lebold et al. [1] developed and presented a non-intrusive torsional vibration method for monitoring small changes in crack growth of Westinghouse 93 reactor coolant pump shaft. This technique is applicable to many types rotating equipment's including centrifugal charging pumps. Ishida et al. [2] presented experimental and theoretical investigation of change in the resonance curve caused by the occurrence of a transverse crack. Sabnavis et al. [3] presented literature review on cracked shaft detection and diagnostics. Davies and Mayes [4] made a research on the effects of a transverse crack on the dynamics of a multi-rotor, multi-bearing system. Dimargonas [5] analyzed continuous shaft with circumferential crack and presented its effect on the torsional dynamic behavior of a shaft. Schmalhorst [6] investigated vibration response of a slender uniform shaft subjected to transverse crack. Green [7] presented global and local asymmetry crack model analyses to identify characteristics of the system that directly attributed to the presence of a transverse crack in the shaft. Huang et al. [8] investigated the dynamic response of a rotating shaft containing a transverse crack. The steady-state response and the stability criteria have been obtained and the effects of crack depth, crack location and rotational speed discussed. Imam et al. [9] developed a comprehensive technology of on-line rotor crack detection and monitoring. The technique, based on the vibration signature analysis approach, can detect incipient transverse rotor cracks in an "on-line mode." Nelson [10] presented a theoretical analysis of the dynamics of a rotor-bearing system with a transverse crack. The rotating assembly has been modeled using finite rotating shaft elements and the presence of a crack is taken into account by a rotating stiffness variation. Ostachowicz [11] presented a method of analysis of the effect of two open cracks upon the frequencies of the natural flexural vibrations in a cantilever beam. Papadopoulos [12] investigated the coupling of longitudinal and bending vibrations of a rotating shaft, due to an open transverse surface crack. The effect of unbalance and gravity has been examined. Tlaisi et al. [13] presented experimental and numerical investigation to identify the presence of a crack in a cylindrical overhung shaft with a propeller at the free end. Modal analysis has been carried out to obtain the shaft response parameters for lateral and torsional vibrations. A lot of research is directed toward the fault diagnosis shaft rotor system. However, only relatively small amount of research work is reported on effects of crack on shaft. Authors found a good potential for the research on cracks on shaft. In the present analysis, experiments have been conducted for the validation of experimental results with numerical. Based on work reported herein, a good correlation between experimental results and numerical investigation can be seen.

3 Experimental Details

The experimental setup for analysis of shaft given in Fig. 1 is designed and fabricated to investigate the vibration characteristic of shaft rotor system. Variable speed motor drives the rotor at above 1440 rpm and consists of a rotor supported on two bearings.

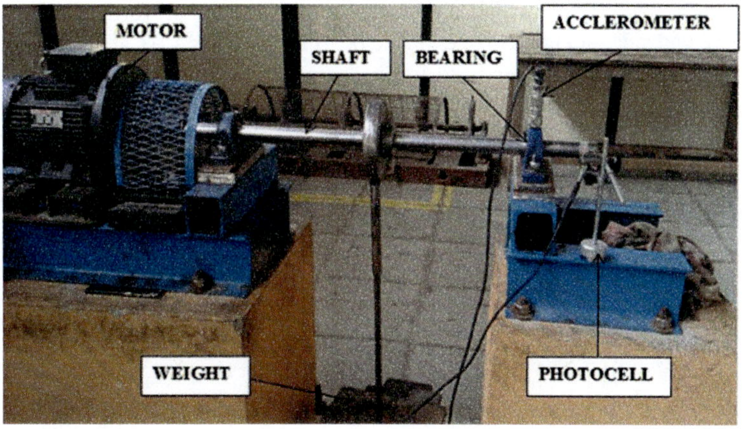

Fig. 1 Experimental setup

The rotor is of diameter 25 mm and length of 850 mm. There is speed control unit and the speed of the rotor may be varied as desired. The whole setup is built on rigid concrete foundation. The motor is connected to the rotor through a flexible jaw-type coupler. The purpose of this jaw-type coupler is to avert the transfer of vibrations of the motor to the shaft rotor system. The shaft is simply supported with rolling bearings at both the ends. A self-aligning bearing (1207K) with adapter sleeve (H207) is selected for the support. The weight is applied at the center of the span. The bearing adapter is used for connecting the weight pan to the central bearing. On the periphery of bearing adapter, a 14 mm diameter hole is tapped with 14 TPI (BSF) tap to fasten the weight pan to the adapter. Accelerometer pickups are kept on bearing housings to measure vibration velocity and photocell for speed measurement.

4 Result and Discussion

Experimental and numerical studies are being presented here considering shaft geometry, material properties, loading, speed with cracks on shaft. In order to validate, experiments are conducted for seven different cases. In light of above comments, the focus of this new investigation is for the accurate prediction of the effects of the cracks on shaft.

4.1 Experimentation Testing Strategy Adopted

Measuring vibration amplitude levels of drive end side and non-drive end side bearings housing with 200 N constant radial load at center of shaft and two speed steps of 800 and 1600 rpm are considered. Taking velocity frequency spectrums at 800 and 1200 rpm with constant load at three directions (horizontal, vertical and axial) of both drive end side and non-drive end side bearings by using FFT analyzer. For each speed, accelerometer has been used for measuring the vibration amplitude. Finding the natural frequencies of shaft by using "Impact Hammer Testing" and comparing the results with natural frequencies found by modal analysis in ANSYS 14. Considering the readings for increasing crack depth in steps of 2 mm up to crack depth of 12 mm to study the effect of crack depth on vibration amplitudes on the support bearings.

4.2 Measurement of Vibration Amplitude Levels Using Vibration Meter

The vibration amplitude levels has been measured in horizontal, vertical and axial direction on both drive end side and non-drive end side bearings by using a vibration meter as shown in Fig. 2. During experimentation, displacement (*D*) has been measured in μm (peak to peak), velocity (*V*) in mm/s (RMS) and acceleration (*A*) in m/s^2 (RMS).

The vibration amplitudes with respect to increasing crack depth at constant load with two speed steps are given in Tables 1, 2, 3, 4, 5 and 6.

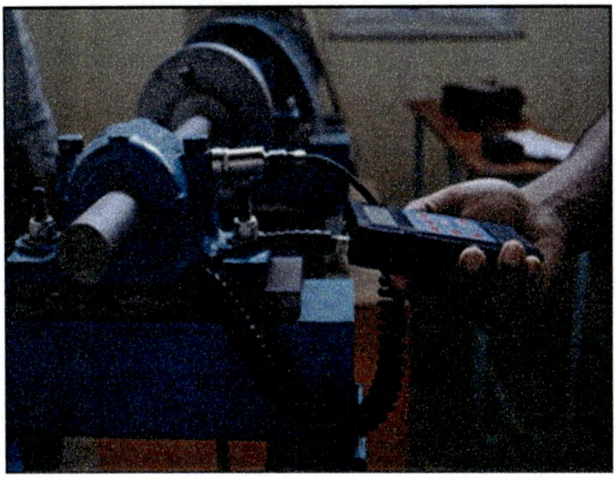

Fig. 2 Vibration measurement using vibration meter

Table 1 Vibration amplitude levels at 800 rpm in horizontal direction

Crack depth	Vibration level at NDE			Vibration level at DE		
	D	*V*	*A*	*D*	*V*	*A*
0	20	1.9	4.2	39	5.1	5.8
2	30	3.6	3.53	29	4.4	3.34
4	36.5	3.35	2.47	49.8	5.99	3.6
6	36.1	4.31	3.11	58.7	6.49	3.76
8	29.5	3.03	2.13	52.7	4.86	2.63
10	28.7	3.15	2.23	58.6	5.5	3
12	34.1	3.26	2.32	67.6	6.24	3.29

Table 2 Vibration amplitude levels at 800 rpm in vertical direction

Crack depth	Vibration level at NDE			Vibration level at DE		
	D	*V*	*A*	*D*	*V*	*A*
0	13	0.6	5.3	8	0.5	3.4
2	19	1.1	0.1	12	0.7	1.18
4	14.5	0.72	0.33	13.2	0.61	0.25
6	23.6	1.34	0.55	19.6	1.23	0.51
8	15.9	0.94	0.36	12.7	0.65	0.19
10	18.8	1.09	0.43	15.5	0.83	0.34
12	26.6	1.39	0.57	20.6	1.07	0.28

Table 3 Vibration amplitude levels at 800 rpm in axial direction

Crack depth	Vibration level at NDE			Vibration level at DE		
	D	*V*	*A*	*D*	*V*	*A*
0	20	2	2.9	27	2.5	3.4
2	19.5	1.3	1.77	12.4	1.7	1.96
4	15.5	1.52	1.31	31	4.32	1.77
6	22.6	1.58	1.31	29.7	2.09	1.53
8	20.3	1.44	1.46	26.9	1.8	1.42
10	25.4	2.09	1.89	37.8	2.61	1.83
12	29.1	2.12	1.81	39.6	2.53	1.57

Table 4 Vibration amplitude levels at 1600 rpm in horizontal direction

Crack depth	Vibration level at NDE			Vibration level at DE		
	D	*V*	*A*	*D*	*V*	*A*
0	54	9.2	19	63	1	15
2	70	11.2	11	76	9	9.3
4	64	14.1	7.9	90	8	2.4
6	82	13.2	9.0	78	1	6.9
8	75	12.9	9.1	53	7	4.9
10	85	13.4	8.8	88	9	5.9
12	90	14.2	8.0	81	9	4.7

4.3 Velocity Frequency Spectrum Using CEMB N500 FFT Analyzer

For the measurement of acceleration, velocity and displacement, the data acquisition system is used. It is CEMB N500, which has 8.0 data acquisition system software. The

Table 5 Vibration amplitude levels at 1600 rpm in vertical direction

Crack depth	Vibration level at NDE			Vibration level at DE		
	D	*V*	*A*	*D*	*V*	*A*
0	26	1.7	9.7	24	1	10
2	42	3.65	4.9	33	2	3.1
4	46	3.2	2.3	40	2	1.1
6	44	3.22	2.7	42	2	1.1
8	54	3.51	1.5	37	2	0.5
10	50	3.37	1.6	36	2	0.6
12	36	2.81	1.8	29	1	0.6

Table 6 Vibration amplitude levels at 1600 rpm in axial direction

Crack depth	Vibration level at NDE			Vibration level at DE		
	D	*V*	*A*	*D*	*V*	*A*
0	28	2.3	7.8	39	3	5.3
2	54	4.7	5.4	79	6	5.59
4	56	8.3	3.7	136	1	10.01
6	56	4.44	4.3	60	5	3.32
8	55.3	2.88	2.5	49	4	2.65
10	66.5	4.61	3.6	58	4	2.84
12	55.6	4.39	2.8	67	6	3.73

vibration signals are measured on the bearing housing of the test bearing by mounting an accelerometer with sensitivity 105.5 mV/g. 43 V data acquisition system has eight analog channels and eight digital channels as shown in Fig. 3. The velocity frequency plot has been recorded with frequency range between 0 and 30,000 rpm. The unit of velocity used is mm/s. From the experimentation, it is determined that the amplitude of vibration is progressive with the velocity of the rotor as shown in Fig. 4a–c.

4.4 Modal Analysis

Natural frequency is the frequency of free vibration of the system. It is constant for a given system and depends on the degree of freedom (DOF). If a system has two DOF, it will have two natural frequencies. As shaft is a continuous system (infinite DOF), it will have infinite natural frequencies. When shaft has a crack, its natural frequencies will change. Thus, finding natural frequency measurement can be used as a tool to detect crack in a shaft. Modal analysis is a method of finding theoretical natural frequencies of a system. Ansys14 has been used to carry out the modal analysis. As

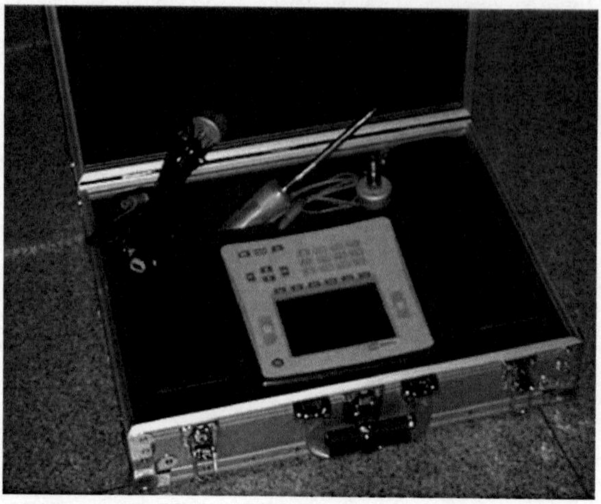

Fig. 3 Two-channel CEMB N500 FFT analyzer

the support bearing in self-aligning, rotation in X- and Y-direction is allowed and all other movements are constrained. The model of shaft in ANSYS 14 is as shown in Fig. 5.

The results of modal analysis as shown in Table 7 give natural frequencies corresponding to different modes and for increasing crack depth in steps of 2 mm up to crack depth of 12 mm.

4.5 Testing Using DEWE-Software 43

The modal analysis carried out in the previous investigation provided the guidelines for setting up the experimental procedure with impact hammer excitation that is described. In addition, an updating procedure has been developed. Impact Hammer Testing is an experimental method of finding the natural frequencies of a system. In this method, the system is hit by a hammer to give an impulse. The result of impact is that the system freely vibrates at its natural frequencies. These natural frequencies have been measured by an accelerometer mounted on the system and a data acquisition system. Figure 6 shows data acquisition system, along with installed software DEWE-soft 43. Figure 7 shows the position of accelerometer on the shaft to find out the natural frequencies.

The frequency response function (FRF) graphs obtained during Impact Hammer Testing with increasing crack depth in steps of 2 mm up to crack depth of 12 mm are as below.

The FRF graphs shown in Fig. 8 give a series of peaks corresponding to the natural frequencies. The natural frequencies corresponding to series of peaks in FRF graphs are shown in Table 8.

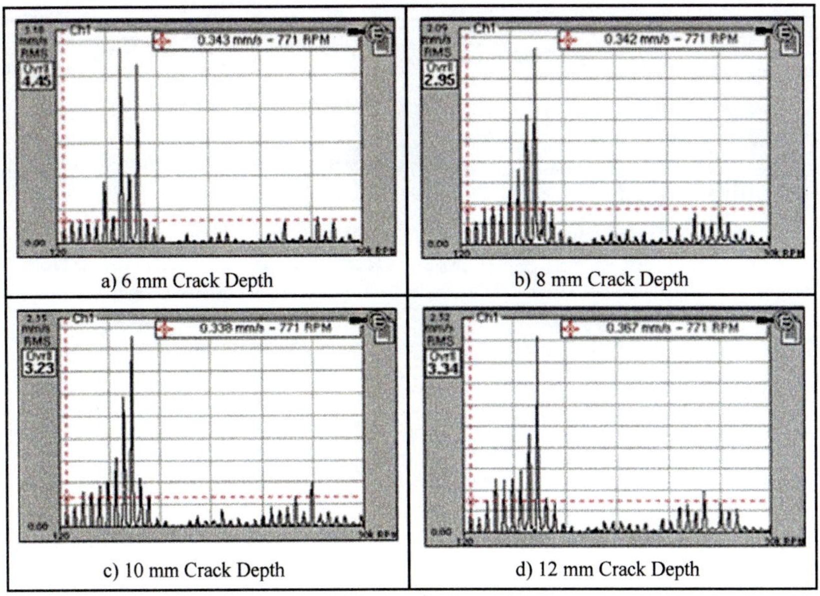

a) 6 mm Crack Depth

b) 8 mm Crack Depth

c) 10 mm Crack Depth

d) 12 mm Crack Depth

(a) Velocity frequency spectrums of non-drive end bearing in horizontal direction at 800 rpm

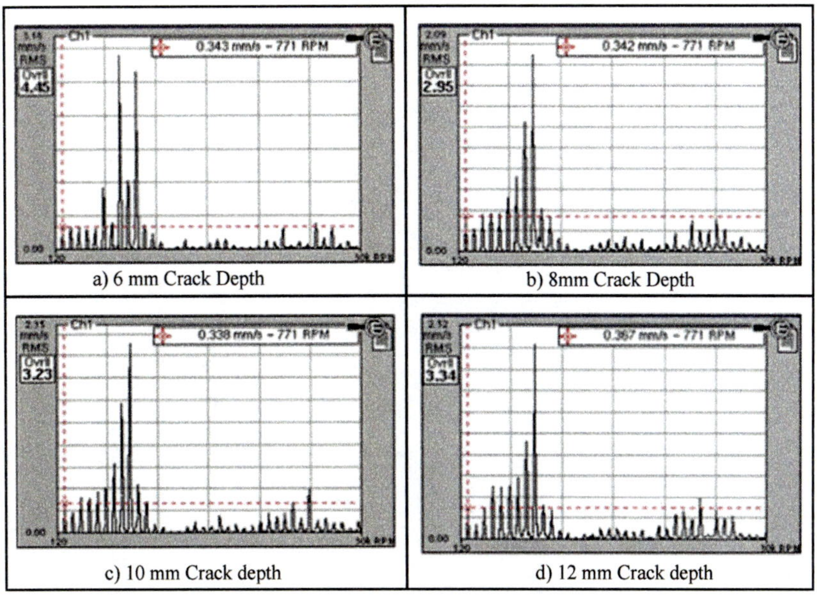

a) 6 mm Crack Depth

b) 8mm Crack Depth

c) 10 mm Crack depth

d) 12 mm Crack depth

(b) Velocity frequency spectrum of non-drive end bearing in vertical direction at 800 rpm

Fig. 4 **a–c** shows vibration amplitude with respect to defect size

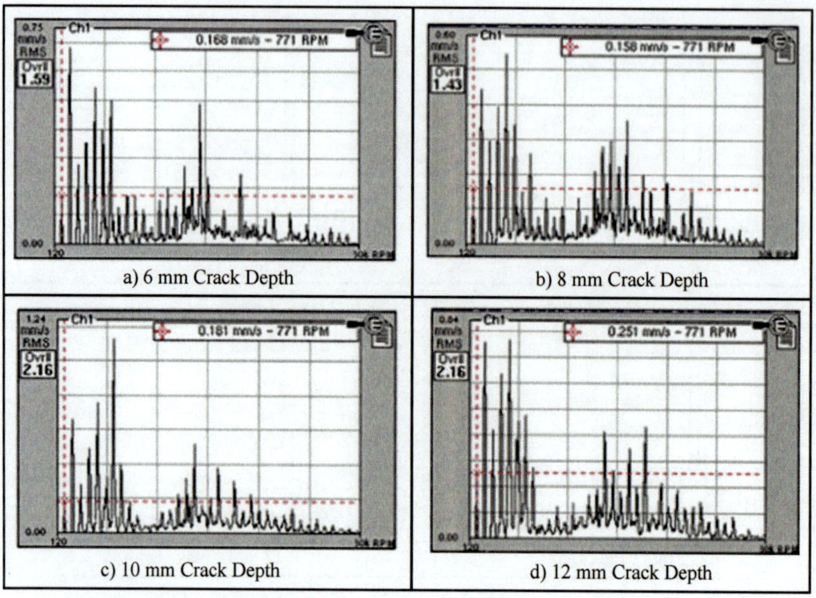

a) 6 mm Crack Depth
b) 8 mm Crack Depth
c) 10 mm Crack Depth
d) 12 mm Crack Depth

(c) Velocity frequency spectrum of non-drive end bearing in axial direction at 800 rpm

Fig. 4 (continued)

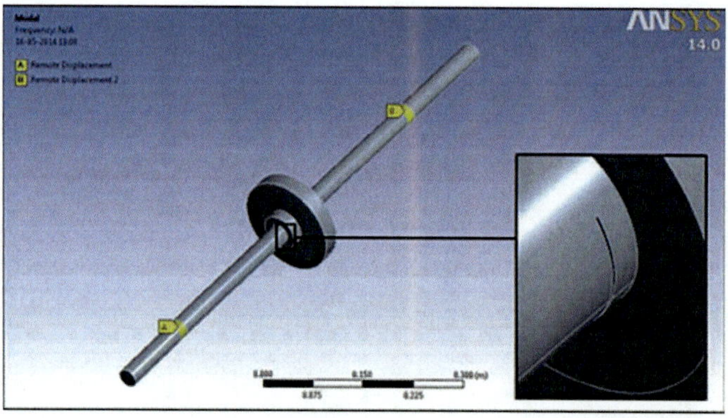

Fig. 5 Shaft model

Shaft vibration and defect frequencies of corresponding defective shafts for variant examine circumstances are measured. Numbers of experiments are conducted with all variables and vibration amplitudes are measured. From the experimentation, it is determined that the amplitude of vibration is progressive with the velocity of the rotor

Table 7 Modal analysis results

Modes	Numerical natural frequencies (Hz)						
	0 mm	2 mm	4 mm	6 mm	8 mm	10 mm	12 mm
1	10	10	10	10	10	98	96
2	15	14	15	14	14	13	13
3	33	33	33	33	33	33	33
4	38	38	38	38	38	38	38
5	39	39	39	39	39	38	38

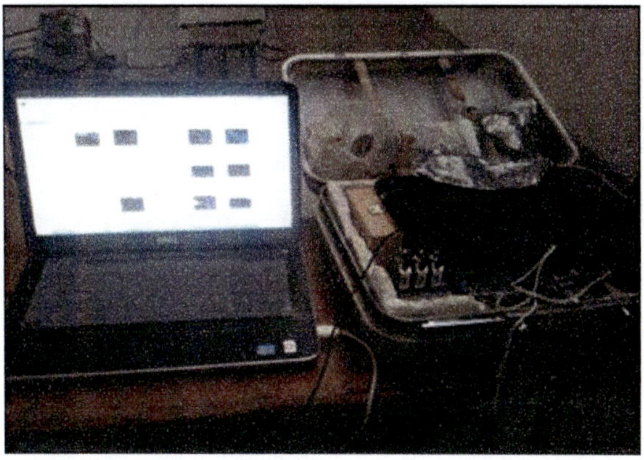

Fig. 6 Data acquisition system with DEWE software

Fig. 7 Position of accelerometer on shaft

as shown in Fig. 9. The leaning of progressive of amplitude with speed may be right as long as the velocity of the rotor is to a lesser extent than natural frequency. The progressive in size of defect the amplitude is ascertained in heighten. The explanation might be defects bring down the stiffness of the shaft which sequentially decreases the natural frequency as shown in Fig. 10. Henceforth, when there is progressive change in defect size, the operative element exceeds toward resonance. By correlating the predominant frequencies with shaft defect frequencies and their harmonics, the number of defects may be predicted.

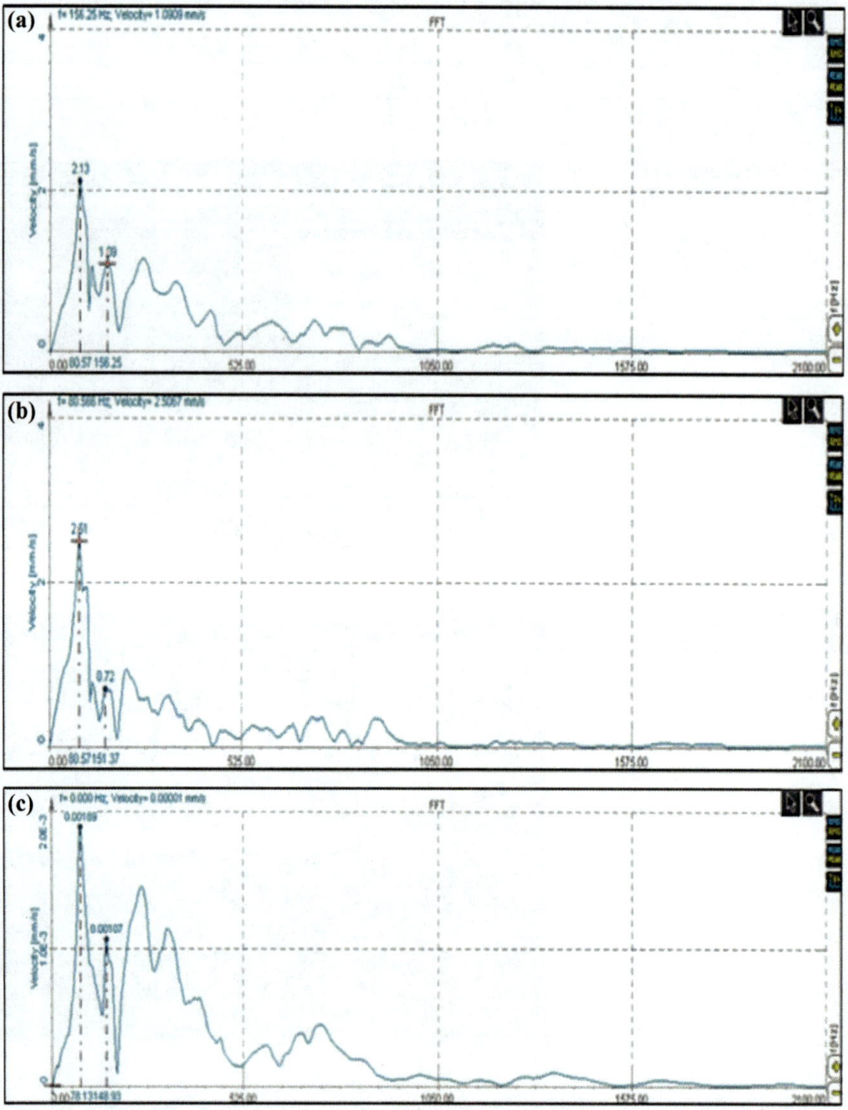

Fig. 8 Natural frequency FRF graphs obtained from DEWE-soft 43, **a** no crack, **b** 2 mm crack, **c** 4 mm crack, **d** 6 mm crack, **e** 8 mm crack, **f** 10 mm crack, **g** 12 mm crack

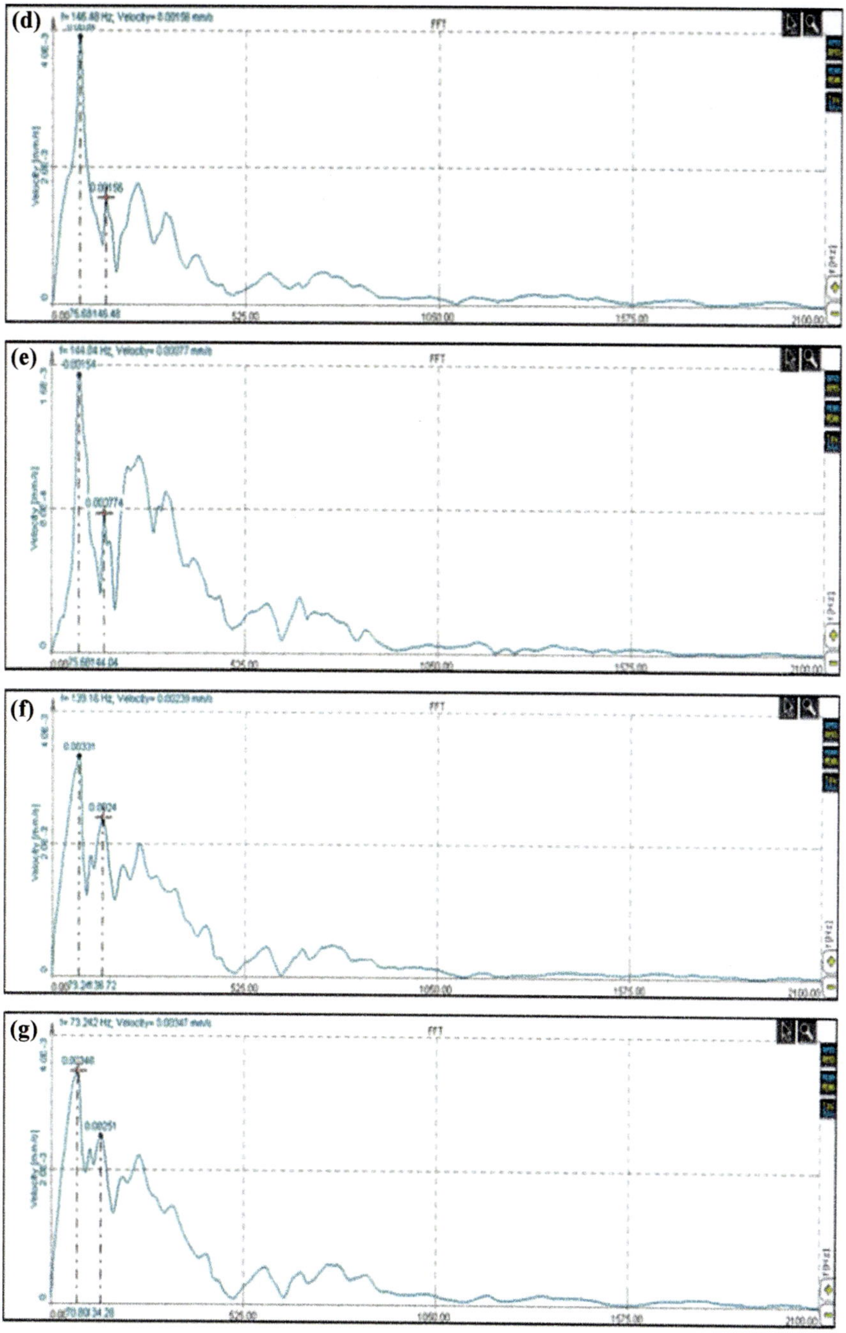

Fig. 8 (continued)

Table 8 Experimental natural frequencies

Modes	Numerical natural frequencies (Hz)						
	0 mm	2 mm	4 mm	6 mm	8 mm	10 mm	12 mm
1	80	80	78	75	75	73	70
2	15	15	14	14	14	13	13

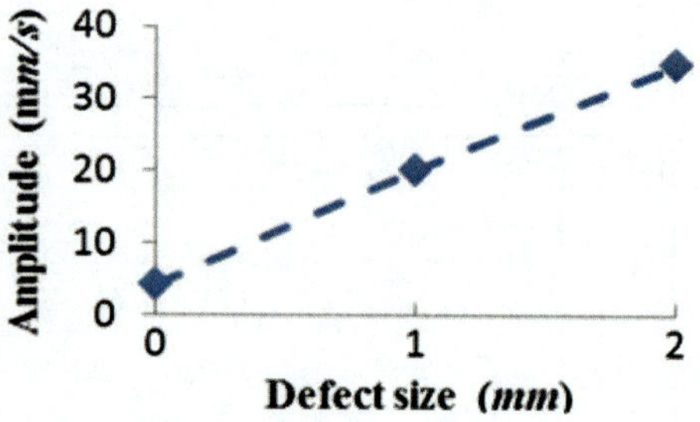

Fig. 9 Influence of defect size on amplitude

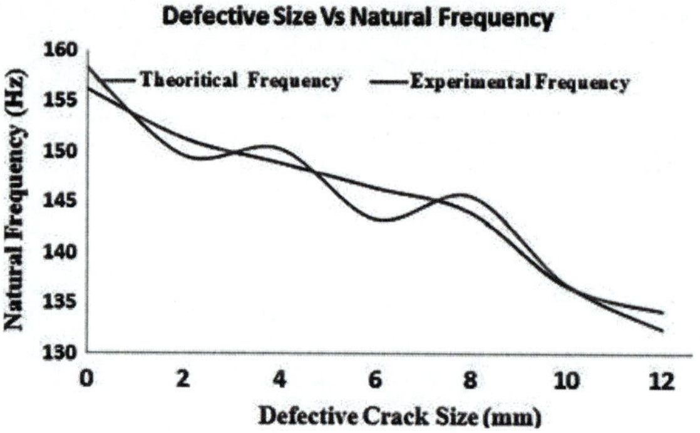

Fig. 10 Frequency spectra

5 Conclusion

Vibration response of defective crack is measured with the help of FFT for various speed. The frequency domain approach is used and absolute values are taken during measurement. The following conclusions are drawn from this research work.

1. The amplitudes at defective frequencies are much larger in occurrence of large defects as contrast to small defects on shaft.
2. Defect size is the most noteworthy factor as of those factors considered, due to raise in defect size vibration level boosts significantly. In addition, natural frequency of shaft also goes on decreasing with the increase in crack depth. Therefore, interface of speed and defect size appears to be more dominating factor for surface defect amplitudes.
3. Experimental and numerical modal examination showed a good fit, which indicates they are closely matched with experimental results. Therefore, we can use

proposed numerical model to predict the vibration of amplitude of healthy and defective shafts with unbalance and looseness.

4. The experimental comparative study revealed that overall vibration level is more as compared to healthy and defective shafts. Predominant peaks are institute to occur at rotational frequency of 3 × rpm of the shaft for all circumstances.

Therefore, the theoretical formulation shows adequate guarantee to set up itself as a contrivance for shaft fault detection.

References

1. Lebold MS, Maynard K, Reichard K, Trethewey M, Bieryla D, Lissenden C, Dobbins D (2004) Using torsional vibration analysis as a synergistic method for crack detection in rotating equipment. In: 2004 IEEE aerospace conference proceedings (IEEE Cat. No. 04TH8720), vol 6, pp 3517–3527
2. Ishida Y, Ikeda T, Yamamoto T, Masuda N (1988) Vibrations of a rotating shaft containing a transverse crack: variations of a resonance curve due to the angular position of an unbalance at the major critical speed. JSME Int J Ser 3 Vib Control Eng Eng Ind 31(1):22–29
3. Sabnavis G, Kirk RG, Kasarda M, Quinn D (2004) Cracked shaft detection and diagnostics: a literature review. Shock Vib Digest 36(4):287
4. Davies WG, Mayes IW (1984) The vibrational behavior of a multi-shaft, multi-bearing system in the presence of a propagating transverse crack. J Vib Acoust Stress Reliab Des 106(1):146–153
5. Dimarogonas A, Massouros G (1981) Torsional vibration of a shaft with a circumferential crack. Eng Fract Mech 15(3–4):439–444
6. Dirr BO, Schmalhorst BK (1988) Crack depth analysis of a rotating shaft by vibration measurement. J Vib Acoust Stress Reliab Des 110(2):158–164
7. Green I, Casey C (2003) Crack detection in a rotor dynamic system by vibration monitoring: Part I—analysis. In: ASME turbo expo 2003, collocated with the 2003 international joint power generation conference. American Society of Mechanical Engineers, pp 569–582
8. Huang SC, Huang YM, Shieh SM (1993) Vibration and stability of a rotating shaft containing a transverse crack. J Sound Vib 162(3):387–401
9. Imam I, Azzaro SH, Bankert RJ, Scheibel J (1989) Development of an on-line rotor crack detection and monitoring system. J Vib Acoust Stress Reliab Des 111(3):241–250
10. Nelson HD, Nataraj C (1986) The dynamics of a rotor system with a cracked shaft. J Vib Acoust Stress Reliab Des 108(2):189–196
11. Ostachowicz WM, Krawczuk M (1991) Analysis of the effect of cracks on the natural frequencies of a cantilever beam. J Sound Vib 150(2):191–201
12. Papadopoulos CA, Dimarogonas AD (1987) Coupled longitudinal and bending vibrations of a rotating shaft with an open crack. J Sound Vib 117(1):81–93
13. Tlaisi A, Akinturk A, Swamidas AS, Haddara MR (2012) Crack detection in shaft using lateral and torsional vibration measurements and analyses. Mech Eng Res 2(2):52

Analysis of Parameters on Bend Force in Nickel-Coated Mild Steel Sheets Through Contour Plot

D. Pritima, B. Stalin, J. Vairamuthu, P. Mallesham, M. Srinivasa Rao, and S. Marichamy

Abstract The contour plot analysis of parameters on bend force is carried out in nickel-coated mild steel sheets using response surface methodology (RSM). The contour plot analysis includes various parameters, namely as punch travel (mm), strain hardening index, holding time (min), punch velocity (mm/s) and width of the sheet (mm) on the bend force. The experimental analysis is achieved through central composite design (CCD). The linear and quadratic polynomial equations are used to predict the mathematical models. Finally, concluded that the proposed models predict the response satisfactorily within the limits and the punch travel is the main significant parameter on bend force. The holding time and strain hardening index provide less effect on bend force.

Keywords Nickel-coated mild steel · Sheet bending · Bend force · Punch travel and RSM

1 Introduction

Bending methods and bending behavior were analyzed in fiber sintered sheet [1]. The developed finite element model was applied to analyze the bend force in cold

D. Pritima
Department of Mechanical Engineering, Vickram College of Engineering, Sivagangai, Tamil Nadu 630561, India

B. Stalin (✉)
Department of Mechanical Engineering, Anna University, Regional Campus Madurai, Madurai, Tamil Nadu 625019, India
e-mail: stalin1312@gmail.com

J. Vairamuthu
Department of Mechanical Engineering, Sethu Institute of Technology, Kariapatti, Virudhunagar, Tamil Nadu 626115, India

P. Mallesham · M. Srinivasa Rao · S. Marichamy
Department of Mechanical Engineering, Sri Indu College of Engineering and Technology, Hyderabad, Telangana 501510, India

A. Arockiarajan et al. (eds.), *Advances in Industrial Automation and Smart Manufacturing*, Lecture Notes in Mechanical Engineering,
https://doi.org/10.1007/978-981-15-4739-3_55

formed stainless steel [2]. The various shapes and angle of bending have been done in stainless steel sheet [3]. The spring-back experimental investigation has been conducted in SPFC 440 sheets by U-die bending process [4].

The cooling during bending process has been analyzed by strip casting and rolling of sheet metals [5]. The Bauschinger effect and behavior of sheet metals were investigated [6]. In RSM, the second-order polynomial-based predominant model was used to optimize the sheet bending parameters [7].

This work describes the contour plot analysis of parameters on bend force in nickel-coated mild steel sheets. The design of experiments was confirmed through central composite approach. The linear and quadratic polynomial equations are used to predict the response.

2 Methodology

2.1 Developing the Design Matrix

The modeling and optimizing the response variables were derived by response surface methodology. The quadratic model and regression model were obtained from the experimental results [8, 9]. The experimental design and bend force results were shown in Table 1. The experiments were designed by central composite design.

2.2 Developing an Empirical Model

The empirical model for bend force of the sheet is a function of punch travel (A), strain hardening index (B), holding time (C), punch velocity (D) and width of the sheet (E) which is mentioned in Eq. 1. The developed model depends on regression coefficients [7].

$$\begin{aligned} \text{Bend force} = & -521.24 - 34.04A + 4725.4B + 1.51C \\ & + 239.01D + 14.78E + 4.73AB + 3.9AC + 9.6AD \\ & + 0.19AE - 224.13BC - 1282.63BD - 61.47BE \\ & + 7.96CD - 0.36CE - 1.84DE \end{aligned} \tag{1}$$

The ANOVA results for bend force are shown in Table 2. It consists of model with correlation with various parameters.

Table 1 The experimental design and bend force results

Std	Run	A: Punch travel (mm)	B: Strain hardening index	C: Holding time (min)	D: Punch velocity (mm/s)	E: Width of the sheet (mm)	Bend force (N)
17	1	15	0.25	5	0.4995	50	190
15	2	15	0.3	3.75	0.4995	50	252
23	3	15	0.25	3.75	0.4995	50	240
25	4	15	0.25	3.75	0.4995	50	240
5	5	20	0.3	2.5	0.166	60	234
21	6	15	0.25	3.75	0.4995	60	234
19	7	15	0.25	3.75	0.166	50	234
26	8	15	0.25	3.75	0.4995	50	200
18	9	15	0.25	3.75	0.166	50	234
9	10	20	0.2	5	0.166	60	234
14	11	15	0.2	3.75	0.4995	50	208
6	12	20	0.2	2.5	0.833	60	254
24	13	15	0.25	3.75	0.4995	50	235
4	14	20	0.3	5	0.166	40	250
16	15	15	0.25	3.75	0.4995	50	236
1	16	20	0.3	2.5	0.833	40	247
13	17	20	0.25	3.75	0.4995	50	210
12	18	10	0.25	3.75	0.4995	50	274
22	19	15	0.25	3.75	0.4995	50	260
3	20	10	0.3	5	0.166	60	210
7	21	10	0.2	5	0.833	60	240
2	22	20	0.2	5	0.833	40	241
11	23	10	0.2	2.5	0.166	40	234
20	24	15	0.25	3.75	0.4995	40	230
10	25	10	0.3	5	0.833	40	230
8	26	10	0.3	2.5	0.833	60	234

3 Contour Plot Analysis

Figure 1a shows the cause of strain hardening index and punch travel on bend. From the figure, it was identified that bend force increases linearly with increase in the strain hardening index. It is also evident that bend force has a maximum when punch travel value is low and the strain hardening index is high. The strain hardening index has a major effect on the bend force [10].

From Fig. 1b, it is confirmed that punch travel and holding time have least effect on bend force. Bend force does not vary much with change in both the punch travel

Table 2 The ANOVA results for bend force

Source	Sum of squares	DF	Mean of square	*F* value	*p*-value Prob > *F*	
Model	5501.978	15	366.7986	1.207773	0.3900	Not significant
A-Punch travel	2048	1	2048	6.743535	0.0266	
B-Strain hardening index	968	1	968	3.187374	0.1045	
C-Holding time	1058	1	1058	3.48372	0.0915	
D-Punch velocity	0	1	0	0	1.0000	
E-Width of the sheet	8	1	8	0.026342	0.8743	
AB	4.281709	1	4.281709	0.014099	0.9078	
AC	1819.389	1	1819.389	5.990778	0.0344	
AD	784.1069	1	784.1069	2.581861	0.1392	
AE	300.6187	1	300.6187	0.98986	0.3432	
BC	600.0502	1	600.0502	1.97581	0.1901	
BD	1398.735	1	1398.735	4.605674	0.0574	
BE	2888.664	1	2888.664	9.511625	0.0116	
CD	33.73698	1	33.73698	0.111087	0.7458	
CE	62.65219	1	62.65219	0.206297	0.6594	
DE	115.8721	1	115.8721	0.381537	0.5506	
Residual	3036.983	10	303.6983			
Lack of fit	1136.983	6	189.4972	0.398941	0.8490	Not significant
Pure error	1900	4	475			
Cor total	8538.962	25				

and holding time. Maximum bend force was induced when the punch travel and holding time are high. Figure 1c shows the interaction of punch velocity and punch travel on the bend force of nickel-coated mild sheet bending. It is observed that the maximum bend force was induced for the lowest punch travel and punch velocity values. Figure 1d shows the interaction of punch velocity and strain hardening index on induced the bend force of sheet bending. It is observed that, for the increase in punch velocity, the increase in bend force is observed. The similar result was obtained for the interaction of the width of the sheet and strain hardening index (Fig. 1e).

From Fig. 1f, it is also evident that the punch velocity and holding time have no effect on the bend force during sheet bending operations. Also, while interacting, the width of the sheet with holding time, the reasonable effect is not observed as seen in

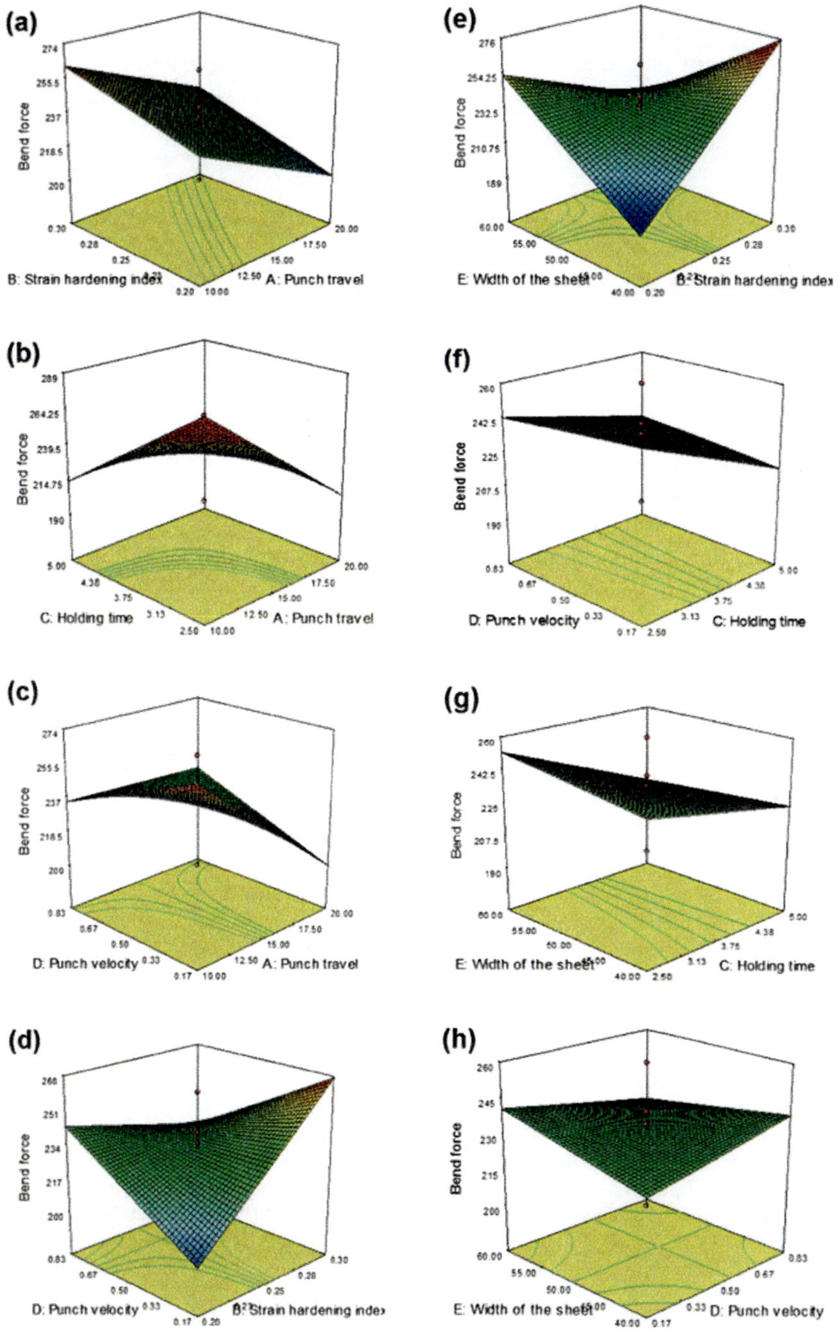

Fig. 1 **a–h** Contour plot–response graphs for bend force

Fig. 1g. The increase in width of the sheet increases the bend force and also increase in punch velocity increases the bend forces and it is obtained from Fig. 1h. The larger amount of force and small amount of die radius required to bend the parts.

4 Conclusions

The below conclusions were drawn from the experimental investigation:

- Sheet bending behavior was analyzed successfully through bend force by using response surface methodology.
- An empirical model was developed to predict bend.
- A maximum bend force of 274 N is obtained under at punch travel is 10 mm, strain hardening index is 0.25, holding time is 3.75 min, punch velocity is 0.4995 mm/s and width of the sheet is 50 mm.
- Punch travel was the for the most part on the bend force.
- Holding time and the strain hardening index have provided less effect on bend force.

References

1. Zhou W, Wang Q, Ling W, He L, Tang Y, Wu F, Liao J, Hui KS, Hui KN (2014) Characterization of three- and four-point bending properties of porous metal fiber sintered sheet. Mater Des 56:522–527
2. Zhou F, Long G (2016) Element interaction of cold-formed stainless steel cross-sections subjected to major axis bending. J Constr Steel Res 118:22–40
3. Gisario A, Barletta M, Venettacci S, Veniali F (2015) External force-assisted Laser Origami (LO) bending: shaping of 3D cubes and edge design of stainless steel chairs. J Manuf Process 18:159–166
4. Leu D-K (2016) A simplified approach for distinguishing between spring-back and spring-go in free U-die bending process of SPFC 440 sheets. Mater Des 94:314–321
5. Wang X, Li H, Liu Z, Liu H, Wang G, Luo Z, Zhan F, Chen S, Lyu L (2016) Effect of cooling rate on bending behavior of 6.5 wt.% Si electrical steel thin sheets fabricated by strip casting and rolling. Mater Charact 111:67–74
6. Zang S, Lee M-G, Sun L, Kim JH (2014) Measurement of the Bauschinger behavior of sheet metals by three-point bending springback test with pre-strained strips. Int J Plast 59:84–107
7. Wang H, Li E, Li GY (2009) Development of hybrid fuzzy regression-based metamodeling technique for optimization of sheet metal forming problems. Mater Des 30:2854–2866
8. Wang H, Li E, Yao LG (2008) Optimization of draw bead design in sheet metal forming based on intelligent sampling by using response surface methodology. J Mater Process Technol 206:45–55
9. Gopalakannan S, Senthilvelan T (2013) Application of response surface method on machining of Al–SiC nano-composites. Measurement 46:2705–2715
10. Eggertsen P-A, Mattiasson K (2009) On the modelling of the bending–unbending behaviour for accurate spring back predictions. Int J Mech Sci 51:547–563

Development of Multiple Regression Models for Wire Spark Erosion Machining of AA2024 Alloy

B. Vishnu Vardhana Naidu, K. C. Varaprasad, N. Manikandan, and J. S. Binoj

Abstract AA2024 is an aluminium alloy, most commonly employed as material for aircraft structural applications. The exceptional properties of these alloys made this material to be suitable for consequent applications. The components used for aerospace applications need very closer tolerances and are more accurate in shape and size. Wire electrical discharge machining (WEDM) is a contemporary approach of material removal which is conceived from the concept of spark erosion machining process which is known as WEDM, predominantly employed for machining of difficult to machine materials and also especially used for making complicated shapes on any electrically conductive material. In this present exploration, an analysis has been performed on WEDM of AA2024 alloy. Pulse on time (μs), pulse off time (μs), flushing pressure (kg/cm^2), wire feed (mm/min) and servo voltage (V) are considered as independent process parameters in this present exploration. Surface roughness (microns) is deemed as desired performance measure which is to be improved. Taguchi's design approach has been adopted for designing and analysing the data attained from the experimentation. An L27 orthogonal array (OA) has been employed to perform the experimental trials. The effects of process variables on desired performance measures such as surface roughness were analysed by Taguchi's response analysis. The significance of input process variables on desired performance measure is examined by ANOVA analysis. Multiple regression analysis has been performed for correlating the relationship amongst the selected independent process variable and preferred performance characteristic. The comparison results proved that the values predicted from the developed regression models were closer with the experimental observations.

Keywords WEDM · AA2024 alloy · Taguchi's methodology · Experimental design · Response analysis · ANOVA · Multiple regression · Single objective optimization

B. Vishnu Vardhana Naidu (✉) · K. C. Varaprasad · N. Manikandan · J. S. Binoj
Micro Machining Research Centre, Department of Mechanical Engineering, Sree Vidyanikethan Engineering College, (Autonomous), Tirupati, Andhra Pradesh 517102, India
e-mail: vishnu234@gmail.com

A. Arockiarajan et al. (eds.), *Advances in Industrial Automation and Smart Manufacturing*, Lecture Notes in Mechanical Engineering,
https://doi.org/10.1007/978-981-15-4739-3_56

1 Introduction

Aluminium alloy is commonly preferred in the manufacturing domain due to its low weight and significantly better strength. The components for aerospace industries extensively employ this as work material [10]. In order to take the advantages of these alloys. The aerospace industries require to make component with more precision and accurate dimension. A negligible amount of error/misalignment in the manufactured components makes consequence in malignant failures. So there is need of giving importance to unconventional machining methods for making these kinds of components with more precision and accuracy [7, 8]. In recent days, the technology of WEDM has been improved in a considerable way for accomplishing the demands in numerous manufacturing domains. In WEDM, the sparks are produced by continuously amongst the tool electrode (wire) and the work piece material that are divided by a distance which makes the possible removal of material in the work sample. The wire action is being controlled for attaining the desired accuracy and shape of the work material. The graphical representation of WEDM is shown in Fig. 1 [3]. WEDM works by making high voltage spark amongst the work material and moving wire with evaporation of work materials. So the WEDM is the appropriate method of material removal for electrically conductive work material irrespective of its hardness.

The potential usage of WEDM method has been recognized as hastily developing method for material removal in recent days and exclusively employed in fabrication of precise and accurate components that are to be used in various engineering applications such as tool fabrication, die making, automobile and aerospace components manufacturing [2, 4].

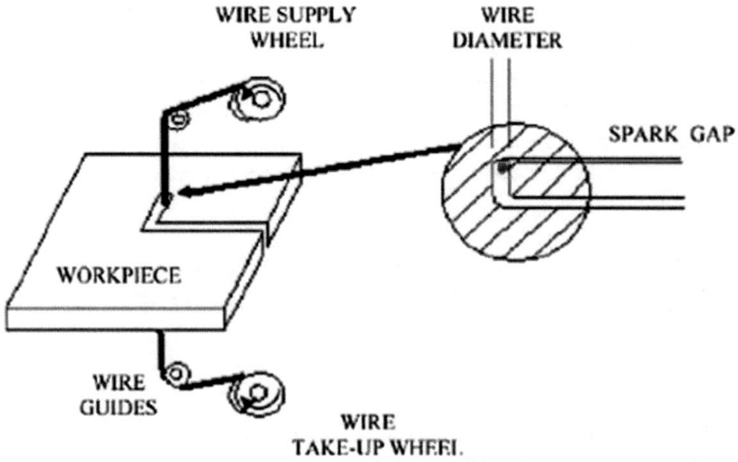

Fig. 1 Schematic of Wire EDM

In recent days, Taguchi's methodology is deemed as a powerful tool for enhancing the rate of production with the help of efficient work plan and ascertaining the better combination of process variables as single-objective optimization problem that makes the possibility to the manufacturer to produce the parts/components with anticipated quality with affordable cost [11]. A multi-objective optimization has been done by hybrid method named as Taguchi-based grey theory in WEDM of various materials and best possible machining parameters were determined for obtaining the better machining performance [1, 9]. A multi-objective optimization model has been developed with various predictive tools for various unconventional machining processes and also the importance of predictive modelling is explained with the help of various advanced predictive techniques. The regression models were developed for desired output measures such as surface roughness and kerf width for optimizing the process variables in WEDM by genetic algorithm method and it is an effective method for predicting the performance measures [13]. The regression models were developed for various machining operations and it is proven that the values predicted from the developed regression models were closer relationship with actual experimental values [5, 6, 8, 12]. The appropriate choice of proper combination of input variable for attaining the improved performance in machining and other accuracy connected activities is fairly perplexing in WEDM process because of presence of numerous process variables. So a careful study is required for ascertaining the best suited combination of process variable to achieve the desired performance improvement in machining.

It is surmised from the available literature that there is an importance requires on investigational study on WEDM of AA2024 alloy, also in development of regression model and prediction of desired performance measures. In this present research article, an exertion has been taken for examining the influence of process variables on surface roughness by Taguchi's method and to develop and predict the desired performance with the help of regression models.

2 Materials and Methods

The setup for experimentation comprised of machining chamber, panel for control and dielectric fluid (deionized water) circulation system. AA2024 has been selected as the work material and it is placed in the vice given inside of the chamber. The selected independent process variables were varied according to the levels and ranges selected. The experimental runs were performed on ELEKTRA make wire-cut electric discharge machine (Model: ULTRACUT 843). Brass is utilized as wire material (terminal) and connected between two rollers. Taguchi's structural approach is utilized for arranging the trial runs. The WEDM parameters and their levels are outlined in Table 1. A L27 OA has been chosen for experimentation. The machining was performed to make a through hole and roughness of the machined surface has been evaluated (Mitutoyo Surface Roughness Tester SJ-210).

Table 1 Machining process parameters

Symbols	Machining parameters	Levels		
		1	2	3
A	Pulse on time (μs)	105	108	111
B	Pulse off time (μs)	54	57	60
C	Flushing pressure (kg/cm^2)	6	9	12
D	Wire feed (mm/min)	2	4	6
E	Servo voltage (V)	20	25	30

3 Results and Discussion

The experimental runs have been performed as per L27 OA for exploring the prominence of independent factors on desired performance characteristic in Wire EDM of AA2024 alloy. An exertion has been made to ascertain the best possible process variables for accomplishing the productive and competent process of Wire EDM. In the Wire EDM process, lower the values of surface roughness is the indicator of superior performance measure. So the surface roughness is considered as smaller the better criterion.

3.1 Influence of Process Parameters on Surface Roughness

The graphical illustration of response graph for roughness of the machined surface during Wire EDM of AA2024 is depicted in Fig. 2. It is perceived from the graph that the values of roughness of the machined surface getting increased with an increase of 'wire feed and servo voltage' and the same has been getting diminished with an increment of 'P_{on}', 'P_{off}' and flushing pressure. The pulse off time is considered as the persuasive process variable for roughness of the machined surface.

Taguchi's analysis has been executed for analysing the roughness of machined surface and the results are presented in Table 2. The optimal combination of process parameters for attaining improved surface finish is *A3B3C3D1E1*. It means that the optimal combination of process parameters for better performance is 'P_{on}'—111 μs; 'P_{off}'—60 μs; flushing pressure—12 kg/cm^2; wire feed—2 mm/min; servo voltage—20 V. Also it is witnessed from the analysis that the 'P_{off}' is the major variable and then it is trailed by pulse on time, servo voltage, wire feed and flushing pressure.

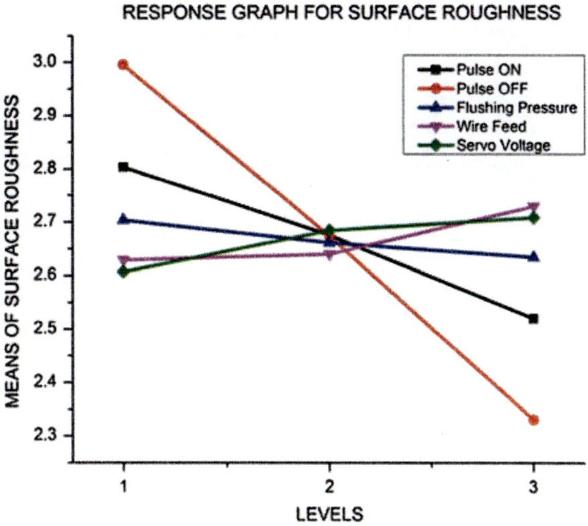

Fig. 2 Response graph for surface roughness

Table 2 Response analysis for surface roughness for WEDM of AA2024 alloy

Levels	Means of surface roughness				
	A	B	C	D	E
1	2.803	2.995	2.703	2.629	2.607
2	2.675	2.673	2.661	2.640	2.684
3	2.520	2.330	2.634	2.729	2.708
Delta	0.282	0.665	0.069	0.1	0.101
Rank	2	1	5	4	3

3.2 *Importance of Process Parameters on Performance Measures (ANOVA)*

The influence of the process variables on desired performance measure is determined by analysis of variance (ANOVA) statistically at 95% confidence level. The ANOVA outcomes are depicted in Table 3. It is witnessed from the analysis that the pulse off time is the preponderant parameter for roughness of machined surface.

Table 3 Analysis of variance for WEDM of AA2024 alloy

Source	DF	Seq SS	Adj SS	Adj MS	*F*	*P*
Surface roughness (mm^3/min)						
Pulse on	2	0.35983	0.35983	0.17992	77.38	0
Pulse off	2	1.99139	1.99139	0.99569	428.21	0
Flushing pressure	2	0.02147	0.02147	0.01073	4.62	0.026
Wire feed	2	0.05427	0.05427	0.02713	11.67	0.001
Servo voltage	2	0.0503	0.0503	0.02515	10.82	0.001
Error	16	0.0372	0.0372	0.00233		

3.3 Development of Multiple Regression Models

Multiple regressions method has been adopted to determine the correlation amongst the selected independent and dependent process variables. The regression models have different models such as linear, quadratic, interaction and full model and have been exhibited in Eqs. (1–4) (for three input parameters):

Linear model:

$$y = \beta_0 + \beta_1 X_1 + \beta_2 X_2 + \beta_3 X_3 \ldots \quad (1)$$

Quadratic model:

$$y = \beta_0 + \beta_1 X_1 + \beta_2 X_2 + \beta_3 X_3 + \beta_4 (X_1)^2 + \beta_5 (X_2)^2 + \beta_6 (X_3)^2 \ldots \quad (2)$$

Interaction model:

$$y = \beta_0 + \beta_1 X_1 + \beta_2 X_2 + \beta_3 X_3 + \beta_4 X_1 X_2 + \beta_5 X_1 X_3 + \beta_6 X_2 X_3 \ldots \quad (3)$$

Second order (Full model):

$$\begin{aligned} y = \beta_0 + \beta_1 X_1 + \beta_2 X_2 + \beta_3 X_3 + \beta_4 (X_1)^2 + \beta_5 (X_2)^2 + \beta_6 (X_3)^2 \\ + \beta_7 X_1 X_2 + \beta_8 X_1 X_3 + \beta_9 X_2 X_3 \ldots \end{aligned} \quad (4)$$

where 'y' is criterion variable and 'X_1', 'X_2' and 'X_3' are predictor variables.
'β_1', 'β_2', 'β_3'……'β_n' are regression coefficients.

3.3.1 Regression Model for Surface Roughness

The empirical relations for interrelating the correlation amongst the preferred performance aspects and the considered process factors have been attained and the preferred regression empirical relation for surface roughness is exposed in Eq. (5).

$$\begin{aligned}\text{SR (microns)} = & -36.8807 + 0.389444\ \text{Pulse ON} \\ & + 0.304889\ \text{Pulse OFF} \\ & + 0.126278\ \text{Flushing Pressure} \\ & + 0.191375\ \text{Wire Feed} \\ & + 0.253578\ \text{Servo Voltage} \\ & - 0.00152469\ \text{Pulse ON} \\ & * \text{Pulse ON} - 0.000522222\ \text{Pulse ON} \\ & * \text{Servo Voltage} - 0.00120988\ \text{Pulse OFF} \\ & * \text{Pulse OFF} - 0.00224444\ \text{Pulse OFF} \\ & * \text{Servo Voltage} - 0.00611728\ \text{Flushing Pressure} \\ & * \text{Flushing Pressure} - 0.000466667\ \text{Flushing Pressure} \\ & * \text{Servo Voltage} - 0.0236389\ \text{Wire Feed} \\ & * \text{Wire Feed} - 0.000208333\ \text{Wire Feed} \\ & * \text{Servo Voltage} - 0.00108222\ \text{Servo Voltage} \\ & * \text{Servo Voltage} \ldots \end{aligned} \tag{5}$$

3.4 *Comparative Analysis of Experimental and Predicted Values by Multiple Regression Equation*

A comparison has been done for the measured values from the conducted trials and the values obtained from the prediction by regression equation are presented in this section.

Comparison of surface roughness with the experimental outcomes and the predicted values by the developed multiple regression model are shown in Fig. 3. From the comparative analysis, it is perceived that the predicted values by the developed regression models were closer with the experimental observations.

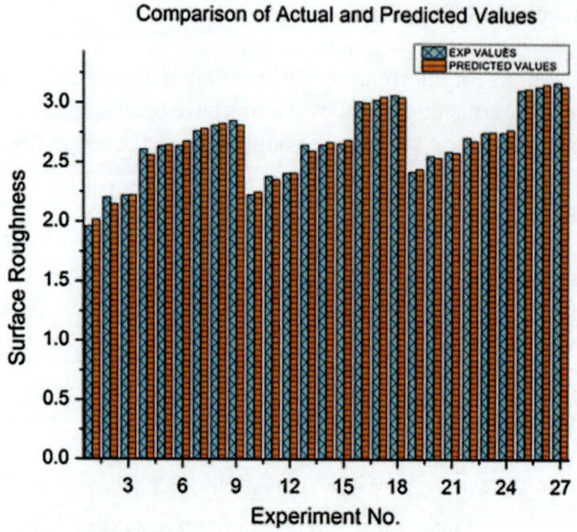

Fig. 3 Comparison of actual and predicted values—surface roughness

4 Conclusions

The experimental analysis presented that the investigations on roughness of the machined surface and development of predictive models for WEDM of AA2024 alloy and the preferred performance characteristics are remarkably influenced by the various dominant input machining variables that are considered in this present analysis. Taguchi's method is employed for designing the experiments and this method has ascertained its competence to be an effectual tool for investigation of the WEDM process. The following were the conclusions obtained from the exploration:

- Taguchi's design of experimental methodology is opted for planning the experimentation. An L_{27}OA has been chosen by considering the input variable, namely pulse on time, pulse off time, flushing pressure, wire feed and servo voltage.
- Taguchi's analysis has been employed to ascertain the best combination of process parameters for attaining the improved machining aspects on considered performance measure.
- The dominance of desired process parameters on the considered performance machining aspects is revealed by Taguchi's single response analysis response graphs.
- ANOVA has been employed to identify the prominence of the independent process parameters on various performance machining characteristics. It is perceived from ANOVA analysis that the pulse off time is the significant process parameter for surface roughness.
- Mathematical models have been evolved to correlate the desired performance measure material removal rate with the various input machining variables. The

comparison results proved that the values envisaged from the developed regression models were closer with the experimental observations.

References

1. Aravind Krishnan S, Samuel GL (2013) Multi-objective optimization of material removal rate and surface roughness in wire electrical discharge turning. Int J Adv Manuf Technol 67(9–12):2021–2032
2. Chalisgaonkar R, Kumar J (2015) Multi-response optimization and modeling of trim cut WEDM operation of commercially Pure Titanium (CPTi) considering multiple user's preferences. Eng Sci Technol Int J 18:125–134
3. El-Hofy Hassan (2005) Advanced machining processes. McGraw-Hill, New York
4. Ho KH, Newman STÃ, Rahimifard S, Allen RD (2004) State of the art in wire electrical discharge machining (WEDM). Int J Mach Tools Manuf 44:1247–1259
5. Jayabal S, Natarajan U (2010) Regression & neuro fuzzy models for prediction of thrust force and torque in drilling of glass fibre reinforced composites. J Sci Ind Res 69(October):741–745
6. Leone C, Addona D, Teti R (2011) Tool wear modelling through regression analysis and intelligent methods for Nickel base alloy machining. CIRP J Manufact Sci Technol 4(3):327–331. https://doi.org/10.1016/j.cirpj.2011.03.009
7. Manikandan N, Arulkirubakaran D, Palanisamy D, Raju R (2019) Influence of wire-EDM textured conventional tungsten carbide inserts in machining of aerospace materials (Ti–6Al–4 V alloy). Mater Manuf Process 34(1):103–111
8. Manikandan N, Binoj JS, Varaprasad KC, Sabari SS, Raju R (2019) Investigations on wire spark erosion machining of Aluminum-l. In: Advances in manufacturing technology, Springer, Singapore, pp 361–369
9. Manikandan N, Kumanan S, Sathiyanarayanan C (2017) Optimisation of electrochemical drilling process using Taguchi method and regression analysis. Int J Mach Mach Mater 19(2):136–159
10. Marcos-Bárcena M, Sebastián-Pérez MA, Contreras-Samper JP, Sánchez-Carrilero M, Sánchez-López M, Sánchez-Sola JM (2005) Study of roundness on cylindrical bars turned of Aluminium–Copper alloys UNS A92024. J Mater Process Technol 162:644–648
11. Ross PJ (1996) Taguchi, techniques for quality engineering. McGraw Hill International Edition, Singapore
12. Sahoo AK, Pradhan S (2013) Modeling and optimization of Al/SiCp MMC machining using Taguchi Approach. Measurement 46(9):3064–3072
13. Selvam MP, Kumar PR (2017) Optimization Kerf width and surface roughness in wirecut electrical discharge machining using brass wire. 21(1):37–55

Numerical Analysis on Gas Turbine Blade of a Nickel-Based Alloy with Titanium Alloy

V. Dhinakaran, R. Surendran, K. P. Manoj Kumar, A. Rahul Kumar, B. K. Nagesha, and M. D. Vijayakumar

Abstract Gas turbine has complicated problems in the turbine blades besides the combustion chamber. The turbine blades are swayed massively due to the high-temperature and high-velocity gas emerging out from the combustion cell and knocking the blades. Principally, the static and thermal loads affect the blades, wherein temperature plays a vital role. Here, we executed a comparative study of the stator blades made using diverse alloys in the gas turbine. The stator blade is modeled with CATIA V5. The finite element analysis is carried out by ABAQUS 2018 to determine the effects of static and thermal loads in the stator blades of the gas turbine. This study has been done to determine which alloy is best, whether titanium alloy or nickel alloy. RR1000, Ti6Al4V and INCONEL 600 are taken into account for the comparison study.

Keywords Static load · Thermal load · Stator blade · Comparing diverse alloy

1 Introduction

The gas turbine is generally a mechanical setup/device which has a wide range of expertise in many fields. After the combustion process, the hot gases with high speed create a rotating effect on the turbine which tends to rotate it in a circular motion. The power is produced by coupling the gas turbine with the generator. Further, power is also used to activate the compressor. The air is compounded by various stages and continuously expand through fixed and moving blades. The high thrust is produced by elevating the working fluid temperature. The important part of the gas turbine in power generation is the stator blade of the gas turbine. The turbine blades play a vital role in extracting power from hot and high-pressure gases of combustion chamber.

V. Dhinakaran (✉) · R. Surendran · K. P. Manoj Kumar · A. Rahul Kumar · M. D. Vijayakumar
Centre for Applied Research, Chennai Institute of Technology, Kundrathur, Chennai, Tamil Nadu 600069, India
e-mail: dhinakaranv@citchennai.net

B. K. Nagesha
Gas Turbine Research Establishment, DRDO, Bengaluru 560093, India

A. Arockiarajan et al. (eds.), *Advances in Industrial Automation and Smart Manufacturing*, Lecture Notes in Mechanical Engineering,
https://doi.org/10.1007/978-981-15-4739-3_57

Due to this effect, the blade wears soon. The operating temperature of the gas turbine is 820–1370 °C and the speed ranges between 3000 and 8000 rotation per minute. To overcome the failures, adequate designs have been implemented in the turbine blades. This paper helps in a better understanding of the strength of a turbine blade by varying different materials. Different materials have been chosen to compare the strength of the turbine blades withstanding static and thermal loads. Hacker [1] has conveyed an exploratory research project to advance the design ability of gas turbine blades with concerning static and thermal loads. Menq [2] reviewed the necessity for the RE process and the coordinate measuring machine features by digitally materializing the current model. Tari [3] examined the significant hurdles in the designing blade of the gas turbine by admitting the effects of turbine blades while cooling after high operating temperatures. Kermanpur [4] discussed the crack proliferation analysis of gas turbine compressor blades of Ti6Al4V and also provided the failure spot. Sindhu and Chikkanna [5] had performed a comparative analysis study of a gas turbine blade with titanium alloy, N 155 alloy and nickel alloy material and spot out the most desirable material among them based on the applicability. Chintala et al. [6] investigated the gas turbine blades in Ansys with various materials and generating a solid model from the FEA results obtained from Ansys. A clear study has been put forth on the selection of material for gas turbine blades. Chintala et al. [7] carried out a finite element analysis for optimizing the number of holes in gas turbine blades to accomplish reliable performance at high temperatures with the assistant of Ansys. Hou et al. [8] distinguished the causes for failures in blades by assessing the static and thermal loads of the gas turbine blades.

2 Material Used

The materials are distinguished by their mechanical properties (Young's modulus, Poisson ratio, yield strength, ultimate strength), ductility and thermal properties since stresses induced in the blades during operation is very high, which can be compensated by stronger materials. This paper has focused on the materials that are dominant and representing efficient material choices for stator blades [4, 9–11]. The World's largest original equipment manufacturers use nickel and titanium alloys materials for the design of the stator blade design and manufacturing. Based on the literature survey and with the knowledge transferred from gas turbine manufacturing industries, the selected materials are Titanium Ti6Al4V [10], INCONEL 600 [10] and RR1000 [10]. The material property knowledge transferred is listed below (Tables 1, 2, 3 and 4).

3 Geometry

Figure 1 blade is considered for finite element analysis. The CAD model is gener-

Table 1 Material property

Property	Units	Ti6Al4V	INCONEL 600	RR1000
Young's modulus (E)	GPa	113.8	207	0.224
Poisson ratio (μ)	–	0.342	0.29	0.24
Density (ρ)	kg/m^3	4430	8430	8300

Table 2 Thermal properties of Ti6Al4V

Thermal conductivity (W/m K)	6.7
Coefficient of thermal expansion E-06 (/K)	8.8

Table 3 Thermal properties of RR1000

Coefficient of thermal expansion E-06 (/K)	Temperature (K)	Thermal conductivity (W/m K)
8.00	293.15	13.5
8.50	323.15	13.8
9.50	373.15	14.5
1.00	473.15	15.3
1.23	573.15	16.8
1.29	673.15	17.6
1.34	773.15	18.7
1.37	873.15	21.5
1.43	973.15	23.1
1.50	1073.15	26
1.64	1173.15	30
1.66	1273.15	34
1.85	1423.15	43.6
1.87	1473.15	

ated by CATIA V5. The latest perforated stator blades with internal cooling lines are considered for modeling. The stator blade consists of leading-edge, trailing edge, suction side and fixture disk. The blade profile is designed to deliver high performance. Originally, air flows through the front edge and leaves out at the tail edge, where the cooling is carried out by the perforated suction side.

4 Finite Element Analysis

Computational mechanics has progressed immensely over the last few decades and today it is the need of the hour for researchers. Structural modeling and analysis are

Table 4 Thermal properties of INCONEL 600

(a)	
Coefficient of thermal expansion E-05 (/K)	**Temperature (K)**
1.33	366.15
1.39	477.15
1.42	589.15
1.46	700.15
(b)	
Thermal conductivity (W/m k)	**Temperature (K)**
14.8	294.15
15.7	366.15
17.4	477.15
19.2	589.15
20.9	700.15

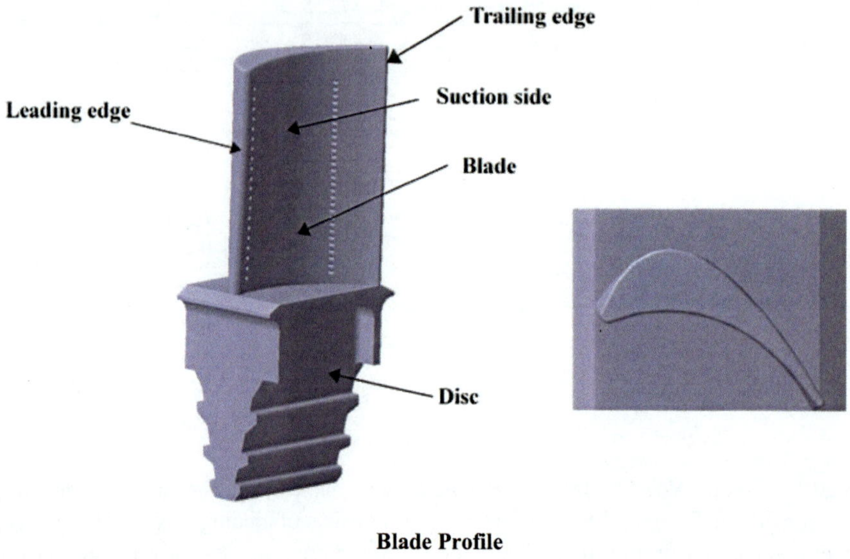

Fig. 1 CAD model of turbine blade

playing a vital role in determining the deflection, induced stress, in response to the applied load and identifying the regions where failures are most likely to occur in the model. The knowledge of the behavior of a structural analysis assists in improving the performance and increasing the efficiency of the design. Static structural analysis is done in ABAQUS to calculate the effect of steady loading conditions on the stator blade of the gas turbine. The deflection obtained for the applied load is determined in

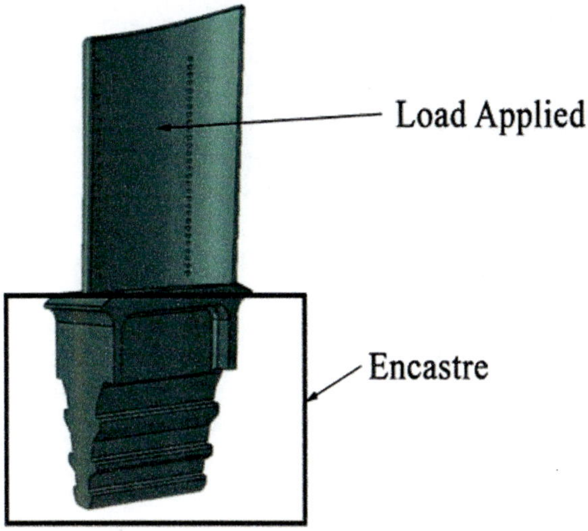

Fig. 2 Loading and boundary conditions

Table 5 Mesh details

Part	Element type	Element shape	Geometric order	No. of elements	No. of nodes
Gas turbine blade	C3D10	Tetrahedral	Quadratic	238,091	370,457

the analysis. The CAD model of the stator blade of a gas turbine engine is imported in ABAQUS. The analysis is done for three selected materials by maintaining the same thermal and static load on the blade. In actual working, the stator blade of the gas turbine is attached to the rotating disk. For the analysis, the root of the stator blade is encased and the load is applied at the reference point coupled with the blade. Three types of loads are subjected to the blade of the gas turbine as discussed by Deepanraj et al. [7], the blades are subjected to the (i) axial, (ii) tangential and the (iii) centrifugal forces. For rotor turbine blade, axial force, $F_a = 16$ N, tangential force, $F_t = 995$ N, centrifugal force, $F_c = 39{,}000$ N. The whole blade is maintained at an operating temperature of 1366.15 K during the analysis which is predefined in ABAQUS. The tetrahedral quadratic element is used for meshing and details of the mesh are given in Table 2 (Fig. 2; Table 5).

5 Results and Discussion

The finite element analysis of the stator blade is performed with diverse materials, i.e., Titanium Ti6Al4V, INCONEL 600 and RR1000 by inducing static and thermal loads with an aim of determining the deflection of the blade. First, the turbine blade design is analyzed with Titanium Ti6Al4V and results are noted.

Figure 3 shows the deflections when it is analyzed with Ti alloy. The maximum deflection obtained is 1.264e−06 m. Then, the analysis is done with INCONEL 600 and deflection is obtained as shown in Fig. 3.

The deflection of the blade is shown in Fig. 4 and the maximum deflection for the given load is 7.085e−07 m.

At last RR1000 is analyzed and from Fig. 5, the maximum deflection obtained is 6.627e−04 m.

The deflections obtained for all the materials are tabulated, and from Table 6, we can see that the Ti alloys blade shows less deflection than the nickel-based alloys. Thus, Ti alloys can be used for the manufacture of the turbine blades which are light in weight, highly stiff, corrosive resistance and have excellent strength. This turbine fan blade is a fatigue-critical model; therefore, it should withstand high strength without destroying its structure. It should possess sufficient ultimate strength and fatigue strength under different operating conditions. The damage in blades occurs due to the storage of high kinetic energy in the rotating blades. Repeated loading of the structure leads to early failure, if it is not properly designed. Though the researchers, scientists and material engineers are working on the development of the material for

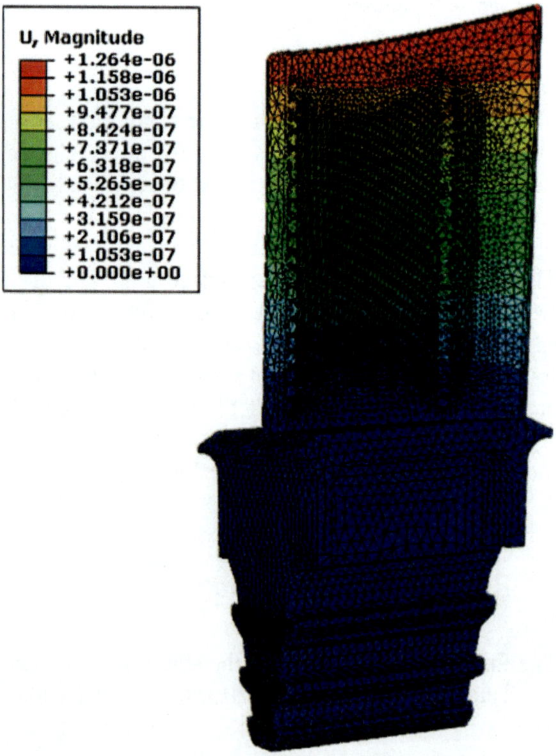

Fig. 3 Deflection of Ti6Al4V

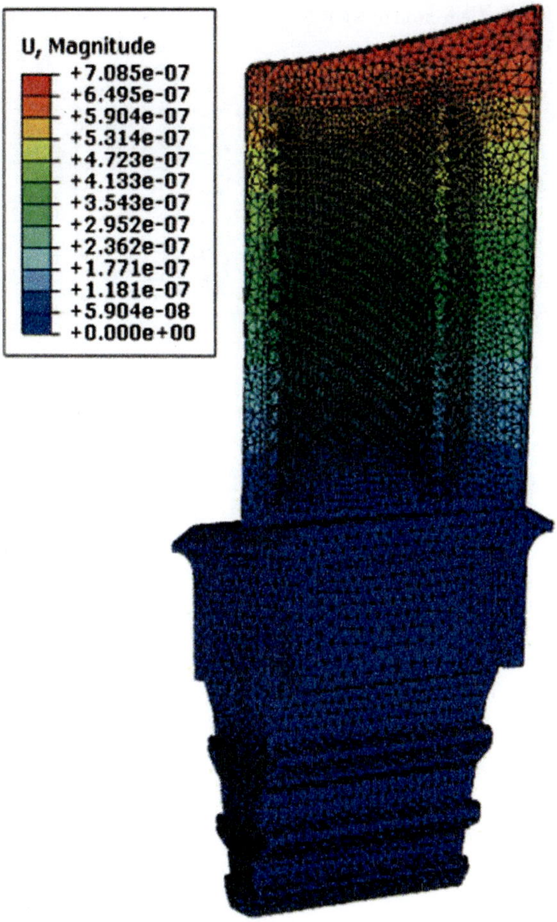

Fig. 4 Deflection of INCONEL 600

the turbine blade with high load resistivity, cost-effective and improved capabilities. The scope is toward the development of the blade manufacturing with high efficiency as a constraint.

6 Conclusion

In this work, the analysis of the blade is done in ABAQUS. The CAD model is generated using CATIA V5 followed by RR1000, Ti6Al4V and INCONEL 600 are taken into account for the comparison study by FEA. It is found that Ti alloys are the most preferred material for manufacturing blades for resisting the axial, tangential and centrifugal load acting on the blade under the operating condition at 1366.66 K due to its overwhelming properties of structural stability when subjected to differential

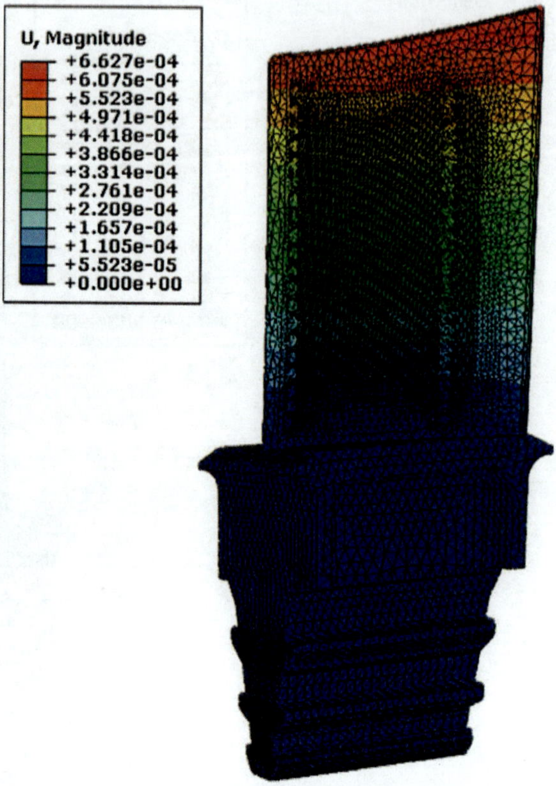

Fig. 5 Deflection of RR100

Table 6 Deflections of all materials

Materials	Deflections (m)
Titanium Ti6Al4V	1.264e−06
INCONEL 600	7.085e−07
RR1000	6.627e−04

loads with increased strength in peak temperatures. The technological innovation leads to a healthy battle among the leading engine manufacturers such as Rolls Royce, Whitney, General Electric, etc. This battle leads the way which is beneficial not only for turbine fan blades but also the overall jet engine development. The persistence and ceaseless efforts of academia, industry and governmental institutions with continuous cooperation will lead to the progressive development of the turbine fan blade.

References

1. Hacker J (2000) Statical analysis of manufacturing deviations and classification methods for probabilistic aerothermal design of turbine blade. Master's thesis. Department of Aeronautics and Astronautics, Aug 2000
2. Menq LC (1994) CMM Feature accessibility and path generation. Int J Prod Res 32(3):597–618
3. Tari I (2000) 3-D simulation of convection in asymmetrically heated turbine blade cooling channels. In: International symposium on heat transfer in gas turbine systems, pp 161–163
4. Kermanpur A, Sepehri Amin H, Ziaei-Rad S, Nourbakhshnia N, Mosaddeghfar M (2008) Failure analysis of Ti6Al4V gas turbine compressor blades. Eng Fail Anal 15(8):1052–1064
5. Sindhu NL, Chikkanna N (2017) Design and analysis of gas turbine blade. Int J Res Appl Sci Eng Technol (IJRASET) 5(VI). IC Value: 45.98. ISSN: 2321-9653
6. Chintala G, Gudimetla P (2014) Optimum material evaluation for gas turbine blade using reverse engineering (RE) and FEA. Procedia Eng 97:1332–1340
7. Deepanraj B, Lawrence P, Sankaranarayanan G (2011) Theoretical analysis of gas turbine blade by finite element method. Sci World 9(9):29–33
8. Hou J, Wicks BJ, Antoniou RA (2002) An investigation of fatigue failures of turbine blades in a gas turbine engine by mechanical analysis. Eng Fail Anal 9(2):201–211
9. Thirugnanasambantham KG, Raju R, Sankaramoorthy T, Velmurugan P, Kannagi A, Reddy MCK, Chandra VR (2018) Degradation mechanism for high-temperature sliding wear in surface-modified In718 superalloy. Cogent Eng 5(1):1–11
10. Dillibabu V, Duraiselvam M, Chandrasekhar U, Raju R (2017) Microstructural studies on laser dissimilar welded Ni and steel alloys for aeronautical turbine applications. Lasers Eng 37(4-6):247–260
11. Wang L, Preuss M, Withers PJ, Baxter G, Wilson P (2005) Energy-input-based finite-element process modeling of inertia welding. Metall Mater Trans B 36(4):513–523

Mechanical and Tribological Properties of Al–Mg–SiC Metal Matrix Composite for Pistons of Two-Stroke Engine

Jacob, V. Dhinakaran, and T. Jagadeesha

Abstract This work is about the tribological and mechanical characterization of aluminium and magnesium metal matrix composite reinforced with silicon carbide (SiC) along with the addition of graphite powder to reduce coefficient of friction and increase lubricity. The Al–Mg MMC is made by squeeze and stir casting method along with the addition of SiC as reinforcement and graphite as a lubricant. Addition of SiC improved the yield strength and tensile strength of the metal matrix composite (MMC). The uniform distribution of the SiC is ensured by fabricating the MMC by squeeze and stirring casting. Magnesium addition to the composite is mainly done to reduce the overall density of the composite. After the fabrication of the MMC, various mechanical tests (like tensile test, hardness test, compression test) and tribological tests (wear test) were conducted to prove the increase in the respective properties. The microstructure and composition of the MMC were analysed using X-ray diffraction and scanning electron microscopy (SEM) techniques. An ANSYS model was also developed at the end to operate a piston with the above composition in the engine operating conditions. Moreover, the mechanical and tribological properties were compared with the original piston material. There was a significant increase in the mechanical properties like tensile strength of the MMC by the addition of SiC as the reinforcement. The hardness value also showed a marginal increase. Tribological property, wear resistance was also increased with the addition of the graphite.

Keywords Composite material · Piston · MMC · Magnesium · SiC

Jacob · T. Jagadeesha (✉)
National Institute of Technology Calicut, Calicut, Kerala 673601, India
e-mail: jagdishsg@nitc.ac.in

V. Dhinakaran
Department of Mechanical Engineering,
Chennai Institute of Technology, Kundrathur, Chennai, Tamil Nadu 600069, India

A. Arockiarajan et al. (eds.), *Advances in Industrial Automation and Smart Manufacturing*, Lecture Notes in Mechanical Engineering,
https://doi.org/10.1007/978-981-15-4739-3_58

1 Introduction

The need for energy is booming up as everyone is looking to lead a comfortable life but the source of energy is limited, most commonly exploited source is fossil fuels and it is depleting at high speed. The time has come where one has to use an energy-efficient system to utilize this source more efficiently in order to utilize it for a prolonged period [1–3].

In the automobile sector, a huge impact can be achieved if the mileage can be increased, to achieve higher mileage most easy solution would be to reduce or decrease the self-weight of vehicle. This gives a breakthrough for application of Al–Mg alloys.

The most dominant matrix material is Al but the emphasis is given to Mg by its virtue for low density, i.e. one-third less than Al. To improved mechanical and chemical properties of Al–Mg, addition of SiC has been beneficial. Al has been used in aero-space and automobile industry as it has a high strength to weight ratio, easy machinability, ductile, low melting and low density. The addition of graphite is done because it is an isotope of carbon which is made up of loosely bonded layers of carbon atoms, which would impart a slippery texture and can be successfully used as a lubricant in stir casting process [4–8].

Al–Mg alloy has engraved its application in various applications mainly in aero-space, marine, automotive and many others. In the current scenario a new Al–Mg (mmc), a combination of Al–Mg would give better results. The proposed work has been an attempt on Al–Mg alloy, developed by stir squeeze casting and martial characterization was done using various modes of testing such as tensile test and wear test. Microstructure was analysed using XRD and SEM. An ANSYS model was prepared to predict the behaviour of a material when made into a piston [9].

2 Experimental Procedure

This section presents how the samples were prepared and how various tests were conducted to measure the desired properties.

2.1 Sample Preparation

Sample of various combinations of Al, Mg, SiC and Graphite were squeeze cast at 100 MPa. Aluminium and magnesium, silicon carbide (1200 mesh) and graphite are weighed and taken in the required proportions as shown in Table 1 are prepared. The aluminium and magnesium ingots are placed in the induction furnace for pre-heating and temperature is raised gradually, i.e. from room temperature to the melting point of magnesium for about an hour. In parallel, a mixture of SiC and graphite is

Table 1 Composition of composite fabricated

Name	Al (%)	Mg (%)	SiC (%)	Graphite (%)
Sample 1	73	20	5	2
Sample 2	83	10	5	2
Original piston	74.5	0.7	–	–

heated to about 400 °C in the pre-heating oven. An inert atmosphere of argon gas is maintained to provide a protective environment which would prevent the burning of magnesium. After melting of metals, a stirrer is introduced in the furnace for achieving homogenous mixture by stirring at 500 rpm and simultaneously mixture of SiC + Graphite is added into the furnace. The stirring is continued for about 15 min with regular up and down motion. Meanwhile, the die and conduit are also pre-heated to 250 °C, in order to prevent solidification. The molten composite is poured into the die and squeezed at a pressure of 100 MPa. The cast is allowed to solidify, after 30 min and the required specimen is obtained. Sample 3 is directly melted in the furnace and squeeze cast is done.

2.2 *Wear Test*

The heat generated due to friction may cause the failure of material; therefore, it is very important to find wear and coefficient of friction. One of the most effective methods of measuring static coefficient of friction and wear rate is using pin-on-disc tribometer. A pin-on-disc tribometer consists of a fixed pin which comes in contact with rotating disc under a known load. Perpendicularity has to be maintained between pin surface and disc surface, to achieve this lapping process is done. Lapping process ensures that the surface is smoother and polished. Coefficient of friction is the ratio of the frictional force to the loading force on the pin.

The sample is mounted on the pin which comes in contact with rotating disc at a constant RPM. A known weight is applied on the pin to create the wear. The coefficient of friction is recorded and from the volume of sample lost the wear rate is calculated.

To investigate the wear, samples are prepared of diameter: 8 mm and length: 25 mm. The disc diameter was selected at 30 mm, with variable loads and speed. All the test samples had undergone lapping to ensure perpendicularity of the test surface. Test is conducted at three different phases namely,

1. Constant load of 10 kg and varying the speed from 200 to 600 rpm in steps of 100 rpm on each sample of the specified size.
2. Constant speed of 400 rpm and varying load from 6 to 14 kg in steps of 2 kg on each sample.
3. Constant speed of 600 rpm and varying load from 6 to 14 kg in steps of 2 kg (Fig. 1).

Fig. 1 Wear test studies done at NIT Calicut

2.3 Microhardness Test

Indentation hardness is used to measures the resistance of a sample to material deformation due to constant compression load from a sharp object. The test used is Vickers hardness test as it is better than other hardness tests because it is independent of the size of the indenter. The basic principle used is to measure the material's ability to resist plastic deformation from a standard source. The unit of the hardness given by the test is Vickers pyramid number (HV) or diamond pyramid hardness (DPH). The hardness of the samples was determined by the testing machine with 40 kg load. The test was conducted on VMT-X digital hardness tester using Al–Mg MMCs of 20 × 20 × 20 mm size. The value of hardness was observed for the variation in the composition of each specimen and the amount of SiC and graphite added. The time for detention for the hardness measurement was 10 s. Three values were taken on different portions for each specimen and the average value is taken.

2.4 Tensile Test

The tensile test was carried out on universal testing machine (UTM) present in NIT Calicut department of mechanical engineering. Ultimate tensile strength of the composite is obtained at the point where the specimen breaks. The specimen samples were prepared as per ASTM B557M-15, with gauge length, diameter, radius of fillet and length of reduced section to be 62.5 mm, 12.5 mm, 9 mm and 75 mm, respectively. The specimen that was made for the UTM is shown in Fig. 2.

Fig. 2 Machine, sample 1: 20% Mg, sample 2: 10% Mg

2.5 Compression Test

A compression test is the opposite of common tensile test. Behaviour of materials under opposing loads is determined by compression test. The given specimen is subjected to compressive loading which leads to crushing and deformation of sample. The compression tests were conducted on a universal testing machine (UTM) present in NIT Calicut department of mechanical engineering, the test was in accordance with American Society for Testing and Material standards E9-09(2018) (ASTM) at room temperature. Short specimens were prepared having length of 25 mm and diameter 30 mm to maintain a *L*/*D* ratio of 0.8.

2.6 Energy-Dispersive X-ray Spectroscopy

This is a method by which different elements present in material can be analysed. This process utilizes a source of X-rays from which beams are incident on the given sample. Each element in the sample has unique atomic structures which produces an identical set of peaks on its electromagnetic spectrum. Thus, the basic principles of this method are grounds on spectroscopy. The accuracy of the EDS method may depend on various factors. Some elements may have overlapping emission peaks which hamper the accuracy in observations. For the smooth functioning of EDS, it is required that the X-ray escapes the specimen. Properties of the specimen, such as density and composition, are determinants of how much X-ray escapes the specimen, and the phenomena of X-ray absorption which leads to inaccurate predictions. In such cases, quantitative corrections methods would be suggested to obtain the proper solution.

3 Results and Discussion

3.1 Tensile Strength Analysis

The engineering stress-strain plots are plotted for all the three samples, using the Instron universal testing machine at NIT Calicut. Figure 3a–c gives the plot for sample 1 (80% Al, 20% Mg), sample 2 (90% Al, 10% Mg) and sample 3 (piston), respectively. The addition of SiC reinforcement has a positive effect on tensile strength and failure strain of the composite. Sample 2 shows to have better tensile and failure strength compare to other two samples.

During the study, it was found that the ultimate tensile strength of the material in sample 2 was 215 MPa and ultimate tensile strength value of 213 MPa for sample 3 (piston). This was due to the addition of SiC nanoparticles with high modulus that transfers the effective load from the matrix to the SiC particles. Moreover, the decrease in grain size leads to an increase in the yield strength of the matrix. The SiC particles also form an obstacle to the movement of dislocations within the matrix that contributes to the increase in tensile strength.

3.2 Compressive Strength Analysis

A compression test is the opposite of common tensile test. Behaviour of materials under opposing loads is determined by compression test. The given specimen is crushed and deformation for different loads is noted. The compression test was conducted on a universal testing machine (UTM) in accordance with American Society for Testing and Material standards (ASTM) at room temperature.

Stress-strain plots for all the three specimens were plotted according to the readings obtained using universal testing machine. The addition of 10% Mg and 5% SiC shows an increase of compressive strength from 514 to 636 MPa. Almost 120 MPa increment in compressive strength is achieved. Increase in the compression strength of MMC is due to the applied compressive load distribution to the strongly bonded SiC reinforcement in Al matrix. The decrease of compressive strength for 20% Mg and 5% SiC reinforced MMC is due to the formation of intermetallic compounds and the segregation of SiC particles in the test specimen (Fig. 4).

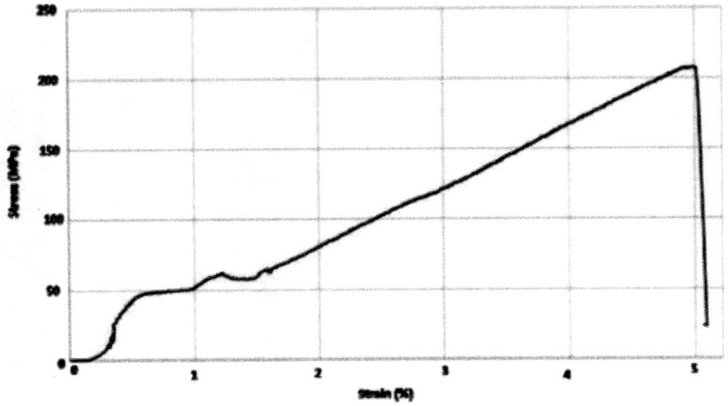

a. Sample 1:20% Mg.

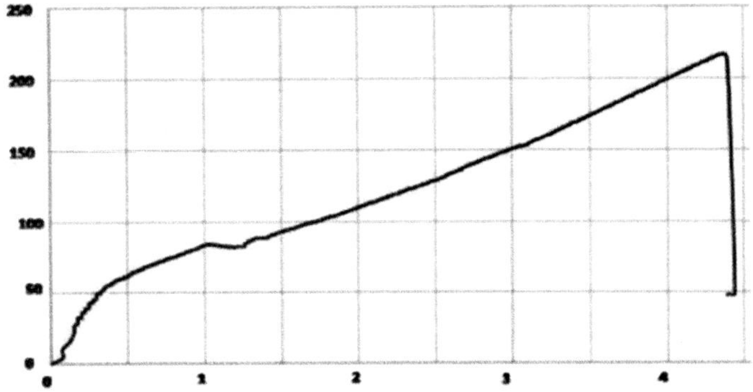

b. Sample 2: 10% Mg.

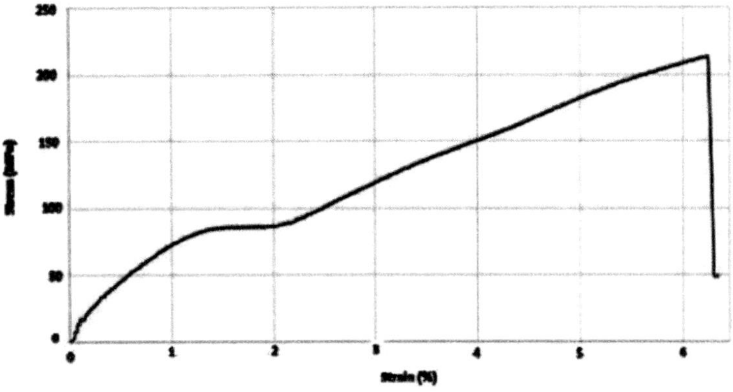

c. Sample 3: original piston

Fig. 3 **a** Sample 1: 20% Mg. **b** Sample 2: 10% Mg. **c** Sample 3: original piston

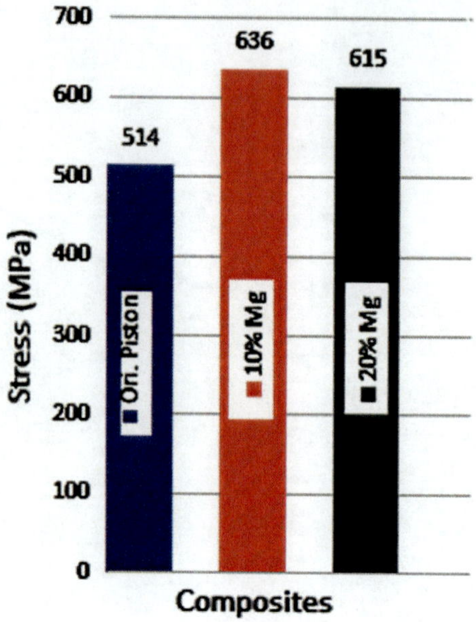

Fig. 4 Wear test studies done at NIT Calicut

3.3 *Wear Test Analysis*

The wear behaviour of the newly casted composite specimens was analysed with focus on two properties—wear rate and coefficient of friction. The measurements were made using a pin-on-disc tribometer and the wear rate was calculated from the weight loss of the pin using the above-mentioned equation. The experiment was repeated for various values of load and speed of rotation of the disc. It was observed that the new composites had a significantly lower wear rate when compared with the older piston material. The drastic decrease in wear rate of the composite can be attributed to the presence of graphite which helps to minimize the wear rate by providing a solid lubricating effect at the interface of composite and the hard surface that is in contact (Fig. 5).

3.4 *Structural Analysis*

In the ANSYS, a three-dimensional model of the piston is modelled. Some unimportant details like oil rings, etc., are omitted to make the analysis simple. Finite mesh is generated on the surface for the analysis. Explosive pressure on the piston head was taken as 6 MPa. A fixed support is given at the gudgeon pin and a friction less support is given on the cylindrical surface to restraint its motion. Figures 6 and 7 show the analysis of samples.

The von Mises stress and strain are given in Table 2.

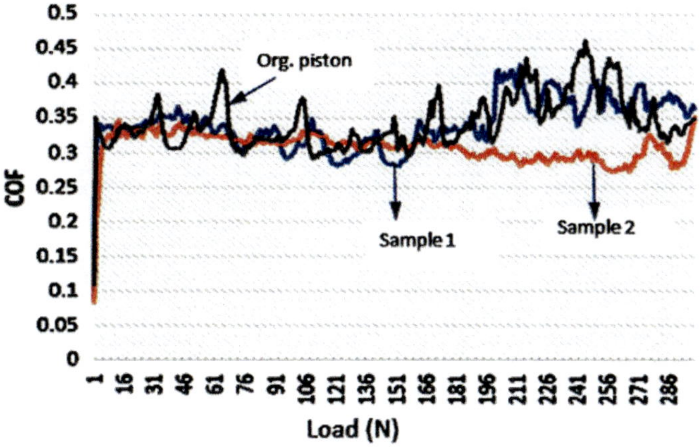

Fig. 5 Original piston, sample 1: 20% Mg, sample 2: 10% Mg

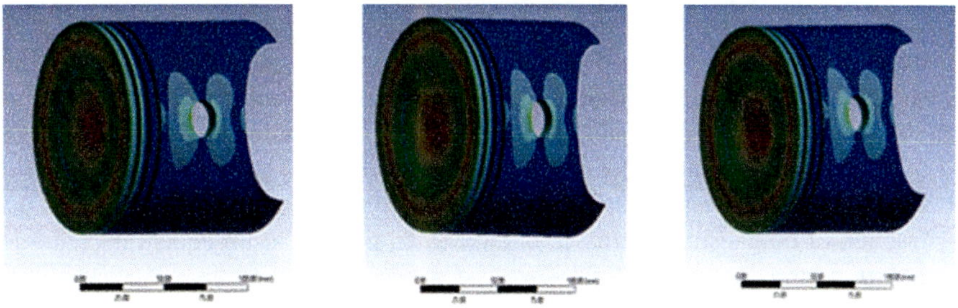

Fig. 6 Sample 1: 20% Mg, sample 2: 10% Mg, original piston

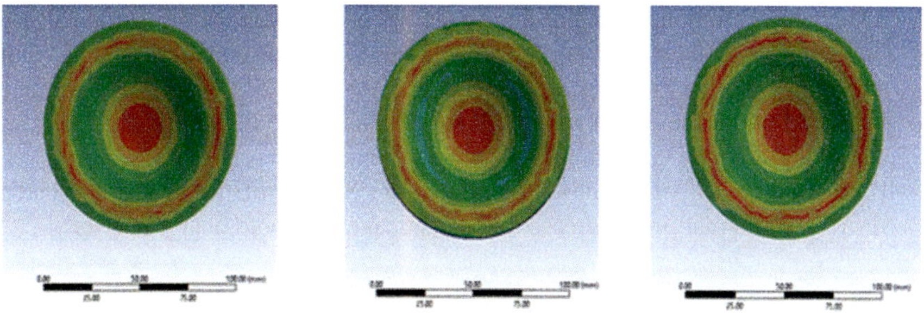

Fig. 7 Sample 1: 20% Mg, sample 2: 10% Mg, original piston

Table 2 Composition of composites prepared

Name	Von Mises stress (MPa)	Strain
Sample 1	254.75	0.00590
Sample 2	250.65	0.00452
Original piston	248.61	0.00472

From Table 2, it is clear that vanishes stress of sample 1 is greater than sample 2 and original piston has least stress of 248.61 MPa. Difference in stress values is not much. Strain of sample 1 is greater than original piston and strain for sample 2 is least.

4 Conclusions

The mechanical properties and tribological characteristics were investigated for new Al–Mg–SiC MMC. The percentages of magnesium in aluminium were varied to prepare, using stir casting fabrication technique. A series of test were conducted to obtain the mechanical properties of the MMC. Tribological properties like wear rate and coefficient of friction were determined using the pin-on-disc test. The major conclusions drawn from the experimental evaluation are:

- Homogenous mixture of Al–Mg was not achieved throughout the samples. In the XRD tests, the presence of aluminium was clearly seen. EDS analysis did shows about 16% of magnesium in the sample but the actual amount added was 20%. This could be the result of localization of different elements which can be attributed to these anomalies.
- Addition of magnesium leads to reduction in the weight of the MMC. As the fraction of magnesium increases in the MMC, there is a subsequent decrease in density.
- Addition of SiC increases the hardness of the MMC.
- The tensile test results show that the sample with 10% magnesium shows greater strength than that with 20% magnesium. Hence, there is a possibility of a superior MMC which might have magnesium fraction between 10% and less than 20%.
- The hardness of the MMC is observed to increase with increase in magnesium percentage.
- The wear rate of the existing piston material is found to be the maximum and it is found the sample containing 20% Mg has better wear properties.
- It was observed a significant reduction in the density of the newly fabricated MMC's with a marginal increase in mechanical and tribological properties.

References

1. Shen MJ, Wang XJ, Ying T, Wu K, Song WJ (2016) Characteristics and mechanical properties of magnesium matrix composites reinforced with micron/submicron/nano SiC particles. J Alloy Compd 686:831–840
2. Yang X, Wang F, Fan Z (2017) Crystallographic study of nucleation in SiC particulate reinforced magnesium matrix composite. J Alloy Compd 706:430–437

3. Qi L, Ju L, Zhou J, Li S, Zhang T, Tian W (2017) Tensile and fatigue behavior of carbon fiber reinforced magnesium composite fabricated by liquid-solid extrusion following vacuum pressure infiltration. J Alloy Compd 721:55–63
4. Tekumalla S, Yang C, Seetharaman S, Wong WL, Goh CS, Shabadi R, Gupta M (2016) Enhancing overall static/dynamic/damping/ignition response of magnesium through the addition of lower amounts (<2%) of yttrium. J Alloy Compd 689:350–358
5. Sharma P, Sharma S, Khanduja D (2015) Production and some properties of Si3N4 reinforced aluminium alloy composites. J Asian Ceram Soc 3(3):352–359
6. Lorimer GW, Yang Z, Li JP, Zhang JX, Lorimer GW, Robson J (2009) Review on research and development of magnesium alloys. Acta Metall Sin 21(5):313–328
7. Manikandan N, Binoj JS, Varaprasad KC, Sabari SS, Raju R (2019) Investigations on wire spark erosion machining of aluminum-based metal matrix composites. In: Lecture notes in mechanical engineering, Advances in manufacturing technology. Springer, Singapore, pp 361–369
8. Lloyd DJ (1989) The solidification microstructure of particulate reinforced aluminium/SiC composites. Compos Sci Technol 35(2):159–179
9. Tahreen N, Zhang DF, Pan FS, Jiang XQ, Li C, Li DY, Chen DL (2014) Influence of yttrium content on phase formation and strain hardening behavior of Mg–Zn–Mn magnesium alloy. J Alloy Compd 615:424–432

Fabrication and Characterization of Al6061 with Boron Carbide Metal Matrix Composites

V. Viswanatha Chari, G. Srinivas Kumar, D. Sai Chaitanya Kishore, and M. Usha Rani

Abstract In recent trends, metal matrix composites (MMC) have attracted the newline potential interest in engineering materials due to their superior mechanical newline properties than the conventional materials. The way of selecting the particular newline properties of the constituent materials is to meet the specific demand. In general, MMCs possess newline increased strength, higher service temperature, decreased part weight and improved newline wear resistance, when compare with materials of monolithic. Aluminium and its alloys are the most widely used AMMC. Matrix reinforced with ceramic particles such as oxides and non-oxides (carbides, nitrides, borides) offers a huge variety of mechanical properties depending on the aluminium and its alloy chemical composition. Among the various ceramics, boron carbide (B_4C) possesses excellent physical and newline mechanical properties like high melting point and hardness, good impact and wear resistance. In the present work, a combination of Al6061 + B_4C metal matrix composites is fabricated with different proportions (1.5, 3, 4.5%) by stir casting technique. Their structural analysis was done by using SEM analysis; wear resistance was found out by using pin-on-disc method; and surface roughness was found out.

Keywords Al6061 · Stir Casting · Pin-on-disc · SEM analysis · Surface roughness

V. Viswanatha Chari (✉) · G. Srinivas Kumar · D. Sai Chaitanya Kishore · M. Usha Rani
Department of Mechanical Engineering, Srinivasa Ramanujan Institute of Technology, Anantapuramu, India
e-mail: visumtech@gmail.com

G. Srinivas Kumar
e-mail: srin.8045@gmail.com

D. Sai Chaitanya Kishore
e-mail: dskishore66@gmail.com

M. Usha Rani
e-mail: usharani.mec@srit.ac.in

A. Arockiarajan et al. (eds.), *Advances in Industrial Automation and Smart Manufacturing*, Lecture Notes in Mechanical Engineering,
https://doi.org/10.1007/978-981-15-4739-3_59

1 Introduction

Composite materials are commonly classified into two levels based on matrix constituent and bade on reinforcement form. Based on matrix, constituent metal matrix is composed of matrix (aluminium, titanium, magnesium, copper) and a dispersed ceramic (oxides, non-oxides) or metallic (lead, tungsten and molybdenum) phase. Based on reinforcement form like particulate composites, fibre-reinforced consists of a matrix reinforced by a dispersed phase in the form of continuous or discontinuous fibres. Aluminium metal matrix composites can be reinforced with various oxides, borides, carbides. But the strength of the AMC's can be improved by combining these materials carefully by addition of quantity, dispersion of reinforcement as well as processing techniques. The variety of manufacturing processes for MMC's such as powered metallurgy route stir casting and vapour deposition are available. The stir casting technique is accepted based on its low-cost processing technique [1–3].

2 Materials

Aluminium alloy Al-6061 has as numerous engineering applications including transportation and construction where the desired mechanical and thermal properties are required like UTS, wear resistance, conductivity. It is mainly suitable for automobile industry applications, due to its highly corrosive resistance. Chemical composition of Al6061 is represented in the tabular column mentioned below.

Al-6i1061 is a precipitated hard aluminium alloy, containing silicon and magnesium has major alloying elements and is originally called as "Alloy 61S"; it is developed in the year 1935 [4]. It possesses good properties of mechanical, exhibits weld ability and is very commonly extruded. In general, it is one of the most widely used aluminium alloy among all of this series (Table 1).

One of the hardest materials in ceramics is boron carbide, ranking third place, which is behind diamond and cubic boron nitride. It has a finite melting point low enough to permit its relatively easy fabrication into shapes. Boron carbide powder is mainly produced by reacting carbon with B_2O_3 in an electric arc furnace, through carbothermal reduction or by gas-phase reactions. For industrial applications like wear parts such as blasting nozzles, wire-drawing dies, powdered metal and ceramic forming dies, thread guides and armour (Table 2).

3 Methodology

Al-6061 is having the chemical composition shown in Table 3. It is melted in a crucible furnace. Graphite is used for making of crucibles. Before composite fabrication, the main considerations to be taken are melting losses of alloy constituents. The composite was fluxed with coverall to prevent dressing. Using of tetrachlorethane, the molten alloy was gasified. The crucible which is modified with the sodium is taken out of the furnace. To molten liquid of matix into semi solid state, by cooling down it to, below the liquidus temperature. During this process, the molten liquid is added by the preheated reinforcements (which is 440 °C for ½ hour) B_4C particulates. The average size B_4C is 10 μm. Molten B_4C and Al6061 particles are stirred manually at regular intervals. After sufficient stirring manually over a period of time, using of automatic mechanical stirring, for about 15 min at a speed of 600 rpm, turn the semi-solid state of liquid into fully liquid state in resistance furnace. Using a thermocouple which is of dipping type, temperature is measured. Finally, after attaining a proper mixing of matrix and reinforcement, it is poured into the preheated cast iron mould in the presence of gravity [5] (Fig. 1).

Table 1 Properties of Al-6061

Physical properties	Density			
	2.700 gm/cm^3			
Mechanical properties	Modulus of elasticity (E)	UTS	Elongation	Poisson's ratio
	68.9 GPa	124–290 MPa	12-25%	0.33
Thermal properties	Melting temperature	Thermal conductivity	Thermal expansion	Specific heat capacity
	585 °C	2.23×10^{-5} K^{-1}	897 J/Kg K	32.5–39.2 Ω*m

Table 2 Properties of B_4C

Colour	Black
Appetence	Crystalline solid
Melting point	2763 °C
Boiling Point	3500 °C
Density	3.52 gm/cm^3
Molecular weight	55.25 g/mol
Thermal conductivity	28 W/m-K@30 °C
Specific gravity	2.1

Table 3 Chemical composition of pure Aluminium alloy (Al-6061)

Element	Mg	Fe	Si	Cu	Mn	V	Ti	Al
% of Wt	1.08	0.17	0.63	0.32	0.52	0.01	0.02	Remainder

Fig. 1 Photograph of composite material Al6061+B_4C

4 Experimentation

4.1 Microstructure

To study the microstructure of fabricated composites, standard metallographic procedure is followed and the samples are prepared. Samples are grinded and then polished by surface grinder and emery paper, respectively. Mirror finish is obtained by polished with alumina powder. To investigate the microstructure of the samples, Keller's reagent is applied to be applied and then with the help of scanning electron microscope (SEM) (Fig. 2).

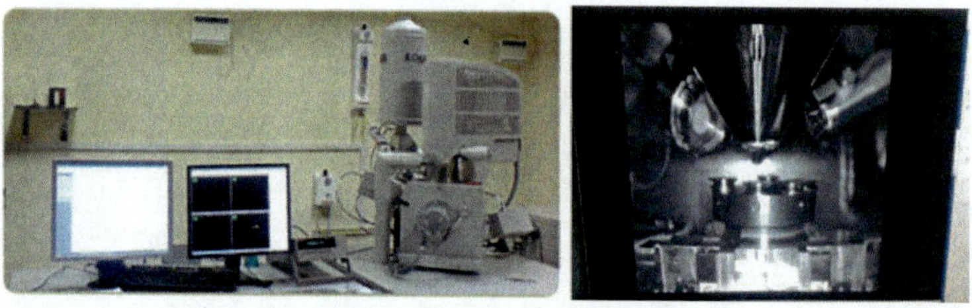

Fig. 2 Photograph of SEM analysis equipment and its inner view

4.2 *Microhardness*

Metal matrix composites of Al with different %wt proportions of various reinforcement (1.5, 3 and 4.5% of B_4C) particulates in Al-6061 as matrix are taken for microhardness test. The test is conducted on Rockwell microhardness tester (Model-RAS) for duration of 20 s with 110 Kg-f, using of 1/16" of ball indenter.

4.3 *Tribological Analysis*

Al6061+B_4C cast composites are made as per the ASTM E8 standard for conducting of wear test to analyse, under dry constant load condition. In this study, flattened specimens of diameter 8 mm and length of 30 mm are made from each composite. The tracking radius on the disc is about 60 mm, on which the pin slides. To get in a fully contact of specimen with a disc, which is rotating, should be pressed against carbon steel disc EN31, having 62 HRC, which is rotating by loading with known weights. Before going for test, the specimens are polished by alumina powder and the rotating disc is by 4000-grade emery paper. Wear resistance under dry sliding condition test is carried out on each cast composite specimens by applying a constant load of 10 N-force and sliding distance of 846 mts at a velocity of 2.82 mts/sec. The test duration is about 5 min at a constant velocity of disc, having 900 RPM. Wear rate, frictional force and coefficient of friction against sliding distance analysis are carried on each specimen [6–12].

5 Results and Discussion

5.1 *Micro structure*

The samples of metallographic were sectioned from the cylindrical cast composites to observe the difference of distribution of B_4C particulates in the Al-6061 matrix; for these samples, etching is done whenever required by the help of 0.5% HF solution and is observed for any developments under EVO® HD scanning electron microscope from Carl Zeiss. All the samples were cast composites which are having different weight fractions (1.5, 3 and 4.5% of B_4C) as reinforcement (Fig. 3, 4 and 5).

To improve the particulate distribution, second stage of mixing is required, which is an automatic; i.e. slurry is heated above the liquidus temperature and then stirred well for 15 min at 500 rpm by using an stirring device which is automatic.

As observed from the above figures, Fig. 5 reveals that the cast composite which is having a wt proportion (4.5% of B_4C) as reinforcement, particle cluster and the absence of B_4C at different locations. Particulates which are added to the molten alloy were observed on the top layer of molten surface, though they have a more value

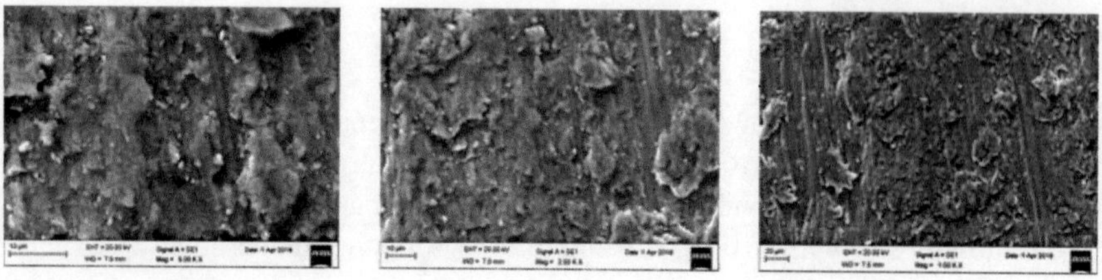

Fig. 3 Micrograph of Al6061+1.5% of B_4C with 5 KX, 2.5 KX and 1 KX

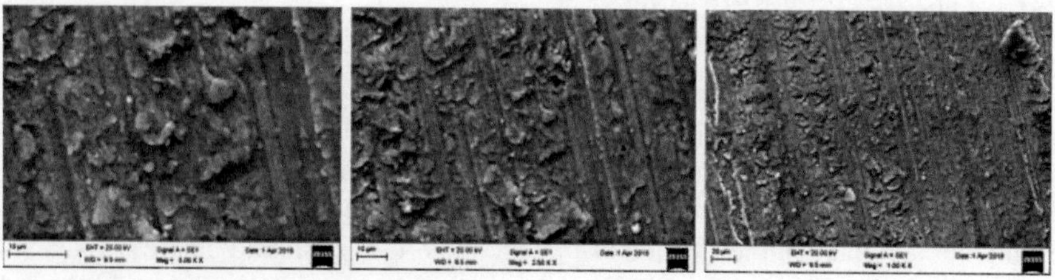

Fig. 4 Micrograph of Al6061+3% of B_4C with 5 KX, 2.5 KX and 1 KX

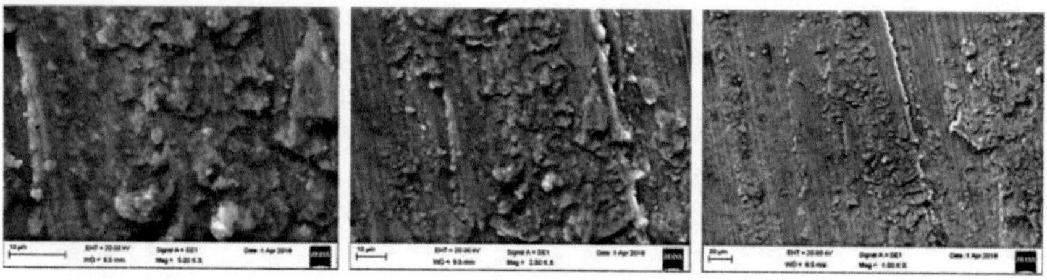

Fig. 5 Micrograph of Al6061+4.5% of B_4C with 5 KX, 2.5 KX and 1 KX

of specific density than the molten alloy. Having higher surface tension and poor wetting properties makes it to float on the surface. Wettability between the ceramics and molten liquid alloy is poor, and its improvement is by applying mechanical force on the particles, by overcoming the surface tension.

From Figs. 3 and 4, it revealed that microstructure of specimens having Wt proportion (1.5 & 3% B_4C) as reinforcement. It clearly observed that the resulting of cluster formation and non-uniform distribution of particulates in the samples.

From Figs. 3, 4 and 5, it revealed that microstructure of specimens having Wt proportion (1.5, 03 and 4.5% B_4C) as reinforcement. It clearly revealed that the uniform dispersion of particulates for all specimens is distributed uniformly, as the increasing order of percentage of reinforcement.

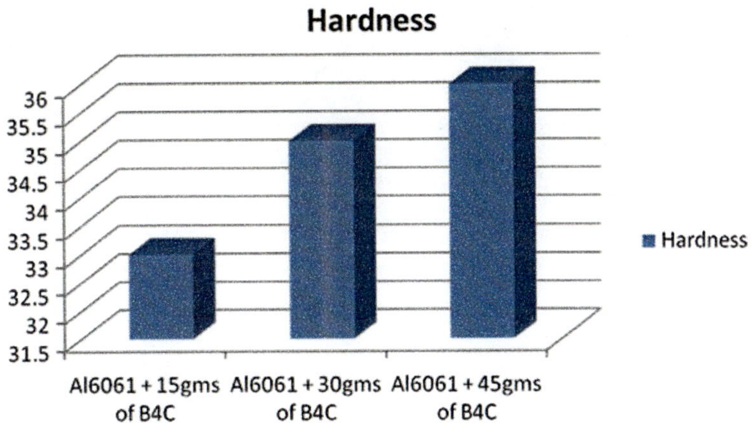

Graph 1 Cast composites hardness

5.2 *Hardness*

Testing has been conducted on all the specimens, with different weight fraction of B_4C (1.5, 03 & 3.5%) as a reinforcement in Al-6061 which are fabricated by stir casting. Average of all the hardness values is taken into account; for each specimen, graph is plotted for the recorded values.

The results as indicated in Graph 1 show that hardness increases as the weight fraction of B_4C increases; here the increasing order of hardness value for each specimen is high, due to the increase of boron content in the cast composites.

5.3 *Wear Analysis*

In order to observe the qualitative proof on the particulate distribution in the matrix, in the ceramic reinforcement (B_4C), wear rate is to be employed. In pin-on-disc testing machine, which is under dry sliding condition, testing was conducted at ambient temperature, and the specimens are loaded against the rotating disc by applying the known weights. The graphs that are obtained by the experimentation would reveal the coefficient of friction, frictional force and wear rate of fabricated composites [4] (Tables 4, 5 and 6).

The wear rate of all specimens (mm^3/Nm) is obtained and is plotted against the sliding distance, as shown in Graph 2 for a constant load of 10 N-force and for different reinforcement of B_4C (1.5, 03 and 4.5%). In Graph 2, it is revealed that at constant load application, wear rate increases, against the sliding distance of all the specimens. Special characteristics were observed in Graph 2 that Al-6061+.4.5% of B_4C has less wear rate among all cast composites, having maximum of 20 mm^3/m at a sliding distance of 846 mts.

Table 4 Wear rate and sliding distance

wt% distance	15 gms of B_4C	30 gms of B_4C	45 gms of B_4C
100	17	2	1
200	25	8	3
300	29	18	6
400	36	22	10
500	42	24	12
600	45	31	15
700	48	32	16
800	51	37	18
846	52	38	20

Table 5 Frictional force and sliding distance

wt% distance	15 gms of B_4C	30 gms of B_4C	45 gms of B_4C
100	3.8	3.6	3.2
200	3.8	3	3.4
300	3.8	2.8	2.6
400	3.5	2.8	2.6
500	3.2	3.1	2.4
600	3.5	2.7	2.4
700	2.8	3.2	2.3
800	2.8	3.3	2.3
846	3.0	3.4	2

Table 6 Frictional coefficient and Sliding Distance

wt% distance	15 gms of B_4C	30 gms of B_4C	45 gms of B_4C
100	0.037	0.036	0.032
200	0.036	0.03	0.034
300	0.037	0.028	0.026
400	0.034	0.028	0.026
500	0.032	0.031	0.025
600	0.03	0.027	0.024
700	0.028	0.032	0.023
800	0.027	0.033	0.023
846	0.031	0.034	0.02

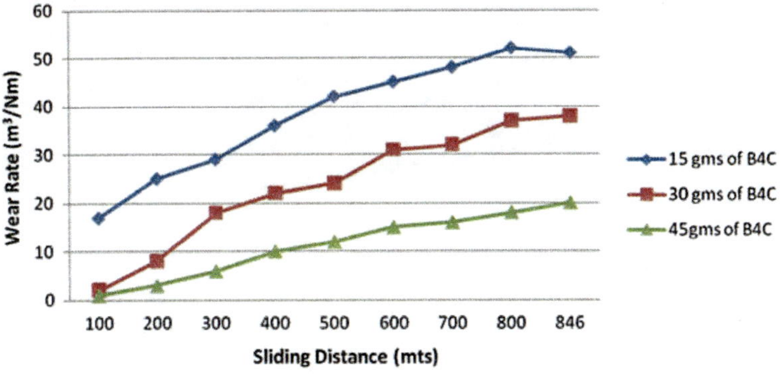

Graph 2 Wear rate versus sliding dist of cast composites

The frictional force of the specimens will give frictional coefficient values directly and is plotted against the sliding distance as shown in Graphs 3 and 4 for a constant load of 10 N for different % wt proportion of reinforcement B_4C. From Graphs 3 and 4, it is shown that coefficient of friction and frictional forces was drastic changes in the cast composites of (1.5 and 03%) B_4C as a reinforcement because of non-uniform particles dispersion in the matrix. It was revealed that, Al6061+4.5% of B_4C has less frictional force and coefficient of friction among all cast composites, due to the increase of carbon content in the cast, having maximum of 3.5 N and 0.035, respectively, over a sliding distance of 846 mts.

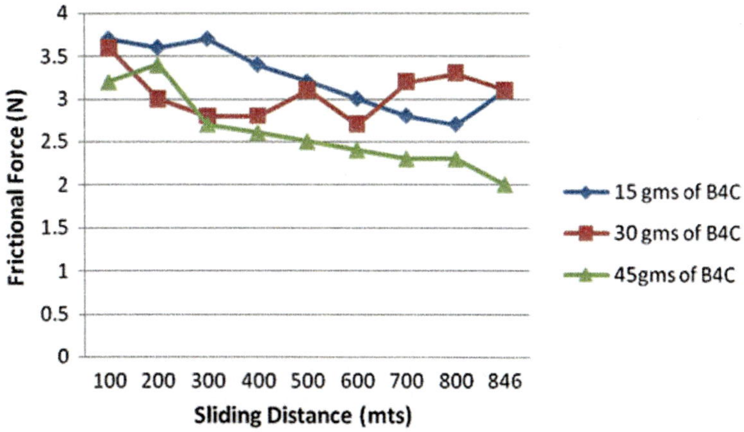

Graph 3 Frictional force versus sliding distance of cast composites

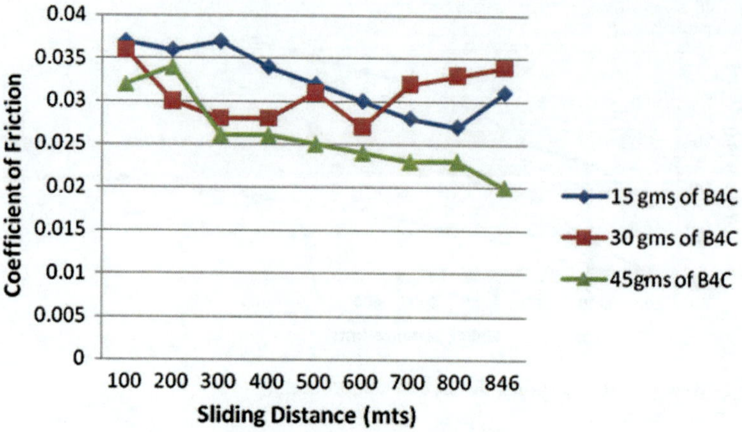

Graph 4 Frictional coefficient and sliding distance of cast composites

6 Conclusions

Here experimental study reveals that:

1. For different % by wt proportion of reinforcement of B_4C particle with Al-6061 as a matrix, B_4C with 4.5% by wt would have lesser frictional force, coefficient of friction and wear rate.
2. B_4C as reinforcements in Al6061 matrix, as a homogenous distribution of particles without agglomeration were achieved rather than B_4C. Strong bonding was observed due to preheating of reinforcements B_4C particulate.
3. Due to the formation of cluster, when B_4C used as reinforcement in the Al6061 matrix which is observed in SEM photograph, the hardness values are higher, but in case of 4.5% wt of B_4C as a reinforcement, due to homogenous distribution of particulates, hardness value is also better. It seems that hardness increases with an increase in B_4C as a reinforcement.
4. The best results have been obtained at 4.5% by wt proportion of B_4C as a reinforcement.

Acknowledgements This work would not have been possible without the support of Srinivasa Ramanujan Institute Of Technology, Anantapuramu, Pulla Reddy Engineering College, Kurnool, and S. V. University, Tirupati, and permitting us to perform the tests in their esteem laboratory.

References

1. Singla M, Deepak Dwivedi D, Singh L, Chawla V (2009) Development of Aluminium based Silicon Carbide particulate metal matrix composite. J Min Mater Character Eng 8(6):455–467

2. Selection and applications—the aluminum association, Inc. Dec 1998
3. Ghosh S, Sahoo P, Sutradhar G (2014) Tribological properties of Al–SiC metal matrix composites:a comparison between sand cast and squeeze cast techniques. J Inst Eng India Ser D 95(2):161–171, July–Dec 2014
4. Pazhouhanfar Y, Eghbali B (2018) Microstructural characterization and mechanical properties of TiB_2 reinforced Al6061 matrix composites produced using stir casting process. Mater Sci Eng, A 710:172–180
5. Srinivas Kumar G, Viswanath Chari V (2017) Comparision of mechanical characterization of Al metal matrix composite and Gray Cast iron of an automobile brake rotor. Int J Eng Innov Res 6(5):2277–5668. ICEMS-2017, ISSN (Online)
6. Kumar N, Gautam G et al (2015) Synthesis and characterization of TiB_2 1Aluminium matrix composites: a review. J Inst Eng India Ser D
7. Agbeleye AA, Esezobor DE et al (2017) Tribological properties of aluminium-clay composites for brake disc rotor applications. J King Saud Univ Sci. http://dx.doi.org/10.1016/j.jksus.2017.09.002
8. Nakanishia Hiroaki, Kakihara Kenji et al (2002) Development of Aluminum metal matrix composites (Al-MMC) brake rotor and pad. JSAE Rev 23:365–370
9. Mohan Krishna SA et al (2015) Research significance, applications and fabrication of hybrid metal matrix composites. IJISET Int J Innov Sci Eng Technol 2(5)
10. Mahajan G, Karve N, Patil U, Kuppan P, Venkatesan K (2015) Analysis of microstructure, hardness and wear of Al–SiC–TiB2 hybrid metal matrix composite. IJST 8(2)
11. Viswanatha Chari V, Srinivas Kumar G (2017) Fabrication and testing of reinforced composites of carbon and agave fibers. IJEIR 6(5)
12. Sai Chaitanya Kishore D, Prahlada Rao K, Mahamani A (2013) Fabrication and characterization of in-situ Al–TiC composite. IJMET 1(4):109–114
13. Mishra AK, Srivastava RK (2017) Wear behaviour of Al-6061/SiC metal matrix composites. Springer. https://doi.org/10.1007/s40032-016-0284-3

Effect of TiB_2 on Mechanical Properties and Microstructural of Aluminium Composite

K. S. Sridhar Raja, J. Hemanandh, J. Mohan Krishna, and R. Muni Sai Preetham

Abstract A356 has its extensive applications in the fabrication of airplane wing parts, turbine blades for jet engine and other complex parts of the airplane which led to lightweight and hard material. TiB_2 is a hard ceramic particle which has high hardness is reinforced in aluminium matrix to increase its mechanical properties. The material was fabricated through liquid metallurgy route by reinforcing different weight percentage of TiB_2 (4 wt%, 8 wt% and 12 wt%). The mechanical properties and morphological studies are analysed. The result reveals that there is an increase in tensile strength, hardness and impact strength of the composite with the increase in reinforcement. SEM morphology shows the bonding between the reinforcing particle and the matrix alloy.

Keywords A356 · TiB_2 · Hardness · Tensile strength · Mechanical properties · SEM

1 Introduction

Aluminium A356 alloy is a versatile material that has been used in many applications due to its greater elongation, high strength and ductility. It is used in some typical applications like automobile brakes, airframe casting, aircraft and missile components, etc. The tensile properties of A356 alloy at high temperatures are influenced by the heat treatment process of the castings, and the duration at the high temperature has good corrosion resistance. TiB_2 is the hardness ceramic particle which has high corrosion resistance with good resistance to oxidation up to 1000 °C and high melting point (>2900 °C).

Sridhar Raja et al. (2015) examined the response of steel slag on the impact strength of the aluminium composite by using liquid casting technique and found out that the microhardness of the composite increases with the increase in weight

K. S. Sridhar Raja (✉) · J. Hemanandh · J. Mohan Krishna · R. Muni Sai Preetham
School of Mechanical Engineering, Sathyabama Institute of Science and Technology, Chennai, India
e-mail: sridhar2raja@gmail.com

A. Arockiarajan et al. (eds.), *Advances in Industrial Automation and Smart Manufacturing*, Lecture Notes in Mechanical Engineering,
https://doi.org/10.1007/978-981-15-4739-3_60

percentage of the slag. The impact strength of the material also reduced due to the non-homogeneous distribution of the particle and porosity [1]. It has been proved that by reinforcing the micro-SiC particles with cast aluminium and with magnesium matrix, the elastic modulus, hardness and tensile strength increase considerably with the increase in reinforcing particles [2–4]. By the addition of hard ceramic particles like B_4C particles with the aluminium matrix, it shows the tensile strength, hardness and impact strength improved drastically. It was found that the particles were distributed homogeneously. To improve the wettability and to remove moisture content, the particles were mixed with K_2TiF_6 during melting [5–8]. Viswanatha et al. (2013) took A356 as a parent material and reinforced with SiCp with different percentage composition using stir casting technique, liquid metallurgy method and found out the uniform distribution of SiC_p was obtained by stir casting technique. With the addition of SiC_p, the hardness of composites increased abruptly, whereas the maximum hardness was obtained for 9% of SiC_p. The inclusion of low wt% of SiC_p to aluminium alloy leads to a decrease in percentage elongation and increase in tensile strength [9]. Hajizamani et al. (2011) reinforced A356 with Al_2O_3 with 10% ZrO_2 nanoparticles employed by stir casting at 850 °C. They found out that the density was increased but it resulted in the increase of yield and ultimate tensile strength and compressive strength increased. Ductility of the aluminium composites was low due to huge porosity content [10].

From the above literature, it was evident that by reinforcing hard ceramic particles the mechanical properties like tensile strength, hardness and impact strength of the material increased. But very few attempts were noticed by reinforcing TiB_2 particles. Titanium diboride has high density (4.52 g/cm^3) along with high hardness and high compressive strength, and it has been used in armour components. An attempt is made in this paper by reinforcing TiB_2 with the cast aluminium A356 alloy which is used in automobile brakes. By varying the weight percentage of TiB_2 (4 wt%, 8 wt% and 12 wt%), the composite was fabricated and its mechanical properties were analysed that is used in the automobile brake pad.

2 Materials and Methods

Commercially available cast A356 aluminium alloy was used in brakes as well as industrial sectors and advanced machinery due to its better wear properties, good castability and good strength. The chemical composition of the A356 aluminium alloy is given in Table 1. To develop lightweighted metal matrix, A356 alloy was reinforced with TiB_2 with different weight percentage (4 wt%, 8 wt% and 12 wt%)

Table 1 Chemical composition of aluminium 356 [1]

Si	Fe	Cu	Mn	Mg	Ni	Zn	Ti	Al
6.58	0.16	0.06	0.06	0.57	0.01	0.01	0.14	Balance

Fig. 1 Stir casting experimental set-up

3 Experimental Procedure

Liquid-state fabrication technique is the most cost economical and simplest method involved in the production of A356 composite. To obtain good mechanical property and good interfacial bonding between the dispersed phase and the matrix, TiB_2 was preheated along with K2TiF6 in the furnace. The A356 aluminium alloy was melted in a crucible at 650 °C. The mixed TiB_2 particle was added slowly, and it was stirred mechanically to get a homogeneous distribution of mixture. The metal was continuously mixed at a speed of 300 RPM and allowed to melt till 730 °C. The mixed metal is then poured into the preheated metal mould to get solidified. With different wt% (4 wt%, 8 wt% and 12 wt%), the material has been fabricated. The experimental set-up for the fabrication of the composite is shown in Fig. 1.

4 Results and Discussion

4.1 Microstructure

Figure 2 shows the microstructure of aluminium A356 metal matrix composite at different weight percentages. It was noticed that the TiB_2 particles are distributed homogeneously with the matrix.

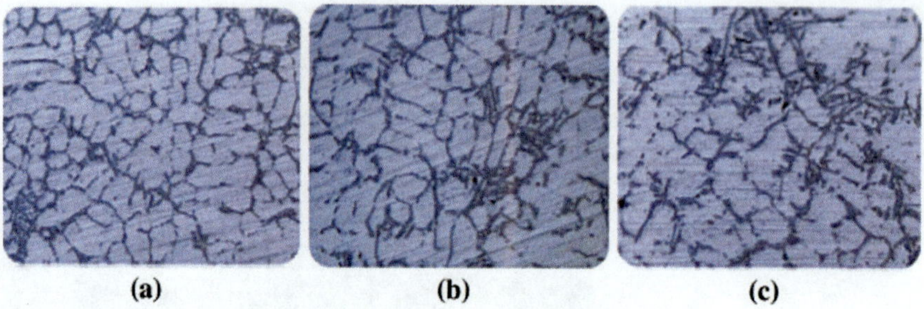

Fig. 2 Microstructure of A356/TiB_2 composite

4.2 Tensile Strength

The fabricated sample was machined according to ASTM E8 standards. Figure 3a shows the fractured tensile specimen. It was noticed from Fig. 3b that the strength of the composite material increased gradually as the reinforcement increased. It was found that the strength improves by 18% when compared with the base aluminium A356 alloy. Figure 3c shows the percentage elongation of the composite. It was

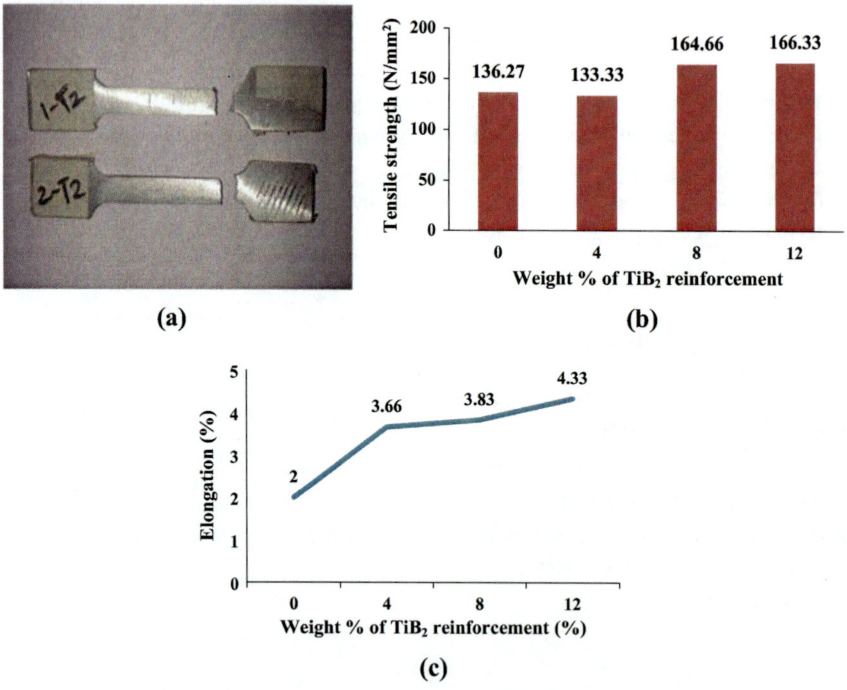

Fig. 3 **a** Fractured tensile strength. **b** Tensile strength of composite. **c** % Elongation

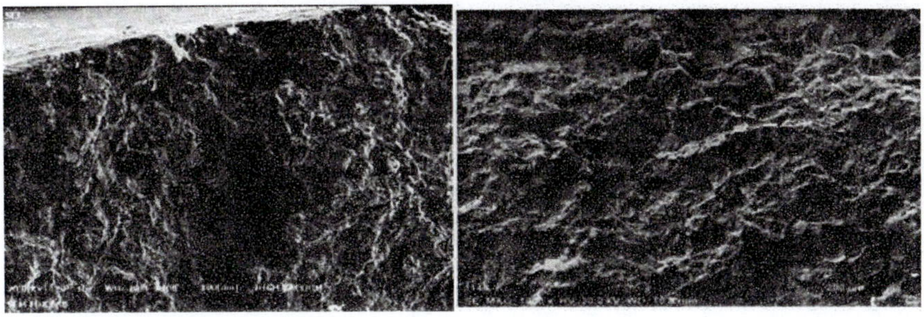

Fig. 4 SEM morphology of A356/TiB_2 composite

observed that the elongation increases rapidly about 54% compared with the A356 alloy.

Figure 4 shows the SEM morphology of A356/TiB_2 composite material. The figure shows that there was a perfect bonding between the alloy and the ceramic particle. It was also observed that there was some agglomeration in some places which cause a reduction in the impact energy and tensile strength of the material.

4.3 *Microhardness*

In order to analyse the morphological behaviour and the grain refinement, the hardness test was analysed as per ASTM standards. To study the effect of reinforced particles and the homogeneous distribution, the microhardness was consistent at different places and the average values were plotted. It was observed from Fig. 5 that by increasing the weight percentage of the TiB_2, the hardness also increased abruptly from 60 to 74 HV for the cast A356 alloy. It was observed that the hardness dropped down to 9.5% at 8 wt% of the reinforced particle, and upon increasing to 12% reinforcement the hardness has improved to 11.5%.

Fig. 5 Hardness of A356/TiB_2 composite

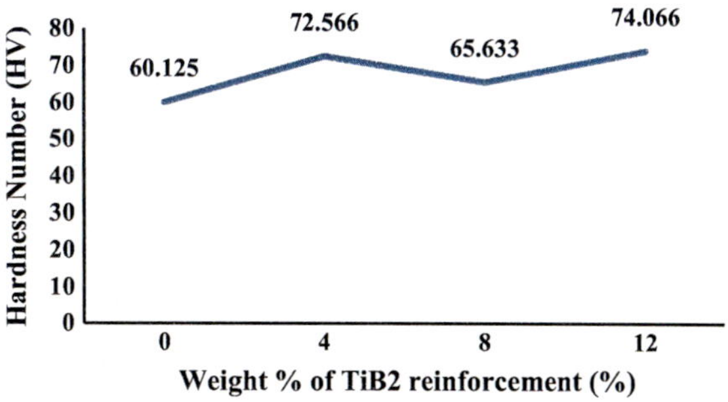

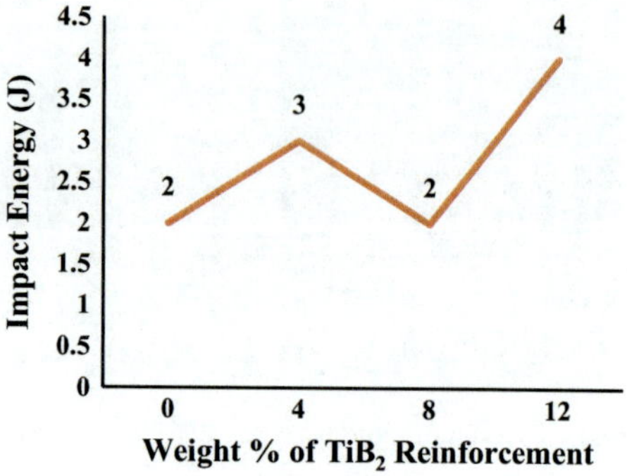

Fig. 6 Impact energy of A356/TiB_2 composite

4.4 Impact Test

The impact test samples were adapted as per ASTM E8-96 standards with the size of 10 × 10 × 55 mm with a V-notch of 2 × 45°. From Fig. 6, it was observed that the energy absorbed by the A356 alloy and the composite material found to be increased with the increase in reinforcing particles. It was also seen that there is a drop in the energy at 8 wt% of TiB_2 particles, due to the agglomeration of the particles.

5 Conclusion

From the above investigation, it was concluded that by reinforcing hard ceramic particle with cast A356 aluminium alloy, the composite was fabricated through liquid metallurgy route. It was observed that the tensile strength and hardness have been increased by 18% and 24% when correlated with the base A356 aluminium alloy. Also, the impact energy gets increased by 50% of the aluminium alloy. The microstructure and the SEM morphology show the distribution of the particle along with the A356 aluminium alloy. It has been proved that as the strength and hardness improved drastically, it can be used as an alternative to the existing automobile brake material.

References

1. Sridhar Raja KS, Bupesh Raja VKKR, Vignesh, Ramana Rao SN (2015) Effect of steel slag on the impact strength of aluminium metal matrix composite. Appl Mech Mater 766–767:240–245
2. Wang XJ, Wang NZ, Wang LY, Hu XS, Wu K, Wang YQ, Huang YD (2014) Processing, microstructure and mechanical properties of micro-SiC particles reinforced magnesium matrix composites fabricated by stir casting assisted by ultrasonic treatment processing. Mater Des 57:638–645

3. Jayakumar A, Rangaraj M (2014) Property analysis of aluminium (LM-25) metal matrix composite. Int J Emerg Technol Adv Eng 4
4. Jebeen J, Moses (2014) A review on effect on properties of Al-SiC composites fabricated by stir casting method. Int J Res Eng Technol 5:283–287
5. Auradi V, Rajesh GL, Kori SA (2014) Processing of B_4C particulate reinforced 6061 aluminium matrix composites by melt stirring involving two step addition. Procedia Mater Sci 6:1068–1076
6. Rebba B, Ramanaiah N (2014) Studies on mechanical properties of 2024 Al–B_4C composites. Adv Mater Manuf Charact 4:42–46
7. Gopal Krishna UB, Sreenivas Rao KV, Vasudeva B (2013) Effect of boron carbide reinforcement on aluminium matrix composites. Int J Metall Mater Sci Eng 3:41–48
8. Rama Rao S (2012) Fabrication and mechanical properties of aluminium-boron carbide composites. Int J Mater Biomater Appl 3:15–18
9. Viswanatha BM, Prasanna Kumar M, Basavarajappa S, Kiran TS (2013) Mechanical property evaluation of A356/SiCp/Gr metal matrix composites. J Eng Sci Technol 8:754–763
10. Hajizamani M, Baharvandi H (2011) Fabrication and studying the mechanical properties of A356 alloy reinforced with Al_2O_3-10% Vol. ZrO_2 nanoparticles through stir casting. Adv Mater Phys Chem 1:26–30

Thermo-Physical Properties of Al_2O_3 and Preparation Technique

S. Baskar and L. Karikalan

Abstract Nanofluid is considered to the next invention of transfer of heat from the fluid which is providing and stimulating the possibility to increase heat transfer characteristics as compared to the base fluid of water. The experimental work of the paper in ethylene glycol (EG)-based Al_2O_3 nanofluid along with SDBS which acts as the surfactant was carried out. There are three various concentrations used as 0.1%, 0.15%, and 0.25%. The particle sizes show that the power on thermal conductivity variant along with volume concentration increases than the base fluid. The measured nanoparticles' sizes are <50 nm. In addition, the fluid thermal conductivity is expected at isothermal process. From the observation, the nanofluid viscosity increases with changeable in the volume concentration. The thermal conductivity of the nanofluid also was improved while increase in volume concentration. The particle sizes show the power on thermal conductivity variant along with volume concentration.

Keywords Nanofluid · Thermal conductivity and viscosity

1 Introduction

Nanomaterial are the materials which are science-based and move toward the nanotechnology. It deals the features of the nanoscale especially their various properties like viscosity, thermal conductivity, density, specific heat and shear stress, etc. The nanoscale size is less than 1/10 of micro-meter in least one dimensioned. The catalytic activities are also revealing innovative method in biomaterials [1]. There are different methods which are available to make the nanoparticles by using ball mill mechanism. In cold conditions, heat transfer processes the secondary refrigerants playing important role [2]. In that, ethylene glycol or brine solution has important

S. Baskar (✉) · L. Karikalan
Department of Automobile Engineering, Vels Institute of Science Technology & Advanced Studies (VISTAS), Chennai, Tamil Nadu 600117, India
e-mail: baskar133.se@velsuniv.ac.in

L. Karikalan
e-mail: karilk2005@yahoo.co.in

A. Arockiarajan et al. (eds.), *Advances in Industrial Automation and Smart Manufacturing*, Lecture Notes in Mechanical Engineering,
https://doi.org/10.1007/978-981-15-4739-3_61

role. The secondary refrigerant carried the heat from the primary source is given, and it is recirculated again and again in order to minimize the primary refrigerants with respect to environmental concerns. It is desirable to the thermal energy by utilizing the ethylene glycol [3]. In commercial and large industrial plants, the secondary refrigerants are widely used. In this case, indirect cooling is attained by secondary refrigerant than the direct cooling. When the temperature requirements of the different parts in a large building are varied, these dissimilar temperatures can be sustained by monitoring the amount of brine solution. The heat is carried by secondary refrigerant by the help of pipes. The use of brines is advisable in order to keep coil and pipe encompassing a toxic refrigerants absent from load place [4].

In order to operate these fluids in secondary refrigeration circuit, an anti-freeze agent should be added in order to avoid freezing particularly in the case of water. However, the additions of the anti-freezing medium in the water severally have a greater influence in thermal transfer properties few as decreased in thermal conductivity, increased in viscosity, etc. In the case of other refrigerants like propanol, ethylene glycol is directly used in the secondary refrigeration circuit without addition of the anti-freeze medium, but they possess the less thermal conductivity that reduces the effectiveness of secondary refrigeration loops.

2 Nanofluids

The modern technology provides the chance to manufacture the nanoparticles which are relatively dissimilar from the close relative material in thermal, mechanical, and electrical properties. The nanofluid expertise is joined with new heat transfer studies in various fields. Generally, the suspended size of the nanoparticles in the nanofluid is less than 100 nm [5]. The size of the particles is large which tends to block the flow in particularly the narrow channel, and the fluid pressure drop is increasing significantly [6]. The nanofluid thermal conductivity is enhanced based on nanoparticle concentrations which is higher particle volume concentrations resulting the higher enhancement.

In recent years, the carbon nanotubes receive much attention among the thermal engineers appropriate to its outstanding thermal, mechanical, and electrical properties. The CNT are classified into two types, namely single-walled nanotubes and multi-walled nanotubes. The single-walled carbon nanotubes (SWCNT) having a diameter in the range of 1 nm [7]. The multi-walled carbon nanotubes (MWCNT) consist of several rolled layer of graphite. The scope of the experimental work is to discover the heat transfer characteristics of nanofluids in favor of heating/cooling applications [8].

3 Properties

The nanoparticles are of immense scientific attention as they are successfully a bridge between bulk materials and molecular structure. A bulk material having physical properties in spite of size is observed.

3.1 Nanofluids Thermal Conductivity

The nanofluids thermal conductivity is playing major role in heat transfer character of heating/cooling applications. The maximum concentrations of nanoparticles are giving greater heat transfer capable in the nanofluid. The experimental work concludes that the nanofluids' thermal conductivity depends on the various factors, namely particle size, volume concentration of the particle, shape of the particle, material, temperature, and base fluid.

3.2 Particle Size and Material

There are various nanoparticle materials which are widely used to prepare the nanofluid preparation. In that, Al_2O_3, TiO_2, Ag, Cu, SiC, TiC, and CuO nanoparticles were regularly utilized in the research. The CNT is used due to its tremendous thermal conductivity. The water is used as base fluids for preparing the nanofluids. According to the Brownian motion of nanoparticle, it will affect the nanofluid thermal conductivity. The volume concentration of nanoparticle is investigated, and most of the researchers reported that rising nanofluid thermal conductivity increases the volume concentration of the nanoparticle. The majority of the works prove that the nanoparticles important parameters affect the thermal conductivity of the fluids. Lee et al. measured the nanofluid thermal conductivity of Al_2O_3 and CuO nanoparticle and establish that the nanofluid with CuO nanoparticle shows good improvement than the others. The nanoparticle size is one more essential fluid parameter of thermal conductivity. They are feasible to create nanoparticle at different sizes in the range between 5 and 99 nm. Eastman et al. accomplished that the nanoparticle sizes are main reason that affects the enhancement of thermal conductivity into the account. The general tendency is that the nanofluid thermal conductivity increases by diminishing the nanoparticle size. It is attained by the mechanism of Brownian motion of the nanoparticle. Though there is a significant quantity of opposing data, reducing thermal conductivity with declining of nanoparticle size is indicated.

3.3 Shape of the Particle

There is two important particles shape is available in the nanofluids study, one is spherical particle in addition to cylindrical particle. The cylindrical particles are frequently having a big length-to-diameter ratio. This type of method is used by Xie et al. for the nanofluids preparation. It was noted that the 4.3 vol.% water-based nanofluid through spherical particle have a thermal conductivity enhancement which is 14.9%, while 4.3 vol.% of the nanofluids among cylindrical particle have thermal conductivity enhancement which is 218%. As a result, it concludes that the cylindrical-shaped nanoparticles are giving superior thermal conductivity enhancement than the spherical-shaped nanoparticles. However, it has to be noticed that the Nanofluid with cylindrical particles decrease the viscosity than those with the spherical nanoparticle. Consequently, the pumping power increases.

4 Results and Discussion

4.1 SEM Image of 0.15 and 0.3 Vol.% Dispersed MWCNT in Base Fluid with 0.20% SDBS with Different Magnification Rates

The scanning electron microscope is used to identify the size of the nanoparticles in the nanofluid. The image is shown in Fig. 1 which is under the 30000× and 50000× magnifications ranges, respectively. The nanoparticles are agglomerated under the

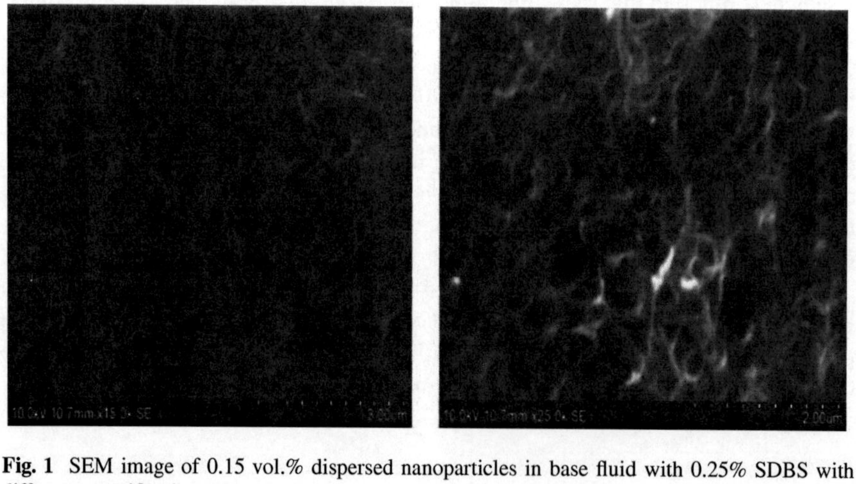

Fig. 1 SEM image of 0.15 vol.% dispersed nanoparticles in base fluid with 0.25% SDBS with different magnification rates

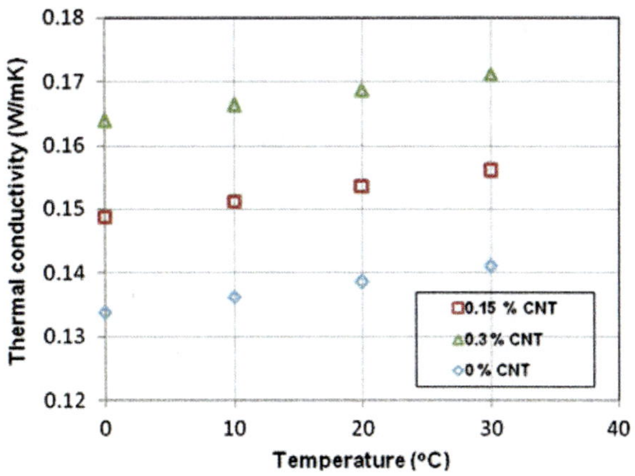

Fig. 2 Temperature versus thermal conductivity

atmospheric conditions which are of the micrometers in sizes. The shape of the individual nanoparticles was close to spherical.

4.2 Temperature Effect on the Thermal Conductivity

Figure 2 illustrate the effect of the nanofluid thermal conductivity at various temperatures. Hence, it is noted that thermal conductivity of nanofluid is increasing with increasing in temperature as well as increasing in concentrations also. The low and high thermal conductivity of fluid is as 0.134 W/mK, 0.149 W/mK, and 0.163 W/mK, respectively.

5 Conclusion

One of the methods to differ the thermo-physical properties and characteristics of fluids is found to be particularly effective in nanoparticles form, because of surface area-to-volume ratio enhancement. The experiment was done at different levels to investigate the effects of the thermo-physical properties and its performance. The experimental work is happening on the effect of various parameters, namely preparation time and nanoparticle size, etc.

References

1. Duangthongsuk W, Wongwises S (2009) Heat transfer enhancement and pressure drop characteristics of TiO_2–water nanofluid in a double-tube counter flow heat exchanger. Int J Heat Mass Transf 52(7–8):2059–2067
2. Farajollahi B, Etemad SG, Hojjat M (2010) Heat transfer of nanofluids in a shell and tube heat exchanger. Int J Heat Mass Transf 53(1–3):12–17
3. Baskar S, Chandrasekaran M, Vinod Kumar T, Vivek P, Ramasubramanian S (2018) Experimental studies on flow and heat transfer characteristics of secondary refrigerant-based CNT nanofluids for cooling applications. Int J Ambient Energy, 1–4
4. Baskar S, Chandrasekaran M, Vinod Kumar T, Vivek P, Karikalan L (2018) Experimental studies on convective heat transfer coefficient of water/ethylene glycol-carbon nanotube nanofluids. Int J Ambient Energy, 1–4
5. Baskar K (2013) Heat transfer characteristics of acetone/water mixture in a tubular heat exchanger with turbulator. In: International Conference on Advanced Nanomaterials and Emerging Engineering Technologies (ICANMEET), pp 627–630
6. Mohammed HA, Bhaskaran G, Shuaib NH, Saidur R (2011) Heat transfer and fluid flow characteristics in microchannels heat exchanger using nanofluids: a review. Renew Sustain Energy Rev 15(3):1502–1512
7. Huminic G, Huminic A (2012) Application of nanofluids in heat exchangers: a review. Renew Sustain Energy Rev 16(8):5625–5638
8. Yang Y, Zhang ZG, Grulke EA, Anderson WB, Wu G (2005) Heat transfer properties of nanoparticle-in-fluid dispersions (nanofluids) in laminar flow. Int J Heat Mass Transf 48(6):1107–1116

Non-destructive Evaluation for Composite Aluminium Composites

I. J. Isaac Premkumar, V. Vijayan, K. Rajaguru, and B. Suresh Kumar

Abstract Aluminium can be used to make parts of aircrafts, and it is less in strength. The combination of aluminium with titanium carbide will increase the strength of the material. By varying the composition of titanium carbide mixture, surface properties of the developed material have been analysed. The characteristics and properties of composite material can be analysed by the non-destructive testing method without breaking the composite. The following tests can be performed to analyse the material are penetrant test and radiographic test. Here, Penetrant test is to identify the surface and radiography test is to identify the sub surface of the material; Different types of aluminium alloys are mixed with different composition of titanium carbide to increase the strength.

Keywords Aluminium · Titanium carbide strength · Non-destructive testing · Penetrant test

1 Introduction

Non-destructive testing can be used to test the material without damaging it [1]. Aluminium characteristic can be changed with respective alloys [2]. Titanium carbide mixture can be mixed with certain percentage aluminium to differentiate the characteristics of the material [3]. Titanium carbide can be mixed with the ratio of 10,20, 30, and 40% with aluminium [4]. Two methods are used to test the composition. Penetrant test can be used to test the surface defect of the material and radiography test can be used to test the internal pores of the material [5]. The interpretation of

I. J. Isaac Premkumar (✉)
Mechanical Engineering Department, JCT College of Engineering and Technology, Coimbatore, Tamil Nadu, India
e-mail: isaacpremkumar007@gmail.com

V. Vijayan · K. Rajaguru · B. Suresh Kumar
Department of Mechanical Engineering, K.Ramakrishnan College of Technology, Samaypuram, Tiruchirappalli, Tamil Nadu 621112, India

A. Arockiarajan et al. (eds.), *Advances in Industrial Automation and Smart Manufacturing*, Lecture Notes in Mechanical Engineering,
https://doi.org/10.1007/978-981-15-4739-3_62

results and formulate accurate decisions based on finding can be made to find the suitability of developed material for specific applications [6–10].

Liquid penetrant test, it can be used to test the surface of the material. The surface smoothness of the material can be visually found out. The penetrant can be applied in both ferrous and non-ferrous material for used to surface detection. LP test can be used in all area of casting, forging and welded objects, etc. Penetrant can be applied in the testing material by applying, spraying, brushings according to the specimen. The dwell time is allowed for capillary action of penetrant to the specimen. Excess penetrant has been removed; the developer is applied on the surface to drawn out the penetrant in the defect and to produce the indication of the surface. Crack within 150 nm can be detected. It was concluded that liquid penetrant testing is more sensitive in detection of surface defect and magnetic inspection is sensitive in sub-surface defects. Finally, ultrasonic method is more sensitive in detection of internal defects. So welds are examined by NDT test [11–23].

Radiography testing is the method of non-destructive testing, where the method is used to identify the defects in the subsurface of the material by using two types of rays, namely x-rays and gamma rays. In this method, the test is conducted thoroughly without damaging the specimen. The result of specimen is executed in two ways, namely by radiograph film and several imaging methods, where more exposure give the metal loss result, where as the less exposure gives the result of excess metal of the specimen to be tested. The specimen is placed between the radiography flim and source of the radiation. The specimen will arrest some radiation. Thicker and more dense area will stop more amount of radiation. The darkness of the flim will vary with respective radiation that reaches the flim through the test object. The applications area aerospace, military defence, offshore, marine, power generation, etc.

2 Materials and Methods

Table 1 shows the composition of the specimens. The specimens are made with varying TiC reinforcement ranging from 10% to 40% in weight proportion. AL 6063 material was the base metal.

Table 1 Composition

Specimen	AL6063 (%)	Tic (%)
1	90	10
2	80	20
3	70	30
4	60	40

3 Result Discussion

The results of the LPT test have been shown in Fig. 1. The porosity has been observed in the specimens throughout the various combinations of aluminium and TiC. The maximum porosity has been found with the specimen containing the maximum aluminium content. So, it is evident that the reinforcement of TiC has increased the hardness of the developed composite with reduced porosity. The radiography test reveals that the heat-carrying capacity has been increased with reinforcement of Al 6063 with TiC as shown in Fig. 2.

4 Conclusion

By the result of penetrant test, the surface smoothness will be identified, the smoothness will be varied by the different composition of aluminium composition. The surface will be rougher at the maximum composition of the titanium carbide. By the result of the radiography test, internal pores and the fatigue falls of the material can be identified. The internal defects will be shown in the X-ray photography results. On analysing these compositions, it has been concluded that the heat withstanding property of aluminium 6063 can be varied with the composition of titanium carbide at different compositions. By analyzing the various composition of aluminium and titanium carbide mixtures, the heat withstanding property can be increased with increasing the composition of titanium carbide.

Fig. 1 LPT test

Fig. 2 Radiography results

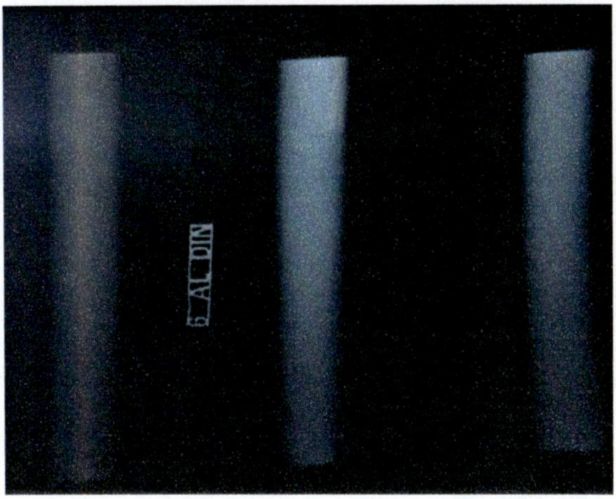

References

1. Cartz, L (1995) Nondestructive testing. ASM International. ISBN 978-0-87170-517-4
2. Hellier C (2003) Handbook of nondestructive evaluation. McGraw-Hill. p 1.1. ISBN 0-07-028121-1
3. Introduction to nondestructive testing. asnt.org
4. Bridges, A (2013) High speed cameras for non-destructive testing, NASA TechBriefs
5. Radiography (RT) Reference Standards, PH Tool. customers.phtool.com
6. Singh S, Goyal A (2007) The origin of echocardiography: a tribute to Inge Edler. Tex Heart Inst J 34:431–8. PMC 2170493. PMID 18172524
7. Ahi, K (2018) A method and system for enhancing the resolution of Terahertz imaging. Measurement
8. Hellier C (2003) Handbook of nondestructive evaluation. McGraw-Hill. p 1.25. ISBN 0-07-028121-1
9. Oldberg T, Christensen R (1999) Erratic measure 4(5)
10. Zolfaghari A, Zolfaghari A, Kolahan F (2018) Reliability and sensitivity of magnetic particle nondestructive testing in detecting surface cracks of welded components. Nondestruct Tes Eval 33(3):290–300
11. Tamizharasan T, Senthil Kumar N, Selvkumar V, Dinesh S (2019) Taguchi's methodology of optimizing turning parameters over chip thickness ratio in machining PM AMMC. SN Appl Sci 1:160
12. Raju R, Sivalingam V, Sun J, Natarajan M, Zhao Y (2019) Experimental and Taguchi-based grey approach of laser metal deposition technique on nickel-based superalloy. Trans Indian Inst Met 72(1):205–214
13. Dinesh S, Rajkumar T, Muthukumarasamy S, Satheesh Kumar G, Kajendrakumar SV, Suresh Kumar B (2019) Analysis and optimization of machining parameters in through feed centerless grinding of high carbon steel. J Mech Eng Technol 9(13):431–441
14. Manikandan N, Raju R, Palanisamy D, Arulkirubakaran D, Kumar S (2018) Investigation on Ti6Al4V laser metal deposition using Taguchi based grey approach. Mater Today: Proc 5(6):14375–14383
15. Parameswaran P, Godwin Antony A, Dinesh S, Radhakrishnan K (2018) Experimental study on mechanical and corrosion characteristics NAB alloy with the addition of chromium. Mater Today Proc 5:8089–8094

16. Jeyaprakash N, Duraiselvam M, Raju R (2018) Modelling of Cr_3C_2-25% NiCr laser alloyed cast iron in high temperature sliding wear condition using response surface methodology. Arch Metall Mater 63(3):1303–1315
17. Dinesh S, Godwin Antony A, Rajaguru K, Vijayan V (2017) Experimental investigation and optimization of material removal rate and surface roughness in centerless grinding of magnesium alloy using grey relational analysis. Mech Mech Eng 21(1):17–28
18. Thirugnanasambantham KG, Raju R, Sankaramoorthy T, Velmurugan P, Kannagi A, Reddy MCK, Chandra VR (2018) Degradation mechanism for high-temperature sliding wear in surface-modified In718 superalloy. Cogent Eng 5(1):1–11
19. Dillibabu V, Duraiselvam M, Chandrasekhar U, Raju R (2017) Microstructural studies on laser dissimilar welded Ni and steel alloys for aeronautical turbine applications. Lasers Eng 37(4–6):247–260
20. Venkatesh R, Vijayan V (2016) Performance evaluation of multipurpose solar heating system. Mech Mech Eng 20(4):359–370
21. Sivalingam V, Sun J, Yang B, Liu K, Raju R (2018) Machining performance and tool wear analysis on cryogenic treated insert during end milling of Ti–6Al–4V alloy. J Manuf Process 36:188–196
22. Dinesh S, Godwin Antony A, Rajaguru K, Vijayan V (2016) Experimental investigation and optimization of material removal rate and surface roughness in CNC turning of EN24 alloy steel. Mech Mech Eng 20(4):451–466
23. Raju R, Manikandan N, Palanisamy D, Arulkirubakaran D, Sambathkumar S, Bhanu Prakash P (2018) Optimization of process parameters in electrical discharge machining of haste alloy C276 using Taguchi's method. Mater Today: Proc 5(6):14432–14439

Investigation on Surface Roughness in Drilling of al/SiC/MoS_2 Metal Matrix Composites

S. Ajith Arul Daniel, A. Parthiban, S. Sivaganesan, and S. Vijay Ananth

Abstract The present examination researches the weight portion and molecule size of silicon carbide (SiC) particulates alongside machining parameters in boring task of Al5059/SiC/MoS_2 half breed metal network composites. The silicon carbide fortifications at various sizes of SiC (10, 20, 40 μm) at different weight divisions (5, 10 and 15%) with aluminium and 2% molybdenum disulphide (MoS_2) are included at steady level. The composites have been created through compo throwing strategy. The trials are completed dependent on Taguchi L_{27} symmetrical exhibit structure. The outcomes decided through Taguchi examination are 5% SiC with 40 μm molecule size, 600 rpm axle speed and 25 mm/min feed have least surface roughness and temperature during boring process.

Keywords Drilling process · Silicon carbide · Compocasting · Surface roughness · ANOVA

1 Introduction

In underway enterprises, aluminium metal matrix composites have a significant impact in late improvements because of their one of a kind mechanical and physical properties [1–4]. In this way, utilities of aluminium composite materials increment and have to a great extent supplanted the customary ferrous material in vehicle, marine and aviation businesses inferable from its wear obstruction, erosion opposition and high quality [5]. By and large, clay particulates are added to aluminium combinations to improve the mechanical quality, wear safe and consumption properties [6]. In any case, the nearness of hard support particles in the materials machining is tranquil troublesome in processing, turning and penetrating process. Because of this condition, it brings about higher instrument wear, poor surface quality and machine vibration [7]. At the long last to achieve the complete shape and size of an item, the material

S. Ajith Arul Daniel (✉) · A. Parthiban · S. Sivaganesan · S. Vijay Ananth
Department of Mechanical Engineering, Vels Institute of Science Technology & Advanced Studies, Chennai, Tamil Nadu 600117, India
e-mail: ajith_danny@yahoo.com

A. Arockiarajan et al. (eds.), *Advances in Industrial Automation and Smart Manufacturing*, Lecture Notes in Mechanical Engineering,
https://doi.org/10.1007/978-981-15-4739-3_63

ought to experience some machining tasks. So the machinability of these advanced materials finds importance in their respective field [8]. To improve the properties of aluminium alloy, the ceramic reinforcements like SiC, Al_2O_3, B_4C and TiC are added to the matrix by various fabrication techniques. Among that stir casting is an effective method for production of these composites, due to low fabrication cost and uniform distribution of reinforcement particles [9]. In case of newly developed aluminium composite material, it must undergo a deep analysis over the input machining parameters and output response variables. Also to maintain good geometrical shapes and for assembling of various parts, accurate drilling is performed [10]. Altunpak et al. [11] carried out drilling process of Al/20%SiC/5%Gr and Al/20%SiC/10%Gr MMC to find the effect of input parameters on cutting force and surface roughness. They reported that increase in feed rate increases surface quality.

2 Experimental Details

2.1 Sample Preparation

Al 5059 is used as lattice material, and its substance combination is shown up in Table 1. Silicon carbide creative particles and molybdenum disulphide were picked as fortresses. SiC particles of three one of a kind sizes (10, 20, 40 μm) (5, 10 and 15%) and weight division of MoS_2 are kept reliable as 2%. The fortifications were preheated in a pot by warming it in a mute heater at 400 °C. The aluminium composite is warmed in a work throwing heater up to a temperature of 700 °C to soften the base combination totally and afterwards chilled off (620 °C) to keep the slurry in semi-strong state [12]. At this stage, the preheated SiC particles and molybdenum disulphide particles were added to the vortex in various mixes. Programmed blending was done with the assistance of mechanical stirrer for around 10 min at mixing pace of 290 rpm. The organizations of created aluminium composites are given in Table 2.

Table 1 Chemical composition of armour aluminium (wt%)

Al	Mg	Mn	Zn	Fe	Si	Zr	Cr	Cu	Ti
91.6	5.33	0.85	0.66	0.47	0.43	0.18	0.20	0.22	0.21

Table 2 Composition of aluminium MMC with hardness

SiC size (μm)	10	10	10	20	20	20	40	40	40
SiC %	5	10	15	5	10	15	5	10	15
Hardness	83	84.52	86.66	78.33	82.33	83.6	75.16	76	78.3

Table 2 demonstrates the hardness of various composites regarding grain size and weight percentage of SiC. Among the AMMCs created, the composites fortified with 15% SiC at 10 μm grain size have the better hardness esteem [6].

2.2 Drilling Experiments

AMMC workpiece with the dimensions of 100 mm × 65 mm × 15 mm is used for the experiments. The cutting tools used for drilling of these AMMCs were uncoated fine-grained high-speed steel (HSS) tool with 6 mm diameter. All cutting tests were carried out under dry condition with spindle speed range of 60–6000 rpm with 802D BMV 40 320D. The surface roughness (Ra) of the milled specimen was calculated by MITUTOYO SJ 210 M portable surface roughness device.

2.3 Microstructure

Microstructure of the created composites was dissected through scanning electron microscope (SEM) so as to distinguish the dispersion of fortification particles on lattice material. The examples were set up according to the required size with the assistance of wire cut EDM. Figure 1a demonstrates the SiC at 5% level at 10 μm and consistent weight level of MoS_2. Similarly, Fig. 1b, c shows the event of SiC at 10% at 20 μm and 15% at 10 μm.

2.4 Process Parameters

Table 3 shows info process parameters picked for the exploratory examination which incorporates cutting rate and feed rate. Levels of each machining parameter are fixed dependent on written works and pilot tests.

3 Experimental Design

3.1 Taguchi S/N Ratio Analysis

The deviation between test esteems and yield esteems, quality misfortune capacity approach was suggested by Taguchi. In Taguchi technique, the reaction factors are dissected as far as signal-to-noise (S/N) proportion, which records the affectability of yield estimated to the clamour factor or wild factor. The correct S/N proportion

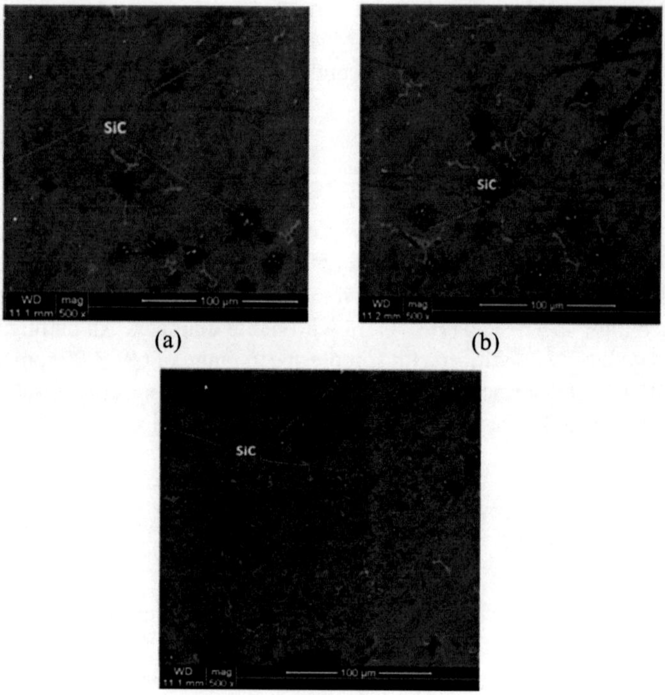

Fig. 1 SEM image of different composites

Table 3 Control parameters and corresponding levels

S. No.	Factors	Unit	Values		
			I	II	III
1	Weight % of SiC	µm	5	10	15
2	Particle size of SiC	%	10	20	40
3	Spindle speed	Rpm	600	1200	1800
4	Feed rate	mm/min	25	50	75

estimation criteria must be picked among three criteria to be specific ‘bigger is better’, ‘normal is better’ and ‘smaller is better’. As the goal is to limit the surface roughness, temperature and cutting power during penetrating process, the ‘smaller is better’ criterion is chosen and the condition is

$$\text{S/N ratio} = -10\log\left(\frac{1}{n}\sum_{i=1}^{n} y_i^2\right) \tag{1}$$

To decide the S/N proportion for material expulsion rate (MRR) which must be improved, the 'bigger is better' criterion is chosen and the condition is

$$\text{S/N ratio} = -10\log\left(\frac{1}{n}\sum_{i=1}^{n}\frac{1}{y_i^2}\right) \quad (2)$$

where 'i' is the number of experiments, and 'y_i' is the experimental results of ith experiments.

4 Results and Discussion

4.1 Signal-to-Noise Ratio Analysis

Table 4 demonstrates L_{27} symmetrical cluster and trial results relating to the info parameters. The S/N proportion worth is resolved for each degree of reaction parameters which recommend the ideal decision, with the end goal that the most influencing or impacting parameters can be recognized by the delta of S/N proportion rank request. The S/N proportion diagram is the simpler portrayal of ideal process parameters.

Figure 2 demonstrates ideally the cutting parameters for surface roughness in boring of this composite which is A1B3C2D1. The level of SiC at 5% and molecule size of SiC at 40 μm cutting pace at 1200 rpm and with least feed rate bring about better surface completion. The feed rate and level of SiC are the most oppressive factor which influences the surface completion of AMMCs. The better surface roughness esteems are accomplished at lower feed rate and higher axle speed which prompts machine vibration and enormous measure of material are expelled, thusly bringing about poor surface of the workpiece. Surface completion extraordinarily relies upon hardness of the composites. Because of the nearness of SiC at 15% and 10 μm molecule size of SiC, the hardness rate is 86.66 BHN at this stage, and the machining of the segment is tranquil troublesome, which results in poor surface completion. Likewise boring the composites at higher cutting velocity of the chip expulsion is simpler which results in better surface completion at increment in axle speed.

4.2 Regression Model Analysis

Equations 3 shows the regression equation generated using Minitab 13 for surface roughness and temperature which have percentage of reinforcement (A), particle size (B), cutting speed (C) and feed rate (D).

$$\text{Surface roughness} = 0.0625556 + 0.0113A - 0.000871429B + 1.51852e$$

Table 4 L_{27} orthogonal array and results

Sl. No	Weight % of SiC	Particle size of SiC	Spindle speed	Feed rate	Surface roughness
1	5	10	600	25	0.212
2	5	10	1200	50	0.224
3	5	10	1800	75	0.329
4	5	20	600	50	0.204
5	5	20	1200	75	0.314
6	5	20	1800	25	0.194
7	5	40	600	75	0.296
8	5	40	1200	25	0.138
9	5	40	1800	50	0.225
10	10	40	600	25	0.22
11	10	40	1200	50	0.254
12	10	40	1800	75	0.337
13	10	10	600	50	0.274
14	10	10	1200	75	0.357
15	10	10	1800	25	0.279
16	10	20	600	75	0.374
17	10	20	1200	25	0.239
18	10	20	1800	50	0.315
19	15	20	600	25	0.294
20	15	20	1200	50	0.341
21	15	20	1800	75	0.428
22	15	40	600	50	0.319
23	15	40	1200	75	0.405
24	15	40	1800	25	0.311
25	15	10	600	75	0.416
26	15	10	1200	25	0.284
27	15	10	1800	50	0.355

$$- 005C + 0.00241111D \quad (3)$$

4.3 *Investigation of Variance for Surface Roughness*

ANOVA is executed to know the criticalness and rate impact of control parameter over the quality highlights of AMMCs shown in Table 5. The complete inconstancy is determined by the aggregate of squared deviations of mean of S/N proportion. In

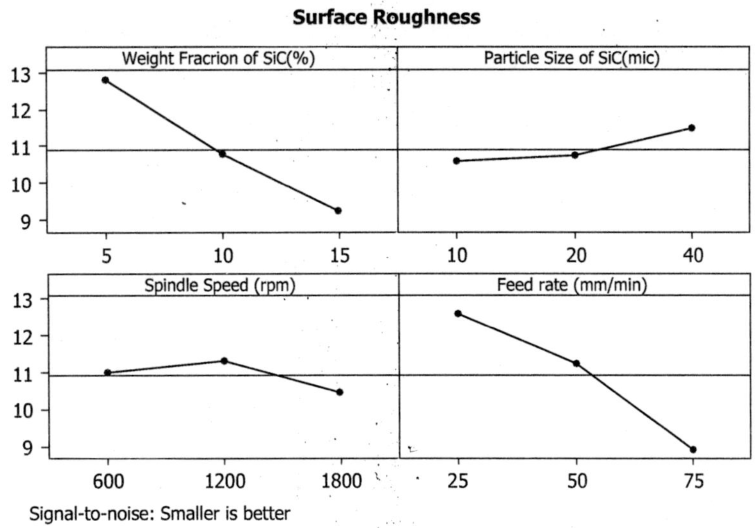

Fig. 2 Mean S/N proportion diagram for surface roughness

Table 5 Investigation of variance for surface roughness

Source	DF	Seq SS	Adj SS	Adj MS	*F*	*P*	%
Weight % of SiC	2	0.057462	0.057462	0.028731	134.6	0.0	42.3
Particle of SiC	2	0.003354	0.003354	0.001677	7.86	0.004	2.4
Spindle speed	2	0.002844	0.002844	0.001422	6.66	0.007	2.6
Feed rate	2	0.068439	0.068439	0.034219	160.3	0.00	50.3
Error	18	0.003841	0.003841	0.00021			2.16
Total	26	0.13594					100

the light of the commitment rate, the most overwhelming viewpoint for penetrating activity is the incremented feed rate (50.34%), weight portion (42.2%), axle speed (2.6%) and molecule of support (2.4%) (SiC) to the composites which impact the surface roughness in boring nature of the AMMCs.

5 Conclusion

The present study and the ideal mix of multi-reaction parameters are distinguished utilizing Taguchi investigation for limiting the surface roughness. By expanding the weight level of SiC, the surface roughness increments. In this manner, the increment in molecule size of SiC prompts diminishes in surface roughness. Anticipated estimation of test information perceived proposed by relapse model is critical of all yield parameters in boring of recently created AMMCs. The ANOVA results additionally suggest that feed rate and weight level of SiC are the most overbearing component which impact the surface roughness of this created AMMCs.

References

1. Rajesh Jesudoss Hynes N, Kumar R (2018) Electrochemical machining of aluminium metal matrix composites. Surf Eng Appl Electrochem 54(4):367–373
2. Rajesh Jesudoss Hynes N, Kumar R, Tharmaraj R, Shenbaga Velu P (2016) Production of aluminium metal matrix composites by liquid processing methods 1728:020558
3. Ajith Arul Daniel S, Pugazhenthi R, Kumar R, Vijayananth S (2019) Multi objective prediction and optimization of control parameters in the milling of aluminium hybrid metal matrix composites using ANN and Taguchi-grey relational analysis. Def Technol
4. Daniel AA, Sakthivel M, Gopal PM, Sudhagar S (2018) Study on tribological behaviour of Al/SiC/MoS_2 hybrid metal matrix composites in high temperature environmental condition. Silicon 10(5):2129–2139
5. Sudhagar S, Sakthivel M, Ashok A, Ajith Arul Daniel S (2017) Investigation of discontinuities in friction stir welding of 2024 aluminum alloy using radiography. Mater Eval 75(8):1032–1037
6. Sahoo AK, Pradhan S (2013) Modeling and optimization of Al/SiCp MMC machining using Taguchi approach. Measurement 46:3064–3072
7. Quigley O, Monaghan J, Reilly PO (1994) Factors affecting the machinability of an Al/SiC metal–matrix composites. J Mater Process Technol 43:21–36
8. El-Gallab M, Sklad M (1998) Machining of Al/SiC particulate metal-matrix composites part I: tool performance. J Mater Process Technol 83:151–158
9. Perng CC, Hwang JR, Doong JL (1993) High strain rate tensile properties of an (Al_2O_3 particles)-(6061-T6 Al alloy) metal-matrix composite. Mater Sci Eng, A 171:213–221
10. Ramulu M, Rao PN, Kao H (2002) Drilling of (Al_2O_3)p/6061 metal matrix composites. J Mater Proc Technol 124:244–254
11. Altunpak Y, Ay M, Aslan S (2012) Drilling of a hybrid Al/SiC/Gr metal matrix composites. Inter J Adv Manuf Technol 60:513–517
12. Daniel AA, Murugesan S, Manojkumar, Sukkasamy S (2017) Dry sliding wear behaviour of aluminium 5059/SiC/MoS_2 hybrid metal matrix composites. Mat Res 20(6):1697–1706

Parametric Effect and Laser Beam Machining of Rhenium Diboride-Based Molybdenum Metal Matrix

Anbarasu Augustine, Joseph Dominic Vijayakumar, S. Paulsingarayar, S. Marichamy, B. Stalin, and V. Dhinakaran

Abstract The applications of super hard material have been increased in aerospace, marine and automobile industries. In the present, the investigation deals with rhenium diboride (ReB_2)-based molybdenum metal matrix which is fabricated through stir casting method. Rhenium diboride is mixed to the molybdenum with various weight percentages such as 3, 5 and 8%. The material properties are evaluated for the sample specimens. Based on the comparison of material properties, molybdenum with 8% of rhenium diboride has been considered for these experimental investigations. Due to high hardness and strength, conventional machining methods are not suitable. Among all the unconventional machining processes, the laser beam machining process is the most suitable for high hardness materials. Various input and output parameters are considered. The responses such as the material removal rate (MRR) and surface roughness (SR) are measured. The effects and contribution of parameters are confirmed through analysis of variance (ANOVA) and Pareto charts.

Keywords Rhenium diboride · Molybdenum metal matrix · Stir casting method · Laser beam machining process and ANOVA

A. Augustine
Department of Mechanical Engineering, Sethu Institute of Technology, Virdhunagar, Tamil Nadu, India

J. D. Vijayakumar · S. Paulsingarayar
Department of Mechanical Engineering, SSM Institute of Engineering & Technology, Dindigul, Tamil Nadu, India

S. Marichamy (✉)
Department of Mechanical Engineering, Sri Indu College of Engineering & Technology, Hyderabad, Telangana 501510, India
e-mail: smari2k5@gmail.com

B. Stalin
Department of Mechanical Engineering, Anna University, Regional Campus Madurai, Madurai, Tamil Nadu 625019, India

V. Dhinakaran
Department of Mechanical Engineering, Chennai Institute of Technology, Kundrathur, Tamil Nadu, India

A. Arockiarajan et al. (eds.), *Advances in Industrial Automation and Smart Manufacturing*, Lecture Notes in Mechanical Engineering,
https://doi.org/10.1007/978-981-15-4739-3_64

1 Introduction

The role of unconventional machining process has rapidly increased in manufacturing industries. The main advantages are good dimensional accuracy, machining very hard materials, excellent surface finish and stress-free process. This process was used for machining precision parts, aerospace, automobile, electronic and medical industries [1, 2]. The laser beam machining was used for machining metal and non-metals. These processes have excellent advantages like more machining accuracy, no thermal damages and very quicker processes. The effect of machining parameters was studied in laser machining process [3].

The characteristics and features were studied in the laser supported machining process [4]. The interactions of input and output parameters were analysed in laser machining on magnesium-based alloy [5]. The laser power was the most important input parameter which affects the quality of machining in the laser cutting process [6]. Laser beam machining parameters were analysed, and experimental investigations were made on mild steel [7]. The molybdenum-based alloys have more wear-resistant properties [8]. The rhenium diboride properties and characteristics were studied in various ambient pressures [9]. Various parameters and their effects were studied in CO_2 laser cutting of aluminium plates [10].

The current work describes about various modules such as fabrication of molybdenum metal matrix by stir casting process, evaluation of material properties after addition of rhenium diboride, laser beam machining process and analysis of significance parameter through ANOVA and Pareto chart.

2 Stir Casting Process

The molybdenum metal matrix is fabricated through stir casting method. The billet of molybdenum is kept under the furnace. The furnace is operated at 1950 °C. The uniform stirrer is applied to the molten metal. The preheated rhenium diboride particles are added with various weight percentages such as 3, 5 and 8%. Again, the stirrer is allowed at 750 rpm for 30 min. The uniform mixture of the molten material is transferred to the mould. The three specimens were prepared and tested the material properties. It is shown in Table 1.

3 Results and Discussion

3.1 Laser Beam Machining Process

Thermal energy is used in laser beam machining process. The high frequencies of a monochromatic laser beams are used to melt the workpiece. AMADA 3015 model

Table 1 Material properties of molybdenum metal matrix

Properties	Hardness (BHN)	Tensile strength (MPa)	Shear strength (MPa)	Impact strength (J)	Density (g/cc)
Sample (1)-3% of ReB_2	304	388	510	12	9.2
Sample (2)-5% of ReB_2	356	434	576	14	10.1
Sample (3)-8% of ReB_2	432	520	620	17	12.7

Table 2 Laser beam machining of molybdenum metal matrix

S. No.	A: cutting speed (m/min)	B: laser power (W)	C: standoff distance (mm)	MRR (mm^3/min)	SR (μm)
1	2	900	0.4	2.12	0.25
2	2	1200	0.6	2.74	0.32
3	2	1500	0.8	3.34	0.42
4	4	900	0.6	2.02	0.16
5	4	1200	0.8	3.56	0.27
6	4	1500	0.4	4.32	0.54
7	6	900	0.8	2.98	0.20
8	6	1200	0.4	4.23	0.58
9	6	1500	0.6	4.77	0.64

laser beam machining is selected for these experimental investigations. The control parameters are measured as per the level of input parameters. The holes (0.5 mm) are produced on surface of the metal matrix. The experimental observations are shown in Table 2. Due to better material properties, the third specimen is considered (8% of ReB_2). The optical non-contact sensor-type tester was used to compute the surface roughness of the drilling holes.

3.2 Analysis of Variance

The machining performance depends on the parameters. The contribution of each parameter has been discussed in, ANOVA, Table 3. The laser power has provided the maximum effect on the material removal rate (63.78%). The high amount of laser power produces more heat. Hence, the rate of material removal is high.

Table 3 Analysis of contribution for each parameter

Parameters	DF	Adjusted sum of square	Adjusted mean of square	*F*-value	*P*-value	Contribution %
Cutting speed	2	2.3894	1.19471	19.37	0.049	31.58
Laser power	2	4.8260	2.41301	39.12	0.025	63.78
Standoff distance	2	0.2274	0.11368	1.84	0.352	03.00
Error	2	0.1234	0.06168			01.64
Total	8	7.5662	–	–	–	100

Pareto chart is to emphasize the most important parameter among the various parameters. The results which are obtained from the ANOVA which are confirmed through Pareto chart analysis are shown in Fig. 1.

Investigation of variance for SR is shown in Table 4. The laser power is the most influential factor on the material removal rate (66.24%). The cutting speed is the second influential parameter.

The ANOVA for SR is validated through Pareto chart, and it is shown in Fig. 2. The interaction of BC, AB and AC has provided less effect on surface roughness.

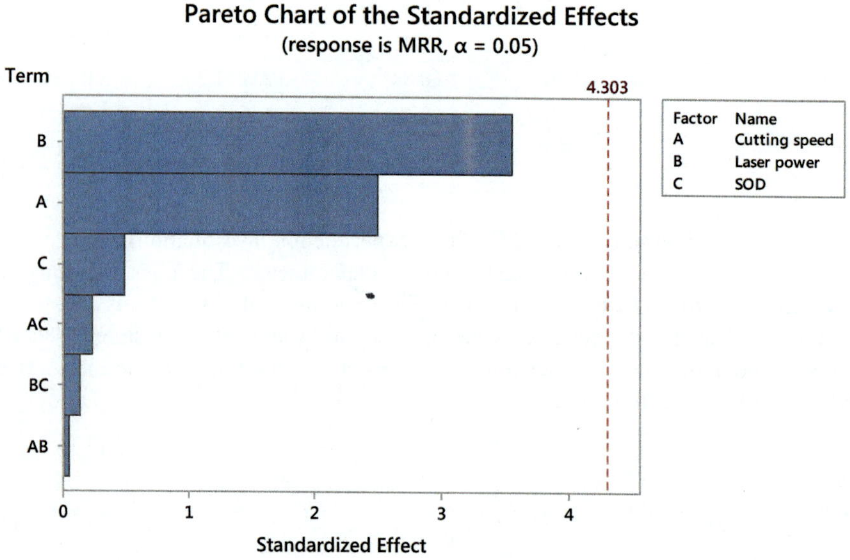

Fig. 1 Pareto chart for the effect of MRR

Table 4 Contribution analysis of parameters for LBM–SR

Source	DF	Sum of square (Adj.)	Mean of square (Adj.)	*F*-value	*P*-value	Percentage %
Cutting speed (A)	2	0.043089	0.021544	19.39	0.049	17.37
Laser power (B)	2	0.164289	0.082144	73.93	0.013	66.24
Standoff distance (C)	2	0.038422	0.019211	17.29	0.055	15.49
Error	2	0.002222	0.001111			00.90
Total	8	0.248022	–	–	–	100

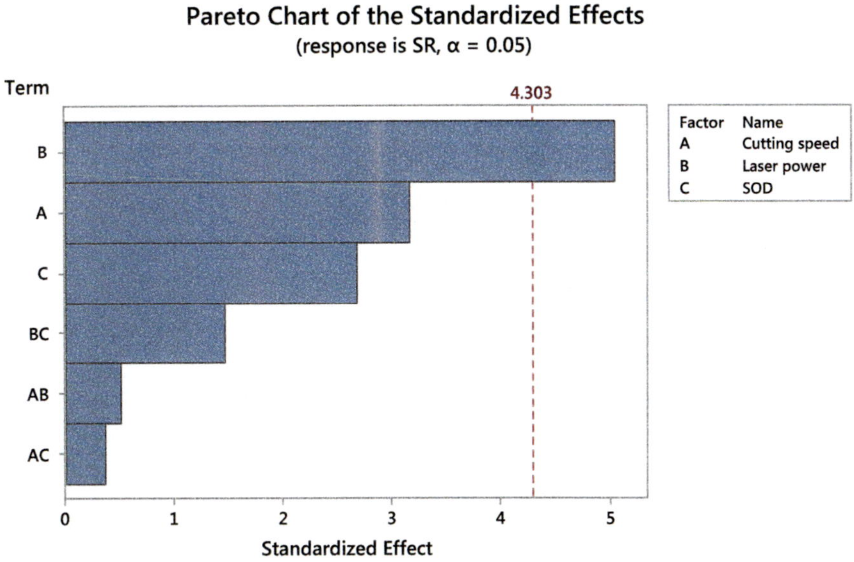

Fig. 2 Pareto chart for effect of SR

4 Evaluation of Wear Properties

One of the most important properties of the material is wear, and it can be defined as the gradual removal of material from the surface due to frictional contact. Its mainly depends on the composition of material. Wear can be observed on any metal by means of mechanical and chemical processes. Pin is made up of molybdenum metal matrix which is rotated on disc. In this experiment, wear rate can be measured according to the change of various input parameters such as applied load (15–25 N), constant track diameter (150 mm), sliding distance (1500–1700 mm) and disc speed (500–900 rpm). During wear test, applied load plays a vital role and minimum wear rate was

Table 5 Experimental results for wear

No.	A: applied load (N)	B: sliding distance (mm)	C: disc speed (rpm)	Wear rate (mm^3/m)
1	15	1500	500	0.0031
2	15	1600	700	0.0024
3	15	1700	900	0.0010
4	20	1500	700	0.0042
5	20	1600	900	0.0035
6	20	1700	500	0.0011
7	25	1500	900	0.0087
8	25	1600	500	0.0063
9	25	1700	700	0.0007

Table 6 Variance analysis for SR

Source	DF	Adj. SS	Adj. MS	*F*-value	*P*-value	Percentage %
Applied load	2	0.000015	0.000008	3.26	0.235	26.31
Sliding distance	2	0.000031	0.000015	6.57	0.132	54.39
Disc speed	2	0.000006	0.000003	1.24	0.446	10.53
Error	2	0.000005	0.000002			08.77
Total	8	0.000057	–	–	–	100

achieved. The grain refinement of material also affects the wear rate. Table 5 shows the experimental results for frictional wear. The minimum wear rate (0.0007 mm^3/m) was achieved at applied load which is 25 N, sliding distance is 1700 mm, and disc speed is 700 rpm. The experimental results for wear are shown in Table 5.

Table 6 shows the analysis of variance for surface roughness. The sliding distance is the most influential parameter which affects the wear rate (54.39%). The least influential parameter is the disc speed (10.53%).

The most influential and least influential parameters are shown in Fig. 3 which affect the wear rate and are confirmed by Pareto chart. The correlation of disc speed and applied load provides less effect on the wear rate.

5 Conclusion

- Through the stir casting process, the molybdenum metal matrix was fabricated.
- By comparison of material properties, the third specimen has the better properties such as hardness (BHN), tensile strength (MPa), impact strength (J) and density (g/cc). So it was considered for these experimental investigations.

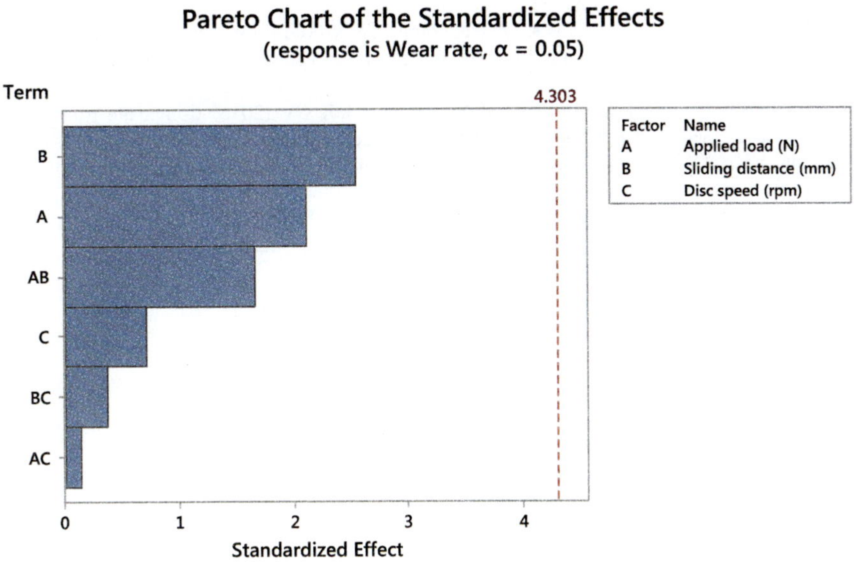

Fig. 3 Pareto chart for effect of wear rate

- The laser beam machining of molybdenum metal matrix was done. The maximum material removal rate is 4.77 mm^3/min, and the minimum surface roughness is 0.16 μm.
- The laser power was the most influential parameter which affects the MRR and SR through analysis of variance and Pareto chart. The standoff distance was the least parameter.
- Through a pin on disc wear test, the minimum wear rate is 0.0007 mm^3/m obtained.

References

1. Dobrzaski LA, Bonek M, Hajduczek E, Klimpel A, Lisiecki A (2005) Comparison of the structures of the hot-work tool steels laser modified surface layers. J Mater Process Technol 164–165:1014–1024
2. Kaczmar JW, Pietrzak K, Wosinski W (2000) The production and application of metal matrix composite materials. J Mater Process Technol 106:58–67
3. Dandekar CR, Shin YC (2010) Laser-assisted machining of a fiber reinforced metal matrix composite. J Manuf Sci Eng 132(6):61
4. Shin YC, Lei S, Rebro P, Rozzi JC (2000) Laser-assisted machining: its potential and future. Mach Technol 11(3):875–885
5. Rashid RAR, Sun Sh, Wang G, Dargusch MS (2013) Experimental investigation of laser assisted machining of AZ91 magnesium alloy. Int J Precis Eng Manuf 14:1263
6. Adelmann B, Hellmann R (2011) Fast laser optimization algorithm. Phys Procedia 12591–12598

7. Srinivasa Raju SVSS, Srinivas K, VenkataRamana M (2005) Parametric investigation of laser cutting and plasma cutting of mild steel E350 material—a comparative study. J Mech Civ Eng 12:1–9
8. Shashkeev KA, Popkov OV, Yurkov YG, Kondrashov SV (2016) Composite tribological materials based on molybdenum disulfide nanoparticles. Russ J Appl Chem 89(4):626–632
9. Chung Hsiu-Ying (2007) Synthesis of ultra-incompressible superhard rhenium diboride at ambient pressure. Science 316:436
10. Senthilkuma V, Periyasamy N, Manigandan A (2015) Parametric investigation of process parameters for laser cutting process. Int J Innov Res Sci Eng Technol 4(5):2773–2779

Microstructure and Mechanical Properties of Two Different Magnesium Alloys Reinforced with Calcium Phosphate Cement Matrix

R. Suresh, N. Shyam Sunder, and Shrishail Kakkeri

Abstract In the present work, an attempt is made to synthesize calcium phosphate cement by using different chemicals such as monetite, calcium carbonate, magnesium hydrogen phosphate tri-hydrate, and magnesium hydroxide in the calculated ratio and it is heat treated in the melting furnace; the sintered powder is then reinforced with magnesium alloy AZ31 and AZ91 series wires. Much effort has been made to enhance their mechanical properties since bending strength of calcium phosphate cement is low due to its brittleness and efforts have been made to increase its strength by adding reinforcement for cement matrix to increase its strength and toughness. The results obtained from this experiment clearly indicates that the workpiece obtained will withstand more amount of load and will have better strength than other workpiece for load bearing, mechanical properties like compressive strength, three-point bending strength were studied for both magnesium alloys reinforced composite at room temperature. Significant increase in mechanical properties was observed for cement with AZ91 reinforcement compared to AZ31 reinforcement. Microstructural analysis reveals that the calcium phosphate biocement contained bigger size granules and smaller size particles to form a dense structure which are evident from SEM microphotographs.

Keywords Load bearing · Calcium phosphate cement · Microstructure · Reinforcement

1 Introduction

Calcium phosphate cement matrix is the alternate material used for bone defects due to its biodegradable in nature without causing any sort of problem to the parent material leading to have a better biocompatibility and high compressive quality. However, their fragile nature and low twisting/rigidity forbid the utilization in burden bearing

R. Suresh (✉) · N. S. Sunder · S. Kakkeri
Department of Mechanical Engineering, Sri Venkateshwara College of Engineering, Bengaluru, Karnataka 562157, India
e-mail: rajappasuresh27@gmail.com

A. Arockiarajan et al. (eds.), *Advances in Industrial Automation and Smart Manufacturing*, Lecture Notes in Mechanical Engineering,
https://doi.org/10.1007/978-981-15-4739-3_65

locales of the skeleton [1] This article highlights support systems of biocements to improve their quality and toughness for an application at burden bearing deformity sides. While porosity decrease depends on the enhancement of an natural concrete property prompting higher quality, the expansion of strands or the production of double setting bond frameworks is extraneous methodology improving quality as well as durability of the lattice. Shockingly, most investigations are committed to the mechanical properties of calcium phosphate biocements just arrangement with one of the displayed methodologies [2].

Considering the benefits of MPCs with brilliant biocompatibility, quick biodegradability, and high starting mechanical quality, it is accepted that the fusing of MPCs into CPCs may result in novel calcium–magnesium phosphate concretes (CMPCs) with improved properties [3]. Calcium-substituted trimagnesium phosphate with the general recipe $Ca_xMg(3-x)(PO4)_2$ ($0 < x < 1.5$) was blended by calcination of powder blends with the suitable stoichiometry and responded with 3.5 M diammonium hydrogenphosphate answer for structuring a cementitious framework of magnesium ammonium phosphate hexahydrate (struvite). The level of ionic substitution was appeared to impact physical concrete properties clinically and appropriate bond definitions with setting time as been ranged from 5 to 15 min and compressive strength of >50 MPa were obtained [4].

A tale philosophy was proposed in the present investigation to consolidate the prevalent support of strands with the advantage of macropores. Huge distance across resorbable strands was utilized to fortify a self-setting concrete to give the required momentary quality and sturdiness, and after that disintegrated to abandon macropores reasonable for bone in growth [5] In this procedures work, a novel bioactive calcium sulfate/magnesium phosphate bond (CSMPC) was manufactured to join the upsides of the calcium sulfate bond and the magnesium phosphate cement [6].

Magnesium calcium biocement with fast setting and degradable qualities was synthezized by utilizing a blend of calcium dihydrogen phosphate and magnesium hydroxide as bond powders. The obtained cement could be effectively solidified furnace and the strength of a compressive quality of 43 MPa in the wake of setting for 1 h is obtained leading to increased in mechanical strength [7]. Fiber-reinforced calcium phosphate is a material where for the most part hydroxyapatite is strengthened with the help of different fibers. The improvement in strength of the material can be done with the orientation of fibers with the matrix [8]. Bone mineral thickness [BMD] for the most part relates with bone quality. BMD testing is utilized to survey bone thickness and analyze osteoporosis. The relationship between CT numbers and mechanical property assessed from cortical bone was observed to be low ($r2 < 0.2$) and cancellous bone was observed to be higher [9].

In present work, an endeavor is made to incorporate calcium phosphate concrete composites with fortifications magnesium combination wires AZ31 and AZ91 for various distances across to contemplate the mechanical properties of the bond.

Table 1 Properties of magnesium wires as indicated by the makers and the references [1]

SN	Magnesium alloy	Wire composition (wt%)	Diameter (mm)	Tensile strength (MPa)	Elongation (%)
1	AZ31	95.4 Mg 3 Al 1 Zn 0.6 Mn	1.6	240	15
2	AZ91	89.6 Mg 9 Al 1 Zn 0.4 Mn	1.6	230	3

2 Experimental Details

2.1 *Materials Used*

The materials used for the present study are magnesium wires AZ31 and AZ91 as reinforcement and calcium phosphate cement as matrix.

Magnesium alloys have hexagonal lattice structure, and deformation of hexagonal lattice is more complicated than cubic latticed metals like aluminum, copper widely used due to its good strength. The composition of two magnesium alloy wires is shown in Table 1. AZ31 and AZ91 are used as the reinforcement which is one of the most widely used because of their strength to weight ratio. It facilitates high strength and a higher specific modulus, good resistance to corrosion.

2.2 *Processing of Cement Mixture*

The reinforcement material used for the present study is aluminum alloys AZ31 and AZ91 wires. Calcium–magnesium phosphate powder with the formula $Ca_{0.75}Mg_{2.25}(PO_4)_2$ was synthesized by mixing the chemicals of monetite ($CaHPO_4$), calcium carbonate ($CaCO_3$), magnesium hydrogen phosphate tri-hydrate ($MgHPO_43H_2O$), and magnesium hydroxide (Mg(OH)2) to 1120 °C for 5 h followed by cooling to room temperature in the furnace only. The organization of various powder blend is appeared in Table 2. The obtained sintered cake was squashed with pestle and mortar and went through a 355 mm strainer. The sintered cake is crushed

Table 2 Structure of powder blends for the planning of $Ca_xMg(3\text{-}x)(PO_4)_2$

SN	$Mg_xCaY(PO_4)_2$	$CaHPO_4$ (mol)	$CaCO_3$ (mol)	$MgHPO_4\_3H_2O$ (mol)	$Mg(OH)_2$ (mol)
1	$Ca_{0.75}Mg_{2.25}(PO_4)_2$	0.5	0.25	1.5	0.75

in ball milling for 4 h with 120 g clumps in a planetary ball factory at 190 rpm This CPC concrete was blended with a 3.5 M diammonium hydrogen phosphate (DAHP) arrangement with a powder/fluid proportion of 3 g/mL on a watch glass dish. The bond glue was filled in silicon elastic molds (6 ~ 7 ~ 45 mm^3) with a spatula. For form tests, AZ31 and AZ91 wire pieces of 45 mm length were included during filling of the mold.

2.3 Testing

The workpiece for compressive test and three-point bending test are set up according to ASTM measures. To research the mechanical conduct of the composites, the compressive tests and three-point bending test were done in the computerized UTM at a test speed of 2 mm/min. Microstructure of arranged example is contemplated by using scanning electron microscope at different magnifications before scanning the cement mixture the polishing is done with the help of emery papers in order to remove dirt on the surface of the workpiece.

3 Results and Discussions

The compressive test results for specimens with two different reinforcement wires are as shown in Table 3; for the specimen with AZ91 reinforcement, the maximum compressive strength is 37.33 MPa compared to AZ31 reinforcement of 36.23 MPa.

3.1 Compressive Stress Observations

It is observed from Fig. 1, when the cement workpiece is subjected into the compressive force in the computerized UTM at a test speed of 2 mm/min, material undergoes a plastic deformation at a peak load of 1304 N for a cross-sectional area of 36 mm^2, and compressive stress obtained for cement mixture reinforced with magnesium alloy AZ31 is 36.23 MPa.

Table 3 Shows the cement with two different magnesium alloy reinforcements

SN	Cement with reinforcement	Cross-section area (mm^2)	Peak load (N)	%Elongation	Strain	Compressive strength (N/mm^2)
1	AZ31	36	1304	2.68	0.994	36.23
2	AZ91	36	1344	4.99	0.998	37.33

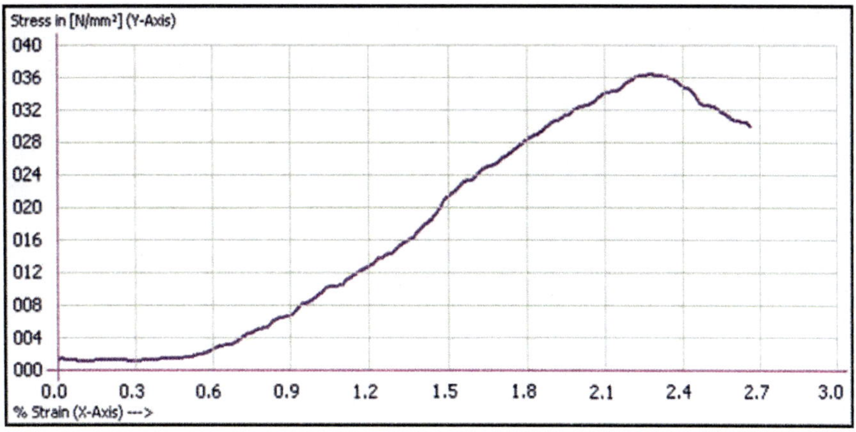

Fig. 1 Shows the variation of stress versus strain for cement with reinforcement magnesium alloy AZ31

It is observed from Fig. 2, when the cement workpiece is subjected to the compressive force in the computerized UTM at a test speed of 2 mm/min, material undergoes a plastic deformation at a peak load of 1344 N for a cross-sectional area of 36 mm^2, and compressive stress obtained for cement mixture reinforced with magnesium alloy AZ91 is 37.33 MPa.

From Table 4, it is observed that the bending strength of cement with reinforcement AZ31 is 38.25 MPa compared with the cement with reinforcement AZ91 is 39.75 MPa. The bending strength is calculated through the formula $\sigma = B_m$ * depth/moment of inertia by submitting the length, width and thickness for rectangular workpiece,

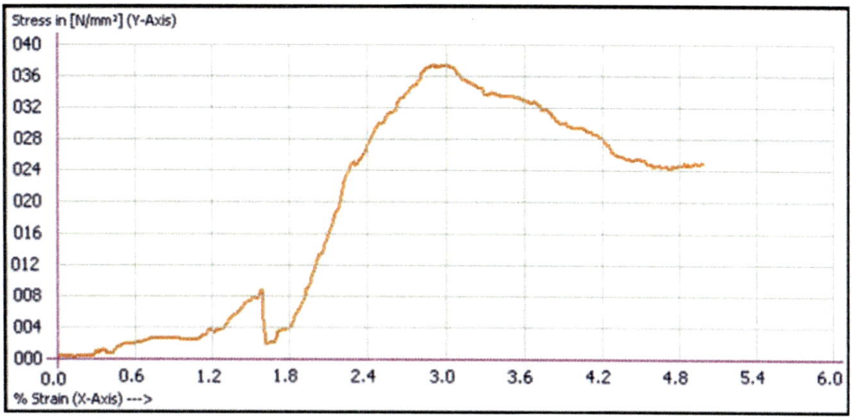

Fig. 2 Shows the variation of stress versus strain for cement with reinforcement magnesium alloy AZ91

Table 4 Shows the cement with two different magnesium alloy reinforcements

SN	Cement with reinforcement	Cross-section area (mm^2)	Peak load (N)	Bending strength (N/mm^2)
1	AZ31	36	102	38.25
2	AZ91	36	106	39.75

and bending strength for two different reinforced cement composite can be obtained as 38.25 MPa and 39.75 MPa, respectively.

3.2 Point Bending Stress Observations

The three-point bending tests was performed in a computerized universal testing machine with a three-point t bend fixture with a speed of 2 mm/min. Figure 3(a) indicates stress vs strain measurement and Fig. 3(b) indicates load vs. displacement for cement. The peak load for cement with reinforcement of magnesium alloy AZ31 is 102N compared with cement with reinforcement of magnesium alloy AZ91 IS 106 N. The bending stresses calculated by using the formula $\sigma = B_m$ * depth/moment of inertia. Magnesium wires show a higher Young's modulus than the calcium phosphate cement.

3.3 Microstructural Studies

Figure 4 represents SEM pictures of the calcium phosphate cement mixture. It was identified that calcium phosphate cement having bigger size molecules shown in Fig. 4(a) and (b) with long arrows. Similarly in the Fig. 4(c) and (d) a small granule like substances shown in small arrows.

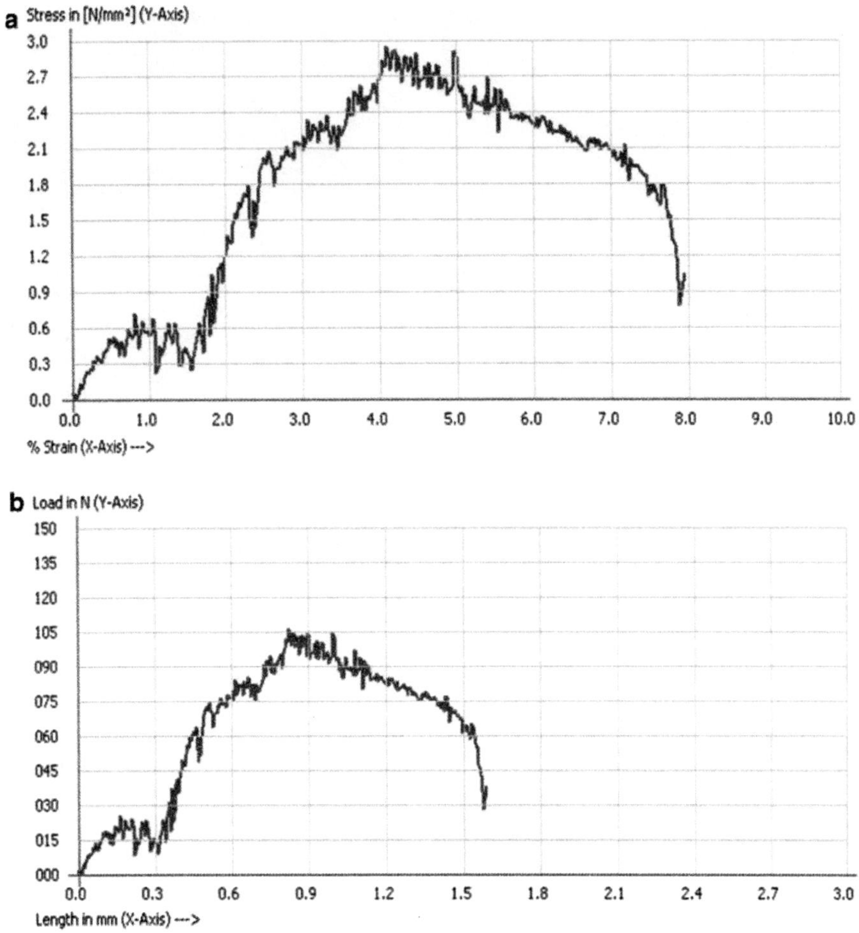

Fig. 3 **a** Stress versus strain diagram for AZ91 cement composite, **b** load versus displacement AZ91 cement composite

4 Conclusions

The results obtained from the present work clearly indicate that the strength of cement is increased and capable of withstanding high amount of load so that it can be replaced with the fractured bone constituent and leading to better strength.

- The chemicals containing monetite, calcium carbonate, magnesium hydrogen phosphate tri-hydrate, and magnesium hydroxide were successfully synthesized in the calculated ratio to obtain calcium–magnesium phosphate cement matrix.
- The compression strength of CPC with AZ31 is 36.23 MPa where as CPC with AZ91D is 37.33 MPa.

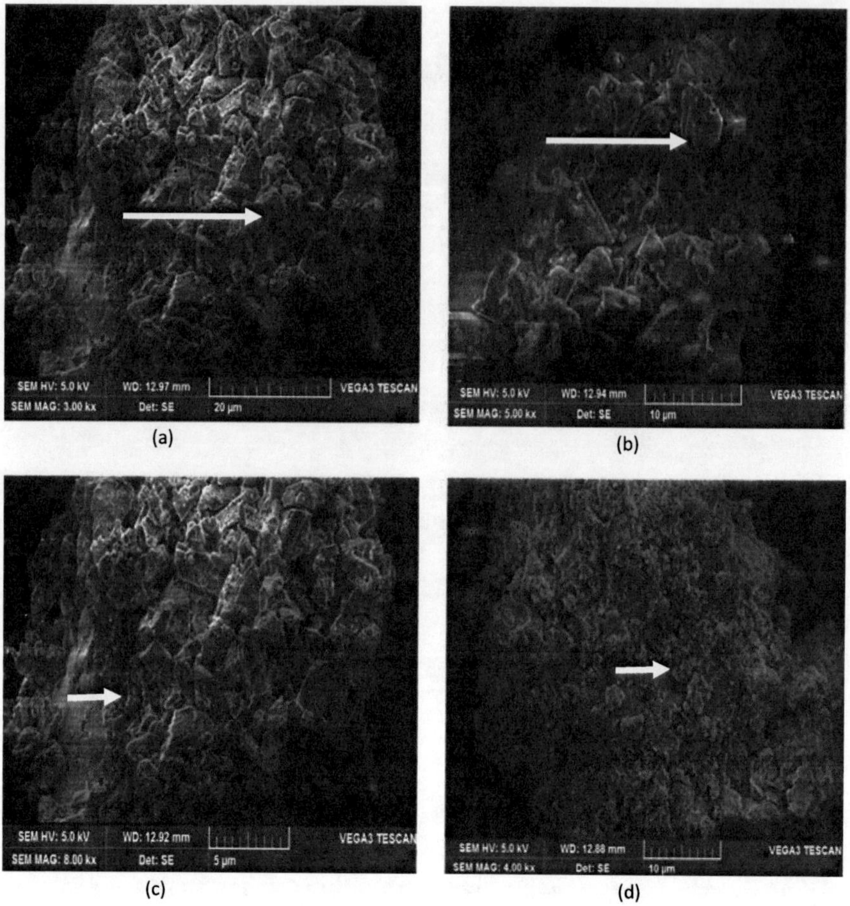

Fig. 4 Scanning electron microscope pictures of the specimen of calcium phosphate cement under various amplifications (Fig(a) 3000 × and Fig(b) 5000×); long arrows indicates magnesium phosphate; (Fig(c) 8000 × and Fig(d) 4000×); short arrows indicates calcium phosphate

- The three-point bending strength of CPC with AZ31 is 38.25 MPa whereas CPC with AZ91D is 39.75 MPa.

References

1. Kruger R, Seitz JM, Ewald A, Bach FW, Groll J (2013) Strong and tough magnesium wire reinforced phosphate cement composites for load bearing bone replacement. J Mech Behav Biomed Mater 20:36–44
2. Geffers M, Groll J, Gbureck U (2015) Reinforcement strategies for load-bearing calcium phosphate biocements. Materials 8:2700–2717. ISSN 1996-1944

3. Li K, Shen Q, Xie Y, Zheng X (2014) Preparation and characterization of calcium-magnesium phosphate cements. Adv Mater Res 1058:83–86
4. Vorndran E, Ewald A, Muller FA, Zorn K, Kufner A, Gbureck U (2011) Formation and properties of magnesium ammonium phosphate hexahydrate biocements in the Ca–Mg–PO_4 system. J Mater Sci: Mater Med 22:429–436
5. Xu HHK, Quinn JB (2002) Calcium phosphate cement containing resorbable fibers for short-term reinforcement and macroporosity. Biomaterials 23:193–202
6. Yang G, Liu J, Li F, Pan Z, Ni X, Shen Y, Huang Q (2014) Bioactive calcium sulfate/magnesium phosphate cement for bone substitute applications. Mater Sci Eng C 35:70–76
7. Jia J, Wei S, Shin J-W, Liu C (2010) Development of magnesium calcium phosphate biocement for bone regeneration. J R Soc Interface 7:1171–1180
8. Krüger R, Groll J (2012) Fiber reinforced calcium phosphate cements on the way to degradable load bearing bone substitutes. Biomaterials 33: 5887–5900
9. Khan SN, Warkhedkar RM RM, Shyam AK Human bone strength evaluation through different mechanical tests. Int J Curr Eng Technol E-ISSN 2277-4106, P-ISSN 2347-5161

Abrasive Water Jet Machining Studies on $AlSi_7$+63%SiC Hybrid Composite

Pruthviraju Garikipati and K. Balamurugan

Abstract To improve the properties of Al/Si_7 composite, 63% SiC (maximum solubility) is added as a secondary reinforcement to form Al/Si_7+63%SiC hybrid matrix composite. Al/Si_7+63% SiC composite is fabricated through stir casting process. The fabricated composite is machined by using abrasive water jet machine (AWJM) and an optimum machining condition is evaluated through gray-related response surface method (G-RSM) by using L_{20} orthogonal array. AWJM are governed with independent parameters such as jet pressure (JP), stand-off distance (SOD), and traverse speed (TS). The dependent responses are taken as material removal rate (MRR) and surface roughness (Ra). The effect of each independent parameter and the output responses is evaluated. The optimum machining condition is predicted through ANOVA. The test experiment is conducted to validate the RSM equation, and it found in the acceptable range.

Keywords Al/si_7+63%SiC hybrid composite · Abrasive water jet machine · G-RSM · ANOVA · Optimization

1 Introduction

The advantages of the composite over the conventional materials make a suitable candidate material in mass-market products, like automobiles, aeronautics, etc. [1, 2]. Aluminum composites are prepared using a ceramic mixture, due to its specific strength and excellent stiffness property even at elevated temperature it gain attention in automotive engines components [3, 4]. The addition of 2 wt% of silica will enhance the ductile property of the alloy with the formation of spheroidization in the eutectic solidification [5]. Microscopy analysis of Al-SiC composite surface reveals that the crack initiation starts at SiC particle-rich zone [6]. Also, it is stated that the addition of SiC as a reinforcement material will improve the machinability through the breakup

P. Garikipati · K. Balamurugan (✉)
Department of Mechanical Engineering, VFSTR (Deemed to be University), Guntur, AP 522213, India
e-mail: kbalan2000@gmail.com

A. Arockiarajan et al. (eds.), *Advances in Industrial Automation and Smart Manufacturing*, Lecture Notes in Mechanical Engineering,
https://doi.org/10.1007/978-981-15-4739-3_66

of the particle. Addition of reinforced particles with a homogeneous distribution will lead to a reduction in porosity and improves bonding strength [7]. Fracture mechanics analysis on Al/SiC is the ductile mode of failure. Fracture analysis on Al/SiC is the ductile mode of failure and this occurs due to deformation, crack propagation along with there inforcement particles and through the voids [8]. The machined surface of Al/SiC composite will have plastic deformation surface and the failure occurs at the interfacial de-bonding regions that lead to an early stage of the fracture [9, 10].

Present work will investigate the AWJM performance on $AlSi_7$+63%SiC hybrid composite. The experimental results are optimized using G-RSM. The role of SiC in the hybrid composite is analyzed, and trial experiments are conducted to validate the experimental results.

2 Material Preparation

The preparation of AlSi7% with 63% SiC is done in an induction furnace by mixing the SiC containing zero impurities with $AlSi_7$ alloy. The temperature of the furnace is reduced to 600 °C which cools the liquid alloy for the incorporation of SiC. The secondary reinforcement (SiC) is pre-oxidized at 900 °C for 2 h and reinforcement abrasive grade with size ranging from 15 to 30 micrometers at the constant rate of SiC is added to the semi-solid matrix alloy with specially designed stirrer semi-solid matrix alloy is mixed in the presence of argon gas productive atmosphere production of the metal matrix and composite is carried out. The melt is heated to 750 °C and the temperature is maintained for 5 min with the continuous stirring condition. The composite slurry has been transferred into preheated iron dies that have the wall thickness of 8 mm. Samples of size 25 cm × 25 cm × 1.2 cm is produced after the removal of the feeder head. Addition of more than 63% of SiC in Al7%Si alloys leads to having improper dispersion and hence the maximum solubility of SiC in Al alloy is resisted with 63% of weight ratio.

3 Experimentations and Observations

The machinability of the fabricated hybrid metal matrix composites that find applications in aerospace and automotive industries is examined using AWJM. Garnet is used as abrasive that has the mesh size of 80. The high-pressure jet passes to 0.67 mm diameter WC nozzle where the entire unit is controlled by PLC. The nozzle is kept perpendicular to the cut material. The experimental setup is shown in Fig. 1.

Among various parameters of AWJM, jet pressure (JP), stand-off distance (SOD), and traverse speed (TS) are chosen parameters with three levels [11]. Through central composite design, theoptimum conditions are predicted. The AWJM parameters are listed in Table 1.

Fig. 1 Abrasive water jet machine and the test sample

Table 1 Machining parameters and its levels

S. No.	Machining parameters	Levels			Units
		1	2	3	
1	JP	220	240	260	bar
2	SOD	1	2	3	mm
3	TS	20	30	40	mm/min

4 Results and Discussion

Gray theory-based response surface methodology (G-RSM)

To identify the optimum condition for the multi-level responses for any machining, gray relational analysis (GRA) uses S/N ratio to obtain a feasible solution for producing kerf surfaces with acceptable level of surface properties [12]. RSM is a statistical tool used to get a quadratic equation for GRG. G-RSM technique is used to predict the optimal parameters condition. Table 2 shows the experimental observations of the hybrid composite in AWJM.

Table 2 Machining data: Al/Si_7+63%SiC composite

Test numbers	Machining parameters			Output responses	
	JP	SOD	TS	MRR	Ra
	(bar)	(mm)	(mm/min)	(g/s)	(μm)
1	220	1	20	0.0042	3.047
2	240	2	30	0.0064	3.592
3	260	3	40	0.0090	3.678
4	240	2	40	0.0086	3.622
5	260	3	20	0.0052	3.596
6	240	2	30	0.0065	3.599
7	260	2	30	0.0065	3.908
8	260	1	40	0.0078	4.415
9	240	3	30	0.0072	3.418
10	240	2	30	0.0064	3.613
11	260	1	20	0.0044	4.445
12	220	3	20	0.0048	3.495
13	220	1	40	0.0070	3.097
14	240	2	20	0.0046	3.606
15	220	2	30	0.0060	3.346
16	240	2	30	0.0063	3.511
17	220	3	40	0.0081	3.386
18	240	2	30	0.0064	3.603
19	240	2	30	0.0065	3.599
20	240	1	30	0.0055	3.733

ANOVA for MRR and Ra of fabricate composite are shown in Table 3. Probability value of MRR and Ra is tabulated as zero. The significant effect of the independent parameters could be identified through the least *P*-value. Further, from the *F*-value, it could be stated that the effect of TS is superior to SOD and JP. The interface between JP × TS has found to have the least significant effect over other interfaces. The effect of SOD on output responses is found to be negligible for MRR, whereas the machining parameters of AWJM, i.e., JP and SOD are having a significant effect on Ra. The influence of TS is found to be very low and has no effect. The above observations can be confirmed through the ANOVA table. The interface effect of JP × SOD has a significant effect than other interface parameters on Ra.

Machining effect: MRR

From Fig. 2a, it is observed that at a lower level of SOD and JP, MRR get considerably decreased. At the middle level of JP and SOD, MRR is increased. Increase in SOD, high-pressure jet accelerates the hard abrasive and tends to increase the width of the jet before it impinges the top surface of the composite material. At the high level

Table 3 ANOVA table: MRR and Ra

Source	DF	Adj SS	Adj MS	*F*-value	*P*-value
ANOVA table: MRR					
Model	6	0.000034	0.000006	117.33	0.000
Linear	3	0.000034	0.000011	233.02	0.000
JP	1	0.000001	0.000001	16.53	0.001
SOD	1	0.000003	0.000003	60.95	0.000
TS	1	0.000030	0.000030	621.58	0.000
2-Way interface	3	0.000000	0.000000	1.64	0.229
JP × SOD	1	0.000000	0.000000	0.17	0.691
JP × TS	1	0.000000	0.000000	2.87	0.114
SOD × TS	1	0.000000	0.000000	1.88	0.193
Error	13	0.000001	0.000000	13.28	0.006
Lack-of-fit	8	0.000001	0.000000		
Pure error	5	0.000000	0.000000		
Total	19				
$R^2 = 98.19\%$, Adjacent $R^2 = 97.35\%$					
Source	DF	Adj SS	Adj MS	*F*-value	*P*-value
ANOVA table: Ra					
Model	6	2.16001	0.36000	85.68	0.000
Linear	3	1.48337	0.49446	117.67	0.000
JP	1	1.34799	1.34799	320.80	0.000
SOD	1	0.13537	0.13537	32.22	0.000
TS	1	0.00001	0.00001	0.00	0.964
2-Way interface	3	0.67664	0.22555	53.68	0.000
JP × SOD	1	0.67483	0.67483	160.60	0.000
JP × TS	1	0.00153	0.00153	0.36	0.557
SOD × TS	1	0.00028	0.00028	0.07	0.800
Error	13	0.05463	0.00420	4.24	0.064
Lack-of-fit	8	0.04760	0.00595		
Pure error	5	0.00702	0.00140		
Total	19				
$R^2 = 97.53\%$, Adjacent $R^2 = 96.40\%$					

operating conditions of JP and SOD abrasives will easily erode away the composite surface that significantly increases the MRR. As the hard abrasives hit the surface of the composite since most of the region in the composite is packed with hard SiC particles leads to the excess removal of materials and increases MRR.

In Fig. 2b, increase in JP and TS shows a considerable increase in MRR. It is believed that at high operating condition, the highly accelerated abrasives grains cut composite with bulk removal of SiC material. SOD has high significance compared to JP. From Fig. 2c, the interfaces between SOD and TS are identified to have the least significant effect. However, with an increase in SOD will influence the width of the water beam that leads to getting high MRR. From Fig. 2d, all the experimental

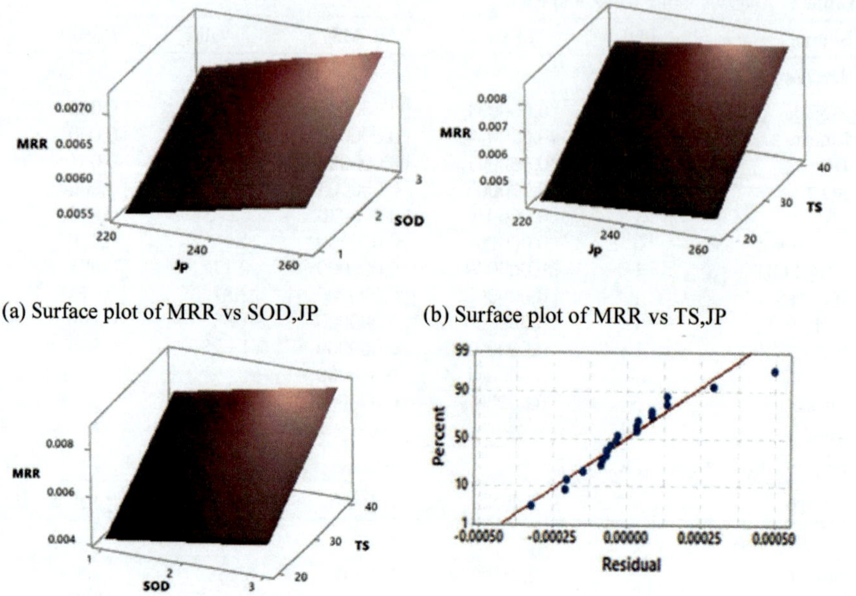

(a) Surface plot of MRR vs SOD,JP

(b) Surface plot of MRR vs TS,JP

(c) Surface plot of MRR vs TS, SOD

(d) Normal Probability of MRR

Fig. 2 Interface plots of the independent parameters on MRR

observations are bound to be in the acceptable level. The optimum machining condition for MRR is identified as with the following conditions: JP = 220 bar, SOD = 1 mm, and TS = 29.2 mm/s.

The predicted equation for MRR is shown in Eq. 1. The residual square of MRR is 98.19% and adjacent of MRR is 97.35% which fits more than 95% of a confidential level.

$$\begin{aligned} \text{MRR} = {} & 0.00283 - 0.000009\ \text{JP} - 0.000156\ \text{SOD} - 0.00006\text{TS} + 0.000002\ \text{JP} \\ & \times \text{SO} + 0.000001\ \text{JP} \times \text{TS} + 0.000011\ \text{SOD} \times \text{TS} \end{aligned} \tag{1}$$

Machining effect: Ra

3D surface plot of Ra is displayed at Fig. 3. The interface effect of JP × SOD on Ra is shown in Fig. 3a. The saddle point could not be obtained from the plot, but the optimum surface can be identified through visual inspection. Figure 3b shows the interface effect of JP and TS on Ra. An increase in the levels of JP and TS will increase the surface wear rate and progress to have excess wear track on the cut surface. This action increases Ra value to a greater extent. Figure 3c shows the interface effect of SOD × TS on Ra and the effects produce by SOD on Ra is superior to TS on Ra. It becomes possible to predict the optimum condition under this working condition. To obtain the acceptable level of Ra, AWJM parameters are to be at the levels as JP of 220 bar, SOD of 1.1 mm, and TS of 21 mm/min. From Fig. 3a–c, it is seen that all considered machining parameters have a major impact on Ra. Figure 3d shows the normal probability plot of the experimental observations. Observations are found to be within the acceptable range.

The equation predicted for Ra is displayed as Eq. 2. From ANOVA table, it is clear that the residual square and adjacent to residual square of Ra is 97.53% and 96.5%, respectively.

$$\begin{aligned} Ra = & -7.07 + 0.04533\ JP + 3.387\ SOD - 0.0153 TS - 0.01452\ JP \times SOD \\ & + 0.000069\ JP \times TS - 0.00059\ SOD \times TS \end{aligned} \tag{2}$$

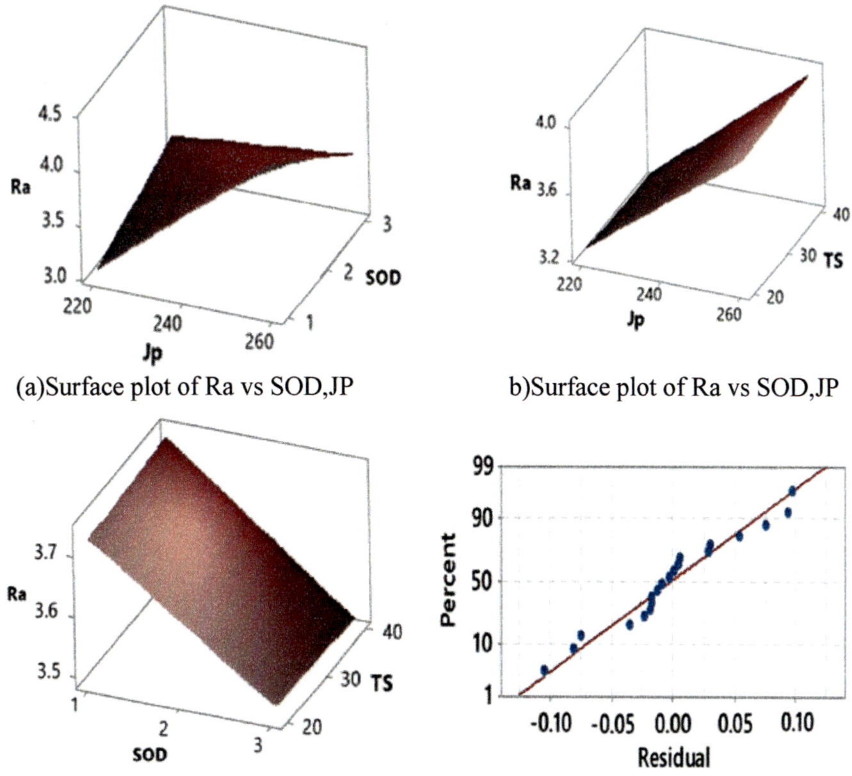

(a)Surface plot of Ra vs SOD,JP b)Surface plot of Ra vs SOD,JP

(c) Surface plot of Ra vs SOD,JP (d) Normal Probability of Ra

Fig. 3 Interface plots of the independent parameters on Ra

5 Experimental Validation

To validate the equations obtained through G-RSM, a trial experiment is performed. The trial machining conditions with the experimental and predicted observations are shown in Table 4. Almost all the predicted observations through the equations are found to be consistent with the experimental observations, and this could be validated through the comparison chart between the experimental and predicted observation and it is shown in Fig. 4.

Table 4 Validation table

Ex. No.	JP	SOD	TS	MRR		Ra	
	(bar)	(mm)	(mm/min)	Experiment	Predicted	Experiment	Predicted
1	245	35	50	0.0125	0.0141	3.392	3.414

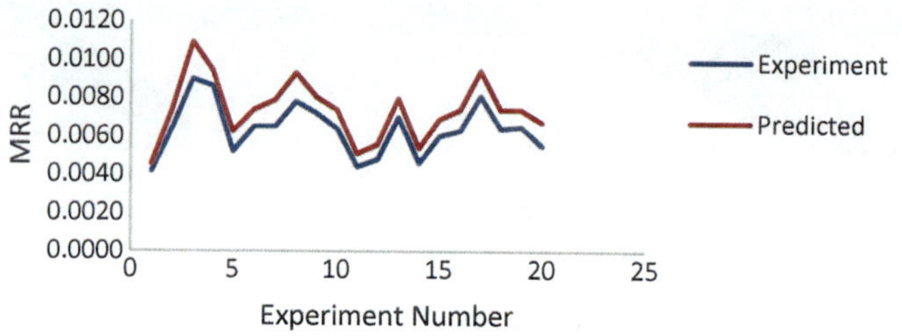

(a) MRR

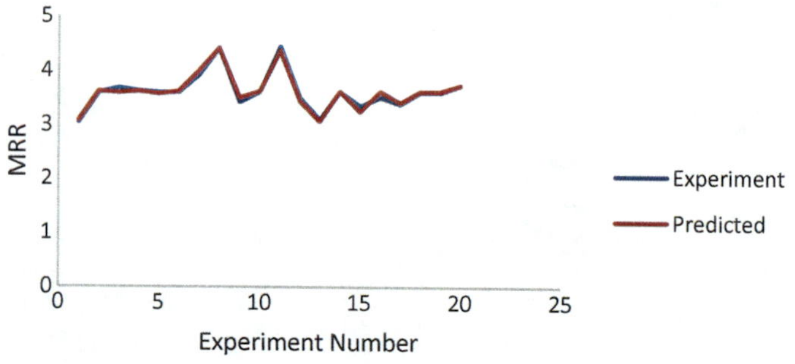

(b) Ra

Fig. 4 Experimental versus predicted observation

6 Conclusions

The parametric investigation on AWJM effects on $AlSi_7$+63%SiC composite is performed and the following conclusion is drawn:

- Jet pressure, stand-off distance, and traverse speed have found to have an important effect over the output responses. The impact of SOD and TS is proved to have greater effect on MRR whereas influence of JP and SOD is higher for Ra.
- Optimized AWJM conditions for MRR is identified as: JP at 20 bar, SOD at 1mmand TS of 29.2 mm/min and to obtain the optimized Ra, the preferable machining condition: JP of 220 bar, SOD of 1.1 mm and TS of 21 mm/s.
- It is observed that the predicted equation best fits the experimental observation and it can be verified by validation table and figure.

References

1. Kaczmar JW, Pietrzak K, Wlosinski W (2000) The production and application of metal matrix composite materials. J Mater Process Technol 106:58–67
2. Valente M, Billi F (2001) Micromechanical modification induced by cyclic thermal stresses on metal matrix composites for automotive applications. Compos: Part B 32:529–533
3. Kelly A, Zweben C (2000) Comprehensive composite materials, vol 3. Elsevier Science Ltd., Oxford
4. Schnabl A, Degischer HP (2003) Thermal cycling creep of a short fiber reinforced aluminium piston alloy. Mater Res Adv Tech 94:743–748
5. Lasagni F, Lasagni A, Holzapfel C, Mucklich F, Degischer HP (2007) Three-dimensional characterization of 'As-Cast' and solution-treated AlSi12(Sr) Alloys by high-resolution FIB tomography. Acta Mater 55:3875–3882
6. Yotte S, Breysse D, Riss J, Ghosh S (2001) Cluster characterization in a metal matrix composite. Mater Charact 46:211–219
7. Lee HS, Yeo JS, Hong SH, Yoon DJ, Na KH (2001) The fabrication process and mechanical properties of SiCp/Al-Si metal matrix composites for automobile air-conditioner compression pistons. J Mater Process Technol 113:202–208
8. Pandey AB, Majumdar BS, Miracle DB (2000) Deformation and fracture of a particle-reinforced aluminum alloy composite: part I. Experiments. Metall Mater Trans A 31:921–936
9. Badini C, La Vecciha M, Giurcanu A, Wenhui J (1997) Damage of 6061/SiCw composite by thermal cycling. J Mater Sci 32:921–930
10. Manikandan N, Binoj JS, Varaprasad KC, Sabari SS, Raju R (2019) Investigations on wire spark Erosion machining of aluminum-based metal matrix composites. Lecture Notes in Mechanical Engineering, Advances in Manufacturing Technology, Springer, Singapore, pp 361–369
11. Balamurugan K, Uthayakumar M, Sankar S, Hareesh US, Warrier KGK (2018) J Aust Ceramic Soc 54:205. https://doi.org/10.1007/s41779-017-0142-7
12. Ghosh D, Doloi B, Das PK (2015) Parametric analysis and optimization on abrasive water jet cutting of silicon nitride ceramics. J Precis Technol 5:294–311

Machining Studies on Various Ply Orientations of Glass Fiber Composite

Jyothhi Yarlagaddaa, Ramakrishna Malkapuram, and K. Balamurugan

Abstract The glass fiber reinforced composite (GFRC) plate of 7 mm thickness, comprising of 11 layers that are stacked at different FPOs likely 0°, 45° and 90° (fiber mats are placed normal to each other) is fabricated using drum winding and hand layup process followed by compression molding at 150 °C for 3–4 h. Fabricated glass fiber reinforced composites are machined by abrasive water jet machine (AWJM) with variable parameters such as jet pressure (JP), stand-off distance (SOD), and traverse speed (TS). Material removal rate (MRR), kerfs angle (KA), and surface profile parameter (Ra) are taken as dependent parameters. Garnet of 80 mesh size is used as erodent material. ANOVA obtained through Grey relation analysis (GRA) infers that JP has superior effect on 90° FPO, followed by SOD on 0° and 45° FPO. The influence of TS is found to have less impact with respect to FPO. JP is found to have high influence at 0° FPO for MRR and 0° 45° for Ra. There is no considerable effect of JP on KA for 45° FPO. SOD has least significance on Ra, but machining at low levels of SOD, irrespective of FPO, observations are noticed for MRR and KA. The effect of TS is high for 90° FPO.

Keywords Fiber ply orientation GFRC · Abrasive water jet · ANOVA · Surface morphology

1 Introduction

Polymer composites replace the commercially available materials due to their lightweight and superior weight-to-stiffness ratio. Aramide et al. [1] stated that combinations of two or more materials that are distinct in nature offer better properties with low specific weight. Erden et al. [2] confirms that fiber strength, modulus, matrix strength, and chemical stability will enhance the mechanical properties of

J. Yarlagaddaa · R. Malkapuram (✉) · K. Balamurugan
Department of Mechanical Engineering, VFSTR (Deemed to be University), Guntur, Andhra Pradesh, India
e-mail: drmramakrishnaa@gmail.com

A. Arockiarajan et al. (eds.), *Advances in Industrial Automation and Smart Manufacturing*, Lecture Notes in Mechanical Engineering,
https://doi.org/10.1007/978-981-15-4739-3_67

the composite. Arikan and Sayman [3] observed that the matrix (epoxy) composites are prone to more damage when compared to thermoplastic matrix composites (polypropylene) because of the inability of the fibers to transfer stress throughout the composite and the damage progresses in impact direction. Nikforooz et al. [4] investigated the thermo-mechanical behavior of continuous E-glass reinforced polyamide composites at below and above glass transition temperatures and found that there is considerable reduction of Young's modulus and interface strength due to the inability of the matrix to transfer load to the fibers. According to Lopez et al. [5] GFRC with organic, polyester, vinyl ester, and epoxy resins have applications in the field such as biomedical, automobile, and aeronautical industry.

Ellyin and Rohrbacher [6] found that the rate of moisture absorption is dependent on the immersion temperatures of glass fiber reinforced epoxy laminates. Wu et al. [7] were able to produce GFRC with proper interfacial adhesion between glass fibers with inter laminar shear strength. Kim et al. [8] studies on the interfacial properties of unidirectional and bidirectional GFRC reveal that unidirectional GFRC exhibit better mechanical and interfacial properties due to its longitudinal fiber pattern. Alam et al. [9] proposed that reinforcement of glass fibers in longitudinal fiber, chopped fiber, or knitted form will improve the mechanical and tribological properties.

Goh et al. [10] investigation on mechanical properties of carbon and GFR thermoplastics fabricated through additive manufacturing reported that these composites exhibit better tensile strength at the proposed manufacturing techniques. Purohit et al. [11] used fly ash as additive material with GFRC with 4%, 6%, and 10% fly ash as filler particles, and the improvement in tensile strength, flexural strength, impact strength is found at 4% fly ash. Gao et al. [12] fabricated E-glass/epoxy plain weave composites and incorporated silica nanoparticles which reported that proper chemical bonding and fiber surface texture at the interface will helps in energy absorption. Zulfli et al. [13] found that the incorporation of 1% MWCNT into the GFRC yields better mechanical properties such as flexural strength and impact strength. Gokuldass and Ramesh [14] investigated the mechanical properties of woven glass fabric and found that there is better fracture toughness for modified epoxy with 11 wt% of nanosilica and the hybrid glass/Kevlar with nanosilica, and micro-rubber-modified epoxy composite possesses very good interfacial bonding than plain composites. Dinesh et al. [15] reported that silane-treated E-glass fiber increases the mechanical and thermal properties of the composite.

Aich et al. [16] stated that the studies on cutting performance and the erosion mechanism of any amorphous materials like borosilicate glass on AWJM will help the process to be satisfactory. Balamurugan et al. [17] stated that AWJM is found to be a suitable technique to cut glass reinforced composite in an eco-friendly manner. Doreswamy et al. [18] studies on the effect of process parameters of AWJM of GFRP confirm that SOD and feed rate have a significant change on kerf taper.

2 Materials and Methods

Commercially available continuous glass fiber, epoxy, and hardener are procured from MS Industries, Kakinada, and Andhra Pradesh. HY556 type hardener and Araldite-LY951 type epoxy are mixed in the ratio of 1:10. Continuous fiber spools are placed at Creel stand, the one end of the fiber from the spool is immersed into the resin tank which is attached to the drum winding machine. The cylindrical sheet obtained by drum winding is cut to mold dimensions and arranged at different FPO's followed by compression molding in hydraulic press at 150 °C for 3 h. Resin is applied to the mold cavity and the fibers are stacked as layer by layer. A gentle load is applied upon placing each layer and then the stacked sample is compressed mold at 150 °C for 3 h. The experimental arrangement of drum winding and the samples are shown in Fig. 1.

The prepared composite is machined using AWJM of Model DIP 6D-2230. AWJM of an orifice diameter 0.25 mm and the WC nozzle of 0.67 mm are used to machine the prepared composite. Garnet is used as abrasive particle. Figure 2 shows the experimental set up of AWJM. Surface roughness of the specimen is measured using SJ-411 surface roughness tester. The parameters used to measure the surface profile roughness on the cut surface are span, measurement range, and probe speed which is 5 mm, 350 μm, and 0.25 mm/s, respectively.

The weight of the composite material before and after machining is determined using Shimadzu made of model AUX220 high-precision weighing balance. MRR can be calculated using Eq. 1:

$$\mathrm{MRR} = \frac{(W_{\mathrm{f}} - W_{\mathrm{i}})}{t} \tag{1}$$

Fig. 1 **a** Drum winding setup. **b** GFRP laminates ply orientation. **c** GFRP laminate of the proposed system

Fig. 2 AWJM experimental setup

where W_i—initial weight of work piece in grams before cutting, W_f—final weight of work piece in grams after cutting, and t—period of trial (s).

3 Results and Discussion

3.1 Developed Model of AWJM for Composite

For the present experimentation work, Taguchi's L27 experimental design is used to design and analyze the significance of control parameters over output responses. JP, TS, and SOD were the selected control variables, and their corresponding level is given in Table 1.

The experimental results obtained after machining are shown in Table 2.

Contribution of control variable in affecting MRR, KA, and Ra are evaluated through ANOVA. The percentage of contribution of three independent variables reveals that the orientation has found to have effect on machining the GFRC and it is given in Tables 3, 4, and 5 for 0°, 45°, and 90°, respectively.

From the observation, it is clear to say that TS has a significant contribution in affecting the output variables. TS have 81.63% contribution on 0° FPO, whereas for 45° and 90°, it is 51.6% and 66.1%, respectively. Followed by TS, SOD shows

Table 1 Selected factor and levels

S. No.	Independent variables	Symbols	Levels			Units
			−1	0	1	
1	Jet pressure	JP	168	216	264	bar
2	Stand-off-distance	SOD	1	2	3	mm
3	Traverse speed	TS	20	30	40	mm/min

Table 2 Experimental observations of GFRC

Ex. No.	JP (bar)	SOD (mm)	TS (mm/min)	MRR (g/s)			KA (°)			Ra (μm)		
				0°	5°	90°	0°	5°	90°	0°	5°	90°
1	168	1	20	71.344	92.4080	104.98	0.001	0.026	0.025	5.12	5.642	5.754
2	168	1	30	107.014	125.9040	104.05	0.006	0.018	0.015	5.017	5.284	6.872
3	168	1	40	136.203	129.9227	143.83	0.016	0.074	0.024	5.732	6.913	8.571
4	168	2	20	84.616	82.9120	102.98	0.010	0.049	0.028	5.62	5.496	6.654
5	168	2	30	120.156	110.0760	103.23	0.012	0.029	0.025	5.478	5.932	8.216
6	168	2	40	170.704	159.5693	161.90	0.003	0.009	0.041	5.599	6.021	6.041
7	168	3	20	88.943	78.4133	101.79	0.012	0.067	0.030	5.95	6.631	7.958
8	168	3	30	124.128	141.8580	111.22	0.004	0.009	0.039	4.981	4.939	7.688
9	168	3	40	164.184	146.7093	155.01	0.008	0.049	0.036	5.667	6.786	7.837
10	216	1	20	72.099	102.9027	109.04	0.005	0.020	0.030	3.752	5.08	5.964
11	216	1	30	110.972	120.6020	107.05	0.008	0.006	0.014	5.031	6.385	7.133
12	216	1	40	144.861	146.1973	187.86	0.032	0.037	0.026	5.683	6.836	8.63
13	216	2	20	160.032	94.2707	106.06	0.018	0.030	0.030	4.831	5.735	6.963
14	216	2	30	125.118	107.9120	106.03	0.022	0.019	0.025	5.117	6.517	8.306
15	216	2	40	168.787	180.4933	219.04	0.005	0.012	0.045	5.307	5.328	6.987
16	216	3	20	98.465	85.5320	105.07	0.019	0.040	0.033	5.748	6.172	7.682
17	216	3	30	136.866	128.5440	113.31	0.004	0.003	0.037	5.021	5.551	6.399
18	216	3	40	160.032	157.1840	201.02	0.018	0.020	0.034	5.656	6.544	7.913
19	264	1	20	78.272	118.3827	118.02	0.009	0.002	0.040	3.558	5.587	6.434
20	264	1	30	110.264	146.7240	112.13	0.011	0.004	0.031	4.472	7.653	7.513

(continued)

Table 2 (continued)

Ex. No.	JP (bar)	SOD (mm)	TS (mm/min)	MRR (g/s)			KA (°)			Ra (µm)		
				0°	5°	90°	0°	5°	90°	0°	5°	90°
21	264	1	40	148.592	164.9253	205.74	0.042	0.018	0.062	4.799	6.022	8.56
22	264	2	20	82.207	101.4573	112.28	0.021	0.019	0.040	4.308	7.062	7.13
23	264	2	30	120.720	125.2540	110.96	0.022	0.005	0.035	4.877	7.151	8.9128
24	264	2	40	172.727	202.5173	254.62	0.003	0.004	0.079	4.743	4.132	7.523
25	264	3	20	97.768	090.3187	110.32	0.022	0.019	0.046	4.755	7.897	8.883
26	264	3	30	147.980	156.1020	112.13	0.007	0.001	0.056	4.184	4.836	6.542
27	264	3	40	155.568	180.8080	228.21	0.013	0.010	0.067	4.403	5.096	8.2905

Table 3 ANOVA of MRR, KA, and Ra for 0°

Parameters	DoF	SS	MS	Contribution Q (%)	*F*-value
JP (bar)	2	0.0017	0.0009	2.0911	0.8027
SOD (mm)	2	0.0112	0.0056	13.6786	5.2507
TS (mm/min)	2	0.0666	0.0333	81.6252	31.3327
Error	2	0.0021	0.001063	2.6051	
Total	8	0.0816		100.0000	

Table 4 NOVA of MRR, KA, and Ra for 45°

Parameters	DoF	SS	MS	Contribution Q (%)	*F*-value
JP (bar)	2	0.0215	0.0107	22.7164	56.6435
SOD (mm)	2	0.0239	0.0119	25.2829	63.0430
TS (mm/min)	2	0.0488	0.0244	51.5997	128.6640
Error	2	0.0004	0.000190	0.4010	
Total	8	0.0945		100.0000	

Table 5 ANOVA of MRR, KA, and Ra for 90°

Parameters	DoF	SS	MS	Contribution Q (%)	*F*-value
JP (bar)	2	0.0103	0.0051	31.9615	20.0089
SOD (mm)	2	0.0001	0.0000	0.2958	0.1852
TS (mm/min)	2	0.0213	0.0107	66.1453	41.4091
Error	2	0.0005	0.000257	1.5974	
Total	8	0.0322		100.0000	

the second highest contribution on 0° and 45° FPO. For 90° FPO, JP has 31.96% contribution in affecting the output performance. From the observation, it is clear to say that FPO has a considerable effect on the output responses.

Figure 1 shows the working of CP-ABE in our proposed system. First, the data user sends a request for a file to the cloud storage. Data owner checks whether the user is an authorized user or not, and based on that, owner may accept or reject the permissions. If the owner accepts the user requests permission, then the user will get the notification. A data user can only download the uploaded files or view the history of the downloaded file. If the user tries to download the uploaded file even though the permission is granted by the data owner, the user can access the file only after satisfying the attributes policy. The attributes have matched the file means it will start downloading. After that, the owner will get the notification of the user who downloaded the file.

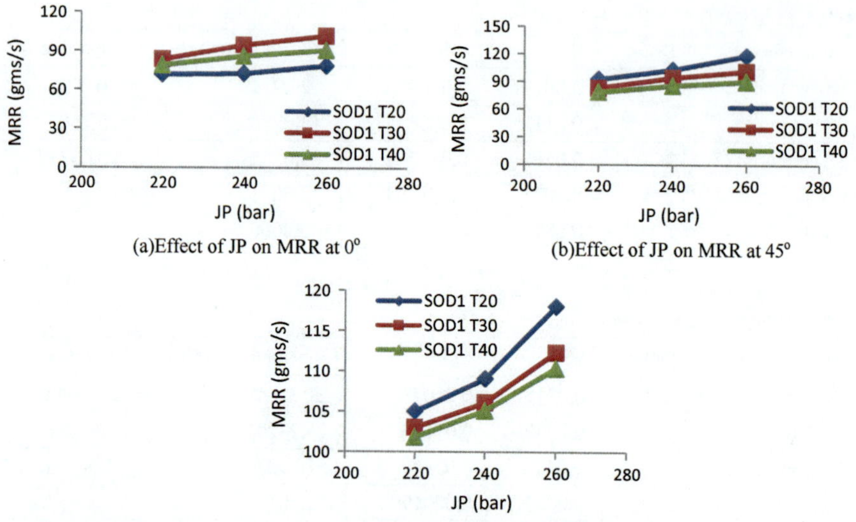

(a)Effect of JP on MRR at 0°

(b)Effect of JP on MRR at 45°

(C)Effect of JP on MRR at 90°

Fig. 3 Effect of JP on MRR

3.2 *Influence Study of JP on MRR, KA, and TS*

3.2.1 Machining Effect on MRR

In AWJM, materials are removed due to the impact of hard abrasive particles by the compressed water jet. Materials are removed due to cutting wear, plastic deformation, and by deformation wear that progressively leads to ductile erosion mechanism. MRR with varied input variables for AWJM on GFRC is shown in Fig. 3. Compared to 45° and 90°, 0° FPO shows variation in MRR when varying the TS.

For 45° and 90° composite, a linear variation in TS is noticed in MRR when varying TS, but at 0° FPO, the MRR is increasing from TS 20 mm/min to 30 mm/min, but at TS 40 mm/min, the MRR drops. This variation in MRR is due to the variation in FPO. At all cutting conditions, maximum MRR is noticed at high JP because of the increased kinetic energy in the erodent particle.

3.2.2 Machining Effect on KA

The KA formed during AWJM is calculated to understand the inclination formed in the cut surface, and the effect of JP on KA is shown in Fig. 4.

In the present work, variation in FPO of GFRC showed the variation in KA when varying the machining variables. KA has been increasing when the TS is increasing from 20 to 40 mm/min. At high TS, the erodent will not be able to penetrate the GFRC throughout the thickness because of the rapid passing of nozzle. This produces the

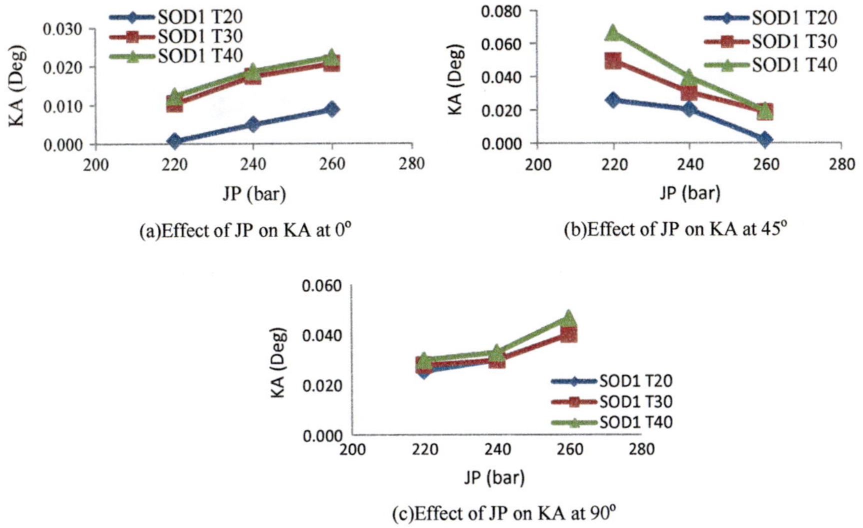

Fig. 4 Effect of JP on KA

shorter kerf at the exit surface and higher kerf at the top cut surface. The shorter kerf at the exit surface is due to reduction of machining period. This tends to increase KA with increased TS. Increase in JP for samples with 0° and 90° FPO, increases the KA, but for the samples with 45° FPO, KA decreases with the increase in JP. At higher JP, the kinetic energy of the water will increase the turbulence, and it creates a water hammering effect. Higher the JP increases the width of the water beam irrespective of SOD. This action tends to have wider slot in the top cut surface and reduced bottom slot width. This leads to the increment in KA. Observation at 45°, FPO shows an increase in JP that reduces KA, and this is because of the formation of narrow cut.

3.2.3 Machining Effect on Ra

From Fig. 5, it is clear that the surface roughness is mostly influenced by TS irrespective of JP and SOD. Surface profile roughness varies with variation in TS. In all the three FPOs, Ra is found to be maximum at high level TS.

At high TS, the nozzle moves more rapidly which forms striation on the cut surface. Also due to this rapid movement the erodent reduces the depth of penetration on the surface. The rapid movement of the nozzle creates brittle fracture on the glass fiber that increases the Ra. Change in levels of JP shows the variation in Ra for 0° and 45° FPOs. Since the nozzle movement is parallel to the fibre orientation that leads to have excess removal of materials. Also increase in JP accelerates the abrasives that progressively lead to have smooth profile roughness. In general, increased JP improves the Ra due to the formation no striations in the cut surface.

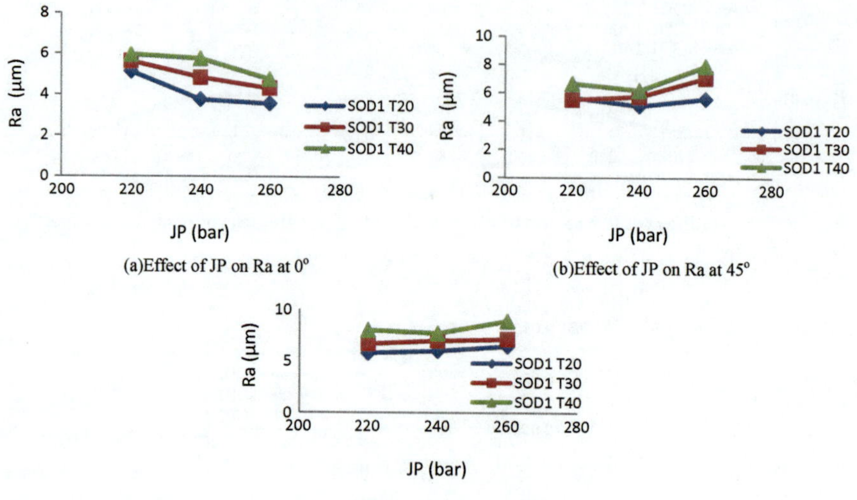

(a)Effect of JP on Ra at 0°

(b)Effect of JP on Ra at 45°

(c)Effect of JP on Ra at 90°

Fig. 5 Effect of JP on Ra

3.3 *Influence Study of SOD on MRR, KA, and TS*

3.3.1 Machining Effect on MRR

The effect of SOD on FPO of GFRC is shown in Fig. 6. GFRC with 45° FPO exhibits similar behavior as other materials when subjected to AWJM. MRR for FPOs at 90° shows no change irrespective of TS and JP.

However, a steady raise is observed, when SOD increases to 2 mm, and the changes noticed are due to the increment in the width of the water beam. The energy transmitted to the abrasive particle through high-pressure water possesses additional energy than the required amount that progressively leads to the increase in MRR. At 1 mm and 2 mm of SOD, for 90° FPO irrespective of TS, MRR has no significant change.

3.3.2 Machining Effect on KA

The influence of the three processing parameters of AWJM on KA is shown in Fig. 7. Higher the TS, larger the KA for AWJM. At 90° FPO, a steady raise of KA is observed at the machining condition of JP 260 bar and TS 30 mm/min. From the observation, it can be stated that the machining sequence of AWJM for 0°, 45°, and 90° FPO, JP and SOD play a vital role on the surface profile. Almost a vertical cut, i.e., KA is equal to zero, is observed at JP of 220 bar and SOD 1 mm.

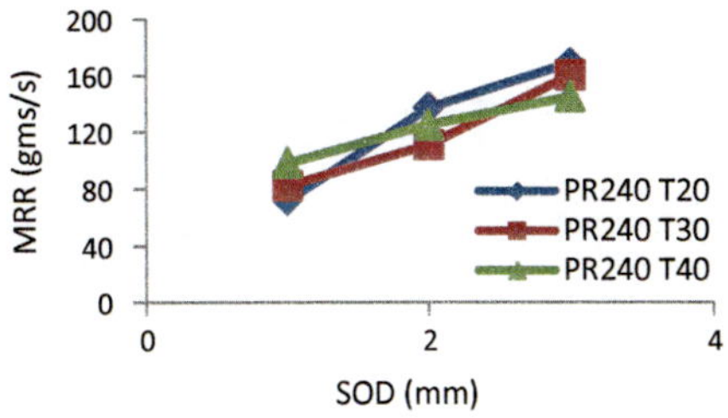

(a)Effect of SOD on MRR at 0°

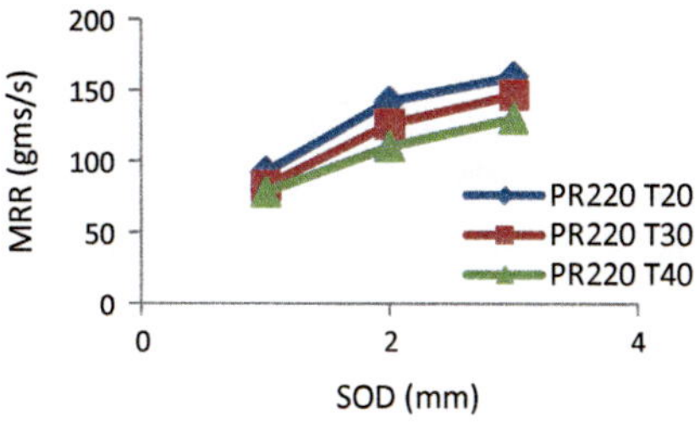

(b)Effect of SOD on MRR at 45°

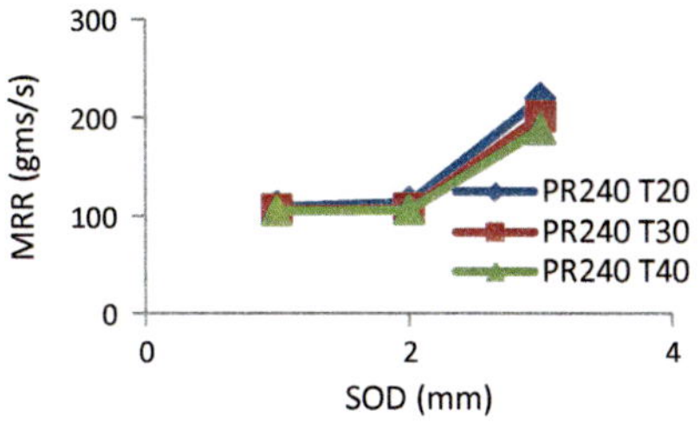

(c)Effect of SOD on MRR at 90°

Fig. 6 Effect of SOD on MRR

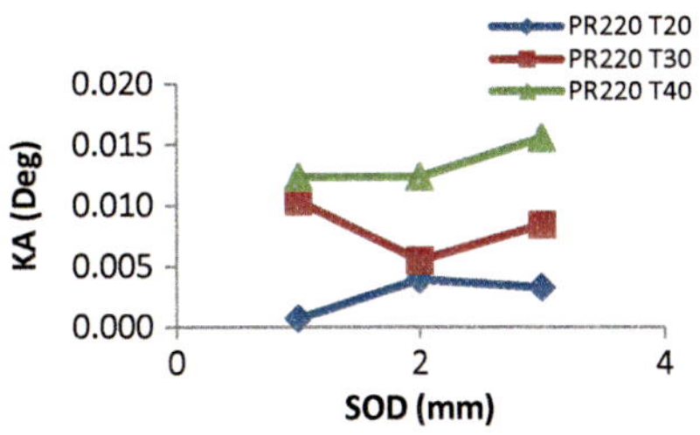

(a) Effect of SOD on KA at 0°

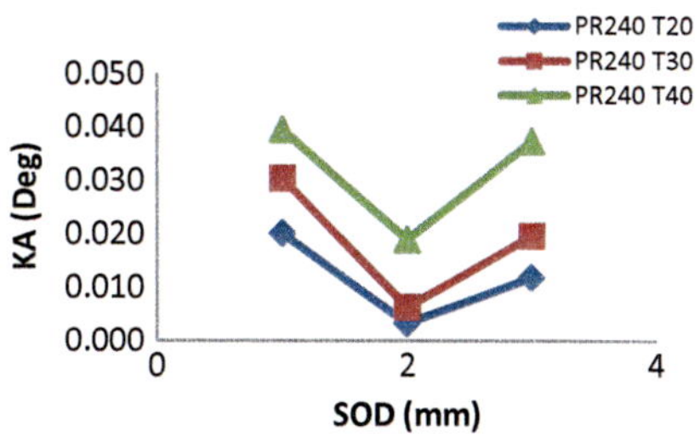

(b) Effect of SOD on KA at 45°

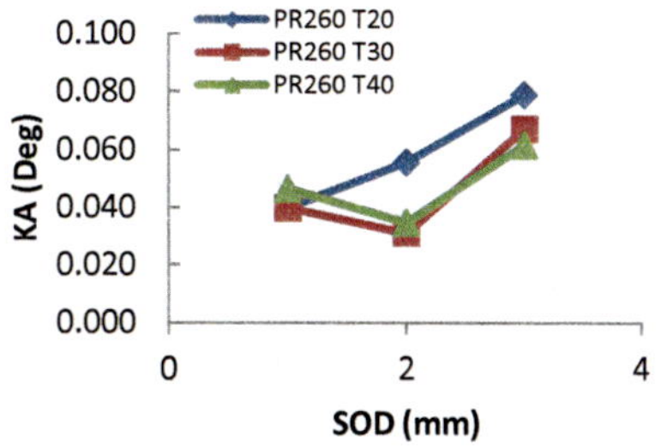

(c) Effect of SOD on KA at 90°

Fig. 7 Effect of SOD on KA

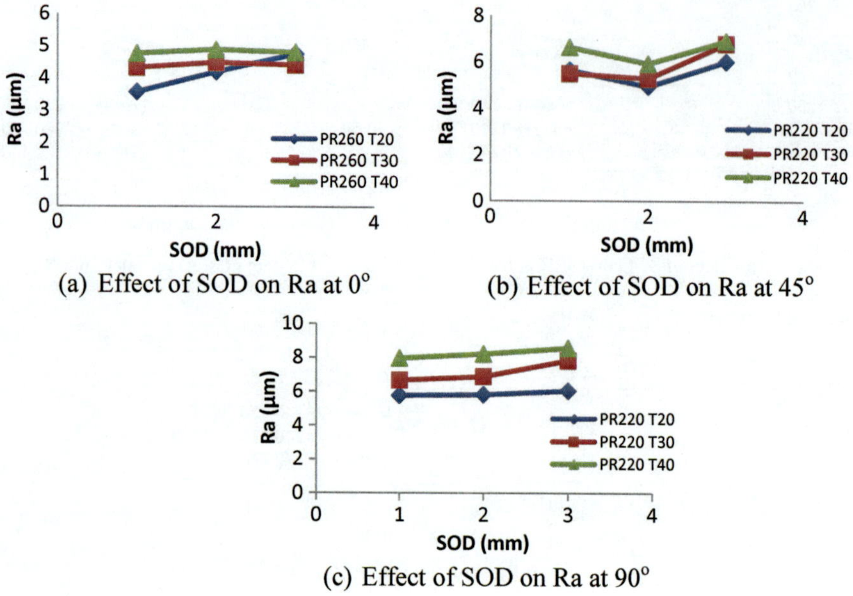

Fig. 8 Effect of SOD on Ra

3.3.3 Machining Effect on Ra

AWJM on various FPOs of GFRC with epoxy resin is shown in Fig. 8. Increase in TS significantly increases the Ra of the composite. All the observations through the image obtained for Ra shows that SOD has no significant effect on the FPO.

3.4 *Influence Study of TS on MRR, KA, and TS*

3.4.1 Machining Effect on MRR

The effect of TS on MRR for different FPO's is shown in Fig. 9. The image obtained at 45° FPO shows the similar behavior as other materials when subjected to AWJM. But at 0° and 90° FPO, the effect of SOD is clearly visible. Particularly at 90° orientation with SOD of 2 and 3 mm, it shows no significant effect irrespective of TS.

3.4.2 Machining Effect on KA

The machining studies on KA at different FPOs of GFRC are shown in Fig. 10. It is observed that the increase in TS with respect to KA tends to increase irrespective of JP and SOD. But at 90° orientation, KA formed on different operating conditions of

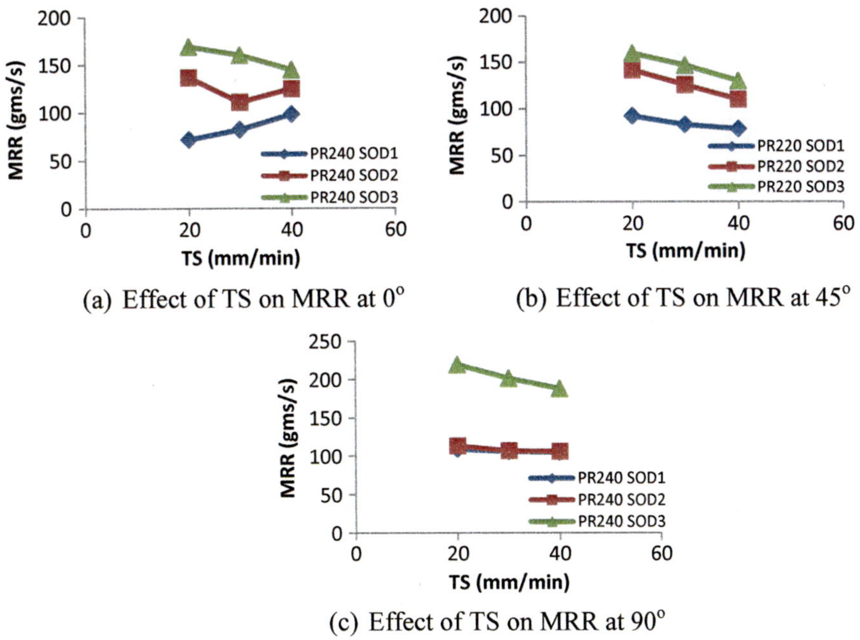

Fig. 9 Effect of TS on MRR

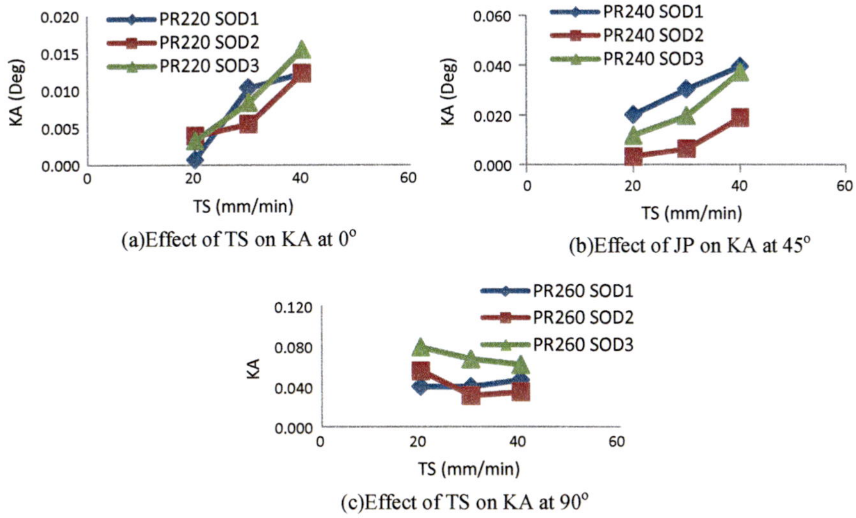

Fig. 10 Effect of TS on KA

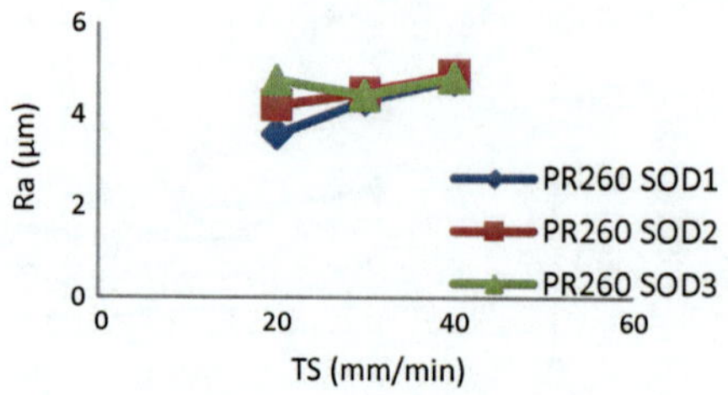

(a)Effect of TS on Ra at 0°

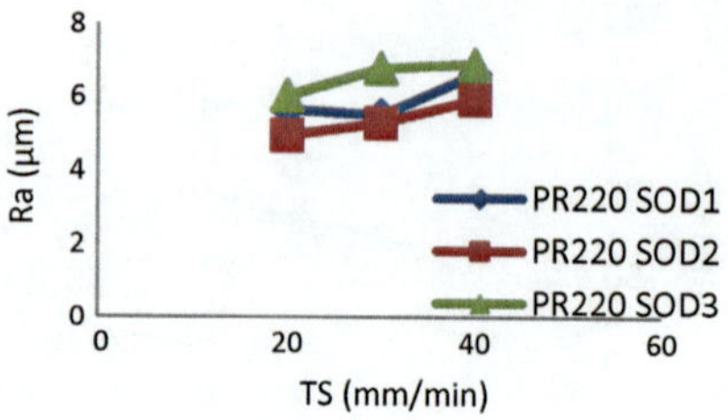

(b)Effect of TS on Ra at 45°

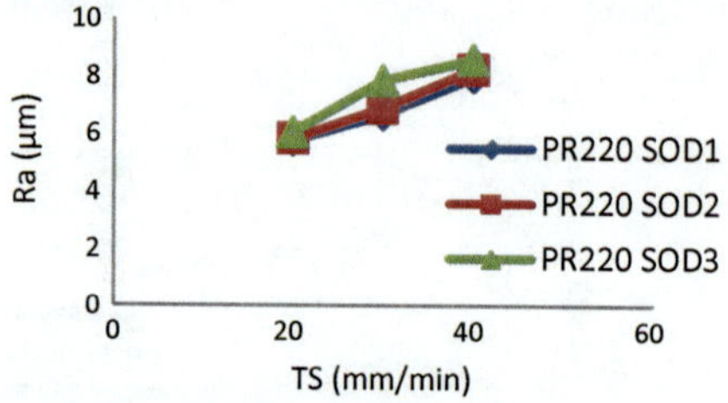

(c)Effect of TS on Ra at 90°

Fig. 11 Effect of TS on Ra

TS is found to be the same. Increase in JP, irrespective of TS is found to have low KA. The highly accelerated abrasive particles clearly cut the surface with less tear mark over the surface. The high-pressure water act as polishing agent, and it smooths the fracture surface created by the GFRC and hence reduces the KA.

3.4.3 Machining Effect on Ra

The influence of TS on Ra for different FPOs of GFRC is shown in Fig. 11. The increase in TS of 0° and 45° FPO shows a negligible amount of change in surface profile roughness. However, there is sudden rise in Ra for 90° FPO with increase in TS, and this is due to the irregular breakage of glass fibers on the machined surface. But, the value of Ra and KA tends to decrease at high operating condition of JP.

3.5 Morphology of Specimens After AWJM

The surface morphology of different ply-oriented GFRC is shown in Fig. 12a–d. It is noticed that the high-accelerated abrasives are hard enough to machine the composites that leads to the formation of striation and results in plastic deformation of the surface. From Fig. 12a, the direction of the motion of the jet is parallel to the fiber from the ply orientation which results in the large amount of material removal. From Fig. 12b, it is well evident that there is fiber pullout and greater disturbance in

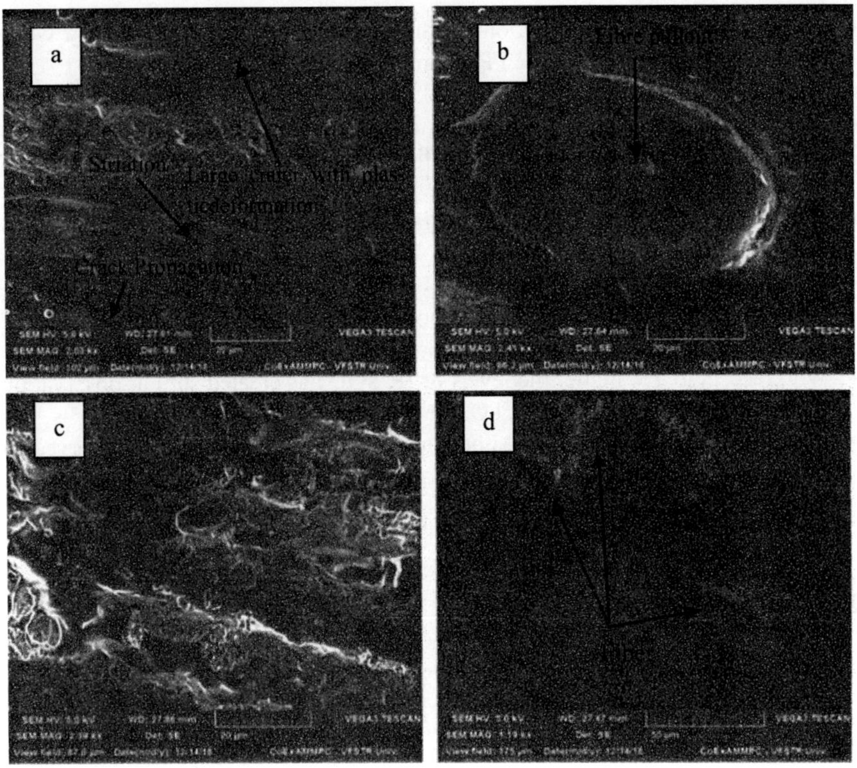

Fig. 12 **a–d** SEM image of GFRC

the plies. This happens due to the inability of the jet to cut the component of a fiber embedded in the matrix. The jet will be deflected and will erode the softer component, i.e., the matrix instead of fiber. This results in the appearance of matrix wash out and protruding fibers because of the load bearing capacity of fiber. Figure 12c shows higher amount of fiber pull out with plowing effect which is the basic machining characteristics of water jet machine in the rough cut zone. It is evident that from Fig. 12d, the energy acquired from the water through abrasives is hard enough to break the glass fiber. However, due to the matrix bonding, the glass fiber has the ability to withstand at high pressure, and this image is obtained at 0°/90° orientation.

4 Conclusion

Glass fiber reinforced epoxy composites with different ply orientations were fabricated using drum winding process. The plate that has the thickness of 7 mm is machined by AWJM with varied parameters, and the observations are as follows:

- TS have significant contribution on the output parameters, particularly while cutting the composite with the FPO parallel to the movement of nozzle.
- The contribution of SOD is superior at 0° and 45° FPO and JP for 90° orientations.
- Material removal happens in the form of ductile erosion mechanism. The fiber with 45° orientation shows the diversified observations compared to 0° and 95° orientations.
- JP has superior effect on 0° FPO for MRR which produces least KA at 45° FPO and have more influence on 0° and 45° for Ra.
- With lesser SOD, MRR and KA have been observed to have less value, and the change in SOD has less significance on Ra, when SOD interaction with TS has found to get influenced with output responses.
- With respect to TS, higher the levels of JP, SOD has created similar observation for MRR and KA, but for 90° FPO, the influence of TS is high on Ra.
- Plastic deformation surface with fiber break, particle pull out has been observed on 0°/90° ply-oriented composite rather than 0° and 90° ply orientation, respectively.

References

1. Aramide FO, Atanda PO, Olorunniwo OO (2012) Mechanical properties of a polyester fibre glass composite. Int J Compos Mater 2:147–151
2. Erden S, Sever K, Seki Y, Sarikanat M (2010) Enhancement of the mechanical properties of glass/polyester composites via matrix modification glass/polyester composite siloxane matrix modification. Fibers Polym 11:732–737
3. Arikan V, Sayman O (2015) Comparative study on repeated impact response of E-glass fiber reinforced polypropylene & epoxy matrix composites. Compos Part B Eng 83:1–6
4. Nikforooz M, Montesano J, Golzar M, Shokrieh MM (2018) Assessment of the thermomechanical performance of continuous glass fiber-reinforced thermoplastic laminates. Polym Testing 67:457–467
5. Lopez FA, Martín MI, Alguacil FJ, Rincón JM, Centeno TA, Romero M (2012) Thermolysis of fibreglass polyester composite and reutilisation of the glass fibre residue to obtain a glass–ceramic material. J Anal Appl Pyrol 93:104–112
6. Ellyin F, Rohrbacher C (2003) The influence of aqueous environment, temperature and cyclic loading on glass-fibre/epoxy composite laminates. J Reinf Plast Compos 22:615–636
7. Wu Z, Li J, Huang C, Li L (2015) Effect of matrix modification on interlaminar shear strength of glass fibre reinforced epoxy composites at cryogenic temperature. Phys Procedia 67:1068–1073
8. Kim JH, Kwon DJ, Shin PS, Beak YM, Park HS, DeVries KL, Park JM (2018) Interfacial properties and permeability of three patterned glass fiber/epoxy composites by VARTM. Compos B Eng 148:61–67
9. Alam S, Habib F, Irfan M, Iqbal W, Khalid K (2010) Effect of orientation of glass fiber on mechanical properties of GRP composites. J Chem Soc Pak 32:265–269
10. Goh GD, Dikshit V, Nagalingam AP, Goh GL, Agarwala S, Sing SL, Wei J, Yeon WY (2018) Characterization of mechanical properties and fracture mode of additively manufactured carbon fiber and glass fiber reinforced thermoplastics. Mater Des 137:79–89
11. Purohit R, Sahu P, Rana RS, Parashar V, Sharma S (2017) Analysis of mechanical properties of fiber glass-epoxy-fly ash composites. Mater Today Proc 4:3102–3109
12. Gao X, Gillespie JW Jr, Jensen RE, Li W, Haque BG, McKnight SH (2015) Effect of fiber surface texture on the mechanical properties of glass fiber reinforced epoxy composite. Compos Part A Appl Sci Manuf 74:10–17

13. Zulfli NM, Bakar AA, Chow WS (2013) Mechanical and water absorption behaviors of carbon nanotube reinforced epoxy/glass fiber laminates. J Reinf Plast Compos 32:1715–1721
14. Gokuldass R, Ramesh R (2018) Mechanical strength behavior of hybrid composites tailored by glass/kevlar fibre-reinforced in nano-silica and micro-rubber blended epoxy. Silicon 1:9
15. Dinesh T, Kadirvel A, Vincent A (2018) Effect of silane modified E-glass fibre/iron (III) oxide reinforcements on UP blended epoxy resin hybrid composite. Silicon 1:2
16. Aich U, Banerjee S, Bandyopadhyay A, Das PK (2014) Abrasive water jet cutting of borosilicate glass. Procedia Mater Sci 6:775–785
17. Balamurugan K, Uthayakumar M, Sankar S, Hareesh US, Warrier KGK (2018) Mathematical modelling on multiple variables in machining LaPO4/Y2O3 composite by abrasive waterjet. Int J Mach Mach Mater 19:426–439
18. Doreswamy D, Shivamurthy B, Anjaiah D, Sharma NY (2015) An investigation of abrasive water jet machining on graphite/glass/epoxy composite. Int J Manuf Eng. https://doi.org/10.1155/2015/627218

Evaluation of Mechanical and Morphological Properties of Hybrid Reinforced Polymer Composites

N. Babu, C. S. Rathnasabapathy, and N. Karunakaran

Abstract This paper deals with development and study of a composite made of natural fibers, namely banana and flax fibers. A polymer matrix composite (polyester resin) using banana and flax fiber as reinforcement was developed. The composites were prepared with different volume ratio of banana and flax fibers. The fabrication process composite is done using a compression molding machine. The composite thus developed is subjected to tensile, impact, flexural loads testing using universal testing machine (UTM) to determine the mechanical properties of this hybrid composite. The composite is observed scanning electron microscope (SEM) to deduce its morphology. Finally, the finite element analysis is done using ANSYS software. The present study indicates that the hybrid composite is found to be stronger when compared to single fiber composites.

Keywords Banana · Flax · SEM · ANSYS · Polyester · UTM

1 Introduction

Natural fiber polymer composite consists of a polymer matrix reinforced with high strength natural fibers rendering them composites are strong and lightweight. The length of fiber, matrix ratio and fiber arrangement influence the physical and mechanical properties of these composites. The strength and the being light make these natural fibers a perfect choice for automotive applications. Composite materials, reinforced with combination two or more types of composite fibers, are called hybrid composites. These hybrid composite materials are found to have a wide range of engineering application in the present scenario. Hybrid composites are attributed by

N. Babu (✉)
ACGCET, Karaikudi, Tamilnadu, India
e-mail: babu.manu11@gmail.com

C. S. Rathnasabapathy · N. Karunakaran
Annamalai University, Tamilnadu, India

A. Arockiarajan et al. (eds.), *Advances in Industrial Automation and Smart Manufacturing*, Lecture Notes in Mechanical Engineering,
https://doi.org/10.1007/978-981-15-4739-3_68

individual component in which there exist balance between their inherent advantages and disadvantages [1–5].

The objective of this study is to determine the mechanical and morphological properties. It is also proposed to carry out a numerical analysis of hybrid composite consisting banana and flax fibers. The analyses are done to evaluate the fitness of the composites in potential areas of application [6–10].

2 Experimental Details

The composite plates are fabricated by using natural fibers like banana and flax fibers with reinforced polymer. The thus fabricated composites are tested for the mechanical properties, morphological properties along with finite element analysis [11, 12].

2.1 Fiber Used

2.1.1 Banana Fiber

Banana plant is a perennial plant which grows in regions with tropical climate. It has leaf sheaths which form the pseudo-stem. Banana, apart from being a major food crop which has prominent health benefits, provides natural fibers. It grows easily and it most commonly found in hot tropical climates. The fibers exhibit excellent mechanical properties, due to high specific strength, comparatively light and are bio-degradable. This makes them a choice of material in variety of applications including automobiles.

Chemical composition of banana fiber:

See Tables 1 and 2.

Table 1 Chemical Composition of Banana Fiber

Fiber	Cellulose (%)	Hemicellulose (%)	Lignin (%)	Moisture (%)
Banana fiber	62–64	19	5	10–11.5

Table 2 Mechanical properties of banana fiber

Fiber	Flexural modulus (GPa)	Tensile strength (MPa)	Elongation at break (%)	Density (g/cm^3)
Banana fiber	4	356	2.6	1.35

2.1.2 Flax Fiber

Banana fiber finds its application in apparels, furniture, paper making, in currency notes. Banana fiber reinforced polypropylenes composites are used for making under floor protection panels in luxurious cars, used in buildings boards and fire resistance boards. Flax fibers are up to 90 cm long and average about 12–16 microns in diameters. The chemical and mechanical properties of flax fiber are given below table.

2.2 Polyester Resin

Polyester resin is the base material for thermoset polyester matrix. It is an unsaturated resin that contains C = C double bonds. It is usually prepared by reaction of glycol and an unsaturated dibasic acid. In order to modify the chemical structure between the cross-linking sites saturated acids, such as isophthalic acid or orthophthalic acid, are also added. Polyester resins can be formulated in a variety ranging from soft and flexible to hard and brittle. The advantages of polyester resin are that they are fast curing, low viscous and cost effective. Though these properties make the release from the molds easier, there is always a difference in shrinkage between the resin and the fibers resulting in uneven depressions (called sink marks) on the molded surface. The principal disadvantage of polyesters over epoxies is their high volumetric shrinkage.

2.3 Fabrication Procedure

In order to create hybrid composite plates, banana fiber and flax were taken in three different volume ratios. The volume of the resin was maintained at 70%. A combination of 20% flax with 10% banana fiber makes the plate 1, 15% flax with 15% banana fiber makes the plate 2 and 10% flax with 20% banana fiber makes the plate 3.

In this study, the composite plates are fabricated using compression molding machines. The molding material is generally preheated and placed in an open, heated mold cavity. Top force is applied to close the mold and the material is pressurized to make contact with the entire surface of the mold. Till the curing of the mold temperature and pressure is maintained.

2.4 Specimen Preparation

The specimen from plate 1 (flax 20%, banana fiber 10% and 70% resin), plate 2 (flax 15%, banana fiber 15% and 70% resin and plate 3 (flax10%, banana20% and 70% resin) and subjected to testing for mechanical and morphological properties (Figs. 1 and 2).

Fig. 1 Tensile specimen 1, 2 and 3

Fig. 2 Flexural specimen 1, 2 and 3

3 Testing of Composite

3.1 Tensile Testing

ASTM std 63810 is adopted to carry out the tensile testing of the composite specimen. The sample is tested using a universal testing machine with maximum load rating 10 kl. The three samples are held in the grip and the deflection corresponding to the load applied is recorded. The break load and the ultimate tensile strength are recorded by applying load until the specimen breaks. A graph is plotted for tensile stress and strain (Figs. 3 and 4).

3.2 Flexural Testing

The flexural test is done as per the ASTM D790 std. The tests are done by using universal testing machine applying loads at the middle of the beam till the specimen bends and eventually fractures. A graph was drawn between the stress and strain for obtaining a projectile.

Fig. 3 Impact specimen 1, 2 and 3

3.3 Impact Testing

The impact test is done as prescribed by the ASTM D790 std. The specimen is impacted by a pendulum arrangement which fractures it on impact. The specimen when breaks absorbs energy and the same is noted.

4 Results and Discussion

4.1 Results of Tensile Testing

The specimens 1, 2 and 3 are tested for tensile properties as discussed in the previous chapter. The mechanical properties thus obtained upon testing are tabled as follows.

It is found on comparison that the specimen 1 has been high peak value when compare to other two specimens. Finally, the specimen 1 has more brittle and good properties compare other two specimens (Figs. 5 and 6).

4.2 Results of Flexural Testing

The specimens 1, 2 and 3 are tested for flexural properties and the values are shown on the following table (Figs. 7 and 8; Tables 3 and 4).

It is found that among the three specimens, specimen 1 exhibits the highest value among the other two. It is inferred that specimen 1 is stronger and less brittle compared to the other two specimens.

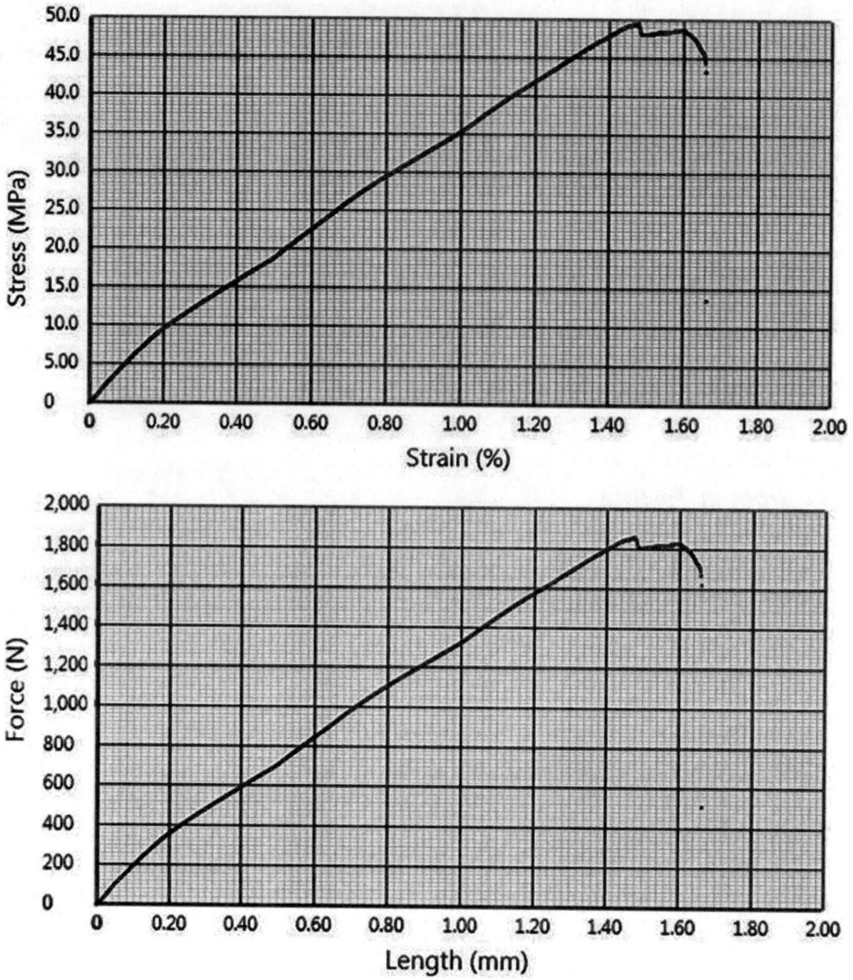

Fig. 4 Flax 20%, Banana 10%—Tensile results

4.3 *Results of the Impact Testing*

The energy absorbed by the specimen during impact equals the energy loss during the impact and it is found that the composite specimen 1 absorbs more energy. Hence, the specimen 1 is found to be stronger while the other two remain less strong. The values observed during the impact strength are tabulated below (Figs. 9 and 10; Tables 5 and 6).

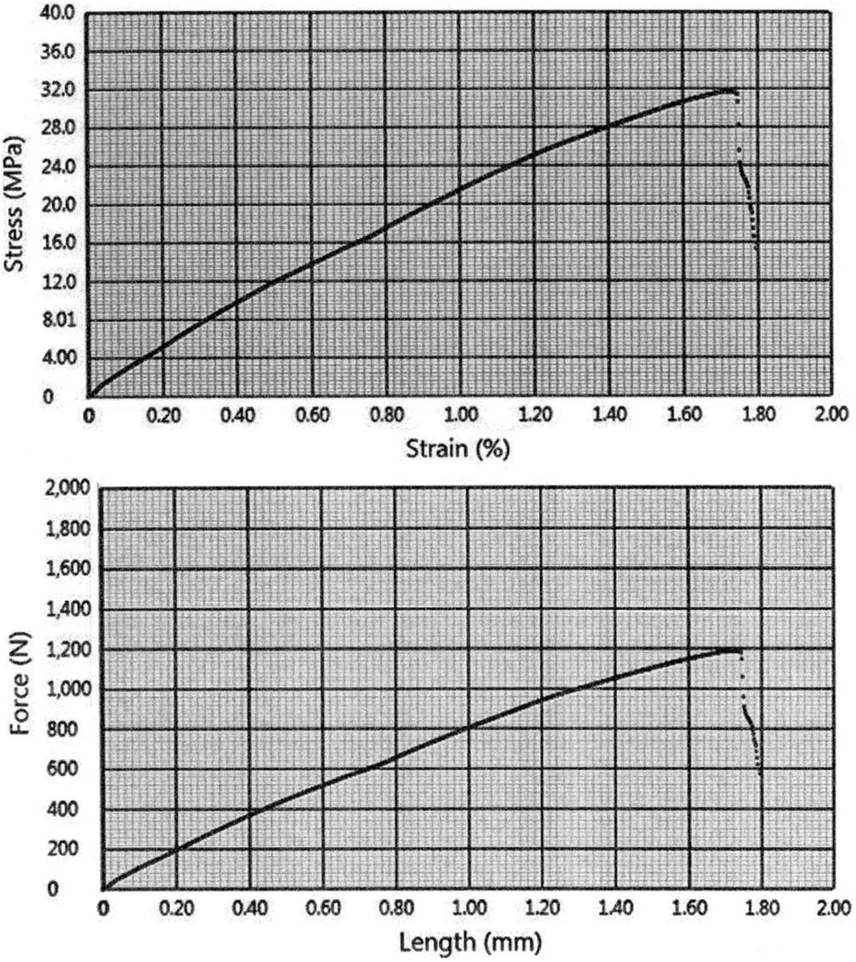

Fig. 5 Specimen 2 Tensile results

4.4 Morphological Properties

The specimen is studied under the scanning electron microscope (SEM) for morphological analysis. The specimens prepared are subjected to test for defects and adhesion bonding between resin and fibers (Figs. 11 and 12; Table 7).

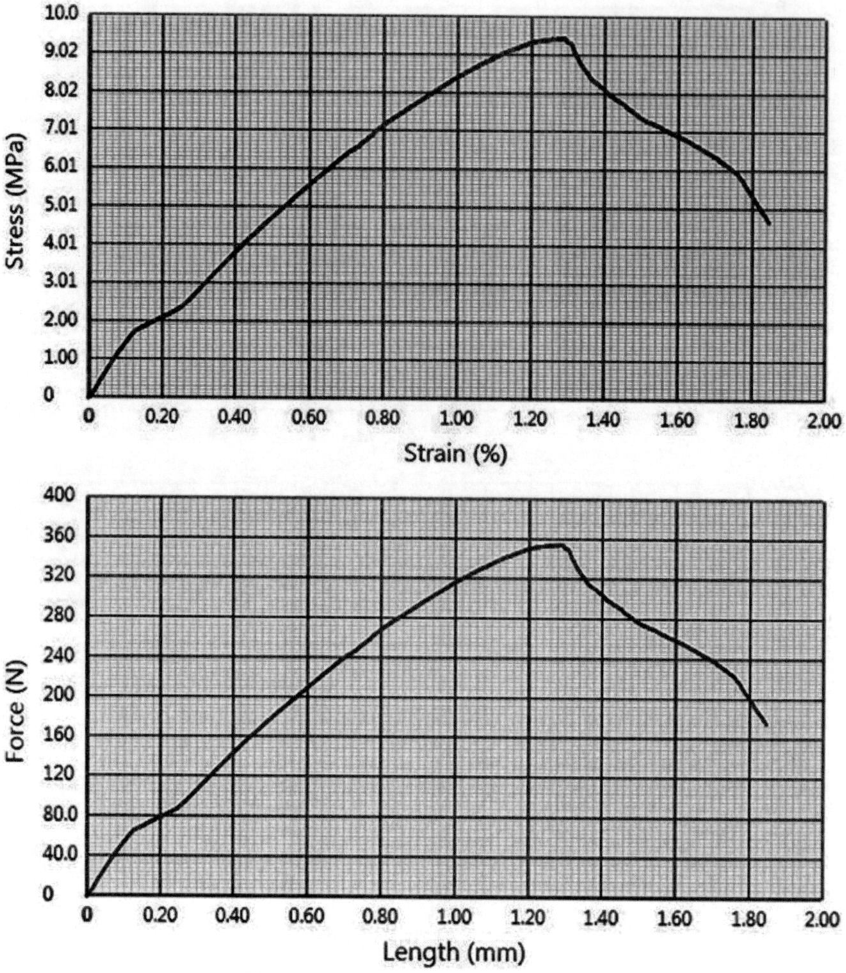

Fig. 6 Specimen 3 Tensile results

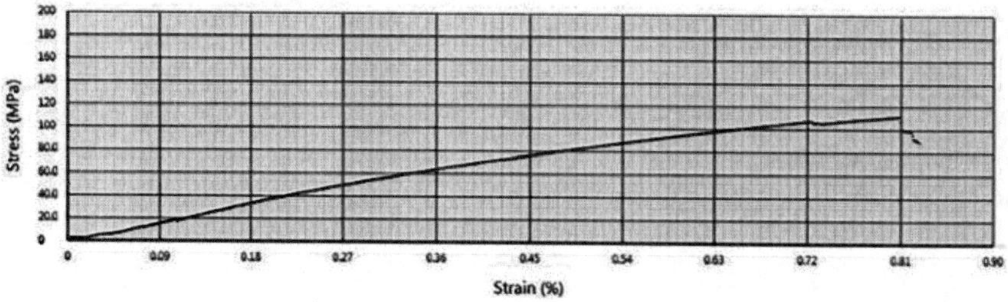

Fig. 7 Specimen 1 flexural results

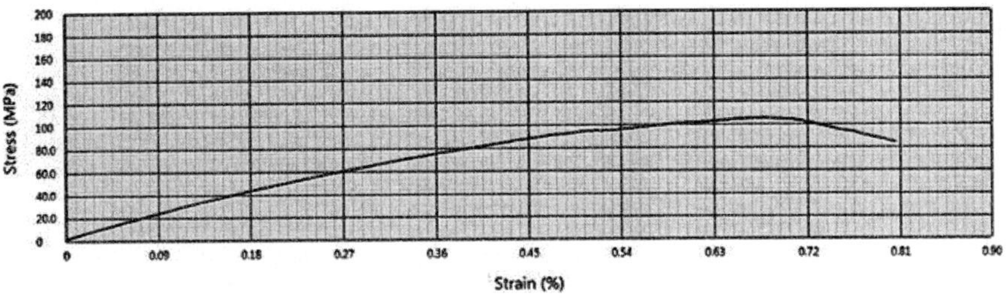

Fig. 8 Specimen 2 flexural results

Table 3 Chemical composition of flax fiber

Fiber	Cellulose (%)	Hemicellulose (%)	Lignin (%)	Pectin (%)
Flax fiber	81	14	3	4

Table 4 Mechanical properties of flax fiber

Fiber	Flexural modulus (GPa)	Tensile strength (MPa)	Elongation at break (%)	Density (g/cm^3)
Flax fiber	50–70	400	1.2–1.6	1.4–1.5

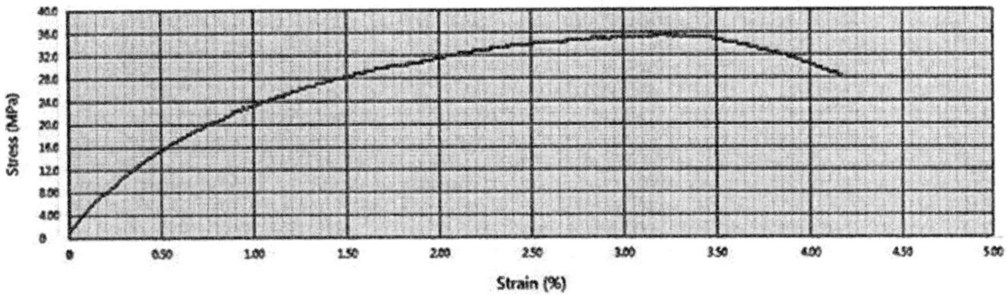

Fig. 9 Specimen 3 flexural results

5 Finite Element Analysis

The finite element analysis is done using ANSYS software. The stress levels experienced across the surface by the composite plate are analyzed. The analysis shows that a varied pattern that ranges from higher stress levels at the test point to lower levels at a point farther away from it. The mesh model is presented in Fig. 13.

Figure 14 shows high stress level indicated by the red zone at the point where the load is applied. Lower stress is represented by the yellow and green zones. The blue zone indicates that the composite plate is free from stress.

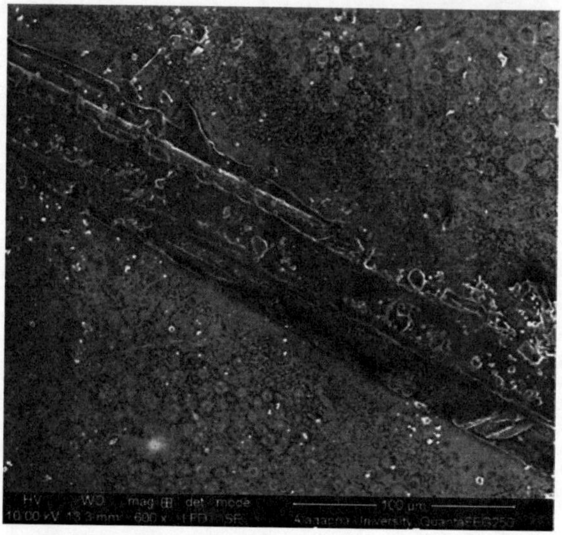

Fig. 10 SEM image of specimen 1

Table 5 Tensile results

Composite specimen	Ultimate force (N)	Ultimate stress (MPa)	Break distance (mm)
Specimen 1	1860	49.6	1.66
Specimen 2	1190	31.7	1.75
Specimen 3	354	9.45	1.85

Table 6 Flexural Results

Composite Specimen	Ultimate Force (N)	Flexural strength (MPa)
Specimen 1	83.67	111.56
Specimen 2	79.33	105.78
Specimen 3	26.67	35.56

6 Conclusions

This paper deals in detail a study of a hybrid natural composite fabricated using banana and flax fiber by compression molding method. It is concluded from the tensile testing that the specimen 1, which withstood a peak load of 49.6 MPa, outperforms the other two specimens 2 & 3 withstanding loads of 31.7 MPa and 9.45 MPa, respectively. It is also evident from the flexural testing that the specimen 1 exhibited a satisfactory performance by withstanding a peak load of 111.56 MPa over the other two specimens whose peak values are 105.78 MPa and 35.55 MPa, respectively. Composite specimen 1 has also proved to possess excellent ability to absorb impact force. This is clear from the results obtained from impact testing. Specimen 1 has

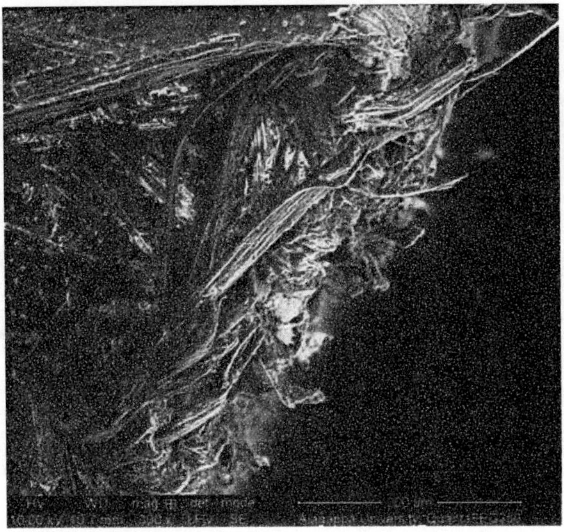

Fig. 11 SEM image of specimen 2

Fig. 12 SEM image of specimen 3

Table 7 Impact results

Composite specimen	Pendulum (J)	Break (J)	Strength-1 (kJ/m^2)	Strength-2 (kJ/m^2)
Specimen 1	27.506	7.0787	188.767	566.300
Specimen 2	27.506	3.6228	96.6081	289.824
Specimen 3	27.506	2.8278	75.4083	226.225

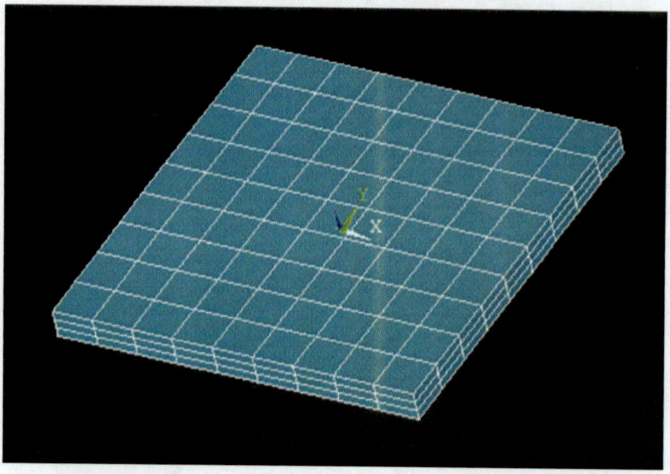

Fig. 13 Meshing diagram

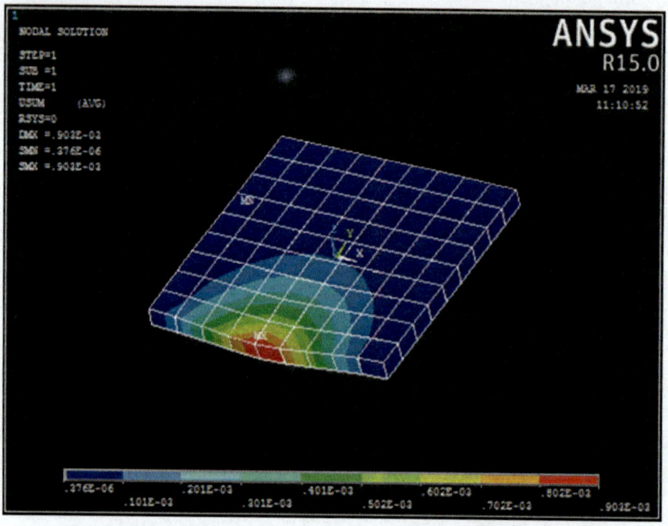

Fig. 14 ANSYS result

the capability to absorb a peak 7.0787 J while the other two specimens 2 and 3 have the capability to absorb attained 3.6228 J and 2.8278 J, respectively. From the SEM results, it is seen few voids are present in the composite specimen. The fiber strength is very much associated with the arrangement and orientation of fibers. There is clear picture bonding between resin and fiber. From the finite element analysis, the amount of stress experienced across the surface of the plate at different places is analyzed by using ANSYS software.

References

1. Rahman M, Das S, Hasan M (2018) Mechanical properties of chemically treated banana and pineapple leaf fiber reinforced hybrid polypropylene composite. In: Advances in materials and processing technology. https://doi.org/10.1080/2374068x.2018.1468972
2. Bakri MKB, Jayamani E, Hamdan S (2016) Processing and characterization of banana fiber/epoxy composites: effect of alkaline treatment
3. Ronald Aseer J, Sankaranarayanasamy K, Jayabalan P, Natarajan R, Priya Dasan K (2013) Morphological, physical and thermal properties of chemically treated banana fiber. J Natural Fibers 10(4):365–380. https://doi.org/10.1080/15440478.2013.824848
4. Srinivasan VS, Rajrndra Boopathi S, Sangeetha D, Vijaya Ramnath B. Evaluation of mechanical and thermal properties of banana and flax based natural fiber composite
5. Ortega Z, Monzon MD, Benitez AN, Kearns M, McCourt M, Hornsby PR (2014) Banana and abaca fiber reinforced plastic composites obtained by rotational molding process, materials and manufacturing processes. 28(8):879/883. https://doi.org/10.1080/10426914.2013.792431
6. Huner U (2017) Effect of chemical surface treatment on flax reinforced epoxy composite. J Natural Fibers. https://doi.org/10.1080/15440478.2017.1369207
7. Chattopadhyay SN, Pan NC, Roy AN, Samanta KK (2018) Pretreatment of jute and banana fibre its effect on blended yarn and fabric. J Natural Fibers. 1080/15440478, 2018, 1469450
8. Saba N, Jawaid M. Epoxy resin based hybrid polymer composite
9. Petel H, Ren G, Hogg PJ, Peijs T (2015) Hemp fibre as alternative to glass fibre in sheet moulding compound. Part-2 impact properties. https://doi.org/10.1179/1743289815y.0000000030
10. Samal SK, Mohanty S, Nayak SK. Banana/glass fiber reinforced polypropylene hybrid composites: fabrication and performance evaluation, polymer-plastic technology and engineering. 48(4):397–414. https://doi.org/10.1080/03602550902725407
11. Pugila D, Biagiotti J, Kenny LM (2005) A review on natural fibre based composite part 2; Application of natural reinforcement in composite material for automobile industry. J Natural Fibers 1:23–65
12. Mohammed L, Ansari M, Pua G et al (2015) A review on natural fiber reinforced polymer composite and its applictions. Int J Polym Sci. Article ID 243947

Experimental Investigation and Characterization of HDPE & LDPE Polymer Composites

S. Ganesan, J. Hemanandh, K. S. Sridhar Raja, and M. Purusothaman

Abstract This paper deals with the preparation and characterization of high- and low-density polyethylene (HDPE & LDPE) composites by mixing with sawdust generated as wastes from wood industries thereby converting wastes to wealthy materials. HDPE & LDPE granules are mixed with sawdust in different formulations by using twin screw extruder. In this extrusion process, granules and sawdust get mixed and extruded in the form of pellets. These pellets are injection molded to get the desired cross-section of specimen. After specimen preparation, they are tested for mechanical properties like tensile, hardness, flexural resistance, and water absorption. The testing results are compared with the properties of existing plastics and virgin HDPE, LDPE to know whether the properties have been increased or decreased and to reveal the application based on mechanical strength of the specimen.

Keywords DPE · LDPE · Sawdust · Pellets

List of Nomenclature

nm	Nanometer
mm	Millimeter
%	Percentage
°C	DegreeCelsius
K	Kelvin
Mpa	Megapascal
N	Newton

S. Ganesan (✉) · J. Hemanandh · K. S. Sridhar Raja · M. Purusothaman
School of Mechanical Engineering, Sathyabama Institute of Science and Technology, Chennai, India
e-mail: gansuma@gmail.com

A. Arockiarajan et al. (eds.), *Advances in Industrial Automation and Smart Manufacturing*, Lecture Notes in Mechanical Engineering,
https://doi.org/10.1007/978-981-15-4739-3_69

1 Introduction

Generation of solid and liquid wastes during manufacturing processes is unavoidable in major industries. Wood sawdust is being considerably used in reinforcing polymer both from a scientific and commercial point of view over the last decades. The various waste generated are classified as given in Table 1. India alone produces, more than 960 MT of solid waste from various industrial and domestic operations and dumping of these wastes in wastelands pose severe environmental threat. Proper recycling of the wastes would certainly save the energy and conserve the resources. Residue management has become a major problem in modern society, especially solid residue with potential toxic effects. The search for innovative solutions for the reuse of solid residue increased in the late twentieth century and has intensified with growing urgency for environmental preservation. Many residue management solutions aim to add value to residue through the development of new materials and processes.

SawDust

Sawdust or wood dust is a by-product of cutting, grinding, drilling, sanding, or otherwise pulverizing wood or any other material with a saw or other tool. It is composed of fine particles of wood. It is also the by-product of certain animals, birds, and insects which live in wood, such as the woodpecker and carpenter ant. It can present a hazard in manufacturing industries, especially in terms of its flammability. A major use of sawdust is for particleboard, coarse sawdust may be used for wood pulp. Sawdust has a variety of other practical uses, including serving as a mulch, as an alternative to clay cat litter, or as a fuel. Until the advent of refrigeration, it was often used in icehouses to keep ice frozen during the summer. It has been used in artistic displays, and as scatter in miniature railroad and other models. It is also sometimes used to soak up liquid spills, allowing the spill to be easily collected or swept aside. As such, it was formerly common on barroom floors. It is used to make Cutler's resin. Mixed with water and frozen, it forms pykrete, a slow-melting, much stronger form of ice. Sawdust is used in the manufacture of charcoal briquettes. The claim for invention of the first commercial charcoal briquettes goes to Henry Ford who created them from the wood scraps and sawdust produced by his automobile factory.

Table 1 Classifications of Wastes

Source	Waste Stream (Examples)
Municipal	Food scrap, office paper, yard waste, plastics, glass, textiles
Hazardous	Petroleum refining residuals, electroplating solvents
Industrial	Automotive industries, coal combustion, pulp, iron scrap, chemical, leather waste, sawdust (wood industries)
Medical	infectious agents, waste human blood, pathological waste
Universal	Batteries, agricultural pesticides, thermostats
Construction	Concrete, asphalt, roofing

Wood sawdust particles reinforced polymer composites; Hossain et al. [1] prepared a composite with both treated and untreated sawdust. Wood sawdust is being considerably used in reinforcing polymer both from a scientific and commercial point of view over the last decades. In this study, three different types of sawdusts were treated with 10% NaOH solution separately. Both chemically treated and untreated sawdust particles reinforcing polymer matrix composites were prepared and their tensile strengths were studied. Experimental results showed that the tensile strength of the composites substantially increases due to chemical treatment of wood particles.

Adhikary et al. [2] investigated the stability, mechanical properties, and the microstructure of wood–plastic composites, which were made using either recycled or virgin high-density polyethylene (HDPE) with wood flour (Pinusradiata) as filler. The post-consumer HDPE was collected from plastics recycling plant and sawdust was obtained from a local sawmill. The tensile and flexural properties of the composites based on recycled HDPE were equivalent to those based on virgin HDPE. Adding maleated polypropylene (MAPP) by 3–5 wt% in the composite formulation significantly improved both the stability and mechanical properties.

Nanoclay reinforced HDPE as a matrix for wood-plastic composites Faruk and Laurent Matuana [3] study was aimed at identifying the best approach of incorporating nanoclay into wood-plastic composites (WPCs) to enhance their mechanical properties. Two different methods of introducing nanoclays into HDPE-based WPCs were examined. The first method involved the reinforcement of HDPE matrix with nanoclay, which was then used as a matrix in the manufacture of the wood-plastic composites (melt blending process). The second method consisted of a direct addition of nanoclay into HDPE/wood-flour composites during conventional dry compounding (direct dry blending process). The mechanical properties of injection Molded WPCs were characterized using flexural, tensile, and dynamic mechanical analysis tests.

In the study of mechanical properties of wood dust reinforced epoxy composite Kumara et al. [4] sundi wood dust reinforced epoxy composite were processed with seven different percent filler weight. The tensile and flexural tests were performed at three different speeds to study the mechanical behavior of the composites. From the observation it was found that the mechanical property increases up to certain filler percentage and then properties gradually decrease. The experimental results support that successful fabrication of sundi wood dust reinforced epoxy composites is possible and that sundi wood dust possesses good filler characteristics as it improves the tensile and flexural properties of the polymericresin.

This paper, Petchwattana and Sirijutaratan [5], aims to utilize the acrylic Core-Shell Rubber (CSR) particles to reduce the brittleness in wood-plastic composites (WPC) prepared from poly(lactic acid) (PLA) and rubber wood sawdust (*Heveabrasiliensis*). Experimental works consisted of two major parts. The first part concentrated on toughening PLA by using CSR particles.

Wood-Polyethylene Composites Using Ethylene-Vinyl Kim et al. [6] described that the sawdust-reinforced linear low-density polyethylene (LLDPE) (1:1) composites were prepared by using ethylene–vinyl alcohol copolymer (EVAL) as adhesion promoter to improve mechanical strength.

Bouafif et al. [7] investigated the effects of fiber variability, size, and content. The effect of fiber size on water uptake is minimal. Increasing fiber load improves the strength and stiffness of the composite but decreases elongation and energy to break. Water uptake increases with increasing fiber content.

2 Materials and Methodology

Sawdust or wood dust is a by-product of cutting, grinding, drilling, sanding, or otherwise pulverizing wood or any other material with a saw or other tool. It is composed of fine particles of wood. It is also the by-product of certain animals, birds, and insects which live in wood, such as the woodpecker and carpenter ant. It can present a hazard in manufacturing industries, especially in terms of its flammability. A major use of sawdust is for particleboard; coarse sawdust may be used for wood pulp. Sawdust has a variety of other practical uses, including serving as a mulch, as an alternative to clay cat litter, or as a fuel. Until the advent of refrigeration, it was often used in icehouses to keep ice frozen during the summer. It has been used in artistic displays, and as scatter in miniature railroad and other models. The claim for invention of the first commercial charcoal briquettes goes to Henry Ford who created them from the wood scraps and sawdust produced by his automobile (Fig. 1).

Reinforcement Phase

Here the teak wood sawdust is taken as reinforcement phase in making the composite. Teak woods are widely used in making furniture and during that process sawdust is generated which has been taken as one kind of waste in preparing the composite. Figure 2 shows the reinforcing phase of composite (Fig. 3).

3 Methodology

The methodology of the present study shown in Fig. 4. Collection of raw material from the wood industry and the HDPE, LDPE granules from the polymer store is the first stage of present study. Collected sawdust wastes have to be pretreatment for removing unwanted particles and to reduce the particle size. Sawdust easily absorbs the moisture in nature. It becomes wet in room temperature which cannot be processed for compounding. So the moisture should be removed from sawdust. First the size of sawdust is reduced by sieving process step by step (from 450 micron to 180 micron) and then the reduced size of sawdust is heated in an electric oven above the room temperature for about 7 to 8 h. Finally the treated sawdust is mixed with

Fig. 1 Sawdust

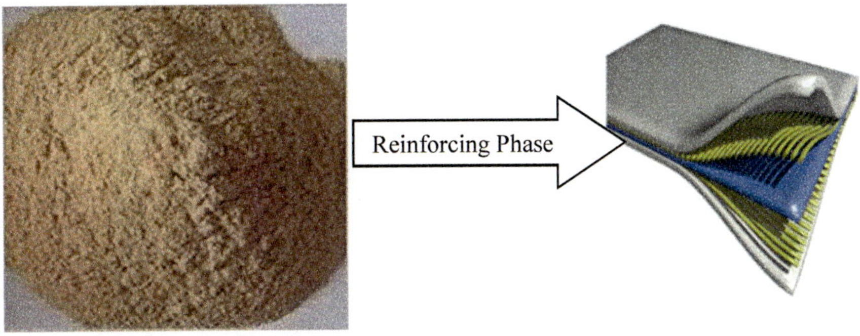

Fig. 2 Teak wood dust as reinforcing phase

HDPE and LDPE granules in different formulations through compounding process in twin screw extruder and then injection molded to get the desired shape of specimens. At last, the specimens are tested for mechanical properties like tensile, hardness, flexural, and water absorption test.

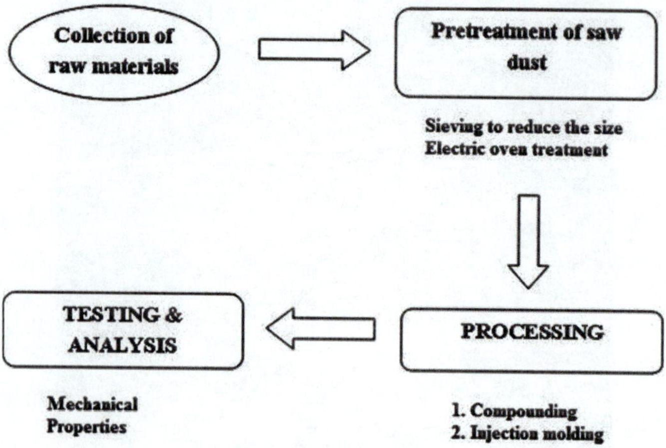

Fig. 3 Methodology

Fig. 4 Sieving of sawdust

Pretreatment of Sawdust

The pretreatment of sawdust includes two following steps:

- Sieving of sawdust
- Electric oven treatment

Sieving of sawdust

In order to reduce the particle size of sawdust, sieving is done. So, that the reduced size of sawdust can be used for processing in twin screw extruder because the hopper of the extruder allows only tiny materials for processing.

Fig. 5 Electric oven treatment of sawdust

This process is carried out in a sieving machine which runs on the application of AC induction motor in which series of sieving plates of different micron size are arranged together. After sieving, it is found that the particle size is reduced from 450 micron to 180 micron. Figure 5 shows the sieving process

Electric oven treatment

The next pretreatment process is the electric oven treatment in which the sawdust is heated above the room temperature for about 3–4 h in order to remove the moisture content present in it because the sawdust naturally absorbs the moisture present in the open atmosphere. So it is heated in a closed chamber and once heated it should be carried out for processing. Figure 5 shows the electric oven treatment.

Formulations

Table 2 shows the three different formulation of HDPE, LDPE granules and sawdust and Fig. 5a shows the pictographs of formulations:

Table 2 Formulations of HDPE, LDPE granules and sawdust

Ingridients	Formulation 1	Formulation 2	Formulation 3 (%)
HDPE	80%	–	40
LDPE	–	80%	40
Sawdust	20%	20%	20

Manual Mixing By SiliconeOil

Before the materials are compounded in twin screw extruder, it is mixed manually in a plastic bag by using silicone oil as binding agent which improves the adhesion between the sawdust and granules before feeding into hopper.

Compounding Process

Compounding is nothing but the mixing process in which the granules are mixed with sawdust uniformly. There are many compounding machines to mix the materials, here the twin screw extruder is used for mixing in which the materials are extruded in wire form and it is made in the form of pellets. Then the pellets are dried and injection molded to get the desired shape. In hopper, the materials are fed and by the action of twin screw it moves from solid phase to melting phase where the granules start to melt and mix with the sawdust.

The heater is preset to the processing temperature of 110–130 °C corresponding to the melting point of HDPE and LDPE. Once the materials came in semi solid form, it is cooled in water so that it turns to solid form by rapid quenching method. The pellets are dried again in electric oven for about 3 h and then it is injection molded for getting the desired shape. Figures 5 and 6 show the twin screw extruder and Fig. 7 shows the pellets.

4 Injection Molding

Injection molding is a manufacturing process for producing parts by injecting material into a mold. Injection molding can be performed with a host of materials, including metals, (for which the process is called die casting), glasses, elastomers, confections, and most commonly thermoplastic and thermosetting polymers. Material for the part is fed into a heated barrel, mixed, and forced into a mold cavity, where it cools and hardens to the configuration of the cavity (Fig. 8).

Injection molding uses a ram or screw-type plunger to force molten plastic material into a mold cavity. This solidifies into a shape that has conformed to the contour of the mold. It is most commonly used to process both thermoplastic and thermosetting polymers, with the volume used of the former being considerably higher. Thermoplastics are prevalent due to characteristics which make them highly suitable for injection molding, such as the ease with which they may be recycled, their versatility allowing them to be used in a wide variety of applications and their ability to soften and flow upon heating. Injection molding consists of high pressure injection of the raw material into a mold which shapes the polymer into the desired shape. Molds can be of a single cavity or multiple cavities. In multiple cavity molds, each cavity can be identical and form the same parts or can be unique and form multiple different geometries during a single cycle. Molds are generally made from tool steels, but stainless steels and aluminum molds are suitable for certain applications (Table 3).

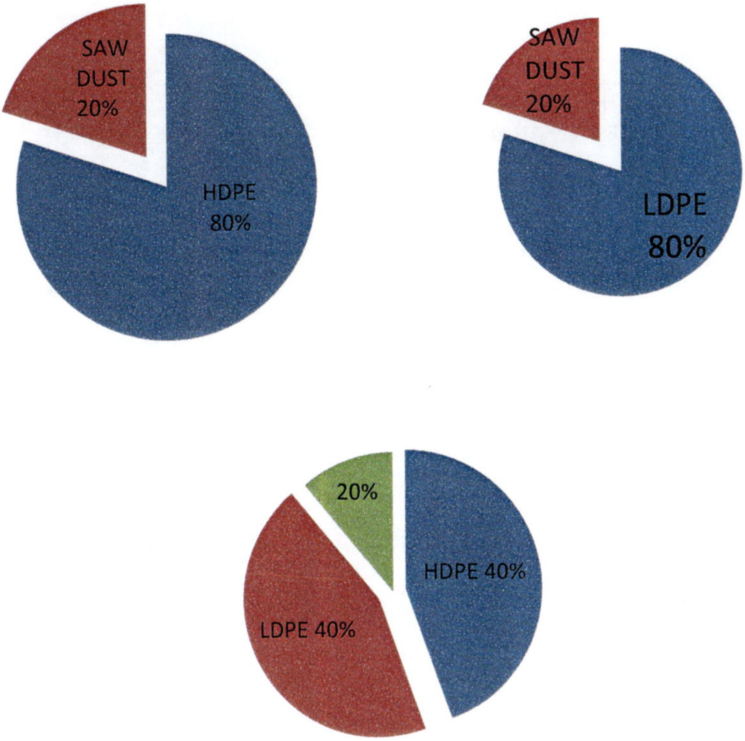

Fig. 6 A pictographs of different formulations

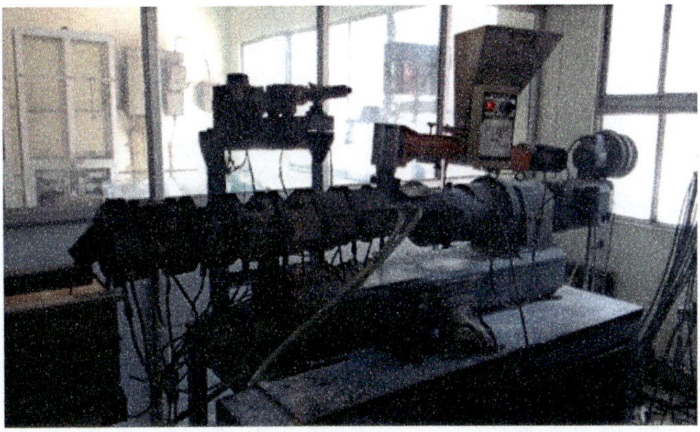

Fig. 7 Twin screw extruder

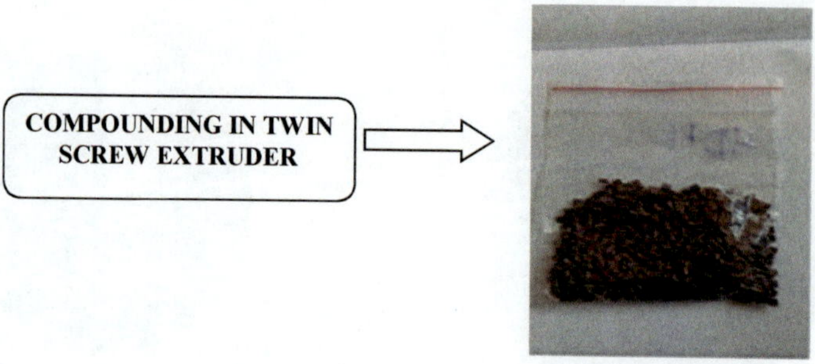

Fig. 8 Pellets extruded in twin screw extruder

Table 3 Tensile properties of prepared composites

Formulation	Specimen (HDPE:SD:LDPE)	Ultimate tensile strength (Mpa)
HSC	80:00:20	18
LSC	00:20:80	7.33
HLSC	40:20:40	11

The pellets as a result of twin extrusion are should be dried because it is water cooled during extrusion process and injection molded at temperature 200 °C to make the final specimens. Figure 9 shows the injection molding process. Figure 10 shows

Fig. 9 Injection molding machine

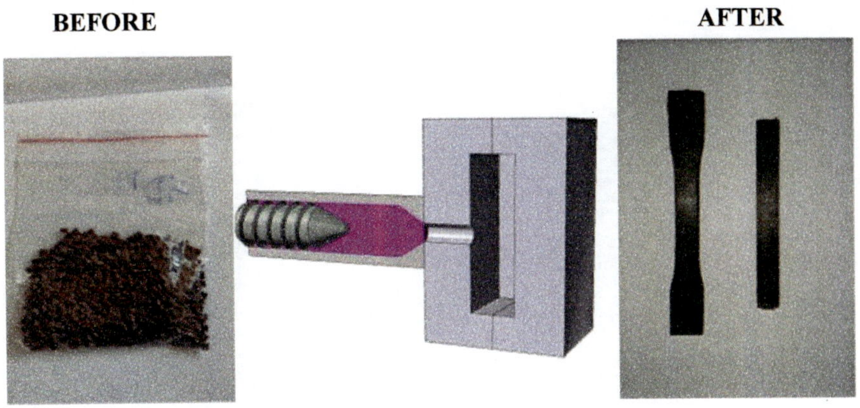

Fig. 10 Specimen preparation by injection molding

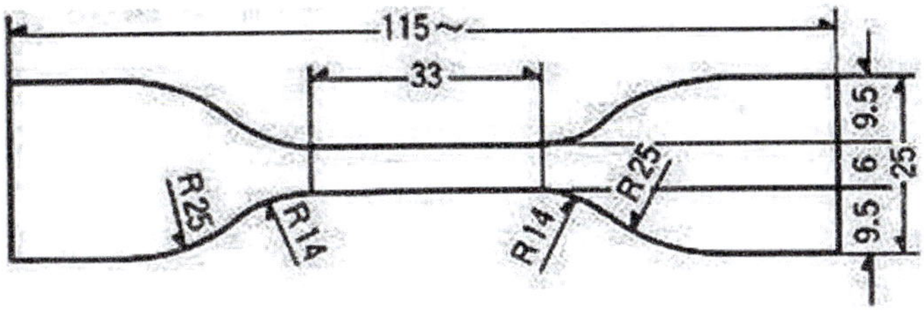

Fig. 11 Tensile test specimen

the injection molding machine and Fig. 11 shows the specimen made by injection molding.

5 Testing

A material testing is performed for a variety of reasons and can be used to collect wide range of information. Testing of raw materials is necessary for checking the quality and standard of incoming material before processing (Figs. 12 and 13).

In this present study, the following tests have been conducted for three samples in each formulation and the average is taken.

- Tensile test
- Hardness
- Flexural test
- Water absorption test.

Specimen Sample

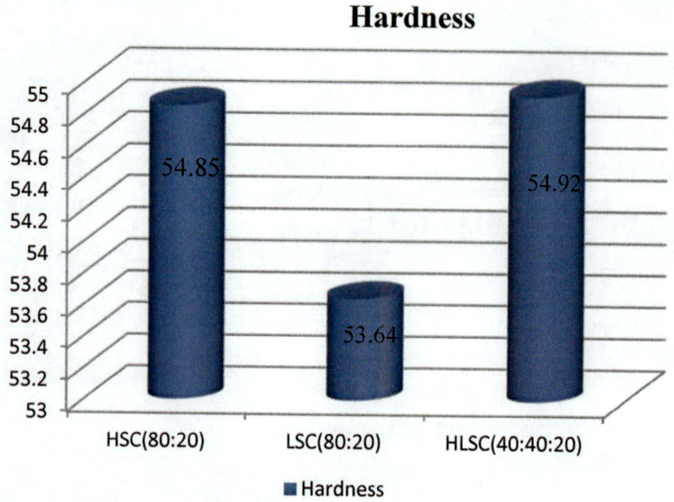

Fig. 12 Hardness values for prepared composites

Figure 9 shows the sample for tensile test.

6 Results and Discussion

The following are the results of various mechanical properties like tensile, hardness, flexural, and water absorption capacity.

Tensile Properties

The variation of the tensile strength as a function of teak wood sawdust is listed in Table 4.

The tensile strength of the 80% HDPE and 20% sawdust content composites was slightly superior to other composites. This may be due to the amount of HDPE content and the type of sawdust. It is also found that reducing the HDPE content decreasing the tensile strength in third formulation. LDPE-sawdust composite shows the least value among all the three formulations. This is due to the fact that LDPE has low density. At the same time, when LDPE is mixed with HDPE, it shows higher tensile strength than LDPE-sawdust composite because the sawdust with HDPE improves the tensile strength. The mechanical properties of the fiber-reinforced polymer composites not only depend on the characteristics of the polymer matrix and the reinforcement fibers but also on the interface relationship between the matrix and the fiber, which is important for the load transfer.

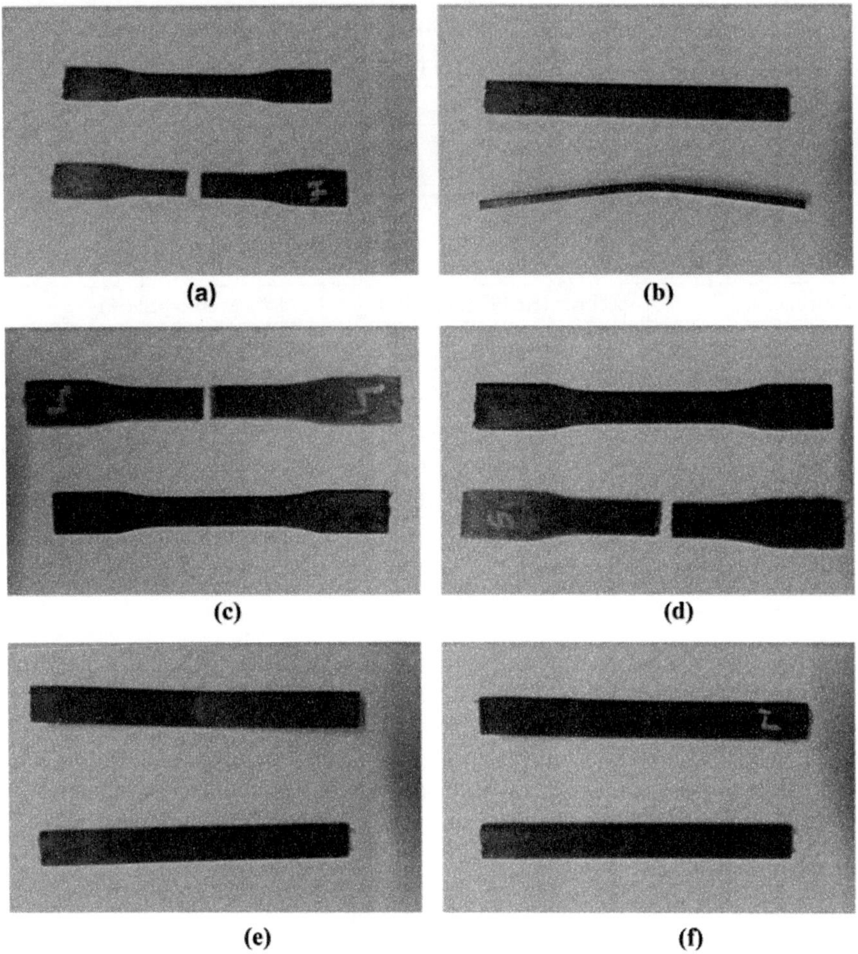

Fig. 13 Tested (Tensile & Flexural) Specimens of prepared composites

Table 4 Flexural properties of prepared composites

Formulation	Specimen (HDPE:SD:LDPE)	Ultimate break load (N)
HSC	80:00:20	51.6
LSC	00:20:80	20
HLSC	40:20:40	35

Flexural Properties

The variation of the flexural strength as a function of teak wood sawdust is listed in Table 5.

Table 5 Hardness of prepared composites

Formulation	Specimen (HDPE:SD:LDPE)	Hardness (Shore D)
HSC	80:00:20	54.85
LSC	00:20:80	53.64
HLSC	40:20:40	54.92

Table 6 Water absorption of prepared composites

Formulation	Specimen (HDPE:SD:LDPE)	Water absorbed (%)
HSC	80:00:20	2.7
LSC	00:20:80	5.26
HLSC	40:20:40	2.7

From the above table, it is found that the HDPE-sawdust composite and HDPE-LDPE-sawdust composites are having the higher flexural strength than LDPE-sawdust composite.

Hardness (ShoreD)

Based on Shore D hardness, it is found that the HDPE-LDPE-sawdust composite shows highest hardness value of 549.2 among the three formulations. It is due to the fact that both HDPE and LDPE improve the hardness in this formulation than the HDPE and LDPE sawdust composite. Also all the three composite show the better results toward hardness above 500. The variation of the hardness as a function of teak wood sawdust is listed in Table 6 and Fig. 9 shows the graphical data.

Water Absorption Test

The water absorption test results are listed in Table 6.

Among the three formulations, LSC has the highest water absorption rate of 5.26% and the other two formulations have the rate of 2.7%.

7 Conclusion

In this study, composites sheets were prepared by a simple, efficient, and scalable technique consists of industrial wood waste with HDPE and LDPE virgin granules. The collected sawdust is sieved, impurities removed and heated. Both are mixed in the various ratio using twin screw extruder and the semi-finished products are obtained and using the injection molding the finished composites specimens are prepared. The prepared composites were tested for finding its properties.

- This work provides possible solution for the management of wood wastes generated from wood industry and municipal wastes.

- In order to remove the presence of impurities content in the wood wastes and to reduce the size, sieving has been carried out and size is reduced to 180micron.
- To remove the moisture content from sawdust, heating of sawdust is carried out in an electric oven for three hours.
- The tensile strength of the 80% HDPE and 20% sawdust content composites were slightly superior to other composites. The mechanical properties of the fiber-reinforced polymer composites not only depend on the characteristics of the polymer matrix and the reinforcement fibers but also on the interface relationship between the matrix and the fiber, which is important for the load transfer.
- The composites with 40% HDPE, 40% LDPE, and 20% sawdust polymer had a great hardness value than other composites.
- Thus the present work offers a valuable solution for recycling and managing of wood wastes from wood factories and furniture makers to usableproducts.
- Here the testing results revealed that the untreated sawdust show less mechanical properties than chemically treated sawdust.

References

1. Hossain MF, Islam MA, Islam MK (2014) Effect of chemical treatment on physical and mechanical properties of wood saw dust particles reinforced polymer matrix composites. Proc Eng 90:39–45
2. Adhikary KB, Pang S, Mark Staiger P (2008) Dimensional stability and mechanical behaviour of wood–plastic composites based on recycled and virgin high-density polyethylene (HDPE), pp 807–815
3. Faruk O, Laurent Matuana M (2008) Reinforced HDPE as a matrix for wood-plastic composites. Nanoclay Compos Sci Technol 68:2073–2077
4. Kumara Rahul, KausikKumarb PrasantaSahooc, Bhowmik S (2014) Study of mechanical properties of wood dust reinforced epoxy composite. Proc Mater Sci 6:551–556
5. Petchwattana N, Sirijutaratan (2014) Mechanical and morphological properties of wood plastic biocomposites prepared from toughened poly(lactic acid) and rubber wood sawdust (heveabrasiliensis). Covavisaruch J Bionic Eng 11:630–637
6. Kim J-P, Yoon T-H, Mun S-P, Rhee J-M, Lee J-S (2006) Wood–polyethylene composites using ethylene–vinyl alcohol copolymer as adhesion promoter. Bioresour Technol 97:494–499
7. Bouafif H, Koubaam A, Perré P, Cloutier A (2009) Effects of fiber characteristics on the physical and mechanical properties of wood plastic composites. Compos Part A 40:1975–1981

Experimental Investigation and Comparing the Mechanical Properties of Glass Fiber-Reinforced Polymers with Carbon Nanotubes and Sawdust

Janarthanam Hemanandh, K. S. Sridhar Raja, S. Ganesan, S. P. Venkatesan, Ashish Kumar Tiwari, and Akshay Sethu

Abstract Glass fiber-reinforced polymers (GFRP) are now a more desirable material to strengthen composite structures stronger, as they are a good resistant to corrosion, having a high strength to weight ratio and quite cheap when compared to other tough polymers. But its frequent use in various infrastructural applications was brought to a hold because of GFRP's limited fatigue life. Here GFRP's prostrate life could be enhanced by adding reinforcements like multi-walled carbon nanotubes and saw dust was fabricating with different weight percentage, and it was further compared for its mechanical properties. It has been proved that multi-walled CNT reinforcement has higher tensile strength, flexural strength and hardness when compared with the sawdust-reinforced particles.

Keywords Carbon nanotubes · Composites · Epoxy resin · Hand layup technique · Saw dust · Glass fiber-reinforced polymers

1 Introduction

The preceding two decades showed that glass fiber-reinforced polymers (GFRPs) gripped a good place as a composite in market as an alternative to other polymer substances. But GFRP was reported to have less fatigue strengths for which it was unable to resist heavy loads because of low stress limits. So if the fatigue strength of the GFRP composites could be improved to a limit keeping in mind of the low-cost factor, it would be used effectively in various potential applications. To overcome the above problems to an extent, carbon nanotubes and saw dust (additive reinforcements) were added with glass fiber to improve its overall mechanical property. Therefore, six different plates were prepared with unique concentrations by hand

J. Hemanandh (✉) · K. S. Sridhar Raja · S. Ganesan · S. P. Venkatesan · A. K. Tiwari · A. Sethu
School of Mechanical Engineering, Sathyabama Institute of Science and Technology, Chennai, Tamil Nadu 600043, India
e-mail: hemanandh123@gmail.com

A. Arockiarajan et al. (eds.), *Advances in Industrial Automation and Smart Manufacturing*, Lecture Notes in Mechanical Engineering,
https://doi.org/10.1007/978-981-15-4739-3_70

layup technique and they were undergone through flexural test, hardness test, impact test, and tensile test.

Genedy et al. [1] demonstrated that the prostrate life of GFRP can be remarkably improved by incorporating MWCNTs during GFRP fabrication, and the results show that the 0.5 wt% and the 1.0 wt%. MWCNTs were able to improve GFRP's enervating life by 1143% and 986%, respectively, compared with neat GFRP. Araújo et al. [2] showed the possibility of reusing the wastes taken from spray-up processing as augmentation in polyester matrix composites with 20, 30, 40, 50 and 60 wt% of recycled fiberglass were prepared by compression molding and compared with pure glass fiber composites. The mechanical properties and water sorption behavior were evaluated. Farhan and Al Omairi [3] aimed to generate a new class of natural fillers-based polymer composites with different types of wood and analyzed their mechanical behavior. The mechanical properties measured include tensile strength, Young's modulus, flexural strength, and impact strength for all specimens at normal conditions. Shukla et al. [4] investigated the mechanical properties by in situ testing at + 70 °C and +100 °C, for GFRP and glass/carbon (G/C hybrid) composites which resulted in considerable loss in inter-laminar shear strength (ILSS) for both the composites as compared to room temperature. The ILSS was nearly equal for both the composite systems, and the effect of fiber hybridization was completely reduced. At low-temperature ex situ conditioning significant reduction in ILSS was observed for both the systems. Further at −60 °C G/C hybrid exhibited 32.4% higher ILSS than GFRP. Their investigation showed that for materials required in low-temperature environmental application, G/C hybrid is a better choice. Sathishkumar et al. [5] did a detail study on glass fiber composites where each type of glass fibers has unique properties and was used for various applications in the form of polymer composites. The mechanical, thermal, water absorption, chemical, and vibration properties of various glass fiber-reinforced polymer composites were reported. Tian et al. [6] applied a special electrochemical anodic oxidation method to modify the surface of carbon fibers with fatty alcohol polyoxyethylene ether phosphate (O3P), triethanolamine (TEA), fatty alcohol polyoxyethylene ether ammonium phosphate (O3PNH4), and ammonium bicarbonate (NH_4HCO_3) used as the electrolyte, respectively. Through the experiment, a uniform catalyst coating was obtained on the surface of CFs after reduction process, which laid the foundation for the growth of uniform and regular CNTs. Mosadeghzad et al. [7] based on recycled PET and the effect of surface treatment and filler loading on the composites prepared a sawdust/UPR composite. This results in an increase of both tensile and flexural moduli with increasing filler contents while the strength decreases. This was overcome by adding the sawdust fillers with 10% sodium hydroxide (NaOH). The tensile, flexural, hardness, and water absorption tests prove that the alkali treatment has strengthened the adhesion between the sawdust and the UPR matrix. Arsh et al. [8] did a detailed study on the mechanical properties of carbon nanotubes and found their remarkably high elastic modulus and tensile strength, making them the most ideal and promising reinforcements in substantially enhancing the mechanical properties of resulting polymer/carbon nanotube composites. Hamidinia and Tajvidi [9] studied the mechanical properties of wood-plastic composites (WPCs) manufactured from sawdust and

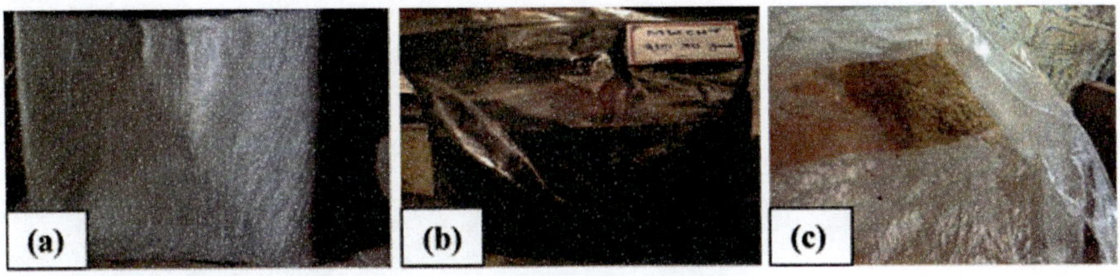

Fig. 1 **a** Glass fibre (chopped strand mat), **b** CNT, **c** sawdust

pure recycled plastics, namely high-density polyethylene (HDPE) and polypropylene (PP). Sawdust was manufactured from beech industrial sawdust by screening to the required particle size and was mixed with different pure or recycled plastics at 50% by weight fiber loading. Flexural and tensile properties as well as the impact strength of the manufactured WPCs were determined according to the suitable standard specifications. Reddy et al. [10] studied the effect of fly ash content on tensile strength and flexural strength of 10 wt% glass fiber epoxy composite. The fly ash content is diverse into steps of 2 g that is 0–10 g when compared to the other composites the composite with 6 g fly ash exhibited better tensile strength.

2 Experimental Details

2.1 Materials and Methods

Choice of raw materials is a prior activity before any fabrication process of composites. The individual properties of each of the raw materials play an important role for defining the overall property of the composite. Glass fiber is used as a reinforcing material along with multi-walled carbon nanotubes (MCNT) and sawdust used as a filler material (Fig. 1).

2.2 Method of Fabrication

2.2.1 Hand Layup Method

It is a simple and economical method of fabrication and processing of composite materials in which after mixing uniformly with resin and hardener, additives are added to the base matrix material with the help of brushes or rollers. Then the fabricated composites are laminated and are left to be cured in the atmosphere (Fig. 2).

Fig. 2 Composite preparation by hand layup

2.3 *Preparations of Composites*

The fiber mat of required size of 500 × 500 mm with zero-degree orientation is placed on the binding sheet on the plywood, and the wax was applied on the plastic sheet. The CNT with different weight ratio proportion (5 g, 10 g, and 15 g) was mixed with the epoxy. The hardener and resin are mixed thoroughly with the ratio of 1:10, and the glass fiber mat is placed on the plastic sheet; it is repeated up to 5 layers and was spread using a roller before and after each glass fiber layer. The fibers and the epoxy layers were subjected to pressure. After fabrication, the GFRP plates were air-cured for 24 h. To disperse the CNT/saw dust in the epoxy, it should be stirred manually in the epoxy resin for 1 min and then mechanically stirred using a magnetic stirrer for 2 h at 80 °C and 800 rpm to insure uniform dispersion of the CNTs/saw dust in the epoxy resin. The samples are fabricated according to ASTM D3171-06 standards. Similarly, by reinforcing sawdust with different weight proportions was fabricated with the same technique (Fig. 3).

3 Results and Discussion

The fabricated composite material is prepared for testing mechanical properties. It depends on the toughness, and flexibility of the material its brittleness has been

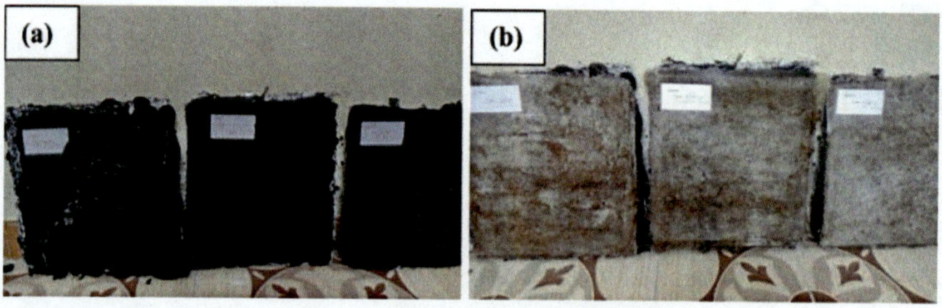

Fig. 3 Fabricated **a** GFRP + MCNT plates, **b** GFRP + sawdust plates

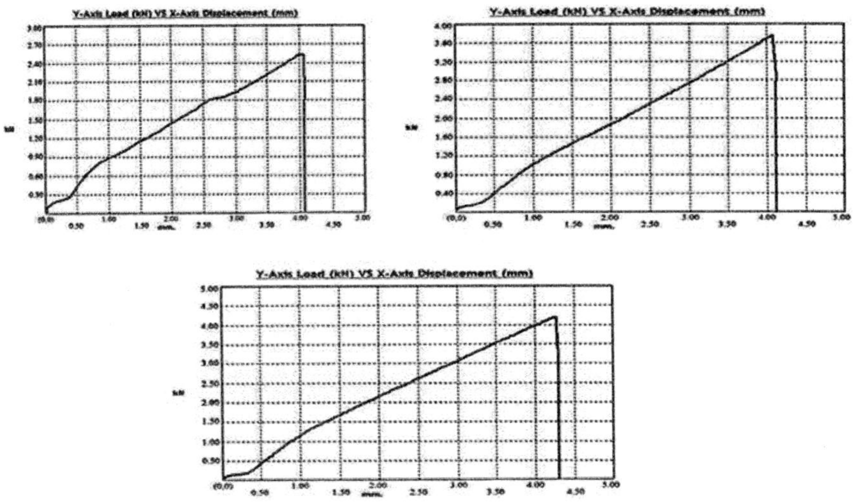

Fig. 4 Tensile load versus displacement graphs when CNT is 5 g, 19 g, 15 g

determined. The prepared material is subjected to mechanical properties like tensile testing, impact testing, and hardness.

3.1 Tensile Test

The tensile test was conducted based on ASTM standards, and its load versus displacement was plotted. For both the composites as shown in Figs. 4 and 5, it was observed from Fig. 4 that when load increases the displacement increases gradually. But in Fig. 5, it can be studied that the displacement has a fluctuation when the load increases. This fluctuation in the displacement is due to non-homogeneous mixture of sawdust particles.

It was observed that both the following additives enhance the overall mechanical properties of composite materials as the results were found satisfactory. From the graph, it was observed that by increasing the reinforcing particle the breaking load increases gradually both in the case of reinforcing CNT and particles of sawdust.

3.2 Flexural Test

In order to study the elasticity modulus of the composite material, three-point bending flexural test has been performed. It helps to determine the stiffness of the material. The result is provided in Table 1. It was noticed that on increasing the weight percentage

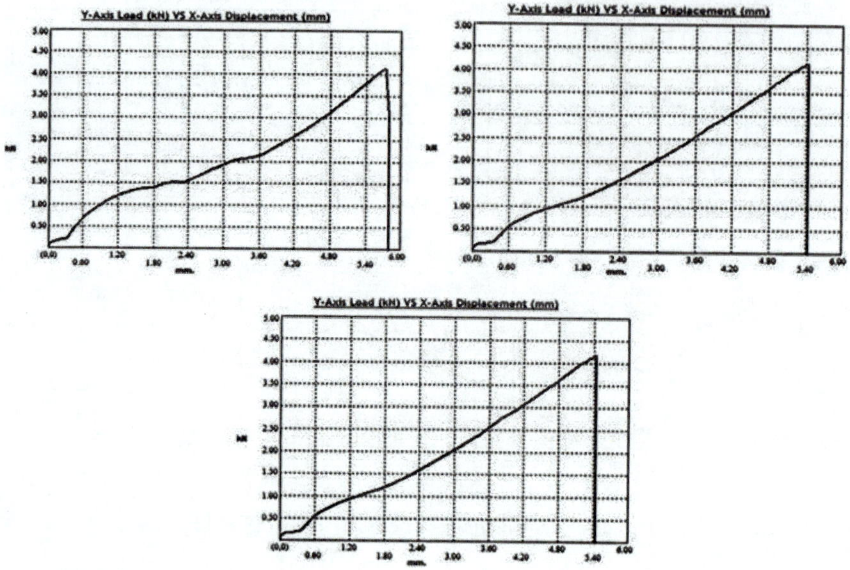

Fig. 5 Tensile load versus displacement graphs when sawdust is 5 g, 10 g, 15 g

Table 1 Flexural test data for CNTs and sawdust composites

Quantity (g)	CNTs (N)	Sawdust (N)
5	165	265
10	235	290
15	335	320

of CNT the flexural strength increased, whereas when the sawdust increases the strength increased gradually and it was plotted in Fig. 6.

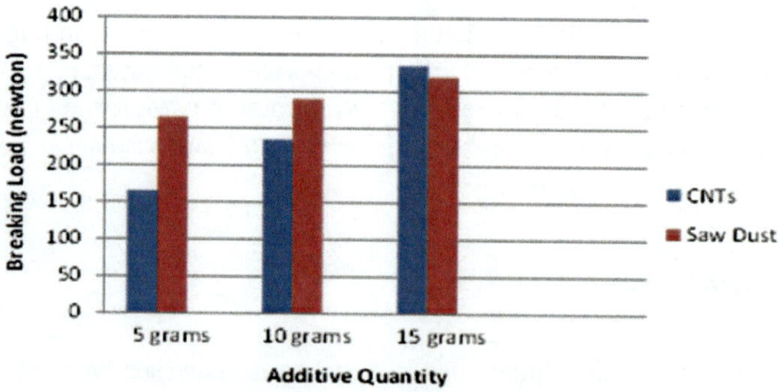

Fig. 6 Tensile strength of the composites

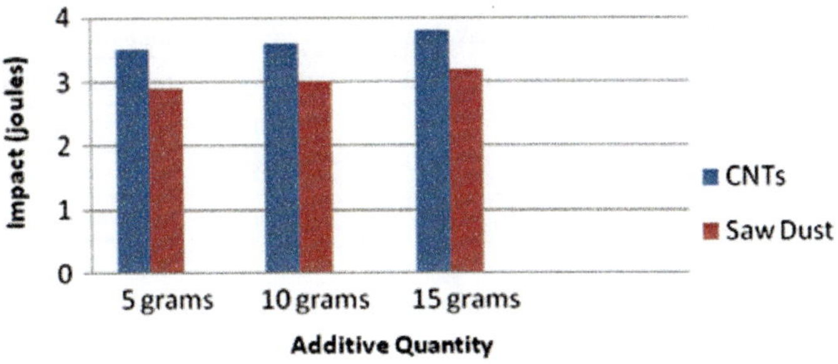

Fig. 7 Impact test of composites

3.3 *Impact Test*

Izod impact test: Impact test was done to measure the impact of resistance and toughness of the materials. From Fig. 7, it was observed that the impact energy was high in carbon nano-tubes-reinforced composites when compared with sawdust-reinforced particles.

3.4 *Hardness Test*

Hardness test is done to measure the hardness of a material in terms of elasticity. The shore hardness was measured at several places to find the elasticity of the material, and the average hardness was plotted in Fig. 8 and tabulated. From Fig. 8, it was observed that both CNT and sawdust-reinforced composite was increased gradually with the increase in weight percentage of the particle. But compared with sawdust reinforcement, CNT-reinforced composite has higher hardness, which processes the elasticity of the material (Table 2).

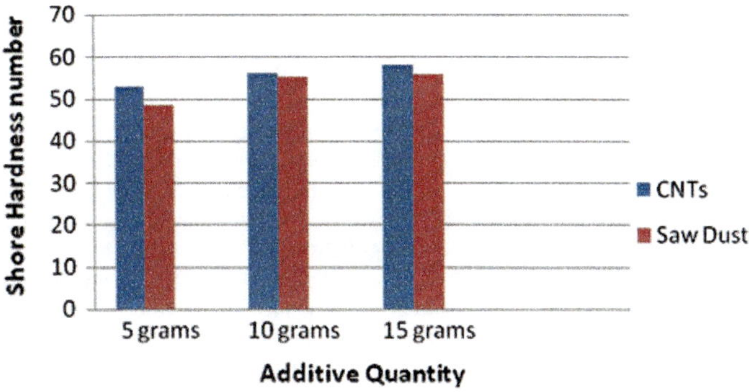

Fig. 8 Hardness of composites

Table 2 Hardness test data for CNTs and sawdust on different compositions

Quantity (g)	CNTs (shore hardness number)		Sawdust (shore hardness number)	
5	48.7	52.87	56.9	48.45
	66.2		58.9	
	29.2		46.7	
	49.7		49.0	
10	48.1	56.02	54.7	55.07
	55.4		59.7	
	59.6		53.8	
	57.2		55.9	
15	48.4	58.1	54.6	55.57
	51.7		59.8	
	56.5		60.3	
	65.7		57.9	

4 Conclusion

GFRP was reported to have less fatigue strengths for which it was unable to resist heavy loads because of low-stress limits. Thus, it was not much effective when used in infrastructural applications. So, to hold this problem to an extent, carbon nanotubes and saw dust were added as additive reinforcements in compositions to GFRP composites. It was observed that both the following additives enhance the overall mechanical properties of composite materials. It has been concluded that both CNT and sawdust-reinforced composite has equal improvement in mechanical properties, but CNT-reinforced composite has little bit higher mechanical properties.

References

1. Genedy M, Daghash S, Soliman E, Taha M (2015) Improving fatigue performance of GFRP composite using carbon nanotubes. Fibers 3:13–29. http://doi.org/10.3390/fib3010013
2. Araújo EM, Araújo KD, Pereira OD, Ribeiro PC, de Melo TJA (2006) Fiberglass wastes/polyester resin composites: mechanical properties and water sorption. Polímeros: Ciência e Tecnologia 16(4):332–335. http://doi.org/10.1590/S0104-14282006000400014
3. Farhan AJ, Al Omairi (2011) Investigation of some mechanical properties of sawdust and chopped reeds/UPE composites. Study of Sawdust: Methyl Methacrylate Polymer Composite Properties, Thesis, 2014, Universiti Teknologi Petronas. Tronoh Perak, Malaysia 1:7–41
4. Shukla MJ, Kumar DS, Mahato KK, Rathore DK, Ray BC (2015) A comparative study of the mechanical performance of glass and glass/carbon hybrid polymer composites at different temperature environments. In IOP Conference Series: Materials Science and Engineering, vol 75, p 012002. http://doi.org/10.1088/1757-899X/75/1/012002
5. Sathishkumar TP, Satheeshkumar S, Naveen J (2014) Glass fiber reinforced polymer composites. J Reinf Plast Comp 33(13):1258–1275. http://doi.org/10.1177/0731684414530790

6. Tian Y, Zheng L, Qin J, Tao Y, Song C, Fan W, Wang Y (2017) Studying the growth of carbon nanotubes on carbon fibers surface under different catalysts and electrochemical treatment conditions. J Fuller Nanotub Carb Nano structures 25(3). http://doi.org/10.1080/1536383X.2016.1273905
7. Mosadeghzad Z, Ahmad I, Daik R, Ramli A, Jalaludin Z (2009) Preparation and properties of acacia sawdust/UPR composite based on recycled PET. Malays Polym J 4(1):30–41
8. Arsh B, Wang Q, Vardhan VK (2004) Mechanical properties of carbon nanotube/polymer composites. 1:9–31
9. Najafi SK, Hamidinia E, Tajvidi M (2006) Mechanical properties of composites from sawdust and recycled plastics. J Appl Polym Sci 100:3641–3645. http://doi.org/10.1002/app.23159
10. Reddy SP, Rao PVCS, Reddy, AC, Parmeswari G (2013) Flexural strength of glass fiber epoxy composites. In International Conference on Advanced Materials and Manufacturing Technologies (AMMT) December 18–20, 2014 JNTUH College of Engineering Hyderabad

Mechanical and Microstructural Characterization of Copper and Carbon Nanotubes Composites

Shridhar Deshpande, D. Amaresh Kumar, C. T. Murali, and Shrishail Kakkeri

Abstract In recent years, copper matrix with carbon nanotubes (CNTS) reinforcement has attracted increasing attention. The distribution of reinforcement in composites is a crucial factor, determines the homogeneity and mechanical properties. By using the powder metallurgy technique various amounts of well dispersed multi-walled carbon nanotubes (MWCNTS) were used to reinforce a copper material. The microstructures of nanocomposites were observed in the optical metallurgical microscope. Wear characterization of the nanocomposites was performed on the pin-on-disk wear testing machine under the different loading conditions. In comparison with copper, addition of MWCNTs decreases the coefficient of friction in cu MWCNT nanocomposites. The results shown in addition to MWCNTs can upgrade the copper and convert it has a wear resistant material. The MWCNT is directly improving the tribological performance and indirectly influencing the microstructure and mechanical properties of nanocomposites by acting as a lubricating medium.

Keywords Carbon nanotubes · Copper powder · Tribological properties

1 Introduction

In present work, metal matrix composite is considered and further, multi-walled carbon nanotubes are being selected as they are easy to produce and cost-effective compared to single-walled carbon nanotubes. In this pure nano-sized copper powder is taken as the matrix material and MWCNTs as reinforcement material with a different weight fraction of reinforcement materials are prepared by using a powder metallurgy technique [1]. The mixing of Cu/CNT composite is done by using ball milling process and the microstructure of Cu/CNT composite is checked in Scanning Electron Microscope (SEM), X-ray diffraction (XRD) analysis techniques. The microstructure image from SEM indicates proper diffusion of CNTs in

S. Deshpande (✉) · D. Amaresh Kumar · C. T. Murali · S. Kakkeri
Department of Mechanical Engineering, Sri Venkateshwara College of Engineering, Bengaluru, Karnataka, India
e-mail: shridhardes@gmail.com

A. Arockiarajan et al. (eds.), *Advances in Industrial Automation and Smart Manufacturing*, Lecture Notes in Mechanical Engineering,
https://doi.org/10.1007/978-981-15-4739-3_71

Table 1 Properties of copper and MWCNTs

Properties	Copper	MWCNTs
Density	7 g/cm^3	1.6–2 g/cm^3
Young's modulus	117 GPa	1.7–2.4 Tpa
Thermal conductivity	394 W/mK	>3000 W/mK
Melting point	1083 °C	35,000 °C

copper surrounding substance of sintered composites. Vickers micro-hardness test is conducted for different percentage of CNT's and Cu composites to determine the hardness values [2]. The tribology properties such as wear, wear resistance, and frictional behavior of each composite specimen are determined by conducting wear analysis on pin-on-disk wear testing machine and compared with the pure copper specimen.

2 Material Selection

The following are the composite material combinations along with their properties are considered for the present project work (Table 1).

3 Methodology

3.1 Fabrication of Nanocomposites

As per the ASTM standards, the nanocomposites were made from copper and MWCNT used as filler materials [3]. By using the powder metallurgy techniques fabrication of nanocomposites was done at room temperature and was sintered at 850 °C temperature for 4 h and then cooled to room temperature (Table 2).

Table 2 Density of different composite with different MWCNT%

Specimen	Copper	Copper + MWCNT 0.5%	Copper + MWCNT 1%	Copper + MWCNT 1.5%
Density (g/cm^3)	7	6.25	5.7	5.27

Table 3 Different temperatures and time is taken for sintering

Temperature in °C	150	250	400	600	850
Time in minutes	45	30	30	30	240

Table 4 Specimen numbers with their compositions

Specimen Number	1	2	3	4
Composition Cu (gm) +MWCNTS (wt%)	50 +0	50 +0.5	50 +1	50 +1.5

3.2 *Experimentation*

In this work, six variations of the reinforcement material with pure copper of and MWCNT, 0.5 wt%, 1 wt%, and 1.5wt% are prepared [4]. The ball milling process is carried out for mixing the copper and MWCNTs of different variation. The Cu/MWCNTs composites of 0.5 wt%, 1 wt%, 1.5wt% are blended thoroughly. Each process is carried for 30 min of duration and the homogeneously blended for good and proper mixing of both powder materials [5] (Table 3).

3.3 *Xrd Test*

XRD analysis was conducted after the ball milling process to determine the mixture of both copper powder and carbon nanotubes [6] (Table 4).

3.4 *Compacting Process*

After ball milling process the blended powder is compacted, by compacting process

3.5 *Density of Composite*

It is defined as the ratio of the weight of the composite material to the volume of the composite material.

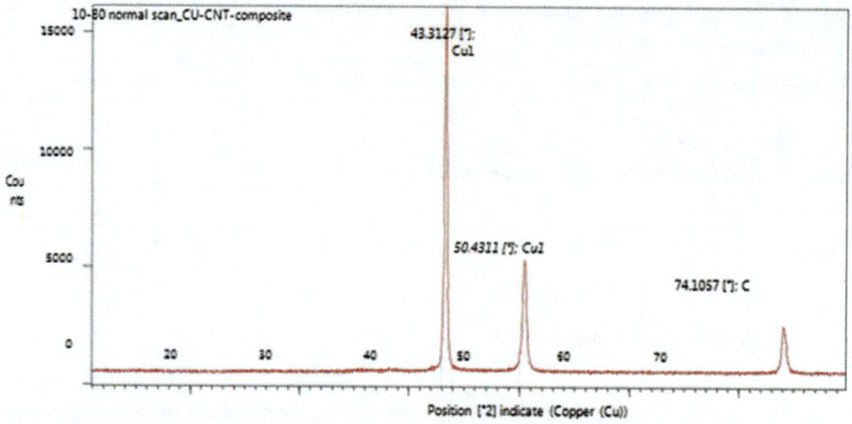

Fig. 1 XRD test pattern for Cu+CNTs1.5% nanocomposites powders

3.6 Sintering Process

After the compacting process, the compacted powders are sintered using heat furnace for the following temperatures and cooled in room temperature.

3.7 Machining and Finishing of Sintered Specimen

Completing the sintering process the sleeves are cooled in room temperature then the outer diameter of sleeves are slowly machined to get the sintered specimen and thereafter further tests are conducted [7].

The microstructure of the entire sintered specimen prepared was also scanned using SEM to know the distribution of multi-walled carbon nanotubes in copper powder and compared with pure copper as shown in Fig. 1.

3.8 Vickers Hardness Test

The hardness of Cu+ MWCNT's nanocomposites and pure copper was measured using a Vickers hardness test machine. Hardness test was conducted in the CML commercial laboratory (Figs. 2, 3, 4, 5, 6, 7 and 8).

The above graph shows the effect of carbon nanotubes as a weight percentage on the hardness of prepared Cu + MWCNTs specimens. It has been observed that the hardness values (HV) of specimen increase with an increase in CNTs and this also indicates that there is a proper distribution of CNTs in the copper matrix [8].

Fig. 2 Compacted specimen

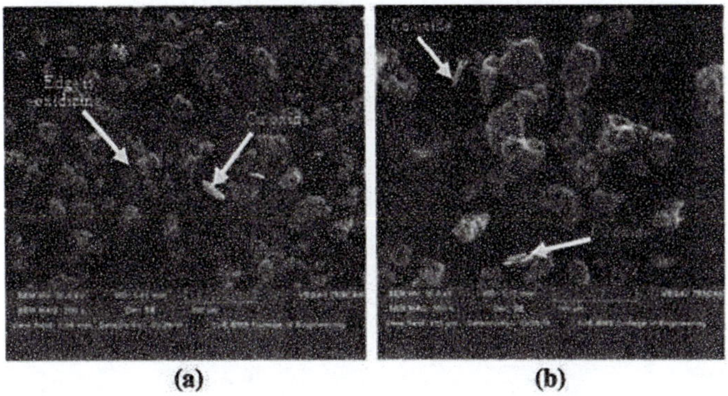

Fig. 3 (**a**, **b**) SEM images Copper powder

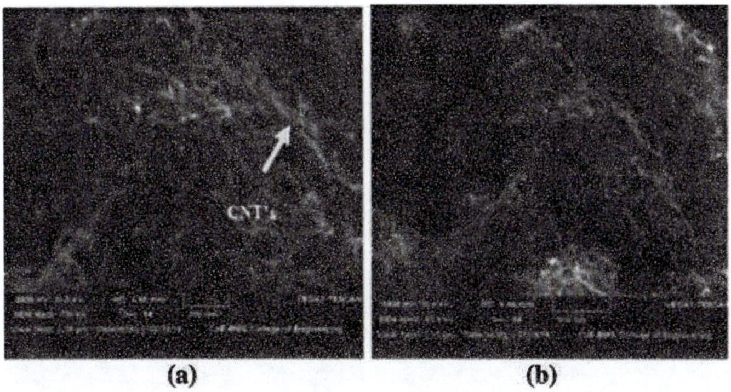

Fig. 4 (**c**, **d**) SEM images of Multi-walled Carbon Nanotubes

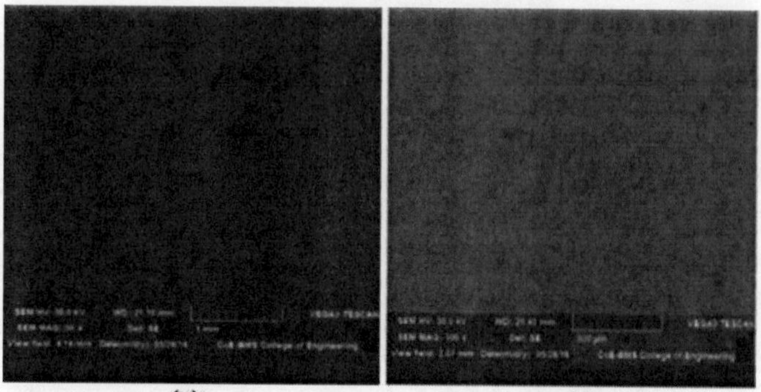

(a) (b)

Fig. 5 SEM images of the Sintered copper specimen (Cu+MWCNT0%)

(a) (b)

Fig. 6 SEM images of Sintered specimen with composition (Cu + MWCNT 0.5%)

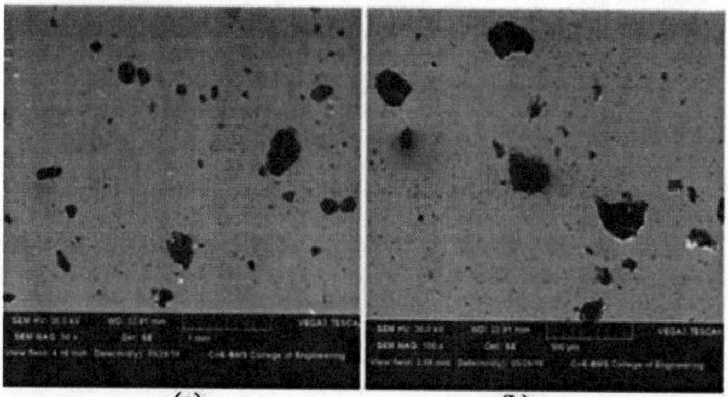

(a) (b)

Fig. 7 SEM images of Sintered specimen with composition (Cu + MWCNT 1%)

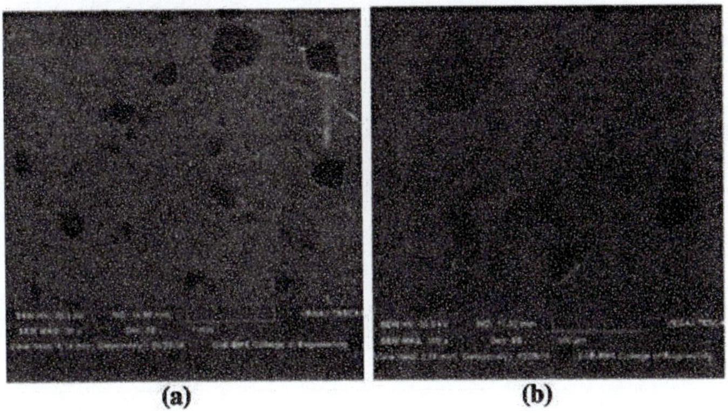

Fig. 8 SEM images of Sintered specimen with composition (Cu + MWCNT 1.5%)

3.9 *Wear Testing Results*

The test was conducted according to ASTM standard G99 obtained under different load and speed conditions [9] for five specimens including pure copper specimen and the following reading was obtained then wear results are tabulated. Specimen: Dia. = 8 mm; Length = 25 mm; Track diameter: 130 mm.

Graphs plotted to wear loss at different load and speed conditions

3.10 *Microstructure of Wear Tested Specimen*

(See Figs. 9, 10, 11, 12, 13, 14 and 15).

Fig. 9 Effect of CNTs weight percentages on the hardness of Cu + MWCNTs composites

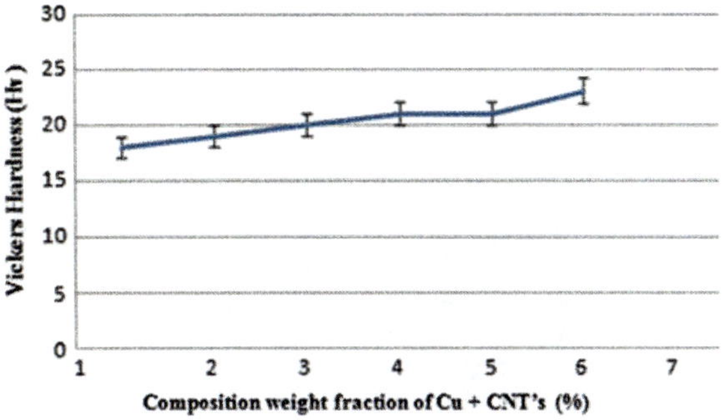

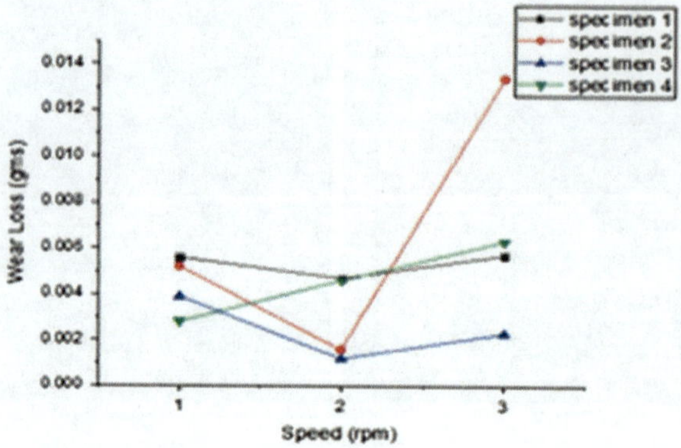

Fig. 10 Above graph shows wear loss versus speed for a constant load of 2 kg

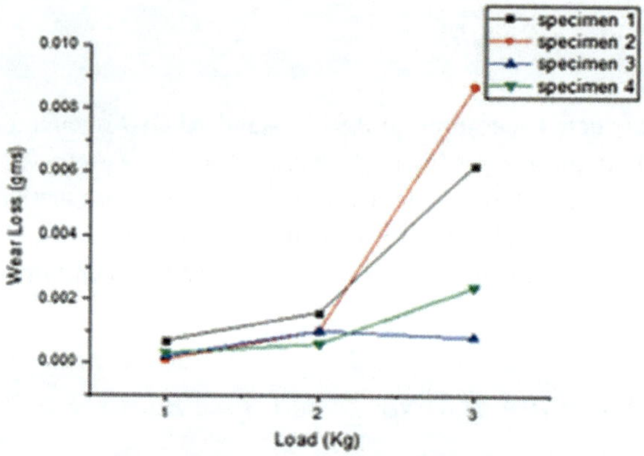

Fig. 11 Above graph shows wear loss versus load for a constant speed of 150 rpm

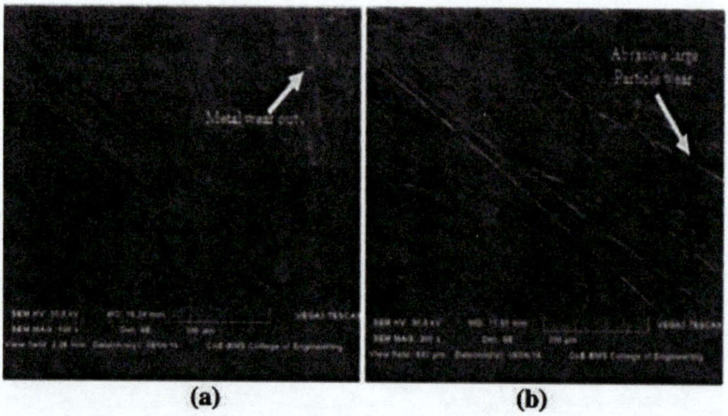

Fig. 12 SEM images of wear tested specimen with composition (Cu + MWCNTs 0%)

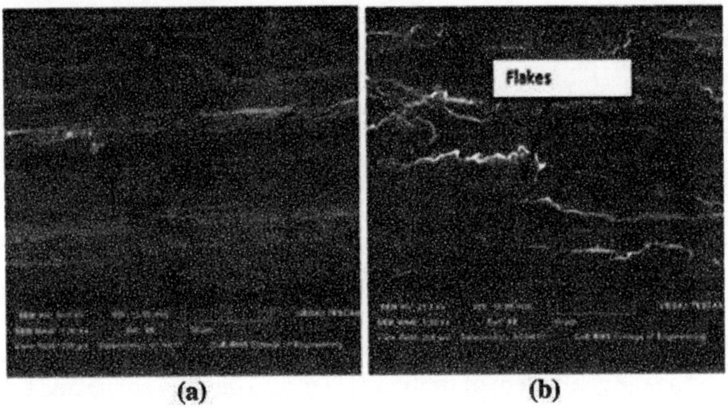

Fig. 13 SEM images of wear tested specimen with composition (Cu + MWCNTs 0.5%)

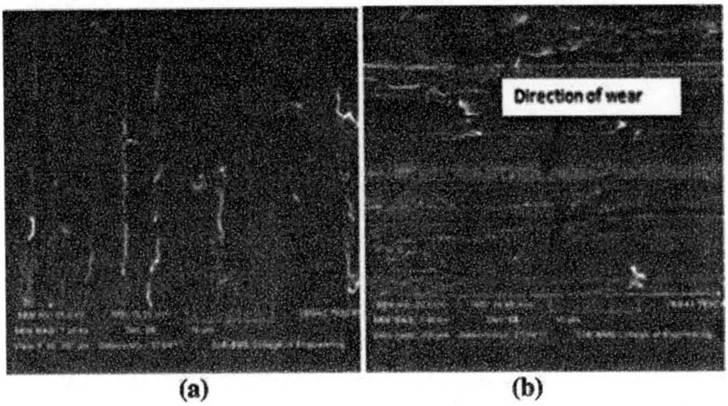

Fig. 14 SEM images of wear tested specimen with composition (Cu + MWCNTs 1%)

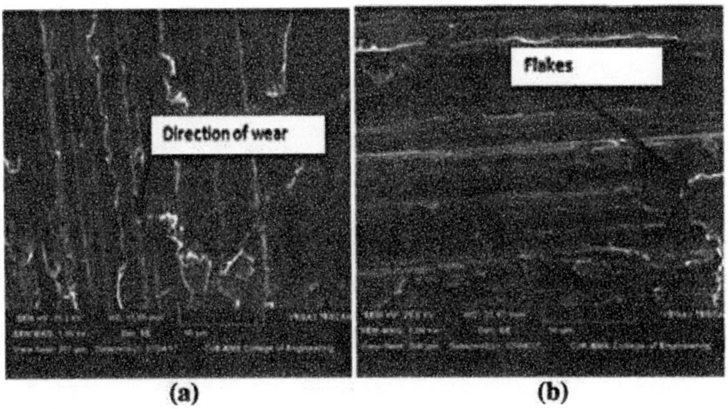

Fig. 15 SEM images of wear tested specimen with composition (Cu + MWCNTs 1.5%)

4 Results and Discussion

4.1 Wear a Test Results Graph that Was Obtained While Performing Tests

Wear test is conducted by using pin on disk method [10]. This method determines wear out of material under dry sliding conditions. Initially, the weight of all specimens which are to be tested is noted down before the mounting of the specimen to specimen holder and disk track is cleaned throughout the experiment. Tribo-meter is used for measuring wear and coefficient of friction [11] in turn directly connected to computer system from which graphs are obtained as shown below.

The wear test was conducted for five specimens including pure copper specimen [12] and following readings were obtained. Test was conducted according to ASTM standard G99 obtained under different load and speed conditions.

Only for the first specimen has shown below and remaining shown in a graph.

4.1.1 Graphs of Pure Copper (1st) Specimen

Figure 16 indicates wear of the pure copper increases quickly and steadily increased and then giving significant high wear rate for 1 kg load and 150 rpm. Figure 17 shows wear of the copper increases quickly and after some period it increases steadily and later it is found slowly decrease in wear rate for 2 kg load at 150 rpm. In (Fig. 18), it is found that wear of the copper increases slowly and then steadily increases and becomes stable after some time. Then wear rate of specimen found stable at 150 rpm for 3 kg load.

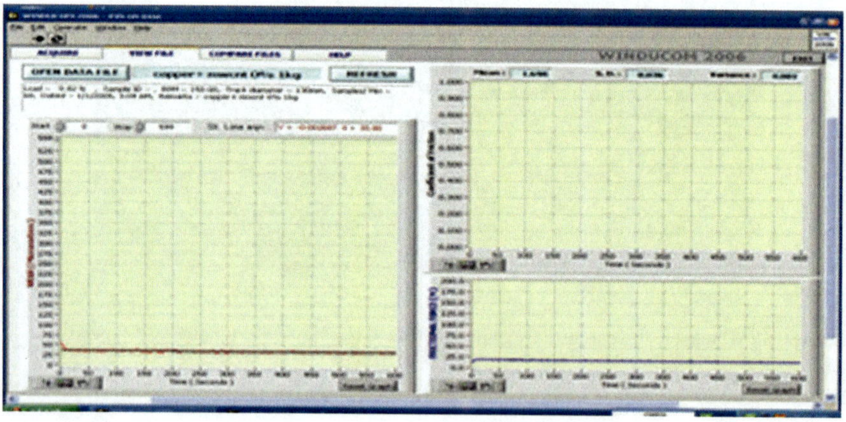

Fig. 16 Wear and friction coefficient graph for pure copper for load 1 kg and with speed 150 rpm

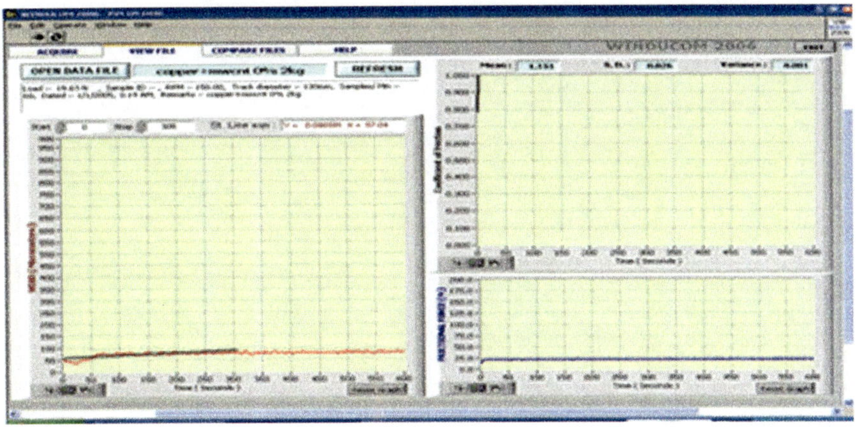

Fig. 17 Wear and friction coefficient graph for pure copper for load 2 kg and with speed 150 rpm

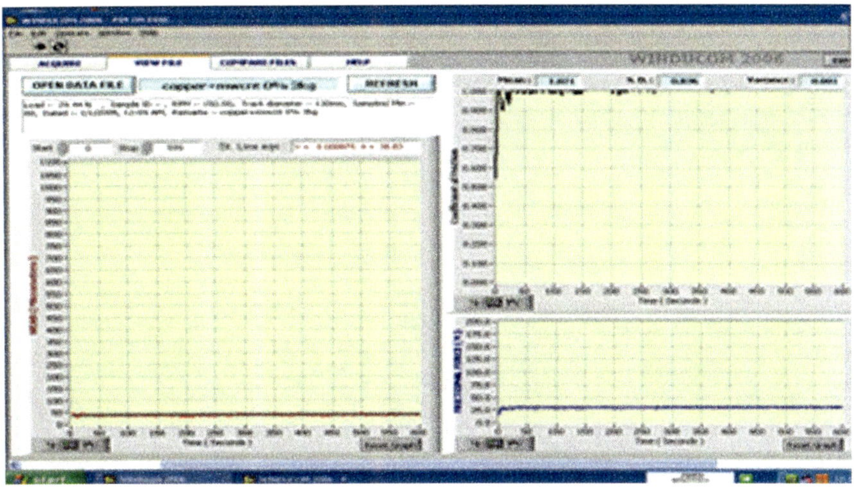

Fig. 18 Wear and friction coefficient graph for pure copper for load 3 kg and with speed 150 rpm

Figure 19 indicates the wear of the copper steadily increases throughout and then it remains same, thus giving constant flow of wear rate at 200 rpm. Figure 20 wear of copper is fluctuating in the initially and stables at the end at 300 rpm. Figure 21 shows the same behavior of wear and then stables at 400 rpm.

For rest of the specimen, the above-said behavior is observed while testing.

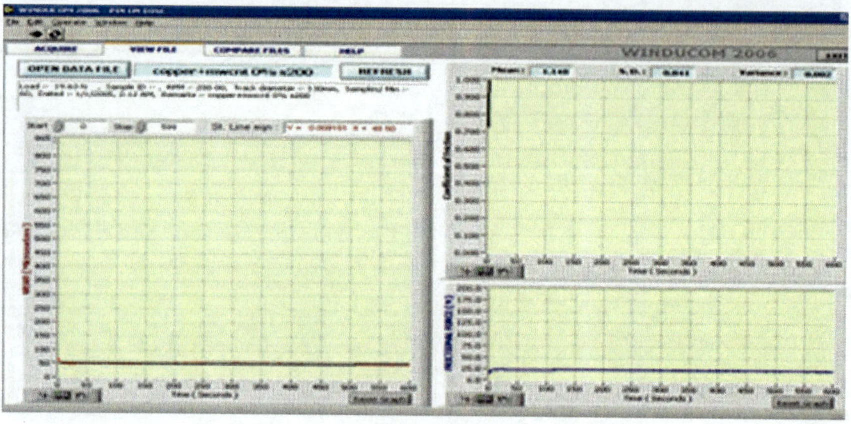

Fig. 19 Wear and friction coefficient graph for pure copper for load 2 kg and with speed 200 rpm

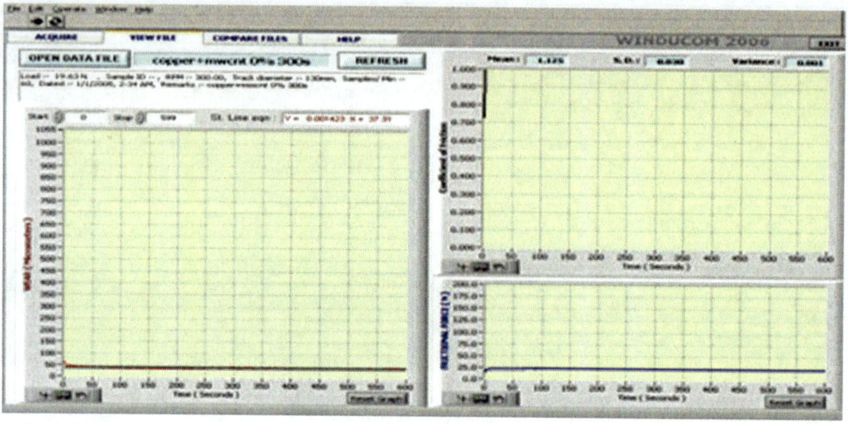

Fig. 20 Wear and friction coefficient graph for pure copper for load 2 kg and with speed 300 rpm

4.1.2 Wear Coefficient Results

Wear coefficient of all wear tested specimen is determined and each specimen is compared to pure copper. Wear coefficients of friction graphs for different percent of CNTs as shown below.

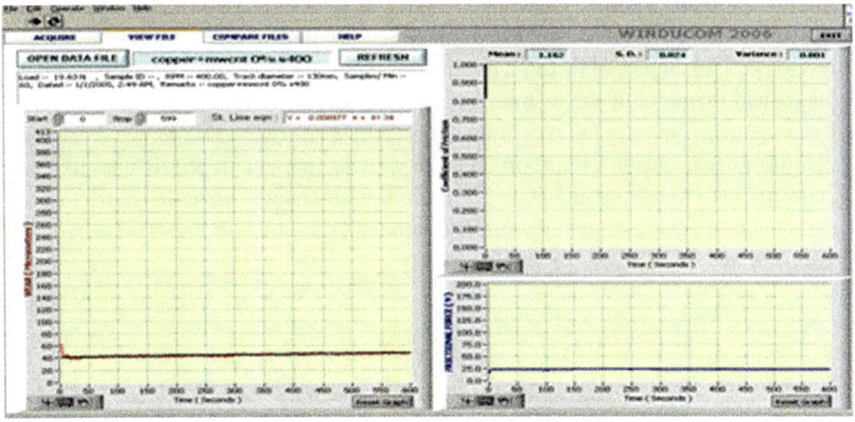

Fig. 21 Wear and friction coefficient graph for pure copper for load 2 kg and with speed 400 rpm

4.2 Wear Rate Results

Wear rate of all wear tested specimen is determined and each specimen is compared to pure copper.

Wear rate graphs for different percentage of CNTs

5 Conclusion

As the density of both copper powder and carbon nanotubes are different therefore powder metallurgy process is carried out to prepare different weight fraction of nanocomposites and thereafter wire EDM process to be followed to obtain the required testing specimens.

(Figures 3a, 4, 5, 6, 7 and 8i) SEM images show the microstructure of test specimens and show that there is even distribution carbon nanotubes in copper matrix and also come across cavities, deep holes, scratches, etc.

Initiating with hardness test, it is experimented that there is slight increase in hardness value of nanocomposites as the percentage of the carbon nanotubes increased as reinforcement's shown in Fig. 1. From the graph, it is shown that an increase of 5% of hardness value for the addition of 1.5% carbon nanotubes compared to the pure copper specimen.

Secondly, wear properties such as adhesive wear are determined by conducting wear test, from wear test conducted the wear rate and coefficient of friction are determined and graphs (Figs. 16, 17, 18 and 19) are obtained. It indicates that as the percentage of carbon nanotubes increases as reinforcement in composites there is decrease in both friction and wear rate.

(Figures 12a, 13, 14 and 15h) SEM images show adhesive wear properties as track direction and formation of flakes. This ensures there is a proper metal to metal contact is happened throughout the testing. This further ensures that carbon nanotubes are the reasons for reducing the wear rate (Figs. 20, 21, 22 and 23).

Uniform distribution of carbon nanotubes in the copper matrix is also likely an enhancement in mechanical and wears properties and hence decreases in friction and wear rate comparing to the pure copper composite specimen. It can be referred from (Figs. 24 and 25).

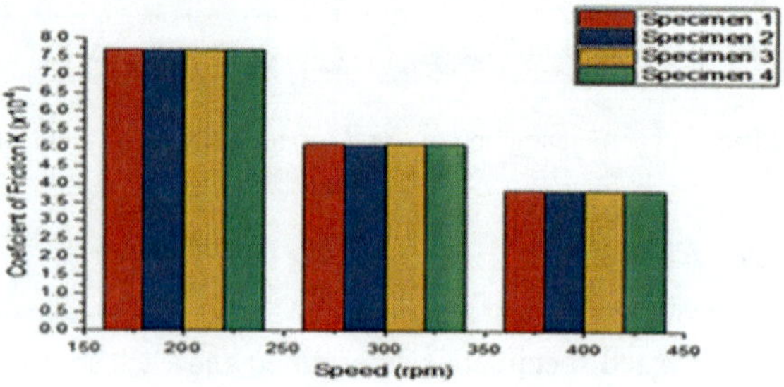

Fig. 22 Coefficient of friction versus speed of constant load with varying speed

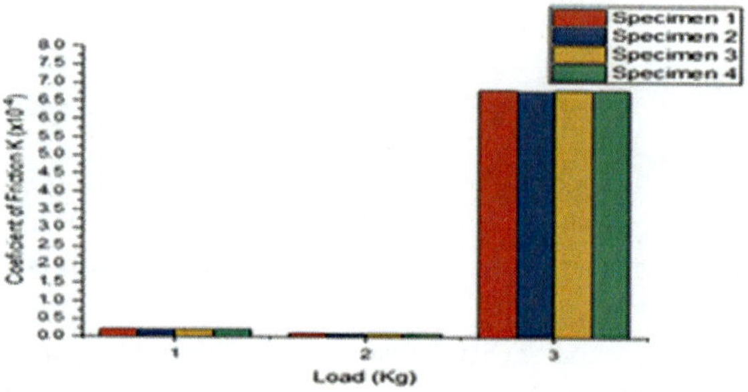

Fig. 23 Coefficient of friction versus load of constant speed with varying load

Fig. 24 Wear rate versus speed of constant load with varying speed

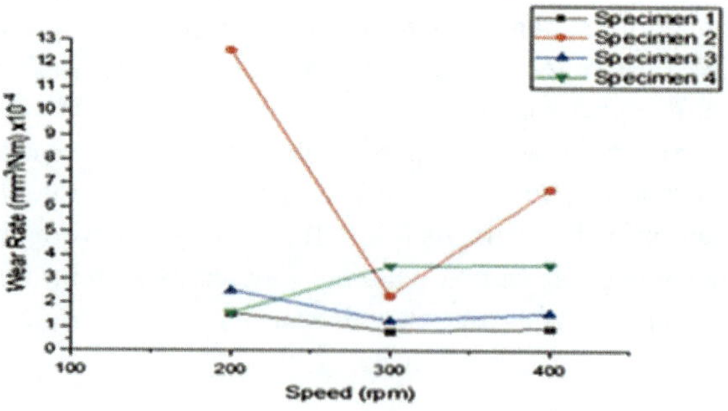

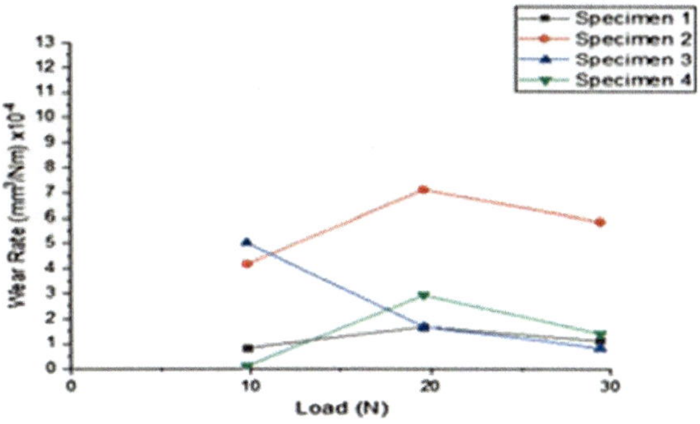

Fig. 25 Wear rate versus load of constant speed with varying load

References

1. Maqbool A, Hussain MNBA, Khalid FAHA, Kim M. Mechanical characterization of copper coated carbon nanotubes reinforced aluminum matrix composites
2. Eiroma K, Jokiniemi J, Auvinen A, Sipiläinen-Malm T, Tapper U Johanna
3. Forsman, Hult E-L, Koskela P, Sarlin J. VTT Technical Research Centre of Finland; Espoo, Finland."Development of Conductive Carbon Coated Copper Nanoparticle Inkjet Fluid"
4. Journal on "Carbon Nanotube-Based Composites-A Review" Minerals & Materials categorization & Engineering, vol 4, no 1, pp 31–46, 2005 jmmce.org
5. Hussain Z, Yusoff M (2013) International Journal of "Effect of Sintering Parameters on Microstructure and Properties of Mechanically Alloyed Copper-Tungsten Carbide Composite Materials", vol 1, no 3, August 2013
6. Rajkumar K, Aravindan S. Tribological performance of microwave sintered copper-CNT composites
7. Tripathi SM, Bholanath TS. Synthesis and Study of Applications of Metal coated Carbon Nanotubes. https://phys.org/news/2006-03-superconductivity-multi-walled-carbon-nanotubes
8. International Journal For Engineering & Technology IJET-IJENS Vol:10 No:04. Finite Element Modeling of Multi-Walled Carbon Nanotubes
9. Daoushb WM, Moa CB, Lima BK, Honga SH, Nama DH. Electrical and mechanical properties of carbon nanotubes reinforced copper nanocomposites fabricated by electroless deposition process
10. Patil RKG, Ramesh C, Chatpalli V. Wear properties of copper and carbon nanotubes. Nanocomposites
11. Bhaumik S, Prabhu S, Vinayagam BK. School of Mechanical Engineering
12. SRM University, Chennai-603203, India. Mechatronics Engineering Department, SRM University, Finite element modeling and analysis of zigzag and armchair type single wall carbon nanotubes

Fabrication, Characterization and Parametric Optimization on Drilling of Heat-Treated A356-SiCp Composites

K. Vinoth Babu, S. Suresh Kumar, P. Ganesan, S. Marichamy, and D. Madan

Abstract This study is dealing with the production, characterization and drilling of aluminum alloy reinforced with SiC particulates. The stir casting route was employed to prepare the composites. The specimen was cut from the prepared composites for the heat treatment. The microstructural analysis was done on the prepared composite which ensures the homogeneous dispersion of SiC particle in the matrix. The incorporation of particulates enhances the mechanical properties of the composite. Based on experimental design, the drilling studies are conducted on the heat-treated specimen and the optimizations of the drilling factors are done through grey relational analysis (GRA). The variables chosen for the experimental work are point angle, fraction of reinforcement, feed and cutting speed. The thrust force and surface finish are considered for an optimization with an objective to minimize both responses.

Keywords Drilling · Heat-treated AMMC · Volume fraction · GRA · ANOVA

1 Introduction

In metal matrix composites (MMCs), one major constituent is metallic matrix and the other constituents may be reinforcements. Aluminum alloy-based composite reinforcing with SiC is used extensively, owing to easy production, higher hardness with lower density and also SiC has higher temperature resistant properties. Owing to these properties, Al-based MMCs are far and wide used in engineering applications. Aluminum metal matrix composites (AMMCs) are fabricated using many processing techniques, but among those techniques stir casting method is cost-effective method

K. Vinoth Babu · P. Ganesan · S. Marichamy · D. Madan
Department of Mechanical Engineering, Sri Indu College of Engineering and Technology, Hyderabad, Telangana 501510, India

S. Suresh Kumar (✉)
Faculty of Mechanical Engineering, Kalasalingam Academy of Research and Education, Srivilliputtur, India
e-mail: sureshme48@gmail.com

A. Arockiarajan et al. (eds.), *Advances in Industrial Automation and Smart Manufacturing*, Lecture Notes in Mechanical Engineering,
https://doi.org/10.1007/978-981-15-4739-3_72

[1–3]. But the improvised properties of composites become the demanding criterion for shaping the composites and convert the engineering components into its final shape [4]. So optimal condition is required for cost-effective machining of MMC component in order to obtain high-quality surface finish with required metal removal [5].

Surface finish depends upon the drill tool materials such as HSS, carbide and coated HSS drills. It is observed that the carbide drills are better than HSS drills while machining Al/17% SiCp MMC [6]. But machining MMC with coated carbide drills is more advantageous than the carbide tools to get quality surface finish, and it is reported by [7]. This is by reason of preventing the development of built-up edge by the coating on the drill. Ramulu et al. [8] noticed that PCD drills provide better surface finish than any other drills but not having the cost advantage. The surface texture depends on the process variables. It is implicit that speed has less influence on surface finish [9] (basavarajappa). The reason is while drilling MMCs, particulates are stuck between surface of the tool and material [10]. While drilling MMCs, the thrust force acts perpendicular to the hole axis and the excessive thrust force retained will lead to damage the workpiece. The factor such as feed, speed, point angle and volume fractions influences the thrust force [11]. It is found that when the feed rate increases, the thrust force is also increased and no predominant changes in feed rate while raising the spindle speed. Thrust force also depends on the drill materials, and the higher thrust force is experienced while drilling composites with coated carbide drills than carbide drills. Thrust force also depends on the reinforcement's volume fraction and it increased with volume. It is increased with the increasing hardness when tested the heat-treated composite material [12]. Tosun et al. [13] discovered the damage in the sub-surface section of the materials and which leads to poor finish. This was happened owing to increased point angle. Taguchi uses orthogonal arrays to define the experimental plans and using this concept, optimization is being done only for single characteristics problem. For dealing with multi-characteristics, GRA can be used effectively by converting all the responses into a single grade. In order to improve the experimental process, the setting of optimized machining parameter is required which improves the quality characteristics [14–16]. With the objective to minimize surface roughness and burr height, the optimal process variables are determined by Tosun [17] using GRA. Haq et al. [18] have attempted to optimize the different process variables while drilling Al-SiC composite material.

From the literature evaluation, it is implicit that many researchers have tried to determine the significance of variable on affecting the machinability responses of composite material. However, the reported research work on machinability of heat-treated composite material and its optimization is limited. The aim of this article is to process the composites through stir casting process, and material characterization is done. By varying the process variables, the drilling is done on the heat-treated A356 alloy-based composite and the optimization of variables also processed [19, 20]. ANOVA is employed to estimate the role of variables affecting the multiple responses. The best parameters' setting is obtained from GRA being validated experimentally.

2 Experimental Procedure

A356 (Al-7Si-0.3 Mg) is a precipitation hardening alloy, and the T6 condition (solution treatment followed by precipitation or age hardening) was used for preparing MMC. Through stir casting route, composite consists of A356 base alloy with different volume ratios of SiC (0%, 10% and 20%) which were fabricated. The alloy was reinforced with SiC of average particle of 23 μm. The prescribed solution treatment is 10 h at 525 °C and age hardening at 165 °C for 8 h. The drilling studies were carried out on the heat-treated MMC specimen size of 50 × 55 × 15 mm in a Hartford CNC vertical machining with Kistler piezoelectric dynamometer which is shown in Figs. 1 and 2. The selected 5 mm drills with varying point angles (90°, 120°, 140°) were coated with TiAlN used throughout the experimental work.

The microstructural features of the AMMC composite as-cast are observed using Leica optical microscope which is presented in Fig. 3. Piezoelectric dynamometer (make: Kistler) is used to measure the thrust force induced. Surface roughness tester (make: Mitutoyo SJ 201) is used to determine the finish on the machined surface. The readings are taken three times at three different positions and averages

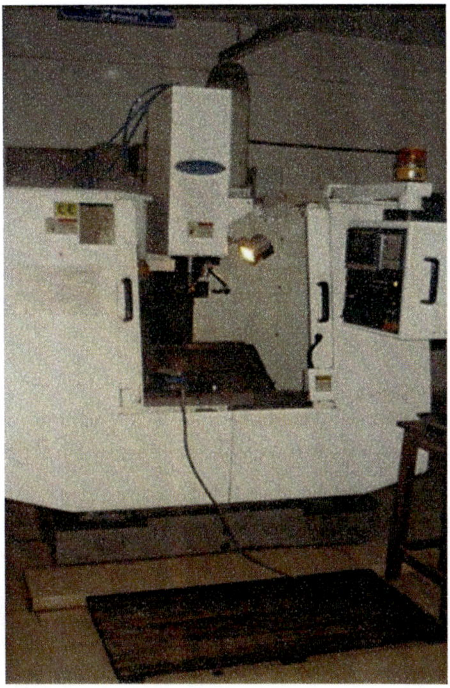

Fig. 1 CNC machining center

Fig. 2 Kisteler dynamometer

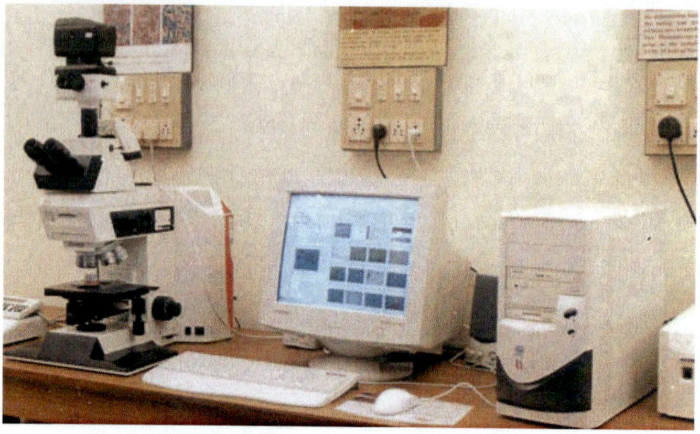

Fig. 3 Leica optical microscopes with Qwin image analyzer

of these readings are noted. Based on Taguchi's L9 orthogonal arrays, the experiments are conducted to reduce the experimental cost of 81 experimental runs (3^4) to 9 experiments.

3 Results and Discussion

3.1 Fabrication and Characterization

The alloy is melted in a crucible at 740 °C and preheated SiC particles are added to the melt via powder addition mechanism with constant stirring at 350 rpm. The mixture is poured into the mold, and the melt is permitted to solidify completely.

Fig. 4 MMC specimens used for drilling studies

The cast is machined to form a required shape. Figure 4 shows the specimens for drilling studies which are made from the produced composite. Figure 5 exhibits the microstructure of the produced composite materials of varying volume fractions.

The micrograph (Fig. 5c) shows the even dispersal of reinforced particulates in the A356 base alloy. The matrix shows the presence of light-colored dendrites as shown in Fig. 5a and the dendrites in damaged/broken stage in Fig. 5b which helps in particulates dispersion. The existence of SiC ceramic particulates in the alloy enhances the hardness of the composites. The hardness observed for AA 356 is 93 BHN, and 10 vol.% and 20 vol.% SiC reinforced matrices are 104 BHN and 118 BHN, respectively. The tensile strength of the alloy and 10 and 20 vol.% reinforced composites are 245 MPa, 264 MPa and 281 MPa, respectively.

3.2 *Machining of MMC*

The drilling is carried out by using carbide tools coated with TiAlN of 5 mm diameter (Fig. 6). Table 1 presents the experimental results observed during drilling process.

3.3 *GRA*

It is a process employed for optimizing the multi-objective machining characteristics. The thrust force and surface roughness are the major drilling characteristics considered for the evaluation. The lower the better level is adopted for both the characteristics. Equation 1 is used to normalize the values [18, 19].

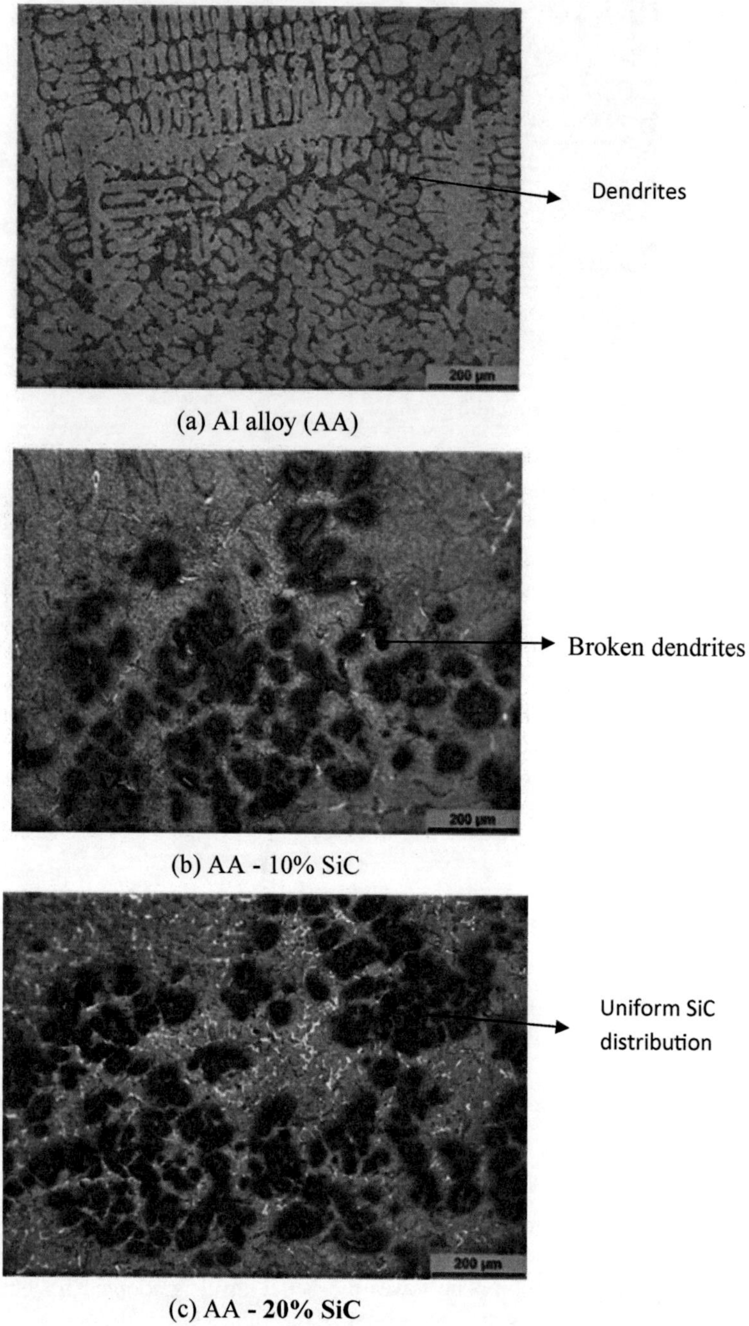

Fig. 5 Optical micrographs

(a) Heat treated alloy (AA)

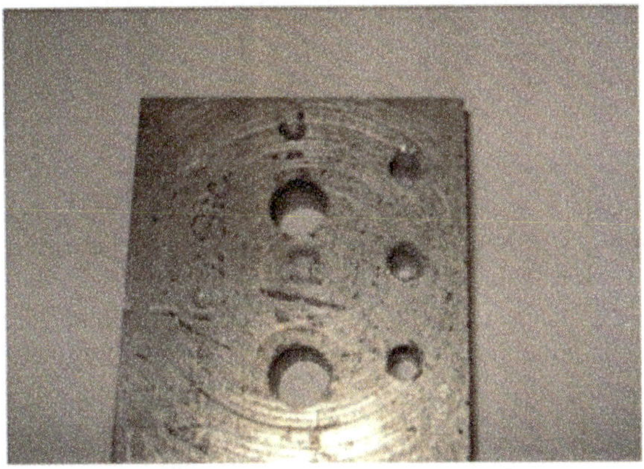

(b) AA – 10 vol. % SiC

(c) AA – 20 vol. % SiC

Fig. 6 Drilled alloy and composites

Table 1 Experimental results

S. No	Speed (S) (rpm)	Feed (F) (mm/rev)	Point angle (PA) (deg)	Volume of SiC (V) (%)	Surface roughness (SR) (μm)	Thrust force (TF) (N)
01	1000	0.05	90	0	4.530	106
02	1000	0.15	120	10	1.740	83.2
03	1000	0.25	140	20	2.060	111.2
04	2000	0.05	120	20	2.420	41.8
05	2000	0.15	140	0	1.530	113.4
06	2000	0.25	90	10	2.450	174.3
07	3000	0.05	140	10	1.540	64.90
08	3000	0.15	90	20	3.396	109.1
09	3000	0.25	120	0	2.123	118.3

$$X_{ij} = \frac{(y_{ij})\max - y_{ij}}{(y_{ij})\max - (y_{ij})\min} \tag{1}$$

Table 2 shows the normalized data of GRA process. Thereafter, grey relational coefficient (GRC) is calculated from the Eq. 2.

Here, x_i^0 is the ideal normalization value for the ith experimental case. Since equal weight is given to all the process variables, the coefficient value considered for the analysis is $\xi = 0.5$.

$$\delta_{ij} = \frac{\min_i \min_j (x_i^0 - x_{ij}) + \varepsilon \max_i \max_j (x_i^0 - x_{ij})}{(x_i^0 - x_{ij}) + \varepsilon \max_i \max_j (x_i^0 - x_{ij})} \tag{2}$$

The average of the GRC of all the variables is considered as grey relational grade which is determined by using the Eq. 3.

Table 2 Normalized values

S. No.	SR	TF
1	0.000	0.515
2	0.930	0.688
3	0.823	0.476
4	0.703	1.000
5	1.000	0.460
6	0.693	0.000
7	0.997	0.826
8	0.378	0.492
9	0.802	0.423

Table 3 GRC and GRG values

Ex. No.	GRC		GRG	Rank
	SR	TF		
1	0.333	0.508	0.421	9
2	0.877	0.615	0.746	3
3	0.739	0.488	0.614	5
4	0.628	1.000	0.814	2
5	1.000	0.481	0.740	4
6	0.620	0.333	0.477	7
7	0.993	0.741	0.867	1
8	0.446	0.496	0.471	8
9	0.717	0.464	0.590	6

$$\alpha_j = \frac{1}{m}\sum_{i=1}^{m} \delta_{ij} \quad (3)$$

where α_j is GRG for the j-th experiment and m denotes the drilling responses. Table 3 shows the values of GRC and GRG.

GRA helps to obtain a single performance grade from multiple response system. Better performance of the experimental work is noted corresponding to higher GRG. Table 4 shows the response values of each process variable.

Figure 7 presents the effect of variables on the multiple responses. From the study, it is found that the best drilling conditions for obtaining minimum response values are speed of 2000 rpm, feed rate at 0.05 and the point angle of 140° with SiC volume fraction of 10%. From Table 5, the combination of all the variables with their levels is identified. Furthermore, a statistical analysis of variance is adopted in the GRA in order to identify the significance of variables which are presented in Table 5. The point angle is the predominant factors (71.238%) on affecting TF and SR followed by the feed rate (14.544%), volume fraction (9.182%) and speed (5.025%). Equation 4 is used to forecast the best condition from the GRG which is shown in Table 6.

Table 4 Response table

Variables	Level 1	Level 2	Level 3	Difference between maximum and minimum	Rank
S	0.5935	0.6769	0.6429	0.0834	4
F	0.7006	0.6525	0.5602	0.1404	2
PA	0.4560	0.7168	0.7405	0.2844	1
V	0.5838	0.6968	0.6328	0.1130	3

Average GRG = 0.6378

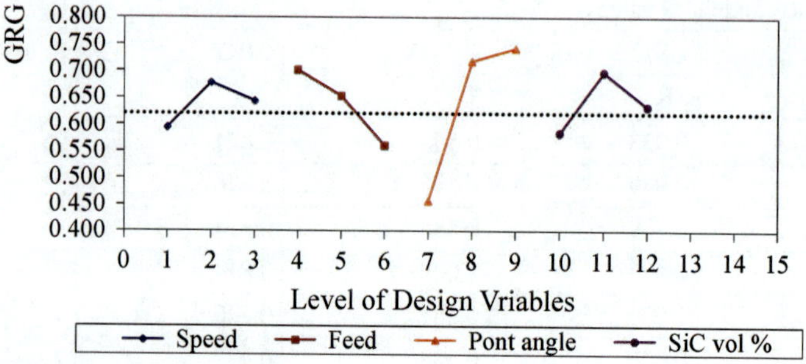

Fig. 7 Effect of machining parameters

Table 5 ANOVA

	DoF	SS	Mean SS	Contribution %	Rank
Speed	2	0.0105	0.0053	5.025	4
Feed rate	2	0.0305	0.0153	14.554	2
Point angle	2	0.1495	0.0747	71.238	1
Vol% of SiC	2	0.0193	0.0096	9.182	3
Total	8	0.2099	0.1049	100	

Table 6 Confirmation test

	Initial condition (rank 1 grey code)	Optimum conditions	
		As predicted	From experiments
Optimum condition (Levels)	A2B1C3D2	A2B1C3D2	A2B1C3D2
Grade	0.867	0.901	0.913
Upgrading in grade		0.034	0.046

$$\widehat{\alpha} = \alpha_m + \sum_{i=1}^{q} (\overline{\alpha}_i - \alpha_m) \tag{4}$$

q—no. of variables

The confirmation test is to be made so as to authorize the findings. It shows good agreement with the predicted values. The notable upgrading in the experimental part for the optimal parameter setting is presented in Table 7.

4 Conclusions

The Al MMC with varying volume fractions of SiC has been produced through stir casting route successfully. The drilling study at varying input conditions was carried out, and the following conclusions were drawn.

- The improved tensile strength and hardness of the composite were observed with increased volume of reinforcement.
- GRA is adopted to identify the optimal conditions for heat-treated Al-SiC composite.
- The optimum conditions obtained are A2B1C3D2 (Speed—2000 rpm, feed rate—0.05, point angle at 140^0 and volume fraction of 10% of SiC.
- The most significant factor affecting the combined responses is the point angle (71.238%) followed by feed rate (14.544%), volume fraction (9.182%) and speed (5.025%).
- Results drawn from grey relational analysis and ANOVA approaches are comparable. The confirmation experiment is also done to confirm the result, and it shows the improvement.

References

1. Zhong Z, Hung NP (2002) Grinding of alumina/aluminum composites. J Mater Process Technol 123:13–17
2. Hocheng H, Lei WT, Hsu HS (1997) Preliminary study of material removal in electrical-discharge machining of SiC/Al. J Mater Process Technol 63:813–818
3. Rajan TPD, Narayan Prabhu K, Pillai RM, Pai BC (2007) Solidification and casting/mould interfacial heat transfer characteristics of aluminum matrix composites. Compos Sci Technol 67:70–78
4. Mueller F, Monaghan J (2001) Non-conventional machining of particle reinforced metal matrix composites. J Mater Process Technol 118:278–285
5. Thirumalai Kumaran S, Uthayakumar M (2014) Investigation on the dry sliding friction and wear behavior of AA6351-SiC-B4C hybrid metal matrix composites. Proc I Mech E Part J J Eng Tribol 228:332–338
6. Muratoglu M (2004) The drilling of Al/SiCp metal–matrix composites. Part II: workpiece surface integrity. Compos Sci Technol 64:1413–1418
7. Basavarajappa S, Chandramohan G, Rao KVN, Radhakrishanan R, Krishnaraj V (2006) Turning of particulate metal matrix composites—review and Discussion. Proc Inst Mech Eng Part B J Eng Manuf 220:1189–1204
8. Ramulu M, Rao PN, Kao H (2002) Drilling of (Al_2O_3)p/6061 metal matrix composites. J Mater Process Technol 124:244–254
9. Basavarajappa S, Chandramohan G, Prabu M, Mukund K, Ashwin M (2007) Drilling of hybrid metal matrix composites-workpiece 106 n surface integrity. Int J Mach Tools Manuf 47:92–96
10. Paulo Davim J, Conceic CA, Antonio (2001) Optimal drilling of particulate metal matrix composites based on experimental and numerical procedures. Int J Mach Tool Manuf 41:21–31
11. Manna A, Bhattacharayya B (2003) A study on machinability of Al/SiC-MMC. J Mater Process Technol 140:711–716

12. Pramanik A, Zhang LC, Arsecularatne JA (2007) An FEM investigation into the behavior of metal matrix composites: tool-particle interaction during orthogonal cutting. Int J Mach Tools Manuf 47:1497–1506
13. Tosun G, Muratoglu M (2004) The drilling of an Al/SiCp metal-matrix composites. Part I: microstructure. Compos Sci Technol 64:299–308
14. Metin K (2011) Modelling the effect of surface roughness factors in the machining of 2024Al/Al_2O_3 particle composites based on orthogonal arrays. Int J Adv Manuf Technol 55:911–920
15. Lin ZC, Ho CY (2003) Analysis and application of grey relation and ANOVA in chemical–mechanical polishing process parameters. Int J Adv Manuf Technol 21:10–14
16. Manikandan N, Binoj JS, Varaprasad KC, Sabari SS, Raju R (2019) Investigations on wire spark erosion machining of aluminum-based metal matrix composites. Lecture Notes in Mechanical Engineering, Advances in Manufacturing Technology, Springer, Singapore, pp 361–369
17. Nihat T (2006) Determination of optimum parameters for multiperformance characteristics in drilling by using grey relational analysis. Int J Adv Manuf Technol 28:450–455
18. Noorul Haq A, Marimuthu P, Jeyapaul R (2008) Multi response optimization of machining parameters of drilling Al/SiC metal matrix composite using grey relational analysis in the Taguchi method. Int J Adv Manuf Technol 37:250–255
19. Palanikumar K, Latha B, Senthilkumar V, Paulo Davim J (2012) Analysis on drilling of Glass Fiber-Reinforced Polymer (GFRP) composites using grey relational analysis. Mater Manuf Process 27:297–305
20. Palanikumar K, Muthukrishnan N, Hariprasad KS (2008) Surface roughness parameters optimization in machining A356/SiC/20% metal matrix composites by PCD tool using response surface methodology and desirability function. Mach Sci Technol 12(4):529–545

A Review on Production of Bioethanol

Rajnish Kumar and Om Prakash

Abstract Rapidly increasing rate of global energy consumption due to increase in population, vehicles, transportation, and modern lifestyle caused consumption of fossil fuel and environmental pollution. Limiting potential of fossil fuel sources has received attention of the investigators to research in the field of renewable biomass-based fuel like ethanol for internal combustion engine. Ethanol is the most commonly researched alcoholic fuel for internal combustion, and it can be produced from sugar containing biomass, starch containing biomass, cellulosic feed stocks, and agricultural residues, etc. The objective of this review is to deal with the recent development in bioethanol production, raw material used for their production, and its production procedure. Generally, there are two technics used such as biological conversion and thermochemical conversion of biomass into bioethanol. Production procedure of bioethanol mainly consists of pretreatments, enzymatic hydrolysis, fermentation, distillation, and filtration. It is found that pretreatment technique of biomass is one of the major challenges for conversion into bioethanol. The selection of pretreatment techniques plays a significant role to increase the efficiency of enzymatic hydrolysis which makes whole process economically feasible. The present review could help researcher to find the renewable resources, and available technics for bioethanol production.

Keywords Bioethanol · Renewable fuel · Biomass · Fossil fuel

1 Introduction

Global energy consumption is increasing rapidly day by day due to increase in population, rate of vehicles in publics and goods transportation, industrialization, and

R. Kumar (✉) · O. Prakash
Mechanical Engineering Department, National Institute of Technology, Patna, India
e-mail: agrajnish@gmail.com

O. Prakash
e-mail: om.prakash@nitp.ac.in

A. Arockiarajan et al. (eds.), *Advances in Industrial Automation and Smart Manufacturing*, Lecture Notes in Mechanical Engineering,
https://doi.org/10.1007/978-981-15-4739-3_73

modern life style. Due to this reason, fuel consumption rate for internal combustion engine increased drastically [1]. As shown in Fig. 1, demand of the fossil fuel increased over the years and it shows a huge difference between oil imports and oil produced in India. To meet these demand of energy consumption, conventional energy sources like petroleum is not bearable. These truths encouraged the researchers to find alternative fuel sources which are renewable as well as sustainable. Bioenergy has got attention of the investigators as alternative renewable energy due to sustainability and availability of its resources. Among the different biofuel resources, alcoholic fuel sources like ethanol are the alternative fuel for internal combustion engine. Biofuel is the main green fuel especially for rural areas [2, 3].

Moreover, conventional fuel sources after burning inside engine produce toxic emissions, greenhouse gas emissions which create global warming and climate alteration. Ethanol is the noble alternative of the non-renewable fuel sources which forms less emissions. The focus to find the alternative fuel was being started after global energy crisis in 1970s [4]. Ethanol is the most researched area as an alternative fuel in the past three decades. Synthesis of ethanol is categorized into two parts which is grain ethanol and biomass ethanol. Ethanol which is produced from sugar crops or starch crops such as corn, barley, sugar beet, cassava, sorghum, millets, rice, and wheat is known as grain ethanol [5]. Moreover, ethanol which is obtained from trees/grasses such as sugarcane, sugarcane tops, bagasse, wheat straw, rice straw, wood chips, and stover's through fermentation process is known as biomass ethanol [6, 7].

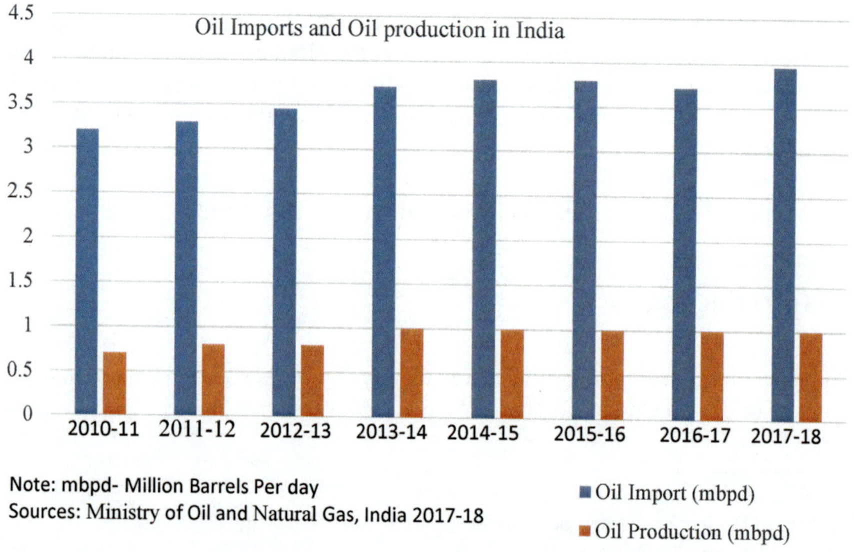

Fig. 1 Crude oil imports and production in India over the year from 2010 to 2018

Table 1 Properties of fuels used in IC engine

Properties	Ethanol	Diesel	Gasoline	Biodiesel
Chemical formula	C_2H_5OH	$C_{12}H_{26}$–$C_{14}H_{30}$	C_7H_{17}	$C_{19}H_{36}O_2$
Molecular weight	46	170–198	101	296
Calorific value (KJ/Kg)	26,700	42,600	43,400	37,100
Octane number	105–110	–	91-98	–
Cetane number	0–5	50	–	46–52
Density (Kg/m^3)	789	846	737	855
Latent heat of vaporization (KJ/Kg)	904	700	350	300
Viscosity (mPa-s)	1.074	3.546		4.51
Flash point (°C)	13	70	43	101
Auto-ignition Temp (°C)	420	256	280	363
Oxygen content (mass%)	34.78	0	0	10.8
Carbon content (mass%)	52.17	84.7–84.74	83.16	77
Hydrogen content (mass%)	13.04	15.3–15.15	16.83	12.16
Sulfur content (mass%)	0	<50		<10

Sources [11–14]

2 Ethanol as IC Engine Fuel

Ethanol is an oxygenated alcoholic biofuel which contains oxygen in their chemical composition (C_2H_5–OH). Extra oxygen molecules provide benefits for complete combustion of ethanol blends during burning [8]. Bioethanol is a flammable, non-toxic, colorless, and polar solvent fuel. It is a good water soluble solvent and it is immiscible with petroleum fuel like diesel, petrol, biodiesel, etc., over a wide range of temperature [9, 10] (Table 1).

3 Materials and Methods for Ethanol Production

3.1 Raw Materials

Currently, majority of ethanol produced through biotechnological process. Agricultural products are used to produce 95% of ethanol in the world [15]. Available resources are categorized into two groups which are grain materials and cellulosic materials or biomass. Ethanol which is obtained from grain materials like sugar crops or starch crops such as corn, barley, sugar beet, cassava, sorghum, millets, rice, and wheat is known as grain ethanol. Likewise, ethanol which is obtained from trees/grasses such as sugarcane, sugarcane tops, bagasse, wheat straw, rice straw, wood chips, and stover is known as biomass ethanol [5] (Table 2).

Table 2 Production process for ethanol as per available resources [5, 12]

Production process	Feed stocks categories	Resource
Biological conversion (Fermentation)	Sugar containing feed stocks	Sugarcane juice, sugar beet juice, sorghum, sugarcane molasses
	Starch containing feed stocks	Sorghum, millets, cassava
	Cellulosic feed stocks	Sugarcane, wheat straw, rice straw, stover, bagasse, wood chip
Thermo-chemical conversion (gasification and synthesis)	Agricultural residues	Wheat straw, rice straw, bagasse, forestry waste materials
	Carbon containing petroleum residues	Petroleum coke

3.2 Methods

Ethanol is a renewable source of energy which can be produced from any biological material that contains sugars or feed stocks that can be converted in sugars [16]. It can be obtained from sugar containing feed stocks and starch containing feed stocks through biochemical conversion of feed stocks which is as fermentation process. Figure 2 explains the detail synthesis procedure of ethanol from lignocellulosic materials like sugar containing feed stocks, starch containing feed stocks, and cellulosic feed stocks through the fermentation process [11]. Agricultural residues, agricultural waste, and cellulosic feed stocks can also be used for the production of ethanol through thermochemical conversion of feed stocks which named as gasification process. The detail synthesis procedure of ethanol from agricultural residues like rice straw, wheat straw, and forestry waste materials, etc., is explained by Fig. 2. Petroleum residues such as coke pit and petroleum coke are used to produce ethanol through gasification process followed by synthesis [12].

Pretreatment is the initial step for conversion of any biomass into ethanol. Pretreatment technique of biomass is one of the major challenges for conversion into bioethanol. Purpose of pretreatment is to increase surface area, remove hemicellulose, and make cellulose easier to get enzymes so that it can rapidly convert into fermentable sugar [17]. Generally, pretreatment method consists of physical, chemical, and thermal processes. Physical pretreatment involves milling, grinding, steam explosion, pressure, and temperature [18]. Chemical pretreatment makes easier conversion of cellulose into fermentable sugar through enzymatic hydrolysis. Usually, alkali treatment, acidic treatment, ammonia treatment, and oxidizing agent treatment are the most promising chemical pretreatments. Enzymatic hydrolysis is the second step for bioethanol production from cellulosic biomass. Enzymes are added to pretreated biomass at low pressure and constant temperature for conversion of cellulose and hemicellulose into fermentable sugar.

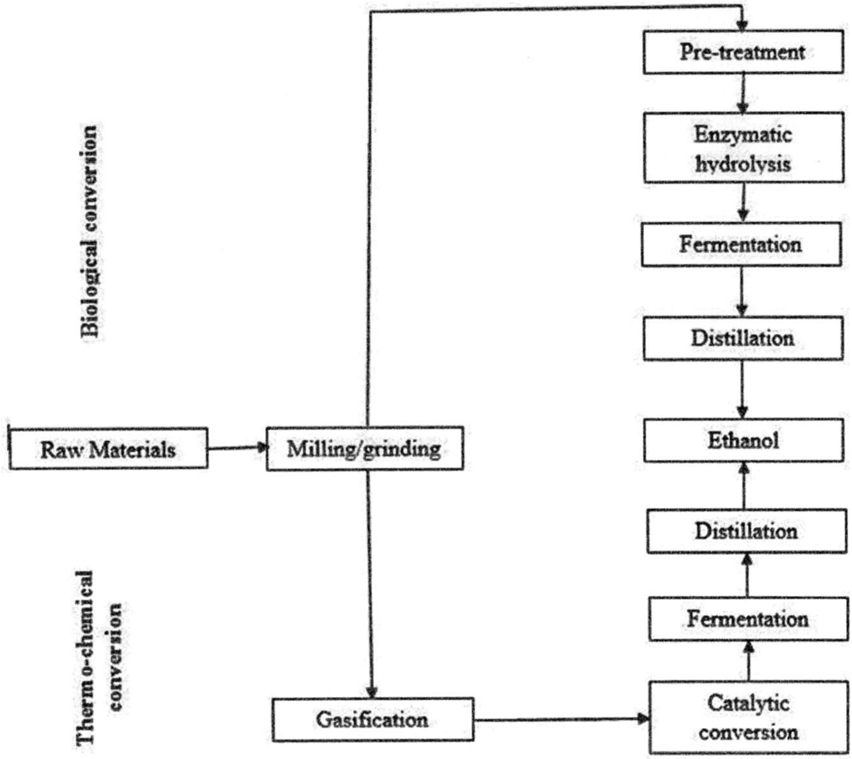

Fig. 2 Detailed production procedure of ethanol

Vlasenko et al. [19] performed for hydrolysis of pretreated rice straw with various acids with concentration (0.5–1% H_2SO_4, 2–3% HCl and 0.5–1% H_3PO_4) and they observed that rice straw pentos converted into monosaccharides solution after 3 h retention time which is suitable for fermentation. Cellulose, hemicellulose and fermentable sugar can be converted into bioethanol by fermentation process. Zhao and Xia [20] carried out fermentation of hydrolysate corn stovers in the presence of the fermentation medium (2 g/L yeast extract, 2.5 g/L KH_2PO_4, 0.25 g/L$CaCl_2$ and 0.25 g/L $MgCl_2$) in a sealed flask at 120 rpm stirrer speed and 30 °C for 72 h and found the ethanol yield 0.393 g/g.

Thammasittirong et al. [21] performed evaluation of ethanol production from sugarcane juice and bagasse both by fermentation method using hydrolysate medium (5 g/L KH_2PO_4, 2 g/L $(NH_4)_2SO_4$, 0.2 g/L $MgSO_4{\cdot}7H_2O$, 1 g/Lpeptone and 5 g/L yeast extract, pH 5.5) and fermentation medium (2% glucose, 1% yeast, and 2% peptone) concluded that the maximum ethanol yield from juice was obtained 2697 kg/ha while the maximum ethanol yield from baggage was obtained 1695.45 kg/ha.

Han et al. [22] studied fermentative ethanol production utilizing waste cake by two-step hydrolysis with α-amylase reported that ethanol produced 46.6 g/L and ethanol yield of 1.13 g/g of reduced sugar and 1.12 g/g of initial dry cake which was maximum ethanol yield among other food waste. An experimental study was done by [23] for acetone, butanol, and ethanol (ABE) production using water algae through fermentation in serum vials media and results obtained that 2.74 g/L of total ABE produced from pretreated algae, 7.27 g/L of total ABE produced from pretreated algae with 1% glucose, 9.74 g/L of total ABE produced from pretreated algae with 2% glucose when xylanase and cellulase enzymes added.

Soam et al. [24] experimentally studied second generation cellulosic ethanol production from rice straw using dilute acid and steam explosion for pretreatment followed by enzymatic hydrolysis and fermentation and analyzed that the obtained ethanol yield was 239 and 253 L by the processing of 1 ton of rice straw and life cycle greenhouse gas emissions was 292 and 288 kg CO_2 per ton of rice straw. Also, results are concluded that cellulosic ethanol production is the sustainable technology for perspective of greenhouse gas emissions and energy efficiency.

4 Conclusion

The utilization of sugar containing, starch containing, agricultural residues, and cellulosic biomass for production of bioethanol is sustainable and better alternative to fossil fuel. Biological conversion of biomass into fermentable sugar after enzymatic hydrolysis is at present the most efficient, attractive technique to the environment point of view. A very balanced combination of pretreatment, hydrolysis, fermentation, and distillation process has to be select for maximum efficiency of conversion.

References

1. Agrawal AK (2007) Biofuels (alcohols and biodiese) applications as fuels for internal combustion engines. Prog Energy Combust Sci 33:233–271
2. Xu K, Lv B, Huo Y, Li C (2017) ScienceDirect toward the lowest energy consumption and emission in biofuel production: combination of ideal reactors and robust hosts. Curr Opin Biotechnol 50:19–24
3. Sakthivel P, Subramanian KA, Mathai R (2018) Indian scenario of ethanol fuel and its utilization in automotive transportation sector. Resour Conserv Recycl 132:102–120
4. Chen H, Shuai S, Wang J (2007) Study on combustion characteristics and PM emission of diesel engines using ester—ethanol—diesel blended fuels. Proc Combust Inst 31:2981–2989
5. Ho M, Woo Y, Han C, Bum Y (2012) Bioresource technology simultaneous production of bio-ethanol and bleached pulp from red algae. Biores Technol 126:198–201
6. Lee OK, Lee EY (2016) Biomass and bioenergy sustainable production of bioethanol from renewable brown Algae biomass. Biomass Bioenergy 92:70–75

7. Momayez F, Karimi K, Sarvari I (2018) Enhancing ethanol and methane production from rice straw by pretreatment with liquid waste from biogas plant. Energy Convers Manage 178:290–298
8. Banugopan VN, Prabhakar S, Annamalai K, Jayaraj S, Kumar SP (2010) Experimental investigation on D.I. diesel engine fuelled by ethanol diesel blend with varying inlet air temperature. Proc Int Conf Front Automob Mech Eng 4:128–134
9. Pandiyan K, Singh A, Singh S, Kumar A, Nain L (2019) Technological interventions for utilization of crop residues and weedy biomass for second generation bio-ethanol production. Renew Energy 132:723–741
10. Parthasarathi R, Gowri S, Saravanan CG (2012) Experimental investigation of performance, emission and combustion characteristics of kirloskar tvi di diesel engine using diesel-ethanol-surfactant as fuel. ARPN J Eng Appl Sci 7(11):1468–1472
11. Jimenez ET, Svoljsak M, Gregorc A, Lisec I, Dorado MP, Kegl B (2011) Physical and chemical properties of ethanol-diesel fuel blends. Fuel 90(2):795–802
12. Binod P, Sindhu R, Singhania RR, Vikram S, Devi L, Nagalakshmi S, Kurien N, Sukumaran RK, Pandey A (2010) Bioethanol production from rice straw: an overview. Biores Technol 101(13):4767–4774
13. Bernhardt W (1977) Future fuels and mixture preparation methods for spark ignition automobile engines. Prog Energy Combust Sci 3:139–150
14. Fang Q, Fang J, Zhuang J, Huang Z (2013) Effects of ethanol-diesel-biodiesel blends on combustion and emissions in premixed low temperature combustion. Appl Therm Eng 54(2):541–548
15. Rosillo-Calle F, Walter A (2006) Global market for bioethanol: historical trends and future prospects. Energ Sustain Dev 10(1):20–32
16. Wi SG, Choi IS, Kim KH, Bae HJ (2013) Bioethanol production from rice straw by popping pretreatment. Biotechnol Biofuel 6:166
17. Zhang H, Zhang P, Ye J, Wu Y, Liu J, Fang W, Xu D, Wang B, Yan L, Zeng G (2018) Comparison of various pretreatments for ethanol production enhancement from solid residue after rumen fluid digestion of rice straw. Biores Technol 247:147–156
18. Comwien J, Boonvithaya N, Chilaluksananukul W, Glinwong C (2015) Direct production of Butanol and Ethanol from cane sugar factory wastewater and cellulosic Ethanol Pilot plant wastewater by Clostridium BeijerinckiiCG1. Energy Procedia 79:556–561
19. Vlasenko EY, Ding H, Labavitch JM, Shoemaker SP (1997) Enzymatic hydrolysis of pretreated rice straw. Bioresour Technol 59(2–3):109–119
20. Zhao J, Xia L (2010) Ethanol production from corn stoverhemicellulosichydrolysate using immobilized recombinant yeast cells. Biochem Eng J 49:28–32
21. Thammasittirong SNR, Chatwachirawong P, Chamduang T, Thammasittirong A (2017) Evaluation of ethanol production from sugar and lignocellulosic part of energy cane. Ind Crops Prod 108:598–603
22. Han W, Xu X, Gao Y, He H, Chen L, Tian X, Hou P (2019) Utilization of waste cake for fermentative ethanol production. Sci Total Environ 673:378–383
23. Ellis JT, Hengge NN, Sims RC, Miller CD (2012) Acetone, Butanol, and Ethanol production from wastewater algae. Biores Technol 111:491–495
24. Soam S, Kapoor M, Kumar R, Borjesson P, Gupta RP, Tuli DK (2016) Global warming potential and energy analysis of second generation ethanol production from rice straw in India. Appl Energy 184:353–364

Fuel from Plastic Waste: A Review

Kundan Kumar Jha, T. T. M. Kannan, and Ashutosh Das

Abstract Plastics are an indispensable part of our today's life, Its specific properties and faster rate of production of plastics are affecting our health; on the other hand, plastics play important role in domestic and industrial applications. In that situation, due to plastic production lots of environmental challenges are arising as if it is not disposed properly. In that condition, waste plastic recycling, regeneration and its utilization are necessary for human life, environment, and nation. The waste plastic to fuel pyrolysis method is commonly used as the process including thermal cracking and vis-breaking. The most commonly used catalysts in the literature for plastic waste pyrolysis include silica-alumina and zeolites. This process is providing as a result of plastic oil, diesel, petrol and other useful fuel which is using as an alternative fuel for engine. The scope of the current research is conversion of plastic waste to useful fuel. It also describes the earlier work on fuel from plastic waste.

Keywords Waste plastics · Pyrolysis · Alternative fuel · Petrol · Diesel

1 Introduction

Plastics are a nonexclusive gathering of engineered or characteristic materials, made out of high-atomic chains whose sole or real component is carbon. In like manner use, the terms plastics, polymers, and pitches are generally comparable. According to a September 2018 report by the Central Pollution Control Board (CPCB), which extrapolated data from 60 major cities, the country generates around 25,940 tonnes of

K. K. Jha (✉) · T. T. M. Kannan
Department of Mechanical Engineering, PRIST Deemed to be University, Thanjavur 613403, India
e-mail: k.k.jhathermal@gmail.com; jhakundan051@gmail.com

T. T. M. Kannan
e-mail: ttmkresearch@gmail.com

A. Das
Centre for Research and Development, PRIST Deemed to be University, Thanjavur 613403, India
e-mail: acadas@gmail.com

A. Arockiarajan et al. (eds.), *Advances in Industrial Automation and Smart Manufacturing*, Lecture Notes in Mechanical Engineering,
https://doi.org/10.1007/978-981-15-4739-3_74

plastic waste a day. About 94% of this comprises thermoplastic, such as polyethylene terephthalate (PET) and polyvinyl chloride (PVC), which is recyclable. A plastic material is any of a substantial gathering of materials comprising entirely or in part of mixes of carbon with oxygen, hydrogen, nitrogen, and other natural or inorganic components which, while strong in the completed state, at some phase in its make is made fluid, and along these lines equipped for being framed into different shapes, most normally through the application, either independently or together, of warmth and weight. Plastics are lightweight, strong, and adaptable, permitting their fuse into a different scope of uses. The use of plastic is building up every day as a result of the qualifications of occupations of it. Along these lines, the plastic waste is moreover augmented step by step. Plastic as a fabricated polymer substitute trademark materials in basically every edge 16 of our life and transform into a key bit of our overall population. Nature has seen a 17 broad heightening in the age of plastics in latest couple of decades and 18 synchronous extended usage of plastic materials. With time, security and durability of plastics have been enhanced constantly, and thus these gatherings of 20 materials are currently considered as an equivalent word for materials being impervious to numerous natural constraints in the twenty-first century. The previous work literature is on hydro-treating and hydro-breaking of waste plastic material for changing over in to fuel for diesel motor. The different investigations and tests are made to comprehend the compound piece of handled pyrolysis fuel from this work in which, the bromine test was increasingly beneficial to show the unsaturated securities in the oil. The value of bromine test as a pointer of unsaturated securities in the oil has been featured.

2 Literature Survey

- Chanashetty and Patil [1] have investigated fuel from waste plastic; they used condenser and reactor for pyrolysis process. They found this method is suitable for large plastic seas problems and helping fuel storage by means of products as diesel, kerosene, and lubricant oil. In this investigation, they used the waste plastic as rigid film, sheet plastic, and expanded foam materials.

Karad and Havalammanavar [2] investigated waste plastic to fuel, petrol, diesel, and kerosene by pyrolysis method on the temperature range 350–500 °C and waste plastic bags, food wrap, vegetable oil bottles, automotive parts garments bags, some carpets refrigerated containers, and they concluded that it saves 1000000 species of oceanic life and green future. Due to eco-friendly, it is involved in Swachh Bharat. Arunkumar and Nataraj [3] explored change of waste plastic into fuel oil within the sight of bentonite as an impetus and utilized materials condenser, reactor strategy is pyrolysis. It gives us outputs that are petrol, diesel, and fuel oil and inputs are PET bottles, shopping bags, plastic packages. They concluded that it provides perfect and green future and fuel efficiency, control of nitrogen, halogen, sulfur which is hazardous for human beings. Mathur et al. [4] investigated extraction of pyrolysis oil from waste plastic where pyrolysis process is using, and the waste plastic inlet is all the types of plastic and outputs are waste oil and diesel fuel. It concludes that the process is carried out grade 5 types of plastic materials, 1.65 l of oil is obtained by using 1.5 kg of plastic. Verma et al. [5] examined engine fuel creation from waste plastic pyrolysis and execution advancement in a CI Motor with diesel mix. Inputs are every one of the sorts of waste plastic and yield plastic pyrolysis oil by utilizing pyrolysis technique. The materials are used: condenser, oil collector, furnace reactor, load controller, gas analyzer, fuel meter, dynamometer, and computer. The conclusion is plastic oil saves 40% blended with diesel without loss of power reduction in exhaust emission. Bezergini et al. [6] investigated alternate diesel from waste plastic by use of pyrolysis plant. The products are diesel fuel, heavy oils, naphtha, kerosene and inlets are bags, bottles, and liquid containers. They got the conclusion from investigation nearly 70–85% less production cost, and this method is promising alternate diesel fuel.

Mustofa Kamal and Zainuri [7] researched green result of fluid fuel from plastic waste by pyrolysis at 900 °C, where yields are fluid fuel and bays are delicate destroyed plastic, drain bottles, cleanser parcels, water funnels receptacles They concluded that in 900 °C of pyrolysis providing more calorific value of the fuel compare to 425 °C, boric acid and cyclopentanone is reduced at 900 °C. Karmina et al. [8] investigated catalytic pyrolysis of waste plastic on the use of stainless steel semi-bath reactor, condenser, and process is thermal pyrolysis inlet as solid waste plastic. Final conclusion high yield (77–88%) of liquid product, the plastic is converted into liquid fuel. Khan et al. [9] investigated pyrolitic waste plastic oil and its diesel blend fuel characterization where the material is used as pyrolysis chamber, sensor wire, storage, condenser valve, the inlet waste plastic is polystyrene, PVC, polypropylene, PE, and outputs are gasoline, liquid fuel, diesel, and oil. They concluded that very small amount of char volatile products. Sharuddin et al. [10] investigated pyrolysis of plastic waste for liquid fuel production as prospection energy resources used materials are chamber, condenser, storage of fuel bath, semi-bath and the method is pyrolysis, thermal cracking. Mostly inlets are all types of plastic and outputs are oil, petroleum, char diesel, and liquid fuel. The conclusion of this investigation is due to thermal treatment; more energy is obtained from plastic comparison of pyrolysis process. More efficient for management uses capacity of landfill needed uses pollution and cost-efficient. The conversion of waste plastic into resources and used catalytic cracking, thermal cracking, and pyrolysis studied. In the research, municipal waste plastic, plastic, rubbers, food containers, packaging, disposal cups, electronic cushioning foams, drainage pipe, carbonated soft drinks bottles, and CD cassette boxes are performed for pyrolysis process to obtain gasoline, liquid, petroleum, gas, paraffin, olefins, naphthalene, and aromatics. The result revealed that the thermal catalytic process is having potential of effective conversion of waste plastic to fuel and concluded that more suitable process on the comparison of land-filling on the conversion of waste plastic into fuel by thermal cracking and pyrolysis method and used aluminum silicates as catalyst; the used plastics are polypropylene, polyethylene, polystyrene for obtaining plastic oil and diesel. They concluded the producing a highly usable commodity that due to its clear burning characteristics. At transformation of waste plastic into low -emissive hydrocarbon fills through synergist depolymerization in another research center was studied, researcher utilized the technique thermo-compound disintegration and pyrolysis process. In this exploration, froth refuse, plastic sacks, plastic holders, plastic jugs used to acquire gas, diesel, and engine oil. Finally, they concluded that it is suitable for society needs. The plastic waste into fuel using pyrolysis process by using grocery bags, garbage bags, squeezable bottles, shrink wrap, stretched films, coating of milk cartoons, toys, and container lids for obtaining plasto-fuel, petrol was studied. The conclusion of this research is good alternatives for diesel and petrol engines and provides sustainable, clean, and green future.

The creation of oil from waste plastics and polythene utilizing pyrolysis and its use in pressure start (CI) Engine they utilized tires, polythene, biomass for getting diesel, polythene oil, synthetic compounds examined. Finally, they concluded that the engine runs satisfactory with 20% blend of oil. A review on waste plastic fuel to get gas, heavy oil, gasoline, petrol oil, diesel oil by using the plastic as polyethylene, polyvinyl chloride, polystyrene with clean and very effective operation which we have left behind over the last decades studied. Design and fabrication of waste plastic oil converter with involving all types of plastics for getting plastic oil and char. In the process, reactor tank was closed and connected with 220 VAC, temperature 380 °C runs for 2 h with 20 min. cracking time. They got average capacity 83.3 ml oil/h. The waste plastic fuel utilized in oil motor by utilizing pyrolysis technique, by utilizing high thickness polyethylene drain, cleanser and oil bottles, toys, compartments utilized outside, parts of plastic sacks, recoil wraps, pieces of clothing packs and holders, refrigerated holders, most jug tops, floor coverings, nourishment wraps and impetus was aluminum silicate, for getting diesel and plastic fuel studies. For testing the 8 ml fuel-run bike of 110 CC, Bajaj Caliber approx. 2 min. finally concluded that in 100 CC Bajaj Discover bike it increases efficiency of bike by 15–20% as compared to petrol used in bike. The creation and investigation of pyrolysis oil from waste plastic in Kolhapur city studied, they utilized manufactured materials for getting pyrolysis oil, diesel, maker gas, and roast. They reasoned that the running expense of pyrolysis plant ranges from 14 to 18 rupees for each liter while the market cost of oil is 40 rupees for every liter. The oil that has been generated by the LDPE plastic by prescribed experimental process shows that the properties are very much comparable to the petrol. Ultimately, pyrolysis of LDPE does not only recover the energy contained in the plastic, but also manage the environment by alternate disposal technique of the waste plastic. They at long last reasoned that warm transformation of city squander plastics yields monetary benefits. From all the above research, discoveries manage pyrolysis technique, vis-breaking, and thermal cracking method for conversion of fuel from plastic waste. Only a few research conducted on plastic waste into fuel by plasma gas method.

3 Conclusion

Hence, on this review the importance of plastic disposal and its uses is completely described. The research on alternate fuel which is the product of waste plastic pyrolysis provides clean environment and also financial support by country. This review likewise serves to the usage of Swachh Bharat Abhiyan of Government of India and executed plastic waste pyrolysis strategy for oil generation as an elective fuel for running the motor. It portrays the long haul point of vitality security and, combined with the natural worries of hazardous waste amassing, is tended to by means of the proposed waste-to-fuel innovation. The present research is focused on the conversion of waste plastic by different methods namely catalytic cracking, vacuum cracking, pyrolysis, and thermal cracking.

References

1. Chanashetty VB, Patil BM (2015) Int J Emerg Technol 6(2):121–128
2. Karad RT, Havalammanavar S (2017) Waste plastic to fuel-petrol, diesel, kerosene 641–545
3. Arunkumar BY, Nataraj CN Conversion of waste plastic into fuel oil in the presence of bentonite as a catalyst. Int Res J Eng Technol 114–119
4. Mathur K, Shubham C, Sunadh H, Aditya P (2016) Extraction of pyrolysis oil from waste plastics. Int Res J Eng Technol
5. Verma A, Raghuvansi A, Quraishi MA, Tirkey JV, Verma C (2018) Engine fuel production from waste plastic pyrolysis (WPO) and performance evaluation in a CI engine with diesel blend. J Mater Environ Sci 6:1712–1721
6. Bezergianni S, Dimitriadis A, Faussone G-C, Karonis D (2017) Alternative diesel from waste plastics. Energies 10:1–12
7. Mustofa Kamal D, Zainuri F (2015) Green product of liquid fuel from plastic waste by pyrolysis at 900 °C. J Energy Power Eng 9:29–34
8. Mileva K, Alesksovski S, Bogeova G (2016) Catalytic pyrolysis of waste plastic into liquid fuel. Sci Pap 57(4):600–604
9. Khan MZH, Sultana M, Al-Mamun MR, Hasan MR (2016) In: Journal of Environmental and Public Health, Pyrolytic Waste Plastic Oil and Its Diesel Blend: Fuel Characterization, vol 10, pp 1–6
10. Sharuddin SDA, Abnisa F, Daud WMAW, KAroua M (2018) Pyrolysis of plastic waste for liquid fuel production as prospective energy resource. In: IOP Conf Ser: Mater Sci Eng 334

Production of Alternate Fuel Using Jatropha Oil and Kerosene—Analysis of Their Performance and Emission Characteristics

J. Hemanandh, S. Ganesan, K. S. Sridhar Raja, V. Aakhash Sivan, and Abdul Maher Khaliq

Abstract The main objective of this experiment is to analyze the emission and performance of Jatropha oil and kerosene blends and to utilize them as a substitute fuel for diesel engines. The hydrotreated Jatropha oil and kerosene were blend in the ratios such as HK10, HK20, and HK30. These fuel blends were tested in the Kirloskar TV1 Engine at different conditions of loads from 0 to 100%. The results were analyzed, and the emission and performance characteristics of the HVO blends were differentiate with pure kerosene and the graphs were plotted.

Keywords Hydrotreated jatropha oil · Diesel engine · Emission and performance characteristics

1 Introduction

In the recent times, there is an increase in the depletion of natural resources and renewable sources of energy worldwide. Because of this depletion, the world has changed into a new form of energy, i.e., alternate energy. As seen, petroleum products are depleting and alternate fuel has shown up in recent days. In addition, the emissions from the fuels play a major role in causing global warming and air pollution. Alternate fuels show greater emission and performance characteristics. Sanjid et al. [1] found that the palm and Jatropha oil blend PBJB10 showed 20.49% reduction in CO emission compared with diesel fuel. This blend showed better results with HC emissions and sound levels. Nalgundwar et al. [2] tested dual biodiesel palm and Jatropha oils with diesel fuel; the blend D70JB15PB15 showed 14.5% CO reduction in the emission characteristics. The blend D90JB5PB5 shows 5.3% increase in NO_x emission. Agarwal et al. [3] found that CO, HC, and CO_2 emissions were less in the B20 engine compared to the diesel engine. With respect to the NO_x parameter, it showed higher rate of emission due to the presence of fuel oxygen. Kathirvelu

J. Hemanandh (✉) · S. Ganesan · K. S. Sridhar Raja · V. Aakhash Sivan · A. M. Khaliq
School of Mechanical Engineering, Sathyabama Institute of Science and Technology, Chennai, Tamil Nadu 600043, India
e-mail: jhemanandh.mech@sathyabama.ac.in

A. Arockiarajan et al. (eds.), *Advances in Industrial Automation and Smart Manufacturing*, Lecture Notes in Mechanical Engineering,
https://doi.org/10.1007/978-981-15-4739-3_75

et al. [4] proved that fish waste and Jatropha oil can be used as alternate fuel for engines without modifying any specifications of the engine, and also, the emissions from the engine showed a better result except NO_x which is higher. Ranucci et al. [5] confirmed that 10% proportion of Jet A1 kerosene and methyl esters showed similar properties compared with the existing aviation fuels. Paula et al. [6] found the presence of higher cylinder pressure and decrease in ignition in delay period resulted in an increased CO_2 and NO_x emission and decreased smoke and [7] proved to increase the ambient temperature to increase the evaporation of Jatropha droplet. Portugal-Pereira et al. [8] showed FT-diesel synthesis results in lesser emissions. Nguyen et al. [9] reveal JHE5 and JHE15 have lesser CO, CO_2, and NO_x emissions, but HC shows rise in the emission characteristics. Nguyen et al. [10] proved NO_x emissions' reduction by 48% and also decrease in [11] found that the blend of formic acid and crude Jatropha oil has greater deoxygenation of 99.5%. Karikalan and Chandrasekaran [12] proves that emissions of CO, smoke, and HC emissions are reduced for Jatropha-Wood Turpentine and Jatropha-Mineral Turpentine blends at 75% load conditions. Karikalan and Chandrasekaran [13] proves that the brake thermal efficiency is equal to diesel fuel in the blend of 80% Jatropha and 20% Turpentine oil at 75% load condition. Murali Krishna et al. [14] showed the increase in the performance level of CE at 27° and 31° BTDC with increase in NO_x emission. The results by Tongroon et al. [15] showed decrease in engine torque by 10% of H-FAME and also a decrease in the exhaust gas emissions. Rahman et al. [16] compared with other blends and JB10 blends show the reduction in the emission of CO and HC by 14 and 16% and the MB10 blend increases by 7% in the NO_x emission. Ashrafur Rahman et al. [17] proved the increase in NO_x emission, brake-specific fuel consumption, and decrease in CO and HC emissions at high idling conditions of Jatropha and palm oil. Acharya et al. [18] found the increase in oxidation stability of B20 and B30 blends, and J50:M50 mixture satisfies En-14214, ASTM D-6751, and IS-15607 standards. Desmira et al. [19] visualize the CO_2, CO, H_2O, and NO presence through FT-IR camera. Senthilkumar and Sankaranarayanan [20] show the smoke emission decrease by 11.4% in PJ20 blend and increase in the NO_x emission in JME blend. Pradhan et al. [21] found that by preheating the Jatropha oil, the CO_2, HC, and NO_x are decreased and the CO emission in PJO blend increase. Madiwale et al. [22] showed improvement in brake power, brake-specific fuel consumption, and brake thermal efficiency at various loading conditions. Sharma et al. [23] found the blend JME80TPO20CNT100. It shows thermal efficiency of about 25.64%. In addition to this, JMETPOCNT and JMETPOCeO2 blends showed decreased NO_x emissions. Chintala et al. [24] prove higher brake thermal efficiency and lesser carbon-based emissions compared with diesel fuel. Jiotode and Agarwal [25] proved that the blend JB20 has low cylinder soot formation under endoscopic visualization and the flame temperature distribution is lower in the biodiesel.

2 Materials and Methods

The Jatropha oil, kerosene, and diesel were purchased from the market. The experiment was conducted as follows: The hydrotreatment was conducted on Jatropha oil. The hydrotreated Jatropha oil and kerosene were blend together in required proportions. The test blends were treated to test in the Kirloskar TV1 engine at different load conditions. The characteristics of the test blends were obtained and analyzed.

3 Experimental Setup

The hydrotreated Jatropha oil blended together with kerosene at different ratios such as HK10, HK20, and HK30 and was subjected to testing in the engine. The blends of hydrotreated Jatropha oil and kerosene were tested in a four-stroke Kirloskar TV1 engine to analyze the emission and performance characteristics. The engine was tested at differing load conditions from 0 to 100% at a constant speed of 1500 rpm. The Eddy current dynamometer was introduced to load the engine. The stipulation of engine is as shown in Table 1. Initially, the engine was let to run at no load condition and the test blends were added in the fuel inlet. The emission and performance characteristics for each sample were analyzed and discussed.

3.1 Table 1 Engine Specifications

See Table 1.

3.2 Table 2 Properties of Oil

See Table 2.

Table 1 Engine specifications

Combustion	Direct injection
Rated power	5.20 kW
Rated speed	1500 rpm
Compression ratio	17.50
Dynamometer	Eddy current

Table 2 Properties of oil

S. No.	Properties	EURO-IV Bharat stage 1460:2005 diesel	ASTM D–751 (IS 5607:2005)	Kerosene	HK10	HK20	HK30	Jatropha
1.	Cetane number	51	–	–	–			
2.	Density at 15 °C kg/m^3	820–845	860–900	800.85	810	820	830	788.8
3.	Kinematic viscosity at 40 °C cst	2–4.5	1.9–6	3.493	3.509	3.524	3.5407	3.65
4.	Flash point °C min	35 °C	130 °C	63.25	65.86	68.46	71.065	89.3
5.	Calorific value kJ/kg	39,000	–	43,781	34,700	34,500	34,200	45,163

4 Results and Discussions

4.1 Carbon Monoxide (CO)

Figure 1 shows the variations in CO emissions of HK and kerosene with load versus CO%. The blend HK10 shows a decrease of 0.5%, HK20 shows a decrease of 5.5%, and HK30 shows a decrease of 26.5%. Thus, it is proved that the blend HK30 shows an optimum result of 26.5%. This is due to the additional oxygen which ensures complete combustion [1].

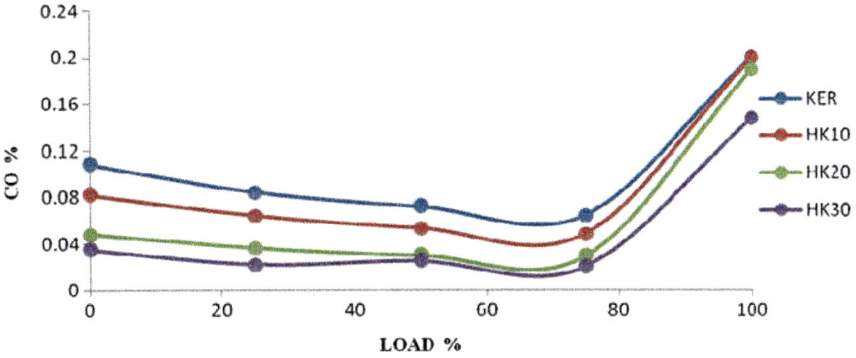

Fig. 1 Variations in CO emissions of HK blends and kerosene

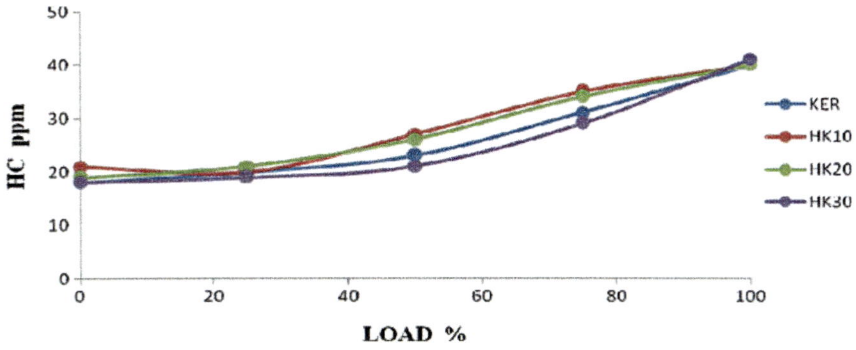

Fig. 2 Percentage of HC emission for HK blends and kerosene

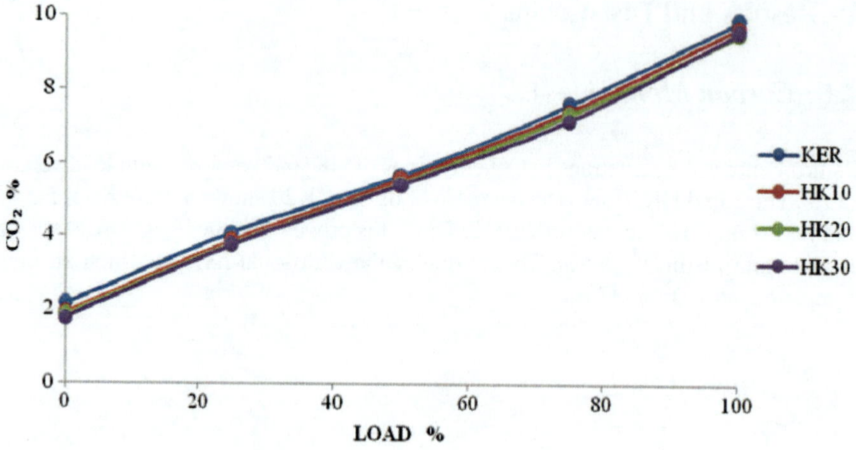

Fig. 3 Comparison of CO_2 emission of HK blends and kerosene

4.2 Hydrocarbons (HC)

Figure 2 shows the variation in HC emissions of HK and kerosene with load versus HC percentage. The blend HK10 shows a decrease of 5%, HK20 shows a decrease of 7.5%, and HK30 shows a decrease of 10%. Thus, it is proven that the blend HK30 shows an optimum result of 10%. This is due to the higher combustion temperatures at higher engine loads [9].

4.3 Carbon Di Oxide (CO_2)

Figure 3 shows the variation of CO_2 emissions of HK and kerosene with load versus CO_2 percentage. The tested blends HK10 show an optimum decrease of 1.93%, HK20 shows a decrease of 7.5%, and HK30 also shows a decrease of 10%. These results prove that HK30 blends show a better yield of 10%. This is due to oxygen content in the fuel [12].

4.4 Nitrogen Oxide (NO_x)

Figure 4 shows the variation of NO_x emissions of HK and kerosene with load versus NO_x percentage. The tested blends HK10 decrease by 2.01%, HK20 decreases by 3.15%, and HK30 blend shows a decrease in 3.33%. This result proves that HK30 blend shows a better result of 3.33%. This is due to the lower burning temperature of the fuel in the cylinder [13].

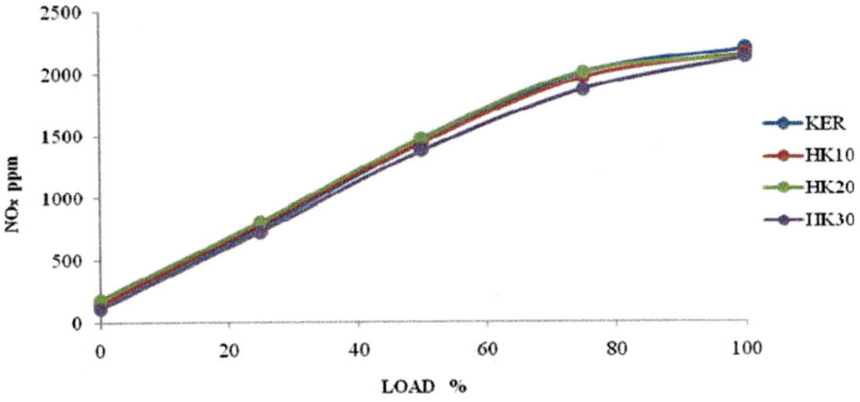

Fig. 4 Comparison of NO_x emissions for HK blends and kerosene

4.5 *Brake Thermal Efficiency (BTE)*

Figure 5 shows the variation of brake thermal efficiency of HK and kerosene with load versus BTHE percentage. The tested blends HK10, HK20, and HK30 showed an optimal increment of 7.07, 6.46, and 10.35%. These results prove that HK30 blend shows an optimum yield of 10.35%. This is due to increase in in-cylinder temperature [2].

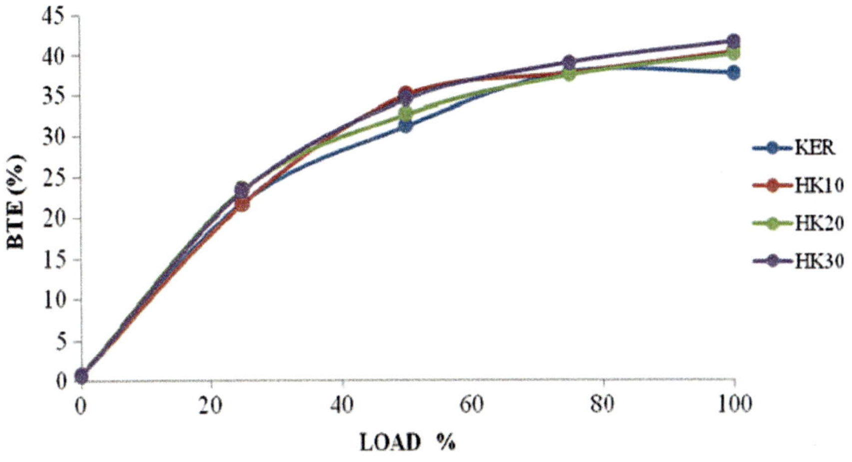

Fig. 5 BTHE comparison for HK blends and kerosene

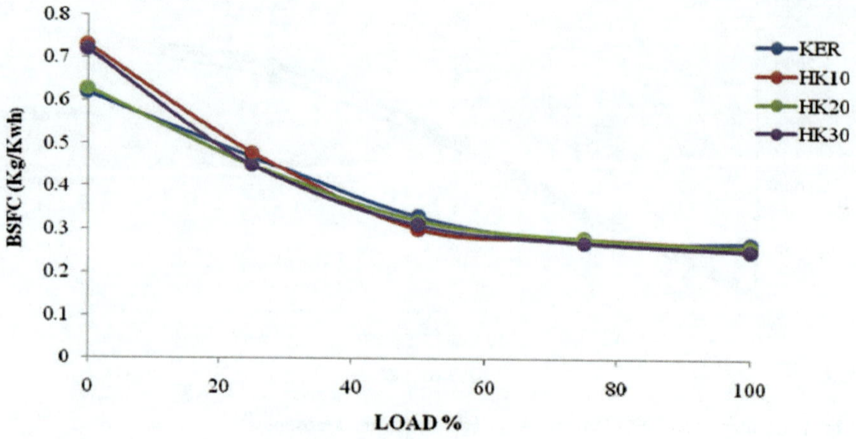

Fig. 6 Variations in BSFC of HK blends and kerosene

4.6 *Brake-Specific Fuel Consumption (BSFC)*

Figure 6 shows the variation of brake-specific fuel consumption of HK and kerosene with load versus BSFC percentage. The tested blends HK10, HK20, and HK30 show a decrement of 3.71, 7.41, and 11.11%. This proves that optimal blend is seen in HK30 with 11.11%. This is due to lower density and viscosity of the fuel [21].

4.7 *Smoke*

Figure 7 shows the variations in smoke of HK and kerosene with load versus smoke percentage. The tested blends HK10, HK20, and HK30 show an optimal increment of 3.97%, 4.46%, and a decrement of 7.95%, respectively. These results prove that HK30 blends show an optimal yield. This variation is due to the quantity of fuel injected into the combustion chamber [10].

5 Conclusions

The objective of this paper is to analyze the emission and the performance characteristics of hydrotreated Jatropha oil and kerosene as an alternate fuel. The analysis of the characteristics is performed, and the results have been studied. The results are as follows:

- The carbon monoxide emission for HK10, HK20, and HK30 decreases by 0.5, 5.5, and 26.5%.

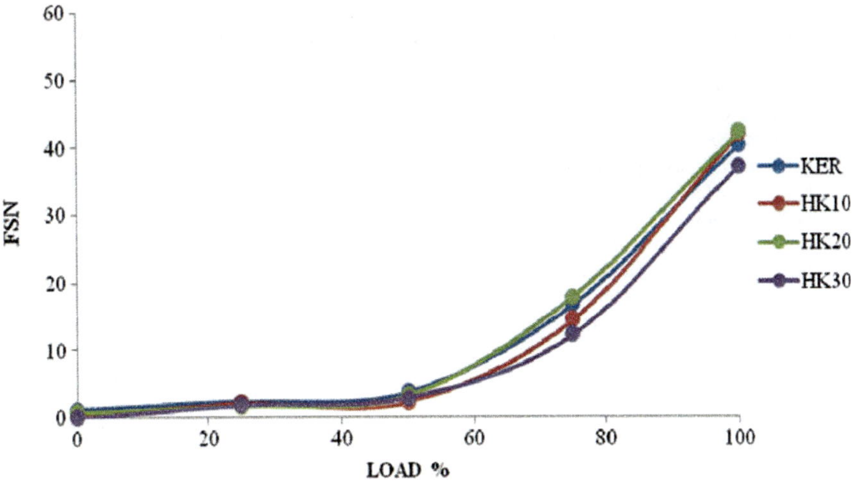

Fig. 7 Variations of smoke of HK blends and kerosene

- The hydrocarbon emission for HK10, HK20, and HK30 decreases by 5, 7.5, and 10%.
- The carbon dioxide emission for HK10, HK20, and HK30 decreases by 1.93, 7.5, and 10%.
- The emission of NO_x for HK10, HK20, and HK30 decreases by 2.01, 3.15, and 3.33%.
- The brake thermal efficiency (BTE) increases for HK10, HK20, and HK30 by 7.07, 6.46, and 10.35%.
- The brake-specific fuel consumption for HK10, HK20, and HK30 decreases by 3.71, 7.41, and 11.11%.
- The smoke emission increases for HK10, HK20 by 3.97%, 4.47% and decreases for HK30 by 7.95%.

References

1. Sanjid A, Masjuki HH, Kalam MA, Ashrafur Rahman SM, Abedin MJ, Palash SM (2014) Production of palm and jatropha based biodiesel and investigation of palm-jatropha combined blend properties, performance, exhaust emission and noise in an unmodified diesel engine. J Clean Prod 65:295–303
2. Nalgundwar A, Paul B, Sharma SK (2016) Comparison of performance and emissions characteristics of DI CI engine fueled with dual biodiesel blends of palm and jatropha. Fuel 176:172–179
3. Agarwal AK, Shrivastava A, Prasad RK (2016) Evaluation of toxic potential of particulates emitted from Jatropha biodiesel fueled engine. Renew Energy 99:564–572

4. Kathirvelu B, Subramanian S, Govindan N, Santhanam S (2017) Emission characteristics of biodiesel obtained from Jatropha seeds and fish wastes in a diesel engine. Sustain Environ Res 27:283–290
5. Ranucci CR, Alves HJ, Monteiro MR, Kugelmeier CL, Bariccatti RA, de Oliveira CR, da Silva EA (2018) Potential alternative aviation fuel from jatropha (Jatropha curcas L.), babassu(Orbignya phalerata) and palm kernel (Elaeis guineensis) as blends with Jet-A1 kerosene. J Clean Prod 185:860–869
6. Paula G, Datta A, Mandal BK (2014) An experimental and numerical investigation of the performance, combustion and emission characteristics of a diesel engine fueled with jatropha biodiesel. Energy Procedia 54:455–467
7. Wanga J, Wang X, Chen H, Jin Z, Xiang K (2018) Experimental study on puffing and evaporation characteristics of jatropha straight vegetable oil (SVO) droplets. Int J Heat Mass Transf 119:392–399
8. Portugal-Pereira J, Nakatani J, Kurisu K, Hanaki K (2016) Life cycle assessment of conventional and optimised Jatropha biodiesel fuels. Renew Energy 86:585–593
9. Nguyen K-B, Dan T, Asano I (2014) Combustion, performance and emission characteristics of direct injection diesel engine fueled by Jatropha hydrogen peroxide emulsion. Energy 74:301–308
10. Nguyen K-B, Dan T, Asano I (2015) Effect of double injection on combustion, performance and emissions of Jatropha water emulsion fueled direct-injection diesel engine. Energy 80:746–755
11. Hwang K-R, Choi I-H, Choi H-Y, Han J-S, Lee K-H, Lee J-S (2016) Bio fuel production from crude Jatropha oil; addition effect of formic acid as an in-situ hydrogen source. Fuel 174:107–113
12. Karikalan L, Chandrasekaran M (2016) Influence of turpentine addition in Jatropha biodiesel on CI engine performance, combustion and exhaust emissions. SAE, Int J Autom Technol 17:697–702
13. Karikalan L, Chandrasekaran M (2017) Performance and pollutants analysis on diesel engine using blends of Jatropha biodiesel and mineral turpentine as fuel. Int J Environ Sci Technol 14:323–330
14. Murali Krishna MVS, Ohm Prakash T, Ushasri P, Janardhan N, Murthy PVK (2016) Experimental investigations on direct injection diesel engine with ceramic coated combustion chamber with carbureted alcohols and crude jatropha oil. Renew Sustain Energy Rev 53:606–628
15. Tongroon M, Suebwong A, Kananont M, Aunchaisri J, Chollacoop N (2017) High quality Jatropha biodiesel (H-FAME) and its application in a common rail diesel engine. Renew Energy Int J 113:660–668
16. Rahman MM, Hassan MHJ, Abul Kalam M, Atabani AE, Memon LA, Ashrafur Rahman SM (2014) Performance and emission analysis of Jatropha curcas and Moringa oleifera methyl ester fuel blends in a multi-cylinder diesel engine. J Clean Prod 65:304–310
17. Ashrafur Rahman SM, Masjuki HH, Kalam MA, Abedin MJ, Sanjid A, Rahman, MM (2014) Assessing idling effects on a compression ignition engine fueled with Jatropha and Palm biodiesel blends. Renew Energy, 68:644–650
18. Acharya N, Nanda P, Panda S, Acharya S (2017) A comparative study of stability characteristics of Mahua and Jatropha biodiesel and its blends. J King Saud Univ—Eng Sci
19. Desmira N, Kuniyuki K, Morita S, Gupta AK (2014) In-situ spectroscopic monitoring of Jatropha oil combustion properties. Renew Energy 63:775–778
20. Senthilkumar P, Sankaranarayanan G (2016) Effect of Jatropha methyl ester on waste plastic oil fueled DI diesel engine. J Energy Inst 89:504–512
21. Pradhan P, Raheman H, Padhee D (2014) Combustion and performance of a diesel engine with preheated Jatropha curcas oil using waste heat from exhaust gas. Fuel 115:527–533
22. Madiwale S, Karthikeyan A, Bhojwani V (2018) Properties investigation and performance analysis of a diesel engine fuelled with Jatropha, soybean, palm and cottonseed biodiesel using ethanol as an additive. Mater Today: Proc 5:657–664
23. Sharma SK, Das RK, Sharma, A (2016) Improvement in the performance and emission characteristics of diesel engine fueled with jatropha methyl ester and tyre pyrolysis oil by addition of nano additives. Springer 38:1907–1920

24. Chintala V, Kumar S, Pandey JK (2017) Assessment of performance, combustion and emission characteristics of a direct injection diesel engine with solar driven Jatropha biomass pyrolysed oil. Energy Convers Manag 148:611–622
25. Jiotode Y, Agarwal AK (2017) Endoscopic combustion characterization of Jatropha biodiesel in a compression ignition engine. Energy, 119:845–851

An Influence on Wear Characteristics of Benthic-Diatom Navicula Sp. Algae Oil by Using Four Ball Tribometer

J. Arunprasad, R. Thirugnanasambantham, R. Rajesh, S. Sugumar, and T. Elango

Abstract In this investigation, as we have seen that the algae oil contains many minerals, we have chosen one of the green algae prokaryotic plants named Benthic-Diatom Navicula Sp. Algae. The Fourier Transform Infrared (FTIR) Spectroscopy analysis is well-known for the Benthic-Diatom Navicula Sp. Algae to Benthic-Diatom Navicula Sp. Algae methyl ester conversion. To convert it into biodiesel formation, the nano additive used is Magnesium oxides (MgO) with a mixer of clean diesel and their blends are B20, B20 + MgO nanoparticle 50 ppm and B20 + MgO nanoparticle 100 ppm were prepared. The MgO nanoparticle is coagulated with the mixer of biodiesel with help of ultrasonicator. When the experiment was undergone, around one hour with the constant temperature at 75 °C, the rotating speed turn with the ball inside the tribometer machine is 1200 rpm at the load of 40 kg per sample of the tests by using four ball tribometer. SEM had analysed worn surfaces of the balls. As a result, wear and friction were significantly lowered when increment of biodiesel with MgO nano-sized particle.

Keywords Benthic-Diatom navicula sp. algae oil · Magnesium oxide (MgO) · Four ball tribometer · SEM/EDX analysis

Nomenclature

DNMSA	Diatom Navicula Sp. Algae oil
DNSME	Diatom Navicula Sp. Algae oil methyl ester
ASTM	American Society for Testing and Materials
KOH	Potassium Hydroxide

J. Arunprasad (✉) · R. Thirugnanasambantham · R. Rajesh · S. Sugumar
Department of Mechanical Engineering, Dhanalakshmi Srinivasan Engineering College, Perambalur, India
e-mail: arunaug21@gmail.com

T. Elango
Department of Mechanical Engineering, Mahendra Institute of Technology, Mallasamudram, India

A. Arockiarajan et al. (eds.), *Advances in Industrial Automation and Smart Manufacturing*, Lecture Notes in Mechanical Engineering,
https://doi.org/10.1007/978-981-15-4739-3_76

B20	80% Petro diesel
B20 + 50 ppm	20% biodiesel + 80% diesel + 50 ppm(MgONano sized)
B20 + 100 ppm	20% biodiesel + 80% diesel + 100 ppm(MgONano sized)
WSD	Wear scar diameter
FTP	Flash temperature parameter
COF	Coefficient of friction
MgO	Magnesium oxide
FFA	Free Fatty Acids
SEM	Scanning Electron Microscope
EDX	Energy Dispersive X ray test

1 Introduction

As introducing the work report, its start to end express that the biofuel which has been used is producing smoothing wear characteristics, good viscosity and so on [1]. The making of a fuel was producing on the thought of inherit the factors such as speed, load and temperature. They are usually a green due to dominance of pigments of chlorophyll A and chlorophyll B. Here, lubricant that we used plays a vital role of decreasing wear and friction in an environment. So the present condition can be used for an extended duration of period. The reform from vegetable oil is an actively generous lubricant on the renewal in the place of petroleum-based lubricant. The popular attribute of biodiesel includes over diesel fuel includes lowering exhaust emissions and toxicity to the environment and human, sustainability and non-flammability level. With the increase of rapid growth and modernisation technological progress, fuel is not only the essential requirement of providing energy but also the lubricity. In the past activity, we prospect the reports and it is confirmed that the effect of nanoparticle scatter in lubricating oil has visible friction and wear reduction between the rotating and sliding areas. Therefore by blending of MgO nanoparticle slightly improves the viscosity of the oil, which is our main concern about to find in this investigation. There is many experiment have been done, the effect of nanoparticle to the synthetic lubricants and vegetable oil, but it was seen for past researchers literatures report that it was very rarely described the content of algae oil-based lubricants. After a component in automobile such as piston, crank angle, piston rings, cylinder, fuels pumps, the lubricity matters a lot for the engine side of view, to extend it in a better performance [2]. Henceforth, it is need to be signing clear crystal that needs to enumerate the properties of lubricant from raw oil and know the density, viscosity and wear properties of four ball tribometer by SEM/EDX analysis, when three balls were sliding and rolling between upper ball having a diameter of ball is 12.7 mm. In this paper content, the algae oil will behold as the supportive fuel of 20% algae oil and 80% diesel oil. After that its wear and friction behaviour is analysed experimentally using the tribometer.

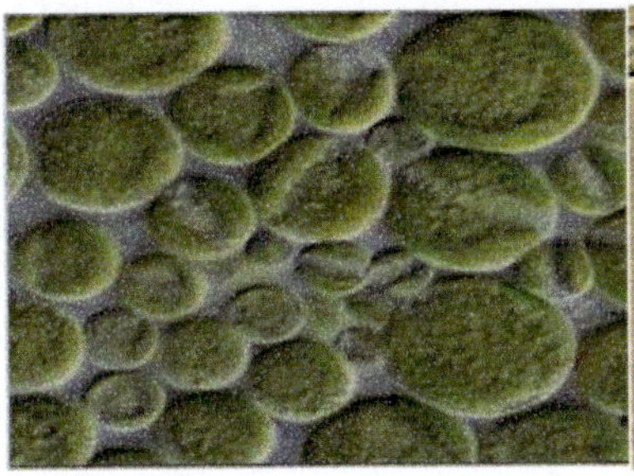

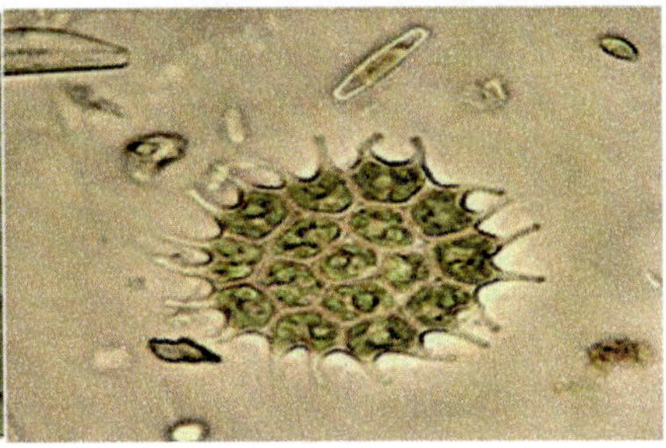

2 Experimental Procedure

2.1 Biodiesel Construction and Designation

For starting the production of biodiesel, it is started from choosing which one is the best for performance, then Benthic-Diatom Navicula Sp. Algae oil has been chosen at last. Henceforth, crude oil has been accumulated from the place where it is available. After some days, information is collected from where it can be produced. Raw oil has been extracted from the species under the standard procedure. After converting the green algae into crude oil, the preparation for upcoming process was needed. So, the process to extract pure biodiesel from algae oil by using transesterification process is reached. Thus, 1000 ml of Benthic-Diatom Navicula Sp. Algae oil was taken in a container with 11gm of potassium hydroxide alkaline catalyst (KOH) 250 ml of methanol was taken in the beaker, and then KOH was poured into the oil beaker. The magnetic stirrer process was used to mix the petrodiesel and biodiesel, and the process went on till it mixed completely, though it takes 45 min to complete it.

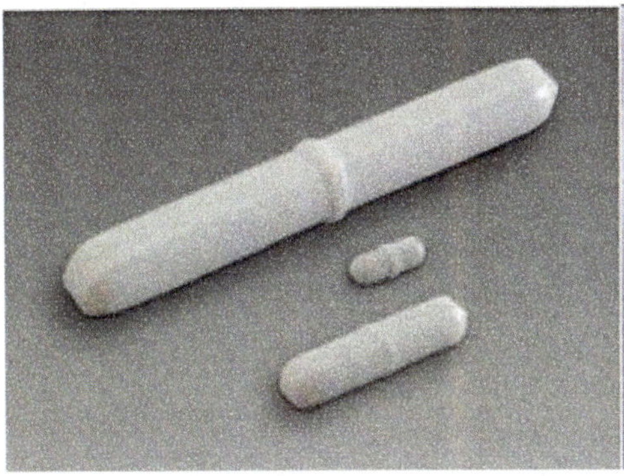

Magnetic stirrer Magnetic stirrer machine.

Magnetic stirrer machine employs a magnetic field around the formation to cause the stir bar immersed in the liquid beaker to stir the liquid quickly and mix completely.

The speed of magnetic stirrer was constantly run inside the beaker with the temperature until it reached 60 °C. The method for removing the glycerine is none transesterification the reaction of the alcohol with an oil, to form a by-product of biodiesel (Tables 1 and 2).

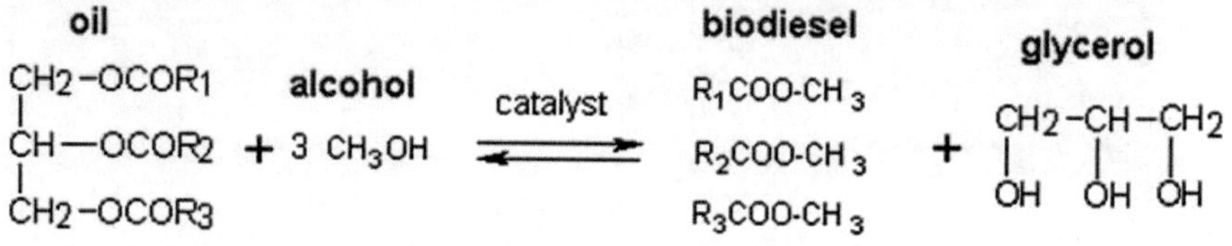

R_1, R_2, R_3 are hydrophobic rest of fatty acids

To execute the further process is not preferable to use edible oil for diesel engine oil directly. There are four methods to reduce the high viscosity of oils to audit

Table 1 a Properties of fuel sample. b The details of MgO nanoparticles

(a)						
Fuel properties	Petrodiesel fuel	Biodiesel	Biodiesel derived from microalgae	B20	B20 + 50 ppm	B20 + 100 ppm
Density (g/cm^3)	0.83–0.90	0.85–0.90	0.85–0.88	0.89	0.88	0.872
Kinematic viscosity (mm^2/s)	1.9–4.1	1.9–6.0	2.0–5.2	3.912	3.603	3.364
Heating value (MJ/kg)	41–45	39.5	38–41	37,300	38,900	40,100
Flashpoint (°C)	>63	>101	116	110	112	113
Pour point (°C)	−15	11.2	−6	−4	−4	−4

(b)	
Item	Specification
Manufacturer	American Elements, USA
Chemical name	Magnesium Nanoparticles
Actual particle size	90–100 nm
Surface area	8–10 m^2/g
Appearance	Black
Melting point	1200 °C
Boiling point	Decomposes
Molecular weight	133.07
Density	6.98 g/cm^3

Table 2 Conditions for the four ball wear test

Test parameters	
Applied load (kg)	40
Rotation (rpm)	1200
Fuel temperature (°C)	75
Test duration (s)	3600
Materials	Chrome alloy steel
Diameter (mm)	12.7
Hardness (HRc)	64
Surface roughness Ra (mm)	0.042

their common diesel engine in operations engine which have been investigated. The methods are pyrolysis, dilution, micro-emulsification and trans-esterification. To check the viscosity of an oil and the process to identify all the properties, trans-esterification process is one the best methods, and its another name is alcoholysis process. By this method, it can be clearly observed that separation can take place and biodiesel can be preserved in the upper place and glycerine comes down in the glass beaker. The rest down part is not in the purpose of work. Then the last part of the trans-esterification process is to heat the biodiesel to 100 °C to convert it into vapour, to remove the water content from it, after that the part becomes ready to use for experiment (Tables 3 and 4).

Table 3 Details of four ball tribometer

Specification	Unit	Description	Accuracy
Model	–	TR 30 H	–
Speed	RPM	300-3000	1
Maximum axial load	N	10,000	
Temperature	°C	AMBIENT to 100	0.5
Scar range	MICRON	100–4000	0.5
Test ball diameter	Mm	12.7	–
Drive motor	KW	1.5	–
Power	V/HZ/VA	380/50/20	0

Table 4 Coefficient of Friction

Samples	Coefficient of Friction
B20 + MgO2 50 PPM	0.07207
B20 + MgO2 100 PPM	0.08064
B20	0.08633

3 Experimental Apparatus

Here, the formulation gets into further preparation, to analyse the tribological method. In this a four ball tribometer TR-30L-IAS, the proficient equipment by DUCOM Instruments, Bengaluru is used to prospect the tremendous working design. Figures 1 and 2 help to demonstrate the working procedure of four balls operated tribometer. It

Fig. 1 Tribometer

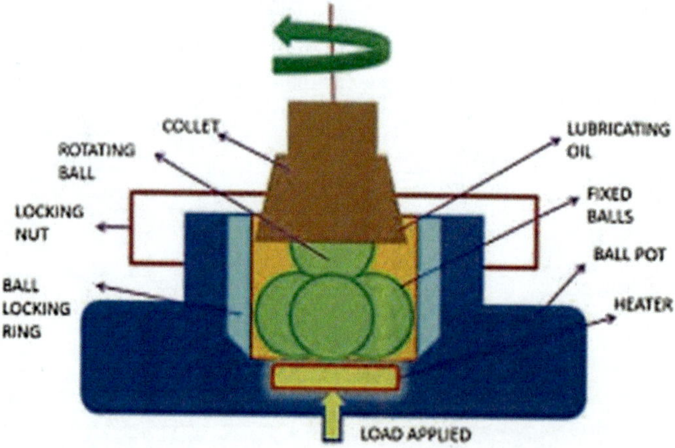

Fig. 2 Schematic of Ball pot

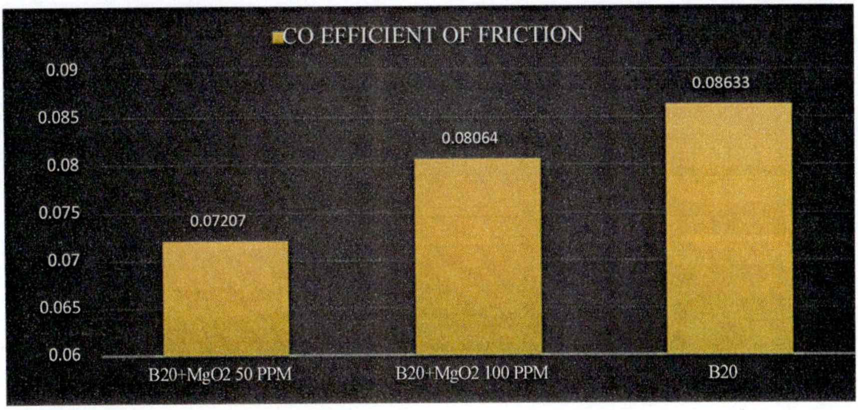

Fig. 3 Coefficient of friction Graph

contains four balls, one of the ball is situated at the top of the other three balls, parallel to the upfront. Three balls dipped the samples in order to get wear phenomena over a top surface of it when it is rotating as per astm norms.

The ball which is fixed on the top collet is meant to press against the top ball to the bottom position. Before that, it is important to wash the ball nicely with acetone to clean the surfaces of the ball. To make it useful for four ball tribometer experiment, the balls are made up of AISI E-52100 chrome alloy steel with a diameter of 12.7 mm and fined polished to grade 25 and 64 HRC as the hardness level. The ball pot is supported with upwards lever holding one ball on a thrust bearing, with a load compound which is fixed with moving lever to measure the normal load on the bottom ball. The four ball element is referred to evaluate the friction, wear and FTP properties of lubricant and oils. The sample is heated up to 75 °C by means of hot coil, ported in a four ball tribometer with a minimum speed of 1200 rpm and a constituent normal load of nearly 394 N. After the process is completed the ball pot is taken and kept into CCD microscope which is used for viewing the images of wear scar in ball. The image viewer software is used here to provide the dimensions of wear diameter and wear image for worn surfaces. The test was projected in the providence with ASTM D 4172. The viscosity of various lubricity oil samples was measured by Redwood Viscometer. The important part is to clean the ball surfaces gently and perfectly, while one ball should be used in a one sample at a time. The experiment went upon one complete hour (Figs. 3 and 4).

3.1 Flash Temperature Behaviour

The FTP states that it is used to evaluate the critical temperature over the lubricant to prepare the ratio between wear scar and diameter [3]. FTP can be experimentally suitable for all given conditions. The equations to evaluate the flash temperature

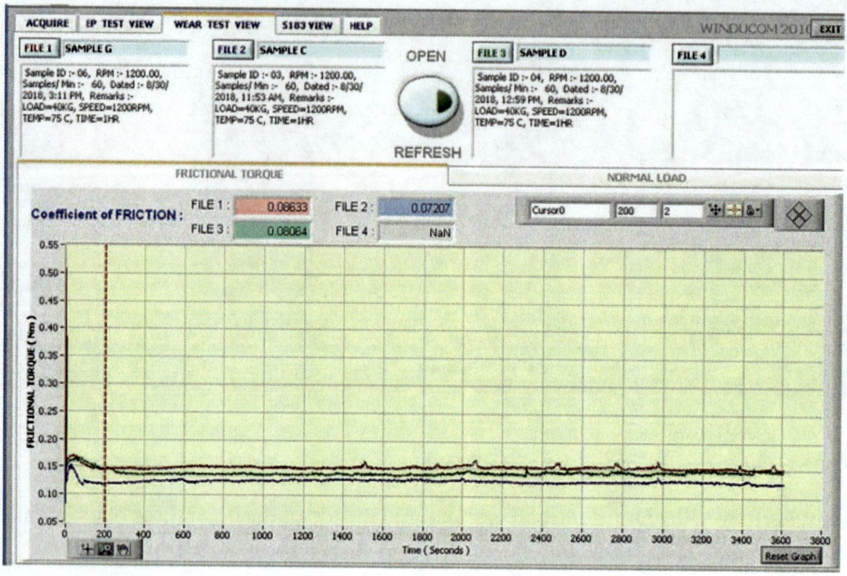

Fig. 4 Frictional Torque graph

parameter upon submitted loads to wear scar diameter are

$$=> \text{FTP} = W/D^{1.4}$$

where,

W—load act on ball (kg)
D—Wear Diameter of the ball (mm).

3.2 *Wear Behaviour*

It is a direct mean to the optical microscope to represent the density of wear, area covered by the wear scared and distances between scar images. The wear evolution is intensely when the ball bearing is wiped,u wear scar is viewed, and when it was affected caught, it was shown on the specimen with wear scar image at forefront. The perfect magnification lens was adjusted for clear focus on wear scar image and visible on image software, to broadcast the lubricant properties on material. The image was viewed synchronously and it was suggested to gather the part of the image which is are ought to be captured and saved for later by print screen button on keyboard. These findings are operationally maintained for bottom part of the ball in the ball pot which was dipped in lubricant. At last, by, the software has measured all the necessary parameters from the wear scar images found (Figs. 5 and 6).

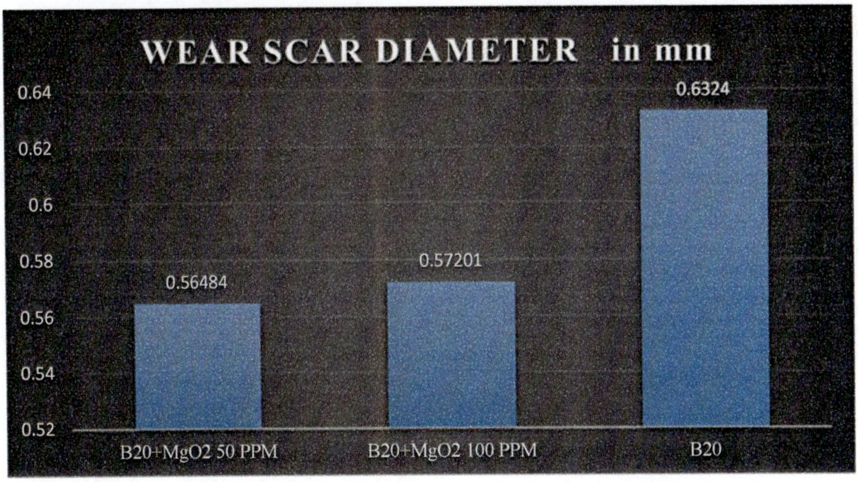

Fig. 5 Wear Scar Diameter Graph

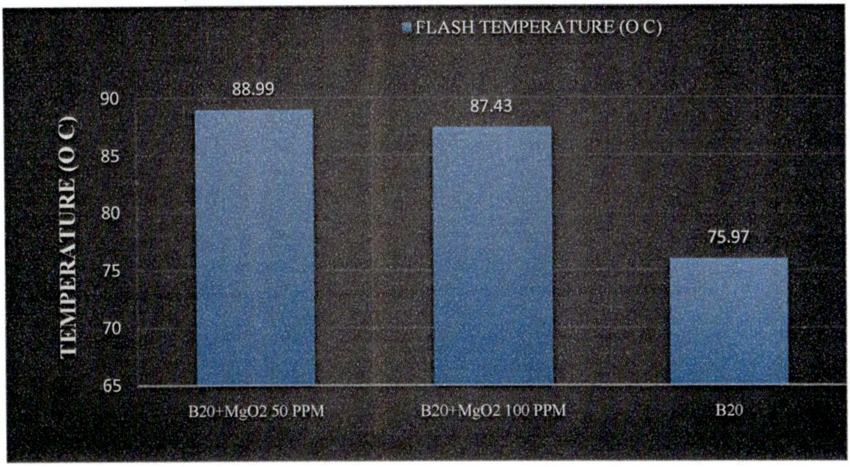

Fig. 6 Flash temperature graph

3.3 Friction Evaluation

In this experiment, load companion is expressed to measure the frictional torque. The frictional torque can be calculated by the coefficient of friction in the product of three times of load to equivalent distance from the centre divided by square root of six (Tables 5 and 6).

Table 5 Wear scar diameter

Samples	B1	B2	B3	Avg Areamm^2	Wear scar diagram mm
B20 + MgO 50 PPM	0.27549	0.21199	0.26425	0.250576667	0.56484
B20 + MgO100 PPM	0.29402	0.24537	0.23157	0.256986667	0.57201
B20	0.31894	0.33132	0.29207	0.31411	0.6324

Table 6 Flash temperature

Sample	Flash Temperature (°C)
B20 + MgO2 50 PPM	88.99
B20 + MgO2 100 PPM	87.43
B20	75.97

$$T = \frac{\mu * 3w * r}{\sqrt{6}}$$

Similarly, coefficient of friction is calculated by frictional torque which equals to the product of square root of six is divided by three times of load (W) and distance from the centre (r).

$$=> u = \frac{T\sqrt{6}}{3wr}$$

where u is the coefficient of friction, w is the load applied, r is the distance from the centre and T is the frictional torque.

4 Results and Discussions

4.1 *Frictional Characteristics*

About the behaviour of frictional coefficient, it is must to say that the zigzag line was unstable throughout every sample test. The period is commonly known as running period. After sometimes, when the force and viscosity of lubricant proceed to be overwhelmed, then the behaviour constant or stabilised known as normal state conditions. The x-axis shows the variation in coefficient of friction and y-axis shows the time. There are two types of conditions to be followed which are eventually decreasing at the graph are (1) running period and (2) normal state conditions (Figs. 7, 8 and 9).

The above contribution in results patched to the fact that at the start, the surface of contact reaches at the steady level. The unsteady state fuel properties increases

Fig. 7 Sample SEM image for B20 +50 ppm MgO

Fig. 8 Sample SEM image for B20 +100 ppm MgO

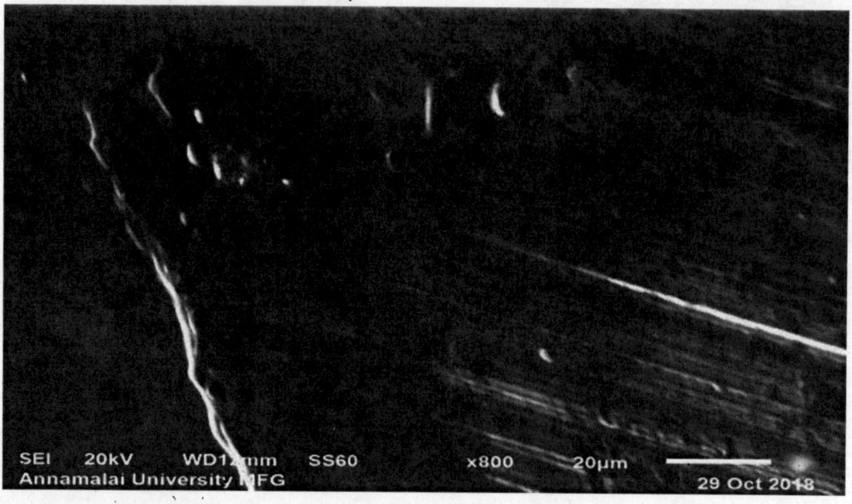

Fig. 9 Sample SEM image for B20

with the time, alike other fuel sublimated blends. This contribution is supposed to be a finding for the presence of ester in biofuel, which provides better scrap or scratch protection characteristics than in hydrocarbon fuel [4]. This result may be characterised as the removal of ester molecules from the biodiesel [5]. The unsteady state friction coefficient is increased with the decrease in decimal values. With this B20 is the highest blended value achieved from the counter list, to achieve the highest proportion in friction coefficient. From the values, the B20 comes first with highest decimal values, then comes B20 + 50 ppm MgO and B20 + 100 ppm MgO. This increasing to decreasing criteria formation does happen because the content of fatty acid present in the sample of biodiesel does not take a part to process lubricant film formation until the operation finishing time takes place, and after some seconds, the formation of thin-film layer is started. It is clearly observed from the friction coefficient graph that B20 makes own position in higher than all other fuel blends. This is supposed to be happen because of long-term oxidation method suppressing in the biodiesel blends.

4.2 Wear Characteristics

The features of wear behaviour confined show the effect of various fuels on wear scar diameter (WSD) under various loads.

The scar diameter of ball increases with an increase of load wear characteristics, with the fact of results that gradually increases the pressure on the metal ball surfaces degenerating the tribological characteristics of biodiesel repercussion of oxidation.

The results showed the compared to B20 reduced wear scar diameter 10.68% and 9.5%.

Compression of the increase of biodiesel concentration should compensate the diameter exceptionally. In these provisions, the biodiesel drops the percentage mark appreciably. This makes can be understandable to the incremental of ester functional groups which shows binding of molecules in lubricant [6]. Actually, biodiesel contains good proposition amount of oxygen it, to reduce the wear and friction between rubbing surfaces that contain acceptance of aliphatic fatty acid Benthic-Diatom Navicula Sp. Algae.

4.3 Flash Temperature Parameter

As we have already discussed, that flash temperature is the temperature which is convinced with lubricant film can be established to strengthen lubricity. A good FTP shows the good lubricity performances and a low possibility of diesel film breakdown. The effect of different samples of biodiesel oil with addition of nano additives responds with increase in every test with different values of such as B20, B20 + 100 ppm, B20 + 50 ppm, and the parameters are 75.97 °C, 87.43 °C and 88.99 °C, respectively.

This extensive condition shows that B20 + 100 ppm exhibits good lubricant performances and its extreme valuable to launch lubricating film that does not break down. Thus, contact surface floated with lubricant oil which surpassed one another and increased the contact surface.

4.4 SEM/EDX Analysis

Although after trans-esterification test, fuel needs to be clean and clear optimisation by filter paper is done and the filter debris was taken to examine using SEM/EDX method. Petrodiesel fuel particles are comparatively bigger than biodiesel blend used fuel [4]. The actual size of the worn surfaces from metal ball partly decreasing when biodiesel quantity increasing with the increase in biodiesel blends.

The particle size measurement of filter debris is extended broadly to a 20 um which provides the adhesive wear characteristics from the breakdown of formation in fuel properties.

It presents the SEM micrograph area surrounded by the wear scar worn surfaces of stationary balls under the pressure load of 40 kg different tested fuels. Hence, the surfaces framework shows that the layer metals from the worn surfaces are prohibited to the sliding direction of the rotating ball. Now, we can that the surface damage is purely greater than 20 um for every usable types of fuel. Therefore, shear strength would increase considerably, in definition of increase flow and ductility, should soften the surface at high temperatures and further reduced viscosity index. Again the mating surface is applied to be adhesive and enhances the quality properties. Based on the obtained results, sometimes wear characteristics suddenly decrease for B20 and constantly increase throughout the duration of oxidation until the half percentage biodiesel blend is produced.

This analysis is made to judge that the B20 +100 ppm MgO will exhibit the compounded lubricating performances in mean of wear and friction, and therefore, it is used to be in automobile engine to enhance the lubricating running engine lifelong.

5 Conclusion

- The below statements present the conclusions for Benthic-Diatom Navicula Sp. Algae oil on this experimental conditions;
- It kept characteristics of other experimented with results out test to validate the presence of ester in biodiesel blend, which gives a desired scrubbing diagram to protection behaviour in relation to petrodiesel fuel properties.
- B20 +100 ppm MgO is the common blend and the most protective behaviour for both wear and viscosity of the fuel, and also linear to the steady time period.
- From the ups and downs of the graph, it is clearly evident that the wear scar diameter and the coefficient of friction decrease and flash temperature increase with the load chosen.
- This increasing field could be equally similar for all blends except B20.

References

1. Horner D (2002) Recent trends in environmental friendly lubricants. J Synthetic Lubrication 18(4):327–347
2. Arumugam S, Sriram G (2012) Effect of bio-lubricant and biodiesel-contaminated lubricant on tribological behavior of cylinder liner—piston ring combination. Tribol Trans 55(4):438–445
3. Anand ON, Chhibber VK (2006) Vegetable oil derivatives: environment-friendly lubricants and fuels. J Synthetic Lubrication 23(2):91–107
4. Arumugam S, Sriram G (2013) Synthesis and characterisation of rapeseed oil bio-lubricant—its effect on wear and frictional behaviour of piston ring–cylinder liner combination. J Eng Tribol 227(1):3–15
5. Wu X, Zhang X, Yang S, Chen H, Wang D (2000) The study of epoxidized rapeseed oil used as a potential biodegradable lubricant. J Am Oil Chem Soc 77(5):561–563
6. Erhan SZ, Asadauskas S (2000) Lubricant basestocks from vegetable oils. Ind Crops Prod 11(2–3):277–282

Experimental Investigation of Performance and Combustion Characteristics of a Diesel Engine Fueled with *Eruca sativa* (Taramira) Biodiesel–Diesel Blend

Mohd Hamid Hussain and C. H. Biradar

Abstract The world's current issues like environmental pollution and fossil fuel inanition made the researchers to discover the alternative to the fossil fuel. Biodiesel has potential in resolving current energy crisis because it is a clean burning diesel and renewable. The present work evaluates the performance and its combustion characteristics of a constant speed diesel engine which is fueled with blends of *Eruca sativa* as biodiesel. Biodiesel is prepared by conventional transesterification from taramira oil. For the blends TB_{10}, TB_{20}, TB_{30}, TB_{40} and TB_{100} the physical–chemical properties (i.e. kinematic viscosity, pour point, density, flash point and calorific value) are measured, and tests were performed on CI engine and are compared with the diesel. The results shows, physical–chemical properties of taramira biodiesel and its blends meet the standard specifications of ASTM D-6751 and EN-14214, performance parameters for TB_{10} and TB_{20} like brake thermal efficiency, mechanical efficiency and indicated mean effective pressure are higher than the diesel, while brake specific fuel consumption is lower. For the combustion characteristics parameters with the use of biodiesel, significant changes are observed and are noteworthy.

Keywords Taramira · Biodiesel · Performance · Combustion · Diesel engine

Abbreviations

BSFC	Brake specific fuel consumption
BTHE	Brake thermal efficiency
CA	Crank angle
CI	Compression ignition
EGT	Exhaust gas temperature

M. H. Hussain (✉)
Department of Mechanical Engineering, PDA College of Engineering, Kalaburagi 585102, India
e-mail: mhhussain095@gmail.com

C. H. Biradar
Department of Automobile Engineering, PDA College of Engineering, Kalaburagi 585102, India

A. Arockiarajan et al. (eds.), *Advances in Industrial Automation and Smart Manufacturing*, Lecture Notes in Mechanical Engineering,
https://doi.org/10.1007/978-981-15-4739-3_77

FAME Fatty acid methyl ester
IMEP Indicated mean effective pressure
TB Taramira biodiesel
TB_{10} 10% Taramira biodiesel
TB_{20} 20% Taramira biodiesel
TB_{30} 30% Taramira biodiesel
TB_{40} 40% Taramira biodiesel
TB_{100} 100% Taramira biodiesel
TDC Top dead center

1 Introduction

The primary fundamental requirement for existence of human is energy, since the last two decades worlds energy demand has encouraged toward searching alternative sources of energy [1]. The alternatives to fossil fuels and commercial sources of renewable energy are biomass, wind and solar energy [2]. Furthermore, the depletion of fossil fuel, elevated and increasing pollution in atmosphere causing numerous concerns of environment like global climate change, air quality made researchers to look for the alternatives to fossil fuels [3, 4]. One of the best ways to cut the worlds dependence on petroleum products is to develop the substitute and renewable fuels like biodiesel [5].

Biodiesel is one of the important biofuel types, which can be produced from vegetable oils and animal fats by means of transesterification technique [6]. Transesterification seems to be the best method for converting straight vegetable oil into biodiesel [7]. In this process, the fatty acids are converted into its corresponding esters with the help of a catalyst [8]. Figure 1 represents the production of biodiesel by transesterification reaction. Biodiesel has many characteristics like it is biodegradable, renewable, green fuel, non-toxic, with excellent lubricity, substantially free from aromatics nature and sulfur [9]. As the biodiesel has characteristics close to base diesel, therefore, it can be mixed in any proportion with diesel homogenously. With the mixing of biodiesel, the diesel engine exhaust emissions get reduced [10]

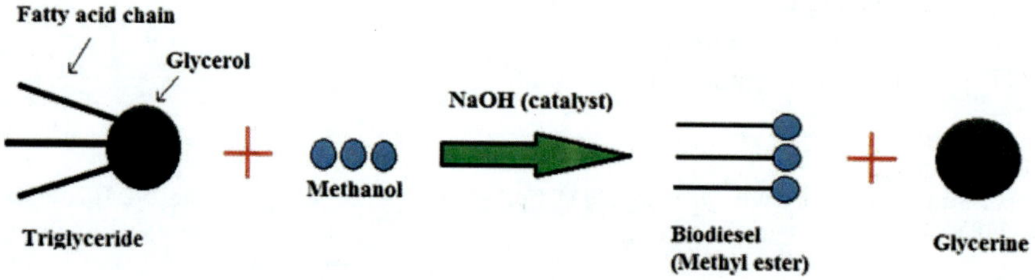

Fig. 1 Transesterification reaction

and it does not add any extra CO_2 to the environment, therefore, it can be considered that the biodiesel is a solution to reduce the pollutant levels in the environment [11].

The production of biodiesel with the use of edible oils raises the concerns of long-term competition between the supply of food and feedstock for biodiesel production [12], therefore use of the non-edible vegetable oils for production of biodiesel is rising globally [13]. *Eruca sativa* commonly known as taramira (in south Asia) and Rocket or Jarjeer (in middle east) [14, 15] belongs to the family Brasssicaceae [16]. It is grown in Europe, North Africa, Middle east countries, India and Pakistan [17], due to its medicinal uses [18] and fourth-generation salad characteristics [19]. In India, taramira is being grown in many states such as Madhya Pradesh, Punjab, Haryana, Rajasthan and Uttar Pradesh. Among these states, Rajasthan is the chief growing state, in terms of production and area [20]. Taramira plant contains non-edible food crops [21] and the oil obtained is non-edible due to its pungent odor [22].

2 Experimentation

2.1 Taramira Biodiesel Production

For the present study, taramira biodiesel is produced by transesterification method. Methanol (1:6 molar ratio) and the catalyst sodium hydroxide (NaOH) (1% w/w) mixture was added to the reaction chamber containing taramira oil at 70 °C with continuous stirring at 1200 rpm. The condenser and thermometer were fitted to the reaction chamber. Reaction has been carried out for 120 min at 65 °C, and then it was transferred to the separating funnel and kept steady for 180 min at room temperature. After settling two layers were formed, top layer identified as biodiesel and bottom layer as glycerin as its density is high [23, 24]. The glycerin was separated out and biodiesel washed with warm water to remove unreacted traces of NaOH, soap formed during reaction [25]. Then, it is dried at 120 °C for 90 min. Finally, the product obtained is taramira biodiesel.

2.2 Properties Testing

The physical–chemical properties of the produced taramira biodiesel and its blends (TB_{10}, TB_{20}, TB_{30}, TB_{40} and TB_{100}) were tested in a certified laboratory by the following American Society for Testing and Material (ASTM) test methods.

2.3 Engine Setup and Test Procedure

The experiments were carried out on a four stroke, compression ignition engine. Speed of the engine 1500 rpm was constant throughout the test and the rated power was 3.5 kW. The loading is provided by eddy current dynamometer connected to engine by universal joint coupling. The technical engine specifications are listed in Table 1. Data from all the measuring instruments and sensors had been collected in data acquisition system and transferred to computer, in which software named ICEnginesoft is installed to analyze the diesel engine performance and combustion characteristics. Figure 2 shows pictorial view of engine setup and Fig. 3 represents line diagram of engine setup. At the beginning engine was run on petro-diesel to create a baseline data and then fueled with blends of taramira biodiesel. The fuel consumption and airflow were measured with the aid of sensors and data acquisition system. The performance and combustion parameters were measured and results were plotted against load for diesel and different blends of taramira biodiesel.

Table 1 Specifications of the engine setup

Parameter	Specifications
Make and model General details	Kirloskar, TV1 Four stroke, single cylinder, constant speed, water cooled diesel engine.
Data acquisition device	NI USB-6210, 16-bit, 250 kS/s
Software	"ICEnginesoft V-9.0" engine performance and combustion analysis software
Rated power	3.5 kW at 1500 rpm
Rated speed	1500 rpm
Compression ratio range	12:1–18:1
Injection variation	0-25°BTDC
Bore × Stroke	87.5 mm × 110 mm
Swept volume	661.45 cc
Dynamometer	Eddy current type, water cooled
Load sensor	Strain gauge load sensor
Temperature sensor	Thermocouple Type-K
Crank angle sensor	Make Kubler, with TDC pulse
Injection	Direct
Calorimeter	Type pipe in pipe 25–250 Lph
Engine cooling supply	40–400 Lph
Pump	Monoblock type

Fig. 2 Engine setup pictorial view

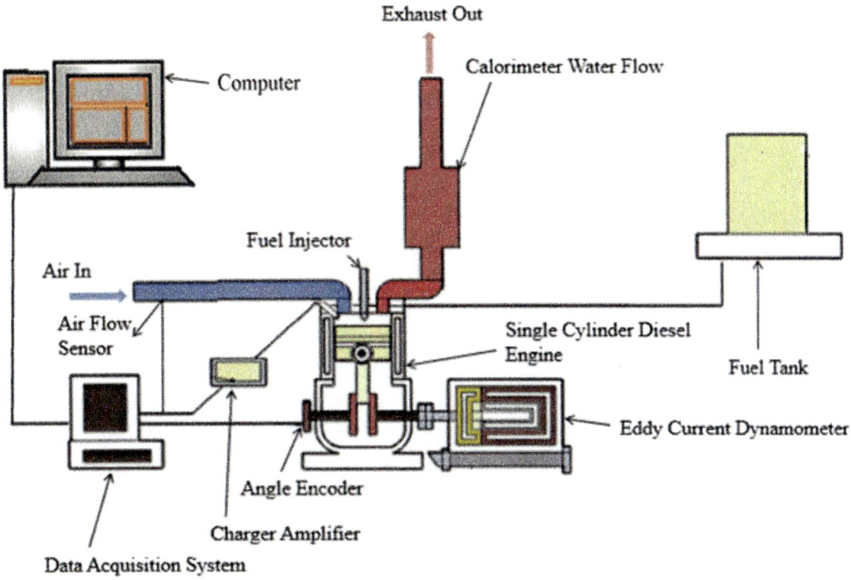

Fig. 3 Engine setup line diagram

Table 2 Instrument ranges, accuracies and uncertainties (%)

Instruments	Range	Accuracy	Uncertainties percentage
Thermocouple (type-K)	0–1000°C	± 1°C	± 0.1
Load indicator	0–50 kg	± 0.1 kg	± 0.2
Speed measuring unit	0–9999 rpm	± 10 rpm	± 0.2
Volumetric fuel flow meter	0–500 mm	± 0.1 mm	± 1
Manometer	0–50 mm	± 1 mm	± 1
Crank angle encoder	0–360°CA	± 1°CA	± 0.2
Pressure pickup	0–25 MPa	± 0.1 MPa	± 0.1

2.4 *Error Analysis*

The precision of the outcomes acquired in the experiment relies on mistakes and uncertainties resulting from the choice, calibration, situation, test planning, atmosphere, observation and reading of instruments. So, instruments were selected according to their percentage of uncertainties and accuracy. Table 2 shows the values of errors for a variety of instruments.

3 Results and Discussion

3.1 *Biodiesel–Diesel Blending Effect on Physical–Chemical Properties*

It is observed from Table 3 that, all the biodiesel blends properties meet the ASTM D-6751 standard specifications. The density has increased as the proportion of taramira biodiesel in the biodiesel–diesel blend increases [28]. Density has the influence on fuel atomization efficiency for airless combustion systems [29]. Kinematic viscosity reflects the resistance to the fluid shear stress [30], for diesel (2.94 mm^2/s) it is minimum whereas for TB_{100} (5.9 mm^2/s) it is near the maximum limit. The volatility of the biodiesel is inversely proportional to the flash point [31]. The growing trend for flash point is observed for all the biodiesel blends, resulting in safe handling and transport. Generally biodiesel has the lower calorific value than that of petro-diesel, this may be due to, higher content of oxygen [32]. The temperature at which fuel loses its flow characteristics is known as pour point, it is considered to be cold flow property because during cold weather condition it establishes the limit for the use of fuels [33].

Table 3 Physical–chemical properties of taramira biodiesel and its blends comparing

Property	ASTM test method	ASTM D-6751 [26]	Petro-diesel [13, 27]	TB_{10}	TB_{20}	TB_{30}	TB_{40}	TB_{100}
Density at 15°C (kg/m^3)	D-127	860-890	842	846	853	861	867	880
Kinematic viscosity at 40°C (mm^2/s)	D-445	1.90-6.0	2.94	3.36	3.89	4.16	4.58	5.9
Flash point (°C)	D-93	Min. 130	68	79.3	85.2	92.8	104	134
Calorific value (kJ/kg)	D-240	Above 37,656	45,389	45,167	44,802	44,417	44,084	43,459
Pour point (°C)	D-2500	−15 to 10	0	−0.16	−0.39	−0.87	−1.31	−2.2

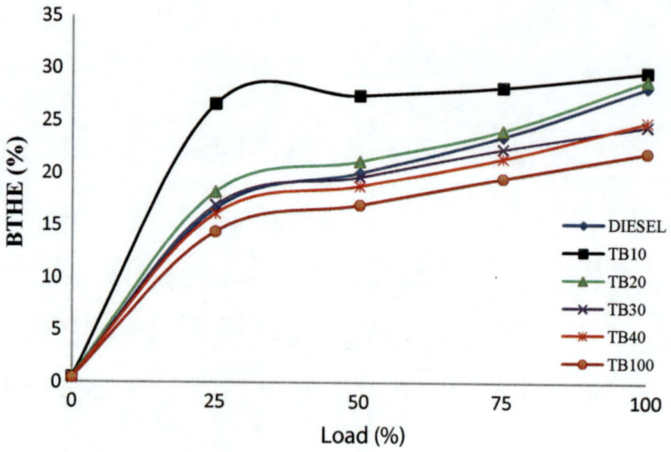

Fig. 4 Variation of BTHE against load

3.2 *Performance*

3.2.1 Brake Thermal Efficiency

BTHE, also known as fuel conversion energy, shows how efficiently the engine turns chemical fuel energy into helpful work. Figure 4 shows the BTHE variation with respect to load for taramira biodiesel and its blends with diesel. From the figure, it can be observed that the BTHE increases as the load increases, at 25% load condition the TB_{10} blend shows the maximum rise in BTHE and it is 60.24% higher compared to the diesel, while during full load 5.17% and 2.44% rise is observed for TB_{10} and TB_{20}, respectively. The increase in BTHE is because of rise in combustion temperature and the peak pressure [34].

3.2.2 Brake Specific Fuel Consumption

The BSFC can be defined as the actual mass of fuel required in an hour to generate 1 kW power. Figure 5 shows the BSFC variation for TB blends and the diesel. It is clear from figure that the BSFC decreases steadily as the engine load increases. The BSFC for TB_{10} and TB_{20} is less than diesel by 10% and 6.67%, respectively, at rated load. It is also found that BSFC rises with the increase in biodiesel concentration in the mixture this could be accredited to the lower calorific value of blends TB_{30}, TB_{40} and TB_{100}, as for more amount of fuel consumed for the same energy output, high cetane number of biodiesel causing change in combustion timing [35].

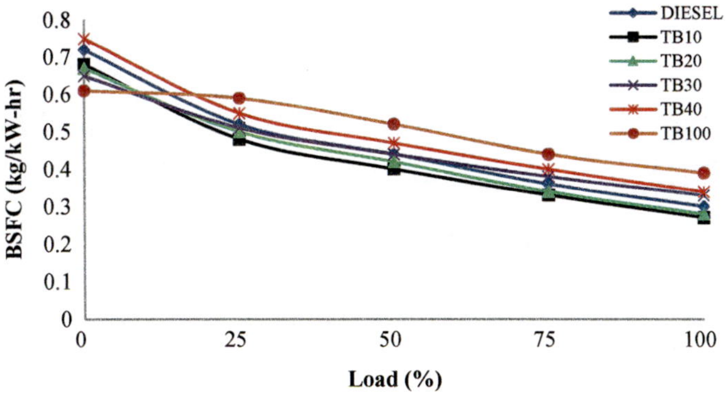

Fig. 5 Variation of BSFC against load

3.2.3 Mechanical Efficiency

Mechanical efficiency is a measure of a machine's efficacy in transforming the device's input power and energy into an output and force motion. Therefore, it shows how successful the engine is transforming the indicated power into useful power [36]. The change in mechanical efficiency with respect to engine load is shown in Fig. 6. From the graph plotted it can be noted that during 25% load on engine TB_{30} shows the maximum efficiency. At rated load condition, the mechanical efficiency for TB_{10} (82.68%) is higher than diesel (80.53%) while the TB_{20} (80.29%) blend is nearly equal to the diesel.

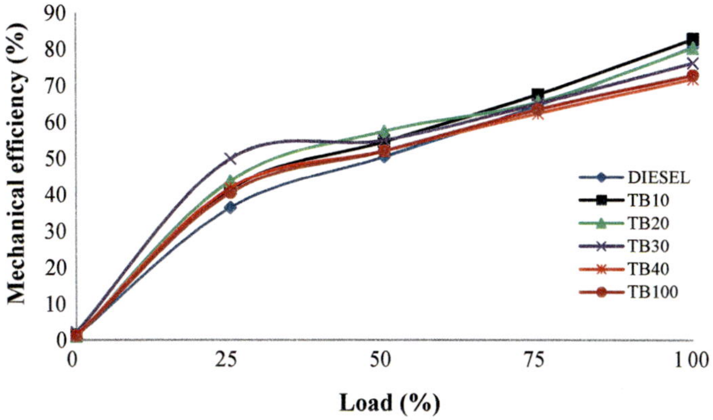

Fig. 6 Variation of mechanical efficiency against load

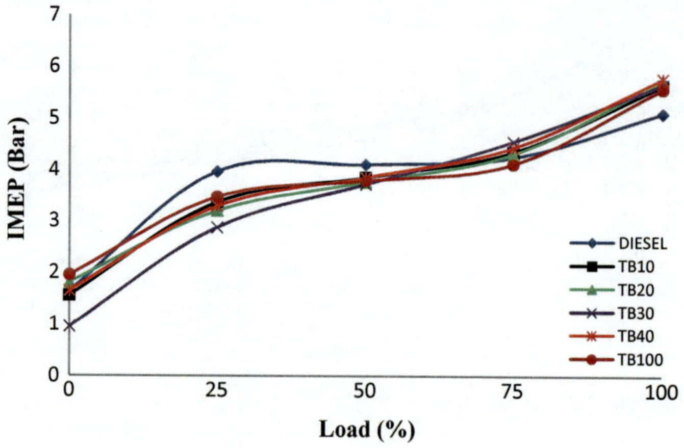

Fig. 7 Variation of IMEP against load

3.2.4 Indicated Mean Effective Pressure

For internal combustion engine, IMEP is the average pressure that is induced in the combustion chamber through the complete thermodynamic cycle [37]. The IMEP variation against load is shown in Fig. 7. It can be observed from the figure that, for taramira biodiesel blends the IMEP found to be lower at lower loads and higher at higher loads than base diesel line, the similar trend is observed with karanja biodiesel blends [38]. The IMEP at rated load for diesel, TB_{10}, TB_{20}, TB_{30}, TB_{40} and TB_{100} are 5.1 bar, 5.64 bar, 5.73 bar, 5.66 bar, 5.77 bar and 5.57 bar, respectively.

3.3 Combustion Characteristics

3.3.1 Exhaust Gas Temperature

Temperature of the exhaust gas is an indicative of the extent of combustion and it will increase with an increase in load on the engine [39]. Figure 8 shows the EGT variation against load on the engine, it is noted from the figure that the EGT rises with raise in percentage of TB in the blend at rated load. This could be due to the improved combustion due to the content of oxygen and also the elevated viscosity of biodiesel blends causing the poor fuel vaporization and atomization, resulting in injected fuel late burning and consequently higher exhaust gas temperature [40]. At rated load, the EGT tested for diesel and TB blends varies from 293.63 to 308.75 °C.

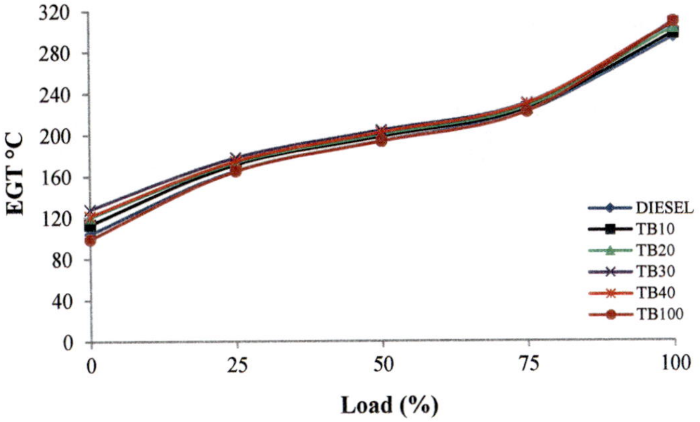

Fig. 8 Variation of EGT against load

3.3.2 Cylinder Pressure

For understanding the combustion process, cylinder pressure measurement is the most significant instrument used to provide adequate data for the assessment of combustion and to determine the burned mass fraction, heat release rate, cylinder pressure-volume, etc. The change in cylinder pressure with respect to °CA for different taramira biodiesel blends at 100% loading is shown in Fig. 9. It can be observed that the TB blends follow the similar trend as that of conventional diesel. For TB blends the cylinder pressure starts increasing significantly 9 °CA before top dead center and continues rise in the pressure is observed up to 10 °CA after TDC. The maximum pressure for TB blends in the cylinder is greater than diesel and ranged

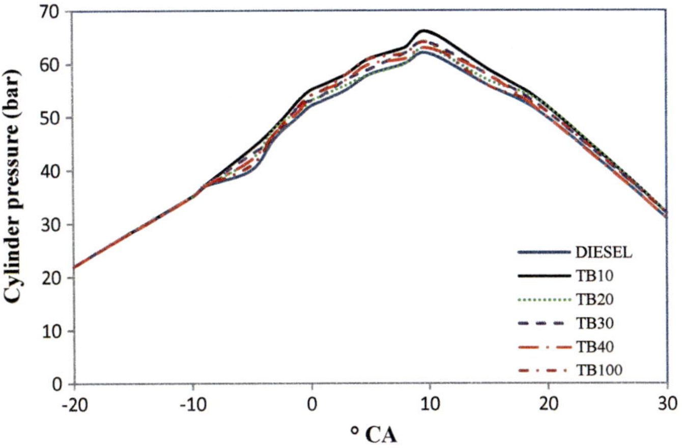

Fig. 9 Variation of cylinder pressure against crank angle

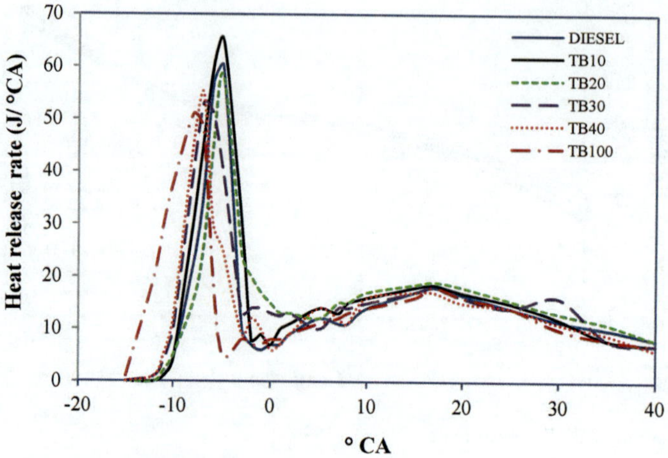

Fig. 10 Variation of heat release rate against crank angle

from 63 to 66 bar for the biodiesel blends, whereas for diesel it is 62 bar. This slightly higher maximum pressure can be accredited to higher viscosity [41] and high cetane no of taramira biodiesel [42].

3.3.3 Heat Releases Rate

The heat release rate curve helps to recognize how much heat energy can be rehabilitated during fuel burning into precious job [43]. It depends upon the physical–chemical properties of the fuel, volume swept in the cylinder by piston in the engine block and cylinder pressure [44]. The variations of heat release rate with respect to crank angle at full load of engine for TB blends and base diesel are plotted in Fig. 10. It can be observed that the combustion starts earlier for the TB and its blends. The peak heat release rate for TB_{10} (65 J/°CA) is higher than petro-diesel (60 J/°CA) and for TB_{20} (58 J/°CA) it is nearly equal to diesel. This could be one of main the reason to increase in brake thermal efficiency [45] of these blends which has been discussed earlier.

4 Conclusions

From the above results obtained for diesel engine performance and combustion characteristics by using taramira biodiesel its blends, the following conclusions can be drawn:

- The physical–chemical properties of taramira oil methyl ester and the blends meet the ASTM standard specifications and are nearer to diesel. The diesel engine can perform successfully with TB and its blends without any engine modifications.
- BTHE is higher and the BSFC is found to be lower for TB_{10} and TB_{20} blends when compared to diesel.
- Mechanical efficiency is higher than diesel for TB_{10}, and it starts decreasing as the proportion of TB in the blend increases. While IMEP for all the TB blends found to be higher than base diesel at higher loads.
- The combustion characteristics analysis showed the rise in EGT and cylinder pressure with the use of TB blends and the heat release rate is found to be maximum for TB_{10}.

Thus, it can be concluded that taramira biodiesel can be used instead of petro-diesel as a substitute and environmental friendly fuel for the diesel engine. The optimum combination of taramira biodiesel and diesel can be determined on the basis of performance and combustion parameters.

References

1. Dwivedi G, Sharma MP (2014) Prospects of biodiesel from Pongamia in India. Renew Sustain Energy Rev 32:114–122. https://doi.org/10.1016/j.rser.2014.01.009
2. Izadyar N, Ong HC, Chong WT, Leong KY (2016) Resource assessment of the renewable energy potential for a remote area: a review. Renew Sustain Energy Rev 62:908–923. https://doi.org/10.1016/j.rser.2016.05.005
3. Amini Z, Ong HC, Harrison MD, Kusumo F, Mazaheri H, Ilham Z (2017) Biodiesel production by lipase-catalyzed transesterification of *Ocimum basilicum* L. (sweet basil) seed oil. Energy Convers Manag 132:82–90. https://doi.org/10.1016/j.enconman.2016.11.017
4. Arumugam A, Ponnusami V (2019) Biodiesel production from Calophyllum inophyllum oil a potential non-edible feedstock: an overview. Renew Energy 131:459–471. https://doi.org/10.1016/j.renene.2018.07.059
5. Keera ST, El Sabagh SM, Taman AR (2018) Castor oil biodiesel production and optimization. Egypt J Pet 27:979–984. https://doi.org/10.1016/j.ejpe.2018.02.007
6. A. E. Biofuel: Sources, Extraction and Determination. Liq Gaseous Solid Biofuels—Convers Tech, InTech (2013) https://doi.org/10.5772/51943
7. Silitonga AS, Atabani AE, Mahlia TMI, Masjuki HH, Badruddin IA, Mekhilef S (2011) A review on prospect of Jatropha curcas for biodiesel in Indonesia. Renew Sustain Energy Rev 15:3733–3756. https://doi.org/10.1016/j.rser.2011.07.011
8. Nantha Gopal K, Thundil Karupparaj R (2014) Effect of pongamia biodiesel on emission and combustion characteristics of di compression ignition engine. Ain Shams Eng J 6:297–305. https://doi.org/10.1016/j.asej.2014.10.001
9. Yan J, Yan Y (2017) Biodiesel production and technologies. Encycl Sustain Technol. Elsevier, pp 261–272. https://doi.org/10.1016/b978-0-12-409548-9.10110-1
10. Venkatesan SP, Kadiresh PN (2016) Influence of an aqueous cerium oxide nanofluid fuel additive on performance and emission characteristics of a compression ignition engine. Int J Ambient Energy 37:64–67. https://doi.org/10.1080/01430750.2014.882863
11. Kumar A, Shukla SK, Tierkey JV (2015) A review of research and policy on using different biodiesel oils as fuel for C.I. Engine. Energy Proc 90:292–304. https://doi.org/10.1016/j.egypro.2016.11.197

12. Balat M (2011) Potential alternatives to edible oils for biodiesel production—a review of current work. Energy Convers Manag 52:1479–1492. https://doi.org/10.1016/j.enconman.2010.10.011
13. Atabani AE, Silitonga AS, Ong HC, Mahlia TMI, Masjuki HH, Badruddin IA et al (2013) Non-edible vegetable oils: a critical evaluation of oil extraction, fatty acid compositions, biodiesel production, characteristics, engine performance and emissions production. Renew Sustain Energy Rev 18:211–245. https://doi.org/10.1016/j.rser.2012.10.013
14. Di Gioia F, Avato P, Serio F, Argentieri MP (2018) Glucosinolate profile of Eruca sativa, Diplotaxis tenuifolia and Diplotaxis erucoides grown in soil and soilless systems. J Food Compos Anal 69:197–204. https://doi.org/10.1016/j.jfca.2018.01.022
15. Jaafar N (2019) Eruca Sativa Linn.: pharmacognostical and pharmacological. Asian J Pharm Clin Res 12:10–17. https://doi.org/10.22159/ajpcr.2019.v12i3.30893
16. Taffner J, Cernava T, Erlacher A, Berg G (2019) Novel insights into plant-associated archaea and their functioning in arugula (Eruca sativa Mill.). J Adv Res. https://doi.org/10.1016/j.jare.2019.04.008
17. Garg G, Sharma V (2014) Eruca sativa (L.): botanical description, crop improvement, and medicinal properties. J Herbs Spices Med Plants 20:171–182. https://doi.org/10.1080/10496475.2013.848254
18. Khoobchandani M, Ojeswi BK, Ganesh N, Srivastava MM, Gabbanini S, Matera R et al (2010) Antimicrobial properties and analytical profile of traditional Eruca sativa seed oil: comparison with various aerial and root plant extracts. Food Chem 120:217–224. https://doi.org/10.1016/j.foodchem.2009.10.011
19. Katsarou D, Omirou M, Liadaki K, Tsikou D, Delis C, Garagounis C et al (2016) Glucosinolate biosynthesis in Eruca sativa. Plant Physiol Biochem 109:452–466. https://doi.org/10.1016/j.plaphy.2016.10.024
20. Garg G, Sharma V (2015) Assessment of fatty acid content and genetic diversity in Eruca sativa (L.) (Taramira) using ISSR markers. Biomass Bioenergy 76:118–129. https://doi.org/10.1016/j.biombioe.2015.03.010
21. Lazzeri L, Errani M, Leoni O, Venturi G (2004) Eruca sativa spp. oleifera: a new non-food crop. Ind Crops Prod 20:67–73. https://doi.org/10.1016/j.indcrop.2002.06.001
22. Orhan I, Kartal M, Sekeroglu N, Esiyok D, Sener B, Ugur A et al (2010) Variations in fatty acid compositions of the seed oil of Eruca sativa Mill. caused by different sowing periods and nitrogen forms. Pharmacogn Mag 6:305. https://doi.org/10.4103/0973-1296.71801
23. Dhanamurugan A, Subramanian R (2013) Performance of single cylinder diesel engine with bael seed biodiesel, vol 72
24. Arun SB, Suresh R, Yatish KV, Omkaresh BR, Channa Keshava Naik N (2017) Use of CaO and Na3PO4 catalysts in the synthesis of biodiesel and investigation of fuel properties. Mater Today Proc 4:11111–11117 (Elsevier Ltd). https://doi.org/10.1016/j.matpr.2017.08.074
25. Ferdous K, Deb A, Ferdous J, Uddin MR, Khan MR, Islam M (2014) Aphanamixis polystachya: a potential non-edible source of biodiesel in Bangladesh. J Chem Eng 28:45–49. https://doi.org/10.3329/jce.v28i1.18111
26. Palash SM, Masjuki HH, Kalam MA, Atabani AE, Rizwanul Fattah IM, Sanjid A (2015) Biodiesel production, characterization, diesel engine performance, and emission characteristics of methyl esters from Aphanamixis polystachya oil of Bangladesh. Energy Convers Manag 91:149–157. https://doi.org/10.1016/j.enconman.2014.12.009
27. Labeckas G, Slavinskas S (2006) The effect of rapeseed oil methyl ester on direct injection Diesel engine performance and exhaust emissions. Energy Convers Manag 47:1954–1967. https://doi.org/10.1016/j.enconman.2005.09.003
28. Saxena P, Jawale S, Joshipura MH (2013) A review on prediction of properties of biodiesel and blends of biodiesel. Proc Eng 51:395–402. https://doi.org/10.1016/j.proeng.2013.01.055
29. Silitonga AS, Masjuki HH, Mahlia TMI, Ong HC, Chong WT, Boosroh MH (2013) Overview properties of biodiesel diesel blends from edible and non-edible feedstock. Renew Sustain Energy Rev 22:346–360. https://doi.org/10.1016/j.rser.2013.01.055
30. Yadav KK, Garg N, Kumar Verma A (2017) Optimization and extraction of oil from mango seed kernel (Mangifera indica) Enhancing Livelihood and Profitability Index of Malihabad

Farmers through Diversified Horti-Enterprise Modules View project Devising technologies for minimization of post harvest. Indian J Agric Sci 7:91–94
31. Atabani AE, Silitonga AS, Badruddin IA, Mahlia TMI, Masjuki HH, Mekhilef S (2012) A comprehensive review on biodiesel as an alternative energy resource and its characteristics. Renew Sustain Energy Rev 16:2070–2093. https://doi.org/10.1016/j.rser.2012.01.003
32. Wakil MA, Kalam MA, Masjuki HH, Atabani AE, Rizwanul Fattah IM (2015) Influence of biodiesel blending on physicochemical properties and importance of mathematical model for predicting the properties of biodiesel blend. Energy Convers Manag 94:51–67. https://doi.org/10.1016/j.enconman.2015.01.043
33. Atabani AE, Mahlia TMI, Masjuki HH, Badruddin IA, Yussof HW, Chong WT et al (2013) A comparative evaluation of physical and chemical properties of biodiesel synthesized from edible and non-edible oils and study on the effect of biodiesel blending. Energy 58:296–304. https://doi.org/10.1016/j.energy.2013.05.040
34. Balakrishnan K, Dey S, Gupta T, Dhaliwal RS, Brauer M, Cohen AJ et al (2019) The impact of air pollution on deaths, disease burden, and life expectancy across the states of India: the Global Burden of Disease Study 2017. Lancet Planet Heal 3:e26–e39. https://doi.org/10.1016/S2542-5196(18)30261-4
35. Gangil S, Singh R, Bhavate P, Bhagat D, Modhera B (2016) Evaluation of engine performance and emission with methyl ester of Karanja oil. Perspect Sci 8:241–243. https://doi.org/10.1016/j.pisc.2016.04.040
36. Srithar K, Arun Balasubramanian K, Pavendan V, Ashok Kumar B (2017) Experimental investigations on mixing of two biodiesels blended with diesel as alternative fuel for diesel engines. J King Saud Univ Eng Sci 29:50–56. https://doi.org/10.1016/j.jksues.2014.04.008
37. Martyr AJ, Plint MA (2012) The combustion process and combustion analysis. Engine Testing 375–406. https://doi.org/10.1016/b978-0-08-096949-7.00015-7
38. Sivaramakrishnan K (2018) Investigation on performance and emission characteristics of a variable compression multi fuel engine fuelled with Karanja biodiesel–diesel blend. Egypt J Pet 27:177–186. https://doi.org/10.1016/j.ejpe.2017.03.001
39. Kathirvelu B, Subramanian S, Govindan N, Santhanam S (2017) Emission characteristics of biodiesel obtained from jatropha seeds and fish wastes in a diesel engine. Sustain Environ Res 27:283–290. https://doi.org/10.1016/j.serj.2017.06.004
40. Kerihuel A, Kumar MS, Bellettre J, Tazerout M (2010) Investigations on a CI Engine using animal fat and its emulsions with water and methanol as fuel. SAE Tech Pap Ser 1. https://doi.org/10.4271/2005-01-1729
41. Aalam CS, Saravanan CG (2017) Effects of nano metal oxide blended Mahua biodiesel on CRDI diesel engine. Ain Shams Eng J 8:689–696. https://doi.org/10.1016/j.asej.2015.09.013
42. Vijay Kumar M, Veeresh Babu A, Ravi Kumar P (2018) The impacts on combustion, performance and emissions of biodiesel by using additives in direct injection diesel engine. Alexandria Eng J 57:509–516. https://doi.org/10.1016/j.aej.2016.12.016
43. Asokan MA, Prabu SS, Kiran P, Bade K, Nekkanti VM, Sai S et al (2019) Performance, combustion and emission characteristics of juliflora biodiesel fuelled DI diesel engine. Energy. https://doi.org/10.1016/j.energy.2019.02.075
44. Yusri IM, Mamat R, Akasyah MK, Jamlos MF, Yusop AF (2019) Evaluation of engine combustion and exhaust emissions characteristics using diesel/butanol blended fuel. Appl Therm Eng. https://doi.org/10.1016/j.applthermaleng.2019.02.028
45. Biradar CH, Subramanian KA, Dastidar MG (2011) Performance improvement and emissions reduction of a DI Diesel Engine for use of Karanja Biodiesel-Diesel Blend (B20) using Diethyl Ether. SAE Tech Pap Ser, vol 1, SAE International. https://doi.org/10.4271/2011-26-0004

Feasibility Studies on Spent Coffee Powder Oil as Alternative to Diesel in CI Engines

S. Sunil, Shrishail Kakkeri, B. S. Chandra Prasad, N. Kapilan, and Shivarudraiah

Abstract In the current work, oil is removed from spent coffee powder utilizing a chemical extraction method. The properties of acquired biodiesel are found out, and it very well may be inferred that the properties of got biodiesel are within ASTM norms. To augment the yield of biodiesel, Taguchi strategy is utilized to improve the response parameters like methanol to oil proportion, catalyst concentration, reaction temperature, reaction time. An improvement of yield up to 15% was found by using Taguchi strategy. Heating value of the produced fuel is competitive with the diesel. The biodiesel blends are tested in a 4S single chamber compression ignition engine to get to the performance and outflow attributes. It is seen that most elevated thermal efficiency is acquired at B20 blend which is 1.33% more than that of diesel at full load and least specific fuel consumption at B20 at full load was 0.15% lesser than diesel. As for emissions, there was a noteworthy decrease in CO, CO_2, and HC while NO_x discharges expanded marginally.

Keywords Biodiesel · Transesterification · Spent coffee oil

1 Introduction

Oil-based goods are the principal form of energy devoured worldwide other than atomic, wind, and coal-based energy sources [1]. India imports about 70–80% of its oil-based fuels from the bay countries which are more frequently under political

S. Sunil (✉) · S. Kakkeri · B. S. Chandra Prasad
Mechanical Engineering Department, Sri Venkateshwara College of Engineering, Bengaluru, Karnataka 562157, India
e-mail: sunilmurthy86@gmail.com

N. Kapilan
Mechanical Engineering Department, Nagarjuna College of Engineering and Technology, Bengaluru, India

Shivarudraiah
Mechanical Engineering Department, University Visvesvaraya College of Engineering, Bengaluru 560001, India

A. Arockiarajan et al. (eds.), *Advances in Industrial Automation and Smart Manufacturing*, Lecture Notes in Mechanical Engineering,
https://doi.org/10.1007/978-981-15-4739-3_78

unsettling influence [2]. Consequently, for India, in the same way as other different countries on the planet, national security for energy and worry for the earth are the standard central purposes for the chase of endless and clean wellsprings of energy [3]. Bio-fuels are the promising alternatives to this hunt. It is a good sign that, today bio-fuels are in effect truly seen from the multidimensional viewpoint of exhausting non-renewable energy source assets, ecological, well-being, and energy security.

Biodiesel as an alternative fuel is winding up progressively significant because of decreasing oil reserves and the natural results of pollution from oil-fuelled engines. Biodiesel, which is produced using sustainable sources, comprises of the basic alkyl esters of unsaturated fatty acids [4]. Significant research has been carried out using seed-borne oils like simarouba, jathropa, pongamia, and so on, even though they give better efficiency and reduced emissions they are not becoming popular due to various factors like seasonal availability, geographical limitations, Government policy matters which restricts the use of land for cultivation of fuel-based crops, higher cost [5]. As a future imminent fuel, biodiesel needs to contend monetarily diesel [6]. One way of making it monetarily competitive is to use less expensive feedstocks which do not possess commercial value and are considered as waste. Some of the examples of such wastes are used vegetable oil, spent coffee powder oil, dairy scum, etc. These feedstocks overcome the above-said issues related to seed-borne oils [7].

In late decades, a huge ascent in coffee production and utilization is observed and thusly an expansion in the coffee squander age. Used coffee powder has no business esteem and is presently discarded as a strong waste or sometimes utilized as manure [8], because of high natural material substance and the nearness of mixes, for example, caffeine, tannins, and polyphenols which can affect the earth and result in the arrival of ozone harming substances into the climate. One approach to deal with this waste is to utilize them as a crude material for biodiesel generation. The utilization of these wastes as a feedstock for biodiesel production helps for the disposal issue, yet in addition, it decreases the generation cost. Therefore, the transfer of spent coffee powder ought to be appropriately overseen. According to the experiments, spent coffee powder comprises of around 18% of oil, which can be changed over to biodiesel by the strategy of transesterification.

2 Materials and Methods

2.1 *Extraction of Oil*

Spent coffee powder was gathered from nearby restaurants (Adyar Anand Bhavan, KIAL Road, and Cafeteria-SVCE). The gathered coffee powder is dried in daylight for 2 days to expel the dampness content. Oil is removed from the dried coffee powder by solvent extraction method utilizing hexane as solvent. From the above strategy, it is found that the oil content in the coffee powder is around 16–18%. The FFA of oil found to be 7.67%.

2.2 *Preparation of Biodiesel*

Oils are made out of triglycerides and free unsaturated fats; FFA assumes a fundamental job during the generation of biodiesel. The FFA over 1% diminishes the yield of the biodiesel [9]. The FFA substance of the raw oil is found by titrating the oil against sodium hydroxide solution utilizing isopropyl alcohol as dissolvable with marker as methyl red. The FFA of the oil is reduced by esterification; it decreases the FFA substance in the oil within the sight of concentrated hydrochloric acid as an impetus and methyl alcohol as a reactant. For 1ltr of base oil, methyl alcohol in the molar ratio of 1:6 and 2.5% v/v of focused sulfuric acid dependent on FFA is utilized. This blend is warmed with consistent mixing at 60 °C for 120 min, and after that it is allowed to settle to overnight in an isolating jar. The FFA substance settles at the top due to low density, and it is expelled from the oil. The FFA substance of the oil is estimated, and the remaining oil is taken for the next stage.

The product of the previous stage is made to respond with 150 ml of methyl alcohol and 1% w/v of impetus potassium hydroxide for 1000 ml of oil and is warmed to 60 °C with consistent stirring for 120 min. The reacted consequence of this stage is made to settle down under gravity. Glycerol and various pollutions which is the bottom layer are disengaged from the methyl esters. The top layer which is required biodiesel is washed a few times with lukewarm water and dried to evacuate the dampness. The yield of biodiesel obtained is around 75–80%.

2.3 *Optimization of Yield*

To augment the yield of biodiesel, Taguchi strategy is utilized to improve the response parameters like methanol to oil proportion, catalyst concentration, reaction temperature, reaction time. In the present work, larger the better case is picked as the yield is to be amplified. This examination is related to four parameters with each at three levels. For this, mix L9 array is appropriate (Table 1).

2.4 *Blending*

Produced spent coffee oil biodiesel is blended with diesel in different extents (10, 20, 30%). For the preparation of 1000 ml B10, mix 100 ml of spent coffee oil biodiesel,

Table 1 Parameters chosen and range

Parameters	Symbol	1	2	3
Molar ratio (oil/methanol)	A	1:6	1:9	1:12
Catalyst concentration (%)	B	0.50	0.75	1.00
Reaction temperature (°C)	C	45	60	75
Reaction time (min)	D	40	80	120

and 900 ml of diesel is blended altogether utilizing magnetic stirrer for 20 min. Likewise, remaining mixes like B20 and B30 are readied.

2.5 Experimental Setup

A 5.2-kW diesel engine is used for the experimentation. Technical specification for the engine is given in Table 2. The engine discharges are estimated by AVL Digas 444N exhaust gas analyser.

Table 2 Engine specifications

Parameters	Specification
Engine make	KIRLOSKAR
Engine type	TV-1
Number of cylinders	Single cylinder
Number of strokes	Four stroke
Rated power	5.2 kW at 1500 rpm
Bore	87.5 mm
Stroke	110 mm
Cubic capacity	662 cc
CR	17.5:1
Rated speed	1500 rpm
Loading	Electrical/eddy current
Engine cooling	Water cooling
Injector pressure	200 bar

Table 3 Fuel properties

Property	Standard	Range	Diesel	B10	B20	B30	B100
Flash point (°C)	ASTM D93	>130	58	66	72	77	133
Kinematic viscosity (cSt)	ASTM D445	1.9–6.0	1.764	2.05	2.14	2.43	3.8
Specific gravity	ASTM D4052	0.87–0.90	0.82	0.831	0.839	0.846	0.874
Ash (% w/w)	IS: 1448 (P4)	0.5 max	Nil	Nil	Nil	Nil	0.01
Calorific value (kJ/kg)	ASTM D240	–	43,000	42,180	41,370	40,540	34,890
Cloud point (°C)	IS: 1448 (P10)	−3 to 12	−9	–	–	–	9
Carbon residue, Rams bottom, %w/w	IS: 1448 (P8)	0.05 max	Nil	Nil	Nil	Nil	0.033

3 Results

3.1 *Fuel Properties*

Various properties of spent coffee oil biodiesel mixes were determined by the ASTM standards and are condensed in Table 3. It will in general be concluded that properties of biodiesel are practically comparable to that of Diesel and are well inside the ASTM benchmarks for biodiesel.

3.2 *Optimization of Yield*

From the above plots, we can presume that the best yield acquired is by molar ratio of A2 (1:9), catalyst concentration of B3 (1.00%), reaction temperature of C2 (60 °C), reaction time of D3 (120 min). The approval trial was completed for the results, and the yield is observed to be 93.25% (Figs. 1 and 2; Table 4).

3.3 *Performance Analysis*

Brake thermal efficiency demonstrates the capacity of the ignition framework to acknowledge the test fuel and gives the way to evaluate how successfully the heat content in the fuel sample was changed over into useful form. It is observed that in case of spent coffee oil biodiesel highest efficiency is obtained at B20 mix which is 1.33% higher at full load as compared to that of diesel. B30 blend is also competitive with diesel (0.82% more at full load). It is observed that the blending of biodiesel with

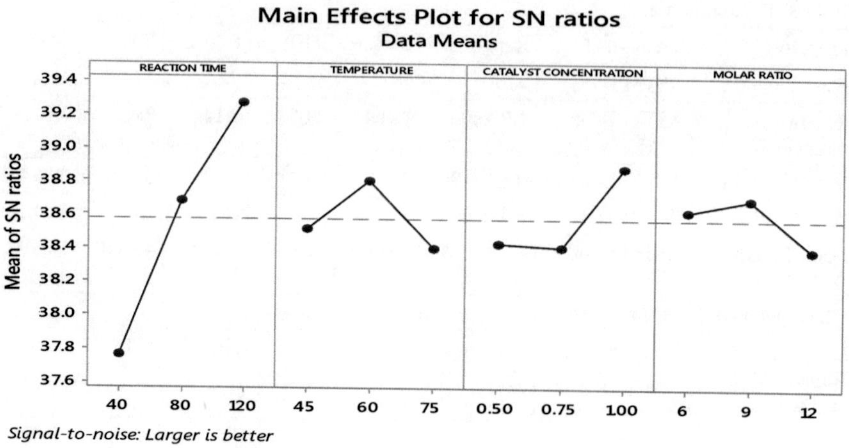

Fig. 1 Interaction plot for yield

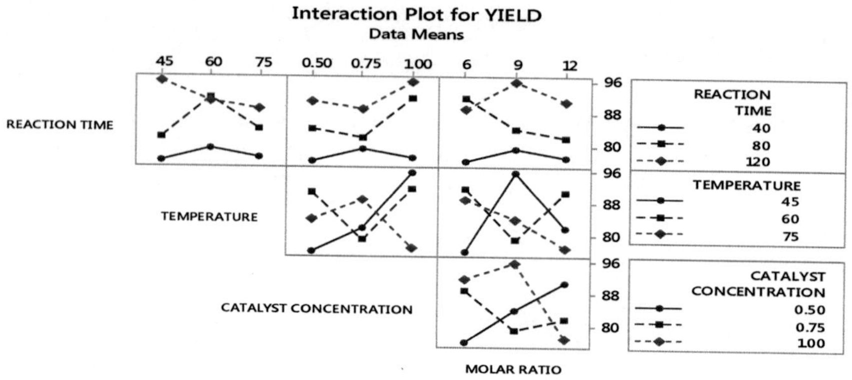

Fig. 2 Main effects plots

Table 4 Response table for means

Level	Reaction time	Reaction temperature	Catalyst conc.	Molar ratio
1	37.77	38.51	38.43	38.63
2	38.68	38.80	38.41	38.69
3	39.27	38.40	38.88	38.40
Delta	1.51	0.40	0.48	0.30
Rank	1	2	3	4

diesel yields good thermal efficiency. The efficiency of engine increases with increase in concentration of biodiesel initially (up to 20% blend) and at higher concentrations it is found to be diminished. The potential purposes behind the increase in efficiency might be because of the extra lubricity given by the biodiesel [10]. The particles of biodiesel fuel contain some measure of oxygen which upgrades the rate of combustion increasing the efficiency [10]. Anyway, increasing the concentration of biodiesel past 30% may not be suggestible as the calorific estimation of the mix decreases quickly which diminishes the efficiency additionally also the viscosity of the blend increases bringing about poor atomization [2] (Fig. 3).

Specific fuel consumption. It is observed that in case of spent coffee oil biodiesel, minimum fuel consumption was at B20 mix at full load which was 0.15% lesser than diesel. In general, between 50 and 90% loads it can be noticed that biodiesel blends have lesser specific fuel consumption as compared to that of diesel. Fuel consumption is higher at lower load, and it diminishes with increment in load. This might be because of higher ignition effectiveness, adequate time, and appropriate mixture formation in the burning chamber [10]. It is seen that the fuel consumption of biodiesel mixes up to B20 was marginally lower than that of diesel. The potential explanations behind decrease in specific fuel utilization might be because of the extra oxygen content and extra lubricity given by the biodiesel [11] (Fig. 4).

Exhaust gas temperature. There is a pointer of the measure of energy discharged during combustion. Higher temperature signifies high energy release [12]. It is observed that in case of spent coffee oil biodiesel highest exhaust gas temperature was 362 °C for B30 blend at full load which is 16.63 °C (4.82%) higher as compared to diesel. In case of B20 blend the exhaust, gas temperature was higher by 3.08%. The explanation behind the expansion in exhaust gas temperature with increment in percentage of biodiesel might be because of the nearness of extra substance of oxygen in the biodiesel which prompts total burning (Fig. 5).

Fig. 3 Brake thermal efficiency

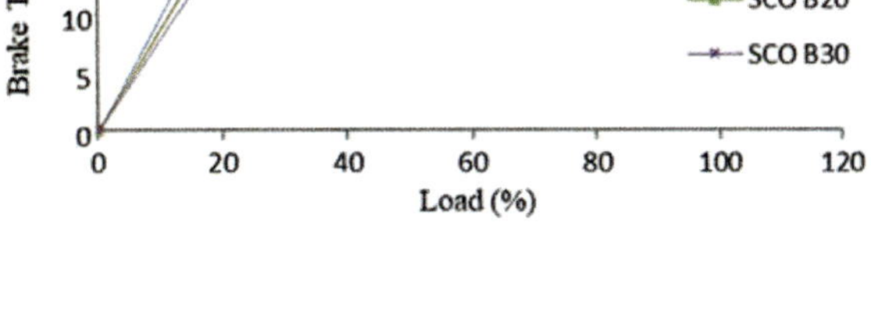

Fig. 4 Brake-specific fuel consumption

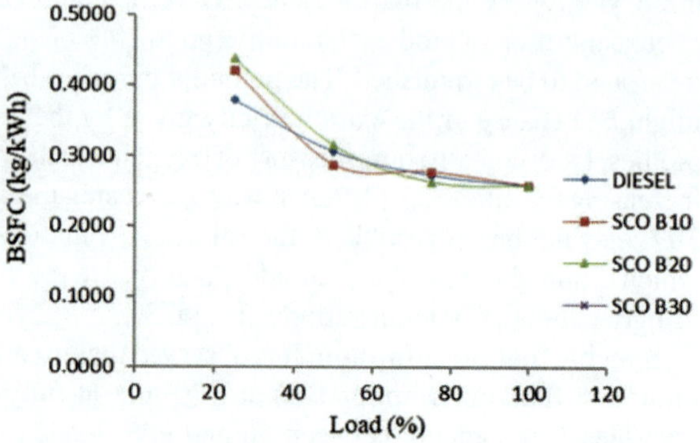

Fig. 5 Exhaust gas temperature

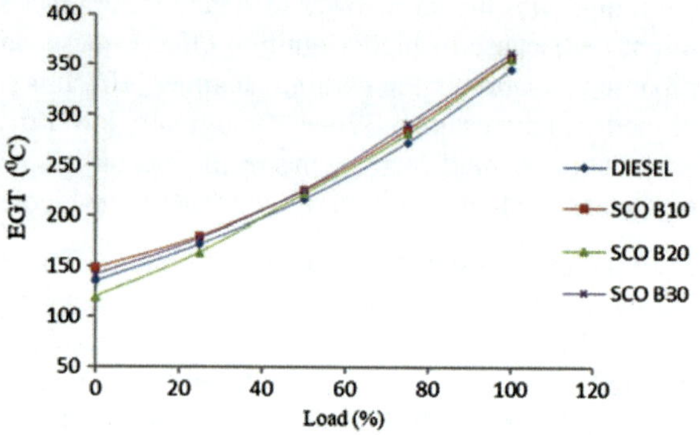

3.4 Emission Analysis

Carbon monoxide. It is a by-product of combustion. The degree of carbon monoxide expands on account of increase in temperature in the chamber, physical and substance properties of the fuel, air–fuel proportion, and short time for combustion. It is seen that the carbon monoxide outflow increments with increment in load. This is run of the mill with all IC engines, since the air–fuel proportion diminishes as the load increases. It is observed that in case of spent coffee oil biodiesel, the carbon monoxide emissions at full load were 20% lower in comparison with diesel in B10 and B20 mixes; whereas for B30 mix, it was 10% lower at full load (Fig. 6).

Carbon dioxide. The formation of carbon dioxide depends on the measure of carbon substance present in the fuel, and it shows how effectively the fuel is burnt inside the burning chamber. Carbon dioxide is a inert gas, and it isn't directly harmful to humans. The carbon dioxide outflow for every one of the blends and diesel goes on increases with load. It is observed that in case of spent coffee oil biodiesel blends the CO_2 emissions at full load was 15.44%, 17.24%, 15.4% higher as compared to that of diesel in B10, B20, and B30 blends, respectively (Fig. 7).

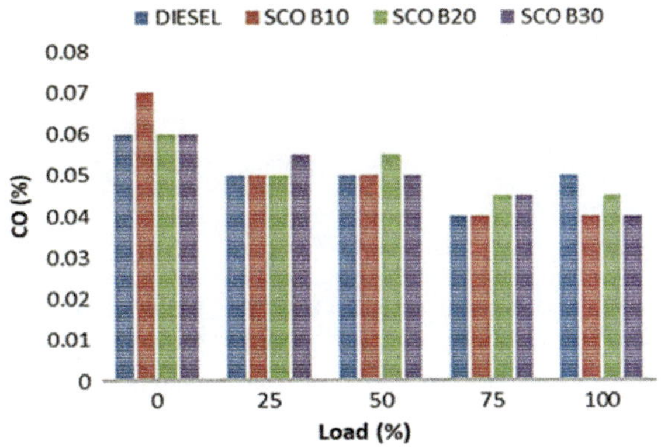

Fig. 6 Carbon monoxide

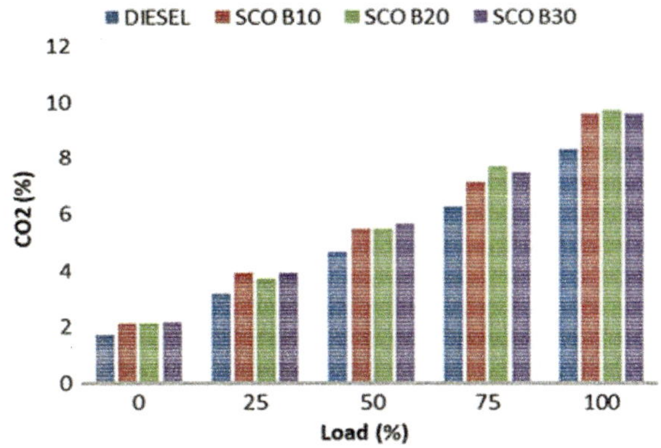

Fig. 7 Carbon dioxide

Hydrocarbon emissions. Increased hydrocarbon discharge demonstrates that there is some measure of unburnt fuel present in the fume's outflows. It is observed that in case of spent coffee oil biodiesel blends the hydrocarbon emission emissions at full load was 17.18%, 23.43%, 20.31% lower as compared to that of diesel in B10, B20, and B30 blends, respectively (Fig. 8).

Oxides of nitrogen in the fumes are a mix of nitric oxide and nitrogen dioxide. High temperatures and accessibility to oxygen are the main fundamental reasons behind the development of NO_x. The oxides of nitrogen emission for all the fuel samples increased with respect to load. It is observed that in case of spent coffee oil biodiesel blends, the NO_x emission emissions at maximum load were 1.21%, 1.97%, 2.93% higher in comparison with diesel in B10, B20, and B30 blends, respectively (Fig. 9).

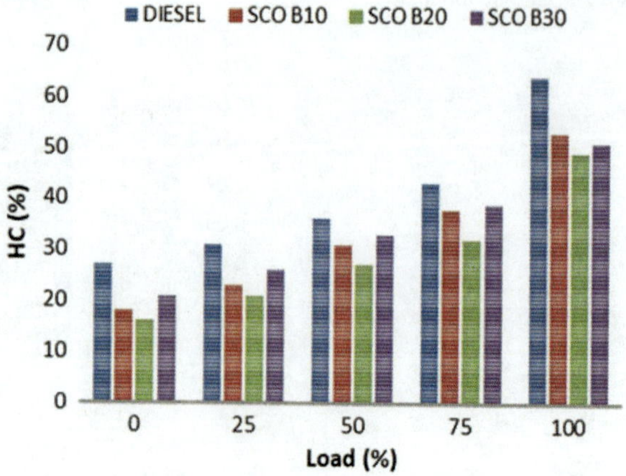

Fig. 8 Hydrocarbon emissions

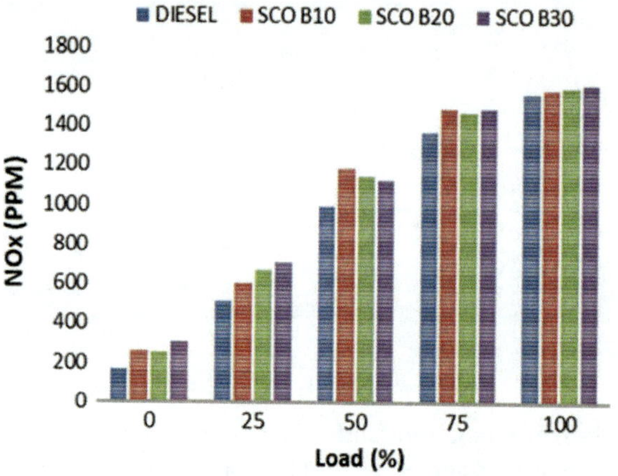

Fig. 9 NO_x emissions

3.5 Combustion Analysis

It is clear that peak pressure increases with increase in load. In all the biodiesel mixes, ignition begins prior in contrast with diesel. It is additionally seen that peak pressure shifts toward TDC with expanding brake power for every one of the mixes. Quite a bit of variety in burning procedure isn't seen among these mixes (Figs. 10 and 11).

4 Conclusions

In the current work, the performance and outflow qualities of a diesel engine fueled with spent coffee oil biodiesel mixes have been talked about and contrasted with that of diesel.

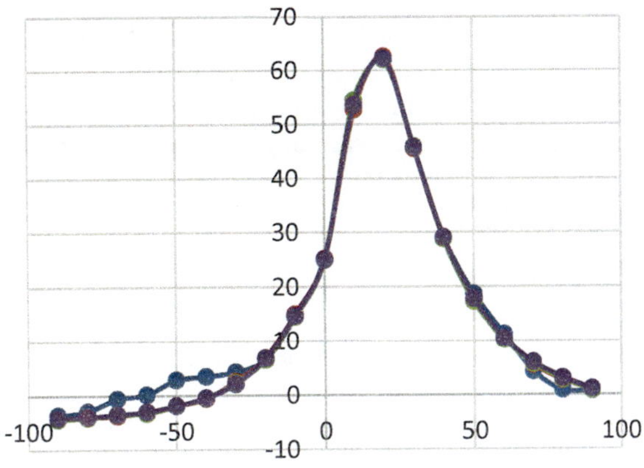

Fig. 10 Pressure versus crank angle diagram (full load)

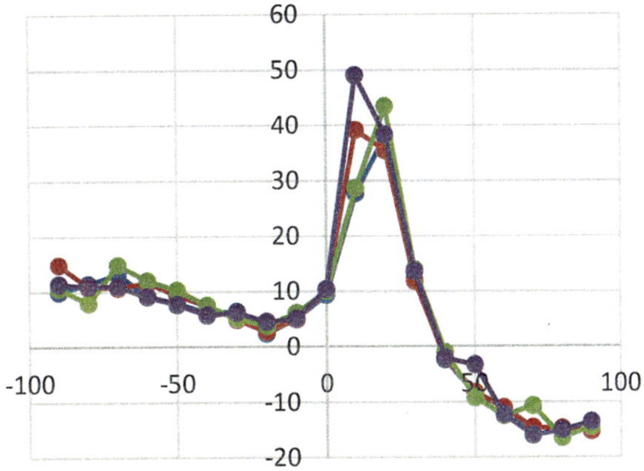

Fig. 11 Heat release rate (full load)

Outcomes of the work are abridged as pursues:

1. The compression ignition engine can function pleasingly on spent coffee oil biodiesel mixes without engine modifications.
2. The deliberate properties of created biodiesel blends met the ASTM D6751 standards.
3. Thermal productivity of biodiesel mixes was in a general sense higher than that of Diesel.
4. It is seen that there is a decrease in CO, HC outflows for spent coffee powder biodiesel blends as compared to diesel. In any case, oxides of nitrogen outflow of spent coffee powder biodiesel mixes are barely higher than diesel.
5. By Taguchi strategy, the best yield acquired is by molar ratio of A2 (1:9), catalyst concentration of B3 (1.00%), reaction temperature of C2 (60 °C), reaction time of D3 (120 min). The approval trial was completed for the results, and the yield is observed to be 93.25%, which resulted in improvement of up to 15% yield as compared to that of crude method.

It may be inferred that spent coffee oil biodiesel mixes could supplement the diesel to some extent so as to help in controlling air contamination, drawing in the accumulation, and reusing of waste to make biodiesel and decrease the dependence on oil-based products without giving up engine performance.

References

1. Xue J et al (2011) Effect of biodiesel on engine performances and emissions. Renew Sustain Energy Rev 15:1098–1116
2. Gopal KN, Pal A, Sharma S, Samanchi C, Sathyanarayanan K, Elango T (2014) Investigation of emissions and combustion characteristics of a CI engine fueled with waste cooking oil methyl ester and diesel blends. Alexandria Eng J 53:281–287
3. Raheman H, Ghadge SV (2007) Performance of compression ignition engine with mahua (*Madhuca indica*) biodiesel. Fuel 86:2568–2573
4. Berchmans HJ, Hirata S (2008) Biodiesel production from crude *Jathropa curacus* L. seed oil with a high content of free fatty acids. Bioresour Technol 99:1716–1721
5. Mofijur M, Masjuki HH (2013) Effect of biodiesel from various feedstocks on combustion characteristics, engine durability and materials compatibility: a review. Renew Sustain Energy Rev 28:441–455
6. Palash SM, Kalam MA (2013) Impacts of biodiesel combustion on NO_x emissions and their reduction approaches. Renew Sustain Energy Rev 23:473–490
7. Bora DK, Baruah DC (2012) Assessment of tree seed oil biodiesel: a comparative review based on biodiesel of a locally available tree seed. Renew Sustain Energy Rev 16:1616–1629
8. https://www.alphainvesco.com/blog/overview-indian-coffee-sector/. Last accessed 2019/07/15
9. Otadi M, Shahrakib A, Goharrokhic M, Bandarchian F (2011) Reduction of free fatty acids of waste oil by acid-catalyzed esterification. Procedia Eng 18:168–174
10. Ramadhas AS, Muraleedharan C, Jayaraj S (2005) Performance and emission evaluation of a diesel engine fueled with methyl esters of rubber seed oil. Renew Energy 30:1789–1800
11. Abedin MJ, Masjuki HH, Kalam MA, Sanjid A, Ashraful AM (2014) Combustion, performance and emission characteristics of low heat rejection engine operating on various biodiesels and vegetable oils. Energy Convers Manag 85:173–189
12. Nwafor OMI (2004) Emission characteristics of diesel engine operating on rapeseed methyl ester. Renew Energy 29:119–129

Emission and Performance Characteristics of Hydrotreated Vegetable Oil and Kerosene as Fuel for Diesel Engines

Hemanandh Janarthanam, S. Ganesan, B. R. S. Aravind, and N. Aman

Abstract The main focus of this work is to study the emission and performance characteristics of hydrotreated vegetable oil (HVO) and kerosene blends and use them as an alternate fuel source for the diesel engines. The HVO and kerosene were blend in the required proportions such as HTB10, HTB20, and HTB30. The HVO blends were tested in a VCR test engine for varied load conditions such as 0%, 25%, 50%, 75%, and 100%. The results were analyzed and the emission and performance characteristics of the HVO blends were compared with diesel.

Keywords Hydrotreated vegetable oil · Kerosene · Diesel engine · Emission and performance characteristics

1 Introduction

Day by day emission levels have been constantly increasing worldwide. Vehicular emissions contribute nitrogen oxides and reactive hydrocarbons into the atmosphere. Carbon dioxide and carbon monoxide are the common pollutants produced by vehicles because of the combustion of fuels like petrol and diesel. Most of the automobiles today use petrol and diesel as fuels and because of that there is a depletion of resources which leads to the development and use of alternate fuels or biofuels. Hydrotreated vegetable oil (HVO) blended biodiesel is a form of renewable diesel that is produced from vegetable fats and oils.

Vojtisek-Lom et al. [1] found that butanol–HVO blends tested in Iveco Tector diesel engine resulted in a 75–78% reduction in elemental carbon. Millo et al. [2] revealed that the importance of blending ultra-low sulfur diesel with 30% by volume of different biofuels tested in an automotive diesel engine resulted in a significant reduction of CO, NO_x, and HC compared to diesel fuel. The study by Buffi et al. [3] states that lower emission concentrations are measured for a specific blend (jet a-1/hrj

H. Janarthanam (✉) · S. Ganesan · B. R. S. Aravind · N. Aman
School of Mechanical Engineering, Sathyabama Institute of Science and Technology, Chennai, Tamil Nadu 600119, India
e-mail: hemanandh123@gmail.com

A. Arockiarajan et al. (eds.), *Advances in Industrial Automation and Smart Manufacturing*, Lecture Notes in Mechanical Engineering,
https://doi.org/10.1007/978-981-15-4739-3_79

80/20 vol%) and higher HRJ concentrations tested in a pressurized swirl combustor resulted in increased CO and NO_x emissions. Hemanandh and Narayanan [4] found out that HVO tested in a CI engine can be used as an alternate fuel for diesel, and emissions can be reduced using HVO when compared with petrol and diesel. Kim et al. [5] showed that iso-HVO and HVO blended diesel has better performance than biodiesel, and they emit less CO than the biodiesel, along with an increase in NO_x at higher loads than those of other conditions after testing them in a 1.5 l DOHC passenger car. Lapuerta et al. [6] suggested some blending strategies of HVO as biofuel and found out that blends with concentrations above 50% cannot be recommended. No et al. [7] found out that the application of HVO from triglyceride-based biomass in C.I engines resulted in substantial reductions in NO_x, CO, PM, and HC. Phimsen et al. [8] found out that the physiochemical properties of the obtained diesel from spent coffee grounds extracted using Soxhlet extractor are almost equal to the specifications of commercial biohydrogenated diesel. Orozco et al. [9] developed a new continuous method using fixed-bed continuous reactors to produce castor oil green diesel which resulted an increase in purity and yield. Singh et al. [10] revealed that hydro-processed renewable diesel can be used as a promising alternative for compression ignition engines. Chau et al. [11] found out that the ignition delay, heat release rate, flame temperature, and so forth tested in a rapid compression expansion machine are corresponding to the HVO fraction and O_2 concentrations in the blend. Plank et al. [12] found out that ECNO of vegetable oils is strongly dominated by combustion chamber temperature and injection pressure compared to conventional diesel fuel. Pinto et al. [13] revealed that olive pomace oil can be systematically converted into biohydrocarbons with a wide range of applications. Srifa et al. [14] found out that the influence of reaction parameters on hydrotreating of palm oil to BHD in an uninterrupted flow fixed-bed reactor over $NiMoS_2/Al_2O_3$ resulted in a product yield of 90%. Rabaev et al. [15] found out that transformation of vegetable oils on Pt/Al_2O_3/SAPO-11 resulted in a 42–48 wt% of jet fuels. Vasquez et al. [16] found out that hydrotreatment of vegetable oils with pd/c catalyst resulted in a 96% yield of jet fuels. Perez-Cisneros et al. [17] found out that the performance of the hydrotreating RD process is more energy-efficient and has higher yields when blends of vegetable oils along with high free fatty acids content and petrol–diesel are premixed and further hydrotreated. Varakin et al. [18] showed that the carbon in CO (Ni)-pmo/C/Al_2O_3 is highly effective in co-hydrotreating of SRGO and hence feedstock. Hwang et al. [19] found that the catalytic deoxygenation of crude jatropha oil experimented in a batch-type reactor with hydrogen in situ is made from 30% formic acid solution that resulted in 97% yield of hydrocarbons and degree of deoxygenation of 99.5%. Han et al. [20] found out that the co-hydrotreatment of tire and vegetable oils resulted in a yield of 90 wt% of hydrotreated oils. Domínguez-Barroso et al. [21] found that the fusion of Pt-Ni/Al_2O_3 + Pd/C with vegetable oil resulted in a product similar in properties with commercial HVO and that can be suitably used as biofuel.

Hagos et al. [22] found that the effects of emulsification fuels and fuel oxygenation are accountable for improving combustion, performance, CO, PM emissions, and reduction of NO_x. Zhang et al. [23] found that the spray tip penetration grew with a decreasing tip velocity and the cone angle increased gradually after a remarkable growth and slight drop, and the prediction of numerical method, when used in conjunction with experimental studies, was proven effective in elucidating the macroscopic characteristics of HVO spray. Szabados and Bereczky [24] found that the comparison of three fuels (fossil fuels, conventional biodiesel, and triglycerides of modified structure) resulted in a moderate improvement in combustion as well as the physiochemical properties of the tested biodiesel which fulfill the European norms and emission result shows that there is an increase in NO_x and reduction in HC. López et al. [25] found out that n-butanol fumigation is a better diesel substitute than hydrous ethanol fumigation. Plazas-González et al. [26] found that hydrotreating of palm oil components with $NiMo/Al_2O_3$ resulted in the production of green diesel with a purity of 97% under optimized conditions. Preuß et al. [27] revealed that the performance of the alcohol–HVO blends and diesel was similar and there was no notable variation in HC, NO_x emission for alcohol–HVO blends and diesel, but there was a limiting in CO and soot emissions for all alcohol blends. Wang et al. [28] proposed the hydrogenation reaction of palm oil with alumina/zsm-5 supported trimetallic catalyst as the optimum catalyst. Diederichs et al. [29] found that the hydro-processed esters and fatty acids (HEFA) process produced lower minimum jet selling prices (MJSP), and 1g and 2g processes involving vegetable and oil and sugar cane juice had significantly higher MJSP.

The main aim of this paper is to study the emissions and performance characteristics of hydrotreated vegetable and kerosene blends and use them as an alternate fuel for diesel.

2 Materials and Methods

The sunflower oil, kerosene, and diesel were obtained from local market. The experiment was conducted in the following phases:

Phase 1: The sunflower oil was hydrotreated.
Phase 2: The HVO and kerosene were blended together and test blends were prepared at required proportions.
Phase 3: The test blends were tested in the VCR test engine for required load conditions.
Phase 4: The emission and performance characteristics of the test blends were analyzed and discussed.

3 Experimental Setup

The HVO blends were tested in 4S, single-cylinder VCR test engine to examine the emission and performance characteristics of the engine for the fuel blends. The engine was examined for different loads at a constant speed of 1500 rpm. The eddy current dynamometer was used to load the engine at different loading conditions. The engine was loaded in steps of 25% from 0 to 100% and the conditions are 0%, 25%, 50%, 75%, and 100%. The HVO was blended with kerosene and diesel at different proportions such as HTB10, HTB20, and HTB30 and tested in the engine. The engine was permitted to run at no load condition in the beginning, and the test blends were added and the emission and performance characteristics for each sample were recorded.

3.1 Engine Specifications

The engine which is used for this experiment with specifications is given in Table 1 (Tables 2 and 3).

Table 1 Specification of DI diesel engine

Combustion	Direct injection
Rated power	3.5 kW
Rated speed	1500 rpm
Compression ratio	12–18:1
Dynamometer	Eddy current

Table 2 Comparison of properties of diesel, biodiesel standards, and HTSFO

S. No	Properties	Euro-IV Bharat stage 1460:2005 diesel fuel	ASTM d-751 (IS 5607:2005)	Refined sunflower oil (feed)	HTSFO (hydrotreated vegetable oil)
1	Calculated cetane index	51	–	35	81
2	Density at 15 °C (kg/m^3)	820–845	860–900	922.5	798.8
3	Viscosity at 40 °C (cSt)	2–4.5	1.9–6	28.3	3.69
4	Flash point (°C/min)	35	130	254	64.5
5	Calorific value (kJ/kg)	39,000	–	38,796	43,754

Source Emission and performance analysis of hydrotreated refined sunflower oil as alternate fuel

Table 3 Comparison of properties of kerosene, HTB10, HTB20, and HTB30

S. No	Properties	HTB10	HTB20	HTB30	Kerosene
1	Density at 15 °C (kg/m^3)	799.005	799.21	799.415	800.85
2	Viscosity at 40 °C (cSt)	3.6703	3.6506	3.6309	3.493
3	Flash point (°C/min)	64.375	64.25	64.135	63.25
4	Calorific value (kJ/kg)	43,950	43,907	43,814	43,781

4 Results and Discussions

4.1 *Carbon Monoxide (CO)*

Figure 1 shows the CO emission for the given test blends in reference to pure diesel. It is found that the test blends HTB10, HTB20, and HTB30 produce reduced CO emission than pure diesel. At full load conditions, the test blend HTB10, HTB20, and HTB30 decreases by 25%, 37%, and 43.2%, respectively. Compared with other blends, the blend HTB30 produces lower CO emission with a percentage of 43.2%. This is due to higher calculated cetane index value enhancing better combustion [4].

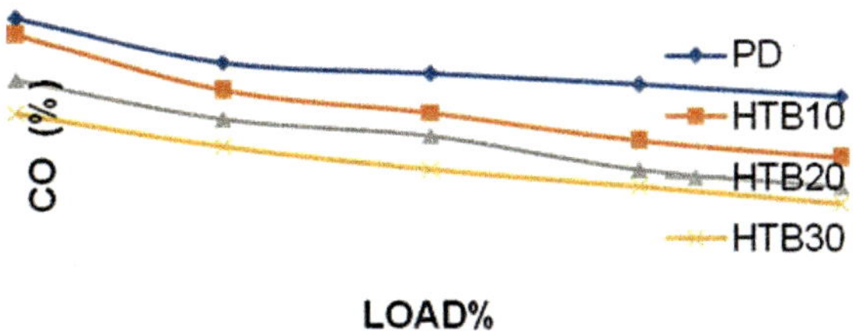

Fig. 1 Variation of CO emissions of HVO blends and diesel

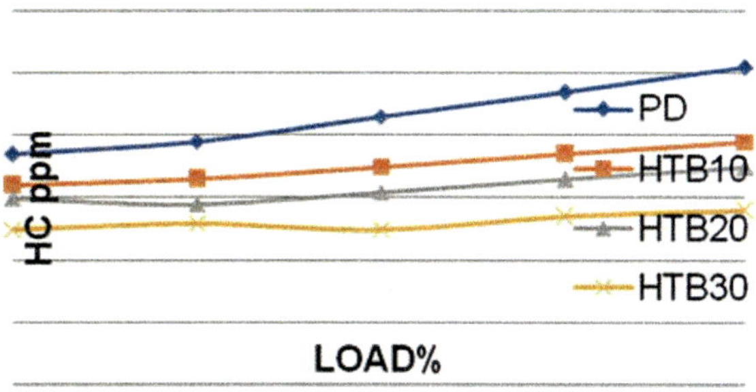

Fig. 2 Percentage of HC emission for HVO blends and diesel

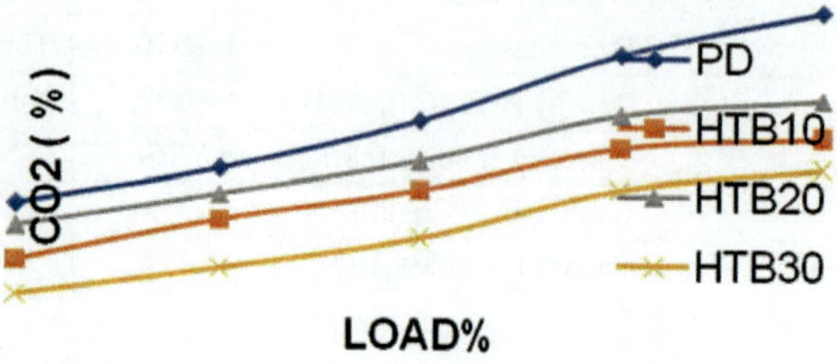

Fig. 3 Comparison of CO_2 emission of HVO blends and diesel

4.2 *Hydrocarbon (HC)*

The HC emission for the following test blends is given in Fig. 2. The hydrocarbon emissions of the test blends are compared with pure diesel for reference. The test results show that the test blends HTB10, HTB20, and HTB30 emit lesser percentage of hydrocarbon compared to pure diesel. At full load conditions, the test blends HTB10, HTB20, and HTB30 decrease by 24%, 31.4%, and 45.1% when compared with pure diesel. From the graph, it can be found out that the HTB30 blend produces lower hydrocarbon emission than other blends when compared with pure diesel. This might be due to an increase in hydrogen content present in the fuel, temperature, and reaction time [4].

4.3 *Carbon Dioxide (CO_2)*

Figure 3 illustrates the emission of carbon dioxide for pure diesel and the test blends. The test results show that the percentage of carbon dioxide emitted by the test blends is lesser than that emitted by pure diesel. From Fig. 3, at full load conditions, it is found that the test blend HTB10 decreases by 35.95%, the blend HTB20 decreases by 25.05%, and HTB30 decreases by 44.75% when compared with pure diesel. It can be concluded that the blend HTB30 produces lower CO_2 emission than pure diesel. The reduction in CO_2 emission might be due to better fuel efficiency [5].

4.4 *Nitrogen Oxides (NO_x)*

Figure 4 illustrates the emission of nitrogen oxides produced by the test blends and PD. PD produces a higher percentage of NO_x, whereas the test blends produce lesser percentage of NO_x. At full load conditions, the test blend HTB10 decreases drastically by 74% compared to pure diesel. Similarly, the blends HTB20 and HTB30 decrease by 43.95% and 52.47%, respectively. From the graph, it can found out that the test blend HTB10 has lower NO_x emission compared to other blends. This may be due to the flame characteristics of the fuel [14].

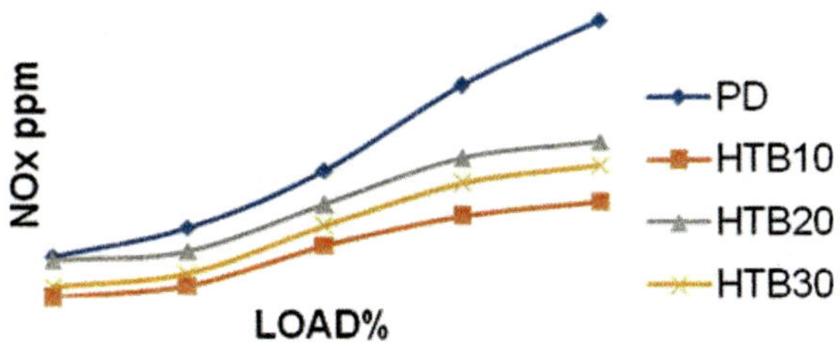

Fig. 4 Comparison of NO_x emissions for HVO blends and diesel

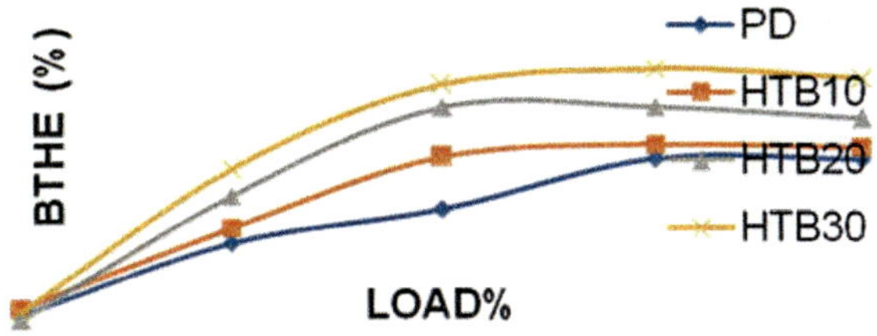

Fig. 5 BTHE comparison for HVO blends and diesel

4.5 Brake Thermal Efficiency (BTHE)

Figure 5 illustrates the brake thermal efficiency (BTHE) of petrol–diesel and test blends. The test blends show a significant increase in BTHE when compared with pure diesel. At full load conditions, the BTHE of blend HTB10 increases by 6.27% than pure diesel. Likewise, the BTHE of the blends HTB20 increases by 18.48% and HTB30 increases by 31.7%, respectively. It can be concluded that the blend HTB10 has higher BTHE than other blends. The increase in BTHE could be due to higher calculated cetane number [10].

4.6 Brake-Specific Fuel Consumption (BSFC)

Figure 6 displays the brake-specific fuel consumption (BSFC) of test blends and pure diesel. The results indicate that the test blend HTB30 has lower specific fuel consumption compared to pure diesel than other blends. At full load conditions, the test blend HTB10 decreases by 18.51%, the blend HTB20 decreases by 40.74%, and the blend HTB30 decreases by 51.85%, respectively, in comparison with pure diesel.

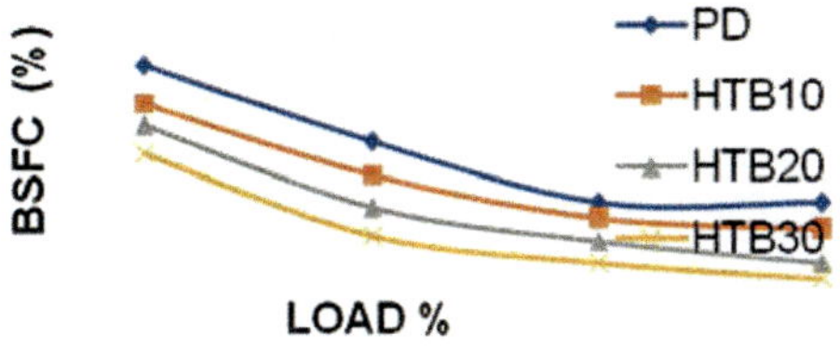

Fig. 6 Variations of BSFC of HVO blends and diesel

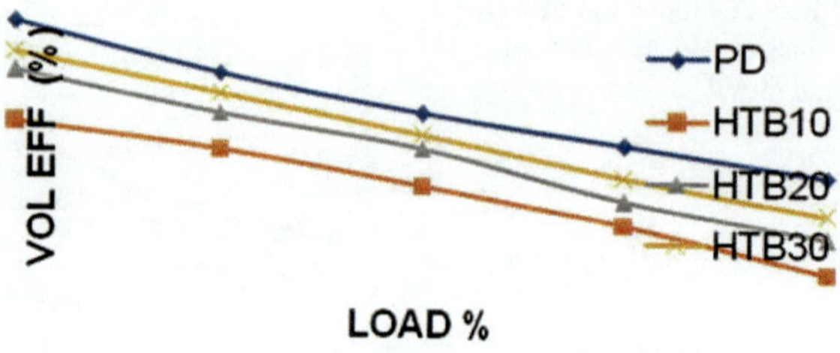

Fig. 7 Comparison of VOL EFF of HVO blends and diesel

The reduction in BSFC might be due to the lower density of the fuel and increase in blending ratios [5].

4.7 *Volumetric Efficiency*

The volumetric efficiency of the test blends is shown in Fig. 7. In comparison with pure diesel, the test blend HTB10 results in lower volumetric efficiency than the other two blends. At full load conditions, the blend HTB10 decreases by 1.16%, HTB20 decreases by 0.74%, and HTB30 decreases by 0.47% when compared with pure diesel. It can be concluded that the blend HTB10 has lower volumetric efficiency than other blends and diesel.

5 Conclusions

The main aim of this paper is to study the emissions and performance characteristics of HVO and kerosene blends and to use them as an alternate fuel for diesel engine. The emission and performance characteristics for the test blends have been studied and results have been obtained. From the obtained results, it is concluded that:

- Comparing the test blends with pure diesel, it is found that the test blend HTB30 yields better CO emission than the other blends with a percentage of 43.2%.
- In comparison with pure diesel, it can be concluded that the blend HTB30 produces lesser HC emission than other two blends with a yield of 45.1%.
- Comparing the test blends with pure diesel, it is found that the CO_2 emission produced by the test blend HTB30 is very less than the other two blends with a percentage of 44.75%.
- In comparison with pure diesel, it can be concluded that the test blend HTB10 produces a low percentage of NO_x than the other blends, with a percentage of 74%.
- The test blends were compared with pure diesel, and it can be concluded that the blend HTB30 produces higher BTHE than other blends with a percentage of 31.7%.

- In comparison with pure diesel, it is found that the test blend HTB30 has minimal brake-specific fuel consumption than the other blend which results in 51.83% lesser specific fuel consumption than pure diesel.
- Comparing the blends with pure diesel, it is found out that the test blend HTB10 has lower volumetric efficiency than other blends with a 1.16% lesser volumetric efficiency than pure diesel.

References

1. Vojtisek-Lom M, Beranak V, Mikuska P, Krumal K, Coufalik P, Sikorova J, Topinka J (2017) Blends of butanol and hydrotreated vegetable oil as a drop-in replacement for diesel engines. Fuel 197:407–421
2. Millo F, Debnath BK, Vlachos T, Ciaravino C, Postrioti L, Buitoni G (2015) Effects of different biofuels blends on performance and emissions of an automotive diesel engine. Fuel 159:614–627
3. Buffi M, Valera-Medina A, Marsh R, Pugh D, Giles A, Runyon J, Chiaramonti D (2017) Emissions characterization tests for hydrotreated renewable jet fuel from used cooking oil and its blends. Appl Energy 201:84–93
4. Hemanandh J, Narayanan KV (2015) Emission and performance analysis of hydrotreated refined sunflower oil as alternate fuel. Alexandria Eng J 54:389–393
5. Kim D, Kim S, Oh S, No S-Y (2014) Engine performance and emission characteristics of hydrotreated vegetable oil in light duty diesel engines. Fuel 125:36–43
6. Lapuerta M, Villajos M, Agudelo JR, Boehman AL (2011) Key properties and blending strategies of hydrotreated vegetable oil as biofuel for diesel engines. Fuel Process Technol 92:2406–2411
7. No S-Y (2014) Application of hydro treated vegetable oil from triglyceride-based biomass to CI engines—a review. Fuel 115:88–96
8. Phimsen S, Kiatkittipong W, Yamada H, Tagawa T, Kiatkittipong K, Laosiripojana N, Assabumrungrat S (2016) Oil extracted from spent coffee grounds for bio-hydro treated diesel production. Energy Convers Manage 126:1028–1036
9. Orozco LM, Echeverri D, Sánchez L, Rios LA (2017) Second-generation green diesel from castor oil: development of a new and efficient continuous-production process. Chem Eng J 322:149–156
10. Singh D, Subramanian KA, Garg MO (2018) Comprehensive review of combustion, performance and emissions characteristics of a compression ignition engine fueled with hydro processed renewable diesel. Renew Sustain Energy Rev 81:2947–2954
11. Chau V, Chinda C, Preechar K, Sato S, Kosaka H (2017) Effects of variable O_2 concentrations and injection pressures on the combustion and emissions characteristics of the petrol-diesel and hydrotreated vegetable oil-based fuels under the simulated diesel engine condition. J Energy Inst 91:1071–1084
12. Plank M, Wachtmeister G, Remmele E, Thuneke K, Emberger P (2017) Ignition characteristics of straight vegetable oils in relation to combustion and injection parameters, as well as their fatty acid composition. Fuel Process Technol 167:271–280
13. Pinto F, Varela FT, Gonçalves M, Netoandré R, Costa P, Mendes B (2014) Production of bio-hydrocarbons by hydrotreating of pomace oil. Fuel 116:84–93
14. Srifa A, Faungnawakij K, Itthibenchapong V, Viriya-Empikul N, Charinpanitkul T, Assabumrungrat S (2014) Production of bio-hydrogenated diesel by catalytic hydrotreating of palm oil over $NiMoS_2$/c-Al_2O_3 catalyst. Bioresour Technol 158:81–90

15. Rabaev M, Landau MV, Vidruk-nehemya R, Koukouliev V, Zarchin R, Herskowitz M (2015) Conversion of vegetable oils on Pt/Al_2O_3/SAPO-11 to diesel and jet fuels containing aromatics. Fuel 161:287–294
16. Vasquez MC, Silva EE, Castillo EF (2017) Hydro treatment of vegetable oils: a review of the technologies and its developments for jet biofuel production. Biomass Bioenergy 105:197–206
17. Perez-Cisneros ES, Sales-Cruz M, Lobo-Oehmichen R, Viveros-Garcıa T (2017) A reactive distillation process for co-hydrotreating of non-edible vegetable oils and petro-diesel blends to produce green diesel fuel. Comput Chem Eng 105:105–122
18. Varakin AN, Salnikova VA, Nikulshina MS, Maslakov KI, Mozhaeva AV, Nikulshin PA (2017) Beneficial role of carbon in Co (Ni) MoS catalysts supported on carbon-coated alumina for co-hydrotreating of sunflower oil with straight-run gas oil. Catal Today 292:110–120
19. Hwang KR, Choi IH, Choi HY, Han JS, Lee KH, Lee JS (2016) Bio fuel production from crude jatropha oil; addition effect of formic acid as an in-situ hydrogen source. Fuel 174:107–113
20. Han Y, Stankovikj F, Garcia-Perez M (2017) Co-hydro treatment of tire pyrolysis oil and vegetable oil for the production of transportation fuels. Fuel Process Technol 159:328–339
21. Domínguez-Barroso MV, Herrera C, Larrubia MA, Alemany LJ (2016) Diesel oil-like hydrocarbon production from vegetable oil in a single process over pt–ni/ Al_2O_3 and pd/c combined catalysts. Fuel Process Technol 148:110–116
22. Hagos FY, Ali OM, Mamata R, Abdullah AA (2017) Effect of emulsification and blending on the oxygenation and substitution of diesel fuel for compression ignition engine. Renew Sustain Energy Rev 75:1281–1294
23. Zhang Z, Lu Y, Ren H, Roskilly AP, Chen L, Smallbone A, Wang Y (2017) Experimental and numerical investigation on the macroscopic characteristics of hydrotreated vegetable oil spray. Energy Proc 142:474–480
24. Szabados G, Bereczky A (2018) Experimental investigation of physicochemical properties of diesel, biodiesel and TBK-biodiesel fuels and combustion and emission analysis in CI internal combustion engine. Renew Energy 121:568–578
25. López A, Cadrazco M, Agudelo AF, Corredor LA, Vélez JA, Agudelo JR (2015) Impact of n-butanol and hydrous ethanol fumigation on the performance and pollutant emissions of an automotive diesel engine. Fuel 153:483–491
26. Plazas-González M, Guerrero-Fajardo CA, Sodré JR (2018) Modelling and simulation of hydrotreating of palm oil components to obtain green diesel. J Cleaner Prod 184:301–308
27. Preuß J, Munch K, Denbratt I (2018) Performance and emissions of long-chain alcohols as drop-in fuels for heavy duty compression ignition engines. Fuel 216:890–897
28. Wang HY, Jiao TT, Li ZX, Li CS, Zhang SJ, Zhang JL (2015) Study on palm oil hydrogenation for clean fuel over ni–mo–w/γ-Al_2O_3–zsm-5 catalyst. Fuel Process Technol 139:91–99
29. Diederichs GW, Mandegari MA, Farzad S, Görgens JF (2016) Techno-economic comparison of biojet fuel production from lignocellulose, vegetable oil and sugar cane juice. Bioresour Technol 216:331–339

Experimental Study on CI Engine by Using Microalgae Oil and Their Blends

P. Moulali, P. Nagaraju, P. Madhuraghava, M. Ravichandra, and S. Neeraja

Abstract The basic objective of this article was to evaluate the effect of microalgae oil (MAO) blends on engine efficiency and pollution characteristics of a single cylinder variable compression direct-injection ignition compression motor running at 1500 rpm. The raw oil was acquired from microalgae manufacturing using a rowing pond with native algal communities. The good MAO was accomplished by trans-esterification technique and mixed with petrol in diverse proportions of B20 (20% MAO–90% diesel), B40 (40% MAO–80% diesel), B60 (60% MAO–70% diesel), B80 (80% MAO–20% diesel), and 100% quantity-based MAO. Their features have been validated based on ASTM standards. Investigational findings show that at B20, brake thermal efficiency, brake specific fuel consumption, and nitrogen oxides are optimal compared to other biodiesel blends.

Keywords Brake thermal efficiency · Microalgae oil · Oxides of nitrogen · Smoke opacity

1 Introduction

Day to day increases the utilization of the fuel consumption and increases the demand of fuel. The fossil fuels are more effected to the environment and also human health. Ali and Hanna [1] evaluated the use of essential oils and refined carbohydrates as gas in diesel engines. The impact of oil preparing techniques such as transesterification, thermal decomposition, dilution, and micro-yeast fermentation on fuel characteristics and its consequent impact on gasoline engine efficiency was addressed. Ikwuagwu et al. [2] separated plant fuel from rubber (Heveabrasiliensis). Methyl ester was structured by transesterification using iron chloride as a liquid with an additional 6-molar quantity of methanol. Melvin et al. [3] manufactured rubber licorice root methyl esters. The acid content of oilseed oil was 35 mg KOH/g. Given the presence

P. Moulali (✉) · P. Nagaraju · P. Madhuraghava · M. Ravichandra · S. Neeraja
Department of Mechanical Engineering, Santhiram Engineering College, Nandyal, Kurnool, AP 518501, India
e-mail: moulalip20@gmail.com

A. Arockiarajan et al. (eds.), *Advances in Industrial Automation and Smart Manufacturing*, Lecture Notes in Mechanical Engineering,
https://doi.org/10.1007/978-981-15-4739-3_80

of elevated FFA, two-stage solvent extraction was used to decrease the FFA quantity of raw rubber seed oil. Atabani et al. [4] generated biofuel from *Camelina sativa*, known as misleading flax. It has a small content of FFA (1.0 wt.%) and elevated concentrations of unsaturated such as linolenic (38.4 wt.%), gondoic (16.8 wt.%), linoleic (14.3 wt.%) and hemp seed oil (14.3 wt.%). The unique properties of camelina oil methyl esters include 3 °C and −4 °C cloud point and pour point values. The direct-injection engine compression qualities fueled with biodiesel included cottonseed oil blends were evaluated by Bharadwaj et al. [5]. The testing evaluation involves three types of biodiesel and diesel seed oil used in cotton plants (10%, 15%, and 20% blends). Shrigiri et al. [6] discovered that brake thermal efficiency at peak loads is lowered by 5.91% and specific fuel consumption. Senthil Kumar et al. [7] in their work on the impact of octane oil resistance on combustion behavior with macroalgae oil-based dual-fuel engine discovered a decrease in combustion length in dual-fuel mode with eucalyptus oil with relatively higher brake fuel efficiency.

2 Biodiesel Preparation

Biofuel can be obtained from vegetable oils or edible oils, non-food vegetable oils or inedible oil, animal fats and surplus or used vegetable oils. Methyl, ethyl, and butyl esters generated through the transesterification of vegetable oils are the most prevalent kinds of biofuels. Therefore, the generally successful application of vegetable oils as biodiesel fuels can be called the ester of vegetable oils.

Microalgae oil manufacturing began from microalgae growing using a gravel pond with indigenous algal populations, with *Desmodesmus* being the primary plant with other species including *Scenedesmus acumunatus*, *Klebsormidium*, *Dictyosphaerium*, and other small fraction trees. *Desmodesmus* as the primary ratio of individuals varies depending on the season but can reach up to 24% of the complete population. The microalgae were collected and sent to a wastewater centrifuge and a tank dryer for a second drainage phase once getting HTL therapy. Algae production is a simple method that disposed of developing the coral by supplying the needed inputs for respiration, harvesting, soil sampling, and extracting oil. The algae cells carry the energy of photons that converts the metal oxides of CO_2 and water into glucose and oxygen. The grains are ultimately transformed within the algae organs into complicated protein, fats, nutrients, and fats. To remove precious lipids, a series of measures must be taken to isolate algae cells and oil [8, 9].

The criterion method begins with centrifuges, filtration, or flocculation techniques to separate the algae biomass from the water broth in the dewatering stage. Centrifuges acquire energy by spinning its soup of algae-water so that water goes away from either the algae bodies. Flocculation involves making algae cells out of location so they can be easily concentrated and removed. When the algae cells are collected, the oil must be divided from the shells. The oil can then be converted to biofuel, jet fuel, ethanol, synthetic fuel, or other substances. The flash points and fire points were measured using flash point tester. The calorific value was evaluated

Table 1 Fuel properties

S. No.	Properties of fuels	Diesel	Microalgae oil
1	Density (g/ml)	0.823	0.8792
2	Calorific valve (Kcals/Kg)	10,865.5	9392
3	Kinematic viscosity at 40 °C	3.01	4.64
4	Cetane number	52	50
5	Flash point °C	51	52
6	Specific gravity	0.837	0.8799
7	Acid number (mg of KOH/gm)	0.4	0.5

Fig. 1 Biodiesel blends

using a fast point test [10, 11]. The amount of Cetane was evaluated using bomb calorimeter. Table 1 shows the fuel characteristics (Fig. 1).

3 Experimental Setup

The experimental setup used for the current experimental investigations comprises of an engine with dynamometer, exhaust meter for exhaust gas analyzer. The test engine used is 4S1C, making Kirloskar's water-cooled, direct-injection ignition compression engine with eddy current dynamometer. This device can be met with greater pressures and can also be used widely in the agricultural and industrial industries. This machine is therefore chosen to carry out studies. Furthermore, needed changes can be produced readily on the piston needed for the current job. The density of the smoke is evaluated with a Bosch smoke meter. A gas analyzer is connected to the engine to evaluate the pollutants in the exhaust gas. This analyzer is used to evaluate three major exhaust emissions, i.e., CO, NO_x, and uHC (Figs. 2, 3 and Table 2).

Fig. 2 Experimental setup

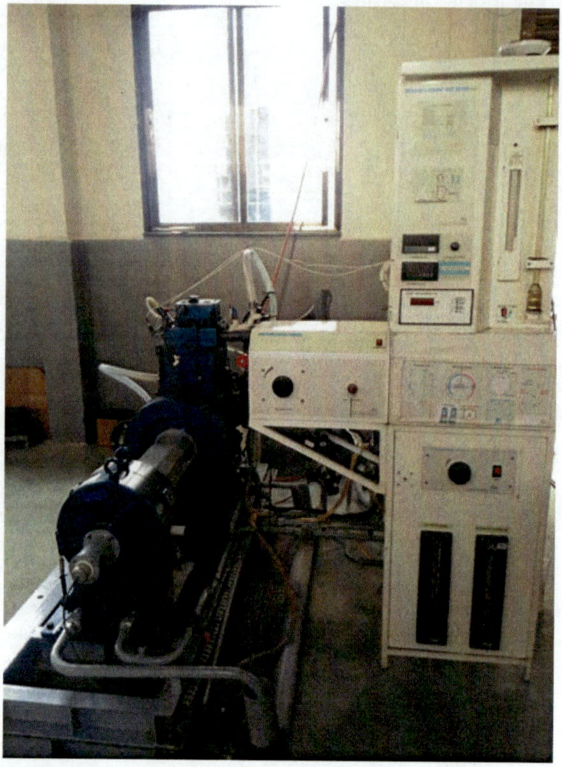

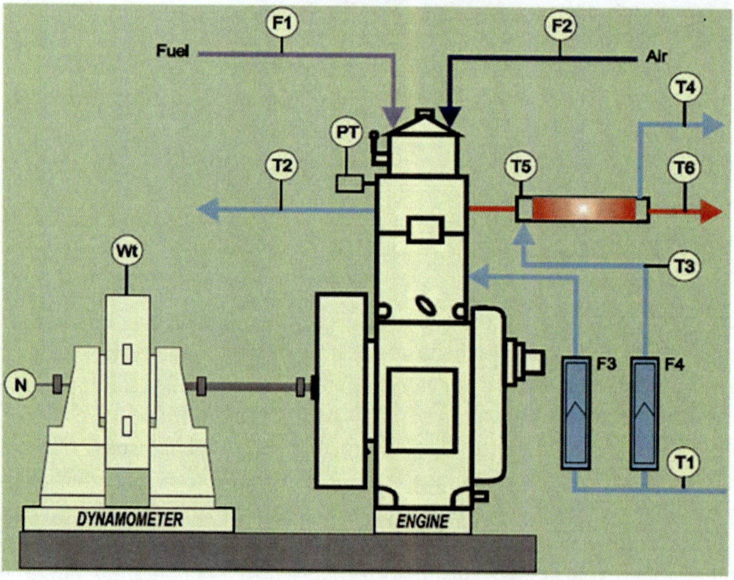

Fig. 3 Schematic diagram

Table 2 Engine Specifications

S. No.	Specification	Ranges
1	Make, type	Kirlosker, 4 S 1 C
2	Stroke × Bore	110 × 87.5 mm
3	Type of cooling	Water cooled
4	Engine speed, rated power speed	1500 rpm, 3.5 Kw
5	Compression ration	12:1 to 18:1
6	Injection point variation	0–25° BTDC
7	Orifice diameter	20 mm
8	Connecting road length	234 mm
9	Type of sensors	K-type
10	Fuel injection pressure	2.5–3 bars

4 Results and Discussion

4.1 Brake Thermal Efficiency (BTE)

Changes in BTE with diesel loads and distinct advertisements for biodiesel fuels are shown in Fig. 4. The diesel fuel brake thermal efficiency showed 27.1% at elevated load and 11.56% at low load. It is observed BTE was increases with increases the load. As shown in Fig. 4, BTE of the methyl ester microalgae oil at 20% of the

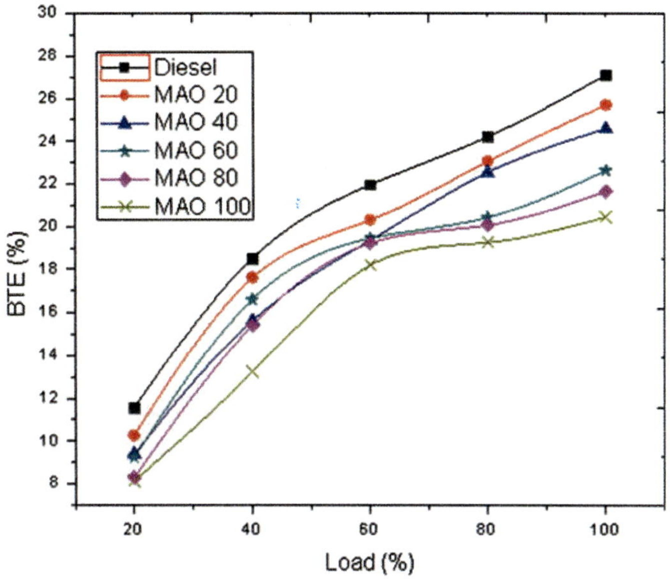

Fig. 4 Changes of brake thermal efficiency

load was observed 10.27%, 9.39%, 9.25%, 8.29%, and 8.13% and high load was observed 25.7%, 24.6%, 22.63%, 21.65%, and 20.45%. With respective fuel blends MAO 20, MAO 40, MAO 60, MAO 80, and MAO 100. It is observed that the brake's thermal efficiency was improved by raising the load. Experiment recognized diesel fuel consists of elevated BTE compared to biodiesel fuel. The primary cause of lowering in heat brake effectiveness is owing to the decreased heating value of biofuel relative with diesel. Brake thermal efficiency of methyl ester microalgae oil at MAO 20 was noted to a maximum extent than other combustion of fuels.

4.2 Brake Specific Fuel Consumption (BSFC)

The comparison of BSFC with load for diesel as well as biodiesel blends as shown in Fig. 5. BSFC of the diesel fuel was recorded 0.81 kg/kWh, 0.5 kg/kWh, 0.41 kg/kWh, 0.34 kg/kWh, and 0.3 kg/kWh with respective 20%, 40%, 60%, 80%, and 100% load. As shown in Fig. 5, BSFC of the methyl ester microalgae oil at 20% of the load was observed 0.89 kg/kWh, 0.91 kg/kWh, 0.98 kg/kWh, 1.02 kg/kWh, and 1.05 kg/kWh and at 100% load was recorded 0.34 kg/kWh, 0.4 kg/kWh, 0.39 kg/kWh, 0.38 kg/kWh, and 0.39 kg/kWh. With respective fuel blends MAO 20, MAO 40, MAO 60, MAO 80, and MAO 100. It is depicted that the SFC was reduced with a rise in the load. It is understood from Fig. 5, the BSFC of the diesel fuel was low compared to biodiesel fuel blends. Biodiesel blend MAO 20 was recorded low brake specific fuel

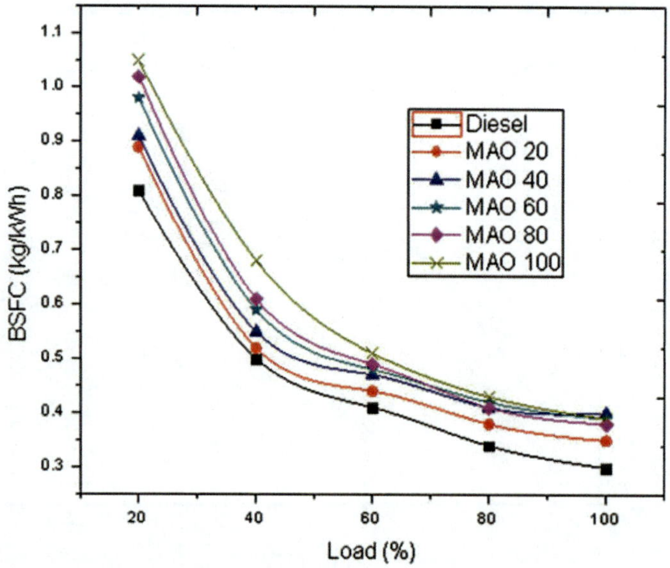

Fig. 5 Variation of brake specific fuel consumption

consumption than remaining blends because of high viscosity and density. Because of the elevated viscosity it generates the lower heating value and consumes a large quantity of fuel.

4.3 Oxides of the Nitrogen (NO_X)

The mixture of nitrogen oxide (NO) and nitrogen dioxide (NO_2) was an oxide of nitrogen emission. Changes of NO_x with diesel load as well as biodiesel blends are shown in Fig. 6. It shows that oxides of the nitrogen increases with increase the load. The values of the oxides of nitrogen emissions were reported 130 ppm, 505 ppm, 691 ppm, 871 ppm, and 978 ppm with respective 20%, 40%, 60%, 80%, and 100% of loads. In biodiesel blends at 100% load was observed 1001 ppm, 1196 ppm, 1210 ppm, 1234 ppm, and 1305 ppm with respective MAO 20, MAO 40, MAO 60, MAO 80, and MAO 100. It is represented that the load increases an oxide of the nitrogen increases. Mainly oxides of nitrogen effected combustion temperature. Hence, the load increase the combustion temperature increases and it leads to increases the oxides of the nitrogen. As shown in Fig. 6, MAO 20 was recorded less oxides of nitrogen than MAO 40, MAO 60, MAO 80, and MAO 100. This is due to less combustion temperature in combustion chamber.

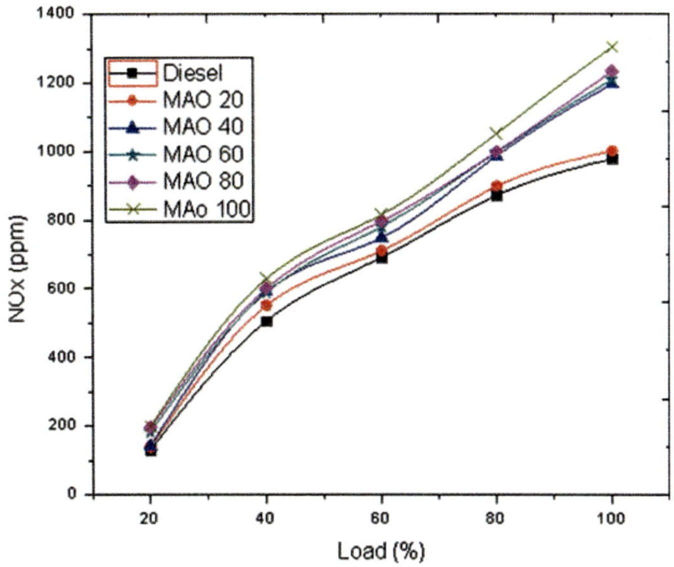

Fig. 6 Variation of oxides nitrogen

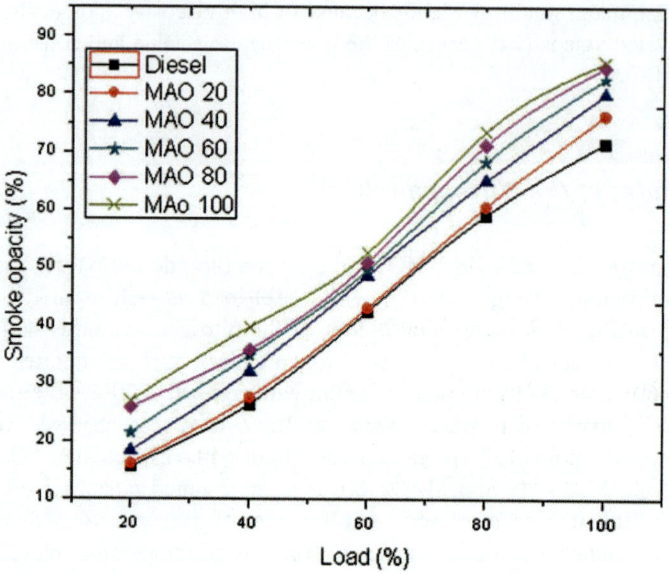

Fig. 7 Variation of smoke opacity

4.4 Smoke Opacity

Variation of smoke opacity with diesel load as well as biofuel mixes as shown in Fig. 7. It demonstrates that the opacity of smoke improves the load. The quality of low load smoke opacity of diesel fuels was 15.5% and complete load was 71.6%. Biofuel smoke opacity at complete load was 73.3%, 79.9%, 82.6%, 84.7%, and 85.5% with MAO 20, MAO 40, MAO 60, MAO 80, and MAO 100, respectively. In this experiment, result shows that MAO20 recorded less smoke opacity than MAO 40, MAO 60, MAO 80, and MAO 100. This is owing to the heavy chemical composition, less oxygen content that even in locally wealthy areas adds to incomplete oil decomposition.

4.5 Hydrocarbons (HC)

Hydrocarbon emissions are primarily caused by incomplete combustion of the fuel. Figure 8 represents the variation of the hydrocarbons with loads for the fuels of the diesel and biodiesel blends. Hydrocarbon emission recorded with diesel fuel was 55 ppm and 156 ppm with respective low load and high load. Hydrocarbon emission observed in the biodiesel blends was 138 ppm, 104 ppm, 115 ppm, 122 ppm, and 82 ppm with respective MAO 20, MAO 40, MAO 60, MAO 80, and MAO 100. Hydrocarbon emission of all blended biodiesel fuels shows less value than diesel

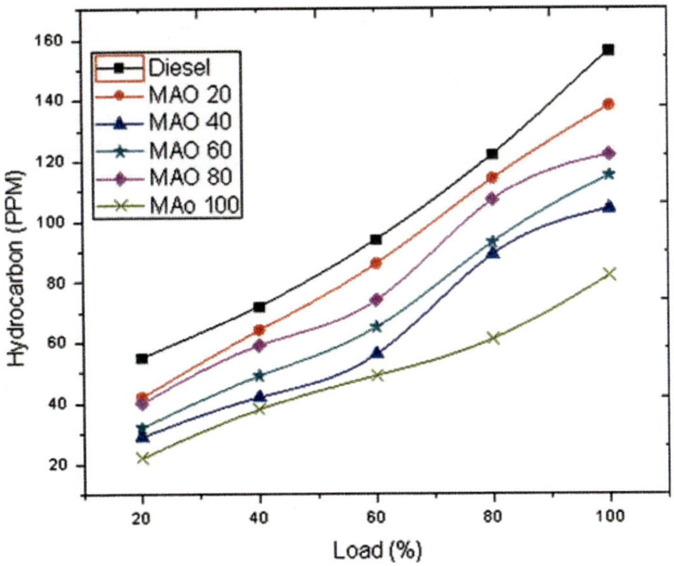

Fig. 8 Variation of hydrocarbon

fuel. Lowest hydrocarbon emissions were noted in pure biofuel diesel. This is prior to the elevated oxygen in pure biofuel, which results in the avoidance of incomplete combustion. Reduces hydrocarbon emissions with pure biofuel fuels.

4.6 Carbon Monoxide (CO) Emissions

Variety of carbon monoxide pollution with fuel load as well as biofuel mixes as shown in Fig. 9. It is shown that carbon monoxide levels are reduced by up to 60% of the load, further rises with increased load. The values of the carbon monoxide emissions were observed 0.107 %vol., 0.1 %vol., 0.097 %vol., 0.123 %vol., and 0.149 %vol. with respective 20%, 40%, 60%, 80%, and 100% load. In biodiesel blends, 60% load was observed minimum carbon monoxide emissions. It was shown that 0.087 %vol., 0.075 %vol., 0.071 %vol., 0.061 %vol., and 0.054 %vol. with respective MAO 20, MAO 40, MAO 60, MAO 80, and MAO 100. Carbon monoxide emissions were shown in Fig. 9 more in diesel fuel than compared with all blended biodiesel fuels. Lowest carbon monoxide emissions were recorded pure biodiesel fuel. High value of the carbon monoxide was observed in biodiesel was MAO20. This would be due to inadequate combination of air-fuel and poor oxidation in heavy loads.

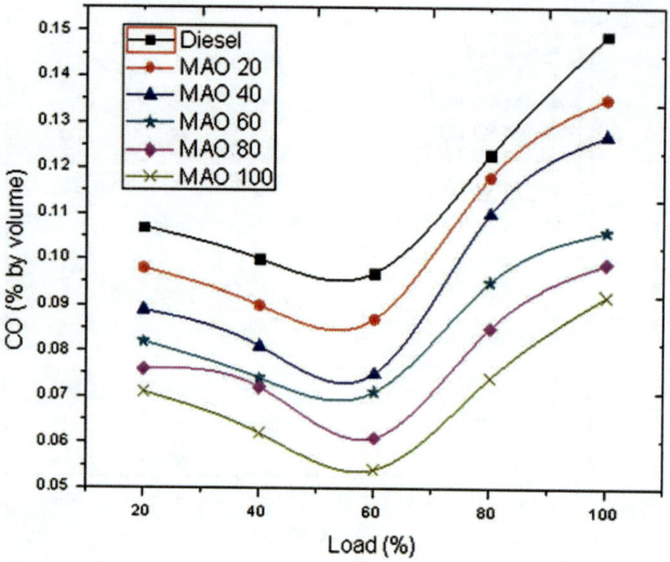

Fig. 9 Variation of carbon monoxide with respective percentage of loads for the microalgae oil blends and diesel fuel

5 Conclusion

The investigational examination into diesel fuel and methyl ester macro algae oil blended MAO 20, MAO 40, MAO 60, MAO 80, and MAO 100. A performance and emission characteristic was evaluated on conventional diesel engine with variation of load percentage at 1500 rpm. Better results were observed at MAO 20 than remaining blends. Details were discussed below.

- MAO 20 brake thermal efficiency at complete load was registered at 27.1%; MAO 20 brake thermal efficiency at complete load was registered at 25.7%.
- BSFC reduces as load increases. 0.3 kg/kWh of full-load diesel fuel BSFC was noted. MAO 20 indicates less specific fuel consumption for brake than other mixes.
- Increases in nitrogen oxides were suggested for biofuel mixes and loads. Nitrogen peak oxides were recorded at complete load. 1001 ppm was noted at complete load MAO 20.
- Smoke opacity of the diesel fuel at complete load condition was noticed 71.6%. MAO 20 was produces less smoke opacity. This is due to less oxygen content that contributes to incomplete fuel oxidation even in locally rich zones.
- Hydrocarbon emission of MAO 20 at full load was reduced 12.5% than diesel fuels. HC emission values of the diesel 156 ppm at 100% load. HC emission of the MAO 20 was recorded 138 ppm

- Carbon monoxide emission of MAO 20 at full was reduced 9.39% than the diesel fuel. MAO 100 was recorded less CO emission than other blends of MAO fuel and diesel fuel.

References

1. Ali Y, Hanna (1994) Alternative diesel fuels from vegetable oils, pp 153–163
2. Ikwuagwu OE, Ononogbu IC, Njoku OU (2000) Production of biodiesel using rubber [*Hevea brasiliensis* (Kunth. Muell.)] seed oil, pp 57–62
3. Melvin Jose DF, Edwin Raj R, Durga Prasad B, Robert Kennedy Z, Mohammed Ibrahim A (2011) A multi-variant approach to optimize process parameters for biodiesel extraction from rubber seed oil, pp 2056–2063
4. Atabani AE et al (2014) A study of production and characterization of Manketti (*Ricinodendron rautonemii*) methyl ester and its blends as a potential biodiesel feedstock. Biofuel Res J 4(2014):139–146
5. Bhardwaj V, Singla V, Gaikwad P, Sharma G (2014) Study of performance characteristics of compression ignition engine fuelled with blends of biodiesel from used cottonseed oil, pp 289–296
6. Shrigiri BM, Hebbel PD, Reddy KHC (2016) Performance, emission and combustion characteristics of a semi-adiabatic diesel engine using cotton seed and neem kernel oil methyl esters, pp 699–706
7. Senthil Kumar M, Nataraj G, Arul Selvan S (2017) A comprehensive assessment on the effect of high octane fuels induction on engine's combustion behaviour of a Mahua oil based dual fuel engine, pp 176–184
8. Hariprasad T et al (2010) Performance and exhaust emissions analysis of a diesel engine using methyl esters of fish oil with artificial neural network aid IACSIT. Int J Eng Technol 2(1). ISSN: 1793-8236
9. Heywood JB (1988) Internal combustion engine fundamentals. McGraw-Hill, United States of America
10. Moulali P et al (2018) Performance and emission characteristics of Homogeneous Charge compression ignition engine with different bio diesel fuels. Int J Eng Technol 7(4.24):157–162
11. Mathan Raj V, Ganapathy Subramanian LR, Manikandaraja G (2017) Experimental study of effect of isobutanol in performance, combustion and emission Characteristics of CI engine fuelled with cotton seed oil blended diesel, pp 1–10

An Investigation on Piston Structural Analysis Related with Experimental Cylinder Pressures Using Different Biodiesel Blend Ratios

I. J. Isaac Premkumar, A. Prabu, V. Vijayan, and S. Dinesh

Abstract The prime objective of this investigation is to ascertain the stress analysis of dissimilar pistons with different compression ratios. The diverse piston geometries of deep cylindrical piston (DCP) and shallow re-entrant piston (SRP) were exclusively designed by the baseline of hemispherical existing piston (HEP) at different compression ratios of 15:1, 17:1, and 16:1, respectively. The combustion characteristics of cylinder pressures were established for various piston geometries using different biodiesel blends of soya bean, rice bran, and pongamia. The engine operating characteristics and load capacity in terms of the pressure with respect to crank angle rotation, air and fuel mixing rate, and turning moments were attained from the experimental investigation. Analysis of piston geometry was done with boundary conditions of the maximum value of the cylinder pressure for pongamia biodiesel blends with diesel. The structural analysis of equivalent stress and total deformation values was plotted against the cylinder pressure values. This analysis predicts the portion of the piston structure that might be damaged on top or bottom skirt. Finally, mechanical thermal stresses developed in the piston were optimized with lesser deformation by finite element analysis (FEM).

Keywords Stress analysis · Equivalent stress · Mechanical stresses

I. J. I. Premkumar (✉)
Mechanical Engineering Department, JCT College of Engineering and Technology, Coimbatore, Tamilnadu, India
e-mail: isaacpremkumar007@gmail.com

A. Prabu
Automobile Engineering Department, SRM University, Chennai, Tamil Nadu, India

V. Vijayan · S. Dinesh
Mechanical Engineering Department, K. Ramakrishnan College of Technology, Samayapuram, Trichy, Tamilnadu, India

A. Arockiarajan et al. (eds.), *Advances in Industrial Automation and Smart Manufacturing*, Lecture Notes in Mechanical Engineering,
https://doi.org/10.1007/978-981-15-4739-3_81

1 Introduction

Piston is the primary part of the engine. The engine, as a prime mover, is the important part of a vehicle. Currently, plethora of research is going on to optimize the engine in terms of manufacturing techniques and other attributes. There are lots of investigation works proposing, new dimensional models, and manufacturing techniques on engine pistons, and this investigational progress has developed with a continuous development over the previous decades [1]. Piston is an important component of the internal combustion engines, due to the role it fulfills, but also due to the complex loads which act upon it. The piston's resistance to the mechanical loads limits the power of the internal combustion engine. Hence, during the designing stage of the piston, there is a need to determine the stresses and deformations due to the mechanical loads, an aspect which was revealed in numerous scientific papers dealing with these issues [2].

Though many experimental investigations were made on piston geometry modifications, there are only a few numerical studies focused on 3-D structural analysis on a different shape of aluminum alloy piston bowl model [3]. In this analysis, effort has been made to contain stress and deformation in FEM analysis by ANSYS software, and the structural analysis of piston geometries was utilized to vary the compression ratios by altering the geometry of the piston model. The modified piston geometries of shallow re-entrant piston (SRP) and deep cylindrical piston (DCP) were compared with baseline geometry of hemispherical existing piston (HEP). The experimental results were considered as the inputs of pressure loads to carry out stress and deformation rates. This paper reveals the 3-D finite element modeling of different aluminum alloy piston with different compression ratios. The optimum piston geometry model was found with optimum characteristics of stress analysis through ANSYS software. The experimental results of combustion cylinder pressure were related to stress analysis and carried out the effects of different piston structure using diesel and biodiesel blends. Maximum factor of safety, minimum equivalent stress, and low deformation rates were carried out for optimum piston model. Our present investigation acts as the practical guidelines for optimum piston geometry found in order to improve engine operating characteristics.

2 Materials and Methods

The real-time low-performing internal combustion engine was used to analyze the cylinder pressure for every piston with soya bean, rice bran, and pongamia biodiesel blends. The combustion results of maximum cylinder pressure were taken as the pressure acting on different geometries of HEP, SRP, and DCP in the range of 6.5, respectively [4]. In the combustion chamber due to explosion of gas, pressure range was applied on the piston head. The pressure force was considered as mechanical load applying on the piston head and the same process was taken as boundary condition

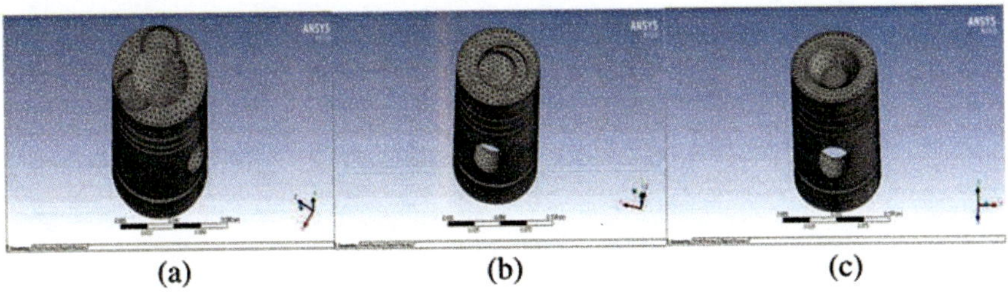

Fig. 1 Mesh images of HEP, SRP, and DCP

Table 1 Mesh details of different piston geometry of DCP, SRP, and HCP

Mesh	DCP	SRP	HEP
Nodes	58,803	49,645	51,873
Elements	33,365	28,314	29,639

in structural analysis. Since the piston moves from TDC to BDC, fixed support had specified at the surface of pinhole [5].

From Fig. 1, it was evident that the parts of structural analysis were attained for different piston geometry designs. The different piston geometry designs were a mesh of DCP, SRP, and HEP that was made up of tetrahedron elements with the following nodes as shown in Table 1. The structural properties of Al 7050 T7451 for baseline piston geometry of HEP were discussed as shown in Table 2.

3 Machining Process

The piston was first machined in center lathe for processing with help of a four-jaw chuck. For the SRP, the piston geometry was altered by making a groove of 4 mm at the top to get eight (8) like shape in the piston head. In DCP, 8 mm was grooved near indentation. After machining, the piston was mounted in vertical milling machine. The milling process was separated into three categories that fact was mill rough, mill semi-finish, and mill finish.

The main purpose of the rough cut stage was to eradicate material in the most proficient mode. In semi-finishing cutting, the process was to make a close to the final shape of the piston and the required surface finish and accurateness were made during final finishing. After the milling process, the grooved surface was cased to obtain accuracy form which is possible by milling process. Subsequently, the surface was wiped by course emery sheet to make surface even and finally fine emery sheet for attaining enhanced swirl and squish progress. The wax was melted and then poured in the combustion chamber in piston and kept undisturbed for 30 min. This method was used to analyze the volume of the piston. The compression ratio of HEP geometry was 16:1. Then, the wax which was poured inside the piston was taken out

Table 2 Structural properties of Al 7050 T7451 for baseline piston geometry of HEP

Material	Elastic modulus (N/mm^2)	Density (Kg/m^3)	Yield strength (MPa)	Tensile strength (MPa)	Poison's ratio	Thermal conductivity (w/(m-k))
Al 7050 T7451	72,000	2830	470	525	0.33	157

Fig. 2 Different piston geometries of HEP, SRP, and DCP

and heated in an electric heater. The molten wax was then poured in a 100 ml beaker and quantity was noted. The quantity of wax for shallow re-entrant piston and deep cylinder piston was 19 ml and 26 ml, respectively. This method was used to found the compression ratio for different shapes of piston geometries as shown in Fig. 2.

4 Results and Discussion

4.1 Structural Analysis of Different Shapes of HEP, SRP, and DCP Geometry

The combustion of gases in the combustion chamber exerts pressure on piston head during power stroke. The ANSYS Workbench was used to inspect the structural characteristics of piston geometries by boundary condition of the pressure force [6]. Consequently whatever the load was applied on piston as a result of power stroke causes damage to piston pin which also induces bending stress [7, 8]. The stress analysis of piston geometries was done under structural boundary conditions which were applied to the finite element model of the pistons [9]. A maximum stress of 8.2473 MPa and minimum of 0.001175 MPa were observed with a total deformation of 0.15 mm which was comparatively good even after more cycle of operation. The equivalent stress and total deformations were found to be good as shown in Fig. 3. The maximum stress of 7.9803 MPa and minimum of 0.0012259 MPa were observed. Total deformation of 0.14 mm was obtained for the applied pressure which was considerable after much number of cycles of operation. The SRP geometry stress values were lower than the other piston geometries as shown in Fig. 4. The maximum stress of 7.8383 MPa and minimum of 0.00088874 MPa were observed. Total deformation of 0.18 mm was obtained for the applied pressure which was considerable after many numbers of cycles of operation. The DCP geometry stress values were higher than the other piston geometries as shown in Fig. 5.

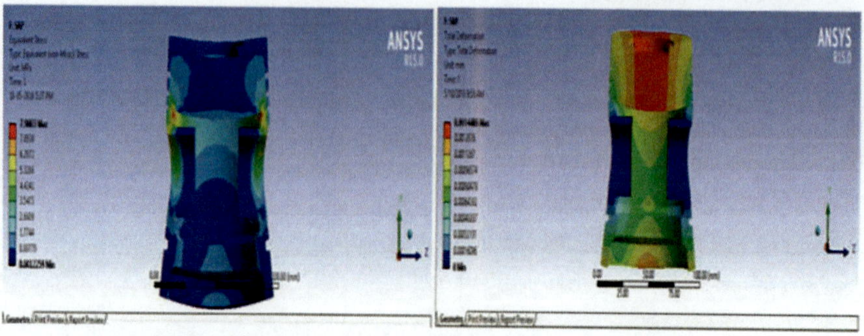

(a) (b)

Fig. 3 **a–b** Equivalent stress and total deformation of HEP geometry model

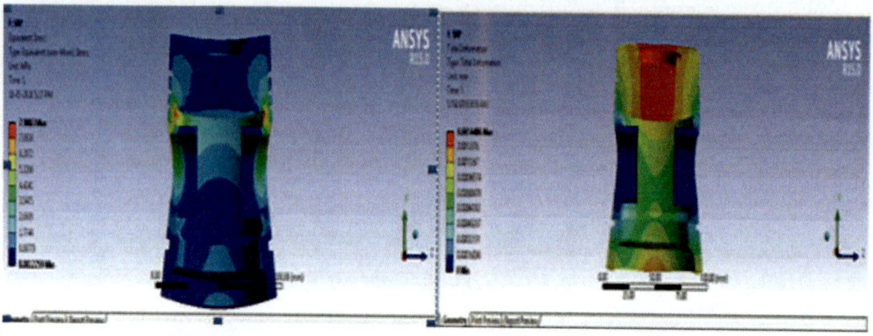

(a) (b)

Fig. 4 **a–b**. Equivalent stress and total deformation of SRP geometry model

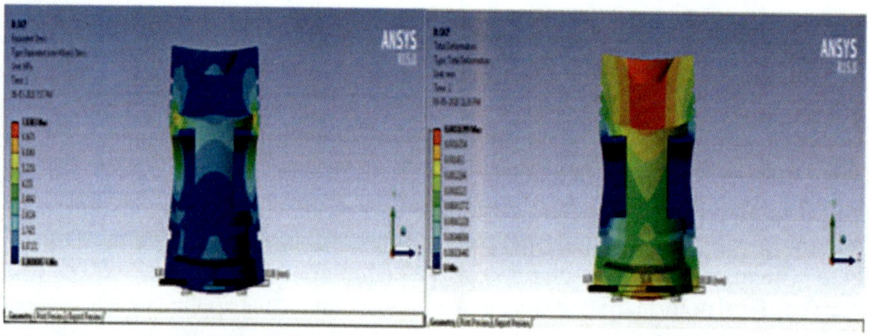

(a) (b)

Fig. 5 **a–b**. Equivalent stress and total deformation of DCP geometry model

Factor of safety of piston head SRP is greater than the other piston heads DCP and HEP. SRP piston head is safe designs when compared to other as it possesses a higher value of 3.4034. The HEP geometry had a minimum factor of safety of 2.9899. In addition, the factor of safety for induced stresses was determined as 3.3067, 2.9899, and 3.4034 for different compression ratios of 15:1, 16:1, and 17:1. From the results, it is very clear that the compression ratio of 17:1 had a higher factor of safety. The factor of safety must be considered for improving the appreciable effects on the stress by increasing the compression ratio [10, 11]. The high compression ratio with SRP piston had given better results of minimum stress and reduced the total deformation rate [12]. The factor of safety is also higher in SRP geometry. Hence, SRP was considered as optimum piston geometry when compared to other piston geometries. The proposed piston head design has reduced the equivalent stress when it is subjected to a maximum pressure [14–16].

4.2 *Combustion Cylinder Pressures Related to Stress Values for Different Piston Geometries*

Stress and deformation of the piston were studied by considering the cylinder pressure with respect to crank angle. The maximum stresses and deformations were observed during the engine operation.

4.2.1 Effect of Combustion Pressure Ranges for Blends of Soya Bean, Rice Bran, and Pongamia Biodiesel with HEP

The maximum equivalent stress of 8.2473 Mpa was developed with B50 soya bean. The equivalent stress of soya bean B10 and B30 demonstrated the results of 7.603 Mpa and 7.8607 Mpa, respectively. The pure diesel and HEP have shown an equivalent stress of 7.3453 Mpa which was lower when compared to all other biodiesel blends with HEP as shown in Fig. 6. Total deformation of 10% soya bean oil with diesel was lesser than other percentages of soya bean oil value of 0.23 mm and it was compared to 50% of the soya bean was 5.17% of value was decreased and 30% of the soya bean was 7.75% of the total deformation value was decreased. This recorded maximum equivalent stress value was comparatively lower of about 7.51 for 30% of rice bran oil and 9.09 of 10% rice bran oil when compared with 50% rice bran oil (8.505 Mpa) as shown in Fig. 7. Total deformation of 10% rice bran oil with diesel was lesser than other percentages of rice bran oil value of 0.21 mm, whereas the 30 and 50% of rice bran oil recorded a maximum piston deformation percentage of about 5.06% and 9.16% than 10% rice bran oil. When the maximum equivalent stress was compared with other blends of pongamia, it was observed that 10% pongamia oil recorded 5.8% lesser equivalent stress and 2.9% lesser value for 30% of pongamia oil as shown in Fig. 8. Total deformation of 10% pongamia oil

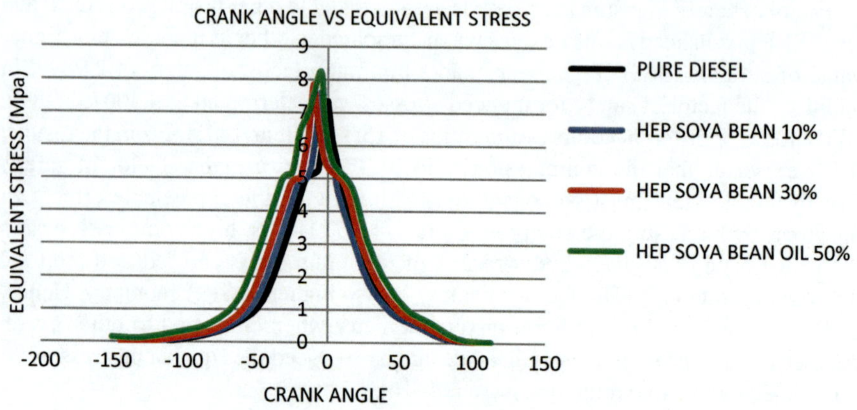

Fig. 6 Crank angle versus equivalent stress for soya bean biodiesel blends with HEP

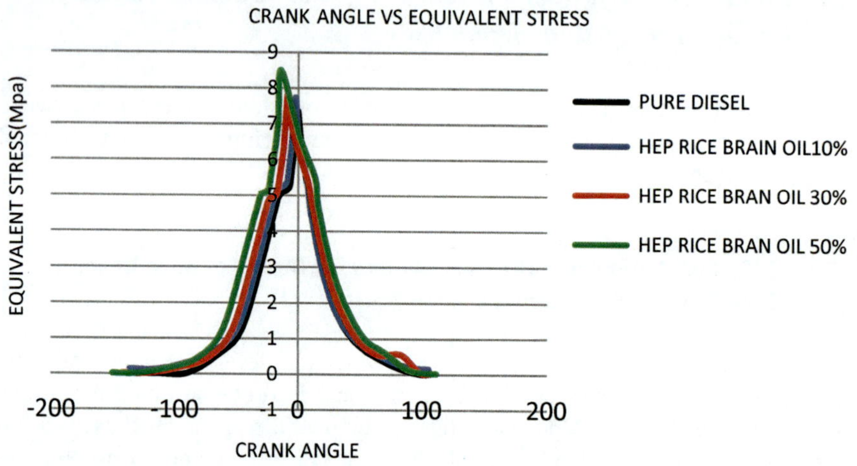

Fig. 7 Crank angle versus equivalent stress for rice bran biodiesel blends with HEP

with diesel was lesser than other percentages of pongamia oil value of 0.24 mm which was 3.025% lower than 30% of pongamia oil. The 50% of pongamia oil has shown 21.1% of higher deformation values compared to other blends. The maximum cylinder pressure was recorded with pongamia B50 which is significantly higher in terms of equivalent stress with higher deformation.

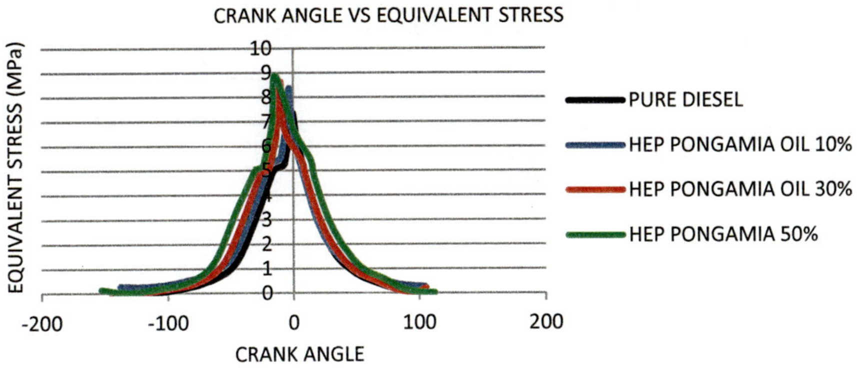

Fig. 8 Crank angle versus equivalent stress for Pongamia biodiesel blends with HEP

4.2.2 Effect of Combustion Pressure Ranges for Blends of Soya Bean, Rice Bran, and Pongamia Biodiesel with SRP

The equivalent stress and total deformation of piston model were compared between soya bean biodiesel blends and diesel with SRP. From Fig. 9, it was observed that the equivalent stress of 50% soya bean oil was comparatively greater than that of diesel. By comparing the equivalent stress effect within different blends of soya bean oil, the maximum equivalent stress was attained with 50% soya bean oil of about 7.9803 Mpa. The obtained value was 4.61% higher than 10% soya bean oil and 3.13% higher than 30% soya bean oil. Total deformation of 10% soya bean oil with diesel was lesser than other percentages of soya bean oil value of 0.19 mm which was 28% lower than 30% of soya bean. The 50% of soya bean showed 29.8% higher value of deformation compared with B10 and B30 blends.

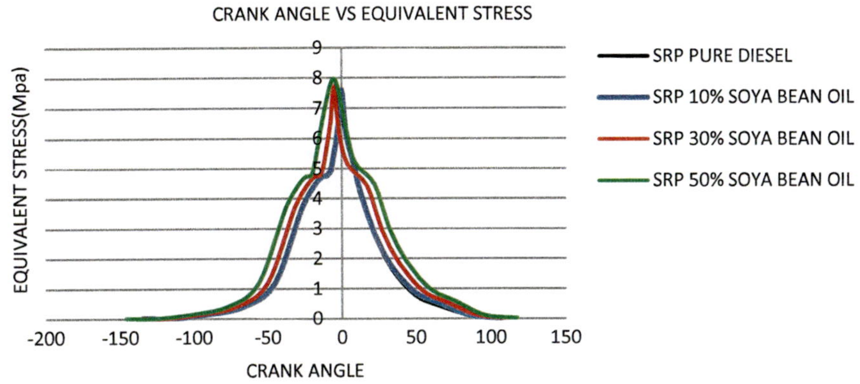

Fig. 9 Crank angle versus equivalent stress for soya bean biodiesel blends with SRP

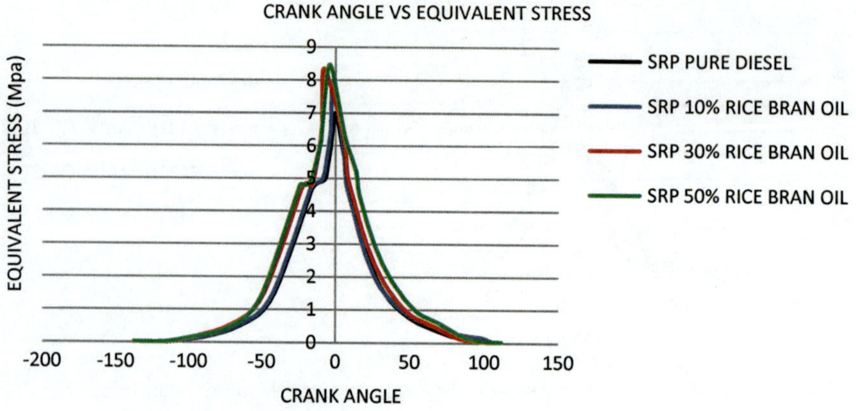

Fig. 10 Crank angle versus equivalent stress for rice bran biodiesel blends with SRP

From Fig. 10, it was observed that the diesel with 50% rice bran oil has shown a higher equivalent stress value of 8.4714 Mpa when compared to other percentages of rice bran oil and diesel. Comparing the effects of stress on total deformation of SRP, the piston total deformation of pure diesel was lower than various percentages of rice bran oil. The total deformation value of lesser for 10% rice bran oil with 0.24 mm which was 4.43% higher than 30% rice bran oil and 6.41% higher than 50% rice bran oil.

The maximum equivalent stress was developed by 50% of pongamia with respect to CAD of 8.717 MPa and it was 4.225 and 1.41% lesser when compared with 10 and 30% pongamia oil, respectively, as shown in Fig. 11. The total deformation value

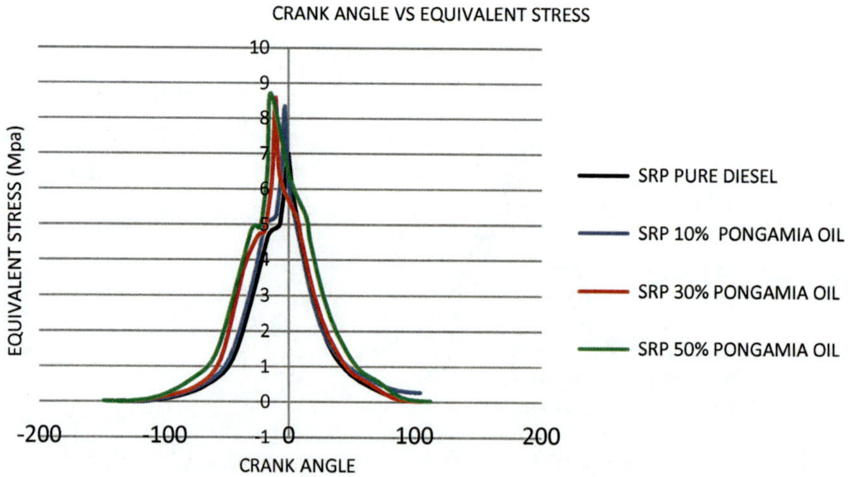

Fig. 11 Crank angle versus equivalent stress for Pongamia biodiesel blends with SRP

was found to be lesser for 10% rice pongamia oil with 0.25 mm which was 3.03% higher than 30% rice bran oil and 4.4% higher than 50% rice bran oil.

The high compression ratio of 17:1 with SRP geometry had better cylinder pressure compared to other piston geometries but with higher equivalent stress and deformation. Most of the investigation reports have revealed that increasing the compression ratio of the engine had produced better power output and also complete combustion process. In this SRP geometry, the combustion starts earlier and a large amount of fuel burned during the premixed combustion which may cause a higher cylinder pressure [13]. Due to the rich oxygen content of biodiesel blends, the cylinder pressure was quite higher than other biodiesel blends and also increases the chance of higher equivalent stress and deformation. The 50% of pongamia biodiesel blend has produced maximum equivalent stress and total deformations of 8.717 Mpa and 0.27 mm which were fairly higher than other biodiesel blends. The increased cylinder pressure allows complete combustion even in the premixed combustion stage.

The SRP geometry had given maximum cylinder pressure by using pongamia B50. This result was evident that maximum stress occurred was 8.717 MPa with a cylinder pressure of 71 bar. This SRP geometry had reached maximum pressure compared to HEP, but it showed vaguely lesser equivalent stress and total deformation. The well design with higher compression ratio of SRP produced better stress and total deformation results and also allows complete combustion as shown in Table 3.

Table 3 Combustion cylinder pressure and stress values of SRP using different biodiesel blends

Fuel and blend ratios	Pressure (bar)	Equivalent stress (MPa)	Maximum principle stress (MPa)	Total deformation (mm)
Diesel	57	7.2453	3.4677	0.0021729
Soya bean 10% blend	62	7.6121	3.7719	0.0023635
Soya bean 30% blend	63	7.7348	3.8327	0.0024016
Soya bean 50% blend	71	8.717	4.3194	0.0027066
Rice bran 10% blend	65	7.9803	3.9544	0.0024779
Rice bran 30% blend	68	8.3487	4.1369	0.0025922
Rice bran 50% blend	69	8.4714	4.1978	0.0026304
Pongamia 10% blend	68	8.3487	4.1369	0.0025922
Pongamia 30% blend	70	8.5942	4.2586	0.0026685
Pongamia 50% blend	71	8.717	4.3194	0.0026766

4.2.3 Effect of Combustion Pressure Ranges for Blends of Soya Bean, Rice Bran, and Pongamia Biodiesel with DCP

The lower stress and total deformations were obtained with various biodiesel blends. Huge variations in cylinder pressure were recorded when compared to HEP and SRP. Maximum cylinder pressure of DCP had produced 61 bar and it showed very lower compared to HEP and SRP. The lower combustion cylinder pressure had given low power output and partial combustion as shown in Table 4.

The minimum equivalent stress developed with respect CAD in the range of 7.3069 MPa for 10% of soya bean biodiesel blend. It was lowered compared to 30% and 50% of the soya bean 7.4397 Mpa and 7.7054, respectively, as shown in Fig. 12. Total deformation of 50% soya bean oil with diesel was 0.22 mm. This value was fairly high when compared to other biodiesel blend cylinder pressures. 10% of soya bean biodiesel was given fairly lower stress and total deformations compared to other diesel and biodiesel blends. In general, increasing the blends of biodiesel has shown an increasing stress and deformation.

The maximum equivalent stress developed with respect to crank value of 7.8383 MPa for 50% of rice bran biodiesel. The diesel and blends of 10 and 30% rice bran values were 7.714 Mpa, 7.5174 Mpa, and 7.5726 Mpa, respectively, as shown in Fig. 13. The diesel with 10% of rice bran oil had given low equivalent stress when

Table 4 Combustion cylinder pressure and stress values of DCP using different biodiesel blends

Fuel and blend ratios	Pressure (bar)	Equivalent stress (MPa)	Maximum principle stress (MPa)	Total deformation (mm)
Diesel	54	7.714	2.6801	0.0020699
Soya bean 10% blend	55	7.3069	2.7298	0.0021082
Soya bean 30% blend	56	7.4397	2.7794	0.0021465
Soya bean 50% blend	58	7.7054	2.8787	0.0022232
Rice bran 10% blend	54	7.5174	2.6801	0.0021699
Rice bran 30% blend	57	7.5726	2.829	0.0021848
Rice bran 50% blend	59	7.8383	2.9283	0.0022615
Pongamia 10% blend	58	7.7054	2.8787	0.0022232
Pongamia 30% blend	59	7.8383	2.9283	0.0022615
Pongamia 50% blend	61	8.104	3.0275	0.0023382

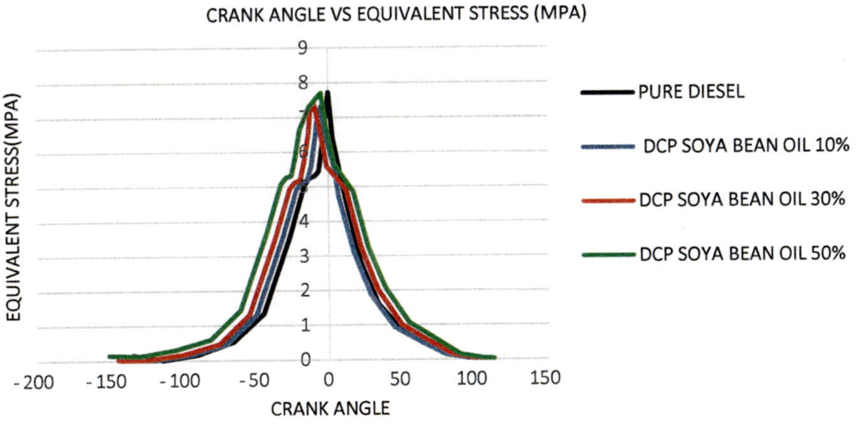

Fig. 12 Crank angle versus equivalent stress for soya bean biodiesel blends with DCP

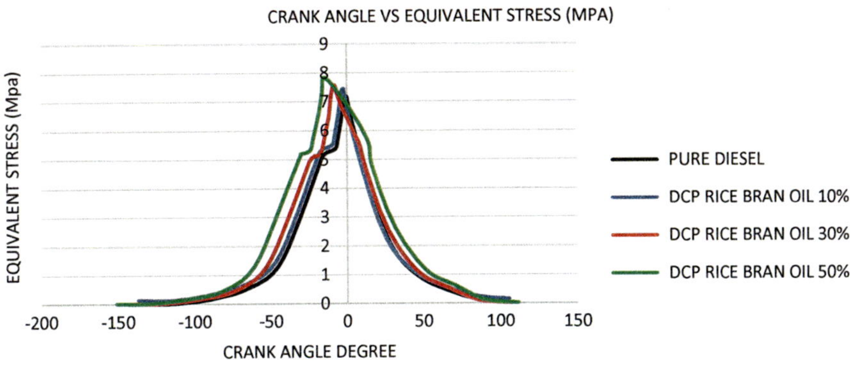

Fig. 13 Crank angle versus equivalent stress for rice bran biodiesel blends with DCP

compared to other biodiesel blends.

Total deformation of 10% rice bran oil was slightly higher than diesel fuel with DCP. The biodiesel blend of 50% rice bran had given 0.22 mm which was higher compared to other blends. The maximum equivalent stress developed with respective crank angle degree for the value of 8.104 MPa for B50 pongamia blends is shown in Fig. 14. The higher cylinder pressure values were given maximum stress with increased total deformation rates. The B10 and B30 pongamia blends were given 7.7054 Mpa and 7.8383 Mpa, respectively, which was increased compared to other biodiesel blends of soya bean and rice bran biodiesels with low compression ratio of DCP. Total deformation of 10% pongamia oil with diesel was lesser than other blends of pongamia. The B50 of pongamia had a maximum deformation value of 0.0023382 which was higher than other blend ratios of B10 and B30.

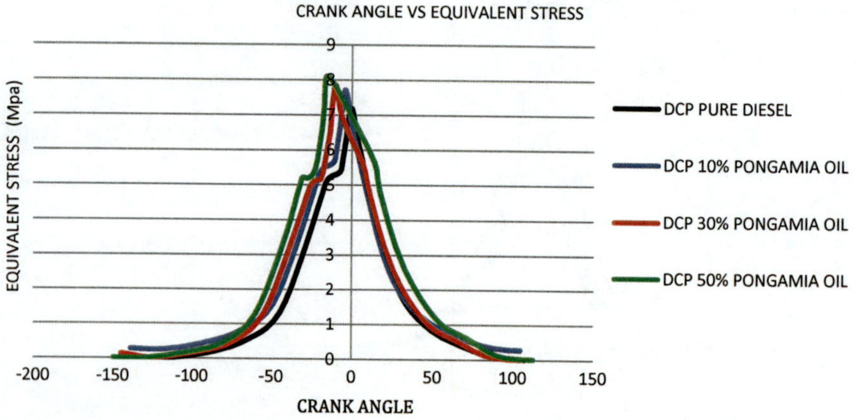

Fig. 14 Crank angle versus equivalent stress for Pongamia biodiesel blends with DCP

DCP cylinder pressure values were lowered compared to other geometries of HEP and SRP and also it has produced slightly lower stress and deformation values. In general, combustion characteristics of cylinder pressure would be lower in low compression ratio of the engine due to lack of oxygen, poor piston bowl shape, and improper air–fuel mixing [3, 13]. The premixed part of the combustion was lower for biodiesels, which results in slightly lowered cylinder pressure. Maximum cylinder pressure was attained in SRP with fairly decreased equivalent stress and total deformation compared to HEP as shown in Table 5. The comparison of structural analysis and also experimental cylinder pressure values-based structural analysis had given better results for SRP when compared to other piston geometries.

Table 5 Maximum equivalent stress and total deformation for various piston geometry models

Various piston geometry	Structural analysis by FEM		Experimental cylinder pressure values-based structural analysis by FEM		
	Maximum equivalent stress (Mpa)	Total deformation (mm)	Pressure (bar)	Maximum equivalent stress	Total deformation (mm)
DCP	7.8383	0.0018399	61	8.104	0.0023381
HEP	8.2473	0.0015612	69	8.8916	0.0027274
SRP	7.9803	0.001448	71	8.717	0.0026766

5 Conclusion

Our experiments on stress and thermal analysis of dissimilar piston geometries at compression ratios of 15:1, 16:1, and 17:1 revealed that the HEP and DCP had maximum stress values compared to SRP due to piston geometry bowl. The maximum equivalent stress value of 8.2473 MPa for HEP was decreased to 7.9803 MPa for SRP due to its well-structured piston design with appropriate compression ratio. The total deformation rate was changed from 0.15 mm of HEP to 0.14 mm of SRP because of enough stiffness in SRP.

The experimental analysis on cylinder pressure for different piston geometries showed that SRP had produced well cylinder pressure values with decreased rate of stress and total deformation when compared to HEP. The experimental cylinder pressure values had given the maximum equivalent stress value of 8.8916 MPa for HEP decreased to 8.717 MPa for SRP.

The decreased value of total deformation and equivalent stress was obtained in SRP geometry. The total deformation was reduced from 0.27 mm of HEP to 0.26 mm of SRP. The factor of safety was 3.4034 for SRP geometry which was greater compared to other piston geometry. So, the SRP geometry was safe under the applied loading conditions. By using ANSYS software in designing parts/assemblies of parts, the designer had the opportunity to analyze different constructive variants in a short time, thereby optimizing the designing work. The results are obtained using two designing methods from the baseline piston, and it was noted that the obtained values had slight a variation with respect to stress.

References

1. Venkatareddy KV, Goud CS (2016) Design and analysis of the piston by using composite materials. Int J Prof Eng Stud 7:154–162
2. Sundaram K, Palanikumar N (2016) Investigation and analysis of Piston by using composite Material. IJARIIE-ISSN (O) 2(2395–4396):1447–1454
3. Buyukkaya E, Cerit M (2007) Thermal analysis of a ceramic coating diesel engine piston using 3-D finite element method. Surf Coat Technol 202:398–402
4. Li CH (1982) Piston thermal deformation and friction considerations. SAE Technical Paper 820086, 1–18
5. Shubham Shrivastava T, Upadhyay S (2016) Thermal analysis of I C engine cylinder block with fins perpendicular to the axis of piston movement. Int J Mech Ind Technol 3:139–149
6. Nikanorova SP, Volkova MP, Gurina VN, Burenkova Y, Derkachenkoa LI, Kardasheva K, Regelb LL, Wilcoxb WR (2005) Structural and mechanical properties of Al–Si alloys obtained by fast cooling of a levitated melt. Mater Sci Eng: A 390:63–69
7. Gawale AA, Shaikh A, Patil V (2012) Nonlinear static finite element analysis and optimization of connecting rod World. J Sci Technol 2:01–04
8. Bhagat AR, Jibhakate YM (2012) Thermal analysis and optimization of I.C. Engine piston using finite element method. Int J Mod Eng Res (IJMER) 2:2919–2921
9. Solanki AB, Patel, HC, Abhishek Y (2014) Makawana design analysis and optimization of hybrid piston for 4 stroke single cylinder 10 HP (7.35 kW) diesel engine. Int J Eng Trends Technol 16:258–262

10. Vaishali R, Khamankar SD (2015) Stress analysis of piston using pressure load and thermal load. Int J Mech Eng (IIJME) 3:1–8
11. Cerit M, Ayhan V, Parlak A, Yasar, Thermal H (2011) Analysis of a partially ceramic coated piston: Effect on cold start HC emission in a spark ignition engine. Appl Therm Eng 31:336–341
12. Pandey H, Chandrakar A, Bhagwat PM (2014) Thermal stress analysis of a speculative IC engine piston using CAE tools. Int J Eng Res Appl 4:60–64
13. Cerit M (2011) Thermo mechanical analysis of a partially ceramic coated piston used in an SI engine. Surf Coat Technol 205:3499–3505
14. Kumar PM, Vijayan V, Suresh Kumar B, Vivek CM, Dinesh S (2018) Computational analysis and optimization of spiral plate heat exchanger. J Appl Fluid Mech 11:121–128
15. Avudaiappan T, Vijayan V, Sundara Pandiyan S, Saravanan M, Dinesh S (2018) Potential flow simulation through lagrangian interpolation meshless method coding. J Appl Fluid Mech 11:129–134
16. Godwin Antony A, Aravind S, Dinesh S, Vijayan V (2017) Analysis and optimization of performance parameters. In Computerized I.C engine using diesel blended with linseed oil & Leishman's solution. Mech Mech Eng 21(2):193–205

Experimental Investigation on the Performance of Tyre Pyrolysis Oil Blended with Palm Stearin Methyl Ester as a Fuel for Single-Cylinder C.I Engine

P. Nagaraju, M. Ravichandra, M L R Chaitanya Lahari, and P H V Sesha Talpa Sai

Abstract Shortage of conventional power sources, ordinary energizes and the unrefined petroleum extracts increases the cost of processed fuel. It is necessary to find out fuel hotspots for viable use in automobile and industrial applications. Research in this area of using oil extracted using pyrolysis process from the waste tyre and mixed with biodiesel is attracting attention. In this survey, a single-cylinder four-stroke conventional diesel engine with waste tyre pyrolysis oil and biodiesel mixtures was performed. Testing on single-barrel four-stroke diesel motor by using biodiesel synthesized with palm stearin oil from waste tyre pyrolysis oil is taken up in this work. Analysis shows the rise in brake thermal efficiency (BTE) to the mix level. It is also noted that pollution such as CO, HC and NOx is limited compared to traditional diesel fuel.

Keywords Biodiesel · Tyre pyrolysis · Emission · Brake thermal efficiency

1 Introduction

Coal, petroleum products, natural gas and other fossil fuels are fast depleting due to increase in vehicular population and industrial usage. Fuels and energy produced using nuclear isotopes such as uranium, thorium and artificial radio-isotopes are good alternatives to fossil fuels but have some limitations. Energy can be produced using methanol, ethanol, butanol and other bio-alcohols and can be stored in electrical form in batteries and fuel cells. Another source of energy production is by hydrogen,

P. Nagaraju (✉) · M. Ravichandra
Department of Mechanical Engineering, Santhiram Engineering College, Nandyal, India
e-mail: nagaraju.pagidyala@gmail.com

M. L. R. C. Lahari
REVA University, Bangalore, India

P. H. V. S. T. Sai
Department of Mechanical Engineering, MRCET, Hyderabad, India

A. Arockiarajan et al. (eds.), *Advances in Industrial Automation and Smart Manufacturing*, Lecture Notes in Mechanical Engineering,
https://doi.org/10.1007/978-981-15-4739-3_82

non-fossil natural gas, biomass and non-fossil methane. Biodiesel is considered to be less harmful to the environment as it is biodegradable.

In India, every year million tyres are being thrown into garbage as scrap after utilizing in automobiles. Direct burning of these scrap tyres emits a lot of dangerous and dangerous gas like CO_2, CO, smoke and NO_x, etc. These waste tyres can be utilized to produce pyrolysis oil and this can be blended to produce biodiesel. Fuel produced by this method can be characterized by their physical and chemical parameters. Chemical and thermophysical properties of the blended diesel are to be evaluated and can be compared with that of raw oil after refining and distillation process. This blended diesel is considered to be useful for better engine performance and increased efficiency. Experimental investigations are needed to make this process still more viable and environmental friendly, which can bring significant economical improvement [1–3].

2 Literature Survey

Ali et al. [4] reviewed the process of pyrolosis and extracted alternative diesel fuels from vegetable oils and their blends in diesel and process of preparation, i.e. transesterification, dilution and micro-emulsion are reviewed. Murugan et al. [5] disclosed emission features and efficiency of a direct injection and single-cylinder diesel engine using 50, 30 and 10% of pyrolysis oil in diesel blends. It was noted that the engine's thermal brake effectiveness improved with enhanced concentration of mixture. Emission gases and smoke content were discovered to be greater at greater loads due to longer ignition delay and elevated aromatic content. Aydin et al. [6] recorded feasibility process for the production of oil from waste tyres using pyrolysis and diesel fuel by multiple desulphurization techniques. Sulphur content was reduced by using CaO, Ca (OH) and NaOH catalysts. 34.25% lower sulphur content was reported using 5% Ca (OH) catalyst compared to diesel. Al-Lal et al. [7] studied about pyrolysis fuels and their desulphurization. Two desulphurization methods are tried in this study and hydrogen was not used in these methods in order to obtain reduced sulphur content. The desulphurization rate reported was 64%. De Marco Rodriguez et al. [8] carried out chemical analysis and studied the behaviour of oil blends of tyre pyrolysis in diesel. Higher portion of aromatics was found in the tyre oil bends and it was observed to be a complex organic mixture. Carbon compounds of five to twenty carbons are present in the mixture. Aromatics percentage varied between 34.7 and 75.6 with operating temperatures from 300 to 700 °C and aliphatics were about 19.8–59.2%.

3 Fuel Preparation and Characteristics

3.1 Preparation of Tyre Pyrolysis Oil

Fixed-bed type reactor was used to carry out pyrolysis process. Temperature was varied between 350 and 400 °C. Condenser and a fractionating column were used in this system to condensate and fractionate the fuel oils. Nitrogen gas was used in an inert environment. Waste tyre was cleaned and cut into small pieces after removing the unwanted iron wires. Figure 1 shows the flow chart for the development of pyrolysis oil

Pyrolytic oil extracted from waste tyre using pyrolysis process was taken and hydro-sulphuric acid of 8% by weight is added. The mixture was stirred using electrical stirrer for 3–4 h. Temperature was preserved at 50 °C. The combination has been permitted to settle for 40 h. By this time, the mixture was formed in two layers. Clear and higher viscous oil was formed on the top and acidic sludge with non-viscous was formed at the bottom. The viscous oil was then activated by using bentonite–calcium oxide (CaO) treatment.

Pyrolytic oil treated with activated bentonite is taken in the ratio of 10–1. Pyrolytic oil and CaO are taken in the ration of 20–1. These are added and mixed in an electrical stirrer for 4 h maintaining 70 °C. The blend can be settled for 24 h and then cleaned utilizing sample fabric. Healed pyrolytic oil was obtained and was distilled using vacuum distillation. Table 1 illustrates the proportion of oil acquired from TPO.

Table 1 shows that the TPO percentage obtained from raw pyrolytic oil after removing sludge initially is 66.9. After vacuum distillation, the percentage is 22.79. The raw TPO sludge carries almost 77.21% of oil.

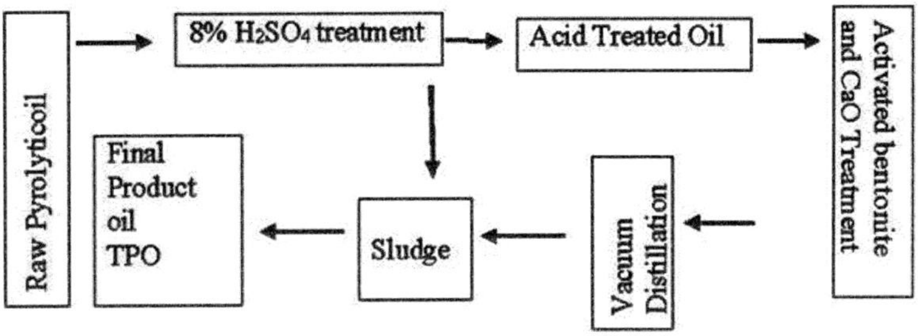

Fig. 1 Preparation of tyre pyrolysis oil—flow chart

Table 1 Percentage of oil obtained from tyre pyrolysis oil

Raw TPO	After removing sludge	Percentage oil	After vacuum distillation	Final percentage of TPO
2150	1440	66.97	490	22.79

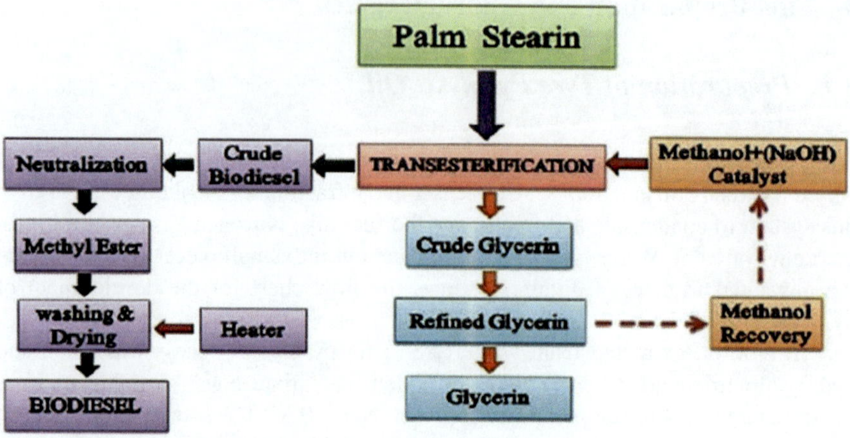

Fig. 2 Flow chart of biodiesel production from palm stearin

3.2 *Preparation of Palm Stearin Methyl Ester*

Methyl esters obtained from vegetable oils with diesel mixtures are proven to be effective appropriate fuels. Ethyl and isopropyl acids of shallow or uncooked palm oil and palm stearin were synthesized utilizing synthetic transesterification responses. The fuel properties were evaluated using different instruments. Alkyl esters had higher viscosities in the range of $4.4 \times 10^{-6} m^2/s$–5.2×10^{-6} m^2/s. Petroleum diesel had viscosity value of 4.0×10^{-6} m^2/s. But the SO_2 emission from alkyl esters was low compared to diesel. Figure 2 shows the flow chart of biodiesel production from palm stearin.

3.3 *Properties of TPO*

(See Table 2).

Table 2 Properties of TPO and diesel fuel

S. No.	Properties	TPO	Diesel
1	Density at 25 °C, Kg/m^3	845.6	820–860
2	Absolute viscosity at 25 °C, cp	1.51	1.91–2.90
3	Pour point, °C	−6	−42–30

Fig. 3 Experimental set-up

4 Experimentation

A vertical air-cooled 1C4S diesel engine is used for experimentation. Compression ratio of 17.5:1 was maintained and 4.4 kW power developed at 1500 rpm. The engine was combined with electric current dynamometer. Figure 3 shows the experimental set-up. Five levels of load were applied, i.e. at 0, 25, 50, 75 and 100%. At all the five load conditions, speed, airflow rate, fuel flow rate, exhaust gas temperature, combustion characteristics and exhaust emissions of HC, CO and smoke were measured. Horiba MEXA-584L model exhaust gas analyser with microprocessor control was used to measure emission levels of CO and HC. Blend percentages of 20, 40 and 50 are experimented in this work.

5 Results and Discussion

Combination of CO and NOx of pedal capacity and difference of smoke density, BTE and BSFC with BP are evaluated.

5.1 Carbon Monoxide Versus BP

The variation of CO emission with BP is revealed in Fig. 4. This can be seen in the CO pollution decreases with enhance in load. CO emission values obtained are

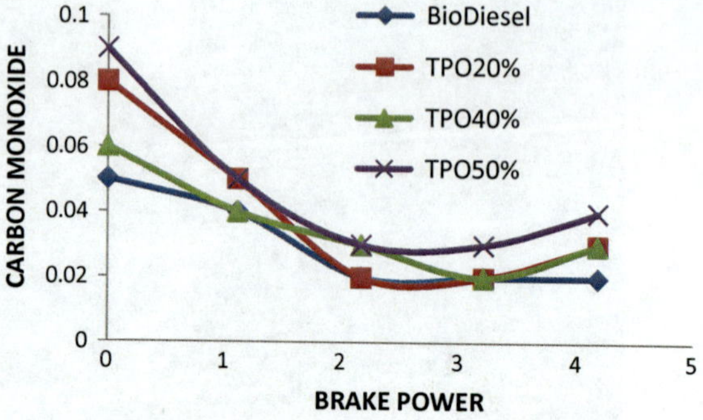

Fig. 4 Variation of carbon monoxide

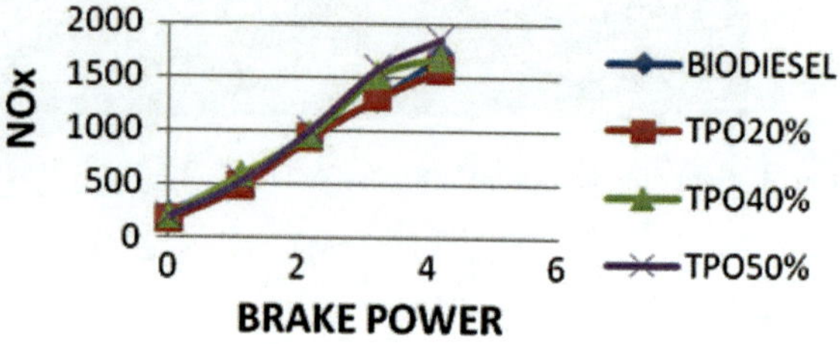

Fig. 5 Variation of No_x with BP

0.03, 0.03, 0.04 and 0.02 at full load conditions for fuels of diesel, TPO20, TPO40, TPO50 and pure biodiesel, respectively. At full load, CO emission using tyre oil blend TPO20 is decreased compared to diesel fuel.

5.2 *NOx Versus BP*

NOx emission values variation with brake power is exhibited in Fig. 5. NOx pollution increases with augment in load. NOx emission values at full freight are 1568, 1692, 1860 and 1701 ppm for fuels of diesel, TPO20, TPO40, TPO50 and pure biodiesel, respectively. The NOx emission, at full load, using tyre oil blend TPO20 is increased compared to the other blends.

5.3 *Smoke Density Versus BP*

It shows smoke density variety with brake strength. As load increases, smoke density reduces in relation to diesel. The emissions of smoke at complete load are 26, 45, 48 and 31 mg/m^3 for diesel, TPO20, TPO40, TPO50 and pure biodiesel fuels, respectively. The smoke density emission using tyre oil blend TPO20 is decreased when compared to the other blends at full load conditions (Fig. 6).

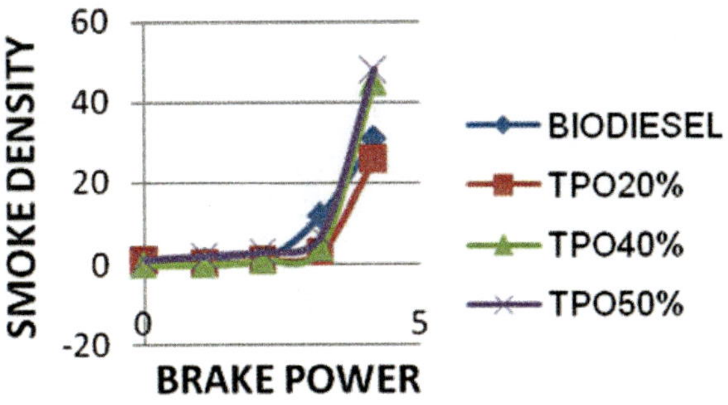

Fig. 6 Variation of smoke density with BP

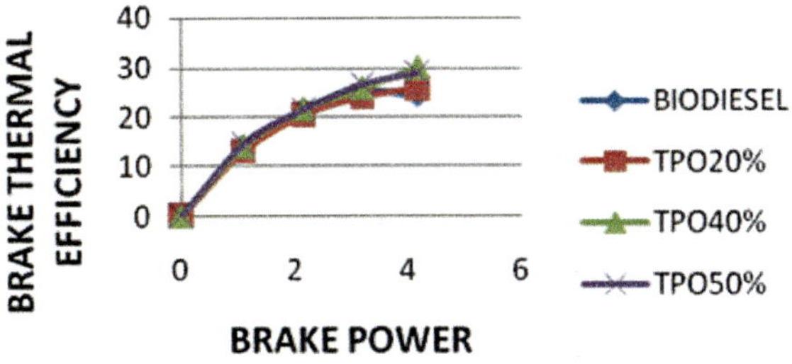

Fig. 7 Difference of brake thermal efficiency with BP

5.4 *Brake Thermal Efficiency Versus BP*

Brake thermal efficiency variation with brake power is shown in Fig. 7. Brake thermal efficiencies obtained at full load are 33.69, 28.1, 26.73, 26.58 and 26.56% for fuels of diesel, TPO20, TPO40, TPO50 and pure biodiesel, respectively. The brake thermal efficiency using tyre oil blend TPO20 is decreased compared to the diesel at full load conditions.

5.5 *BSFC Versus BP*

Figure 8 displays variations in specific fuel consumption with brake energy. The emissions are decreasing as the load rises. The SFC levels at steaming pile are 0.33, 0.53, 0.30, 0.31, 0.30, 0.31 and 0.31 kg/kw-hr for diesel, TPO20, TPO40, TPO50 and pure biofuel successively. At max power environments, the SFC using tyre oil blend TPO20 is decreased compared to the diesel.

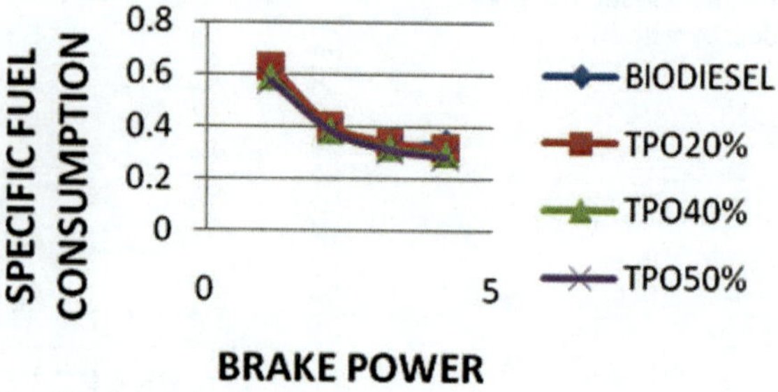

Fig. 8 Change of specific fuel consumption with BP

6 Conclusion

Experimentation is conducted to examine and evaluate the effect of palm stearin methyl ester (PSME) fuel injection pressure, exhaust gas quality and stricter emission levels that used a single-cylinder constant CI engine compressor ratio. The by-product of palm derived oils (palm stearin) could be useful alternate for producing biodiesel with less harm to environment. PSME of about 98.10% yield was obtained from refined, bleached and deodourized (RBD) palm stearin. Brake thermal efficiency and rise in specific fuel efficiency are decreased across the engine installation using PSME relative to diesel fuels. Decline in CO, HC emissions and smoke density is noted for PSME throughout engine operation relative to baseline operation. It is observed that better performance and emission characteristics of the engine can be obtained with TPO addition to PSME. It can be shown that direct injection pressure plays a crucial role in emission reduction and improving engine performance PSME-TPO fuel is best suited for CI engine.

References

1. Reisman JI (1997) Air emissions from scrap tyre combustion. Tech Rep EPA-600/R-97-115
2. Williams PT, Besler S (1995) Pyrolysis-thermogravimetric analysis of tyres and tyre components. Fuel 74(9):1277–1283
3. Mastral AM, Murillo R, Callen MS, Garcia T (2000) Optimisation of scrap automotive tyres recycling into valuable liquid fuels. Resour Conserv Rcycling 29(4):263–272
4. Ali Y (1994) Alternative diesel fuels from vegetable oils. Bioresour Technol 50:153–163
5. Murgan S, Ramaswamy M, Nagaranja C (2008) The use of tyre pyrolysis oil in dieselengines. Waste Manag 28(12):2743–2749
6. Aydin H, Ilkilic C (2012) Optimization of fuel production from waste vehicle tyres by pyrolysis and resembling to diesel fuel by various desulfurization methods. Fuel 102:605–612
7. Al-Lal AM, Bolonio D, Llamas A, Lapuerta M, Canoira L (2015) Desulfurization of pyrolysis fuels obtained from water: lube oils, tyres and plastics. Fuel 150:208–216
8. de Marco Rodriguez I, Laresgoiti MF, Cabrero MA, Torres A, Chomon MJ, Caballero B (2001) Pyrolysis of scrap tyres. Fuel Process Technol 72(1):9–22

Experimental Investigation on Mixing of Waste Plastic and Tyre Pyrolysis Oil Blends with Diesel in a Single-Cylinder Four-Stroke Diesel Engine

M. Mohammed Riyaz and M. P. Rangaiah

Abstract Energy consumption in the form of fossil fuels has increased continuously. Because of its lightweight, simple carrying and low price, the modern world utilizes many products with waste plastic and waste tyre as a main source. As a result of this, there is an issue of dumping plastic and tyres into areas of land due to the drastic rise in these reserves each year, municipal waste posses many environmental problems. The latest trend toward converting waste into quality fuels is waste to energy. This present article introduces the use of waste plastic pyrolysis oil and waste tyre pyrolysis oil as an alternative fuel for testing the four-stroke single-cylinder VCR water-cooled diesel engine. Performance and emissions characteristics for mixing waste plastic and waste tyre pyrolysis oil together with diesel mixtures with different blending ratios such as D100, B5-D95, B10-D90, B15-D85, B20-D80, B25-D75, and B30-D70 at a speed of 1500 rpm with loads of 2, 4, 6, 8, and 10 kg at rated brake power are investigated. Compared to diesel and all other blends, the B20-D80 blend found decent effects. For B20-D80 blend, brake thermal efficiency has increased compared to pure diesel, and brake-specific fuel consumption is reduced. With enhanced blend ratios, CO emissions were increased, NO_x reduced with increase in blend ratios, and UHC emissions and smoke are decreased than diesel fuel.

Keywords Waste plastic and tyre pyrolysis oil · Alternative fuels · CI engine · Pyrolysis · Performance and emissions

M. Mohammed Riyaz (✉)
Advanced Internal Combustion Engines, Department of Mechanical Engineering, JNTUA College of Engineering, Ananthapuramu, India
e-mail: mohammedriyaz7050@gmail.com

M. P. Rangaiah
Department of Mechanical Engineering, JNTUA College of Engineering, Ananthapuramu, India
e-mail: mpranga9@gmail.com

A. Arockiarajan et al. (eds.), *Advances in Industrial Automation and Smart Manufacturing*, Lecture Notes in Mechanical Engineering,
https://doi.org/10.1007/978-981-15-4739-3_83

Nomenclature

WPPO	Waste plastic pyrolysis oil
WTPO	Waste tyre pyrolysis oil
CO	Carbon monoxide
UHC	Unburnt hydrocarbons
CO_2	Carbon dioxide
NO_x	Oxides of nitrogen
DI	Direct ignition
CI	Compression ignition
PPM	Parts per million

1 Introduction

Energy, environment, and economy are the main contributors to the country's development. Daily, fossil fuels are depleting, resulting in rising prices of petroleum products around the globe. This shows the need for a cheaper alternative fuel to meet the transportation needs and electric power generation in the rural area. On the other hand, waste product generated with growth of population and its management plays essential role to maintain economic and healthy country. Among many types of solid waste, plastic waste is a major problem as its non-biodegradable in nature. Due to their lightweight, durability, non-perishability, quicker production rate, and flexibility in design and commodity supply, plastic use has become a vital component of today's world [1, 2]. Venkatesan [1] reported that, based on the experimental study into the combustion and performance features of a four-stroke compression ignition engine driven with diesel and waste plastic oil blends, the BTE showed small differences at low load, but noticeable changes were observed at complete load which might be due to increased calorific value of WPPO blends, and mechanical efficiency was observed to have minimal impact on all fuel mixtures at all loads, but BSFC was reported to be higher for all non-load fuel mixtures, and fuel consumption declined with load increases. Varma [2] researched the experimental inquiry on the addition of waste plastic oil resulting in increased brake thermal efficiency (BTE) and decreased BSFC. Muhammed Qasim and Ansari [3] conducted experimental study on a diesel engine operated with fuel blends extracted from a combination of Pakistani waste tyre oil and soybean oil biodiesel revealed that all the fuel blends tested showed reduced BTE, smoke emissions, but increased NO_x emissions. Ventaka subbaiah and Durga Prasad [4] have noticed that the greater deposition of plastic waste in the earth's pyrolysis cycle leads to a better transformation of plastic to effective hydrocarbon fuels such as petrol and diesel. The cost of manufacturing per liter of fuel reduces when compared to crude oil refining when performing this method effectively. This method showed an alternative way of generating fuel that could satisfy the worldwide demand for fuel due to enhanced fuel demand. The plastic oil blend [B20] generates the same power as diesel. Plastic oil indicates slightly greater emission values than diesel when absorbing the emission trait graphs. It is feasible to distinguish better fuel quality from the alteration in manufacturing such as pressure heating of the acquired crude type of fuel. If manufacturing is carried out effectively with defined parameters, quality fuel can be produced, and this will take the place of standard diesel. Mani [5] conducted the test on performance, emission, and combustion characteristics of a DI diesel engine using waste plastic oil and reviles that engine fueled with waste plastic oil exhibits greater heat efficiency up to 75% of the rated energy and smoke emission decreases by about 40%, but CO and NO_x emissions increase when the engine is operating at complete load circumstances. percentnoted that the brake thermal efficiency of the engine was 3–4% smaller for PPO90 and 2–3% lower for PPO70 compared to diesel, and all measured emissions (NO_x, UHC, CO, and CO_2) were greater for PPO70 and PPO90 compared to all measured emissions (PPO70 and PPO90). Subramanyam [6] researched the performance investigation of diesel engine using waste plastic pyrolysis oil and diesel blends that waste to energy is the current trend that will concentrate on different research fields. Waste plastic oil blend's (WPO20) brake thermal efficiency BTE of 26.24% at 3.67 kw bp was higher than diesel (24.85%). Specific fuel consumption was slightly greater than diesel fuel consumption. Plastic oil's exhaust gas temperature with diesel blends is growing owing to longer plastic oil ignition delay. Carbon monoxide emissions increase with increased brake power under load circumstances at 3.67 kw. The CO emissions for the blends like WPO15, WPO20, WPO30, WPO40, and WPO50 are 3.69, 3.63, 3.72, 3.96, 4.18, and 4.18%. Carbon dioxide emissions from plastic oil blends with diesel rise significantly at WPO20 brake power with 2.94 kw lower than diesel emissions owing to full fuel combustion. Hydrocarbon emission plastic oil at all brake powers blends more than diesel emission. The emissions of 3.67 kw bp hc were 1606, 1610, 1609, 1614, 1616, and 1618 ppm for diesel, WPO15, WPO20, WPO30, WPO40, and WPO50. Paramar [7] operates on emission analysis of CI engine using tyre pyrolysis oil and diesel blend suggesting that CO emissions are increased with TPO mix in exhaust emissions. The D75T25 mix provides the required impact. The emission of HC, CO, and CO_2 decreases with the TPO blend and the optimum values of the D95T5 blend. With the rise in separate TPO blend D75T25, the emission of NO_x and O_2 provides the desired impact than other blends and diesel. Senthil Kumar [9] investigated a DI naturally aspirated engine using tire pyrolysis oil–diesel blends as a biodiesel and realized that pyrolysis blends' brake thermal efficiencies are lesser than diesel. This may be due to reduced heating value and reduced pyrolysis combustion. Due to reduced viscosity, the rise in thermal efficiency for 50% compared to 75% can be ascribed to improved fuel atomization. Reduction in thermal efficiency by approximately 14% is greater than that of the other pyrolysis oil blends which is eventually closer to blending the distilled method to fix the distinct aims, and NO_x values for 50 and 75% of the pyrolysis oil are lower than diesel, and CO and HC emissions for diesel can be high owing to the existence of unsaturated oil. Smoke is greater than diesel for tyre pyrolysis blends. Mohammad Younus [10] researched the performance and emission attributes of diesel engine fueled with tyre pyrolysis oil and diesel blends with additives and reported that BTE is improved for all blends compared to standard diesel fuel, and BSFC is lowered for all blends compared to diesel fuel. CO, CO_2, and HC emissions are considerably reduced with the blends, and lastly, the blend TPE 20 demonstrates better efficiency compared to other blends (TP10, TP20, TP30, TPE10, and TPE30) and diesel. Pawan conducted his research work on the performance and emission characteristics of tyre pyrolysis oil (TPO) blend with diesel for weird composition and revealed that the engine's brake thermal efficiency decreases with increased concentration of TPO blend than diesel. BTE for the operation of diesel is at varying loads. In case of TPO 10% blending, it is 22.47%. In the event of a 20% mix, it is 15.83%. It is 18.55 for 30% mix, and the TPO–DF mix demonstrates greater BSFC value than diesel because of reduced TPO–DF blend calorific value. As the TPO–DF's BSFC load increases, it reduces. TPO–DF's mechanical efficiency improves as the load rises. The highest efficiency achieved is 21.64% for 10%. In the event of a 20% blend, it is 53.2%. And it is 49.7% in 30%. At greater loads, the CO emission is greater and grows with the blend proportion (TPO–diesel). Similarly, the HC emission is large for greater loads. The emission concentration is high in all load circumstances for the CO_2 emission mix at 30%.

2 Production Process of WPPO and WTPO

2.1 Pyrolysis

In the lack of oxygen, pyrolysis is usually described as the monitored heating method of a material. The macro-molecular structures of polymers are split into lower molecules and sometimes monomer units in plastics and tyre pyrolysis. Further degradation of these subsequent molecules is dependent on a number of different conditions including (and not restricted to) temperature, residence time, catalyst presence, and other process circumstances. In a cylindrical chamber or reactor, plastic waste is continually handled. At 300–500 °C, the plastic and tyre oil is pyrolized.

2.1.1 Condenser

It cools the reactor's entire heated vapor. The cold water to pass through its exterior region enters and exits the outlet through the inlet. This is used for vapor cooling. The gaseous hydrocarbons are condensed to about 30–35 °C at a temperature of around 35 °C.

2.1.2 Reactor

It is a 1000 mm long cylindrical stainless steel container with an inner diameter of 300 mm, an outer diameter of 320 mm secured on one end and an outlet tube on the other. Inside, the reactor is put into the entire cylindrical container. External heating is carried out from below the container and inside the reactor using the raw material such as coal, wood, and coke. The reactor for lagging is produced of stainless steel, mild steel, and clay. The reactor is heated at about 1450 °C and higher temperatures.

2.2 *Process Description*

The method of thermal cracking is used to convert waste plastic and tyre into liquid fuel. For this specific experiment, only one sort of waste plastic is chosen, i.e., low-density polyethylene. Waste plastic is collected and cleaned using liquid soap and water as strong smooth forms. Washed plastics and tyre are cut to a size of 3–5 cm to fit conservatively into the reactor. The experiment is performed with no vacuum applied during the thermal cracking process under a closed system. In a batch process system, we used polyethylene plastics of low density due to the relatively low conversion temperatures for these plastics. At the beginning of melting, the waste plastics and tyre, heat is applied from 100 °C, the melted waste plastic and tyre transform into liquid slurry when temperature increases gradually. When the temperature rises to 570 °C, liquid slurry becomes vapor, and then, the vapor passes through a condenser unit. We obtain liquid fuel at the end of this phase. Between 100 and 250 °C, about 20–30% of the fuel is gathered, and then, the next 40% is obtained when raised to 325 °C, and the output is lastly finished when retained at 400 °C. Plastic and tyre pieces are not immediately broken down during the thermal cracking phase because plastics have distinct chain hydrocarbons. Only the straight chain hydrocarbon is applied in the starting mode heat. The temperature profile that slowly breaks this plastic and tyre carbon bond is further improved. The long chains' molecules are dissolved step by step with the temperature rise.

3 Engine Setup and Procedure

For experimental testing, Kirloskar single-cylinder water-cooled variable compression ratio diesel engine is used. To apply loads to the engine, the Eddy current dynamometer is connected to the flywheel. To inject the petrol, a 200 bar injection pressure is retained (Fig. 1).

The pressure of the cylinder is evaluated by the piezo-sensor installed on the head of the engine cylinder and the angle of the crank encoded on the fly wheel. The conventional engine is supplied with 0° to 25° BDC injection point variation. The HC, CO, CO_2, UBHC and NO_x emissions are evaluated using the fire gas analyzer

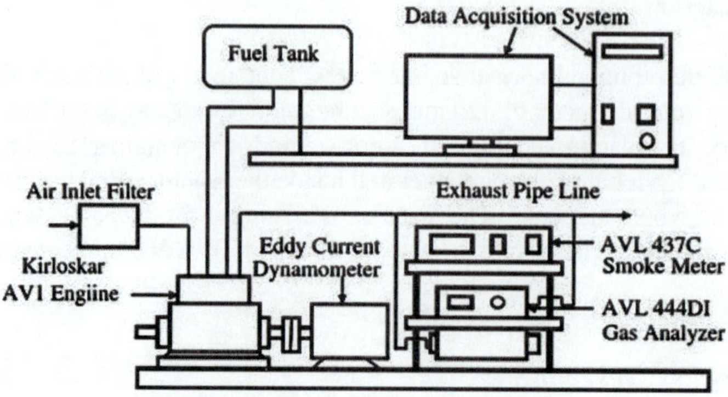

Fig. 1 Experimental setup of test engine

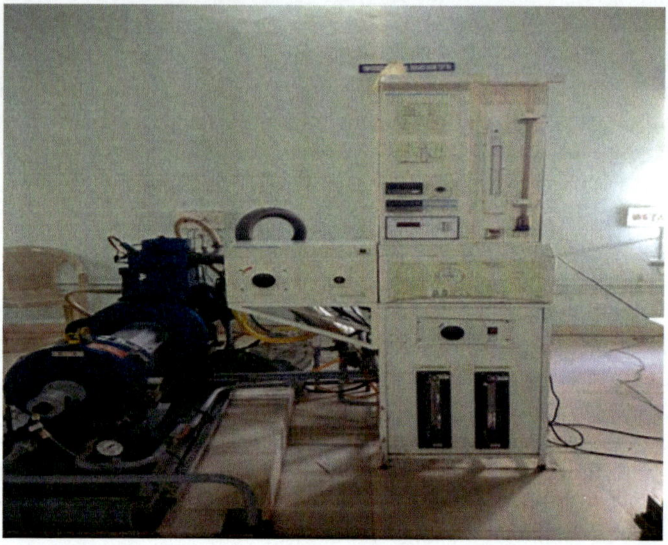

Fig. 2 Experimental setup of CI engine

AVL-DIGAS 444. AVL smoke meter is used to measure the opacity of the smoke (Fig. 2; Table 1).

3.1 Test Procedure

This research focuses on the use of WPPO and WTPO as fuel in the conventional diesel direct injection (DI) engine. Some of these two biodiesel characteristics are closer to diesel fuel. As a result, WPPO and WTPO biodiesel are selected as an

Table 1 Test engine specifications

Engine make	VCR engine test setup single-cylinder, four-stroke, diesel with EGR (computerized) engine
Type	Kirloskar type single-cylinder, four-stroke diesel, water-cooled engine.
Bore and stroke	87.5 and 110 mm
Compression ratio	12–18
Fuel	Diesel and petrol
Rated brake power and speed	3.5 KW and 1500 rpm
Dynamometer	Eddy current, water cooled with loading unit
Ignition system	Compression ignition
Connecting rod length	234 mm
Software	"IC Engine software"

Table 2 Blending proportions table

Blends	Blends proportions
B5	2.5% WPPO + 2.5% WTPO + 95% Diesel
B10	5% WPPO + 5% WTPO + 90% Diesel
B15	7.5% WPPO + 7.5% WTPO + 85% Diesel
B20	10% WPPO + 10% WTPO + 80% Diesel
B25	12.5% WPPO + 12.5% WTPO + 75% Diesel
B30	15% WPPO + 15% WTPO + 70% Diesel

alternative fuel, and the highest possible diesel substitution is explored in a DI diesel engine. However, the combination of these two biodiesel mixed with diesel has a reduced amount of cetane number. We used these WPPO and WTPO oil by reducing its viscosity. Also engine will be made to run with blends of diesel and WPPO and WTPO biodiesel consisting of 5, 10, 15, 20, 25, and 30% into the diesel fuel (Table 2).

4 Properties of Diesel, WPPO, WTPO, and Its Blends

See Tables 3 and 4.

Table 3 Physical properties of diesel, waste plastic pyrolysis oil, and waste tyre pyrolysis oil

S. No.	Fuel	Density (kg/m^3) at 30 °C	Kinematic viscosity (cst) at 30 °C	Specific gravity	Flash point °C	Fire point °C	Calorific value kj/kg	Cetane number
1	Diesel	840	2.02	0.84	43	45	42,000	55
2	WPPO	830.55	2.04	0.830	37	40	44,340	52
3	WTPO	920.35	2.23	0.92	38	41	42,600	58

Table 4 Physical properties of biodiesel mixtures

S. No.	Fuel	Density (kg/m^3) at 30 °C	Kinematic viscosity (cst) at 30 °C	Specific gravity	Flash point °C	Fire point °C	Calorific value kj/kg
1	B5-D95	843	1.98	0.843	35	37	42,750
2	B10-D90	841	1.332	0.841	37	40	43,100
3	B15-D85	841.5	1.296	0.8415	37	40	43,480
4	B20-D80	841	1.248	0.841	37	40	43,500
5	B25-D75	841	1.480	0.841	37	41	43,930
6	B30-D70	841.1	1.4006	0.8411	37	42	44,300

5 Results and Discussion

5.1 Performance Analysis

5.1.1 Brake-Specific Fuel Consumption (BSFC)

The differences in brake-specific fuel consumption with regard to the brake power of distinct blends are shown in Fig. 3. The plot reveals that the BSFC is diminishing as the load rises. The BSFC obtained are 0.39, 0.366, 0.361, 0.36, 0.34, 0.36, and 0.38 kg/kw h for diesel fuels, B5-D95, B10-D90, B15-D85, B20-D80, B25-D75, and B30-D70, respectively, under maximum load conditions. BSFC's percentage variation of all diesel blends at different load circumstances is declining at 0.938, 0.925, 0.923, 0.871, 0.924, and 0.974%. Increased BSFC was noted at low load, which is due to greater aromatic compounds with enhanced viscosity and volatility. This results in inadequate air–fuel mixture, particularly during lean operations. For the B20-D80 mix, which is 0.34 kg/kw h, the minimum BSFC is acquired, i.e., 0.871% reduces with 10 kg than diesel fuel. The BSFC for B20-D80 is achieved lower compared to all oil blends, but for B5-D95 and B30-D70, the BSFC is closer to diesel.

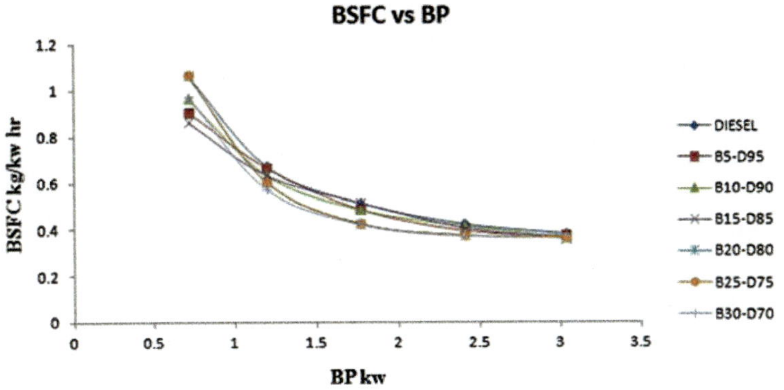

Fig. 3 Comparison of BSFC with BP

5.1.2 Brake Thermal Efficiency (BTE)

The brake thermal efficiency variations regarding brake power is shown in Fig. 4. The brake thermal efficiency generally improves with increased load. As the load increases, the brake power increases and reduces the bsfc which improves the brake thermal efficiency. At low loads, it was hard to atomize WPPO and WTPO mixed fuel particles, leading in lower brake thermal efficiency. Higher BTE was noted for enhanced concentration of WPPO and WTPO at part and complete load conditions, which is due to better preparing of air–fuel mixture at high loads. All the fuel blends showed a rising trend in brake thermal efficiency with load. Brake thermal efficiency is 22.08, 23.45, 23.68, 23.18, 24.73, 23.78, and 22.67% for fuel blends such as diesel, B5-D95, B10-D90, B15-D85, B20-D80, B25-D75, and B30-D70, respectively, from graphs at 10 kg load conditions. The percentage variation of brake thermal efficiency of all diesel blends at different load conditions is improved by 1.062, 1.072, 1.049,

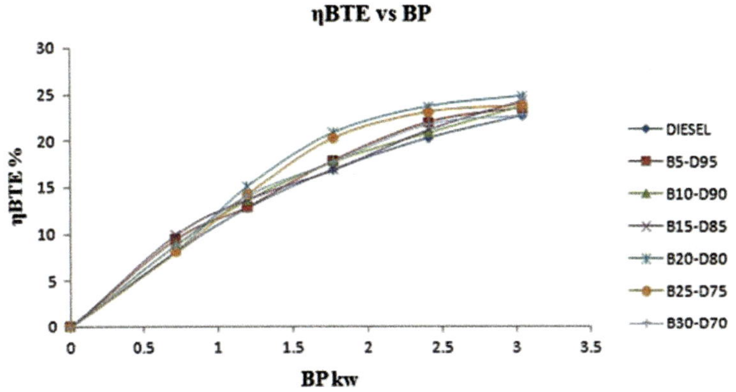

Fig. 4 Comparison of ηBT versus BP

1.12, 1.076, and 1.02%. The maximum brake thermal efficiency for the B20-D80 blend is 24.73%, i.e., 1.12% higher than diesel fuel, and blends such as B25-D75 and B30-D70 reduced as compared to B20-D80 owing to a rise in the proportion of WPPO and WTPO in diesel fuel.

5.1.3 Mechanical Efficiency (Mechη)

Mechanical efficiency is the measure of the engine's useful work with fuel and energy input. The graph 5 demonstrates mechanical efficiency variations for all fuel mixtures under different load conditions, and it is noted that with increasing loads for pure diesel to all blends, there is an increase in mechanical efficiency. At 10 kg load conditions, the mechanical efficiency values for fuel blends such as diesel, B5-D95, B10-D90, B15-D85, B20-D80, B25-D75, and B30-D70 were 52.50, 53.47, 53.52, 53.71, 54.10, 54.32, and 56.94%. The percentage variation in the mechanical efficiency of all blends at different load circumstances is improved by 1.018, 1.019, 1.023, 1.030, 1.034, and 1.08%. The mechanical efficiency is improved from the diesel fuel to the blend B30-D70 due to rise in the mixing ratios. Higher ignition delay results in increased premixed combustion and lower fuel atomization at elevated loads. The maximum mechanical efficiency is achieved by 56.94% for the blend B30-D70, i.e., 1.08% higher than diesel (Fig. 5).

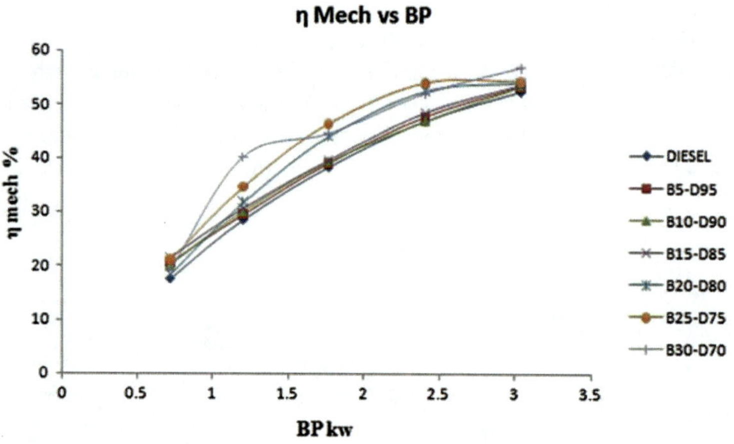

Fig. 5 Comparison of η mech with BP

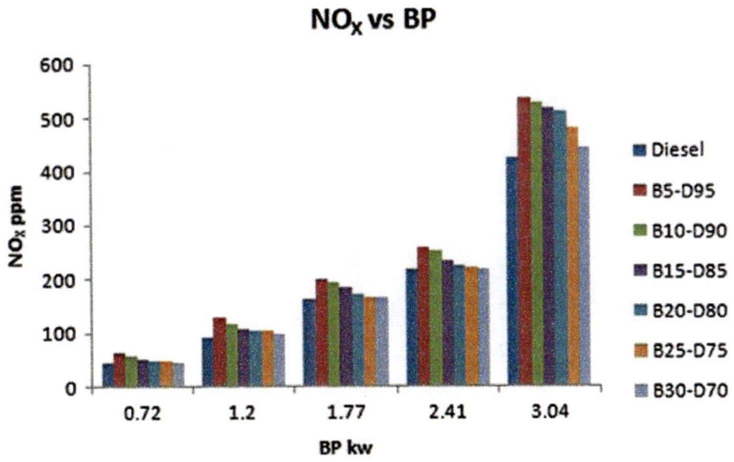

Fig. 6 Comparison of NO_x emissions versus BP

5.2 *Emission Analysis*

5.2.1 Oxides of Nitrogen Emissions (NO_x)

Figure 6 shows the variation of oxides of nitrogen with distinct blends at varying brake power. The increase in temperature causes the formation of nitrogen oxides in the exhaust. But in this case, the engine temperature is equivalent to normal diesel for B15-D85 and B20-D80, and the blend B30-D70 is nearer to diesel. This is because pyrolyzed fuel contains more oxygen content than standard diesel fuel, although the engine temperature is less. This provides excess oxygen, resulting in surplus nitrogen oxide release, and enhanced delay in the ignition of WPPO, promoting premixed combustion by providing more time for fuel to be injected for ignition, may also be another reason to increase NO_x. That is the reason we received that much NO_x emissions. It is portrayed from the graph that all blends such as B5-D95, B10-D90, B15-D85, B20-D80, B25-D75, and B30-D70 yield greater NO_x emissions. Among them, B25-D75 and B30-D70 produce NO_x emissions, i.e., 479 and 442 ppm, which is much closer to diesel fuel, and the B5-D95 blend provides maximum NO_x emissions, i.e., 534 ppm.

5.2.2 Carbon Monoxide Emissions

Carbon monoxide (CO) is a slightly less dense, colorless, odorless, and tasteless gas than air. Figure 7 compares the emissions of carbon monoxide (CO) to the brake power. In practice, when there is incomplete fuel burning, carbon monoxide can be seen in the exhaust. For the obtained pyrolysis oil, there is a presence of aromatics in both WPPO and WTPO. This element does not exhibit for complete

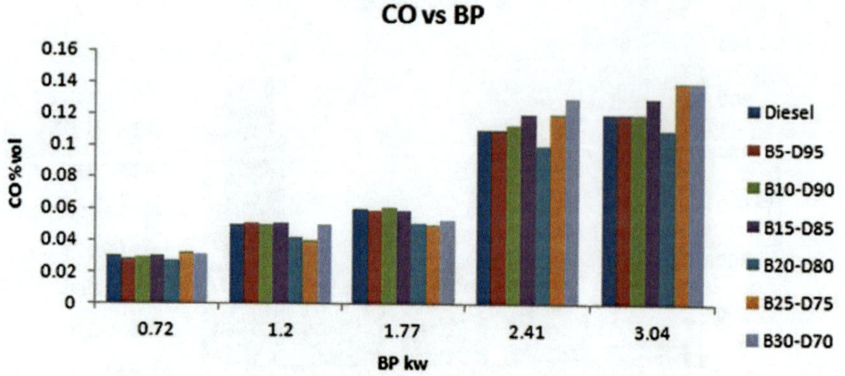

Fig. 7 Comparison of CO emissions versus BP

combustion. From the graph at different load conditions, the blends B25-D75 and B30-D70 give higher CO emissions, and B20-D80 emits lesser CO% by volume compared to all blends and diesel because of having required quantity of oxygen for complete combustion. The CO emissions obtained are 0.12%, 0.12%, 0.12%, 0.13%, 0.11%, 0.14%, and 0.14% for the blends like diesel, B5-D95, B10-D90, B15-D85, B20-D80, B25-D75, and B30-D70, respectively.

5.2.3 Carbon Dioxide Emissions

The variation of carbon dioxide with different blends with varying brake power is represented graphically in Fig. 8. Complete combustion occurs when adequate supply of oxygen makes carbon dioxide which is seen in engine exhaust. The oils we used, WPPO and WTPO, may have excess oxygen. This leaves excess oxygen that eventually releases large amounts of exhaust carbon dioxide. In reality, carbon dioxide in the diesel engine was seen as pollution accompanied lived green house gas. From the graph at different load conditions, the blend B30-D70 and diesel emit lesser co_2 emissions of about 5.84 and 6.1%, whereas the blend B20-D80 gives 6.62% of CO_2 emissions due to more amount of oxygen present in the fuel which makes to burn the fuel completely than other fuel mixtures.

5.2.4 Unburnt Hydrocarbons Emissions (UHC)

Figure 9 shows the variation of unburned hydrocarbons with distinct blends at varying brake power. It is generally possible to see hydrocarbons in the exhaust when the fuel has a greater amount of cetane number, and there is not enough time to perform combustion process. Compared to standard diesel, the cetane number of WPPO and WTPO pyrolyzed oil is lower, and hence, the level of combustion rises and

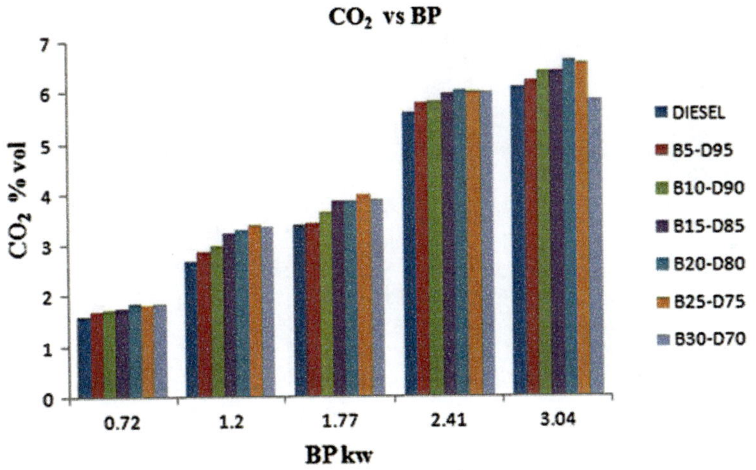

Fig. 8 Comparison of CO_2 emissions versus BP

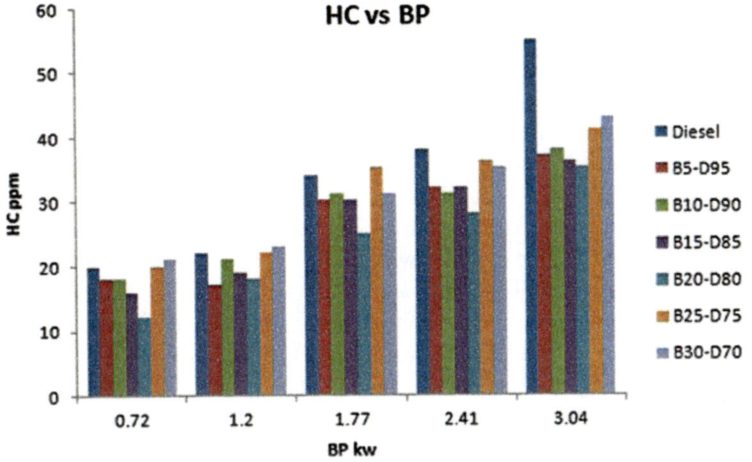

Fig. 9 Comparison of HC emissions versus BP

hydrocarbon emissions reduces. From the graph, it is noted that all blends such as B5-D95, B10-D90, B15-D85, B20-D80, B25-D75, and B30-D70 produce lower UHC emissions. Among them, B15-D85 and B20-D80 give lesser UHC emissions, i.e., 36 and 35 ppm.

5.2.5 Smoke Opacity

Opacity of smoke rises with load increases. Figure 10 demonstrates a graph plotted

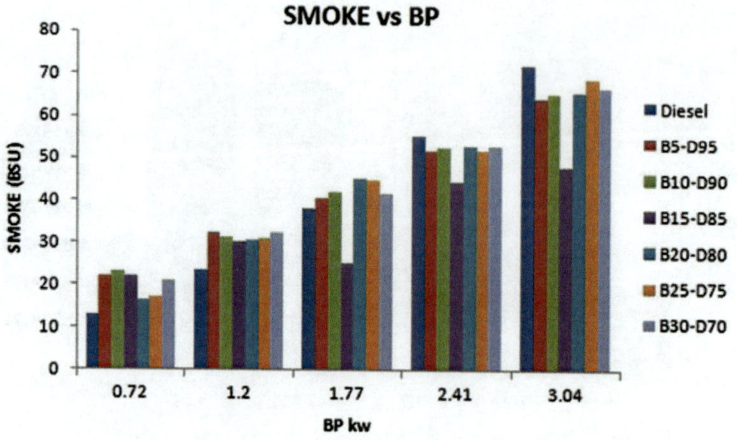

Fig. 10 Comparison of Smoke opacity versus BP

between the opacity of smoke and the engine brake power fueled with mixtures, i.e., diesel, B5-D95, B10-D90, B15-D85, B20-D80, B25-D75, and B30-D70. In general, smoke opacity of diesel fuel is greater than other fuels. Due to more oxygen content, smoke opacity reduces with a rise in the percentage of the blends. The minimum opacity of smoke is for the blend B15-D85 (47.8BSU) while considering the fuel and the maximum in the blend B25-D75 (68.6BSU). Opacity of smoke in diesel fuel is noted at 71.8 BSU. It can be observed that the amount of smoke is lesser than diesel for all blends. The reason for this decreased smoke is the availability of oxygen within the engine cylinder with premixed and homogeneous charge well before the beginning of combustion. The few reasons for decreased smoke are higher combustion temperature, prolonged combustion duration, and rapid flame propagation. However, there was a noticeable white smoke emission at higher load owing to lack of adequate oxygen and abnormal combustion.

6 Conclusions

The following conclusions were drawn based on the experimental investigation on performance and emission characteristics of a four-stroke single-cylinder diesel engine fueled with straight diesel and blends of WPPO and WTPO with diesel in a blending ratios of D100, B5-D95, B10-D90, B15-D85, B20-D80, B25-D75, and B30-D70, respectively.

1. The physio-chemical characteristics of WPPO and WTPO such as kinematic viscosity, density, dynamic viscosity, flash point, fire point, and calorific value and its diesel blends were found.

2. The engine's brake thermal efficiency improves as the blending proportions increase from B5-D95 to B20-D80 and subsequently reduces for B25-D75 to B30-D70. The maximum brake thermal efficiency of the B20-D80 blend is 24.73% and the minimum for diesel (D100) approximately 22.08%.
3. Brake-specific fuel consumption reduces with load which increases from B5-D95 to B20-D80 in all blending ratios and increased later for B25-D75 to B30-D70. At the B20-D80 blend, the minimum BSFC is 0.34 kg/kw h, and a maximum value of 0.39 kg/kw h is obtained.
4. The mechanical efficiency increased with increase in fuel blends from diesel, B5-D95 to B30-D70. The maximum mechanical efficiency is obtained at blend B30-D70 is 56.94%, and minimum values are obtained at diesel fuel of about 52.50%.
5. The CO emission decreases with increase in the blending proportions, whereas at higher loads, the CO emission is increased for the blends B25-D75 and B30-D70, i.e., 0.14 ppm. The minimum CO emissions occurred at B20-D80 blend is about 0.11 ppm.
6. Because of presence of oxygen content in both WPPO and WTPO, the CO_2 emissions are more for all blending proportions than the diesel fuel. B20-D80 blend gives 6.62% of CO_2 emissions.
7. The UHC emissions decrease with increase in the blending proportions. The blends B20-D80 and B15-D85 give lesser UHC emissions of about 35 and 36 ppm than other fuel blends.
8. The increase in temperature causes the formation of nitrogen oxides in the exhaust. But in this case, the engine temperature is more than standard diesel for B15-D85 and B20-D80, and the blend B30-D70 is nearer to diesel. Among all, the blends B25-D75 and B30-D70 yield NO_x emissions, i.e., 479 and 442 ppm, which is much nearer to diesel fuel, and the B5-D95 blend yields maximum NO_x emissions, i.e., 534 ppm.
9. In general, smoke opacity of diesel fuel is greater than other fuels. Due to more oxygen content, smoke opacity reduces with a rise in the percentage of the mixture. For the blends B15-D85 and B20-D80 of approximately 47.8 BSU and 65.4 BSU, the minimum smoke opacity is acquired while considering the fuel. The diesel fuel smoke opacity is noted at 71.8 BSU. The B20-D80 blend provides less opacity smoke than diesel.

As a result, the B20-D80 is chosen as an optimum mixture owing to better results and can be used as a replacement for diesel fuel.

References

1. Venkatesan H, Sivamani S, Bhutoria K, Vora HH (2017) Experimental study on combustion and performance characteristics in a DI CI engine fuelled with blends of waste plastic oil. Alex Eng J. 1110-0168

2. Verma A, Raghuvansi A, Quraishi MA, Tirkey JV, Verma C Engine fuel production from waste plastic pyrolysis (WPO) and performance evaluation in a CI engine with diesel blend. J Mater Environ Sci. ISSN: 2028;2508 CODEN:JMESCN
3. Qasim M, Ansari TM, Hussain M (2017) Experimental investigations on a diesel engine operated with fuel blends derived from a mixture of Pakistani waste tyre oil and waste soybean oil biodiesel. Received: 5 May 2017. Accepted: 2 October 2017# Springer, GmbH Germany
4. Venkata Subbaiah M, Sunil KuarReddy S, Durga Prasad B (2018) Performance and emission characteristics of diesel engine fuelled with waste plastic bio-oil/diesel. In: 2nd international conference on emerging trends in engineering, sciences and management-2018 (ICEESM'18)
5. Mani M, Subash C, Nagarajan G (2009) Performance, emission and combustion characteristics of a DI diesel engine using waste plastic oil. Appl Therm Eng 29:2738–2744
6. Vasukumar N, Subramanyam D, Tammineni L (2017) Performance investigation of diesel engine using waste plastic pyrolysis oil and diesel blends. IJESC 7(11)
7. Paramar MD, Patel SC, Patel TM, Patel P (2016). Emission analysis of C.I engine using tyre pyrolysis oil and diesel blend. IJRTER-2016 (2455-1457)
8. Bhatt Prathmesh M, Patel Paresh D (2012) Suitability of tyre pyrolysis oil (TPO) as an alternative fuel for internal combustion engine. IJAERS I(IV):61–65
9. Senthil Kumar R, Prabu M. (2014) Experimental investigation of a di diesel engine using tyre pyrolysis oil-diesel blends as a biodiesel 5(2). ISSN 0976-6340
10. Mohammad Younus SK, Ranjith Kumar V, Hanumantha Rao YV (2013) Performance and emissions characteristics of diesel engine fueled with tyre pyrolysis oil&diesel blends with additives 2(5):32–37. www.ijesi.org

Effect of Exhaust Gas Recirculation and Cerium Oxide on Tire Pyrolysis Oil Blends

M. Ravichandra, P. Nagaraju, P. Moulali, and K. Nagaraju

Abstract Sustainable fuels are required to use ever-increasing consumption of fossil fuels and large-scale environmental decline. Authors around the globe are working with a multitude of alternative fuels to meet future energy requirements. Biofuel is one of the most successful renewable fuels owing to emissions of CO, HC, and CO_2. In the case of biodiesel, moreover, pollution of NOx is expanded in the CI engine. NOx emissions are one of the main causes of health problems, ozone depletion, and climate change. Therefore, there is a need to decrease engine emissions by finding appropriate methods or using alternative fuels. By using CI engine, we performed experiments on tire pyrolysis oil blends B5 (5% TPO and 95% diesel fuel), B10 (10% TPO and 90% diesel fuel), and B20 (20% TPO and 80% diesel fuel) at multiple loads and found B20 to be less fuel consumption, excellent brake heat effectiveness and not up to norms. More study carried out using exhaust gas recirculation (EGR) for B20 and noted excellent effectiveness and emissions. Eventually performed by placing the element (cerium oxide) for the B20 and noted that less pollution.

Keywords Compression ignition engine · Tire pyrolysis oil · Exhaust gas recirculation · Cerium oxide

1 Introduction

Owing to its very elevated expansion rate and intrinsic lean flame, the combustion engine has the largest thermal effectiveness (engine efficiency) of any useful inner or external diesel engine, which allows surplus air to dissipate heat. Compared to two-stroke non-direct-injection gasoline engines, a tiny efficiency loss is also prevented as unburned fuel is not visible at valve mixing and therefore no fuel flows straight from intake/injection to discharge. Low-speed diesel vehicles (as used in boats and other vehicles where general engine weight is comparatively insignificant) can have

M. Ravichandra (✉) · P. Nagaraju · P. Moulali · K. Nagaraju
Department of Mechanical Engineering, Santhiram Engineering College, Nandyal, Kurnool, Andhra Pradesh 518501, India
e-mail: ravichandra.me@srecnandyal.edu.in

A. Arockiarajan et al. (eds.), *Advances in Industrial Automation and Smart Manufacturing*, Lecture Notes in Mechanical Engineering,
https://doi.org/10.1007/978-981-15-4739-3_84

a thermal efficiency exceeding 50%. Abd-Alla [1] The objective of this job is to evaluate the ability of EGR to decrease exhaust pollution, especially NO_X emissions, and to determine the scope of implementation of this method. A thorough assessment of past and current EGR outcomes will be implemented on the pollution and efficiency of diesel engines, spark ignition motors, and duel petrol motors. Gürü et al. [2] Cerium oxide being a rare-earth alloy with dual-valance state presence has outstanding catalytic activity owing to its oxygen buffering capacity, particularly in the nano-sized shape. This results in a concurrent drop and oxidation of nitrogen dioxide and hydrocarbon pollution from the diesel engine, successively, when used as an element in the fuel. The current research explores the impact of cerium oxide nanoparticles on diesel engine efficiency and emissions. Cerium oxide nanoparticles were extracted by chemical technique and analysis methods such as TEM, EDS, and XRD was used. Friedrich [3] Having completed the first IEA-report on the worldwide production of biodiesel production in December 1997 as an attempt to provide knowledge about biodiesel development at an early stage, this new report is updating the first report after years of intensive activity to get commercial biodiesel production established in the fuel markets. Graham [4] The recirculation of exhaust (EGR) is efficient in reducing nitrogen oxides (NO_x) from diesel motors because it reduces the flame intensity and the quantity of oxygen of the working fluid in the exhaust stream. Akers [5] Cloud value is evaluated using a UV–vis spectrophotometer. An embedded undergraduate laboratory exercise concerning the synthesis and description of a diesel fuel replacement, biodiesel, is provided. Biofuel is presently a common subject in the field of green chemistry as it is a non-petroleum oil that can be effectively manufactured from unused vegetable oil. It needs no engine changes as a fuel and outcomes in reduced pollution including toxins and carbon dioxide. De Menezes et al. [6] Diethylene glycol dimethyl ether (Diglyme) is used as an additive in this research to enhance oil characteristics. The physicochemical characteristics and harmful emissions of mixed gas and conventional diesel were researched. Then the exhaust pollution of the heavy-duty diesel engine was experimentally assessed for the sole combination of diesel and diglyme-diesel petrol. Jesila and Murugan [7] The pyrolysis process, pyrolysis reactors, their product yield, distinctive tire evaluation like proximate assessment, elemental analysis and pyrolysis method are explored in this study. The recycle tire pyrolysis begins at 250 °C and is finished at 550 °C. The existence of a solvent generates lighter oil with a dramatic rise in the concentration of single-ring aromatics. For the inquiry, Sharma and Murugan [8] performed five distinct blends of varying TPO, ranging from 10 to 50% at volume-based steps of 10%. Interestingly, the engine's combustion and pollution behavior deviated in the mix after 20% TPO [9]. There was a decrease in effectiveness at complete load with 30, 40, and 50% TPO in the mix. The engine's combustion, efficiency, and pollution behavior fueled by JMETPO blends are contrasted and described in this document with that of diesel and JME action. Sajith et al. [10] as gas reserves are gradually depleted and fuel energy needs an increasing economy; so in order to fill the gap among demand and supply, we need to figure out the alternative source of oil supply. Biofuel is one of the potential sources for reducing CO, HC, and SO_x pollution, but owing to a higher oxygen content in biofuel nearly 16% higher than neat diesel, the damaging emissions of NO_x have lately been introduced as a fuel additive to minimize this NO_x emissions of metal oxide nanoparticles. Ghosh and Dutta [11, 12] EGR is a technique of decreasing combustion engine emissions [13]. The concept is based on the exhaust gas's thermodynamic properties, decreasing combustion temperature and thus decreasing nitrogen oxide emissions. Ravi Kumar et al. [14] A diesel engine's efficiency is affected by multiple variables such as compression ratio, fuel air ratio, and velocity. Variable compression techniques are used in IC motors to improve fuel efficiency under varying loads. Kumar and Sharma [15] In this evaluation research, we find that only 20% of each biodiesel (defined as MB20, PB20, and JB20) was studied in a diesel engine, since open literature suggests the potential use of up to 20% biofuel in a diesel engine without engine change [16].

2 Preparation of Tire Pyrolysis Oil

Tire pyrolysis is one of the durable techniques of producing oil from waste products such as run-out tires. Oil production from waste tires is a series of steps engaged. It starts with an 8–10 tone capacity reactor. It is 16 mm thick and 8 feet long. In stage 1, the waste tires are warmed up to 70 °C in the containment vessel. The thermal content of fossil fuels such as a mixture of coal, wood, and waste was used. At such a heat, hydrogen grew from the heated tires. In stage 2, owing to the change in temperature, the oil gasses in the reactor core grew and the gasses slowly emerged from the top. Ducts are given to meet the definition at the top of the reactor. Pressure controls can be used to show pressure created within the reactor because both hydrogen gas and petroleum fumes are extremely combustible. Much oil is generated in the heart of the reactor itself and is captured individually in the ships. The reactor fumes are sent to the transducers for cooling. Fume's condensation consists of amount of steps; there are four phases in the sector. In stage 3, fume's condensation takes place. An oil is ultimately carried to ambient temperature since four phases of condensation. In phase one from 1500c it is carried to 120 °C, in phase one it is carried to 90 °C from 120 °C, in phase one it is carried to 60 °C from 90 °C, in phase it is carried to 30 °C from 60 °C. All in all, it took 18 h of one batch processing, in which 12 h of manufacturing and 6 h of evaporation. The reactor is left in the tires with carbon residue or ash and the steel radial cables. After cooling the reactor, the ash thus generated and steel are separated and distinct requests are sent out. Research studies indicate that 40–45% tire pyrolysis oil, 30–35% carbon black, 20% steel can be retrieved from the method. The TPO collected and financed properties in accordance with the ASTM norms. As shown in Table 1, the diesel fuel and tire pyrolysis oil characteristics were reported (Table 2).

Table 1 Properties of fuels

S. No.	Properties of fuels	Diesel	Tire Pyrolysis Oil (TPO)
1	Density (Kg/m^3)	825	908
2	Calorific valve (MJ/Kg)	42.4	39.7
3	Kinematic viscosity at 40 °C	3.04	3.7
4	Cetane number	54	48
5	Flash point °C	56	47
6	Fire Point °C	56	53
7	Boiling point	180–360	100–360

Table 2 Properties of blends

Fuels properties	B10	B20
Density (Kg/m^3)	885	891
Calorific valve (MJ/Kg)	35.75	34.12
Kinematic viscosity at 40 °C	5.4	5.62
Specific gravity	0.885	0.897

3 Experimental Setup

As shown in the experiment schematic diagram in Fig. 1, it consists of a 1 cylinder with 4 strokes, water-cooled. Fuel injection system engine with eddy current load engine was used to perform the study. A U-tube manometer with an orifice meter was used to measure the water flow and fuel flow supplied to quantify the amount of air supply to the engine. To quantify the angle of the crank, AVL 365C was coupled

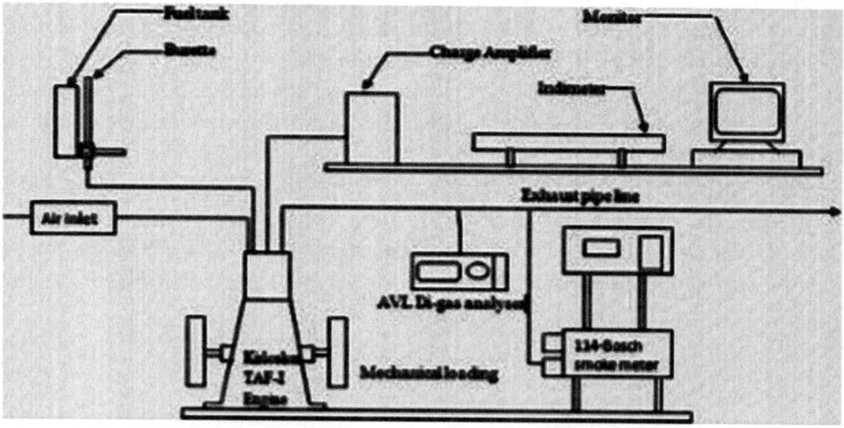

Fig. 1 Schematic diagram for the experimental

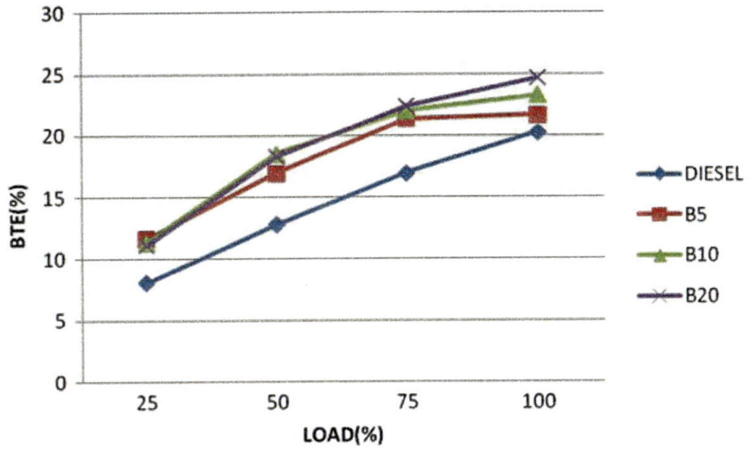

Fig. 2 Variation of BTE with load

with the crank of the engine. The exhaust gas pollution such as NO_x, HC, CO_2, O_2, and CO were recorded using HG540 5 gas analyzer, and AVL 437 smoke meter was used to record the smoke saturation of the waste heat.

4 Results and Discussion

4.1 Performance and Emission Characteristics of TPO

4.1.1 Brake Thermal Efficiency

Figure 2 demonstrates the variation in brake heat effectiveness (BTE) with the proportion of load for diesel and TPO 5, TPO 10, and TPO 20, respectively. Results are showing that the BTE increases with increase in the load. The experimental values were observed 8.06, 12.06,16.95, 20.27% for the diesel fuel; 11.64, 16.94, 21.33, 21.66% for the TPO5; 11.27,18.52, 22.01, 23.27% for the TPO10; and 11.14, 18.31, 22.38, 24.69% for the TPO20 with respective 25, 50, 75, 100% load. Optimum BTE is noted at complete load in TPO20. This is due to greater calorific value and low viscosity.

4.1.2 Indicated Thermal Efficiency

Figure 3 demonstrates the difference in the specified thermal efficiency (ITE) for diesel and TPO 5, TPO 10, and TPO 20, respectively. Results are showing that the ITE decreases with increase in the load. The experimental values are observed 46.17,

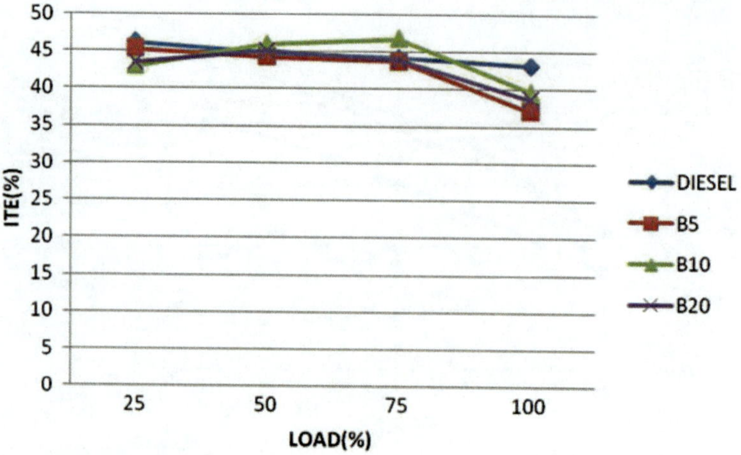

Fig. 3 Changes of ITE with load

44.84, 44.22, 43.2% for the diesel fuel; 45.29, 44.27, 43.76, 37.14% for the TPO5; 42.97, 45.94, 46.74, 39.65% for the TPO10; and43.50, 45.09, 43.88, 38.71% for the TPO20 with respective 25, 50, 75, 100% load. Maximum ITE is observed in TPO10 at 75% of load.

4.1.3 Mechanical Efficiency

Figure 4 indicates the mechanical effectiveness (ME) difference with the proportion of the load for diesel and TPO 5, TPO 10, and TPO 20, respectively. Results are

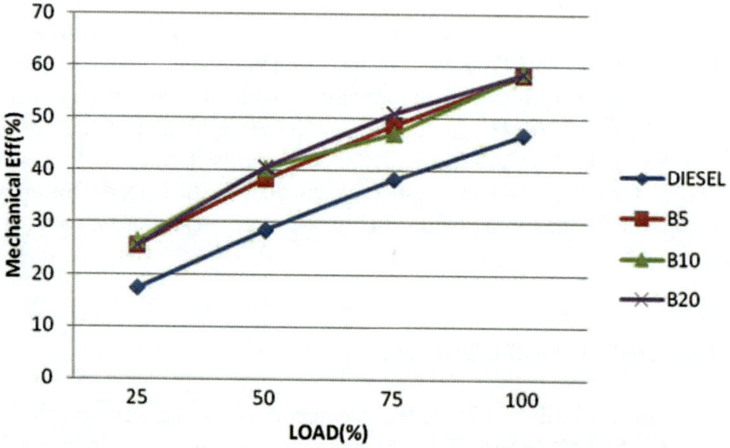

Fig. 4 Changes of ME with load

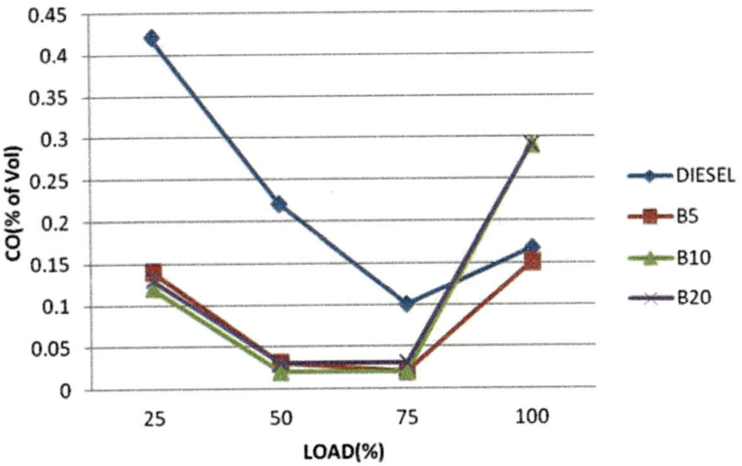

Fig. 5 Changes of CO with load

showing that the ME increases with increase in the load. The experimental values are observed 17.45, 28.56,38.32, 46.93% for the diesel fuel; 25.70, 38.25, 48.74, 58.32% for the TPO5; 26.46, 40.31, 47.09, 58.69% for the TPO10; and 25.61, 40.60, 51.01, 58.61% for the TPO20 with respective 25, 50, 75, 100% load. Maximum mechanical efficiency is observed in TPO10 at full load.

4.1.4 Carbon Monoxide

Figure 5 indicates the variety of CO with the corresponding proportion of diesel and TPO 5, TPO 10, and TPO 20. Results are showing that the carbon monoxide increases with increase in the load. The experimental values are observed 0.42, 0.22, 0.1, 0.167% of volume for the diesel fuel; 4.79, 5.38, 6.31, 63.94% of volume for the TPO5; 4.54, 5.55, 5.99,7.24% of volume for the TPO10; and 4.30, 5.44, 6.29, 7.35% of volume for the TPO20 with respective 25, 50, 75, 100% load.

4.1.5 Brake-Specific Fuel Consumption

Figure 6 demonstrates the change in brake-specific fuel consumption with the corresponding proportion of diesel and TPO 5, TPO 10, and TPO 20 load. Results are showing that the BSFC decreases with increase in the load. The experimental values are observed 1.06, 0.67, 0.51,0.42 kg/kwh for the diesel fuel; 0.74, 0.51, 0.4, 0.4 kg/kwh for the TPO5; 0.77, 0.47, 0.38, 0.38 kg/kwh for the TPO10; and 0.75, 0.46, 0.39, 0.37 kg/kwh for the TPO20 with respective 25, 50, 75, 100% load.

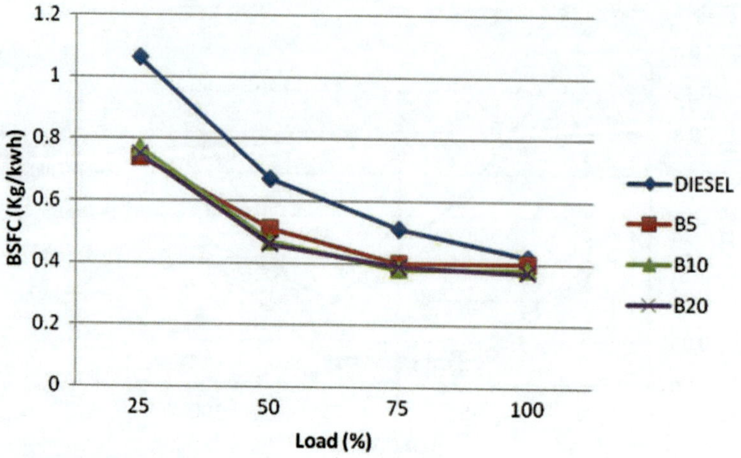

Fig. 6 Changes of BSFC with load

4.1.6 Oxides of Nitrogen

Figure 7 indicates the NO_X variety with the corresponding proportion of diesel, TPO5, TPO10, and TPO20 load. Results are showing that the NO_X increases at TPO20 with increase in the load. The experimental values are observed 186 ppm, 499, 791, and 1435 ppm for the diesel; 168, 485, 786, and 1358 ppm for the TPO5; 170, 426, 802, and 1372 ppm for TPO10; and 172, 418, 815, and 1386 ppm for TPO20 with respective 25, 50, 75, 100% load.

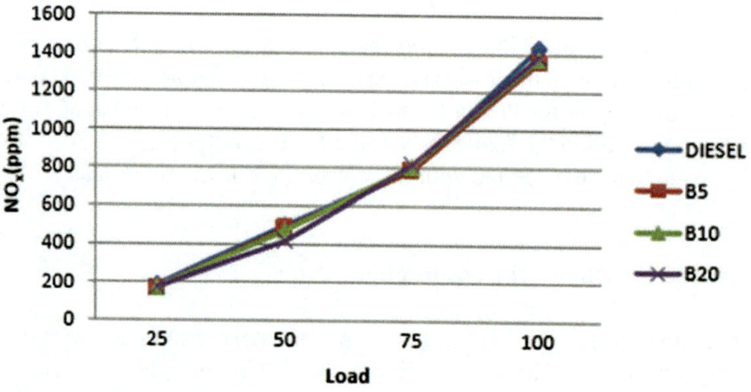

Fig. 7 Changes of NOx with load

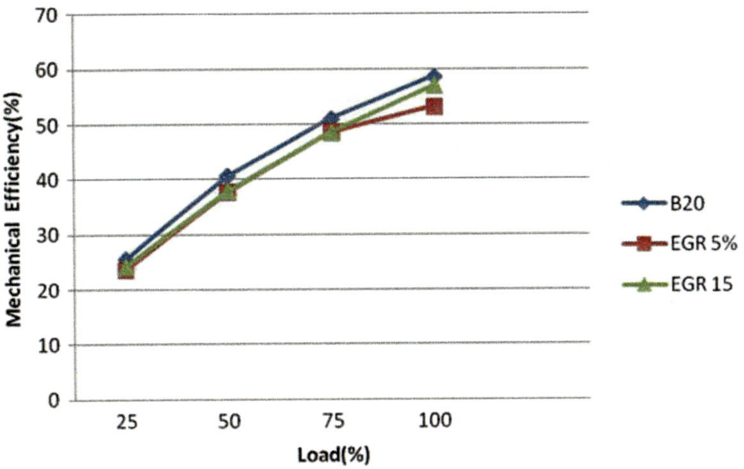

Fig. 8 Changes of ME with load

4.2 *Exhaust Gas Recirculation Performance and Emission Properties*

4.2.1 Mechanical Efficiency

Figure 8 demonstrates the change in mechanical efficiency for TPO20, EGR5 percent, and EGR15 percent, respectively. Results are showing that the ME increases with increase in the load. The experimental values are observed 25.61, 40.6,51.01, 58.61% for the TPO20; 23.57, 37.59, 48.57, 53.15% for the EGR5%; and 24.27, 37.85, 48.49, 56.92% for the EGR15% with respective 25, 50, 75, 100% load. Maximum ME is observed in TPO20 at full load.

4.2.2 Brake Thermal Efficiency

Figure 9 demonstrates the change in engine heat effectiveness (BTE) with the corresponding proportion of load for TPO20, EGR5 percent, and EGR15 percent. Results are showing that the BTE increases with increase in the load. The experimental values are observed 11.14, 18.31,22.38, 22.69% for the TPO20; 10.94, 16.44, 20.29, 21.39% for the EGR5%; and 11.2, 18.58, 21.75, 21.77% for the EGR15% with respective 25, 50, 75, 100% load. Maximum BTE is observed in TPO20 at full load.

4.2.3 Indicated Thermal Efficiency

Figure 10 demonstrates the change in the specified thermal efficiency for TPO20,

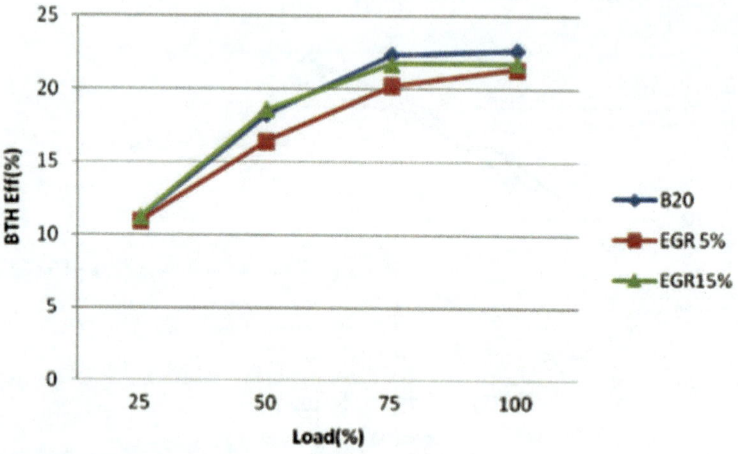

Fig. 9 Changes of BTE

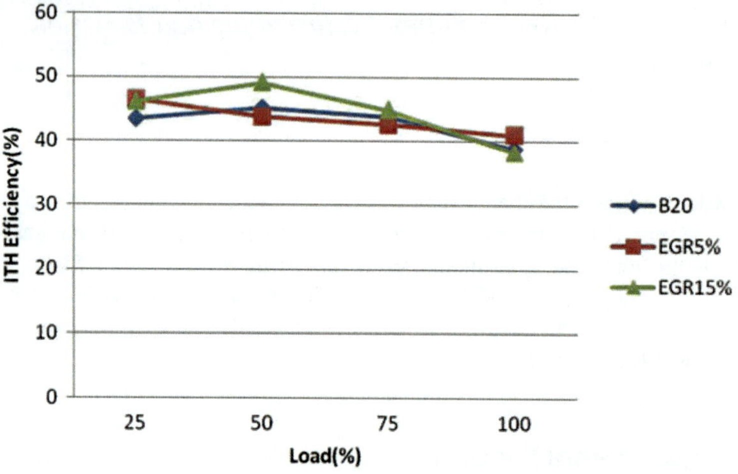

Fig. 10 Changes of ITE

EGR5 percent, and EGR15 percent respectively. Results are showing that the ITE decreases with increase in the load. The experimental values are observed 43.5, 45.09, 43.88, 38.71% for the TPO20; 46.43, 43.73, 42.61, 41.02% for the EGR5%; and 46.19, 49.08, 44.86, 38.25% for the EGR15% with respective 25, 50, 75, 100% load. Maximum ITE is observed in EGR15% at 50% of load

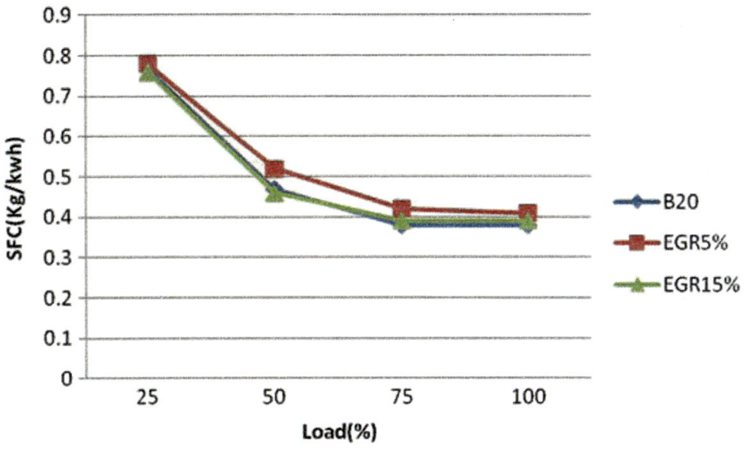

Fig. 11 Changes of BSFC

4.2.4 Brake-Specific Fuel Consumption

Figure 11 demonstrates the change in brake-specific fuel consumption with the corresponding proportion of TPO20, EGR5 percent, and EGR15 percent. Results are showing that the BSFC decreases with increase in the load. The experimental values are observed 0.77, 0.47,0.38, 0.38 kg/kwh for the TPO20; 0.78, 0.52, 0.42, 0.41 kg/kwh for the EGR5%; and 0.76, 0.46,0.39, 0.39 kg/kwh for the EGR15% with respective 25, 50, 75, 100% load.

4.2.5 Carbon Dioxide

Figure 12 demonstrates the carbon dioxide difference with the corresponding proportion of TPO20, EGR5 percent, and EGR15 percent. As results are showing that the CO increases at TPO20 and EGR5% with increase in the load and decrease at EGR15%. The experimental values are observed 1, 1, 1.6, 1.9% of volume for the TPO20; 0.3, 0.2, 0.2, 0.4% of volume for the EGR5%; and 0.3, 0.3, 0.3, 0.2% of volume for the EGR15% with respective 25, 50, 75, 100% load.

4.2.6 Oxides of Nitrogen

Figure 13 demonstrates the NOX variety with the corresponding proportion of TPO20, EGR5 percent, and EGR15 percent. Results are showing that the NOX increases at TPO20 with increase in the load. The experimental values are observed 172, 418, 815, and 1386 ppm for the TPO20; 145, 387, 419, and 938 ppm for EGR5%; 133, 267,390, and 719 ppm for EGR15% with respective 25, 50, 75, 100% load.

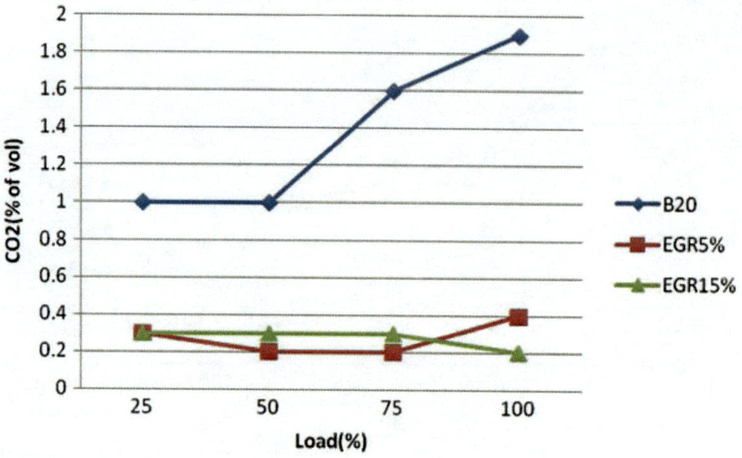

Fig. 12 Changes of CO_2

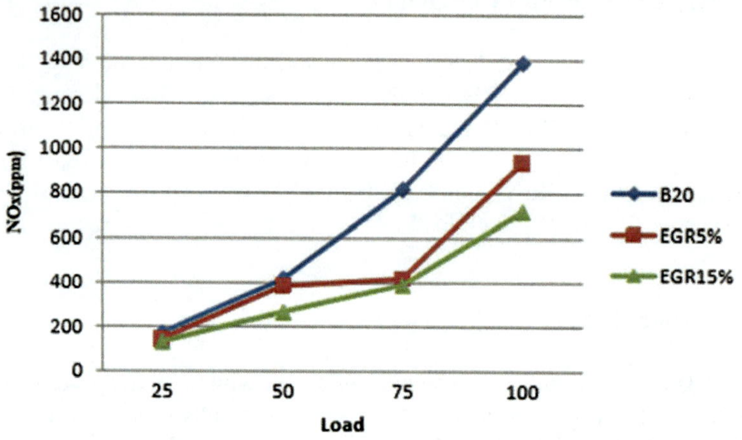

Fig. 13 Changes of NO_x

4.3 Performance, Combustion with Ceo Additives

4.3.1 Mechanical Efficiency

Figure 14 indicates the mechanical effectiveness (ME) difference with the corresponding proportion of the load for TPO20, EGR15 percent, and cerium oxide. Results are showing that the ME increases with increase in the load. The experimental values are observed 25.61, 40.6,51.01, 58.61% for the TPO20; 24.27, 37.85, 48.49, 56.92% for the EGR15%; and 26.87, 38.09, 50.33, 55.84% for the cerium

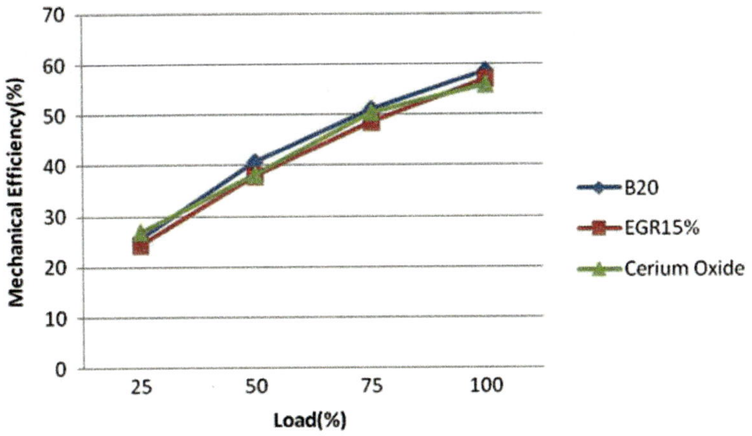

Fig. 14 Changes of Mechanical Efficiency

oxide with respective 25, 50, 75, 100% load. Maximum mechanical efficiency is observed in TPO20 at full load.

4.3.2 Brake Thermal Efficiency

Figure 15 demonstrates the change in brake heat effectiveness (BTE) with the corresponding proportion of TPO20, EGR15 percent, and cerium oxide load. Results are showing that the BTE increases with increase in the load. The experimental values are observed 11.14,18.31, 22.38, 24.69% for the TPO20; 11.21, 18.58, 21.75, 21.77% for

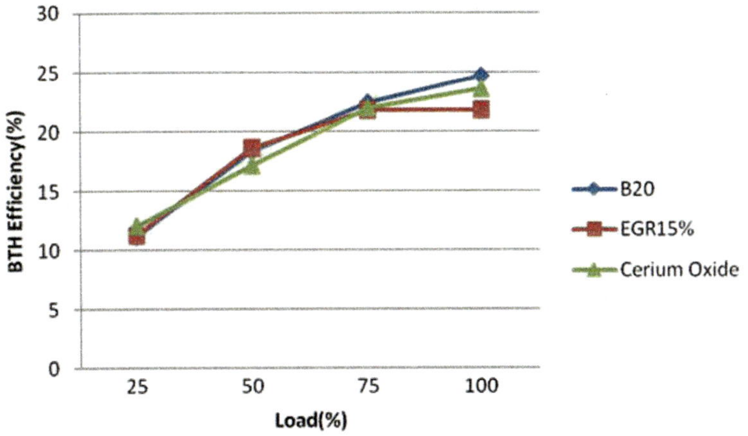

Fig. 15 Changes of brake thermal efficiency

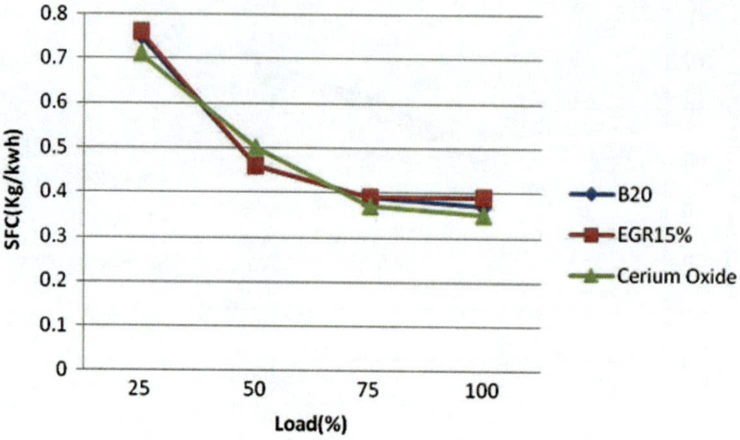

Fig. 16 Changes of specific fuel consumption

the EGR15%; and 12.01, 17.1, 21.93, 23.55% for the cerium oxide with respective 25, 50, 75, 100% load. Maximum BTE is observed in TPO20 at full load.

4.3.3 Brake-Specific Fuel Consumption

Figure 16 demonstrates the change in brake-specific fuel consumption (BSFC) with the corresponding proportion of TPO20, EGR15 percent, and cerium oxide load. Results are showing that the BSFC decreases with increase in the load. The experimental values are observed 0.75, 0.46, 0.39, 0.37 kg/kwh for the TPO20; 0.76, 0.46, 0.39, 0.39 kg/kwh for the EGR15%; and 0.71, 0.5, 0.37, 0.35 kg/kwh for the cerium oxide with respective 25, 50, 75, 100% load.

4.3.4 Carbon Dioxide

Figure 17 shows the carbon dioxide variation with the respective percentage of the load for TPO20, EGR15 percent, and cerium oxide. As results are showing that the CO increases at TPO20 and cerium oxide with increase in the load and decrease in EGR15%. The experimental values are observed 1, 1, 1.6, 1.9% of volume for the TPO20; 0.3, 0.3, 0.3, 0.2% of volume for the EGR15%; and 1.2, 1.5, 1.7, 2% of volume for the cerium oxide with respective 25, 50, 75, 100% load.

4.3.5 Oxides of Nitrogen

Figure 18 indicates the NOX variation with the corresponding proportion of load for cerium oxide, 15 percent EGR, and TPO20. As the figure represents the percentage

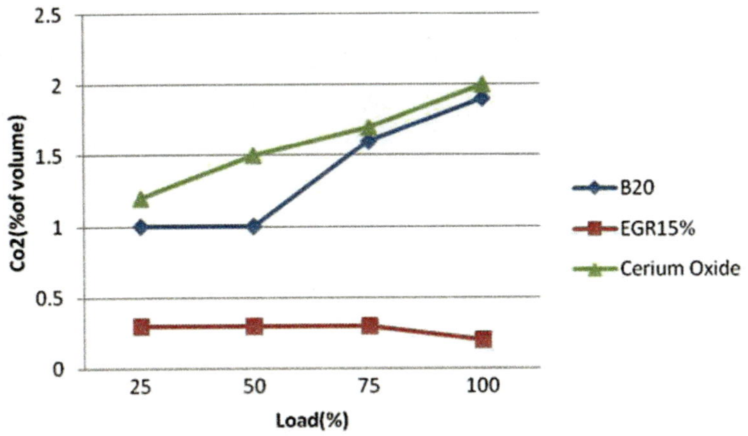

Fig. 17 Changes of CO_2

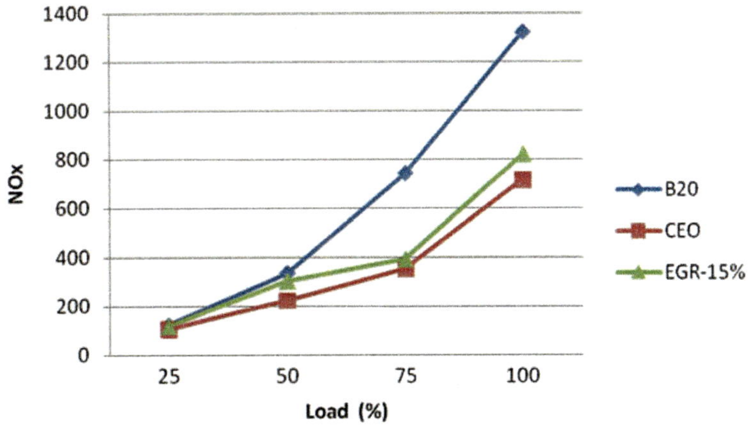

Fig. 18 Changes of NO_X

of the load and also raises the oxides of nitrogen. Results are showing that the NO_X increases at TPO20 with increase in the load. The experimental values are observed 172, 418, 815, and 1386 ppm for the TPO20 and 128, 251, 345, and 692 ppm with respective 25, 50, 75, 100% load.

4.3.6 B20 With Additive (Cerium Oxide)

As shown in Fig. 19a–c, the changes of the cylinder pressure, mass fraction burned, and net heat release with crank angle on CI engine for 20% bio-fuel blend, EGR 15%, and cerium oxide. The start of combustion is started at 345 °CA BTDC. The

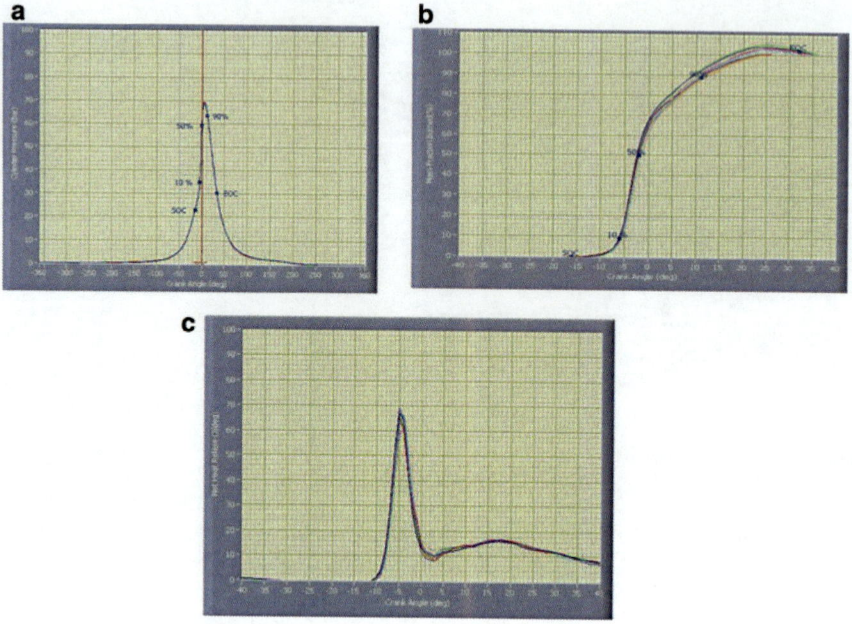

Fig. 19 **a** Variation of cylinder pressure **b** Variation of mass fraction **c** Variation of net heat release

cylinder pressure is recorded 23 bar, maximum cylinder pressure is recorded is 68 bar at 10°CA, ATDC, and end of the combustion is observed at a cylinder pressure of 65 bar. The mass fraction burned at the end of the combustion is 100%.

5 Conclusion

The experimental was studied on CI engine for tire pyrolysis oil (TPO) blended and variation of EGR with added additives of cerium oxide

- At B20 (TPO20) biodiesel blend was observed with increased performance and reduction of emissions as compared to other blends B5, B10.
- Emissions are further reduced with exhaust gas recirculation, at 15% of EGR was to 15–20%.
- The performance and emission features of biodiesel fuel are more impacted by cerium oxide additives.

References

1. Abd-Alla GH (2002) Using exhaust gas recirculation internal combustion engines: a review. J Energy Convers Manage 43(8):1027–1042
2. Gürü M, Karakaya U, Altiparmak D, Alicilar A (2002) Improvement of Diesel fuel properties by using additives. Energy Convers Manag 43(8):1021–1025
3. Friedrich S (2004) A world wide review of the commercial production of biodiesel—a technological, economic and ecological investigation based on case studies. Des Institutes fürTechnologie und NachhaltigesProduktmanagement, Vienna
4. Graham TR (2004) Diesel engine exhaust gas recirculation: review on advanced and novel concept. J Prog Energy Combust Sci 45:883–900
5. Akers MS et al (2006) Determination of the heat of combustion of biodiesel using bomb calorimetry. J Chem Educ 83(2), Feb 2006
6. De Menezes EW, Da Silva R, Cataluña R, Ortega RJC (2006) Effect of ethers and ether/ethanol additives on the physicochemical properties of diesel fuel and on engine tests. Fuel 85(5–6):815–822
7. Jesila J, Murugan S (2007) Distilled Tyre Pyrolysis Oil as an alternate fuel for CI engine. SilnikiSpalinowe 266–273
8. Sharma A, Murugan S (2007) Investigations on the behaviour of a DI diesel engine fuelled with jatropha methyl ester (JME) &Tyre pyrolysis oil(TPO) blends. Fuel 108:699–708
9. Agarwal AK, Rajamanoharan K (2009) Experimental investigationsof performance and emissions of Karanja oil and its blends in a singlecylinder agricultural diesel engine. Appl Energy 86(1)
10. Sajith V, Sobhan CB, Peterson GP (2010) Experimental investigations on the effects of cerium oxide nanoparticle fuel additives on biodiesel. Adv Mech Eng
11. Hirst SM et al. (2011) Bio-distribution and in vivo antioxidant effects of cerium oxide nanoparticles in mice. Environ Toxicol. Article first published online 26 May 2011
12. Ghosh S, Dutta D (2012) Performance and emission parameters analysis of gasoline engine with exhaust gas recirculation. Int J Eng Res Dev
13. Awad S, Varuvel EG, Loubar K, Tazerout M (2013) Single zone combustion modeling of biodiesel from wastes in diesel engine. Fuel 106:558–68
14. Ravi Kumar N, Sekhar YMC, Adinarayana S (2013) Effects of compression ratio and EGR on performance, combustion and emissions of Di injection diesel engine. Int J Appl Sci Eng (11)1:41–49
15. Kumar K, Sharma MP (2016) Performance and emission characteristics of a diesel engine fuelled with biodiesel blends 6(2)
16. Hurakalli R et al. (2015) Effect of exhaust gas recirculation on the performance and emission of two stroke spark ignition Engine. JSIR 4

Production of Paraffinic Fuel by Hydrotreatment of Waste Sunflower Cooking Oil Using Nobel Catalyst

Hemanandh Janarthanam, S. Ganesan, S. P. Venkatesan, A. M. Rakesh, and B. Sathish Kumar Reddy

Abstract The work reported here is about the combustion characteristics, performance, and emission analysis of the hydrotreated waste cooking oil as fuel in vertical, 4.3 kW, DI 4S diesel engine with compression ratio of 17.5:1, at a constant speed of 1500 rpm. The waste sunflower cooking oil was hydrotreated in the high-pressured trickle bed reactor. The HR538 type of ($Ni\text{-}Mo/Al_2O_3$) was used as catalyst. The waste sunflower cooking oil was treated with catalyst at 6 MPa and at a temperature of 633 K. The two fuels prepared for testing are 25% hydrotreated waste cooking oil fuel blended with petro diesel (B25) and (B100) which is 100% neat hydrotreated fuel. At different loading conditions, these fuels were tested in the diesel engine. The emission and performance of the hydrotreated waste cooking oil showed significant decrease in CO, HC, NO_x, smoke, brake specific fuel consumption and increase in brake thermal efficiency when compared with petro diesel.

Keywords HR 538 catalyst · Hydrotreated waste cooking oil · Decarboxylation · Combustion · Performance and emissions

Nomenclature

WCO	Waste cooking oil
WCOB40	40% of waste cooking oil blended with 60% of petroleum diesel
WCOB20	20% of waste cooking oil blended with 80% of petroleum diesel
WCOB5	5% of waste cooking oil blended with 95% of petroleum diesel
HVO	Hydrotreated vegetable oil
HTWCO	Hydrotreated waste cooking oil
HTWCO B25	25% hydrotreated waste cooking oil + 75% petro diesel fuel
HTWCO B100	100% hydrotreated waste cooking oil

H. Janarthanam (✉) · S. Ganesan · S. P. Venkatesan · A. M. Rakesh · B. Sathish Kumar Reddy
School of Mechanical Engineering, Sathyabama Institute of Science and Technology, Chennai, Tamil Nadu 600119, India
e-mail: hemanandh123@gmail.com; foreversri14@gmail.com

A. Arockiarajan et al. (eds.), *Advances in Industrial Automation and Smart Manufacturing*, Lecture Notes in Mechanical Engineering,
https://doi.org/10.1007/978-981-15-4739-3_85

PD	Petro diesel
BIS	Bureau of Indian Standards
LHSV	Hourly space velocity
CO	Carbon monoxide
HC	Hydrocarbon
CO_2	Carbon dioxide
NO_x	Nitrogen oxides
FSN	Filter smoke number
BSFC	Brake specific fuel consumption
BTE	Brake thermal efficiency
aTDC	After top dead center
bTDC	Before top dead center

1 Introduction

Many commercial kitchens heat the oil to the maximum making it unfit for cooking. Triglycerides form due to the excessive heating of the oil. These oils harm human health and pollute the environment. The systematic disposal of waste cooking oil can be done by converting it into biodiesel as fuel for diesel engines.

Huber et al. [1] found that hydrotreating the sunflower oil produced more alkenes, and it could increase the reaction chemistry under optimum operating conditions. Simacek et al. [2] experimented the rapeseed oil at higher temperatures and pressures with three hydrorefining catalyst and produced C_{17}–C_{18} alkenes which was same as diesel fuel. Solymosi et al. [3] investigated by converting the waste cooking oil into diesel fuel using heterogeneous catalyst with LHSV of 1 h^{-1} at a temperature and pressure of 320–380 °C and 20–80 bar. This produced the high cetane number and lower aromatic and sulfur content. Bezergianni et al. [4] conducted experiment with waste cooking oil using hydrotreatment process at different pressure, LHSV, and temperature which resulted in the removal of sulfur, nitrogen, and oxygen more than 90%. Sotelo-Boyas et al. [5] hydrotreated rapeseed oil using multiple catalyst at different pressure and temperature with LHSV of 3 h^{-1}. The high yield of hydrocarbon was obtained using the NiMo/γ-Al_2O_3. Bezergianni et al. [6] studied the removal of heteroatom and saturation in double bonds in the waste cooking oil using hydrotreatment process. By changing the different operating temperatures resulting in lower temperature suitable for diesel production and higher temperature, they could lower temperature so as to be suitable for diesel production and higher temperature for gasoline production. Guzman et al. [7] found that the crude palm oil when hydroprocessed with a NiMo/γ-Al_2O_3 catalyst at 40–90 bar could produce paraffin and high cetane index. Kousoulidou et al. [8] found that the waste cooking oil when hydrotreated using catalyst resulted in elevation of NO_x and depletion in CO, hydrocarbons, and CO_2. Liu et al. [9] reported about the removable of oxygen and CO_2 in hydrotreated jatropha oil using sulfided Ni–Mo/SiO_2–Al_2O_3 as catalyst.

Liu et al. [10] experimented with the jatropha oil using the Ni-Mo/SiO_2–Al_2O_3 as catalyst at LHSV of 7.6 h, 4 MPa pressure, temperature of 350 °C, with H_2/oil ratio of 800, to convert it into the diesel fuel using hydrotreatment process. Sebos et al. [11] attempted to convert the refined cotton seed oil into hydrotreated fuel using CoMo/Al_2O_3 catalyst. 100% conversion of fuel was obtained at temperature of 305–345 °C, 30 bar, and LHSV of 5 h^{-1}–25 h^{-1}. Jindal et al. [12] experimented with the jatropha and karanja methyl esters and reported decrease in NO_x, HC, and smoke and higher power when compared to petro diesel. But, the brake thermal efficiency was same as petro diesel. Saravanan et al. [13] reported that crude rice bran oil when blended with diesel resulted in 25% blend giving reduction in CO, unburnt hydrocarbon, and heat release while increasing peak pressure. Sugiyama et al. [14] studied emission and combustion analysis of hydrotreated vegetable oil by spray analyzer, engine dynamometer, and vehicle test that resulted in decrease in particulate matter and hydrocarbon due to high cetane value and absence of aromatics. Kalam et al. [15] conducted experiment with the waste coconut oil and palm oil methyl esters in diesel engine. They prepared two fuels C5 and B5 and compared them with ordinary diesel. The result revealed that decrease in CO, HC, and brake thermal efficiency for C5 and B5 while increase in NO_x for B5 blend. Hwang et al. [16] established that the waste cooking oil could power the common rail diesel engine by differing the injection pressure and injection timing. The result showed increase in NO_x in all conditions. The CO, HC, and smoke were decreased a high injection pressure. Roy et al. [17] prepared three types of canola biofuel and tested them in DI diesel engine. Used canola and pure canola showed the similar release and performance of DI diesel engine. The HC and CO emission were less for canola blends. Mumtaz et al. [18] conducted experiment on waste cooking oil using NOVOZYME-435, with $NaOCH_3$ catalyst. The CO, PM, and NO_x were decreased for WCOB40, WCOB20, and WCOB5, respectively. Nantha Gopal et al. [19] found depletion in CO, HC, and smoke and elevation in specific fuel consumption and NO_x when waste cooking oil was tested in the diesel engine. Nayak and Pattanaik [20] revealed that when dimethyl carbonate additive was added in different proportions in mahua oil and used as fuel, it could reduce CO, HC, smoke, brake specific fuel consumption, and NO_x but increased brake thermal efficiency and engine temperature with increase in additive. Jindal et al. [21] investigated on the blends of methyl esters of waste cooking oil with diesel, and these blends were blended with n-butanol at 10 and 20%. The result showed that there was increase in HC, BSFC, NO_x, and that CO decreased for 10% n-butanol.

In this study, the properties of hydrotreated waste cooking oil were better than properties of petro diesel. The aim of the experiment is to evaluate the performance, emission, and combustion characteristics of diesel engine using hydrotreated waste cooking oil fuel.

2 Materials and Methods

Different types of waste cooking oil were procured from Sathyabama University's central cooking facility. The central cooking facility uses 18,000 L per month. Around 3500 L of waste cooking oil is generated per month. These oils were collected and filtered using a filter to remove solid food particles and impurities in the oil. Stage 1: The three different waste vegetable oils were thoroughly mixed. Stage 2: The waste cooking properties were tested and compared with those of petro diesel. Stage 3: The waste cooking oil was hydrotreated and blended with diesel in various proportions.

2.1 Production of Hydrotreated Oil Using Waste Cooking Oil

The hydrotreated vegetable oil was produced from the trickle bed, High Pressure Cata Test Pilot plant, Vinci technologies, France. The schematic diagram of the High Pressure Cata Test Pilot Plant is shown in Fig. 1. The filtered waste cooking oil was stored in the source tank. The oil was pumped, preheated, and mixed with hydrogen before it reached the reactor. The heterogeneous sulfided HR538 type of nickel–molybdenum/aluminum oxide (Ni-Mo/Al_2O_3) was used as catalyst. The reactor temperature was maintained at 350 °C, with pressure of 6 MPa, H2/HC ratio of 300 L/L, and at LHSV of 2 h^{-1}. The LHSV was maintained for proper wetting of the catalyst.

The hydrocracking is the process which converts the heavier molecules into lighter molecules. The hydrocarbons are produced by cracking the carbon bonds within the scale of boiling point of the diesel. This can be produced by feeding the hydrogen at high temperature and heterogeneous catalyst at high temperature. The triglycerides

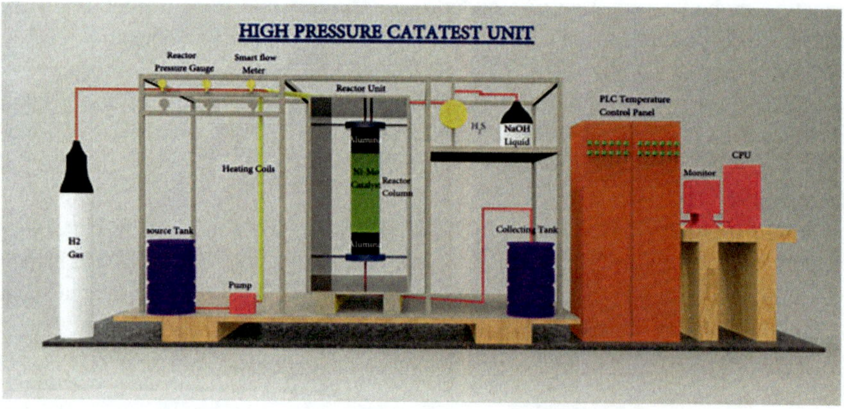

Fig. 1 Schematic diagram of the hydrotreatment plant

molecules are divided into separate chains, and hydrocarbons similar to petro diesel fuel are produced.

The long chain hydrocarbons are cracked into short chain hydrocarbon using catalyst. There are three methods used to crack the hydrocarbon chain. The main reactions that take place inside the reactor are decarboxylation, decarbonylation, denitrification, and hydrodeoxygenation. The carboxyl group (–COOH) is removed in the decarboxylation process, and it forms the carbondioxide (CO_2), or carbon monoxide is removed.

The process of separating (C=O) from the carbonyl group is called decarbonylation. This forms the n-alkanes and water. With the help of isomerization catalyst, the n-alkanes are converted to iso-alkanes. The n-alkanes are converted into iso-alkanes with isomerization catalyst. The triglycerides are converted into propane. In hydrodeoxygenation, the oxygen molecules are eliminated in the triglyceride along with hydrogen and form n-paraffins and water. The paraffins are in the range of C15–C18. The deoxygenation absorbs more hydrogen than decarboxylation reaction.

In hydrotreatment catalytic process, the heteroatoms such as nitrogen, sulfur, oxygen, and metals are removed, and also, the process eliminates the aromatics and olefins. In the catalytic reaction, the carbon and sulfur bonds are broken, and hydrogen sulfide is formed. The hydrogen sulfide gas is collected in the NaOH liquid. The colorless NaOH liquid turns black.

The samples of the waste cooking oil and hydrotreated waste cooking oil are shown in Figs. 2 and 3. The blend of HTWCO B25 with petro diesel is prepared for testing in the diesel engine to study the emission and performance of the green diesel.

The properties of WCO, fatty acid methyl esters of waste cooking oil, and hydrotreated waste cooking oil are determined using standard ASTM D–751 (IS

Fig. 2 Waste cooking oil (WCO)

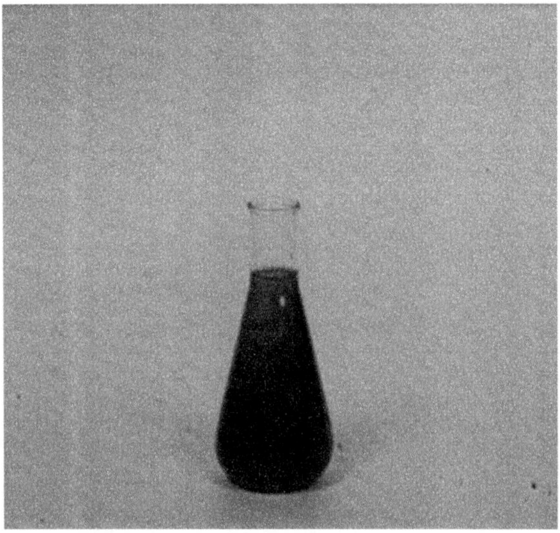

Fig. 3 Hydrotreated waste cooking oil (HTWCO)

5607:2005) methods as given in Table 1. The properties of the WCO and HTWCO are given in Tables 1 and 2. The cold flow properties of HTWCO are improved by

Table 1 Comparison of properties of petro diesel, WCO, and HTWCO

S. No.	Properties	Euro-IV Bharat stage 1460:2005 diesel	ASTM D–751 (IS 5607:2005)	WCO	HTWCO
1.	Cetane number	51	–	37	84
2.	Density at 15 °C kg/m^3	820–845	860–900	924.9	788.8
3.	Kinematic viscosity at 40 °C cst	2–4.5	1.9–6	32.8	3.85
4.	Flash point °C min	35 °C	130 °C	256	94.2
5.	Calorific value kJ/kg	39,000	–	36,132	43,829

Table 2 Specifications of test engine

Type: Kirloskar oil engine, vertical, 4S, single-acting, high-speed C.I diesel engine	
Combustion	Direct injection
Rated power	4.3 kW
Rated speed	1500 rpm
Compression ratio	17.5:1
Fuel injection pressure	210 bar
Dynamometer	Eddy current

Table 3 Ultimate analysis of test results

S. No.	Composition	WCO (%)	HTWCO (%)
1.	Carbon	52.56	89.93
2.	Hydrogen	4.56	9.18
3.	Nitrogen	4.42	0.02
4.	Sulfur	0.98	0.01
5.	Oxygen	4.12	0.8

using the hydroisomerization catalyst to form green diesel. The increase in cetane number was found. There was increase in calculated cetane index, flash point, and the calorific value, but decrease in density and kinematic viscosity. Table 3 shows the composition of carbon, hydrogen, nitrogen, sulfur, and oxygen content in the fuel which was found by ultimate analysis test.

3 Experimental Set Up

To test the hydrotreated waste cooking oil, vertical stationary four-stroke Kirloskar DI diesel engine with a constant speed of 1500 rpm was used. The specifications of the Kirloskar DI diesel engine are given in Table 2. The schematic diagram of the experimental set up is shown in Fig. 4. Initially, the diesel engine was started with diesel fuel and permitted to run at zero load condition for 15 min for each proportion before applying the load. The alternator was coupled with the engine, and the load was varied (0, 25, 50, 75, and 100%). After the engine was warmed, the different blends of hydrotreated test fuels such as HTWCOB25 and HTWCOB100 were tested. The load applied to the engine was gradually increased step by step from 25% up to 100% at a constant speed of 1500 rpm. The emission parameters (CO_2, NO_x, CO, HC, and smoke) were measured using the AVL DIGAS 444 analyzer and AVL smoke meter 415, respectively. The average of three fuel tests was recorded and compared with petro diesel.

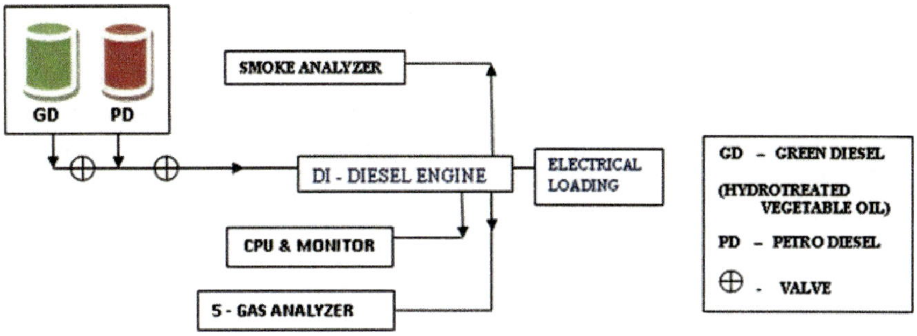

Fig. 4 Schematic diagram of the experimental set up

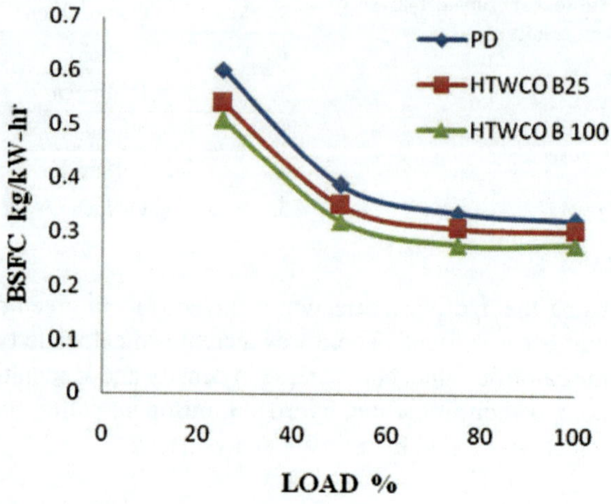

Fig. 5 Variation of BSFC with respect to PD, HTWCO B25, and HTWCO B100

4 Results and Discussion

4.1 Performance Analysis

4.1.1 Brake-Specific Fuel Consumption (BSFC)

The comparison of the BSFC values and the load trend is shown in Fig. 5. The figure shows that when the engine is fueled with hydrotreated fuel, the value of the BSFC decreases slightly in comparison with petro diesel. The BSFC decreases with HTWCO B25 and HTWCO B100 as load increases. At maximum load condition for both HTWCO B25 and HTWCO B100, there is a decrease in BSFC by 12.5% and 25%, respectively. The engine performance depends on the fuel properties and the spray characteristics. The lower kinematic viscosity of the HTWCO fuel and better fuel injection results in decrease of BSFC in comparison with respect to petro diesel.

4.1.2 Brake Thermal Efficiency (BTE)

Figure 6 depicts the brake thermal efficiency of the HTWCO B25 and HTWCO B100 at constant speed. The BTE increased gradually for test fuels compared with petro diesel. At maximum load condition, there is an increase of 11.14 and 27.48% for HTWCOB25 and HTWCOB100, respectively, with respect to petro diesel. This is because of the higher calorific value of the hydrotreated waste cooking oil.

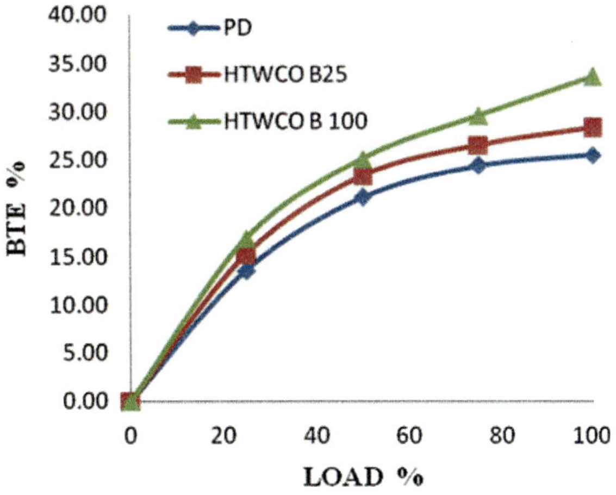

Fig. 6 Variation of BTE with respect to PD, HTWCO B25, and HTWCO B100

4.2 *Emission Analysis*

4.2.1 Carbon Monooxide (CO)

The variation of CO for PD, HTWCO B25, and HTWCO B100 is shown in Fig. 7. Due to the absence of the carboxyl group (–COOH) in the decarboxylation process, the carbon monoxide gets removed. The results show a decrease of 42.8%, and 58.33% for HTWCOB25 and HTWCOB100, respectively, with respect to petro diesel. This decreased value can be attributed to higher cetane number in the hydrotreated fuel that improved the combustion of the fuel.

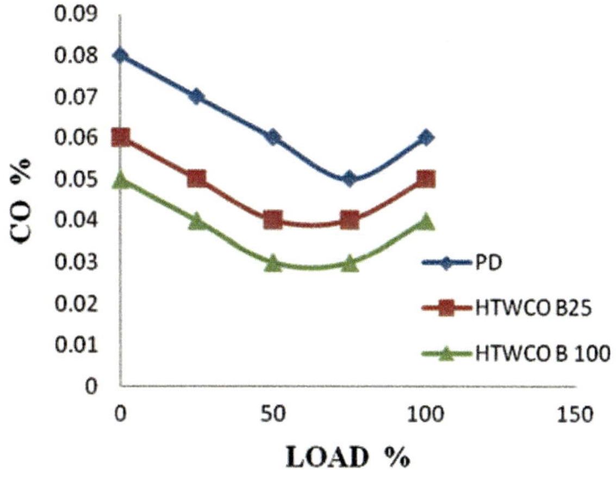

Fig. 7 Variation of CO with respect to PD, HTWCO B25, and HTWCO B100

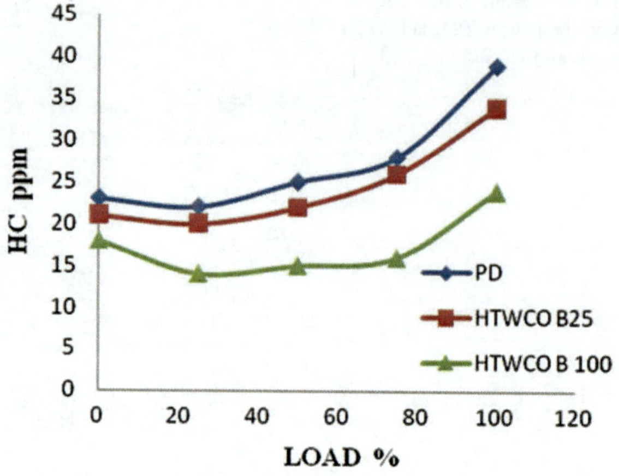

Fig. 8 Variation of HC with respect to PD, HTWCO B25, and HTWCO B100

4.2.2 Hydrocarbon (HC)

The emission of HC with respect to load is shown in Fig. 8. At the initial load condition, the decrease in HC was observed for HTWCO B25 and HTWCO B100 in comparison with PD. As the load increases, there is a decrease of 13.51% and 40.5% for HTWCOB25 and HTWCOB100, respectively, with respect to petro diesel. The reduction of HC in hydrotreated waste cooking oil could be due to the removal of oxygen during deoxygenation process. This improved the combustion and reduced the HC.

4.2.3 Nitrogen Oxide (NO_x)

Figure 9 shows the variation of NO_x with respect to PD, HTWCO B25, and HTWCO B100. The NO_x emission of the HTWCO B25 was similar to that of petro diesel up to 50% of load and thereafter decreased. At full load condition, the NO_x decreased by 9.1% with respect to petro diesel. At initial load, HTWCOB100 NO_x emission is slightly lower than petro diesel and HTWCOB25. This decreases to 18.18% difference at maximum load condition. This reduction in NO_x with respect to petro diesel for HTWCO B25, and HTWCO B100 is due to the denitrification in the waste cooking oil during hydrotreatment process and also due to the shorter ignition delay.

4.2.4 Smoke

The smoke emission of the PD, HTWCO B25, and HTWCO B100 is shown in Fig. 10. The smoke emission is measured in terms of FSN (Filter Smoke Number). At full load condition, the decrease in HTWCOB25 and HTWCO B100 is by 13.51% and

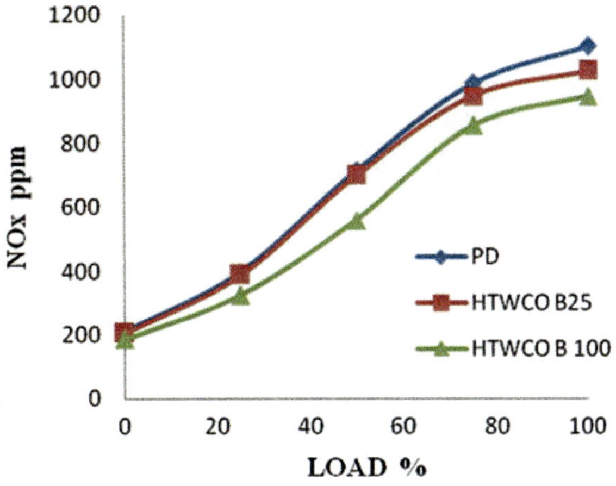

Fig. 9 Variation of NO_x with respect to PD, HTWCO B25, and HTWCO B100

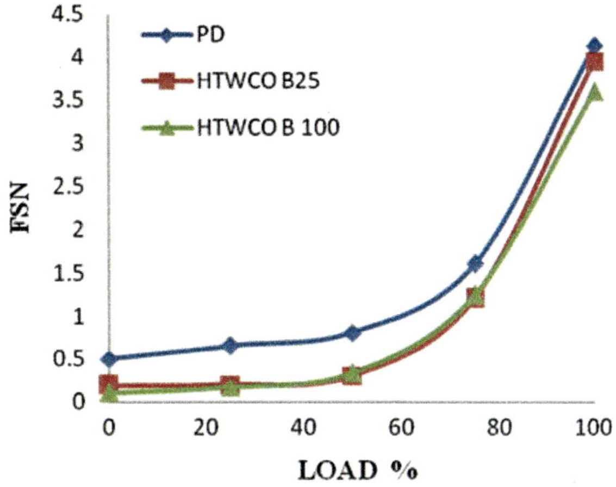

Fig. 10 Variation of FSN with respect to PD, HTWCO B25, and HTWCO B100

40.5%, respectively, with petro diesel. This decrease in smoke is due to the absence of aromatics in the fuel and the presence of higher hydrogen content in fuel.

4.2.5 Pressure—Crank Angle Diagram

The combustion process of the PD, HTWCO B25, and HTWCO B100 is shown in Fig. 11. The combustion process of the petro diesel and the hydrotreated fuel has similar characteristics. In first phase of the combustion, the fuel mixes with air forming the rich fuel zone. Since the presence of hydrogen in the hydrotreated fuel enhances the shorter ignition delay, it improves the combustion. The maximum pressure for the PD was 68.614 bat at 8 ATDC. The maximum pressures for HTWCO

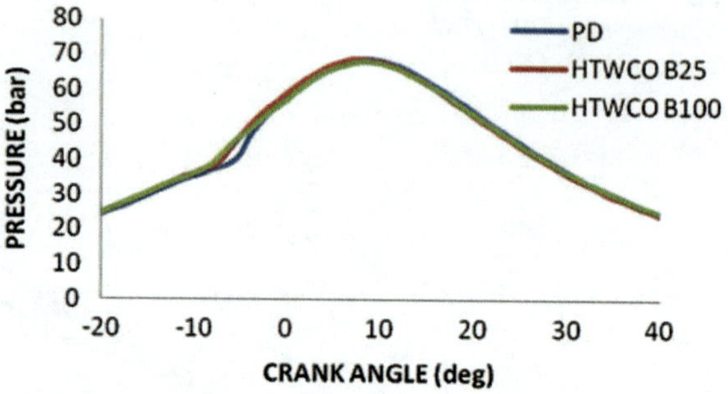

Fig. 11 Variation of pressure–Crank angle with respect to PD, HTWCO B25, and HTWCO B100

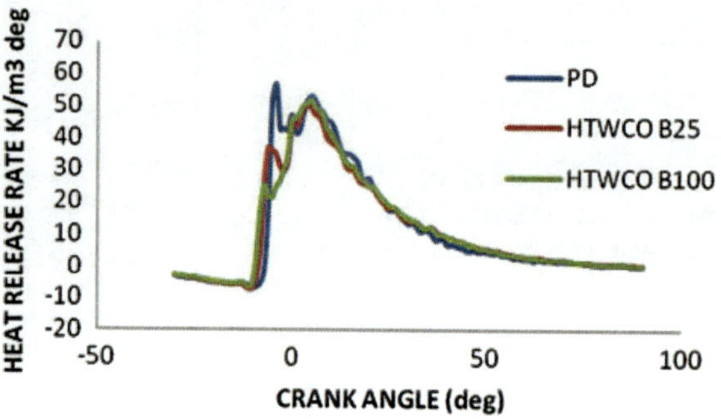

Fig. 12 Variation of heat release rate with respect to PD, HTWCO B25, and HTWCO B100

B25 and HTWCO B100 were 68.749 bar at 8 ATDC and 67.699 bar at 8 ATDC, respectively. This could be because of higher cetane index.

4.2.6 Heat Release Rate (HRR)

The maximum heat release rate of the PD, HTWCO B25, and HTWCO B100 is shown Fig. 12. The maximum HRR for PD, HTWCO B25, and HTWCO B100 were 56.674 kJ/m^3 deg at 4 BTDC, 49.612 kJ/m^3 deg at 4 ATDC, and 51.727 kJ/m^3 deg at 5 ATDC. The decrease in heat release rate was observed in the HTWCO B25 and HTWCO B100. This was because of the higher calorific value of the fuel and the absence of oxygen in fuel. It was also achieved because of lean fuel mixture.

5 Conclusions

In the latest study, the waste cooking was hydrotreated at a constant temperature of 350 °C, pressure 6 MPa, H_2/HC ratio of 300 L/L, and at LHSV of 2 h^{-1}. The performance, emission, and combustion characteristics of petro diesel and hydrotreated fuel extracted from waste cooking oil and its blend were compared.

The results are mentioned below with respect to petro diesel and full load condition

1. Decrease in BSFC for HTB25 and HTB100 of 12.5% and 25.5%, respectively
2. Significant reduction for CO, HC, NO_x, and smoke for HTWCO B25 and HTWCO B100 fuel when compared to petro diesel at full load condition.
3. The increase in brake thermal efficiency for HTWCO B25 and HTWCO B100 by 11.14% and 27.48%, respectively.
4. The high cetane index value improved the combustion of the hydrotreated fuel.

Thus, the above findings help us to conclude that 100% hydrotreated waste cooking oil can be used as fuel directly without any modification in the engine, leading to lesser emissions and better performance of the diesel engine.

Acknowledgements The authors would like to thank the Management of Sathyabama University and Chennai Petroleum Corporation Limited for granting permission to perform the performance and emission studies in thermal engineering laboratory and at the hydrotreatment pilot plant.

References

1. Huber GW, O'Connor P, Corma A (2007) Appl Catal A-Gen 329:120
2. Simacek P, Kubicka D, Sebor G Pospisil M (2009) Fuel 88:456
3. Solymosi P, Varga G, Hancsok J (2014) Chem Eng Trans 39:1387
4. Bezergianni S, Dimitriadis A, Kalogianni A, Knudsen KG (2011) Ind Eng Chem Res 50:3874
5. Sotelo-Boyas R, Liu Y, Minowa T (2011) Ind Eng Chem Res 50:2791
6. Bezergianni S, Dimitriadis A, Kalogianni A, Pilavachi PA (2010), Bioresour Technol 101:6651
7. Guzman A, Torres JE, Prada LP, Nunez ML (2010) Catal Today 156:38
8. Kousoulidou M, Dimaratos A, Karvountzis-Kontakiotis A, Samaras Z (2010) J Energy Eng-ASCE 09
9. Liu Y, Sotelo-Boya´s R, Murata K, Minowa T, Sakanishi K (2009) Chem Lett 38:552
10. Liu Y, Sotelo-Boy R, Murata K, Minowa T, Sakanishi K (2011) Energy Fuel 25:4675
11. Sebos I, Matsoukas A, Apostolopoulos V, Papayannako N (2009) Fuel 88:145
12. Jindal S, Nandwana BP, Rathore NS (2010) Energy Fuels 24:1565
13. Saravanan S, Nagarajan G, Lakshmi Narayana Rao G, Sampath S (2007) Energy Sustain Dev 11:83
14. Sugiyama K, Goto I, Kitano K, Mogi K, Honkanen M (2011) SAE Int J Fuels Lubr 5:205
15. Kalam MA, Masjuki HH, Jayed MH, Liaquat AM (2011) Energ 36:397
16. Hwang J, Qi D, Jung Y, Bae C (2014) Renew Energ 63:9
17. Roy MM, Wang W, Bujold J (2013) Appl Energ 106:198
18. Mumtaz MW, Adnan A, Mahmood Z, Mukhtar H, Malik MF, Qureshi FA, Raza A (2012) Int J Green Energy 9:68
19. Nantha Gopal K, Pal A, Sharma S, Samanchi C, Sathyanarayanan K, Elango T (2014) AEJ 53:281
20. Nayak SK, Pattanaik BP (2014) Energy Procedia 14
21. Jindal M, Rosha P, Mahla SM, Dhir A (2015) RSC ADV 5:33863

Experimental Investigation on HCCI Engine with Changes of the Air Intake Temperature

P. Moulali, T. Hari Prasad, and B. Durga Prasad

Abstract The submitted work introduces the impact of intake air temperature on the level of homogeneous charge compression ignition (HCCI) performance and emission. Using port fuel injection (PFI), the standard CI engine is altered into a HCCI engine. The performance and emission concentrations of standard CI engine and HCCI engine at distinct intake air temperatures (ambient, 100 °C, 120 °C and 140 °C) are compared at a steady engine speed of 1500 rpm. Nitrogen oxides (NO_x) and smoke opacity are decreased while operating the HCCI engine at ambient air intake temperature relative to the CI engine. NO_x emissions of the HCCI engine are further reduced with the increase of the intake air temperature up to 120 °C. Hence, the hydrocarbon (HC) emissions of HCCI engine reduce at ambient intake air temperature but the carbon monoxides (CO) emissions of HCCI engine are more compared to CI engine.

Keywords Brake thermal efficiency (BTE) · Intake air temperature · Oxides of nitrogen port fuel injection · HCCI engine

1 Introduction

Internal combustion (IC) engines play a crucial role in human lives, as day to day, the usages of the IC engines are increasing constantly. The IC engine became the integral parts of the society effectively since the nineteenth century. Nevertheless, various studies since then indicated the significant increase in the air pollution. Spark ignition engines have low compression ratio. Therefore, SI engine has low brake

P. Moulali (✉)
Research Scholar, JNTUA, Ananthapuramu, AP 515002, India
e-mail: moulalip20@gmail.com

T. Hari Prasad
Sree vidyanekathan Engineering College, Tirupati, India

B. Durga Prasad
JNTUA, Ananthapuramu, India

A. Arockiarajan et al. (eds.), *Advances in Industrial Automation and Smart Manufacturing*, Lecture Notes in Mechanical Engineering,
https://doi.org/10.1007/978-981-15-4739-3_86

thermal efficiency and low emissions. Though the compression ratio of CI engine is high, the engine produces harmful emissions [1, 2]. It generates more thermal break effectiveness than the SI engine. This situation encouraged the scientists to sound for fresh engines based on combustion technology such as the homogeneous charging compression ignition (HCCI) engine [3]. The port fuel injection (PFI) system is used in the HCCI engine to prepare the air–fuel mixture, like the spark ignition (SI) engine, but it is a self-ignition mechanism. HCCI engine combustion operates close to the constant volume process. But, the internal combustion (IC) engines release hazardous emissions [4]. Novel technology is introduced, and the active thermo-atmospheric combustion (ATAC) [5, 6] is a new combustion principle in two-stroke engine. The HCCI engine emissions are reduced with the use of different premixed fuels and cooled EGR [7, 8].

The peak pressure is reduced by 21% when the swirl ratio of inside cylinder is 4 with 30% EGR compared to a swirling ratio of 1–0% EGR [9]. Different fuels were used to control the emissions of biogas fuelled HCCI engine and biodiesel fuelled HCCI engine without changing its brake power [10, 11].

The impact of intake air temperature on emissions of gasoline HCCI particulate matter (PM) was described in the literature [6, 12]. It is noted that, with the rise of intake air temperature, the PM emissions were lowered. The rise in EGR outcomes in less NO and less PM. It is also noted that with the rise in intake air temperature, the smoke emissions were lowered [6, 13]. The experiment was conducted to control the smoke emission levels using turbocharged multi-fuel injection (MFI) engine with butane and diesel. It is noticed that the exhaust emissions are smoke, NO and CO_2 were reduced significantly [14]. It was found that the optimal value of fuel quantity was 1.5 mg and offset angle was 9.7° [15].

An experiment result shows that NO and smoke emissions could be reduced. With reform EGR (REGR), it is observed that even at full load operation the premixed combustion rate was improved [7, 16]. It is noticed that the NO and PM emissions were reduced up to 25% EGR. Further increases in the 25% EGR, the emissions were increased which reduced the performance on HCCI engine [17].

2 Experimental Set-up

A single-cylinder, four-stroke, water-cooled, direct Injection engine with eddy current load engine is used to perform the experiment and investigate the CI engine's combustion performance and emission qualities for diesel fuel with brake power. The experimental photo graphic views are shown in Fig. 1. Further, the experimental study is performed by modifying the single-cylinder diesel engine with extra fuel injector in the HCCI engine (injecting the fuel at outside of the combustion chamber to prepare the homogeneous charge compression) and air preheater to supplement the intake air temperature for mixing the fuel at outside of the engine. PE3SP ECU is installed for controlling the amount of the fuel supply based on the changes of

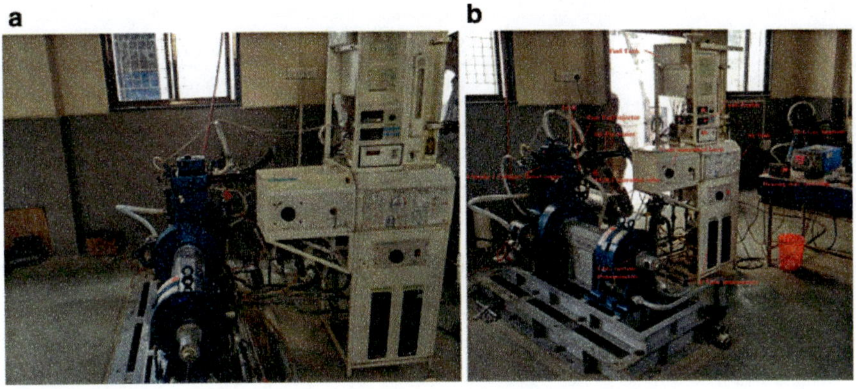

Fig. 1 Photo view of the **a** CI engine experimental set-up, **b** HCCI engine experimental set-up

brake power and air intake temperature [18–20]. PE3SP and air preheater are shown in Fig. 1b.

3 Experimental Procedure

The water flow rate is maintained at 120 lph (litre per hour) before starting the experiment. The HCCI engine is operated on no load condition to warm up engine until the air temperature reaches the saturated condition. The experiment is carried out on CI engine mode for the diesel fuel with respective brake power. The experiment is performed on HCCI engine with distinct ambient, 100 °C, 120 °C and 140 °C intake air temperatures for brake-powered diesel fuel. The experiment is conducted completely at constant engine speed of 1500 rpm and observed the performance, combustion and emission analysis [21–23]. The experimental details are in Table 1 (Table 2).

Table 1 Engine parameters

S. No.	Specification	Ranges
1	Model	TV1
2	Compression ration	12:1 to 18:1
3	Stroke x Bore x length	110 × 87.5 × 234 mm
4	Power	3.5 kW
5	Cooling type	Water cooled
6	Inlet value open	4.5°BTDC
7	Inlet value close	35.5°ABDC
8	Exhaust value open	35.5°BBDC

Table 2 Specification of gas analyser

S. No.	Specification	Measuring range	Resolution
1	HC	0–10,000 ppm	1 ppm
2	CO_2	0–20%	0.01%
3	CO	0.000–9.999%	0.001%
4	O_2	0–25%	0.01%
5	NO_x	0–5000 ppm	1 ppm
6	Measuring range	0–100% opacity in %	

4 Results and Discussion

4.1 Brake Thermal Efficiency (BTE)

Figure 2 indicates BTE changes in brake power with regard to the different ambient, 100 °C, 120 °C and 140 °C intake air temperatures in the HCCI engine and CI engine. BTE rises with BP rise. It is identified that the BTE of the HCCI engine at ambient temperature condition is lower than conventional CI engine of 27.3%. Because the diesel fuel does not mix at low temperature, it is difficult for the preparation of homogeneous mixture. Thus, it reduces the BTE of HCCI engine at ambient temperature. However, with the rise in intake air temperature up to 120 °C, the HCCI engine's brake heat efficiency is improved. This is due to the preparing of a more homogeneous mixture of air fuel for combustion and fuel atomization. It is found later that

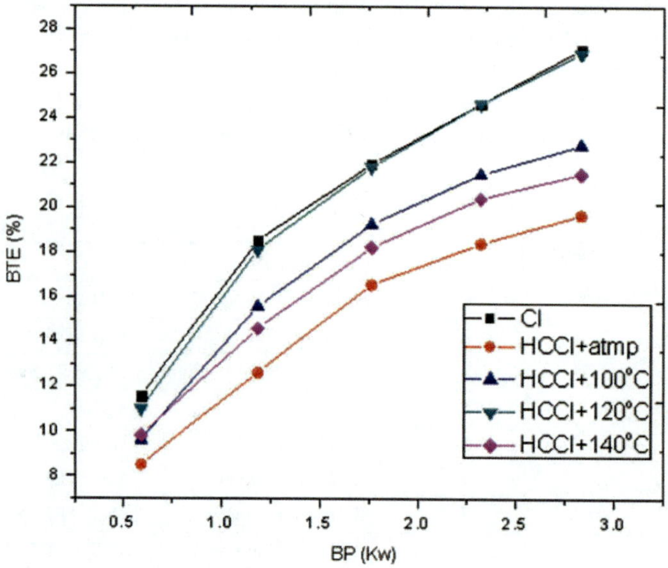

Fig. 2 BP versus BTE

with the rise in intake air temperature from 120 °C, the BTE reduces outside the engine cylinder owing to misfire in the combustion chamber and fuel evaporation. The BTE of the HCCI engine is increased by 26.8%, 15.24% and 20% at intake air temperature of 120 °C compared to HCCI engine with intake air temperature of ambient, 100 °C and 140 °C, respectively, at full load condition. It is discovered that the CI and HCCI engine brake heat effectiveness are equivalent to 120 °C air intake temperature.

4.2 *Brake-Specific Fuel Consumption (BSFC)*

Figure 3 illustrates the BSFC brake modifications with regard to the CI and HCCI engine brake energy changes. It is observed that with a rise in brake power for the CI and HCCI engine, the BSFC reduces. Experimental results demonstrate that the BSFC is more in the HCCI engine than in the CI engine. It is also noted that the HCCI engine's BSFC with intake air temperature 120 °C is lowered by 22 per cent compared to the ambient air intake temperature. This is because of good preparation of homogeneous mixture preparation at outside the engine cylinder, and this results in reducing of ignition delay and better combustion at lean mixture. The BSFC values of HCCI engine are 0.5 kg/kWh, 0.42 kg/kWh, 0.39 kg/kWh and 0.5 kg/kWh at ambient air intake temperature, 100 °C, 120 °C and 140 °C, respectively, at a brake power of 2.83 kW.

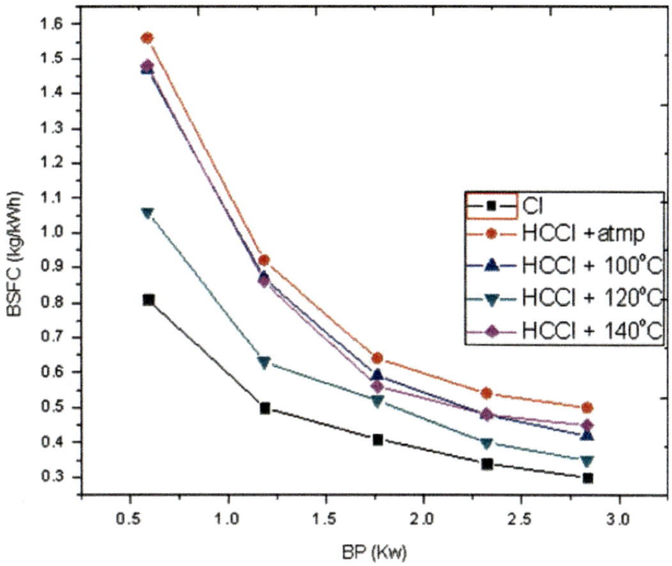

Fig. 3 BP versus BSFC

4.3 *Emission Characteristics*

CI and HCCI engine exhaust emission features are evaluated with regard to brake power. To measure the angle of the crank, AVL 365C is connected to the crank of the engine. Using the HG-540-5 gas analyser, the exhaust gas emissions are HC, NO_x, CO_2, O_2 and CO. The AVL 437 smoke metre is used to record the smoke opacity of the exhaust gas. The AVL 437 smoke metre is used to record the exhaust gas's smoke opacity.

4.3.1 Oxides of Nitrogen (NO_x)

Figure 4 shows the changes of the NO_x with respect to brake power of the CI and HCCI engine. The change of oxides of nitrogen is recorded at various intake temperatures of ambient, 100 °C, 120 °C and 140 °C. The oxides of nitrogen are reduced by 21.06% when HCCI engine is operated at atmospheric intake air temperature than that of CI engine. It is noticed that with increasing the intake air temperature up to 120 °C, the NO_x emissions are reduced. However, the continuous rise in the consumption of air temperature from 120 °C increases the nitrogen oxides. The oxides of nitrogen of the HCCI engine are decreased by 21.24%, 1.15% and 14.9% at intake air temperature of 120 °C compared to HCCI engine with intake air temperature of ambient, 100 °C and 140 °C, respectively. The oxides of nitrogen of the HCCI engine are reduced because

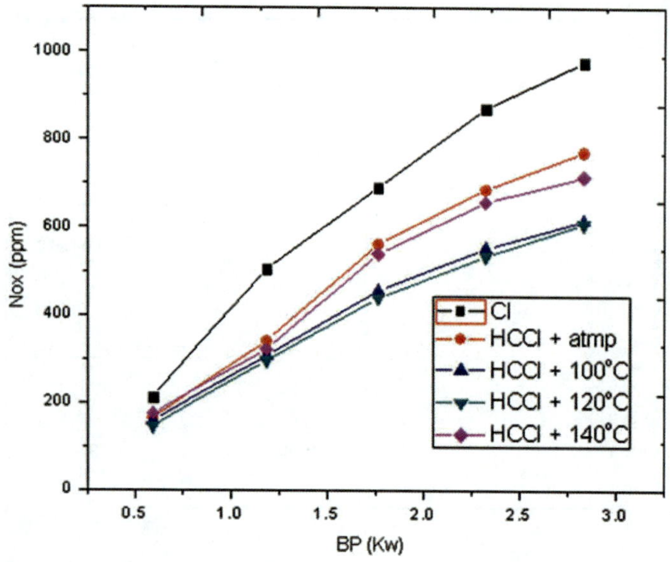

Fig. 4 BP versus NO_x

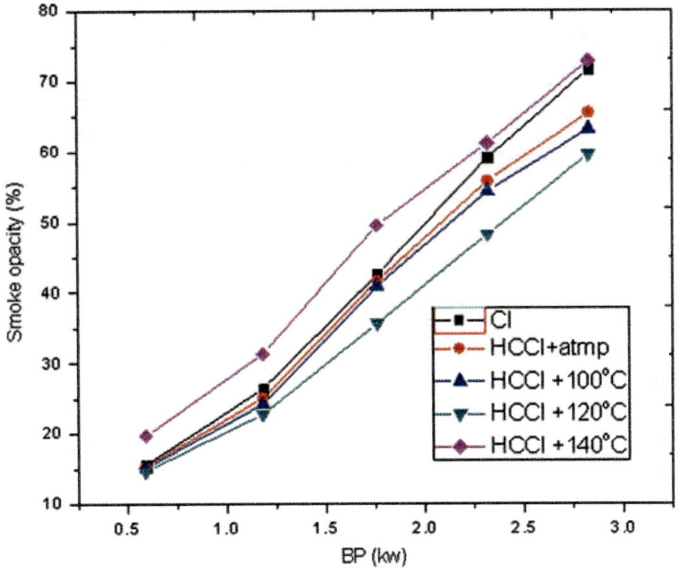

Fig. 5 BP versus smoke opacity

of low combustion temperature inside the cylinder. Since, CI engine combustion was observed high temperature its increases the oxides of the nitrogen.

4.3.2 Smoke Opacity

Figure 5 illustrates improvements in smoke opacity with regard to CI and HCCI engine brake strength with diesel fuel. Opacity of smoke rises with increased loads for CI and HCCI engine. It is observed that the smoke opacity of the HCCI engine is less than that of the CI engine. It is reduced by 8.51% compared to CI engine at an ambient intake air temperature. Due to homogeneous mixture preparing at the external port, the beginning of combustion (SOC) and ignition delay are decreased in HCCI engine. Experimental values of the smoke opacity are recorded as 15.5%, 26.4%, 42.5%, 59.1%, 71.6% for the CI engine and 15.3%, 25.1%, 41.6%, 55.8%, 65.5% for the HCCI engine at load of 20%, 40%, 60%, 80%, 100%, respectively, at an air intake temperature. Lowest smoke opacity is observed at an intake air temperature of 120 °C.

4.3.3 Hydrocarbons (HC)

The HCCI engine runs at low combustion process. The low combustion process avoids the complete combustion of charge or some of charge might leave from the cylinder without burning. Hence, it increases the unburned hydrocarbons on the exhaust. An experimental outcome demonstrates that hydrocarbon emissions increase with higher intake air temperatures up to 120 °C and further decreases HC emissions. HC emissions are reduced by 1.56% in HCCI engine at ambient intake temperature compared to CI engine at low load. Hydrocarbon emissions of CI engine at low load are recorded as 156 ppm, whereas the HC emissions of HCCI engine are 154 ppm, 166 ppm, 172 ppm and 156 ppm at intake air temperature of ambient, 100 °C, 120 °C and 140 °C, respectively.

4.3.4 Carbon Monoxide (CO) Emissions

The experimental findings indicate that the CO emissions are higher in HCCI engine compared to standard diesel engine. It is observed that the CO emissions are reduced up to medium load and after that the CO emissions are increased. The CO emissions are released during combustion and rely on the temperature of combustion. A low vapour zone like the fluid flow close to the cylinder wall triggers the development of CO in the HCCI mode engine. The CO emissions are released during the combustion and depend on the combustion temperature. Another variable was the regionally rich mixture, which during the fuel injection system continuously raises the CO concentration in the exhaust owing to absence of oxygen. A low combustion region such as boundary layer near the cylinder wall causes the CO creation in HCCI mode engine. Another factor is the locally rich mixture, which steadily increases the CO concentration in the exhaust due to lack of oxygen during the combustion process. Experimental values of CO for the CI engine are 0.107 % vol., 0.1 % vol., 0.097 % vol., 0.123 % vol. and 0.149 % vol., whereas for HCCI engine the CO values are 0.581 % vol., 0.472 % vol., 0.415 % vol., 0.69 6% vol. and 0.884 % vol. This is discovered that less CO emissions are registered by the CI engine relative to the HCCI engine. This is due to the elevated temperature of combustion and adequate oxygen for wealthy combustion of fuel (Figs. 6 and 7).

5 Conclusion

Compression ignition (CI) engine performance and emission features and homogeneous loaded compression ignition (HCCI) using diesel fuel are researched.

- Brake thermal efficiency of HCCI is reduced by 17.9% at intake air temperature of 120 °C than ambient temperature. The BTE efficiency of HCCI engine at 120 °C and CI engine is equal for all the loads.
- Brake-specific fuel consumption was reduced by 22% while HCCI engine running intake air temperature is 120 °C than atmospheric air intake temperature.
- Oxides of nitrogen were reduced by 21.06% while running HCCI engine at atmospheric intake air temperature than the CI engine. HCCI engine at intake air

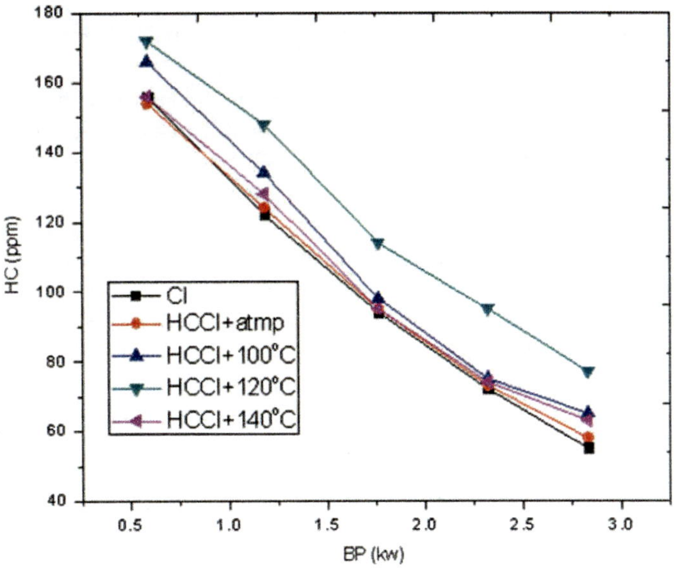

Fig. 6 BP versus HC

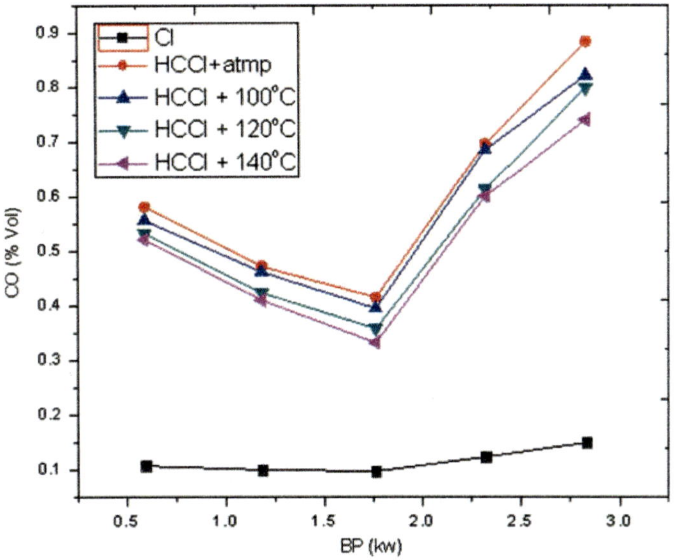

Fig. 7 BP versus CO

temperature 120 °C reduces the oxides of the nitrogen 21.24% and 14.9% than intake temperature atmosphere, 140 °C, respectively.

- Smoke opacity was reduced by 8.51% while running HCCI engine at intake air temperature was atmospheric than compared with CI engine. Further it reduced smoke opacity by 16.76% while running HCCI engine at intake air temperature is 120 °C than base CI engine.
- Hydrocarbon emission was reduced by 1.56% while running HCCI engine atmospheric condition than the conventional CI engine at low load.
- An experimental result of carbon monoxide emission was reduced up to medium load after that increases. This is because of insufficient fuel air mixture at full load. The value of the carbon monoxide CI engine was noticed 0.097% vol. And HCCI engine was 0.11% vol. At 60% of the load.

References

1. Yao M, Zheng Z, Liu H (2009) Progress and recent trends in homogeneous charge compression ignition (HCCI) engines. Prog Energy Combust Sci 35:398–437
2. Heywood JB (1988) Internal combustion engine fundamental. McGrawhill, New York
3. Kim DS, Kim MY, Lee CS (2007) Combustion and emission characteristics of a partial homogeneous charge compression ignition engine when using two-stage injection. Combust Sci Technol 179(3):531–551. https://doi.org/10.1080/00102200600671914
4. Zhang C, Zhang C, Xue L, Li Y (2017) Combustion characteristics and operation range of a RCCI combustion engine fueled with Direct injection n-heptane and pipe injection. Energy 125:439–448
5. Maurya RK, Agarwal AK (2011) Experimental investigation on the effect of inlet air temperature and air–fuel ratio on cycle to cycle changes of HCCI combustion and performance parameter. Appl Energy 88:1153–1163
6. Misztal, Xu H et al (2009) Influence of inlet air temperature on gasoline HCCI particulate emission. Combust Sci Tech 181:695–709
7. Tsolakis A, Megaritis A (2005) Partially premixed charge compression ignition engine with on-board H-2 production by exhaust gas fuel reforming of diesel and diesel. Int J Hydrogen Energy 30:731–45
8. Kim D, Lee C (2006) Improved emission characteristics of HCCI engine by various premixed fuels and cooled EGR. Fuel 85(5–6):695–704
9. Mohamed Ibrahim M, VarunaNarasimhan J, Ramesh A (2015) Comparison of the predominantly premixed charge compression ignition and the dual fuel modes of operation with biogas and diesel as fuels. Energy 89:990–1000
10. Ramesh A, Swami Nathan S, Mallikarjuna JM (2010) An experimental study of the biogas–diesel HCCI mode of engine operation. Eng Convers Manage 51:1347–1353
11. Ganesh D, Nagarajan G (2012) Performance and emission analysis on mixed-mode homogeneous charge compression ignition (HCCI) combustion of biodiesel fuel with external mixture formation. SAE International. https://doi.org/10.4271/2011-01-2450
12. Singh AP (2017) Partially homogeneous charge compression Ignition Engine development for low volatility fuels. ACS Publications. https://doi.org/10.1021/acs.energyfuels.6b02832
13. Gowthaman, Sathiyagnanam AP (2016) The effect ofinlet charge temperature on a diesel-fuelled homogeneous charge compression ignition engine. Int J Ambient Energy 38(2): 590–596

14. Kim Y, Kim K, Lee K (2008) A study on the combustion and emission characteristics of diesel fuel blended with ethanol in an HCCI engine. SAE Technical Paper 2008-32-0026. https://doi.org/10.4271/2008-32-0026
15. Yadav J, Ramesh A (2018) Injection strategies for reducing smoke and improving the performance of a butanol-diesel common rail dual fuel engine. Appl Energy 212:1–12
16. Qiang F, Junhua F, Zhuang J et al (2012) Influences of pilot injection and exhaust gas recirculation (EGR) on combustion and emissions in a HCCI-DI combustion engine. Appl Therm Eng 48:97–104
17. Gowthaman S, Sathiyagnanam AP (2017) The effect of exhaust gas recirculation on performance and emission characteristics of HCCI engine. Int J Ambient Energy 38(2):178–185. https://doi.org/10.1080/01430750.2015.1074616
18. Onishi S, Jo SH, Shoda K, Jo PD, Kato S (1979) Active thermoatmosphere combustion (ATAC)—a new combustion process for internal combustion engines. SAE Tech Pap Ser
19. Gentili R, Frigo S, Tognotti L, Habert P, Lavy J (1997) Experimental study on ATAC (active thermo-atmosphere combustion) in a two-stroke gasoline engine. SAE Tech Pap Ser
20. Aoyama T, Hattori Y, Mizuta J, Sato Y (1996) An experimental study on premixed-charge compression ignition gasoline engine. SAE Tech Pap Ser
21. Xingcai L, Libin J, Junjun M, Zhen H (2007) Experimental study on the cycle-by-cycle changes of homogeneous charge compression ignition combustion using primary reference fuels and their mixtures. Proc Inst Mech Eng—Part D J Automob Eng 221:859–866
22. Singh AP, Agarwal AK (2012) An experimental investigation of combustion, emissions and performance of a diesel fuelled HCCI engine. SAE Tech Pap Ser
23. Tanaka S et al (2003) Two-stage ignition in HCCI combustion and HCCI control by fuels and additives. Combust Flame 132(1–2):219–239

Experimental Investigation on the Performance of Refrigerator with Nitrile Rubber and Glass Wool as Insulating Material

K. Kiran Kumar and B. Chandra Mohan Reddy

Abstract The refrigeration is the science of producing and maintaining the temperature below that of the surrounding atmosphere. Insulating material is main component in the refrigeration system. The work of insulating material in household refrigeration system is to restrict the transfer of heat from system to surroundings. Hence, the coefficient of performance of refrigeration system depends on the insulating material. So, the selection of suitable insulating material for household refrigeration system is an important task in order to increase the coefficient of performance of refrigeration system. In this paper, experimental investigation is carried out on the performance of household refrigeration system with nitrile rubber and glass wool against polyurethane foam (PUF). This paper propose, suitable insulating material for the household refrigeration system and which material have high tendency to restrict heat transfer and smart temperature holding capability and smart cop and low energy consumption of the refrigerator.

Keywords Household refrigeration system · Nitrile rubber · Glass wool · Performance of glass wool · Performance of nitrile rubber

1 Introduction

The refrigeration is process of removing heat from a substance under controlled conditions. Maintaining the low temperature is carried out by the number of subsystems. Insulating material is main component in the refrigeration system. The work of insulating material in household refrigeration system is to restrict the transfer of heat from system to surroundings. Hence, the coefficient of performance of refrigeration system depends on the insulating material. So, the selection of suitable insulating

K. K. Kumar (✉)
Department of Mechanical Engineering, Srinivasa Ramanujan Institute of Technology, Rotarypuram, Anantapur 515701, Andhra Pradesh, India
e-mail: kiranshalivahana6@gmail.com

B. C. M. Reddy
Department of Mechanical Engineering, JNT, Anantapur, Andrapradesh, India

A. Arockiarajan et al. (eds.), *Advances in Industrial Automation and Smart Manufacturing*, Lecture Notes in Mechanical Engineering,
https://doi.org/10.1007/978-981-15-4739-3_87

material for household refrigeration system is an important task in order to increase the coefficient of performance of refrigeration system [1, 2].

2 Literature Review

Importance of insulating material *source:* ASHRAE, 1981 in this paper, the efficiency of the refrigerator can be enhanced by the proper insulating materials. Depending up on the application, the type of the insulating material also varies. Nitrile rubber is the one of best outcome materials to improve the performance of the refrigerator [3].

Domkundnw ararora Properties of insulating material Author based on data from ASHRAE, 1981 in this context we observe that the many insulating materials are available for reduce the cool effect from the cooling cabin. This context also gives the idea about the properties for good insulating material depending upon the situation [4].

3 Experimental Setup and Procedure

The experimental setup consists of 80 L capacity of household refrigerator, designed to work with R-134a refrigerant that has the four main components compressor, condenser, capillary tube and evaporator coil. The experimental setup consists of the conventional vapor-compression refrigerating machine with proper instrumentation to measure temperature at evaporator section and to measure energy required by the system.

An energy meter is used to measure the energy input to the compressor motor, and temperature indicator is used to measure the temperature inside the evaporator. The schematic diagram of the experimental setup is shown in Fig. 1. The insulating materials are bonded to the evaporator on the outer surface in order not to reduce storage capacity. In this experimental setup, study of heat distribution and temperature holding capacity of evaporate with three different types of insulating materials they are PUF, glass wool and nitrile rubber. During the first step, an insulating material cabin is constructed with the help of polyurethane foam (PUF), and then, the capacity of transfer of heat and temperature holding capacity of PUF has been studied. In the second step instead of polyurethane foam, an insulating material cabin is constructed with the help of glass wool, and then, the capacity of transfer of heat and temperature holding capacity of glass wool has been studied. Similarly, in the third step instead of glass wool, an insulating cabin is constructed with the help of nitrile rubber, and then, the capacity of transfer of heat and temperature holding capacity of nitrile rubber has been studied (Tables 1 and 2).

Procedure

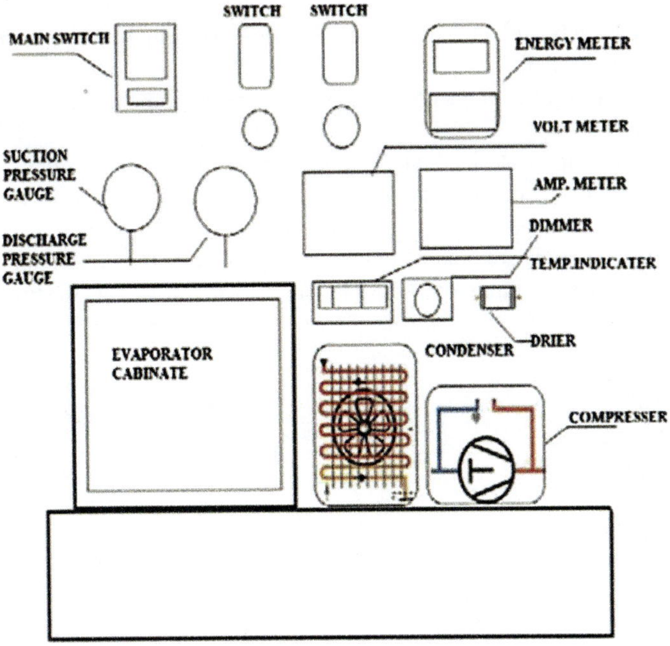

Fig. 1 .

Table 1 Evaporator box Specifications

Particulates	Specifications (cm)
Length	33
Height	12
Width	25
Thickness	1.5

Table 2 Specifications of refrigerator

Refrigerator model	ALLWYN refrigerator 80 L
Ice making time	2 h 30 min
Voltage range	220 + 10% V 50 Hz AC
Rated energy consumption	2.00 kWh/24 h max
Rated gross volume	80 L
Rated general storage volume	72 L
Fuse rating	5A
Refrigerant	R-134a

1. The household refrigerator is selected, which is working on vapor compression refrigeration system
2. At the inlet and outlet of the compressor pressure and temperature, indicators are installed.
3. With the help of pressurized nitrogen, flushing of the system is done
4. The vapor compression refrigeration system is charged with R134a refrigerant.
5. With the help of soap solution, leakage test is done
6. Insulating materials are bonded on the outer surface of the evaporator.
7. After starting the test unit, pressure and temperatures are recorded (Figs. 2, 3 and 4).

Fig. 2 Refrigerator setup when pressure and temperature gauges are fixed

Fig. 3 Evaporator cabin bonded with glass wool

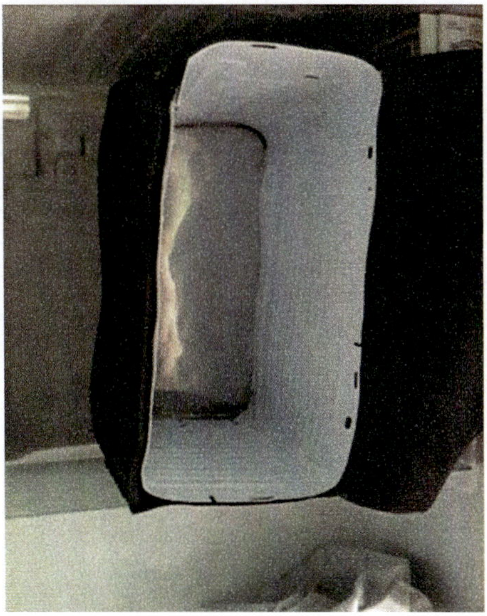

Fig. 4 Evaporator cabin bonded with nitrile rubber

4 Results and Discussion

By performing this experiment, the following results were obtained (Tables 3 and 4; Chart 1).

The above graph shows that

1. Refrigeration effect of the system increases with nitrile rubber (190 kJ/kg) as insulating material compared with PUF (177 kJ/kg) and glass wool (175 kJ/kg).

Table 3 .

Parameter	Existing insulating material (PUF)	Glass wool as insulating material	Nitrile rubber as insulating material
Refrigerating effect (kJ/kg)	177	175	190
Compressor work (kJ/kg)	66	76	65
COP	2.68	2.30	2.92

Table 4 .

Parameter	Existing insulating material (PUF)	Glass wool as insulating material	Nitrile rubber as insulating material
Energy consumption	0.11 kw/hr	0.11 kw/hr	0.09 kw/hr

Chart 1 Refrigeration effect and compression work comparison

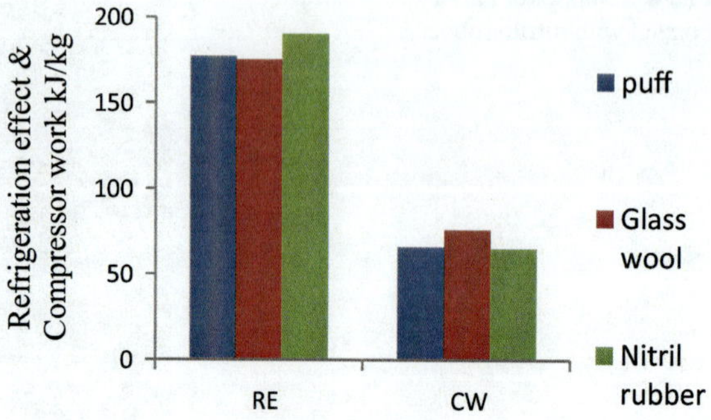

Chart 2 COP comparison

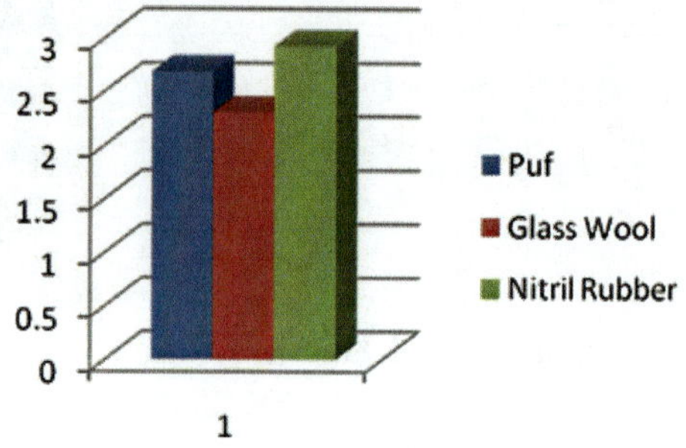

2. The compressor work is decreased by using nitrile rubber because of its high insulating capacity compared with PUF and glass wool (Chart 2).

The cop of the VCR system is enhanced by using the nitrile rubber. Due to the unique properties, nitrile rubber is able to reduce the heat losses from the cooling cabin. Apart from this, PUF shows good results compared with glass wool.

5 Conclusion

In this present investigation, glass wool, polyurethane foam (PUF) and nitrile rubber such three different types of insulating material are used. From the experimental investigation, it is found that the temperature holding capacity of nitrile rubber is more than that of glass wool and polyurethane foam (PUF). If nitrile rubber used instead of glass wool and PUF, the energy consumption of household refrigerator is reduced. And it found that nitrile rubber has a better coefficient of performance than glass wool and PUF.

References

1. Domkundnwar A (1981) Properties of insulating material: source: prepared by Author based on data from ASHRAE
2. Importance of insulating material *source:* ASHRAE 1981
3. Nandi DN (2004) Hand book on insulation and refractory. edn 2004
4. American Society of Heating (2001) Refrigeration and Air-Conditioning Engineers, inc www.ashre.org. Reprinted by permission from ASHRAE 2001 Hand book fundamentals

Enhancement of Energy Efficiency Using Environmentally Benign Refrigerant Blends in Vapour Compression Refrigeration System

P. Elumalai, R. Vijayan, V. Subburam, and S. Maniraj

Abstract This study involves theoretical investigations made on a simple vapour compression system working with R152a/R1270/R600 refrigerant mixture. The properties of the mixtures along with that of R22 were obtained from REFPROP 7.0 software for the operating temperature ranging from 0 to 60 °C. CYCLE_D software has also been used for finding the proposed refrigerant mixture's performance. Test results from theoretical study presented that the coefficient of performance of the HFCs and HC refrigerant mixtures being 2.1% higher than that of R22. Compressor power of the mixtures was 1.8% less than that of R22 at 7 °C evaporator temperature and 53 °C condenser temperature. Mixture of the refrigerants showed higher mass flow rate than that of R22, and also its deviation was 36% from R22. Compressor shell outlet temperature of the refrigerant mixtures was 9.6% lower than R22 at 7 °C evaporator temperature and 53 °C condenser temperature. In this study, the selected ternary refrigerant mixture was found to provide better energy efficiency, and hence, it can be used as a suitable replacement for HCFC 22 in residential air conditioners.

Keywords Refrigerant · Environment · Ozone depletion · Condenser · Evaporator · R22

P. Elumalai (✉)
Department of Mechanical Engineering, Nehru Institute of Engineering and Technology, Coimbatore, Tamilnadu, India
e-mail: eluammu@gmail.com

R. Vijayan
Department of Mechanical Engineering, Government College of Engineering, Salem, Tamilnadu, India

V. Subburam · S. Maniraj
Department of Mechanical Engineering, Paavai Engineering College, Namakkal, Tamilnadu, India

A. Arockiarajan et al. (eds.), *Advances in Industrial Automation and Smart Manufacturing*, Lecture Notes in Mechanical Engineering,
https://doi.org/10.1007/978-981-15-4739-3_88

1 Introduction

R22 is being mainly used in the heat pumps and air-conditioners for the past two decades, and it has the leading sales percentage when compared with other refrigerants. R22 has less ozone depleting substance comparing with other CFCs. However, it contains chlorine which is an ozone depleting substance. Shaika et al. [1] have stated that it becomes essential to use naturally eco-friendly refrigerants in the process of developing alternative refrigerants to fulfil the provisions of Kyoto and Montreal protocols. The eco-friendly hydrocarbon refrigerants are flammable in nature, and hence, its leakage from system should be avoided for safety purpose. Theoretical thermodynamic studies have been carried out on refrigerants like R32, R290, R600, R600a, R1270, R134a and R152a as alternative working fluids for R12, R22 and R134a [1]. Mota-Babiloni et al. [2] have reported that the working substances commonly used in refrigerators and air conditioners like R134a, R410A and R404A are going to be phased out by EU regulations (No.517/2014) as their ODP and GWP values are high due to extended use.R32, R125, R134a and R152a are HFCs and R1234yf are HFOs by composition. The greenhouse effect causing the global warming has become a serious global issue now. As a remedy, the Kyoto protocol has given outlines to relieve hydrofluorocarbons (HFCs) which are classified as one of the greenhouse gases [2]. The study of Lontsi et al. [3] is the combined effect of injection/compression cycles of refrigeration proposed for the production of cold required for freezing and refrigeration. The performance of the system has been analysed by using R290, R152a and R134a as refrigerants for their cold production temperature level and the impact of temperature on environment [3]. Yoo et al. [4] have stated that currently all the countries around the globe are taking steps to prevent ozone depletion and global warming. A refrigerant with zero value of ODP like R134a finds higher usage in automotive air conditioning system, but its high GWP is a dangerous factor. Another refrigerants, R152a, has zero ODP, and GWP is suggested as good alternative for R134a [4]. He et al. [5] have used a mixture of HFC152a/HFC125 (1,1difluoroethane/pentafluoroethane) refrigerants in domestic refrigerator and also calculated the performance of refrigeration cycle. The theoretical analysis carried out on HFC152a/HFC125 showed that the composition of HFC152a with a mass fraction of 0.85 produces same performance as that of R22 [5].

Wu et al. [6] in their attempt to find an alternative for R22 tested the mixtures of R152a/R125/R32 with different mass proportions. NIST REFPROP 7.0 software was used for developing the refrigerant performance and thermodynamic performance of the proposed refrigerant mixture and R22 for comparative analysis. New mixture was utilized in the domestic refrigerator which actually designed for R22. The results revealed that the new mixture could be a promising substitute for R22 [6]. Park et al. [7] have tested hydrocarbon refrigerants R290 and R1270 in heat pump of one ton of refrigeration capacity with a hermetically sealed compressor using water/glycol mixture, and water have secondary heat transfer working fluids. The experiments were conducted in different operating conditions and the coefficient of performance and other parameters obtained [7].

Calm et al. (2004) have reported that the highly toxic HFC refrigerants could not maintain a choice for long term to the society. The other alternatives available are R290 and R1270 which are hydrocarbon refrigerants with favourable thermodynamic and transport properties and also categorized as zero ODP and GWP refrigerants [8]. Corberan et al. [9] have stated that though R290 and R1270 have low ODP and GWP, their high flammability nature may lead to dangerous accidents, and because of this problem, international societies have taken severe steps to prohibit their use. But their eco-friendly nature has brought them back into the process [9]. Ahamed et al. [10] have reported that developing countries like Malaysia are using halogenated refrigerants as they possess better thermo-physical and thermo dynamic properties. This type of refrigerant runs in vapour compression refrigeration system and has high ODP and GWP. The alternate for halogenated refrigerants were investigated, and pure butane and isobutane that gave high COP than R134a were suggested [10].

Arcaklioglu et al. [11] analysed vapour compression system working with halogenated refrigerants and stated that during phase changes under constant pressure, change occurs in condenser and evaporator temperatures. Disadvantage of the system is that temperature gliding requires to be assumed. Usage of pure hydrocarbons is suggested to solve this problem [11]. Fatouh et al. [12] have suggested the use of HC mixtures as refrigerant in domestic refrigerators instead of R134a. Simulation study was used to predict the performance of R134a, propane(R290), commercial butane (R600) and propane/n-butane/R600a refrigerants with different composition by mass in domestic refrigerators. The simulation results showed R600 as having low COP and high operating pressure [12]. The present research work involves investigation of R152a/R1270/R600 mixtures with different mass proportions in vapour compression refrigeration system.

2 Refrigerant Mixture Selection

The proposed HFC and HC ternary mixtures with different percentage of composition to investigate for replacement of R22 are mentioned below.

- M1 = R152a-60%/R1270-20%/R600-20%
- M2 = R152a-20%/R1270-20%/R600-60%
- M3 = R152a-10%/R1270-10%/R600-80%
- M4 = R152a-10%/R1270-80%/R600-10%
- M5 = R152a-80%/R1270-10%/R600-10%.

Among the selected combination of refrigerant mixtures, M5 is found to be the optimum mixture with low GWP, high COP and temperature glide also is within the limit. Table 1 below presents the COP and temperature glide obtained for mixtures containing different proportions of HCFC 22 at −15 °C of evaporator temperature and 30 °C of condenser temperature [13].

Table 1 Selected refrigerant mixture with various proportions corresponding to temperature glide and COP

Refrigerant mixtures	R152a (%)	R1270 (%)	R600 (%)	COP	Temperature glide (°C)	
					T_{eva}	T_{cond}
M1	60	20	20	3.23	3	3.6
M2	20	20	60	3.28	7.3	9.6
M3	10	10	80	3.27	4.7	7.6
M4	10	80	10	3.18	4.2	3.9
M5	80	10	10	3.25	3.4	3.6

From the table, the temperature glide of refrigerant mixtures for both T_{eva} and T_{cond} should have the acceptable limits less than 12 °C at a given operating conditions. Since they are non-azeotropic (refrigerant) mixtures, they exhibit a temperature variation during constant pressure phase change. Table 1 inferred that M5 provides the best combinations of high COP and low-temperature glide which is favourable for getting good energy efficiency. Figure 1 represents Vapour Compression Refrigeration System in which the working fluid as refrigerant undergoes phase changes, is one of the refrigeration cycles and is the best suitable method for air-conditioning applications.

GWP of blend can be calculated [14] (from IPCC 2013),

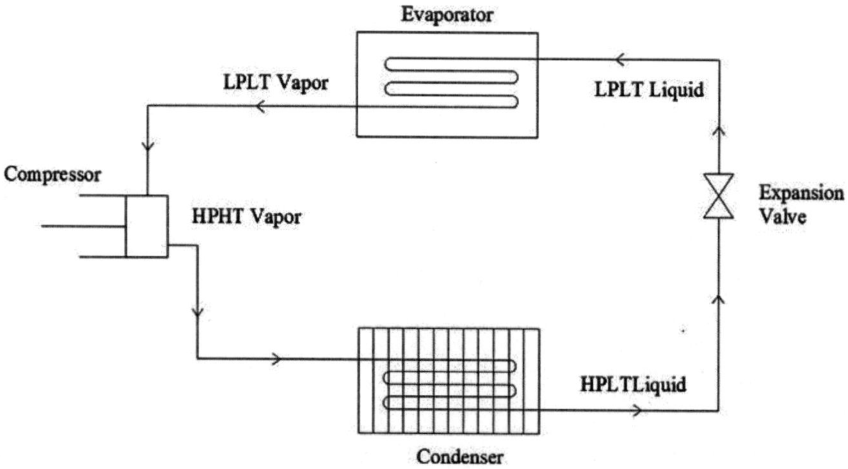

Fig. 1 Schematic diagram of Vapour Compression Refrigeration (VCR) system

$$\text{GWP of blend(M5)} = \text{Proportion by \% mass of component A} \times \text{GWP of A} + \\ \text{Proportion by \% mass of component B} \times \text{GWP of B} + \\ \text{Proportion by \% mass of component C} \times \text{GWP of C}$$

where

$$\begin{aligned} \text{Component A} &= \text{R152a},\ \text{GWP of A} = 124,\ \text{Mass percentage} = 80\% \\ \text{Component B} &= \text{R1270},\ \text{GWP of B} = 2,\ \text{Mass percentage} = 10\% \\ \text{Component C} &= \text{R600},\ \text{GWP of C} = 3,\ \text{Mass percentage} = 10\% \\ &= (0.8 \times 124) + (0.1 \times 2) + (0.8 \times 3) \\ &= 101.8 \end{aligned}$$

2.1 Simulation

The performance studies of the proposed refrigerant mixtures in vapour compression system are done using CYCLE_D software. The software can simulate required conditions of operations and produce performance parameters of the given refrigerants of single component and multi-component blends. The basic system for simulation comprises components such as compressor, compressor suction line, discharge line, condenser, expansion valve and evaporator. REFPROP 7.0 software designed and developed by NIST, USA, is used to study the required properties of the proposed refrigerants mixtures along with the existing R22 refrigerant. The database of the software can produce thermo-physical property models for a variety of industrially applied fluids and fluid mixtures an accepted standard.

3 Results and Discussion

The theoretical study carried out on the performance of environmentally benign refrigerant mixtures with different proportion of HFC and HC blends involves measurement of various performance parameters such as pressure ratio, discharge temperature, compressor power, mass flow rate, refrigeration effect and COP which are listed below in Table 2.

The performance parameters under different operating conditions are discussed below.

3.1 Coefficient of Performance

Figure 2 reveals how the condenser temperature influences the co-efficient of performance of the proposed mixtures of R152a/R1270/R600 with five different mass proportions and that of R22. For increase in condenser temperature, the refrigerant effect decreases which in turn lead to decrease the cooling capacity. In course with these changes, the increase in pressure ratio of compressor causes the required compressor work to increase. The combined influence of these two input parameters decreases the coefficient of performance as can be seen from Fig. 2. Figure 3 shows the percentage deviation of COP of the test mixtures from R22 which is set as base for comparison. It can be noticed from Fig. 3 that the coefficient of performance of M5 is 2.10% higher than that of R22 due to its highest cooling capacity at 53 °C condenser temperature and 7 °C evaporator temperature.

Table 2 Summary of test results refrigerants various mass percentage

Performance parameters	R22	R152a-60 R1270-20 R600-20	R152a-20 R1270-20 R600-60	R152a-10 R1270-10 R600-80	R152a-10 R1270-80 R600-10	R152a-80 R1270-10 R600-10
COP	3.24	3.39	3.31	3.4	3.13	3.31
Compressor power, kW	1.08	1.07	1.03	1.03	1.12	1.06
Pressure ratio	3.35	3.61	3.89	4.1	3.21	3.67
Mass flow rate, kg/s	0.025	0.016	0.013	0.013	0.014	0.016
Discharge temperature (°C)	86.2	76.7	73.4	77.9	75.4	68.2
Refrigeration effect, kJ	140.08	225.43	269.01	273.99	244.15	217.11

Fig. 2 Variation of COP with condenser temperature

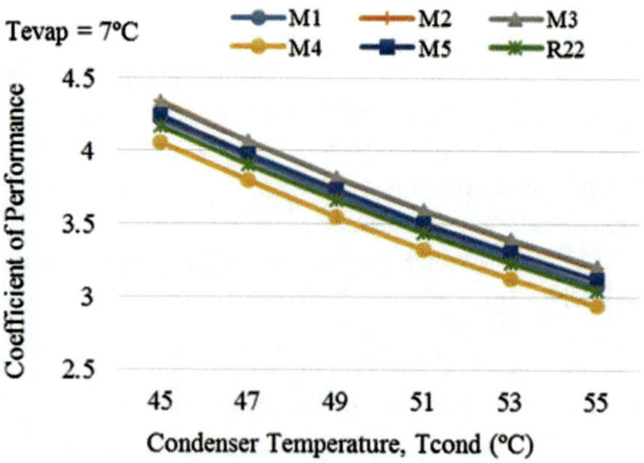

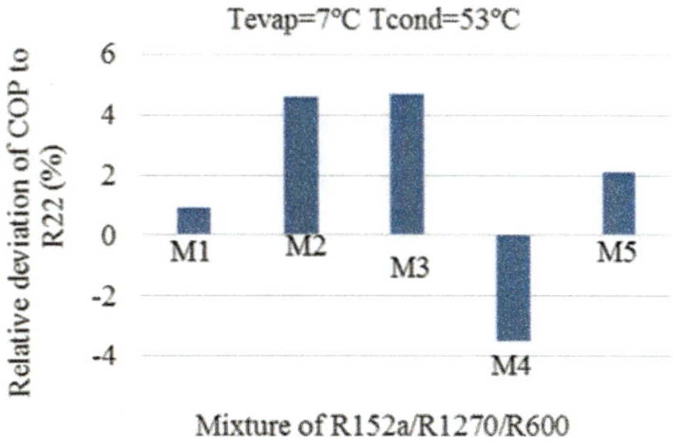

Fig. 3 Deviation of COP compared with R22

3.2 *Pressure Ratio*

Figure 4 represents the effect of condenser temperature on pressure ratio of the proposed refrigerant mixtures under test along with existing R22. The increase in T_{cond} causes in increasing pressure ratio across the hermitically sealed compressor which is due to increase in discharge pressure of compressor in direct relation with the condenser temperature. Figure 5 is plotted to show the deviation of pressure ratio of the test mixtures when compared to R22 at the operating condition of 7 °C evaporator temperature and 53 °C condenser temperature. The figure reveals that the pressure ratio of M5 mixture is 8.5% higher than that of R22 because of its lowest suction pressure.

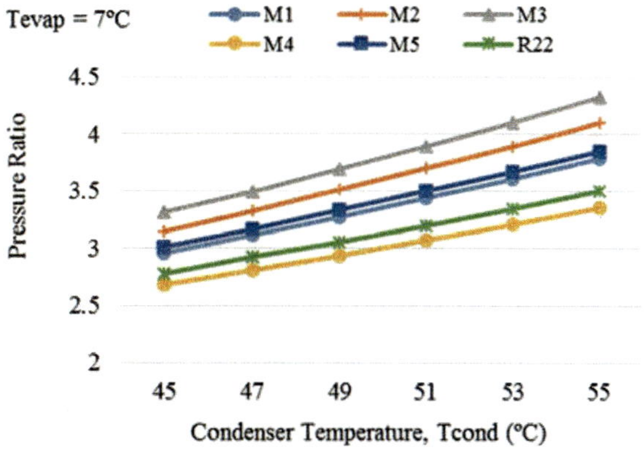

Fig. 4 Variation of pressure ratio with condenser temperature

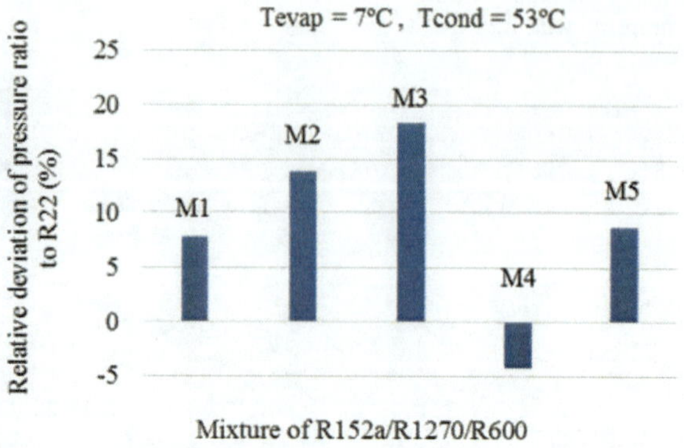

Fig. 5 Deviation of pressure ratio compared with R22

3.3 *Discharge Temperature*

Figure 6 reveals the influence of condenser temperature on discharge temperature for the proposed mixtures and also R22 with which test mixtures are compared. The graph clearly indicates that increase in condenser temperature increases with discharge temperature of compressor. Figure 7 provides the relative deviation in percentage of the test mixtures M1to M5 (with different mass proportions of R152a/R1270/R600) from the base mixture R22 at the operating temperature of 7 °C evaporator temperature and 53 °C condenser temperature. It can be observed that the refrigerant mixture of M5 depicts the highest deviation of 26.3% from R22, and it is lower than the discharge temperature of HCFC 22. The lower compressor discharge temperature of M5 is favourable property in terms of refrigerant stability, lubricant characteristics and prolonging the operating life of the compressor (Devotta et al.).

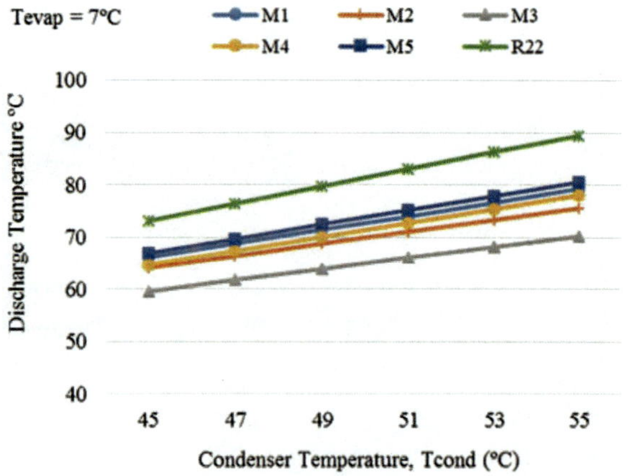

Fig. 6 Variation of compressor discharge temperature with condenser temperature

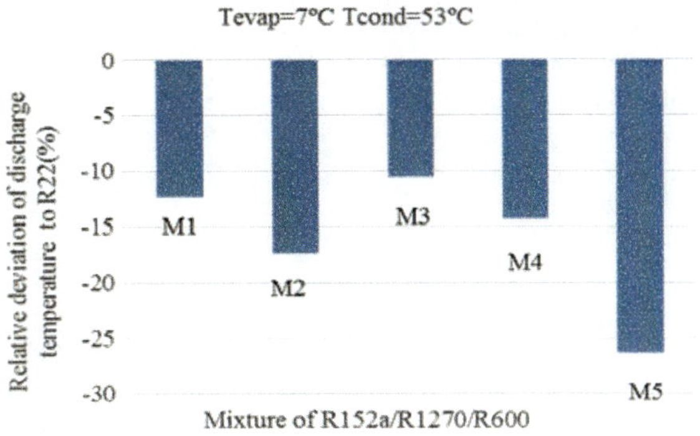

Fig. 7 Deviation of compressor discharge temperature compared with R22

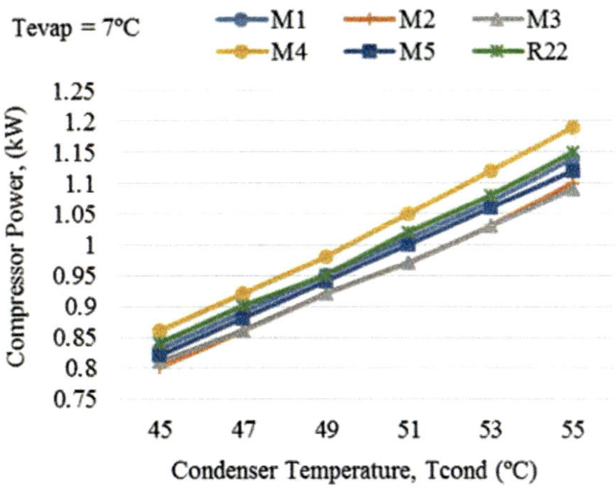

Fig. 8 Variation of compressor power with condenser temperature

3.4 *Compressor Power*

Figure 8 shows the influence of condenser temperature on condenser power to cause variations of test mixtures M1to M5 and base mixture R22. When the condenser temperature is increased, it leads to increase in pressure ratio which in turn increases the compressor power. Figure 9 represents the deviation percentage of compressor power of the proposed mixtures M1to M5 from that of base mixture R22 at the test conditions of 7 °C evaporator temperature and 53 °C condenser temperature. It can be observed that the graph that M5 is having 1.8% less compressor power compared to R22, whereas M4 shows 3.5% higher compressor power.

3.5 *Mass Flow Rate*

Figure 10 shows the effect of mass flow rate of proposed refrigerant mixtures and

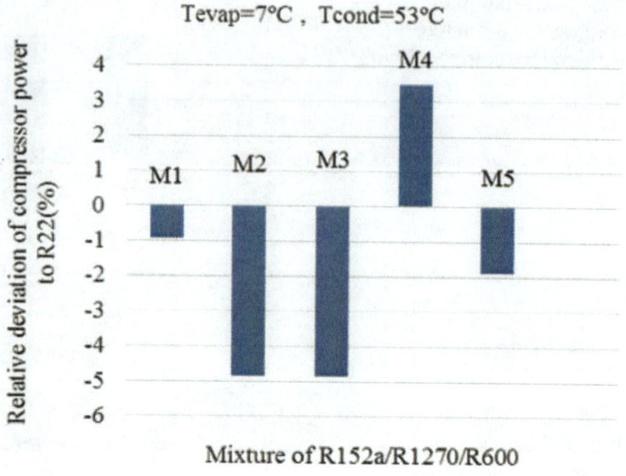

Fig. 9 Deviation of compressor power compared with R22

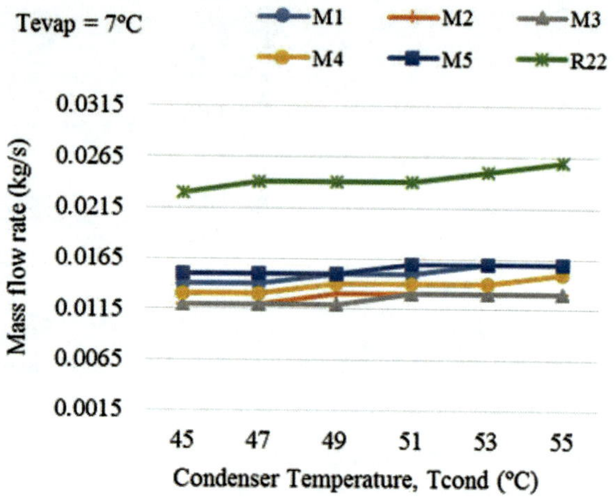

Fig. 10 Variation of mass flow rate with condenser temperature

base refrigerant with respect to the changes in condenser temperature for the test mixtures and base refrigerant. The pattern of variation shows that the mass flow rate of refrigerant mixtures increases with the increase in condenser temperature. The results show that the mass flow rate of HCFC 22 is the highest among the all refrigerants tested. Figure 11 shows the deviation in mass flow rate of the proposed mixtures (five different mass proportions of R152a/R1270/R600 designated as M1, M2, M3, M4 and M5) from that of R22 as base refrigerant at specified condition of 7 °C evaporator temperature and 53 °C condenser temperature. The bar plot clearly shows that the mass flow rate of M5 is 56.25% lower than that of refrigerant HCFC 22.

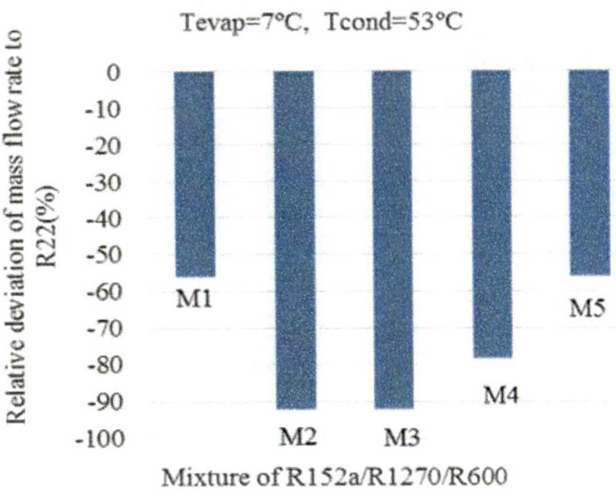

Fig. 11 Deviation of refrigerant mass flow rate compared with R22

3.6 *Refrigerating Effect*

Figure 12 shows the influence of condenser temperature to cause in variation in refrigerant effect of the test mixtures and R22. The effect of increase in condenser temperature is that of decrease in cooling capacity caused by the decrease in evaporation latent heat. Figure 13 represents the relative deviation of refrigeration effect of the test mixtures in comparison to R22 at specified condenser (53 °C) and evaporator (7 °C) temperature. The graph depicts that of M5 having a deviation of 35.4% higher cooling effect than R22. It can also be observed that the deviation of M3 is the highest among the test mixtures and its cooling effect is 48.87% higher than that of R22 (Fig. 13).

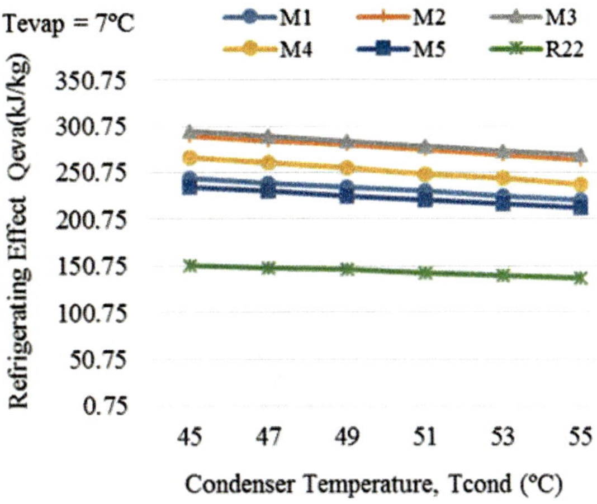

Fig. 12 Variation of refrigerating effect (cooling) with condenser temperature

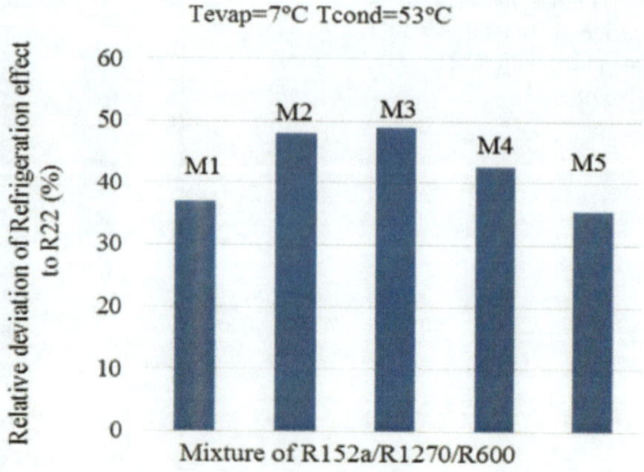

Fig. 13 Deviation of Refrigeration effect compared with R22

4 Conclusion

The main objective of this study is to identify an eco-friendly refrigerant mixture with desirable performance parameters to replace R22 in vapour compression refrigeration system. The ternary refrigerants R152a/R1270/R600 with five different mass proportions are taken for study, and their thermal performances are compared with that of R22 to find the optimal combination among the test mixtures. CYCLE _D software is used for simulation and analysis. The performance characteristics of the test fluids along with R22 are measured for study and the major conclusions are given below.

1. The coefficient of performance(COP) of M1 to M3 and M5 were found to be higher than that of R22 by 4.7%, 4.6%,2.1%, and 0.9%, respectively, and M4 gave 3.51% lower COP than that of R22.
2. The pressure ratio of M1 to M3, M5 was higher than that of HCFC 22 by 7.8%, 13.8%, 18.3% and 8.71%, respectively, and that of M4 was 4.3% lower than R22.
3. The discharge temperature of the compressor for M1, M2, M3, M4, M5 were found to be lower than that of R22 by 12.3%, 17.04%, 10.6%, 14.3%, and 26.3%, respectively.
4. The compressor power of M4 was 3.5% higher than that of R22, whereas mixture (M1, M2, M3, M5) were found to be lower by 0.93%, 4.85%, 4.85%, 1.88%, respectively.

The theoretical investigation of the thermal performance of the new ternary mixtures of R152a/R1270/R600 has given good results compared to the existing R22 mixture. From the study, it can be inferred that these mixtures can be a suitable alternatives to R22 in air-conditioning and heat pump applications. Among the test mixtures, M5 mixture is found to have optimal composition as it has shown better energy efficiency than other mixture.

References

1. Shaika SV, Babub PTA (2017) Thermodynamic performance analysis of eco friendly refrigerant mixtures to replace R22 used in air conditioning applications. Energy Procedia 109:56–63
2. Mota-Babiloni A, Navarro-Esbrí J, Barragán-Cervera Á, Molés F, Peris B (2015) Analysis of new HFC/HFO mixtures as alternatives of high GWP refrigerants in refrigeration and HVAC systems. Int J Refrig (2015)
3. Lontsi F, Hamandjoda O, Mayi OTS, Kemajou A (2016) Development and performance analysis of a multi-temperature combined compression/ejection refrigeration cycle using environment friendly refrigerants. Int J Refrig
4. Yoo SY, Lee DW (2009) Experimental study on performance of automotive air conditioning system using R-152a refrigerant. Int J Automot Technol 10(3):313–320
5. He MG, Li TC, Liu ZG, Zhang Y (2005) Testing of the mixing refrigerants HFC152a/HFC125 in domestic refrigerator. Appl Therm Eng 1169–1181:25
6. Wu J, Chu Y, Jing H, Liu Z (2009) Performance of mixture refrigerant R152a/R125/R32 in domestic air-conditioner. Int J Refrig 32:1049–1057
7. Park KJ, Jung D (2008) Performance of R290 and R1270 for R22 applications with evaporator and condenser temperature variation. J Mech Sci Technol 22:532–537
8. Wu JH, Yang LD, Hou J (2012) Experimental performance study of a small wall room air conditioner retrofitted with R290 and R1270. Int J Refrig 35 (2012)
9. Corberan JM, Segurado J, Colbourne D, Gonzalvez J (2008) Review of standards for the use of hydrocarbons refrigerants in A/C, heat pump and refrigeration equipment. Int J Refrig 31(4):748–756
10. Ahamed JU, Saidur R, Masjuki HH (2010) Thermodynamic performance analysis of R-600 and R600a as refrigerant. Eng e-Trans 5(1):11–18. (ISSN 1823-6379)
11. Arcaklioglu E, Cavosuglu A, Erisen A (2005) An algorithmic approach towards finding better refrigerant substitutes of CFCs in terms of the second law of thermodynamics. Energy Convers Manag 46:1595–1611
12. Fatouh M, El Kafafy M (2016) Essessment of propane/ commercial butane mixtures as possible alternatives to R134a in domestic refrigerators. Energy Convers Manag 47:2644–2658
13. Khurmi RS, Guptha (2009) A text book of Refrigeration and Air Conditioning. S. Chand Publishers, New Delhi
14. Climate Change (2013) The physical science basis. Contribution of working group I to the fifth Assessment Report of the Intergovernmental panel on Climate Change (IPCC). Cambridge University Press, Cambridge, United Kingdom and New York, NY, USA, p 1535

Investigation of Performance of Vapor Compression Refrigeration System with Forced Air-Cooling Condenser With R600a and Hydrocarbon Mixture (R290 + R600a) as Refrigerant

Mulla Irfan Ahmad, S. M. Jameel Basha, V. Viswanatha Chari, and G. Srinivas Kumar

Abstract The environmental impacts like global warming and ozone depletion have become a challenge for R & AC Industry. Even though the refrigerant R134a is better in performance, it is categorized under hydrofluorocarbon which has more GWP and poor heat transfer rates. So, the refrigerant is likely to be phased out; therefore, there is a need to go for alternative refrigerant. The other refrigerants may be hydrocarbons and mixture of hydrocarbons. Hydrocarbons are called natural refrigerants because they occur in the earth's material cycle, as a by-product of natural gas production, or in oil refineries. They are non-toxic as well as ozone-friendly. Also in domestic refrigerator, condenser is the heat exchanger which condenses the refrigerant to liquid phase by losing latent heat. It exchanges the heat with atmosphere through natural convection. The heat transfer coefficient for the natural circulating air is less than the forced circulating air. In order to increase heat transfer rate, forced convection is used. In this work, a cabin is modeled by using aluminum sheet. Performance of 220L capacity refrigerator equipped with forced convection apparatus working on hydrocarbons will be analyzed.

Keywords Al-6061 · Stir casting · Wear behavior · SEM analysis · Brake rotor · R134a · Hydrocarbons · R600a · Forced convenction

M. I. Ahmad (✉) · V. Viswanatha Chari · G. Srinivas Kumar
Department of Mechanical Engineering, Srinivasa Ramanujam Institute of Technology, Anantapuramu, India
e-mail: irfan0338@gmail.com

V. Viswanatha Chari
e-mail: visumtech@gmail.com

G. Srinivas Kumar
e-mail: srin.8045@gmail.com

S. M. Jameel Basha
Sri Krishnadevaraya Engineering College, Gooty, India
e-mail: jameeljunaid@gmail.com

A. Arockiarajan et al. (eds.), *Advances in Industrial Automation and Smart Manufacturing*, Lecture Notes in Mechanical Engineering,
https://doi.org/10.1007/978-981-15-4739-3_89

1 Introduction

In today's world, refrigeration is very important for both domestic and industrial purposes. Eighty percentage of domestic refrigerator used in India works on R134a as refrigerant. R134a has good thermodynamic properties and thermophysical properties. But the main disadvantage of R134a is having high GWP index of 1300. It is likely to be phased out. The alternative may be hydrocarbons [1–3]. Hydrocarbons occur as a by-product in earth's material cycles. So, they are called as "natural refrigerants." In the early 1920s, they were used as refrigerants. In the early 1950s, slowly they were replaced by halogenated hydrocarbons as they are non-toxic and nonflammable and they are referred to as "safety refrigerants." Meanwhile, it was found that these halogenated hydrocarbons are damaging the ozone layer, as well as leading to global warming which resulted in banning. By using natural refrigerant as a refrigerant, which can counter or reduce the global warming effect. The major components in refrigerant system are compressor, condenser, expansion valve and evaporator. The condenser is a heat exchanging device in domestic refrigerator in which the refrigerants get condensed by losing latent heat at constant pressure. The mode of heat transfer between the condenser and ambient air is convection. It is divided into two types, natural convection and forced convection depending on how the fluid flow is initiated. In case of natural convection, fluid motion occurs naturally due to density difference or pressure differences or buoyancy effect. The cooler fluid will be having high density when compared to hot fluid, so cooler fluid falls down. In case of forced convection, external forces initiate the flow like pump or fan. The fluid motion enhances heat transfer; i.e., higher the velocity higher the heat transfer rate [4]. The expression for rate of heat transfer is given by Newton's law of cooling.

$$q_{\text{convection}} = h(T_s - T_a)\left(\text{W/m}^2\right)$$
$$Q_{\text{convection}} = hA(T_s - T_a)(\text{w})$$

2 Methodology

Figure 1 shows the line diagram of the present work and how the experimental setup was developed. First, the experimental setup was developed from the refrigerator capacity of 220L having R134a compressor. The experimental setup has four temperature-measuring points and four pressure-measuring points. The temperature is measured using digital thermocouples, whereas pressure is measured by using low- and high-pressure gauges. Figure 3 shows the cabin in which induced fans and exhaust fans are attached. The induced fans pump cold air toward the condenser which is covered by the cabin, and the exhaust fans blow out the hot air from the cabin. The main purpose for the arrangement of fans was to increase the heat transfer rate from the condenser to the environment, due to which refrigerant gets sub-cooled

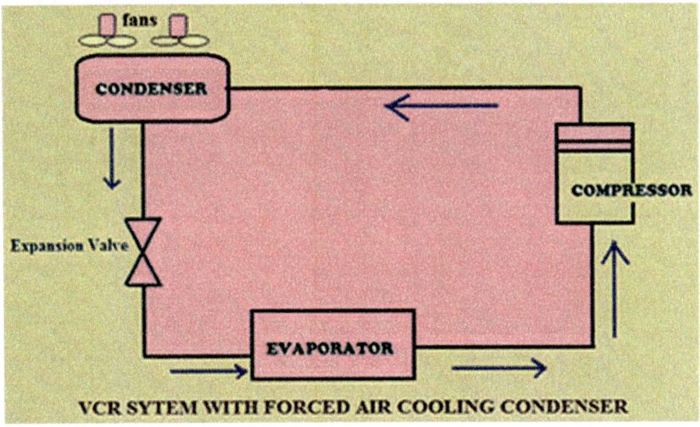

Fig. 1 Line diagram of experimentation

[5, 6]. As a result, the refrigeration effect increases. Further advantage of this fan arrangement is it reduces the pressure ratio and creates turbulence inside the cabin which in turn increases the heat transfer [4]. As a result, power consumption of the compressor decreases and the life duration of the compressor increases. The cabin is modeling by using NX CAD software (Fig. 2).

Fig. 2 Experimental setup without modification

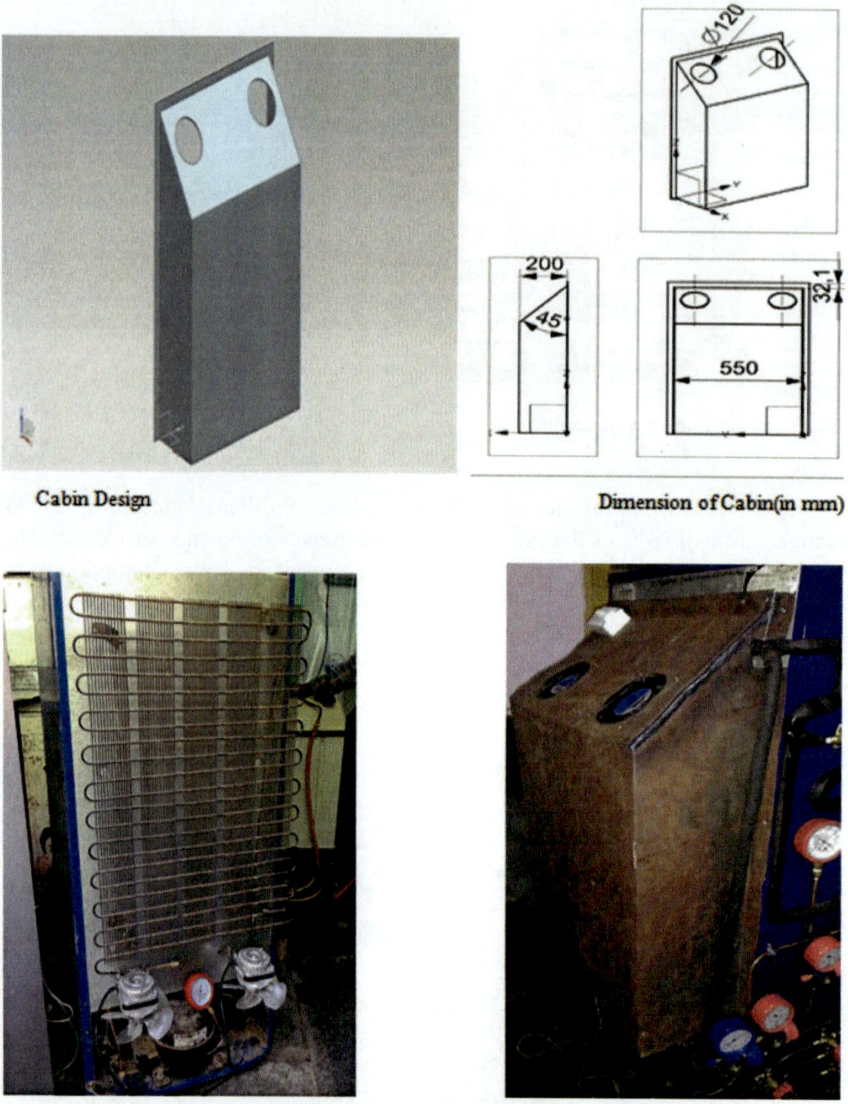

Fig. 3 Total experimental setup

3 Experimentation

Now for the experimental setup, the vacuuming and evacuation were done by another compressor which is up to 25 in/hg. After that, nitrogen gas was filled in the system and then the soap bubble test was conducted, to check whether there are any leakages. After that, again vacuum is created. The main purpose of this was to remove moisture, air and oil. Now initially the refrigerant R134a (135 g) is charged into the system by using charging unit, initial reading of thermocouples and pressure gauges is noted, and the readings were noted for every 5 °C decrease in the evaporator temperature. Then after taking the readings, the power supply to system is stopped. When the evaporator temperature reaches the ambient temperature, the power supply to system was given and also to the fan setup. Same steps are followed, and various readings were taken. After this, the R134a refrigerant was replaced with hydrocarbon mixture (R290 + R600a (60 + 40)). This mixture is charged (60 g) with charging unit after creating the vacuum. The same steps are followed, and various readings were taken with fan and without fan. After taking the readings, the R134a is replaced with the R600a. An amount of 60 g R600a refrigerant is charged into the compressor after vacuuming and evacuation [7, 8]. Then, various readings were taken without and with fan following the same procedure.

4 Results and Discussion

The above graphs clearly indicate that the time taken to reach the evaporator temperature to −100 °C is minimum for R600a refrigerant with forced convection, then followed by forced convection cooling with hydrocarbon mixture and then followed by R134a with forced convection. In case of non-forced cooling, the order is R600a, hydrocarbon mixture and R134a. So, the figure states time taken for achieving cooling is minimum with forced convective cooling at the condenser (Figs. 4 and 5).

The above graph clearly states that the amount of power consumed in the forced convection-cooled condenser is less when compared with natural air-cooled condenser. It clearly also states that the power consumption is minimum for the R600a, then followed by hydrocarbon mixture and R134a.

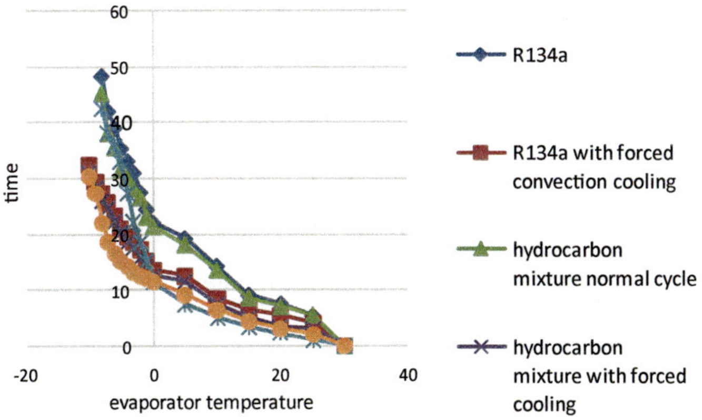

Fig. 4 Evaporator temperature versus time

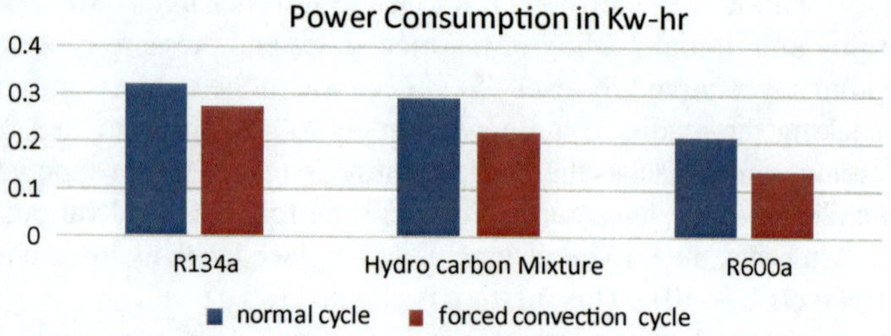

Fig. 5 Power consumed by the system when it is operated for two hours

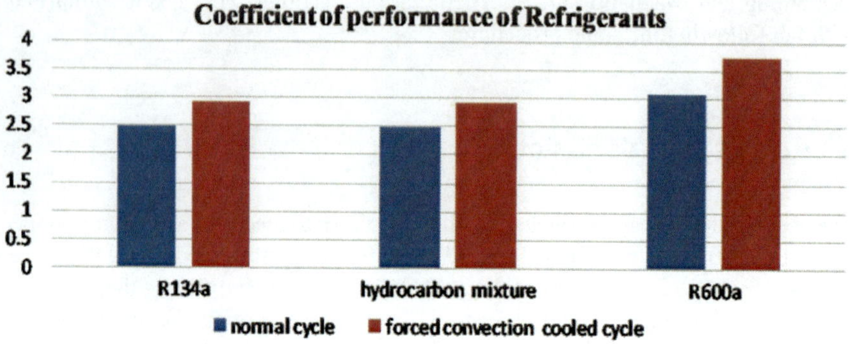

Fig. 6 Coefficient of performance of refrigerants

Figure 6 Coefficient of performance for different refrigerants without and with forced cooling

The above graph clearly indicates that coefficient of performance is better for forced convection-cooled cycle. And the coefficient of performance is better for R600a with forced convection-cooled condenser, then followed by the hydrocarbon mixture and R134a, respectively.

5 Conclusions

Here, experimental study reveals that:

From the results obtained, the following conclusions are made:

The modification from air-cooled condenser to forced air-cooled condenser of refrigerator has the following effects:

1. The refrigeration effect is increased by 7% in case of R600a and R 134a and 6.7% in case of hydrocarbon mixture.

2. The pressure ratio is decreased by 25.47% in case of R600a, 18.3% in case of hydrocarbon mixture and 15.14% in case of R134a due to which the life period of the compressor increases.
3. The power consumption of the system decreased by 33.33% in case of R600a, 24.1% in case of hydrocarbon mixture and only 15.6% in case of R134a.
4. The coefficient of performance of system with forced air-cooled condenser is increased by 21.31%, and in case of hydrocarbon mixture it is found to be 18.33%, whereas in case of R134a it is found to be just 15.14.

From the above results, it can be concluded that refrigeration system with R600a as refrigerant along with forced convection air-cooled condenser is having better performance.

References

1. Dalkilic AS, Wongwises S (2010) A performance comparison of vapour compression refrigeration system using various alternative refrigerants. Int Commun Heat Mass Transf 37:1340–1349
2. Yu CC, Teng TP (2014) Retrofit assessment of refrigerator using hydrocarbons refrigerants. Appl Therm Eng 66:507–518
3. Mahajan RY, Borikar SA (2014) Performance evaluation of domestic refrigerator using HC-12a refrigerant as an alternative to R12 and R134a. Int J Eng Sci 3(10):26–37
4. Singh AP, Kumar V, Pandey SK, Yadav M, Ram K, Upadhayay SK, Jain PA, Mishra VK (2015) Performance analysis of domestic refrigerator with forced and natural convection. Pelagia Res Libr Adv Appl Sci Res 6(7):216–223
5. Hu SS, Huang BJ (2005) Study of a high efficiency residential split water-cooled air conditioner. Appl Therm Eng 25:1599–1613
6. Slama RB (2009) Water-heater coupled with the refrigerator to develop the heat of the condenser. Int Renew Energy Congr Sousse Tunisia
7. Lee MY, Lee DY, Kim Y (2008) Performance characteristics of a small-capacity direct cool refrigerator using R290/R600a (55/45). Int J Refrig 31:734–741
8. Mohanraj M, Jayarai S, Muraleedharan C, Chandrasekhar P (2009) Experimental investigation of R290/R600a mixture as an alternative to R134a in a domestic refrigerator. Int J Therm Sci 48:1036–1042
9. Abdul Qureshi M, Bhatt S, Comparative analysis of cop using R134a & R600a refrigerant in domestic refrigerator at the steady state condition

Development of Multi-functioning Organic Waste Shredding Machine for Natural Compost

E. Pavankumar, S. Baskaran, R. Prithivirajan, S. Vinoth Kumar, and A. Karpagaraj

Abstract The organic waste such as food, flowers, vegetables and fruits and garden waste such as dry leaves, wet leaves and gross are not often effectively recycled according to municipal waste management statistics in India. Nowadays, the recycling of organic waste is very important for creating better and good environment. There are three conventional recycling techniques in India, i.e., organic waste converter (OWC), bio-energy plant and natural composting. In conventional recycling process, the waste is converted into manure and gas by standard processes like shredding and squeezing. In this work, it is proposed to design and fabricate the shredding machine by integrating these processes in a single unit, thereby reducing processing time and increasing the production rate. Such a machine would make the process easier, safer, faster and more economical. It can be used in hostels, restaurants, schools, colleges and vegetables market yards.

Keywords Organic waste · Shredding machine · Natural compost

E. Pavankumar · S. Baskaran (✉) · R. Prithivirajan
Department of Mechanical Engineering, Madanapalle Institute of Technology & Science, Madanapalle, Andhra Pradesh, India
e-mail: baskaranslvm@gmail.com

E. Pavankumar
e-mail: epavankumaar@gmail.com

R. Prithivirajan
e-mail: prithivi72@gmail.com

S. Vinoth Kumar
School of Mechanical Engineering, Shandong University, 250061 Jinan, China
e-mail: svkceg@gmail.com

A. Karpagaraj
Department of Mechanical Engineering, National Institute of Technology, Patna, India
e-mail: akraj@nitp.ac.in

A. Arockiarajan et al. (eds.), *Advances in Industrial Automation and Smart Manufacturing*, Lecture Notes in Mechanical Engineering,
https://doi.org/10.1007/978-981-15-4739-3_90

1 Introduction

In recent years in India, the accumulation of municipal solid waste (MSW) is very high in the country due to increase in population. It has emerged as a severe problem concern with environmental and aesthetic issues for recycling or removing completely the above stated waste. According to the Central Pollution Control Board (CPCB) statistics, out of total municipal solid waste generated in India during 2015, only 15–20% amount of waste is recycled or treated for reuse. The major problems and aspects which are impacting proper management of MSW are limited collection at the doorstep, reluctance in public to take ownership, unavailability of adequate funds, lack of access to proper technology and unscientific disposal of MSW at dump sites. Municipal solid waste includes household, commercial, industrial, construction and demolition waste in either solid or semi-solid form excluding industrial hazardous wastes but including treated biomedical wastes. It also includes wastes from hotels and restaurants, sanitation residue and waste from streets. Nowadays, processing of organic waste has become popular for producing natural compost and to attain more beneficial from byproduct of compost as a renewable energy resource such as biogas. Natural compost can be produced from the organic waste such as agriculture waste, garden waste (i.e., dry leaves, wet leaves and gross), garbage waste such as vegetable market waste, fruit market waste and fish market by shredding and squeezing operations. Shredding is the process of cutting the large pieces into small size pieces, thereby increasing the biological decomposition easily. In the conventional organic waste recycling process, shredding and squeezing processes are performed by physically separated units to convert organic waste into manure and gas. These recycling processes can reduce the amount and size of the waste into very small pieces which enhance the decomposition of wastes quickly and effectively.

The literature review is carried out based on various existing shredder machines and the following important topics such as organic waste, shredder, squeezer, design and development and fabrication of machines. Hande and Deshpande [1] developed the methodology to fabricate portable organic waste chopping machine for producing the compost. Pavankumar et al. [2] have designed and fabricated the economic organic waste shredding machine to shred agriculture waste, kitchen debris, cooked foods, etc. Sanjay Kumar and Hemanth Kumar [3] developed agricultural waste shredder machine for producing vermin compost from chopping of coconut leaves and areca leaves. The performance of the shredder was measured by accounting total quantity of powder chopped for different time durations. Tegegne [4] developed the dual shaft multi-blade shredder for shredding the plastic wastes for recycling purpose. The developed shredder showed the satisfactory performance and produced 11 kg of plastic pieces per hour. Amiebenomo et al. [5] designed and developed the breadfruit shredder machine and reported that press weight has a significant effect on the time to shred and invariably the shredding efficiency of the machine. Etoamaihe et al. [6] has developed a reciprocating motion cassava shredding machine and reported that the shredding efficiency mainly depends on the size of the shredding aperture of the machine. The machine produced the maximum shredding efficiency of 92%

at shred aperture 3 mm with shredding speed of 975 rpm. Sara [7] developed a shredder attachment for landfilling compactor machine with consideration of safety and environmental requirements. In another study, Fitzgerald [8] developed a pre-shredding machine to shred municipal solid wastes prior to disposal. Jibrin et al. [9] designed and developed crop residue crushing machine to enhance agricultural production in an economic manner. In this study, by keeping in mind that integrating shredding and squeezing operations in a single machine, a portable organic waste shredder is developed and evaluated for its performance. This machine will be an alternative to the existing shredder, and squeezer machines perform multi-function in single machine, thereby reducing the processing time and the final cost of the machine.

2 Methodology

Figure 1 shows the methodology for executing the present work. Based on the literature review of organic waste, existing shredding machine, design and development, the problem has been identified for the present work. After collecting the all related information, the objectives are framed. Before fabrication, the individual components are designed by using the CATIA V5 software. The raw materials are selected

Fig. 1 Methodology

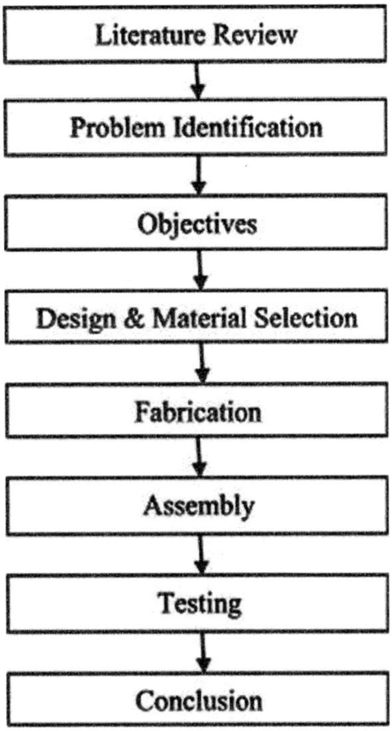

Table 1 List of parts and type of material

Components	Type of material
Shredder and squeezer shaft	Stainless steel
Shredder and squeezer body	Mild steel
Shredder and squeezer housing	Stainless steel
Shredder blade	Mild steel
Squeezer blade	Mild steel
Frame	Mild steel bar

based on the properties required such as strength, hardness and availability to fabricate the shredding machine. Based on the detailed design of individual components of shredding machine, the fabrication work is started by using cutting, welding, grinding, painting, etc. Finally, the individual components are assembled together for achieving desired function. To evaluate the performance of the fabricated machine, the testing is carried out by using the selected organic wastes and reported.

3 Materials Selection

The materials for fabricating shredder were selected by considering the cost, availability, mechanical properties, fabrication methods and environmental factors. Table 1 depicts the materials used for manufacturing different parts of the shredder. The main parts of the shredder are shafts, blades, housing and frames.

4 Design Considerations

Before performing design analysis, certain mechanical and operational factors were considered for successful design, fabrication and operations of the shredding machine. The important mechanical factors are strength and rigidity of the construction, shredder should accommodate different speeds, the hopper should accommodate different sizes of organic waste, the machine should be coated with nontoxic corrosion-resistant paints to avoid rust and finally the vibration and noise level of the machine should be low. The important operational factors are: Shredder should ensure uniformity of shreds produced and the capacity of the machine should be high enough to achieve fast processing of organic wastes.

5 Design Analysis of Major Components

Based on the standard formulas, torque, force, power, length of the belt and speed of pulley, shafts were calculated to fabricate the components of multi-function organic waste shredder machine; some of these parameters are

1. Power need for shredding (for motor selection)
2. Force at each stage
3. Pulley calculations such as length and speed
4. Torque and V-belt calculations
5. Average shredding speed (i.e., 350 rpm).

5.1 Torque

The following formulas are used for calculating the torque, power and force.

$$T = F \times r$$

$$\text{Electric power (P)} = T \times \omega$$

$$T = (60 \times P)/(2\pi \times n)$$

$$\omega = 2\pi N/60$$

where

- T Torque transmitted by the shaft (N m)
- P Transmitting power (kW)
- n Speed of transmission shaft (rpm)
- ω Angular velocity (rad/s)
- N Shredding speed (rpm)
- F Shredding force (N)
- r Radius of the shredding shaft (m).

Assumptions: Power = 1 hp = 746 W = 0.746 kW, $n = 1440$ rpm, $N = 350$ rpm, Torque produced by the motor:

$$T = (60 \times 746)/(2 \times 3.14 \times 1440)$$
$$T = 4.9495\,\text{N m}$$

Now, angular speed of motor shaft,

$$\omega = 2\pi n/60 = 2 \times 3.14 \times 1440/60$$
$$= 150.72 \text{ rad/s}.$$

5.2 Pulley

Speed of pulleys: The expression for diameter of driven pulley or shaft pulley is given by

$$D_s = D_m \times N_m/N_s,$$

where

D_s Diameter of shaft pulley (mm)
D_m Diameter of motor pulley (mm)
N_m Speed of electric motor (rpm)
N_s Speed of shaft (rpm).

Speed of squeezer shaft:

Diameter of shaft pulley (D_s) = 200 mm
Diameter of electric motor pulley (D_m) = 75 mm
Speed of electric motor (N_m) = 1440 rpm

Speed of driven shaft:

$$D_s = D_m \times N_m/N_s$$
$$N_s = 75 \times 1440/200$$

Speed of the driven shaft (squeezer shaft), $N_s = 540$ rpm.
Now, angular speed of squeezer shaft

$$\omega_{\text{squz}} = 2\pi N_s/60$$
$$= 2 \times 3.14 \times 540/60$$

Angular speed of squeezer shaft $\omega_{\text{squz}} = 56.62$ rad/s.

Speed of shredder shaft:

Diameter of shaft pulley (D_s) = 200 mm
Diameter of electric motor pulley (D_m) = 100 mm

Speed of electric motor = 540 rpm (i.e., shredder shaft is driven by the squeezer shaft).

Speed of driven shaft:

$$D_s = D_m \times N_m/N_s$$
$$N_s = 100 \times 540/200$$

Speed of the driven shaft (shredder shaft) $N_s = 275$ rpm.
Now, angular speed of shredder shaft

$$\omega_{\text{shred}} = 2\pi N_s/60$$
$$= 2 \times 3.14 \times 275/60$$

Angular speed of shredder shaft $\omega_{\text{shred}} = 28.783$ rad/s.
Torque produced: During an operation

At squeezer shaft:

$$\text{Power} = \text{Torque}\,(\tau) \times \text{speed}$$
$$P = \tau \times \omega$$
$$\tau = P/\omega$$
$$\tau = P/(2pN_s/60)$$
$$= (P \times 60)/(2pN_s)$$
$$= (746 \times 60)/(2 \times 3.14 \times 540)$$

Torque producing by squeezer shaft = 13.1988 Nm.
At shredder shaft: $P = \tau \times \omega$

$$\tau = P/\omega$$
$$\omega = P/(2pN_s/60)$$
$$= (P \times 60)/(2pN_s)$$
$$= (746 \times 60)/(2 \times 3.14 \times 275)$$

Torque producing by shredder shaft = 25.9176 N m.

Force generated:
At squeezer shaft:
We know that $\tau = \text{F} \times r$

$$F = \tau/r \text{ where radius of squeezer shaft } r = 60\,\text{mm}$$
$$= 13.198/0.06$$

Squeezing force at squeezer shaft $F_{\text{squz}} = 219.982$ N.

At shredder shaft:

We know that $\tau = F \times r$

$$F = \tau / r \text{ where radius of squeezer shaft } r = 70\,\text{mm}$$
$$= 25.9178/0.07$$

Shredding force at shredder shaft $\text{Fs}_{\text{hred}} = 370.254$ N.

5.3 Belt

The expression for the V-belt length is

$$\frac{\pi}{2}(D_S + D_m) + 2C + \left(\frac{D_S - D_m}{4C}\right)^2$$

where $C =$ Centre distance between two pulleys.

Motor–squeezer:

$$L = \frac{\pi}{2}(20 + 7.5) + 2 \times 57 + \left(\frac{20 - 7.5}{4 \times 57}\right)^2$$

The belt length between motor shaft to squeezer shaft $L = 157.86$ cm.

Squeezer–shredder:

$$L = \frac{\pi}{2}(20 + 10) + 2 \times 36 + \left(\frac{20 - 10}{4 \times 36}\right)^2$$

The belt length between squeezer shaft to shredder shaft L = 110.79 cm.

6 Design of Major Components

After detailed design analysis of individual components, the drawings of the shredding machine are drawn by CATIA V5 software. Figure 2 shows the main components that are base frame, housing, shredder shaft with blades, hopper and assembled view of shredding machine.

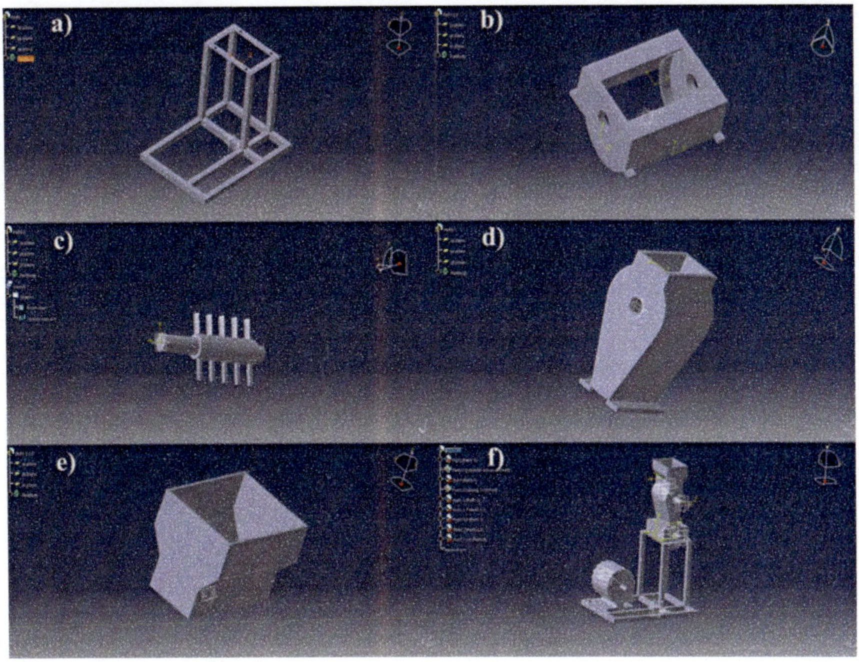

Fig. 2 Design drawings of shredding machine **a** base frame, **b** housing, **c** shredder shaft with blades, **d** shredder housing, **e** hopper, **f** assembled view

7 Fabrication of Machine Parts

Prior to the fabrication of the machine, the design drawings are thoroughly checked with the dimensions. During fabrication, the following points should be taken care such as avoiding very sharp edges, joints should be tight, ensuring minimum production cost and providing lubrication holes. Fabrication process involves converting the raw materials into components and also evaluating the performance. The steps involved in fabrication process are selection of raw materials, selection of fabrication process and fabrication of parts and assembly. The main components are body frame, squeezer shaft, shredder shaft, cutting blades, hopper, mesh and housing. These parts are fabricated by using various operations such as welding, cutting, grinding, turning and drilling. Figure 3 shows the individual parts fabricated by different operations and finally fabricated shredding machine.

8 Performance Evaluation

Finally, the developed organic waste shredding machine is subjected to performance evaluation test. The shredder machine performance was measured using total quantity

Fig. 3 Fabrication of shredding machine by various operations **a** turning, **b** grinding, **c** welding, **d** fabricated shredding machine

Table 2 Performance test results

S. No	Type of waste	Initial weight (kg)	Time for shredding and crushing in sec	Final weight of waste after shredding (kg)	Loss in weight (kg)
1	Food waste	5	300	5.00	0.00
2	Garden waste	5	600	4.90	0.10
3	Vegetable waste	5	450	4.95	0.05

of different organic wastes shredded for different time durations. It converts the solid organic wastes which are too hard, to cut into small pieces for decomposing and easy digest. The output of the wastes is good for producing natural compost, and it acts as a good biofertilizer for cultivation. By growing the plants using this type of natural compost can decrease the usage of chemical fertilizers which are harmful to health.

The fabricated shredding machine is mainly aimed to incorporate the mechanisms such as shredding and squeezing within a single unit to reduce investment on machinery and space occupying. Irrespective of input size of different organic wastes (i.e., vegetables, fruits), the expected output size of shredded organic waste was obtained as 10-25 mm. Different types of organic wastes were used for evaluating the performance of the machine. They are food waste from homes, hotels and restaurants, vegetables and fruits waste from markets and garden waste. A stopwatch and weighing balance have provided for measuring shredding time and mass of the waste. The procedure for performance evaluation is given as follows: (1) Each type of waste is weighed before the experimentation.. (2) Shredded and crushed waste weighed again after the shredding operation. (3) The respective time to completely shred each type of waste is measured with the stopwatch. Table 2 shows the test results of different organic wastes shredded by developed machine approximately. Figure 4 shows the performance of shredding machine at various stages. The machine processed about maximum 90 kg of waste (from home and hotels) within one hour, and it can perform the waste according to size, and hardness of different types of waste is available.

9 Shredding Efficiency of the Shredding Machine

The shredding efficiency was calculated considering mass of organic waste before and after shredding. From Table 2, average mass of waste before shredding is 5000 g, and average mass of waste after shredding is 4950 g. Therefore, shredding efficiency can be calculated as follows

Fig. 4 Performance evaluation of shredding machine **a** feeding the vegetable waste, **b** and **c** during shredding, **d** shredded vegetable waste

$$
\begin{aligned}
\text{Shredding efficiency} \\
&= (\text{Avg. mass of initial weight of waste/Avg. mass of final weight of waste}) \times 100 \\
&= (4950/5000) \times 100 \\
&= 99\%
\end{aligned}
$$

10 Conclusions

The multi-function shredded machine was successfully designed, fabricated and tested for evaluating its performance. This machine is better option to shred any kind of organic waste with low cost. The machine is designed taking into consideration various demands of farmers and other customers. This machine is ideal equipment for shredding the organic waste to obtain organic compost and can be used in the OWC machine, biogas plant and natural composting process. This machine is integrated with shredding and squeezing within a single unit, i.e., multi-function organic waste shredder machine. So, the cost of the equipment is very less.

11 Recommendations

The purpose of this multi-function shredding machine is mainly focused on shredding the organic waste in affordable economic conditions of farmers and other users, thereby improving their economy. The developed shredding machine can be used in a commercial way for providing organic compost to various other farmers. Hence, it is recommended this portable multi-function shredding machine to farmers and other users to produce the organic compost. Due to the low cost of fabrication, the machine can be adopted by small-scale producers.

References

1. Hande AS, Deshpande AA (2014) Methodology for design & fabrication of portable organic waste chopping machine to obtain compost—a review. Int J Innov Res Sci Techn 1(7):132–135
2. Pavankumar SB, Sachin KR, Shankar R, Thyagaraja B, Madhusudhan T (2018) Design and fabrication of organic waste shredding machine. Int J Eng Sci Invention 7(6):26–31
3. Sanjay Kumar IM, Hemanth Kumar TR (2015) Design and development of agricultural waste shredder machine. Int J Innov Sci Eng Technol 2(10):164–172
4. Tegegne A, Tsegaye A, Ambaye E, Mebrhatu R (2019) Development of dual shaft multi blade waste plastic shredder for recycling purpose. Int J Res Sci Innov 6(1):49–55
5. Amiebenomo SO, Omorodion II, Igbesi FC (2013) Development and performance evaluation of an African breadfruit shredder machine. Int J Eng Res Technol 2(3):1–11
6. Etoamaihe UJ, Iwe MO (2014) Development and performance evaluation of a reciprocating motion cassava shredder. Int J Eng Sci 3(4):6–15
7. Sara E (2015) Design of a shredder attachment for a waste compactor. B.Tech Thesis, Department of mechanical engineering Vilnius Gediminas Technical University
8. Fitzgerald GC (2009) Technical and economic analysis of pre-shredding municipal solid wastes prior to disposal. Thesis, Department of Earth and Environmental Engineering. Columbia University
9. Jibrin MU, Amonye MC, Akonyi NS, Oyeleran OA (2013) Design and development of a crop residue crushing machine. Int J Eng Inventions 2(8):28–34

A Technological Assessment of the Ocean Wave Energy Converters

D. Madan, P. Rathnakumar, S. Marichamy, P. Ganesan, K. Vinothbabu, and B. Stalin

Abstract Amid growing concern about global warming, renewable energy sources are considered more important now. Energy is extracted from many renewable sources. Among this all, the ocean wave is one of the largest sources. Harvesting energy from this source has been one of the interesting and great challenges to researchers for many years. There are plenty of methods accepted to harvest electrical energy from sea waves. However, in reality, technical and economical intricacies are issues in harvesting energy from the ocean. Effective use and sustainability of wave energy converters (WECs) depend on competence in dealing with circumstances in maximum energy extraction, their configuration in the rough seasonal waves and addressing economic issues. This article gives a complete overview of WECs in comparison with all other ocean energy converters. It will help researchers to speed up their ongoing research on developing ocean WECs. The aim of this evaluation was to serve the information and describe the significance of ocean WECs. Another objective was to broaden the search of results gained in other fields in the WEC and to stimulate research on topics that have not been in the focus so far.

Keywords Renewable energy · Ocean energy · Wave energy · Energy converter and challenges

---

D. Madan · S. Marichamy (✉) · P. Ganesan · K. Vinothbabu
Department of Mechanical Engineering, Sri Indu College of Engineering and Technology, Hyderabad, Telangana, India
e-mail: smari2k5@gmail.com

P. Rathnakumar
Department of Mechanical Engineering, Navodaya Institute of Technology, Raichur, Karnataka, India

B. Stalin
Department of Mechanical Engineering, Anna University, Regional Campus Madurai, Madurai, Tamil Nadu 625019, India

A. Arockiarajan et al. (eds.), *Advances in Industrial Automation and Smart Manufacturing*, Lecture Notes in Mechanical Engineering,
https://doi.org/10.1007/978-981-15-4739-3_91

1 Introduction

An effect to harness energy from the ocean wave was scientifically and technologically taken in the eighteenth century. A large number of technological developments accompanied by innovative concepts have been made to extract the energy from the ocean wave because the ocean wave energy could desirably become an important component of future energy supply. The most welcome advance would be a large-scale affordable means of electricity storage. Without this, even a combination of affordable ocean-wave-driven energy and electricity generation could not provide a reliable base-load supply. The research ahead is daunting. Of course, it poses challenge too because the expectations of energy in future are high. They combine the anticipation of continued supply improvement in access, reliability and affordability in affluent countries, not only with the necessity of increasing energy consumption in the world but also with harnessing, distributing and converting the marine energy flows in ways compatible with maintenance of local and global environmental quality. In the course of time, a proliferation of wave energy technologies has come one after another. It is not an impossible option. So, thousands of different methodologies in energy capture from the ocean wave have been perpetually registered patent-wise in the world.

The first patent was documented in the year 1799 by Girard and Son in France [1]. Being pioneers, their technology of converting the wave into energy certainly inspired many inventors in 1980. A large number of patents were filed on many kinds of techniques to harness the energy from ocean waves [2]. International Energy Agency, being a source of information, updates every stage of development in the wave energy research. A few authors have strongly proposed their respective classifications upon the selection criteria and authenticated property. Interest and attitude of inventors at the global level mount day by day. Their protocols and guidelines give a clear picture to the fresh researchers and ocean energy device developers.

George Hagerman, the great oceanographic expert, classified the wave energy conversions based upon the most well-known principles. Figure 1 represents a classification of the wave energy converters (WECs) mooted by him on the basis of the motion with energy absorption (surge, heave and pitch), structure (fixed or flexible), types of reaction forces (inertial reference, anchored support and fixed structure) and types of working fluid (oil, water and air). It is still useful in reviewing the WECs [3]. Harris classified that the WECs on their conversion concept are based on the location and operation of principles (Fig. 2) [4]. Falcão perceived devices of various wave energy conversions and classified them by means of working principles, and therefore, the devices were considered to be a prototype and development stage (Fig. 3).

Even though a significant development in these devices and its commercialization has been prompted in the recent period, their view is not focused on the particular shore devices in the assessment. Anyhow, the different kinds of mooring possibilities and their design constraints for the WECs are hereby discussed at length.

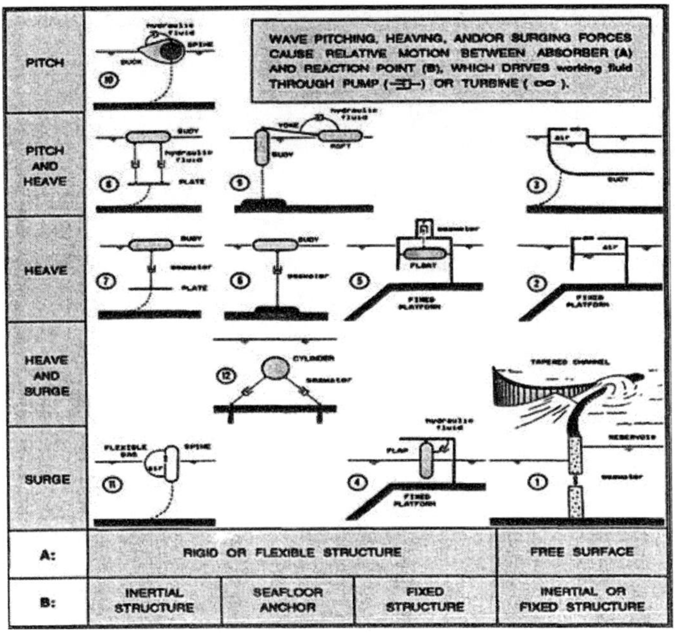

Fig. 1 Categorization of wave energy conversion principles by Hagerman [3]

2 Classification of WECs upon Locations

According to the location, the energy conversion devices are categorized as onshore devices, near-shore devices and offshore devices.

2.1 On Shore

The device is situated on the seashore, which has a lot of advantages of being a utility network as close as to the grid. However, the maintenance of the system is simple and easy to operate. As the waves attenuate passing over the shallow region, they have not much probability and damage at extreme condition. When the waves are broken or scatted, the shallow water gets accelerated towards the shore and makes a great impact upon the structures of the device. Hence, conditions and careful construction are to be needed. Locating devices in the shoreline commonly damages the coastal circumstances abruptly. The devices could seldom be designed for the mass manufacturing due to the factors such as the nature of location, site-specific requirement and environment.

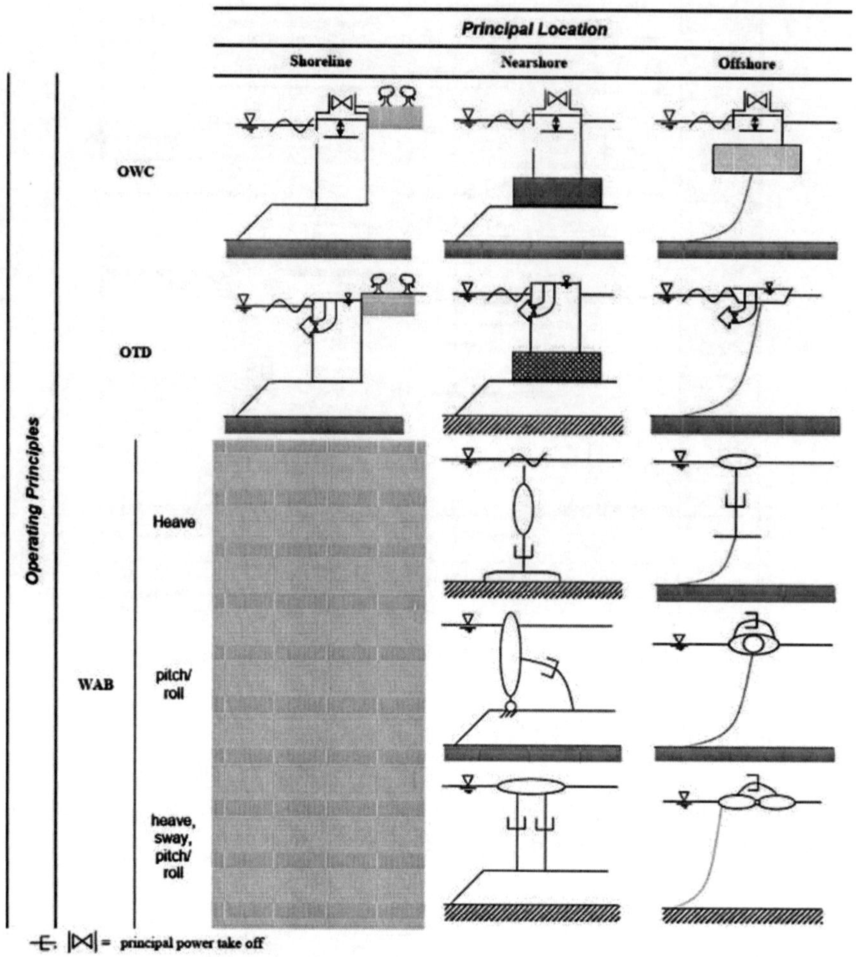

Fig. 2 WEC devices based on operating principles and principal locations [4]

2.2 Near Shore

Energy harvesting devices are generally attached to the seabed located near the shore. Usually, energy conversion takes place at the shore. Shallow water leads to wave with inadequate power. It restricts fielding potential also.

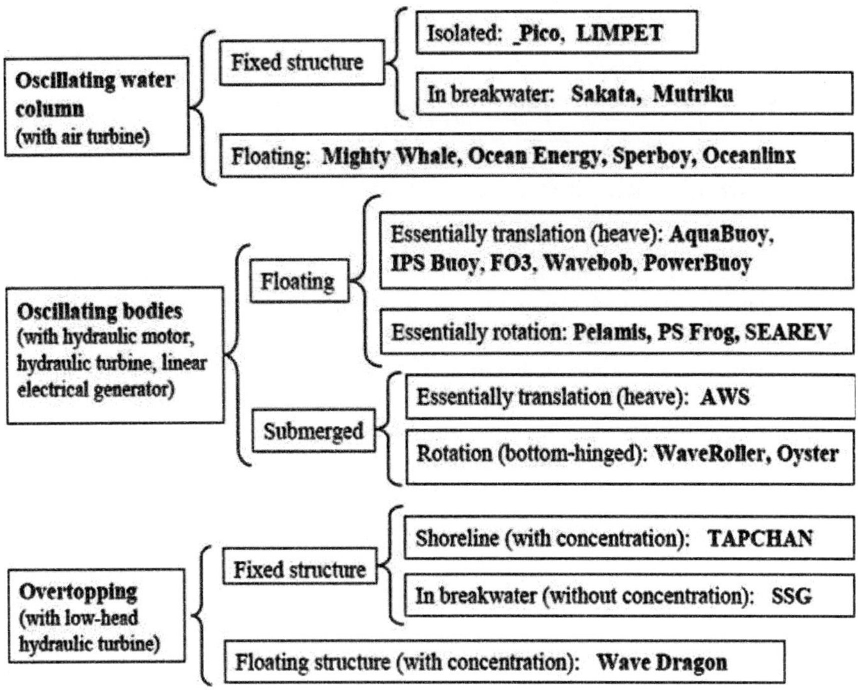

Fig. 3 Classifications of wave energy converters [2]

2.3 Offshore

The device is located in the deep water. The profound water is not clearly meant for the scientific purpose. So, it may be either ten meters or more than 40 m too. Definition is still vague. All in all, the advantage of this method is too difficult to construct and maintain the devices at the offshore based upon the reason of being great wave height and surplus energy in the deep water. Anyhow, cost incurred in transmitting electricity from the shore to the grid is too high. Besides, the submerged devices seem to be inefficient. The devices have to be designed to stay alive at the extreme conditions. Additionally, there is a separate cost for the construction too. The offshores are categorized into two divisions such as floating type and resting on seabed. Floating devices reduce the cost of construction, whereas the devices on the seabed may be expensive.

3 Classification of WECs upon the Operating Principles

The WECs are classified based on the following three category principles.

a. Oscillating water column
b. Overtopping devices
c. Wave-activated bodies.

3.1 Oscillating Water Column (OWC)

An OWC (Fig. 4a, b) contains an air chamber, which is open at the bottom. It seems to be so partially submerged under the seawater line that can make an air chamber above the seawater surface. On the top surface of the chamber, there is a turbine incorporated with electricity generating unit. As soon as the waves approach the devices, water enters into the chamber and gives much force on the air inside the chamber. The compressed air blows in the atmosphere through the turbine driven by an electrical generation. The movement water column falls down, and air returns to the chamber in the same way as long as turbine rotates. The repeated action of the waves generates electricity with little fluctuation.

A low-pressure well turbine is used in the application as it runs in the unidirectional rotation of backward and forward directions of the air flow through the open air. The noise which roars for the production and operation is a great challenge to the OWC. According to the structure, it is grouped into two categories such as fixed-structure OWC and floating-structure OWC.

Most of fixed-structure OWC (Fig. 4c) devices are situated either on the shoreline or near shore. Generally, the devices stand at the bottom of the sea or may have been fixed to the rocky cliff. To maintain and install these devices in easiest way is

Fig. 4 **a**, **b** OWC device operational principle (courtesy: open EI), **c** The LIMPET OWC, installed on the Isle of Islay, Scotland (courtesy: Wavegen), **d** backward bent duct buoy-type OWC in Galway Bay, Ireland, (courtesy: Ocean energy)

Fig. 5 **a** Wave energy conversion plant at India [5], **b** 125 kW rated power OWC established in 1990 at Trivandrum, India [6]

highly advisable and advantageous. Floating-structure OWC (Fig. 4d) is so advantageous that it is possible to attain higher frequencies of harvested electricity than the frequency of waves through these devices. It is too much difficult to maintain and install. Besides, it should be operated under the water cables.

In 1982, an OWC was tested in Indian coast with OWC device (Fig. 5a) and utilized for desalination application. In 1990, an OWC prototype (Fig. 5b) was installed in India (Vizhinjam, Kerala, India, 1990) [7].

3.2 Overtopping Devices (OTD)

An overtopping device called a terminator (Fig. 6a) is as much similar as conventional low-head hydroelectric power unit convert wave energy into potential energy. The device usually captures the seawater waves in a reservoir and releases it through the low-head turbine. An example is cited in Fig. 6b [8]. The device known as wave

Fig. 6 **a** OTD operational principle, **b** wave dragon prototype at test site in Nissum Bredning

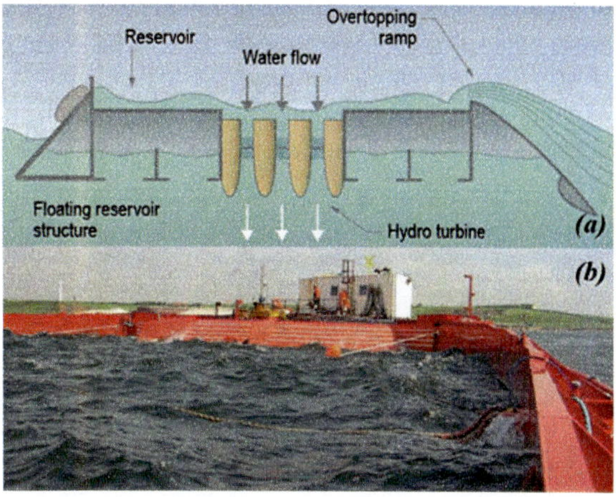

dragon contains a pair of big and curved reflectors which turn waves into a central receiving reservoir part where the waves flow above the slope or perhaps may be at the top of the reservoir and releases low-head water through turbines. Both controlling and stability which are essential for the wave direction pose a challenge to the devices.

OTDs are categorized into two types such as fixed-structure OTS (It works in areas with a tiny tidal range) and floating-structure OTD (It is applicable to the coastal regions as much close to a rocky coastline).

3.3 Wave-Activated Bodies (WAB)

Offshore devices called as third-generation energy devices which produce energy by way of the relative motion between waves and oscillating bodies, which may be floated or submerged. Despite, they may exploit much powerful wave regime present in the deep water. Undoubtedly, the third-generation system is more complicated than the first-generation system because OBSs are intricated with problems in mooring accessing to the maintenance and laying underwater cables. On the other hand, some recent systems seem to be close to the demo-stage.

The WABs are grouped as single-body heaving buoy, two-body heaving buoy, fully submerged heaving systems, pitching devices, over-hanging and bottom-hinged system and multi-body systems.

These devices consist of small point absorbing device (horizontal size negligible in comparison with incident wavelength), attenuation-large floating structure mingled with main wave direction and oscillated waves (Pelamis and Anaconda), termination-large device entangled with wave crest (Salter's Duck) and sea floor mounted device work with different pressures. It is one of the references which denote the oscillating wave absorbing body. Energy is taken through the power take-off system with resistant motion of the body. The reference may be either fixed or flexible (Fig. 7). However, it can be in the position of damping platter or by another body. Activation of buoys may be translational or rotational. Certainly, wave energy in offshore is greater than near shore because there could be less energy loss in the offshore waves. Hence, these devices are more effective than other WECs.

Single-body heaving buoy. Single-body heaving buoy reacts against a pivot positioned on the seabed. This system is conceived as a point absorber. Experts in the

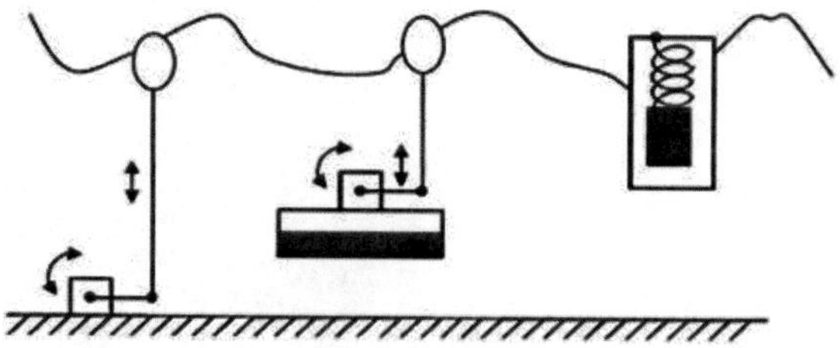

Fig. 7 Reaction mechanisms in various types for point absorber

nineteenth century carried out the early scientific research on point absorbers. J. Falnes and K. Budal, C. C. Mei, D. V. Evans and J. N. Newman were a few of these pioneers. Despite their publication regarding the basic theory of power extraction from the waves and deduced expressions for maximum power absorption, they hold views that the point absorbers absorb more than 100% power than is incident on its width. The dimensions are smaller than their wavelengths.

In 1980, the device named G-1T was tested in Tokyo Bay. It has a wedge-shaped buoy of 1.8 m × 1.21 m rectangular platform at the water level and 1.2 m draft. By and large, vertical motion of the device was guided by fixed steel structure. The hydraulic ram circuit was used as PTO.

In 1983, the device named Norwegian buoy (Fig. 8a) was tested in Trondheim Fjord [9], having spherical floater performed heave oscillations of a strut which is connected to an anchor on the seabed. The body erected with an air turbine is a phase controlled by latching which is executed by locking the spot of the body twice per each half-wave cycle allowing for the excitation force on the buoy and maximum buoy speed to coincide, to achieve maximum power capture a condition that should be fulfilled.

In 1990, the above-mentioned device with some modifications was tested in Denmark. The buoy was coupled to a fixed structure at the bottom through a spring-loaded cable. A relative motion between the float and seabed structure that activates the PTO system contains piston pump supplies pressurized water into the turbine.

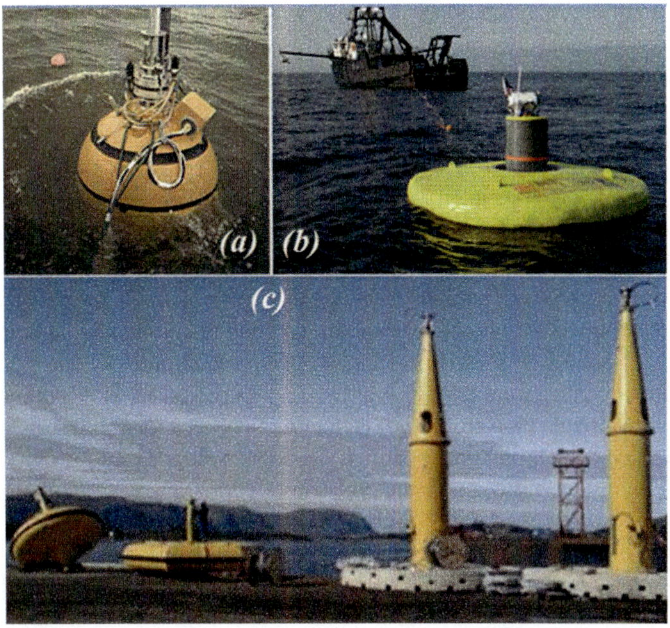

Fig. 8 **a** Norwegian heaving buoy in Trondheim Fjord, 1983. **b** L-10 wave energy converter with linear electrical generator. **c** Swedish heaving buoy by means of linear electrical generator [A]

In 2008, a new system (Fig. 8b) developed by Oregon State University, USA, in which a heaving buoy drives a magnetic linear electrical generator. It has an angular-shaped buoy (3.5 m radius) and a deep draught spar (6.7 m length) and by a cable, the spar is taut-moored to the seabed. The buoy is a one-dimensional heave related to a spar. The relative motion between the buoy and spar could be changed into electricity by a linear generator permanently [6].

The taut-moored buoy concept (Fig. 8c) has been developed by Uppasala University, Sweden. In which, instead of Piston pump a linear electrical generator directly positioned on the sea floor. A generator associated with buoy located at the surface of the ocean and acts as PTO [10].

Apart from this, some experimental level devices have been proposed. The device consisting of a floating buoy directly connects a vertical movable rack that alone makes a unique drive assembly to keep continuous rotation from downward and upward motion of the float. Another device possesses of steel container, arm, counter mass and mechanical PTO. The arm pivoted at its mid and subjected to an inverted steel container and balancing counter mass are at each end. When incidental wave passes the container, the arm gets imbalanced. Due to this imbalance, the arm oscillates and drives the PTO system [11]. Another offshore floating device called Aegir Dynamo that contains of a spar anchored with the seafloor and unidirectional gearbox containing float moving along the spar. A conventional electrical generating unit coupled with the device is to produce electrical energy [12]. Another direct drive WEC consists of ball and screw arrangement and converts up and down motion of buoy into unidirectional rotation. The device was tested in wave flume. The technical feasibility becomes accessible and viable [13].

Two-body heaving system. The conception of a single floating has a few difficulties in reaction between the seabed and floating body due to the same distance between them. The energy can be converted into power from the different relative oscillating motion between two bodies.

In 1985, the concept of the bipartite absorber of a two-body heaving system was introduced. It is made of two floaters. The outer structure being acted as reference having low resources but being an inner, it acts as a resonating absorber. The concept about the device was adopted in the wave bob: the mass of the inner body got increased by connectivity to a fully submerged body positioned at the bottom level. In 1978, the IPS buoy (Fig. 9a) was created by Sven. A. Noren, Sweden, and refined by the company inter project service, in which a buoy linked to a submerged open vertical tub, which had a piston. The relative motion between the piston and floater tube system drives the PTO mechanism. In 1981, some inventors introduced the same with modifications to solve the problem of end-stops. An end-stop applies to all moving body converters with rigid connections to PTOs. An end-stop is nothing but a central part of the tube where the piston slides and limits its stroke.

Aqua buoy is an inter project service heaving buoy. In 2000, it was developed with a combined concept of IPS with a pair of tube pumps to make high-pressure overflowing water. The model of the same was tested in coast of Oregon, Pacific Ocean, in 2007.

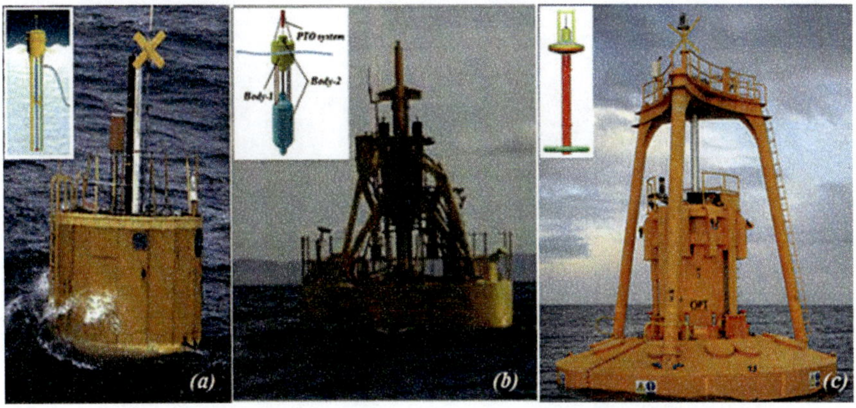

Fig. 9 **a** IPS buoy OWEC (W4P), **b** wave bob (Courtesy: Wave bob Ltd.), **c** the power buoy (Courtesy: Ocean Power Technologies)

A few years ago, 1:4 scale model of the wave bob was tested in the sheltered waters of Ireland Galway Bay (Fig. 9b). It contained two co-axial symmetric buoys: the relative motion between these buoys was converted into electricity energy through a hydraulic ram and pump system. In addition, the co-axial submerged body linked with inner buoy located underwater, which is to increase the inertia.

In 2008, WEC named a power buoy (Fig. 9c) developed by the American Company Ocean Power Technologies, and 40 kW prototype was installed at the coast of Santoña, Spain. A relative motion acted between disc-shaped floated and a submerged cylindrical body which was converted into electrical energy by means of hydraulic P.T.O. The floater terminated at the bottom activates a horizontal damper plate so as to raise the inertia through the mass of water surrounded by it.

Fully submerged heaving systems. Archimedes wave swing (AWS) was developed in Holland (Fig. 10a). There was an upper part called as an oscillating floater and the lower part called a basement. The upper part which pushes down under the crest moves up under the trough. A linear generator resists this motion with the interior air pressure as a spring. A prototype that was built rated 2 MW. In spite of ineffective two trials to sink it into a position of the Portugal northern coast, it was effectively installed and tested in 2004 [6].

In 2011, 80 kW rated another device (fully submerged) heave 7-m-diameter body named CETO was tested in Australia (Fig. 10b). The vertical movement of the rigid-moored buoy brings concerning high-pressure water pump. The PTO is positioned on the shore.

In 1970, David V. Evans, University of Bristol, invented with a concept regarding linear water theory. The device includes a circular cylinder called as Bristol cylinder whose shape looks a submerged horizontally and axis parallel to the crest of the incoming regular wave. The cylinder centre is fixed in a circle of small radius. To develop this project, the British Wave Energy Program funded from 1975 to 1982.

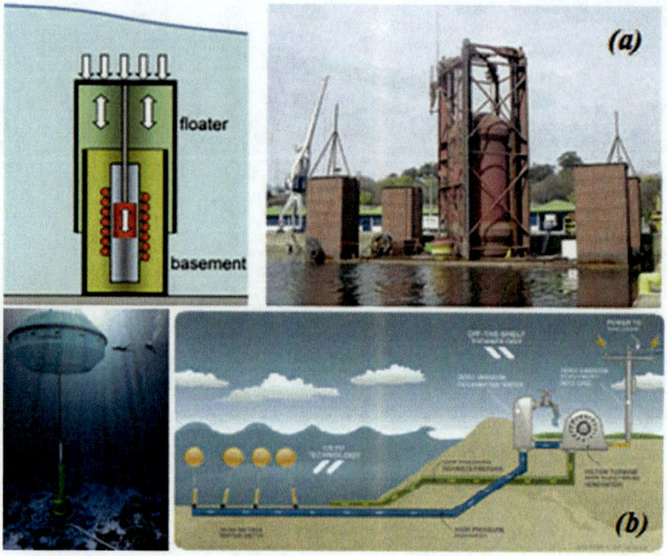

Fig. 10 **a** AWS scheme and prototype **b** CETO wave energy converter and schematic diagram (courtesy: CWE Ltd.)

Pitching Devices. Among the WAB systems, the conversion of energy is based on relative rotation. The nodding duck created and introduced by Stephen. H. Salter was approved as an excellent offshore device in the years 1970 and 1980. The advanced versions' appearance in others' research was put into practice. The first version had a ducks string mounted on a long spire in alignment with the direction of wave crest with a hydraulic–electric PTO system as reflected in Fig. 11a.

Later on Salter modified and projected a solo duck in which the structure of orientation against which the nodding duck (duck in movable condition) reacts is illuminated by a gyroscope in Fig. 11b; the concept of duck introduced by him had not become operational and successful.

The Pelamis is a snake-like slack-moored articulated structure. It composes of four cylindrical sections connected by hinged joints. It is in alignment with the wave direction. The hydraulic rams control the wave-activated motion of these joints, which pump pressurized oil through hydraulic motor that drives electrical generators.

The wave-stimulated motion of joints activates the hydraulic rams, which pumps oil through hydraulic motors driving electrical generators. A set of three Pelamis devices was installed in Portuguese northern coast in 2008 (Fig. 11c), making it the first worldwide grid-connected wave farm.

Over-Hanging and Bottom-Hinged System. It is purely based on the reversed pendulum, which is hinged at the seabed concept. Stephen Salter invented the mace. It has a buoy with symmetrical with respect to a universal joint at the sea vertical axis. It swings about that joint at the sea bottom. It is possible for the power take off reaction to the seabed through cables only. The wave-activated oscillating motion

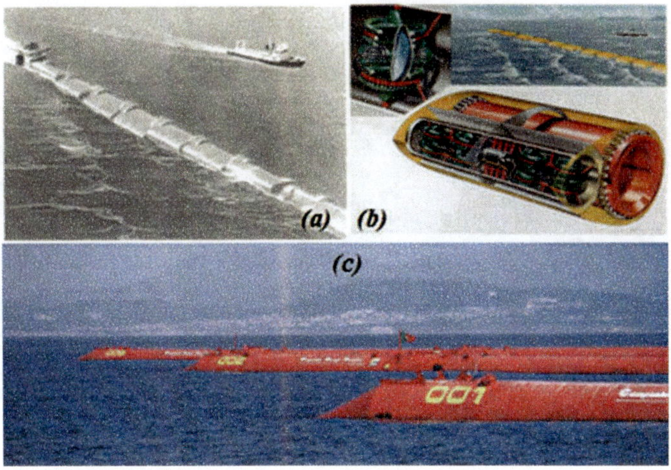

Fig. 11 **a** Salter duck, **b** the Duck version equipped with gyroscopes (Courtesy: Univ. of Edinburgh), **c** the 3 units × 750 kW Pelamis wave farm in calm sea off northern Portugal (Courtesy of Barros)

of the buoy is converted into electricity energy through means of a hydraulic PTO device. A suitable example for this system is Oyster developed by Aquamarine Power Ltd. (Fig. 12).

Multi-body System. It is a multi-body system in which energy can be turned into different oscillations between two bodies. The device, which has many numbers of floating bodies, response and energy are collected by common frame and PTO system. The wave makes the buoys to swing and pumps the oil in the hydraulic PTO circuit.

Fig. 12 Oyster prototype being assembled (courtesy: Aquamarine Power)

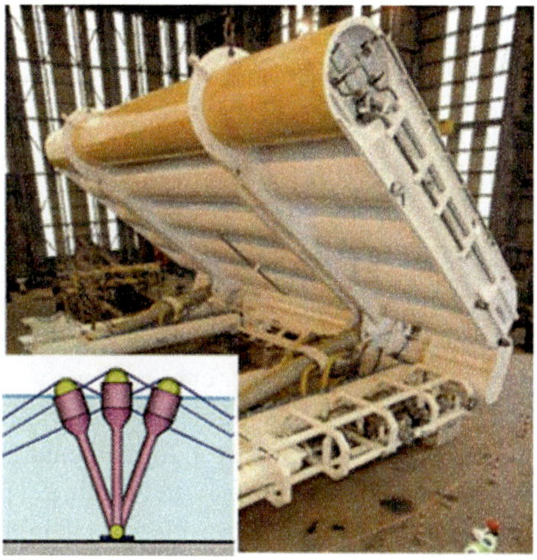

Fig. 13 **a** Multi-body device "Buldra" tested in the southern coast of Norway. **b** Wave Star deployed in Nissum Bredning, Denmark

Examples of many body systems are as follows: in 2000, a 1:20 scale model was tested in Trondheim, Norway. In mid-2000 s, a device called as "Buldra" 1:3 scale multi-body device was tested in Norway southern coast (Fig. 13a). A 24-m-long, 5.5 kW capacity, 10 buoys on each side, 1:10 scale model was deployed in 2006 in Nissum Bredning, Denmark (Fig. 13b). For a couple of years, it was productively tested with grid connection.

The WEC for electricity generation for the past few years has been progressive, but utilization of energy in the other parts of the world has also been developing rapidly parallel. Testing under the red sea condition is a final phase of modelling WECs. In each and every system, optimal wave energy absorption includes the same type of resources which has a generating as well as the size of the structure connected to the wavelength.

4 Challenges

4.1 Challenges for Technical Viability

There are many factors influencing the design of a WEC. Some of the important factors are reviewed. When compared to other renewable sources such as solar and wind, energy concentration is spatially higher in wave energy sources. However, wave energy possesses a higher energy density and is more persistent and predictable than wind energy. The development of a WEC from concept stage to commercial has not reached to a significant level. Implementation of a WEC in a realistic scenario is more complicated than that of a wind turbine. The geometry of a WEC is associated with wave parameter such as wavelength, height, and period. The prototypes are tested in laboratories under some specific wave climate only, but the real wave is different. The resonance of an energy conversion mechanism such as front-end interface of WEC with the natural wave is not a predictable one. In addition, the power transfer unit of the WEC is designed for a certain range, but it should withstand rough weather condition also. Owing to environment load variations in wave, current, and wind, designing optimized mooring systems is difficult.

4.2 Challenges for Commercial Viability

Although the reliability of the a WEC is high compared to other renewable sources, it is commercially expensive because of the initial investment, transportation of devices, offshore maintenance, installation difficulty, cost of equipment and special materials used. Considering huge expenses, investors are not giving much attention on this kind of investments. Therefore, we expect a substantial monetary support from governments.

5 Conclusions

Ocean energy is clean and renewable. The ocean possesses energy in many forms such as tidal energy, tidal current, ocean current, ocean thermal energy, salinity gradient and ocean wave energy. Wave energy is the most promising resource among all other resources in the ocean in which many successful projects and a lot of research activities are going on. Using energy from this kind of sources is an important move towards meeting proposed renewable energy targets. Although the ocean is the largest energy resource, the usage and utilization are very less compared to other resources. Considering complication in erection, maintenance and implementation power plants, along with economic cost involved from the conceptual stage to the commercial stage, it requires huge expenses. Therefore, necessary research on the possibilities available to reduce the cost of investment and technical issues needs to be undertaken. This article has given an updated review of WEC technologies along with a summary of different types of device documented and assessed, and reports published in the past.

References

1. Mc Cullen P, Clément A, Falcão A, Fiorentino A, Gardner F, Hammarlund K, Thorpe T (2002) Wave energy in Europe: current status and perspectives. Renew Sustain Energy Rev 6(5):405–431
2. De Falcão O (2010) Wave energy utilization: a review of the technologies. Renew Sustain Energy Rev 14(3):899–918
3. Hals J (2010) Modelling and phase control of wave-energy converters. Norwegian University of Science and Technology, Trondheim

4. Harris RE, Johanning L, Wolfram J (2004) Mooring systems for wave energy converters: a review of design issues and choices 180–189
5. Sundar V, Moan T, Hals J (2010) Conceptual design of OWC wave energy converters combined with breakwater structures. OMAE2010 3:479–489
6. Falcão AFO (2014) Modeling of wave energy conversion. University of Lisbon
7. Madan D, Rajendran M, Prabhakaran M, Anandan K, Sagadevan S (2015) Renewable energy scenario in Tamil Nadu. Teri Inf Dig Energy Environ 14(4):481–492
8. Kempener R, Neumann F (2014) Wave energy: technology brief. International Renewable Energy Agency Ocean Energy Technology Brief 4
9. Falnes J (2004) Ocean waves and oscillating systems. Cambridge University Press
10. Babarit A, Hals J, Muliawan MJ, Kurniawan A, Moan T, Krokstad JSL (2012) Numerical benchmarking study of a selection of wave energy converters. Renew Energy 41:44–63
11. Madan D, Rajendran M, Amarkarthik A, Prabhakaran M, Sagadevan S (2015) Experimental investigation on impact of container geometry on efficiency of non-buoyancy body type wave energy converter. Indian J Sci Technol 8(29):1–5
12. Al-Habaibeh A, Su D, Mc Cague J, Knight A (2010) An innovative approach for energy generation from waves. Energy Convers Manag 51(8):1664–1668
13. Agamloh EB, Wallace AK, Von Jouanne A (2008) A novel direct drive ocean wave energy extraction concept with contact-less force transmission system. Renew Energy 33(3):520–529

A Secure Data Sharing Using IDSS CP-ABE in Cloud Storage

T. P. Ezhilarasi, N. Sudheer Kumar, T. P. Latchoumi, and N. Balayesu

Abstract In day-to-day life, cloud computing plays an important role in storing and retrieving the huge volumes of data. Data's which are taken from social media (Whatsapp, Facebook, Twitter etc.), Organizaitons (MNC companies), Business data (Sensex, Forbes Global, Forbes cloud, TS Historical bus cycles etc.,) and other datas from our day to day life records are not enough memory to store in Big Data. While the storage size is increasing, the security problem also increases simultaneously. Therefore, there is a security issue in cloud computing. There are large numbers of studies conducted to improve security in cloud computing. However, most of the survey is not applicable for cloud providing security in mobile devices because of less power consuming and computing resources. The issue in the existing mobile application is no proper security of data on cloud storage. In this paper, we propose a lightweight data sharing scheme for mobile cloud computing. It accesses a control from the ciphertext-policy attribute-based encryption technology which is used in the basic environment of cloud and makes some changes in the structure of the access control tree that is suitable for mobile cloud computing. This technology in the mobile cloud will remove the major portions in the access control tree, and it transforms by using the external proxy servers. Additionally, it is used to reduce the cost of user revocation. The experimental results of our LDSS system can efficiently reduce the mobile device overhead.

Keywords Cloud computing · Mobile computing · Data storage · Data security · Intrusion detection system · Encryption technique

T. P. Ezhilarasi
Department of Computer Science and Engineering, RAAK College of Engineering and Technology, Puducherry, India

N. Sudheer Kumar · T. P. Latchoumi (✉) · N. Balayesu
Department of Computer Science and Engineering, VFSTR (Deemed to Be University), Guntur, India
e-mail: tplatchoumi@gmail.com

A. Arockiarajan et al. (eds.), *Advances in Industrial Automation and Smart Manufacturing*, Lecture Notes in Mechanical Engineering,
https://doi.org/10.1007/978-981-15-4739-3_92

1 Introduction

Cloud computing is the one which utilizes both hardware and software assets for processing the conveyed administration over a system through the Internet. Service provided in the cloud storage allows the data owner to obtain their data into the cloud and provides the data accessing service to the authorized users. The data owner and the cloud server both are not in the similar trusted domains, i.e., cloud server is partially trusted that cannot depend on providing the access policy. Traditional methods generally need the data owner to do the encryption of data and send the decryption keys to the authorized users to address the above challenge. There is an issue in high overhead and the difficulty in managing the key by the data owner. This paper [1] stated the designing access to the controlled framework in the cloud systems to achieve fine access control.

People are creating the information and the endeavors that are to be put away/recovered the information quickly. The data owners are highly interested in outsourcing their complex local data to be managed for economic savings and increase flexibility. However, sensitive data should be encrypted first and then by using traditional utilization of data service to search the data in the plaintext. As the demand for many users and the outsourced files of data in the cloud environment is a challenging issue. It is very difficult to solve issues like practical performance requirements, the usability of the system, and high-level searching experiences of the users. This paper [2] discussed how security is provided in the cloud data and solved the similarity searching problems over the data in the cloud. The mechanism designed for exploiting the storage of data as well as searching the documents using similarity metric distance called trie traverse search index. This proves the guarantee of privacy-preserving under the treatment of hard security. This mechanism further enriches the entire application spectrum. And it shows how the proposed method supports the fuzzy searching technique for all types of users with inconsistencies on input. This method is implemented in the Amazon cloud platform with the real set of data to demonstrate the validity of the proposed system.

Due to the inadequacy of revocation and decryption of the ciphertext, the existing ciphertext-policy attribute-based encryption (CP-ABE) technique cannot be directly applied in constructing the multi-authority storage system through data access control, where the users might hold the multiple authorities. This paper [3] stated that data access control for multi-authority cloud storage (DAC-MACS) is secure and an effective data access scheme for revocation and decryption of ciphertext. And precisely, they modeled a new multi-authority CP-ABE technique for efficient decryption and revocation to achieve both backward and forward security. New cryptographic scheme Attribute-Based Proxy Re-Encryption (ABPRE) is extended from traditional re-encryption method to increase the access control capability of the user. Here, users can freely re-encrypt the ciphertext with one access polity and another user with the different access policy. This paper [4] proposed the scheme that proven to choose the plaintext in a secure manner by selecting the structure and secure the master key with random oracles. In this, they developed another key delegating ability and discussed issues like stronger security in the applications.

2 Background

Most of the organizations have more sales moving across the countries, and for that, collecting the data (selling, buying, remaining stock, etc.) is a very critical problem for developing their business [5]. All these people should move all over in olden days, and now, Internet connectivity is addressing this issue. Data synchronization with their respective corporate server is used to solve the issue which is addressed by sales data collection through the smartphone (offline) whenever the access required.

2.1 Cloud Computing

Cloud computing name is derived from the symbol of cloud-shaped which represents the general use of designing the complex infrastructure [6]. It assigns remote services for user data, software and computation. It consists of both hardware and software resources that make third party services consistently maintain the availability of internet connections. Thus, it provides access control to all innovative applications and high-end computer servers in the networks. The objective is to apply this on high-performance computing (HPC) and supercomputing that is utilized by military application to perform the researches over trillions of computations per second and in consumer-oriented applications like financial portfolios (delivering the personalized data) and computer games (to deliver the large immersive data storage).

Cloud computing always uses the huge groups of servers through networks running at very low cost. Consumer technology has dedicated the networks for processing of data over them. So that it shares the infrastructure of Information Technology (IT) containing many systems that are connected [7]. To maximize the security in cloud computing, power virtualization technique is used.

3 Existing System

Generally, access control approach is dividing into four categories. (1) Simple ciphertext access control, (2) hierarchical access control, (3) attribute-based encryption (ABE) access control and (4) homomorphic encryption (HE) access control. In the existing system, no model is for mobile storage in the cloud environment.

Issues in the existing system are:

- In many organizations, there is an issue in data privacy for many data owners.

- The access control mechanism is either not convenient or not enough which is provided by the cloud service provider (CSP).
- Not meet all the data owners' requirements.
- In mobile devices, the availability of a large amount of data storage of computational resources is not enough.
- Revocation cost is very high.
- Accordingly, the important issue in the existing system is data security.

The current solutions to cloud computing could not solve all the existing problems. There is no proper solution to solve the secure data problem while sharing of data in the mobile cloud environment [8–11]. This admiration of cloud computing helps us to store or retrieve data from anywhere at any time. So, we considered in which specific cloud computing the data are accessed efficiently and conveniently for mobile devices.

4 Proposed System

In the proposed system, we used the third category of access control method, i.e., ABE with novel modifications which have higher computational cryptographic operations to the provider of the cloud, and it reduces the mobile user's total communication cost.

We proposed an algorithm of LDSS for mobile storage environment for CP-ABE called LDSS-CP-ABE to provide an access control efficiently over ciphertext. Proxy servers are used for both encryption and decryption methods. Because of using this proxy server, it highly reduces the computational overhead on the mobile device's client side.

Figure 1 shows the working of CP-ABE in our proposed system. First, the data user sends a request for a file to the cloud storage. Data owner checks whether the user is an authorized user or not, and based on that, owner may accept or reject the permissions. If the owner accepts the user requests permission, then the user will get the notification. A data user can only download the uploaded files or view the history of the downloaded file. If the user tries to download the uploaded file even though the permission is granted by the data owner, the user can access the file only after satisfying the attributes policy. The attributes have matched the file means it will start downloading. After that, the owner will get the notification of the user who downloaded the file.

4.1 Data-Flow Diagram

Data-flow diagram (DFD) is also called a bubble chart. It is a graphical representation of a system in terms of input, processing of data and output generation. It is the most

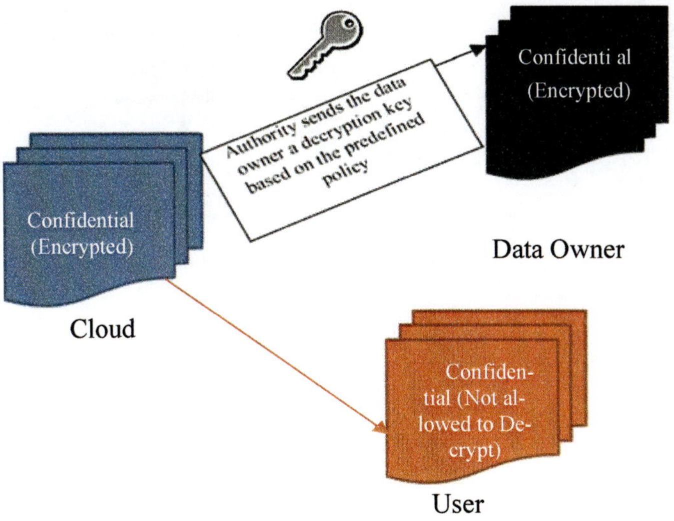

Fig. 1 CP-ABE architecture of the proposed system

important modeling tools for system components. It tells about the processing of the system, how an external entity interacting with the systems, how the transformation of data moves from input to output and flow of information. It is used any level of system abstraction that increased the flow of information and detail of functional requirements. Figure 2 represents the DFD of data owner responsibilities.

The cloud server is to check first whether the user is an authorized data owner or not. If he/she is an authorized data owner, then allow them to login, providing the decrypt key for accessing the data, uploading of files, delete the files, etc.

Figure 3 represents the DFD of user responsibilities. The data owner is to check first whether the user is an authorized user or not. If he/she is an authorized user, then allow them to login, providing the attribute key request for accessing the data like view the uploaded files and download the files and view the downloaded file history. And Fig. 4 shows the cloud owner responsibilities.

4.2 UML Diagrams

Use case diagram (behavioral diagram) is to present the graphical representation of the system functionality in terms of actors, their goals and dependencies between the use cases. In this, Fig. 5 represents the use case diagram of the proposed system which contains four actors (data owner, cloud, trust authority and the data user), their responsibilities in the proposed system and any dependencies between any of the use cases.

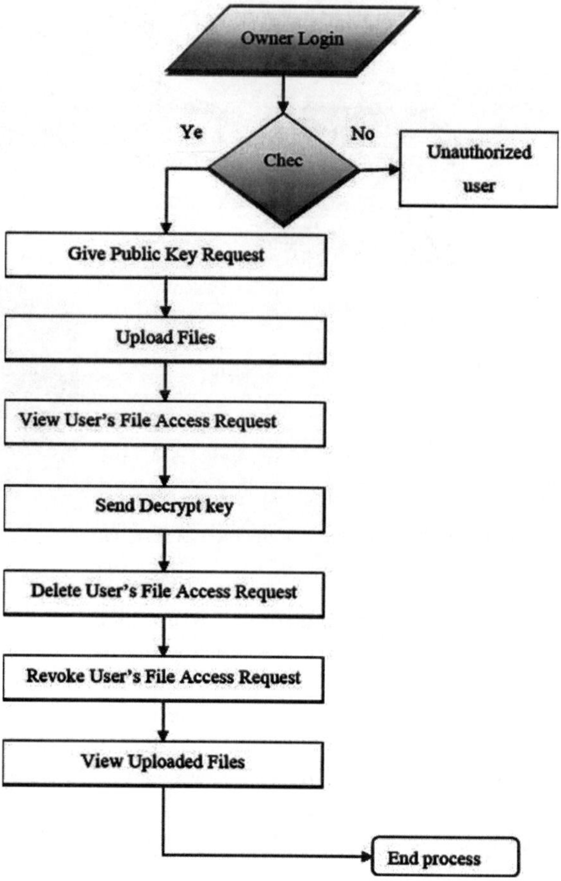

Fig. 2 DFD of data owner responsibilities

Sequence diagram (interaction diagram) also called as event diagrams or event scenarios or timing diagrams shows that how the system process operates with one another and in which order the system is working. Figure 6 shows the working order of the proposed system.

4.3 Algorithm

The proposed algorithm is mainly specified by four steps.

Step 1: Setup—there is no input in the setup, but it considers the parameter and generates the public key (PK) and master key (MK).

Step 2: Encryption—it encrypts the public key (PK), message (M) and access (A). It converts the plain text (PT) into ciphertext (CT). The user whose key matches with these policies can get permission to access the information.

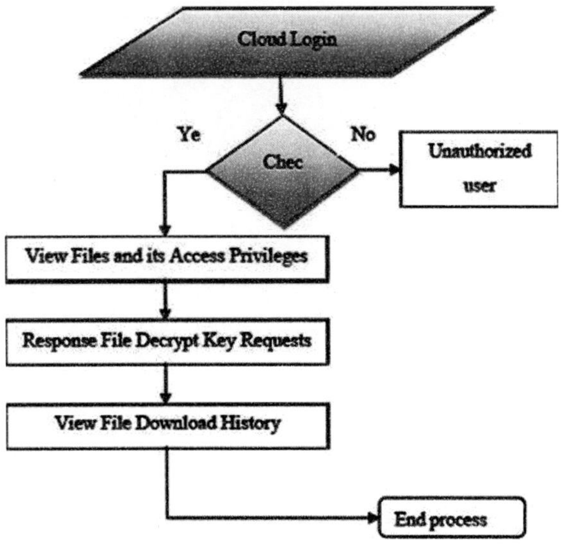

Fig. 3 DFD of user responsibilities

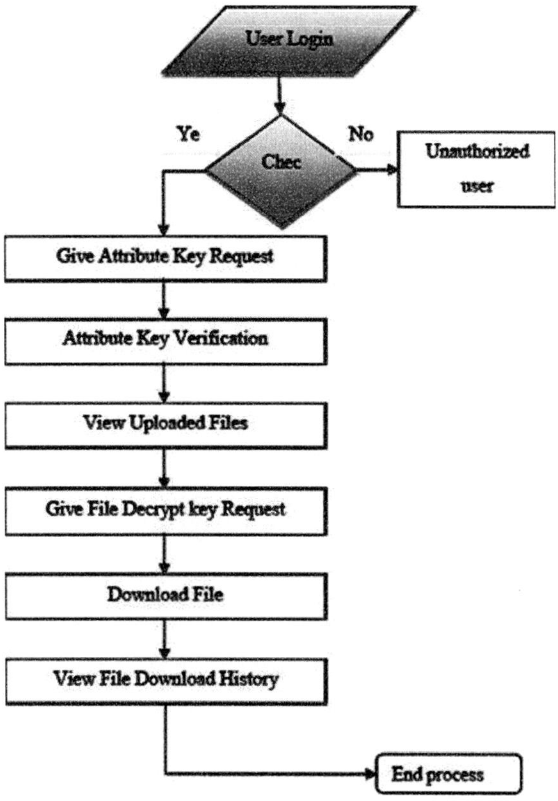

Fig. 4 DFD of cloud owner responsibilities

Fig. 5 Use case diagram of the proposed system

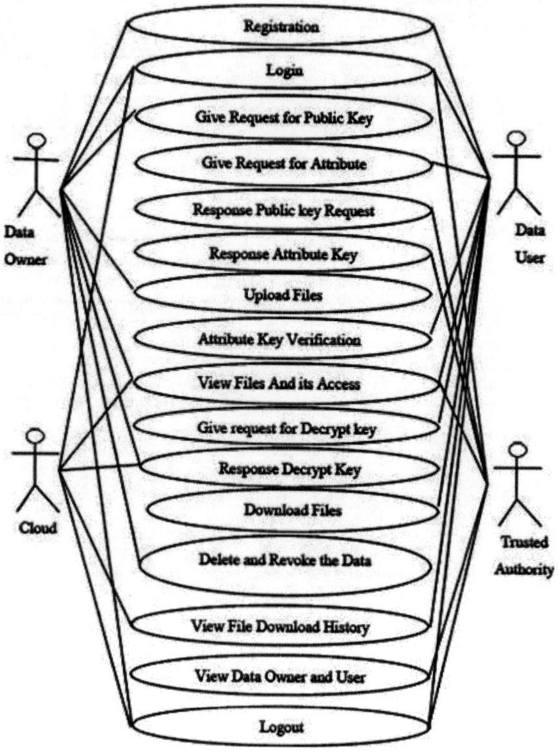

Step 3: Private Key Generation—master key (MK) and the attributes such as input generate the private key (SK).

Step 4: Decryption—it takes the set of attributes public key (PK), ciphertext (CT) and private key (SK). If these attributes satisfy the policy, then only the data owner can convert the ciphertext (CT) into plain text (PT) (i.e., decryption) and get the original message (M).

5 Implementation and Results

5.1 Screen Shots

Data owner registration is shown in Fig. 7. For that registration, the data owner must provide all the details like name, contact number, mail id, address, etc.

Public key (PK) request from the user is shown in Fig. 8. Data users need to request the public key for accessing the data from the cloud.

Trusted authentication is shown in Fig. 9. This is the one to check the authorized user or owner or not.

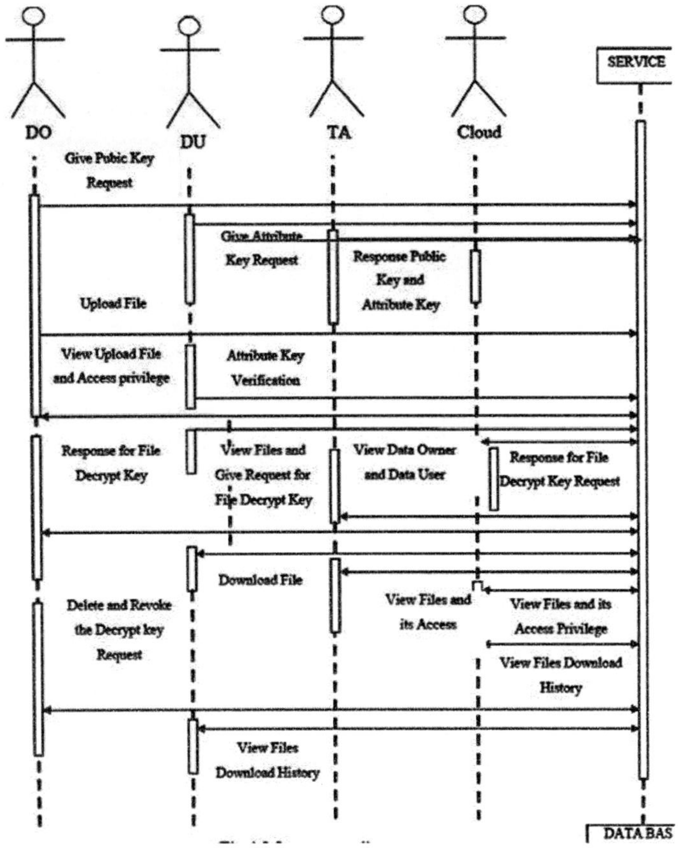

Fig. 6 Sequence diagram of the proposed system

The data owner has the rights to upload the files after getting the private key (SK) as shown in Fig. 10.

5.2 *Test Case*

The objective of testing is to find errors. It is the technique of finding all possibilities of error. Testing uses the numerous sorting of checks. In each check, it reports a demand for testing. It is used to determine the usual occurring of errors with minimum time and effort. A successful test determines the undiscovered error also. And the good test case is all possibility of discovering an error. Testing is for improving the quality and the standards of the system. And test cases of all forms are included in Table 1.

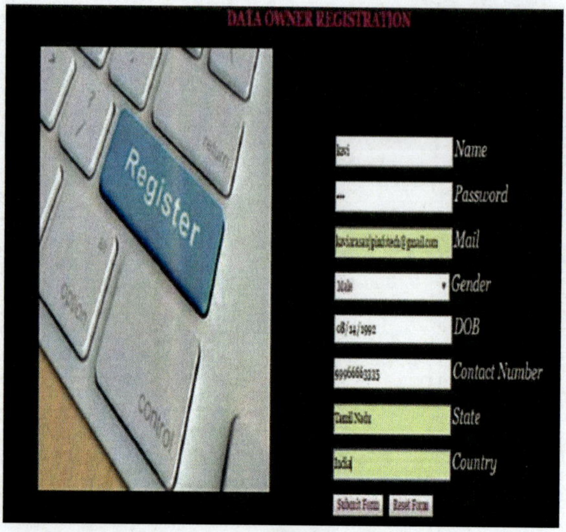

Fig. 7 Data owner registration

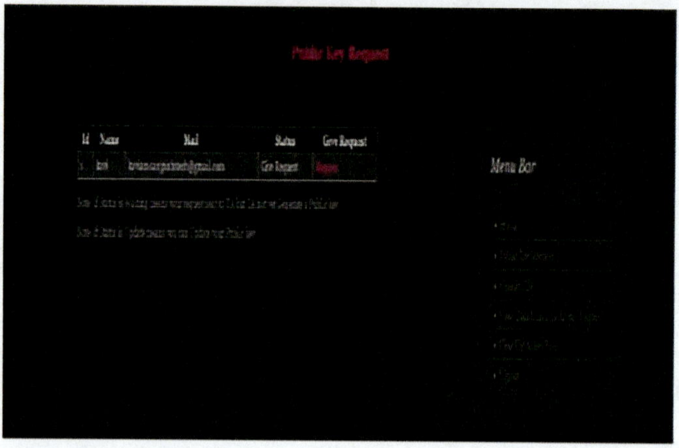

Fig. 8 Public key request

5.3 Results

The current application is developed for Android mobile phones using the Android framework. It can be used for Android mobile phones when it is not connected to network or Wi-Fi. This allows the user to easily capture the information when there are in roaming or offline as well as they can carry the customer data with them wherever they go. Merits of this application are mentioned below.

i. The welcome screen is displayed.
ii. Checking the authentication of the user while login.

Fig. 9 Trusted authentications

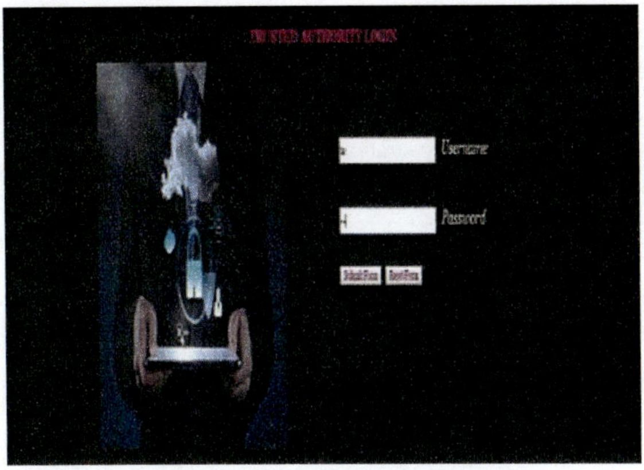

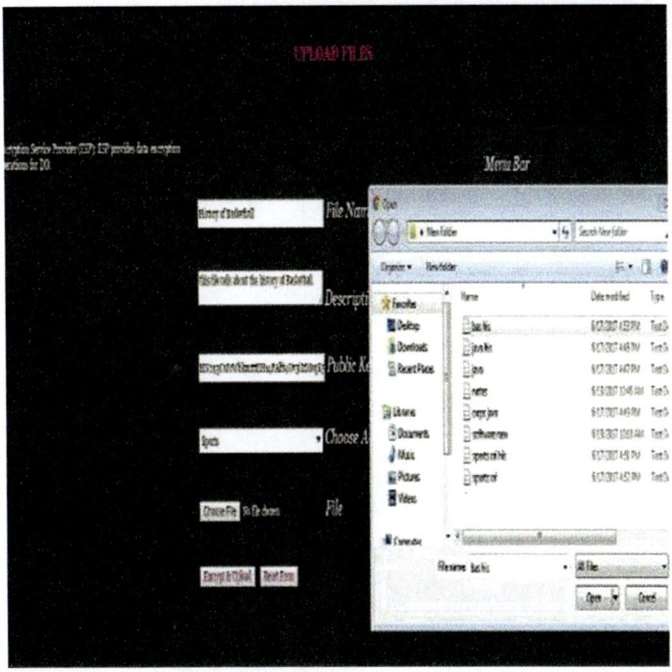

Fig. 10 Data owner is uploading the files in the cloud server

iii. Choose the online or offline mode on the home screen.
iv. A menu bar tab is to switch between the online, offline, home screen and the recent activity.
v. In the online mode, the data owner can list, create, update, delete and download the files.
vi. Users can access the data in the offline mode.

Table 1 Test cases

Test case id	Test case description	Actual value	Entered value	Status
1	Register user details in the registration page	Fill all the fields in the page	All the fields filled	Pass
2	Give user name in the text box	User name must be given in alphabets	User name given in alphabets and numeric values	Fail
3	Password to be entered in the password box	Password must be given correctly	Password entered wrongly	Fail
4	Phone number must be entered in phone number box during registration	Phone number must be given in 10 digits	Phone number given in 10 digits	Pass
5	Validating the functionality of browse button	System should select the corresponding file	Selected the file what we expected	Pass

vii. Data owners also can create, update and delete the files in offline mode.
viii. It saves the client visits both in offline and online modes.

The proposed system reduces the client-side overhead in which it introduces a minimal additional cost on the server side. This system benefits in implementing security while sharing the data on mobile devices. It results in better performance when compared with the existing ABE schemes.

6 Conclusion and Future Work

Many studies in cloud computing are based on the access control ABE scheme. Traditional ABE is not suitable for access control in mobile cloud storage because the issue is about only limited resources. We propose an LDSS-CP-ABE to overcome the issue in limited resources, huge storage of data and security on accessing the data. We introduced the proxy server to solve the above-mentioned issues. The experimental results ensure the data privacy and reduce the overhead on users' side in the mobile storage. In the future, we will model the new encryption approaches for data integrity. And, to increase the potential in mobile cloud, we will do the ciphertext for data sharing that is in order.

References

1. Gentry C, Halevi S (2011) Implementing gentry's fully-homomorphic encryption scheme. In: Annual international conference on the theory and applications of cryptographic techniques. Springer, Berlin, Heidelberg, pp 129–148

2. Kwon H, Kim D, Hahn C, Hur J (2017) Secure authentication using ciphertext policy attribute-based encryption in mobile multi-hop networks. Multim Tools Appl 76(19):19507–19521
3. Zhang Y, Dong Z, Xiaofeng C, Jin L, Hui L (2016) Efficient attribute-based data sharing in mobile clouds. Pervasive Mob Comput 28:135–149
4. Skillen A, Mohammad M (2013) On implementing deniable storage encryption for mobile devices
5. Jiang, Y, Willy S, Yi M, Fuchun G (2018) Ciphertext-policy attribute-based encryption against key-delegation abuse in fog computing. Future Gener Comput Syst 78:720–729
6. Wang Q, Li P, Hu X, Jianfei S, Zhiguang Q (2017) Ciphertext-policy attribute-based encryption with delegated equality test in cloud computing. IEEE Access 6:760–771
7. Yang, K, Xiaohua J, Kui R (2013) Attribute-based fine-grained access control with efficient revocation in cloud storage systems. In: Proceedings of the 8th ACM SIGSAC symposium on Information, computer and communications security. ACM, pp 523–528
8. Han J, Susilo W, Mu Y, Zhou J, Au MH (2014) Improving privacy and security in decentralized ciphertext-policy attribute-based encryption. IEEE Trans Inf Forensics Secur 10(3):665–678
9. Zhou Z, Huang D, Wang Z (2013) Efficient privacy-preserving ciphertext-policy attribute based-encryption and broadcast encryption. IEEE Trans Comput 64(1):126–138
10. Jayakumar L, Janakiraman S, Latchoumi TP, Shanthoshini (2017) Dynamic virtual server for optimized web service interaction. Int J Pure Appl Math 117(19):371–377
11. Latchoumi TP, Jayakumar L, Ezhilarasi TP (2017) Detecting generic objects by using an adaptation of spatial features. J Adv Res Dyn Control Syst 1401–1408

Mechanical Characterization of Palmyra- and S Glass Fibre-Reinforced Hybrid Polymer Composites

Navuluri Padma Sravya and S. Sivaganesan

Abstract The need of composite materials is intensive in last few decades as the reinforced polymer composites can be processed easily in many constraints and there are many advantages of using Palmyra fibre. Here, the Palmyra fibre is combined with glass fibre as matrix epoxy resin is taken to improve the properties. In the present work, Palmyra fibre is reinforced with glass fibre and epoxy resin in different ratios by hand lay-up method and is processed and different tests are conducted. This includes evaluation of tensile strength, flexural strength, compression strength and impact strength has been studied and discussed. The characterization of composites reveals that different ratios of Palmyra palm and glass fibre are having significant effect on the mechanical properties of composites. Tensile strength, tensile modulus, flexural strength, flexural modulus, compression strength, compression modulus and impact strength are found to be improved for all treated samples when compared to untreated samples. Flexural and tensile properties are improved due to removal of dust particles on surface of fibre; hence, cell walls of fibre are opened to wet matrix interface.

Keywords Palmyra fibre · Epoxy resin · Hand lay-up method · Characterization · Treated and untreated samples

1 Introduction

Composites are classified into two types by the geometry of the reinforcement and by the type of matrix [1]. According to the type, reinforcement can be classified as particulate composites, flake composites, fibre composites. Fibres are generally classified into two groups: synthetic fibres and natural fibres. The fibres which can be obtained from natural sources such as plants and animals are referred to as natural fibre [2]. The fibres whose structural properties and chemical compositions are significantly modified during the formation and are manufactured artificially are referred

N. P. Sravya (✉) · S. Sivaganesan
Department of Mechanical Engineering, VISTAS, Chennai, India
e-mail: navuluri.sravya@gmail.com

A. Arockiarajan et al. (eds.), *Advances in Industrial Automation and Smart Manufacturing*, Lecture Notes in Mechanical Engineering,
https://doi.org/10.1007/978-981-15-4739-3_93

to as synthetic fibre [3]. The most common advanced composites are polymer matrix composites consisting of polymer reinforced by thin-diameter fibres [4]. The hand layup method is used for the fabrication of polymer reinforced short length palmyra fibres. Initially the composites are made into plate shape by using compression moulding process. The sound absorption test is performed by using impedence test tube method [5]. Composite plates with the fibrous waste material content varying from 30 to 70% by weights were studied. It was observed that the incorporation of pfw in the matrix imparted poor reinforcement effect in tensile. Moisture absorption studies on pfw/glass fibre polyester composites were carried out separately and observed that the composite absorbed 9.23% of moisture in 166 h, 5.57% of moisture in 154 h and 3.5% of moisture in 175 h for the composites with 50% pfw/10% glass fibre, 40% pfw/20% glass fibre and 30% pfw/30% glass fibre, respectively. These Palmyra fibre waste and glass fibre hybrid composites could be moulded into standard size and used as partition boards, panel for windows and cupboards, etc., which need further attention on effect of moisture absorption on properties [6].

2 Experimental Set-Up

Glass moulds are made by different dimensions of ASTM standard which are used to prepare epoxy/glass fibre nanocomposites. For making the composite for testing the tensile strength and flexural strength, the glass moulds prepare as per the ASTM standards (ASTM D 638 and D 618). The glass moulds have the dimensions of (130 × 130 × 3) mm.

For making the composite for testing the impact strength of fibrous composite by conducting the Izod and Charpy test, prepare a glass mould as per the ASTM standard (ASTM D 256) having dimensions of (100 × 100 × 13) mm. For testing the compression strength of composite fibre, make a composite and prepare a mould as per the ASTM standards (ASTM D 695). Glass moulds have dimensions of (10 × 10 × 10) mm (Fig. 1).

2.1 Preparation of Composites

By the rule of hybrid mixtures, composites are prepared by hand lay-up technique. [7]. Initially by ASTM standards with 13 × 13 × 0.3 cm^3, the glass moulds are prepared. For the removal of moisture, the filler is kept in oven for half an hour. Before taking out from the oven, the surfaces of the mould are sprayed with polyvinyl alcohol. For curing, at least the mould should be kept aside at least for 24 h [8]. The casting is taken out from the mould using spatulas and knives [9]. According to ASTM standard sizes, the composites are cut for testing. On universal testing machine (3369) tests are performed. The average of results for each tested composition of at least 5 samples is listed. This technique is suitable for low volume production of product

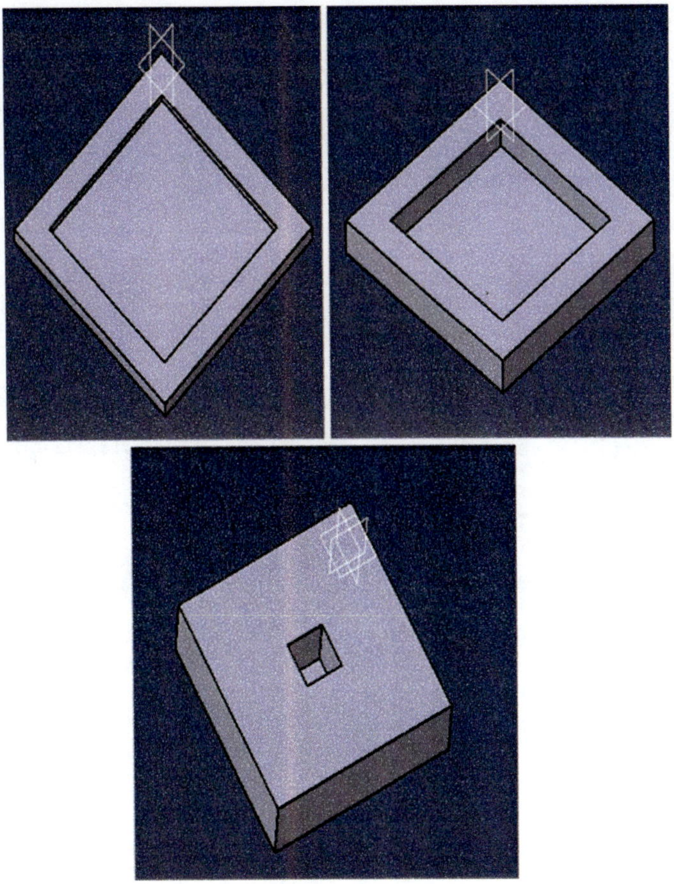

Fig. 1 Tensile, flexural, impact and compression mould

[10]. Reinforcement fibres can be cut and laid in the mould. The amount of resin and direction of the fibres depend on the designer. To make the reaction fast, a catalyst is used. The fibres with the resin are impregnated with the use of a brush or a roller. The quality of saturation and the amount of resin controlling responsibility are taken by the lay-up technician [11].

Characterization of Palmyra-reinforced polymer composites includes tensile, flexural, compression, impact tests. The combination system is given below (Table 1).

As per ASTM standards, samples are subjected to appropriate tests after fabrication (Fig. 2).

Table 1 Combinations of the samples

Sample No.	Epoxy in %	Palmyra in %	Glass fibre in %
1	90	10(25)	10(75)
2	90	10(50)	10(50)
3	90	10(75)	10(25)
4	90	10(100)	10(0)

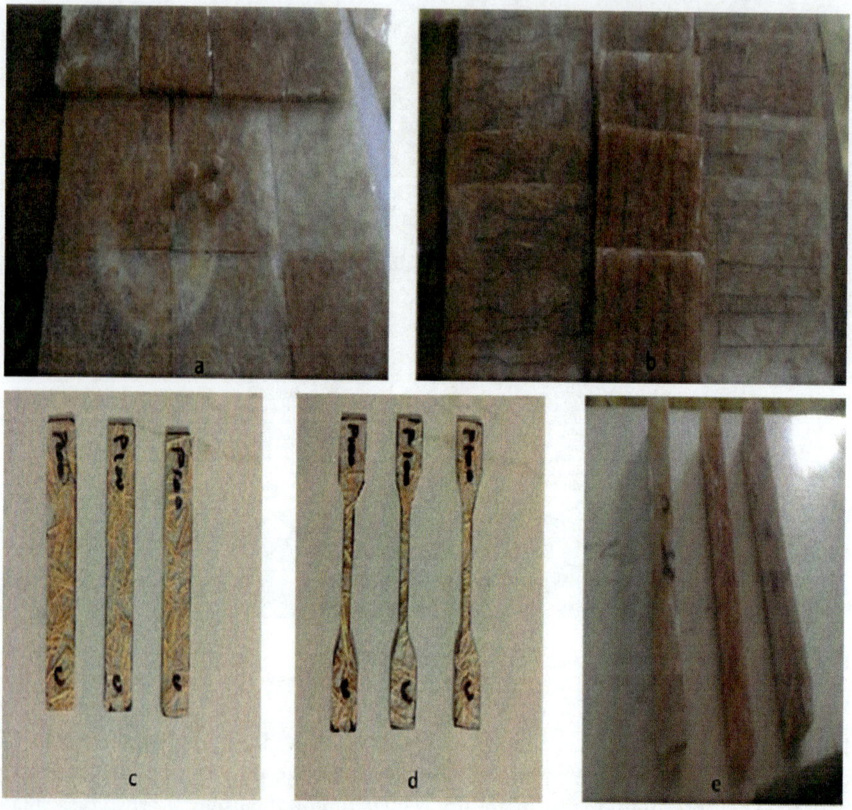

Fig. 2 **a**. Workpieces after taking from the glass moulds. **b**. Workpieces in the glass moulds after taking out of furnace. **c**, **d**, **e**. Testing samples

3 Results and Discussion

The results of tests which are conducted for various characterizations are listed. Evaluation of mechanical properties has been included. The interpretation of comparison and results of samples are also presented [12]. The different ratios of Palmyra and glass fibre are characterized. Different ratios of Palmyra are presented in the table below.

Table 2 Tensile strength of treated and untreated fibred composites

Sample No.	Tensile strength (Mpa)	
	Untreated	Treated
1	19.63	23.52
2	24.45	28.03
3	13.23	16.52
4	15.10	20.76

3.1 Tensile Strength

From the above results observed tensile strength of all treated samples is showing higher performance. Tensile strength of treated sample is showing higher performance when compared to other samples. Treated sample-2 shows 12.7% more than the untreated sample-2. Removal of substances which are waxy and the surfaces with impurities of the fibre during chemical treatment provides higher performance (Table 2).

3.2 Tensile Modulus

From the above results observed tensile modulus of all treated samples is showing higher performance. Tensile modulus of treated epoxy, 10(50) % Palmyra and 10(50) % glass fibre (sample-2) is showing higher performance when compared to other samples. Treated sample-2 shows 29.08% more than the untreated sample-2. Removal of substances which are waxy and the surfaces with impurities of the fibre during chemical treatment provides higher performance (Table 3; Fig. 3).

Table 3 Tensile modulus of treated and untreated fibred composites

Sample No.	Tensile modulus (Mpa)	
	Untreated	Treated
1	2333.17	3232.97
2	2607.18	3676.45
3	1650.21	2628.83
4	2407.45	3030.15

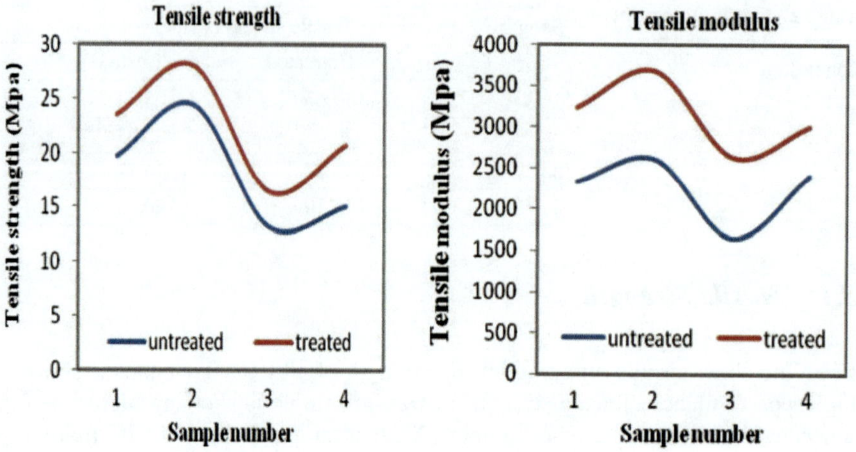

Fig. 3 Sample numbers versus tensile strength and tensile modulus

Table 4 Flexural strength of treated and untreated fibred composites

Sample No.	Flexural strength (Mpa)	
	Untreated	Treated
1	51.33	71.87
2	71.29	95.83
3	64.73	87.86
4	57.39	62.23

3.3 Flexural Strength

From the above results observed flexural strength of all treated samples is showing higher performance. Flexural strength of treated sample-2 is showing higher performance when compared to other samples. Treated sample-2 shows 23.8% more than the untreated sample-2. Removal of substances which are waxy and the surfaces with impurities of the fibre during chemical treatment provides higher performance (Table 4).

3.4 Flexural Modulus

From the above results observed flexural modulus of all treated samples is showing higher performance when compared to untreated samples. Flexural modulus of treated sample-2 is showing higher performance when compared to other samples. Treated sample-2 shows 29.8% more than the untreated sample-2. Removal of

Table 5 Flexural modulus of treated and untreated fibred composites

Sample No.	Flexural modulus (Mpa)	
	Untreated	Treated
1	4448.31	6115.84
2	5509.77	7864.64
3	4399.95	5969.59
4	3797.97	4900.8

Table 6 Compression strength of treated and untreated fibred composites

Sample No.	Compression strength (Mpa)	
	Untreated	Treated
1	31.15	47.36
2	48.37	61.85
3	29.98	40.43
4	34.34	49.60

substances which are waxy and the surfaces with impurities of the fibre during chemical treatment provides higher performance (Table 5).

3.5 *Compression Strength*

From the above results observed compression strength of all treated samples is showing higher performance. Compression strength of treated sample-2 is showing higher performance when compared to other samples. Treated sample-2 shows 19.7% more than the untreated sample-2. Removal of substances which are waxy and the surfaces with impurities of the fibre during chemical treatment provides higher performance (Table 6).

3.6 *Compression Modulus*

From the above results observed compression modulus of all treated samples is showing higher performance when compared to untreated samples. Compression modulus of treated sample-2 is showing higher performance when compared to other samples. Treated sample-2 shows 29.7% more than the untreated sample-2. Removal of substances which are waxy and the surfaces with impurities of the fibre during chemical treatment provides higher performance (Table 7).

Table 7 Compression modulus of treated and untreated fibred composites

Sample No.	Compression modulus (Mpa)	
	Untreated	Treated
1	1760.73	2828.25
2	2893.73	4172.78
3	2198.65	3285.35
4	1589.75	2246.18

Table 8 Impact strength of treated and untreated fibred composites

Sample No.	Energy absorbed (joules)	
	Untreated	Treated
1	3.0	4.1
2	4.3	5.6
3	2.1	3.8
4	0.8	1.77

3.7 Impact Strength

From the above results observed impact strength of all treated samples is showing higher performance than untreated samples. Treated impact strength of sample-2 is showing higher performance when compared to other samples. Treated sample-2 shows 23.2% more than the untreated sample-2. Removal of substances which are waxy and the surfaces with impurities of the fibre during chemical treatment provides higher performance (Table 8; Figs. 4, 5 and 6).

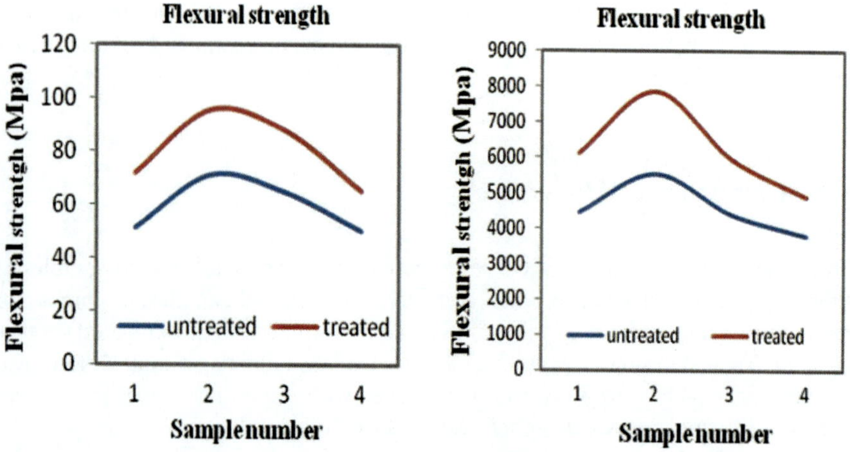

Fig. 4 Sample number versus flexural strength and flexural modulus

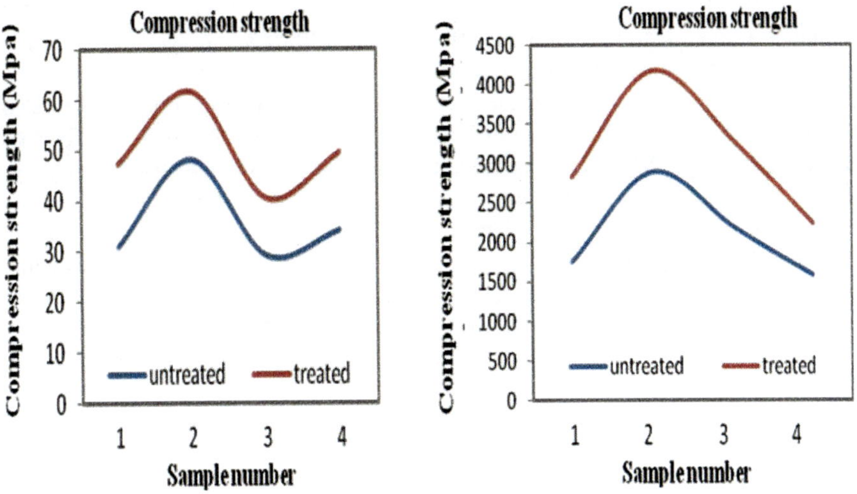

Fig. 5 Sample numbers versus compression strength and compression modulus

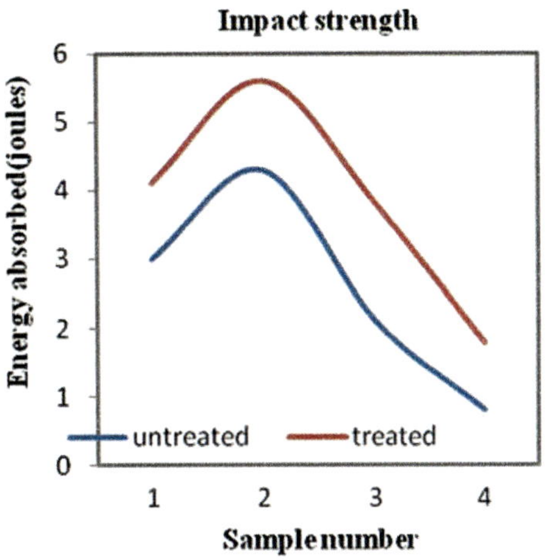

Fig. 6 Sample numbers versus energy absorbed

4 Conclusion

The mechanical properties considered are found to be improved for all treated samples when compared to untreated samples. Tensile strength of treated sample-2 which is of ratio 90% epoxy, 10(50) % Palmyra and 10(50) % glass fibre is showing higher performance when compared to other samples. Treated sample-2 shows 12.7% more than the untreated sample-2. Removal of substances which are waxy and the surfaces with impurities of the fibre during chemical treatment provides higher performance. Tensile modulus of treated sample-2 is showing higher performance when compared to other samples. Treated sample-2 shows 29.08% more than the untreated sample-2. Flexural strength of treated 90% epoxy, 10(50) % Palmyra and 10(50) % glass fibre is showing higher performance when compared to other samples. Treated sample-2 shows 23.8% more than the untreated sample-2. Flexural modulus of treated sample-2 is showing higher performance when compared to other samples. Treated sample-2 shows 29.8% more than the untreated sample-2. Compression strength of treated 90% epoxy, 10(50)% Palmyra and 10(50)% glass fibre is showing higher performance when compared to other samples. Treated sample-2 shows 19.7% more than the untreated sample-2. Compression modulus of treated sample-2 is showing higher performance when compared to other samples. Treated sample-2 shows 29.7% more than the untreated sample-2. Impact strength of treated 90% epoxy, 10(50)% Palmyra and 10(50)% glass fibre is showing higher performance when compared to other samples. Treated sample-2 shows 23.2% more than the untreated sample-2. Flexural and tensile properties are improved due to removal of dust particles on the surface of the fibre; hence, cell walls of fibre are opened to wet matrix interface.

References

1. Ashok kumar M, Ramachandrareddy G, Siva bharathi Y, Venkata Naidu S, Naga Prasad Naidu V (2010) Frictional coefficient, hardness, impact strength, and chemical resistance of reinforced sisal glass fibre epoxy hybrid composites. J Compos Mater 44(26):3195–31202. https://doi.org/10.1177/0021998310371551
2. M. Ashok Kumar, T. Maruthi Chowdary, K. Chandra Sekhar Balaji, E. Dhanunjaya Goud, S. Nagaraju, K. Shaik Ahmmed, B. Raja Sekhar. "Utilisation of CF/Sawdust Reinforced Epoxy Hybrid Composites on Mechanical Properties", International Letters of Chemistry, Physics and Astronomy, 2015
3. Velmurugan R, Manikandan V (2007) Mechanical properties of palmyra/glass fiber hybrid composites. Applied Science and Manufacturing, Composites Part A
4. Santhosh J (2014) Study of properties of banana fiber reinforced composites. Int J Res Eng Technol 2014
5. Vishnu Vardhini KJ, Murugan R, Surjit R. Effect of alkali and enzymatic treatments of banana 9
6. Ramachandra Reddy G, Ashok Kumar M, Karthikeyan N, Mahaboob Basha S. (2015) Tamarind fruit fiber and glass fiber reinforced polyester composites. Mech Adv Mater Struct
7. Montazer, M. Influences of different enzymatic treatment on denim garment. Appl Biochem Biotechnol 09/01/2009
8. Lu Y (2007) Fabrication, morphology and properties of soybean oil-based composites reinforced with continuous glass fibers. Macromol Mater Eng
9. Ishak ZAM, Bakar AA (1995) An investigation on the potential of rice husk ash as fillers for epoxidized natural rubber (ENR). Eur Polym J
10. Narendiranath Babu T, Dogra S, Chauhan M, Singh R et al (2018) A review on Aloe/Kapok/Palmyra/Corn Fiber/Vetivernatural fibers for biomedical applications. Mater Today: Proc
11. Jawaid M, Cellulosic/synthetic fibre reinforced polymer hybrid composites: A review. Carbohydr Polym 20110801
12. Biswas S, Kindo S, Patnaik A (2011) Effect of fiber length on mechanical behavior of coir fiber reinforced epoxy composites. Fiber Polym